ONE CENTURY OF THE DISCOVERY OF ARSENICOSIS IN LATIN AMERICA (1914–2014)

Arsenic in the Environment – Proceedings

Series Editors

Jochen Bundschuh
Faculty of Engineering and Surveying and National Centre for Engineering in Agriculture (NCEA), The University of Southern Queensland, Toowoomba, Australia

Prosun Bhattacharya
KTH-International Groundwater Arsenic Research Group, Department of Sustainable Development, Environmental Science and Engineering, KTH Royal Institute of Technology, Stockholm, Sweden

ISSN: 2154-6568

PROCEEDINGS OF THE 5TH INTERNATIONAL CONGRESS ON ARSENIC IN THE ENVIRONMENT, BUENOS AIRES, ARGENTINA, 11–16 MAY 2014

One Century of the Discovery of Arsenicosis in Latin America (1914–2014)

As 2014

Editors

Marta I. Litter
Gerencia Química, Centro Atómico Constituyentes, Comisión Nacional de Energía Atómica (CNEA), Buenos Aires, Argentina

Hugo B. Nicolli
Instituto de Geoquímica (INGEOQUI) and Consejo Nacional de Investigaciones Científicas y Técnicas (CONICET), San Miguel, Buenos Aires, Argentina

Martin Meichtry & Natalia Quici
Gerencia Química, Centro Atómico Constituyentes, Comisión Nacional de Energía Atómica (CNEA), Buenos Aires, Argentina

Jochen Bundschuh
Faculty of Health, Engineering and Sciences & National Centre for Agriculture, The University of Southern Queensland, Toowoomba, Australia

Prosun Bhattacharya
KTH-International Groundwater Arsenic Research Group, Department of Sustainable Development, Environmental Science and Engineering, KTH Royal Institute of Technology, Stockholm, Sweden

Ravi Naidu
Cooperative Research Centre for Contamination Assessment and Remediation of the Environment (CRC-CARE), University of South Australia, Mawson Lakes, SA, Australia

CRC Press
Taylor & Francis Group
Boca Raton London New York

CRC Press is an imprint of the
Taylor & Francis Group, an **informa** business
A BALKEMA BOOK

ISGSD

International Society of Groundwater for Sustainable Development

Cover photo

The picturesque Copahue volcano with a cascade. Classified as a stratovolcano, it is located at the border between Argentina and Chile. The word Copahue means "sulphur waters" in Mapuche. There are nine volcanic craters in the region, forming a linear belt of approximately 2 km. The 300 m wide eastern summit crater lake contains briny, acidic water with a pH ranging between 0.18 and 0.30. It is an important source of arsenic in the region. The geology of Copahue is characterized by stratified sedimentary and volcanic rocks ranging in age from Eocene to Pliocene. Eruptions from the crater lake during the past century have resulted in the deposition of a huge pile of pyroclastic sediments with sulfur fragments. The present day volcano became active since 1.2 Ma towards the east of Copahue. The modern caldera formed between 0.6 and 0.4 Ma BP, with dimensions of 20 km by 15 km and generated large pyroclastic flows, extending up to 37 km from the volcano. This elongated shield volcano has experienced at least six eruptions within the Holocene, the most recent during December 2012.

(*See also: http://www.geotimes.org/nov07/article.html?id=Travels1107.html*)

CRC Press
Taylor & Francis Group
6000 Broken Sound Parkway NW, Suite 300
Boca Raton, FL 33487-2742

First issued in paperback 2018

CRC Press/Balkema is an imprint of the Taylor & Francis Group, an informa business

Typeset by V Publishing Solutions Pvt Ltd., Chennai, India

ISBN-13: 978-1-138-00141-1 (hbk)
ISBN-13: 978-1-138-37263-4 (pbk)

Published by: CRC Press/Balkema
P.O. Box 11320, 2301 EH Leiden, The Netherlands
e-mail: Pub.NL@taylorandfrancis.com
www.crcpress.com – www.taylorandfrancis.com

Visit the Taylor & Francis Web site at
http://www.taylorandfrancis.com

and the CRC Press Web site at
http://www.crcpress.com

One Century of the Discovery of Arsenicosis in Latin America (1914–2014) – Litter, Nicolli, Meichtry, Quici, Bundschuh, Bhattacharya & Naidu (Eds)
© 2014 Taylor & Francis Group, London, ISBN 978-1-138-00141-1

Table of contents

Section 4: Arsenic remediation and removal technologies

4.1 Technologies based on adsorption and co-precipitation

One Century of the Discovery of Arsenicosis in Latin America (1914–2014) –
Litter, Nicolli, Meichtry, Quici, Bundschuh, Bhattacharya & Naidu (Eds)
© 2014 Taylor & Francis Group, London, ISBN 978-1-138-00141-1

About the book series

Although arsenic has been known as a 'silent toxin' since ancient times, and the contamination of drinking water resources by geogenic arsenic was described in different locations around the world, almost a century back in Argentina in 1914—it was only two decades ago that it received overwhelming worldwide public attention. As a consequence of the biggest arsenic calamity in the world, which was detected more than twenty years back in West Bengal, India and other parts of Southeast Asia, there has been an exponential rise in scientific interest that has triggered high quality research. Since then, arsenic contamination (predominantly of geogenic origin) of drinking water resources, soils, plants and air, the propagation of arsenic in the food chain, the chronic affects of arsenic ingestion by humans, and their toxicological and related public health consequences, have been described in many parts of the world, and every year, even more new countries or regions are discovered to have elevated levels of arsenic in environmental matrices.

Arsenic is found as a drinking water contaminant, in many regions all around the world, in both developing as well as industrialized countries. However, addressing the problem requires different approaches which take into account, the differential economic and social conditions in both country groups. It has been estimated that 200 million people worldwide are at risk from drinking water containing high concentrations of As, a number which is expected to further increase due to the recent lowering of the limits of arsenic concentration in drinking water to 10 μg/L, which has already been adopted by many countries, and some authorities are even considering decreasing this value further.

The book series "Arsenic in the Environment—Proceedings" is an inter- and multidisciplinary source of information, making an effort to link the occurrence of geogenic arsenic in different environments and the potential contamination of ground- and surface water, soil and air and their effect on the human society. The series fulfills the growing interest in the worldwide arsenic issue, which is being accompanied by stronger regulations on the permissible Maximum Contaminant Levels (MCL) of arsenic in drinking water and food, which are being adopted not only by the industrialized countries, but increasingly by developing countries.

Consequently, we see the book series *Arsenic in the Environment-Proceedings* with the outcomes of the international congress series Arsenic in the Environment, which we organize biannually in different parts of the world, as a regular update on the latest developments of arsenic research. It is further a platform to present the results from other from international or regional congresses or other scientific events. This Proceedings series acts as an ideal complement to the books of the series *Arsenic in the Environment*, which includes authored or edited books from world-leading scientists on their specific field of arsenic research, giving a comprehensive information base. Supported by a strong multi-disciplinary editorial board, book proposals and manuscripts are peer reviewed and evaluated. Both of the two series will be open for any person, scientific association, society or scientific network, for the submission of new book projects.

We have an ambition to establish an international, multi- and interdisciplinary source of knowledge and a platform for arsenic research oriented to the direct solution of problems with considerable social impact and relevance rather than simply focusing on cutting edge and breakthrough research in physical, chemical, toxicological and medical sciences. It shall form a consolidated source of information on the worldwide occurrences of arsenic, which otherwise is dispersed and often hard to access. It will also have role in increasing the awareness and knowledge of the arsenic problem among administrators, policy makers and company executives and improving international and bilateral cooperation on arsenic contamination and its effects.

Both of the book series cover all fields of research concerning arsenic in the environment and aims to present an integrated approach from its occurrence in rocks and mobilization into the ground- and surface water, soil and air, its transport therein, and the pathways of arsenic introduction into the food chain including uptake by humans. Human arsenic exposure, arsenic bioavailability, metabolism and toxicology are treated together with related public health effects and risk assessments in order to better manage the contaminated land and aquatic environments and to reduce human arsenic exposure. Arsenic removal technologies and other methodologies to mitigate the arsenic problem are addressed not only from the technological perspective, but also from an economic and social point of view. Only such inter- and multidisciplinary approaches will allow a case-specific selection of optimal mitigation measures for each specific arsenic problem and provide the local population with arsenic-safe drinking water, food, and air.

Jochen Bundschuh
Prosun Bhattacharya
(*Series Editors*)

One Century of the Discovery of Arsenicosis in Latin America (1914–2014) – Litter, Nicolli, Meichtry, Quici, Bundschuh, Bhattacharya & Naidu (Eds)
© 2014 Taylor & Francis Group, London, ISBN 978-1-138-00141-1

Obituary

Arun Bilash Mukherjee, D.Sc.
Formerly Senior Research Scientist, Environmental Sciences,
Department of Biological and Environmental Sciences, University of Helsinki, Finland
1938–2013

We dedicate this Volume of Proceedings of the the International Congress of Arsenic in the Environment (As 2014) to our colleague, friend and coworker Dr. Arun Bilash Mukherjee, who had been an active researcher and scientist working in the field of arsenic in the environment for a period of more than 13 years. Born in Faridpur, East Bengal (now Bangladesh) in the year 1938, he had his early education in Faridpur. He completed his degree of Bachelor of Science in Metallurgy from University of Calcutta (Kolkata), India and soon after he started his professional career as a metallurgist and worked in well known companies in the metal industry in Finland and in the copper industry in Finland, India, and Zaire (presently known as the Democratic Republic of Congo). His interest in metals brought him back to Finland in 1976, where he continued with higher studies starting with a M.Sc. in Process Metallurgy from the Helsinki University of Technology, Finland followed by the degree of Licentiate in Technology from the Department of Forest Products and he received Doctor of Science (D.Sc.) from the University of Helsinki, Finland in 1994.

His research interests included environmental biogeochemistry of trace elements, emission inventories, soil remediation, waste management and recycling, and fate of trace elements in coal and coal combustion by-products and groundwater arsenic problems in the developing countries. He had conducted several national and international investigations into trace elements and chemicals especially for the developing countries. Dr. Mukherjee had published approximately 80 papers, articles, and book chapters in peer reviewed journals, conference proceedings, series and symposia. He is also co-author of the book entitled Trace Elements from Soil to Humans, published by Springer-Verlag, Germany in Spring 2007.

Since 2000, he shifted his research interest toward the global problem of arsenic in groundwater, its occurrence, fate and management for drinking water supplies. He has published a number of articles on groundwater arsenic in Bangladesh in collaboration with the KTH-International Groundwater Arsenic Research Group at the KTH Royal Institute of Technology, Stockholm, Sweden. Through the collaborative network of the KTH-International Groundwater Arsenic Research Group, he has co-edited a number of books including "*Arsenic in Soil and Groundwater Environment: Biogeochemical Interactions, Health Impacts and Remediation*" published in the Elsevier Series "Trace Elements in the Environment" (2007); "*Groundwater and Sustainable Development: Problems, Perspectives and Challenges*" (2008); and "Natural arsenic in groundwaters of Latin America—Occurrence, health impact and remediation (2009), the first volume of the Book Series "Arsenic in the Environment" published by CRC Press/Balkema,The Netherlands.

He had been active and energetic throughout his academic career. In 2011 he retired, followed by his illness and he passed away on 30th August 2013 in Helsinki, Finland. The International Society of Groundwater for Sustainable Development will always remember his contributions towards the advancement of knowledge on arsenic in environmental systems.

One Century of the Discovery of Arsenicosis in Latin America (1914–2014) –
Litter, Nicolli, Meichtry, Quici, Bundschuh, Bhattacharya & Naidu (Eds)
© 2014 Taylor & Francis Group, London, ISBN 978-1-138-00141-1

Organizers

ORGANIZERS OF BIANNUAL CONGRESS SERIES: ARSENIC IN THE ENVIRONMENT

Jochen Bundschuh
International Society of Groundwater for Sustainable Development (ISGSD), Stockholm, Sweden
University of Southern Queensland (USQ), Toowomba, QLD, Australia

Prosun Bhattacharya
KTH-International Groundwater Arsenic Research Group, Department of Sustainable Development, Environmental Science and Engineering, KTH Royal Institute of Technology, Stockholm, Sweden
International Society of Groundwater for Sustainable Development (ISGSD), Stockholm, Sweden

LOCAL ORGANIZING COMMITTEE

Marta Irene Litter
Gerencia Química, Comisión Nacional de Energía Atómica, San Martín, Provincia de Buenos Aires, Argentina
Consejo Nacional de Investigaciones Científicas y Técnicas (CONICET), Universidad de San Martin, San Martín, Argentina

Hugo B. Nicolli
Instituto de Geoquímica (INGEOQUI) and Consejo Nacional de Investigaciones Científicas y Técnicas (CONICET), San Miguel, Provincia de Buenos Aires, Argentina

Ana María Ingallinella
Centro de Ingenieria Sanitaria, Universidad Nacional de Rosario, Provincia de Santa Fé, Argentina

Alicia Fernández Cirelli
Alejo Pérez Carrera
Facultad de Veterinaria, Universidad de Buenos Aires, Argentina
Consejo Nacional de Investigaciones Científicas y Técnicas (CONICET), Universidad de San Martin, San Martín, Argentina

Silvia Farias
Comisión Nacional de Energía Atómica, Argentina

MEMBERS

Guillermina Bongiovanni
Instituto Multidisciplinario de Investigación y Desarrollo de la Patagonia Norte (IDEPA), Neuquén-Cinco Saltos, Argentina
Consejo Nacional de Investigaciones Cientificas y Técnicas (CONICET), Universidad Nacional del Comahue, Neuquén, Argentina

Cristina Pérez Coll
Universidad Nacional de San Martín, Buenos Aires, Argentina

Edda Villaamil
Julio Navoni
Facultad de Farmacia y Bioquímica, Universidad de Buenos Aires, Argentina

Alejandra Volpedo
Facultad de Veterinaria, Universidad de Buenos Aires, Argentina
Consejo Nacional de Investigaciones Cientificas y Técnicas (CONICET), Universidad Nacional del Comahue, Neuquén, Argentina

Rubén Fernández
Centro de Ingenieria Sanitaria, Universidad Nacional de Rosario, Argentina

Susana García
Ricardo Benitez
Silvia Rivero
Ministry of Health, Government of Argentina, Argentina

Leonardo Pflüger
Universidad Centro de Altos Estudios en Ciencias Exactas (CAECE)

YOUTH LOCAL ORGANIZING COMMITTEE

Natalia Quici
Martín Meichtry
Comision Nacional de Energía Atómica, Consejo Nacional de Investigaciones Cientificas y Técnicas (CONICET), Universidad Tecnológica Nacional, San Martín, Prov. de Buenos Aires, Argentina

Verónica Sotomayor
Instituto Multidisciplinario de Investigación y Desarrollo de la Patagonia Norte (IDEPA), Neuquén, Argentina
Consejo Nacional de Investigaciones Cientificas y Técnicas (CONICET), Universidad Nacional del Comahue, Neuquén, Argentina

Cynthia Corroto
Agua y Saneamientos Argentinos S.A, Argentina
Facultad de Veterinaria, Universidad de Buenos Aires, Argentina

Fernanda Vazquez
Facultad de Veterinaria, Universidad de Buenos Aires, Argentina
Consejo Nacional de Investigaciones Cientificas y Técnicas (CONICET), Universidad Nacional del Comahue, Neuquén, Argentina

CONGRESS EVENT ORGANIZER

Maria Isabel Cambón

One Century of the Discovery of Arsenicosis in Latin America (1914–2014) – Litter, Nicolli, Meichtry, Quici, Bundschuh, Bhattacharya & Naidu (Eds)
© 2014 Taylor & Francis Group, London, ISBN 978-1-138-00141-1

Scientific committee

B. Figueiredo: *UNICAMP, Campinas, SP, Brazil*
R.B. Finkelman: *Department of Geosciences, University of Texas at Dallas, Richardson, Texas, USA*
A. Fiúza: *University of Porto, Porto, Portugal*
S. García: *Ministry of Health, Buenos Aires, Argentina*
S.E. Garrido: *Mexican Institute of Water Technology, Jiutepec, Mor., Mexico*
M. Gasparon: *The University of Queensland, Australia*
A.K. Giri: *CSIR-Indian Institute of Chemical Biology, Kolkata, India*
W. Goessler: *University of Graz, Austria*
D.N. Guha Mazumder: *DNGM Research Foundation, Kolkata, India*
L.R. Guimaraes Guilherme: *Federal University of Lavras, Lavras, M.G., Brazil*
X. Guo: *Peking University, Peking, PR China*
B. Hendry: *Cape Peninsula University of Technology, Cape Town, South Africa*
M. Hernández: *National University of La Plata, La Plata, Argentina*
J. Hoinkis: *Karlsruhe University of Applied Sciences, Karlsruhe, Germany*
C. Hopenhayn: *University of Kentucky, Lexington, KY, USA*
Q. Hu: *University of Texas at Arlington, USA*
M.F. Hughes: *Environmental Protection Agency, Research Triangle Park, NC, USA*
A.M. Ingallinella: *Centro de Ingeniería Sanitaria (CIS), Facultad de Ciencias Exactas, Ingeniería y Agrimensura, Universidad Nacional de Rosario, Rosario, Provincia de Santa Fe, Argentina*
G. Jacks: *Department of Sustainable Development, Environmental Sciences and Engineering, KTH Royal Institute of Technology, Stockholm, Sweden*
J.-S. Jean: *National Cheng Kung University, Tainan, Taiwan*
C.-J. Chen: *Academia China, Taipéi, Taiwan*
N. Kabay: *Chemical Engineering Department, Engineering Faculty, Ege University, Izmir, Turkey*
I.B. Karadjova: *Faculty of Chemistry, University of Sofia, Sofia, Bulgaria*
A. Karczewska: *Institute of Soil Sciences and Environmental Protection, Wroclaw University of Environmental and Life Sciences, Poland*
D.B. Kent: *US Geological Survey, Menlo Park, CA, USA*
N.I. Khan: *The Australian National University, Canberra, Australia*
K.-W. Kim: *Department of Environmental Science and Engineering, Gwangju Institute of Science and Technology, Gwangju, South Korea*
W. Klimecki: *Department of Pharmacology and Toxicology, University of Arizona, Tucson, Arizona, USA*
M. Labas: *FICH-UNL-INTEC(UNL-CONICET), Santa Fe, Argentina*
R. Leyva Ramos: *Autonomous University of San Luis Potosí, Luis Potosí, México*
M.I. Litter: *Comisión Nacional de Energía Atómica, and Universidad de Gral. San Martín, San Martín, Argentina*
D.L. López: *Ohio University, Athens, Ohio, USA*
L. Ma: *Universidad de Florida, Florida, USA*
M. Mallavarapu: *University of South Australia, Mawson Lakes, Australia*
N. Mañay: *De la República University, Montevideo, Uruguay*
J. Matschullat: *Interdisciplinary Environmental Research Centre (IÖZ), TU Bergakademie Freiberg, Freiberg, Germany*
A. Mukherjee: *Department of Geology and Geophysics, Indian Institute of Technology (IIT), Kharagpur, India*
R. Naidu: *CRC Care, University of South Australia, Mawson Lake Campus, Mawson Lake, SA, Australia*
J.A. Navoni: *University of Buenos Aires, Buenos Aires, Argentina*
J.C. Ng: *National Research Centre for Environmental Toxicology, The University of Queensland, Brisbane, Australia*
H.B. Nicolli: *Instituto de Geoquímica (INGEOQUI), San Miguel, Provincia de Buenos Aires, Argentina and Consejo Nacional de Investigaciones Científicas y Técnicas (CONICET), Argentina*
B.N. Noller: *The University of Queensland, Australia*
D.K. Nordstrom: *U.S. Geological Survey, Menlo Park, CA, USA*
G. Owens: *Mawson Institute, University of South Australia, Australia*
P. Pastén González: *Pontificia Universidad Católica de Chile, Chile*
C.A. Pérez: *LNLS-Brazilian Synchrotron Light Source Laboratory, Campinas, SP, Brazil*
A. Pérez Carrera: *University of Buenos Aires, Buenos Aires, Argentina*
C. Pérez Coll: *National San Martín University, San Martín, Argentina*
B. Petrusevski: *UNESCO-IHE, Institute for Water Education, Delft, The Netherlands*
L. Pflüger: *Secretary of Environment, Buenos Aires, Argentina*
G.E. Pizarro Puccio: *Pontificia Universidad Católica de Chile, Chile*

B. Planer-Friedrich: *University Bayreuth, Bayreuth, Germany*
D.A. Polya: *School of Earth, Atmospheric and Environmental Sciences, The University of Manchester, UK*
T. Pradeep: *Indian Institute of Technology Madras, Chennai, India*
I. Queralt: *Institute of Earth Sciences Jaume Almera, CSIC, Spain*
J. Quintanilla: *Institute of Chemical Research, Universidad Mayor de San Andrés, La Paz, Bolivia*
M. Rahman: *University of South Australia, Mawson Lakes, SA, Australia*
AL. Ramanathan: *School of Environmental Science, Jawaharlal Nehru University, New Delhi, India*
G. Román Ross: *Senior Consultant and Project Manager, Barcelona, Spain*
A.M. Sancha: *Department of Civil Engineering, University of Chile, Santiago de Chile, Chile*
E. Scarlato: *University of Buenos Aires, Buenos Aires, Argentina*
C. Schulz: *La Pampa National University, Argentina*
O. Selinus: *Geological Survey of Sweden (SGU), Uppsala, Sweden; Linneaus University, Kalmar, Sweden*
A. SenGupta: *Lehigh University, Bethlehem, PA, USA*
V.K. Sharma: *Florida Institute of Technology, Florida, USA*
A. Shraim: *The University of Queensland, Brisbane, Australia*
M. Sillanpää: *Lappeenranta University of Technology, Lappeenranta, Finland*
P.L. Smedley: *British Geological Survey, Keyworth, UK*
A.H. Smith: *University of California, Berkeley, CA, USA*
M. Stýblo: *University of North Carolina, Chapel Hill, USA*
C. Tsakiroglou: *Foundation for Research and Technology, Hellas, Greece*
M. Vahter: *Karolinska Institutet Stockholm, Sweden*
J.W. Vargas de Mello: *Federal University of Viçosa, Viçosa, MG, Brazil*
D. Velez: *Institute of Agrochemistry and Food Technology, Valencia, Spain*
E. Villaamil: *University of Buenos Aires, Buenos Aires, Argentina*
Y. Zheng: *Lamont-Doherty Earth Observatory, Columbia University, New York, USA*
Y.-G. Zhu: *Chinese Academy of Sciences, Research Centre for Eco-Environmental Sciences, Beijing, China*

One Century of the Discovery of Arsenicosis in Latin America (1914–2014) –
Litter, Nicolli, Meichtry, Quici, Bundschuh, Bhattacharya & Naidu (Eds)
© 2014 Taylor & Francis Group, London, ISBN 978-1-138-00141-1

Foreword (President, KTH)

KTH
VETENSKAP
OCH KONST

ROYAL INSTITUTE OF TECHNOLOGY

PRESIDENT

Arsenic is a natural or anthropogenic contaminant in many areas around the globe, where human subsistence is at risk. It is considered as a class 1 carcinogen, and its presence in groundwater has emerged as a major environmental calamity in several parts of the world. In several regions of the world especially in different countries of Asia such as Bangladesh, Cambodia, China, India, Nepal, Pakistan, Taiwan, Thailand and Vietnam, the situation of arsenic toxicity is alarming and severe health problems are reported amongst the inhabitants relying on groundwater as drinking water. However recent investigations have shown that the arsenic problem in many Latin American countries is of the same order of significance. The use of arsenic contaminated groundwater for irrigation and its bioavailability to food crops and ingestion by humans and livestock through the food chain has presented additional pathways for arsenic exposure. The widespread discovery of arsenic in Asia has paved the way to the discovery of the presence of this element in different environmental compartments as a "silent" toxin globally. New areas with elevated arsenic occurrences are reported in groundwater exceeding the maximal contamination levels set by the WHO and other national and international regulatory organizations are identified each year.

Since the beginning of the 21st century, there has been a remarkable increase in interest in the field of arsenic. Many research councils and international donor organizations have provided significant support to local and international research teams to assess the extent of the problem and the strategies needed to minimize the risk of arsenic exposure among the population. As a consequence, there has been a radical increase in the number of scientific publications that give a holistic overview on the dynamics of arsenic in the soil and water environment, and its impact on human health.

The review on arsenic conducted by the WHO/FAO Joint Expert Committee (JEC) on Food Additives nearly four year back resulted in withdrawal of the Provisional Tolerable Weekly Intake (PTWI) due to several identified gaps, particularly related to accurate measurement of dietary and other exposure pathways as well as the speciation of arsenic and bioavailability that account for the total daily intake of arsenic. Long-term exposure to arsenic is related to non-specific pathological irreversible effects and has significant social and economic impacts. Thus, arsenic in environment is clearly a concern that needs an inter- and multi-disciplinary and cross-disciplinary platform of research including hydrogeology and hydrogeochemistry, environmental sciences, food and nutrition, toxicology, health and medical sciences, remediation technologies and social sciences.

Since the first International Congress on "Arsenic in the Environment" at the UNAM, Mexico City in 2006, there has been an overwhelming response from the scientific community engaged with multidisciplinary facets of arsenic research to participate and present their research findings on this platform. I am happy to see that this congress has taken a form of biennial congress series with rotating venues at different continents globally and is providing a common platform for sharing knowledge and experience on multidisciplinary issues on arsenic occurrences in groundwater and other environmental compartments on a global scale for identifying the risks and design innovative approaches for the assessment and management of arsenic in the environment, and reducing the health impacts.

The following three events namely the 2nd International Congress (As 2008), with the theme "Arsenic from Nature to Humans" (Valencia, Spain) and the 3rd International Congress (As 2010) with the theme "Arsenic in Geosphere and Human Diseases" (Tainan, Taiwan) and 4th International Congress on Arsenic in the Environment (As 2012) with a theme "Understanding the Geological and Medical Interface" (Cairns, Australia) had been successfully organized and participated by a leading scientific community around the globe. The upcoming 5th International Congress on Arsenic in the Environment (As 2014) is envisioned with a theme "One Century of the Discovery of Arsenicosis in Latin America (1914–2014)" to be organized in Buenos Aires, Argentina between 11th and 16th May, 2014, with an aim to provide another international, multi- and interdisciplinary discussion platform for the presentation of cutting edge research on the hydrogeological, geochemical, mining, toxicological, medical, water treatment and other related social issues on environmental arsenic by bringing together scientific, medical, engineering and regulatory professionals.

I feel proud to write this foreword to this Volume of Arsenic in the Environment—Proceedings Series, that contains the extended abstracts of the presentations to be made during the forthcoming 5th International Congress on Arsenic in the Environment-As 2014. The present volume "One Century of the Discovery of Arsenicosis in Latin America (1914-2014)" being published as a new volume of the book series "Arsenic in the Environment-Proceedings under the auspices of the International Society of Groundwater for Sustainable Development (ISGSD), will be an important updated contribution, comprising a large number of over 200 extended abstracts submitted by various researchers, health workers, technologists, students, legislators, and decision makers around the world that would be discussed during the conference. Apart from exchanging ideas, and discovering common interests, the scientific community involved in this this specialized field needs to carry out researches, which not only address academic interests but also contribute to the societal needs through prevention or reduction of exposure to arsenic and its toxic effects in millions of exposed people throughout the world.

I deeply appreciate the efforts of the International Organizers from KTH-International Groundwater Arsenic Research Group, KTH Royal Institute of Technology and the National Commission of Atomic Energy (CNEA) and the University of Buenos Aires Argentina and the entire editorial team for their efforts to bring together for their untiring work with this volume and hope that the book will serve a broad purpose of improving knowledge required for the management of arsenic in the environment for protecting human health.

Professor Dr. Peter Gudmundson
President
KTH Royal Institute of Technology,
Stockholm, Sweden
January, 2014

One Century of the Discovery of Arsenicosis in Latin America (1914–2014) –
Litter, Nicolli, Meichtry, Quici, Bundschuh, Bhattacharya & Naidu (Eds)
© 2014 Taylor & Francis Group, London, ISBN 978-1-138-00141-1

Foreword (President, USQ)

UNIVERSITY OF SOUTHERN QUEENSLAND AUSTRALIA

The University of Southern Queensland has great pleasure in co-organising the International Congress "As 2014: One Century of the Discovery of Arsenicosis in Latin America (1914–2014)".

Worldwide more than 200 million people currently suffer from either arsenic contamination of water resources or other geoenvironments. Evidence also suggests the incidence of arsenic contamination of drinking and irrigation water has also doubled in the last ten years with reports of contamination from over 75 countries. Hence, arsenic is an increasing global problem which will require global solutions. Research into the occurrence, mobility and bioavailability of arsenic in different environments including aquifers, soils, sediments as well as the food chain, will all be increasingly important.

The University of Southern Queensland provides education and research services to the local, regional and global communities. A particular focus is the conduct of applied research to support sustainable development and fulfilling lives. USQ has an emerging geochemical and groundwater research capability supported by existing internationally recognised expertise across seven key research centres. The National Centre for Engineering in Agriculture and the Australian Centre for Sustainable Catchments are both actively involved in research to utilise waste streams, mitigate the impacts of mining, and rehabilitate contaminated and degraded landscapes while the Centre for Rural and Remote Area Health conducts health related research in rural and remote communities. Much of this research is conducted in collaboration with international partners and is directed towards safeguarding environments while contributing to the sustainable utilisation of our mineral, energy and water resources to optimise the economic, social and ecological benefits for society.

I congratulate the organisers for their success and making progress with the 5th international conference of the series "Arsenic in the Environment" at the University of Buenos Aires in Argentina and and acknowledge the collaborative and cooperative efforts of the Royal Institute of Technology (KTH) in co-organising this event. I also hope that these proceedings will serve as a lasting record of our improving knowledge base to better manage groundwater and geoenvironments as well as protect communities.

Jan Thomas
Vice-Chancellor and President
The University of Southern Queensland,
Toowoomba, Australia
January, 2014

One Century of the Discovery of Arsenicosis in Latin America (1914–2014) –
Litter, Nicolli, Meichtry, Quici, Bundschuh, Bhattacharya & Naidu (Eds)

Foreword (President, CNEA)

Arsenic has been an issue of concern in last decades because of the serious incidence of the element on the human health arising from the ingestion of water with small amounts of arsenic for prolonged periods. Contamination by arsenic is a serious public health problem at global level in our planet, due to the carcinogenic and neurotoxic power of the element. Around 100 million people are at risk in Asia and 14 million people are potentially affected in Latin America, with high impact in Argentina.

Arsenic intoxication in Argentina was first mentioned in the medical literature in 1913 by Dr. Goyenechea, followed by Dr. Ayerza, who related some clinical manifestations in patients in the locality of Bell Ville (Province of Córdoba) to the consumption of groundwater with high concentrations of arsenic.

Since 2006, four events of the International Congress on "Arsenic in the Environment" were held in different countries (Mexico 2006, Spain 2008, Taiwan 2010 and Australia 2012). A wide range of topics related to the problem of arsenic have been addressed, ranging from geological distribution, health effects, advances in analytical measurements and methods for removal and mitigation of pollution by arsenic and its chemical species. The four editions of the Congress counted with the assistance of researchers and professionals from many countries worldwide, especially the most affected by the problem. Prominent speakers made very important contributions to the advance of the knowledge on the arsenic state of art.

In the 4th International Congress of Arsenic in the Environment, carried out in Australia in May 2012, Argentina has been proposed and elected to host the following Congress in 2014. For this reason, the organization of the 5th International Congress in the Latin American region is considered relevant, with the title "As2014, One Century of the Discovery of Arsenicosis in America (1914–2014)". This new edition of the Congress intends to invite and integrate researchers of a wide age range, involved in different study areas in an open forum, with the aim to strengthen relations between academia, industry, research laboratories, government agencies and the private sector. An optimal atmosphere allows sharing the interchange of knowledge, discoveries and discussions about the problem of arsenic in the environment. Therefore, and without doubts, the organization of As2014 has relevant national and regional positive consequences.

The National Atomic Energy Commission (CNEA) is proud to organize this event, which is in close agreement with the basic principles of environmental policy of the Institution, which promotes a responsible attitude in the care of the environment, the conservation of the natural resources and the prevention of the pollution for CNEA workers and for the whole society. We congratulate the As2014 Local Organizing Committee for the almost 300 contributions from more than 1000 contributors belonging to different countries of America, Asia, Africa, Europe and Oceania.

Norma Boero
President
National Atomic Energy Commission (CNEA),
Buenos Aires, Argentina
February, 2014

One Century of the Discovery of Arsenicosis in Latin America (1914–2014) –
Litter, Nicolli, Meichtry, Quici, Bundschuh, Bhattacharya & Naidu (Eds)
© 2014 Taylor & Francis Group, London, ISBN 978-1-138-00141-1

Editors' foreword

Arsenic is an element that has attracted people's attention since ancient times. It used to be considered as a powerful poison, and it is now in the news as a result of the largest environmental calamity in Asia. This has led to the rediscovery of arsenic despite it always exists in the environment as a "silent" toxin in the world. Every year, new areas with arsenic in environments exceeding the maximal contamination levels set by international organizations are identified.

Arsenic is a natural or anthropogenic contaminant in areas where human subsistence is at risk. It comprises a large number of chemical species with a wide variation in toxicity. Arsenic induces non-specific pathological effects which are difficult to reverse and result in serious social impact. It is thus essential to carry out cross-border and interdisciplinary studies. Scientific interest in arsenic has increased remarkably since the last century. By using the ISI Web of Knowledge as a search tool, there were 40,620 published articles on arsenic in its database for the period 1950–2013. Among them, 25,498 have been published in the 21st century. The increase in scientific publications has been spectacular in areas such as environmental sciences, food, metabolism, health and remediation. Arsenic is an old toxin offering unparalleled opportunities for research.

The First International Congress on *Arsenic in the Environment* (As 2006), was held in Mexico City in 2006 with the theme of "*Natural Arsenic in Groundwater of Latin America*". Following the success of this conference, the Congress has been organized biennially, continuing with the theme of the congress held in Mexico and also including toxicology, health effects, medical aspects, social sciences and policy issues with rotating venues in different continents. The following International Congress on Arsenic in the Environment, As 2008 was held in Valencia, Spain (May 2008) with the theme of "Arsenic from Nature to Humans", As 2010 in Tainan, Taiwan (May 2010), with the theme of "Arsenic in Geosphere and Human Diseases" and As 2012 in Cairns, Australia (July, 2012) with a theme "Understanding the Geological and Medical Interface of Arsenic". All these congresses have provided participants with an up-to-date, global overview of the arsenic research with multi- and interdisciplinary perspectives.

The forthcoming fifth International Congress on Arsenic in the Environment (As 2014) in Buenos Aires, the Capital of Argentina between 11–16 May, 2014 provides yet another exciting opportunity to the scientists to present their state of the art research results both on platform and posters, exchange ideas, discover and discuss common research interests.. The event in Argentina is also significant, since 2014 marks the completion of a century of the discovery of problems related to groundwater arsenic contamination and the related arsenicosis disease in different parts of the country. Besides the academic interests, the scientific community involved in arsenic research needs to find the practical solutions to fullfil the societal needs for the prevention or reduction of arsenic exposure and its toxic effects in millions of exposed people throughout the world.

We have received a large number of over 300 extended abstracts which were submitted from researchers, health workers, technologists, students, legislators, government officials to the 3rd international congress *Arsenic in the Environment* with the theme *Arsenic in Geosphere and Human Diseases.*

The topics to be covered in the Congress As 2014 have been grouped under the five general thematic areas:

Theme 1: Arsenic in environmental matrices (air, water and soil)
Theme 2: Arsenic in food
Theme 3: Arsenic and health
Theme 4: Removal technologies
Theme 5: Mitigation management and policy

We thank the members of the international scientific committee, for their dedicated efforts on reviewing the extended abstracts. Further, we thank the sponsors of the Congress from Argentina: the National Atomic Energy Commission (CNEA), the National Research and Technology Council (CONICET), The National University of San Martín (UNSAM), The National University of Buenos Aires (UBA), The Argentine Foundation of Nanotechnology, Aguas Santafesinas S.A., the Secretary of Water from Santiago del Estero Province, the Secretary of Environment and Sustainable Development of Argentina, AIDIS-Argentina, ProH2O GEH Wasserchemie, Germany, SATIA, and some other national and international institutions, for the successful organization of the event. We are extremely grateful to the CRC-CARE (University of South Australia), The DN Guha Mazumder Foundation, Kolkata, India, Third World Academy of Sciences (TWAS) and the Organisation for the Prohibition of Chemical Weapons (OPCW-decision pending) for their generous support.

The International Organizers Prosun Bhattacharya and Jochen Bundschuh would like to thank the Local Organizing Committee for the support to organize the 5th International International Congress, Arsenic in the Environment (As 2014), the Swedish International Development Cooperation Agency (Sida) for the support within the framework of the Sustainable Arsenic Mitigation—SASMIT (Sida Contribution 73000854) and the KTH Royal Institute of Technology for supporting the activities of the KTH-International Groundwater Arsenic Research Group at the Department of Sustainable Development, Environmental Sciences and Engineering, Stockholm on global networking and advocacy for sustainability of safe drinking water supplies.

Jochen Bundschuh would thank the International Soceity of Groundwater for Sustainable Development (ISGSD) and the University of Southern Queensland, Australia for participation and final editorial work for the volume.

We would like to acknowledge also Mrs. María Cambón, for her assistance during the compilation of the abstracts published in this volume.

Lastly, the editors thank Janjaap Blom and Lukas Goosen of the CRC Press/Taylor and Francis (A.A. Balkema) Publishers, The Netherlands for their patience and skill for the final production of this volume.

Marta I. Litter
Hugo B. Nicolli
Martin Meichtry
Natalia Quici
Jochen Bundschuh
Prosun Bhattacharya
Ravi Naidu
(*Editors*)

One Century of the Discovery of Arsenicosis in Latin America (1914–2014) – Litter, Nicolli, Meichtry, Quici, Bundschuh, Bhattacharya & Naidu (Eds)

List of contributors

Abate, S.: *Fundación Barrera Zoofitosanitaria Patagónica (Funbapa), Viedma, Rio Negro, Argentina; Universidad Nacional de Rio Negro, Argentina*
Abimbola, A.F.: *Department of Geology, University of Ibadan, Ibadan, Nigeria*
Abou Hamdan, W.: *Thermochemistry Laboratory, Department of Chemistry, University of Surrey, UK*
Abrahão, W.A.P.: *Soil Department, Federal University of Viçosa, Viçosa, MG, Brazil; National Institute of Science and Technology INCT-Acqua, Belo Horizonte, MG, Brazil*
Abreu, C.A.: *Instituto Agronômico (IAC), Campinas (SP), Brasil*
Acarapi, J.: *Centro de Investigaciones del Hombre en el Desierto, CIHDE, Arica, Chile*
Achear, H.: *Servicio Geológico Minero Argentino, Instituto de Tecnología Minera, Argentina*
Agostini, E.: *Departamento de Biología Molecular, FCEFQyN, Universidad Nacional de Río Cuarto, Río Cuarto, Córdoba, Argentina*
Aguilar Madrid, G.: *Unidad de Investigación y Salud en el Trabajo, Instituto Mexicano del Seguro Social, México DF, México*
Aguilera-Alvarado, A.F.: *Departamento de Ingeniería Química, División de Ciencias Naturales y Exactas, Campus Guanajuato, Universidad de Guanajuato, Guanajuato, México*
Aguirre, M.E.: *Departamento de Química, Facultad de Ciencias Exactas y Naturales, Universidad Nacional de Mar del Plata, Provincia de Buenos Aires, Argentina*
Ahmad, A.: *Hoofd Ingenieursbureau, Brabant Water, BC 's-Hertogenbosch, The Netherlands; KTH-International Groundwater Arsenic Research Group, Department of Sustainable Development, Environmental Science and Engineering, KTH Royal Institute of Technology, Stockholm, Sweden*
Ahmad, S.A.: *Department of Occupational and Environmental Health, Bangladesh*
Ahmed, A.: *U-Chicago Research Bangladesh, Ltd., Dhaka, Bangladesh; Columbia University Arsenic Research Project, Dhaka, Bangladesh*
Ahmed, K.M.: *Department of Geology, University of Dhaka, Curzon Hall Campus, Dhaka, Bangladesh*
Ahmed, S.: *Institute of Environmental Medicine, Karolinska Institutet, Stockholm, Sweden; International Center for Diarrhoeal Disease in Bangladesh, Dhaka, Bangladesh*
Ahsan, H.: *The University of Chicago, Chicago, IL, USA*
Ahuja, S.: *Ahuja Consulting, Rutledge Court, Calabash, NC, USA*
Airasca, A.: *Universidad Tecnológica Nacional, Facultad Regional Bahía Blanca, Provincia de Buenos Aires, Argentina*
Aitken, J.: *School of Chemistry, The University of Sydney, New South Wales, Australia*
Ajees, A.A.: *Herbert Wertheim College of Medicine, Florida International University, Miami, USA*
Al Hakawati, N.: *Thermochemistry Laboratory, Department of Chemistry, University of Surrey, UK; Department of Chemistry, Lebanese American University, Lebanon*
Al Rawahi, W.A.P.: *ICP-MS Facility, Department of Chemistry, University of Surrey, Guildford, UK*
Alam, A.M.S.: *Department of Chemistry, University of Dhaka, Dhaka, Bangladesh*
Alaniz, E.: *Carrera de Gestión Ambiental, Universidad Blas Pascal, Córdoba, Argentina*
Alarcón-Herrera, M.T.P.: *Advanced Materials Research Center (CIMAV), Chihuahua, Mexico*
Alauddin, M.: *Department of Chemistry, Wagner College, Staten Island, New York, USA*
Alava, P.: *National Exposure Research Laboratory, NERL, RTP, NC, USA*
Alegre, J.: *Laboratorio de Investigación y Desarrollo, UNCAUS, P.R.. Saenz Peña, Chaco, Argentina*
Alfaro-Barbosa, J.M.: *Facultad de Ciencias Químicas, Universidad Autónoma de Nuevo León, Nuevo León, México*
Ali, M.A.: *Faculty of Animal Husbandry, Bangladesh Agricultural University, Mymensingh, Bangladesh*
Ali, Md.: *Mahavir Cancer Institute and Research Centre, Patna, Bihar, India*
Alimohammadi, M.: *Center for Solid Waste Research, Institute for Environmental Research, Tehran University of Medical Sciences, Tehran, Iran*

Alonso, D.L.: *Universidad Nacional de Colombia, Bogotá, Colombia*
Alshana, U.: *Gazi University, Ankara, Turkey*
Altun, B.: *Gazi University, Ankara, Turkey*
Alvarez Gonçalvez, C.V.: *Instituto de Investigaciones en Producción Animal (INPA), CONICET, Facultad de Ciencias Veterinarias, Universidad de Buenos Aires, Buenos Aires, Argentina*
Álvarez, M. del P.: *CONICET, Argentina*
Álvarez, P.: *Instituto Nacional de Tecnología Industrial (INTI), San Martín, Buenos Aires, Argentina*
Alves, M.: *Centro de Desenvolvimento Sustentável, Universidade de Brasília (CDS/UnB), Brazil; Community of Practice in Ecohealth, Latin America and Caribbean (CoPEH-LAC), Brasília, DF, Brazil*
Amaral, V.P.: *Bolsista Iniciação Científica CDTN/CNEN, Belo Horizonte, MG, Brazil*
Amaro, A.S.: *CONICYT Regional, CIDERH, Water Resources Research and Development Center, University Arturo Prat, Iquique, Chile*
Amaro, S.: *Center of Research and Development of Water Resources, Faculty of Health Sciences, University Arturo Prat, Tarapacá, Chile*
Ameer, S.S.: *Department of Laboratory Medicine, Section of Occupational and Environmental Medicine, Lund University, Lund, Sweden*
Amini, M.: *Eawag, Swiss Federal Institute of Aquatic Science and Technology, Dübendorf, Switzerland*
Amiotti, N.: *Departamento de Agronomía, UNS, Argentina; CERZOS-CONICET, Argentina*
Ancelet, T.: *GNS Science, Lower Hutt, New Zealand*
Andrade Figueiredo, J.: *PPGeoc, Pernambuco Federal University, Brazil*
Angeles, G.: *Departamento de Geografia, Universidad Nacional del Sur, Provincia de Buenos Aires, Argentina*
Annaduzzaman, M.: *KTH-International Groundwater Arsenic Research Group, Department of Sustainable Development, Environmental Science and Engineering, KTH Royal Institute of Technology, Stockholm, Sweden*
Ansone, L.: *Department of Environmental Science, University of Latvia, Riga, Latvia*
Anticó, E.: *Department of Química, University of Girona, Girona, Spain*
Aparicio, J.: *Coordinación de Tecnología Hidrológica, Instituto Mexicano de Tecnología del Agua, IMTA, Jiutepec, Morelos, México*
Araújo, A.S.: *Kinross Gold Corporation, Paracatu, MG, Brazil*
Araya, J.A.: *Instituto Tecnológico de Costa Rica, Cartago, Costa Rica*
Arbelaez, J.: *Dow Water and Process Solutions, Saudi Arabia*
Arcagni, M.: *Laboratorio de Análisis por Activación Neutrónica, Centro Atómico Bariloche, CNEA, Bariloche, Argentina*
Arce, L.: *Life Quality Unit, Defensoría de los Habitantes, Costa Rica, USA*
Arenas, M.J.: *Centro de Investigaciones del Hombre en el Desierto, CIHDE, Arica, Chile*
Armas, A.: *Cooperativa Obras Sanitarias de Venado Tuerto, Venado Tuerto, Santa Fe, Argentina*
Armendariz, A.L.: *Departamento de Biología Molecular, FCEFQyN, Universidad Nacional de Río Cuarto, Río Cuarto, Córdoba, Argentina*
Armienta, M.A.: *Instituto de Geofísica, Universidad Nacional Autónoma de México, México*
Arreola Mendoza, L.: *Departamento de Biociencias e Ingeniería CIIEMAD-IPN, México DF, México*
Arribére, M.A.: *Laboratorio de Análisis por Activación Neutrónica, Centro Atómico Bariloche, CNEA, Bariloche, Argentina*
Ascari, J.: *Bolsista de Pós-Doutorado CDTN/CNEN, Belo Horizonte, MG, Brazil*
Asgari, A.R.: *School of Public Health, Tehran University of Medical Sciences, Tehran, Iran; Center for Solid Waste Research, Institute for Environmental Research, Tehran University of Medical Sciences, Tehran, Iran*
Asik, E.: *Gazi University, Ankara, Turkey*
Assis, I.R.: *Soil Department, Federal University of Viçosa, Viçosa, MG, Brazil*
Atabey, E.: *General Directorate of Mineral Research and Exploration, Ankara, Turkey*
Ataman, O.Y.: *Middle East Technical University, Ankara, Turkey*
Auge, M.: *Buenos Aires, Argentina*
Auler, L.M.L.A.: *Centro de Desenvolvimento da Tecnologia Nuclear (CDTN/CNEN), Belo Horizonte, MG, Brazil*
Avalos, C.: *MIT Sloan School of Management, USA*
Avalos-Borja, M.: *Centro de Nanociencias y Nanotecnología-UNAM, Ensenada, IPICyT—Instituto Potosino de Investigación Científica y Tecnológica, División de Materiales Avanzados, San Luis Potosí, SLP, México*
Avigliano, E: *Instituto de Investigaciones en Producción Animal (INPA-CONICET-UBA), Facultad de Ciencias Veterinarias, Universidad de Buenos Aires, Buenos Aires, Argentina*

Ayllon-Vergara, J.C.: *Hospital Español, México DF, México*
Ayora Ibañez, C.: *Instituto de Diagnóstico Ambiental y Estudios del Agua, CSIC, Barcelona, Spain*
Azevedo, A.A.: *Universidade Federal de Viçosa, Viçosa, MG, Brazil*
Babu, Y.R.: *CSIR-National Geophysical Research Institute, Uppal Road, Hyderabad, India*
Baeza Reyes, A.: *Universidad Nacional Autónoma de México, México*
Baeza Terrazas, F.: *Colegio de Médicos Cirujanos y Homeópatas del Estado de Chihuahua, A.C., Nayarit, México*
Bahr, C.: *GEH Wasserchemie GmbH & Co., KG, Osnabrück, Germany*
Baisch, P.: *Federal University of Rio Grande, Brazil*
Baker, A.J.M.: *School of Botany, The University of Melbourne, Australia*
Balaji, T.: *CSIR-National Geophysical Research Institute, Uppal Road, Hyderabad, India*
Balestrasse, K.: *Instituto de Investigaciones en Biociencias Agrícolas y Ambientales (INBA, CONICET/UBA), Facultad de Agronomía, Universidad de Buenos Aires, Buenos Aires, Argentina*
Ballinas, M.L.: *Facultad de Ciencias Químicas, Universidad Autónoma de Chihuahua, Chihuahua, México*
Balverdi, C.: *LABTRA, Facultad de Bioq., Química y Farmacia, UNT, Tucumán, Argentina*
Balverdi, P.: *Laboratorio de Análisis de Trazas (LABTRA), Facultad de Bioq., Química y Farmacia, UNT, Tucumán, Argentina*
Bao, P.: *State Key Laboratory of Urban and Regional Ecology, Research Center for Eco-Environmental Sciences, Chinese Academy of Sciences, Beijing, China*
Barbosa Jr., F.: *Faculdade de Ciências Farmacêuticas de Ribeirão Preto, Universidade de São Paulo, Ribeirao Preto, SP, Brazil*
Barbosa, F.A.R.: *Laboratório de Limnologia Ecotoxicologia e Ecologia Aquática, Universidade Federal de Minas Gerias, Brazil*
Barcelos, G.R.M.: *Faculdade de Ciências Farmacêuticas de Ribeirão Preto, Universidade de São Paulo, São Paulo, Brazil*
Barg, G.: *Faculty of Psychology, Catholic University of Uruguay, Uruguay*
Barone, F.: *Facultad de Cs. Exactas y Naturales, Universidad de Buenos Aires, Argentina*
Barrera, R.: *Departamento de Química, Facultad de Ciencias, Universidad Nacional de Colombia, Bogotá, Colombia*
Barrera-Hernández, A.: *Centro de Investigación y de Estudios Avanzados del IPN, Departamento de Toxicología, México DF, México*
Barrionuevo, L.: *AISA IONIC S.A., Buenos Aires, Argentina*
Basanta, B.: *Facultad de Cs. Exactas y Naturales, Universidad de Buenos Aires, Argentina*
Bassani, R.: *JLA Argentina S.A., Bv. Italia, Gral. Cabrera, Córdoba, Argentina*
Basualdo, J.P.: *Laboratorio de Biodiversidad, Ultraestructura y Ecofisiología de Microalgas, Instituto de Biodiversidad y Biología Experimental Aplicada, CONICET, Universidad de Buenos Aires, Buenos Aires, Argentina*
Batista, B.L.: *Universidade Federal do ABC, Santo André, SP, Brazil; Faculdade de Ciências Farmacêuticas de Ribeirão Preto, Universidade de São Paulo, Brazil*
Battaglia-Brunet, F.: *BRGM, ISTO, UMR 7327, Orléans, France*
Bauda, P.: *LIEC, CNRS-UMR 7360, Université de Lorraine, Metz, France*
Bayzidur R.: *The School of Public Health and Community Medicine, Faculty of Medicine, The University of New South Wales, Sydney, NSW, Australia*
Befani, R.: *Facultad de Ciencias Agropecuarias, Universidad Nacional de Entre Ríos, Paraná, ER, Argentina*
Beldoménico, H.: *Programa de Investigación y Análisis de Residuos y Contaminantes Químicos PRINARC Facultad de Ingeniería Química, Universidad Nacional del Litoral, Santa Fe, Argentina*
Benjumea-Flórez, C.F.: *Santo Tomas University, Bucaramanga, Colombia*
Beone, G.M.: *Institute of Agricultural and Environmental Chemistry, Università Cattolica del Sacro Cuore, Piacenza, Italy*
Berardozzi, E.: *Departamento de Hidráulica, Facultad de Ingeniería, UNLP, Argentina*
Berg, M.: *Eawag, Swiss Federal Institute of Aquatic Science and Technology, Dübendorf, Switzerland*
Bernardos, J.: *Faculty of Exact and Natural Sciences, UNLPam, Argentina*
Bertolino, L.C.: *Centro de Tecnologia Mineral (CETEM), Cidade Universitária, Rio de Janeiro, Brasil*
Bhattacharya, P.: *KTH-International Groundwater Arsenic Research Group, Department of Sustainable Development, Environmental Science and Engineering, KTH Royal Institute of Technology, Stockholm, Sweden*
Bhowmick, S. *Department of Chemistry, University of Kalyani, Kalyani, West Bengal, India; Faculty of Sciences, University of Girona, Campus de Montilivi, Girona, Spain*

Bia, G.: *Centro de Investigaciones en Ciencias de la Tierra (CICTERRA), CONICET and Universidad Nacional de Córdoba, Argentina*

Bidone, E.D.: *Departamento de Geoquímica, Instituto de Química, Universidade Federal Fluminense (UFF), Brasil; Curso de Pós-graduação em Geoquímica Ambiental, Universidade Federal Fluminense (UFF), Brazil*

Bilir, N.: *Hacettepe University, Ankara, Turkey*

Billib, M.: *Institute of Water Resources Management, Hydrology and Agricultural Hydraulic Engineering—WAWI, Leibniz University of Hannover, Germany*

Bissacot, L.C.G.: *Yamana Gold Corporation, São Paulo, Brazil*

Biswas, A.: *DNGM Research Foundation, Kolkata, India*

Biswas, N.: *Civil and Environmental Engineering, University of Windsor, Windsor, Ontario, Canada*

Blanc, G.: *UMR CNRS, EPOC, University of Bordeaux, France*

Blanco Coariti, E.: *Instituto de Investigaciones Químicas, Universidad Mayor de San Andrés, La Paz, Bolivia*

Blanco, M. del C.: *Departamento de Agronomía, Universidad Nacional del Sur, Bahía Blanca, Argentina*

Blanco, R.: *Environmental Management Unit, Caja Costarricense de Seguro Social, Universidad de Costa Rica, Costa Rica*

Blanes, P.S.: *Área Química General, FCBioyF, IQUIR-CONICET, Universidad Nacional de Rosario, Argentina*

Boeykens, S: *Laboratorio de Química de Sistemas Heterogéneos (LaQuiSiHe), Departamento de Química, Universidad de Buenos Aires, Buenos Aires, Argentina; Facultad de Ingeniería, Universidad de Buenos Aires, Buenos Aires, Argentina*

Boglione, R.M.: *Facultad Regional Rafaela, Universidad Tecnológica Nacional, Rafaela, Santa Fe, Argentina*

Boischio, A.: *Sustainable Development and Health Equity, Pan American Health Organization, USA*

Bonaventura, M.M.: *Instituto de Biología y Medicina Experimental (IByME, CONICET), CABA, Buenos Aires, Argentina; Facultad de Ingeniería, Universidad de Buenos Aires (UBA), CABA, Argentina*

Bongiovanni, G.A.: *CONICET-Universidad Nacional del Comahue (IDEPA), Neuquén-Cinco Saltos, Argentina; IDEPA (Multidisciplinary Institute of Scientific Research and Development from North Patagonia), (CONICET-CCT COMAHUE, National University of Comahue), Neuquén y Cinco Saltos, Argentina*

Bonilla, C.A.: *Department of Hydraulic and Environmental Engineering, Pontificia Universidad Católica de Chile, Santiago, Chile*

Bonis, M.L.: *Laboratoire Sols et Environnement (LSE), Université de Lorraine (INPL (ENSAIA)/INRA), Vandoeuvre-lès-Nancy Cedex, France*

Bonomi, H.: *Facultad de Cs. Exactas y Naturales, Universidad de Buenos Aires, Argentina*

Boochs, P.W.: *Institute of Water Resources Management, Hydrology and Agricultural Hydraulic Engineering—WAWI, Leibniz University of Hannover, Germany*

Borgnino, L.: *Centro de Investigaciones en Ciencias de la Tierra (CICTERRA), CONICET, and FCEFyN Universidad Nacional de Córdoba, Córdoba, Argentina*

Borneo, G.: *AISA IONIC S.A., Buenos Aires, Argentina*

Boschetti, G.: *Facultad de Ciencias Agropecuarias, Universidad Nacional de Entre Ríos, Paraná, ER, Argentina*

Bose, N.: *A.N. College, Patna, India*

Bottini, R.: *Instituto de Biología Agrícola de Mendoza (IBAM), CONICET, Mendoza, Argentina*

Boubinova, R.: *Institute of Chemical Technology, Prague, Czech Republic*

Boudaghi Malidareh, H.: *School of Public Health, Tehran University of Medical Sciences, Tehran, Iran*

Boudaghi Malidareh, P.: *Antibiotic Sazi IRAN Co., (ASICO), Mazandaran, Iran; Islamic Azad University, Ghaemshahr, Iran*

Boudaghi Malidareh, Z.: *Mazandaran Province Remedy and Social Security, Iran*

Bourguignon, N.S.: *Instituto de Biología y Medicina Experimental (IByME, CONICET), CABA, Argentina*

Bovi Mitre, G.: *Grupo INQA, Facultad de Ciencias Agrarias, Universidad Nacional de Jujuy, Argentina; Grupo INQA, Facultad de Ciencias Agrarias, San Salvador de Jujuy, Argentina*

Bradham, K.: *National Exposure Research Laboratory, NERL, RTP, NC, USA*

Braeuer, S.: *Institute of Chemistry, Analytical Chemistry, University of Graz, Austria*

Brandi, R.J.: *INTEC (UNL-CONICET), Güemes, Santa Fe, Argentina; FICH (UNL), Ciudad Universitaria, Santa Fe, Argentina*

Brendova, K.: *Department of Agroenvironmental Chemistry and Plant Nutrition, Czech University of Life Sciences, Prague, Czech Republic*

Bretzler, A.: *Eawag, Swiss Federal Institute of Aquatic Science and Technology, Dübendorf, Switzerland*
Bril, H.: *University of Limoges, GRESE, Limoges Cedex, France*
Broberg, K.: *Department of Laboratory Medicine, Section of Occupational and Environmental Medicine, Lund University, Lund, Sweden; Institute of Environmental Medicine, Karolinska Institutet, Stockholm, Sweden*
Bruno, M.: *Departamento de Saneamiento Básico, Dirección General de Salud Ambiental, Sistema Provincial de Salud, S.M. de Tucumán, Argentina*
Brusa, L.: *Programa de Investigación y Análisis de Residuos y Contaminantes Químicos, PRINARC, Facultad de Ingeniería Química, Universidad Nacional del Litoral, Santa Fe, Argentina*
Brusa, M.: *Departamento de Química, Facultad de Ciencias Exactas y Naturales, Universidad Nacional de Mar del Plata, Provincia de Buenos Aires, Argentina*
Buchhamer, E.E.: *Departamento de Química Analítica, UNCAus, Pcia, R. Sáenz Peña, Chaco, Argentina*
Buitrón, E.B.: *Ministerio de Obras y Servicios Públicos, Ministerio de Salud & Secretaría de Recursos Hídricos, La Pampa, Argentina*
Bundschuh, J.: *Faculty of Health, Engineering and Sciences and National Centre for Agriculture, University of Southern Queensland, Toowoomba, Australia; KTH-International Groundwater Arsenic Research Group, Division of Land and Water Resources Engineering, Department of Sustainable Development, Environmental Science and Engineering, Royal Institute of Technology (KTH), Teknikringen, Stockholm, Sweden*
Burgaz, S.: *Gazi University, Ankara, Turkey*
Burgess, W.G.: *Department of Earth Sciences, University College London, London, UK*
Burguete-García, A.I.: *Instituto Nacional de Salud Pública, Cuernavaca, Morelos, México*
Burló, F.: *Departamento de Tecnología Agroalimentaria, Universidad Miguel Hernández, Ctra. Beniel, Orihuela, Alicante, Spain*
Buse, J.B.: *University of North Carolina at Chapel Hill, North Carolina, USA*
Bustingorri, C.: *Instituto de Investigaciones en Biociencias Agrícolas y Ambientales (INBA, CONICET/UBA), Facultad de Agronomía, Universidad de Buenos Aires, Buenos Aires, Argentina*
Buzek, F.: *Czech Geological Survey, Prague, Czech Republic*
Cabanillas-Vidosa, I.: *JLA Argentina S.A., Bv. Italia, Gral. Cabrera, Córdoba, Argentina*
Caille, N.: *Rothamsted Research, Harpenden, Hertfordshire, UK*
Cakmak Demircigil, G.: *Gazi University, Ankara, Turkey*
Camerotto Andreani, P.A.: *Instituto de Investigación e Ingeniería Ambiental, UNSAM, Buenos Aires, Argentina; Departamento de Ciencias Geológicas, Facultad de Ciencias Exactas y Naturales, UBA, CABA, Argentina*
Campaña, D.H.: *Facultad Regional Bahía Blanca, Universidad Tecnológica Nacional, Provincia de Buenos Aires, Argentina*
Campbell, L.M.: *Environmental Science, Saint Mary's University, Canada*
Campins, M.: *National Technological University-Mar del Plata (UTN-MDP), Buenos Aires, Argentina*
Campos, N.V.: *Universidade Federal de Viçosa, Viçosa, MG, Brazil*
Cano-Rodriguez, I.: *Departamento de Ingeniería Química, División de Ciencias Naturales y Exactas, Campus Guanajuato, Universidad de Guanajuato, Guanajuato, México*
Cano-Rodriguez, S.: *Departamento de Química, Facultad de Ciencias, Universidad Nacional de Colombia, Bogotá, Colombia*
Cantoni, M.: *Agronomic Institute of Campinas (IAC), Campinas, São Paulo, Brazil*
Cao, Y.S.: *School of Water Resources and Environment, China University of Geosciences, Beijing, China*
Carbonell Barrachina, A.A.: *Departamento de Tecnología Agroalimentaria, Universidad Miguel Hernández, Ctra. Beniel, Orihuela, Alicante, Spain*
Carlotto, N.: *Facultad de Cs. Exactas y Naturales, Universidad de Buenos Aires, Argentina*
Carneiro, M.C.: *Centro de Tecnologia Mineral (CETEM), Cidade Universitária, Rio de Janeiro, Brasil*
Carrasco, C.: *Center of Research and Development of Water Resources, Faculty of Health Sciences, University Arturo Prat, Tarapacá, Chile*
Carrier, M.: *CNRS, University of Bordeaux, ICMCB, IPB-ENSCBP, Pessac, France*
Carro Perez, M.E.: *Universidad Nacional de Córdoba, Córdoba, CONICET, Argentina*
Cartwright, J.R.: *Thermochemistry Laboratory, Department of Chemistry, University of Surrey, UK*
Casiot, C.: *HSM, CNRS-UMR 5569, Université de Montpellier, Montpellier, France*
Castaño-Iglesias, C.: *Universidad Miguel Hernández, Departamento de Tecnología Agroalimentaria, Ctra. Beniel, Orihuela, Alicante, Spain*

Castellanos, W.: *Departamento de Saneamiento Básico, Dirección General de Salud Ambiental, Sistema Provincial de Salud, S.M. de Tucumán, Argentina*

Castilhos, Z.C.: *Center for Mineral Technology, CETEM, Ministry of Science, Technology and Innovation, RJ, Brazil; Instituto de Química, Departamento de Geoquímica, Universidade Federal Fluminense (UFF), Centro, Niterói, RJ, Brasil*

Castillo, E.: *Universidad Nacional de Colombia, Bogotá, Colombia*

Castillo, F.: *IPICyT—Instituto Potosino de Investigación Científica y Tecnológica, División de Materiales Avanzados, Camino a la Presa San José, SLP, México*

Castro Grijalba, A.: *Laboratory of Analytical Chemistry for Research and Development (QUIANID), ICB-UNCuyo, Mendoza, Argentina*

Cattaneo, F.: *Instituto Nacional de Tecnología Agropecuaria, Estación Experimental Agropecuaria Concepción del Uruguay (NTA EEA), Provincia de Entre Ríos, Argentina*

Cattani, I.: *Institute of Agricultural and Environmental Chemistry, Università Cattolica del Sacro Cuore, Piacenza, Italy*

Cebrián, M.E.: *Centro de Investigación y Estudios Avanzados, Departamento de Toxicología, México DF, México; College of Pharmacy, University of Arizona, Tucson, AZ, USA*

Celebi, C.R.: *Akpol Medical Center, Ankara, Turkey*

Cesar, R.G.: *Departamento de Geoquímica, Instituto de Química, Universidade Federal Fluminense (UFF), Centro, Niterói, RJ, Brasil*

Chakrabarti, R.: *Adelphi, Berlin, Germany*

Chandrasekhar, A.K.: *Department of Earth Sciences, Indian Institute of Technology Bombay, Mumbai, India*

Chandrasekharam, D.: *Department of Earth Sciences, Indian Institute of Technology Bombay, Mumbai, India*

Chatterjee, D.: *Department of Chemistry, University of Kalyani, Kalyani, West Bengal, India*

Chen, B.: *P S Analytical Ltd., Orpington, Kent, UK*

Chen, C.-J.: *Genomics Research Center, Academia Sinica, Taipei, Taiwan*

Chen, Y.: *New York University, New York, USA*

Chen, Z.: *Research Center for Eco-Environmental Sciences, Chinese Academy of Sciences, Beijing, China; Lancaster Environment Centre, Lancaster University, Lancaster, UK*

Chen, Z.-S.: *Department of Agricultural Chemistry, National Taiwan University, Taipei, Taiwan*

Cheng, W.F.: *Graduate Institute of Oncology, National Taiwan University, Taipei, Taiwan*

Cheng, X.: *New York University, New York, USA*

Cherchiet, L.: *Gerencia Química, Comisión Nacional de Energía Atómica, San Martín, Provincia de Buenos Aires, Argentina*

Chiocchio, V.M.: *Cátedra de Microbiología Agrícola, Facultad de Agronomía, UBA, Buenos Aires, Argentina; Instituto de Investigaciones en Biociencias Agrícolas y Ambientales, INBA, (CONICET-UBA), Ciudad Autónoma de Buenos Aires, Argentina*

Chiou, Hung-Yi: *School of Public Health, Taipei Medical University, Taipei, Taiwan*

Chiu, H.W.: *Department of Environmental and Occupational Health, National Cheng Kung University, Medical College, Tainan, Taiwan*

Chong, G.: *Departamento de Ciencias Geológicas, Universidad Católica del Norte, Antofagasta, Chile*

Choque, D.: *Grupo INQA, Facultad de Ciencias Agrarias, Universidad Nacional de Jujuy, Argentina*

Choudhury, R.: *Department of Civil Engineering, Indian Institute of Technology Guwahati, Assam, India*

Ciminelli, V.S.T.: *Departamento de Engenharia Metalúrgica e de Materiais, Escola de Engenharia, Universidade Federal de Minas Gerais, Belo Horizonte, Brazil; National Institutes of Science and Technology Acqua, Universidade Federal de Minas Gerais—UFMG, Belo Horizonte, Brazil*

Cirello, D.: *Servicio Geológico Minero Argentino, Instituto de Tecnología Minera, Argentina*

Coelhan, M.: *Research Center Weihenstephan for Brewing and Food Quality, Technische Universität München, Freising, Germany*

Coelho, R.M.: *Instituto Agronômico (IAC), Campinas (SP), Brasil*

Cohen, M.: *Toxicología Ambiental, Departamento de Ciencias Marinas, Instituto de Investigaciones Marinas y Costeras, Facultad de Ciencias Exactas y Naturales, Universidad Nacional de Mar del Plata, Mar del Plata, Argentina*

Colazo, J.: *Instituto Nacional de Tecnología Agropecuaria, Estación Experimental Agropecuaria Concepción del Uruguay (NTA EEA), Provincia de Entre Ríos, Argentina*

Concha, G.: *National Food Administration, Toxicology Division, Uppsala, Sweden*

Conforti, V.: *Facultad de Ciencias Exactas y Naturales, Universidad de Buenos Aires, Argentina; IBBEA-CONICET, Buenos Aires, Argentina*

Cordeiro Silva, G.: *INCT-Acqua, Belo Horizonte, Brazil; Departamento de Engenharia Metalúrgica e de Materiais, Escola de Engenharia, Universidade Federal de Minas Gerais, Belo Horizonte, Brazil*
Córdova-Villegas, L.G.: *Advanced Materials Research Center (CIMAV), Chihuahua, Mexico*
Corelli, M.: *JLA Argentina S.A., Bv. Italia, Gral. Cabrera, Córdoba, Argentina*
Cornejo, L.: *Laboratorio de Investigaciones Medioambientales de Zonas Áridas, LIMZA, Escuela Universitaria de Ingeniería Industrial, Informática y Sistemas, EUIIIS, Universidad de Tarapacá, Arica, Chile; Centro de Investigaciones del Hombre en el Desierto, CIHDE, Arica, Chile*
Corns, W.T.: *P S Analytical Ltd., Orpington, Kent, UK*
Cortes González, T.: *Salud Ambiental, Facultad de Medicina de la Universidad Autónoma de Coahuila, México*
Cortés, S.: *Department of Public Health, Faculty of Medicine, Pontifical Catholic University of Chile, Chile*
Cortés-Palacios, L.: *Facultad de Zootecnia y Ecología, Universidad Autónoma de Chihuahua, Chihuahua, México*
Costa, S.P.: *GEOBIOTEC, University of Aveiro, Campus Universitário de Santiago, Aveiro, Portugal*
Courtin-Nomade, A.: *University of Limoges, GRESE, Limoges Cedex, France*
Coynel, A.: *UMR CNRS, EPOC, University of Bordeaux, France*
Crapez, M.A.C.: *Curso de Pós-graduação em Biologia Marinha, Universidade Federal Fluminense (UFF), Brazil*
Cremonezzi, D.: *Universidad Nacional de Córdoba, Córdoba, Argentina*
Crubellati, R.: *Servicio Geológico Minero Argentino, Instituto de Tecnología Minera, Argentina*
Cumbal, L.: *Centro de Nanociencia y Nanotecnologia, Universidad de las Fuerzas Armadas, Quito, Ecuador*
Cunningham, A.: *York College of City University of New York, Jamaica, New York, USA*
Custo, G.: *Gerencia Química, Comisión Nacional de Energía Atómica, San Martín, Provincia de Buenos Aires, Argentina*
Cuzzio, G.: *National Technological University-Mar del Plata (UTN-MDP), Buenos Aires, Argentina*
da Silva, L.I.D.: *Centro de Tecnologia Mineral (CETEM), Cidade Universitária, Rio de Janeiro, Brasil*
Daga, R.: *Laboratorio de Análisis por Activación Neutrónica, Centro Atómico Bariloche, CNEA, Argentina; CONICET, Argentina*
Dalmaso, M.: *Faculty of Exact and Natural Sciences, UNLPam, Argentina*
Dang, T.H.: *UMR CNRS, EPOC, University of Bordeaux, France; Faculty of Chemistry, BaRia-VungTau University, Vietnam*
Danil de Namor, A.F.: *Thermochemistry Laboratory, Department of Chemistry, University of Surrey, UK; Instituto Nacional de Tecnología Industrial, Argentina*
Daraei, H.: *Kurdistan Environmental Health Research Center, Kurdistan University of Medical Sciences, Kurdistan, Iran*
Das, S.: *Department of Earth Sciences, National Cheng Kung University, Tainan, Taiwan*
Datta, S.: *Department of Geology, Kansas State University, Manhattan, KS, USA*
Davy, P.K.: *GNS Science, Lower Hutt, New Zealand*
de Albuquerque Menor, E.: *Department of Geology, Pernambuco Federal University, Brazil*
De Capitani, E.M.: *Poison Control Center, School of Medicine, State University of Campinas, Campinas, SP, Brazil*
de Carvalho, F.: *Dow Water and Process Solutions, Saudi Arabia*
de Jesus, I.M.: *Evandro Chagas Institute (IEC/MS), Belém, Pará, Brazil*
de la Peña Torres, A.: *Departamento de Ingeniería Química, División de Ciencias Naturales y Exactas, Campus Guanajuato, Universidad de Guanajuato, Guanajuato, México*
de Mello, J.W.V.: *Soils Department, Federal University of Viçosa, Viçosa, MG, Brazil; National Institute of Science and Technology, INCT-Acqua, Belo Horizonte, MG, Brazil*
De Mello, W.Z.: *Departamento de Geoquímica, Instituto de Química, Universidade Federal Fluminense (UFF), Centro, Niterói, RJ, Brasil; Environmental Geochemistry Department, Universidade Federal Fluminense, Niterói, RJ, Brazil*
de Moraes, M.L.B.: *Soils Department, Federal University of Viçosa, Viçosa, MG, Brazil; National Institute of Science and Technology, INCT-Acqua, Belo Horizonte, MG, Brazil*
De Pietri, D.E.: *Dirección Nacional de Determinantes de la Salud e Investigación, Ministerio de Salud de la Nación, Buenos Aires, Argentina; Departamento de Ingeniería Civil, Facultad Regional Buenos Aires, Universidad Tecnológica Nacional, Argentina*
de Titto, E.: *Dirección Nacional de Determinantes de la Salud e Investigación, Ministerio de Salud de la Nación, Buenos Aires, Argentina*

De Vizcaya-Ruiz, A.: *Centro de Investigación y de Estudios Avanzados del IPN, Departamento de Toxicología, México DF, México*
Deb, D.: *DNGM Research Foundation, Kolkata, India*
Del Razo, L.M.: *Centro de Investigación y de Estudios Avanzados del IPN, Departamento de Toxicología, México DF, Mexico*
Delgado-Caballero, M.R.: *Centro de Investigación en Materiales Avanzados (CIMAV), Chihuahua, México*
Demergasso, C.: *Centro de Biotecnología, Universidad Católica del Norte, Antofagasta, Chile*
Deowan, S.A.: *Karlsruhe University of Applied Sciences, Germany*
Derderián, N.: *Licenciatura en Gestión Ambiental, Universidad CAECE, Buenos Aires, Argentina*
Devin, S.: *LIEC, CNRS-UMR 7360, Université de Lorraine, Metz, France*
Dhar, R.K.: *York College of City University of New York, Jamaica, New York, USA*
Di, C.H.: *School of Medicine, Hangzhou Normal University, Hangzhou, Zhejiang, China*
Diacomanolis, V.: *National Research Centre for Environmental Toxicology (Entox), The University of Queensland, Brisbane, Queensland, Australia*
Dias, A.C.: *GEOBIOTEC, University of Aveiro, Campus Universitário de Santiago, Aveiro, Portugal*
Dias, L.E.: *Soil Department, Federal University of Viçosa, Viçosa, MG, Brazil*
Dias, L.E.: *Soil Department, Federal University of Viçosa, Viçosa, MG, Brazil*
Díaz, E.: *Facultad de Ciencias Agropecuarias, Universidad Nacional de Entre Ríos, Paraná, ER, Argentina*
Díaz, M.P.: *Escuela de Nutrición e Instituto de Investigaciones en Ciencias de la Salud, CONICET, Facultad de Ciencias Médicas, Universidad Nacional de Córdoba, Córdoba, Argentina*
Díaz, S.L.: *CONICET, Argentina*
Dietrich, S.: *Instituto de Hidrología de Llanuras, CONICET, Azul, Argentina*
Dietze, V.: *Deutscher Wetterdienst, Zentrum für Medizin-Meteorologische Forschung, Freiburg, Germany*
Dóczi, R.: *Technical University, Budapest, Hungary*
Dollard, M.A.: *LIEC, CNRS-UMR 7360, Université de Lorraine, Metz, France*
Domingo, E.J.: *Departamento de Ingeniería Civil, Facultad Regional Buenos Aires, Universidad Tecnológica Nacional, Argentina*
Domingos, L.M.: *Center for Mineral Technology, CETEM, Ministry of Science, Technology and Innovation, RJ, Brazil*
Donna, F.: *Instituto Argentino de Investigaciones de las Zonas Áridas (IADIZA-CCT CONICET), Mendoza, Argentina*
Donselaar, M.E.: *Delft University of Technology, Delft, The Netherlands*
Dopp, E.: *Institute of Hygiene and Occupational Medicine, University Hospital Essen, Essen, Germany*
Doronila, A.I.: *School of Botany, The University of Melbourne, Australia*
Dorr, F.: *Facultad de Cs. Exactas y Naturales, Universidad de Buenos Aires, Argentina*
dos Santos Afonso, M.: *INQUIMAE/DQIAQF, Facultad de Ciencias Exactas, Universidad de Buenos Aires, Buenos Aires, Argentina*
Doušová, B.: *Institute of Chemical Technology in Prague, Czech Republic*
Dragana Jovanovic, C.D.: *Institute of Public Health of Serbia "Dr Milan Jovanovic Batut", Belgrade, Serbia*
Drobná, Z.: *University of North Carolina at Chapel Hill, North Carolina, USA*
Duarte Portocarrero, E.: *Laboratorio de Salud Pública, Secretaria Distrital de Salud de Bogotá, Colombia*
Duarte, G.: *Escola de Engenharia, Departamento de Engenharia Metalúrgica e de Materiais, Universidade Federal de Minas Gerais, Belo Horizonte, Brazil*
Duarte-Sustaita, J.J.: *Universidad Juarez del Estado de Durango, Gomez Palacio, Durango, México*
Durán, A.: *Departamento de Saneamiento Básico, Dirección General de Salud Ambiental, Sistema Provincial de Salud, S.M. de Tucumán, Argentina*
Durán, E.: *Departamento de Saneamiento Básico, Dirección General de Salud Ambiental, Sistema Provincial de Salud, S.M. de Tucumán, Argentina*
Echevarria, G.: *Laboratoire Sols et Environnement (LSE), Université de Lorraine (INPL (ENSAIA)/INRA), Vandoeuvre-lès-Nancy Cedex, France*
Egler, S.: *Centre of Mineral Technology, Rio de Janeiro, Brazil*
Elorza, C.G.: *Ministerio de Salud, La Pampa, Argentina*
Engström, K.: *Department of Laboratory Medicine, Section of Occupational and Environmental Medicine, Lund University, Lund, Sweden*
Ersoz, M.: *Advanced Technology Research and Application Center, Selcuk University, Turkey*
Ertas, N.: *Gazi University, Ankara, Turkey*

Escauriaza, C.E.: *Department of Hydraulic and Environmental Engineering, Pontificia Universidad Católica de Chile, Santiago, Chile*
Escudero, L.: *Centro de Investigación Científica y Tecnológica para la Minería (CICITEM), Antofagasta, Chile*
Escudero, L.B.: *Chemistry Institute of San Luis, San Luis, Argentina*
España, C.: *Departamento de Química, Facultad de Ciencias, Universidad Nacional de Colombia, Bogotá, Colombia*
Esper, J.: *Kinross Gold Corporation, Paracatu, MG, Brazil*
Espinosa Fematt, J.A.: *Universidad Juárez del Estado de Durango, Gómez Palacio, Durango, México*
Espósito, M.: *Departamento de Agronomía, UNS, Argentina*
Etcheber, H.E.L.: *UMR CNRS, EPOC, University of Bordeaux, France*
Evangelou, E.: *Institute of Soil Mapping and Classification, Larissa, Greece*
Ewy, W.: *ProH2O S.A., Buenos Aires, Argentina*
Faial, K.C.F.: *Evandro Chagas Institute, Ministry of Health, Belém, PA, Brazil*
Faial, K.R.F.: *Evandro Chagas Institute, Ministry of Health, Belém, PA, Brazil*
Falnoga, I.: *Department of Environmental Sciences, Jožef Stefan Institute, Ljubljana, Slovenia*
Farías, S.S.: *Departamento de Química Analítica, Gerencia Química, Comisión Nacional de Energía Atómica, Buenos Aires, Argentina*
Farnfield, H.: *ICP-MS Facility, Department of Chemistry, University of Surrey, Guildford, UK*
Favas, P.J.C.: *University of Trás-os-Montes e Alto Douro, UTAD, School of Life Sciences and the Environment, Quinta de Prados, Vila Real, Portugal*
Fernandez Cirelli, A.: *Instituto de Investigaciones en Producción Animal (INPA), CONICET, Facultad de Ciencias Veterinarias, Universidad de Buenos Aires, Buenos Aires, Argentina*
Fernandez, C.: *Instituto de Ingeniería Biomédica, Universidad de Buenos Aires, Buenos Aires, Argentina*
Fernández, M.I.: *Facultad de Bioquímica, Química y Farmacia, Universidad Nacional de Tucumán, Tucumán, Argentina*
Fernández, R.G.: *Centro de Ingeniería Sanitaria, Facultad de Ciencias Exactas, Ingeniería y Agrimensura, Universidad Nacional de Rosario, Argentina*
Ferral, A.E.: *Carrera de Gestión Ambiental, Universidad Blas Pascal, Córdoba, Argentina*
Ferreccio, C.: *Department of Public Health, Faculty of Medicine, Pontifical Catholic University of Chile, Chile*
Ferreira da Silva, E.: *GEOBIOTEC, GeoBiosciences, Geotechnologies and Geoengineering Research Center, University of Aveiro, Portugal*
Ferreira da Silva, E.: *GEOBIOTEC, Universidade de Aveiro, Campus de Santiago, Aveiro, Portugal*
Ferreira, A.P.: *Fundação Oswaldo Cruz, FIOCRUZ, Ministry of Health, Rio de Janeiro, Brazil*
Ferreira, C.A.: *Pós Graduação em Ciência e Tecnologia das Radiações Minerais e Materiais, Centro de Desenvolvimento da Tecnologia Nuclear (CDTN/CNEN), Belo Horizonte, MG, Brazil*
Ferreira, M.: *Departamento de Geoquímica, Instituto de Química, Universidade Federal Fluminense (UFF), Centro, Niterói, RJ, Brasil*
Ferreira, O.: *Kinross Gold Corporation, Paracatu, MG, Brazil*
Ferrero, M.A.: *Planta Piloto de Procesos Industriales Microbiológicos (PROIMI-CONICET), Tucumán, Argentina*
Fiedler, F.: *Karlsruhe University of Applied Sciences, Germany*
Fiúza, A.: *Faculty of Engineering, CIGAR, University of Porto, Portugal*
Fletcher, T.D.: *The University of Melbourne, Melbourne, VIC, Australia*
Fontanella, M.C.: *Institute of Agricultural and Environmental Chemistry, Università Cattolica del Sacro Cuore, Piacenza, Italy*
Fontàs, C.: *Department of Química, University of Girona, Girona, Spain*
Ford, S.: *Department of Geology, Kansas State University, Manhattan, KS, USA*
Fort, F.: *Licenciatura en Gestión Ambiental, Universidad CAECE, Buenos Aires, Argentina*
Foster, S.: *Ecochemistry Laboratory, Institute for Applied Ecology, University of Canberra, Bruce, Australia*
Francesconi, K.A.: *Institute of Chemistry, Analytical Chemistry, Graz, Austria*
Francisca, F.M.: *Universidad Nacional de Córdoba, Córdoba, CONICET, Argentina*
Franco, M.W.: *Laboratório de Limnologia Ecotoxicologia e Ecologia Aquática, Universidade Federal de Minas Gerias, Brazil*
Frangie, M.S.: *Instituto Nacional de Tecnología Industrial, San Martín, Buenos Aires, Argentina*
Fry, R.: *University of North Carolina at Chapel Hill, North Carolina, USA*

Funes Pinter, M.I.: *Instituto de Biología Agrícola de Mendoza (IBAM), CONICET, Mendoza, Argentina*
Futuro, A.: *Faculty of Engineering, CIGAR, University of Porto, Portugal*
Gaiero, D.: *Centro de Investigaciones en Ciencias de la Tierra (CICTERRA), CONICET and Universidad Nacional de Córdoba, Argentina*
Gailer, J.: *Department of Chemistry, University of Calgary, Calgary, Canada*
Gait, N.del Valle: *Nuevo Hospital San Roque, Córdoba, Argentina*
Galende, M.: *INRA, UMR BIOGECO, University of Bordeaux, Talence, France; INRA Cestas, France*
Galindez, J.M.: *Departamento de Hidráulica, Facultad de Ingeniería, UNLP, Argentina*
Galzenati, V.J.: *Laboratorio de Biodiversidad, Ultraestructura y Ecofisiología de Microalgas, Instituto de Biodiversidad y Biología Experimental Aplicada, CONICET, Universidad de Buenos Aires, Buenos Aires, Argentina*
Gamiño-Arroyo, Z.: *Departamento de Ingeniería Química, División de Ciencias Naturales y Exactas, Campus Guanajuato, Universidad de Guanajuato, Guanajuato, México*
Gandolfi, A.J.: *Department of Pharmacology and Toxicology, College of Pharmacy, University of Arizona, Tucson, AZ, USA; Departamento de Toxicología, Centro de Investigación y Estudios Avanzados, México D.F., México*
Gao, X.B.: *China University of Geosciences, Wuhan, China*
Garay, F.: *INFIQC, Departamento de Físico Química, Facultad de Ciencias Químicas, Universidad Nacional de Córdoba. Pabellón Argentina, Ciudad Universitaria, Córdoba, Argentina*
Garcés Mendoza, I.D.: *Universidad Autónoma Metropolitana, Azcapotzalco, México*
García Aróstegui, J.L.: *Instituto Geológico y Minero de España, Madrid, España*
García Einschlag, F.S.: *Departamento de Química, Facultad de Ciencias Exactas, Instituto de Investigaciones Fisicoquímica Teóricas y Aplicadas(INIFTA), UNLP, Argentina*
García, J.I.: *Laboratorio de Química de Sistemas Heterogénos (LaQuíSiHe), Departamento de Química, Universidad de Buenos Aires, Argentina*
García, J.J.: *Laboratorio de Química de Sistemas Heterogéneos (LaQuiSiHe), Departamento de Química, Universidad de Buenos Aires, Argentina*
García, M.G.: *Centro de Investigaciones en Ciencias de la Tierra (CICTERRA), CONICET, and Universidad Nacional de Córdoba, Córdoba, Argentina*
Garcia-Martínez, A.: *Instituto Nacional de Salud Pública, Cuernavaca, Morelos, México*
García-Salgado, S.: *School of Civil Engineering, Technical University of Madrid, Madrid, Spain*
Garcia-Vargas, G.G.: *Universidad Juárez del Estado de Durango, Gomez Palacio, Durango, México; Facultad de Medicina, UJED, Gómez Palacio, México*
Gardon, J.: *IRD—Hydrosciences Montpellier (HSM), France*
Garrido, S.E.: *Instituto Mexicano de Tecnología del Agua, Jiutepec, Morelos, Mexico*
Gasparon, M.: *School of Earth Sciences, The University of Queensland, Brisbane, Australia; National Institutes of Science and Technology: Acqua, Universidade Federal de Minas Gerais—UFMG, Belo Horizonte, Brazil*
Gbadebo, A.M.: *Department of Environmental Management and Toxicology, Federal University of Agriculture, Abeokuta, Ogun State, Nigeria*
George, G.N.: *Department of Geological Sciences, University of Saskatchewan, Saskatchewan, Canada*
German, M.: *Lehigh University, Bethlehem, USA*
Gerpe, M.: *Toxicología Ambiental, Departamento de Ciencias Marinas, Instituto de Investigaciones Marinas y Costeras, Facultad de Ciencias Exactas y Naturales, Universidad Nacional de Mar del Plata, Mar del Plata, Argentina; Consejo Nacional de Investigaciones Científicas y Técnicas (CONICET), Buenos Aires, Argentina*
Ghorbel, S.: *University of Limoges, GRESE, Limoges Cedex, France*
Ghose, A.: *DNGM Research Foundation, Kolkata, India*
Ghosh, A.K.: *A.N. College, Patna, India*
Ghosh, S.A.: *Anugrah Narayan College, Patna, Bihar, India*
Gianni, M.: *Fundación Barrera Zoofitosanitaria Patagónica (Funbapa), Viedma, Rio Negro, Argentina*
Gimenez, M.: *Facultad de Cs. Exactas y Naturales, Universidad de Buenos Aires, Argentina*
Giménez, M.C.: *Departamento Química Analítica, Universidad Nacional del Chaco Austral, Argentina; Laboratorio de Química Analítica, Univ. Nacional del Chaco Austral. P.R. Sáenz Peña, Chaco. Argentina*
Giordano, Rafael: *Net Log Service, Córdoba, Argentina*
Giorgi, A.D.N.: *Universidad Nacional de Luján, Buenos Aires, Argentina*
Giri, A.K.: *Molecular and Human Genetics Division, CSIR-Indian Institute of Chemical Biology, Calcutta, India*
Glasauer, S.: *School of Environmental Sciences, University of Guelph, Guelph, ON, Canada*

Gobbi, A.: *Laboratorio de Química de Sistemas Heterogénos (LaQuíSiHe), Departamento de Química, Universidad de Buenos Aires, Argentina*
Goessler, W.: *Institute of Chemistry, Analytical Chemistry, University of Graz, Austria*
Goldaracena, V.: *Nuevo Hospital San Roque, Córdoba, Argentina*
Gomez, M.L.: *Instituto Argentino de Investigaciones de las Zonas Áridas (IADIZA-CCT CONICET), Mendoza, Argentina*
Gómez, R.: *Grupo ECOCATAL, Departmento de Química, Universidad Autónoma Metropolitana, Iztapalapa, México*
Gómez, V.: *Instituto Tecnológico de Zacatepec, Morelos, México*
Gonnelli, C.: *Institute of Agricultural and Environmental Chemistry, Università Cattolica del Sacro Cuore, Piacenza, Italy; Dipartimento di Biologia, Università degli Studi di Firenze, Firenze, Italy*
Gonsebatt, M.E.: *Department de Medicina Genómica y Toxicología Ambiental, Instituto de Investigaciones Biomédicas, Universidad Nacional Autónoma de México, México*
González, A.: *Centro de Ingeniería Sanitaria, Facultad de Ciencias Exactas, Ingeniería y Agrimensura, Universidad Nacional de Rosario, Argentina*
Gonzalez, C.R.: *Pontificia Universidad Católica, Santiago, Chile*
Gonzalez, L.: *Servicio Geológico Minero Argentino, Instituto de Tecnología Minera, Argentina*
González, N.: *Cátedra de Hidogeología, Facultad de Ciencias Naturales y Museo, Universidad Nacional de La Plata, La Plata, Argentina*
Gonzalez-Horta, C.: *Facultad de Ciencias Químicas, Universidad Autónoma de Chihuahua, Chihuahua, México*
Goso, G.: *Departamento de Evolución de Cuencas, Facultad de Ciencias, Universidad de la República (UDELAR), Montevideo, Uruguay*
Graieb, O.J.: *Universidad Tecnológica Nacional/CEDIA, Tucumán, Argentina*
Grande, A.: *Facultad de Cs. Exactas y Naturales, Universidad de Buenos Aires, Argentina*
Graziano, J.H.: *Department of Environmental Health Sciences, Mailman School of Public Health, Columbia University, New York, USA*
Green-Ruiz, C.R.: *Unidad Académica Mazatlán, Instituto de Ciencias del Mar y Limnología, Universidad Nacional Autónoma de México*
Grefalda, B.: *School of Earth Sciences, The University of Queensland, Brisbane, Australia*
Grefalda, B.: *Escola de Engenharia, Departamento de Engenharia Metalúrgica e de Materiais, Universidade Federal de Minas Gerais, Belo Horizonte, Brazil*
Grela, M.A.: *Departamento de Química, Facultad de Ciencias Exactas y Naturales, Universidad Nacional de Mar del Plata, Provincia de Buenos Aires, Argentina*
Griffa, C.A.: *Facultad Regional Rafaela, Universidad Tecnológica Nacional, Rafaela, Santa Fe, Argentina*
Groenendijk, M: *Hoofd Ingenieursbureau, Brabant Water, BC 's-Hertogenbosch, The Netherlands*
Grosbois, C.: *University F. Rabelais of Tours, Tours, France*
Großmann, B.: *Karlsruhe University of Applied Sciences, Germany*
Guallar, E.: *Department of Environmental Health Sciences, BSPH, Johns Hopkins University, Baltimore, USA*
Guarda, V.L.M.: *Department of Pharmacy, Federal University of Ouro Preto, Ouro Preto, Minas Gerais, Brazil*
Guasch, H.: *Universitat de Girona, Cataluña, Spain*
Guedes, F.A.: *Laboratório de Limnologia Ecotoxicologia e Ecologia Aquática, Universidade Federal de Minas Gerias, Brazil*
Guha Mazumder, D.N.: *DNGM Research Foundation, Kolkata, India*
Guha Mazumder, R.N.: *DNGM Research Foundation, Kolkata, India*
Guido, M.: *Faculty of Chemistry, University of the Republic of Uruguay, Uruguay*
Guilherme, L.R.G.: *Department of Soil Science, Federal University of Lavras (UFLA), Lavras, Minas Gerais, Brazil*
Guimarães, M.: *LNEG, National Laboratory for Geology and Energy, S. Mamede de Infesta, Portugal*
Gukelberger, E.: *Karlsruhe University of Applied Sciences, Germany*
Guo, H.M.: *School of Water Resources and Environment, China University of Geosciences, Beijing, China*
Guo, S.L.: *Ministry of Education Key Laboratory of Xinjiang Endemic and Ethnic Disease, Shihezi University, China*
Gupta, V.: *National Institute of Industrial Engineering (NITIE), Mumbai, India*
Gutierrez, C.: *Coordinación de Tecnología Hidrológica, Instituto Mexicano de Tecnología del Agua—IMTA, Jiutepec, Morelos, México*

Gutiérrez-Torres, D.S.: *Facultad de Ciencias Químicas, Universidad Autónoma de Chihuahua, Chihuahua, México*
Gutierrez-Valtierra, M.P.: *Departamento de Ingeniería Química, División de Ciencias Naturales y Exactas, Campus Guanajuato, Universidad de Guanajuato, Guanajuato, México*
Guzmán-Mar, J.L.: *Facultad de Ciencias Químicas, Universidad Autónoma de Nuevo León, Nuevo León, México*
Haertig, C.: *Environmental Geochemistry Group, University of Bayreuth, Germany*
Hahn-Tomer, S.: *Helmholtz Centre for Environmental Research, UFZ, Department of Environmental Microbiology, Leipzig, Saxony, Germany*
Halder, D.: *Department of Chemistry, University of Kalyani, Kalyani, West Bengal, India*
Harari, F.: *Institute of Environmental Medicine, Karolinska Institutet, Stockholm, Sweden*
Harini, P.: *CSIR-National Geophysical Research Institute, Hyderabad, India*
Harris, H.H: *School of Chemistry and Physics, Adelaide University, Adelaide, South Australia, Australia*
Hasan, M.A.: *Department of Geology, University of Dhaka, Dhaka, Bangladesh*
Hasan, R.: *Columbia University Arsenic Research Project, Dhaka, Bangladesh*
Hénault, C.: *INRA, Research Center of Orléans, Soil Science Unit, Orléans, France*
Hernández, A.: *Instituto Nacional de Tecnología Industrial (INTI), San Martín, Buenos Aires, Argentina*
Hernández, A.: *Grup de Mutagènesi, Departament de Genètica i de Microbiologia, Facultat de Biociències, Universitat Autònoma de Barcelona, Bellaterra, Spain; CIBER Epidemiología y Salud Pública, ISCIII, Madrid, Spain*
Hernández, M.A.: *Cátedra de Hidogeología, Facultad de Ciencias Naturales y Museo, Universidad Nacional de La Plata, La Plata, Argentina*
Hernández-Alcaraz, C.: *Instituto Nacional de Salud Pública, Cuernavaca, Morelos, México*
Hernández-Castellanos, E.: *Centro de Investigación y de Estudios Avanzados del IPN, Departamento de Toxicología, México DF, México*
Hernández-Cerón, R.: *CIBER Epidemiología y Salud Pública, ISCIII, Madrid, Spain*
Hernández-Ramirez, A.: *Facultad de Ciencias Químicas, Universidad Autónoma de Nuevo León, Nuevo León, México*
Hernández-Ramírez, R.U.: *Instituto Nacional de Salud Pública, Cuernavaca, Morelos, México*
Herrera, A.V.: *Instituto Nacional de Salud Pública, Cuernavaca, Morelos, México*
Herrera, V.: *Center of Research and Development of Water Resources, Faculty of Health Sciences, University Arturo Prat, Tarapacá, Chile*
Herrmann, M.: *Karlsruhe University of Applied Sciences, Germany*
Hettiarachchi, G.: *Kansas State University, Manhattan, KS, USA*
Hidalgo Estévez, M.C.: *Departamento de Geología, Universidad de Jaén, Jaén, España*
Hilbe, N.: *Programa de Investigación y Análisis de Residuos y Contaminantes Químicos, PRINARC, Facultad de Ingeniería Química, Universidad Nacional del Litoral, Santa Fe, Argentina*
Hinojosa-Reyes, L.: *Facultad de Ciencias Químicas, Universidad Autónoma de Nuevo León, Nuevo León, México*
Hinrichsen, S.: *Environmental Geochemistry Group, University of Bayreuth, Germany*
Hisarli, N.D.: *Gazi University, Ankara, Turkey*
Hoeft McCann, S.: *United States Geological Survey, Menlo Park, California, USA*
Hoinkis, J.: *Karlsruhe University of Applied Sciences, Germany*
Honma, T.: *Niigata Agricultural Research Institute, Nagaoka, Japan*
Hossain, M.: *KTH-International Groundwater Arsenic Research Group, Department of Land and Water Resources Engineering, Royal Institute of Technology (KTH), Teknikringen, Stockholm, Sweden*
Hosseini, S.A.: *School of Public Health, Tehran University of Medical Sciences, Tehran, Iran*
Hosseini, S.S.: *School of Public Health, Tehran University of Medical Sciences, Tehran, Iran*
Houdková, P.: *Institute of Chemical Technology in Prague, Czech Republic*
Hryczyñski, E.: *Laboratorio de Investigación y Desarrollo, UNCAUS, P.R. Saenz Peña, Chaco, Argentina*
Hsiao, B.-Y.: *Genomics Research Center, Academia Sinica, Taipei, Taiwan*
Hsu, L.-I: *Genomics Research Center, Academia Sinica, Taipei, Taiwan*
Hsy, L.-I.: *Genomics Research Center, Academia Sinica, Taipei, Taiwan*
Hu, Q.H.: *China University of Geosciences, Wuhan, China; The University of Texas at Arlington, Texas, USA*
Huang, L.: *Centre for Mined Land Rehabilitation (CMLR), Sustainable Minerals Institute (SMI), The University of Queensland, Brisbane, Queensland, Australia*
Huang, T.-H.: *Department of Agricultural Chemistry, National Taiwan University, Taipei, Taiwan*

Huerta-Rosas, B.: *Departamento de Ingeniería Química, División de Ciencias Naturales y Exactas, Campus Guanajuato, Universidad de Guanajuato, Guanajuato, México*
Iglesias, M.: *Faculty of Sciences, University of Girona, Campus de Montilivi, Girona, Spain*
Illa, A.: *Ingesco, Terrassa (Barcelona), Spain*
Inácio, M.: *GEOBIOTEC, GeoBiosciences, Geotechnologies and Geoengineering Research Center, University of Aveiro, Portugal*
Ingallinella, A.M.: *Centro de Ingeniería Sanitaria, Facultad de Ciencias Exactas, Ingeniería y Agrimensura, Universidad Nacional de Rosario, Argentina*
Invernizzi, R.: *Nuclear Chemistry Department, National Atomic Energy Commission of Argentina, Buenos Aires, Argentina*
Iriel, A.: *Instituto de Investigación en Producción Animal (INPA-UBA-CONICET), Facultad de Ciencias Veterinarias, Buenos Aires, Argentina*
Ishida-Gutiérrez, M.C.: *Facultad de Ciencias Químicas, Universidad Autónoma de Chihuahua, Chihuahua, México*
Islam, M.A.: *Department of Chemistry, University of Dhaka, Dhaka, Bangladesh*
Islam, Md.S.: *Hokkaido University, Sapporo, Hokkaido, Japan*
Islam, T.: *Columbia University Arsenic Research Project, Dhaka, Bangladesh*
Islam, T.: *U-Chicago Research Bangladesh, Ltd., Dhaka, Bangladesh*
Jaćimović, R.: *Department of Environmental Sciences, Jožef Stefan Institute, Ljubljana, Slovenia*
Jacks, G.: *KTH-International Groundwater Arsenic Research Group, Department of Sustainable Development, Environmental Science and Engineering, KTH Royal Institute of Technology, Stockholm, Sweden*
Jakovljevic, B.: *Institute of Hygiene and Medical Ecology, School of Medicine, Belgrade, Serbia*
Jamieson, H.E.: *Department of Geological Engineering, Queen's University, Kingston, ON, Canada*
Jasmine, F.: *The University of Chicago, Chicago, IL, USA*
Jean, J.-S.: *Department of Earth Sciences, National Cheng Kung University, Tainan, Taiwan*
Jesus, I.M: *Evandro Chagas Institute, Ministry of Health, Belém, PA, Brazil*
Jia, Y.: *State Key Laboratory of Biogeology and Environmental Geology, China University of Geosciences, Beijing, P.R. China; School of Water Resources and Environment, China University of Geosciences, Beijing, China*
Jiang, J.Y.: *New York University, New York, USA*
Jigyasu, D.K.: *Centre of Advanced Study in Geology, University of Lucknow, Lucknow, India*
Jing, C.: *Research Center for Eco-Environmental Sciences, Chinese Academy of Sciences, Beijing, China*
Johannesson, K.: *Department of Earth and Environmental Science, Tulane University, New Orleans, LA, USA*
Johnson, C.A.: *Eawag, Swiss Federal Institute of Aquatic Science and Technology, Dübendorf, Switzerland*
Joulian, C.: *BRGM, ISTO, UMR 7327, BP, Orléans, France*
Juárez, Á.: *Facultad de Ciencias Exactas y Naturales, Universidad de Buenos Aires, Argentina; IBBEA-CONICET, Buenos Aires, Argentina*
Juncos, R.: *Laboratorio de Análisis por Activación Neutrónica, Centro Atómico Bariloche, CNEA, Bariloche, Argentina*
Kader, M.: *Centre for Environmental Risk Assessment and Remediation, University of South Australia, Mawson Lakes, SA, Australia; Cooperative Research Centre for Contamination Assessment and Remediation of the Environment (CRC CARE), University of South Australia, Mawson Lakes, SA, Australia*
Kadioglu, E.: *Gazi University, Ankara, Turkey*
Kaegi, R.: *Eawag, Swiss Federal Institute of Aquatic Science and Technology, Switzerland*
Kaneko (Kadokura), A.: *Niigata Agricultural Research Institute, Nagaoka, Japan*
Kar, S.: *Department of Earth Sciences, National Cheng Kung University, Tainan, Taiwan*
Karim, M.R.: *Department of Population Dynamics, National Institute of Preventive and Social Medicine (NIPSOM) Mohakhali, Dhaka, Bangladesh*
Kechit, F.: *INRA, UMR BIOGECO 1202, University of Bordeaux, Talence, France; INRA Cestas, France*
Khamseh Safa, Z.: *Macquarie University, Sydney, NSW, Australia*
Khan, M.A.S.: *Faculty of Animal Husbandry, Bangladesh Agricultural University, Mymensingh, Bangladesh*
Khan, N.I.: *Fenner School of Environment and Society, The Australian National University, Canberra, Australia*
Kibria, M.G.: *Department of Geology, Kansas State University, Manhattan, KS, USA*
Kibriya, M.G.: *The University of Chicago, Chicago, IL, USA*
Kihl, A.: *Ragn Sells AB, Stockholm, Sweden*
Kippler, M.: *Institute of Environmental Medicine, Karolinska Institutet, Stockholm, Sweden*
Kirk, M.F.: *Kansas State University, Kansas, USA*

Kirschbaum, A.: *Museo de Ciencias Naturales, Instituto de Bio y Geociencias del NOA, Universidad Nacional de Salta, Argentina*
Klavins, M.: *Department of Environmental Science, University of Latvia, Riga, Latvia*
Klimecki, W.: *College of Pharmacy, University of Arizona, Tucson, AZ. USA*
Knezevic, T.: *Institute of Public Health of Serbia "Dr Milan Jovanovic Batut", Belgrade, Serbia*
Kohan, M.: *National Exposure Research Laboratory, NERL, RTP, NC, USA*
Kohfahl, C.: *Instituto Geológico y Minero de España, Madrid, España*
Kolev, S.D.: *School of Chemistry, Faculty of Science, The University of Melbourne, Australia*
Komárek, M.: *Department of Environmental Geosciences, Faculty of Environmental Sciences, Czech University of Life Sciences, Prague, Czech Republic*
Kong, S.Q.: *China University of Geosciences, Wuhan, China*
Kordas, K.: *School of Social and Community Medicine, University of Bristol, UK*
Korfali, S.: *Department of Chemistry, Lebanese American University, Lebanon*
Kovács, B.: *University Debrecen, Debrecen, Hungary*
Krapp, A.: *Universidad Nacional de Córdoba, Córdoba, Argentina*
Krehel, A.: *Department of Geological and Environmental Science, Hope College, Holland, MI, USA*
Krejcova, S.: *Institute of Chemical Technology, Prague, Czech Republic*
Krikowa, F.: *Ecochemistry Laboratory, Institute for Applied Ecology, University of Canberra, Bruce, ACT, Australia*
Krishnamohan, M.: *National Research Centre for Environmental Toxicology, The University of Queensland, Brisbane, Australia*
Kumar, A.: *Mahavir Cancer Institute and Research Centre, Patna, Bihar, India*
Kumar, M.: *School of Environmental Sciences, Jawaharlal Nehru University, New Delhi, India*
Kumar, R.: *Mahavir Cancer Institute and Research Centre, Patna, Bihar, India; A.N. College, Patna, India*
Kumpiene, J.: *Luleå University of Technology, Luleå, Sweden*
Kurasaki, M.: *Hokkaido University, Sapporo, Hokkaido, Japan*
Kuvar, R.: *Centre of Advanced Study in Geology, University of Lucknow, Lucknow, India*
La Porte, P.F.: *Department of Medicine, Emory University School of Medicine, Atlanta, Georgia, USA*
Lahiri, S.: *Saha Institute of Nuclear Physics, Kolkata, India*
Lamb, D.T.: *Centre for Environmental Risk Assessment and Remediation, University of South Australia, Mawson Lakes, SA, Australia; Cooperative Research Centre for Contamination Assessment and Remediation of the Environment(CRC CARE), University of South Australia, Mawson Lakes, SA, Australia*
Lamela, P.A.: *IDEPA Multidisciplinary Institute of Scientific Research and Development from North Patagonia, (CONICET-CCT COMAHUE), National University of Comahue, Neuquén y Cinco Saltos, Argentina*
Lan, L.E.: *Departamento de Ingeniería Química, Facultad Regional Buenos Aires, Universidad Tecnológica Nacional, Argentina; UDB-Química, Facultad Regional Buenos Aires, Universidad Tecnológica Nacional, Argentina*
Lan, V.M.: *Center for Environmental Technology and Sustainable Development, Hanoi University of Science, Vietnam*
Landi, S.M.: *Instituto Nacional de Metrologia, Qualidade e Tecnologia, Divisão de Metrologia de Materiais, RJ, Brasil*
Laniyan, T.A: *Department of Earth Sciences, Olabisi Onabanjo University, Ago-Iwoye, Ogun State, Nigeria*
Lantz, R.C.: *Department of Pharmacology and Toxicology, College of Pharmacy, University of Arizona, Tucson, USA*
Lara, J.A.: *Planta Piloto de Procesos Industriales Microbiológicos (PROIMI-CONICET), Tucumán, Argentina*
Larralde, A.L.: *INQUIMAE, DQIADF, Facultad de Ciencias Exactas y Naturales, Universidad de Buenos Aires, Buenos Aires, Argentina*
Lavado, R.S.: *Instituto de Investigaciones en Biociencias Agrícolas y Ambientales, INBA, (CONICET-UBA), Ciudad Autónoma de Buenos Aires, Argentina*
Lavezzari, E.: *Departamento de Ingeniería Química, Facultad Regional Buenos Aires, Universidad Tecnológica Nacional, Argentina; UDB-Química, Facultad Regional Buenos Aires, Universidad Tecnológica Nacional, Argentina*
Lazarova, Z.: *Department of Health and Environment, AIT—Austrian Institute of Technology, Vienna, Austria*
Le, L.A.: *Institute of Chemistry, Vietnam Academy of Science and Technology, Hanoï, Vietnam*
Leal, P.R.: *Departamento de Ciencias Geológicas, Facultad de Ciencias Exactas y Naturales, UBA, CABA, Argentina*

Lei, M.: *College of Resource and Environment, Hunan Agricultural University, Changsha, P.R. China*
Leiva, E.A.: *Departamento de Ingeniería Hidráulica y Ambiental, Pontificia Universidad Católica de Chile, Santiago, Chile*
Leiva, E.D.: *Department of Hydraulic and Environmental Engineering, Pontificia Universidad Católica de Chile, Santiago, Chile*
Leiva, S.: *National Technological University-Faculty of Trenque Lauquen (UTN-FRTL), Buenos Aires, Argentina*
Lescano, M.R.: *INTEC (UNL-CONICET), Güemes, Santa Fe, Argentina*
Lescure, T.: *BRGM, ISTO, UMR 7327, Orléans, France; LIEC UMR 7360, University of Lorraine, Bridoux Campus, Metz, France*
Levy, D.: *Columbia University, New York, USA*
Levy, I.K.: *Gerencia Química, Comisión Nacional de Energía Atómica, San Martín, Provincia de Buenos Aires, Argentina*
Lhotka, M.: *Institute of Chemical Technology, Prague, Czech Republic*
Li, J.: *College of Water Sciences, Beijing Normal University, Beijing, China; Engineering Research Center of Groundwater Pollution Control and Remediation, Ministry of Education, Beijing, China*
Liberman, C.: *Instituto Nacional de Tecnología Agropecuaria, Estación Experimental Agropecuaria Concepción del Uruguay (NTA EEA), Provincia de Entre Ríos, Argentina*
Libertun, C.: *Instituto de Biología y Medicina Experimental (IByME, CONICET), CABA, Buenos Aires, Argentina*
Lictevout, E.: *Faculty of Health Sciences, CIDERH, Water Resources Research and Development Center, University Arturo Prat, Iquique, Chile*
Lienqueo, H.: *Centro de Investigaciones del Hombre en el Desierto, CIHDE, Arica, Chile*
Lima, A.: *Dow Water and Process Solutions, Saudi Arabia*
Lima, C.A.: *Independent Risk Assessor*
Lima, M.O.: *Evandro Chagas Institute, Ministry of Health, Belém, PA, Brazil*
Limón Pacheco, J.H.: *Department de Medicina Genómica y Toxicología Ambiental, Instituto de Investigaciones Biomédicas, Universidad Nacional Autónoma de México, México*
Litter, M.I.: *Gerencia Química, Comisión Nacional de Energía Atómica, San Martín, Provincia de Buenos Aires, Argentina*
Liu, M.L.: *New York University, New York, USA*
Liu, S.: *Research Center for Eco-Environmental Sciences, Chinese Academy of Sciences, Beijing, China*
Liu, Y.J.: *Centre for Mined Land Rehabilitation (CMLR), Sustainable Minerals Institute (SMI), The University of Queensland, Brisbane, Queensland, Australia*
Livore, A.B.: *EEA INTA, Concepción del Uruguay, Entre Ríos, Argentina*
Lizama Allende, K.: *Universidad de Chile, Santiago, Chile*
Llano, J.: *INQUIMAE/DQIAQF, Facultad de Ciencias Exactas y Naturales, UBA, CABA, Argentina*
Lohmayer, R.: *Environmental Geochemistry Group, University of Bayreuth, Germany*
Londonio, J.A.: *Departamento de Química Analítica, Gerencia Química, Comisión Nacional de Energía Atómica, Argentina*
Loomis, D.: *International Agency for Research of Cancer, Lyon Cedex, France*
López Callejas, R.: *Universidad Autónoma Metropolitana, Azcapotzalco, México*
López Paraguay, M.Z.: *Centro de Investigación en Materiales Avanzados (CIMAV), Chihuahua, México*
López Pasquali, C.E.: *Universidad Nacional de Santiago del Estero, Santiago del Estero, Argentina*
Lopez, A.: *Instituto Nacional de Salud, Ministerio de Salud, San Salvador, El Salvador*
López, A.R.: *Departamento de Ingeniería Química, Facultad Regional Buenos Aires, Universidad Tecnológica Nacional, Argentina; Departamento de Ingeniería Civil, Facultad Regional Buenos Aires, Universidad Tecnológica Nacional, Argentina*
Lopez, D.L.: *Department of Geological Sciences, Ohio University, Athens, Ohio, USA*
López, W.G.: *Laboratorio de Investigación y Desarrollo, UNCAUS, P.R. Saenz Peña, Chaco, Argentina*
López-Carrillo, L.: *Instituto Nacional de Salud Pública, Cuernavaca, Morelos, México*
Loppinet-Serani, A.: *CNRS, University of Bordeaux, ICMCB, IPB-ENSCBP, Pessac, France*
Lord, G.: *ICP-MS Facility, Department of Chemistry, University of Surrey, Guildford, UK*
Loureiro, M.E.: *Universidade Federal de Viçosa, Viçosa, MG, Brazil*
Lujan, J.C.: *Universidad Tecnológica Nacional/CEDIA, Tucumán, Argentina*
Luo, J.: *State Key Lab of Pollution Control and Resource Reuse, School of Environment, Nanjing University, China*

Luo, Z.: *Key Laboratory of Urban Environment and Health, Institute of Urban Environment, Chinese Academy of Sciences, Xiamen, China*
Lux Lantos, V.: *Instituto de Biología y Medicina Experimental (IByME, CONICET), CABA, Argentina*
Ma, L.Q.: *State Key Lab of Pollution Control and Resource Reuse, School of Environment, Nanjing University, China*
Machovič, V.: *Institute of Chemical Technology in Prague, Czech Republic*
Magalhães, S.M.S.: *Faculdade de Farmácia, Universidade Federal de Minas Gerais, Brazil*
Mahanta, C.: *Department of Civil Engineering, Indian Institute of Technology Guwahati, Assam, India*
Maher, W.A.: *Ecochemistry Laboratory, Institute for Applied Ecology, University of Canberra, Bruce, ACT, Australia*
Mahvi, A.H.: *School of Public Health, Tehran University of Medical Sciences, Tehran, Iran; Center for Solid Waste Research, Institute for Environmental Research, Tehran University of Medical Sciences, Tehran, Iran; National Institute of Health Research, Tehran University of Medical Sciences, Tehran, Iran*
Maidar, T.: *Central Hospital of Mongolia, Mongolia*
Maiti, M.: *Rorkee University, Rorkee, India*
Maizel, D.: *Planta Piloto de Procesos Industriales Microbiológicos (PROIMI-CONICET), Tucumán, Argentina*
Makino, T.: *National Institute for Agro-Environmental Sciences, Tsukuba, Japan*
Maleki, A.: *Kurdistan Environmental Health Research Center, Kurdistan University of Medical Sciences, Kurdistan, Iran*
Mañay, N.: *Faculty of Chemistry, University of the Republic of Uruguay (UDELAR), Uruguay*
Manrique Carrera, R.M.: *Licenciatura en Gestión Ambiental, Universidad CAECE, Buenos Aires, Argentina*
Marapakala, K.: *Herbert Wertheim College of Medicine, Florida International University, Miami, USA*
Marasco, L.: *Facultad de Cs. Exactas y Naturales, Universidad de Buenos Aires, Argentina*
Marchand, L.: *INRA, UMR BIOGECO, University of Bordeaux, Talence, France; INRA Cestas, France*
Marchi, M.C.: *INQUIMAE, Facultad de Ciencias Exactas, Universidad de Buenos Aires, Buenos Aires, Argentina; Centro de Microscopías Avanzadas, Facultad de Ciencias Exactas, Universidad de Buenos Aires, Buenos Aires, Argentina*
Marcilla, A.L.: *British Geological Survey, Keyworth, Nottingham, UK*
Marcilla, A.L.: *Patagonia BBS, General Roca, Río Negro, Argentina*
Marcó, L.: *U. Centroccidental Lisandro Alvarado, Lara, Venezuela*
Marco-Brown, J.L.: *Instituto de Investigación e Ingeniería Ambiental (3IA), Escuela de Ciencia y Tecnología, UNSAM, San Martín, Argentina*
Marcos, R.: *Grup de Mutagènesi, Departament de Genètica i de Microbiologia, Facultat de Biociències, Universitat Autònoma de Barcelona, Bellaterra, Spain; CIBER Epidemiología y Salud Pública, ISCIII, Madrid, Spain*
Markwitz, A.: *GNS Science, Lower Hutt, New Zealand*
Martín Domínguez, I.R.: *Centro de Investigación en Materiales Avanzados (CIMAV), Chihuahua, México*
Martínez-Fernández, D.: *Department of Environmental Geosciences, Faculty of Environmental Sciences, Czech University of Life Sciences, Kamycka, Prague, Czech Republic*
Martínez-Villegas, N.: *IPICyT—Instituto Potosino de Investigación Científica y Tecnológica, División de Geociencias Aplicadas, San Luis Potosí, SLP, México*
Martinis, E.M.: *QUIANID, Instituto de Ciencias Básicas, Universidad Nacional de Cuyo, Padre Jorge Contreras, Parque General San Martín, Mendoza, Argentina; Consejo Nacional de Investigaciones Científicas y Técnicas (CONICET), Argentina*
Masetto, N.: *Dow Water and Process Solutions, Saudi Arabia*
Matos, A.: *Kinross Gold Corporation, Paracatu, MG, Brazil*
Matos, J.A.: *Instituto de Química, Departamento de Geoquímica, Universidade Federal Fluminense (UFF), Centro, Niterói, RJ, Brasil*
Mauricci, J.J.: *Membrane Technical Managers at Cooperativa Obras Sanitarias de Venado Tuerto, Santa Fe, Argentina*
Maury, A.M.: *Facultad de Ingeniería, Universidad de Buenos Aires, Buenos Aires, Argentina*
May-Ix, L.A.: *Grupo ECOCATAL, Departmento de Química, Universidad Autónoma Metropolitana, Iztapalapa, México*
Mazej, D.: *Department of Environmental Sciences, Jožef Stefan Institute, Ljubljana, Slovenia*
Mazumdar, A.: *School of Water Resources Engineering, Jadavpur University, Kolkata, India*
McCarthy, D.T.: *Monash University, Melbourne, VIC, Australia*

McDonald, K.J.: *Department of Ecosystem Science and Management, University of Wyoming, Wyoming, USA*
McRae, C.: *Macquarie University, Sydney, NSW, Australia*
Megharaj, M.: *Centre for Environmental Risk Assessment and Remediation, University of South Australia, Mawson Lakes, SA, Australia; Cooperative Research Centre for Contamination Assessment and Remediation of the Environment (CRC CARE), University of South Australia, Mawson Lakes, SA, Australia*
Meichtry, J.M.: *Departamento de Ingeniería Química, Facultad Regional Buenos Aires, Universidad Tecnológica Nacional, Argentina*
Mejia, R.: *Instituto Nacional de Salud, Ministerio de Salud, San Salvador, El Salvador*
Melgoza-Castillo, A.: *Facultad de Zootecnia y Ecología, Universidad Autónoma de Chihuahua, Chihuahua, México*
Mello De Capitani, E.: *School of Medicine, State University of Campinas, Campinas, SP, Brazil*
Mello, J.W.V.: *Universidade Federal de Viçosa, Viçosa, MG, Brazil*
Melo, M.C.: *Soils Department, Federal University of Viçosa, Viçosa, MG, Brazil; National Institute of Science and Technology, INCT-Acqua, Belo Horizonte, MG, Brazil*
Mench, M.: *INRA, UMR BIOGECO, University of Bordeaux, Talence, France; INRA Cestas, France*
Mendez Guimaraes, E.: *Laboratório de Raios X, Universidade de Brasilia (UnB), Brasil*
Mendez, M.A.: *University of North Carolina at Chapel Hill, North Carolina, USA*
Menezes, M.A.B.C.: *Centro de Desenvolvimento da Tecnologia Nuclear (CDTN/CNEN), Belo Horizonte, MG, Brazil*
Mertens, F.: *Centro de Desenvolvimento Sustentável, Universidade de Brasília (CDS/UnB), Brazil; Community of Practice in Ecohealth, Latin America and Caribbean (CoPEH-LAC), Brasília/DF, Brazil*
Mesdaghinia, A.R.: *School of Public Health, Tehran University of Medical Sciences, Tehran, Iran; Center for Solid Waste Research, Institute for Environmental Research, Tehran University of Medical Sciences, Tehran, Iran; National Institute of Health Research, Tehran University of Medical Sciences, Tehran, Iran*
Michálková, Z.: *Department of Environmental Geosciences, Faculty of Environmental Sciences, Czech University of Life Sciences, Prague, Czech Republic*
Mildiner, S.: *Facultad de Cs. Exactas y Naturales, Universidad de Buenos Aires, Argentina*
Miqueles, E.X.: *Laboratorio Nacional de Luz Sincrotron (LNLS), Campinas, Brasil*
Mirabelli, G.A.: *Instituto Bac Spinoza, Córdoba, Argentina*
Mirlean, N.: *Federal University of Rio Grande, Brazil*
Mizrahi, M.: *Instituto de Investigaciones Fisicoquímicas Teóricas y Aplicadas, La Plata, Argentina*
Mladenov, N.: *Department of Civil Engineering, Kansas State University, Manhattan, KS, USA*
Molina, F.V.: *INQUIMAE, Facultad de Ciencias Exactas y Naturales, Universidad de Buenos Aires, Argentina*
Molina, R.: *Population Record of Tumors, La Pampa Province, Argentina*
Molinari, B.L.: *National Atomic Energy Commission (CNEA), National Scientific and Research Council (CONICET), Buenos Aires, Argentina*
Mondino, C.: *Nuevo Hospital San Roque, Córdoba, Argentina*
Monteiro, M.I.C.: *Center for Mineral Technology, Ministry of Science, Technology and Innovation, Rio de Janeiro, Brazil*
Montenegro, A.: *INQUIMAE, Facultad de Ciencias Exactas y Naturales, Universidad de Buenos Aires, Argentina*
Montini, L.: *Departamento de Saneamiento Básico, Dirección General de Salud Ambiental, Sistema Provincial de Salud, S. M. de Tucumán, Argentina*
Morales, I.: *Postgraduate Program in Earth Sciences, UNAM, Mexico*
Morales, V.: *Pontificia Universidad Católica, Santiago, Chile*
Morisio, Y.: *Departamento de Química Analítica, Gerencia Química, Comisión Nacional de Energía Atómica, Buenos Aires, Argentina*
Morrison, L.: *Earth and Ocean Sciences, School of Natural Sciences and Ryan Institute, National University of Ireland, Galway, Ireland*
Moschione, E.: *National Technological University-Mar del Plata (UTN-MDP), Buenos Aires, Argentina*
Mugrabi, F.I.: *Departamento de Ingeniería Química, Facultad Regional Buenos Aires, Universidad Tecnológica Nacional, Argentina*
Mukherjee, A.: *Department of Geology and Geophysics, Indian Institute of Technology-Kharagpur, West Bengal, India*
Munera-Picazo, S.: *Departamento de Tecnología Agroalimentaria, Universidad Miguel Hernández, Ctra. Beniel, Orihuela, Alicante, Spain*

Muñoz, E.: *Departamento de Química, Facultad de Ciencias, Universidad Nacional de Colombia, Bogotá, Colombia*
Muñoz, S.E.: *Escuela de Nutrición e Instituto de Investigaciones en Ciencias de la Salud, CONICET, Facultad de Ciencias Médicas, Universidad Nacional de Córdoba, Córdoba, Argentina*
Murao, S.: *Sans Frontiere Progres NGO, Mongolia; Institute for Geo-Resources and Environment, AIST, Japan*
Murray, J.: *Museo de Ciencias Naturales, Instituto de Bio y Geociencias del NOA, Universidad Nacional de Salta, Argentina*
Nadra, A.D.: *Departamento de Química Biológica, IQUIBICEN-CONICET, Facultad de Cs. Exactas y Naturales, Universidad de Buenos Aires, Argentina*
Naidu, R.: *Centre for Environmental Risk Assessment and Remediation, University of South Australia, Mawson Lakes, SA, Australia; Cooperative Research Centre for Contamination Assessment and Remediation of the Environment (CRC CARE), University of South Australia, Mawson Lakes, SA, Australia*
Najmanova, J.: *Department of Agroenvironmental Chemistry and Plant Nutrition, Czech University of Life Sciences, Prague, Czech Republic*
Nakamura, K.: *National Institute for Agro-Environmental Sciences, Tsukuba, Japan*
Namavar, S.: *National Iranian Oil Products Distribution Company (NIOPDC), Yasuj, Kohgiluyeh Boyer Ahmad, Iran; School of Public Health, Tehran University of Medical Sciences, Tehran, Iran*
Nandan, M.J.: *CSIR-National Geophysical Research Institute, Uppal Road, Hyderabad, India*
Nannavecchia, P.: *Facultad de Ciencias Exactas y Naturales, Universidad de Buenos Aires, Argentina*
Nasr, H.: *Sustainable Development and Health Equity, Pan American Health Organization, USA*
Nath, A.: *Mahavir Cancer Institute and Research Centre, Patna, Bihar, India*
navas-Aien, A.: *Department of Environmental Health Sciences, BSPH, Johns Hopkins University, Baltimore, USA*
Navoni, J.A.: *Cátedra de Toxicología y Química Legal, FFyB, UBA CABA, Buenos Aires, Argentina*
Navratilova, J.: *Department of Nutrition, University of North Chapel Hill, CH, NC, USA*
Nazmara, Sh.: *Center for Solid Waste Research, Institute for Environmental Research, Tehran University of Medical Sciences, Tehran, Iran*
Neal, A.: *Department of Geology, Kansas State University, Manhattan, KS, USA*
Neidhardt, H.: *Eawag, Swiss Federal Institute of Aquatic Science and Technology, Switzerland*
Neves, O.: *CERENA, Centre for Natural Resources and the Environment, Instituto Superior Técnico, University of Lisbon, Portugal*
Neves, P.C.: *Centro Nacional de Pesquisa em Arroz e Feijão, EMBRAPA, Brazil*
Ng, J.C.: *National Research Centre for Environmental Toxicology (Entox), The University of Queensland, Brisbane, Queensland, Australia; CRC for Contamination Assessment and Remediation of the Environment, Mawson Lakes, Adelaide, South Australia, Australia; INCT-Acqua, Belo Horizonte, Brazil*
Niclis, C.: *Escuela de Nutrición e Instituto de Investigaciones en Ciencias de la Salud, CONICET, Facultad de Ciencias Médicas, Universidad Nacional de Córdoba, Córdoba, Argentina*
Nicolli, H.B: *CONICET, Argentina; Instituto de Geoquímica, Argentina*
Niero, L.: *Waste Science and Technology, Luleå University of Technology, Sweden; Padua University, Italy*
Nieto Moreno, N.: *Facultad de Cs. Exactas y Naturales, Universidad de Buenos Aires, Argentina*
Nigrele, J.P.: *Licenciatura en Gestión Ambiental, Universidad CAECE, Buenos Aires, Argentina*
Nogueira, F.: *INCT-Acqua, Belo Horizonte, Brazil; Departamento de Engenharia Metalúrgica e de Materiais, Escola de Engenharia, Universidade Federal de Minas Gerais, Belo Horizonte, Brazil*
Noller, B.N.: *Centre for Mined Land Rehabilitation, The University of Queensland, Brisbane, Queensland, Australia*
Nordstrom, D.K.: *US Geological Survey, Boulder, CO, USA*
Nriagu, J.: *Department of Environmental Health Sciences, School of Public Health, University of Michigan, Ann Arbor, MI, USA*
O'Reilly, J.: *ICP-MS Facility, Department of Chemistry, University of Surrey, Guildford, UK*
Ocampo Gómez, G.: *Salud Ambiental, Facultad de Medicina de la Universidad Autónoma de Coahuila, México*
Ohba, H.: *Niigata Agricultural Research Institute, Nagaoka, Japan*
Ojeda, G.: *Departamento de Saneamiento Básico, Dirección General de Salud Ambiental, Sistema Provincial de Salud, S.M. de Tucumán, Argentina*
Oláh, Z.: *Technical University, Budapest, Hungary*
Olivas Calderón, E.H.: *Universidad Juárez del Estado de Durango, Gómez Palacio, Durango, México*
Oliveira, A.M.: *Escola de Engenharia, Departamento de Engenharia Metalúrgica e de Materiais, Universidade Federal de Minas Gerais, Belo Horizonte, Brazil; INCT-Acqua, Belo Horizonte, Brazil*

Olivieri, V.: *Cooperativa Obras Sanitarias de Venado Tuerto, Venado Tuerto, Santa Fe, Argentina*
Olmos, V.: *Cátedra de Toxicología y Química Legal, Facultad de Farmacia y Bioquímica, Universidad de Buenos Aires, Argentina*
Ono, F.B.: *Department of Soil Science, Federal University of Lavras (UFLA), Lavras, Minas Gerais, Brazil*
Opio, F.K.: *Queen's University, Canada*
Orange, D.: *IRD, BIOEMCO, IWMI Office, SFRI, Hanoï, Vietnam*
Orantes, C.: *Instituto Nacional de Salud, Ministerio de Salud, San Salvador, El Salvador*
Oremland, R.: *United States Geological Survey, Menlo Park, California, USA*
Ormachea Muñoz, M.: *Instituto de Investigaciones Químicas, Universidad Mayor de San Andrés, La Paz, Bolivia; KTH-International Groundwater Arsenic Research Group, Department of Sustainable Development, Environmental Science and Engineering, Royal Institute of Technology, Stockholm, Sweden*
Orsini, M.P.: *Instituto Nacional de Tecnología Industrial, San Martín, Buenos Aires, Argentina*
Ortiz, M.: *Gerencia Química, Comisión Nacional de Energía Atómica, San Martín, Provincia de Buenos Aires, Argentina*
Ortolani, V.: *Quality Control Management, "Ente Regulador de Servicios Sanitarios", Province of Santa Fe, Argentina*
Osicka, R.M.: *Departamento Química Analítica, Universidad Nacional del Chaco Austral, Argentina*
Osorio-Yáñez, C.: *Centro de Investigación y de Estudios Avanzados del IPN, Departamento de Toxicología, México DF, México*
Osterwalder, E.: *Helmholtz Centre for Environmental Research, Department of Environmental Microbiology, UFZ, Leipzig, Germany*
Ozols, A.: *Instituto de Ingeniería Biomédica, Universidad de Buenos Aires, Buenos Aires, Argentina*
Pabón Reyes, D.C.: *Gerencia Química, CAC-CNEA, San Martín, Argentina*
Pacini, V.A.: *Centro de Ingeniería Sanitaria, Facultad de Ciencias Exactas, Ingeniería y Agrimensura, Universidad Nacional de Rosario, Argentina*
Pagnout, C.: *LIEC, CNRS-UMR 7360, Université de Lorraine, Metz, France*
Pal, S.: *School of Water Resources Engineering, Jadavpur University, Kolkata, India*
Palmieri, H.E.L.: *Centro de Desenvolvimento da Tecnologia Nuclear (CDTN/CNEN), Belo Horizonte, MG, Brazil*
Palmieri, M.A.: *Biodiversity and Experimental Biology Department, School of Exact and Natural Sciences, University of Buenos Aires, Buenos Aires, Argentina*
Pandolpho, L.V.R.A.B.: *Department of Soil, Federal University of Vicosa, Vicosa, Minas Gerais, Brazil*
Panigatti, M.C.: *Facultad Regional Rafaela, Universidad Tecnológica Nacional, Rafaela, Santa Fe, Argentina*
Panique Lazcano, D.R.: *Universidad Nacional de Córdoba, Córdoba, Argentina*
Panozzo, J.: *Facultad de Ciencias Agropecuarias, Universidad Nacional de Entre Ríos, Paraná, ER, Argentina*
Paoloni, J.D.: *CONICET, Argentina*
Paredes, J.: *Posgrado en Ingeniería UNAM, Jiutepec, Morelos, Mexico*
Parsons, P.: *New York State Department of Health, Albany NY, USA*
Parvez, F.: *Department of Environmental Health Sciences, Mailman School of Public Health, Columbia University, New York, USA*
Paschoali, T.: *Escola de Engenharia, Departamento de Engenharia Metalúrgica e de Materiais, Universidade Federal de Minas Gerais, Belo Horizonte, Brazil; INCT-Acqua, Belo Horizonte, Brazil*
Pasten, P.A.: *Department of Hydraulic and Environmental Engineering, Pontificia Universidad Católica de Chile, Santiago, Chile*
Patchineelam, S.: *Environmental Geochemistry Department, Fluminense Federal University, Niteroi, RJ, Brazil*
Patinha, C.: *GEOBIOTEC, Universidade de Aveiro, Campus de Santiago, Aveiro, Portugal*
Patiño Reyes, N.: *Laboratorio de Salud Pública, Secretaria Distrital de Salud de Bogotá, Universidad Nacional de Colombia, Colombia*
Patop, I.: *Facultad de Cs. Exactas y Naturales, Universidad de Buenos Aires, Argentina*
Paula, E.S.: *Faculdade de Ciências Farmacêuticas de Ribeirão Preto, Universidade de São Paulo, São Paulo, Brazil*
Paul-Brutus, R.: *U-Chicago Research Bangladesh, Ltd., Dhaka, Bangladesh*
Paulelli, A.C.C.: *Faculdade de Ciências Farmacêuticas de Ribeirão Preto, Universidade de São Paulo, São Paulo, Brazil*
Pavlikova, D.: *Department of Agroenvironmental Chemistry and Plant Nutrition, Czech University of Life Sciences, Prague, Czech Republic*

Peacey, J.: *Queen's University, Canada*
Pellizzari, E.E.: *Laboratorio de Microbiología, Universidad Nacional del Chaco Austral, P.R. Sáenz Peña, Chaco, Argentina*
Peng, L.: *College of Resource and Environment, Hunan Agricultural University, Changsha, PR China*
Pereira, V.: *GEOBIOTEC, GeoBiosciences, Geotechnologies and Geoengineering Research Center, University of Aveiro, Portugal*
Pereyra, S.: *Instituto de Altos Estudios Sociales, Universidad Nacional de San Martín, Campus Miguelete, San Martín, Provincia de Buenos Aires, Argentina*
Pérez Carrera, A.: *Instituto de Investigaciones en Producción Animal (INPA), CONICET, Facultad de Ciencias Veterinarias, Universidad de Buenos Aires, Buenos Aires, Argentina*
Pérez Coll, C.S.: *Instituto de Investigación e Ingeniería Ambiental, UNSAM, San Martín, Argentina*
Pérez, A.L.: *INQUIMAE/DQIAQF, Facultad de Ciencias Exactas, UBA, Buenos Aires, Argentina*
Perez, C.A.: *Laboratório Nacional de Luz Síncrotron (LNLS), Campinas, Brazil*
Pérez, R.D.: *School of Mathematics, Astronomy and Physics, (CONICET-CCT CORDOBA), National University of Cordoba, Cordoba, Argentina*
Pérez, Roberto D.: *National University of Córdoba, Córdoba, Argentina*
Persson, I.: *Department of Chemistry, Swedish University of Agricultural Sciences, Uppsala, Sweden*
Petrusevski, B.: *UNESCO-IHE Institute for Water Education, Delft, The Netherlands*
Picado, M.: *La Voz de Bagaces Association, Costa Rica, USA*
Picco, P.: *National Technological University-Faculty of Trenque Lauquen (UTN-FRTL), Buenos Aires, Argentina*
Piccoli, P.: *Instituto de Biología Agrícola de Mendoza (IBAM), CONICET, Mendoza, Argentina*
Pidustwa, V.: *Quality Control Management, "Ente Regulador de Servicios Sanitarios", Province of Santa Fe, Argentina*
Pineda-Chacón, G.: *Centro de Investigación en Materiales Avanzados (CIMAV), Chihuahua, México*
Pinheiro, B.: *Centre of Mineral Technology, Rio de Janeiro, Brazil*
Pistón, M.: *Facultad de Química, Química Analítica, DEC, Universidad de la República (UDELAR), Montevideo, Uruguay*
Pizarro, G.E.: *Department of Hydraulic and Environmental Engineering, Pontificia Universidad Católica de Chile, Santiago, Chile*
Plá, R.: *Nuclear Chemistry Department, National Atomic Energy Commission of Argentina, Buenos Aires, Argentina*
Planer-Friedrich, B.: *Environmental Geochemistry Group, University of Bayreuth, Germany*
Pope, J.: *CRL Energy Ltd., Christchurch, New Zealand*
Pradeep, T.: *DST Unit of Nanoscience and Thematic Unit of Excellence, Department of Chemistry, Indian Institute of Technology Madras, Chennai, India*
Pratas, J.: *Faculty of Sciences and Technology, Department of Earth Sciences, University of Coimbra, Coimbra, Portugal*
Pratts, P.B.: *Ministerio de Obras y Servicios Públicos, Ministerio de Salud & Secretaría de Recursos Hídricos, La Pampa, Argentina*
Promige, M.A.L.: *Faculty of Animal Husbandry, Bangladesh Agricultural University, Mymensingh, Bangladesh*
Puente-Valenzuela, C.O.: *Universidad Juárez del Estado de Durango, Gomez Palacio, Durango, México*
Qin, P.: *College of Resource and Environment, Hunan Agricultural University, Changsha, PR China*
Queirolo, E.I.: *Center for Research, Catholic University of Uruguay, Uruguay*
Queralt, Ignasi: *Jaume Almera Institute, (ICTJA-CSIC), Barcelona, Spain*
Quevedo, H.: *Centro de Ingeniería Sanitaria, Facultad de Ciencias Exactas, Ingeniería y Agrimensura, Universidad Nacional de Rosario, Argentina*
Quici, N.: *Gerencia Química, Comisión Nacional de Energía Atómica, San Martín, Provincia de Buenos Aires, Argentina*
Quino, I.: *Instituto de Investigaciones Químicas (IIQ), Universidad Mayor de San Andrés, La Paz, Bolivia*
Quiñones, O.: *Aquatech Internacional SA de CV, México*
Quintanilla, J.: *Instituto de Investigaciones Químicas (IIQ), Universidad Mayor de San Andrés, La Paz, Bolivia*
Quintero, C.: *Facultad de Ciencias Agropecuarias, Universidad Nacional de Entre Ríos, Paraná, ER, Argentina*
Quinteros, E.: *Instituto Nacional de Salud, Ministerio de Salud, San Salvador, El Salvador*
Quinteros, L.: *Nuevo Hospital San Roque, Córdoba, Argentina*

Quiquinto, A.: *Grupo INQA, Facultad de Ciencias Agrarias, Universidad Nacional de Jujuy, Argentina*
Raber, G.: *Institute of Chemistry, Analytical Chemistry, Graz, Austria*
Rahman, M.A.: *Department of Chemistry, University of Dhaka, Dhaka, Bangladesh*
Rakibuz-Zaman, M.: *U-Chicago Research Bangladesh, Ltd., Dhaka, Bangladesh*
Rakotoarisoa, O.: *University of Limoges, GRESE, Limoges Cedex, France*
Ramanathan, AL.: *School of Environmental Sciences, Jawaharlal Nehru University, New Delhi, India*
Ramila, C.dP.: *Department of Hydraulic and Environmental Engineering, Pontificia Universidad Católica de Chile, Santiago, Chile*
Ramírez Alvarado, J.A.: *Posgrado en Ingeniería Universidad Nacional Autónoma de México, Campus Morelos-IMTA, México*
Ramirez, A.: *Departamento de Tecnología Agroalimentaria, Universidad Miguel Hernández, Ctra. Beniel, Orihuela, Alicante, Spain*
Ramos Chávez, L.: *Departamento de Medicina Genómica y Toxicología Ambiental, Instituto de Investigaciones Biomédicas, Universidad Nacional Autónoma de México, México*
Ramos Ramos, O.E.: *Instituto de Investigaciones Químicas (IIQ), Universidad Mayor de San Andrés, La Paz, Bolivia; KTH-International Groundwater Arsenic Research Group, Division of Land and Water Resources Engineering, Department of Sustainable Development, Environmental Science and Engineering, Royal Institute of Technology (KTH), Teknikringen, Stockholm, Sweden*
Ramos, C.: *Gerencia Química, Comisión Nacional de Energía Atómica, San Martín, Provincia de Buenos Aires, Argentina*
Ramos, R.G.: *Dow Water and Process Solutions, Saudi Arabia*
Raqib, R.: *International Center for Diarrhoeal Disease in Bangladesh, Dhaka, Bangladesh*
Rastelli, S.E.: *Centro de Investigación y Desarrollo de Pinturas (CIDEPINT, CICPBA, CCT La Plata CONICET), La Plata, Argentina; Facultad de Ciencias Naturales y Museo, Universidad Nacional de La Plata, Argentina*
Raychowdhury, N.: *Department of Geology and Geophysics, Indian Institute of Technology-Kharagpur, WB, India*
Recio Vega, R.: *Salud Ambiental, Facultad de Medicina de la Universidad Autónoma de Coahuila, México*
Reddy, K.J.: *Department of Ecosystem Science and Management, University of Wyoming, Wyoming, USA*
Regan, J.M.: *Department of Civil and Environmental Engineering, The Pennsylvania State University, University Park, PA, USA*
Reina, F.D.: *Departamento de Ingeniería Química, Facultad Regional Buenos Aires, Universidad Tecnológica Nacional, Argentina*
Reis, A.P.: *GEOBIOTEC, University of Aveiro, Campus Universitário de Santiago, Aveiro, Portugal*
Ren, Y.: *College of Resource and Environment, Hunan Agricultural University, Changsha, PR China*
Requejo, F.: *Instituto de Investigaciones Fisicoquímicas Teóricas y Aplicadas, La Plata, Argentina*
Resnick, C.: *Department of Environmental Health Sciences, BSPH, Johns Hopkins University, Baltimore, USA*
Resnizky, S.M.: *Nuclear Chemistry Department, National Atomic Energy Commission of Argentina, Buenos Aires, Argentina*
Rial, E.: *Fundación Barrera Zoofitosanitaria Patagónica (Funbapa), Viedma, Rio Negro, Argentina*
Ribeiro Guevara, S.: *Laboratorio de Análisis por Activación Neutrónica, Centro Atómico Bariloche, CNEA, Bariloche, Argentina*
Ribó, A.: *Instituto Nacional de Salud, Ministerio de Salud, San Salvador, El Salvador*
Ríos de Molina, M. del C.: *Facultad de Ciencias Exactas y Naturales, Universidad de Buenos Aires, Argentina; IQUIBICEN-CONICET, Buenos Aires, Argentina*
Ríos, P.: *National Direction of Epidemiological Surveillance, Ministry of Health, Ecuador*
Risser, T.: *LIEC, CNRS-UMR 7360, Université de Lorraine, Metz, France*
Rivera-Hernández, J.R.: *Posgrado en Ciencias del Mar y Limnología, Universidad Nacional Autónoma de México, Sin., México*
Rizzo, A.: *Laboratorio de Análisis por Activación Neutrónica, Centro Atómico Bariloche, CNEA, Bariloche, Argentina; CONICET, Argentina*
Robles, A.D.: *Toxicología Ambiental, Departamento de Ciencias Marinas, Instituto de Investigaciones Marinas y Costeras, Facultad de Ciencias Exactas y Naturales, Universidad Nacional de Mar del Plata, Mar del Plata, Argentina; Consejo Nacional de Investigaciones Científicas y Técnicas (CONICET), Buenos Aires, Argentina*
Rodrigues, G.: *Centro de Ciências Naturais e Humanas, Universidade Federal do ABC, Brazil*

Rodriguez Castro, M.C.: *Universidad Nacional de Luján, Buenos Aires, Argentina; CONICET, Argentina*
Rodriguez, A.I.: *INQUIMAE/DQIAQF, Facultad de Ciencias Exactas y Naturales, UBA, CABA, Argentina*
Rodriguez, A.V.: *INQUIMAE/DQIAQF, Facultad de Ciencias Exactas y Naturales, UBA, CABA, Argentina*
Rodríguez, D.: *Instituto de Biología y Medicina Experimental (IByME, CONICET), CABA, Buenos Aires, Argentina*
Rodriguez, D.J.: *Departamento de Ciencias Básicas, Universidad Nacional de Lujan, Luján, Provincia de Buenos Aires, Argentina*
Rodriguez, I.: *Geophysics Institute, Universidad Nacional Autónoma de México, México*
Rodríguez, R.: *Geophysics Institute, Universidad Nacional Autónoma de México, México*
Rodriguez-Lado, L.: *Soil Science and Agricultural Chemistry, University of Santiago de Compostela, Spain*
Rojas, J.: *Departamento de Química, Facultad de Ciencias, Universidad Nacional de Colombia, Bogotá, Colombia*
Rojas, P.: *Instituto Tecnológico de Costa Rica, Cartago, Costa Rica*
Roldán, C.S.: *F.A.C.A. (UNComa), Cinco Saltos, Río Negro, Argentina*
Román, M.D.: *Escuela de Nutrición e Instituto de Investigaciones en Ciencias de la Salud, CONICET, Facultad de Ciencias Médicas, Universidad Nacional de Córdoba, Córdoba, Argentina*
Romero, C.: *Facultad de Medicina, Universidad Nacional de Tucumán, Tucumán, Argentina*
Romero, L.: *Universidad Politécnica de Morelos, Jiutepec, Morelos, Mexico*
Romero, L.G.: *Instituto Tecnológico de Costa Rica, Cartago, Costa Rica*
Romero, M.B.: *Toxicología Ambiental, Departamento de Ciencias Marinas, Instituto de Investigaciones Marinas y Costeras, Facultad de Ciencias Exactas y Naturales, Universidad Nacional de Mar del Plata, Mar del Plata, Argentina; Consejo Nacional de Investigaciones Científicas y Técnicas (CONICET), Buenos Aires, Argentina*
Ronco, A.M.: *Instituto de Nutrición y Tecnología de los Alimentos (INTA), University of Chile, Chile*
Rosales, B.M.: *Centro de Investigación y Desarrollo de Pinturas (CIDEPINT, CICPBA, CCT La Plata CONICET), La Plata, Argentina*
Rosas-Castor, J.M.: *Facultad de Ciencias Químicas, Universidad Autónoma de Nuevo León, Nuevo León, México*
Rosen, B.P.: *Herbert Wertheim College of Medicine, Florida International University, Miami, USA*
Rosso, E.: *JLA Argentina S.A., Bv. Italia, Gral. Cabrera, Córdoba, Argentina*
Rothenberg, S.J.: *Instituto Nacional de Salud Pública, Cuernavaca, México*
Rousselle, P.: *LIEC, CNRS-UMR 7360, Université de Lorraine, Metz, France*
Roy, S.: *U-Chicago Research Bangladesh, Ltd., Dhaka, Bangladesh*
Rozo-Correa, C.E.: *Santo Tomas University, Bucaramanga, Colombia*
Rubio-Andrade, M.: *Facultad de Medicina, UJED, Gómez Palacio, México*
Rubio-Campos, B.E.: *Departamento de Ingeniería Química, División de Ciencias Naturales y Exactas, Campus Guanajuato, Universidad de Guanajuato, Guanajuato, México*
Ruiz-Ruiz, E.: *Facultad de Ciencias Químicas, Universidad Autónoma de Nuevo León, Nuevo León, México*
Sabadini-Santos, E.: *Curso de Pós-graduação em Geoquímica Ambiental, Universidade Federal Fluminense (UFF), Brazil*
Saha, C.: *Department of Clinical and Experimental Pharmacology, School of Tropical Medicine, Kolkata, India*
Sales, A.M.: *LABTRA, Facultad de Bioq., Química y Farmacia, UNT, Tucumán, Argentina*
Salomón, M.V.: *Instituto de Biología Agrícola de Mendoza (IBAM), CONICET, Mendoza, Argentina*
Sampayo-Reyes, Adriana: *Centro de Investigación Biomédica del Noreste, Instituto Mexicano del Seguro Social (IMSS), Monterrey, Mexico*
San Román, E.: *INQUIMAE/DQIAyQF, Facultad de Ciencias Exactas y Naturales, Universidad de Buenos Aires, Argentina*
Sanchez-Palacios, J.T.: *School of Botany, The University of Melbourne, Australia*
Sánchez-Peña, L.C.: *Centro de Investigación y de Estudios Avanzados del IPN, Departamento de Toxicología, México DF, México*
Sánchez-Ramirez, B.: *Facultad de Ciencias Químicas, Universidad Autónoma de Chihuahua, Chihuahua, México*
Sandrini, R.: *JLA Argentina S.A., Bv. Italia, Gral. Cabrera, Córdoba, Argentina*
Sanguinetti, G.S.: *Centro de Ingeniería Sanitaria, Facultad de Ciencias Exactas, Ingeniería y Agrimensura, Universidad Nacional de Rosario, Argentina*

Sankar, M.S.: *Department of Geology, Kansas State University, Manhattan, KS, USA*
Santos Pontes, B.M.: *PPG Minas, Pernambuco Federal University, Brazil*
Santos, M.C.B.: *Departamento de Geoquímica, Instituto de Química, Universidade Federal Fluminense (UFF), Centro, Niterói, RJ, Brasil*
Santos, V.S.: *Faculdade de Ciências Farmacêuticas de Ribeirão Preto, Universidade de São Paulo, São Paulo, Brazil*
Saralegui, A.: *Laboratorio de Química de Sistemas Heterogéneos (LaQuiSiHe), Departamento de Química, Universidad de Buenos Aires, Buenos Aires, Argentina*
Sarmiento Tagle, M.: *Carrera de Gestión Ambiental, Universidad Blas Pascal, Córdoba, Argentina*
Schenone, N.F.: *Instituto de Investigaciones en Producción Animal (INPA-CONICET-UBA), Facultad de Ciencias Veterinarias, Universidad de Buenos Aires, Buenos Aires, Argentina*
Schierano, M.C.: *Facultad Regional Rafaela, Universidad Tecnológica Nacional, Rafaela, Santa Fe, Argentina*
Schmidt, E.: *Departamento de Agronomía, UNS, Argentina*
Schmidt, S.: *Karlsruhe University of Applied Sciences, Germany*
Schulz, C.: *Faculty of Exact and Natural Sciences, UNLPam, Argentina*
Sekhar, B.M.V.S.: *CSIR-National Geophysical Research Institute, Hyderabad, India*
Selim, H.M.: *School of Plant, Environmental and Soil Sciences, Louisiana State University, Baton Rouge, LA*
SenGupta, A.: *Department of Civil and Environmental Engineering, Lehigh University, Bethlehem, USA*
Sequeira, M.: *CERZOS-CONICET, Argentina; Departamento de Ingeniería, UNS, Argentina*
Serce, H.: *Urgup Hospital, Nevsehir, Turkey*
Sereno, L.L.: *Ministerio de Obras y Servicios Públicos, Ministerio de Salud & Secretaría de Recursos Hídricos, La Pampa, Argentina*
Servant, R.E.: *Departamento de Química Analítica, Gerencia Química, Comisión Nacional de Energía Atómica, Argentina*
Shahina: *Centre of Advanced Study in Geology, University of Lucknow, Lucknow, India*
Shakya, S.K.: *Environment and Public Health Organization, New Baneshwor, Kathmandu, Nepal*
Shamsudduha, M.: *Institute for Risk and Disaster Reduction, University College London, London, UK*
Sharifi, R.: *Shiraz University, Shiraz, Iran*
Shekoohiyan, S.: *School of Public Health, Tehran University of Medical Sciences, Tehran, Iran*
Shen, H.: *Key Lab of Urban Environment and Health, Institute of Urban Environment, Chinese Academy of Science, Xiamen, China*
Shengheng, H.: *Institute of Technology of Cambodia, Phnom Penh, Cambodia*
Siddique, M.N.E.A.: *Department of Chemistry, University of Dhaka, Dhaka, Bangladesh*
Siegfried, K.: *Helmholtz Centre for Environmental Research, UFZ, Department of Environmental Microbiology, Leipzig, Saxony, Germany*
Sierra-Campos, E.: *Universidad Juárez del Estado de Durango, Gomez Palacio, Durango, México*
Sigrist, M.: *Programa de Investigación y Análisis de Residuos y Contaminantes Químicos, PRINARC, Facultad de Ingeniería Química, Universidad Nacional del Litoral, Santa Fe, Argentina*
Silbergeld, E.K.: *Department of Environmental Health Sciences, BSPH, Johns Hopkins University, Baltimore, USA*
Sileo, E.E.: *INQUIMAE, DQIADF, Facultad de Ciencias Exactas y Naturales, Universidad de Buenos Aires, Buenos Aires, Argentina*
Silva, G.C.: *Escola de Engenharia, Departamento de Engenharia Metalúrgica e de Materiais, Universidade Federal de Minas Gerais, Belo Horizonte, Brazil; INCT-Acqua, Belo Horizonte, Brazil; National Institutes of Science and Technology Acqua, Universidade Federal de Minas Gerais—UFMG, Belo Horizonte, Brazil*
Silva, J.: *EMBRAPA Vegetables, Brasília, Federal District, Brazil*
Silva, L.do N.: *Center for Mineral Technology, Ministry of Science, Technology and Innovation, Rio de Janeiro, Brazil; Polytechnic School, Universidade Federal do Rio de Janeiro, Brazil*
Silva, R.O.: *Instituto Agronômico (IAC), Campinas (SP), Brasil*
Silva, R.S.V.: *Departamento de Geoquímica, Instituto de Química, Universidade Federal Fluminense (UFF), Centro, Niterói, RJ, Brasil*
Silva, S.A.: *Universidade Federal de Viçosa, Viçosa, MG, Brazil*
Silva, S.C.: *Soil Department, Federal University of Viçosa, Viçosa, MG, Brazil*
Silva, V.: *Membrane Technical Managers at Cooperativa Obras Sanitarias de Venado Tuerto, Santa Fe, Argentina*

Simonnot, M.O.: *Laboratoire Réactions et Génie des Procédés (LRGP), Université de Lorraine-CNRS, Nancy Cedex, France*
Singh, I.B.: *Centre of Advanced Study in Geology, University of Lucknow, Lucknow, India*
Singh, J.K.: *Mahavir Cancer Institute and Research Centre, Patna, Bihar, India*
Singh, M.: *Centre of Advanced Study in Geology, University of Lucknow, Lucknow, India*
Singh, S.: *Department of Earth Sciences, Indian Institute of Technology, Roorkee, India*
Slavkovich, V.: *Columbia University, New York, USA; Department of Environmental Health Sciences, Mailman School of Public Health, Columbia University, New York, USA*
Šlejkovec, Z.: *Department of Environmental Sciences, Jožef Stefan Institute, Ljubljana, Slovenia*
Slokar, Y.M.: *Unesco-IHE Institute for Water Education, Delft, The Netherlands*
Smichowski, P.: *Departamento de Química Analítica, Gerencia Química, Comisión Nacional de Energía Atómica, Buenos Aires, Argentina*
Sohm, B.: *LIEC, CNRS-UMR 7360, Université de Lorraine, Metz, France*
Solari, C.A.: *Laboratorio de Biodiversidad, Ultraestructura y Ecofisiología de Microalgas, Instituto de Biodiversidad y Biología Experimental Aplicada, CONICET, Universidad de Buenos Aires, Buenos Aires, Argentina*
Soria, A.G.: *Facultad de Bioquímica, Química y Farmacia, Universidad Nacional de Tucumán, Tucumán, Argentina*
Soria, E.: *Laboratorio de Farmacotecnia, UNCAUS, P.R. Sáenz Peña, Chaco, Argentina*
Soriano-Pérez, S.H.: *Departamento de Ciencias Químicas, Universidad Autónoma de San Luis Potosí, SLP, México*
Soro, E.M.: *Instituto Nacional de Tecnología Agropecuaria, Estación Experimental Agropecuaria Concepción del Uruguay (NTA EEA), Provincia de Entre Ríos, Argentina*
Sosa, S.: *Facultad de Cs. Exactas y Naturales, Universidad de Buenos Aires, Argentina*
Sotomayor, V.: *IDEPA (CONICET-UNComa), Neuquén, Argentina*
Sousa, A.J.: *CERENA, Technical Superior Institute, Lisbon, Portugal*
Souza, J.M.O.: *Faculdade de Ciências Farmacêuticas de Ribeirão Preto, Universidade de São Paulo, São Paulo, Brazil*
Spagnoletti, F.N.: *Cátedra de Microbiología Agrícola, Facultad de Agronomía, UBA, Buenos Aires, Argentina; Instituto de Investigaciones en Biociencias Agrícolas y Ambientales, INBA, (CONICET-UBA), Ciudad Autónoma de Buenos Aires, Argentina*
Spallholz, J.E.: *Division of Nutritional Sciences, Texas Tech University, Lubbock, Texas, USA*
Sridhar, M.K.C.: *Department of Environmental Health Sciences, Faculty of Public Health, University of Ibadan, Ibadan, Nigeria*
Stekolchik, E.: *Department of Chemistry, Wagner College, Staten Island, New York, USA*
Stengel, C.: *Eawag, Swiss Federal Institute of Aquatic Science and Technology, Switzerland*
Steuerwald, A.J.: *New York State Department of Health, Albany NY, USA*
Stibilj, V.: *Department of Environmental Sciences, Jožef Stefan Institute, Ljubljana, Slovenia*
Stockwell, P.B.: *P S Analytical Ltd., Orpington, Kent, UK*
Stýblo, M.: *Department of Nutrition, University of North Chapel Hill, CH, NC, USA*
Suess, E.: *Institute of Biogeochemistry and Pollution Dynamics, ETH Zürich, Switzerland*
Sun, G.-X.: *State Key Laboratory of Urban and Regional Ecology, Research Center for Eco-Environmental Sciences, Chinese Academy of Sciences, Beijing, China*
Szakova, J.: *Department of Agroenvironmental Chemistry and Plant Nutrition, Czech University of Life Sciences, Prague, Czech Republic*
Szücs, Z.: *Institute for Nuclear Research of the Hungarian Academy of Sciences, Debrecen, Hungary*
Taga, R.: *Centre for Mined Land Rehabilitation, The University of Queensland, Brisbane, Queensland, Australia; National Research Centre for Environmental Toxicology (Entox), The University of Queensland, Brisbane, Queensland, Australia*
Talano, M.A.: *Departamento de Biología Molecular, FCEFQyN, Universidad Nacional de Río Cuarto, Río Cuarto, Córdoba, Argentina*
Tan, X.H.: *School of Medicine, Hangzhou Normal University, Hangzhou, Zhejiang, China*
Tasic, M.: *Water Supply Company of Subotica, Serbia*
Távora, R.: *Centro de Desenvolvimento Sustentável, Universidade de Brasília (CDS/UnB), Brazil; Community of Practice in Ecohealth, Latin America and Caribbean (CoPEH-LAC), Brasília/DF, Brazil*
Temporetti, C.: *Facultad de Ciencias Agropecuarias, UNER, Entre Ríos, Argentina*
Thakur, B.K.: *National Institute of Industrial Engineering (NITIE), Mumbai, India*
Thambidurai, P.: *Department of Earth Sciences, Indian Institute of Technology Bombay, Mumbai, India*

Thao, M.T.P.: *Center for Environmental Technology and Sustainable Development, Hanoi University of Science, Vietnam*
Thiry, Y.: *Agence Nationale Pour la Gestion des Déchets Radioactifs (Andra), Châtenay Malabry, France*
Thomas, D.J.: *National Exposure Research Laboratory, NERL, RTP, NC, USA*
Tlustos, P.: *Department of Agroenvironmental Chemistry and Plant Nutrition, Czech University of Life Sciences, Prague, Czech Republic*
Tobar, N.E.: *Instituto de Investigaciones en Biociencias Agrícolas y Ambientales, INBA, (CONICET-UBA), Ciudad Autónoma de Buenos Aires, Argentina*
Tojo, N.: *Departamento de Ingeniería Civil, Facultad Regional Buenos Aires, Universidad Tecnológica Nacional, Argentina; Facultad Regional Buenos Aires, UTN, Argentina*
Torres, E.: *Instituto de Diagnóstico Ambiental y Estudios del Agua, CSIC, Barcelona, Spain*
Trang, P.T.K.: *Center for Environmental Technology and Sustainable Development, Hanoi University of Science, Vietnam*
Travaglio, M.: *Fundación Barrera Zoofitosanitaria Patagónica (Funbapa), Viedma, Rio Negro, Argentina*
Travar, I.: *Luleå University of Technology, Luleå, Sweden; Ragn Sells AB, Stockholm, Sweden*
Trinelli, M.A.: *INQUIMAE/DQIAQF, Facultad de Ciencias Exactas, UBA, Buenos Aires, Argentina; Instituto de Investigación e Ingeniería Ambiental (3iA), UNSAM, Buenos Aires, Argentina; Departamento de Ciencias Geológicas, Facultad de Ciencias Exactas y Naturales, UBA, CABA, Argentina*
Trompetter, W.J.: *GNS Science, Lower Hutt, New Zealand*
Trovatto, M.M.: *Cátedra de Hidogeología, Facultad de Ciencias Naturales y Museo, Universidad Nacional de La Plata, La Plata, Argentina*
Trupti, G.: *Department of Earth Sciences, Indian Institute of Technology Bombay, Mumbai, India*
Tsadilas, C.D.: *Institute of Soil Mapping and Classification, Larissa, Greece*
Tsakiroglou, C.D.: *Foundation for Research and Technology Hellas, Institute of Chemical Engineering Sciences, Platani, Patras, Greece*
Tseng, Y.-C.: *Department of Environmental and Occupational Health, National Cheng Kung University, Medical College, Tainan, Taiwan*
Tsitouras, A.: *Institute of Soil Mapping and Classification, Larissa, Greece*
Tufo, A.E.: *INQUIMAE, DQIADF, Facultad de Ciencias Exactas y Naturales, Universidad de Buenos Aires, Buenos Aires, Argentina*
Tumenbayar, B.: *Sans Frontiere Progres NGO, Mongolia; Institute for Geo-Resources and Environment, AIST, Japan*
Tuncer, A.M.: *Hacettepe University, Ankara, Turkey*
Tziouvalekas, M.: *Institute of Soil Mapping and Classification, Larissa, Greece*
Uramgaa, J.: *Central Hospital of Mongolia, Mongolia*
Vaca Mier, M.: *Universidad Autónoma Metropolitana, Azcapotzalco, México*
Vahter, M.: *Institute of Environmental Medicine, Section for Metals and Health, Karolinska Institutet, Stockholm, Sweden*
Valdovinos Flores, C.: *Departamento de Medicina Genómica y Toxicología Ambiental, Instituto de Investigaciones Biomédicas, Universidad Nacional Autónoma de México, México*
Valiente, L.: *Instituto Nacional de Tecnologia Industrial, Argentina*
Valles-Aragón, M.C.: *Advanced Materials Research Center, CIMAV, Chihuahua, Mexico*
Valverde, J.: *Instituto Tecnológico de Costa Rica, Cartago, Costa Rica*
van de Wetering, S.: *Hoofd Ingenieursbureau, Brabant Water, BC 's-Hertogenbosch, The Netherlands*
van Ommen, C.: *Water Supply Company Vitens, The Netherlands*
van Paassen, J.: *Water Supply Company Vitens, The Netherlands*
Vangronsveld, J.: *Environmental Biology, Hasselt University, Diepenbeek, Belgium*
Vargas, I.T.: *Department of Hydraulic and Environmental Engineering, Pontificia Universidad Católica de Chile, Santiago, Chile*
Vargas, M.J.: *Instituto Tecnológico de Costa Rica, Cartago, Costa Rica*
Vasavi, G.V.: *CSIR-National Geophysical Research Institute, Hyderabad, India*
Vattino, L.: *Facultad de Cs. Exactas y Naturales, Universidad de Buenos Aires, Argentina*
Vázquez, C.: *Comisión Nacional de Energía Atómica y Facultad de Ingeniería, Universidad de Buenos Aires, Buenos Aires, Argentina*
Vázquez, H.P.: *Quality Control Management, "Ente Regulador de Servicios Sanitarios", Province of Santa Fe, Argentina*
Vega, A.S.: *Departamento de Ingeniería Hidráulica y Ambiental, Pontificia Universidad Católica de Chile, Santiago, Chile*

Vega, M.: *Department of Geology, Kansas State University, Manhattan, KS, USA*
Veloso, R.W.: *Department of Soils, Federal University of Viçosa, Viçosa, MG, Brazil*
Viana, I.B.: *Universidade Federal de Viçosa, Viçosa, MG, Brazil*
Viera, M.R.: *Centro de Investigación y Desarrollo de Pinturas (CIDEPINT, CICPBA, CCT La Plata CONICET), La Plata, Argentina; Facultad de Ciencias Exactas, Universidad Nacional de La Plata, Argentina*
Viet, P.H.: *Center for Environmental Technology and Sustainable Development, Hanoi University of Science, Vietnam*
Vignale, F.: *Facultad de Cs. Exactas y Naturales, Universidad de Buenos Aires, Argentina*
Vilca, P.: *Centro de Investigaciones del Hombre en el Desierto, CIHDE, Arica, Chile*
Villaamil Lepori, E.C.: *Cátedra de Toxicología y Química Legal, Facultad de Farmacia y Bioquímica, Universidad de Buenos Aires, Argentina*
Villagra Cocco, A.: *Nuevo Hospital San Roque, Córdoba, Argentina*
Villalobos, M.: *Departamento de Geología, Universidad Nacional Autónoma de México, Ciudad Universitaria, México D.F., México*
Villarroel, L.: *Department of Public Health, Faculty of Medicine, Pontifical Catholic University of Chile, Chile*
Viniegra-Morales, D.: *Colegio de Médicos Cirujanos y Homeópatas del Estado de Chihuahua, A.C., Nayarit, México*
Vítková, M.: *Department of Environmental Geosciences, Faculty of Environmental Sciences, Czech University of Life Sciences, Prague, Czech Republic*
Vodopivez, C.L.: *Argentinean Antarctic Institute (IAA), Buenos Aires, Argentina*
Volpedo, A.V.: *Instituto de Investigaciones en Producción Animal (INPA-CONICET-UBA), Facultad de Ciencias Veterinarias, Universidad de Buenos Aires, Buenos Aires, Argentina*
von Brömssen, M.: *Ramböll Sweden AB, Stockholm, Sweden; KTH-International Groundwater Arsenic Research Group, Department of Land and Water Resources Engineering, Royal Institute of Technology (KTH), Teknikringen, Stockholm, Sweden*
Wang, J.P.: *National Research Centre for Environmental Toxicology, The University of Queensland, Brisbane, Australia*
Wang, J.S.: *College of Water Sciences, Beijing Normal University, Beijing, China; Engineering Research Center of Groundwater Pollution Control and Remediation, Ministry of Education, Beijing, China*
Wang, P.-P.: *State Key Laboratory of Urban and Regional Ecology, Research Center for Eco-Environmental Sciences, Chinese Academy of Sciences, Beijing, China*
Wang, X.B.: *School of Medicine, Hangzhou Normal University, Hangzhou, Zhejiang, China*
Wang, Y.-H.: *Division of Urology, Department of Surgery, Taipei Medical University, Shuang Ho Hospital, New Taipei City, Taiwan; Graduate Institute of Clinical Medicine, College of Medicine, Taipei Medical University, Taipei, Taiwan*
Wang, Y.-J.: *Department of Environmental and Occupational Health, National Cheng Kung University, Medical College, Tainan, Taiwan*
Wang, Y.X.: *China University of Geosciences, Wuhan, China*
Wang, Z.: *Key Laboratory of Urban Environment and Health, Institute of Urban Environment, Chinese Academy of Sciences, Xiamen, China*
Ward, N.I.: *ICP-MS Facility, Department of Chemistry, University of Surrey, Guildford, UK*
Waring, J.: *Ecochemistry Laboratory, Institute for Applied Ecology, University of Canberra, Bruce, Australia*
Watts, M.J.: *British Geological Survey, Keyworth, Nottingham, UK; Patagonia BBS, General Roca, Río Negro, Argentina*
Weaver, V.: *Department of Environmental Health Sciences, BSPH, Johns Hopkins University, Baltimore, USA*
Webb, O.A.: *Thermochemistry Laboratory, Department of Chemistry, University of Surrey, UK*
Weinzettel, P.A.: *Instituto de Hidrología de Llanuras, Comisión de Investigaciones Científicas (CIC), Azul, Argentina*
Williams, P.N.: *The University of Nottingham, Malaysia Campus, Semenyih, Malaysia; Lancaster Environment Centre, Lancaster University, Lancaster, UK*
Wilson, N.: *Golder Associates (NZ) Limited, Takapuna, New Zealand*
Winkel, L.H.E.: *Eawag, Swiss Federal Institute of Aquatic Science and Technology, Switzerland*
Woodrow, I.E.: *School of Botany, The University of Melbourne, Australia*
Wragg, J.: *British Geological Survey, Keyworth, Nottingham, UK*

Wu, F.: *New York University, New York, USA; Departments of Environmental Medicine, New York University School of Medicine, New York, USA*

Wu, M.M.: *Graduate Institute of Oncology, National Taiwan University, Taipei, Taiwan*

Wu, Y.: *State Key Laboratory of Biogeology and Environmental Geology, China University of Geosciences, Beijing, P.R. China*

Wuilloud, R.G.: *Laboratory of Analytical Chemistry for Research and Development (QUIANID), ICB-UNCuyo, Mendoza, Argentina; Consejo Nacional de Investigaciones Científicas y Técnicas (CONICET), Argentina*

Xu, W.: *Key Lab of Urban Environment and Health, Institute of Urban Environment, Chinese Academy of Science, Xiamen, China*

Xu, Y.G.: *Xiyuan Hospital, China Academy of Traditional Chinese Medicine, Beijing, China*

Xue, X.-M.: *Key Lab of Urban Environment and Health, Institute of Urban Environment, Chinese Academy of Sciences, Xiamen, China*

Yan, C.: *Key Laboratory of Urban Environment and Health, Institute of Urban Environment, Chinese Academy of Sciences, Xiamen, China*

Yang, J.: *College of Water Sciences, Beijing Normal University, Beijing, China; Engineering Research Center of Groundwater Pollution Control and Remediation, Ministry of Education, Beijing, China*

Yang, L.: *Ministry of Education Key Laboratory of Xinjiang Endemic and Ethnic Disease, Shihezi University, China; School of Medicine, Hangzhou Normal University, Hangzhou, Zhejiang, China*

Yang, T.-Y.: *Molecular and Genomic Epidemiology Center, China Medical University Hospital, China Medical University, Taichung, Taiwan*

Ye, J.: *Key Lab of Urban Environment and Health, Institute of Urban Environment, Chinese Academy of Sciences, Xiamen, China*

Yin, X.-X.: *State Key Laboratory of Urban and Regional Ecology, Research Center for Eco-Environmental Sciences, Chinese Academy of Sciences, Beijing, China*

Yoshinaga, M.: *Herbert Wertheim College of Medicine, Florida International University, Miami, USA*

Yousefi, N.: *School of Public Health, Tehran University of Medical Sciences, Tehran, Iran; Center for Solid Waste Research, Institute for Environmental Research, Tehran University of Medical Sciences, Tehran, Iran*

Yue, W.F.: *College of Water Sciences, Beijing Normal University, Beijing, China; Engineering Research Center of Groundwater Pollution Control and Remediation, Ministry of Education, Beijing, China*

Yunesian, M.: *Center for Solid Waste Research, Institute for Environmental Research, Tehran University of Medical Sciences, Tehran, Iran*

Yunus, M.: *Columbia University Arsenic Research Project, Dhaka, Bangladesh*

Zacaron, R.: *Centre of Mineral Technology, Rio de Janeiro, Brazil*

Zahid, A.: *Ground Water Hydrology, Bangladesh Water Development Board, Dhaka, Bangladesh*

Zalazar, C.S.: *INTEC (UNL-CONICET), Güemes, Santa Fe, Argentina; FICH (UNL), Ciudad Universitaria, Santa Fe, Argentina*

Zamboni, W.Z.: *Environmental Geochemistry Department, Fluminense Federal University, Niteroi, RJ, Brazil*

Zamero, M.: *Facultad de Ciencias Agropecuarias, Universidad Nacional de Entre Ríos, Paraná, ER, Argentina*

Zamoiski, R.: *Department of Environmental Health Sciences, BSPH, Johns Hopkins University, Baltimore, USA*

Zamora, A.: *National Technological University-Mar del Plata (UTN-MDP), Buenos Aires, Argentina*

Zdrenka, R.: *Institute of Hygiene and Occupational Medicine, University Hospital Essen, Essen, Germany*

Zeng, Q.: *College of Resource and Environment, Hunan Agricultural University, Changsha, PR China*

Zhang, D.: *State Key Laboratory of Biogeology and Environmental Geology, China University of Geosciences, Beijing, P.R. China; School of Water Resources and Environment, China University of Geosciences, Beijing, P.R. China*

Zhang, H.: *Lancaster Environment Centre, Lancaster University, Lancaster, UK*

Zhang, J.: *Key Lab of Urban Environment and Health, Institute of Urban Environment, Chinese Academy of Science, Xiamen, China*

Zhang, S.-Y.: *State Key Laboratory of Urban and Regional Ecology, Research Center for Eco-Environmental Sciences, Chinese Academy of Sciences, Beijing, China*

Zhao, D.: *State Key Lab of Pollution Control and Resource Reuse, School of Environment, Nanjing University, China*

Zhao, F.-J.: *Rothamsted Research, Harpenden, Hertfordshire, UK*

Zhao, K.: *School of Water Resources and Environment, China University of Geosciences, Beijing, China*

Zhao, L.: *Centre for Mined Land Rehabilitation (CMLR), Sustainable Minerals Institute (SMI), The University of Queensland, Brisbane, Queensland, Australia*
Zheng, Y.: *Queens College of City University of New York, Flushing, New York, USA; Lamont-Doherty Earth Observatory of Columbia University, Palisades, New York, USA*
Zhou, X.Q.: *School of Water Resources and Environment, China University of Geosciences, Beijing, China*
Zhu, Y.-G.: *State Key Laboratory of Urban and Regional Ecology, Research Center for Eco-Environmental Sciences, Chinese Academy of Sciences, Beijing, China; Key Lab of Urban Environment and Health, Institute of Urban Environment, Chinese Academy of Sciences, Xiamen, China*
Zuo, R.: *College of Water Sciences, Beijing Normal University, Beijing, China; Engineering Research Center of Groundwater Pollution Control and Remediation, Ministry of Education, Beijing, China*

Plenary presentations

One Century of the Discovery of Arsenicosis in Latin America (1914–2014) – Litter, Nicolli, Meichtry, Quici, Bundschuh, Bhattacharya & Naidu (Eds)
© 2014 Taylor & Francis Group, London, ISBN 978-1-138-00141-1

Arsenic in mining: Sources and stability

V.S.T. Ciminelli
Department of Metallurgical and Materials Engineering, Universidade Federal de Minas Gerais, Belo Horizonte, Brazil
National Institute of Science and Technology on Mineral Resources, Water and Biodiversity, INCT-Acqua, Brazil

ABSTRACT Mining and metallurgical activities involve the disposal of arsenic residues, in amounts that are expected to increase due to population and industrial growth combined with mining of low-grade ores. Key questions arise: What are the conditions that will ensure the structural and chemical stability of the wastes and tailings materials now and in the foreseeable future? How can one rigorously assess the potential, long-lasting risks associated to disposal of arsenic residues related to the mineral sector? In addition to the traditional tests for classification of residues, other tools are needed and increasingly used. These tools include assessment of bioaccessibility and bioavailability to better evaluate the risks to human health, and advanced analytical techniques to understand at a molecular level the mechanisms of arsenic fixation in the environment. This information is expected to provide the technical and scientific basis for the design of better strategies for As fixation in mining tailings.

1 INTRODUCTION

Mining and metallurgical activities are sources of arsenic wastes, thus creating potential sources for contamination. Therefore, the long-term stability of the mining residues should be well understood and controlled in order to prevent further As release into the environment. The weathering of As-bearing rocks may release arsenic to water and soil (Smedley & Kinniburgh, 2002). Arsenical residues may be also eroded and spread as fallout dust. The long-term stability of arsenic compounds is a function of several parameters, including site characteristics, particle size and crystallinity, presence/absence of oxygen, complexing agents, among others (Riveros *et al.*, 2001). In a review focusing on the stability of As residues under industrial conditions, Harris (2000) stated that "the situation regarding the long-term stability is perhaps not as clear as it appeared to be previously". Ritcey (2005) pointed out the existence of few works focusing on the stability of arsenical residues produced under industrial conditions and exposed to the conditions found in a disposal environment. We shall demonstrate the important advances that took place in the last decade in understanding the stability of arsenic wastes generated from mining and metallurgical operations. The advances were supported by the use of advanced analytical tools, combined with multidisciplinary efforts that led to a better understanding of the geological, environmental, biological and medical interfaces of arsenic. This knowledge can, in turn, be used to develop improved, long-term storage or disposal options.

2 ARSENIC WASTES IN MINING AND METALLURGY

Arsenic is one of the contaminants of concern frequently found in wastes from base metals (e.g. copper), uranium and gold mining and metallurgy. Waste rock generated during mining, tailings from mineral processing and metallurgy, exposed material in the final pit are examples of As sources. In these cases, the main arsenic-bearing minerals are arsenopyrite (FeAsS), arsenian pyrite ($Fe(As,S)_2$, scorodite ($FeAsO_4.2H_2O$), amorphous or crystalline ferric arsenates (Nordstrom, 2012). Arsenopyrite is typically found in the final mine pit, in the flotation and hydrometallurgical tailings, and in waste rock material. Scorodite is found in the wastes and tailings of weathered orebody; it is a product of the hydrothermal oxidation of sulfides, for instance, in the pretreatment of refractory gold ores. In regions of former artisanal mining, arsenic is usually found as co-precipitates or adsorbed on the sediments.

In order to comply with increasingly stringent environmental legislation, arsenic removal from aqueous and gaseous effluents from processing As-bearing ores often becomes necessary, thus creating a second category of residues to be disposed of. Coprecipitation methods using iron salts are the predominant methods in effluent treatment in copper and gold industry. From these methods, the arsenical ferrihydrite process at ambient temperatures and the hydrothermal precipitation of iron arsenates are the most accepted methods for As fixation. The arsenical ferrihydrite process comprises both precipitation of amorphous ferric arsenates and adsorption onto iron oxyhydroxides (Pantuzzo *et al.*, 2010). Large amounts of residues with variable arsenic content (e.g. 3–8% As in arsenical ferrihydrite process) are produced in the effluent treatment, thus requiring large areas for waste disposal. Precipitation with lime or lime combined with ferric salts follows in importance as industrial practices. Nevertheless, it is well established that the disposal of calcium arsenates should be avoided due to the well known instability in the presence of carbon dioxide or carbonate (Robins, 1983). Conventional processing of precious and base metal concentrates containing arsenopyrite by roasting will produce fumes with volatile and soluble arsenic trioxide. The gaseous effluent containing As_2O_3 is usually quenched in water. Arsenic is dissolved, further oxidized to As(V), precipitated as calcium or ferric arsenate, separated by sedimentation and filtration and finally disposed of.

3 STABILITY OF ARSENIC WASTES

Stability Eh-pH diagrams can summarize the thermodynamic stability of the various As phases usually found in the mining areas (Lu & Zhu, 2011). Under acidic conditions, inorganic, soluble arsenite (predominantly $H_3 AsO_3$ under reducing conditions) is more toxic, and more poorly adsorbed on mineral sorbents present in soils and sediments than arsenate ($H_3 AsO_4$). The chemical stability of arsenopyrite is favored under reducing conditions, absence of oxygen and circumneutral pH, such as the conditions provided by a water column in a tailings dams. Sulfide oxidation and dissolution kinetics may be enhanced by microbial reactions under acidic, aerated conditions. Chemical oxidation is favoured under alkaline conditions. Wet/dry cycles and intermitent exposure to oxygen of arsenopyrite wastes, as often found in the "beaches" of tailings dams, should be avoided, as they will favor arsenic dissolution and mobilization. Conversely, scorodite is the stable phase under oxidizing conditions. As pH increases and iron oxy-hydroxides precipitates, arsenic fixation will occur via adsorption and co-precipitation. All these reactions are reported to occur in deposits of tailings (Bissacot & Ciminelli, 2014). Remobilization is expected to occur under strong alkaline conditions, due to the formation of soluble iron and arsenate species. Under reducing conditions, the ferric oxy(hydroxides) undergo reduction and dissolution, with the subsequent release of arsenic as well.

The disposal of solid residues should comply with local regulations. The solids are usually treated by extraction methods to measure the degree of target elements or mobilization of substances under specified dissolution conditions, such as those of the Toxicity Characteristic Leaching Procedure—TCLP (USEPA, 2008) and its variations. In addition to the extraction conditions and the mechanism of dissolution of a given mineral phase, factors such as degree of crystallinity, particle size distribution, and particle morphology are known to affect the dissolution behavior. In this context, the usefulness of the EPA TCLP method 1312 or SPLP method 1311 (USEPA, 2008) to classify mineral processing wastes, in special sulfide tailings, has been argued. Nevertheless, some general conclusions can be drawn from short-term extraction tests applied to arsenical residues. Amorphous ferric arsenate phases, typical of ferrihydrite process, are shown to be more soluble than the crystalline arsenates obtained under hydrothermal conditions. Arsenic release from arsenical ferrihydrite decreases with the increase of Fe/As molar ratio. Dissolved arsenic from calcium arsenates is reported more than 1000 times greater than that released from crystalline and amorphous arsenates (Krause & Ettel, 1989; Swash & Monhemius, 1995a,b). Ladeira & Ciminelli (2004) showed the importance of arsenic oxidation state, the nature of sorbent phase and solution composition in the As leached in TCLP tests. While only 1–2% max. of As(V) was released from As-sorbed on goethite, gibbsite and on enriched Fe- and Al oxisol, leaching of the As(III) reached 32%, the highest values corresponding to the solutions containing sulfate ions. Oxisol and goethite were superior to gibbsite with respect to As immobilization. In another investigation, particle size, morphology, ageing and morphology were shown to have important effects on dissolution behavior in scorodite TCLP-leachability. Arsenic dissolution from precipitated scorodite in TCLP tests was shown to decrease from 13.6 to 0.1 mg L^{-1} as precipitation time increased. After ageing, scorodite produced under atmospheric conditions showed soluble TCLP-arsenic concentrations similar to those from scorodite synthesized under hydrothermal conditions. Round shaped scorodite particles showed an arsenic leachability higher than that of plate-like shaped scorodite for particles with the same specific surface area (Caetano *et al.*, 2009). The use of organic covers was shown to affect the geochemical stability of As tailings (Paktunc, 2013). The reducing conditions imposed by organic cover materials led to the formation of readily soluble secondary minerals. Arsenic concentrations in the TCLP leachate was shown to gradually increase from less than 0.085 to 13 mg L^{-1} and Fe from 28 to 179 mg L^{-1} towards the biosolids. This finding reinforces the need of in-depth investigations prior to the reutilization of mining sites as a closure option. The need to better assess the risks posed by a given substance, has led to other approaches, described in the following paragraphs.

4 BIOACCESSIBILITY AND ARSENIC SPECIATION

In vivo and *in vitro* bioavailability and bioaccessibility tests are used as the main indicators of potential risk that a chemical poses to environment and human health and, therefore, have become useful tools to determine As exposure. Main exposure occurs through ingestion of As-contaminated water and food crops. Unintentional ingestion of contaminated soils is also a source to be evaluated in mining sites (Naidu, 2012).

Arsenic occurs in the environment in various oxidation states (–3, –1, 0, +3, and +5). The mobility and toxicity of dissolved As vary depending on the aqueous speciation. In general, the oxidized forms of inorganic As (+3, +5) are considered more toxic than the organic compounds or the reduced forms found in the natural sulfide minerals (Smedley & Kinniburg, 2002; Brown 1999).

Inorganic arsenic in water is considered highly bioavailable (70–90%) through absorption in the gastrointestinal tract (Naidu, 2012; IARC, 2004). Naidu (2012) summarizes the findings from various investigations using different animal models to assess As bioavailability in As-contaminated soils and sites, such as smelter soils, mine sites, gossan soils, and former railway corridors, where contamination from arsenical pesticides were the main source. The bioaccessibility depends strongly on the As speciation and follows in general the overall trend indicated by TCLP tests. Arsenic(V) and As(III) (e.g. Ca ferric arsenate, arsenolite (As_2O_3), amorphous ferric arsenates, As-bearing ferric (oxy)hydroxides, in decreasing order, are more bioaccessible than As(0) and As (–1) (e.g. As-rich pyrite (FeS_2), arsenopyrite (FeAsS)). Crystalline scorodite ($FeAsO_4 \cdot 2H_2O$) show low bioaccessibility (Plumlee *et al.*, 2006). In a recent study, Toujagueza *et al.* (2013) found bioaccessible As ranging from 0.65 to 40.5% in mining tailings containing up to 35,372 mg As kg^{-1}. The authors ascribed maximum As bioaccessibility to the presence of goethite and amorphous Fe arsenate and low bioaccessibility to arsenic in arsenopyrite and scorodite. Smith *et al.* (2009) showed an increase in arsenic bioaccessibility with decreasing particle size, thus highlighting the importance of this parameter in assessing risk in contaminated environments.

The characterization of the As-bearing phases by traditional analytical techniques is not trivial (Parviainen *et al.*, 2012). Therefore, a convincing association of the results obtained from extraction methods, such as TCLP and bioaccessibility, and As-sources may prove challenging. In view of these difficulties, synchrotron-based analytical techniques have been increasingly applied to investigate As deportment, speciation, and bonding characteristics (Essilfie-Dughan *et al.*, 2013), with clear advances in our understanding of bioaccessibility and stability of mine tailings. This conclusion is supported by some results discussed in the previous paragraph. Bulk X-ray absorption spectroscopy has helped to identify the molecular environment of As in various matrices for more than a decade (Foster *et al.*, 1998). Micro-X-ray absorption spectroscopy (μ-XAS) (Paktunc *et al.*, 2004; Kwong *et al.*, 2007), micro-X-ray diffraction (μ-XRD) (DeSisto *et al.*, 2011), micro-X-ray fluorescence (μ-XRF), and combinations of these methods (Corriveau *et al.*, 2011, Landrot *et al.*, 2012, Toujagueza *et al.*, 2013) allowed one to discern various features of the different constituents in a complex matrix. A combination of synchrotron-based techniques with theoretical molecular modeling, or with other spectroscopic techniques, improved our understanding of the mechanisms of arsenic fixation in typical substrates found in the environment (Ladeira *et al.*, 2001; Müller *et al.*, 2010; Duarte *et al.*, 2012). The combination of conventional with sophisticated analytical tools has helped to advance in the assessment of the potential, long-lasting risks associated to disposal of arsenic residues generated by the mineral sector.

5 CONCLUSIONS

The feasibility and competitiveness of the mineral industry are increasingly dependent on the social approval. The mineral sector is driven to positively respond to all stakeholders concerns, and to demonstrate the safety of the industrial operations. Advanced analytical tools, combined with multidisciplinary efforts have improved our understanding of the stability of arsenic wastes. This information, in turn, helps to establish the conditions to improve the structural and chemical stability of mining wastes and tailings now and in the foreseeable future.

ACKNOWLEDGEMENTS

The author thanks the support of the Brazilian agencies (CNPq, CAPES and FAPEMIG). The collaboration with Massimo Gasparon, Jack Ng, Igor Vasconcelos and Helio Duarte is gratefully acknowledged. Special thanks to Dr. Gabriela Cordeiro Silva for her assistance in the preparation of this manuscript.

REFERENCES

Bissacot, L.C.G. & Ciminelli, V.S.T. 2014. Arsenic Mobility at Alkaline Conditions in a Gold Mine Tailings Dam. (This Volume).

Brown, G., Foster, A.L. & Ostergren, J.D. 1999. Mineral surfaces and bioavailabilty of heavy metals: a molecular-scale perspective. *Proceedings of the* National Academy *of Sciences* 96: 3388–3395.

Bundschuh, J., Litter, M.I., Parvez, F., Roman-Ross, G., Nicolli, H.B., Jean, J.S., Liu, C.W., Lopez, D., Armienta, M.A., Guilherme, L.R.G., Cuevas, A.G., Cornejo, L., Cumbal, L. & Toujaguez, R. 2012. One century of arsenic exposure in Latin America: a review of history and occurrence from 14 countries. *Sci. Total Environ.* 429: 2–35.

Caetano, M.L., Ciminelli, V.S.T., Rocha, S.D.F. Spitale, M.C. & Caldeira, C.L. 2009. Batch and Continuous Precipitation of Scorodite from Diluted Industrial Solutions. *Hydrometallurgy* 95: 44–52.

Corriveau, M.C., Jamieson, H.E., Parsons, M.B. & Hall, G.E.M. 2011. Mineralogical characterization of arsenic in gold mine tailings from three sites in Nova Scotia. *Geochem. Explor. Environ. Anal.* 11: 179–192.

DeSisto, S.L., Jamieson, H.E. & Parsons, M.B. 2011. Influence of hardpan layers on arsenic mobility in historical gold mine tailings. *Appl. Geochem.* 26: 2004–2018.

Duarte, G., Ciminelli, V.S.T., Dantas, M.S.S., Duarte, H.A., Vasconcelos, I.F., Oliveira, A.F. & Osseo-Asare, K. 2012. As(III) immobilization on gibbsite: Investigation of the complexation mechanism by combining EXAFS analyses and DFT calculations. *Geochimica et Cosmochimica Acta* 83: 205–216.

Essilfie-Dughan, J., Hendry, M.J., Warner, J. & Kotzer, T. 2013. Arsenic and iron speciation in uranium mine tailings using X-ray absorption spectroscopy. *Applied Geochemistry* 28:11–18.

Foster, A.L., Brown, Jr. G.E., Tingle, T.N. & Parks, G.A. 1998. Quantitative As speciation in mine tailings using X-ray absorption spectroscopy. *American Mineralogist* 83:553–568.

Harris, G.B. 2000. The removal and stabilization of arsenic from aqueous process solutions: past, present and future. In: Young, C. (Ed.), *Minor Elements 2000. Processing and Environmental Aspects of As, Sb, Se, Te and Bi.* SME, Littleton, Co, USA, p. 3–20.

IARC 2004. Arsenic in drinking water, in: IARC Monographs on the Evaluation of Carcinogenic Risks to Humans, some Drinking-Water Disinfectants and Contaminants, Including Arsenic, vol.84, WHO – International Agency for Research on Cancer, Lyon, 2001, p. 39–270.

Kwong, Y.T.J., Beauchemin, S., Hossain, M.F. & Gould, W.D. 2007. Transformation and mobilization of arsenic in the historic cobalt mining camp, Ontario, Canada. *J. Geochem. Explor*. 92: 133–150.

Krause, E. & Ettel, V.A. 1989. Solubilities and stabilities of ferric arsenate compounds. *Hydrometallurgy* 22: 311–337.

Ladeira, A.C.Q. & Ciminelli, V.S.T. 2004. Adsorption and desorption of arsenic on an oxisol and its constituents. *Water Research* 38: 2087–2094.

Ladeira, A.C.Q., Ciminelli, V.S.T., Duarte, H.A. & Alves, M.C.M. 2001. Mechanism of anion retention from EXAFS and density functional calculations: arsenic(V) adsorbed on gibbsite. *Geochimica et Cosmochimica Acta* 65: 1211–1217.

Landrot, G., Tappero, R., Webb, S.M. & Sparks, D.L. 2012. Arsenic and chromium speciation in an urban contaminated soil. *Chemosphere* 88:1196–120.

Lu, P. & Zhu, C. 2011. Arsenic Eh-pH diagrams at 25 °C and 1 bar. *Environmental Earth Science* 62:1673–1683.

Müller, K., Willscher, S., Dantas, M.S.S. & Ciminelli, V.S.T. 2010. A comparative study of As(III) and As(V) in aqueous solutions and adsorbed on iron oxyhydroxides by Raman Spectroscopy. *Water Research* 44: 5660–5672.

Naidu, R. 2012. Bioavailability and bioaccessibility of arsenic for ecological and human health risk assessment: The geological and health interface. In: J. Ng, B. Noller, R. Naidu, J. Bundschuh & P. Bhattacharya (eds.), *Understanding the Geological and medical Interface of Arsenic, As 2012 - Arsenic in the environment Proceedings*, p. 3–10.

Nordstrom, D.K. 2012. Arsenic in the geosphere meets the anthroposphere. In: J. Ng, B. Noller, R. Naidu, J. Bundschuh & P. Bhattacharya (eds.), *Understanding the Geological and medical Interface of Arsenic, As 2012 - Arsenic in the environment Proceedings*, p. 15–19.

Paktunc, D. 2013. Mobilization of arsenic from mine tailings through reductive dissolution of goethite influenced by organic cover. *Applied Geochemistry* 36: 49–56.

Paktunc, D., Foster, A., Heald, S. & Laflamme, G. 2004. Speciation and characterization of arsenic in gold ores and cyanidation tailings using X-ray absorption spectroscopy. *Geochimica et Cosmochimica Acta* 68: 969–983.

Pantuzzo, F.L. & Ciminelli, V.S.T. 2010. Arsenic association and stability in long-term disposed arsenic residues. *Water Research* 44: 5631–5640.

Parviainen, A., Lindsay, M.B.J., Pérez-López, R., Gibson, B.D., Ptacek, C.J., Blowes, D.W. & Loukola-Ruskeeniemi, K. 2012. Arsenic attenuation in tailings at a former Cu–W–As mine, SW Finland. *Applied Geochemistry* 27: 2289–2299.

Plumlee, G.S., Morman, S.A. & Ziegler, T.L. 2006. The toxicological geochemistry of earth materials: An overview of processes and the interdisciplinary methods used to understand them. *Med. Mineral. Geochem.* 64: 5–57.

Ritcey, G.M. 2005. Tailings management in gold plants. *Hydrometallurgy* 78: 3–20.

Riveros, P.A., Dutrizac, J.E. & Spencer, P. 2001. Arsenic disposal practices in the metallurgical industry. *Canadian Metallurgical Quarterly* 40(4):395–420.

Robins, R.G. 1983. The stabilities of arsenic(V) and arsenic (III) compounds in aqueous metal extraction systems. In: Osseo-Asare, K., Miller, J.D.(Eds.), Hydrometallurgy, Research, Development and Plant Practice. TMS-AIM, USA, p. 404–406.

Smedley, P.L. & Kinniburgh, D.G. 2002. A review of the source, behaviour and distribution of arsenic in natural waters. *Applied Geochemistry* 17: 517–568.

Smith, E., Weber, J. & Juhasz, A.L. 2009. Arsenic distribution and bioaccessibility across particle fractions in historically soils. *Environmental Geochemistry and Health* 31: 85–92.

Swash, P.M. & Monhemius, A.J. 1995a. The disposal of arsenical wastes: technologies and environmental considerations. In: N.J. Roberts (ed.), *International Minerals & Metals Technology*. Sterling. Pubs. Ltd., London, p. 121–125.

Swash, P.M. & Monhemius, A.J. 1995b. Synthesis, characterization and solubility testing of solids in the Ca-Fe-AsO_4 system. In: T.P. Hynes & M.C. Blanchette (eds.), *Sudbury '95 Mining and the Environment*. CANMET, Ottawa, Canada, p. 17–28.

Toujagueza, R., Ono, F.B., Martins, V., Cabrera, P.P., Blanco, A.V., Bundschuh, J. & Guilherme, L.R.G. 2013. Arsenic bioaccessibility in gold mine tailings of Delita, Cuba, *Journal of Hazardous Materials* (*online*).

USEPA—U.S. Environmental Protection Agency, Washington, DC. 2008. Test Methods for Evaluating Solid Waste, Physical/Chemical Methods. Doc. no. SW–846. 3rd Ed. http://www.epa.gov/epawaste/hazard/testmethods/ sw846/

One Century of the Discovery of Arsenicosis in Latin America (1914–2014) – Litter, Nicolli, Meichtry, Quici, Bundschuh, Bhattacharya & Naidu (Eds)
© 2014 Taylor & Francis Group, London, ISBN 978-1-138-00141-1

Prospective evaluation of arsenic-related health effects in children

M. Vahter
Institute of Environmental Medicine, Karolinska Institutet, Stockholm, Sweden

S. Ahmed
Institute of Environmental Medicine, Karolinska Institutet, Stockholm, Sweden
International Center for Diarrhoeal Disease in Bangladesh, Dhaka, Bangladesh

M. Kippler
Institute of Environmental Medicine, Karolinska Institutet, Stockholm, Sweden

R. Raqib
International Center for Diarrhoeal Disease in Bangladesh, Dhaka, Bangladesh

K. Broberg
Institute of Environmental Medicine, Karolinska Institutet, Stockholm, Sweden
Occupational and Environmental Medicine, Lund University, Lund, Sweden

ABSTRACT: Our on-going longitudinal mother-child cohort study (N~2,000) in rural Bangladesh, an area with highly variable arsenic concentrations in well-water, aims at evaluating potential health effects of prenatal and childhood exposure to arsenic through water and food, especially rice. Urinary arsenic metabolites were measured twice during pregnancy and in children at 1.5, 5, and 10 years of age. We assessed fetal and child growth and morbidity (infectious diseases), which all seemed to be adversely affected by arsenic. After the first year in life, concurrent exposure was more influential on health and development than prenatal exposure. Arsenic-related impairment of child development (IQ), observed at 5 years of age, was gender-dependent. Evaluated modes of action included oxidative stress, epigenetic modifications, inflammation, and impaired immune function, the latter associated with increased morbidity and mortality. Undernourished children were generally more susceptible to arsenic toxicity, especially immune effects, but children's arsenic methylation efficiency had little influence.

1 INTRODUCTION

Inorganic arsenic, a potent toxicant and carcinogen, is frequently present in drinking water. Recent studies showed that exposure through food may also be common, especially rice, including rice-based infant foods (Meharg *et al.*, 2008, Ljung *et al.*, 2011). While adverse health effects in adults are well-documented, consequences of early-life exposure are not well researched. This is in spite of the fact that arsenic is known to readily pass the placenta and that the exposure to arsenic through water and food provides a higher dose of arsenic per kilo body weight, compared to adults. The overall aim of our on-going studies in rural Bangladesh is to evaluate prospectively potential health effects of prenatal and childhood arsenic exposure. More than 95% of the inhabitants in the study area, 50 km southeast of the capital Dhaka, use drinking water from local wells, the arsenic concentrations of which vary considerably (Vahter *et al.*, 2006). Because the diet is mainly rice-based, it also causes elevated exposure to arsenic, independent of the arsenic concentration in the drinking water (Vahter *et al.*, 2006, Gardner *et al.*, 2011).

Maternal arsenic exposure during pregnancy was found to be associated with lower size at birth (Rahman *et al.*, 2009, Kippler *et al.*, 2012) and increased risk of infant mortality (Rahman *et al.*, 2007, Rahman *et al.*, 2010) and morbidity, particularly lower respiratory tract infections and diarrhea (Raqib *et al.*, 2009, Rahman *et al.*, 2011). We also observed that arsenic affected immune markers in placenta (Ahmed *et al.*, 2011), and impaired infants' thymus size (Moore *et al.*, 2009) and function (Ahmed *et al.*, 2012). These findings indicate that arsenic-exposure in early childhood is immunosuppressive, as previously shown in older Mexican children (Soto-Pena *et al.*, 2006).

The results prompted us to continue to follow the children's health and development and to elucidate influential factors. This report summarizes the results in children up to 10 years of age, with focus on exposure pattern over time and effects on child growth and morbidity.

2 METHODS

Our longitudinal mother-child cohort (N~2,300) was initiated nested in a large, population-based food and micronutrient supplementation trial in pregnancy, in which the children's growth, health and development were followed until 5 years of age (Persson *et al.*, 2012). We have then followed the children of the arsenic cohort at 10 years of age. For evaluation of prenatal exposure, we measured arsenic metabolites in maternal urine (by AAS) twice in pregnancy (gestational weeks 8 and 30) (Vahter *et al.* 2006). Arsenic metabolites in urine of about 2,000 of their children were measured at 1.5, 5 and 10 of age, using HPLC on line with hydride generation and ICP-MS (Fangstrom *et al.*, 2009, Gardner *et al.*, 2011). Child anthropometry and development (motor and cognitive function and behavior) were assessed in about 2,000 children at 0.7, 1.5, 5 and 10 years of age; child development by problem-solving tests, Bayley Scales of Infant Development-II, and Wechsler Preschool and Primary Scale of Intelligence (IQ). Immune function in the children was further evaluated by measuring delayed type hypersensitivity using PPD skin test and response to vaccination against rubella (MMR vaccine). Also, oxidative DNA damage (urinary 8-oxo-7,8-dihydro-2'-deoxyguanosine, 8-oxodG), plasma cytokines, and modification of DNA methylation were evaluated.

The impact of arsenic on the different outcomes was evaluated using multivariable-adjusted regression analysis, with urinary arsenic as continuous or categorized variable, depending on linearity of the associations. We considered multiple potential confounders and influential factors, including family characteristics, nutrition, and co-exposures. We also stratified for socio-economic status, BMI, and gender. Polymorphisms in genes potentially related to arsenic metabolism and toxicity, in particular arsenic(+III oxidation state) methyltransferase (AS3MT), DNA-methyl-transferase 1a and 3b (DNMT1a and DNMT3b, respectively), phosphatidylethanolamine N-methyltransferase (PEMT), and betaine-homocysteine methyltransferase (BHMT), were genotyped with Sequenom, and AS3MT expression was measured by quantitative real-time polymerase chain reaction using TaqMan expression assays.

3 RESULTS AND DISCUSSION

3.1 *Child growth and development*

The effects of prenatal arsenic (mean 80 μg L^{-1}, total range 1–1,470 μg L^{-1} in maternal urine) on infant growth and development were unexpectedly weak (Hamadani *et al.*, 2010, Saha *et al.*, 2011), considering that arsenic is readily passing from the mother's blood to the fetus (Concha *et al.*, 1998). Possibly this was, at least partly, related to the observed increased arsenic methylation efficiency in early pregnancy leading to lower percentage of the highly toxic MMA metabolite in urine (Gardner *et al.*, 2012). Also, most children were prominently breast-fed for several months (Khan *et al.*, 2013), which efficiently protects against arsenic exposure (Fangstrom *et al.*, 2008). However, the continued childhood exposure to arsenic through water and food, leading to about 35 μg L^{-1} in urine at 1.5 years, 50 μg L^{-1} at 5 years, and 60 μg L^{-1} at 10 years, seems to affect the children's health more than the prenatal exposure. At the age of 21 months, the girls in the fourth quintile of concurrent urinary concentrations of arsenic metabolites (26–46 μg L^{-1}) were almost 300 g lighter and 0.7 cm shorter than those in the first quintile (<16 μg L^{-1}), and had adjusted odds ratios (95% confidence interval) for underweight and stunting of 1.57 (95% confidence interval (CI) 1.02–2.40) and 1.58 (CI:1.05–2.37), respectively (Saha *et al.*, 2012). Arsenic-related effects on child growth and development was followed-up at 5 years. In the longitudinal analysis, multivariable-adjusted attributable differences in children's weight at 5 years were −0.33 kg (CI: −0.60, −0.06) comparing those with highest and lowest arsenic exposure (≥ 95th percentile and ≤5th percentile) (Gardner *et al.*, 2013). The associations were more pronounced in girls than in boys. Similarly, arsenic exposure was significantly inversely associated with child IQ at 5 years, in particular verbal IQ (Hamadani *et al.*, 2011). Concurrent exposure at 5 years of age seemed to have a larger influence on IQ than the prenatal exposure (arsenic in maternal urine during pregnancy), and again, the effect was more prominent in girls than in boys. An increase in the girls' urine of 100 μg L^{-1} was associated with 1–3 points lower IQ (especially verbal IQ). This may not be a severe effect for the individual child, but clearly of concern, considering the number of children exposed in Bangladesh and elsewhere.

We also found that the children's concurrent arsenic exposure was associated with increased risk of infections, in particular respiratory infections, at 5 years of age (Gardner *et al.*, manuscript). Again, we found that the children's exposure had stronger impact on their health than the prenatal exposure, but we found no major gender difference. Arsenic exposure in the children was also associated with increased risk of not responding to the provided antigen in the PPD-test at 5 years of age, which provides further support that arsenic affected the children's cell-mediated immunity. In the long run, the observed developmental immunotoxicity of arsenic may have consequences for multiple immune-based diseases (infections and chronic diseases), and possibly various cancers.

3.2 *Modes of action and susceptibility factors*

Prenatal arsenic exposure was associated with in-creased inflammation (pro-inflammatory cytokines) and decreased number of immune cells in placenta (Ahmed *et al.*, 2011). These effects seemed to be mediated, at least partly, via oxidative stress. Arsenic exposure was also inversely associated with sjTRECs measures in cord blood mononuclear cells, indicating that the prenatal arsenic exposure impaired thymic function in the newborn child (Ahmed *et al.*, 2012), in line with the previously observed reduced thymus size (Moore *et al.*, 2009). The impaired thymus output seemed to be mediated by oxidative stress and apoptosis of thymocytes, as indicated by an increased level of 8-oxodG in cord-blood, as well as decreased expression of antioxidant-related genes and increased expression of apoptosis-related genes. The observed effects may explain the previously indicated arsenic-related increase in infant morbidity (Rahman *et al.*, 2011). In general, undernourished children seemed to be more susceptible to arsenic, possibly due to their lower anti-oxidant defense.

In our recent evaluation of the impact of early-life arsenic exposure on the early-life epigenetic status, we found that early prenatal arsenic exposure (maternal urinary arsenic in gestational week 8, but not at gestational week 30) was associated with decreased cord blood DNA methylation, particularly in newborn boys (Broberg *et al.*, manuscript). In the boys, 372 (74%) of the 500 CpG sites with strongest association with maternal urinary arsenic showed lower methylation with increasing urinary arsenic (Spearman correlation >–0.62), but in girls only 207 (41%) showed inverse correlation. Pathway analysis showed overrepresentation of affected cancer-related genes in boys, but not in girls. The observed associations between arsenic exposure and DNA methylation might reflect interference with the *de novo* DNA methylation in early fetal life.

One major susceptibility factor for arsenic-related toxicity is the efficiency of arsenic metabolism. Methylation of inorganic arsenic, which takes place mainly in the liver, to MMA generally increases toxicity, while addition of a second methyl group, DMA is associated with decreased toxicity. We evaluated the impact of polymorphisms in five methyltransferase genes on arsenic metabolism and found that six AS3MT polymorphisms were significantly associated with arsenic metabolite pattern in urine ($p \leq 0.01$). The most frequent AS3MT haplotype was associated with a higher percentage of MMA in maternal urine. Four polymorphisms in the DNMT genes were also associated with metabolite patterns (Engstrom *et al.*, 2011). Noncoding AS3MT polymorphisms affected gene expression of AS3MT in umbilical cord blood, demonstrating that one functional impact of the AS3MT polymorphisms may be altered levels of gene expression (Engstrom *et al.*, 2013).

4 CONCLUSIONS

The major effect of prenatal arsenic exposure in infancy seems to be impaired immune function and arsenic-related increased morbidity. Important modes of action seem to be oxidative stress and apoptosis of immune cells. However, at 5 years of age, the concurrent exposure seemed highly influential. Effects on infant growth and cognitive development were unexpectedly small, though statistically significant. Effects on child growth and development were seen at 2–5 years of age, and seemed to be more influenced by concurrent exposure than by prenatal exposure.

Taken together, our findings indicate that early-life arsenic exposure may have major public health impact, in particular when considering the many children exposed to arsenic worldwide. The difficulties to mitigate arsenic in well water are of great concern, as the effects of the continued arsenic exposure on the children seemed to be important for child health. While the commonly observed skin effects of arsenic are more frequent in men than in women (Lindberg *et al.*, 2010), girls seem to be more susceptible to certain of the early-life effects of arsenic than boys. However, the arsenic-related modification of DNA methylation in cord blood, which might have long-term consequences, seemed to be more prominent in boys. The more long-term health effects of early-life arsenic exposure remain to be investigated. Studies performed in northern Chile reported elevated risks of both cancer and non-cancer effects in adults who had been exposed to arsenic very early in life (Dauphiné *et al.*, 2011; Smith *et al.*, 2012).

ACKNOWLEDGEMENTS

We acknowledge the MINIMat food micronutrient supplementation trial (PIs L.A. Persson, Uppsala University, and S. El Arifeen, icddr,b, Dhaka).

For the nested arsenic projects, we acknowledge the fruitful collaboration with Dr. J. Hamadani, and Dr. F. Tofail, and their collaborators. At the Karolinska Institute, we acknowledge the analytical laboratory personnel M. Grandér, H. Nordqvist, and B. Palm.

REFERENCES

Ahmed, S., Ahsan, K.B., Kippler, M., Mily, A., Wagatsuma, Y., Hoque, A.M., Ngom, P.T., El Arifeen, S., Raqib, R. & Vahter, M. 2012. In utero arsenic exposure is associated with impaired thymic function in newborns possibly via oxidative stress and apoptosis. *Toxicol. Sci.* 129: 305–314.

Ahmed, S., Mahabbat-e Khoda, S., Rekha, R.S., Gardner, R.M., Ameer, S.S., Moore, S., Ekstrom, E.C., Vahter, M. & Raqib, R. 2011. Arsenic-associated oxidative stress, inflammation, and immune disruption in human placenta and cord blood. *Environ. Health Perspect.* 119: 258–264.

Concha, G., Vogler, G., Lezcano, D., Nermell, B. & Vahter, M. 1998. Exposure to inorganic arsenic metabolites during early human development. *Toxicol. Sci.* 44: 185–190.

Dauphiné, D.C., Ferreccio, C., Guntur, S., Yuan, Y., Hammond, S.K., Balmes, J., Smith, A.H. & Steinmaus, C. 2011. Lung function in adults following in utero and childhood exposure to arsenic in drinking water: preliminary findings. *Int. Arch. Occup. Environ. Health* 84(6): 591–600.

Engstrom, K., Vahter, M., Mlakar, S.J., Concha, G., Nermell, B., Raqib, R., Cardozo, A. & Broberg, K. 2011. Polymorphisms in arsenic(+III oxidation state) methyltransferase (AS3MT) predict gene expression of AS3MT as well as arsenic metabolism. *Environ. Health Perspect.* 119: 182–188.

Engstrom, K.S., Hossain, M.B., Lauss, M., Ahmed, S., Raqib, R, Vahter, M. & Broberg, K. 2013. Efficient Arsenic Metabolism—The AS3MT Haplotype Is Associated with DNA Methylation and Expression of Multiple Genes Around AS3MT. *Plos One* 8:e53732.

Fangstrom, B., Hamadani, J., Nermell, B., Grander, M., Palm, B. & Vahter, M. 2009. Impaired arsenic metabolism in children during weaning. *Toxicol. Appl. Pharmacol.* 239: 208–214.

Fangstrom, B., Moore, S., Nermell, B., Kuenstl, L., Goessler, W., Grander, M., Kabir, I., Palm, B., Arifeen, S.E. & Vahter, M. 2008. Breast-feeding Protects against Arsenic Exposure in Bangladeshi Infants. *Environ. Health Perspect.* 116: 963–969.

Gardner, R., Hamadani, J., Grander, M., Tofail, F., Nermell, B., Palm, B., Kippler, M. & Vahter, M. 2011. Persistent exposure to arsenic via drinking water in rural Bangladesh despite major mitigation efforts. *Am. J. Public Health.* 101 Suppl 1:S333–338.

Gardner, R.M., Engstrom, K., Bottai, M., Hoque, W.A., Raqib, R., Broberg, K. & Vahter, M. 2012. Pregnancy and the methyltransferase genotype independently influence the arsenic methylation phenotype. *Pharmacogenet. Genomics* 22: 508–516.

Gardner, R.M., Kippler, M., Tofail, F., Bottai, M., Hamadani, J., Grander, M., Nermell, B., Palm, B., Rasmussen, K.M. & Vahter, M. 2013. Environmental exposure to metals and children's growth to age 5 years: a prospective cohort study. *Am. J. Epidemiol.* 177: 1356–1367.

Hamadani, J., Tofail, F., Nermell, B., Gardner, R., Shiraji, S., Bottai, M., Arifeen, S., Huda, S. & Vahter, M. 2011. Critical windows of exposure for arsenic-associated impairment of cognitive function in pre-school girls and boys: a population-based cohort study. *Int. J. Epidemiol.* 40: 1593–1604.

Hamadani, J.D., Grantham-McGregor, S.M., Tofail, F., Nermell, B., Fangstrom, B., Huda, S.N., Yesmin, S., Rahman, M., Vera-Hernandez, M., Arifeen, S.E. & Vahter, M. 2010. Pre- and postnatal arsenic exposure and child development at 18 months of age: a cohort study in rural Bangladesh. *Int. J. Epidemiol.* 39: 1206–1216.

Khan, A.I., Hawkesworth, S., Ekstrom, E.C., Arifeen, S., Moore, S.E., Frongillo, E.A., Yunus, M., Persson, L.A. & Kabir, I. 2013. Effects of exclusive breastfeeding intervention on child growth and body composition: the MINIMat trial, Bangladesh. *Acta Paediatr.* 102: 815–823.

Kippler, M., Wagatsuma, Y., Rahman, A., Nermell, B., Persson, L.A., Raqib, R. & Vahter, M. 2012. Environmental exposure to arsenic and cadmium during pregnancy and fetal size: a longitudinal study in rural Bangladesh. *Reprod. Toxicol.* 34: 504–511.

Lindberg, A.L., Rahman, M., Persson, L.A. & Vahter, M. 2008. The risk of arsenic induced skin lesions in Bangladeshi men and women is affected by arsenic metabolism and the age at first exposure. *Toxicol. Appl. Pharmacol.* 230(1): 9–16.

Ljung, K., Palm, B., Grandér, M. & Vahter, M. 2011. Highconcentrations of essential and toxic elements in infant formula and infant foods–A matter of concern. *Food Chemistry* 127: 943–951.

Meharg, A.A., Sun, G., Williams, P.N., Adomako, E., Deacon, C., Zhu, Y.G., Feldmann, J. & Raab, A. 2008. Inorganic arsenic levels in baby rice are of concern. *Environ. Pollut.* 152: 746–749.

Moore, S., Prentice, A.M., Wagatsuma, Y., Fulford, A.J.C., Collinson, A.C., Raqib, R., Vahter, M., Persson, L.Å. & Arifeen, S.E. 2009. Early-life nutritional and environmental determinants of thymic size in infants born in rural Bangladesh. *Acta Pediat.* 98(7): 1168–75.

Persson, L.A., Arifeen, S., Ekstrom, E.C., Rasmussen, K.M., Frongillo, E.A. & Yunus, M. 2012. Effects of prenatal micronutrient and early food supplementation on maternal hemoglobin, birth weight, and infant mortality among children in Bangladesh: the MINIMat randomized trial. *JAMA* 307:2050–2059.

Rahman, A., Persson, L.A., Nermell, B., El Arifeen, S., Ekstrom, E.C., Smith, A.H. & Vahter, M. 2010. Arsenic exposure and risk of spontaneous abortion, stillbirth, and infant mortality. *Epidemiology* 21: 797–804.

Rahman, A., Vahter, M., Ekstrom, E.C. & Persson, L.A. 2011. Arsenic exposure in pregnancy increases the risk of lower respiratory tract infection and diarrhea during infancy in Bangladesh. *Environ. Health Perspect.* 119: 719–724.

Rahman, A., Vahter, M., Ekstrom, E.C., Rahman, M., Golam Mustafa, A.H., Wahed, M.A., Yunus, M. &

Persson, L.A. 2007. Association of arsenic exposure during pregnancy with fetal loss and infant death: a cohort study in Bangladesh. *Am. J. Epidemiol.* 165: 1389–1396.

Rahman, A., Vahter, M., Smith, A.H., Nermell, B., Yunus, M., El Arifeen, S., Persson, L.A. & Ekstrom, E.C. 2009. Arsenic exposure during pregnancy and size at birth: a prospective cohort study in Bangladesh. *Am. J. Epidemiol.* 169: 304–312.

Raqib, R., Ahmed, S., Sultana, R., Wagatsuma, Y., Mondal, D., Hoque, A.M., Nermell, B., Yunus, M., Roy, S., Persson, L.A., Arifeen, S.E., Moore, S. & Vahter, M. 2009. Effects of inutero arsenic exposure on childimmunity and morbidity in rural Bangladesh. *Toxicol. Lett.* 185:197–202.

Saha, K.K., Engstrom, A., Hamadani, J.D., Tofail, F., Rasmussen, K.M. & Vahter, M. 2012. Pre- and postnatal arsenic exposure and body size to two years of age: a cohort study in rural Bangladesh. *Environ. Health Perspect.* 120(8): 1208–14.

Smith, A.H., Marshall, G., Liaw, J., Yuan, Y., Ferreccio, C. & Steinmaus, C. 2012. Mortality in young adults following in utero and childhood exposure to arsenic in drinking water. *Environ. Health Perspect.* 120(11): 1527–1531.

Soto-Pena, G.A., Luna, A.L., Acosta-Saavedra, L., Conde, P., Lopez-Carrillo, L., Cebrian, M.E., Bastida, M., Calderon-Aranda, E.S. & Vega, L. 2006. Assessment of lymphocyte subpopulations and cytokine secretion in children exposed to arsenic. *Faseb J.* 20: 779–781.

Vahter, M., Li, L., Nermell, B., Rahman, A., Arifeen, S.E., Rahman, M., Persson, L.A. & Ekström, E.C. 2006. Arsenic exposure in pregnancy - a population based study in Matlab, Bangladesh. *J. Health Popul. Nutr.* 24(2): 236–245.

One Century of the Discovery of Arsenicosis in Latin America (1914–2014) –
Litter, Nicolli, Meichtry, Quici, Bundschuh, Bhattacharya & Naidu (Eds)
© 2014 Taylor & Francis Group, London, ISBN 978-1-138-00141-1

Affordable point-of-use water purification using nanomaterials

T. Pradeep
DST Unit of Nanoscience and Thematic Unit of Excellence, Department of Chemistry, Indian Institute of Technology Madras, Chennai, India

ABSTRACT: Creation of affordable materials for constant release of silver ions in water is one of the most promising ways to provide microbially safe drinking water for all. Combining the capacity of diverse nanocomposites to scavenge toxic species such as arsenic, lead, and other contaminants along with the above capability can result in affordable, all-inclusive drinking water purifiers that can function without electricity. Here we show that such constant release materials can be synthesized in a simple and effective fashion in water itself without the use of electrical power. These materials have been used to develop an affordable water purifier to deliver clean drinking water at US $2.5/y per family. We are in the process of implementing such solutions in arsenic affected areas of India. In about 18 months, we would have implemented our technology for solving the problems of 600,000 people. Experiences from these efforts will be discussed.

1 INTRODUCTION

Safe drinking water is a significant, but simple indicator of development. Its availability at point of use can save over 2 million human lives (of the 3.575 million deaths caused by water, sanitation, and hygiene issues, 42.6% are due to diarrhea alone: 3.575 million × 0.426 = 1.523 million lives), can avoid over 2 billion diarrheal infections, and can contribute over U$S 4 billion to the global gross domestic product (formula used: Σ (number of deaths attributed to diarrhea in each country × corresponding country's per capita gross domestic product). Considering the challenges associated with traditional disinfectants, solutions based on state-of-the-art science and technology hold the key for safe drinking water and novel approaches are being looked at. It has been long known that silver, especially in the nanoparticle form, is an effective disinfectant and works for a wide spectrum of bacteria and viruses. Numerous approaches are available for the synthesis of biocidal silver nanoparticles or colloids, including the use of matrices. The biocidal property of silver nanoparticles, usually in the size range of 10–20 nm, is attributed to the release of trace quantities of silver ions in water, which, although being sufficient for microorganism killing, does not exhibit toxicity to humans. Although a number of silver-based biocidal compositions have been synthesized, those have not been able to reach the masses in large volumes (e.g., silver nanoparticle loaded ceramic candles). Massive deployment has been hampered due to the following reasons: (a) Drinking water contains many species (e.g., inorganic ions and organics) that anchor on the surface of nanoparticles, making sustained silver ion release difficult; (b) suitable anchoring substrates that limit the scaling of nanoparticle surfaces while simultaneously preventing their release into water are not available; and (c) continued retention of the nanoparticles in the matrix is difficult.

2 METHODS/EXPERIMENTAL

In this work, we demonstrate a unique family of nanocrystalline metal oxyhydroxide-chitosan granular composite materials prepared at near room temperature through an aqueous route. The origin of crystallinity in the composition is attributed to abundant O and –OH functional groups on chitosan, which help in the crystallization of metal oxyhydroxide and also ensure strong covalent binding of the nanoparticle surface to the matrix. X-ray Photoelectron Spectroscopy (XPS) confirms that the composition is rich with surface hydroxyl groups. Using hyperspectral imaging, the absence of nanoparticle leaching in the water was confirmed. Further, a unique scheme to reactivate the silver nanoparticle surface is used for continual antimicrobial activity in drinking water. Several other composites have been developed that can scavenge other contaminants in water. We demonstrate an affordable water purification device based on such composites developed over several years and undergoing field trials in India, as a potential solution for widespread eradication of the waterborne disease burden as well as to remove heavy metal contamination.

3 RESULTS AND DISCUSSION

The antimicrobial composition consists of an aluminum oxyhydroxide–chitosan composite (BM) with silver particles of 10–20 nm diameter embedded in it and is capable of sustained release of silver ions [40 ± 10 parts per billion (ppb)] in natural drinking water over an extended volume of water passing through it, to achieve effective removal of microorganisms. The antimicrobial composite (referred to as Ag-BM) is unique as it is made in water at near room temperature, using a biopolymer, and dried in ambient conditions to obtain water-insoluble granules, yielding Na_2SO_4 as the major by-product (>90%), thereby making it a green synthesis. The concentration of silver ion leached into drinking water from the prepared composite at relevant temperatures (5–35 °C) is significantly less than the maximum permissible limit of 100 ppb (secondary standard, US Environmental Protection Agency), thereby requiring no secondary filtration to remove excess silver ions. This controlled release at temperatures of relevance to drinking-water applications over extended periods is an important advantage of the composite. The details mechanism and the characterization of the composite is given in the figure below.

The material has been used to develop prototypes which has been used for field trials with success. Similar composite structure for iron oxyhydroxide nanostructures have been used to create an arsenic filter. Products based on this composite is being introduced in the arsenic affected areas.

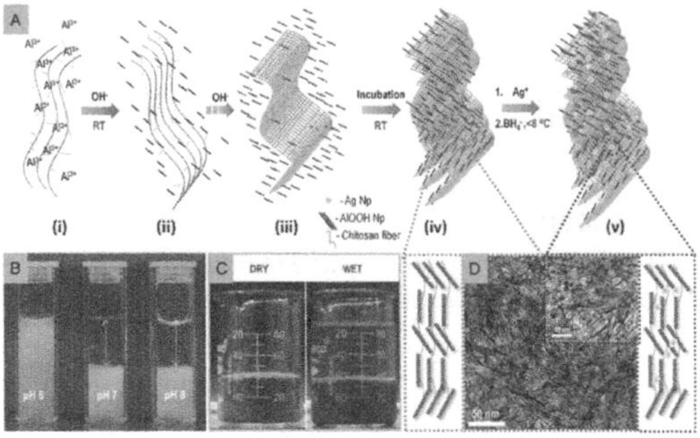

Figure 1. Mechanism for the preparation of composite and origin of its physical strength in water due to network structure. (A) Mechanistic scheme for the formation of Ag-BM composite, as learned through various experiments. (i–v) (i) Al^{3+} complexes with chitosan solution; (ii and iii) alkali treatment leads to the formation of aluminum hydroxide nanoparticles and random coiled chitosan network; (iv) aluminum hydroxide nanoparticles bind to chitosan network, possibly through covalent sharing of oxygen, leading to the formation of aluminum oxyhydroxide; and (v) silver nanoparticles form on the aluminum oxyhydroxide–chitosan network. (B) Photographs of the system during synthesis. Presence of aluminum hydroxide in the supernatant is clearly visible below pH 6 whereas bound aluminum hydroxide settles at pH 7 and pH 8, leading to a clear supernatant. (C) Photographs of the composite granules and of the same in water to illustrate that the material is stable in water. (D) Graphical representation and corresponding TEM images showing the aluminum oxyhydroxide-chitosan network without (green box) and with (red box) embedded silver nanoparticles. Adapted with permission from PNAS.

4 CONCLUSIONS

In their entirety, the proposed device and materials present a compelling solution for achieving the United Nations millennium development goal of sustainable access to safe drinking water. We believe that frugal science based on nanotechnology can make a lasting impact on society. There are over 200 million households in India. Dissemination of this technology in various forms such as cartridges, sachets, etc., can generate large employment opportunities in the villages. The production of composites and water filter devices and their deployment and servicing can contribute to the local economy. Various modifications of the composite with different compositions have been developed with comparable performances.

ACKNOWLEDGEMENTS

The author acknowledges Nanoscience and Nano Mission of the Department of Science and Technology (DST) for supporting the research program on nanomaterials. This work was originally published in PNAS (2013).

REFERENCES

(2004) The World Factbook 2004 (Central Intelligence Agency, Washington, DC).

(2009) Nanotechnology Applications for Clean Water, eds Savage N, et al. (William Andrew Publication, Norwich, NY).

Clasen TF, Haller L (2008) Water quality interventions to prevent diarrhoea: Cost and cost-effectiveness (WHO, Geneva).

Pradeep T, Anshup (2009) Noble metal nanoparticles for water purification: A critical review. Thin Solid Films 517(24):6441–6478.

Prüss-Ustün A, Bos R, Gore F, Bartram J (2008) Safer Water, Better Health (WHO, Geneva).

Shannon MA, et al. (2008) Science and technology for water purification in the coming decades. Nature 452(7185):301–310.

Section 1: Arsenic in environmental matrices (air, water, soil and biological matrices)

1.1 Geological and hydrogeological controls of arsenic in groundwater and other aquatic systems

One Century of the Discovery of Arsenicosis in Latin America (1914–2014) – Litter, Nicolli, Meichtry, Quici, Bundschuh, Bhattacharya & Naidu (Eds)
© 2014 Taylor & Francis Group, London, ISBN 978-1-138-00141-1

Hydrogeochemical controls on arsenic occurrence in Bangladesh aquifer

R.K. Dhar & A. Cunningham
York College of City University of New York, Jamaica, New York, USA

Y. Zheng
Queens College of City University of New York, Flushing, New York, USA
Lamont-Doherty Earth Observatory of Columbia University, Palisades, New York, USA

ABSTRACT: A hydrogeochemical study was conducted in Bangladesh to compare and contrast groundwater chemical compositions in the shallow Holocene and deep Pleistocene aquifers. Groundwaters from several multilevel monitoring wells, as well as from a significant number of existing private wells were sampled and examined for obtaining detailed physicochemical properties. Groundwater is HCO_3 type, although the shallow water is Ca-Mg rich while the deep water is Na rich. Shallow wells with elevated As concentrations (>50 μg/L) are all reducing with negative ORP values between −50 & −200 mV, with predominantly As(III). Sediment characterization shows the elevated As(III) and Fe(II) concentrations in shallow aquifers. PHREEQC calculations indicate super-saturation of Fe and P with respect to siderite, vivianite and hydroxyapatite at shallow depth. Shallow waters with elevated As also contain high level of organic nutrients. Laboratory column experiments are in progress to understand the As mobility and retention in chemically amended groundwaters and sediments.

1 INTRODUCTION

The occurrence of elevated As concentrations in groundwater has been linked to reducing groundwaters (BGS & DPHE, 2001). However, detailed hydrogeochemical processes that contribute to mobilization of As (and Fe) remain elusive. Mobilization of As caused by the reduction of As-bearing Fe-oxyhydroxides is a widely accepted explanation for enrichment of As in groundwater of Bangladesh. However, what processes lead to the reducing conditions in Bangladesh aquifers remain unclear. Organic matter that derived either from decomposition of peat layers (McArthur *et al.*, 2004) or from inflow of superficial organic carbon (Harvey *et al.*, 2002) have both been proposed to explain the reducing conditions encountered in the aquifers there.

A detailed hydrogeochemical study was conducted on groundwaters from six nests of monitoring wells installed in Araihazar, Bangladesh that capture nearly the entire dynamic range of groundwater As concentrations encountered in Bangladesh. Wells are installed to a Holocene shallow aquifer (5–30 m), a Holocene deep aquifer (>30 m) and a Pleistocene deep aquifer (>30 m) that are typical of high yielding aquifers used throughout Bangladesh. A wide range of hydrochemical parameters were measured in these representative samples to examine the systematics of As mobilization with major components of the groundwater, as well as redox conditions and nutrient characteristics in the shallow aquifer and deep aquifers, respectively.

2 METHODS

2.1 *Location of the nests of monitoring wells*

A total of 6 nests of 37 monitoring wells (5–91 m) (Table 1) were installed in 25 km² of study area to capture the wide spectrum of arsenic distribution.

At all sites except site B, shallow Holocene aquifer composed mostly of gray sand, are separated from the deep Pleistocene aquifer composed of orange/light brown sand aquifer by at least one fine sediment layer of clay or silty clay. The thickness of each section varies considerably from site to site (Dhar *et al.*, 2008).

2.2 *Sampling protocols and sample analysis*

Groundwaters were sampled from monitoring wells in January 2003 after 15–30 minutes of pumping, and it was allowed conductivity, temperature, pH and ORP readings to stabilize in a flow cell equipped with multi-probes of conductivity, temperature, pH, ORP. Selected existing private wells at Site A, spanning a range of As concentrations from <5 μg/L to 860 μg/L (van Geen *et al.*, 2003), were also sampled similarly by removing the heads

Table 1. Monitoring wells location.

Well nest site	Village name	Geographic coordinates	Depth range (m)
A	Dari Satyabandi	23.785°N, 90.603°E	7–91
B	Baylakandi	23.780°N, 90.639°E	7–91
C	Bhuyan Para	23.790°N, 90.611°E	5–53
E	Hatkhola Para	23.790°N, 90.616°E	5–38
F	Laskardi	23.774°N, 90.605°E	6–58
G	Laskardi(Bilbari)	23.774°N, 90.601°E	6–52

of the wells in January 2001. Samples (filtered & not filtered) were collected in replicate into pre-cleaned HPDE/glass bottles for trace metals, major ions and DOC (dissolved organic carbon).

Dissolved O_2, alkalinity, dissolved Fe(II), dissolved NH_4^+ and inorganic As species separation were measured or conducted on site using field colorimetric methods and anion exchange column, respectively. Concentrations of 33 dissolved trace metals and major ions including As, P, Fe, Mn, S, Ca^{2+}, Mg^{2+}, Na^+, K^+, SiO_2 and separated As(III) were measured in filtered-acidified groundwater by high-resolution inductively-coupled plasma mass spectrometry (HR ICP-MS). Laboratory results were within 5% of certified values. Dissolved F^-, Cl^-, Br^-, NO_2^-, NO_3^-, PO_4^{3-}, SO_4^{2-} were measured by ion chromatography (DIONEX-500 IC system). Dissolved reactive phosphate (DRP) was also determined using the classic molybdate-blue colorimetric method, modified also to determine As (Dhar *et al.*, 2004). The Shimadzu (TOC-V) instrument was used to measure total organic carbon (TOC) and total water-borne nitrogen (TN).

2.3 *Geochemical calculation*

PHREEQC (version 2.32) was used to investigate precipitation reactions based on groundwater chemical composition data of all monitoring wells. For comparison and validation, MINTEQA (version 2.0) was also used to calculate saturation indices for 11 common minerals including siderite ($FeCO_3$), rhodochrosite ($MnCO_3$), vivianite ($Fe_3(PO_4)_2$, $8H_2O$), calcite ($CaCO_3$) and dolomite ($CaMg(CO_3)_2$). Saturation Index SI is defined as:

$$SI = Log\,(IAP/K_{sp})$$

where IAP is the Ion Activity Product and K_{sp} is the solubility product for a given temperature. When SI is zero, or IAP = K_{sp}, the water is at thermodynamical equilibrium with respect to a mineral. The positive values of SI indicate supersaturation, whereas the negative values of SI suggest undersaturation. The PHREEQC code also calculates the anion/cation balance for each water composition. The anion/cation balance is calculated using the equation:

$$\text{Ion balance (\%)} = \frac{[\Sigma\,\text{cations(eq/L)} - \Sigma\,\text{anions (eq/L)}]}{[\Sigma\,\text{cations(eq/L)} + \Sigma\,\text{anions (eq/L)}]} \times 100$$

For simple groundwater compositions, the anion/cation balance should not exceed a few percent. Evolution of groundwaters was also determined by constructing a Piper diagram with major cations and anions.

2.4 *Column experiment*

A series of column experiments have been under investigation at laboratory ambient conditions. The columns were packed with different media such as beach sand (Fisher, S25-10 sea washed), iron coated beach sands, zerovalent iron (HePure) and their combinations to assess the sorption capacity of arsenic. Groundwaters from upper glacial aquifers (UGA) in southeast Queens of New York City with limited phosphate were spiked with 3 mg L^{-1} of inorganic As (As(III) + As(V)) to conduct all the experiments. Contaminated waters have been pumped through the columns against gravity at a flow rate of ~0.1 mL/min using a 12 channel Gilson Minipuls Evolution peristaltic pump. Same phosphate limited groundwater spiked with organics nutrients and other redox sensitive elements such as Fe, Mn will be pumped through the arsenic loaded columns. These column experiments were performed to understand the sorption, mobilization and retention of arsenic in aquifers of various chemical conditions.

3 RESULTS AND DISCUSSION

3.1 *Chemistry of shallow aquifers*

Groundwaters were nearly neutral at all sites (6.17–7.01). Most of the groundwaters from shallow aquifers were reducing with negative ORP values. Groundwaters at all sites were consistently getting more reducing as they approach to deeper layers above the less permeable clay. ORP ranged from −20 mV to −209 mV except in the shallowest well at Site F, which showed a positive ORP value (112 mV).

Conductivities of groundwaters from shallow aquifers did not show any obvious depth trends at most sites. Interestingly, the average conductivity of the groundwater increased in the sequence C, F, E, G, A, B. Approximately, 25 to 50% of the major cations in most shallow groundwaters were Ca or Mg, except in two wells at F and B sites. Concentrations of Na and K were low at all six sites. The most important anion in the shallow groundwater was HCO_3^-, which accounted for 70% or more of the anion contents. Perhaps the most significant feature was that all shallow waters were depleted in SO_4^{2-}. However, there is a detectable trend in low level of SO_4^{2-} at all sites, where concentrations decreased systematically with depth. Nitrate concentrations also decreased with depth in near surface environment only at a few sites. The concentrations of F^-, NO_2^-, Br^- were not detectable (<0.1 mg/L).

Consistent with negative ORP values from all shallow monitoring wells except F-6 m, most of As was in the form of As(III), determined by anion exchange separation. The proportion of As(III) generally increased with depth and approached 100% at F, A and B sites.

Fe and Mn concentrations were characterized more by the differences among the sites than by the gradient with depth at each individual site.

Total Organic Carbon (TOC), Total Nitrogen (TN), NH_4^+ and P generally increased with depth, and sometimes the profiles resembled that of As. TN concentrations were measured for the waters from the four sites. TN concentration for C, F, E and G sites were 0.7 ± 0.33 mg/L, 0.8 ± 0.87 mg/L, 1.3 ± 1.2 mg/L, 2.9 ± 2.0 mg/L, respectively. NH_4^+ concentrations for C, F, E, G, A and B sites were 0.7 ± 0.31 mg/L, 0.8 ± 1.1 mg/L, 1.4 ± 1.6 μg/L, 3.7 ± 2.5 μg/L, 4.2 ± 3.3 mg/L (except A-7 m; 18.9 mg/L), and 3.6 ± 5.1 mg/L, respectively. Therefore, most of the total nitrogen is ammonia-N. Average P concentrations at all sites were relatively high (20 ± 2.6 μmole/L).

3.2 *Chemistry of deep aquifers*

Groundwaters from deep Holocene and Pleistocene aquifers are neutral with pH values ranging from 6.17–6.94. Consistent with the negative ORP values of about −160 mV for the two deep groundwaters associated with grey sediments in the Holocene deep aquifer at Site B, arsenic is nearly 100% As(III) for B-41 m and B-53 m. The deepest well, B-91 m, also with a negative ORP value of −134 mV, has no detectable As(III), but total As was only 1 μg/L. The ORP values of waters from the Pleistocene aquifers were relatively less negative, especially at A-91 m, with values reaching +26 mV. At A, C, F sites, with <2 μg/L total As, As(III) was not detectable (<0.1 μg/L). However, As(III) was detectable at G and E sites, with all As in the form of As(III).

Compared to the shallow aquifer at each site, the conductivities of deep aquifer water were ~50% lower at sites A & B, were similar at Sites C and E, and were ~3 times higher at Sites F & G. Groundwater from the Holocene deep aquifer from Site B was Ca-depleted (<5%), with the proportion of Na increasing and the proportion of Mg decreasing as the groundwater As level decreased from 21 μg/L (41 m), to 15 μg/L (53 m) and finally to 1 μg/L (91 m). The anion in the 3 wells from the Holocene deep aquifer at Site B is by HCO_3^- with limited Cl^- at neutral pH of 7.01 ± 0.12. Groundwaters from other sites were dominated by HCO_3^- and again were nearly depleted in SO_4^{2-}, although at Sites F and G, the proportions of Cl were close to 50%.

The concentrations of Fe and Mn in the deep aquifers were generally several times lower than those of the shallow aquifer at each site, but Mn concentration exceeded the WHO guideline value in several waters.

The concentrations of TOC in the deep Pleistocene aquifers are generally lower than those of the shallow aquifer at most of the sites. The concentrations of TN in the deep Pleistocene aquifers are much lower than those of the shallow aquifer at all sites, except Site E.

3.3 *PHREQQC*

Almost all groundwaters were undersaturated with respect to evaporated minerals such as halite (NaCl), gypsum ($CaSO_4 \cdot 2H_2O$) and several carbonate minerals, including calcite ($CaCO_3$), dolomite ($CaMg(CO_3)_2$) and, in most cases, rhodochrosite ($MnCO_3$). Most groundwaters were found to be supersaturated with respect to kaolinite ($Al_2SiO_5(OH)_4$), a secondary clay mineral. Groundwaters associated with grey sediments at all sites were supersaturated with respect to siderite ($FeCO_3$) and vivianite ($Fe_3(PO_4)_2.8H_2O$), including Holocene deep aquifer waters at Site B except the deepest well at 91 m at Site B and wells at Site F (6–19 m). On the contrary, Pleistocene deep aquifer waters at all sites except Site F were found to be undersaturated with respect to siderite ($FeCO_3$).

3.4 *Discussion*

Elevated arsenic concentrations were found when waters displayed negative ORP values, between about −50 mV and −200 mV, consistent with As mobilization under reducing conditions observed throughout Bangladesh. In the shallow aquifer, only two wells out of 41 wells sampled displayed positive ORP values. The concentration of As was 0.5 μg/L at F-6 m with an ORP value of +112 mV.

The concentration of As was 9 μg/L for W4068 with an ORP value of +58 mV. A reducing condition, however, may not necessarily be the only indicator for As occurrence in shallow aquifers. This has been observed in various wells. Despite showing very negative ORP, groundwaters from Sites C, F and E showed ~3 times lower As than those from Sites A and B. For example, groundwaters from the same depth (~14 m), from two sites, displayed similar ORP values, about −160 mV. However, the concentrations of As were 73 μg/L and 517 μg/L at Site C and A, respectively. This suggests that other properties, unique to each site, also contribute to As mobilization. At least, one such property is known to be different between the two sites: the phosphate-extractable sediment As concentration. The P-extractable As were found at separate boreholes (~14 m depth) as ~0.3 mg/kg and 1.7 mg/kg at Site C and Site A, respectively (Zheng *et al.*, 2005).

Several nutrients including total organic compounds, important for microbial activities, were systematically higher in the shallow aquifer waters than in the deep aquifer waters. This difference, however, is especially great for ammonia-N and, not surprisingly, total nitrogen at all sites. However, it is difficult at this point to determine which nutrient might be limiting microbial activity in the deep aquifer. It is worth pointing out that most nutrients, TOC, ammonia and P increase with depth at Sites C, F and E. Tritium-helium dating of groundwater at these three sites also shows a depth gradient, with increasing age with depth. On the first order, the increase of ammonia concentrations with depth appears to result from remineralization of organic matter. The increase of P, however, could be complicated because both remineralization of organic matter as well as any processes that release As from sediment will also likely release P. The increase of TOC concentrations may also be related to microbial remineralization of organic matter, except that it is not known how labile these TOCs are.

Regardless of how and why many of the nutrients become elevated in the shallow aquifer water, our data demonstrate that understanding of whether microbial activity is the key of As mobilization, i.e, it is nutrient limited or not in the deep aquifer, is crucial. This is because if infiltration or leakage of shallow aquifer water to the deep aquifer occurs, it will significantly modify the biogeochemical reactions in the deep aquifer by introducing more As, more nutrients, perhaps more labile carbon and a different microbial community.

Although reductive dissolution of Fe-oxides has been widely accepted as the first order control on As mobilization in the Bangladesh aquifer, the lack of correlation between groundwater As and Fe has remained a puzzling feature. Several explanations have been proposed. Iron reduction without dissolution was proposed a potential mechanism to decouple As and Fe release (van Geen *et al.*, 2004). Reductive dissolution of As-bearing Fe oxides near the surface and then transport to depths was proposed as an alternative mechanism to explain the significantly decoupled redox state and groundwater As distribution (Polizzotto *et al.*, 2005). The mineral precipitation calculations show that groundwater Fe concentrations are regulated by precipitation reactions of siderite and vivianite in the shallow aquifer. Siderite has been identified by XRD studies of the Holocene aquifer sediment in our study area and at other sites in Bangladesh (Ahmed *et al.*, 2004). The fact that groundwater Fe was supersaturated with respect to siderite and vivianite at five sites in Araihazar suggests that processes releasing Fe are recent and probably ongoing because solutions can remain at a supersaturated state forever.

While As and P release were coupled in shallow groundwater greater than 5 years old (Dhar *et al.*, 2008), phosphate was also found to be supersaturated with respect to phosphate minerals. The results suggest that processes releasing P to the groundwater are recent and probably ongoing. Processes releasing P to groundwater include remineralization of organic matter and reductive dissolution or reduction of Fe oxides.

4 CONCLUSIONS

This study confirms that reducing conditions are key to the mobilization of As in Bangladesh shallow aquifers, where most of the As species exist as arsenite. However, As concentrations are significantly decoupled from Fe and P, in part because Fe and P concentrations are regulated by several mineral precipitation reactions. The supersaturation of Fe and P with respect to their minerals, however, indicates that processes releasing them are recent or even ongoing, and this implies that mobilization of As is also an ongoing process in the subsurface.

The Holocene shallow and deep aquifers are of Ca-Mg-HCO_3 type while the Pleistocene deep aquifer is of Na-K-HCO_3 type. Sulfate is a very minor component of all groundwaters. In addition, there is a remarkable difference in several nutrients, especially between the Holocene shallow aquifer and the Pleistocene deep aquifer. Future studies of the implication of obvious contrast in nutrients concentration on As mobility in both aquifers and viability of deep aquifer as a drinking water source are much needed.

ACKNOWLEDGEMENTS

This study was supported by PSC-CUNY Research grant and USEPA-NIEHS/Superfund Basic Research Program grant 1 P42 ES10349.

REFERENCES

Ahmed, K.M., Bhattacharya, P., Hasan, M.A., Akhter, S.H., Mahbub Alam, S.M., Hossain Bhuyian, M.A., Imam, M.B., Khan, A.A. & Sracek, O. 2004. Arsenic enrichment in groundwater of the alluvial aquifers in Bangladesh: an overview. *Applied Geochemistry* 19:181–200.

BGS & DPHE. 2001. Arsenic contamination of groundwater in Bangladesh. In D. G. Kinniburgh & P. L. Smedley (eds), *British Geological survey Technical Report WC/00/19*. Keyworth, UK.

Dhar, R.K., Zheng, Y., Rubenstone, P. & van Geen, A. 2004. A rapid colorimetric method for measuring arsenic concentration in groundwater. *Analytica Chimica Acta* 526:203–209.

Dhar, R.K., Zheng, Y., Stute, M., van Geen, A., Cheng, Z., Shanewaz, M., Shamsudduha, M., Hoque, M.A., Rahman, M.W. & Ahmed, K.M. 2008. Temporal variability of groundwater chemistry in shallow and deep aquifers of Araihazar, Bangladesh. *Journal of Contaminant Hydrology* 99(1–4):97–111

Harvey, C.F., Swartz, C.H. & Badruzzaman, A.B.M. 2002. Arsenic mobility and groundwater extraction in Bangladesh. *Science* 2998:1602–1606.

McArthur, J.M., Banerjee, D.M., Hudson-Edwards, K.A., Mishra, R., Purohit, R., Ravenscroft, P., Cronin, A., Howarth, R.J., Chatterjee, A., Talukder, T., Lowry, D., Houghton, S. & Chadha, D.K. 2004. Natural organic matter in sedimentary basins and its relation to arsenic in anoxic groundwater: the example of West Bengal and its worldwide implications. *Applied Geochemistry* 19:1255–1293.

Polizzotto, M.L., Harvey, C.F., Sutton, S.R. & Fendorf, S. 2005. Processes conducive to the release and transport of arsenic into aquifers of Bangladesh. *PNAS* 102:18819–18823.

van Geen, A., Rose, J., Thoral, S., Garnier, J.M., Zheng, Y. & Bottero, J.Y. 2004. Decoupling of As and Fe release to Bangladesh groundwater under reducing conditions. Part II: Evidence from sediment incubations. *Geochimica Cosmochimica Acta* 68:3475–3486.

van Geen, A., Zheng, Y., Versteeg, V., Stute, S., Horneman, A., Dhar, R., Steckler, M., Gelman, A., Small, A, Ahsan, H., Graziano, J., Hussain, I. & Ahmed, K.M. 2003. Spatial variability of arsenic in 6000 contiguous tube wells of Araihazar, Bangladesh. *Water Resources Research* 39(5): 1140 (HWC 3:1–16).

Zheng, Y., van Geen, A., Stute, M., Dhar, R., Mo, Z., Cheng, Z., Horneman, A., Gavrieli, I., Simpson, H.J. & Versteeg, R. 2005. Geochemical and hydrogeological contrasts between shallow and deeper aquifers in two villages of Araihazar, Bangladesh: Implications for deeper aquifers as drinking water sources. *Geochimica Cosmochimica Acta* 69(22): 5203–5218.

One Century of the Discovery of Arsenicosis in Latin America (1914–2014) –
Litter, Nicolli, Meichtry, Quici, Bundschuh, Bhattacharya & Naidu (Eds)
 ISBN 978-1-138-00141-1

Tectonic-sourced groundwater arsenic in Andean foreland of Argentina: Insight from flow path modeling

A. Mukherjee & N. Raychowdhury
Department of Geology and Geophysics, Indian Institute of Technology-Kharagpur, West Bengal, India

P. Bhattacharya
KTH-International Groundwater Arsenic Research Group, Department of Sustainable Development, Environmental Science and Engineering, KTH Royal Institute of Technology, Stockholm, Sweden

J. Bundschuh
Faculty of Health, Engineering and Sciences & National Centre for Agriculture, University of Southern Queensland, Toowoomba, Australia

K. Johannesson
Department of Earth and Environmental Science, Tulane University, New Orleans, LA, USA

ABSTRACT: The groundwater arsenic enriched Chaco-Pampean plain of Argentina is located in the active foreland of continental arc dominated Andean orogenic belt. Rhyolitic volcanic glass fragments are a major component of the aeolian-fluvial aquifer sediments, which is dotted with many hot springs that are related to the palaeo-igneous extrusion in the vicinity. Several Salinas in the areas may have originated because of the tectonic evolution of the region. Hydrogeochemical analyses, thermodynamic mixing diagrams and flow path modeling analyses of groundwater samples collected from the Santiago del Estero province suggest that predominant evolutionary processes of the groundwater include chemical weathering with monosialitization silicate of weathering and evaporate dissolution. Anorthite, albite and As-enriched volcanic glass seems to contribute to the major dissolution phases. Subsequently, co-introduced oxyions mobilized the solid-phase As to groundwater by competitive ion exchanged. Further liberation might have taken place by counter-ion activity due to transition of the Ca-rich to Na-rich groundwater due to groundwater mixing with recharged brackish surface water from Salinas or by evaporative concentration due to the prevailing arid climate.

1 INTRODUCTION

Argentina is the first country in Latin America to have reported groundwater arsenic (As). The As enriched Chaco-Pampean plains of Argentina covers an area of about 10^6 km^2 that includes the study area in Santiago Del Estero (Bundschuh *et al.* 2004; Bhattacharya *et al.*, 2006). The quality of usable groundwater resources in the area is poor due to high salinity and elevated concentrations of As and other toxic trace elements. More than 1.2 million residents of this region are largely dependent on this questionable quality groundwater for drinking and domestic purposes. The study area in Rio Dulce river watershed has a semi-arid climate with mean annual precipitation of ~530 mm, and vegetation comprises low bushes and grasslands (Bejarano and Nordberg, 2003). Groundwater table depth ranges from 1 to 9 m below ground level with direction of groundwater flow from NW to SE (Bhattacharya *et al.*, 2006). Several previous studies suggested that the source of the As in the Argentine groundwater are in the Cenozoic loess sediments that constitute much of the major aquifers (e.g. Smedley *et al.*, 2004) and volcanic glass fragments, intermittently present in these sediments (e.g. Bhattacharya *et al.*, 2006; Nicolli *et al.*, 2010). These sediments have been hypothesized to be source of the groundwater As that mobilize under oxidized hydrogeochemical conditions.

Tectonically, the Chaco-Pampean plain is located in the Andean foreland basin around the piedmont of the Central Andean orogenic system. The orogeny of the Central Andes was caused by subductions that occurred in the absence of major plate collisions. The Andes developed by convergence and subsequent subduction of oceanic

Nazca plate below the continental South American plate. Since Cretaceous, extrusive volcanic activity along the Central Andes resulted in igneous rock like andesites, basalts, rhyolites and volcanic glass and ash. The ash beds commonly include quartz, feldspar and biotite existing in a glassy matrix. The volcanic ash layers are discontinuous, with variable thickness.

2 METHODOLOGY

Groundwater samples (n = 60) were collected from different hand pumped and open dug wells in the study area (Figure 1) and analyzed for field analytes along with major, minor and trace solutes using standard hydrogeochemical techniques as explained in Bhattacharya *et al.* (2006). Minimal reaction-flow path (inverse) models were developed for pairs of wells located along possible flowpaths, by using hydrogeochemical code PHREEQC (see Raychowdhury *et al.*, 2013 for details).

The models were generated following Mukherjee and Fryar (2008) and Johannesson and Tang (2009) for testing two major hypotheses: (i) chemical weathering of volcanic glass within the aquifer sediments are a major source of cations and dissolved silica to study area groundwaters; and (ii) mixing of fresh groundwater with recharged brackish surface water from local Salinas that might contribute dissolved solutes to the local groundwaters and/or high evaporation. A total of 20 different simulations successfully converged within the acceptable uncertainty range that tested the scenarios for I) mixing of fresh groundwaters with recharged brackish waters (mixing reactions), II) without any mixing, III) only ion exchange and IV) both mixing and reactions. The simulations included mole-balance, alkalinity-balance, electron-balance, water-balance and charge-balance equations to solve a) mixing fraction of each input members and b) mole transfers of minerals and gases. The selected pairs of wells used as end members are only based on their distinctiveness in hydrochemistry, and they are located up gradient-down gradient to each other along suggested general flow direction. The simulations evaluated groundwater evolution through silicate weathering (quartz, amorphous silica, feldspars, clay minerals, volcanic glass etc.), evaporite dissolution (halite and thenardite), reversible cation exchange reactions (Ca^{2+}/Mg^{2+}, Ca^{2+}/Na^{+}, Ca^{2+}/K^{+}, Na^{+}/Mg^{2+}, and Na^{+}/K^{+}) along with some oxide phases or combination of these. Volcanic glass dissolution was defined in the simulations as:

$$(Na_{0.0422}\,K_{0.0325}\,Ca_{0.0023}\,Mg_{0.0001}\,Fe_{0.0029}\,Al_{0.0789}\,Si_{0.419}\,O_{1.0})\text{ glass} + H_2O + 0.325CO_2 \rightarrow 0.0442\,Na^{+} + 0.0325\,K^{+} + 0.0023Ca^{+2} + 0.0001\,Mg^{+2} + 0.0029\,Fe^{+2} + 0.0789\,Al^{+3} + 0.419H_4SiO_{4+}\,0.324HCO_3^{-}$$

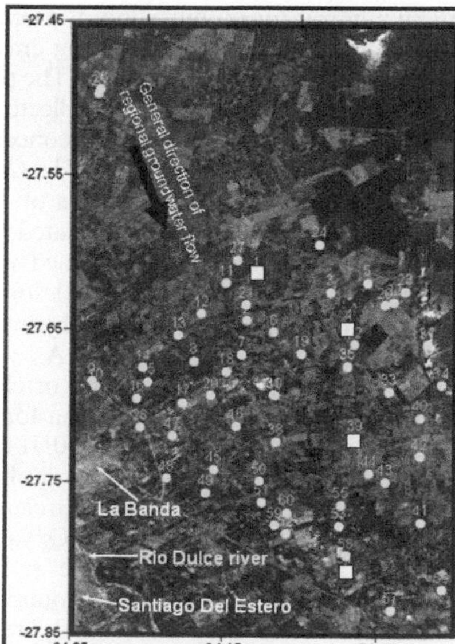

Figure 1. Geographic location of groundwater samples collected and used in this study.

3 RESULTS AND DISCUSSION

3.1 *Insights from hydrogeochemistry*

Chemical analyses of groundwater samples show extensive solute chemistry variability and demonstrate the evolutionary processes (Raychowdhury *et al.*, 2013) (Figure 2). Bicarbonate and Cl^{-} concentrations are widely variable, ranging from 112 to 1069 mg/L. The relatively high pH, alkalinity, and chloride concentrations are suspected to be due to the effect of evaporative concentration in the characteristic arid, oxidizing environment of the study area. Sulfate concentrations are very high with median of 248 mg/L. The combination of high SO_4^{2-} and Cl^{-} concentrations reflect the influence of recharge from local *salinas* and salt flats located in the vicinity of the study area. The *salinas* are probably evaporated Pre- and syn-Andean orogenic inland sea remnants. While, the potentially marine sourced monovalent Na and K concentrations of the sampled groundwater range up to 2557 and 110 mg/L, respectively, generally non-marine sourced, divalent cations such as Ca^{2+} (median: 74.9 mg/L) and Mg^{2+} (median: 14.6 mg/L) have relatively low concentrations. Oxyanion-forming metals and metalloids that are common in aquifers with volcanogenic minerals

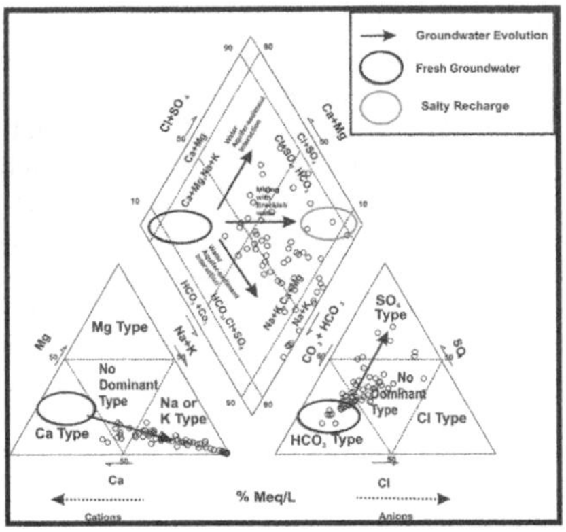

Figure 2. Piper diagram showing the evolutionary mechanisms for the groundwater samples collected in this study.

and oxidative conditions (e.g. As, Mo, and V) are present in exceptionally high concentrations, with As up to ~7.5 mg/L and a median of 0.076 mg/L. Groundwater V and Mo concentrations range up to ~1.9 mg/L (median: 0.06 mg/L) and 1.4 mg/L (median: 0.02 mg/L). Arsenate (As(V)) dominates the groundwater As species.

In the silicate-aluminosilicate stability diagrams, most of the groundwater samples cluster around the K-feldspar, kaolinite, and muscovite tri-junction, where all three minerals are stable. The groundwater samples are supersaturated with respect to quartz and undersaturated with respect to amorphous silica. The samples plot close to the kaolinite-Na-smectite boundary line. The stability diagrams strongly suggest that both kaolinite and smectite may form by albite weathering in arid condition. These calculations suggest a monosiallitization type of chemical weathering predominates in the study region. During monosiallitization weathering, most solutes are mobilized from solid mineral phases to the groundwater, with only remnant Si in secondary clay minerals like kaolinites.

The simulation results of several flow path models worked with Si as non-conservative solutes suggesting dissolution input and/or precipitation of silicates and glasses along the flow paths. Several models also evaporite dissolution leading higher concentrations of Na^+, Cl^-, and SO_4^{2-} in several locations thus influencing the evolution of groundwater with recharge influx from local Salinas or brackish geothermal spring water or evaporative enrichment of groundwater. Only few simulations could be explained by ion exchange reactions. However, majority of the models suggested input from volcanic glass dissolution, ion exchange reactions and mixing with evaporite saturated waters along the flow paths.

3.2 *Hypothesis of tectonic-sourced arsenic and its mobilization*

The primary source of the As in the aquifers of the Andean foreland, including the Chaco-Pampean groundwater is hypothesized to be the volcanic glass that form an important constituent within the aquifer sediment. The results of the hydrogeochemical modeling demonstrate that weathering of the volcanic materials is thermodynamically favorable. Such volcanogenic sources, in association with hydrothermal fluids, may contain deep subsurface As transported to the surface. In addition to glass dissolution, a major geogenic source of As in the study area and its vicinity may include abundant mineralized, hydrothermal zones and hot springs that are present in and around the study area. Tectonically, the As-bearing hydrothermal fluids likely reflect recycled products of precipitated sulfides of Fe, Cu, Zn and/or As-bearing Fe(III) oxides/oxyhydroxides. Meteoric waters also infiltrate up to a 7–10 km depth, become heated, scavenge As from their hot surroundings, and then flow toward the surface to produce geysers and hot springs. Hence, many of these hot springs may be conceptualized to be potential pathways of As transport from magmatic sources to the Earth's surface (Webster and Nordstrom, 2003). It is hypothesized that hot spring waters would subsequently recharge to the groundwater system and possibly contribute to surface waters, subsequently mixing and/or recharging to groundwaters of meteoric origin. Hence, such a mechanism may partially explain the presence of As-enriched groundwaters in the aerated loess aquifers of the Chaco-Pampean plain.

Groundwater pH has been considered as one of the main controls for mobilization of arsenic in groundwater, especially in arid, oxidizing environmental conditions (Smedley *et al.*, 2005). The pH of Chaco-Pampean plain groundwaters collected for this study range from 6.6 to 8.9, and As concentrations positively correlate with pH ($\rho = 0.51$). One of the primary mechanisms of mobilization of As in the Chaco-Pampean plain may also be related to the dissolution of the volcanic glass contained within the aquifer sediments. The competitive desorption of As from hydrous metal oxides and hydroxides may result in high concentrations of As within groundwater and results to a moderate correlation of Si with As ($\rho = 0.34$). Other oxyanion-forming e.g. V is strongly correlated with As ($\rho = 0.913$) and this may indicate similar sources and/or mobilization and transport processes for both trace elements in Chaco-Pampean plain groundwaters. Oxyanions like PO_4^{3-} and HCO_3^- show moderate relationships with As in the Chaco-Pampean groundwater ($\rho = 0.38$ and 0.43 for PO_4^{3-} and HCO_3^-, respectively). These observations imply that the mobilization of As is strongly favored by ionic competition for mineral surface sites. Higher ionic strength enhances inner sphere surface complexation of As,

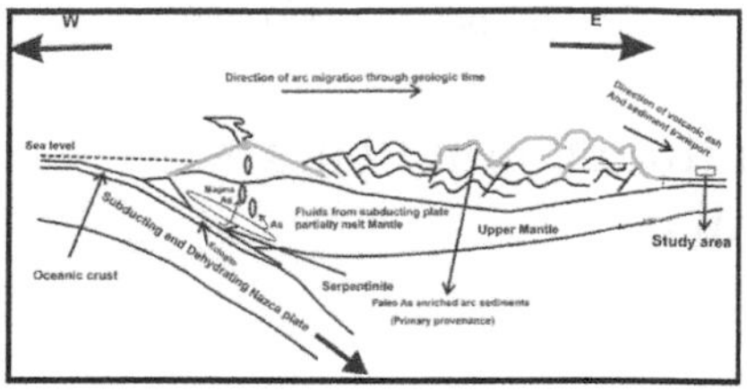

Figure 3. Hypothetical model of tectonic sourced As in aquifers of the Andean foreland (Raychowdhury *et al.*, 2013).

but decreases outer sphere surface complexation, causing sorbed As mobilization (Scanlon *et al.*, 2009). Such a process might take place with change from Ca to Na-rich water. The $Na/Ca^{0.5}$ is > 1 for all the sampled groundwaters of the Chaco-Pampean plain and they are strongly correlated with As concentrations ($\rho = 0.72$).

Also, As exhibits a moderate positive relationship with Na ($\rho = 0.54$), but an inverse relationship with Ca concentrations ($\rho = -0.66$) in the Chaco-Pampean groundwater samples. The strong correlation of $Na/Ca^{0.5}$ with As in the Chaco-Pampean plain groundwaters indicates that the counter-ion effect (Scanlon *et al.*, 2009) may be an important mechanism responsible for As mobilization in these groundwaters.

4 CONCLUSIONS

The origin of arsenic in the Chaco-Pampean plain of Argentina may be hypothesized to be tectonically controlled and is transported to the surface aquifers by extrusive volcanism or hydrothermal fluids ejecting through structurally controlled hot springs. Volcanisms in the Andean arc lead to deposition of volcanic ash layers of different geological ages. This volcanic ash beds constitute major parts of the study area aquifer. Rhyolitic glass present in these ash beds and silicate rocks in the volcanics of the Andean range undergoes hydrolytic dissolution leading by monosialitization weathering to influx of major and trace oxy-anions to the groundwater (e.g. As, V, Mo, etc.) and lead to precipitation of secondary clays. Solutes are available from recharge and subsequent mixing of As-laden hydrothermal fluids and hot spring water with groundwater. These As-rich hydrothermal fluids approach the surface, and become diluted with aerated groundwater. Subsequently, these oxidized As species adsorb onto surfaces of metallic oxides-hydroxides by process of outer surface complexation. However, co-liberated competitive ions with similar ionic radii (e.g. Si, V) due to chemical weathering processes, compete for surface sites of aquifer matrix, and displaces As from solid to dissolve phase. This phenomenon is further enhanced by counter-ion effect due to increase of ionic strength of the groundwater, caused by change Ca-rich groundwater to Na-rich groundwater. Such hydrogeochemical evolution might have been caused by enhanced Na influx from silicate weathering, mixing with brackish water or evaporative concentration in semi-arid environment.

REFERENCES

Bejarano, G. & Nordberg, E., 2003. *Mobilisation of arsenic in the Rıo Dulce alluvial cone, Santiago del Estero Province, Argentina.* Master Thesis, Dept. of Land and Water. Research. Engineering., KTH, Stockholm, Sweden, TRITA-LWR-EX-03-06; 40 pp.

Bhattacharya, P., Claesson, M., Bundschuh, J., Sracek, O., Fagerberg, J., Jacks, G., Martin, R.A., Storniolo, A. del R. & Thir, M., 2006. Distribution and mobility of arsenic in the Rıo Dulce alluvial aquifers in Santiago del Estero Province, Argentina. *Science of Total Environment*. 358:97–120.

Bundschuh, J., Farias, B., Martin, R., Storniolo, A., Bhattacharya, P. & Cortes, J., 2004. Groundwater arsenic in the Chaco–Pampean Plain, Argentina: case study from Robles County, Santiago del Estero Province. *Applied Geochemistry,* 19(2):231– 43.

Johannesson, K.H. & Tang. J. 2009. Conservative behavior of arsenic and other oxyanion-forming trace elements in an oxic groundwater flow system. *Journal of Hydrology,* 378:13–28.

Mukherjee, A. & Fryar, A.E., 2008. Deeper groundwater chemistry and geochemical modeling of the arsenic affected western Bengal basin, West Bengal, India. *Applied. Geochemistry*, 23: 863–892.

Nicolli, H.B, Bundschuh, J, García, J.W, Falcón, C.M. & Jean, J.S., 2010. Sources and controls for the mobility of arsenic in oxidizing groundwaters from loess-type sediments in arid/semi-arid dry climates—Evidence from the Chaco-Pampean plain (Argentina). *Water Research,* 44:5589–5604.

Raychowdhury, N., Mukherjee, A., Bhattacharya, P., Johanneson, K., Bundschuh, Sifuentes, G., Nordberg, E., Martin, R. & Storniolo, A., 2013. Provenance and fate of arsenic and other solutes in the Chaco-Pampean plain of the Andean foreland, Argentina: from perspectives of hydrogeochemical modeling and regional tectonic setting. *Journal of Hydrology*, dx.doi.org/10.1016/j.jhydrol.2013.07.003

Scanlon, B.R., Nicot, J.P., Reedy, R.C., Kurtzman, D., Mukherjee, A. & Nordstrom, D.K., 2009. Elevated naturally occurring arsenic in a semiarid oxidizing system, Southern High Plains aquifer, Texas, USA. *Applied Geochemistry*, 24: 2061–2071.

Smedley, P.L., Kinniburgh, D.G., Macdonald, D.M.J., Nicolli. H.B., Barros. A.J., Tullio. J.O., Pearce. J.M. & Alonso. M.S., 2005. Arsenic associations in sediments from the loess aquifer of La Pampa, Argentina. *Applied Geochemistry*, 20: 989–1016.

Webster, J.G. & Nordstrom, D.K., 2003. Geothermal arsenic, in Arsenic in Ground Water, A.H., Welch and K.G. Stollenwerk (eds), Kluwer Academic Publishers, Boston, pp. 101–25.

One Century of the Discovery of Arsenicosis in Latin America (1914–2014) –
Litter, Nicolli, Meichtry, Quici, Bundschuh, Bhattacharya & Naidu (Eds)
 ISBN 978-1-138-00141-1

Dissolution and entrapment of arsenic in fluvial point bars—case study of the Ganges River, Bihar, India

M.E. Donselaar
Delft University of Technology, Delft, The Netherlands

ABSTRACT: Spatially variable arsenic concentrations in shallow aquifers in Holocene Ganges River deposits in Bihar (India) suggest geological conditioning of the propagation of arsenic-contaminated water in terms of stratigraphy and sediment type. High arsenic concentrations are confined to sandy point bars in abandoned river bends. The arsenic is geogenic and occurs in solid state in clay minerals and hydrated iron-arsenic-oxide quartz grain coatings. Microbial respiration in a redox-controlled environment triggers the reductive dissolution of arsenic. Analysis of two 50-m-deep cored boreholes shows that lithofacies heterogeneity and inherent permeability contrast in stacked point-bar deposits and associated clay plugs is the key to the dissolution and stratigraphic entrapment of arsenic. An electromagnetic survey visualized the shape, size and spatial distribution of the stratigraphic entrapment in the shallow sub surface.

1 INTRODUCTION

Arsenic-contaminated groundwater causes a widespread, serious health risk affecting millions of people world-wide. In India, it was only discovered in 2002 that groundwater in shallow aquifers of Holocene Ganges River deposits in Uttar Pradesh (UP) and Bihar is highly arsenic-contaminated (Chakraborti *et al.*, 2003). Arsenic concentrations in Bihar reach 1,800 $\mu g\ L^{-1}$, far in excess of the WHO guidelines for safe drinking water of 10 $\mu g\ L^{-1}$ (WHO, 1993). An extensive governmental arsenic inventory campaign in Bihar aims to map the extent and magnitude of the contamination (Saha, 2009). To date, while the inventory is not complete, it is estimated that 25% of the 103.8 million population of Bihar is exposed on a daily basis to arsenic-contaminated drinking and irrigation water.

The arsenic contamination has a geogenic origin. Pyrite-bearing shale from the Proterozoic Vindhyan Range, arsenic-copper mineralization in the Bundelkhand Granite in UP, and the gold belt of the Son Valley are potential sources of arsenic (Shah, 2010). Upon weathering, the arsenic is transported in solid state by rivers. Arsenic in Holocene Ganges River sediments is associated with hydrated iron-oxide coatings on quartz and clay minerals (Shah, 2008). The arsenic is subsequently released to the groundwater in a redox-controlled environment (Singh *et al.*, 2010). Microbial respiration triggers the reductive dissolution of iron and arsenic.

Water wells tapping from Holocene deposits in the affected areas show a spatial variability in arsenic concentration in three dimensions. Low arsenic concentrations occur in areas that are well-flushed by groundwater flow due to high-hydraulic head (Shah, 2010). Concentrations increase in a poorly-flushed subsurface environment. Recent studies indicate a relationship between spatial variability of arsenic concentration, and stratigraphy and sediment type (Shah, 2008, 2010; Saha, 2009). The aim of the present paper is to analyze the role of the spatial variability of fluvial lithofacies in the propagation and trapping of the arsenic-contaminated water flux in the subsurface. Focus is on the stratigraphic entrapment in permeable point bar sediment.

2 METHODS

The study is based on the analysis of borehole data and an electro-magnetic survey of Holocene Ganges River deposits in Bihar. Two 50-m-deep wells were drilled in a fluvial point bar and the fringing, sediment-filled abandoned river bend, or clay plug (Figure 1). Core recovery was about 80%, and the cores were accurately depth-constrained. Gamma-ray and deep resistivity logs were run in both boreholes. A Transient Electromagnetic (TEM) survey was performed in the area between the two wells to obtain a detailed depth image of the point bar lithofacies distribution in the shallow subsurface.

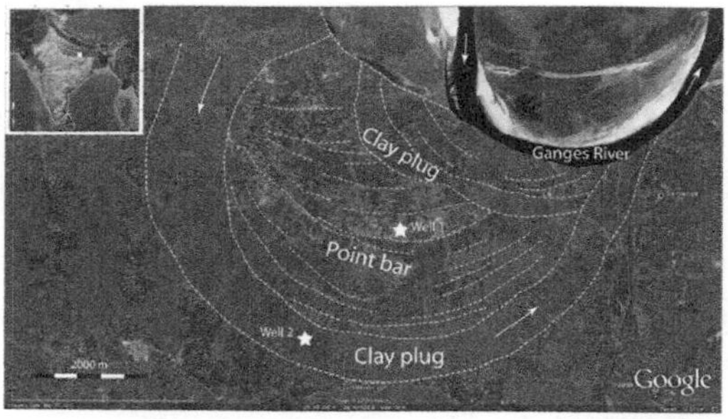

Figure 1. Study area with interpreted morphology of the point bar and clay plug. The main source of drinking water of the villages comes from multiple, shallow hand pump wells which tap from the permeable point-bar sand. Insert: map of India with location of study area (white dot).

A 3-D time-resistivity image was constructed by repeating the measurements in a horizontal grid. Time-to-depth conversion was obtained by validation of the time section with the resistivity logs in the adjacent boreholes.

3 RESULTS AND DISCUSSION

Core analysis reveals that the stratigraphic succession in both wells is subdivided in two sequences with a sharp break at ~28 m depth. The lower sequence consists of gravel layers and coarse-grained gravelly sand. Permeability is very high, to the point that the drilling mud has completely invaded the core. This sequence is interpreted as formed by shallow braided rivers. The mineralogy suggests that the source area of the rivers was to the south, on the stable Indian Craton. The upper sequence consists of medium- to fine-grained, laminated sand, silt and organic-matter containing clay, organized in three fining-upward units with a thickness of 5 to 12 m. The units formed by vertical stacking of successive generations of Ganges River point bar sediment. The top of the sequence in Well 02 is a 12-m-thick succession of silt and black clay, rich in organic carbon. The succession formed as clay-plug fill of the oxbow-lake that envelopes the point bar sand (Figure 1). The 12 m thickness of the point-bar units and clay plug equals the depth of the present-day Ganges River just north of the well locations. The boundary between both sequences is interpreted as a sequence boundary which marks the southward shift of the Ganges River belt to this area, with truncation of the underlying braided river deposits.

The sequence boundary shows up in the TEM survey as a sharp change from high (above) to low resistivity (below). The inclined point-bar to clay plug interface has a marked resistivity contrast. Lateral lithofacies change of inclined point-bar sand and clay layers is beyond the resolution of the method.

Arsenic concentration measurements in the boreholes and hand pump wells show high but variable concentrations in the stacked point-bar sequence. The sequence boundary is characterized by a sharp peak in arsenic concentration, whereas in the lower, braided river sequence the concentrations drop. It is interpreted that a free-moving groundwater flux is present in the highly permeable gravel and gravelly-sand below the sequence boundary. The flux effectively flushes the permeable sediment, hence the low arsenic concentration. Downward percolation of arsenic-enriched water to the sequence boundary accumulates at the top of the free-moving groundwater flux; hence the peak in arsenic concentration. The assumption has to be further corroborated by detailed measurements and tracer tests.

4 CONCLUSIONS

Arsenic contamination in the shallow aquifer domain of Ganges River deposits in Bihar (India) is characterized by large spatial variability of concentration levels. The arsenic is of geogenic origin and its occurrence in groundwater is the result of dissolution of Fe-As oxides in a redox-controlled environment. The permeability contrast between low-permeable clay plug and high-permeable point-bar sand creates a stratigraphic trap in which groundwater with dissolved arsenic accumulates. Core analysis of two 50-m-deep boreholes in a point bar and juxtaposed clay plug show superposition of two fluvial sequences separated by a sequence boundary at 28 m depth. The lower sequence consists of stacked, high-permeable braided river gravel and coarse-grained gravelly sand; the upper sequence is made up of 5–12 m thick stacked point-bar units and associated organic matter-rich clay plug sediment. The lateral continuity of the sequence boundary and the overall shape of the permeable fluvial deposits are observed as resistivity contrasts in an electro-magnetic survey. Measured arsenic concentrations in the boreholes and hand pump wells show high but variable levels in the stacked point-bar sequence and low levels in the underlying braided river sequence. A sharp peak in arsenic concentration at the sequence boundary is interpreted by the permeability contrast between the two fluvial sequences.

ACKNOWLEDGEMENTS

The author gratefully acknowledges financial support for this study from the European Union Erasmus Mundus—EURINDIA—Lot 13 program. The author is indebted to colleagues at A.N College, Patna, India (A.G. Bhatt, N. Bose and A.K. Ghosh) and TU Delft (J. Bruining, G. Schaepman) for their assistance with the data acquisition and analysis.

REFERENCES

Acharyya, S.K., Lahiri, S., Raymahashay, B.C. & Bhowmik, A. 2000. Arsenic toxicity of groundwater in parts of the Bengal basin in India and Bangladesh: the role of Quaternary stratigraphy and Holocene sea-level fluctuation. *Environmental Geology* 39: 1127–1137.

Chakraborti, D., Mukherjee, S.C., Pati, S., Sengupta, M.K., Rahaman, M.M., Chowdhury, U.K., Lodh, D., Chanda, C.R., Chakraborti, A.K. & Basu, G.K. 2003. Arsenic groundwater contamination in Middle Ganges Plain, Bihar, India: A future danger? *Environmental Health Perspective* 111: 1194–1201.

Saha, D. 2009. Arsenic groundwater contamination in parts of Middle Ganges Plain, Bihar. *Current Science* 97: 753–755.

Shah, B.A. 2008. Role of Quaternary stratigraphy on arsenic-contaminated groundwater from parts of Middle Ganges Plain, UP–Bihar, India. *Environmental Geology* 53: 1553–1561.

Shah, B.A. 2010. Arsenic-contaminated groundwater in Holocene sediments from parts of Middle Ganges Plain, Uttar Pradesh, India. *Current Science* 98: 1359–1365.

Singh, M., Singh, A.K., Swati, Srivastava, N., Singh, S. & Chowdhary, A.K. 2010. Arsenic mobility in fluvial environment of the Ganges Plain, northern India. *Environmental Earth Science* 59: 1703–1715.

WHO 1993. *Guideline for drinking water quality. Recommendations, second ed.*, vol 1. WHO, Geneva.

One Century of the Discovery of Arsenicosis in Latin America (1914–2014) – Litter, Nicolli, Meichtry, Quici, Bundschuh, Bhattacharya & Naidu (Eds)
 ISBN 978-1-138-00141-1

Geochemical processes controlling the mobilization of arsenic distribution in the Ganga Alluvial Plain

D.K. Jigyasu, R. Kuvar, Shahina, M. Singh & I.B. Singh
Centre of Advanced Study in Geology, University of Lucknow, Lucknow, India

S. Singh
Department of Earth Sciences, Indian Institute of Technology, Roorkee, India

ABSTRACT: A total of 145 water and 48 sediment samples from the Gomati River Basin were analyzed to understand the role of geochemical process in elevated arsenic concentrations and its mobilization in the Ganga Alluvial Plain. The EDS study of biotite grain suggests that chemical weathering process releases various elements, including arsenic along with iron. In the river water, arsenic concentrations increase with the rise in temperature and reduction in discharge from the winter to the summer season. The elevated arsenic concentrations in the groundwater and river water are influenced by the anthropogenic inputs of phosphorous and labile organic carbon through sewage disposal.

1 INTRODUCTION

The elevated levels of arsenic in water threaten human health in large areas globally and its origin is intensively discussed by researchers worldwide. There remains considerable debate about the key control of arsenic concentrations in water (Benner, 2010; Neumann, *et al.*, 2009). Arsenic toxicity in the Ganga Alluvial Plain is common, influencing large population; however, the exact processes of arsenic mobilization are not yet properly understood.

The source of arsenic in the Ganga Alluvial Plain is geogenic, where arsenic is mobilized into water and food chain from the sediments. The human population in this high population density region is exposed to arsenic toxicity by using surface and ground water for drinking, cooking and irrigation. Here, we are presenting the data on arsenic distribution in sediments and water of the Gomati River Basin and mobilization processes leading to elevated arsenic concentration in water.

2 STUDY AREA

The Gomati River (a tributary of the Ganga River) drains about 30,437 km^2 of the northern part of Ganga Alluvial Plain. The basin experiences a warm and humid subtropical climate with four prominent seasons; the winter, the summer, the monsoon and the post-monsoon. Geologically, the basin is made up of unconsolidated sediments characterized by mica-rich silty sand. The Gomati River receives its sediment and water supply within the alluvial plain. The sediments of the Gomati River are essentially fine sand, composed of mostly quartz, feldspar and mica. Organic debris and small clay pellets are common. Mica content is very high in coarser fractions. Biogeochemical processes may be prominent in the basin due to large annual temperature variation (5–40 °C), monsoon precipitation and fine grained nature of the sediments.

3 METHODS AND EXPERIMENT

The alluvial plain sediments ($n = 16$), bedload sediments ($n = 15$), suspended load sediments ($n = 17$), river water samples ($n = 36$) and shallow groundwater samples ($n = 109$) were collected from the Gomati River Basin. All sediment and water (filtered <0.45 μm) samples were analyzed for their arsenic concentration by INAA and ICP-MS, respectively. Biotite grains from the Gomati River sediments were studied by EDS analyzer.

4 RESULTS AND DISCUSSION

4.1 *Arsenic in alluvial plain and river sediments*

Arsenic concentrations vary in the range of 3.0–19.6 ppm in the alluvial plain sediments, 0.1–4.4 ppm in the bedload sediments and 0.7–14.9 ppm in the suspended load sediments (Singh *et al.*, 2012). The geogenic source of arsenic is the alluvial plain sediments. The average arsenic concentration in the bedload sediments (1.4 ppm) and suspended load (5.3 ppm) is lower than that in the alluvial plain sediments (10.4 ppm). We propose that arsenic is released from the mica dominated unconsolidated alluvial plain sediments by chemical weathering. Anthropogenic inputs of organic rich waste material into the river system accelerate the arsenic release processes.

4.2 *Chemical weathering of biotite*

Arsenic concentration is high in the Gomati River sediments and most of it is present in mica minerals. Arsenic is considered to be derived from biotite in the Ganga Delta region (Seddique *et al.*, 2008). The EDS image of the margins of biotite grain of the Gomati River sediments shows the effect of physical and chemical alteration. The alteration is strong along the margin of the biotite grain indicated by depletion in various elements (Figure 1). The EDS chemical data indicate that the process of biotite mineral dissolution release aluminum, manganese, titanium, potassium and iron along with arsenic. The chemical weathering process of biotite releases these elements as follows (Dong *et al.*, 1998):

$$K_{0.65}(Al_{1.10}Ti_{0.15}Fe_{0.50}Mg_{0.55})(Si_{3.20}Al_{0.80})O_{10}(OH)_2 + 8H^+ = (Al_{1.71}Fe_{0.31}Ti_{0.03})Si_2O_5(OH)_4 + 0.65K^+ + 0.19Al^{3+} + 0.55Mg^{2+} + 0.19Fe^{2+} + 0.12Ti^{4+} + 1.2Si^{4+} + 3H_2O$$

4.3 *Arsenic in river water and groundwater*

Arsenic concentration in the Gomati River water varies between 0.53 to 5.7 ppb in different seasons attaining highest values during the summer season. High temperature causes the enormous growth of microbial species during the summer season, which helps the mobilization of arsenic from the river sediments to the river water (Oremland and Stolz, 2003).

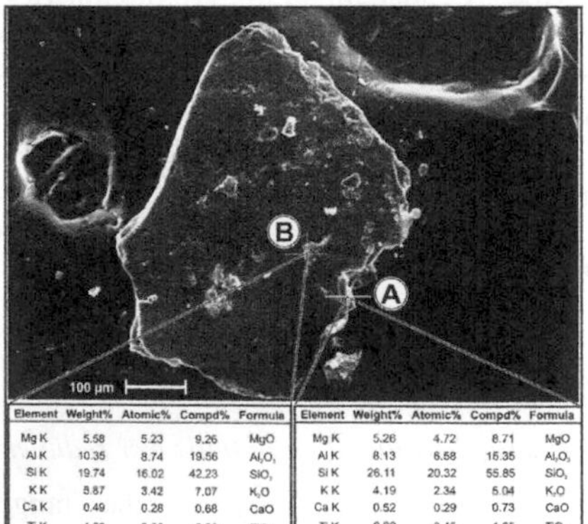

Element	Weight%	Atomic%	Compd%	Formula
Mg K	5.58	5.23	9.26	MgO
Al K	10.35	8.74	19.56	Al_2O_3
Si K	19.74	16.02	42.23	SiO_2
K K	5.87	3.42	7.07	K_2O
Ca K	0.49	0.28	0.68	CaO
Ti K	1.38	0.66	2.30	TiO_2
Fe K	14.69	5.99	18.90	FeO
O K	41.90	59.67		
Totals	100.00			

Element	Weight%	Atomic%	Compd%	Formula
Mg K	5.26	4.72	8.71	MgO
Al K	8.13	6.58	15.35	Al_2O_3
Si K	26.11	20.32	55.85	SiO_2
K K	4.19	2.34	5.04	K_2O
Ca K	0.52	0.29	0.73	CaO
Ti K	0.99	0.45	1.65	TiO_2
Fe K	9.84	3.85	12.66	FeO
O K	44.97	61.45		
Totals	100.00			

Figure 1. Scanning electron micrograph of a biotite flake from the Gomati River Sediments. It shows weathered margins (A) and unweathered surface (B) along with chemical data tables expressed in weight%, atomic%, compound% and formula.

In the Gomati River Basin, arsenic concentration in the shallow groundwater varies from 0.2 to 4.5 ppb. Groundwater samples, with high phosphorous content (<40 ppb), show significant positive correlation ($r^2 = 0.89$, $n = 10$) with arsenic concentrations. High phosphorous contents are attributed to the migration of untreated sewage waste into the shallow aquifer. Anthropogenic activities accelerate the natural geochemical processes by providing labile organic carbon through agricultural activity and the disposal of organic matter rich liquid and solid waste material in the sediment and water of the Gomati River Basin.

5 CONCLUSIONS

Arsenic in the Ganga Alluvial Plain is derived from a geogenic source. It is released during the chemical weathering of alluvial plain sediments, in particular from the biotite mineral. Additionally, anthropogenic activities, causing reducing oxygen deficient conditions and providing phosphorous and labile organic carbon, play a significant role in accelerating the mobilization of arsenic into water from sediments of the Ganga Alluvial Plain.

ACKNOWLEDGEMENTS

DKJ and RK are thankful to University Grants Commission, New Delhi for the financial support. IBS is thankful to INSA, New Delhi for the financial support. The Geological Survey of India, northern region Lucknow, is gratefully acknowledged for the EDS analysis of biotite.

REFERENCES

Benner, S. 2010. Anthropogenic arsenic. *Nature Geoscience 3*: 5–6.

Dong, H., Peacor, D.R. & Murphy, S.F. 1998. TEM study of progressive alteration of igneous biotite to kaolinite throughout a weathered soil, *Geochimica et Cosmochimica Acta*, 62: 1881–1887.

Neumann, R.B., Ashfaque, K.N., Badruzzaman, A.B.M., Ali, M.A., Shoemaker, J.K. & Harvey, C.F. 2009. Anthropogenic influences on groundwater arsenic concentrations in Bangladesh. *Nature Geoscience,* 1–7.

Oremland, R.S. & Stolz, J.F. 2003. The Ecology of Arsenic. *Science*, 300: 939–944.

Seddique, A.A., Masuda, H., Mitamura, M., Shinoda, K., Yamanaka, T., Itai, T., Maruoka, T., Uesugi, K., Ahmed, K.M. & Biswas, D.K. 2008. Arsenic release from biotite into a Holocene groundwater aquifer in Bangladesh. *Applied Geochemistry,* 23: 2236–2248.

Singh, M., Srivastava, A., Shinde, A.D., Acharya, R., Reddy, A.V.R. & Singh, I.B. 2012. Study of arsenic (As) mobilization in the Ganga Alluvial Plain using neutron activation analysis. *Journal of Radioanalytical and Nuclear Chemistry,* 294: 241–246.

One Century of the Discovery of Arsenicosis in Latin America (1914–2014) – Litter, Nicolli, Meichtry, Quici, Bundschuh, Bhattacharya & Naidu (Eds)

Arsenic mobilization in alluvial aquifers of the Brahmaputra floodplains: Hydrochemical and mineralogical evidences

R. Choudhury & C. Mahanta
Department of Civil Engineering, Indian Institute of Technology Guwahati, Assam, India

ABSTRACT: To better understand the sources and mobilization processes responsible for arsenic enrichment in groundwater in the Brahmaputra floodplains in Assam, India, where recent cases of arsenic contamination have been reported, hydrochemical characteristics of the groundwater and the mineralogical features of the aquifer sediments were studied. The aqueous arsenic levels are strongly depth-dependent in the study area and the high arsenic concentrations are found at depths between 15 ft. and 70 ft., with a maximum up to 60.68 µg/L. Geochemical studies using X-ray diffraction and scanning electron microscopy demonstrate that iron oxides/oxyhydroxides with chlorite, illite, quartz and feldspar are the dominant arsenic bearing phases in the sediments. *R*esults of hydrogeochemical analysis coupled with the sediment mineral characterization indicates that the reductive dissolution of iron oxyhydroxides is the possible major process of arsenic mobilization in the study area.

1 INTRODUCTION

Elevated As concentrations in water, soil and sediments is a major public health concern in many parts of the world, due to both natural and anthropogenic sources (Cullen & Reimer, 1988; Smedley & Kinniburgh, 2002; Smith *et al.*, 2000). Yet, the factors that affect As mobilization from solid phase into the aqueous phase remain only partially understood (Blute *et al.*, 2009).

The objective of this study was to understand the geochemical processes responsible for arsenic mobilization based on hydrochemical and mineralogical evidences.

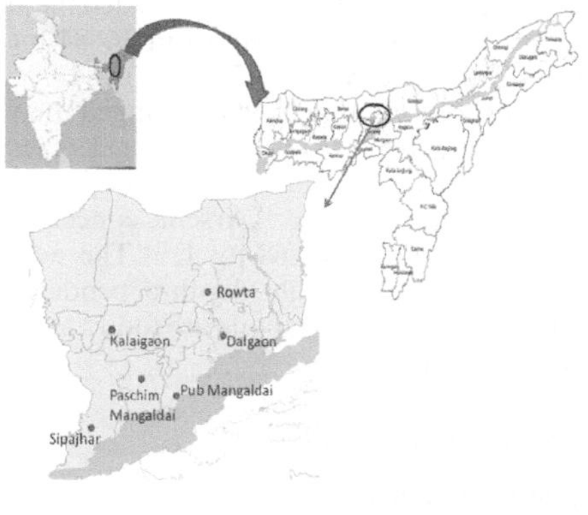

Figure 1. Map showing the study area.

2 METHODOLOGY

2.1 *Sample collection*

Evaluating the results of the arsenic screening and surveillance program (2006–2011), the high arsenic contaminated areas were deciphered, and drilling around a few of these highly arsenic contaminated wells was proposed. For sediment sample collection, a borehole was drilled around a few highly arsenic contaminated wells. Drilling was done by hand flapper technique as described by Horneman et al., (2004). The subsurface sediments samples were collected from every 10 ft. up to a depth of 80 ft.

Apart from sediment samples collected from the drilled wells, groundwater samples from 18 tube well sources located in the vicinity of the drilled well sites were also collected. Standard procedures of groundwater sample collection and preservation was followed. Prior to sampling, the tube well was pumped for 10 minutes to remove groundwater stored in the well. In-situ measurements for physical parameters viz. pH and electrical conductivity of the samples were done using portable pHmeter (pH Testr 10-Eutech) and a conductivity meter (EC Testr 11-Eutech).

2.2 *Groundwater and sediment analysis*

A Varian 55 Atomic Absorption Spectrophotometer (AAS) was used for determining the major cations viz. calcium (Ca^{2+}), sodium (Na^{+}), magnesium (Mg^{2+}), potassium (K^{+}), manganese (Mn) and iron (Fe). While arsenic concentrations were measured using Hydride Generation Technique in AAS with Vapor Generation Assembly (VGA), the major anions like chloride (Cl^-), phosphate (PO_4^{3-}) and sulfate (SO_4^{2-}) were analyzed in unacidified water samples. All the samples were analyzed following APHA (1999). Piper trilinear diagram to evaluate the geochemistry of groundwater of the study area was plotted with the help of GWW software.

Mineralogical quantification of the sediment samples was studied combining X-ray diffraction (Model XRD 3003TT, Seifert, Make; Rich Seifert & Co., Ahrensburg, Germany), SEM coupled with an energy dispersive X-ray analysis.

3 RESULTS AND DISCUSSION

3.1 *Groundwater geochemistry*

The pH of groundwater was of acidic to nearly neutral nature, with values ranging between 6.6 and 7 (Table 1). Low Eh values in the study area as reported by Gustav & Daniel (2006) in the Darrang district reveal the reducing condition of the aquifers in the study area. The concentration of major cations in groundwater were Ca^{2+} (3.57–13.0 mg L^{-1}), Na^+ (7.2–27.4 mg L^{-1}), K^+ (1.02–7.91 mg L^{-1}) and Mg^{2+} (1.2–23.8 mg L^{-1}). HCO_3^- dominated the anion chemistry with values ranging from 36–84 mg L^{-1} while Cl^- varied between 3.95–15.20 mg L^{-1}. Concentrations of SO_4^{2-} and PO_4^{3-} ranged between 2.48–15.05 mg L^{-1} and 0.14–3.95 mg L^{-1} respectively, while NO_3^- concentrations showed a narrow range of 0.01 to 2.80 mg L^{-1}. Arsenic concentration ranged between 0.73–60.68 µg L^{-1}. The aqueous arsenic levels are strongly depth-dependent and the high arsenic concentrations are found at depths between 15 ft. and 70 ft.

3.2 *Sediment mineralogy*

Qualitative sediment mineralogical investigations using XRD revealed the bulk mineralogical compositions to be dominated by quartz, phyllosilicates, feldspars and clay minerals along with iron oxides/oxyhydroxide minerals viz. goethite, hematite and jacobsite. SEM/EDS analysis profiles of the aquifer sediments showed that the mineralogy was dominated by aluminosilicates of Na, K, Mg as well as that of Fe and Mn. SEM/EDS analysis shows the presence of an arsenic peak even at 10 ft. depth (Figure 2).

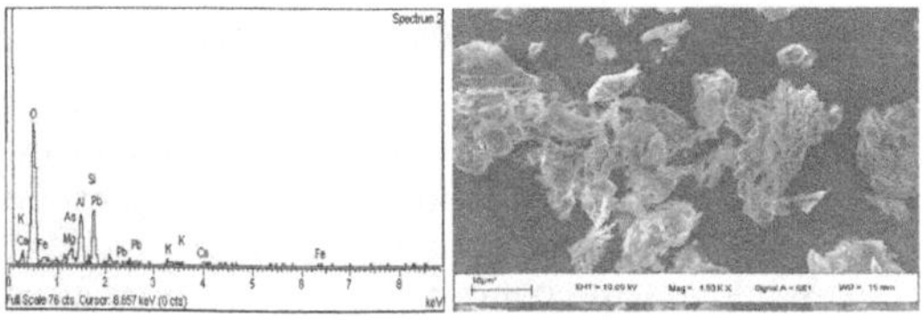

Figure 2. SEM/EDS image of sediment sample at 10 ft. depth.

4 CONCLUSIONS

The chemical characteristics and nature of As release into the groundwater from aquifer sediments were investigated to improve our understanding of the solid and dissolved arsenic fate in the aquifer under study. Arsenic concentration in the groundwater samples were above the WHO guidelines of 0.01 mg L^{-1} and ranged between 0.73 – 60.68 mg L^{-1}. Low concentrations of NO_3^- and SO_4^{2-} coupled with high concentrations of HCO_3^- also supported strong reducing conditions prevailing in the target aquifers. The reducing conditions and a general relationship between dissolved As, NO_3^-, SO_4^{2-} and HCO_3^- are consistent with the mechanism of reductive dissolution of Fe(III) oxides/oxyhydroxides via respiration of organic matter, with concomitant release of arsenic, as has been indicated elsewhere in other case studies (Bhattacharya *et al.*, 1997; Nickson *et al.*, 2000; Welch *et al.*, 2000).

With no major industrial activities in the study area, arsenic contamination in the region is indicated to be of geogenic origin, and the arsenic-rich phases in the aquifer sediments, therefore, remain the only likely source of arsenic in groundwater.

REFERENCES

APHA. Standard Methods for the Examination of Water and Wastewater, 1999, 20th edition, American Public Health Association, American Water Works Association, Water Environment Federation.

Bhattacharya, P., Chatterjee, D. & Jacks, G. 1997. Occurrence of arsenic contaminated groundwater in alluvial aquifers from delta plains, Eastern India: options for safe drinking water supply. *Water Resource Development* 13: 79–92.

Blute, N.K., Jay, J.A., Swartz, C.H., Brabander, D.J. & Hemond, H.F. 2009. Aqueous and solid phase arsenic speciation in the sediments of a contaminated wetland and riverbed. *Applied Geochemistry* 24: 346–358.

Cullen, W.R. & Reimer, K.J. 1989. Arsenic speciation in environment. *Chemical Review* 89: 713–764.

Enmark, G. & Nordborg, D. 2007. *Arsenic in the groundwater of the Brahmaputra floodplains, Assam, India-Source, distribution and release, mechanisms*. Retrieved from the url: www2.lwr.kth.se/Publikationer/

Horneman, A., van Green, A., Kent, D.V., Mathe, P.E., Zheng, Y., Dhar, R.K., O'Connel, S., Hoque, M.A., Aziz, Z., Shamsudduha, M., Seddique, A.A. & Ahmed, K.M. 2004. Decoupling of As and Fe release to Bangladesh groundwater under reducing conditions. Part I: Evidence from sediment profile. *Geochim. Cosmochim. Acta* 68: 3459–3475.

Nickson, R.T., McArthur, J.M., Ravenscroft, P., Burgess, W.S. & Ahmed, K.M. 2000. Mechanism of arsenic poisoning of groundwater in Bangladesh and West Bengal. *Applied Geochemistry* 15: 403–413.

Welch, A.H., Westjohn, D.B., Helsel, D.R. & Wanty, R.B. 2000. Arsenic in ground water of the United States: occurrence and geochemistry. *Ground Water* 38: 589–604.

One Century of the Discovery of Arsenicosis in Latin America (1914–2014) – Litter, Nicolli, Meichtry, Quici, Bundschuh, Bhattacharya & Naidu (Eds)
© 2014 Taylor & Francis Group, London, ISBN 978-1-138-00141-1

Occurrence and lithostratigraphic control of arsenic concentrations of the groundwater of Barak valley (Assam), North Eastern India

P. Thambidurai, D. Chandrasekharam, A.K. Chandrashekhar & G. Trupti
Department of Earth Sciences, Indian Institute of Technology Bombay, Mumbai, India

ABSTRACT: Arsenic (As) contamination in groundwater is emerging in Barak valley, Assam, North Eastern India. The pyritiferous silt, coal, and siderite with high concentration of As (2.5 to 810 mg/kg), derived from Tipam formation of Mio-Pliocene age, are considered the source for As in groundwater in the region. Although the high spatial variability of As concentration in groundwater ranges from 12 to 310 μg/L within the valley, the stratigraphic formation and groundwater residence time are the main likely determinants of the aquifer arsenic concentrations.

1 INTRODUCTION

High arsenic (As) concentration of groundwater in sedimentary basin is a globally major problem of health concern. The presence of As in groundwater has been reported extensively in recent years from different parts of the world including a few from Asia (Smedley and Kinniburgh, 2002). In India, As is reported from several states including West Bengal, Bihar, Uttar Pradesh, Jharkhand, etc. Its abundance has been primarily reported from the Bengal Delta of West Bengal (Chandrasekharam *et al.*, 2001; Farooq *et al.*, 2011). More recently, studies show that the problem of As contamination is emerging in many North Eastern (NE) states of India including Assam, Manipur, Mizoram, etc. (Singh, 2004; Thambidurai *et al.*, 2012). Arsenic in groundwater is often associated with geologic sources; however, a specific source of As has not yet been clearly identified. The main objective of the present study is the elucidation of the distribution of arsenic in stratigraphic sequence sediments and groundwater zone from Barak valley, Assam.

2 MATERIAL AND METHODS

The study area is situated in the southern region of Assam, parts of northern Mizoram and eastern Manipur, and it lies between Eastern Bangladesh and Western Manipur, NE India. Representative rock samples were collected from different formations of the Barak valley and northern part of Mizoram as well as western border of Manipur. Groundwater samples were collected from the study area for major and trace element analysis. The bulk mineral composition of the sediment samples were analyzed by X-ray diffraction analysis (XRD—Rigaku, Japan). Total concentration of arsenic in sediment samples were measured by the powder press method using wavelength dispersive X-ray fluorescence spectrometer (PW-XRF—Phillips), with standard curves based on Canadian soil standard (CCRMP). The contents of major elements and trace elements in sediment samples were determined by XRF.

3 RESULTS AND DISCUSSION

3.1 *Sediment and water chemistry*

The sediments in this area reveal that the stratigraphic sequences consist of Bhuban formation overlain by Bokabil subgroup, and Tipam groups of Mio-Pliocene age, which contain shale, ferruginous sandstone, clay and sand, silt and clay with gravel and occasional intercalation of coal beds, and also Holocene of Dihing and alluvial formation. XRD studies on silt, coal and iron-coated blackened mineral in clay samples reveal the presence of pyrite and siderite, respectively (Figure 1a & b). XRF results indicate that As ranges are 2.5 to 810 mg/kg in the sediments. The pyritiferous silt has 810 mg/kg and coal contains 418 mg/kg, while clay contains 50 mg/kg. The SEM images show (Figure 2) well developed euhedral pyrite crystals is about 25 to 40 μm in diameter and rhomboid forms were also observed. These are distributed throughout the entire thickness of the silt and coal patches of Tiapm. Siderite occurs as spheroids, and the presence of both suggests diagenetic origin from Fe, S and As-rich groundwater. However, it is felt that the water table fluctuation

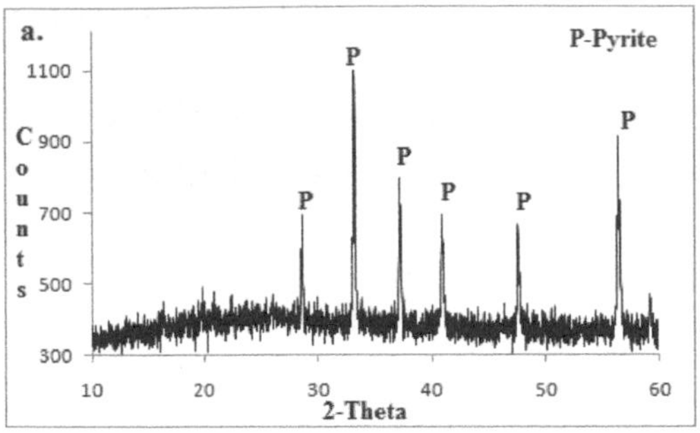

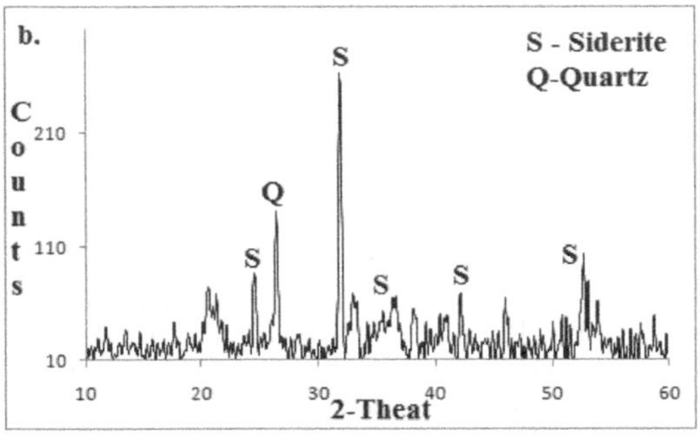

Figure 1. a) XRD pattern of coal concretions b) XRD pattern of siderite.

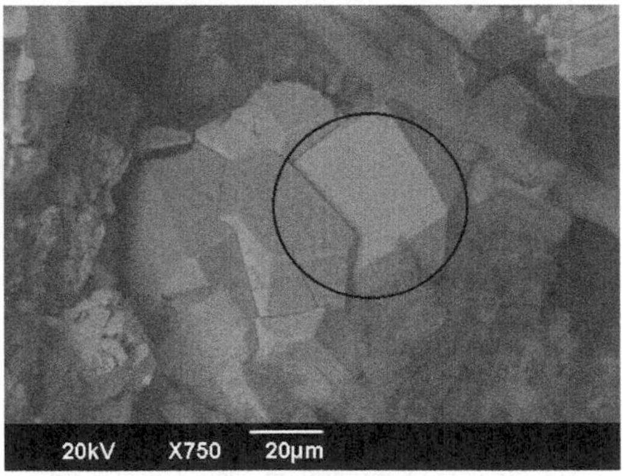

Figure 2. SEM image of pyrite.

may also cause repeated cycles of oxidation and reduction in the aquifer sediments.

The water chemistry results show that the groundwater arsenic content seems to correlate well with paleo-channel and/or floodplain sediments. Groundwater from the youngest sediment in the valley contains high As up to 310 μg/L at Banskandi, about 20 km eastern side of Silchar, whereas the groundwater directly trapped in Tipam formation at Bairabi, and Jiribam gives concentration of As is 12 μg/L. Initially, we expected more As in groundwater in hilly area according to the bulk As concentration present in the sediments, but overall results show an increase in arsenic in the lower part of valley than in the hill ranges, reflecting that arsenic release is due to the increase of the groundwater residence time. This adds further confidence to the relationship between sediment age and the groundwater arsenic concentration.

The As concentration in the alluvial areas perhaps owes its source to the Tipam rock and later enrichment in groundwater (Thambidurai *et al.*, 2013). Moreover, As contamination occurs on either side of Barak and Dhaleswari rivers, which originate from western Manipur and Mizoram, respectively. The water flow system of the valley area on both directions seems to consist of a series of local flow systems due to the N-S direction anticline inter-fingering of older rock formation with alluvium. Therefore, it is not possible for high As water to flow from one location to another because of the wedge shape aquifer increasing rapidly in thickness from hills to valley side. In the flood plains of major rivers of this area, migration of meanders gives a complex distribution of older and younger sediments. Furthermore, a high variability of spatial distribution of arsenic in the groundwater is observed in the floodplain areas. The poor correlation between dissolved iron and arsenic may be due to the loss of Fe though precipitation of siderite (Bhattacharya *et al.*, 2009), and/or involvement of other geochemical processes.

4 CONCLUSIONS

Arsenic concentration in groundwater of the Barak valley ranges from 12 to 310 μg/L, and sediments contain from 2.5 to 810 mg/kg. The poor correlation between As and Fe in groundwater is possibly due to the precipitation of dissolved Fe as siderite solids ($FeCO_3$). The pyritiferous coal, silt and siderite concretions from Tipam formation of Mio-Pliocene age rocks show anomalous contents of As; this is considered the most important contamination source in the groundwater. The water table fluctuations result in repeated cycles of oxidation and reduction and along with residence time of groundwater in the aquifer contribute to As mobilization in groundwater. The source of the As-bearing sediments of the Barak valley (Assam) is thus presently considered to be from Tipam rocks of the highly elevated Mizoram and Manipur hills ranges.

REFERENCES

Bhattacharya, P., Aziz Hasan, M., Sracek, O., Smith, E., Matin Ahmed, K., von Brömssen, M., Huq, I. &

Naidu, R. 2009. Groundwater chemistry and arsenic mobilization in the Holocene flood plains in south-central Bangladesh. *Environ. Geochem. Health*, 31:23–43.
Chandrasekharam, D., Karmakar., J., Berner, Z. & Stueben, D, Arsenic contamination in groundwater, Murshidabad district, West Bengal. WRI-10th Procedia, A.A. Balkema Pub. Com. 2001: 1051–1054.
Singh, A.K., 2004. Arsenic contamination in the groundwater of North Eastern India: Proceedings on national seminar on hydrology, Roorkee. 22–24.
Smedley, P.L. & Kinniburgh, D.G. 2002. A review of the source, behavior and distribution of arsenic in natural waters. *Appl. Geochem.*, 17(5): 517–568.
Thambidurai, P., Chandrashekhar, A.K. & Chandrasekharam, D., 2013. Geochemical signature of arsenic contaminated groundwater in Barak valley—Assam, and neighborhood, northeastern India: WRI-14th Procedia Earth and Planetary Sci., 7: 834–837
Thambidurai, P., Chandrasekharam, D., Chandrashekhar, A.K., & Farooq, S.H. 2012. Arsenic contamination in groundwater of Surma basin of Assam and Mizoram, North Eastern India: Understanding the Geological and Medical Interface of Arsenic—As 2012. Jul 2012: 47–49: Proceeding 4th International Congress on Arsenic in the Environment, J.C. Ng., B.N. Noller, R. Naidu, J. Bundschuh & P. Bhattacharya (eds), Taylor & Francis Gr., London.

One Century of the Discovery of Arsenicosis in Latin America (1914–2014) –
Litter, Nicolli, Meichtry, Quici, Bundschuh, Bhattacharya & Naidu (Eds)
 ISBN 978-1-138-00141-1

New arsenic province and major source in Mongolia

B. Tumenbayar & S. Murao
Sans Frontiere Progres NGO, Mongolia
Institute for Geo-Resources and Environment, AIST, Japan

T. Maidar & J. Uramgaa
Central Hospital of Mongolia, Mongolia

ABSTRACT: New discovered arsenic polluted areas are located in the southeast part of Mongolia, and have good territorial correlation with China. The major primary source of arsenic are: (1) Gold and polymetalic (Zn, Pb) mineralization associated with arsenic anomalies. Most Au and Zn, Pb deposits have high content of arsenopyrite, which is easily soluble under certain conditions. And well known source of arsenic in water; (2) Cretaceous sediments. High content of arsenic were recorded in the Cretaceous brown coal. Lacustrine sediments accumulated in the continental Cretaceous depression for 100 million years and periodical arsenic-rich volcanic activity are good sources of arsenic.

1 INTRODUCTION

In Mongolia, the arsenic concentration in wells ranges up to 0.015 mg L^{-1} and, in river water, up to 0.160 mg L^{-1}. Almost 100 out of 911 samples showed higher values than the WHO Standard (0.01 mg L^{-1}) (Narantuya *et al.*, 2004). High levels of arsenic (3.2 mg L^{-1}) in Mongolian villagers' hair were found by our survey (Murao *et al.*, 2004; Murao *et al.*, 2011) and there is a pressing need to capture a holistic picture about how widely arsenic contamination is distributed in the nation.

2 SURVEY

According to our field investigation on well waters and rivers near the Gold provinces and the comparison of geochemical anomalies all around Mongolia, main arsenic concentrated areas are located in the southeast part of Mongolia (Figure 1).

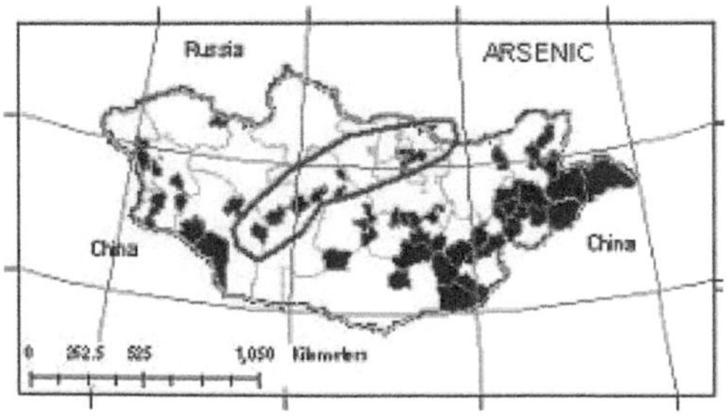

Figure 1. Arsenic contamination areas of well waters in Mongolia.

Arsenic anomalies have been also found in the river water close to the famous gold deposits Boroo (14.0 µg L^{-1}) and Gatsuurt (31.6 µg L^{-1}) in North Mongolia. The high concentration is indicated within the Bayankhongor gold belt in the central and southwest part of Mongolia (within circle in Figure 1).

Measurement of arsenic in brown coal from 14 deposits of eastern Mongolia shows that 10 deposits have high As contents (50 to 216 mg L^{-1}).

3 DISCUSSION

Distribution of arsenic contaminated area and brown coal deposits were similar (Figure 2).

According to our research, the main arsenic polluted areas are located in the southeast part of Mongolia, having good territorial correlation with China (Figure 3).

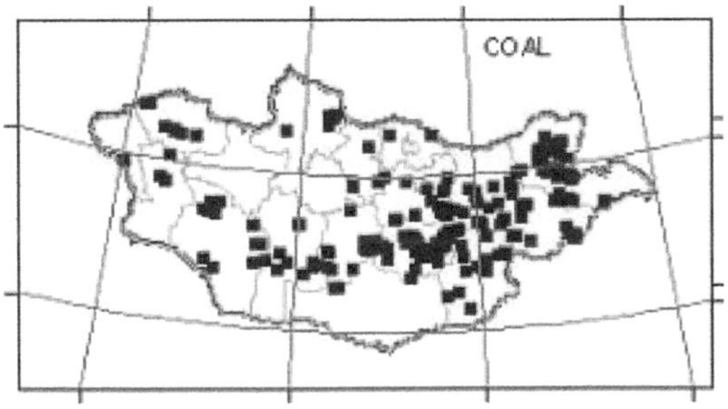

Figure 2. Distribution of brown coal deposits in Mongolia.

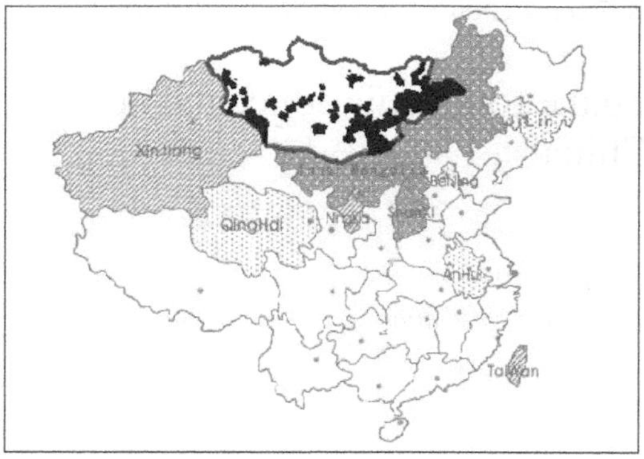

Figure 3. Territorial correlation of Arsenic anomalies in Mongolia and China. (Data of China by Sun (2004)).

4 CONCLUSIONS

Two kinds of rock type seems to be the major sources of arsenic:

1. Gold and polymetalic (Zn, Pb) mineralization associated with arsenic anomalies. Most Au, Zn and Pb deposits have high content of arsenopyrite, which is easily soluble under certain conditions and well known sources of arsenic concentration in water.
2. Cretaceous sediments. Lacustrine sediments accumulated in the continental Cretaceous depression during 100 million years and periodical arsenic rich volcanic activity were good sources. They gave geochemical conditions for arsenic solubility in groundwater.

ACKNOWLEDGEMENTS

Authors express their thanks to Mr. Batjargal (Central Geological Laboratory (Mongolia) for analyses and Dr. K. Tsukada and Dr. K. Yamamoto from Nagoya University (Japan) for cooperation.

REFERENCES

Murao, S., Tumenbayar, B., Sera, K., Futatsugawa, S. & Waza, T. 2004. Finding of high level arsenic for Mongolian villagers' hair. *Int. J. PIXE* 14(3–4): 125–131.

Murao, S., Sera, K., Tumenbayar, B., Saijaa, N. & Uramgaa, J. 2011, High level of Arsenic reaffirmed for human hairs in Mongolia. *Int. J. PIXE* 21(3–4): 119–124.

Narantuya, L., Bolormaa, I., Burmaa, B. et.al. Survey report on Arsenic determination in Mongolia. 2004 *UNICEF*: 74 pp.

Sun, G. 2004. Arsenic contamination and arsenicosis in China. *Toxicology and Applied Pharmacology* 198: 268–271.

One Century of the Discovery of Arsenicosis in Latin America (1914–2014) – Litter, Nicolli, Meichtry, Quici, Bundschuh, Bhattacharya & Naidu (Eds)
© 2014 Taylor & Francis Group, London, ISBN 978-1-138-00141-1

The influence of the Wenchuan earthquake to arsenic concentration in groundwater of Sichuan province, China

J. Yang, J.S. Wang, W.F. Yue, J. Li & R. Zuo
College of Water Sciences, Beijing Normal University, Beijing, China
Engineering Research Center of Groundwater Pollution Control and Remediation, Ministry of Education, Beijing, China

ABSTRACT: In order to develop the impact of earthquakes to As content in groundwater, a comprehensive evaluation was conducted after the Wenchuan earthquake in the Sichuan Province, China. This evaluation focused on the temporal and spatial variation of As in groundwater of Deyang, An town, and Guangyang city. The results showed that As concentration sharply increased hundreds of times just after the earthquake and decreased to usual levels as the time went by. The temporal As pollution showed that more attention should be paid to the safety of water supply after earthquakes. The samples in Guangyuan city, especially in Beicheng area, had the highest As content (0.460 mg/L, 24th May, 2008), though it was the farthest point from the epicenter. The impact of earthquakes to groundwater quality was complex, and there might be several influencing factors except distance, such as hydrogeological conditions and background values of metal in soils and rocks.

1 INTRODUCTION

The Wenchuan earthquake of May 12, 2008, with a Richter magnitude scale of 8.0, caused devastating geotechnical and geoenvironmental issues in the central region of the Sichuan province, China. After an earthquake, the quality estimation of groundwater source becomes very urgent and important. Arsenic (As) is classified as one of the most toxic and carcinogenic chemical elements. As a normal indicator of groundwater quality, high concentrations of As in groundwater source may threaten residents' health and constitute a high-priority public health problem. In addition, As is a naturally occurring contaminant in groundwater, found largely as the result of minerals associated with previous volcanic activity dissolving from weathered rocks, ash and soils (Christodoulidou *et al.*, 2012). However, the presence of arsenic can also be attributed to anthropogenic activities. Releasing from iron oxides appears to be the most common cause of elevated As concentration in groundwater (Welch *et al.*, 2000). The main objective of this study was to investigate the temporal variation of As concentration in groundwater after an earthquake in short term.

2 METHODS

2.1 *Study area*

The Wenchuan earthquake occurred along the Longmen-Shan thrust fault belt, at the eastern margin of the Tibetan Plateau, and impacted directly the range of 10 km. 27% groundwater sources of Sichuan province were distributed in this earthquake region. In this study, five groundwater sources were chosen to collect groundwater samples for analysis: Luojiang water plant in Deyang city (G1), old town water plant in An town (G2), and Dongba (G3), Nanhe (G4), Beicheng (G5) water plants in Guangyuan city. The daily water supply capacity of these water plants was 1×10^4–4×10^4 m^3.

2.2 *Sample collection and analysis*

Groundwater samples corresponded to loose rock pore water; the lithology of aquifer was Q3, Q4 cobblestone. These samples were collected from monitoring wells of water plants. At each sampling site, three water samples were collected. The samples were collected in 1000 mL capacity polystyrene bottles and brought to the laboratory where they were filtered. After that, samples were kept refrigerated for further analysis. Arsenic concentrations in groundwater samples were determined by Inductively Coupled Plasma Atomic Emission Spectrometry (ICP-AES, Ultima, France) with a detection limit of 0.005 mg/L. All chemicals used in the experiment were of analytical grade, and deionized water was used for all procedures.

3 RESULTS AND DISCUSSION

3.1 *Temporal variation of As in groundwater*

As shown in Table 1, the As concentrations of groundwater before earthquake, twelve days after earthquake, twenty days after earthquake, and five months after earthquake were measured. The concentration fluctuations before and after Wenchuan earthquake were obvious. Before the earthquake, the As content was very low (0.0002–0.0003 mg/L) and all below the best level of groundwater environmental quality standards in China (< 0.005 mg/L). Immediately after the earthquake (24th May, 2008), the concentration of As increased significantly and was 510–2300 times higher than the 2007 values. Then, after another seven days (1st June, 2008), the As content decreased, but all of the values still belonged to the worst level of groundwater environmental quality standards in China (> 0.05 mg/L). After five months of the earthquake, the contents almost fell to the usual 2007 level.

The dates certified the sudden increase of As concentration in groundwater after the Wenchuan earthquake. In Anton's research (2013), shaking of earthquake caused structural damage or destruction of building and bridges, damage to infrastructure affecting water supply, wastewater disposal, power and communications. The As^{3+} and As^{5+} species may be mainly released from soil and rock, especially from soil and groundwater contamination by the leakage of toxic chemicals by the sudden disturbance of flow distribution (Wang and Mulligan, 2006). Therefore, As concentration in groundwater varied with the influence of earthquake on the time scale. After the earthquake, the flow became stable and the As contents went down to a normal and safety level.

Table 1. Variation of As concentration in groundwater (mg/L).

	Date			
Sample	2007	24th May 2008	1st June 2008	Oct. 2008
G1	0.0003	0.182	0.161	0.0003
G2	0.0003	0.153	0.152	0.0004
G3	0.0002	0.370	0.220	0.0002
G4	0.0002	0.200	0.200	0.0002
G5	0.0002	0.460	0.340	0.0002

3.2 *Spatial variation of As in groundwater*

The variation of As concentration in different sampling points was discrepant (Table 1). The increasing As of Guangyang city was the highest, especially in Beicheng water plant (0.458 mg/kg), and it was relatively lower in An town and Deyang city.

There may be three reasons for this spatial difference. First, the distance of sampling points to the epicenter of Wenchuan county and the precise earthquake force to groundwater flow field were different. Secondly, the hydrogeological conditions were very complex and differed from area to area. Thirdly, the background values of As in soils of Deyang, An town, and Guangyang city were also different, being 7.30, 7.50, and 7.12 mg/kg, respectively. The epicentral distance of Deyang, An town, and Guangyang city were 90, 98, and 245 km, respectively. There was no direct correlation between the As content and the epicentral distance. The hydrologic responses to earthquakes were many-sided. The confining layers and movement of groundwater create significant artesian pressure in the deeper aquifers (Cubrinovski *et al.*, 2010). Wang and Mulligan showed similar results, attributed to the fact that As had different valence states and speciation in different kinds of soil and rock (2006). As a result, the mechanisms that led to liquefaction and lateral spreading of As may be complex.

4 CONCLUSIONS

A systematic comparison about As contents in groundwater was conducted between different periods after the Wenchuan earthquake. The dates showed that the impact of the earthquake to the As content was obvious in short-term after the earthquake, and it decreased as the time went by. According to the spatial variation of As, the impact of earthquake cannot be estimated only by the epicentral distance. It may be influenced by many reasons, such as the hydrogeological conditions and background values in soil or rock. In conclusion, more attention and management should be paid to the temporally As pollution after an earthquake, especially in drinking groundwater sources.

ACKNOWLEDGEMENTS

This study was supported by the study project of Environmental Safety Assessment and Response

Measures of Wenchuan Earthquake and the Fundamental Research Funds for the Central Universities. The authors would like to thank colleagues at the Beijing Normal University who provided extensive help and support for this study.

REFERENCES

Anton, K.G., Nicholas, F.D.W., Simon, C.C. & Jari, P. 2013. Kaipio. Groundwater responses to the recent Canterbury earthquakes: a comparison. *Journal of Hydrology* 504: 171–181.

Christodoulidou, M., Charalambous, C., Aletrari, M., Nicolaidou, K.P., Petronda A. & Ward N.I. 2012. Arsenic concentrations in groundwaters of Cyprus. *Journal of Hydrology* 468–469: 94–100.

Cubrinovski, M., Bray, J.D., Taylor, M., Giorgini, S., Bradley, B.A., Wotherspoon L. & Zupan J. 2011. Soil liquefaction effects in the central business district during the February Christchurch Earthquake. *Seismological Research Letters*, 82 (6): 893–904.

Wang, S. & Mulligan, C.N. 2006. Occurrence of arsenic contamination in Canada: Sources, behaviour and distribution. *Science of the Total Environment.* 366: 701–721.

Welch, A.H., Westjohn, D.B., Helsel, D.R. & Wanty, R.B. 2000. Arsenic in ground water of the United States: occurrence and geochemistry. *Ground Water* 38, 589–604.

One Century of the Discovery of Arsenicosis in Latin America (1914–2014) – Litter, Nicolli, Meichtry, Quici, Bundschuh, Bhattacharya & Naidu (Eds)
 ISBN 978-1-138-00141-1

Chemical and isotopic characteristics of high arsenic groundwater in the Songnen basin, China

H.M. Guo
State Key Laboratory of Biogeology and Environmental Geology, China University of Geosciences, Beijing, P.R. China
School of Water Resources and Environment, China University of Geosciences, Beijing, P.R. China

Y. Wu
State Key Laboratory of Biogeology and Environmental Geology, China University of Geosciences, Beijing, P.R. China

D. Zhang
State Key Laboratory of Biogeology and Environmental Geology, China University of Geosciences, Beijing, P.R. China
School of Water Resources and Environment, China University of Geosciences, Beijing, P.R. China

P. Ni
State Key Laboratory of Biogeology and Environmental Geology, China University of Geosciences, Beijing, P.R. China

Q. Guo
School of Water Resources and Environment, China University of Geosciences, Beijing, P.R. China

ABSTRACT: High As groundwater was generally of Na-HCO_3 type in the Songnen basin, which has relative low ORP and neutral-weak alkaline pH. Along the groundwater flow path, As concentration showed an increasing trend. High As groundwater was mainly distributed in the low-lying areas. Vertically, at depths between 20 and 30 m, 60 and 90 m, groundwater usually had high As concentration. Distribution of groundwater As was dependent of redox potential, hydrogeological settings, mineral precipitation, and mineralization of organic carbon. Isotopic results showed that dissimilatory Fe(III) reduction coupled with organic carbon degradation would be related to As mobilization in the aquifers. Although both Fe and As were released during these redox processes, sulfide mineral precipitation would be the sink of Fe and As in groundwater, which resulted in the fact that no good correlation exists between dissolved Fe and As.

1 INTRODUCTION

High As groundwater has been widely found in the inland basins of the northwest of China, including the Datong basin, the Hetao basin, the Taiyuan basin, the Yinchuan basin, the Songnen basin (Guo *et al.*, 2011). Although natural occurrence of high As groundwater was recognized in the late 19th century, little is known about geochemical and isotopic characteristics of those high As groundwater. Bian *et al.* (2012) observed that high As groundwater mainly occurred in the low-lying areas of the basin. They found that arsenic in groundwater is closely related to the geological background and tectonic movement. High As groundwater was mainly hosted in the aquifers at depths between 20 and 100 m (Tang *et al.*, 2010). In this paper, we investigated geochemical and isotopic characteristics of those high As groundwater and their implication for As mobilization in the aquifers.

2 METHODS AND MATERIALS

The Songnen basin is located in the west of Jilin province, which is a Meso-Cenozoic fault basin. Thick alluvial-lacustrine sediments occur in the basin, with the thickness up to 5000 m. Several aquifer systems have been recognized, including porous unconfined aquifer (Shallow aquifers), porous confined aquifers (Mid-deep aquifers), the pore-fracture aquifers of upper Neogene Da'an and Taikang group (Deep aquifer I) and the Cretaceous fracture-pore aquifers (Deep aquifer II). This study mainly focused on the shallow aquifer and mid-deep aquifers.

Groundwater samples were taken in five field campaigns in July 2013. Parameters, including water temperature, pH, and E_h, were measured at the time of groundwater sampling. Alkalinity was measured by the Gran titration methods at the time of sampling day. Concentrations of major cations and trace elements were determined by ICP-AES and ICP-MS, respectively. Anions were analyzed by IC. Arsenic species were determined by HPLC-AFS. D and ^{18}O were analyzed by i2010 (Picarro), and ^{13}C of dissolved organic carbon by iTOC (Picarro).

3 RESULTS AND DISCUSSION

Groundwater was neutral to weak alkaline with pH between 7.2 and 9.0. Bicarbonate was the major anion, with concentrations between 263 and 1290 mg/L and average of 628 mg/L. Sodium was the major cation, with concentrations between 25 and 509 mg/L. Groundwater was generally of Na-Mg/Ca-HCO_3. Groundwater had F^- concentration between 0.35 and 2.77 mg/L, with average of 1.27 mg/L. Of all samples, 72.4% contained F^- concentration greater than 1.0 mg/L, which is the drinking water standard for F^-.

Groundwater ORP ranged between −125 and 67.6 mV, showing weak-moderate reducing conditions. Dissolved organic carbon has concentration range between 4.1 and 32.8 mg/L. In these environments, concentrations of NO_3^- and SO_4^{2-} were relatively low, with averages of 4.35 and 69.7 mg/L. Relatively high NH_4-N and S^{2-} concentrations were observed in the groundwater. Arsenic concentrations ranged between <0.1 and 338 μg/L. Of all samples, 85.1% had As concentration greater than 85.1 μg/L. Arsenic(III) was the major As species, with concentration between 2.0 and 265 μg/L.

Groundwater had $\delta^{18}O$ values between −11.3‰ and −2.73‰, and δD values between −90‰ and −43.3‰. The ^{18}O and D lay around the local meteoric water line, showing their precipitation origins. Shallow groundwater generally had higher values of $\delta^{18}O$ and δD than deep groundwater. It indicated that shallow groundwater experienced more intensive evaporation than deep one. Values of $\delta^{13}C$ in dissolved organic carbon ranged between −26‰ and 1.4‰. Shallow groundwater had $\delta^{13}C$ between −24.2‰ and −11.5‰, deep groundwater between −26‰ and 1.4‰. The greater range of $\delta^{13}C$ values indicated that more intensive microbial activities occurred in deep groundwater.

High As groundwater mainly occurred in the southeast of the study area, which is the local discharge area of groundwater flow system. From the recharge area to the discharge area, As concentration showed an increasing trend. In the areas with lower hydraulic gradients, groundwater As was generally higher. Vertically, high As groundwater was mainly found at depths between 20 and 30 m, and 60 and 90 m, in shallow aquifer and deep aquifer, respectively.

Distribution of groundwater As was dependent of redox potential, hydrogeological settings, mineral precipitation, and mineralization of organic carbon. High As concentration was mainly observed in groundwater with ORP <0 mV. It indicated that As was mobilized in reducing conditions, where Fe/Mn oxides were reduced. It supported the fact that high As groundwater generally had high Fe and Mn concentrations. It was suggested that reductive dissolution of Fe oxides should be the major mechanisms for As mobilization in the aquifers, which is the most accepted mechanism for genesis of high As groundwater in Southeast Asia (Nickson *et al.*, 1998; Islam *et al.*, 2004; Guo *et al.*, 2013). Approximately along the groundwater flow path, groundwater As concentration showed an increasing trend. It may be related to the higher flow rate and lower residence time in the recharge area than the discharge area. The higher hydraulic gradient in the recharge area (3.8×10^{-4}) than in the discharge area (2.1×10^{-4}) provided evidence for this implication. Besides, lithologic characteristics (e.g., high organic matter content) may play a role in developing anaerobic conditions (McArthur *et al.*, 2004).

Not all groundwaters with low ORP values had high As concentration. Groundwaters with low ORP and high S^{2-} concentration usually had low As concentration, indicating that sulfide mineral (pyrite) precipitation would be the sink of groundwater As (Guo *et al.*, 2013a; *Guo et al.*, 2013b). Spatial distributions of groundwater and $\delta^{13}C$ showed that high $\delta^{13}C$ values corresponded to high As concentration. Since high $\delta^{13}C$ values of DOC indicated that microbial degradation of DOC occurred in high-As groundwater aquifers, microbe-mediated redox processes would be related to As mobilization in the aquifers, especially dissimilatory Fe(III) reduction (Islam *et al.*, 2004).

ACKNOWLEDGEMENTS

The study has been financially supported by National Natural Science Foundation of China (Nos. 41222020 and 41172224), the program of China Geology Survey (12120113103700), the Program for New Century Excellent Talents in

University (No. NCET-07–0770), and the Chinese Universities Scientific Fund (No. 2652013028).

REFERENCES

Bian, J.M., Tang, J., Zhang, L.S., Ma, H.Y. & Zhao, J. 2012. Arsenic distribution and geological factors in the western Jilin province, China. *J. Geochem. Explor*. 112: 347–356.

Guo, H.M., Zhang, B., Li, Y., Berner, Z., Tang, X.H. & Norra, S. 2011. Hydrogeological and biogeochemical constrains of As mobilization in shallow aquifers from the Hetao basin, Inner Mongolia. *Environ. Pollut*. 159: 876–883.

Guo, H.M., Liu, C., Lu, H., Wanty, R., Wang, J. & Zhou, Y.Z., 2013a. Pathways of coupled arsenic and iron cycling in high arsenic groundwater of the Hetao basin, Inner Mongolia, China: An iron isotope approach. *Geochim. Cosmochim. Acta* 112: 130–145.

Guo, H.M., Zhang, Y., Jia, Y.F., Zhao, K. & Li, Y. 2013b. Spatial and temporal evolutions of groundwater arsenic approximately along the flow path in the Hetao basin, Inner Mongolia. *Chinese Science Bulletin* 58(25): 3070–3079.

Islam, F.S., Gault, A.G., Boothman, C., Polya, D.A., Charnock, J.M., Chatterjee, D. & Lloyd, J.R. 2004. Role of metal-reducing bacteria in arsenic release from Bengal delta sediments. *Nature* 430: 68–71.

McArthur, J.M., Banerjee, D.M., Hudson-Edwards, K.A., Mishra, R., Purohit, R., Ravenscroft, P., Cronin, A., Howarth, R.J., Chatterjee, A., Talukder, T., Lowry, D., Houghton, S., Chadha, D.K. 2004. Natural organic matter in sedimentary basins and its relation to arsenic in anoxic ground water: The example of West Bengal and its worldwide implications. *Appl. Geochem.* 19: 1255–1293.

Nickson, R., McArthur, J., Burgess, W., Ahmed, M., Ravenscroft, P. & Rahman, M. 1998. Arsenic poisoning of Bangladesh groundwater. *Nature*, 395: 338.

Tang, J., Bian, J.M., Li, Z.Y. & Wang, C.Y. 2010. Inverse geochemical modeling of high arsenic groundwater: a case study of the arsenic endemic area in western Jilin Province. *Geol. China* 37(3): 754–759. (In Chinese with English Abstract).

One Century of the Discovery of Arsenicosis in Latin America (1914–2014) –
Litter, Nicolli, Meichtry, Quici, Bundschuh, Bhattacharya & Naidu (Eds)
© 2014 Taylor & Francis Group, London, ISBN 978-1-138-00141-1

Arsenic and trace elements in natural fountains of towns in the Iron Quadrangle, Brazil

C.A. Ferreira
Pós Graduação em Ciência e Tecnologia das Radiações Minerais e Materiais/Centro de Desenvolvimento da Tecnologia Nuclear (CDTN/CNEN), Belo Horizonte, MG, Brazil

H.E.L. Palmieri & M.A.B.C. Menezes
Centro de Desenvolvimento da Tecnologia Nuclear (CDTN/CNEN), Belo Horizonte, MG, Brazil

ABSTRACT: The concentrations of the trace elements Li, Al, V, Cr, Mn, Co, Ni, Cu, Zn, As, Rb, Sr, Cd, Ba, Pb, Sb and the physiochemical parameters—EC, TDS and turbidity—were measured in 37 fountains of 18 towns in the Iron Quadrangle, Brazil. The minimum and maximum pH values found are not within the guidelines recommended for drinking water by the WHO and the Brazilian legislation. However, in only one fountain, the level of arsenic was above the limit established by the WHO. In this region, occurrence of As has been found closely associated with sulfide-rich gold-bearing rocks. In addition, in only one fountain, the value for Pb exceeded the recommend by the WHO. For all the other trace elements analyzed, the concentrations determined were within the allowed limit established by the Brazilian Ministry of Health and the WHO.

1 INTRODUCTION

The Iron Quadrangle (IQ, Minas Gerais State, Brazil) is an area of great prominence in the Brasilian scenario, in reason of its mineral resources and natural beauty, important in the Brazilian history and expressive of economic and social development mainly in the beginning of the 20th century with the discovery of great iron ore reserves. Not only extensive iron deposits but also hydrothermal gold deposits are found in this region.

The Iron Quadrangle, with an area of approximately 7200 km^2, is located in the mid-south of the state of Minas Gerais (MG), between the basins of the Velhas river (sub-basin of the San Francisco river) and of the Piracicaba river (sub-basin of the Doce river).

Previous studies have shown high levels of arsenic (Deschamps *et al.*, 2002; Palmieri, 2006) in aquatic and terrestrial environments of different regions of the IQ. According to Deschamps *et al.* (2002), such high amounts of arsenic in water, soils, and sediments are related to both natural causes as well as to past and recent mining activities.

Due to the presence of vast mineral resources in this region and the intense exploration of these resources, the present study elaborated a diagnosis on the quality of the natural waters destined to human consumption in 37 fountains of 18 towns in the Iron Quadrangle. In the past, these fountains played an essential and strategic role in supplying these towns with potable water. Even today, this water is used by both the local population and visitors who trust its quality. The popular idea that natural spring water is "clean" is misleading since many toxic elements may be naturally present.

Weathering of minerals within the rocks is an important source of dissolved ions in groundwater (Marghade *et al.*, 2012). Geochemical processes, occurring within the groundwater and their reaction with aquifer materials, are responsible for changes in groundwater chemistry (Rao & Rao, 2009). Urbanization and related anthropogenic activities also affect groundwater quality.

2 MATERIALS AND METHODS

Fountain water samples were collected in towns spread in the IQ region, in August 2011. After collection, the samples were filtered through Millipore 0.45 μm filters and immediately acidified with nitric acid (pH < 2).

Total dissolved solids (TDS), conductivity (EC), pH and turbidity were conducted *in situ* on the samples using the portable meters Myron L Ultrameter II and a Turbidimeter 2100Q Hach, respectively. The trace elements (Li, Al, V, Cr, Mn, Co, Ni, Cu, Zn, As, Rb, Sr, Cd, Ba, Pb and Sb) were carried out using a Perkin Elmer ELAN DRC-e ICP-MS spectrometer equipped with an auto sampler (AS-93plus).

Solutions, standards, and dilutions were prepared with ultra-pure water (18.2 MΩ.cm), and ultrapure HNO_3 69.5% (w/w) (Fluka) was used for the preparation of all standard solutions and the preservation and dilution of the samples.

Multi-element Standard 3 PerkinElmer N9301720 solutions containing Ag, Al, As, Ba, Be, Bi, Cd, Co,

Cr, Cs, Cu, Ga, In, Li, Mn, Ni, Pb, Rb, Se, Sr, Tl, U, V, Zn (10.0 mg L^{-1}) were used for preparing the calibration curves, and all the solutions and samples were prepared in 1% HNO_3 for ICP-MS.

3 RESULTS AND DISCUSSION

The range of the analytical results of the fountain water samples collected is given in Table 1. The pH values ranged from 4.5 to 8.7, with a mean of 6.2. The minimum and maximum pH values are not within the guidelines recommended for drinking water by the WHO (2004) and the Brazilian legislation (Brazilian Ministry of Health Decree -518-03/2004). Though pH has no direct effect on human health, it shows close relationships with some other chemical constituents of water. It was observed that fountains with pH below 5 presented higher concentrations of the trace elements, due to the fact that acidity enhances the dissolution of some trace elements whereas high pH values facilitate leaching of others, such as Mo (Brady, 1984).

All the other results of the physicochemical parameters, EC, TDS and turbidity, are within the range established by the WHO (2004) and the Brazilian legislation (Brazilian Ministry of Health Decree -518-03/2004), as shown in Table 1. In only one fountain in the town of Ouro Preto/MG, the level of arsenic was above the limit of 10 µg L^{-1} established by WHO (2004). In this region of the Iron Quadrangle, As has been found closely associated with sulfide-rich gold-bearing rocks (Mello *et al.*, 2006). Also, only in one fountain in the town of Santa Bárbara, the Pb value found (26.1 µg L^{-1}) exceeded the value recommend by the WHO (2004).

For all the other analyzed trace elements, the concentrations found are within the permissible limit established by Brazilian legislation (Ministry of Health 518-03/2004) and WHO-2004.

Table 1. pH, EC (µS cm^{-1}), TDS (mg L^{-1}), turbidity (NTU), and range of concentrations of As and trace elements (µg L^{-1}) for 37 samples collected in 18 towns in the IQ region.

Parameters	Range	Brazilian values (2004)	WHO guideline values (2004)
pH	4.5–8.7	6.0–9.5	6.5–8.0
EC	3.3–324	–	1500
TDS	2.3–233	1 000	<600
Turbidity	0.06–3.03	5 NTU	5 NTU
Li	0.04–7.47	–	–
Al	<0.2–213	200	200
V	<0.1–2.98	–	
Cr	<0.12–7.80	50	50
Mn	0.20–259	100	500
Co	0.01–5.56	–	–
Ni	0.14–17.6	–	20
Cu	0.07–203	2000	2000
Zn	<0.5–381	5000	3000
As	<0.06–12.0	10	10
Rb	0.04–4.90	–	–
Sr	0.30–322	–	–
Cd	<0.01–0.18	5	3
Ba	5.00–608	700	700
Pb	<0.01–26.1	10	10
Sb	0.01–0.65	5	5

4 CONCLUSION

The minimum and maximum pH found are not within the guidelines recommended for drinking water by WHO and Brazilian legislation. In only one fountain the level of arsenic was above the limit established by WHO. In this region of the Iron Quadrangle. As has been found closely associated with sulfide-rich gold-bearing rocks. Also, in only one fountain in the town of Santa Bárbara the value for Pb found exceeded the value recommended by WHO. For all the other trace elements analyzed the concentrations found are within the permissible limit established by the WHO (2004) and the Brazilian legislation (Brazilian Ministry of Health Decree -518-03/2004).

ACKNOWLEDGEMENTS

We thank CDTN/CNEN and FAPEMIG (Fundação de Amparo à Pesquisa do Estado de Minas Gerais) for their financial support.

REFERENCES

Brady, N.C. 1984. *Nature and properties of soils. 8th Edition*. Macmillan Publishers Co. INC. New York.

Brazilian Ministry of Health Decree-518-03/2004. Guideline values for drinking water quality.

Deschamps, E., Ciminelli, V.S.T., Lange, F.T., Matschullat, J., Raue, B. & Schmidt, H. 2006. Soil and sediment geochemistry of the Iron Quadrangle. Brazil: The case of arsenic. *Journal of Soils and Sediments* 2: 216–222.

Marghade, D., Malpe, D.B. & Zade, A.B. 2012. Major íon chemistry of shallow groundwater of a fast growing city of Central India. *Environ. Monit. Assess.* 184: 2405–2418.

Mello, J.W.V. 2006. Mineralogy and arsenic mobility in arsenic-rich Brazilian soils and sediments. *Journal of Soils and Sediments* 6: 9–19.

Palmieri, H.E.L. Distribuição, Especiação e Transferência de Hg e As para a Biota em Áreas do Sudeste do Quadrilátero Ferrífero. MG. *Ph.D Thesis*. Federal University of Ouro Preto. Minas Gerais. Brasil. 169 pp.

Perkin Elmer. U.S. EPA Method 2008 for the Analysis of Drinking Waters and Wastewaters.

Rao, N.S. & Rao, P.S. 2009. Major ion chemistry of groundwater in a river basin: A study from India. *Environmental Geology* 61(4): 757–775.

WHO. 2004. *Guidelines for drinking water quality V.1 Recommendations*. Geneva, Switzerland.

One Century of the Discovery of Arsenicosis in Latin America (1914–2014) – Litter, Nicolli, Meichtry, Quici, Bundschuh, Bhattacharya & Naidu (Eds)
© 2014 Taylor & Francis Group, London, ISBN 978-1-138-00141-1

Geostatistical analysis for the delineation of potential groundwater As contamination

C.A. Johnson, A. Bretzler, M. Amini, L.H.E. Winkel & M. Berg
Eawag, Swiss Federal Institute of Aquatic Science and Technology, Dübendorf, Switzerland

L. Rodriguez-Lado
Soil Science and Agricultural Chemistry, University of Santiago de Compostela, Spain

ABSTRACT: Arsenic-contaminated drinking water poses a health threat to millions of households, worldwide. With an incomplete knowledge of where such contamination may occur, modeling large scale geogenic groundwater contamination is a useful tool. Here we describe a procedure that is able to take into account geochemical expertise, statistical techniques, available datasets of measurements and available geospatial information (geology, soil properties, climate, topography) to develop hazard maps for arsenic.

1 INTRODUCTION

Arsenic (As)-enriched groundwater is used as drinking water by millions of households on a global scale (Smedley & Kinniburgh, 2013). Some of the first incidences of chronic arsenic poisoning (arsenicosis) from groundwater consumption were reported in Taiwan, Chile, Argentina and Mexico, but the scale of this health threat was recognised in the mid-1990s after the discovery of a large population suffering from groundwater derived arsenic poisoning in the Bengal Delta. To date, regions affected by arsenic enrichment in groundwater resources are known to be widespread in Asian countries but also others in Africa, Europe, and the USA. However, many countries have not carried out detailed arsenic surveys and the worldwide scale of As-affected regions therefore still remains largely unknown.

Modelling of arsenic contamination potential can be a useful tool in pinpointing regions at risk and recognising potentially safe groundwater resources, both laterally and at aquifer depth. This is particularly important for low-income countries. Visual outputs of the modelling procedure, such as hazard maps, are very effective in awareness creation and represent a first step in initiating arsenic mitigation activities.

It is known that the natural enrichment of arsenic in groundwater systems is related to different types of environmental conditions (apart from geothermal and volcanic origin): i) sedimentary aquifers under strongly reducing conditions, where arsenic is predominantly present in its reduced state As(III) and its release related to reductive dissolution of As-bearing iron (hydr)oxides in sediments, ii) non-reducing but alkaline environments in closed basins in arid and semi-arid regions, where high pH leads to alkaline desorption of As from mineral oxides (Smedley & Kinniburgh, 2013), iii) fractured bedrock aquifers of usually metamorphic geology with high pH and low dissolved oxygen. This geochemical/hydrological understanding can be used to predict areas at risk of arsenic enrichment. The underlying principle and challenge is to find proxies that are indicative of the environments where arsenic release is most prominent. These can be surface geological, geochemical or climatic raster-based data, backed up with measured As data points. Using statistical procedures, the factors that best correlate with As-affected areas are determined. Once these factors have been established and verified with actual measured As concentrations, predictions can also be made for areas where As measurements are not yet available.

2 METHODOLOGY

Measured data points of groundwater arsenic concentrations are necessary for model calibration and validation. They are aggregated to a suitable spatial scale. Geographic Information Systems (GIS) are used for storing, manipulating and analyzing all types of geographical data and are an integral part of the modeling procedure. The following geospatial layers are required:

- **Geology**. Chemically reducing conditions are associated with young Holocene sedimentary environments. Elevated As concentrations are often associated with geothermal environments indicated, for example, by geothermal springs.

- **Soil properties**. Soil conditions such as pH, organic carbon content and drainage properties are proxies for conditions in subsurface aquifers.
- **Topography**. Digital Elevation Models (DEM), parameters such as slopes and the Topographic Wetness Index (TWI) can be extracted.
- **Climate**. Temperature, precipitation and evapotranspiration are parameters that point to aridity or humidity of a region and will influence groundwater composition and residence times.

Should elevated As concentrations be due to different (hydro)geochemical processes in the area selected for modeling, it is necessary to delineate process regions. This is an iterative process in which the relative significance of variables for a given release process is statistically analyzed. The greater the significance, the higher will be the rank in the delineation process. It is worth mentioning that, during the process of clustering, statisticians and geochemists need to work together closely.

The selection of the modeling procedure depends on the scale and the quality and spatial distribution of input data (Amini *et al.*, 2010). Neural networks can be applied on all scales, but need a comparatively large number of measurements to have a good model fit. Logistic regression can also be applied where data point density is low. Kriging, an interpolation method, is only suitable for regional and local scale modeling if data points are positioned reasonably close together and ideally are evenly distributed.

3 RESULTS AND DISCUSSION

Models developed at different scales are shown in Figure 1. In the case of the global model, two conditions defined as reducing and high-pH/oxidizing were defined. An evapotranspiration/precipitation index was used to separate humid and arid conditions. Sub-soil (30–100 cm) pH and organic carbon contents were used to account for groundwater pH and organic carbon content. The relative importance of variables was determined to be the sequence climate, geology followed by soil parameters.

A regional model of SE Asia was developed by Winkel *et al.* (2008). Though only 1 process region needed to be considered and maps of soil parameters were available, the effort lays in the digitization of very different national geological maps in order to understand depositional environments. The model was able to correctly predict As contamination in Sumatra.

With data on aquifer composition, 3D models can be constructed, as shown by Winkel *et al.*

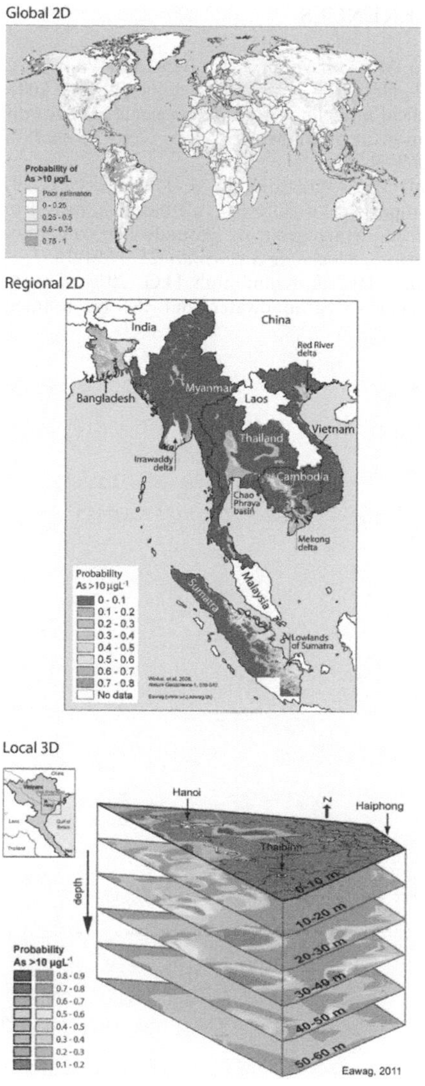

Figure 1. Predicted As groundwater contamination from a global to local scale.

(2011) for the Red River Delta in Vietnam. Using three-dimensional geological data, arsenic risk was determined at different depths in 10 m steps. This kind of model can be very useful for water supply planning and when drilling new drinking water wells. For the majority of situations, 2D have to suffice. Winkel *et al.* (2011) were able to show that while their 3D model correctly classified 74% of their data, the corresponding 2D model of the same area classified 65% correctly.

The key challenge for future developments needs high-resolution geological information, both in terms of scale and classification.

REFERENCES

Amini, M., Abbaspour, K.C., Berg, M., Winkel, L., Hug, S.J., Hoehn, E., Yang, H. & Johnson, C.A. 2008. Statistical modeling of global geogenic arsenic contamination in groundwater. *Environ. Sci Technol.* 42(10): 3669–3675.

Amini, M., Abbaspour, K. & Johnson, C.A. 2010. A comparison of different rule-based statistical models for modeling geogenic groundwater contamination. *Environ. Modelling & Software* 25: 1650–1657.

Smedley, P.L. & Kinniburgh D.G. 2013. Chapter 11: Arsenic in groundwater and the environment. In O. Selinus, B. Alloway, J.A. Centeno, R.B. Finkelman, R. Fuge, U. Lindh & P.L. Smedley (eds,) *Essentials of Medical Geology, Second Edition.* Springer.

Winkel, L., Berg, M., Amini, M., Hug, S.J. & Johnson, C.A. 2008. Predicting groundwater arsenic contamination in Southeast Asia from surface parameters. *Nature Geoscience* 1: 536–542.

Winkel, L.H.E., Trang, P.T.K., Lan, V.M., Stengel, C., Amini, M., Ha, N.T., Viet, P.H. & Berg, M. 2011. Arsenic pollution of groundwater in Vietnam exacerbated by deep aquifer exploitation for more than a century. *PNAS*, 108(4): 1246–1251.

One Century of the Discovery of Arsenicosis in Latin America (1914–2014) – Litter, Nicolli, Meichtry, Quici, Bundschuh, Bhattacharya & Naidu (Eds)
© 2014 Taylor & Francis Group, London, ISBN 978-1-138-00141-1

Integrated study of arsenic contamination in different matrices and targets in La Matanza, Buenos Aires province, Argentina

C. Vázquez
Comisión Nacional de Energía Atómica y Facultad de Ingeniería, Universidad de Buenos Aires, Buenos Aires, Argentina

L. Marcó
U. Centroccidental Lisandro Alvarado, Lara, Venezuela

M.C. Rodríguez Castro
CONICET, Argentina

S. Boeykens & A.M. Maury
Facultad de Ingeniería, Universidad de Buenos Aires, Buenos Aires, Argentina

ABSTRACT: The aim of this study was to investigate exposure to arsenic in La Matanza inhabitants, Argentina. In order to establish a full view of arsenic exposure in the area, several matrices and targets were investigated. As matrices, water and soil samples were analyzed. As targets, canine and human hair were selected. The results of the present study demonstrate chronic contamination in dog hair at the study area. These results concur with arsenic levels found in soil and water but not in human hair.

1 INTRODUCTION

Arsenic (As) is found in the environment due to human activities as well as natural geological processes. In Argentina, the origin of As is mainly natural, and related to different geological processes. During the Quaternary period, accumulation of eolic and fluvial materials was produced in arid environmental conditions. In this region, volcanic material is an important component of loessic and sandy eolic deposits (Smedley *et al.*, 1988). The physical and chemical characteristics of the hydrogeological system are mainly determined by the form of accumulation of materials and the nature of loessic deposits.

The presence of As in water constitutes a serious world health concern due to its toxicity (Abernathy *et al.*, 2003; McClintock *et al.*, 2012). Depending on the intensity and duration of exposure, this element can be acutely lethal or may have a wide range of health effects in humans and animals.

The concern in Argentina about As and its influence on human health dates back to the previous century (Besuschio *et al.*, 1980). The disease adscribed to As contamination was later called 'chronic regional endemic hydroarsenism' (HACRE, 'hidroarsenicismo crónico regional endémico', in Spanish), produced by the consumption of water with high As levels.

In our study, we focused to La Matanza district. In order to establish a full view of As exposure in the area, several matrices and targets were analyzed. As matrices, water and soil samples were analyzed. As targets, canine and human hair were chosen because As tends to bind with sulfhydryl groups in keratin of hair (Karagas *et al.*, 2000). Canine hair was chosen because dogs share the same environment as humans, developing many of the same diseases (Felsburg, 2002). Also, both dogs and humans share drinking water, but, unlike humans, who ingest waters from different sources, dogs consume mainly groundwater.

The aim of this study was to investigate acute and chronic exposure to As in La Matanza inhabitants.

2 METHODS

2.1 *Study site*

The surveyed area was Los Alamos, an 80 km^2 neighborhood that belongs to the Matanza district, located 19 miles (31 km) away from the capital city, in the Buenos Aires province, Argentina (Figure 1). Previous studies from this group showed that the selected area has elevated As levels in groundwater, which is the main source of drinking water for the local population (Vázquez *et al.*, 2012). In this study, 100% of the samples was above the guidelines provided by the WHO (1984).

2.2 *Samples*

A total of twelve water and soil samples were taken from wells and backyards belonging to neighbors of the area. Water samples were filtered through 47 mm GF/C Whatman filters, acidified with HNO_3 (Merck) to pH 2 and stored in bottles until As determination by Total Reflection X Ray Fluorescence (TXRF) instrumentation. For the measurement, 2 µL of the sample were placed on a quartz reflector and dried. Soil samples were dried and pressed for analysis with a wavelength X ray fluorescence spectrometer.

A total of fifty samples of dog hair, coming from male and female dogs, kept as companion animals, were collected. Hair human samples of eighteen neighbors of the area were also collected. In both cases, a lock of hair was taken from the lower neck area, cut with stainless steel ethanol-clean scissors. Samples were stored in clean plastic bags until sample digestion. Sample digestion was performed before the analysis. Dog hair was analyzed by TXRF after digestion using a domestic microwave oven by adding 30% v/v H_2O_2 followed by 65% v/v HNO_3. Human hair was analyzed by hydride generation atomic absorption spectroscopy after digestion. All the procedures were validated using standard reference material: IAEA H4 (animal muscle); IAEA A2-74 (animal blood) and SRM 1577 (bovine liver).

3 RESULTS AND DISCUSSION

In water samples, all the values were above those allowed for human consumption by the WHO (10 µg L^{-1}) (Table 1). In the case of human hair

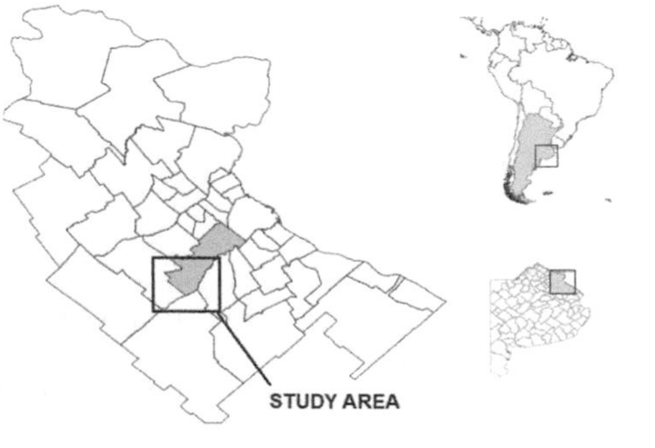

Figure 1. Schematic map of the area of study located in the Buenos Aires province, Argentina.

Table 1. Average, standard deviation and number of samples of As levels in the different matrices and targets. B.r.v. stands for "below reference value".

	AVG	SD	*n*
Water (µg L^{-1})	57	9	12
Soil (µg g^{-1})	6.2	3.3	10
Human hair (µg g^{-1} DW)	b.r.v.	–	18
Dog hair (µg g^{-1} DW)	24	2	50

samples, As concentrations were below the reference value for non exposed professionally individuals (1 µg g^{-1} DW) (WHO, 1984; Sassone *et al.*, 1994). Soil samples had low content of As. In the case of dog hair, no reference values are available. Nevertheless, values found were higher than those found in dogs coming from Eastern Europe (Kozak *et al.*, 2002.).

4 CONCLUSIONS

The results of the present study demonstrate chronic contamination in dog hair at Los Álamos neighborhood. These results concur with As levels found in soil and water but not in human hair. Dog hair may serve as an alert for local population concerning As exposure risks. Additional studies are needed to demonstrate the same chronic effects in the local human population. Although As concentration in dog hair is useful for the detection of exposure, its use as an indicator of the degree of exposure on an individual basis must be carefully considered.

ACKNOWLEDGEMENTS

This work was supported by IAEA Research Contract 15998 and Programa Nacional de Voluntariado Universitario 2009, 2010 and 2011.

REFERENCES

Abernathy, C.O., Thomas, D.J. & Calderon, R.L. 2003. Health Effects and Risk Assessment of Arsenic. *The journal of Nutrition* 133(10): 1536–1538.

Besuschio, S.C, Perez Desanzo, A.C. & Croci, M. 1980. Epidemiological associations between arsenic and cancer in Argentina. *Biol. Trace Elem. Res.* 2: 41–55.

Felsburg, P.J. 2002. Overview of immune system development in the dog: comparison with humans. *Hum. Exp. Toxicol.* 21: 487–492.

Karagas, R.M., Tosteson, T.D. & Blum, J. 2000. Measurement of low levels of arsenic exposure: a comparison of water and toenail concentrations. *Am. J. Epidemiol.* 152: 84–90.

Kozak, M., Kralova, F., Sviatko, P., Bilek, J. & Bugarsky, A. 2002. Study of the content of heavy metals related to environmental load in urban areas in Slovakia, *Bratisl. Lek. Listy.* 103: 231–237.
McClintock, T.R., Chen, Y., Bundschuh, J., Oliver, J.T., Navoni, J., Olmos, V. & Parvez, F. 2012. Arsenic exposure in Latin America: biomarkers, risk assessments and related health effects. *Sci. Tot. Environ.* 429: 76–91.
Neiger, R.D. & Osweiler, G.D. 1992. Arsenic Concentrations in Tissues and Body Fluids of Dogs on Chronic Low-Level Dietary Sodium Arsenite. *J. Vet. Diag. Invest.* 4: 334–337.
Neiger, R.D. & Osweiler, G.D. 2011. Arsenic concetrations in tissues and body fluids of dogs on chronic low-level dietary sodium arsenite. *J. Vet. Diagn* 14(1): 57–61.
Rashed, N. & Soltan, M.E. 2005. Animal hair as biological indicator for heavy metal pollution in urban and rural areas. *Environ. Monit. Assess.* 110(1–3): 41–53.
Sassone, A.H., Sinelli, M., Fernández de la Puente, G, Sarasino, C., López, C. & Roses, O.E. 1994. *Acta Toxicológica Argentina* 2: 17.
Smedley, P.L., Nicolli, H.B., Barros, A.J. & Tullio, J. 1998. Origin and mobility of arsenic in groundwater from Pampean Plain, Argentine. *WRI-9 9th Int. Symp. on Water Rock Interactions*, Taupo, New Zealand pp. 275–278. Balkema, Rotterdam.
Vázquez, C., Palacios, O., Boeykens, S., Saralegui, A., Maury, A.M., Caracciolo, N., Alvarez, L., Botbol, L., Glinka, L., Lunati, C., Macri, D., Rodríguez Castro, M.C., Visacovsky, A. 2012. Problemática Ambiental en el Área Metropolitana de Buenos Aires: Estado de Situación y Alternativas de Solución en Virrey del Pino, La Matanza. *Ciencia* 7(25): 31–38.
WHO (World Health Organization). 1984. *Guidelines for drinking-water quality. WHO chronicle,* 4th ed. 38: 104–108.

One Century of the Discovery of Arsenicosis in Latin America (1914–2014) – Litter, Nicolli, Meichtry, Quici, Bundschuh, Bhattacharya & Naidu (Eds)
 ISBN 978-1-138-00141-1

Arsenic distribution pattern in surface and groundwater matrix of southwestern Nigeria

A.M. Gbadebo
Department of Environmental Management and Toxicology, Federal University of Agriculture, Abeokuta, Ogun State, Nigeria

ABSTRACT: Increasing concentrations of arsenic in different environmental matrices and its health effects has been an issue of major concern for centuries. This study investigated the levels of arsenic in different sources of drinking water from southwestern Nigeria. Surface and groundwater samples were analyzed for different physico-chemical parameters and arsenic content using standard procedures. The results indicated that arsenic concentration ranged from 0.0001 to 7.63 mg L^{-1} for surface water and from 0.05 to 1.5 mg L^{-1} for groundwater samples. Almost 95% of the water samples contain arsenic concentrations above the WHO value of 0.001 mg L^{-1}, thereby implying that most of the water sources in Nigeria have high arsenic concentration. This calls for urgent attention to avoid arsenic poisoning in the nation.

1 INTRODUCTION

Drinking water is one of the powerful determinants of health status of man in his environment as his survival depends on its quality (WHO, 2010). Water pollution is a growing hazard in many developing countries like Nigeria. Present concern is much related to the presence of chemical pollutants in water. One of these pollutants is arsenic, which has cumulative toxic properties, carcinogenic potential and adverse health effects on acute and chronic exposure (Park, 2001). Arsenic occurs in nature: it is present, primarily, in rocks, soils, and waters matrices. In turn, nearby biota are exposed to arsenic, and certain arsenic compounds tend to accumulate in plant leaves and animal tissue through various anthropogenic activities (Brooke, 2002). Many nations are in quest for unpolluted water resources as a means of significant reduction of waterborne diseases. However, much of the earth's crust is saturated with arsenic, which filters into the surface and groundwater and may have health implications when used for drinking purposes (Memon, 2002). It has therefore become mandatory to critically analyze the water sources in southwestern Nigeria for assessing the arsenic levels. maximum allowable limits A total of eighty-five (85) water samples were collected for this analysis, i.e. (55) surface water comprising (33) spring water samples (from Okeigbo, Ikogosi-Ekiti, Ilaro), 12 rain water samples (from Abeokuta) and 10 reservoir water samples (from Abuja). In addition, 30 well water samples were collected from different geological terrains of southwestern Nigeria. All the water samples were analyzed for different physico-chemical parameters such as temperature (using a Celsius thermometer), electrical conductivity (using a WTWLF 90 portable conductivity meter), etc. The anions were analyzed in the non-acidified water samples using gravimetry, phenoldisulphonic acid absorption, turbidimetric and spectrophotometric procedures, while arsenic and other cations in the acidified water samples were measured using the ICP-MS method at the Activation Laboratory in Canada.

2 RESULTS AND DISCUSSION

2.1 *Surface water*

Surface water data collected were subjected to Descriptive Statistical Analysis and it was observed that 80% of the water samples were heavily contaminated with arsenic, with concentrations above the maximum Allowable Limit (0.01 mg L^{-1}) stipulated by the World Health Organization (Fig. 1). The As concentration trend in the surface water samples was in the order: reservoir > rain water > spring water. It was also observed that the As content was in the order: Okeigbo Spring > Ekiti Spring > Ilaro Spring. The Fe concentration in the surface water samples generally ranges from 0.0032–433.7 mg L^{-1}.

The pH of the surface water samples generally ranged from 5.8–8.29, thereby making the water samples to range between slightly acidic to slight alkaline in nature, while the electrical conductivity ranged 37–72.36 µs cm^{-1}, which implies low

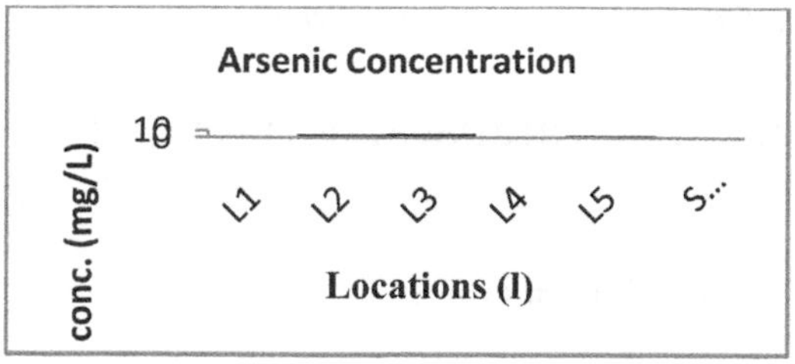

Figure 1. Arsenic concentration in the surface water samples.

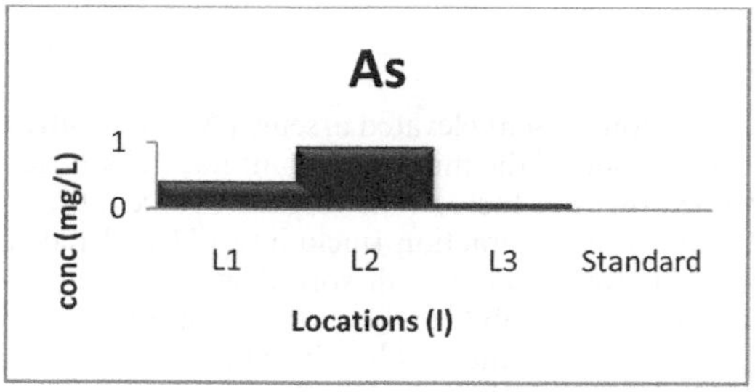

Figure 2. Arsenic concentration in the groundwater samples.

mineralization. The result also proves that the surface waters are highly polluted, probably because of their proximity to various anthropogenic activities. Results further explain that the anions were below the maximum allowable limits stipulated by the WHO.

2.2 *Groundwater*

The results show that pH of the groundwater samples ranged 4.52–6.53 and 6.5–7.80 in Lagos and Ayetoro, which are in the sedimentary geological terrain, while the pH ranges 6.73–8.35 in Abeokuta, which is in the basement complex geological terrain. The electrical conductivity ranges 768.23–471 $\mu s\ cm^{-1}$ in Lagos and Ayetoro and 394.5 $\mu s\ cm^{-1}$ in Abeokuta. Similarly, the analyzed anions in all the locations were below the maximum allowable limits. Fe ranged 0.3–0.49 mg L^{-1} and 90–442 mg L^{-1} in Lagos and Ayetoro, while 20–70 mg L^{-1} in Abeokuta. Arsenic concentrations ranged 0.1–0.9 mg L^{-1} and 0.03–0.15 mg L^{-1} in Lagos and Ayetoro, and 0.2–1.5 mg L^{-1} in Abeokuta, higher than the maximum allowable WHO limits of 0.01 mg L^{-1} (Figure 2).

3 CONCLUSIONS

The high concentration of arsenic in both the surface and groundwater samples implies a great danger to the health of the population if it is not treated immediately. Cumulative health effects of arsenic consumption in the water may result in arsenic related diseases such as skin cancer. This can be avoided by implementing arsenic removal using appropriate techniques before consumption.

ACKNOWLEDGEMENTS

The author wish to express his appreciation to the undergraduate and postgraduate students who have assisted in the fieldwork of this research.

REFERENCES

Brooke, A. 2002. Assessment of arsenic levels in surface water of the upper Yakima river valley, Kittitas county, Washington, Research & Extension Regional Water Quality Conference.

Memon, M., Sommor, M.S. & Puno, H.K. 2002. Status of water bodies and their effect on human health in district Tharparkar. *J. Appl. Sci.* 2: 386–389.

Park, K. 2001. *Park's Textbook of Preventive and Social Medicine*, 16th edn. B.D. Bhanot Publishers, Jabalpur, India, p. 483, 497.

WHO. 2010. Water for Health. Guidelines For Drinking – water Quality. Incorporating First and Second Addendum Addendum to 3rd Ed. Vol. 1 Recommendations. World Health Organization, Geneva, Switzerland.

One Century of the Discovery of Arsenicosis in Latin America (1914–2014) –
Litter, Nicolli, Meichtry, Quici, Bundschuh, Bhattacharya & Naidu (Eds)
© 2014 Taylor & Francis Group, London, ISBN 978-1-138-00141-1

Spatial distribution of arsenic in the region of Tarapacá, northern Chile

A.S. Amaro
CONICYT Regional/CIDERH, Water Resources Research and Development Center, University Arturo Prat, Iquique, Chile

B.C. Venecia Herrera & E. Lictevout
Faculty of Health Sciences, CIDERH, Water Resources Research and Development Center, University Arturo Prat, Iquique, Chile

ABSTRACT: Surface and groundwater in the Tarapacá region present elevated arsenic (As) concentrations, with high spatial heterogeneity. The sulfide minerals are one of the most important natural sources of As in groundwater, being their presence mainly controlled by three factors: the original source (atmosphere, rocks, minerals and/or water), the processes of water-rock interaction (including pH and redox environment) and groundwater flow conditions. Arsenic generally occurs as dissolved species, the predominant oxidation states being As(III) and As(V). As(III) is more mobile and also the most toxic form. The rural population of the region is supplied by untreated resources and it is highly exposed to adverse effects of arsenic. These conditions require a geohydrologic study with the purpose of interpreting mobility, distribution and accumulation of As in the water. The aim of this project is to analyze the concentrations of arsenic in the region of Tarapacá.

1 INTRODUCTION

At least four million people in South American drink water contaminated with arsenic (As). In the 70's, more than 3% of the population of Chile was exposed to contaminated water, making this problem a major concern of health public (CYTED, 2008).

Arsenic is present in water, mainly in groundwater, as geogenic As associated with the volcanism in the Andes. As from these sources moves into the environment (surface water and groundwater, soils, etc.) by natural dissolution, weathering of rocks or mining activities. Other sources of As of minor importance and highly localized, are artificial activities, (like electrolytic processes for metal production) and agricultural products (for example, the use of pesticides containing As).

Currently, in the north of Chile, some people continue drinking water with arsenic concentrations far above the limits established by the 409/1. Of 2005 Chilean norm (INN, 2005), which has set a maximum level of arsenic in drinking water of 10 μg L^{-1} for 2015, in accordance with the limits set by the WHO (1993) and the European Union (Directive 98/83/EC; EU, 1998).

2 METHODS/EXPERIMENTAL

2.1 *Study area*

To provide a regional image of the distribution of As in groundwater used for human consumption in Tarapacá, northern Chile, results from 41 samples of rural potable water and rivers were obtained in September 2013 (black points in Figure 1).

Since 2005, the contents of arsenic in drinking water in the region (gray points in Figure 1) have been systematically monitored.

2.2 *Methods*

The methodologies require strict controls of sampling and transportation. In laboratory, standardized methodologies for water analysis have been used.

The samples were collected in three polyethylene containers, of wide mouth with tip and back cover, which were refrigerated to be transported to the laboratory. The measurement of temperature (°C), pH, dissolved oxygen (mg L^{-1}) and electrical conductivity (μS cm^{-1}) were carried out *in situ*, when collecting the samples, using a multiparameter probe (WTW).

The samples were filtered by 0.45 μm and pH, total dissolved solids (TDS), sodium (Na^+), potassium (K^+), calcium (Ca^{2+}), magnesium (Mg^{2+}), chloride (Cl^-), sulfate (SO_4^{2-}), bicarbonate (HCO_3^-) and carbonate (CO_3^{2-}) were analyzed.

Total arsenic quantification was performed with HG-AAS.

3 RESULTS AND DISCUSSION

The analyzed rural potable water of Tarapacá presented an arsenic content ranging from very low values

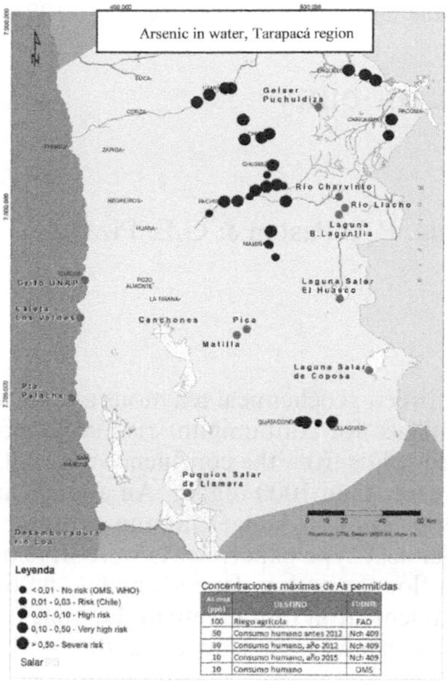

Figure 1. Distribution of arsenic in waters of Tarapacá region.

(0.1 μg L^{-1}) in Guatacondo, in the north, to high concentrations (345.85 μg L^{-1}) in Camiña, at the south. The highest values, exceeding the limit established by law, were observed in Quebrada de Camiña.

The values of boron (B) did not present a regional variability and also no correlation was observed with the arsenic values. The lowest value was observed in Soga (0.01 μg L^{-1}) and the highest one in Pachica (5.33 μg L^{-1}).

pH values fluctuated between 6.66 and 8.03, and the waters can be considered slightly alkaline; the most acid values were recorded in Quebrada de Camiña. The electric conductivity showed very heterogeneous values, ranging from 185.9 to 2876 μS cm^{-1}.

With respect to the monitored rivers, the highest values of arsenic observed (415.0 μg L^{-1}) have been measured in Pachica, and no correlation with boron was registered.

Taking into account the results obtained between 2005 and 2011, it can be said that the waters of rivers and springs of the endorreic basins of the salt flats in the high Andes present arsenic concentrations between 10 and 60 μg L^{-1}, but their concentrations can reach values between 80.000 and 100.000 μg L^{-1} in salt lakes and brines. The As concentrations in the waters of the aquifers of Pica and Matilla did not exceed 30 μg L^{-1}, while in sectors with similar geothermal activity, for example Puchuldiza, the concentrations were higher than 5.000 μg L^{-1} (Lictevout *et al.*, 2013).

In Puquios Salar de Llamara, the arsenic contents varied between 2.000 and 8.000 μg L^{-1}. The waters at the outfall of the Loa river are contaminated with arsenic with values between 1200 and 1700 μg L^{-1} (Lictevout *et al.*, 2013).

In the region of Tarapacá, the arsenic concentration in aquifers and rivers has been a source of concern for regional and national authorities for several decades, because concentrations above 30 μg L^{-1} in freshwater and about 1000 μg L^{-1} in evaporated water have been reported. However, salinity and boron content in water did not show correlation with the concentration of arsenic, as it would be expected (Romero *et al.*, 2003); this may be due to the heterogeneity observed in the different basins of the region. In consequence, it can be said that thc geology of the region, especially the volcanic sediments, the existence of salt lakes, the predominance of closed basins, the thermal areas and anthropogenic activities as copper mining are the main sources of these metalloid in water.

With advances in modern analytical methods used for the detection of As at low concentrations and the introduction of new national limits for drinking water (10 μg L^{-1}), several areas having until now levels of As below the limit, will be classified within the range of unsafe concentrations; this will cause a marked increase in the number of people exposed in the near future.

4 CONCLUSIONS

Groundwater is the main source of supply in the Tarapacá region. In rural waters of the region, arsenic concentrations below 30 μg L^{-1} have been reported; however, normal concentrations vary between 50 and 500 μg L^{-1}.

REFERENCES

CYTED 2008. IBEROARSEN. Distribución del arsénico en las regions Ibérica e Iberoamericana (Distribution of arsenic in Iberian and Latin American regions). Editors: J. Bundschuh; A. Pérez Carrera; M.I. Litter. Argentina. ISBN: 13 978-84-96023-61-1.

EU (European Union) 1998. Council Directive 98/83/EC on the quality of water intended for human consumption.

INN (National Standardization Institute) 2005. Norma Chilena 409/1. Of 2005, Agua Potable Parte 1: Requisitos (Chilean norm 409/1. Of 2005, Drinking water part 1: Requirements).

Lictevout, E., Mass, C., Córdoba, D., Herrera, V. & Payano, R. 2013. Diagnóstico y Sistematización de los Recursos Hídricos de la región de Tarapacá (Diagnostic and systematization of Water Resources in the region of Tarapacá). CIDERH-UNAP, Iquique, Chile.

Romero, L., Alonso, H., Campano, P., Fanfani, L., Cidu, R., Dadea, C., Keegan, T., Thornton, I. & Farrago, M. 2003. Arsenic enrichment in waters and sediments of the río Loa (Second region, Chile). *Applied Geochemistry* 18: 1399–1416.

One Century of the Discovery of Arsenicosis in Latin America (1914–2014) – Litter, Nicolli, Meichtry, Quici, Bundschuh, Bhattacharya & Naidu (Eds)
 ISBN 978-1-138-00141-1

Chemical-hydrodynamic control of arsenic mobility at a river confluence

P.A. Guerra, C. Gonzalez, C.E. Escauriaza, C.A. Bonilla, P.A. Pasten & G.E. Pizarro
Pontificia Universidad Católica de Chile, Santiago, Chile

ABSTRACT: A better understanding of the interactions between geochemical reactions and the hydrodynamics of a river confluence may lead to innovative strategies for contaminant risk evaluation and management. To better understand these processes, suspended solids from the confluence were collected, finding arsenic-rich suspended solids with concentrations greater than 1000 mg/kg. An average arsenic/iron ratio of 44 mg arsenic/g iron was estimated with very little variation, suggesting the most important process in arsenic fate is the precipitation of iron oxides. Jar test type experiments were developed at different mixing ratios and mixing velocities, concluding that floc formation depends mainly on the proportion in which waters are mixed (pH and concentration). Arsenic/iron ratios were in the same order of magnitude as those collected on field, therefore constituting a constant in arsenic fate at the confluence.

1 INTRODUCTION

River confluences are natural reactors where key processes controlling contaminants at different scales occur (Schemel et al. 2000, Sánchez et al. 2005). Better understanding of geochemical reactions and hydrodynamics of a river confluence may provide innovative contaminant risk evaluation and management strategies.

The Lluta River is a precious water resource for the extremely arid regions of Arica y Parinacota in northern Chile. Although the construction of a dam is being planned to increase water availability, the likely accumulation of arsenic-rich iron oxy/hydroxides in the sediments of the dam has prompted careful analysis from environmental authorities and the local scientific community. The main source of arsenic-rich fine particles is the confluence of the Caracarani and the Azufre Rivers, located at 3986 m and approximately 160 km northeast of Arica, near to the village of Humapalca. Its distinction as the main source of arsenic-rich sediments makes it an ideal natural laboratory for studying the interactions between geochemical reactions and hydrodynamics.

Both rivers present similar flow rates during the year, with floods occurring during the so-called "Bolivian Winter." The Azufre River and Caracarani River oscillate between 45 to 245 L/s and 170 to 640 L/s, respectively. The Azufre River is born near the Tacora Volcano and an abandoned sulfur mine and has extreme chemical characteristics, with a pH <2, conductivity > 10 mS/cm, total arsenic concentrations >2 mg/L, and total iron concentrations that vary between 35 and 125 mg/L. The Caracarani River has pH 8, conductivity of approximately 1.5 mS/cm, total arsenic concentration <0.1 mg/L and total iron concentration.

Acoustic Doppler velocimetry conducted at the confluence showed spatially variable turbulence intensities in both the stream-wise and across-stream directions that produce heterogeneous distributions of particulate material on the river bed (Escauriaza et al. 2011). The pH, electrical conductivity and turbidity were recorded in several transects downstream from the confluence; these revealed highly variable spatial distributions throughout the first 300 m.

2 EXPERIMENTAL

2.1 *Field work*

After the Caracarani River and the Azufre River mix, characteristic light-brown fine particulate materials are formed. These were sampled as suspended solids by filtration at 0.45 µm and later analyzed by total x-ray fluorescence. Coatings on riverbed rocks were measured by portable x-ray fluorescence, then washed with hydrochloric acid solution to compare coated and bare surfaces.

2.2 *Laboratory studies*

To understand the chemical and hydrodynamic interactions controlling the fate of arsenic at the confluence, we reproduced the mixing of waters from both rivers at different mixing ratios (Azufre/Caracarani 0.3 and 0.2) and velocities (10, 50 and 80 rpm) by using 20-minute jar-test experiments. The resulting solutions were filtered at 0.45 μm to analyze the aqueous phase by total x-ray fluorescence, optical spectrometry emission inductively coupled plasma and ionic chromatography. The solids retained on the 0.45 μm membranes were analyzed by portable x-ray fluorescence.

3 RESULTS

3.1 *Characterization of suspended solids sampled downstream from the confluence*

The suspended solids have fluctuating concentrations of iron and arsenic (24 ± 22 g/kg and 1130 ± 1210 mg/kg, respectively). Nine samples taken in different seasons were analyzed, but maintain a relatively constant arsenic/iron ratio of 44 ± 5.7. Riverbed sediments of particle size <2 mm have much lower concentrations of these elements (17±3.0 g/kg and 68 ± 54 mg/kg. Six samples taken in different seasons were analyzed). These results suggest that the formation of iron oxides at the confluence controls arsenic transport. These solids apparently settle quickly after their formation, as arsenic-rich coatings were found on riverbed rocks, reaching up to 1157 mg/kg of arsenic. Upwards of 90% of the arsenic was removed by applying a solution of hydrochloric acid to the rock surface.

3.2 *Results of jar test experiments*

Despite the high variability in the amount of produced suspended solids, a greater amount of solids tended to be formed with higher pH (or lower mixing ratio). The formation of iron oxides occurs

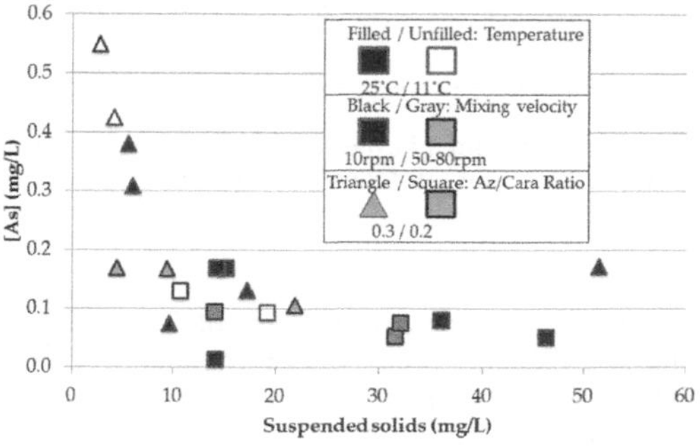

Figure 1. Arsenic removal is greater when there is more solid produced.

at pH 3 at the concentrations of our experiments. Lower As concentrations are achieved with greater production of solids or floccules once each experiment has finished, showing a tendency of the solid to become saturated in arsenic (Figure 1). Furthermore, arsenic/iron ratios in all the experiments are within an order of magnitude of the suspended solids found on field. However, solids produced at a lower pH (or mixing ratio of 0.3) are more concentrated in arsenic, showing better affinity to iron.

4 CONCLUSIONS

Arsenic-rich solids are found downstream from the Caracarani River-Azufre River confluence in suspended materials, floccules, and rock coatings. The confluence is a natural reactor where chemical and hydrodynamic conditions allow for the formation of arsenic-scavenging iron oxides.

Jar tests show the formation of suspended solids is the key to removing arsenic from the system. Experimental arsenic/iron ratio values were within an order of magnitude of the field measurements, showing the importance of suspended solids in the fate of arsenic. The jar test experiments conclude that despite the different mixing velocities tested, it is the chemical conditions given by the mixing ratio that determine the formation of arsenic-adsorbent iron-oxides in form of floccules.

ACKNOWLEDGEMENTS

Research was funded by Fondecyt 1100943. Travel funds from project 1110440 are acknowledged.

REFERENCES

Schemel L.E., Kimball B.A. & Bencala K.E. 2000. Colloid formation and metal transport through mixing zones affected by acid mine drainage near Silverton, Colorado. *Applied Geochemistry* 15: 1003–1018.

Sánchez J., López E., Santofimia E., Reyes J. & Martín J. 2005. The impact of acid mine drainage on the water quality of the Odiel River (Huelva, Spain): Evolution of precipitate mineralogy and aqueous geochemistry along the Concepción-Tintillo segment. *Water Air and Soil Pollution* 173: 121–149.

Escauriaza, C. Gonzalez, C., Guerra, P., Pasten, P. & Pizarro, G. 2011. Turbulent mixing in a river confluence controls the fate of contaminated sediments. *American Physical Society, 64th Division of Fluid Dynamics Annual Meeting, November 20–22 2011.* Baltimore, Maryland, United States.

One Century of the Discovery of Arsenicosis in Latin America (1914–2014) –
Litter, Nicolli, Meichtry, Quici, Bundschuh, Bhattacharya & Naidu (Eds)
 ISBN 978-1-138-00141-1

Speciation of arsenic in a saline aquatic ecosystem in northern Chile

V. Herrera
Faculty of Health Sciences, Center of Research and Development of Water Resources, Tarapacá, University Arturo Prat, Tarapacá, Chile

A.S. Amaro & C. Carrasco
Center of Research and Development of Water Resources, Tarapacá, Faculty of Health Sciences, University Arturo Prat, Tarapacá, Chile

ABSTRACT: In order to interpret the mobility of arsenic (As) in the solid phase (functional and operational speciation), geochemical and physicochemical measurements were performed on sediments of "permanent lagoons" of the high Andes salt flats and Loa river estuary in northern Chile. The ecosystems where the metalloid concentrations in water (1.8 to 80 mg L^{-1}) and sediment (40 to 870 mg/kg) are high, present toxicity. The high mobility of As is controlled by the acidity, calcium carbonate content (10–58%) and redox conditions of shallow saline water column in the estuary or salt-brine water in the permanent lagoon.

1 INTRODUCTION

In northern Chile, the volcanic sediments, the closed lake basins and the thermal areas are the main sources of arsenic (Lictevout *et al.*, 2013). There are specific mechanisms responsible for mobility of some chemical species in the environment that makes common species can be easily mobilized, depending on the pH, redox conditions, temperature and composition of the solutions (Smedley & Kinniburgh, 2002). In the 90s, within the European community, programs for improvement and harmonization of the quality of speciation studies of trace metals extracted from the solid phase were implemented, in order to standardize results (Das *et al.*, 1995).

The aim of this work is to determine the mobility of arsenic in contaminated aquatic ecosystems in northern Chile as permanent lagoons associated with high Andean salt flats and the Loa river estuary by previous geochemical and physicochemical characterization of matrices. Simple and sequential extractions from sediments with solutions especially selected, allowed estimating the mobility of arsenic under environmental conditions.

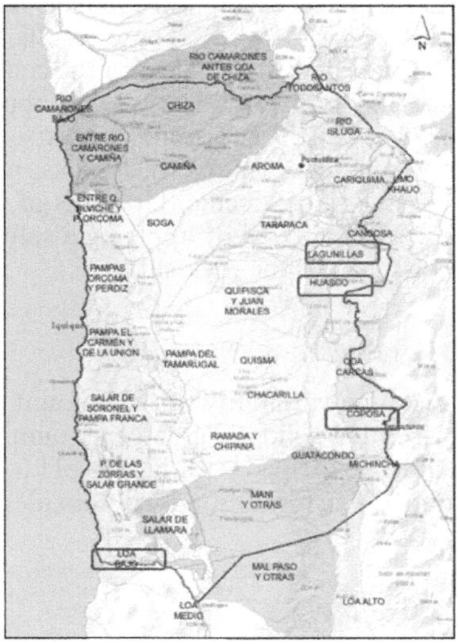

Figure 1. Study area in the region of Tarapacá, northern Chile: Salt flats of Lagunilla, Huasco and Coposa and Loa river estuary.

2 METHODS/EXPERIMENTAL

The study area (Figure 1) considered the estuary of the final part of the sector belonging to the "low Loa" of the Loa river basin, with a total of 4 sites. In the place, located near the Bolivia border, about 3800 m altitude, and with endorrheic basins surrounding the Chilean Altiplano, samples were taken at two sites in permanent lagoons of Pampa Lagunilla salt flats, at three sites of the lagoon of the Huasco salt flats and at two sites of the lagoon of the Coposa salt flats, during the dry season (November) and post-rain (April) of 2009 and 2010.

The sediments were dried, homogenized and sieved to 1 mm to determine the concentration of easily oxidized organic carbon by potentiometric titration with dichromate and calcium carbonate.

The 63 μm sediments were subjected to digestive processes on Teflon pumps with a HCl-HNO_3 (3:1) mixture at 170 °C for 3 h. The same treatment was applied to certified reference materials Estuarine Sediment (SRM 1646) of the National BSS Bureau of Standards (Washington DC).

Functional and operational speciation was developed using the modified protocol of BCR (reducible and oxidable) for single and double extractions with different reagents recommended in literature, such as water, acetic acid buffer, sodium acetate, oxalic acid, ammonium oxalate, acetic acid solutions and diluted sulfuric acid. The extract was kept at 4 °C until the quantification of arsenic by absorption spectroscopy with hydride generation (HG-AAS).

3 RESULTS AND DISCUSSION

The permanent lagoons of the salt flats in study concentrated arsenic in the evaporated water. The waters of Huasco and Coposa are alkaline and vary from salted to brines. Sodium is present as the predominant cation, and the predominant anions are chloride and sulfate. In the salt flats of the Pampa Lagunilla lagoon, the waters are classified as sodium sulfate and less saline (*dulces-saladas*). The sediments of these lagoons have the particularity of containing high concentrations of calcium carbonate (42–68%), which are independent of the fluctuations in water composition that cover them, particularly those of Lagunilla, which presents significant content of precipitated sulfates (30%) and relatively lower contents of calcium carbonate (15–20%). All sediments have a low organic matter content (1–3%) and are enriched in arsenic. The maximum value determined in the sediments was 268 ± 8 mg kg^{-1} in the dry season in the lagoon of Huasco and 70 ± 5 mg L^{-1} in the waters. The evaluation of the physicochemical characteristics and the presence of trace elements in the sediments, together with the determination of the major ions in water are essential to select and propose reagents for functional and operational speciation of arsenic in sediments and, thus, for understanding the mobility in these ecosystems.

In order to compare the results of extractions, the experimental conditions with different extracts were kept constant. The results of the removal percentage with extracting solutions of acetic acid-acetate buffer in sediments (2–11%) show that arsenic is quite available and mobile, which is attributable to Na_2HAsO_4 associated with the phase of carbonate and/or coprecipitated as $CaHAsO_4$. There is also a substantial fraction adsorbed on amorphous Fe and Al oxides present in the sediments, and the results of removal rate with oxalate buffer were higher than 50% in Huasco, but lower (37%) in Coposa and Lagunilla (21%).

There is a good linear correlation between the concentration of arsenic extracted with selected solutions and the complete extraction. When the concentration of iron in the sediment increases, the mobility of arsenic increases under acidic and reducing conditions. Extraction with 0.43 M acetic acid and 0.25 M sulfuric acid was almost 100% with an estimated overall mobility.

The waters of the river estuary of Loa are alkaline, saline and have a reducing character inversely proportional to dissolved oxygen concentrations; they are of sodium-chloride type, and have high levels of sulfate and arsenic (1.7 ± 0.4 mg L^{-1}). These estuarine sediments also exhibit a significant content of calcium carbonate (30–40%) and are heavily contaminated with arsenic, peaking at 823 ± 13 mg kg^{-1}, coinciding with the zone of lower turbulence. The metalloid is mobilized under reducing conditions (54% removed with hydroxylamine chloride) and remobilized under oxidizing conditions (43% removed with hydrogen peroxide), and the residual fraction retains 0.3–1.6%. Arsenic in sediments is easily interchangeable: 20–29% is removed with acetic acid (0.11 M).

We should note that the optimal extraction involves a protocol that allows complete and selectively dissolving the component of the solid phase without removing or changing another phase. However, the effect of the extractant acidity in aqueous medium induces removal of other fractions susceptible to changes of pH.

4 CONCLUSIONS

The waters of the high Andes transport and deposit, in a complex way, various elements associated with the volcanic activity of the lacustrine basin and the evaporitic sediment of the permanent lagoons as a concentrator matrix of arsenic of volcanic origin. The mobility of arsenic is complete (100% removal) under acid pH change and high under reducing conditions and acid pH (60% removal).

The waters and saline and alkaline sediments of Loa river estuary are heavily contaminated with arsenic, what is easily interchangeable and bioavailable. In addition, with a significant change in the acidity of these ecosystems, arsenic could be removed in its totality.

REFERENCES

Das, A., Chakraborty, R., Cervera, M. & De la Guardia, M. 1995. Review: Metal speciation in solid matrices. *Talanta* 42: 1007–1030.

Lictevout, E., Mass, C., Córdoba, D., Herrera, V. & Payano, R. 2013. Diagnóstico y Sistematización de los Recursos Hídricos de la región de Tarapacá. CIDERH-UNAP, Iquique, Chile.

Smedley, P. & Kinniburg, D. 2002. A review of the source, behavior and distribution of arsenic in natural waters. *Appl. Geochem.* 17: 568–571.

One Century of the Discovery of Arsenicosis in Latin America (1914–2014) – Litter, Nicolli, Meichtry, Quici, Bundschuh, Bhattacharya & Naidu (Eds)
 ISBN 978-1-138-00141-1

Trophodynamics of arsenic in Nahuel Huapi Lake: Effect of Puyehue-Cordón Caulle volcanic eruption

R. Juncos, A. Rizzo, M. Arcagni, M.A. Arribére & S. Ribeiro Guevara
Laboratorio de Análisis por Activación Neutrónica, Centro Atómico Bariloche, CNEA, Bariloche, Argentina

L.M. Campbell
Environmental Science, Saint Mary's University, Canada

ABSTRACT: Volcanic eruptions are important natural sources of arsenic (As) to the ecosystems. The Nahuel Huapi Lake, in northern Patagonia, was impacted by the eruption of the Puyehue-Cordón Caulle volcanic complex on June 4th, 2011. We investigated food web transfer patterns of As in the most affected shore of the lake (Brazo Rincón) and the effect of volcanic deposition on the As concentrations of the biota. Food web As biodilution was observed, with the highest As concentrations in littoral snail *Chilina* sp. After the volcanic eruption, significant differences in As concentrations were found only for a few species, including two decapods, a snail and a small fish. The impact of volcanic deposition is associated with increased As in most benthic organisms, reflecting the importance of sediments as a key pathway of As to the food web.

1 INTRODUCTION

Volcanic eruptions are important natural sources of arsenic (As) to aquatic ecosystems. Aquatic organisms can accumulate, retain, and transform arsenic species inside their bodies when exposed through their diet and intake through other routes/sources such as water, soil, particles, etc. (Rahman *et al.*, 2012). Therefore, As has the potential to be transferred through food webs to higher trophic levels, and finally to humans by fish consumption.

The Nahuel Huapi Lake was affected by a massive pyroclastic material deposition after the eruption of the Puyehue–Cordón Caulle volcanic complex (CVPCC) in June 4th, 2011. This situation provides a natural experiment for studying the trophic dynamics of As in a deep oligotrophic lake under the influence of volcanic activity. The goal of this work is to study food web transfer patterns of As in Nahuel Huapi Lake and evaluate the effect of a volcanic eruption on the As contents of the biota.

2 METHODS

2.1 *Study area*

The Nahuel Huapi Lake (40°55'S, 71°30'W) is the largest natural lake of northern Patagonia (surface area: 557 km^2, maximum depth: 464 m). Brazo Rincón (BR) is the northernmost branch of the lake, closest to volcano location. It was the most affected by ash deposition after the eruption, and thus selected for the purposes of this work.

2.2 *Sample preparation, analytical procedures, and data analysis*

Different organisms were sampled from BR before (summer-autumn 2011) and in two moments after (winter-spring 2011 and summer-autumn 2012) the eruption (Table 1). Plankton samples were collected via vertical tows performed with three different mesh size nets. Benthic macroinvertebrates were caught either with baited traps or by hand along the shoreline. The most abundant fish species were captured setting gillnets (for big fish) or seine nets and traps (for small fish). Samples were homogenized and lyophilized. For trophic level determinations, stable nitrogen isotopes were measured via DELTAplusXP continuous flow stable isotope ratio mass spectrometer. Total As concentrations ([As]) were determined by Instrumental Neutron Activation Analysis (INAA). Variations of [As] with trophic positions were assessed through regression analysis (δ^{15} N vs. [As] log10 transformed). Arsenic differences between organisms before and after eruption were assessed using Student's t-test.

3 RESULTS AND DISCUSSION

3.1 *Arsenic trophodynamics*

Arsenic tends to decrease with increasing trophic level in BR before the eruption. Logarithm of [As] was negatively and significantly regressed with $\delta^{15}N$ ($r^2 = 0.34$, $p < 0.05$), suggesting food web biodilution (Figure 1; Table 1). This is consistent with previous work in Moreno Lake in Patagonia (Revenga *et al.*, 2012), and with results in other freshwater ecosystems (Chen *et al.*, 2000).

3.2 *Arsenic distribution in biota before volcanic eruption*

Varying [As] were observed in different organisms from BR (Table 1). These values were close to those found in similar lakes (Revenga *et al.*, 2012), and below values found in contaminated lakes (Chen *et al.*, 2000). Arsenic concentrations significantly differed among the plankton fractions, phytoplankton having the highest values although with high variability, while zooplankton, link to fish (small puyen) in the pelagic food web, having the lowest values.

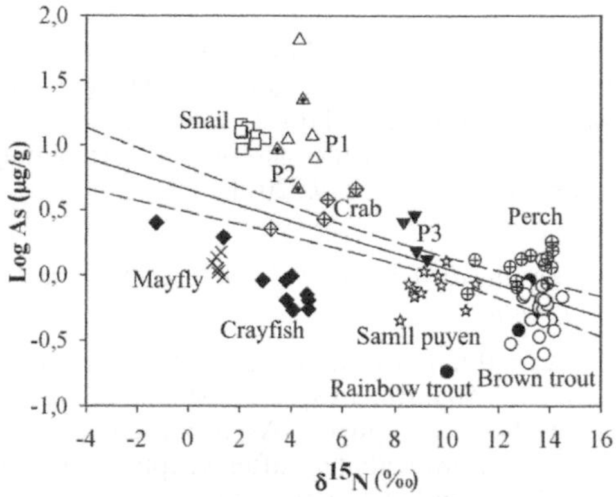

Figure 1. Relationship between $\delta^{15}N$ and Log As in organisms from Brazo Rincón, Nahuel Huapi lake before CVPCC eruption. Regression line and confidence intervals are shown. P1: phytoplankton, P2: mixed plankton, P3: zooplankton.

Among invertebrates, the benthic feeder snail (*Chilina* sp.) showed the highest As concentrations, both in muscle and hepatic tissue, suggesting an important As input to the food web through

Table 1. Mean values and standard deviation of δ15 N and [As] for summer-autumn, 2011 (Su-Au'11) before CVPCC eruption, and mean [As] for winter-spring, 2011 (Wi-Sp'11) and summer autumn, 2012 (Su-Au'12), after CVPCC eruption, for different components of Brazo Rincón biota.

	Before eruption		After eruption	
	$\delta^{15}N$	As (µg g^{-1})	As (µg g^{-1})	
Organism	(‰)	Su-Au'11	Wi-Sp'11	Su-Au'12
Plankton				
Phytoplankton	4.5 ± 0.5	23.8 ± 27.4	–	–
Mixed plankton	4.7 ± 1.3	10.1 ± 8.5	5.1 ± 1.6	7.6
Zooplankton	8.8 ± 0.4	2.1 ± 0.7	3.12 ± 1.3	2.46
Macroinvertebrates				
Mayfly[a]	1.2 ± 0.1	1.2 ± 0.2	<2	2.27 ± 1.8
Snail[b]	2.4 ± 0.3	11.9 ± 1.7	12.3 ± 1.7	13.4 ± 2.2
Snail[c]		17.3 ± 2.1	14.8 ± 0.8	17.5 ± 2.9
Crab[b]	5.1 ± 1.2	4.1 ± 0.7	7.9 ± 4.1	3.7 ± 1.7
Crab[c]		3.5 ± 1.9	5.0 ± 2.1	7.7 ± 1.2
Crayfish[b]	4.1 ± 0.6	1.0 ± 1.9	1.6 ± 1.4	1.0 ± 0.4
Crayfish[c]		3.7 ± 1.1	1.9 ± 0.8	2.1 ± 1.2
Fish				
Small puyen[a]	9.4 ± 0.9	0.8 ± 0.2	1.6 ± 0.3	0.4 ± 0.1
Creole perch[b]	13.1 ± 1.1	1.2 ± 0.3	–	0.6
Creole perch[c]		3.8 ± 0.9	–	3.4
Rainbow trout[b]	12.7 ± 1.5	0.5 ± 0.3	–	0.4 ± 0.2
Rainbow trout[c]		0.5 ± 0.1	–	0.6 ± 0.3
Brown trout[b]	13.8 ± 0.4	0.6 ± 0.2	–	0.6 ± 0.4
Brown trout[c]		0.4 ± 0.2	–	0.5 ± 0.2

[a] Whole organism; [b] muscle tissue; [c] hepatic tissue.

a littoral pathway. Arsenic concentrations in fish were low. Creole perch showed the highest values (liver > muscle) among predator fish. This could be explained by differential diets, perch eating mainly benthic organisms while salmonid eating mainly small puyen (Juncos *et al.*, 2011).

3.3 *Arsenic distribution in biota after volcanic eruption*

Significant differences were found only for the *Aegla* sp. crab, the *Samastacus* sp. crayfish, the *Chilina* sp. snail, and the small puyen. [As] increased in muscle of *Aegla* sp. immediately after eruption (winter-spring 2011), and in hepatic tissue in summer 2011. Lower [As] were found in hepatic tissues of *Samastacus* sp. and *Chilina* sp. in summer-autumn 2012 (Table 1). Small puyen exhibited higher [As] immediately after eruption but lower concentrations the following summer. Changes in food availability and therefore in As availability, caused by habitat disruption after ash deposition, could explain some of the differences found. However, biota developmental stage, seasonal variability, and also As contents in water and sediments, must be taken into consideration for better descriptions.

4 CONCLUSIONS

Arsenic levels in biota from Nahuel Huapi Lake are similar to those found in uncontaminated systems. Besides high [As] in lower trophic levels, biodilution processes resulted in low [As] in fish, therefore not representing a risk to human health. The impact of volcanic products hardly shows up in most benthic organisms (e.g., *Chilina* sp., *Samastacus* sp. and *Aegla* sp.), reflecting the importance of sediments as inputs pathways of As to the food web.

ACKNOWLEDGEMENTS

This work was supported by the projects PICT 33838, Agencia Nacional de Promoción Científica y Tecnológica (Argentina), and IAEA TCP ARG7007.

REFERENCES

Chen, C.Y., Stemberg, R.S., Klaue, B., Blum, J.D., Pickhardt, P.C. & Folt, C.L. Accumulation of heavy metals in food web components across a gradient of lakes. *Limnology and Oceanography* 45(7):1525–1536.

Juncos, R., Beauchamp, D.A. & Vigliano, P.H. 2013. Modeling prey consumption by native and nonnative piscivorous fishes: implications for competition and impacts on shared prey in an ultraoligotrophic lake in Patagonia. *Transactions of the American Fisheries Society* 142:268–281.

Rahman, M.A., Hasegawa, H. & Lim, R.P. 2012. Bioaccumulation, biotransformation and trophic transfer of arsenic in the aquatic food chain. *Environmental Research* 116:118–135.

Revenga, J., Campbell, L.M., Arribére, M.A. & Ribeiro Guevara, S. 2012. Arsenic, cobalt and chromium food web biodilution in a Patagonia mountain lake. *Ecotoxicology and Environmental Safety* 81:1–10.

One Century of the Discovery of Arsenicosis in Latin America (1914–2014) – Litter, Nicolli, Meichtry, Quici, Bundschuh, Bhattacharya & Naidu (Eds)
© 2014 Taylor & Francis Group, London, ISBN 978-1-138-00141-1

Arsenic in different compartments of northwest Patagonian lakes, Argentina

A. Rizzo, R. Daga & R. Juncos
Laboratorio de Análisis por Activación Neutrónica, Centro Atómico Bariloche, CNEA, Argentina
CONICET, Argentina

M. Arcagni, M.A. Arribére & S. Ribeiro Guevara
Laboratorio de Análisis por Activación Neutrónica, Centro Atómico Bariloche, CNEA, Argentina

ABSTRACT: We report arsenic concentrations ([As]) in the principal components of the planktonic and benthic food web of Northern Patagonia Andes Range ultraoligotrophic lakes, Argentina. In addition, [As] were also determined in muscle of four fish species occupying the highest trophic position in six lakes belonging to Nahuel Huapi and Los Alerces National Parks. We also present [As] measured in sedimentary sequences from lakes located in both parks, with special attention to layers related to tephra deposits and layers from the second half of the 20th century.

1 INTRODUCTION

Arsenic (As) is a highly ubiquitous and potentially toxic element. Natural and anthropic sources are the cause of high arsenic concentrations in freshwaters, causing harmful effects to organisms by direct or indirect ingestion through the food chain pathways. Northern Patagonia is a region characterized by a great diversity of watersheds and glacial ultraoligotrophic lakes. Arsenic was studied in lakes from Nahuel Huapi (NHNP) and Los Alerces (LANP) National Parks. Due to its isolation and lack of industrial activities, these parks have been historically protected from direct anthropogenic contamination, and although for the past 40 years the population has increased in Bariloche city, reaching almost 140,000 inhabitants, the population density in the area remains low. This enabled us to analyze [As] in uncontaminated freshwater systems, but with inputs from natural sources due to the proximity of both national parks to the Southern Volcanic Zone (SVZ). The present work reports [As] determined in the principal food web components and As concentrations in sediment sequences from Nahuel Huapi, Moreno, Espejo Chico and Traful (NHNP), and Rivadavia, and Futalaufquen (LANP) lakes, reviewing previous works (Ribeiro Guevara *et al.*, 2004, 2005; Arribére *et al.*, 2010), and including new data.

2 METHODS

2.1 *Study site*

The studied lakes are located in NHNP (40°15'–41°34'S; 71°40'–72°54'W) and LANP, (42°35'–43°11'S; 71°35'–72°09'W), on the eastern side of the Andes Range, North Patagonia, close to the SVZ. The studied lakes, of glacial origin, are ultraoligotrophic, warm monomictic and are surrounded by native forest.

2.2 *Sample preparation and analytical procedures*

Large fish such as the introduced salmonids *Oncorhynchus mykiss* (Rainbow Trout), *Salmo trutta* (Brown Trout), and *Salvelinus fontinalis* (Brook Trout), and the native *Percichthys trucha* (Creole Perch) were captured using gill net gangs. Plankton samples were collected with nets of three different mesh sizes (10 (P1), 50 (P2), and 200 μm (P3)). Benthic macroinvertebrates such as insect larvae of Plecoptera (stonefly), Odonata (dragonfly), Ephemeroptera (mayfly), Trichoptera (caddisfly); gastropods *Chilina* sp., bivalves *Diplodon chilensis*, and hirudeans (leeches) were handpicked from submerged logs and stones, and around macrophytes; biofilm was obtained by scraping submerged stones. [As] were determined in muscle from fish and large macroinvertebrates and in whole body in the rest of the samples. Short sedimentary cores were recovered with a messenger-activated gravity type corer (Ribeiro Guevara *et al.*, 2005). Total [As] in biota and sediments were determined by Instrumental Neutron Activation Analysis (INAA) on freeze-dried samples (Arribére *et al.*, 2010).

3 RESULTS AND DISCUSSION

3.1 *Arsenic in plankton and in the benthic food web*

Total [As] in the invertebrate community vary within 2 orders of magnitude, from 0.43 to 5.70 μg g^{-1} dry weight (DW) in insect larvae, hirudeans and zooplankton (P3), to concentrations up to 39 μg g^{-1} DW in *D. chilensis*.

The high [As] observed in benthic (biofilm) and pelagic (P1 and P2) producers (Figure 1) show that [As] tends to decrease with increasing trophic level, as was observed in other studies (Rosso *et al.*, 2013).

Overall, comparing [As] distribution in the different species among lakes (Figure 1), it is clear that organisms from Moreno East, with longstanding and more populated human settlements, tend to have higher [As] than their counterparts in the less populated lakes.

3.2 *Arsenic in fish*

Total [As] determined in muscle from the four fish species occupying the highest trophic position are reported in Table 1 for the studied lakes. Total [As] in fish muscle do not show a clear distribution pattern among species, and it vary in a wide range, from less than 0.04 µg g^{-1} DW in Rainbow Trout from Lake Futalaufquen, to 1.62 µg g^{-1} DW in the same species from Lake Moreno East. These values were below the [As] at which negative effects appear in fish from polluted systems (2 to 5 µg g^{-1} wet weight), but are considerably higher than [As] found in fish from inhabited pristine aquatic ecosystems (0.005 to 0.201 µg g^{-1} DW) (Rosso et al. 2013).

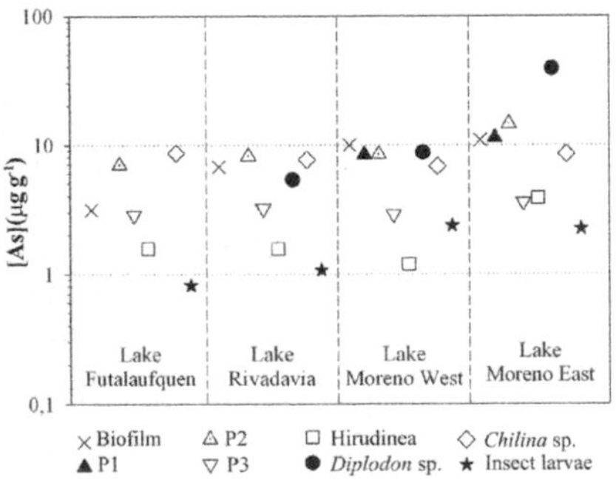

Figure 1. Arsenic concentrations in food web components in three lakes studied.

Table 1. Arsenic concentrations in muscle in four fish species.

	As (µg g^{-1} dry weight)			
Lake	Creole Perch	Brook trout	Rainbow trout	Brown trout
Traful	<1	<0.5	0.65	0.20–0.42
Espejo Chico	0.24	<0.7	0.25	0.33–1.09
Nahuel Huapi	0.36–0.66	<0.7	0.2–<1	0.17–0.62
Moreno West (MW)	0.26–0.56	0.12–0.93	0.12–0.41	<0.7
Moreno East (ME)	0.14–0.86	0.10–0.95	0.16–1.62	–
Rivadavia	0.17–0.53	0.20	0.06–0.26	0.15–0.23
Futalaufquen	0.21–0.48	0.13	0.04–0.17	0.14–0.21

3.3 *Arsenic in sediment cores*

In general, no evidence of a relevant [As] increase was observed in recent years in the sedimentary sequences (4.37 to 21.5 µg g^{-1}), and the higher [As] in sediment in the uppermost layers of some cores (74 to 250 µg g^{-1}) were associated with diffusive diagenetic processes, adsorbed onto particulate surfaces of Mn and Fe oxides (Ribeiro Guevara *et al.*, 2005). However, new data obtained from lake sequences recovered recently showed [As] increased in sediment associated with recent tephras. This was observed in sedimentary sequences from Futalaufquen Lake after the 2008 Chaitén eruption (45.5 µg g^{-1}), and in Nahuel Huapi Lake after the 2011 Cordón Caulle eruption (64.2 µg g^{-1}). In both cases, [As] increased up to 5 times compared with the previous ones immediately below the recent tephras deposited, as possible indicators of As derived from the volcanic activity.

4 CONCLUSIONS

Total [As] tends to decrease with increasing trophic level, with higher [As] in organisms sampled near places with more populated human settlements. [As] in fish were below toxic values. The proximity to the SVZ and the possible effects of volcanic eruptions on lake systems, together with increasing human population and its activities, makes necessary further understanding of processes determining As bioavailability from sediments as main reservoirs of elements.

ACKNOWLEDGEMENTS

This work was partially funded by the projects PICT 2005-33838, PICT 2006-1051, and PICT 2007-00393, Agencia Nacional de Promoción Científica y Tecnológica (Argentina), and by IAEA TCP ARG7007.

REFERENCES

Arribére, M.A., Campbell, L.M., Rizzo, A.P., Arcagni, M., Revenga, J. & Ribeiro Guevara, S. 2010. Trace elements in plankton, benthic organisms, and forage fish of Lake Moreno, Northern Patagonia, Argentina. *Water Air Soil Pollut.* 212: 167–182.

Ribeiro Guevara, S., Bubach, D.F., Vigliano, P.H., Lippolt, G. & Arribére, M.A. 2004. Heavy metals and other trace elements in native mussel Diplodon chilensis from Northern Patagonian lakes, Argentina. *Biol. Trace Elem. Res.* 102: 245–63.

Ribeiro Guevara, S., Rizzo, A., Sánchez, R. & Arribére, M.A. 2005. Heavy metal inputs in Northern Patagonia lakes from short sediment cores analysis. *J. Radioanal. Nucl. Chem.* 265: 481–493.

Rosso, J.J., Schenone, N.F., Pérez Carrera, A. & Fernández Cirelli, A. 2013. Concentration of arsenic in water, sediments and fish species from naturally contaminated rivers. *Environ Geochem Health* 35: 201–214.

One Century of the Discovery of Arsenicosis in Latin America (1914–2014) – Litter, Nicolli, Meichtry, Quici, Bundschuh, Bhattacharya & Naidu (Eds)
© 2014 Taylor & Francis Group, London, ISBN 978-1-138-00141-1

Arsenic, selenium and lead contamination from the waters in surface Itapessoca catchment, northeastern Brazil

B.M. Santos Pontes
PPG Minas, Pernambuco Federal University, Brazil

E. de Albuquerque Menor
Department Geology, Pernambuco Federal University, Brazil

J. Andrade Figueiredo
PPGeoc, Pernambuco Federal University, Brazil

ABSTRACT: Anomalous concentrations of arsenic, selenium and lead, from surface water are presented in this study, focusing the Itapessoca catchment situated at northeastern Brazil. This coastal catchment shows a low demographic occupation and low industrial development, but extensive sugar cane cultivation, shrimp farm, fish ponds, and a cement industry. Results pointed, as the major pollutant contaminations in arsenic, the shrimp farm (110x the recommended USEPA limit) and the fish ponds/cement industry (29x the recommended USEPA limit).These contaminations spread up to 3 km downstream the contaminant focus. The agricultural activities are responsible for moderate anomalous contaminations, not surpassing 2.5x the recommended USEPA limit. Evidence of contamination in arsenic and selenium found in this study calls into question the exploitation of seafood and shrimp in the estuarine compartment of the Itapessoca catchment, requiring investigation of potential health problems for human consumers.

1 INTRODUCTION

Arsenic in the environment appears largely disseminated since this element shows diversified chemical species. Because its toxicity and carcinogenic properties, it poses a significant human concern (Tsai *et al.*, 2003). High dissolved-As concentrations are primarily associated with groundwater (Kar *et al.*, 2011) forcing large populations worldwide to be in risk of toxic exposure from drinking water (Liu *et al.*, 2004). Additionally, adverse health effects have been reported by intake of fisheries from aquacultural ponds were waters are As-contaminated (Ling *et al.*, 2005). Kar *et al.* (2011) shows that As in pond waters and sediments are important exposure media due to its bioaccumulation in fish and shrimp.

Arsenic concentrations in natural waters are large, ranging from <0.5 to >5000 $\mu g\ L^{-1}$ (Smedley & Kinniburgh, 2002). Surface waters with "abnormal" concentrations are a serious environmental problem upon the transfer of this contamination to groundwater, especially when these reservoirs are exploited for drinking water. USA river baselines show low As concentrations (< 2.5 $\mu g\ L^{-1}$) if their courses crosses rural areas or domains without industrial activities (Waslenchuk, 1979); several European catchments display identical conditions (Pettine *et al.*, 1992).

This study makes a hydrogeochemical approach about expected environmental As-contamination provoked by sugar cane cultivations, shrimp farm, fish ponds, and a cement industry, all of them included in a tropical catchment (Itapessoca River), situated at Pernambuco State, northeastern Brazil. This catchment is characterized by general low population density and by intensive exploitation of fisheries and sea foods. Surface waters (30 cm depth) were collected from some stations located in the fluvial and estuarine courses of this river, and As content was analyzed in samples of this water.

2 METHODS

Water samples were filtered through 0.45 μm memebranes, stabilized at acidic conditions (pH ≤ 2), cooled and maintained at 4 ^{o}C before being sent to Actlab's (Canada) for chemical analyses. Water samples were analyzed with Perkin Elmer Sciex ELAN 9000 ICP/MS.

Sampling was performed in dry season, into a time interval of 35 days, always near minimum low tidal, with medium sea level between 0–0.1 m. The water pH range during sampling was 6.6–7.1 in the fluvial course and 7.3–7.6 in the estuarine compartment; BOD_5 was ≤ 2.0 in all cases.

3 RESULTS AND DISCUSSION

The chemical As–Se–Pb data (Table 1) were compared with standards for drinking water and brackish water from USEPA (2002).

The analytical results pointed out two main focuses of contamination (Figure 1). The first one is linked to a shrimp farm, and responsible for anomalous concentrations of arsenic and selenium; the second one, to the set cement industry/fish ponds, and responsible for high Se-As-Pb concentrations. A third focus refers to an isolated Pb anomalous value linked to a headwater of the catchment, near an area crossed by highways.

The main arsenic contamination (110x the USEPA pattern) spreads from the shrimp farm, decreasing its concentration to downstream waters. However, despite a dilution effect at the aquatic media, the arsenic concentration still 10x the USEPA limit 3 km away from the shrimp farm. In contrast, some regional use of fertilizers and pesticides in the agricultural activities resulted in relatively low As-anomalies reaching 2.5x the USEPA tolerance patterns. Selenium and arsenic (respectively 2–11x and 4.5–28x the USEPA pattern) show contaminated values at least 3 km far away from the focus composed by cement industry/fish ponds.

Table 1. Summary of chemical results—surface waters: Itapessoca catchment.

Sampling stations		Arsenic ($\mu g\ L^{-1}$)	Lead ($\mu g\ L^{-1}$)	Selenium ($\mu g\ L^{-1}$)
Fluvial domain	B1	0.12	12	0.9
	B2	0.11	0.29	1.7
	B3	0.18	0.82	0.9
	B4	0.35	1.5	0.3
	B5	0.15	0.81	0.1
	B13	0.25	0.4	0.25
Estuarine domain	B6	11.28	0.4	106
	B7	ND	0.56	2
	B8	ND	7.42	25.2
	B9	0.30	20.2	21
	B10	1.75	1.13	104
	B11	4.07	0.48	111
	B14	15.51	ND	108
USEPA		0.14	10	10

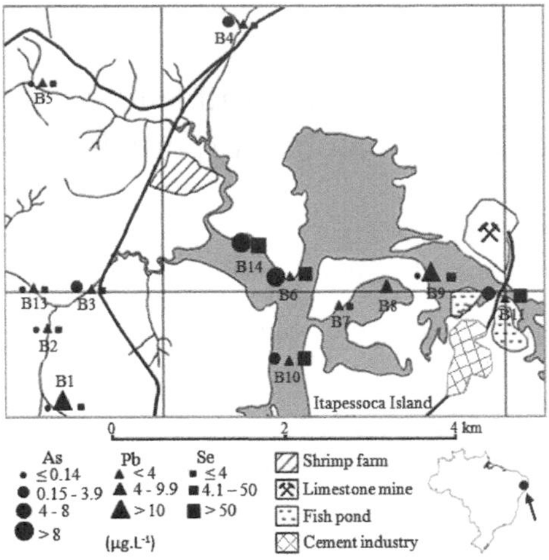

Figure 1. As-Se-Pb concentrations in surface waters, Itapessoca River, NE, Brazil.

4 CONCLUSIONS

Degradation of coastal environments through aquatic contamination has been recognized worldwide due to persistent and accumulative threat from pathogenic chemical species. The contemporary problems linked to environmental contaminations are nearly all of anthropic origin. In this context, the recorded As-Se-Pb data from surface waters of the Itapessoca coastal catchment show that shrimp farms and fish ponds—an investment very profitable and multiplicative in tropical countries—are among the most important anthropic sources of arsenic contamination. In the investigated catchment, plumes of As-contaminated waters from shrimp farms and from fish ponds can reach more than 3 km downstream from the contaminant focus. In contrast, the use of fertilizers and pesticides in local agricultural activities show low anomalies in arsenic not surpassing 2.5x the USEPA tolerance patterns. Lead contamination can be a consequence of the resurgence of groundwaters from areas crossed by highways.

Evidence of contamination of arsenic and selenium found in this study calls into question the exploitation of seafood and shrimp in the estuarine compartment of the Itapessoca catchment, requiring investigation of potential health problems for human consumers.

REFERENCES

Kar, S., Maity, J.P., Jean, J.S., Liu, C.C., Nath, B., Yang, H.J. & Bundschuh, J. 2010. Arsenic-enriched aquifers: Occurrences and mobilization of arsenic in groundwater of

Ganges Delta Plain, Barasat, West Bengal, India. *App. Geochem.* 25: 1805–1814.

Kar, S., Maity, J.P., Jean, J.S., Liu, C.C., Liu, C.W., Bundschuh, J. & Lu, H.Y. 2011. Health risks for human intake of aquacultural fish: Arsenic bioaccumulation and contamination. *J. Environ. Sci. Health* 46: 1266–1273.

Levink, K., Steinnes, E. & Pappas, A.C. 1978. Contents of some heavy metal in Norwegian rivers. *Nord. Hydrology* 9: 197–206.

Ling, M.P., Liao, C.M., Tsai, J.W. & Chen, B.C. 2005. A PBPK/PD modeling-based approach can assess arsenic bioaccumulation in farmed tilapia (*Oreochromis mossambicus*) and human health risks. *Integr. Environ. Assess. Manag.* 1: 40–54.

Liu, C.W., Jang, C.S. & Liao, C.M. 2004. Evaluation of arsenic contamination potential using indicator kringing in the Yun-Lin aquifer (Taiwan). *Sci. Tot. Environ.* 321: 173–188.

Pettine, M., Camusso, M. & Martinotti, W. 1992. Dissolved and particulate transport of arsenic and chromium in the Po River, Italy. *Sci. Tot. Environ.* 119: 253–280.

Tsai, S.Y., Chou, H.Y., The, H.W., Chen, C.M. & Chen, C.J. 2003. The effects of chronic arsenic exposure from drinking water on the neurobehavioral development in adolescence. *Neurotoxicology* 24: 747–753.

USEPA. 2002. National recommended water quality criteria. *U.S. Environmental Protection Agency—Office of Water,* EPA-822-R-02-047.

Waslenchuk, D.G. 1979. The geochemical controls on arsenic concentrations in southeastern United States rivers. *Chemical Geology*, 24: 315–325.

One Century of the Discovery of Arsenicosis in Latin America (1914–2014) – Litter, Nicolli, Meichtry, Quici, Bundschuh, Bhattacharya & Naidu (Eds)

Diurnal geochemical changes at Champagne Pool, Waiotapu, New Zealand

J. Pope
CRL Energy Ltd., Christchurch, New Zealand

N. Wilson
Golder Associates (NZ) Limited, Takapuna, New Zealand

ABSTRACT: Champagne Pool discharges reduced neutral chloride fluids in the Waiotapu geothermal area of New Zealand. Diurnal geochemical cycles strongly influence the concentration and speciation of trace elements including arsenic in the Champagne Pool discharge zone. Arsenic, antimony, and thallium are enriched in sulfide precipitates that form within the spring and are transported onto the surrounding sinter area. When the fluid cools, oxygen is introduced by cyanobacteria and algae photosynthesis. The input of oxygen and the resulting redox shift destabilizes the sulfide precipitates and releases increased concentrations of arsenic, antimony, and thallium downstream during the day. We present trace element concentrations from within Champagne Pool and from ~150 m downstream of the spring that demonstrate strong diurnal cycles in concentration. Arsenic speciation within Champagne Pool also varies on a diurnal cycle, with maxima of oxy-arsenite and dithioarsenate, compared to trithioarsenate, occurring during the day. The diurnal trend within Champagne Pool appears to be the product of photochemical processes because photosynthesizing micro-organisms cannot survive in the spring. Downstream of the spring daytime maxima of arsenate compared to arsenite occurs. Diurnal trends in the concentration of trace elements in the discharge are likely to relate to complex trends changes in sulfur species concentrations rather than the direct oxidation of sulfide precipitates by dissolved oxygen.

1 INTRODUCTION

Champagne Pool is a spring that discharges ~20 L/s of reduced, neutral chloride fluid at ~70°C in the Waiotapu geothermal area of New Zealand. The spring occupies a roughly circular 70 m diameter hydrothermal eruption crater and is famous for deposition of gold rich sulfide precipitates in the sinter zone surrounding the pool (Weisberg 1969). These precipitates form a suspended floc within the spring and are transported onto the sinter zone around the spring in solid form. Champagne pool is a modern and active analogue for sinter style hydrothermal ore deposits (Hedenquist 1983, 1985).

The geothermal fluid discharged at the surface is related to the reservoir fluid through boiling with little conductive cooling or dilution. The fluid contains elevated concentrations of typical conservative geothermal solutes including Cl, B, Li, Na and others (Hedenquist 1991). The fluid contains dissolved gases, hydrogen sulfide and carbon dioxide (Giggenbach 2004). Trace element concentrations are also enriched and include arsenic, antimony, thallium (Pope *et al.*, 2004) and gold (Pope *et al.*, 2005).

The geochemistry of the discharge zone at Champagne Pool changes systematically throughout the diurnal cycle. These changes relate to oxygen input by organisms that photosynthesize including cyano-bacteria and algae (Jones 1999, Renault, 1999). The oxygen destabilizes sulfide precipitates releasing elevated concentrations of arsenic, antimony, and thallium and other trace elements during the day. The sulfide oxidation processes also releases, higher concentrations of sulfate and decreases pH during the day compared to night in the Champagne Pool drainage zone. At night these sulfide precipitates remain stable within the Champagne Pool discharge zone. Other photochemical or biogeochemical reactions are also possible in the catchment because many of the solid and dissolved species have multiple oxidation states.

We present additional data on diurnal geochemical processes at Champagne Pool including previously unpublished trace element data indicating cycles throughout a diurnal sampling period as well as speciation information for arsenic and sulfur throughout a dawn to dusk sampling period (Wilson 2009).

2 METHODS

Samples for trace element analysis were collected every 4 hours throughout a diurnal cycle at Champagne Pool and about 150 m downstream of the pool in 1999. At the time sampling was completed daylight hours were from about 7 am to 9 pm. Subsequently, samples were collected at a 5 sites within the catchment zone every two hours (Pope 2004). Dawn to dusk sampling along the same drainage channel was conducted in June 2007 (Wilson, 2009), and additional samples from Champagne Pool were collected in January 2011. The discharge was dry in January 2011so no downstream samples could be collected.

Filtered (0.45 μm) and non-filtered samples were collected into acid washed, sample rinsed containers in 1999 and 2007, in January 2011samples were filtered to 0.20 μm. Samples were not acid preserved because mild acidification ($pH < 2$) commonly used for sample preservation lead to formation of precipitates in the samples. Measurement of pH and electrical conductivity were made at the time of sampling as were sulfide concentrations in June 2007 and January 2011 (methylene blue method).

Trace element analysis was completed by ICP-MS. Arsenic speciation for samples collected in June 2007 were measured using HG-AAS; As speciation in samples collected in January 2011 were measured by AEC-ICP-MS.

3 RESULTS

Concentrations of dissolved components within and downstream of Champagne Pool are similar in magnitude to those published previously (Gigenbach 1994, Pope 2004, Pope 2005). During the diurnal sampling programme completed in 1999 concentration of trace elements at Champagne Pool did not vary systematically throughout the diurnal cycle. At the sample site 150 m downstream of Champagne Pool, concentrations of several trace elements including arsenic were elevated during the daylight hours compared to night (Figure 1a,b & c).

Dissolved arsenic concentrations in Champagne Pool are dominated by arsenite (As(III)) and thio-arsenate species (Figure 2) with a midday minimum of trithioarsenate and maximum of dithioarsenate and oxy-arsenite. In samples collected ~25 m downstream of the discharge in June 2007, the relative abundance of oxidised arsenic species increased until early afternoon as observed by arsenate abundance (Figure 3) with similar increase in total dissolved arsenic (Figure 1). The ratio of arsenic: lithium (Figure 4) had a similar trend to total concentration of arsenic over this time (Wilson 2009).

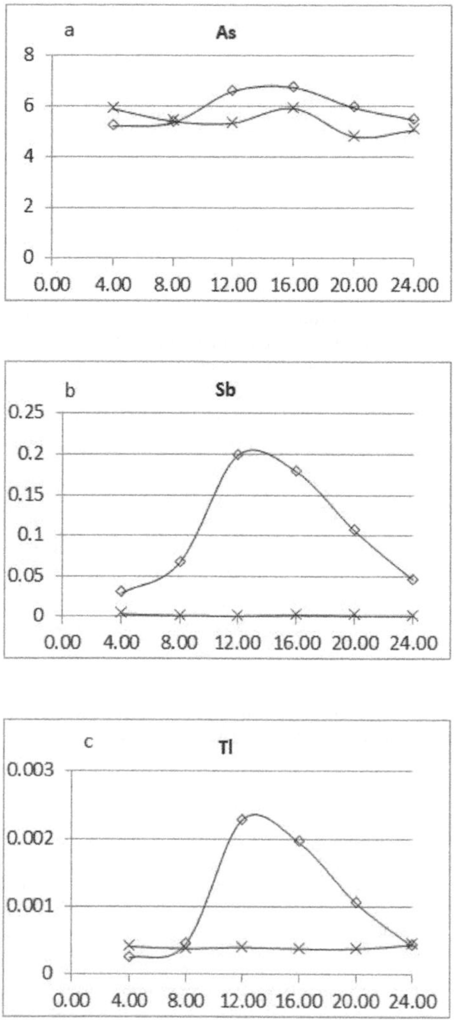

Figure 1a,b & c. Arsenic (As), antimony (Sb) and thallium (Tl) concentrations (mg/L) within Champagne Pool (×) and about 150 m downstream (◊) throughout a diurnal cycle.

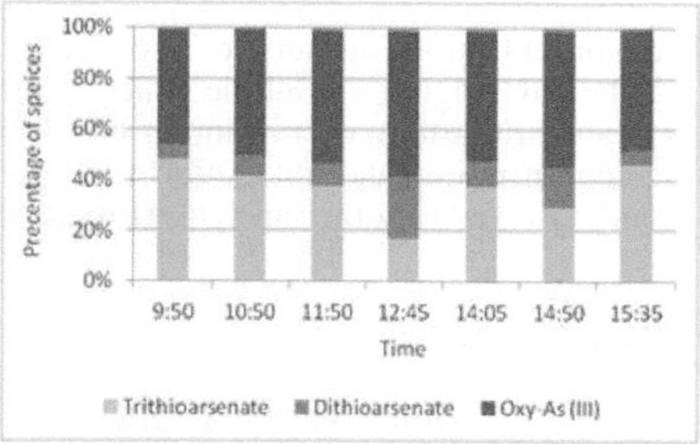

Figure 2. Arsenic speciation within Champagne Pool from dawn to dusk.

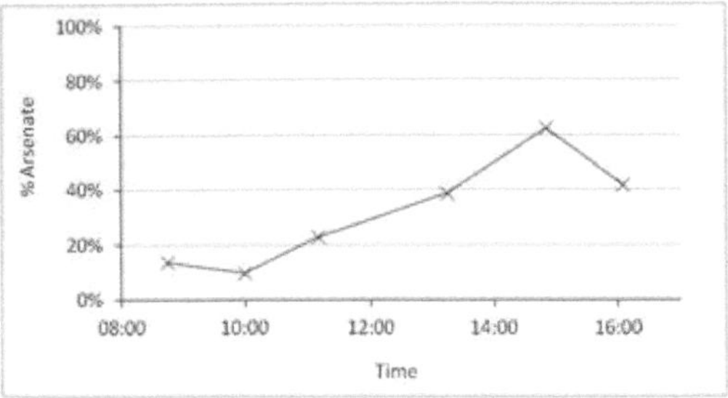

Figure 3. Abundance of arsenate downstream of Champagne Pool between dawn and dusk.

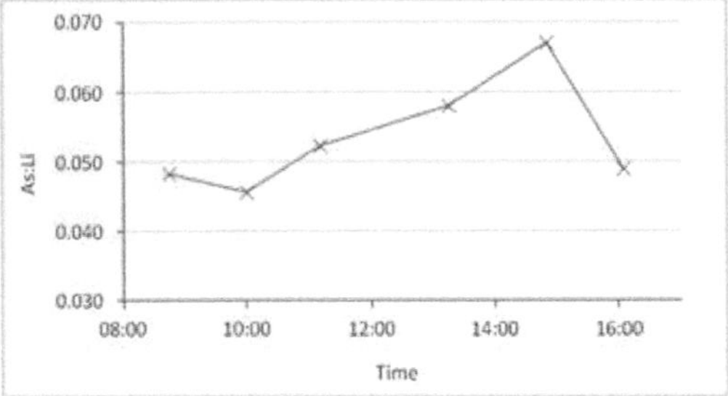

Figure 4. Arsenic: lithium (As:Li) ratio downstream of Champagne Pool between dawn and dusk.

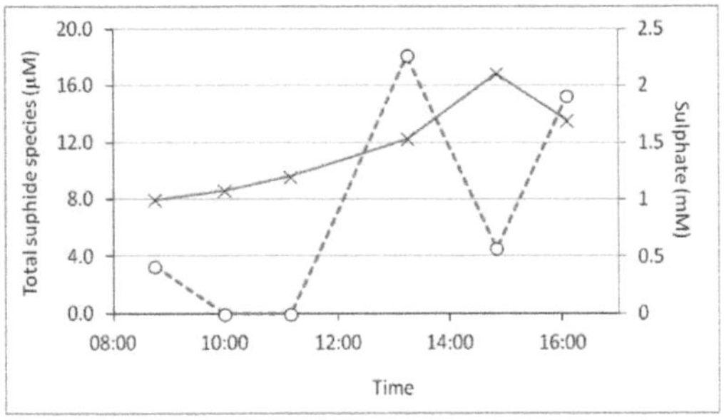

Figure 5. Sulfide and sulfate concentrations downstream of Champagne Pool between dawn and dusk.

The concentration of sulfate also increased during the day (Figure 5). Sulfide concentrations reached an early afternoon maximum but overall sulfide concentrations did not exhibit the systematic trend observed for other analytes (Figure 5).

4 DISCUSSION

Neutral chloride geothermal fluids discharged at Champagne Pool are enriched in trace elements. The trace elements are transported by the geothermal fluid onto a sinter zone around the pool in dissolved form and in solid form as a sulfide precipitate floc.

Within the spring, dissolved absolute concentrations of trace elements do not change in in a systematic manner during diurnal cycle. However arsenic speciation changes slightly with maximum oxy-arsenite and dithioarsenate during daylight hours and most direct sunlight.

There is no evidence that daytime variations in arsenic concentrations are a function of physical processes such as evaporation. If evaporation is a principal mechanism, the ratio of arsenic to conservative elements such as lithium would be stable. Instead arsenic:lithium ratios have similar daytime changes to total concentrations.

We propose that on the sinter terrace downstream of the pool biological input of dissolved oxygen by cyano-bacteria and algae destabilise sulfide precipitates and release elevated concentrations of trace elements during the day. Trace elements that are most enriched in the sulfide precipitate, including arsenic, antimony and thallium have the largest daytime increases in concentration. Other trace elements, sulfate, pH, dissolved oxygen and alkalinity also have diurnal changes (Pope 2004).

There is little evidence for direct photosynthetically-driven bacterial transformations (Wilson 2009). Rather, these increases in concentration may be caused by changes in sulfur species equilibrium that occur with oxygen input into an otherwise reducing environment.

The shift from a reducing to an oxidizing system is also reflected in changes in arsenic species distribution. Champagne Pool is a mix of reduced oxy-arsenic species (arsenite) and thioarsenate species, which can only form in the presence of dissolved sulfide (Planer-Friedrich *et al.*, 2007). Downstream, the relative contribution of these species, which are all measured as As(III) using HG-AAS (Planer-Friedrich & Wallschläger 2009), decrease as total arsenic concentrations increase. The oxidation of these arsenic species increases overall arsenic mobility within the system, and means night time re-precipitation of arsenic mobilized during the day is less likely than if these species had not oxidised.

5 SUMMARY

We present previously un-published diurnal concentration trends for arsenic, antimony and thallium in the Champagne Pool drainage zone. These results support previous interpretations of diurnal changes in concentration downstream of Champagne Pool. Addition of oxygen by photosynthesis is a driving factor in these processes.

In addition, new diurnal geochemical trends have been identified, including daytime maxima in arsenite and dithioarsenate compared to trithioarsenate within Champagne Pool. These changes cannot be caused by photosynthesis, because cyanobacteria and algae cannot survive at 70°C; these microbes can only be present on the sinter around the pool at lower temperatures. It is likely that the trend observed within Champagne Pool is caused by direct photochemical processes, rather than biotic processes.

There is an increase in arsenate compared to arsenite during the day downstream of Champagne Pool and a similar trend in the concentration of sulfate downstream of the spring. Changes in arsenic and sulfur speciation identified in this study indicate that the reaction mechanisms causing diurnal concentration variations are likely to be more complex that simple oxidation of sulfide precipitates by dissolved oxygen. Instead the oxidation processes are likely to be stepwise with different partially oxidized species and some of which could be significant in the mobilization of arsenic and other trace elements in the stream system draining the geothermal area.

REFERENCES

Giggenbach, W.F., Sheppard, D.S., Robinson, B.W., Stewart, M.K., and Lyon, G.L., 1994, Geochemical structure and position of the Waiotapu Geothermal Field, New Zealand: Geothermics, v. 23, p. 599–644.

Hedenquist, J.W., 1983, Waiotapu, New Zealand: The geochemical evolution and mineralisation of an active hydrothermal system. [Ph.D thesis], University of Auckland.

Hedenquist, J., 1991, Boiling and dilution in the shallow portion of the Waiotapu geothermal system, New Zealand.: Geochimica et Cosmochimica Acta, v. 55, p. 2753–2765.

Hedenquist, J.W., and Henley, R.W., 1985, Hydrothermal eruptions in the Waiotapu geothermal system, New Zealand: Their origin, associated breccias, and relation to precious metal mineralisation: Economic Geology, v. 80, p. 1640–1669.

Jones, B., Renaut, R.W., and Rosen, M.R., 1997, Vertical zonation of biota in microstromatolites associated with hotsprings, North Island, New Zealand.: Palaios, v. 12, p. 220–236.

Planer-Friedrich B, Wallschläger D. 2009. A critical investigation of hydride generation-based arsenic speciation in sulfidic waters. Environmental Science & Technology 43: 5007–5013.

Planer-Friedrich B, London J, McCleskey R.B, Nordstrom DK, Wallschläger D. 2007. Thioarsenates in geothermal waters of Yellowstone National Park determination, preservation, and geochemical importance. Environmental Science & Technology 41: 5245–5251.

Pope, J.G.M., D.; Clark, M.W.; Brown, K.L. (2004). Diurnal variations in the chemistry of geothermal fluids after discharge, Champagne Pool, Waiotapu, New Zealand. Chemical Geology 203: 253–272.

Pope, J., K.L. Brown, et al. (2005). Gold Concentrations in Springs at Waiotapu, New Zealand: Implications for Precious Metal Deposition in Geothermal Systems. Economic Geology 100: 677–687.

Renaut, R.W., Jones, B., and Rosen, M.R., 1999, Role of microbes in the formation of gold-bearing siliceous sinter at Champagne Pool, Waiotapu, New Zealand, Geological Society of America 1999 annual meeting, Volume 31, Geological Society of America, p. 392.

Weissberg, B.G., 1969, Gold-silver ore grade precipitates from New Zealand geothermal waters: Economic Geology, v. 64, p. 95–108.

Wilson N.J. 2009. The behaviour of antimony in geothermal systems and their receiving environments. Unpublished thesis, Univeristy of Auckland, Auckland. 235 p.

One Century of the Discovery of Arsenicosis in Latin America (1914–2014) – Litter, Nicolli, Meichtry, Quici, Bundschuh, Bhattacharya & Naidu (Eds)
 ISBN 978-1-138-00141-1

Arsenic occurrence in Brazilian tropical beaches

N. Mirlean & P. Baisch
Federal University of Rio Grande, Brazil

ABSTRACT: High concentrations of As exceeding the environmentally acceptable thresholds were found in the beach sands of Brazilian coast in the states of Bahia and Espirito Santo. The high As concentration is related with calcium carbonate and iron in the beach sands. Calcareous bioclasts participate in As retention in marine environments and its consequent accumulation in beach deposits. Presumably, carbonate bioclast gains the high metalloid content in the surface oxic horizon of sediment during the early diagenetic redistribution of Fe and As. Cliffs and terraces of Barreiras-group sandstones in the near-shore zone of Bahia and Espirito Santo play an important role in the development of the calcareous algae *Corallina panizzoi*. The skeletal parts of this algal species constitute the principal carbonate material that supports the enrichment of the beaches with As.

1 INTRODUCTION

The Brazilian environmental act on marine sediment quality control (CONAMA, 2012) has increased the interest on arsenic distribution in the coastal zone of Brazil, principally due to the environmental limitations that are imposed on dredging operations in port areas. Our pilot assessment of the beach- and near-shore-sediment compositions in Santa Cruz—ES, Brazil in 2008 revealed high levels of As (i.e., above the probable effect level threshold) (Mirlean *et al.*, 2011). We supposed that the peculiarities of As geochemistry and accumulation in the surf sediments of this part of Espirito Santo state may be present in the whole coastline of the state and adjacent state of Bahia due to the analogous geological, geochemical, and tropical biological patterns throughout this region. The general objective of this study is the determination of As distribution in beach sediments along the 1,200,000 km segment of Brazilian coastline. The specific objective is demonstrating the involvement of calcareous bioclast in beach sands in the enrichment of As through diagenetic redistributions of chemical elements.

2 METHODS/EXPERIMENTAL

2.1 *Studied area and sampling*

The study area—the coastline of Brazil between 11°30′ S and 21°17′ S—included the coastlines of the two Brazilian states: Bahia and Espirito Santo. The studied coastline mainly consists of continuous strings of sandy beaches without significant human occupancy or industrial activities. All of the samples used in the current study were collected in 2011–2012. Sand samples were collected from 27 beaches along the coastline of the study area from the surface layer using a plastic shovel and were placed in plastic bags. The sampling points were selected in "clean" beaches, away from the influence of industry and densely populated areas.

2.2 *Chemical analyses*

Arsenic was determined by electrothermal atomic absorption spectrometry that employed a Perkin-Elmer 800 instrument equipped with a Zeeman background corrector using pyrolytically-coated tubes with a platform. To use a higher ashing temperature, $Pd(NO_3)_2$ and $Mg(NO_3)_2$ matrix modifiers were used. Calcium and Fe levels were determined using flame (acetylene—air) atomic-absorption spectrometry. The addition of lanthanum chloride solution was performed for Ca determination to suppress the interference of elements forming stable oxyanions, according to EPA-7140 method (USEPA, 1986). The detection limits were calculated according to IUPAC recommendations (1994).

3 RESULTS AND DISCUSSION

3.1 *As, Fe and Ca concentrations in beach sands*

The total As concentration in the sands of the studied beaches ranged from 1.1 mg kg^{-1} to 119.6 mg kg^{-1}. The average As content in the beach sediments was 18.1 mg kg^{-1}. The Threshold Effect

Level [TEL]: 19 mg kg^{-1} for As was exceeded in 33% of the total studied beach-sand samples, and in two cases, the Probable Effect Level [PEL]: 70 mg kg^{-1} was also exceeded. About 60% of the sands in the studied beaches had a low (< 0.5%) concentration of iron, and it surpassed 1% in only four samples. The $CaCO_3$ content in the beach deposits exhibited a broad range from very low values up to about 20%. The interrelation among the three studied elements demonstrated very strong positive correlation between As and Ca ($R = 0.86$, $n = 27$, $p < 0.01$), a low significant Fe/As correlation ($R = 0.39$, $n = 27$, $p < 0.01$) and between Ca and Fe ($R = 0.35$, $n = 27$, $p < 0.01$).

The distribution of As in the beach sands along the studied Brazilian coastline demonstrated the presence of segments with different metalloid concentrations. The northern beaches of Bahia state had the lowest As concentrations, generally less than 7 mg kg^{-1}. The beaches of the central Bahia states displayed relatively elevated concentrations of As; in some sampling points, these concentrations approximated or exceeded the TEL of the metalloid. The next segment of the Espirito Santo state coastline showed high and very high concentrations of As in beach sands, which sometimes exceeded the TEL by several fold and approximated or even surpassed the PEL for the studied metalloid.

3.2 *The geochemical process of beach sands enrichment by As*

The enrichment of beach sands with As appears to be a successive result of the diagenetic redistribution of this metalloid in near-shore sediment. We suppose that calcareous clastic material, which contains the highest As concentrations in beach sands, is the material that previously composed the upper oxidized layer of near-shore sediment. This clastic material forms in the shelf zone because of the death of calcareous organisms. Because calcareous detritus is lighter than quartz, the skeleton particles of calcareous algae and corals compose the uppermost stratum of the local sediments. The deposition of Fe(III) hydroxides occurs in this stratum, and these materials in turn adsorb the As(III) and/or As(V) ions diffusing upward from the suboxic sediment horizon. The surface of calcareous particles serves as the substrate for the fixation of Fe(III) hydroxides enriched by As. This process is clearly observed in Espirito Santo state where the beach sands, which are predominantly composed of calcareous algae detritus, are yellow to brown. The distribution of detritus of calcareous algae *Corallina panizzoi* in studied beaches is in high correlation with As concentration in beach sands ($R = 0.77$, $n = 27$, $p < 0.01$). Carbonate particles from the upper horizon of near-shore sediments, due to their light weight, are transported over time to the coast by waves and form As-rich beaches.

This mechanism of beach-sand enrichment with As, is most clearly evident in the southern part of the studied Brazilian coastline and is not apparent in the northern portion of Bahia state coast line, where the beach sands are also rich in calcareous detritus but do not exhibit high As concentrations. The carbonate material in the beach sands from the northern part of Bahia state consists of coral detritus and is predominantly uncolored by Fe(III) hydroxides. Most likely, this clastic material has not been in contact with the suboxic/oxic zone of marine sediment, where it could be significantly enriched with Fe and As. Normally, in coral-reef zones, the upper sediment horizon, which is composed of coarse-grained coral detritus, is quite thick, and Fe(III) hydroxide deposition occurs in the lower horizon, which does not participate in the transport of material to the beach. In addition to the difference in the oxic zone, the differences in the properties of the carbonate material of corals and carbonate algae, such as texture, density, porosity, resistance to attrition, etc., might be responsible for Fe/As accumulation and retention. We also assume that calcium carbonate itself may be directly involved in As retention (Romero *et al.*, 2004; Mirlean *et al.*, 2011). These aspects of the As distribution along coasts that are rich in calcareous detritus is the topic of a separate study.

4 CONCLUSION

Beach sediments in various segments of the coast of Bahia and Espirito Santo states are anomalously enriched with arsenic, exceeding the legislative environmental thresholds. Generally, the As concentration in beach sediments increases from the northern part of Bahia state to the south, achieving its maximum in the beaches of Espirito Santo state. The As content in beach sands correlates well with carbonate. We believe that the enrichment of beach sands with As is a successive result of the diagenetic redistribution of metalloid in near-shore sediments. The calcareous material is a depositor and transporter of iron hydroxides enriched in As from the oxic layer of surface sediments toward a beach. This process occurs most obviously due to the involvement of calcareous algae clastic material. Considering the large geographical scale of the described As enrichment phenomenon, it is possible to assume the existence of similar As-rich beaches in other parts of the world.

ACKNOWLEDGEMENTS

The research was funded by CNPq-Brazilian National Research Council through grant 470541/2010-5.

REFERENCES

CONAMA Ministério do Meio Ambiente, Conselho Nacional do Meio Ambiente. *Resolução nº 454, de 01 de novembro de 2012* (in Portuguese).

IUPAC, 1994. Analytical Methods Committee. *Analyst* 119: 16–32.

Mirlean, N., Baisch, P., Travassos, M.P. & Nassar, C. 2011. Calcareous algae bioclast contribution to sediment enrichment by arsenic on the Brazilian subtropical coast. *Geo-Marine Letters* 31: 65–73.

Romero, F.M., Armienta, A. & Carillo-Chavez, A. 2004. Arsenic sorption by carbonate-rich aquifer, a control on arsenic mobility at Zimapán, Mexico. *Archives of Environmental Contamination and Toxicology* 47: 1–13.

USEPA 1986. Test Methods for Evaluating Solid Waste. *National Technical Information Service*, Springfield.

One Century of the Discovery of Arsenicosis in Latin America (1914–2014) – Litter, Nicolli, Meichtry, Quici, Bundschuh, Bhattacharya & Naidu (Eds)

Hourly concentrations of arsenic associated with Particulate Matter

T. Ancelet, P.K. Davy & W.J. Trompetter
GNS Science, Lower Hutt, New Zealand

ABSTRACT: During field studies where Particulate Matter (PM) samples were collected on hourly time-scales, a surprising result was that concentrations of arsenic associated with PM showed distinct diurnal cycles, with peak concentrations occurring in the evening and the morning. PM emissions from the combustion of copper chrome arsenate-treated timber were identified as the source of elevated arsenic concentrations. Here we present how arsenic concentrations at a number of urban locations throughout New Zealand vary on hourly time-scales and the effect that short, extremely high arsenic concentration events can have on meeting health-based guideline values.

1 INTRODUCTION

Particulate matter concentrations in New Zealand urban environments have been shown to have distinct diurnal cycles, independent of community size or population (Trompetter *et al.*, 2010), with peak PM_{10} concentrations occurring between 10 pm–midnight and 8–11 am. To understand the sources and factors contributing to these elevated PM concentrations, we undertook intensive monitoring campaigns in a number of urban locations throughout New Zealand where PM samples were collected on an hourly time-scale. Wood combustion for home heating was identified as the dominant PM source in each location (Ancelet *et al.*, 2012; Ancelet *et al.*, 2013). Interestingly, in each location, concentrations of arsenic associated with PM were high and this work presents the source of elevated arsenic concentrations, how arsenic concentrations vary over a day and how extremely high concentration, short duration arsenic pollution events can impact on daily and annual average arsenic concentrations.

2 EXPERIMENTAL

2.1 *Particulate Matter sampling*

Ambient air monitoring was conducted in four urban areas across New Zealand between 2010 and 2012 (Figure 1). Full descriptions of the sampling setups have been reported (Ancelet et al., 2012; 2013). Briefly, in each urban area, hourly PM samples were collected using a Streaker sampler (PIXE International Corporation, USA) and associated hourly PM masses were determined using beta attenuation monitors (E-BAM; Met One Instruments, Inc.). In total, more than 13,000 samples were collected.

2.2 *Sample analysis and source identification*

Ion Beam Analysis (IBA) techniques were used to measure the concentrations of elements with atomic numbers above neon in the hourly PM samples. Full details of the analysis methodologies have been reported (Trompetter *et al.*, 2005). Black carbon, an indicator of combustion, was measured using light reflection (Ancelet *et al.*, 2011).

Identification of the sources contributing to PM was performed using Positive Matrix Factorization (PMF). With PMF, sources are constrained to have

Figure 1. Location of sampling sites throughout New Zealand.

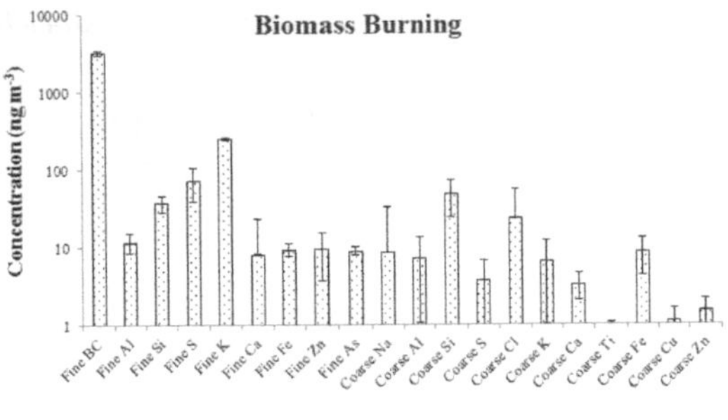

Figure 2. Typical wood combustion source profile showing the association of arsenic.

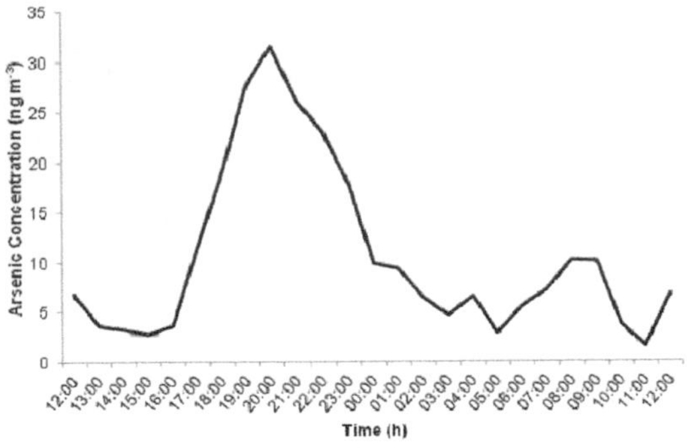

Figure 3. Average hourly arsenic concentrations.

non-negative species concentrations, no sample can have a negative source contribution and error estimates for each observed point are used as point-by-point weights. This is a distinct advantage of PMF, since it can accommodate missing or below detection limit data that is a common feature of environmental monitoring (Song *et al.*, 2001). Data screening and source apportionments were performed as reported (Ancelet *et al.*, 2012).

3 RESULTS AND DISCUSSION

Each of the four urban areas studied were found to have a number of PM sources, including wood combustion for home heating, vehicles, sea salt and soil. Critically, in each urban location, elevated arsenic concentrations were associated with the wood combustion source, indicating that at least some homes were using Copper Chrome Arsenate (CCA)-treated timber as fuel for domestic heating purposes. CCA-treated timber is no longer commonly used in other parts of the world, but it is frequently used for construction purposes in New Zealand to prevent fungal and insect attacks. Figure 2 presents a typical wood combustion source profile.

In each urban location, hourly average arsenic concentrations during peak pollution events were in excess of 200 ng/m^3. It is likely that the use of CCA-treated timber as a fuel for domestic heating is intermittent and opportunistic, because it would be prohibitively expensive to burn only treated timber.

Average hourly concentrations of arsenic in each urban area displayed the same diurnal pattern as PM concentrations, with a large peak in the evening and smaller peak in the morning, as shown in Figure 3. The potential health effects associated with this type of exposure to arsenic are unknown.

4 CONCLUSIONS

We have identified that the combustion of CCA-treated timber is responsible for elevated arsenic concentrations in a number of New Zealand urban environments. Peak hourly arsenic concentrations were regularly above 200 ng/m^3 and the diurnal cycles of arsenic concentrations were identical to those of PM. This is the first example of arsenic concentrations associated with PM on an hourly time-scale.

ACKNOWLEDGEMENTS

The authors thank the local councils involved in these studies (Greater Wellington Regional Council, Otago Regional Council, Nelson City Council and Auckland Council). Bruce Crothers, Ed Hutchinson and Chris Purcell are thanked for their support in sampling, setup and filter analysis.

REFERENCES

Ancelet, T., Davy, P.K., Mitchell, T., Trompetter, W.J., Markwitz, A. & Weatherburn, D.C. 2012. *Environmental Science and Technology* 46: 4767–4774.

Ancelet, T., Davy, P.K., Trompetter, W.J., Markwitz, A. & Weatherburn, D.C. 2011. Carbonaceous aerosols in an urban tunnel. *Atmospheric Environment* 45: 4463–4469.

Ancelet, T., Davy, P.K., Trompetter, W.J., Markwitz, A. & Weatherburn, D.C. 2013. Particulate matter sources on an hourly time-scale in a rural community during the winter. *Journal of the Air and Waste Management Association*, in press. DOI: 10.1080/10962247.2013.813414.

Song, X.H., Polissar, A.V. & Hopke, P.K. 2001. Sources of fine particle composition in the northeastern US. *Atmospheric Environment* 35: 5277–5286.

Trompetter, W.J., Davy, P.K. & Marktiz, A. 2010. Influence of environmental conditions on carbonaceous particle concentrations within New Zealand. *Journal of Aerosol Science* 41(1): 134–142.

Trompetter, W.J., Markwitz, A. & Davy, P.K. 2005. Air particulate research capability at the New Zealand Ion Beam Analysis Facility using PIXE and IBA techniques. *International Journal of PIXE* 15: 249–255.

1.2 Controls of solid phase chemistry and sorption mechanisms

One Century of the Discovery of Arsenicosis in Latin America (1914–2014) –
Litter, Nicolli, Meichtry, Quici, Bundschuh, Bhattacharya & Naidu (Eds)
© 2014 Taylor & Francis Group, London, ISBN 978-1-138-00141-1

Arsenic in the solid phase of aquifers of the Pampean region (Argentina): Implications in the natural contamination of groundwaters

M.C. Blanco
Departamento de Agronomía, Universidad Nacional del Sur, Bahía Blanca, Argentina

ABSTRACT: Geoavailability of arsenic (As) in the solid phase of loess soils-sediments and transference up to excessive (As > 10 μg L^{-1}) in the Pampean aquifers puts at risk the population health when used as drinking water. These aspects connect transversally hydrology, sedimentology and pedology with medical geology, aiming to explain the distribution of HACRE. Origin of As is explained considering As contents of volcanic glass and of minerals of volcanic origin hosted in loess sediments. Total As is normal or slightly higher than average in the earth crust, but yielded elevated As in groundwaters. Geopedogenetic complexities of soils-sediments and partition of As in the particle size fractions are relevant to As source and its transference to waters. Weathering of As-bearers and concentration in the aqueous phase is controlled by local geochemistry. The clay fraction behaves as a source or a sink of arsenic, and evaporation also concentrates As in solution.

1 INTRODUCTION

Geoavailability of arsenic in the solid phase of loess and loess derived sediments and the subsequent incorporation into the aqueous phase of the Pampean aquifers, in where it eventually reaches unacceptable concentrations, puts at risk the health of people that solely utilize affected groundwaters as drinking water. These aspects connect transversally hydrology, sedimentology and pedology with medical geology, aiming to explain the relationship between natural geological factors and population's health as well as the geographical distribution of certain medical problems as hydroarsenicism.

The Chaco-Pampean region, about 1,000,000 km^2, is the most populated geographic region of Argentina that contributes to socioeconomic development owing to the quality of edaphic and hydric resources and agroindustry activity. At the north, the region comprises areas of the Chaco plain (east of Salta, Chaco and Formosa provinces, and the north of Santiago del Estero and Santa Fe provinces) and the Pampa plain, which continues towards the central and southern areas of south of Santiago del Estero Province, eastern plain of Tucumán Province, central and southern plain of Santa Fe Province, a large south and southeast plain of Córdoba Province, southeast of San Luis Province, and Buenos Aires and La Pampa provinces.

In some wide areas, water availability and its quality could be the main limitation to economic growth due to high salinity, hardness and to an excess of As, F and other trace elements, which turn waters unsuitable for human and cattle consumption or irrigation without previous treatment. In this region, water quality problems may affect human health particularly linked to the daily consumption of drinking waters with high As and/or As-F concentrations exceeding the WHO provisional Guideline Value (GV), which are 10 μg L^{-1} for As and 1.5 mg L^{-1} for F, respectively. Consequences related to As exposure via drinking water are skin-pigmentation disorders and the increasing incidence of several types of neoplasms as well as lung, liver and bladder cancer, and several cardiovascular, renal, hematological and even neurological disorders. These health problems are regionally known as HACRE (Hidroarsenicismo Crónico Regional Endémico, Endemic Regional Chronic Hydroarsenicism).

In addition, high F concentrations in drinking water (>1,500 μg L^{-1}) cause both dental and bone fluorosis in rural and periurban inhabitants without access to a central water supply. Hydroarsenism and fluorosis must be considered a public health issue because they are high frequency events. Estimate of potentially affected population at the Chaco-Pampean plain reaches from 2 to 8 million inhabitants considering the 10 μg L^{-1} limit (Bundschuh *et al.*, 2008). However, generalized and complete studies on the sources of As, F and associated trace elements, their mobility and their concentration in surface waters and shallow aquifers are still lacking, particularly for groundwaters close to surface where the highest As and F concentrations are found.

Populations at risk can be separated in urban, where the problem mostly is solved by installing relatively expensive treatment plants and rural and periurban, usually without a possibility to apply a treatment or consume potable waters from alternative sources. Although significant information about the most compromised areas is available, the distribution of As and F and their effects on humans, cattle and crops are far from being completely understood. Nevertheless, great advances in knowledge with respect to the presence of As and associated trace elements in groundwaters had been achieved since the early 20th century. Argentina was the first country in Latin America from where As occurrence in groundwater has been reported. More intense and frequent draught events affected the Pampean region during the 1990–2000 decade, particularly at their SSW extremes. Consequently, the demand of surface or groundwater provision sources for human consumption, irrigation and husbandry grew considerably, leading to identification of new areas susceptible to suffer HACRE after a long latency period.

Nowadays, remarkable efforts of several scientific groups are devoted to identify natural and anthropic As-sources to waters, to quantify the As content and its partition into solid phase, to understand the processes of As mobility and entrance in solution as well as the spatial variability and monitoring in the Chaco-Pampean region and other Latin-American countries. At present, the hydroarsenicism problem is focused by the national, province and local press, this being an advance to aware populations at risk on negative effects of HACRE. In addition, preventive medical campaigns were driven aiming to a reduction of the consequences on individual health and demanding technical solutions for a decrease of As concentrations up to acceptable levels in drinking waters (Sociedad Argentina de Dermatología, 2013).

Despite the high social and economic relevance acquired by the As problem, some sectors of the population and even of the scientific community do not give enough credit to the public health consequences of an excess of As in drinking waters, arguing that HACRE morphologies are not frequently observed, in particular, in the southern extreme of the Chaco-Pampean area. The same is valid for F and fluorosis pathologies. It is highly probable that the number of affected people is underestimated because population in some parts of rural areas does not attend to medical consultation. In addition, robust statistics referred to figures of HACRE at the Pampean region, indicative of total number of individuals identified by sex, age, nutritional status and doses correlated to hydrochemistry of water sources are still missing in the country.

The objective of this contribution is to analyze the mineralogy of the solid phase of the loess soils-sediments and their relationship with groundwaters, aiming to understand the arsenic distribution in the phreatic aquifers of the Pampean region.

2 RESULTS AND DISCUSSION

The Chaco-Pampean plain is modeled by erosion-sedimentation processes related to climate changes of great and low magnitude, particularly those that occurred during the Quaternary and mainly in the Holocene, which shaped the river basins, influenced their hydric regimes (dominant dry climates in arid/semiarid zones referred to present day and paleoclimates), the aquifer hydrochemistry and, consequently, the As and other trace elements concentrations in groundwater. At the northern parts of the Chaco-Pampean plain, surface waters and shallow aquifers have decreasing As concentrations along groundwater flowpaths which start at the volcanic and hydrogeothermal areas of the Andes Ranges. Here, the origin of As, associated to B and F, is explained by past and recent volcanic activity and high temperature geothermal processes (As ranges in groundwater between 470–770 $\mu g\ L^{-1}$ and in geothermal sources between 50–9,900 $\mu g\ L^{-1}$).

Surface and subsurface runoff derived from the areas adjacent to the Andes Ranges discharge in endorreic basins of hyperarid areas where As and the associated trace-elements are concentrated up to an excess. At the eastern areas of the Pampean region, adjacent to the Paraná river, As values in surface and groundwater are relatively low and acceptable for human consumption. The origin of groundwater As in the shallow aquifers at the meridional sections of the Pampean region is more controversial and still heavily discussed. Generally, it is explained by taking into account the As contents of volcanic glass and minerals of volcanic origin hosted in loess sediments, the volcanic glass dissolution and the leaching of As from sediments after volcanic glass dissolution, which all contribute to the concentration of As dissolved in groundwater. The presence of volcanic glass is also associated to distinct volcanic ash layers interbeded into the pedo-sedimentary loess sequences hosting the aquifers. The hypothesis of volcanic glass dissolution is still considered as a main primary process (see e.g., Nicolli *et al.*, 1989).

Although the loess mineral suite of the Pampean Formation does contain volcanic glass shards (frequently 25–50% reaching 63% in exceptional cases), layers of volcanic ash are not ubiquitous in the entire southern Pampean region. Nevertheless, a relationship between volcanic glass and high arsenic concentrations in groundwater is generally accepted. Other authors state a reciprocal relationship between As in shallow groundwaters and the loess mineral association, which besides volcanic glass contemplate other minerals of volcanic origin that include As in their crystalline structure yielding As contamination of hydric resources (Blanco *et al.*, 2006).

Despite the similarity in mineralogical composition between the loess of the Pampean Formation and the post Pampean section, quantitative differences merge from a pulsatile sedimentation derived from a similar source located at the Andes Ranges and the *Patagonia Extrandina*. The sand fraction of loess sediments in the interfluve includes quartz, feldspars, plagioclase, volcanic glass, phytollites, biotite, muscovite, lithic fragments, opaques, scarce tourmaline, zircon, and rutile, olivine, epidote, augite, hypersthene, lithic fragments, green hornblende, brown hornblende and scarce biotite. Illite and interstratified illite-smectite and chlorite-smectite and quartz, microcline, pyroxene and amphibole are identified in the <2 μm fraction. Arsenic contents in the loess ranges from 6.4 mg kg^{-1} in the Tertiary loess of the Pampa Formation to 22 mg kg^{-1} in the Holocene loess-derived alluvial sediments that filled up river basins, being normal or slightly higher than the average content in the earth's crust.

In addition, a comparison between the Pampean and post-Pampean loess sediments, the alluvial sediment interbasins and the different intrabasin landforms, demonstrated vertical and lateral heterogeneity and a grouping of sediments according to their mineral composition, and their As content was determined by local sedimentary sorting along transport.

Partition of As in the different loess particle size fractions is a critical factor related to As source, to its geoavailability in the sedimentary sequences and to its transference to groundwaters (Blanco *et al.*, 2010). Arsenic and other trace element concentrations above the guideline value in shallow groundwaters could be associated to: i) normal or slightly higher As contents in the Aeolian loess and loess derived alluvial sequences; ii) either a low or a high volcanic glass content (As: 6.8–10.40 mg kg^{-1}) in the loess mineral phase; iii) the presence of volcanic ash layers. The rocks of Ventania Hilly system formed mainly by metaquartzites and schists with very low As and other trace elements and do not contribute, significantly, to arsenic provision to groundwaters. On the other hand, lithologies of the Pampean Hills, constituted by calcareous and acidic plutonic–metamorphic units of Precambrian age as crystalline schists, granites and marbles, may be contributive As sources to the aquifers at the central part of the Chaco-Pampean plain.

Generally, As and associate trace elements are adsorbed onto the surface of Fe and Al oxides and oxihydroxides (hematite, goethite, $Fe(OH)_3$(a), magnetite and gibbsite), restricting the mobility of these complex species and, consequently, regulating their distribution in groundwaters. However, sorption is limited in bicarbonated groundwaters; at a high pH, desorption occurs increasing the concentration of As and associated trace elements mainly in the groundwaters close to surface (Nicolli *et al.*, 2007). In general terms, the aquifers of the Chaco-Pampean plain are included in thick sequences of several hundred meters composed by clastic sediments forming a multilayer interconnected aquifer system overlying a crystalline impervious basement, which is faulted in ascended and descended blocks that behave as aquifuge. At the base, the hydrological system comprises thick marine pelitic sediments of the Paraná Formation (green clays from Miocene: aquiclude) hosting very saline waters of Na-sulfate-chloride type superposed, at a regional level, by very thick and heterogeneous deposits of Aeolian loess and loess-like sediments from the Pampa Formation (Plio-Pleistocene). Towards the northeast of Argentina, aquifer levels are also included in sediments of the Puelches Formation, which is not certainly identified towards the south and southwestern areas of the Chaco-Pampean plain. The Puelches aquifer is recharged by rainfall and from the Pampean aquitard, extending into the central and septentrional Pampean areas. Hydrochemical differences between the Puelches and the Pampean aquifers are remarkable.

New areas with As > 10 μg L^{-1} in groundwaters of the southern Pampa suppose productive limitations and risk for health of the periurban and rural population without access to tapped water. Geopedogenetic complexities expressed in the spatial variability of soils-sediments are reflected in the As geoavailability. As provision source and the accumulation processes in waters were investigated. A geochemical characterization of loess soils and sediments (Aeolian) and loess-derived (alluvial and lacustrine) hosting the shallow aquifers (Pampeano Formation and Post-Pampean) gave As in the range 6.4–22 mg kg^{-1} in the Bahía Blanca region, 4.3–7.8 mg kg^{-1} (sand fraction) and 8.9–29.8 mg kg^{-1} (clay fraction) in areas of Coronel Dorrego district.

Parallely, the phreatic aquifer was surveyed in a radii of 200 km centered in Bahía Blanca to determine As and other oligoelements. Moreover, a detailed survey was performed in groundwaters of Coronel Dorrego and Bahía Blanca, this last extended to surface waters and the deep thermal aquifer, to detect the time and spatial variability of arsenic. In the recharge, As concentrations are acceptable (<10 μg L^{-1}) and do not vary in time. Arsenic had a marked seasonal variability at the discharge areas in Atlantic coast, reaching excessive levels between 30 μg L^{-1} (wet season) and 180 μg L^{-1} (dry season) in zones of Coronel Dorrego. Coastal areas adjacent to Bahía Blanca had elevated As in surface waters (10–130 μg L^{-1}) and in phreatic aquifers (7–302 μg L^{-1}) and acceptable levels in the deep thermal aquifer. Excessive As limitates water quality for human consumption and agriculture and cattle production in more than 80% and 90% of the samples of Coronel Dorrego and Bahía Blanca districts.

3 CONCLUSIONS

The partition of arsenic in the soil coarse and fine grain size fractions and in water is critically related to its source and its capacity to be conserved in or driven from the aquifer.

Arsenic is first removed from the minerals of the loess psammitic fraction (50–2,000 μm) in view of the occurrence of mineral weathering reactions, mainly hydrolysis of silicates and carbonate dissolution controlled by the local geochemical factors: pH, Eh, alkalinity, salinity, predominant anions and cations, and competition with other ions.

All reactions are regulated by silicate hydrolysis and carbonate equilibria. As noted above, secondary $CaCO_3$ concretions in the sediment (calcretes) and also as nodules, layers or cement were found. Dissolution of $CaCO_3$ is an important process which generates high pH values due to CO_2 consumption and correspondingly lower pCO_2 values. Solubilization processes of volcanic glass and intensive leaching of minerals hosted in loess sediments during the rainy period summer period in a subtropical to temperate subhumid and semiarid climate give rise to an efficient weathering that affects water-rock interactions and contribute to an increasing amount of As to the aqueous phase.

The subsequent complex behavior of arsenic is affected by processes occurring in the fine fraction as are adsorption–desorption, codissolution—coprecipitation, competition for adsorption sites, and oxide-reduction. Desorption processes and the increase of arsenic concentration in the aquifers are controlled by groundwater pH. Hydrogeochemical evolution to alkaline pH can favor As desorption.

The ion exchange properties of the pelitic fraction provide the means to explore mechanisms of adsorption and desorption of arsenic ions in groundwater. The clay fraction can apparently behave either as a source or as a sink of arsenic. Evaporation is another mechanism that accounts to concentrate As in solution above the guide value. The population usually drinks As-affected phreatic waters and, in most cases, they are not aware of the consequences of long-term consumption of potentially toxic waters. Future research should focus on improving the systematic examination of the spatial variability of geological and edaphic materials and their linkage to hydrogeochemical characteristics with a view to filling the knowledge gap regarding arsenic speciation processes in the groundwater of the region.

However, the strongest efforts should be devoted to developing processes, mechanisms and accessible technologies to lower As up to acceptable levels in waters.

ACKNOWLEDGEMENTS

The author thanks to SeCyT-UNS and CONICET (Argentina) for funding the respective research projects to investigate the As problem.

REFERENCES

Blanco M. del C., Paoloni, J.D., Morrás, H., Fiorentino, C.E. & Sequeira, M.E. 2006. Content and distribution of arsenic in soils, sediments and groundwater environments of the Southern Pampa region, Argentina. *J. of Env. Toxicology.* DOI 19.1002/tox.20219.

Blanco M. del C, Paoloni, J.D., Morrás, H.J.M, Sequeira, M.E., Amiotti, N.M., Bravo, O.A., Díaz, S. & Espósito, M. 2010. Partition of arsenic in soils-sediments and the origin of natural groundwater contamination in the southern Pampa region (Argentina). *Environ. Earth Sci.* DOI 10.1007/s12665.011-1433-x.

Bundschuh, J., Nicolli, H.B., Blanco, M. del C., Blarasin, M., Farías, S., Cumbal, L., Cornejo, L., Acarapi, J., Lienqueo, H., Arenas, M., Guérèquiz, R., Battacharya, P., García, M.E., Quintanilla, J., Deschamps, E., Viola, Z., Castro de Esparza, M.L., Rodríguez, J., Pérez Carrera, A. & Fernádez Cirelli, A. 2008. Distribución de arsénico en la región sudamericana. In: J. Bundschuh, A. Pérez Carrera & M.I. Litter (eds), *Distribución del arsénico en las regiones Ibérica e Iberoamericana*: 137–185 CYTED, IBEROARSEN.

Nicolli, H.B, Suriano, J.M., Gómez Peral, M.A., Ferpozzi, L.H. & Baleani, O.A. 1989. Groundwater contamination with arsenic and other trace elements in an area of the Pampa, Province of Córdoba, Argentina. *Environ. Geol. Water Sci.* 14(1):3–16.

Nicolli, H.B., Tineo, A., García, J.W., Falcón, C.M., Merino, M.H. & Etchichury, M.C. 2007. Arsenic contamination source of groundwater from the Salí basin, Argentina. *Water-Rock Interaction* 2: 1237–1240.

One Century of the Discovery of Arsenicosis in Latin America (1914–2014) – Litter, Nicolli, Meichtry, Quici, Bundschuh, Bhattacharya & Naidu (Eds)

Arsenic adsorption on iron mineral phases under reducing conditions: Results from an *in-situ* field experiment

H. Neidhardt, M. Berg, C. Stengel, L.H.E. Winkel & R. Kaegi
Eawag, Swiss Federal Institute of Aquatic Science and Technology, Switzerland

P.T.K. Trang, V.T.M. Lan, M.T.P. Thao & P.H. Viet
Center for Environmental Technology and Sustainable Development, Hanoi University of Science, Vietnam

ABSTRACT: Intrusion of As-rich groundwater into previously As-free aquifers due to extensive groundwater extraction is an increasing cause of concern, especially in Southeast Asia. We studied *in-situ* As adsorption processes by exposing synthetic Fe-mineral coated sand, as well as original sediment material, to high As groundwater under reducing conditions. Samples were placed in three monitoring wells located in a well-investigated study area (Van Phuc, Vietnam). The amount of adsorbed As reflected an As adsorption potential in the following order: hematite sand > ferrihydrite sand > original Pleistocene sand > goethite sand, and increased proportionally to the concentration of dissolved As in the respective groundwater. Results further indicate transformation of the exposed Fe-minerals, which is presumably controlled by biogeochemical processes. This alteration is expected to cause pronounced changes in the As adsorption potential of the Fe-minerals over time, which is subject to ongoing research.

1 INTRODUCTION

In many Asian aquifer systems, previously As-free aquifers are subject to an intrusion of As-enriched groundwater due to an increased water abstraction, threatening precious groundwater resources (Neidhardt *et al.*, 2013, Winkel *et al.*, 2011). This is currently observed at the village of Van Phuc, which is located on the river banks of the Red River about 10 km southeast of Hanoi. Here, groundwater enriched in dissolved As_{III}, Fe_{II}, NH_4^+, PO_4^{3-} and HCO_3^- laterally flows from a Holocene aquifer into a Pleistocene aquifer, where groundwater is mostly free of As (Van Geen *et al.*, 2013). The Pleistocene sediments are rich in Fe_{III}-minerals, which presumably determine the As mobility in this specific transition zone. Here, the As retardation behavior of the respective sediments is believed to be influenced by (1) competitive adsorption of As and replacement by competing anions (Postma *et al.*, 2007) and (2) transformation of the Fe-minerals induced by a high concentration of dissolved Fe_{II} and/or microbially mediated processes (Radloff *et al.*, 2011). The aim of this study is to investigate the short-term effects of these processes along a sharp redox transition zone in Van Phuc, using original aquifer sediments and synthetic Fe-minerals in an *in-situ* experiment.

2 METHODS

Preparation of synthetic hematite, goethite and ferrihydrite coated sand (coated on cristobalite, SiO_2) was done after Schwertmann & Cornell (2000). Fe-minerals were verified by XRD measurements. Additionally, original Pleistocene orange sand was used that has been previously obtained from the study site. Samples were freeze-dried and sieved to separate the target fraction of 0.2 to 0.5 mm grain size. The *in-situ* exposure of the samples to local groundwater was done via three monitoring wells representing different hydrochemical conditions. One well (AMS-12) was located within young, organic-rich sediments with strongly reducing conditions, where As is likely mobilized. Two other wells (AMS-1 and -5) were located in a transition zone with a similar groundwater composition, but the As mobilization, respective retardation, is exhausted as suggested by long-term monitoring data.

Before introducing the samples, field parameters were recorded and groundwater was sampled for the analysis of major and principal ions (ICP-MS, IC), As speciation (As_V filter cartridges) and Fe speciation (phenantroline method). Samples were then inserted in range of the well screenings by using custom made sample holders covered with nylon mesh (mesh size: 0.2 mm). Positions were checked with a borehole camera system. Samples were exposed for seven days to record the short-term effects. The elemental composition was determined from oven-dried sample material by means of acid microwave digestion (HNO_3 65%, H_2O_2 30%) and ICP-MS. In addition, SEM images were taken to record changes of the mineral surfaces. The SEM was further equipped with an EDS system to check the elemental composition of the samples on spot.

3 RESULTS AND DISCUSSION

3.1 *Groundwater composition*

Results of the groundwater analysis revealed that local aquifers are strongly reducing. Groundwater belongs to the Ca-Mg-HCO_3-type with a near-neutral pH value (6.7 to 6.9) and a low ORP (−160 to −178 mV). Concentrations of dissolved As (174 to 471 µg/L; As_{III} >91.4%), Fe (8.84 to 13.3 mg/L; Fe_{II} >88.9%) and PO_4^{3-} (0.58 to 1.00 mg/L) are clearly elevated compared to local As-free groundwater.

3.2 *Arsenic adsorption and alteration of iron mineral samples*

In the following results from the short-term (seven days) exposed samples are presented, while long-term samples will remain exposed for six months. After removal of the samples, a partial reductive dissolution of ferrihydrite and the formation of a blackish precipitate were already recognized by the naked eye. SEM images of Fe-mineral surfaces indicated further alterations of the exposed samples (Figure 1).

The elemental composition of the exposed mineral samples reflected an As adsorption potential in the following order: hematite coated sand > ferrihydrite coated sand > original Pleistocene sand > goethite coated sand (fig. 2). The As adsorption of all samples appeared to be linear, although groundwater

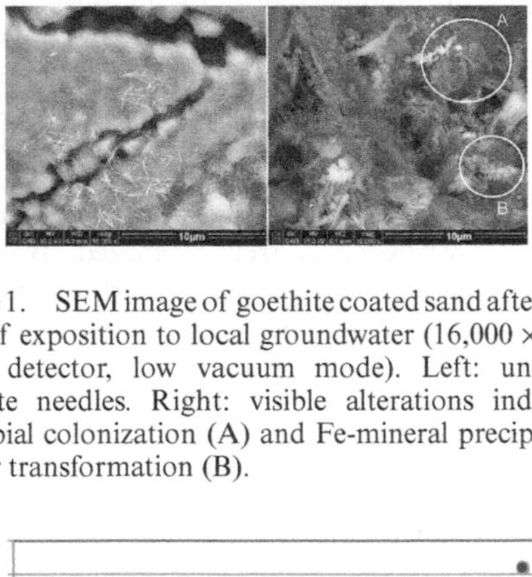

Figure 1. SEM image of goethite coated sand after seven days of exposition to local groundwater (16,000 × mag., GDA detector, low vacuum mode). Left: unaltered goethite needles. Right: visible alterations indicating microbial colonization (A) and Fe-mineral precipitation and/or transformation (B).

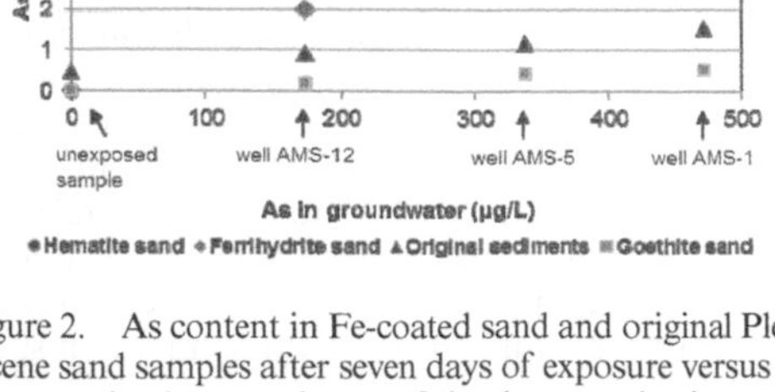

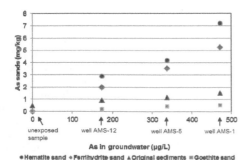

Figure 2. As content in Fe-coated sand and original Pleistocene sand samples after seven days of exposure versus As concentration in groundwater of the three monitoring wells.

contained different concentrations of compounds that are known to compete with As for binding sites (e.g., PO_4^{3-}, SiO_4^{4-}, HCO_3^-, DOC). Solely goethite coated sand seems to have reached a certain saturation with increasing As concentration in the water.

4 CONCLUSIONS

Results demonstrate that fast and visible changes of the Fe-minerals have occurred during the contact with the Fe_{II}-rich groundwater under strongly reducing conditions. We therefore consider the *in-situ* exposure of synthetic Fe-minerals as well as original sediment samples as a promising method to investigate ongoing As adsorption processes at a local transition zone, where groundwater with high As_{III} and Fe_{II} concentrations is advected into less reducing Pleistocene aquifer sediments. Further and more detailed characterization of the long-term exposed samples will help to better understand the mechanisms and consequences of Fe-mineral alteration and its effects on the As adsorption behavior under natural conditions.

ACKNOWLEDGEMENTS

The authors thank the German Research Foundation for financial support (Project NE-1852/1-1) and CETASD members for supporting the fieldwork.

REFERENCES

Neidhardt, H., Berner, Z., Freikowski, D., Biswas, A., Winter, J., Chatterjee, D. & Norra, S. 2013. Influences of groundwater abstraction on the distribution of dissolved As in shallow aquifers of West Bengal, India. *Journal of Hazardous Materials,* 262: 941–950.

Postma, D., Larsen, F., Hue, N.T.M., Duc, M.T., Viet, P.H., Nhan, P.Q. & Jessen, S. 2007. Arsenic in groundwater of the Red River floodplain, Vietnam: Controlling geochemical processes and reactive transport modeling. *Geochimica et Cosmochimica Acta,* 71: 5054–5071.

Radloff, K.A., Zheng, Y., Michael, H.A., Stute, M., Bostick, B.C., Mihajlov, I., Bounds, M., Huq, M.R., Choudhury, I., Rahman, M.W., Schlosser, P., Ahmed, K.M. & Van Geen, A. 2011. Arsenic migration to deep groundwater in Bangladesh influenced by adsorption and water demand. *Nature Geoscience,* 4: 793–798.

Schwertmann, U. & Cornell, R.M. 2000. Iron Oxides in the Laboratory: Preparation and Characterization. Second Edition. Weinheim: Wiley-VCH.

Van Geen, A., Bostick, B.C., Trang, P.T.K., Lan, V.M., Mai, N.-N., Manh, P.D., Viet, P.H., Radloff, K., Aziz, Z., Mey, J.L., Stahl, M.O., Harvey, C.F., Oates, P., Weinman, B., Stengel, C., Frei, F., Kipfer, R. & Berg, M. 2013. Retardation of arsenic transport through a Pleistocene aquifer. *Nature,* 501: 7466.

Winkel, L.H.E., Trang, P.T.K., Lan, V.M., Stengel, C., Amini, M., Ha, N.T., Viet, P.H. & Berg, M. 2011. Arsenic pollution of groundwater in Vietnam exacerbated by deep aquifer exploitation for more than a century. *PNAS,* 108(4): 1246–1251.

One Century of the Discovery of Arsenicosis in Latin America (1914–2014) –
Litter, Nicolli, Meichtry, Quici, Bundschuh, Bhattacharya & Naidu (Eds)
 ISBN 978-1-138-00141-1

Arsenic and associated trace elements in the solid phase and their interrelationships with the aqueous phase in loessic aquifers of the Southern Pampa, Argentina

S.L. Díaz & J.D. Paoloni
CONICET, Argentina

H.B. Nicolli
CONICET, Argentina
Instituto de Geoguímica, Argentina

E. Schmidt, M. del C. Blanco & M. Espósito
Departamento de Agronomía, UNS, Argentina

N. Amiotti
Departamento de Agronomía, UNS, Argentina
CERZOS-CONICET, Argentina

M. Sequeira
CERZOS-CONICET, Argentina
Departamento de Ingeniería, UNS, Argentina

ABSTRACT: In the Southern Pampa, excess levels of As are due to its geoavailability within loess sediments-soils sequences hosting the phreatic aquifers. We analyzed total contents of As and associated trace elements in the loess solid phase, including spatial distribution of As, F, B and V in hydric resources. The interrelationships between solid and aqueous As contents were studied in El Divisorio Creek Basin applying statistics (Principal Components Analysis, PCA). Total As in soils ranges 5.1–20.7 mg kg^{-1}, being lower in the valley slopes and appeared irregularly distributed in the alluvial plain. Alluvial activity determines intrabasin distribution of As-bearing minerals as reflected in As contents in the solid phase. Variability of As in waters is explained by the local factors controlling water residence time in the aquifers, by processes of silicate hydrolysis, oxidation-reduction, sorption-desorption and formation of complex ions in the fine fraction (<2 μm).

1 INTRODUCTION

Independently of their natural or anthropogenic origin, utilization of As toxic waters (OMS, USEPA, CAA; As >10 μg L^{-1}) has an effect on human health and becomes an important soil productivity limitation due to the association of As with another oligoelements (F, V, B). Excess of As in the phreatic aquifers of the Pampa Region (Argentina) is related to its geoavailability in Holocene loess type sediments and loess derived parent materials. These sediments integrate hosting pedosedimentary sequences which, in its upper levels, include the most productive soils of the region.

The objectives of the work are: i) to analyze geoavailability and spatial variability of As and other trace elements associated (Ba, Br, Co, Cr, Fe y Na) in the solid phase, ii) to investigate the spatial distribution of As, F, B and V concentrations in groundwater, iii) to establish the interrelationships between both phases within the El Divisorio Creek (piedmont of the Positive of Ventania Domain, Argentina).

2 METHODS

Samples (n: 36) were collected in seven soil-sediments profiles, three in the upper basin (S) and four in the basin lowlands (P). Soil profiles were referred as: S1, top of a topographic height; S2, valley slope; S3, soil terrace; P1, interfluves; P2, valley slope; P3, terrace and P4, alluvial basin. In the solid phase, were determined As, Ba, Br, Co, Cr, Fe and Na by induced neutron activation analysis (Actlabs, Canada). In water surface samples (n: 9) and groundwater (n: 37), As (Hydride Generator and ICP-OES), B and V (ICP-MS) and F (specific electrode) were determined. The results were studied by statistical procedures using Principal Components Analyses (PCA) methodology.

3 RESULTS AND DISCUSSION

In soils, total As contents were normal in the range 5.1 to 20.7 mg kg^{-1}, both extreme values reached in soils of the middle-lower basin of El Divisorio Creek. P4 had a more irregular vertical distribution for As content (9.3–20.7 mg kg^{-1}) according to alluvial dynamics. In valley slopes from both basin sections, soils yielded similar mean As content (S2: 7.0 mg kg^{-1}, P2: 7.8 mg kg^{-1}). Remobilization and sourcing of materials in unstable landforms within the valley could control the intrabasin distribution of As-bearing minerals, which is then reflected in the As content determined in the solid phase. The origin and latter mobility of available As towards the aquifer, as well as its spatial distribution within the landscape, is controlled by: a) the lithology and geochemistry of loess and loess-derived sediments-soils, b) aeolian and alluvial reworking processes, c) the hydrochemistry and d) the water residence time in the aquifers at a local scale (Blanco *et al.*, 2006; Xie & Naidu 2006). Arsenic bioavailability depends mainly on the following variables: a neutral to alkaline soil reaction, clay content and clay mineralogy, as well as sorption and desorption processes in every landform, which regulate the entrance of As ions to the aqueous phase, its mobility and the spatial variability of As concentrations within the aquifer. The correlation between PC1 and the studied variables demonstrates positive loads for Br (0.673) and Co (0.868) and negatively for Na (–0.769). PC2 has a high and positive relationship with As (0.814) and Cr (0.686) in soils of the alluvial plain. The PC1-PC2 biplot (Figure 1) shows that the solid phase of the soils of the upper basin had a higher content of Na and Ba and a lower content of As, Fe, Co, Br and Cr with respect to those of the middle-lower basin.

A 97.3% of phreatic waters had As exceeding the guide value ranging 10 to 110 µg L^{-1} and coexisting with other trace elements as B (120–1,420 µg L^{-1}), F (20–488 µg L^{-1}) and V (40–800 µg L^{-1}). Although high concentrations were determined in the whole basin, the highest levels indicative of worsening of the water quality were detected in areas of the middle–lower basin. The PC1-PC2 analyses explain 87.5% of total variability of contaminants in the hydric resource.

The distribution of every studied element proves a strong heterogeneity in hydrochemistry

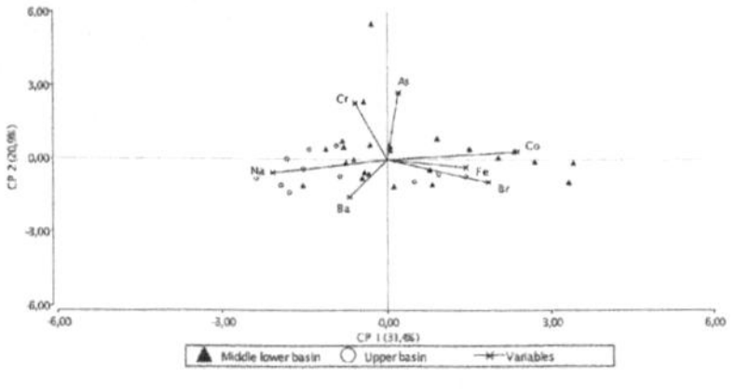

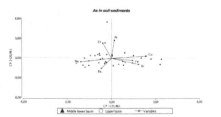

Figure 1. PC1-PC2 biplot in soil-sediments.

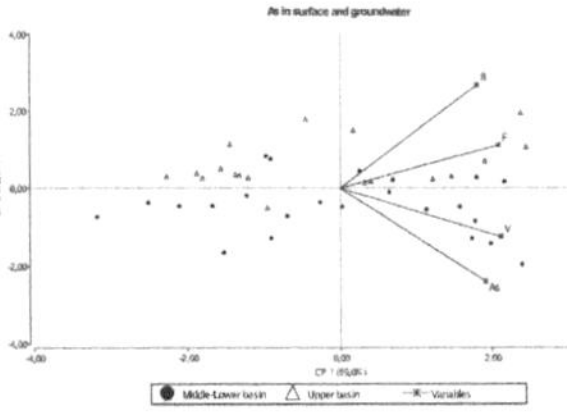

Figure 2. PC1-PC2 biplot in waters.

of groundwater. The biplot (Figure 2) showed that water having a relative better quality owing to a lower concentration for every element belong to the upper basin, some zones of the middle basin and small areas of the lower basin.

4 CONCLUSIONS

Erosion and sedimentation processes control the spatial distribution of loess sediments and loess derived soils associated to diverse landforms in the El Divisorio Creek. The interrelationships aquifer/sediment-loess derived soil explain the As geographic variability in shallow groundwater and the toxicity associated to levels of arsenic >10 µg L^{-1} (WHO). Variability of mean As contents in the solid phase is attributed to alluvial activity that causes intrabasin mobilization and deposition of sediments carrying As-bearing minerals. Spatial and time variability of As concentration and the associated groundwater toxicity could be attributed to a longer residence time of water contacting aquifers' lithology and to processes such as silicate hydrolysis, redox reactions that scavenge As from the solids, adsorption-desorption and formation of complex ions in the fine fraction (<2 µm) leading to unacceptable levels at certain landscape positions conducive to a heterogeneous and patchy spatial pattern.

ACKNOWLEDGMENTS

Authors thanks to SeCyT-UNS, CONICET and ANPCYT (Argentina) for funding the respective research projects to investigate the As problem in the southern Pampean region (Argentina).

REFERENCES

Blanco, M. del C., Paoloni, J.D., Morrás, H., Fiorentino, C.E. & Sequeira, M.E. 2006. Content and distribution of arsenic in soils, sediments and groundwater environments of the Southern Pampa region, Argentina. *J. of Env. Toxicology.* DOI 19.1002/tox.20219.

Xie, Z.M. & Naidu, R. 2006. Factors influencing bioavailability of arsenic to crops. In R. Naidu, E. Smith, G. Owens, P. Bhattacharya & P. Nadebaum (eds.), *Managing arsenic in the environment, from soil to human health.* CSIRO Publishing. 223–234.

One Century of the Discovery of Arsenicosis in Latin America (1914–2014) –
Litter, Nicolli, Meichtry, Quici, Bundschuh, Bhattacharya & Naidu (Eds)
© 2014 Taylor & Francis Group, London, ISBN 978-1-138-00141-1

Identification of the As-bearing phases in fresh volcanic Andean ashes

G. Bia, L. Borgnino, M.G. García & D. Gaiero
Centro de Investigaciones en Ciencias de la Tierra (CICTERRA), CONICET and Universidad Nacional de Córdoba, Córdoba, Argentina

ABSTRACT: The sources and dynamics of arsenic in fresh volcanic ashes collected during the eruptions of Hudson (1991) and Puyehue (2011) volcanoes have been studied. The chemical and mineralogical compositions of both volcanic ashes were analyzed by ICP/OES, DRX, and SEM-EDS. Batch experiments were conducted to evaluate the kinetics of the arsenic release under variable pH values. Results indicate that the release is enhanced under both acidic and alkaline conditions. Besides, the positive linear trends found between Fe and As concentrations in the leachates ($p = 0.0107$ milliQ water; $p = 0.0022$ pH 3; and $p = 0.0081$ pH 10) suggest that arsenopyrite or/and its alteration product scorodite is an important As-bearing phase present in the ash samples. This phase would be the main responsible of the As release under acidic conditions. At higher pH, other mechanisms are involved: desorption from Fe (hydr)oxide coatings (at neutral to slightly alkaline conditions) and dissolution of both arsenopyrite and volcanic glass at pH higher than 9.

1 INTRODUCTION

Large regions of Argentina are affected by the deposition of great amounts of volcanic ashes, a geological material responsible for the natural contamination that affects water reservoirs and soils. A number of studies assign the high concentrations of As measured in groundwaters of the Chaco-Pampean plain to the alteration of volcanic glass and rock fragments spread in the loessic sediments that blanket the entire region (e.g., Nicolli *et al.*, 2012). Most of these conclusions were achieved on the basis of water geochemistry analysis, but little is known about the As concentrations and As-bearing phases present in loess and in the volcanic particles themselves. Some authors suggested that natural contaminants (such as As, V, F and Hg) in volcanic glass are associated with thin coatings of soluble salts and secondary Fe-bearing phases (oxides and sulfides) precipitated onto the particles surface (i.e., Delmelle *et al.*, 2007; Borgnino *et al.*, 2013).

The aim of this work is to identify the As-bearing phases present in two fresh Andean volcanic ashes, in order to assess the mechanisms that control the release of this natural contaminant to the water.

2 METHODS AND EXPERIMENTAL

2.1 *Sampling and chemical and mineralogical analyses*

Two volcanic ash samples were collected immediately after the eruptions of the Hudson (1991) and Puyehue (2011) volcanoes in nearby regions to them. The bulk chemical composition was determined by ICP/OES after acid digestion. Minerals present in the samples were identified by X-ray Diffraction (XRD) and SEM/EDS measurements. The chemical composition of the near-surface region was determined by X-ray photoelectron spectroscopy (XPS).

2.2 *Arsenic release experiments in volcanic ash*

Batch experiments were performed in order to determine the kinetics of the As release. The experiments we carried out under different pH conditions. Three suspensions were prepared with 1.0000 g of dry ash and 20 mL of milli Q water (pH 6.5), HNO_3 10^{-3} M (pH 3.0) and NaOH 10^{-3} M (pH 10.0), respectively. The pH value was kept constant by adding either 0.1 M HNO_3 or NaOH solutions. Aliquots of the suspension were withdrawn after 1, 24, 72, and 168 hours of the experiment start, and filtered through a 0.45 μm cellulose membrane filter. Total As and Fe contents were analyzed in acidified (1% HNO_3) dilutions by ICP-MS. Detection limits were 0.22 μg L^{-1} and 6 μg L^{-1} for As and Fe, respectively.

3 RESULTS AND DISCUSSION

3.1 *Chemical and mineralogical characteristics of volcanic ashes*

Hudson and Puyehue volcanic ashes show andesitic and dacitic composition respectively. Frequent glass grains are observed by SEM and the obtained XRD patterns are typical of amorphous compounds. The chemical characterization of single grain minerals in the Hudson ash sample allowed to identify the presence of Ti and Fe (hydr)oxides, gypsum, illite and chromite associated with altered volcanic glass. Some anhedral crystals of pyrite were also found. In the sample of Puyehue, primary minerals such as andesine, quartz, and cristobalite were recognized, and, using mapping, some minority phases were identified such as gypsum, Fe (hydr)oxides and pyrite.

3.2 *Kinetics experiments*

Figure 1a illustrates the results obtained from the kinetics experiments performed with suspensions of the Hudson ashes in milli Q water, HNO_3 and NaOH solutions. As observed, the kinetic process involves two stages: a first step, which occurs during the first hour, followed by a second much slower step that involves a gradual release of As that reaches equilibrium after 150 h. The Puyehue ashes followed the same trend (not shown).

Figure 1b shows the relationship between As and Fe measured in the leachates of the Hudson volcanic ash. The observed trends, that are identical to those determined in experiments performed with the Puyehue sample, suggest that these two elements have a common source or a similar mechanism of release. Under acidic and neutral conditions, the As/Fe ratio is markedly higher (Figure 1b) than under alkaline conditions. This behavior could explain the presence of different arsenic sources, more likely pyrite grains and

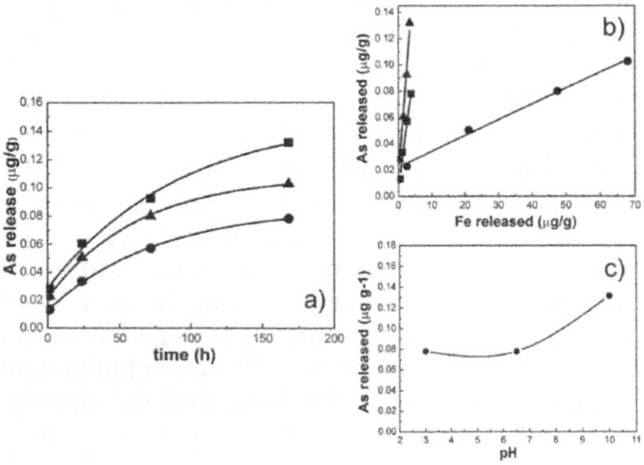

Figure 1. (a) Kinetics curves showing the release of As with time, (b) relationship between As and Fe, in leachates of the Hudson ashes. Circles: milliQ water; triangles: pH 3; squares: pH 10 (c) As released as a function of pH.

Fe (hydr)oxide coatings detected by SEM/EDS. Arsenopyrite can dissolve under both acidic and alkaline pH conditions. Besides, at high pH the dissolution of amorphous volcanic glass could not be ruled out, as its dissolution is favored ($pH > 9$). The positive linear trend observed between Na and As concentrations in the leachates may support this last hypothesis. Taking into account that the composition of the ashes is dominated by Na silicates, the measured increase in Na concentration could be a consequence of the volcanic glass dissolution. Another process that may release As to the solution at high pH is desorption from Fe (hydr)oxides that were identified by SEM/EDS analysis, as thin coatings associated with altered volcanic glass.

In view of this results, dissolution of arsenopyrite predominates at acidic pH, while at higher pH (>8) desorption from Fe (hydr)oxide coatings is probably the main mechanism of As release. Highly alkaline conditions ($pH > 9$) ignite the dissolution of the volcanic glass.

It is important to mention that no individual As minerals were detected by either XRD or SEM-EDS. The detection of low levels of As by this last method is not adequate when Mg-bearing minerals are also present, due to the overlap of peaks. XPS is a surface-sensitive technique that permits to determine the chemical composition of the nearsurface region (2–10 nm). The results obtained here may indicate that when the ash samples are in contact for 90 min with acidic and alkaline solutions, the proportion of As atoms in the nearsurface region is lower than in the untreated sample. This trend agrees with the results obtained in the kinetic experiments above mentioned.

4 CONCLUSIONS

The enhanced release of arsenic at both, acidic and alkaline pH, and the positive linear trends between Fe and As concentrations ($p = 0.0107$ milliQ water; $p = 0.0022$ pH 3; and $p = 0.0081$ pH 10) suggest that arsenopyrite or/and its alteration product scorodite ($FeAsO_4{\cdot}2H_2O$) is an important As-bearing phase present in the ash samples. This phase would be the main responsible for the As release under acidic conditions. At higher pH, other mechanisms are involved: desorption from Fe (hydr)oxide coatings (at neutral to slightly alkaline conditions) and dissolution of both arsenopyrite and volcanic glass. The latter strongly dissolves at pH higher than 9, which explains the increased release of As observed at alkaline pH values.

ACKNOWLEDGEMENTS

Authors wish to acknowledge the assistance of CONICET and UNC for the support facilities used in this investigation. G. Bia acknowledges a doctoral fellowship from CONICET.

REFERENCES

Borgnino, L., Garcia, M.G., Bia, G., Stupar, Y., Le Coustumer, Ph. & Depetris, P.J. 2013. Mechanisms of fluoride release in sediments of Argentina's central region. *Science of the Total Environment* 443: 245–255.

Delmelle, P., Lambert, M., Dufrene, Y., Gerin, P. & Oskarsson, N. 2007. Gas/aerosol-ash interaction in volcanic plumes: new insights from surface analyses of fine ash particles. *Earth Planet. Sci. Lett.* 259: 159–170.

Nicolli, H.B., Bundschuh, J., Blanco, M. del C., Tujchneider, O.C., Panarello, H.O., Dapeña, C. & Rusansky, J.E. 2012. Arsenic and associated trace-elements in groundwater from the Chaco-Pampean plain, Argentina: Results from 100 years of research. *Science of the Total Environment* 429: 36–56.

One Century of the Discovery of Arsenicosis in Latin America (1914–2014) – Litter, Nicolli, Meichtry, Quici, Bundschuh, Bhattacharya & Naidu (Eds)
© 2014 Taylor & Francis Group, London, ISBN 978-1-138-00141-1

Arsenic attenuation in aqueous phase is linked with stabilization onto Fe minerals in a high Andean watershed

E.D. Leiva, C.dP. Ramila, I.T. Vargas, C.E. Escauriaza, C.A. Bonilla, G.E. Pizarro & P.A. Pasten
Department of Hydraulic and Environmental Engineering, Pontificia Universidad Católica de Chile, Santiago, Chile

J.M. Regan
Department of Civil and Environmental Engineering, The Pennsylvania State University, University Park, PA, USA

ABSTRACT: We investigated the relevance of the interaction between Fe and As in iron-rich sediments for As attenuation in surface waters of the Chilean Altiplano, under climatic conditions of extreme altitude (>4,000 masl) and aridity (<310 mm year^{-1}). Our results show that As concentrations in aqueous phase were attenuated (>70%) due to stabilization processes in Fe-rich sediments. Total As content in Fe sediment profiles exceeds 0.9 g kg^{-1} and in the surface is higher than 4 g kg^{-1}. As and Fe show perfect correlation in the Fe sediment profiles indicating a close connection between Fe minerals and As mobilization. Altogether, these results indicate that As attenuation is controlled by stabilization processes onto Fe minerals on the surface and at depth, thus controlling the discharge of As-rich runoff to the surface waters.

1 INTRODUCTION

Arsenic (As) is a ubiquitous and toxic trace metalloid, which is widely distributed in natural environments (Smedley & Kinniburgh, 2002). The different As species exhibit variation in their solubility, toxicity, transport, and bioavailability. As(V) interacts more strongly than As(III) with Fe(III)-oxyhydroxides (Oremland & Stolz, 2003), abundant minerals with high binding affinity toward different metals present in sediments, soils and mine waste. In recent years, the behavior of As in nature has been extensively studied (Nordstrom, 2002; Oremland & Stolz, 2003), but the biogeochemical mechanisms governing mobilization and stabilization of As in fluvial systems are still unclear.

The Chilean Altiplano has water scarcity and the mining activity promotes the release of contaminants (toxic metals and nonmetals) in this area. In the Azufre River sub-basin, the release of contaminants from natural and anthropogenic sources negatively impact the quality of rivers and surface waters. Initial results of field analysis show that in the upper section of this sub-basin, the As concentrations varies between 1.0 to 3.5 mg L^{-1}, being observed high concentrations of As in hydrothermal springs (>0.7 mg L^{-1}) and in Fe-rich sediments (>4 g kg^{-1}).

Interaction between iron (Fe) minerals and As together with microbial reactions can have a significant role in the As load of runoff. This research investigates the processes involved in the attenuation and stabilization of As in solid phase and, consequently, in the fate of As in this system, contributing in the development of future remediation strategies.

2 METHODS

2.1 *Study site: Upper section of the Lluta River Watershed*

Lluta River Watershed (LRW) is located in the XV Region of Arica and Parinacota, in the north of Chile (18°00′-18°30′S and 70°20′–69°22′W). In the upper section of the LRW (Azufre River sub-basin), two hot surface runoff channels, which drain from the same hot spring and are associated with Fe minerals, were chosen for a transect sampling. Each transect contained four sampling points for hydrogeochemical and analysis of sediment profiles.

2.2 *Geochemical analyses*

Water samples from the hydrothermal transects were taken at the sampling points and were analyzed for total As and Fe. The on-site water analyses included temperature, pH, dissolved oxygen (DO) and conductivity. For solid phase profile analyses, the layers of sediment profiles were dried at 40 °C and digested for total As, Fe, S, Zn, Mn and Pb analyses. Elemental analysis was done by TXRF and ICP.

3 RESULTS AND DISCUSSION

3.1 *Arsenic concentration decreases downflow in the transects due to natural attenuation*

AsD concentrations (Figure 1) shown a significant decrease in few meters after the hydrothermal source (from 0.8 to 0.19 mg L^{-1} for transect 1 and from 0.8 to 0.17 mg L^{-1} for transect 2), demonstrating a clear attenuation of AsD concentrations.

Complementarily to the attenuation observed in aqueous phase, the As concentrations in the solid phase of the riverbed are increased from the hydrothermal source. Likewise, pH decrease is observed along the transects. This is relevant because As(V) is adsorbed more efficiently onto Fe(III)-oxyhydroxides between pH 4–7 (Dixit & Hering, 2003). In addition, the analyses of the presence of As(III) and As(V) in the sampling points of the transects shows a significant AsD oxidation downstream the source (over 90%). Thus, the As oxidation together with the As attenuation data and pH measurements support the idea that As oxidation favors natural As attenuation, through adsorption processes onto Fe minerals.

We previously demonstrated that the oxidation of As(III) is controlled by biological activity mediated by enzymatic activity AroA-like (Leiva *el al.*, 2014). Thus, the microbial As(III) oxidation promotes the attenuation of AsD concentrations.

3.2 *Arsenic and iron concentrations are positively correlated, in deep sediments*

To reveal more precisely the processes controlling the As distribution we performed analysis of

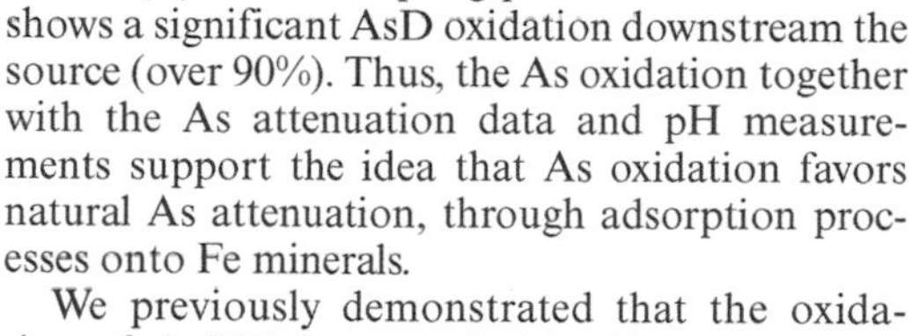

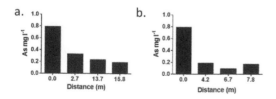

Figure 1. Dissolved As concentrations in sampling points of transect 1 (a) and transect 2 (b).

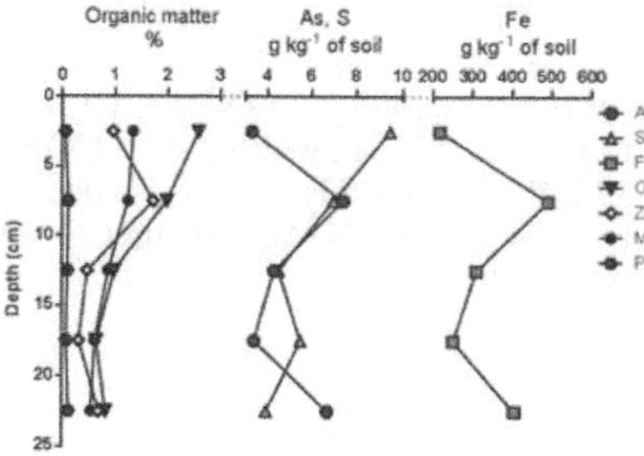

Figure 2. Vertical profile (25 depth) cm of total As, Fe, S, organic matter, Zn, Mn and Pb contents in sediment samples located in the hydrothermal source.

sediment profiles in the sampling points of the hydrothermal transects. The results of the profiles presented in Figure 2 (Representative profile) show that there are high concentrations of As, Fe and S in the solid phase.

The total As content of sediments in of the Azufre River sub-basin ranges between 0.9 and 7.4 g kg^{-1}. In all sediment profiles, the positive correlation between As and Fe concentrations indicates that these elements have a similar pattern in their concentrations at depth and strongly suggests that the environmental fate of these elements are linked.

Fe oxides are ubiquitous in soils and sediments and are particularly important for the As mobilization, because can being capable of containing high concentrations of As.

The relationships between As and Fe in sediment profiles suggest once again that As is mainly stabilized onto Fe-minerals.

4 CONCLUSIONS

Our results demonstrate that the natural attenuation of As is controlled by microbial oxidation reactions, pH decrease and stabilization onto Fe-minerals. In addition, the significant correlation between Fe and As indicates that mainly Fe-minerals control the mobilization of As by adsorption onto preformed Fe minerals or by co-precipitation of As with Fe in the formation of these minerals. Fe sediments are deposits of high concentrations of As, which can accumulate in deeper areas. These findings contribute to the understanding of the biogeochemical mechanisms of As(III) oxidation and its relevance in the fate, transport, and availability of As in surface water flows for fluvial systems rich in As.

ACKNOWLEDGEMENTS

The authors acknowledge FONDECYT 1100943/2010, FONDECYT 1130936/2013, CONICYT 24121233/2012. This study was also partially supported by CORFO 09CN14-5709 and CONICYT/FONDAP 15110020.

REFERENCES

Dixit, S. & Hering, J.G. 2003. Comparison of arsenic(V) and arsenic(III) sorption onto iron oxide minerals: implications for arsenic mobility. *Environmental Science and Technology* 37: 4182–4189.

Leiva, E.D, Ramila, C.d.P., Vargas, I.T., Escauriaza, C.R., Bonilla, C.A., Regan, J.M., Pizarro, G.E. & Pasten, P.A. 2014. Natural Attenuation Process by Microbial Oxidation of Arsenic in a High Andean Watershed. *Science of the Total Environment* 466–467: 490–502.

Nordstrom, D.K. 2002. Public health: enhanced: worldwide occurrences of arsenic in groundwater. *Science* 296: 2143–2145.

Oremland, R.S & Stolz, J.F. 2003. The ecology of arsenic. *Science* 300: 939–944.

Smedley, P.L & Kinniburgh, D.G. 2002. A review of the source, behaviour and distribution of arsenic in natural waters. *Applied Geochemistry* 17: 517–568.

One Century of the Discovery of Arsenicosis in Latin America (1914–2014) – Litter, Nicolli, Meichtry, Quici, Bundschuh, Bhattacharya & Naidu (Eds)
© 2014 Taylor & Francis Group, London, ISBN 978-1-138-00141-1

Arsenic mobilization in the unsaturated zone

S. Dietrich
Instituto de Hidrología de Llanuras, CONICET, Azul, Argentina

E. Torres & C. Ayora Ibañez
Instituto de Diagnóstico Ambiental y Estudios del Agua, CSIC, Barcelona, Spain

P.A. Weinzettel
Instituto de Hidrología de Llanuras, Comisión de Investigaciones Científicas (CIC), Azul, Argentina

ABSTRACT: This work presents the results of laboratory experiments carried out on unsaturated zone samples obtained from the Azul creek basin, Buenos Aires province, Argentina. The objective of the experiments was to determine the arsenic (As) distribution in the different mineral phases along the soil profile. Lixiviation experiments were performed to detect those zones with higher As concentration in the solids. Then, a sequential extraction procedure was accomplished to determine the retention of As in the different mineral phases of the sediments. The lixiviation experiments showed a more As-enriched upper part of the soil profile and that As concentration decreased with depth. The sequential extraction procedure revealed that zones with higher As concentration are related to desorption from Fe(III) oxide and hydroxides.

1 INTRODUCTION

High As concentrations are a common feature in many shallow aquifers in Argentina. In such cases where unsaturated zone sediments contain important amount of As, under certain geochemical conditions, the element can be mobilized from the sediments and transported into the groundwater.

Therefore, the objectives of this work were to analyze the distribution of As along the soil profile and to elucidate the As-bearing mineral phases.

2 MATERIALS AND METHODS

2.1 *Study site*

The study site is located in Azul city, Buenos Aires province, Argentina (36° 46' S; 59° 53' W). The soil, which extends from the surface to a depth of 120 cm, has been classified as a petrocalcic Paleudol (Soil Survey Staff, 1999). At the top of soil profile lies a loamy A horizon, with moderate to strong granular structure (0–18 cm depth). Below, a Bt horizon, clay textured with strong and firm coarse columns (18–43 cm depth), lies, followed by a BCk horizon, with a silty clay texture (43–66 cm depth). Finally, there is a petrocalcic horizon (140 cm depth), very firm and massive, with important amounts of calcium carbonate. From 140 cm depth downwards, the unsaturated zone is composed of silty loessoid sediments, also with calcium carbonate.

Arsenic concentration in shallow groundwater is 0.032 mg L^{-1}.

2.2 *Batch lixiviation experiments*

Batch lixiviation experiments were performed on 10 samples obtained from depths between 50 and 170 cm. Each sample represents 10 cm length of the soil profile. Distilled water was brought into contact with a known mass of sediment and stirred for six hours at room temperature. Previous experiments, which extended during a week, demonstrated that As release from sediments occur in the first two hours. A soil/water ratio of 1:5, with 20 g of sediments, was used. The resultant solution was separated from sediments by centrifugation. Major and trace elements were analyzed by ICP-AES and ICP-MS, respectively.

2.3 *Sequential extractions*

Sequential extractions are a widely used methodology to study elements speciation in sediments. A modified procedure of Torres and Auleda (2013) was used to determine the distribution of As in the mineral phases. The procedure consists of five steps. In each step, a different mineral phase is dissolved and the elements are analyzed in the resulting solution. Step 1: dissolution of water soluble minerals, such as secondary sulfates. Samples are brought into contact with distilled water for one hour. Step 2: removal of adsorbed and interchangeable ions and calcite. A 1 M NH_4-acetate solution is added to the sediments of step 1. Step 3: Fe(III) dissolution of poorly crystallized oxide-hydroxides. A 0.2 M NH_4-oxalate mixed with 0.2 M oxalic acid is added to the remaining sediments of step 2. This step must be performed in the darkness. Step 4: dissolution of crystallized Fe(III) oxides, such us goethite, ferrihydrite and hematite. The same solution as step 2 is added to step 3 sediments but the solution temperature is maintained at 80 °C. Step 5: dissolution of silicates and residual phases. The sediments from step 4 are digested with a solution of HCl, HNO_3 and HF. The slurry is heated up to 250 °C.

3 RESULTS AND DISCUSSION

3.1 *Batch lixiviation experiments*

The results of batch lixiviation experiments are shown in Figure 1. The major As lixiviation occurred in the upper part of the soil profile whereas an important decrease in As lixiviation is observed from 55 cm up to 90 cm depth. The As concentration remained almost similar from 90 cm depth downwards.

Since major As release was accomplished within the first six hours, these results suggest that adsorption is the main process that controls As retention by sediments.

3.2 *Sequential extractions*

The results of the sequential extractions experiments are shown in Figure 2. The bars indicate the relative amount of As released in each extraction step for each depth interval. The total amount of As delivered to the solution (mg kg^{-1}) along the first four step appears above the bars. Since the silicate dissolution is a slow process, which hardly may incorporate important amount of As to the infiltration water, the results of the fifth step were excluded from the chart.

For the three upper depth intervals, the greater amount of As delivered to the solution is related to poorly crystalline and crystallized Fe(III) oxides. Both comprise almost the 80% of the As. These results suggests that adsorption/desorption from Fe(III) oxides are the main processes that control

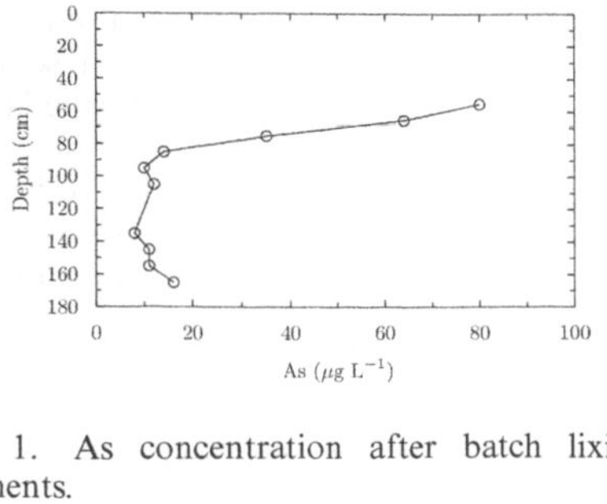

Figure 1. As concentration after batch lixiviation experiments.

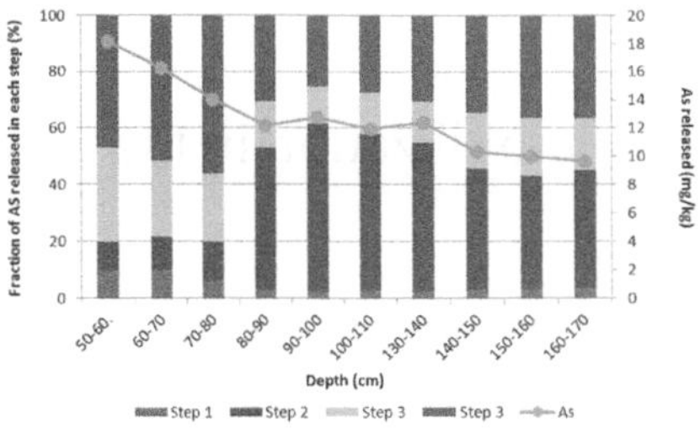

Figure 2. As release in each of the first four steps of the sequential extraction procedure. Percentages are relative to these total amounts. The line indicates the total amount of As delivered in each depth interval (mg/kg).

the As retention/release from sediments. Such evidences are consistent with the literature (Smedley and Kinniburgh, 2002).

In contrast, deeper samples showed that calcite dissolution is the main process that delivered As. Sediments in this part of the unsaturated zone contain high proportion of calcium carbonate (Section 2.1). However, the amounts delivered are lesser than in the upper part, especially between 140 and 170 cm depth.

4 CONCLUSIONS

Batch lixiviation experiments and sequential extractions were performed on samples coming from the unsaturated zone.

Lixiviation experiments showed that greater amounts of As were released in the upper part of the unsaturated zone.

Sequential extractions indicated that desorption of As from Fe(III) oxides was the main process controlling the As concentration in the solution in the upper part of the profile.

In deeper zones, calcite dissolution may exert an important mechanism regulating As concentration.

ACKNOWLEDGEMENTS

Support for this work came from Instituto de Hidrología de Llanuras and ANPCyT (PICT-1988/06), Argentina.

REFERENCES

Smedley, P.L. & Kinniburgh, D.G. 2002. A review of the source, behaviour and distribution of arsenic in natural water. *Appl. Geochem.* 17: 517–568.

Soil Survey Staff. 1999. Soil Taxonomy. A Basic System of Soil Classification for making and Interpreting Soil Surveys. Agric. Handbook No. 436, 2nd. Edition. NRCS-USDA. US Govern. Printing Office Washington, D.C.

Torres, E. & Auleda, M. 2013. A sequential extraction procedure for sediments affected by acid mine drainage. *J. Geochem. Explor.* 128: 35–41.

One Century of the Discovery of Arsenicosis in Latin America (1914–2014) –
Litter, Nicolli, Meichtry, Quici, Bundschuh, Bhattacharya & Naidu (Eds)
 ISBN 978-1-138-00141-1

Groundwater-sediment sorption mechanisms and role of organic matter in controlling arsenic release into aquifer sediments of Murshidabad area (Bengal basin), India

S. Datta
Department of Geology, Kansas State University, Manhattan, KS, USA

K. Johannesson
Department of Earth and Environmental Sciences, Tulane University, New Orleans, LA, USA

N. Mladenov
Department of Civil Engineering, Kansas State University, Manhattan, KS, USA

M.S. Sankar, S. Ford, M. Vega, A. Neal & M.G. Kibria
Department of Geology, Kansas State University, Manhattan, KS, USA

A. Krehel
Department of Geological and Environmental Science, Hope College, Holland, MI, USA

G. Hettiarachchi
Department of Agronomy, Kansas State University, Manhattan, KS, USA

ABSTRACT: The current study attempts to delineate the role of dissolved organic matter in the release of sediment-bound arsenic from shallow aquifer sediments in the Bengal basin. Water samples and sediment cores were collected from 4 sites along the east and west banks of the river Bhagirathi (Nabagram and Kandi: low As, western part; Beldanga and Hariharpara: high As, eastern part). Fluorescence components were of three types: Humic like (A&C), Protein or Tyrosine Like (B) or Tryptophan protein or Phenol like (T). DOM characterization indicated that microbial proteins (Tyrosine (B) and Tryptophan (T)) are dominant in the low As areas, while humic DOM (A and C) were more prevalent in the high As groundwaters. Cl/Br molar ratio of high As wells were low compared to low As wells. The results imply that OM classification plays a vital role in releasing arsenic from aquifer sediments to shallow groundwaters in the Bengal basin.

1 INTRODUCTION

Arsenic groundwater contamination in the Bengal basin is one of the greatest environmental calamities in history. As research progresses, more and more sites along the Indo-Gangetic alluvial plain are being identified with high content of arsenic, resulting in a serious need for further investigation of the processes behind arsenic release and mobilization (Datta et al. 2011; McArthur et al. 2004). The current study attempts to understand the role of dissolved organic matter (DOM) in the retention of dissolved arsenic, as well as to characterize various fractions of these DOM throughout the area of study. The study area encompasses the east and west banks of the river Bhagirathi and the tributary of the river Ganges (flowing N-S transect through the district of Murshidabad). The western portion of the river Bhagirathi is predominantly older alluvium of Pleistocene to Holocene age that is characterized by oxidized ferruginous sand, silt and clay with caliche (Bhagirathi formation). The eastern side of the Bhagirathi is occupied by the Bhagirathi Ganges formation, which consists of newer alluvium, or the Arambag formation of Holocene age, which is mostly sand, silt and clay and also ferruginous components. The organic rich sediments of interest are situated between 20–50 m depth within this aquifer. Four areas were of particular interest and are characterized by the extent of their arsenic content: 1) Nabagram and Kandi (Low As; western part) and 2) Beldanga and Hariharpara (High As; eastern part).

2 METHODS/EXPERIMENTAL

Water samples were collected from 39 hand pumped tube wells (~90–110 ft.), 9 ponds and 6 irrigation wells (~60–80 ft.) along with 4 sediment cores (~150 ft.) from these sites. Characterization of DOC was done using analog fluorescence spectrophotometer Aqualog (Horiba). Total organic carbon (TOC) and total organic nitrogen content of the water samples were analyzed using TOC-LCSH. Fluorescence components of West Bengal waters are identified through the analysis of the Excitation Emission Matrices (EEM) and are found to be: 1. Humic like (A&C), excitation wavelength 330–350 nm and emission wavelength at 420–480 nm & 380–480 nm. 2. Protein like or Tyrosine like (B), excitation wavelength 270–280 nm and emission wavelength at 300–320 nm. 3. Tryptophane like or Protein like or Phenol like (T) with excitation wavelength 270–280 nm and emission wavelength at 320–350 nm (Parlanti et al. 2000).

3 RESULTS AND DISCUSSION

Hydrogeochemical field analyses indicated a strong contrast between the west and east portions of the river Bhagirathi. High As (10–1263 µg/L) and low Mn (0.1–1.3 mg/L) was detected in the eastern portion within Beldanga and Hariharpara, while low As (0–15 µg/L) and high Mn (0.2–4.2 mg/L) were detected in Nabagram and Kandi (west) (Figure 1). Total Fe and Fe^{2+} values at higher in high As areas and subsequently are of low concentrations in waters with low As. Hydrochemical variations between these zones parallel strongly with the difference in sediment appearance and composition.

Dissolved organic matter (DOM) characterization studies indicated that microbial proteins (Tyrosine (B) and Tryptophan (T)) are dominant in the low arsenic areas, while humic DOM (A and C) were more prevalent in the high arsenic groundwaters. DOC was found to be positively correlated with arsenic from depth-wise analyses, and the respective values for the high and low arsenic areas were 1.5–3.2 mg/L and 0.5–1.3 mg/L. Cl/Br molar ratio of high As wells were low compared to the low As wells. Sequential extraction results revealed that majority

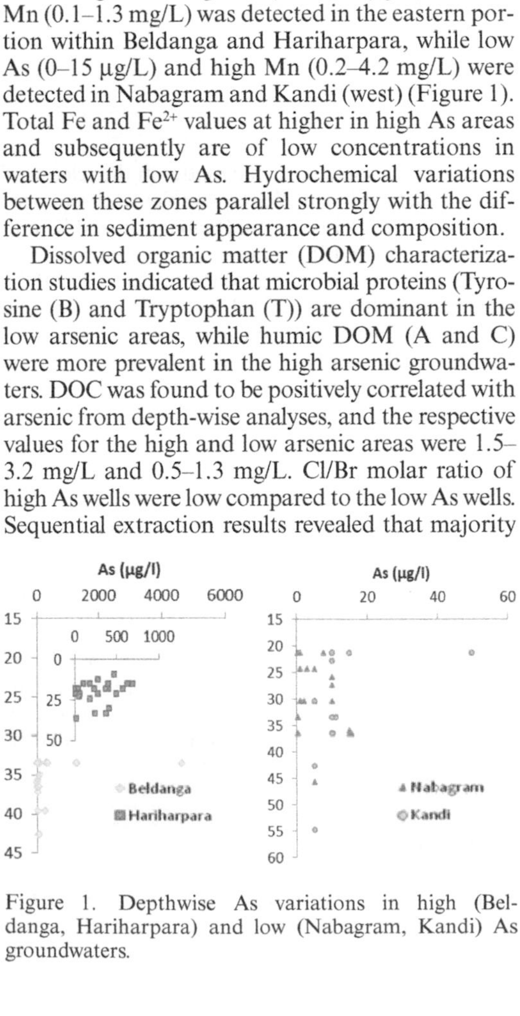

Figure 1. Depthwise As variations in high (Beldanga, Hariharpara) and low (Nabagram, Kandi) As groundwaters.

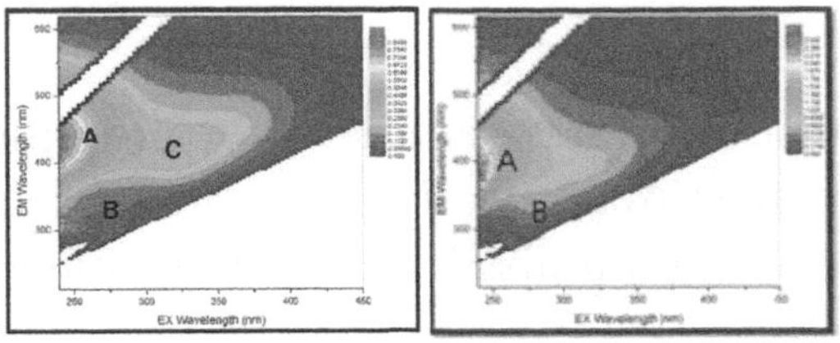

Figure 2. EEM for tubewell water samples TW-126 HK (left, Hariharpara) showing presence of Humic components 'A', 'C' and a Bacterial component 'B', and TW-135 KHN (right, Kandi) showing the presence of Humic component 'A' and a Bacterial component 'B'.

of the sediment bound As is present in residual solid phases (>40%) and relatively less bioavailable.

Microbial mediated reductive dissolution of FeOOH in the presence of organic matter is the major mechanism by which sediment bound As (<50 m depth) is released into the groundwater (McArthur et al. 2004). The darker organic matter rich sediments (OM both sediment bound and anthropogenically derived) existing at the depth range 20–50 m with reducing environment persisting in both high and low As areas are possible reasons for elevated levels of As in this region.

DOC and TON of the high As waters range from 3.2 to 2.4 mg/L and 2.9 mg/L to 1.65 mg/L (respectively) whereas in the low As area the ranges are 1.2 mg/L to 0.6 mg/L and 0.08 mg/L to 0.3 mg/L. Our analysis of the Excitation Emission Matrix (EEM) reveals that there is a difference in fluorescence components of high As and low As areas. The surface waters (pond) and the shallow ground water (tube wells) of high As areas have 2 humic components (A&C) and a protein component (B). However the Humic component (C) of shallow groundwater of high As area is not very strong. Some of the tube wells even do not show humic component 'C'. Deep tube wells (irrigation water) of the high As areas show very mild signatures of Humic component (A&C) but shows appreciable protein like peak B. EEM of most of the surface waters and the shallow ground waters of low As areas show one of the humic components (A) and a protein component (B) (Figure 2). However, the deep groundwater shows very mild signatures of humic components (A or C) and had good signature of Tryptophane like peak (T). Fluorescence index (FI), freshness index and SUVA values of the high As areas are 1.5–1.8, 0.8–0.9, and 2.2–3.1 mg/L m respectively; and for the low As areas: 1.6–2.1, 0.84–1.1, 0.1–3.7 mg/L m.

4 CONCLUSIONS

The results of the current study imply that, besides sediment type and water chemistry, organic matter classification plays a vital role in releasing arsenic from aquifer sediments to shallow groundwaters in the Bengal basin. It is thereby thought that microbial mediated reductive dissolution of goethite, in coincidence with organic matter, is the primary mechanism by which As is released from sediments and into the shallow groundwater. Therefore, much consideration must be paid to the darker, organic rich sediments between 20 and 50 m depth for possible sources of elevated As in the study area.

REFERENCES

Datta, S., Neal, A.W., Mohajerin, T.J., Ocheltree, T., Rosenheim, B.E., White, C.D. & Johannesson, K.H. 2011. Perennial ponds are not an important source of water or dissolved organic matter to groundwaters with high arsenic concentrations in West Bengal, India. *Geophysical Research Letters.* 38: L20404.

McArthur, J.M., Banerjee, D.M., Hudson-Edwards, K.A., Mishra, R., Purohit, R., Ravenscroft, P. & Cronin, A. 2004. Natural organic matter in sedimentary basins and its relation to arsenic in anoxic groundwater: the example of West Bengal and its worldwide applications. *Applied Geochemistry.* 19: 1255–1293.

Parlanti, E., Worz, K., Geoffrey, L. & Lamotte, M. 2000. Dissolved organic matter fluorescence spectroscopy as a tool to estimate biological activity in a coastal zone submitted to anthropogenic inputs. *Organic Geochemistry.* 31(12): 1765–1781.

One Century of the Discovery of Arsenicosis in Latin America (1914–2014) – Litter, Nicolli, Meichtry, Quici, Bundschuh, Bhattacharya & Naidu (Eds)
 ISBN 978-1-138-00141-1

Contamination of arsenic in the Red River watershed (China/Vietnam): Distribution, source and flux

T.H. Dang
University of Bordeaux, UMR CNRS, EPOC, France
Faculty of Chemistry, BaRia-VungTau University, Vietnam

A. Coynel, G. Blanc & H. Etcheber
University of Bordeaux, UMR CNRS, EPOC, France

C. Grosbois
University of Tours, GéHCO, France

D. Orange
IRD, BIOEMCO, IWMI Office, SFRI, Hanoï, Vietnam

L.A. Le
Institute of Chemistry, Vietnam Academy of Science and Technology, Hanoï, Vietnam

ABSTRACT: This study is based on high resolution dataset of hydrological and arsenic analyses (dissolved and particulate phases) during 2008–2009 at five stations from upstream to downstream Red River (China/Vietnam), combined with two sampling campaigns covering the whole Vietnamese watershed at low and high water levels. The result showed that the dissolved and particulate As concentrations in the Red River were higher than world average and strongly decrease from upstream to downstream. In addition, the temporal variation of As concentrations is related to hydrological conditions, suggesting a dilution effect/change in As source(s) with the hydrology. Multidimensional statistical analyses combined with As-maps show that the highest As concentrations are originated from the upstream catchment (in China). The mineralogical characterization of As-bearing phases of stream sediment showed that As is mainly trapped in the particulate fraction of Fe oxyhydroxides. Finally, the Red River contributed 0.3% (particulate) and 1.6% (dissolved) of global As fluxes.

1 INTRODUCTION

The Red River (China/Vietnam, A = 155 000 km^2, Fig. 1) is a typical humid tropics river in term of water discharge and sediment load and strongly affected by human activities (Dang *et al.*, 2010). The Red River Delta is one of the most populous areas in the world, and the demand for clean water in this area has rising since 20 years ago. However, high As concentrations [As] in groundwater (up to 3050 μg L^{-1}, Berg *et al.*, 2007), i.e. more than 300 times the WHO drinking water standard of 10 μg L^{-1}) and drinking water have posed a serious health threat to millions of people. Despite the known As pollution affecting the groundwater in the Red River Delta, little information is available on particulate and dissolved [As] in the fluvial system.

This study is based on a hydrogeochemical monitoring with a high temporal and spatial resolution using daily water discharges, daily suspended particulate matter concentrations and As analyses (weekly to monthly in dissolved and particulate phases) at five permanent observation stations from upstream to downstream Red River during 2008–2009 (Figure 1).

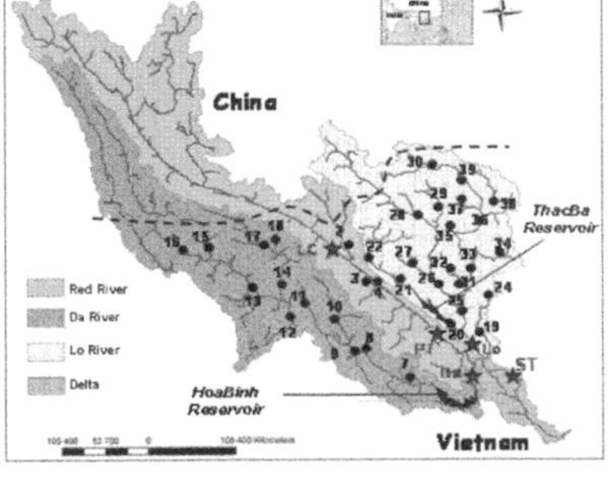

Figure 1. Map of the Red River basin and the study site. The black line represents the China/Vietnam frontier.

In addition, two snapshot campaigns covering the whole Vietnamese watershed was performed in forty sites at low (March-April 2008) and high (September 2008) water levels (Figure 1). The objectives of this study are to (i) determine temporal As variability and quantify annual As flux; and (ii) assess the spatial distribution of As concentrations and identify potential As sources.

2 RESULTS AND DISCUSSION

2.1 *Temporal variation of arsenic concentrations and arsenic fluxes of the Red River*

During 2008–2009, the dissolved As concentrations measured in the Red River at the SonTay station varied between 2.17 and 4.02 μg L^{-1}, and its values varied between 26.6 and 127 mg kg^{-1} for particulate phase. The average dissolved and particulate As concentrations were 3.25 μg L^{-1} and 77.8 mg kg^{-1}, respectively. The world average values were 0.62 μg L^{-1} for dissolved and 36 mg kg^{-1} for particulate As concentrations (Viers *et al.*, 2009). In addition, a negative correlation was observed between As concentrations and water discharges (Figure 2), suggesting a dilution effect for the dissolved phase and a change in punctual As source(s) with hydrology (e.g. mining/industrial point sources) and/or grain size-related effect for the particulate phase.

Annual As fluxes transported by the Red River were estimated at 361 t/yr and 1650 t/yr in dissolved and particulate phases, respectively. Considering the As flux carried by major world rivers (Viers *et al.*, 2009) the contribution of the Red River at the global scale was estimated at 0.3% for particulate and 1.6% for dissolved As fluxes.

2.2 *Spatial variation of arsenic concentrations*

The summary of the statistical parameters describes the variability of [As] measured at five

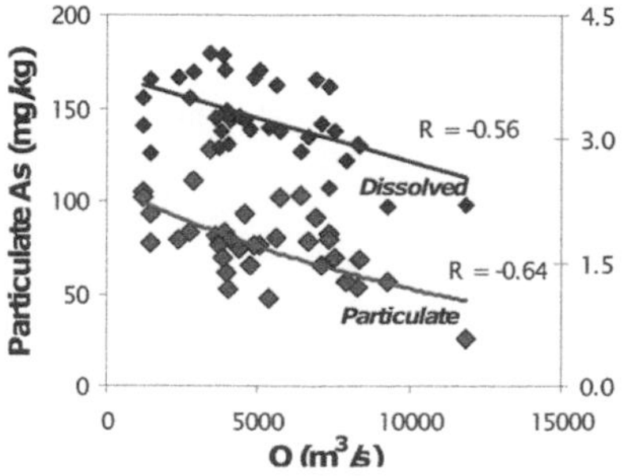

Figure 2. Relationship between dissolved and particulate As concentrations and water discharges (Q) in the Red River at ST (outlet of the watershed) during the 2008–2009 period.

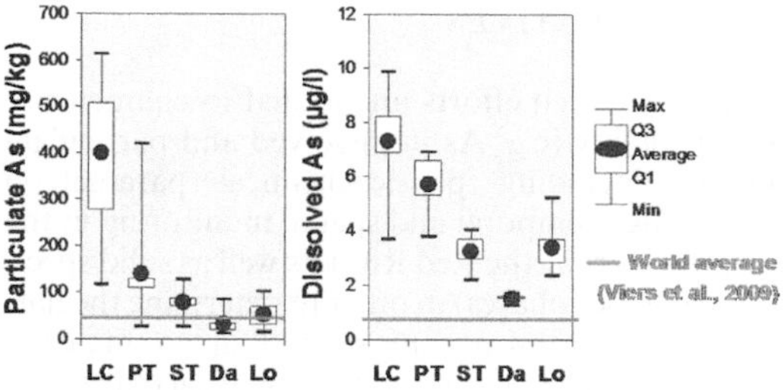

Figure 3. Statistic distribution of particulate and dissolved [As] in the Red River during the 2008–2009 period.

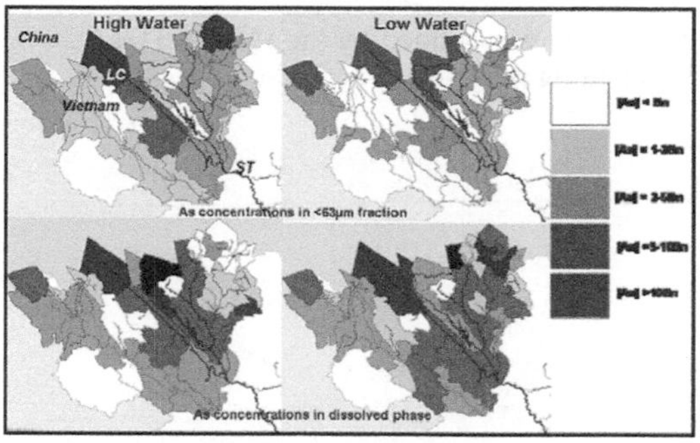

Figure 4. Spatial distribution of [As] in the Red River watershed.

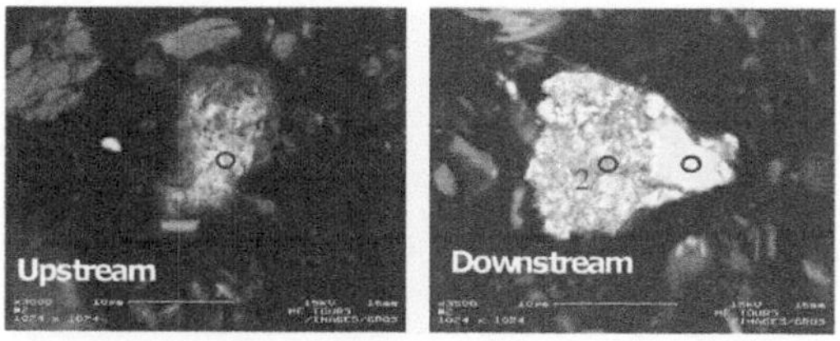

Figure 5. BSE images of Fe-oxyhydroxides (up to 1% wt. As).

strategic sites in the Red River watershed during 2008–2009 (Figure 3) showing that the highest [As] are observed at the LC site (i.e. in China); its values strongly decrease from upstream to downstream.

The [As] from 40 sites in the Red River watershed were used to establish the maps (GIS tool, Arciew®, Figure 4). The high As anomalies are observed in the upstream Red River in both high and low water levels and may be due to natural and/or anthropogenic sources in the upstream part (Figure 4).

2.3 *Mineralogical characterization*

Furthermore, micro-scale characterization of As bearing phases by SEM and EPMA suggested that As is part of highly hydrated/hydroxylated products with variable Fe:As composition (ferrihydrite-type) (Figure 5).

3 PERSPECTIVES

Future research efforts are needed to characterize water quality (e.g. As in dissolved and particulate phases including physicochemical parameters) with a high temporal and spatial monitoring in the Chinese part of the Red River as well as solid speciation (carrier phases) in order to determine the natural and/or anthropogenic sources and evaluate the bioavailability and the eco-toxicological effects.

ACKNOWLEDGEMENTS

This study was funded by the INSU (ST/EC2CO) River-Sông program.

REFERENCES

Berg, M., Stengel, C., Trang, P.T.K., Pham, H.V., Sampton, M.L. & Leng, M. 2007. Magnitude of arsenic pollution in the Mékong and Red River Deltas-Cambodia and Vietnam, *Science of the Total Environment* 327: 413–425.

Dang, T.H., Coynel A., Orange, D., Blanc, G., Etcheber, H., & Le, L.A. 2010. Long-term monitoring (1960–2008) of the river-sediment transport in the Red River Watershed (Vietnam): temporal variability and dam-reservoir impact. *Science of the Total Environment* 408: 4654–4664.

Viers, J., Dupréa, B. & Gaillardet, J. 2009. Chemical composition of suspended sediments in World Rivers: New insights from a new database. *Science of the total environment* 407: 853–868.

One Century of the Discovery of Arsenicosis in Latin America (1914–2014) – Litter, Nicolli, Meichtry, Quici, Bundschuh, Bhattacharya & Naidu (Eds)
© 2014 Taylor & Francis Group, London, ISBN 978-1-138-00141-1

Hydrogeochemistry and microbial geochemistry on different depth aquifer sediments from Matlab, Bangladesh

M.G. Kibria, M.F. Kirk & S. Datta
Department of Geology, Kansas State University, Manhattan, Kansas, USA

P. Bhattacharya, M. Hossain, M. von Brömssen & G. Jacks
KTH-International Groundwater Arsenic Research Group, Department of Sustainable Development, Environmental Science and Engineering, KTH Royal Institute of Technology, Stockholm, Sweden

K.M. Ahmed
Department of Geology, University of Dhaka, Dhaka, Bangladesh

ABSTRACT: Arsenic (As) poses the greatest hazard towards drinking water quality in Bangladesh. Tubewell drinking water is one of the main sources for household based water options in rural Bangladesh. Our study area is in Matlab Upazila, in Bangladesh. The overall objective of this research and the SASMIT project is to develop a community based initiative for sustainable As mitigation by developing a sediment color based tool for the local drillers prioritizing on the hydrogeological and biogeochemical investigations. For this purpose we analyzed different depth colored sediments and water for find out the sustainable low Arsenic contaminated aquifer.

1 INTRODUCTION

Tubewell drinking water is one of the main sources for household based water options in rural Bangladesh. Targeting shallow and low-As aquifers in Bangladesh has been a difficult task. It is argued that groundwater from reddish/light gray sediments, within shallower depth (400 ft) are As safe. Different researchers suggested that the colors of the sediments could be used as a simple tool by the local drillers to target As-safe aquifers (Biswas *et al.*, 2011; von Brömssen *et al.*, 2009). This study indicated that it is possible to assess the relative risk of elevated concentrations of As in aquifers if the color of the sediments are known which would help the local drillers to target safe aquifers for drilling new tubewells. Thus following a simplified use of this sediment color concept, a sustainable mitigation approach can be established in Matlab Upazila, and other areas in Bangladesh with similar geological settings, as well as elsewhere in the world, for improving the safe water coverage based on the initiatives of the local drillers.

2 METHODS

2.1 *Hydrogeochemical methods*

Water samples from 10 piezometer nested tube wells (average ~30–800 ft depth, each nest covers 6 different depths) were collected. Pore water samples were collected from undisturbed core samples by using rhizon samplers (www.rizosphere.com). Groundwater was analyzed for understanding the variation of As and other elements like Mn, Fe, phosphate (PO_4^{3-}), sulfate (SO_4^{2-}), Dissolved oxygen (DO), nitrate (NO_3^{2-}), ammonium (NH_4^+), chloride (Cl^-) were analyzed via Ion Chromatography (IC), various in situ probes and UV-Vis Spectrophometer. The cations (As, Ca^{2+}, Mg^{2+}, Na^+, K^+) were measured using ICP-MS. The water quality parameters like pH, alkalinity, Total Dissolved Solids (TDS), resistivity and conductivity were compared to groundwater As concentrations. Stable isotope ($\delta^{18}O$, dD) compositions of groundwater were studied to understand the recharge processes for these shallow and intermediate depths aquifers.

2.2 *Solid phase chemistry and geomicrobiology*

We collected undisturbed core samples from different depths, within aquifers. From North Matlab site the following depth samples were collected: 30, 90, 150, 210, 265, 310, 330 and 365 ft. For analyzing microbial activity on the different depth aquifers and aquitard samples, we collected undisturbed samples in 15 ml Centrifuge tubes. Once retrieved, samples were preserved under anaerobic condition immediately scraping out with a sterilized spatula from inside the core. Samples were later capped and frozen all-through before analyses. DNA was recovered from 8 different depths from the entire length of the South Matlab core. Via sequencing of DNA of these sediments, we utilized to identify microbial community composition for a complete depth profile of a core. Based on the concentration levels of As in the sediments, sequential extractions was performed and thereafter analysis by ICP-OES to understand the control of sediment fractions in housing As. Synchrotron aided µXANES and µXRD studies were conducted for solid state As speciation (As^{3+} and As^{5+}) in different depth core samples at Brookhaven National laboratory (BNL) on multiple beamlines to understand the spot mineralogy and oxidation states of arsenic as present in these sediments.

3 RESULTS AND DISCUSSION

3.1 *Hydrogeochemical results*

EC is almost same in all the samples from shallow and deeper depths, which is within 500–1500 µS/cm. Eh varies with depth in the shallow and intermediate depths and is within –40 to +80 mV. DOC is higher in all the shallow aquifer waters but it decreases with the depth and it shows very low values in the deeper depths. The major source of higher DOC in shallow aquifers may originate from different types of organic matters. Shallow to intermediate wells show less Cl concentration 0–900 mg L^{-1} and the deep tubewell concentration is 800–1500 mg L^{-1}. The presence of high chloride in the deeper aquifer could be due to high evaporation rate or by sea water influence. Only few samples from shallower depth show high nitrate concentration 15–30 mg L^{-1} where as others are very low within <5 mg L^{-1} range. The HCO_3^- ranges are 150–600 mg L^{-1} in shallow depth and 50–200 mg L^{-1} in deeper aquifers (Figure 1).

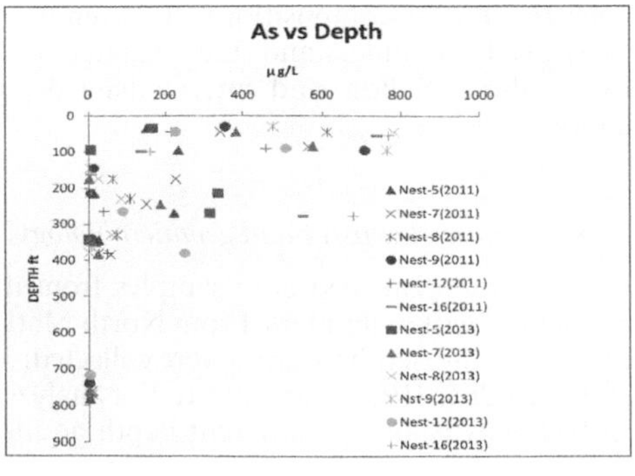

Figure 1. Changing of As concentration with time in the Matlab groundwater.

Shallow depth tube wells are mainly fine to very fine grained sediments and these aquifer sediments are mainly gray to dark gray colored. Intermediate depth sediments are medium to coarse grained sediments which are mainly light gray to brownish gray color according to Munsell (www.munsell.com) color chart. We also find that most of the similar bacterial community relates similar sediment grain sizes and colors. Our phylogenetic tree shows bacterial communities are related with sediment color, grain size and As contamination in water. No sulfate reducing bacteria is present in our samples. Most of the gray and dark gray color sediment shows As(III) and As(V) speciation from bulk XANES analyses. But light gray color sediment does not show any As speciation.

3.2 *Isotopic analyses*

The Local Meteoric Water Line (LMWL) drawn from Mukherjee *et al.* (2007). The results indicate rain water is the major source of recharge to the of Matlab groundwater. The data shows that both shallow and deep groundwater of the areas are occurring together indicating rainwater as the major source. The similarity in isotopic signatures of both shallow and deep groundwaters in the entire nest and their position towards LMWL indicates monsoonal rain as the main recharge to these areas.

4 CONCLUSIONS

Our study characterizes the different sediments from shallow and intermediate aquifers of Matlab Upazila, describe the lithofacies and the genesis of sediments and establish the relationships between aqueous and solid phase geochemistry and color of sediments which is a prerequisite for further studies on the sustainability of the oxidized low-As aquifers for drinking water supplies. This will allow to establish the relation between solid and aqueous phase geochemistry and thereby to understand the mineralogical and microbial controls on the nature and kinetics of arsenic release and other contaminants into the associated groundwater. If we succeed, it will forward this study towards a broader scale regional remediation project that incorporates the enquiry of efficiency of sediment color as an easy tool for identifying safe aquifers in major As risk prone areas.

ACKNOWLEDGEMENTS

Sida-SASMIT Contribution No. 7500085, NASA Kansas Space Grant Consortium, Geological

Society of America (GSA) Graduate Student Research Grant, NSF Hydrology No. 1054971.

REFERENCES

Biswas, A., Nath. B, Bhattacharya, P., Halder, D., Kundu, A. K., Mandal. U., Mukherjee, A., Chatterjee, D. & Jacks, G. 2011. Testing tubewell platform color as a rapid screening tool for arsenic and manganese in drinking water wells. *Environmental Science & Technology* 46: 1434–1440.

Mukherjee, A., Fryar, A. & Rowe, H. D. 2007. Regional-scale stable isotopic signatures of recharge and deep groundwater in the arsenic affected areas of West Bengal, India. *Journal of Hydrology*. 334: 151–161.

von Brömssen, M., Larsson, S. H., Bhattacharya, P., Hasan, M. A., Ahmed, K.M., Jakariya, M., Sikder, M. A. Sracek, O., Doušová, B., Patriarca, C., Thunvik, R. & Jacks, G. 2008. Geochemical characterisation of shallow aquifer sediments of Matlab Upazila, Southeastern Bangladesh—Implications for targeting low-As aquifers. *Journal of Contaminant Hydrology* 99: 137–149.

*One Century of the Discovery of Arsenicosis in Latin America (1914–2014) –
Litter, Nicolli, Meichtry, Quici, Bundschuh, Bhattacharya & Naidu (Eds)*
 ISBN 978-1-138-00141-1

Study of arsenic in groundwater tube well and borehole sediments of Padma-Jamuna Meghna deltas subjected to contamination of arsenic with its mobilization in Bangladesh

M.N.E.A. Siddique, M.A. Islam, M.A. Rahman & A.M.S. Alam
Department of Chemistry, University of Dhaka, Dhaka, Bangladesh

ABSTRACT: A huge population is at health risk from exposure to As-contaminated groundwater in Bangladesh. Water samples were collected from ground tube wells from certain arsenic affected locations, and silt and borehole sediments were also collected from the Padma, Jamuna and Meghna river bed. Arsenic was determined in both water and digested sediment samples by Ag-DDTC modified UV visible method. Most of the tube well waters are contaminated with arsenic of above 0.05 mg L^{-1} and the range of arsenic in borehole sediments (1–6 m depth) of above rivers is 2.71–9.69 mg kg^{-1}. Arsenic concentration varies with depth of the tube wall and depth of borehole near river bed. It has been observed that those tube wells are contaminated with arsenic in places where higher amounts of arsenic were found in areas of the borehole sediments.

1 INTRODUCTION

Arsenic contamination of groundwater in large areas of Bangladesh and West Bengal, India has received much attention and is considered as one of the worst environmental disasters in the world. The presence of high concentrations of arsenic in groundwater is not generally dependent on the concentration of arsenic in soils. The constituents and environmental conditions of the soil have a greater influence on arsenic speciation and mobility than the total concentration in soils. Arsenic is harming developing countries like India and Bangladesh more vehemently owing to obvious reasons. Groundwater poisoning by arsenic, positioned in the top rank as a carcinogenic agent, has created a great health concern, affecting at least 70 million people in Bangladesh and in West-Bengal. In Bangladesh, 61 out of 64 districts are contaminated with arsenic. Some districts are highly affected and some are less contaminated. It has been reported that higher amounts of arsenic were found in groundwater near the river bank districts of Padma, Jamuna and Meghna. The mass populations of those areas are drinking tube well water because it is free from pathogenic microorganism. Unfortunately, arsenic was found in tube well water beyond the acceptable limit (0.05 mg/L). The groundwater is coming from about 100 m depth. High amounts of arsenic have been found in this level. These arsenic-adsorbed solid phases are transported, entrapped and deposited at the sediment-water interface in the sediments. The aim of this work is to see the levels of contamination of As in the sediments of the Padma, Jamuna and Meghna deltas in Bangladesh.

2 EXPERIMENTAL

2.1 *Sampling sites*

Hajigong thana was considered as a sampling location for drinking tube well water. Hajigonj thana is in the downstream of the Padma-Jamuna-Meghna belt. The Hajigong is in the southern part of Bangladesh where the three rivers meet together. About 141 tube water samples from 29 villages were collected. Samples were collected in polypropylene containers and preserved with HNO_3 (pH 2–3) for further arsenic analysis. Five locations from Padma river, seven locations from Jamuna river, and five locations from Surma river and Kushiara along the Meghna river were considered for collecting sediment samples with a normal digging procedure. The silts (mesh size 5 micron) were separated from the borehole sediments. These samples were dried in oven at 110 °C for removing moisture.

2.2 *Digestion procedure of sediments*

The sediment samples were digested following the HNO_3 and $HClO_4$ digestion method as follows. Both acids were analytical reagent grade. An accurately weighed amount (0.1 g) of the sample was taken in a Teflon acid bomb. 3.0 mL of nitric acid

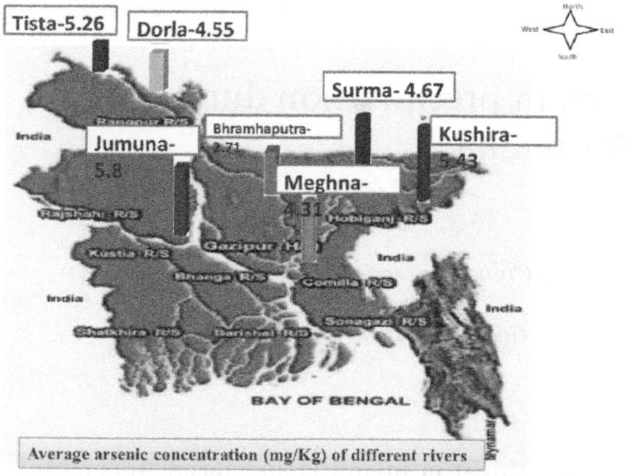

Figure 1. Average as concentration of Jumuna Delta & Meghna Delta.

and 2.0 mL of perchloric acid were added to the sample. Then, the acid bomb was placed in an oven for heating at 200 °C for two hours.

2.3 *Spectrophotometric method of analysis*

Arsine (AsH_3) generation followed by complexation with silver-diethyldithiocarbamate (Ag-SCSN $(C_2H_5)_2$) solution was used. In the method, arsenic reacts with a solution of Ag-DDTC, complexes with morpholine in chloroform to form a soluble red complex, and absorbance is measured at the maximum (535 nm).

$$\text{Ag-DDTC} + AsH_3 \rightarrow \text{As-DDTC (red color)}$$

3 RESULTS AND DISCUSSION

Among the 141-tube wells water samples of Hajigong Thana, the mean, median, mode, standard deviation, variance and range were 0.507, 0.423, 0.540, 0.389, 0.152 and 2.52 mg L^{-1}, respectively. Karl Pearson's coefficient of skewness of this distribution is 1.912, which implies that it is a negative asymmetrical skewed distribution. In this distribution, the mean (0.507) is pulled toward the low valued items. There are also some values that are much smaller than the majority. This indicates that about 60% of the values lie between 0.080 and 0.507 mg L^{-1}, and 40% of the data of this distribution is higher (0.51 to 2.60 mg L^{-1}).

Arsenic in borehole sediments varied from 2.31 to 84.25 mg kg^{-1} along the Padma river, varied from 2.91 to 7.89 mg kg^{-1} along the Jumuna river and from 2.08 to 5.84 mg kg^{-1} along the Meghna river.

The average amount of arsenic was found 9.69 mg kg^{-1} in sediments of the Padma river, 5.61 mg kg^{-1} in sediments of the Jumuna river and 4.28 mg kg^{-1} in sediments of the Meghna river, indicating that the highest arsenic content was found in sediments of Padma River. The tube wells extract water from <100 m, which is used for drinking. It was found that most (about 95%) tube wells are contaminated with arsenic. This observation indicates that there are some positive correlations of arsenic in borehole sediment with arsenic found in tube well waters in this location.

4 CONCLUSIONS

Arsenic was found in borehole sediments of locations from Padma, Jamuna, Surma and Meghna rivers in Bangladesh, which indicates that arsenic is mobilized from upstream to downstream and finally leached to groundwater level. As a result, large arsenic amounts were found in tube well water samples. Most of the people of the region are affected with arsenicosis (melanosis and keratosis) coming from drinking tube well water. Therefore drinking ground water is not the solution of arsenic problem. The government of Bangladesh should take initiatives to use surface water for drinking after proper treatment.

ACKNOWLEDGEMENTS

Authors greatly acknowledge the Ministry of Education, Government of the People's Republic of Bangladesh for financial grant to carry out this research work.

REFERENCES

Anawar, H.M., Akai, J. & Komaki, K., 2003. Geochemical occurrence of arsenic in groundwater of Bangladesh: sources and mobilization processes. *J. Geochem. Explor.* 77: 109–131.

Ferdousi, F.K., Rahman, M.A., Siddique, M.N.A., Alam, A.M.S. 2008. Environmental Impact of Arsenic on water, soil and food chain in Hajigonj, Bangladesh: A case study. *Dhaka Univ. J. Sci.* 56 (1); 107–111.

One Century of the Discovery of Arsenicosis in Latin America (1914–2014) – Litter, Nicolli, Meichtry, Quici, Bundschuh, Bhattacharya & Naidu (Eds)
ISBN 978-1-138-00141-1

Organic matter decomposition and calcium precipitation during As release in the Hetao Basin, Inner Mongolia

Y. Jia & H.M. Guo
State Key Laboratory of Biogeology and Environmental Geology, China University of Geosciences, Beijing, P.R. China
School of Water Resources and Environment, China University of Geosciences, Beijing, China

ABSTRACT: High As groundwater poses a great healthy threat in inland Hetao Basin, Inner Mongolia. The evolution of high As groundwater lead to the occurrence of groundwater with high Fe, NH_4^+ and low SO_4^{2-}, which is the same as the river delta area like Bangladesh. In the Hetao basin, the high As groundwater normally has low Ca^{2+} concentration. With the analysis of saturation index of calcite ($SI_{calcite}$) and related chemical components, microbial degradation of Dissolved Organic Matter (DOC), which was widely believed to play an important role in releasing As, was also found to promote the accumulation of HCO_3^- in groundwater and further sequester Ca^{2+} as calcite. Groundwater flow path was selected to see the scale of this Ca^{2+} sequestration and to calculate consumed DOC during the release of As. This provides some insight into the importance of DOC in As cycling in groundwater.

1 INTRODUCTION

High As groundwater was reported to occur in two different kinds of areas, river deltas and inland basins. The former groundwater is characterized with low Na^+, SO_4^{2-}, neutral pH and high Ca^{2+}, while the latter always possesses high Na^+, pH and low Ca^{2+} (Jia & Guo, 2013). This chemical variation is largely determined by climate and hydrological conditions, which also lead to the difference in mobilizing As. High Total Dissolved Solid (TDS) in inland basin groundwater was frequently found, which may promote the precipitation of major elements like Ca and Mg and accumulation of Na. The objectives of this study are to find (1) how the evolution of high As groundwater influence the fate of Ca in groundwater and (2) the implication of this influence.

2 MATERIALS AND METHODS

2.1 *Study area*

The Hetao Plain of Inner Mongolia lies on the north of the Yellow River and south of Yin Mountains, with annual precipitation from 130 to 220 mm and evaporation from 2000 to 2500 mm. With the low relief in the middle of basin, groundwater flows from the north to the middle basin as well as from the south to basin. It is located in a fault basin formed at the end of the Jurassic with fine clastic sediments mainly deposited in an inland lake. Grey sediment is always found in aquifers. There is seldom any river but rather some discharged ponds and irrigation channels distribute in the basin. The specific study area here is an area of approximately 25 km^2 right from the alluvial fan to the flat plain.

2.2 *Sampling and methods of analysis*

One hundred and five groundwater samples were collected from wells at depths >50 m in the first semi-confined aquifer, which include 13 samples approximately along the groundwater flow path. Parameters, including pH, Eh, and alkalinity were measured in field using a multiparameter portable meter (HANNA, HI 9828), while concentrations of S^{2-}, Fe^{2+}, and NH_4-N were determined using a portable spectrophotometer (HACH, DR2800) in field. Major anions were determined using ion chromatography (DX-120, Dionex), major cations by inductively coupled plasma atomic emission spectroscopy (iCAP 6300, Thermo), As and trace elements by ICP-MS (Agilent). DOC in groundwater was determined by total carbon analyzer (TOC-Vwp, Shimadzu).

3 RESULTS AND DISCUSSION

3.1 *Negative correlation between As and Ca in groundwater*

Groundwater overall presents moderate reductive condition with Oxidation-Reduction Potential (ORP) ranges from -224 to 151 mV and average value of −91.6 mV. TDS differs from 252 to 1897 mg L^{-1} with average value of 565 mg L^{-1} while pH varies from 6.67 to 8.31 with average of 7.67.

In low As groundwater, Ca/Cl meq ratio can be higher than 2.28, while approximately 0.5 was found in high As groundwater (Figure 1a). This means that in the process of As enrichment, Ca tends to precipitate. Low Ca/Cl ratio also accompany with the high HCO_3^- concentration and high $SI_{calcite}$, indicating the high HCO_3^- causes groundwater being over saturated with respect to calcite and precipitation of calcite (Figure 1b, c). Near linear relationship between HCO_3^- and NH_4-N demonstrates that the increase in HCO_3^- was mainly caused by degradation of DOC, which was proved to play a key factor controlling As mobilization not only in this study area but also in Bangladesh (Guo *et al.*, 2008; Fendorf *et al.*, 2010). Besides high DOC content were always found in groundwater with high As and HCO_3^- and low Ca/Cl ratio. So, it is obvious that DOC is a controlling factor in the processes of As release and Ca immobilization.

3.2 *Ca variation along groundwater flow path*

In order to see Ca immobilization during As enrichment, 13 samples approximately along the groundwater flow path were selected, which roughly starts from alluvial fan to the flat plain with the distance of around 5 km. From position 2.2 km to the end, groundwaters are over saturated with respected to calcite. However, the significant rise of HCO_3^- was observed along with the great increase in As concentration (Figure 2). This shows strong degradation of DOC occurs. However, an increase trend of DOC was observed along flow path, which may be explained by continuous dissolution of DOC from sediment since in the downward of flow path fine sediment always have high content of organic matter.

From 1.0 km, the begin of As rise (under detection), to 4.5 km, the maximum of As concentration (342 μg/L), consumed DOC was roughly estimated in this path. The equation (1) below was used to connect the consumed DOC and eliminated Ca in this process, which means that one mole Ca^{2+} precipitation equals one mole consumed dissolved organic carbon (CH_2O). The difference in Ca/Cl ratio

$$CH_2O \rightarrow CO_2 \rightarrow HCO_3^-$$
$$CO_3^- + Ca^{2+} \rightarrow CaCO_3 \quad (1)$$

between the initial in 1 km and in 4.5 km multiply Cl^- concentration comes out the lost Ca^{2+}, which is 741 mg L^{-1} (18.5 mmol), so the consumed CH_2O is also 18.5 mmol, which equals to 222 mg L^{-1} organic carbon. This seems to be a great amount of DOC, which is much higher than DOC concentration in groundwater. This is a rough result which may be inaccurate due to some unexpected factors. The further application of PHREEQC to calculate all possible chemical processes, including reduction process related to DOC, inorganic process like Ca precipitation step by step will give us more valuable information on roles of DOC in As mobilization and Ca^{2+} precipitation along the flow path and predict evolution of these high As groundwater.

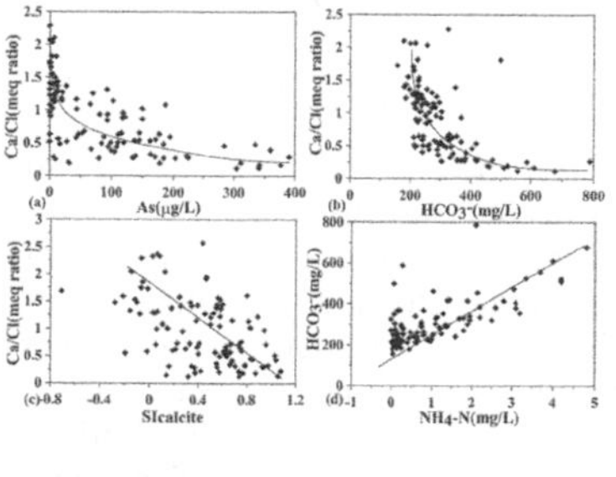

Figure 1. Plots of Ca/Cl meq ratio and As (a), Ca/Cl meq ratio and HCO_3^- (b), Ca/Cl meq ratio and $SI_{calcite}$ (c), and HCO_3^- and NH_4-N (d).

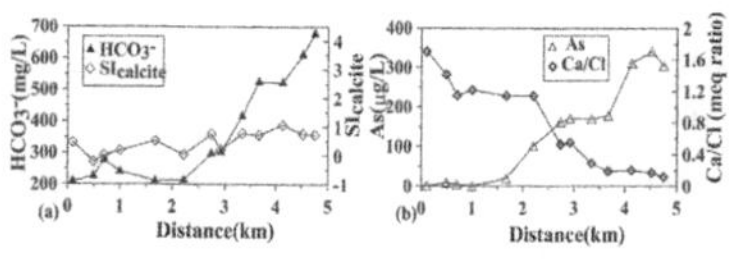

Figure 2. Plots of HCO_3^-, $SI_{calcite}$ (a) and As, Ca/Cl meq ratio (b) approximately along the flow path. Arsenic data were cited from Jia *et al.*, 2013 (in preparation).

4 CONCLUSIONS

The degradation of DOC in reducing aquifers promotes the release of As as well as Ca^{2+} precipitation in the study area. With the lost Ca^{2+}, a roughly consumed organic matter can be estimated. The detailed model simulating chemical processes including redox processes like reduction of Fe(III) and SO_4^{2-} and inorganic processes like Ca dissolution and precipitation with location and time step along the flow path will present As response to these processes and maybe help to predict the fate of As along flow path and in the future.

ACKNOWLEDGEMENTS

The study is financially supported by the National Natural Science Foundation of China (Nos. 41222020 and 41172224).

REFERENCES

Fendorf, S. Michael, H. A. van Geen, A. 2010. Spatial and Temporal Variations of Groundwater Arsenic in South and Southeast Asia. *Science* 328(5982): 1123–1127.

Guo, H.M., Yang, S.Z, Tang, X.H., Li, Y. & Shen, Z.L. 2008. Groundwater geochemistry and its implications for arsenic mobilization in shallow aquifers of the Hetao Basin, Inner Mongolia. *Science of Total Environment* 393 (1): 131–144.

Jia, Y.F. & Guo, H.M. 2013. Hot topics and trends in studies of high arsenic groundwater. *Advances in earth science* 58(1): 51–61. (In Chinese with English abstract)

Jia, Y.F., Guo, H.M., Jiang, Y.X., Wu, Y. & Zhou, Y.Z. 2013. Hydrogeochemical zonation and its implication for arsenic mobilization in deep groundwaters near alluvial fans in the Hetao Basin, Inner Mongolia (in preparation).

One Century of the Discovery of Arsenicosis in Latin America (1914–2014) – Litter, Nicolli, Meichtry, Quici, Bundschuh, Bhattacharya & Naidu (Eds)
© 2014 Taylor & Francis Group, London, ISBN 978-1-138-00141-1

Evaluation of arsenic and its controlling factors in aquifer sands of district Samastipur, Bihar, India

M. Kumar & AL. Ramanathan
School of Environmental Sciences, Jawaharlal Nehru University, New Delhi, India

P. Bhattacharya
KTH-International Groundwater Arsenic Research Group, Department of Sustainable Development, Environmental Science and Engineering, KTH Royal Institute of Technology, Stockholm, Sweden

ABSTRACT: A set of 96 water samples from shallow tubewells and 14 sediment samples from Samastipur district to know the level of arsenic (As) and its controlling factors were analyzed. Groundwater samples were collected from tubewells of different depths, and a wide range of concentrations in the range 0.19–135 μg L^{-1} was found. Scanning electron micrograph study of the sediments shows intense chemical weathering. Present study also support reductive dissolution of FeOOH triggered by organic matter oxidation as an As mobility factor in the aquifer sands.

1 INTRODUCTION

Arsenic (As) pollution in India was reported in 1976 in four states: Punjab, Haryana, Himachal Pradesh and Uttar Pradesh of Northern India (Datta & Kaul, 1976). Lower Ganga plain of West Bengal was revealed in 1984 (Garai *et al.*, 1984). Chakraborti *et al.* (2003) reported As in middle Ganga plain of Uttar Pradesh and Bihar. Up to date, As has been reported in many states of India predominantly in alluvial plains. There are some studies done on controlling factors and mobilization mechanisms, which suggest that microbially mediated oxidation of organic matter triggers reductive dissolution of FeOOH in alluvial plain (Bhattacharya *et al.*, 1997; Nickson *et al.*, 2000; Chauhan *et al.*, 2009; Kumar *et al.*, 2010). However, there are very few studies in middle Gangetic plain.

India stands second on most populated countries bearing 17.5% of the population on a limited area of 2.4%. The Gangetic plain of Uttar Pradesh and Bihar shows large area with high As pollution in water. Due to highly fertile and most populated areas, the calamity occurring is also high. This study focuses on As mobility in the aquifer sands in Samastipur district of Bihar, India, taking into account the water-sediment interaction.

2 METHODS AND EXPERIMENTAL

2.1 *Study area*

The study was conducted in the Samastipur district of Bihar state. The area of Samastipur district is 2904 km^2 and its population, reported to be 4.25 million (Census, 2011), is segmented in 20 community development blocks. The district is linked with Budhi Gandak in the northeast side and Ganga with the southwest side. The district is peneplain, intersected by numerous streams like Budhi Gandak, Baya Nadi and Balan Nadi, etc. The southwestern monsoon brings much needed rainfall, and nearly 70% of the precipitation occurs during the months of July to September. The mean annual temperature lies between 24–26 °C (Kumar *et al.*, 2010). Seven blocks, affected by high As content (Dal Singh Sarai, Vidhyapati Nagar, Mohiuddin Nagar, Mohanpur, Patory, Ujiyarpur, Sarai Ranjan), from the Samastipur district of Bihar, were selected for the proposed research work.

2.2 *Sample collection and analysis*

Ninety six-tubewell water samples from various depth, collected during pre-monsoon season (June, 2013), along with one core sediment of 40 m were analyzed for major cations and other trace elements. An Agilent 7500c (Agilent Technologies) inductively coupled plasma mass spectroscopy was used to determine the amount of As and other elements in water and sediment samples. Scanning Electron Microscope-Energy Dispersive X-ray diffraction (SEM-EDS) was used for sediment analysis.

2.3 *Analysis of Standard Reference Materials (SRMs)*

Standard Reference Materials (SRMs) 1640 (trace elements in natural water) from the National Institute of Standards and Technology (NIST) and 2711 (Montana soil) were used to verify the results for a and other elements in water and sediment.

3 RESULTS AND DISCUSSION

3.1 *Distribution of arsenic in groundwater of Samastipur district*

The range of As concentration in Samastipur district varies from 0.9–135 µg L^{-1}. Among the 96 analyzed samples, 22% exceed WHO & BIS (Bureau of Indian Standards) permissible limit of 10 µg L^{-1} for drinking. Arsenic concentration is high in tubewells logged in 10–30 m depth; below 30 m, As is within WHO & BIS permissible limit. Greater than 56% of the samples have higher concentrations of As(III) than those of As(V). Iron concentration was relatively less with a median value of 259 µg L^{-1}. Arsenic shows poor correlation with iron due to precipitation (Figure 1).

The correlation matrix shows a very good relationship with HCO_3^-, TOC (Total Organic Carbon), NO_2^- and NH_4^+, indicating reducing environmental conditions and microbial degradation as factor for As mobilization.

3.2 *Distribution of arsenic in the aquifer sands of the Samastipur district*

Concentration of As varies from 15.6 to 19.7 mg kg^{-1} at 9.1 m. Low concentration of As (7–4 mg kg^{-1}) prevails at 12–15 m and then increases at up to 13 mg kg^{-1} at 18 m and further decreases (4–5 mg kg^{-1}) up to 33.5 m.

Deposition of the weathered material brought from the Himalaya, which undergoes further chemical weathering in the alluvial basin, and trigger mobilization of several anions and cations (Singh *et al.*, 2004; Chauhan *et al.*, 2009). Iron concentration was less due to precipitation of the minerals. Present study area shows intense chemical weathering influences As mobilization (Figure 2).

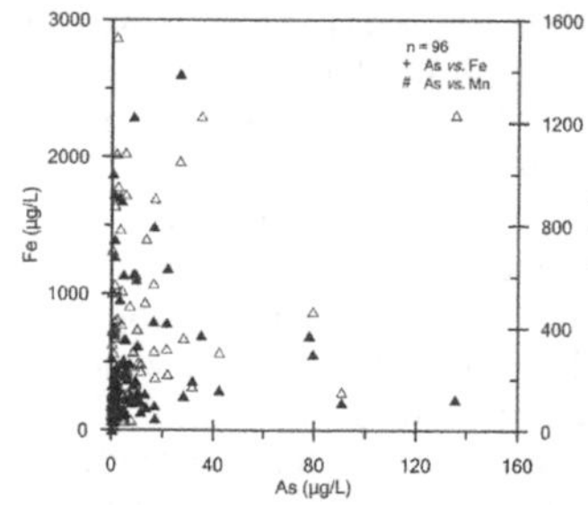

Figure 1. Arsenic shows poor correlation with iron.

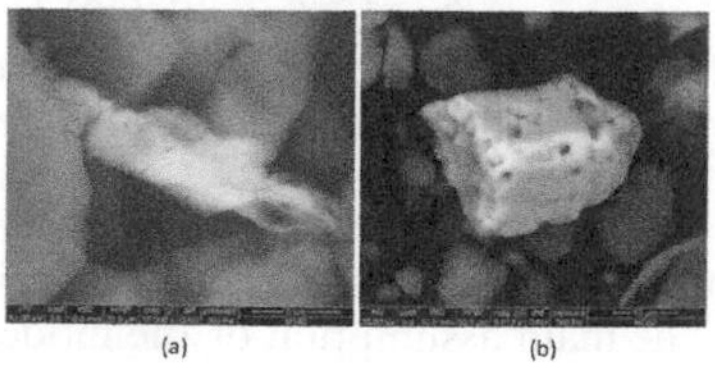

Figure 2. Scanning electron micrograph showing evidence of weathering (a) Fe-oxide precipitated (b) highly weathered zircon.

4 CONCLUSIONS

There is spatial variation in As concentration all over the study area. Release of As from aquifer sand of Samastipur supports reductive dissolution of FeOOH triggered by intense chemical weathering and precipitation of iron minerals. Concentration of As decreases in deeper aquifers.

ACKNOWLEDGEMENTS

Authors are thankful to Center for Environmental Risk Assessment and Remediation, University of South Australia for providing lab facilities and to Crawford fund for providing financial support. KTH-Sida and Linnaeus Palme program (KTH-JNU) also acknowledged for partial support.

REFERENCES

Acharyya, S.K. & Shah, B.A. 2007. Groundwater arsenic contamination affecting different geologic domains in India--a review: influence of geological setting, fluvial geomorphology and Quaternary stratigraphy. *J. Environ. Sci. Health A Tox. Hazard Subst. Environ. Eng.* 42(12): 1795–1805.

Bhattacharya, P., Chatterjee, D. & Jacks, G., 1997. Occurrence of arsenic-contaminated groundwater in alluvial aquifers from the Bengal Delta Plain, Eastern India: options for a safe drinking water supply. *Int. J. Water Resource Development* 13: 79–92.

Chakraborti, D., Mukherjee, S.C., Pati, S., Sengupta, M.K., Rahman, M.M. & Chowdhury, U.K. 2003. Arsenic groundwater contamination in Middle Ganga Plain, Bihar, India: a future danger. *Environ. Health Perspect.* 111(9): 1194–201.

Chauhan, V.S., Nickson, R.T, Chauhan, D., Iyengar, L. & Sankararamakrishnan, N. 2009. Ground water geochemistry of Ballia district, Uttar Pradesh, India and mechanism of arsenic release. *Chemosphere* 75(1): 83–91.

Datta, D.V. & Kaul, M.K. 1976. Arsenic content of tubewell water in villages in northern India. A concept of arsenicosis. *J. Assoc. Phys. India* 24: 599–604.

Garai, R., Chakraborti, A.K., Dey, S.B. & Saha, K.C. 1984. Chronic arsenic poisoning from tubewell water. *J. Ind. Med. Assoc.* 82(1): 34–35.

Kumar, P. Kumar, M., Ramanathan, A.L. & Tsujimuri, M. 2010. Tracing the factors responsible for arsenic enrichment in groundwater of the middle Gangetic Plain, India: a source identification perspective. *Environ. Geochem. Health* 32(2): 129–146.

Nickson, R.T., McArthur, J.M., Ravenscroft, P., Burgess, W.G. & Ahmed, K.M. 2000. Mechanism of arsenic release to groundwater, Bangladesh and West Bengal. *Applied Geochemistry* 15: 403–413.

Singh, B. 2004. Late Quaternary history of the Ganga Plain. *J. Geol. Soc. India* 64: 431–454.

One Century of the Discovery of Arsenicosis in Latin America (1914–2014) –
Litter, Nicolli, Meichtry, Quici, Bundschuh, Bhattacharya & Naidu (Eds)
© 2014 Taylor & Francis Group, London, ISBN 978-1-138-00141-1

Modeling the fate of arsenic in a fluvial confluence: A case study

P.A. Guerra, C.R. Gonzalez, C.E. Escauriaza, V. Morales, G.E. Pizarro & P.A. Pasten
Pontificia Universidad Católica, Santiago, Chile

ABSTRACT: Complex chemical, hydrological and hydrodynamic interactions occur at fluvial confluences receiving the contribution of streams with high arsenic concentrations. Unraveling this complexity may lead to control strategies towards more sustainable drinking water supply, or reduced environmental risks in arsenic-laden watersheds. A confluence affected by acid drainage in the Lluta River watershed, Northern Chile, was used as a study model. We coupled geochemical and computational fluid dynamic simulations to model the 3D profiles of water chemistry and arsenic sorbed to particles formed at the confluence. The model revealed that distinct mixing patterns control the arsenic load and the fate of these reactive particles. Further control of the confluence mixing conditions may lead to enhance natural conditions for arsenic attenuation for removal from the bulk water.

1 INTRODUCTION

The complex interactions among chemical, hydrological and hydrodynamic processes make fluvial confluences natural reactors that are capable of controlling the fate and transport of contaminants through drainage networks. These processes are notorious in confluences affected by Acid Mine Drainage (AMD), where particles of iron (Fe) and aluminum (Al) oxides and oxyhydroxides form (Sarmiento et al. 2009). These particles sorb toxic trace elements like arsenic (As) from solution, providing a natural attenuation mechanism for dissolved As.

Our observations of the Azufre River—Caracarani River confluence located in the Chilean Altiplano (17°50’21” S, 69°42’26” W) show the complexity of arsenic transport. While the mixing ratio between the two streams drives chemical reactions, the hydrodynamics characterized by 3D velocity fields confirm that the river downstream is far from being a completely mixed reactor, as particles and contaminants are heterogeneously distributed. In this work, we seek to better understand chemical-hydrodynamic interactions at the confluence by coupling geochemical and 3D computational fluid dynamics simulations. We aim to obtain and analyze 3D profiles of water chemistry and arsenic sorption on particles formed at the confluence to elucidate how the flow mixing patterns determine the chemical and physical fate of arsenic.

The performance of the Azufre River-Caracarani River confluence may have noticeable impacts on the suitability and sustainability of using the Lluta River as a drinking water source for urban and rural communities in this extremely arid environment.

2 METHODS

2.1 *Hydrodynamic—geochemical modeling*

Computational Fluid Dynamics (CFD) simulations coupled with an advection-diffusion model were previously carried out upon a geometrically simplified version of this confluence. The parameter computed through a three dimensional grid to show the heterogeneous mixture was w_{Azufre} (Q_{Azufre}/Q_{Total}). The global mixing ration modeled was 0.3 L/L. The length of simulation (field scale) was 114 m.

The geochemical model consisted of a mixture of water from both rivers. Water qualities are indicated in Table 1. For the mixture, the formation of amorphous $Fe(OH)_3$ and $Al(OH)_3$ and the sorption of arsenate (As(V)) onto these phases was calculated. The PHREEQC code was used along with the wateq.4f database, which contains the surface complexation models of Dzombak & Morel (1990) for hydrous ferrous oxides (HFOs). The Karamalidis & Dzombak (2010) model was used for arsenic sorption on gibbsite (as a proxy for $Al(OH)_3$). These calculations were performed for a range of w_{Azufre} from 0 to 1 L/L (with 0.01 L/L intervals). The results were incorporated to the existing hydrodynamic model, obtaining three dimensional profiles of pH, reactive phases and dissolved arsenic. The simulations of w_{Azufre} are an average of approximately 40 seconds of hydrodynamic simulations. The main assumption of the model is that reactions occur within this time frame.

Table 1. Water quality considered for geochemical simulations.

Parameter	Azufre river	Caracarani river
Flowrate (L/s)	45–245	170–641
pH	2.2	8.7
Temperature (°C)	10	10
pe	9.0	9.0
Fe^{3+} (mg/L)	47	0.57
Al^{3+} (mg/L)	156	0.7
As(V) (mg/L)	2.1	0.05
SO_4^{-2} (mg/L)	2450	477
Cl^- (mg/L)	920	190
Na^+ (mg/L)	333	170
K^+ (mg/L)	88	20
Mg^{2+} (mg/L)	159	55
Ca^{2+} (mg/L)	233	120

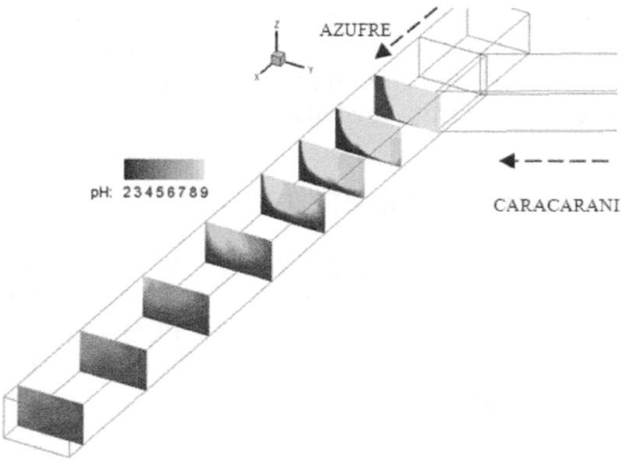

Figure 1. pH profile after the confluence. There is a transition from a clear interface between the two rivers towards complete mixture.

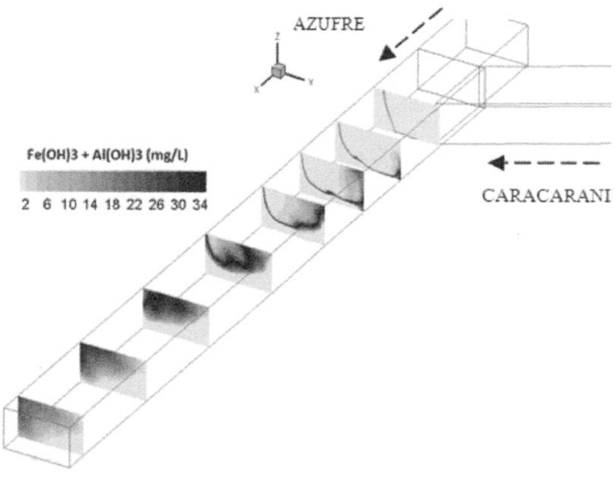

Figure 2. Profile of $Fe(OH)_3$ + $Al(OH)_3$. Particles start to form at an interface. Maximum particle concentration is reached several meters after the junction point.

3 RESULTS AND DISCUSSION

3.1 *3D distribution of pH*

Profiles of pH contribute to identify the reactive post-confluence zones. Confluence hydrodynamics reveals mixing patterns of how the streams blend in to each other. Figure 1 shows two mixing zones. The first zone occurs where a clear interface between the Azufre River (AMD tributary) and the Caracarani River is observed. The interface is a layer between these two tributaries. The pH of this interface fluctuates between 4 and 5. The second mixing zone is a transition zone towards complete mixture. In waters affected by acid streams, a mixture of reactive hydroxides are formed; $Al(OH)_3$ precipitates at a pH > 4.5 and $Fe(OH)_3$ at a pH > 3. Through this transition zone, there is a wider "reactive" pH band. The influence of the highly acidic Azufre River results in a complete mixture of pH 3. Complete mixture is not presented in the model.

3.2 *3D distribution of reactive phases and sorption of As(V)*

Figure 2 shows the distribution of the concentration of $Fe(OH)_3$(s) and $Al(OH)_3$(s) particles formed after the confluence. A mixture of these phases forms at the interface (pH 4–5). However, maximum concentration is reached several meters after the confluence. Approximately 50 m after the junction point, a wide band of particles >30 mg/L is observed. As expected, this band tends to disappear towards the zone of complete mixture zone (pH = 3). These phases dissolve when complete mixture is reached, consequently releasing As from their surfaces.

Field observations (Guerra et al. 2012) reported particle settling areas and arsenic rich rock coatings downstream the confluence area. Furthermore, field measurements indicated that the reactive mixing and settling of As-bearing particles downstream from the confluence resulted in the removal of ~50% of the arsenic flux in the bulk water. This is a relevant discrepancy from the hydrochemical model, as it reveals an important mechanism in the removal of As from the water column.

4 CONCLUSIONS

Arsenic is attenuated by reactions triggered by distinct mixing patterns and by the later settling of these particles. Despite the limitations of the model (e.g., kinetics for As sorption, precipitation/dissolution and particle aggregation are not included), this integrated approach helps to visualize mixing scenarios that are not obvious for the fate and transport of particles. Further control and design of the confluence mixing conditions may lead to enhance natural conditions for arsenic attenuation and removal from the bulk water.

ACKNOWLEDGEMENTS

This work was financed by CONICYT/ FONDAP 15110020 project and CONICYT/ FONDECYT 130936 project. The CONICYT-PCHA/Doctorado Nacional/2010-21100173scholarship is also acknowledged.

REFERENCES

Dzombak, D.A. & Morel, F.M. 1990. *Surface Complexation Modeling: Hydrous Ferric Oxide.*. New Jersey: Wiley.

Guerra, P.A, Simonson, K. & Pasten, P. 2012. The fate of arsenic in sediments formed at a river confluence affected by acid mine drainage. *AGU Fall Meeting; 3–7 December 2012.* San Francisco: American Geophysical Union.

Karamalidis, A. & Dzombak, D. 2010. *Surface Complexation Modeling: Gibbsite.* New Jersey: Wiley.

Sarmiento, A., Nieto, J., Olías, M. & Cánovas, C. 2009. Hydrochemical characteristics and seasonal influence on the pollution by acid mine drainage in the Odiel river Basin. *Applied Geochemistry* 24(4): 697–714.

One Century of the Discovery of Arsenicosis in Latin America (1914–2014) – Litter, Nicolli, Meichtry, Quici, Bundschuh, Bhattacharya & Naidu (Eds)
© 2014 Taylor & Francis Group, London, ISBN 978-1-138-00141-1

Arsenic in soils and sediments from Paracatu, MG, Brazil

E.D. Bidone
Departamento de Geoquímica, Instituto de Química, Universidade Federal Fluminense (UFF), Centro, Niterói, RJ, Brasil

Z.C. Castilhos
Departamento de Geoquímica, Instituto de Química, Universidade Federal Fluminense (UFF), Centro, Niterói, RJ, Brasil
Centro de Tecnologia Mineral (CETEM), Cidade Universitária, Rio de Janeiro, Brasil

M.C.B. Santos, R.S.V. Silva & R.G. Cesar
Departamento de Geoquímica, Instituto de Química, Universidade Federal Fluminense (UFF), Centro, Niterói, RJ, Brasil

L.C. Bertolino
Centro de Tecnologia Mineral (CETEM), Cidade Universitária, Rio de Janeiro, Brasil

ABSTRACT: Soils and sediments are mainly of concern in this work for two reasons: firstly, because a Brazilian quality criteria for soils and sediments is available, and secondly, due to the geochemical behavior of As in these media, which might allow its transference to the drinking water, the major environmental route of exposure to the human population. This work evaluated As levels in these matrixes and made considerations on its geochemical distribution subsequently site-specific environmental and geological gold mining characteristics. 70% of soil samples and 90% of sediment samples are above the legal values established by Brazilian legislation for As. The As levels are influenced by the distance from the ore body and the deactivated gold mining activities.

1 INTRODUCTION

According to Möller *et al.* (2001), "Morro do Ouro" gold ore deposit contains Au grade from 0.4 to 0.6 g Au t^{-1} and As levels >4,000ppm; followed by the oxidized minerals and the phyllites host rocks, that show decreased Au grade and As concentration (from 0,4 to 0,5 g Au t^{-1} and >1,000ppmAs; and 0,15 g Au t^{-1} and <1,000 mg L^{-1} As, respectively). The Morro do Ouro gold is associated primarily with arsenopyrite (FeAsS). Anomalous gold and sulphide mineralizationt is localized within a 120–140 meter thick high strain zone that dips (20°) to the SW and is traceable for over 6 km along a NE-SW trend, and more than 3 km in width (Henderson, 2006). Soils in Paracatu are Latosol type—a highly weathered and well-drained soil. The objective of this work was to investigate the As levels in soils and sediments and the process controlling As mobility. This study is part of the environmental and health assessment conducted by Brazilian research institutions under the general coordination of CETEM.

2 METHODS

2.1 *Sampling*

The sediments (20) and soils (21) samples were collected during the dry season (September-October), including sampling points outside the gold mining area along the three watershed which are sub-basins of Paracatu River watershed: (i) Córrego Rico—its spring is located inside the open pit mine site and at its medium segment is located at Paracatu Municipality; (ii) Ribeirão Entre-Ribeiros—it is downstream of tailing dam and potentially receives its effluents; (iii) Rio Escuro—outside the direct influence of gold mining, it is considered as a reference area and it is the main drinking water source for urban population of Paracatu. Samples were dried and sieved: 1,7 mm (Total) and 0,075 mm (silt-clay) fractions. As levels were analyzed by ICP-MS (Model 42 IC-MS, Perkin Elmer MS). The limit of detection was 0.5 mg kg^{-1}.

3 RESULTS AND DISCUSSION

3.1 *Results*

I. *Soils*. a) Soils showed ~15% of <0,075 mm fraction for all samples studied; b) As levels (mean ± sd) resulted as follows: (i) Córrego Rico – 462.62 ± 572.11 mg L^{-1} (n = 8). It was found a negative gradient of As levels from up to the downstream. Whereas As in soils from a sampling point close to (2 Km) the gold mining open pit (upstream) resulted in 1,752.90ppm, samples far from (70 Km) the mine (downstream) showed 5.4 mg L^{-1}. (ii) Ribeirão Entre-Ribeiros – 215.37 ± 371.99 (n = 9). The highest As levels (>1,131.50ppm and 467.3ppm) were found in the abandoned open pits and small scale gold mining sites (SSGM), respectively. As level downstream (7 Km) of the tailing dam was 81.5ppm. (iii) Rio Escuro – 11.68 ± 7.35ppm (n = 4). The As level in soils close to the drinking water supply to Paracatu city was 16.3ppm. II. *Sediments*. a) Sediments showed distinct percentage of <0,075 mm among the watershed studied, varying from 7 ± 4% (Córrego Rico) to 19 ± 22% (Ribeirão Entre-Ribeiros). b) As levels (mean ± sd) resulted as follows: (i) Córrego Rico– 1,616.43 ± 1,720.84 (n = 7). A negative gradient of As levels from upstream (4,297.20 mg L^{-1}, 2 Km far from open pit) to downstream (291.80 mg L^{-1}, 70 Km far from mining site) was found once more. (ii) Ribeirão Entre-Ribeiros – 1,114.50 ± 1,158.00 (n = 10). The highest As levels (1,638.20ppm and 3,216.80 mg L^{-1}) were found in the abandoned open pits and SSGM sites, respectively. Close to 7 km downstream from the tailing dam, the As level reaches 159.00 mg L^{-1}. (iii) Rio Escuro – 40.77 ± 56.41 mg L^{-1} (n = 3). The As level in sediments at the drinking water supply site to Paracatu city is 12.8ppm. III. *Statistics*. a) The mean of As in sediments and soils, considering all samples resulted in 1,207.36 ± 1408,00 mg L^{-1} and 280.30 ± 467.50 mg L^{-1}, respectively. As concentration in sediments are higher than in soils (Student's t-test: p = 0.01, α = 0.05, n = 18) and there is a significant correlation between them (Pearson's r = 0,499, α = 0,05, n = 18). b) As levels in sediments, as well as in soils, from impacted watershed (Córrego Rico e Ribeirão Entre-Ribeiros) are similar. c) As levels in sand (<1,7 mm > 0,075 mm) and in silt-clay (<0,075 mm) fractions of soils and sediments showed no differences. However, the two size fractions were significantly correlated (α = 0,01).

3.2 *Discussion*

The normal range of As in soils is 1 to 40ppm, but much higher levels (more than 1000 mg L^{-1}) can be localized in the neighborhood of gold (Au) mining plants. Natural concentrations of As in sediments are typically lower than 10 mg L^{-1}, but they can vary widely around the world, e.g., from about 0.1 to 4,000 mg L^{-1} (ATSDR 2007). I. *As in soils*. High As levels as 1000 mg L^{-1} are observed at watersheds located in the same direction of the gold ore body (Rico and Entre-Ribeiros). Over geological time, the ore body and phyllites host have been submitted to intense weathering processes, thus resulting in mineralogical and chemical changes (e.g., arsenopyrite and/or other As sulphites oxidation, As releases and consenquent sorption by Fe and Al oxides and/or other weathering materials) and erosions, originating secondary gold deposits (eluvial deposits, alluvium and laterites) mined by SSGM in the past. The decreasing gradient of As levels from upstream to downstream might be related to the transference sharply from the ore body to the rocks and soils. II. *As in soils compared to the Brazilian legal criteria*. The Resolution n. 420/2009 of "Conselho Nacional do Meio Ambiente"(CONAMA) determines quality criteria for soil (dry weight, dw) for distinct land uses: 15ppm As for prevention; 35ppm for agriculture areas, 55ppm for residencial areas and 150ppm for industrial areas. (i) Córrego Rico—While 75% of samples showed results above the value for industrial areas, 25% showed results below the prevention value. (ii) Ribeirão Entre-Ribeiros – 22% of samples are above the value for industrial areas whereas 33% is below the prevention value (iii) Rio Escuro – 100% of samples showed As levels close or below the prevention criterion, including samples from the watershed, which is located the drinking water source to Paracatu population. III. *As in sediments*. As levels in sediments may be considered high. The spatial distribution profile of As in sediments is similar to that found in soils. IV. *As in sediments compared to the Brazilian legal criteria*. The Resolution CONAMA n. 454/2012 determines quality criteria for sediments (dw). As in freshwater sediments: 5.9 mg L^{-1} As (Level 1, L1 – low probability of adverse effects level on biota) and 17.0 mg L^{-1} As (Level 2, L2 – higher probability of adverse effect level on biota). (i) Córrego Rico (n = 7)- 86% of samples resulted >L2 and 14% between L1 and L2. (ii) Ribeirão Entre-Ribeiros (n = 10) - 90% of samples >L2 and 10%<L1. (iii) Rio Escuro (n = 3) - 2 samples <L1 and 1 sample >L2. At the drinking water source supply to Paracatu population, As concentration is 12,8 mg L^{-1}. V. *Soils and sediment relationship and granulometry*. The reasons why As levels in sediments are higher than in soils are still under study by mineralogical analysis and sequential extractions. One could hypothesize that the retention of sediments, especially at low energy for transport freshwater segments (e.g., swamps and abandoned SSGM open pits), overall during the dry season, may increase the As levels in sediment samples. This argument is supported by similar spatial gradient of As levels found for soils and sediments, alongside the most impacted watersheds. The analogous distribution of As between sand and silt-clay fractions of soils and sediments may be resulted are due to the fact that arsenate is mainly adsorbed by Fe an Al oxides, which may originate fine grains, several sized aggregates, coating on mineral grains and association with weathered aluminosilicates.

4 CONCLUSIONS

Soils and sediments showed high As levels (>1,000 mg L^{-1}). 70% of soil and 90% of sediment samples showed As concentrations above the Brazilian quality criteria. As levels found in sediments were 4 times higher than in soils. Additionally, they are correlated and showed decreasing concentration as the mining site distance increases, but there are also associations with abandoned SSGM open pits and low energy freshwater environment. Additional studies are being conducted to better understand the behavior As considering the specific site conditions.

REFERENCES

ATSDR, US Agency for toxic substances and Disease Registry 2007. Toxicological profile for arsenic. Atlanta, Georgia.

Henderson, RD. 2006. Paracatu Mine Technical Report. p 179.

Möller, J.C., Batelochi, M., Akiti, Y., Maxwell, S. & Borges, A.L. 2001. Geologia e caracterização dos recursos minerais de Morro do Ouro, Paracatu, Minas Gerais. In C.P. Pinto & M.A. Martins-Neto (eds), *Bacia do São Francisco: Geologia e Recursos Naturais, p. 199–234 – SBG/MG – Belo Horizonte.*

One Century of the Discovery of Arsenicosis in Latin America (1914–2014) – Litter, Nicolli, Meichtry, Quici, Bundschuh, Bhattacharya & Naidu (Eds)
© 2014 Taylor & Francis Group, London, ISBN 978-1-138-00141-1

Arsenic in atmospheric particulate matter in Paracatu, MG, Brazil

W.Z. de Mello & J.A. Matos
Departamento de Geoquímica, Instituto de Química, Universidade Federal Fluminense (UFF), Centro, Niterói, RJ, Brasil

Z.C. Castilhos
Departamento de Geoquímica, Instituto de Química, Universidade Federal Fluminense (UFF), Centro, Niterói, RJ, Brasil
Centro de Tecnologia Mineral (CETEM), Cidade Universitária, Rio de Janeiro, Brasil

L.I.D. da Silva, M.C. Carneiro & M.I.C. Monteiro
Centro de Tecnologia Mineral (CETEM), Cidade Universitária, Rio de Janeiro, Brasil

ABSTRACT: Concentrations of As were determined in atmospheric suspended particulate matter at eight air monitoring stations installed in the surroundings of an open pit gold mine in Paracatu (state of Minas Gerais, Brazil). The concentrations of As associated with total suspended particulates varied from < 0.64 to 18.8 ng m^{-3} (mean = 5.7 ± 4.0 ng m^{-3}). Concentrations of As varied spatially and seasonally, controlled by origins of soil-derived dust, prevailing wind direction (northeast), wind speed and seasonal distribution of precipitation.

1 INTRODUCTION

Arsenic (As) is an element found naturally in the lithosphere, hydrosphere, biosphere and atmosphere. However, it is one of the most toxic elements to humans, whose exposure usually occurs through food, water and air. Background levels for As in air are typically 0.2–1.5 ng m^{-3} for rural and 0.5–3 ng m^{-3} for urban areas (WHO, 2001). The city of Paracatu (MG) is potentially exposed to air pollution caused by As associated with atmospheric particulate matter (PM) arising from operational activities related to a nearby open pit gold mining. The objective of this study was to investigate the concentrations of As associated with atmospheric PM and the major processes controlling their spatial and temporal variations in Paracatu. This study is part of the environmental and health assessment conducted by Brazilian research institutions under the general coordination of CETEM.

2 METHODS

2.1 *Study area*

The municipality of Paracatu (8,229.6 km^2) is located in the northwest border of Minas Gerais (MG) state (southeast region of Brazil) and has a population of *ca.* 90,000 inhabitants, of which about 95% live in the Pacaratu city. The economy is centered on cattle raising, agriculture (mainly soybean, corn, rice and beans) and gold mining. The gold-mine operations in Morro do Ouro, a low-grade gold deposit located close to the northern border of Paracatu city, began in the late 80 s. Nowadays, it is the largest open pit gold mine in the world. Local mean annual precipitation is *ca.* 1,424 mm (for the period 2004–2012). The region has distinct April-September dry (290 mm) and October-March rainy (1,134 mm) seasons. Mean annual temperature is 21 °C with mean monthly values varying from 17 to 24 °C. The prevailing wind direction is northeast (NE), with monthly mean wind speed varying from 2.8 to 4.4 m s^{-1}, with maximum in the dry season.

2.2 *Sampling and analysis*

Atmospheric PM samples, collected from May 2011 to June 2012, by the gold mining company were requested by the Municipality of Paracatu for determination of As. Sample collection, on glass fiber filters, was carried out at eight air quality monitoring stations, installed in the surroundings of Morro do Ouro (within *ca.* 5 km radius), which belong to an air monitoring network of the gold mining company established in the area. Each station has a total suspended particulate (TSP) high-volume sampler and one of them has a PM10 (PM with aerodynamic diameters less than 10 μm).

TSP and PM10 samples were collected for a 24-h sampling period. For the determination of As, one 1″ × 8″ strip of each filter was inserted in a 50-mL polypropylene tube and ultrasonically extracted (at 70 °C for 3 h) with 10 mL of a mixture of HCl (16.75%) and HNO_3 (5.55%). The resulting solution was allowed to cool and then completed to 20 mL with ultra-pure water. The next steps were centrifugation, filtration through a 0.45 μm pore size Nylon membrane, and analysis by ICP-OES. Recovery efficiencies were in the range of 80–120%. The limit of detection (LoD) was 7.5 μg As L^{-1} (in solution) and 0.64 ng As m^{-3} (in air). Concentrations of As in blank glass fiber filters were below the LoD.

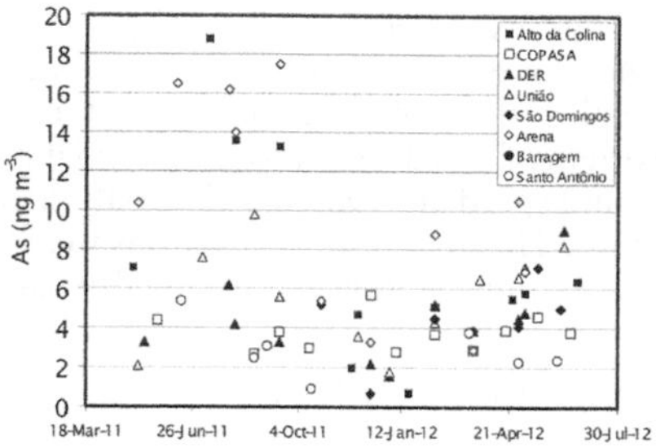

Figure 2. Seasonal variation of concentrations of As associated with total suspended particulates measured at 8 air quality monitoring stations in Paracatu (MG).

3 RESULTS AND DISCUSSION

3.1 *Concentrations*

Arsenic content was determined in 112 TSP samples, corresponding to 14 samples from each of the eight air monitoring stations (one sample for each month), of which 42 samples (38%) were below the LD. Concentrations of As in TSP measured at the eight stations varied from < 0.64 to 18.8 ng m^{-3}, with mean (± standard deviation) and median values of 5.7 ± 4.0 ng m^{-3} and 4.6 ng m^{-3}, respectively. The highest concentration was found at the Alto da Colina station. At the Arena station, where both TSP and PM10 were sampled, of the 14 samples provided from each of these sample categories, 6 of them were collected simultaneously and exhibited As concentration values above the LoD. The average PM10/TSP As concentration ratio (n = 6) was 0.61 (median = 0.65), varying from 0.36 to 0.84.

3.2 *Spatial variation*

Figure 1 shows the mean concentrations of As associated with TSP measured at the eight air monitoring stations. The mean concentrations of As varied spatially from < 0.64 to 10.2 ng m^{-3}.

The lowest mean concentrations of As were found at Barragem and Santo Antonio stations, both located upwind of the gold mine of Morro do Ouro. Concentrations of As at Barragem station were all (n = 14) < 0.64 ng m^{-3}. The highest mean concentrations of As were found at the Arena and Alto da Colina stations, both located downwind and near (0.5–1 km) to the current active mining area of Morro do Ouro. Interestingly, Santo Antonio station exhibited the highest mean TSP concentration (51 μg m^{-3}; geometric mean = 38 μg m^{-3}) and the lowest mean mass concentration of As in the TSP (2.6 μg As g^{-1} TSP). The Arena station had the second highest mean TSP concentration (48 μg m^{-3}; geometric mean = 36 μg m^{-3}) and the highest mean mass concentration of As in the TSP (6.7 μg As g^{-1} TSP). Therefore, the highest concentrations of As found at the sites located nearest to the southwest border of the current active mining area of Morro do Ouro are attributed to emissions of dusts from operational activities, the higher concentrations of As in the dust, and the predominant NE winds.

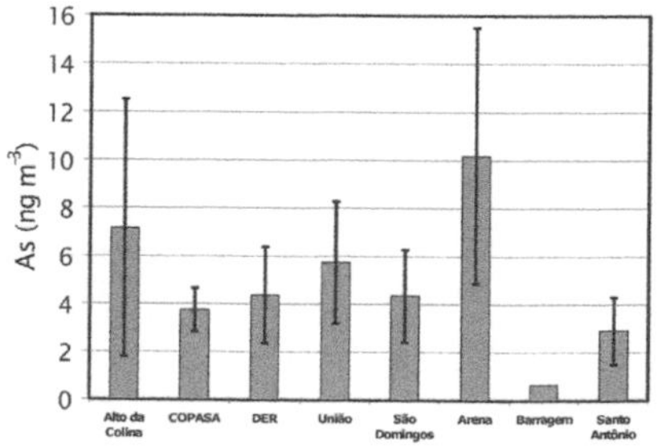

Figure 1. Mean (± standard deviation) concentrations of As associated with total suspended particulates at eight air quality monitoring stations in Paracatu (MG). At Barragem station, all concentrations were below the LoD (< 0.64 ng m^{-3}).

3.3 *Seasonal variation*

Concentrations of As associated with TSP varied seasonally. During the study period, the highest concentrations occurred between May and September, and the lowest between October and April (Figure 2). These periods overlap respectively the dry and rainy season pattern for the study region. Accordingly, during the rainy season precipitation plays an important role mitigating dust dispersion from the mining area. Besides, mean wind speed increases progressively from 2.9 m s^{-1} in April to 4.2 m s^{-1} in September, decreasing subsequently.

4 CONCLUSIONS

Mean concentration of As in atmospheric TSP was 5.7 ± 4.0 ng m^{-3}, varying from < 0.64 to 18.8 ng m^{-3}.

Concentrations varied spatially and seasonally by more than an order of magnitude, controlled by a combination of anthropogenic and natural processes (origin of soil dusts, prevailing wind direction (NE), wind speed and seasonal distribution of precipitation). The highest concentrations were found at Alto da Colina station and during dry season.

ACKNOWLEDGEMENTS

To Brazilian Agencies CAPES and CNPq.

REFERENCE

World Health Organization (WHO), 2001. Arsenic and Arsenic Compounds – Environmental Health Criteria 224. 2nd Edition. WHO, Geneva.

One Century of the Discovery of Arsenicosis in Latin America (1914–2014) – Litter, Nicolli, Meichtry, Quici, Bundschuh, Bhattacharya & Naidu (Eds)
© 2014 Taylor & Francis Group, London, ISBN 978-1-138-00141-1

The influence of nutrients on arsenic adsorption onto nanomaghemite

D. Martínez-Fernández, M. Vítková, Z. Michálková & M. Komárek
Department of Environmental Geosciences, Faculty of Environmental Sciences, Czech University of Life Sciences Prague, Prague, Czech Republic

ABSTRACT: The effects of the iron nano-oxide maghemite (Fe_2O_3) on the bioavailability of trace elements (As, Al, Cu, Fe, Mn, Mg, Zn) were tested by a set of adsorption experiments, performed with a soil solution with/without the nanomaghemite (NM) and increasing concentrations of nutrients (K, N and P). For this purpose, KNO_3, NH_4NO_3 and KH_2PO_3 were added in different proportions to a soil solution obtained from an As-contaminated soil, to create a set of combination of them during NM-nutrients interaction, and study the changes in the availability of the elements with emphasis on the competition between contaminant and nutrients for the sorption sites. NM was able to decrease As concentration in a soil solution, but among the nutrients studied, the presence of P reduced the effectiveness of the NM for As adsorption. The concentration of Ca, Mg, Mn and Na were also affected by K, N and P in the soil solution probably because of competition for the sorption sites in the NM.

1 INTRODUCTION

Bioremediation of contaminated soils with metals and metalloids through the establishment of a vegetation cover (phytostabilization) is a viable alternative for their recovery and conservation of surrounding areas and groundwater. Conditioning the soil is a key factor for the survival and the growth of plants, with nanomaghemite (NM) being a promising material for the stabilization of inorganic pollutants (Mueller and Nowack, 2010). Due to their reactivity and relatively large specific surface area (tens to hundreds of $m^2 g^{-1}$), iron nano oxides (particle size of 1–100 nm) are important scavengers of contaminants (Waychunas *et al.*, 2005), especially for As (Kim *et al.*, 2012), being a good choice for contaminated soils because of their oxidative properties (Komárek *et al.*, 2013). Since low nutrient content is one of the most-limiting problems for growing plants in contaminated soils (Walker *et al.*, 2007), there is a need to investigate if the fertilization might reduce the effectiveness of the NM.

The effect of NM on the bioavailability of Trace Elements (TEs) was tested by a set of adsorption experiments, performed with a soil solution with/without NM and increasing concentrations of nutrients, with emphasis on the competition between contaminant and nutrients for the sorption sites.

2 METHODS/EXPERIMENTAL

2.1 *Soil and NM*

Mokrsko is an area with increased As levels located in central Czech Republic (1588 mg As kg^{-1} soil). Iron nano-oxides made of maghemite (Fe_2O_3) (< 50 nm nanopowder; Sigma Aldrich) were tested as immobilizing agents for TEs. From the contaminated soil, 10 g were mixed with 100 ml of $CaCl_2$ 0.01 mol l^{-1} for the extraction of the available TEs in the soil fraction (Quevauviller, 1998).

2.2 *Experiment design and procedure*

In order to add nutrients to the soil solution, common hydroponic solutions were used. In that way, changes in pH were avoided, and it was possible to study the effect of those compounds on a hypothetical addition with NM to the soil. MES buffer was also used (1.5 mmol L^{-1}). KNO_3 and NH_4NO_3 and KH_2PO_4 were used to add K, N and P respectively. Since the compounds contain some element in the same formula, a set of combinations were designed to study their individual effect. Three concentrations (0, 3 and 6 mmol L^{-1}) of each nutrient were obtained taking into account that KNO_3 involves the addition of K and N, KH_2PO_4 of K and P, but NH_4NO_3 involves the addition of a double amount of N. The software Visual MINTEQ was used to guarantee that the forms in the solution would be more than 98.8% available in all cases (K^+, NO_3^-, NH_4^+ and PO_4^{-3}). In tubes of 10 mL, the different combinations were added separately to the diluted 1:8 soil solution, and then, 1 mL of 10% w/v of the NM was added to obtain a final concentration of 1% w/v in the soil-nutrients-solution. A control without NM was used to compare the effect of the presence of NM in the equilibrium. After agitation for 24 h in the darkness, pH was determined, and the samples were centrifuged and filtered by

Table 1. Concentration of TEs in the equilibrium with the treatment of NM and the three different nutrients. Mean values denoted by the same letter in a column do not differ significantly according to Tukey's test ($p > 0.05$); ns not significant; *$p < 0.05$; **$p < 0.01$; ***$p < 0.001$. When necessary, values were transformed to satisfy normality and variance homogeneity tests.

		μg L^{-1}						mg L^{-1}			
Treatment	pH	Al	As	Cd	Cu	Fe	Zn	Ca	Mg	Mn	Na
Control	4.66	15.76	25.60 a	2.73	2.93	1.03	58.57	47.66 a	1.47 a	0.125 a	0.75 b
NM+ 0 mmol l^{-1}K	4.65	16.23	18.03 b	2.80	3.27	10.87	31.80	46.16 b	1.42 b	0.121 b	0.78 b
NM+ 3 mmol l^{-1}K	4.68	17.20	17.17 b	2.63	3.03	–	22.30	43.16 c	1.30 c	0.112 c	0.88 a
NM+ 6 mmol l^{-1}K	4.67	18.17	16.47 b	2.60	3.33	–	18.43	41.77 d	1.25 d	0.108 d	0.94 a
ANOVA	ns	ns	*	ns	ns	ns	ns	***	***	***	***
Control	4.66	15.76	25.60	2.73	2.93	1.03	58.57	47.66 a	1.47 a	0.125 a	0.75
NM+ 0 mmol l^{-1}N	4.66	16.23	18.30	2.57	2.77	19.10	20.87	47.54 a	1.46 a	0.124 ab	0.79
NM+ 3 mmol l^{-1}N	4.63	17.50	20.20	2.67	2.57	–	32.00	46.86 ab	1.44 ab	0.123 ab	0.76
NM+ 6 mmol l^{-1}N	4.65	16.23	18.03	2.80	3.27	10.87	31.80	46.16 b	1.42 b	0.121 b	0.78
ANOVA	ns	ns	ns	ns	ns	ns	ns	**	**	*	ns
Control	4.66	15.76 ab	25.60 a	2.73 ab	2.93	1.03 b	58.57	47.66 a	1.47 a	0.125 a	0.75 d
NM+ 0 mmol l^{-1}P	4.67	18.16 a	16.46 b	2.60 b	3.33	0.16 b	18.43	41.77 b	1.25 b	0.108 b	0.93 c
NM+ 3 mmol l^{-1}P	4.64	16.40 a	20.46 ab	2.56 b	3.23	2.30 b	17.37	41.35 b	1.25 b	0.107 b	1.84 d
NM+ 6 mmol l^{-1}P	4.61	13.50 b	21.40 ab	2.86 a	3.63	6.10 a	19.53	40.10 c	1.22 c	0.105 c	2.73 a
ANOVA	ns	***	**	*	ns	***	ns	***	***	***	***

0.45 μm nylon filters. Although the size of the NM was smaller than the loop size of the filters, they were suitable for filtration due to the aggregation of the NM. After a 1:2 dilution, all the samples were acidified with HNO_3 to 2% before the analysis by ICP-OES (Varian, VistaPro, Australia).

2.3 *Statistical analysis*

ANOVA was used to determine the significance of treatment effects, and differences were determined using Tukey's test (performed with IBM SPSS Stadistics 19 software).

3 RESULTS AND DISCUSSION

The addition of NM to the As-contaminated soil solution decreased the concentration of available As, Ca, Mg and Mn (Table 1). The presence of P in equilibrium was the only responsible to reduce the As adsorption capacity of NM. According to Chowdhury and Yanful (2010), the competition of AsO_4^{-3} and PO_4^{-3} for the sorption site in contaminated water reduced the effectiveness of maghemite NM, which agrees with our results. Although the concentration of Zn was no affected separately, a significant reduction of the available Zn was detected when the data were treated to compare the control with the NM without any nutrient (from 58.5 to 23.96 μg Zn L^{-1}; $p < 0.05$; data not shown). The concentration of Cu was no affected by the presence of NM or the nutrients. Ca, Mg and Mn concentrations showed a tendency to be lower in the presence of NM and nutrients. The opposite behavior was detected for Na in the presence of K and P, but without differences for N. Although the total concentration of TEs in the soil was high (22.383, 389, 21, 23.782, 499, 26 and 93 mg kg^{-1} for Al, As, Cu, Fe, Mn, Pb and Zn, respectively), the concentrations of the available elements in the soil solution was no high enough to show important changes caused by the NM interaction. Previous experiences with the same NM showed that the concentration of available As could be reduced more than 48%.

4 CONCLUSIONS

The effectiveness of the NM (iron nano-oxides (Fe_2O_3)) for adsorption of TEs is affected by the concentration of nutrients in the medium. NM were able to reduce As concentration in a soil solution but, among the nutrients studied, the presence of P reduced the As adsorption capacity, probably because of As-P surface interference. Competition for the adsorption sites must happen between the elements Ca, Mg, Mn and Na with the nutrients K, N and P.

ACKNOWLEDGEMENTS

The authors thank to Hana Šillerová for the ICP analysis and the European project Postdok ČZU (ESF/MŠMT CZ.1.07/2.3.00/30.0040).

REFERENCES

Chowdhury, S.R. & Yanful, E.K. 2010. Arsenic and chromium removal by mixed magnetite-maghemite nanoparticles and the effect of phosphate on removal. *Journal of Environmental Management* 91(11), 2238–2247.

Kim, K.R., Lee, B.T., Kim, K.W. 2012. Arsenic stabilization in mine tailings using nano-sized magnetite and zero valent iron with the enhancement of mobility by surface coating. *Journal of Geochemical Exploration* 113, 124–129.

Komárek, M., Vaněk, A., Ettler, V. 2013. Chemical stabilization of metals and arsenic in contaminated soils using oxides - A review. *Environmental Pollution* 172, 9–22.

Mueller, N.C. & Nowack, B. 2010. Nanoparticles for remediation: solving big problems with little particles. *Elements* 6, 395–400.

Quevauviller, Ph. 1998. Operationally defined extraction procedures for soil and sediment analysis I. Standardization. *Trends in analytical chemistry* 17, 289–298.

Walker, D.J., Bernal, M.P., Correal, E. 2007. The influence of heavy metals and mineral nutrient supply on *Bituminaria bituminosa*. *Water, Air, and Soil Pollution* 184, 335–345.

Waychunas, G.A., Kim, C.S., Banfield, J.F. 2005. Nanoparticulate iron oxide minerals in soils and sediments: unique properties and contaminant scavenging mechanisms. *Journal of Nanoparticle Research*, 7, (4–5) 409–433.

1.3 Groundwater quality in Latin America

One Century of the Discovery of Arsenicosis in Latin America (1914–2014) – Litter, Nicolli, Meichtry, Quici, Bundschuh, Bhattacharya & Naidu (Eds)
© 2014 Taylor & Francis Group, London, ISBN 978-1-138-00141-1

Arsenic in the groundwater of the Buenos Aires province, Argentina

M. Auge
Consultant hydrogeologist, Buenos Aires, Argentina

ABSTRACT: A regional map of 1:3.000.000 was elaborated with the lines of threshold of 0.05 and 0.1 mg L^{-1} of dissolved As, in the groundwater of the Buenos Aires province. The map shows that 87% of the provincial territory (267,000 km^2) has groundwater that exceeds the standard of 0.05 mg L^{-1} recommended for drinking water and only 13% (40,000 km^2) of its total area (307,000 km^2) account with underground water suitable for human consumption. Fortunately, around 91% of the total population of the province of Buenos Aires is placed in areas where As in groundwater is less than 0.05 mg L^{-1} and, therefore, the water is drinkable according to the existing regulations.

1 GENERALITIES

The province of Buenos Aires, with 307,000 km^2, is the largest province in the continental territory of Argentina (2,790,000 km^2). It is also the most populous, with 15.6 million people, out of a total of 40 million, and it is the one that generates the highest GDP (35%) of Argentina (425,000 million USD/year).

About 37% of the Buenos Aires province population is supplied with groundwater and the rest is supplied with surface water, coming especially from Paraná and De la Plata rivers. Around 87% (267,000 km^2) of the total provincial area (307,000 km^2) has groundwater with arsenic concentration exceeding the current standard for drinking water (50 μg L^{-1}). However, in this region, only lives the 9% (1.37 millions) of the total population of the province (15.6 millions).

The Buenos Aires province is part of the south end of the Great Chaco-Pampean plain, which occupies 1 million km^2 in Argentina (Figure 1) and therefore topographic slopes are very scarce (10^{-3} to 10^{-5}). The plain is interrupted by two hill cords of low heights (Tandilia and Ventania), which occupy only 5% of the province. The climate is temperate with an average annual temperature ranging from 17 °C in the NE to 15 °C in the SW, while the average annual rainfall also decreased from 1,000 mm in the NE to 350 mm in the SW. In most areas, the soils are suitable for crops and natural pastures; these conditions, together with the factors mentioned above, determine a great regional potential for crops and livestock production.

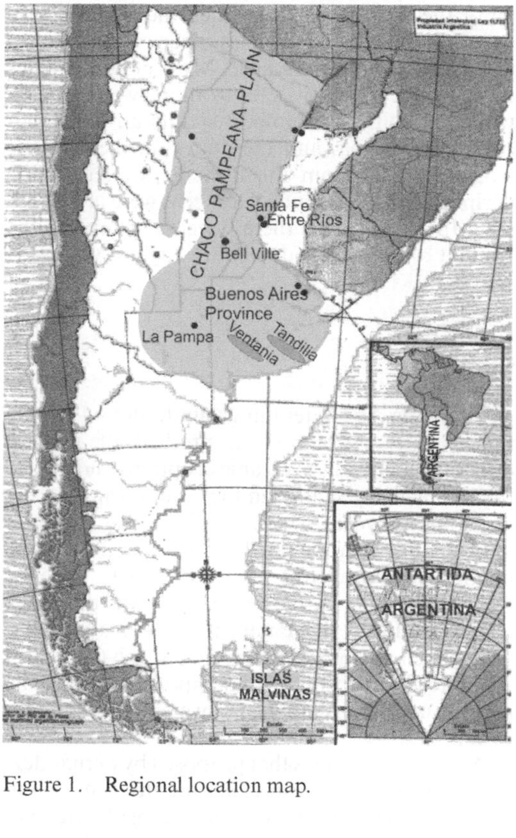

Figure 1. Regional location map.

2 INTRODUCTION

Arsenic is one of the elements that due to its high toxicity, has a significant limitation on the water potability. Prolonged consumption of water with elevated As concentrations produces severe damage to the human body, leading to a condition known as chronic endemic regional hydroarsenicism (HACRE in Spanish).

After ingestion of water with arsenic, the element is absorbed by the blood and accumulates preferentially in lungs, liver, kidneys, skin, teeth, hair and nails. Characteristic disorders resulting from chronic exposure are: thickening of palms and soles, increased pigmentation of the skin and development of skin, lung and larynx cancer. Arsenic intoxication in Argentina was first mentioned in the medical literature in 1913 by Dr. Mario Goyenechea in Rosario, specifying the origin from the ingestion of arsenical water by two patients from the Bell Ville city in the province of Córdoba (Goyenechea, 1913). Therefore, the clinical manifestations were called "Bell Ville disease".

Part of this work paper is based in a previous one (Auge *et al.*, 2013).

3 METHODOLOGY

For the mapping of total As content in groundwater of the Buenos Aires province, the results of 640 chemical analysis of samples taken at 74 counties and 159 towns was available. Most of the samples came from cooperatives for drinking water supply, which exploits the Pampeano aquifer, the dunes of Sandy Pampa, and the dunes on the Atlantic Coast. Results from the Puelche aquifer, in the periphery of the Buenos Aires city, were also available. In cases where there were variations in the content of different samples, the highest value was selected.

4 ORIGIN

Most of the As content in groundwater of Argentina has a natural origin, product of mineral dissolution linked to volcanic eruptions and hydrothermal activity, mainly in the Andes Range in the last 5 million years, which remains today, although in a much more attenuated form (Auge, 2009). The main transport agent from the Andes to the east, reaching the Chaco-Pampean plain was the wind, which produced the Pampeano Loess accumulation, which are intercalated volcanic ash (tuff) with volcanic glass (obsidian) and appears as one of the main sources of As in groundwater (Nicolli *et al.*, 1989).

A different origin is that proposed by Fernández-Turiel *et al.* (2005), especially for the NW region of Argentina and N of Chile, considering that the main contribution to arsenic was produced by transport through a big paleo-river network, from the Andes to the E and W, respectively.

Other minor sources of As, but that also can deteriorate water quality locally, are those related to mining activities, production and use of pesticides, glass manufacturing, electronics and foundries.

5 MOBILITY

The mobility of As depends essentially on the redox conditions and the pH of the water. In oxidizing medium, which is prevalent in shallow groundwater, arsenic is dissolved as As(V). Under reducing conditions, arsenic is predominantly present as As(III), which is the most toxic form. The microbiological activity also influences the mobility of As: if it is intense, it can substantially modify the redox conditions of the environment (Ormeland *et al.*, 2004).

6 POTABILITY

The potability standards, as well as the quality of most foods, have been evolving towards stricter thresholds over time. Arsenic is no an exception and, thus, Decree 6553/74 of the Buenos Aires province, which stated as desirable value < 0.01, acceptable 0.01 and tolerable 0.1 mg L^{-1}, was amended in 1996, by Law 11820, currently in effect, which established only a tolerable limit of 0.05 mg L^{-1} (50μg L^{-1}). In this regard, the World Health Organization (2008) sets as reference value for As 0.01 mg L^{-1}, while clarifying that it is a provisional guideline value "in view of the scientific uncertainties".

In Argentina, there are great differences between rules regarding the content of As in drinking water and this is verified also among neighboring provinces. For example, compared to the 0.05 mg L^{-1} of the Buenos Aires province, La Pampa takes 0.15, Santa Fe 0.1, Córdoba and Río Negro 0.05, and Entre Ríos 0.01 mg L^{-1}, although in no case representative toxicological studies were made.

7 AREAL DISTRIBUTION

Figure 2 shows the spatial distribution of total As content of the most exploited aquifers for drinking water in the Buenos Aires province. It shows a marked predominance of areal concentrations that exceed the tolerable limit, because the 87% of the province (267,000 km^2) has values greater than 0.05 mg L^{-1} and only 13% (40,000 km^2) has lower contents. This condition is fortunately not repeated when analyzing the number of people affected, because sites with high population density are located in areas with levels less than 0.05 mg L^{-1}. In this regard, three groundwater regions, suitable for human consumption according to current regulations, have been identified:

1. That located in the NE sector of the province, which occupies 16,600 km^2 and extends from the SE (sampling point-Bavio) to the NW

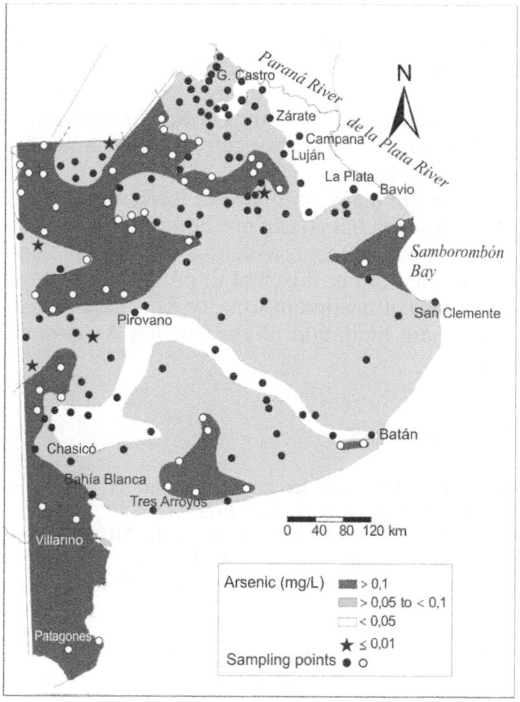

Figure 2. Arsenic concentration in groundwater of the Buenos Aires province.

(G. Castro). In this region, the majority of groundwater provided by cooperatives and companies is extracted from the Puelche aquifer. It is the most densely populated part of the province, as it is home of the 24 districts of the periphery of Buenos Aires city (Conurbano Bonaerense), with 9.9 million inhabitants (INDEC, 2012) and other major cities like La Plata, Zárate, Campana and Luján. The entire population of this region reached 12.3 million in 2010, representing 79% of the total population of the Buenos Aires province (15.6 millions). Here, the Puelche aquifer has As concentrations below 0.05 mg L^{-1}, but only 20% of the total drinking water has subterranean origin, because the rest 80% comes from Paraná and De la Plata rivers, with even less concentrations than the Puelche aquifer.

2. That located near the central and southwest part of the province, has a boomerang form and extends from SE to NW, from Batán until Pirovano and from there, to the SW, to the vicinity of Chasicó. This region, which occupies 21,700 km^2 and is home of 1.5 million people, largely coincides with hilly and interhilly environments of Tandilia and Ventania ranges. In this region, Pampeano is the most exploited aquifer for human supply.
3. The belt that matches the neighboring dunes to the Atlantic Coast, stretching from San Clemente to the N to the vicinity of Bahía Blanca in the SW, occupies only 1,700 km^2 and has 400,000 permanent inhabitants, although, during the summer, the population typically triplicates due to tourism. In this region, groundwater is captured in the free aquifer of the dunes, or in one semiconfined, underlying previous one.

Finally, the population situated in regions with less than 0.05 mg L^{-1} of As in groundwater, is 14.2 million, or 91% of the province total (15.6 millions).

Regarding the regions with more than 0.05 mg L^{-1} of As in water, the region with contents between 0.05 and 0.1 mg L^{-1} stands out for its size (about 178,000 km^2), but it has only 470,000 inhabitants (3% of the total population), and it is supplied primarily by the Pampeano aquifer. The areas exceeding 0.1 mg L^{-1}, are distributed in patches in: the vicinity of Samborombón Bay (5,600 km^2), the NW (44,200 km^2), the S (Tres Arroyos-7,600 km^2) and stick to S of Bahía Blanca (Villarino-Patagones), and lobular-shaped N to Adolfo Alsina (31,600 km^2). Together this region covers about 89,000 km^2 and 880,000 inhabitants (6% of the province) live there.

8 TREATMENT

There are many processes to lower the content of As in water. The most appropriate choice depends on several factors such as the volume of water to be treated and the initial concentration of As, the speed required for the treatment, the number of users, the type and amount of waste generated by the treatment, the cost of treatment and technology and the economic means available for the use. The most frequently used treatments are the following:

1. *Coagulation-precipitation-filtration.* This is the most widely used technique in the world, and uses metal salts as coagulants.
2. *Adsorption.* The most effective adsorbents are clays, especially the ferruginous ones (laterite). A decrease greater than 95% has been reached at pilot level with this adsorbent.
3. *Ion exchange.* Sulfate resins are used for removing As(V) and nitrate resins for As(III).
4. *Reverse osmosis.* It is based on the use of semipermeable membranes that permit the passage of water but retains much of the substances in solution, in this case, As.

In any case, all the processes used to lower the arsenic content in water, end in an residual effluent (liquid, solid, or semi-solid) that introduces a high risk to human health and the environment. Therefore, it is vital to provide conditions of maximum security for disposal. This requirement is the main drawback faced by arsenic removal plants.

9 CONCLUSIONS

Most of the As incorporated into the aquifers that are used for human supply in the Buenos Aires province and the rest of the country, has natural origin and comes from the dissolution of minerals associated with volcanic eruptions and hydrothermal activity, mainly from the Mountains Range, in the last 5 million years. The main transport agent from the Andes to the east, reaching the Chaco-Pampean plain, was the wind, which produced the Pampeano Loess accumulation, in which volcanic ash (tuff) is intercalated with glass of the same source (obsidian), and it appears as one of the main sources of As in groundwater.

Considering the potability regulations related to dissolved As in groundwater of the Buenos Aires province, which establishes a tolerable limit of 0.05 mg L^{-1}, there is a marked predominance of areal concentrations above that value. Thus, 87% of the province (267,000 km^2) has groundwater with more than 0.05 mg L^{-1} and only 13% (40,000 km^2) has lower contents. This condition is fortunately not repeated when analyzing the number of people potentially affected, because the most densely populated sites are located within areas with levels less than 0.05 mg L^{-1}. This is true for 91% of the total population of the province and therefore, only the remaining 9% live in regions with more than 0.05 mg L^{-1} of dissolved total As in groundwater.

At the country level, there are significant differences between rules regarding the content of As in drinking water in the different parts and this is verified also among neighboring provinces. Thus, compared with the 0.05 mg L^{-1} value for the province of Buenos Aires, La Pampa takes 0.15, Santa Fe 0.1, Córdoba and Río Negro 0.05, and Entre Ríos 0.01 mg L^{-1}. However, no toxicological studies based on a representative number of inhabitants and over a time period has been made to determine with greater certainty the potability limit regarding As.

There are many processes to lower the content of As in water, but all end in a liquid, solid, or semisolid residual effluents with high risk to health and environment; thus, it is vital to provide disposal in conditions of maximum security. This requirement is the main limitation of the existing As removal plants.

REFERENCES

Auge, M. 2009. Arsénico en el agua subterránea. *Fed. Méd. de la Prov. de Buenos Aires* 1:15.

Auge, M., Espinosa Viale, G. & Sierra, L. 2013. Arsénico en el agua subterránea de la Provincia de Buenos Aires. In *VIII Congreso Argentino de Hidrogeología. T II: 58–63, La Plata.*

Fernández Turiel, J., Galindo, G., Parada, M., Gimeno, D., García-Vallés, M. & Saavedra, J. 2005. Estado actual del conocimiento sobre Arsénico en el agua de Argentina y Chile: origen, movilidad y tratamiento. In *Taller de Arsénico en aguas, origen, movilidad y tratamiento: 1–22. IV Congreso Argentino de Hidrogeología. Río Cuarto - Argentina.*

Goyenechea, M. 1913. Sobre la nueva enfermedad descubierta en Bell Ville. *Rev. Méd. de Rosario* 7:48.

INDEC. 2012. Censo Nacional de Población, Hogares y Viviendas 2010. Ser. B, (2), T 1:1–371 and Ser. B, (2), T 2:1–406. Buenos Aires.

Nicolli, H., Suriano, J., Gómez Peral, M., Ferpozzi, L. & Baleani, O. 1989. Groundwater contamination with arsenic and other trace-elements in an area of the Pampa, Province of Cordoba, Argentina. *Environm. Geol. Water Sci.* 14 (1): 3–16.

Ormeland, R., Stolz, J. & Hollibaugh, J. 2004. The microbial arsenic cycle in Mono Lake, California. *Microbiol. Ecol.* 48 (1): 15–27.

World Health Organization. 2008. Guidelines for Drinking-water Quality. Ed. 3: 1–398.

One Century of the Discovery of Arsenicosis in Latin America (1914–2014) – Litter, Nicolli, Meichtry, Quici, Bundschuh, Bhattacharya & Naidu (Eds)
 ISBN 978-1-138-00141-1

Water quality data and presence of arsenic in groundwater in Trenque Lauquen County, province of Buenos Aires (Argentina)

E. Moschione, G. Cuzzio, M. Campins & A. Zamora
National Technological University-Mar del Plata (UTN-MDP), Buenos Aires, Argentina

P. Picco & S. Leiva
National Technological University-Faculty of Trenque Lauquen (UTN-FRTL), Argentina

ABSTRACT: A study on the presence of arsenic in groundwater is of great importance to guarantee the safe access to drinking water for the population. In this paper, 33 wells were evaluated in Trenque Lauquen County, north-west of the Buenos Aires province, looking for information about the state of the hydric resource referred to this natural pollutant. The results showed 66% of the wells with arsenic levels above the provisional reference value, and 34% with concentrations from 10 to 47.7 µg/L. This finding, already observed in other regions where high As levels in the waters are not appropriate for drinking, opens a discussion showing the necessity of a deep epidemiological assessment for setting a permanent guide value for water quality and health effects on local population.

1 INTRODUCTION

The presence of high arsenic levels in groundwater is a very well-known public health problem reported in countries such as Argentina, Brazil, Chile, China, India, Mexico and Taiwan (Galetovic Garabantes, 2003). A chronic exposure to arsenic due to water consumption may cause dermal lesions, hyperpigmentation, peripheral neuropathy, skin, bladder and lung cancers, and peripheral vascular disease (WHO, Arsenic in Drinking Water, 2011).

In Argentina the most affected area is the Chaco-Pampean plain, where the arsenic content in groundwater is attributable to volcanic glass and ashes in the loessic sediments of the region (Bocanegra, 2002). Although there are many studies relating the presence of arsenic to regional chronic endemic hydroarsenicism (HACRE), there is a lack of data about water quality and possible effects of the pollutant in the local populations of the Buenos Aires province (Nonna, 2006). In this study, data of water quality of Trenque Lauquen County are presented and arsenic is investigated in wells for different uses.

2 METHODOLOGY

2.1 *Study area*

Trenque Lauquen County (Buenos Aires province) has a population of 43,000 inhabitants (INDEC, 2010) and a surface of 5,500 km² (Figure 1). The zone belongs to the hydrogeological province named Arid-Chaco-Pampean plain and the north-west hydrogeological region of Buenos Aries province, including the *Postpampeano, Pampeano, Araucano* and *Formación Paraná* hydrogeological units (Bocanegra, 2012).

2.2 *Field data*

A total of 33 wells were selected and localized by GPS in Trenque Lauquen County, and a map was constructed in Google earth to show the sampling area. Field activities and sampling were carried out in March and April 2013.

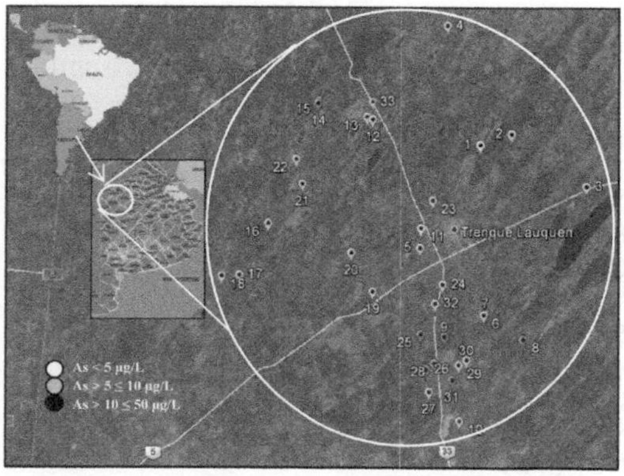

Figure 1. Study area and sampling locations.

2.3 *Laboratory analysis and data management*

Water samples were sent to the Laboratory of UTN-UA-MDP for the chemical analysis of 18 parameters using standardized test methods (APHA, AWWA, WEF, 2012). The presence of arsenic was specially investigated by the EPA 7061A (AAE-hydride generation) test method in the Laboratory Group of Environmental Studies (GEMA) SRL. Data were statistically evaluated and software Aquachem 4.0 was applied (Calmbach, and Waterloo Hydrogeologic INC, 1998).

3 RESULTS AND DISCUSSION

3.1 *Water quality in the study area*

Water quality data obtained in this study were found to be very different within each other. In Table 1, minimum, maximum and average concentrations for the main water-quality parameters are presented.

This heterogeneous concentration distribution is characteristic of the hydrogeology and management conditions of the study area. There are wells with salinity (total dissolved solids) between 11677 and 356 mg L^{-1}, and sodium and chloride concentrations from 3125 to 27 mg L^{-1} and 3869 to 6 mg L^{-1}, respectively, these ions being the most variable ones.

As can be seen in Figures 2 and 3, water composition is HCO_3^--Na^+, or Cl^--Na^+ in most wells, with variable concentrations of Ca^{2+}, Mg^{2+} and SO_4^{2-}.

3.2 *Arsenic in Trenque Lauquen County*

Arsenic concentrations detected in the study area were below 50 µg L^{-1} in all samples. Out of 22 from 33 wells evaluated, the pollutant was not detected (<5 µg L^{-1}), while 11 wells had detectable values (Figure 1). Two of these last wells showed concentrations of 8 and 10 µg L^{-1} and the other nine samples had values between 11 and 47.7 µg L^{-1} (the maximum value observed).

Table 1. Main water quality parameters.

Parameter	Max	Min	Average
Conductivity (µS cm^{-1})	15000	397	3910
pH	8.44	6.74	7.75
Total dissolved solids (mg L^{-1})	11677	356	2792
Na (mg L^{-1})	3125	27	707
K (mg L^{-1})	100	6	31
Ca (mg L^{-1})	302	21	91
Mg (mg L^{-1})	364	14	85
Hardness (mg $CaCO_3$ L^{-1})	2253	127	579
Alkalinity (mg $CaCO_3$ L^{-1})	1010	167	415
Cl^- (mg L^{-1})	3869	6	802
SO_4^{2-} (mg L^{-1})	1870	7	379
NO_3^- (mg L^{-1})	433	9	106
As (µg L^{-1})	47.7	<5	7

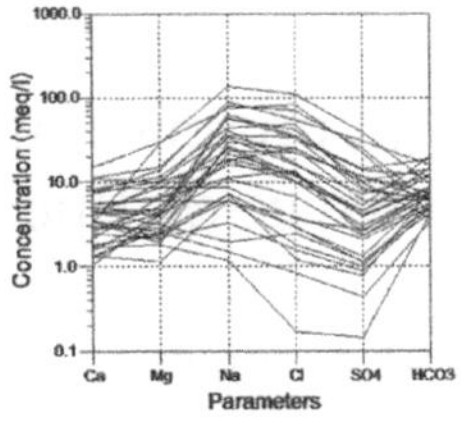

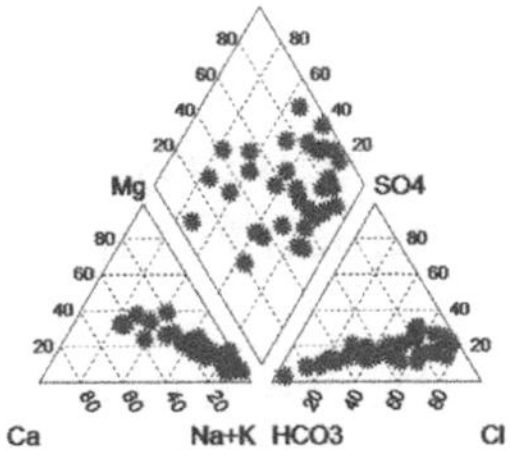

Figures 2 and 3. Schoeller and Piper graphs.

3.3 *Results and current reference values*

The maximum allowed value for As in drinking water according to the Argentine Food Code was 50 µg L^{-1} until 2007. In last updates of 2007 and 2012, 10 µg L^{-1} was defined as reference concentration, but the need of new and deep epidemiological studies to confirm this value were initiated (CAA, 2012). The WHO also recommends this provisional value, due to the necessity of optimizing standardized analytical methods to detect these low arsenic levels, risk studies in local populations, and a further development of water technologies to remove the pollutant to desired concentrations (WHO 2011).

4 CONCLUSION

Chemical composition of groundwater in Trenque Lauquen County is very variable, with high salinities in many cases. The main water quality problems in this area are related with the concentrations of total solids, chloride, sulfate, nitrate and hardness, which are not appropriate for human consumption and also for cattle or agriculture.

Most of the analyzed samples presented undetectable arsenic values (<5 µg L^{-1}), while those with higher concentrations where always below 50 µg L^{-1}.

REFERENCES

APHA, AWWA, WEF, 2012. *Standard Methods for the Examination of Water and Wastewater*. 22nd. Ed.

Argentine Food Code, Cap. XII. 2012.

Bocanegra, E., Moschione, E., Picco, P., Leiva, S. & Zamora, A. 2012. Hydrogeochemical tools applied to the evaluation of Trenque Lauquen groundwater, in Buenos Aires Province, Argentina. In *XI*

Latin-American Congress of Hydrogeochemistry and Hydrogeology. IV Colombian Congress, ALHSUD-ACH. Colombia.

Bocanegra, E., Martinez, D. & Massone, H. 2002. Arsenic in ground water: Its impact in health. *Ground Water and Human Development*. 21–27. ISBN 987-544-063-9.

Calmbach and Waterloo Hydrogeologic Inc. 1998. Aquachem Aqueous Geochemical Analysis, Plotting and Modeling.

Galetovic Garabantes, A. & de Fernicola, N. 2003. Arsénico en el agua de bebida: un problema de salud pública. *Brazilian Journal of Pharmaceutical Sciences* 39(4): 365–372.

INDEC. (2010). National Census www.censo2010.indec.gov.ar

Nonna, S. 2006. *Epidemiología del Hidroarsenicismo Crónico Regional Endémico en la República Argentina*. Ministry of Health, Secretary of environment and sustainable development, Argentine Toxicological Association.

WHO. 2011. Guidelines for Drinking-water Quality 4th Ed. ISBN 978 92 4 154815 1.

WHO. 2011. Arsenic in Drinking Water. Background document for development of WHO Guidelines for Drinking-water Quality. WHO/SDE/WSH/03.04/75/Rev/1.

One Century of the Discovery of Arsenicosis in Latin America (1914–2014) – Litter, Nicolli, Meichtry, Quici, Bundschuh, Bhattacharya & Naidu (Eds)
 ISBN 978-1-138-00141-1

Groundwater contamination with arsenic, Región Lagunera, México

P.W. Boochs & M. Billib
Institute of Water Resources Management, Hydrology and Agricultural Hydraulic Engineering—WAWI, Leibniz University of Hannover, Germany

C. Gutiérrez & J. Aparicio
Coordinación de Tecnología Hidrológica, Instituto Mexicano de Tecnología del Agua—IMTA, Jiutepec, Morelos, México

ABSTRACT: The groundwater of the Región Lagunera (México), which represents the main source of drinking water for more than 2 million people, shows high arsenic concentrations. Large areas are far above the Mexican standard of 25 μg L^{-1}. The aquifer is strongly overexploited and the groundwater levels in the center have declined more than 100 m in less than 50 years. The drawdown has caused probably the dissolution and migration of geogenic arsenic from former geothermal activities. Data analysis shows that the arsenic concentration is correlated to groundwater age: "older" water has higher arsenic content than "younger" water. The process of the genesis of the arsenic pollution implicates that the highest content is in the bottom of the aquifer where the oldest water can be found. Possibilities to provide drinking water of good quality for the region are artificial recharge or the use of surface water and/or arsenic treatment, for example subterranean immobilization.

1 INTRODUCTION

Since 1962, Mexican agencies have reported elevated arsenic concentrations in extensive areas of the unconfined alluvial aquifer of the Región Lagunera, which has caused adverse health effects in both people and animals. Arsenic is naturally occurring and the measured concentrations are in the range of 5 to 750 μg L^{-1} (IMTA, 1990; Boochs *et al.*, 2007). Large areas of the alluvial aquifer have arsenic concentrations that are quite above the Mexican standard of 25 μg L^{-1} for drinking water.

Actually, there are more than 4,800 pumping wells in the area, mainly used for irrigation. The total volume of water pumped from the aquifer reached up to 1,221.8 Mm^3/year and is at least two times greater than the recharge (CNA, 2002). These conditions caused a drawdown of more than 100 m in the center of the aquifer within the last 50 years.

The Mayrán and Viesca Lagoons run dry and the groundwater quality deteriorated. The drawdown induced the dissolution and migration of the geogenic arsenic within the aquifer, which is endangering the water supply of the population and the agricultural production in the area.

2 INVESTIGATION AREA

2.1 *Groundwater table and discharge*

At the beginning of the 19th century, there were only dugs and shallow wells of low abstraction capacity. The water table was close to the surface over the whole basin. During the period 1950–1970, several agricultural wells for irrigation purposes were built, and in the 1960's abstraction reached 1,300 Mm^3 per year. Furthermore, in 1960 the Nazas River was canalized and the natural infiltration from the river was nearly stopped. In consequence, the groundwater levels drawdown dramatically.

2.2 *Arsenic in the Lagunera Aquifer*

According to the measurement of IMTA (1990) the arsenic concentrations ranged from 3 to 443 μg L^{-1}. The highest arsenic concentrations were found in the areas of the former lagoons located at the north eastern part of the basin and decreased towards the riverbeds. In general, areas with high arsenic concentration also have high levels of sulfates, fluorides, chlorides, sodium, boron and lithium, above their respective regulated standards. The study included 95 wells distributed over the whole main aquifer, 41% of which showed As concentrations above the Mexican drinking water standard.

Recent measurements of CNA and WAWI in 2010show that the arsenic concentration compared to 1990 has increased, especially at the northern part of the region, where the levels reached up to 750 µg L^{-1} (Billib *et al.*, 2012).

3 RESULTS AND DISCUSSION

The presence of arsenic has been related to several potential sources (IMTA, 1990): hydrothermal activity, use of arsenical pesticides, mining activities and sedimentary origin. The data analysis showed that arsenic is naturally occurring and that its most probable source is due to extinct, intrusive hydrothermal activity combined with a sedimentary process.

This assumption was based on the following: (1) high arsenic concentrations are close to extrusive igneous rocks; (2) high arsenic concentrations are observed in areas with high concentrations of boron, lithium, chloride and fluoride (probably they were the result of magmatic cooling); (3) some wells located at the southeast and northwest of the Region Lagunera registered high temperatures, strong smell of hydrogen sulfide (H_2S) and elevated arsenic concentrations; and (4) high arsenic concentrations are associated with the "oldest" waters of the aquifer.

This was verified by measurements of the groundwater age, carried out by IMTA (1990). The data show that "older" water has higher arsenic concentration than "younger" water (Figure 1).

Furthermore, SIMAS (2013) (waterworks of Torreon) observed from 2004 to 2010 in 20 of 75 drinking water wells a significant increase of the arsenic concentration (from a mean of 20 to 35 g/L). In the same period, the static water level of the wells decreased in mean of about 28 m. Figure 2 shows that the greater the depth to groundwater, the higher the arsenic concentration.

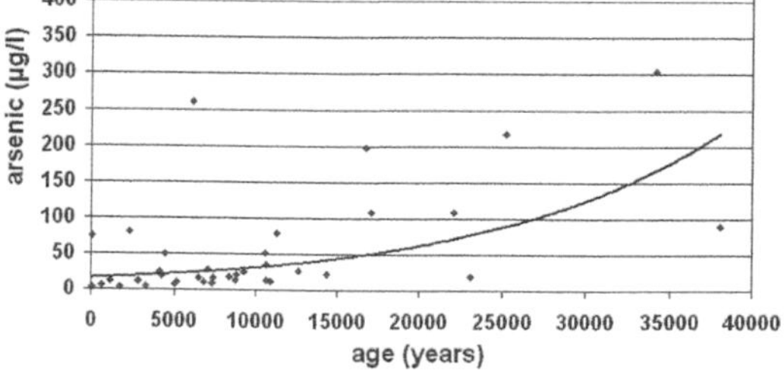

Figure 1. As concentration vs. groundwater age (IMTA, 1990).

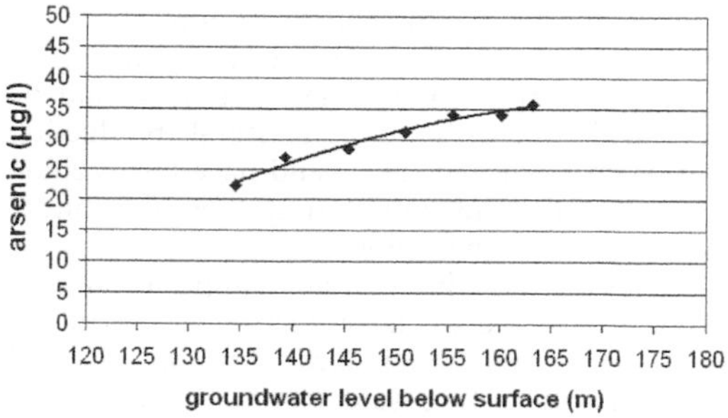

Figure 2. As concentration vs. groundwater level (SIMAS, 2013).

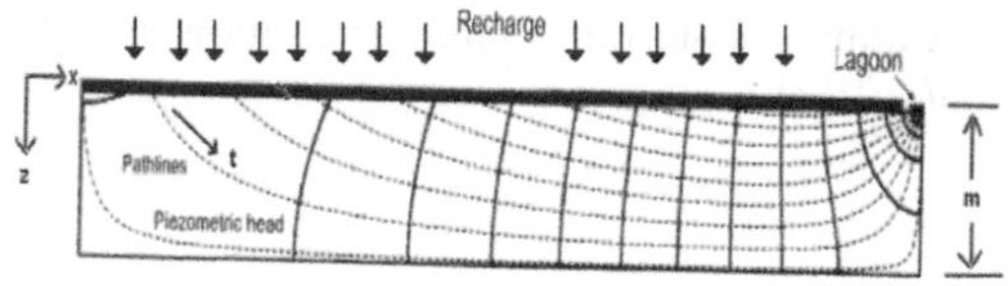

Figure 3. Schematic diagram of piezometric head and path lines.

The correlation between arsenic content and groundwater age resp. the depth of the groundwater level can be illustrated by the schematic diagram in Figure 3. Shown is a vertical crossection of the aquifer. The groundwater recharge is usually by precipitation or/and infiltration from rivers.

The "young" water infiltrates into the underground and along the path lines it moves. Over time, the water moved deeper and with increasing length of path lines the water get "older".

The intensive groundwater abstraction mainly for irrigation and the canalization of the river Nazas, which stopped the infiltration of river water into the groundwater, induced a drawdown of the groundwater level of about 100 m in the center of the area within the past 100 years. Increasingly "older" water was pumped from deeper layers of the aquifer with higher arsenic content.

In addition, the use of arsenic-rich groundwater for irrigation introduce *per se* a further increase of the arsenic content by "leaching" of surplus water.

4 CONCLUSIONS

It can be assumed that, under the current conditions of over-exploitation of the aquifers in the Región Lagunera, the groundwater levels will continue to decrease and consequently the arsenic concentration will continue to rise.

In order to overcome this situation, the intensive groundwater withdrawals have to be reduced.

Because the most significant part is used for irrigation, the technology and the efficiency of irrigation should be improved. Furthermore, it should be examined whether the agricultural production can be changed, but this would have far-reaching consequences and implications for the future development of the region.

To ensure the drinking water supply, the appropriate wells have to be protected against inflow of arsenic groundwater. Near Torreón, artificial recharge with surface water can reduce the arsenic migration towards the municipal drinking water wells. A pilot project was carried out in 2000 by IMTA and CNA. Other possibilities to supply drinking water of good quality are the use of surface water from reservoirs or arsenic treatment, for example subterranean immobilization (Boochs *et al.*, 2012).

REFERENCES

Billib, M., Boochs, P.W, Aparicio, J. & Gutiérrez, C. 2012. Hydro-geochemical Investigation of the Arsenic in the Comarca Lagunera. *Report MEX 08-009, Hannover, Germany.*

Boochs, P.W, Billib, M., Aparicio, J. & Gutiérrez, C. 2007. Management of the arsenic groundwater system Comarca Lagunera. *Proc. AGU Joint Assembly, Acapulco, Mexico.*

Boochs, P.W, Billib, M. & Krüger, T. 2012. In-situ Immobilisation of Arsenic (in German), *Wasser und Abfall,* 14, No. 09.

CNA, 2002. Estudio COAH-DGO, *Comisión Nacional del Agua, Torreón, México.*

IMTA, 1990. Estudio Hidrogeoquímico e isotópico del acuífero granular de la Comarca Lagunera. *Jiutepec, México.*

SIMAS, 2013. Informal report, *Sistema Municipal de Aguas y Saneamiento, Torreón, México.*

One Century of the Discovery of Arsenicosis in Latin America (1914–2014) – Litter, Nicolli, Meichtry, Quici, Bundschuh, Bhattacharya & Naidu (Eds)
© 2014 Taylor & Francis Group, London, ISBN 978-1-138-00141-1

Heavy metal concentration in arsenic contaminated groundwater of the Chaco Province, Argentina

M.C. Giménez & R.M. Osicka
Departamento Química Analítica, Universidad Nacional del Chaco Austral, Argentina

P.S. Blanes
Área Química General, FCBioyF, IQUIR-CONICET, Universidad Nacional de Rosario, Argentina

Y. Morisio & S.S. Farías
Gerencia Química, Comisión Nacional de Energía Atómica, Buenos Aires, Argentina

ABSTRACT: The occurrence of arsenic and other trace elements have been investigated in the groundwater of the central region of the Chaco Province, Northern Argentina. The arsenic concentrations samples ranged between 0.7 to 1990 μg L^{-1}; 91% (n = 45) exceeded the 10 μg L^{-1} WHO (World Health Organization) provisional standard limits for drinking water. A significant correlation between As and F was detected ($r^2 = 0.50$). In addition, 16, 78, 31, 13 and 4.5% of groundwater samples had, respectively Al, B, Fe, Mn and Sb exceeding CAA (Código Alimentario Argentino) guideline values. The values corresponding to Ag, Be, Cr, Ni, Pb, Se, Zn were found below the quantification limit. The presence of As and HM in the groundwater of this region may be due to local geochemical conditions that facilitate their transfer to water. This represents an important potential public health issue.

1 INTRODUCTION

For at least three decades, the occurrence of high concentrations of Arsenic (As) and other trace elements in drinking water has been recognized as a great public health concern in several parts of the world. In particular, elevated As concentrations have been reported in groundwater of many different regions of Argentina including La Pampa, Buenos Aires, Chaco, Cordoba, Santiago del Estero, Santa Fe and Tucumán Provinces (Nicolli *et al.*, 2012).

This work examines the occurrence of As and 23 Heavy Metals (HM) in groundwater of the area of Comandante Fernández, in the central region of the Chaco Province, northern Argentine. The relationships between As occurrence and these chemical components are discussed.

2 EXPERIMENTAL

2.1 *Location of the study area and sampling*

The study area is located in Department Comandante Fernández, Chaco Province, northern Argentina. It covers an area of 1500 km^2, with a population of approximately 0.10 million people. This area is located within the geologic province known as the Chaco Pampean. The aquifers are hosted in superposed sequences of aeolian and fluvial sediments of the Tertiary and Quaternary ages. As there are no superficial sources of groundwater, recharge of the phreatic aquifer is produced from rainfall, while deep aquifer recharge occurs from west to east and northwest–southeast and it is supposed to be located in the higher parts of the sub-Andean outcrops (Larroza and Fariña, 2005). Forty five groundwater samples were collected during May–October 2010 (a dry season), in rural and suburban areas of the Comandante Fernández Department (Chaco Province).

2.2 *Groundwater analysis*

Arsenic concentration was determined by HG-AAS (hydride generation atomic absorption spectroscopy). Total concentrations of 23 other trace elements (Ag, Al, B, Ba, Be, Cd, Co, Cr, Cu, Fe, Mn, Mo, Ni, P, Pb, S, Sb, Se, Si, Sr, Ti, V and Zn) were also simultaneously determined by inductively coupled plasma–optical emission spectrometry (ICP–OES). This equipment was linearly calibrated from 1 to 100 μg L^{-1} with custom certified standard solution. Accuracy of the procedure was verified by analyzing a certified reference material, NIST SRM 1643 c. All statistical analysis of data was performed using STATGRAPHICS 5.1.

3 RESULTS AND DISCUSSION

3.1 *Physicochemical parameters*

The pH values varied from 6.50 to 8.94 (mean of 7.54). EC values varied from 0.69 and 24.9 mS/cm (mean of 5.27 mS/cm). This value of EC may be due to contributions from Na^+, HCO_3^-, Cl^- and SO_4^{2-}. A Piper diagram of major ion compositions (not show), indicated that bicarbonate (135 to 1868 mg L^{-1}) and chloride (14.4 to 8193 mg L^{-1}) were the dominant anions followed by sulfate (18 to 3485 mg L^{-1}), and that sodium was the prevalent cation (132 to 5857 mg L^{-1}).

3.2 *Total arsenic and fluoride*

While the presence of As in the groundwater of this region of Argentina is natural, the local geology and rainfall have been shown to have a major impact on the variations of As concentration in groundwater.

The distribution of fluoride (*F*) concentrations was heterogeneous and similar to those of As. A positive correlation was observed between As and *F* ($r^2 = 0.50$), with 31% of the samples exceeding the WHO guideline value of 1.5 mg L^{-1}. 47% samples had *F* concentrations > 1 mg L^{-1} recommended by CAA (2007).

3.3 *Other trace elements*

Amongst the 23 target analytes, concentrations also exceeded WHO recommended drinking water limits for B, Ba, Cd, and Sb. Table 1 shows the percentage of groundwater samples exceeding these WHO and CAA guidelines.

Boron concentration ranged from 156 to 219 10 μg L^{-1}. The salinity, EC, type of sediments and pH may be the crucial factors that determine boron mobility in the groundwater system. The correlation (as r^2) is moderate but clearly existing between As with B, Mo and V: 0.11, 0.20 and 0.42, respectively, showing their common origin in the volcanic glasses.

Some samples have Fe, Al, and Mn concentrations above CAA guideline value. The processes of dissolution and release from oxides and oxyhydroxides, mainly Al, Mn, and Fe, may control presence and mobility of As and *F* in groundwater. The co-contamination of As and *F* observed in many parts of Chaco-Pampean plain is associated with the desorption from oxyhydroxides by the pH increase.

Other elemental concentrations (Co, Cu, Ni, S, Se, Sn, Sr, V and Zn) were very low. Although analyzed for many elements (Ag, Be, Cr, Pb, Se, Ti) were not found in quantifiable amounts and were thus not reported.

Table 1. Risk–based drinking water criteria and the percentage of groundwater samples exceeding these criteria.

	Risk-based drinking water criteria (μg L^{-1})		Percentage of groundwater samples ($n = 45$) exceeding drinking water criteria	
Element	WHO*	CAA**	WHO	CAA
As	10***	50	91	73
Al	none	200	NA	16
B	2400	500	78	78
Ba	700	none	2	NA
Cd	3	5	13	2
Cr	50***	50	0	0
Cu	2000	1000	0	0
Fe	none	300	NA	31
F	1500	1000	31	47
Mn	none	100	NA	16
Ni	70	20	0	0
Pb	10***	50	0	0
Sb	20	20	4.5	4.5
Se	10***	10	0	0
Zn	none	5000	NA	0

NA, Not Applicable.
*The WHO has not established risk–based drinking water criteria for Al, Ag, Be, Bi, Co, Fe, Mn, Mo, Si, Sr, Ti, V and Zn.
**CAA, Código Alimentario Argentino.
*** Provisional guideline value.

The presence of As and HM in the samples collected from this region may be due to local geochemical conditions that facilitate the transfer of naturally occurring As and HM from soil and sediment to the water. High pH and high concentration of HCO_3^- would facilitate the dissolution of volcanic glass; thus, trace elements may enter groundwater cycles, forming anionic complexes in alkaline solutions and become mobile.

4 CONCLUSION

The population in the Department Comandante Fernández, Chaco Province, may potentially be exposed to the chronic toxicological effects of both hydroarsenicism and fluorosis, increasing the risks of contracting other diseases derived from them. Since the groundwater studied here constitutes the principal source of drinking water, mitigation efforts should not be limited to As, as health risks from other toxic elements present in the drinking water must also be addressed in this region.

ACKNOWLEDGEMENTS

This work was supported with funds from PI 36/00005 UNCAus (Univ. Nac. del Chaco Austral, Argentina).

REFERENCES

CAA (Código Alimentario Argentino). 2007. http://www.anmat.gov.ar/CODIGOA/Capitulo_XII_Agua_2007–05.pdf. (accessed 04.13).

Comisión Nacional de Alimentos Acta N° 93, Reunión Ordinaria 30/11 y 01/12–/2011 (Prórroga Art. 982 y 983 del CAA).

Larroza F. & Fariña L.S. Caracterización hidrogeológica del sistema acuífero Yrenda (SAY) en Paraguay: recurso compartido con Argentina y Bolivia, IV Congreso Argentino de Hidrogeología, Córdoba, Argentina, Tomo II, 2005, pp. 125–134.

Nicolli H.B., Bundschuh J., Blanco M.C., *et.al.* 2012. Arsenic and associated trace-elements in groundwater from the Chaco-Pampean plain, Argentina: Results from 100 years of research, Sci. Total Environ. 429: pp. 36–56.

World Health Organization (WHO). Recommendations. Vol. 1. In Guidelines for Drinking Water Quality. 4th Ed.; Geneva, Switzerland, 2011.

One Century of the Discovery of Arsenicosis in Latin America (1914–2014) – Litter, Nicolli, Meichtry, Quici, Bundschuh, Bhattacharya & Naidu (Eds)
© 2014 Taylor & Francis Group, London, ISBN 978-1-138-00141-1

Study of the concentration of arsenic in drinking water from the Province of Tucumán, Argentina

G. Ojeda, A. Durán, E. Durán, M. Bruno, W. Castellanos & L. Montini
Departamento de Saneamiento Básico, Dirección General de Salud Ambiental, Sistema Provincial de Salud, S.M. de Tucumán, Argentina

ABSTRACT: The aim of this study was to evaluate the content of Arsenic (As) in drinking water in the province of Tucumán, Argentina. We analyzed 687 water samples in 192 locations, representing 86.4% of the total population. The As content was determined by using the semiquantitative test Merckquant®. The results were grouped into four categories of As concentration: 67.7% of sites (130) had a concentration <2.5–10 µg/L, 14.1% (27) between >10–<25 µg/L, 13.5% (26) between 25–49 µg/L, and 4.7% (9) recorded higher values ≥50 µg/L. There was a population of 105.805 inhabitants exposed to As in the water network, greater than 10 µg/L and less than 50 µg/L. The study will enable to supplement the information for the review of As concentration limits.

1 INTRODUCTION

The province of Tucumán is located in the northwest of Argentina, between 26° and 28° south latitude, and the meridians of 64° 30′ and 66° 30′ west longitude, covering an area of 22.524 km^2. It is divided into 17 departments with a population of 1.448.188 inhabitants (Census 2010). It is the smallest province in the country and one of the most densely populated (64.3 inhabitants/km^2).

The Argentine Food Code (CAA, Código Alimentario Argentino, 2007) provided a permitted limit of As in drinking water of 50 µg/L up to 2004. From that year the WHO (2011), and from 2007, the CAA, recommend a maximum concentration of 10 µg/L.

Epidemiological studies help to answer the needs of information and provide support to decision-makers in the adoption of acceptable limits in Water Quality Standards whose parameters influencing health (Corey *et al.*, 2005).

The Sanitation Department of Tucumán has conducted surveys since 1982 in a discontinuous way (Vapiano and Davolio, 1983; Castellanos and Ojeda, 2008) and there is no updated information on arsenic in the water network in the province of Tucumán.

This work aims to evaluate the distribution of As in drinking water network in the province of Tucuman, developing a risk map for to provide information to initiate epidemiological studies that allow authorities and agencies to revise the control limits of As concentration in drinking water.

2 METHODS/EXPERIMENTAL

For this study, 192 localities were surveyed from 2008 to 2013, and 687 samples of water for human consumption from the distribution network were collected. In each locality, surveyed samples were taken at least from a user of the service, an educational place and a health care center. Sampling sites were georeferenced with a Garmin GPS.

The population of the surveyed localities reached 86.4% (1.250.992) of the total inhabitants in the Province.

Samples were transported to the laboratory respecting the chain of custody, and the As contents were semiquantitatively determined using the Merckoquant® test.

For the distribution map of the surveyed localities, based on categories of As concentration in tap water, the ArcView Gis 3.2 program was used.

3 RESULTS AND DISCUSSION

Table 1 shows the percentage of localities surveyed according to categories of As content established for this study. Figure 1 shows the distribution of As in water network in the towns of the province of Tucumán.

Table 2 shows the results of the population of the surveyed towns that are exposed to different categories of As contents.

To evaluate the results, the recommendations proposed by the WHO (2011) were taken into

Table 1. Categories of As content (µg/L) in tap water in Tucumán. Years 2008–2013.

Categories As (µg/L)	Localities	Percentage (%)
<2.5–10	130	67.7
>10–<25	27	14.1
25–49	26	13.5
≥50	9	4.7
Total	192	100.0

Figure 1. Map of As distribution in water network in the province of Tucuman. Years 2008–2013.

Table 2. Population exposed in the categories of As content (µg/L) in Tucumán. Years 2008–2013.

Categories As (µg/L)	Population	Percentage (%)
<2.5–10	1136433	90.8
>10–<25	56825	4.5
25–49	48980	3.9
≥50	8754	0.7
Total	1250992	100.0

account. It was observed that 67.7% of surveyed sites (130) had a concentration of As in drinking water within the acceptable range (from <2.5–10 µg/L), representing 90.8% of the population of the province (1.136.433 inhabitants). Similarly, 14.1% (27) had concentrations >10–<25 µg/L and 13.5% (26) between 25–49 µg/L; these two categories correspond to the range of As concentration that WHO and the CAA propose to deepen their study to establish the maximum allowable limit.

Also noteworthy is that nine localities, whose population reached 8754 inhabitants, had concentrations in the water network above the limit established in Argentina (≥50 µg/L). They include a town in the northwest from which there was no previous records.

4 CONCLUSION

The towns with concentrations between >10–<50 µg/L of As correspond to the southeast of the Province, with a population of 105.805 inhabitants exposed. Further study is recommended as the effects of arsenic in drinking water at levels above 50 µg/L are very well know. The present investigations provide the basis to initiate epidemiological studies that will be useful to assess health effects in the range >10–<50 µg/L, and to define the maximum allowable limit compatible with acceptable risks to health.

REFERENCES

Castellanos, W. & Ojeda, G. Determinación de arsénico en agua de consumo. Provincia de Tucumán. Dirección General de Salud Ambiental. Departamento de Saneamiento Básico. SI.PRO.SA. Años 2008–2009. (Data not published).

Código Alimentario Argentino (CAA). 2007. Capítulo XII, bebidas hídricas, agua y agua gasificada.

Corey, G., Tomasini, R. & Pagura, J. 2005. Estudio epidemiológico de la exposición al arsénico a través del consumo de agua. *Programa de fortalecimiento del ENRESS.* Provincia de Santa Fe. Argentina.

INDEC. 2010. Censo Nacional de Población, Hogares y Viviendas (Argentina).

Viapiano, J. & Davolio, F. 1983. Presencia de arsénico en el agua para el consumo humano. *6° Congreso Argentino de Saneamiento*. Salta.

WHO. 2011 *World Health Organization. Guidelines for drinking-water quality*-4th ed. Geneve.

One Century of the Discovery of Arsenicosis in Latin America (1914–2014) –
Litter, Nicolli, Meichtry, Quici, Bundschuh, Bhattacharya & Naidu (Eds)
 ISBN 978-1-138-00141-1

Arsenic in drinking and environmental water in the province of Río Negro, Argentina

S. Abate
Fundación Barrera Zoofitosanitaria Patagónica (Funbapa), Viedma, Rio Negro, Argentina
Universidad Nacional de Rio Negro, Argentina

M. Travaglio, M. Gianni & E. Rial
Fundación Barrera Zoofitosanitaria Patagónica (Funbapa), Viedma, Rio Negro, Argentina

ABSTRACT: Inorganic arsenic (As) is naturally present at high levels in different sources of water in many countries including Argentina, and is highly toxic. People are exposed to chronic arsenic poisoning by the consumption of water or contaminated food. One important action to control this problem is the prevention of exposure, for what it is necessary to know the level of As in different sources of water and food. In Argentina, there are huge areas where people do not know the level of arsenic in water used to drink or for agricultural uses. The aim of this study is to know the level of As in water from different origins (groundwater, surface and household distribution water). Samples were taken in the Río Negro province between 2012 and 2013. This information contributes to complete the As map of Argentina, of great importance in order to take preventive measures in public health and agrifood production.

1 INTRODUCTION

The availability of water for human consumption and safe agrifood production is a critical aspect for Public Health.

Arsenic (As) is a natural pollutant in different sources of water (mainly groundwater) of many countries, including Argentina. High levels of As in water imply a risk for public health through water consumption as well as agrifoods exposed to contaminated water. Poisoning in humans is frequently associated to this, and can lead to degenerative problems like cancer.

One of the most important actions to avoid As intoxication is the control of water and food with levels above the maximum accepted limits; for this, it is necessary to know the levels of this metal in different water sources. Accordingly, since many years ago, Argentina counts with a map with the distribution of As in water, even though there are still many areas where the As level remains unknown. In Figure 1, the map of the Secretary of Sustainable Development of Argentina (October 3rd, 2013) does not reveal data of As concentration in the Río Negro Province.

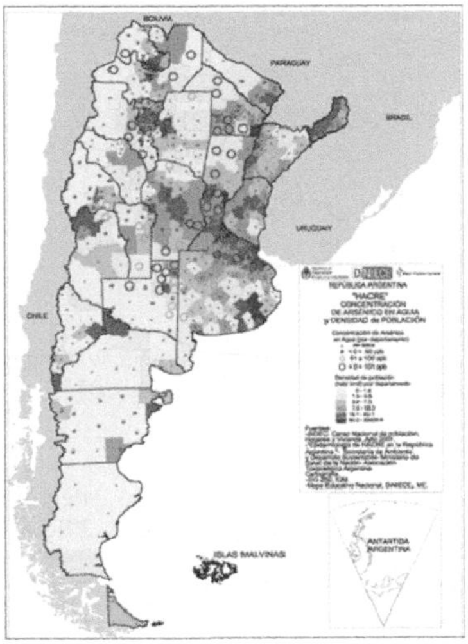

Figure 1. Actual map of As distribution in water, in Argentina.

2 METHODS/EXPERIMENTAL

2.1 *Samples*

133 samples of water were taken (surface $n = 22$; household distribution $n = 11$, groundwater $n = 102$) in different sites of the Rio Negro Province between 2012 and 2013. Samples were put in clean devices and sent to the Laboratory according Standard Methods (APHA).

2.2 *Methods*

Samples were processed by a colorimetric method (Hach: arsenic test kit) as strategy of screening (Abate, 2012): samples with values of As above 10 µg L^{-1} received a second evaluation by Atomic Absorption

Atomic with Graphite Furnace (AAS-GF) according to Standard Methods (APHA), using an UNICAM equipment, model Solaar with current calibration.

3 RESULTS AND DISCUSSION

Results of our studies are shown in Figure 2.

Levels of As higher than 10 μg L^{-1} were found in household distribution water in Valcheta, Los Menucos, Aguada de Guerra, Sierra Colorada and El Cuy. Considering recent reports from the National University of Rio Negro about dental injuries compatible with chronic consumption of toxic levels of As in indigenous inhabitants of these areas (personal communication), this situation should be considered by human health authorities in order to adopt prevention measures, such as installing arsenic removal systems—either centralized or domestic—and to ensure the appropriate disposal of the removed arsenic. There are an increasing number of effective and low-cost options for removing arsenic from small or household supplies. It is possible to use treated surface water, rain water, or to do new and deeper explorations of groundwater with levels of As lower than 10 μg L^{-1} according to the Código Alimentario Argentino (CAA).

As levels in surface water from Negro and Chico rivers were below 10 μg L^{-1} in every site under study,

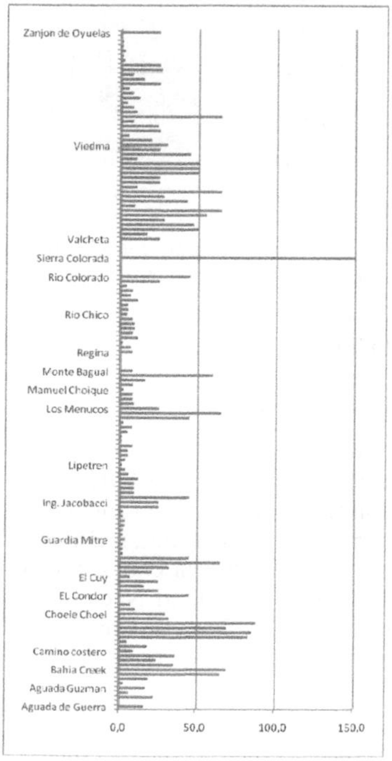

Figure 2. Levels of As in 133 samples of water from different locations in Rio Negro, Argentina.

suggesting no concern of anthropic contamination levels. These low As levels confirm that contamination due to the Puyehue volcano ashes generated in the 2012 eruption does not constitute a risk to Public Health to date.

As levels in groundwater were compatible with livestock production in Rio Colorado, Choele Choel, Seaside Road and Viedma. These locations have the largest cattle stock in the Río Negro province. Considering the variability of results observed in Seaside Road and Viedma (range: 2–88 μg L^{-1}), and the low number of samples processed in Choele Choel and Rio Colorado, it is necessary to increase the number of samples in order to guarantee that As levels are not a limiting factor for cattle production in these areas. This goal is relevant, given the need to increase reproductive rates of cattle in the region, considering that it has lost more than 50% of the cattle stock after five years of severe drought, and the inability to incorporate animals from other provinces given the new restrictions on admission of animals transmitting foot and mouth disease virus.

4 CONCLUSION

The decrease in surface water sources safe for consumption (increased evaporation due to higher average temperatures, irrigation use, and different kinds of pollution by industrial development and mismanagement of wastes) on one hand, and the need to increase agricultural food production to meet the growing global demand on the other, determines the increasingly frequent use of groundwater.

High levels of As found in groundwater from various areas of the province of Rio Negro highlight the need to check the map of the As of Argentina. Therefore, education and mitigation measures can be implemented, aimed at reducing the likelihood of consumer poisoning by this chronic toxic.

ACKNOWLEDGEMENTS

To the Ford Foundation and to National University of Rio Negro for partial financial support of this study.

REFERENCES

APHA-American Public Health Association. 2012. Standard Methods for the Examination of Water and Wastes, Washington DC, 22nd edition.

Código Alimentario Argentino. *Cap XII. Bebidas analcohólicas, bebidas hídricas, agua y agua gasificada*.

Travaglio, M. & Abate, S. 2012. Validación de un método rápido de determinación de As en agua, para aplicarlo en screening de muestras ambientales que permita clasificarlas en rangos de concentración de As según diferentes criterios de aptitud para consumo humano. *E-ices 8, 8a Encuentro del International Center for Earth Sciencies, acta de resúmenes p. 92.*

One Century of the Discovery of Arsenicosis in Latin America (1914–2014) –
Litter, Nicolli, Meichtry, Quici, Bundschuh, Bhattacharya & Naidu (Eds)

Geological and hydrogeochemical characteristics of an aquifer with arsenic, can they help to define the As origin?

I. Morales
Postgraduate Program in Earth Sciences, UNAM, Mexico

R. Rodriguez & M.A. Armienta
Geophysics Institute, UNAM, Mexico

ABSTRACT: In central Mexico, there are many regions where groundwater contains As over the national standard for drinking water. In El Bajio, Guanajuato, the presence of arsenic could be linked to igneous rocks of different origin, to the local geothermal activity and to the presence of deep faults and fractures through which hot water rises. In the area, there are geothermal wells, which have surface temperatures between 35–50 °C. Isotopic studies and geothermometers indicate that water is recirculated mixed with irrigation water. Arsenic can be originated in deep rocks contained in metal sulfides or to another mineral.

1 INTRODUCTION

High arsenic As groundwater concentrations represent a risk for the population due to their toxicity (Armienta & Segovia, 2008). In El Bajio, Guanajuatense, Central Mexico, As has been also reported. In Juventino Rosas, the As concentrations in some wells are greater than the Mexican standard for drinking water (Rodríguez & Morales, 2013). The natural origin of As in the aquifer is related to the local geology and physico-chemical processes that control their availability and mobility in water (Smedley & Kinninbuerg, 2002). Likewise, geothermal activity observed in the site could be linked to the high concentrations, because of the hot water depth; in combination with metal ion concentrations and pH, this allows alteration of minerals, including the tendency to precipitation-solubilization of certain minerals and sorption-desorption elements (Ellis, 1979).

2 AREA OF STUDY

2.1 *Location and geology*

The municipality is located in the TransMexican Volcanic Belt, in El Bajio Guanajuatense, Central Mexico. The site is located between geological units composed of fractured volcanic rocks of different composition and lacustrine sediments. Most wells have a depth between 150 and 250 m. The faults and fractures observed in the site are product from compressional and extensional events from Paleocene and Oligocene-Miocene, respectively (Figure 1).

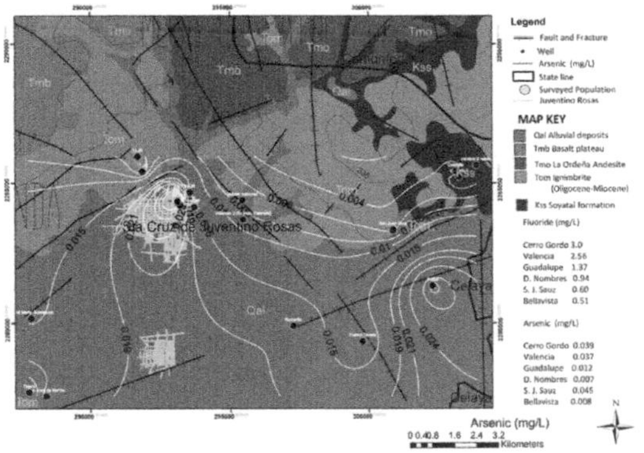

Figure 1. Geology and arsenic concentration in Juventino Rosas, El Bajio Guanajuatense, Central Mexico.

3 METHODOLOGY

3.1 *Groundwater sampling*

A groundwater sampling was carried out in Juventino Rosas in the period 2010–2013. Sampling was realized following international and national standards. Chemical determinations, including Fluorine, were done at the Analytical Lab of the Geophysics Institute of the Universidad Nacional Autonoma de Mexico, UNAM. Some urban and agriculture wells were sampled, at least three times.

3.2 *Geothermometers*

The application of geothermometers is a tool used for the exploration and development of geothermal

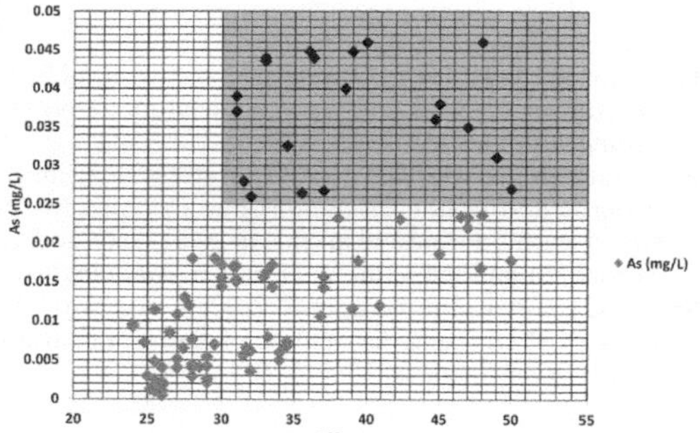

Figure 2. Arsenic concentration in mg/L with respect to the temperature of well water.

resources and reserves, estimation of temperature using isotopic and chemical composition of groundwater. There are two groups of geothermometers: 1) chemical and 2) isotopic. The first uses the content of major cations in water (Na, Ca, K, Mg), SiO_2 and relative abundance of gaseous components. The second group occupies isotope concentrations of deuterium and ^{18}O in groundwater and in geothermal environments to identify its source (Giggenbach, 1988). The first group is based on the chemical balance and the temperature dependence between water and minerals. It is assumed that the water bodies preserve a chemical composition during its ascent from the aquifer to the surface. This assumption is not always true because the interaction with cooler water or another aquifer water influences the final composition, although mineralogical alteration studies in geothermal conditions have shown the presence of equilibrium in different geothermal fields.

In most geothermal systems, unusual high concentrations of lithium, cesium, fluoride, boron and arsenic exist. These elements have been taken as indicators of the presence of magmatic flows in geothermal waters (Ellis, 1979).

4 RESULTS AND DISCUSSION

4.1 *Results*

In Juventino Rosas, high As and F concentrations in urban and agriculture wells were detected. This water is the only one source to supply the population.

Figure 2 shows the behavior of arsenic with respect to the temperature increase. The data show that in wells with hot water high concentrations of arsenic exist, and that several of the obtained values

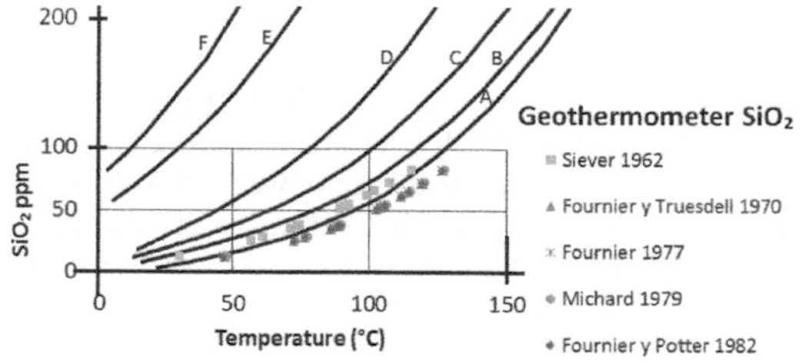

Figure 3. Geothermometry in JR and solubility curves of silica. A) silica-quartz. B) silica-quartz. C) silica chalcedony. D) cristobalite silica-α. E) Silica-β cristobalite (Fournier, 1977). F) silica amorphous silica.

exceed the limit allowed by the Mexican standard NOM-127, mainly at temperatures above 30 °C.

Figure 3 shows the application of different silica geothermometers in collected samples from J. Rosas. The mineral phase that controls the solubility of silica is quartz, and the temperature of collected samples are between 50 °C and 150 °C, which represents an environment of low-medium geothermal temperature.

4.2 *Discussion*

In Juventino Rosas, the As concentrations in some wells are greater than the Mexican standard for drinking water. Arsenic can be found in rocks, soils, and surface and groundwater. Its availability depends on different geological and anthropogenic processes (Smedley & Kinninbuerg, 2002). Its toxicity depends on the oxidation state, the chemical structure and the solubility. The metal sulfides are the most important minerals that increase arsenic content in groundwater flows. In geothermal environments, elements such as fluoride, chloride and boron dissolved from rocks exist, even before a significant hydrothermal alteration occurs in the rock.

In this area, arsenic could be associated to geothermal water, to the hydrogeology, local tectonic (faults and fractures) and mineralogy from volcanic rocks. Through the faults and fractures, geothermal waters containing dissolved arsenic and fluoride. could be circulating.

5 CONCLUSIONS

High concentrations of arsenic and fluoride in well water of El Bajio, Guanajuato could be related to geothermal activity, low-medium temperature, faults and fractures present in the site and to hydrogeochemical processes occurring in the aquifer.

ACKNOWLEDGEMENTS

The research was financed by the PAPIIT UNAM grant, Num IN102113. CMAPAS Salamanca facilitates the access to the urban wells. CMAPAJR Juventino Rosas people helped in the groundwater sampling. To A. Aguayo N. Ceniceros and O. Cruz for chemical analyses.

REFERENCES

Armienta, M.A. & Segovia, N. 2008 Arsenic and fluoride in the groundwater of Mexico. *Environ. Geochem. Health,* 30: 345–353.

Ellis, A.J. 1979. Chemical geothermometry in geothermal system. *Chem. Geol.* 25: 219–226.

Giggenbach, W.F. 1988. Geothermal solute equilibria. Derivation of Na-K-Mg-Ca geoindicators. *Geochim. Cosmochim. Acta* 52: 2749–2765.

Rodriguez, R. & Morales, I. 2013. Water management patterns and health affectations due to groundwater supply with fluorine and arsenic in a village of central Mexico. Submitted *to Int. J. Environ. Contamination.*

Smedley, P.L & Kinninburg, D.G. 2002. A review of the source, behavior and distribution of arsenic in natural waters. *Appl Geochem*, 17: 517–568.

One Century of the Discovery of Arsenicosis in Latin America (1914–2014) –
Litter, Nicolli, Meichtry, Quici, Bundschuh, Bhattacharya & Naidu (Eds)
 ISBN 978-1-138-00141-1

Evaluation of inorganic arsenic species in drinking water of the southeastern of the Buenos Aires province, Argentina

A.D. Robles
Toxicología Ambiental, Departamento de Ciencias Marinas, Instituto de Investigaciones Marinas y Costeras, Facultad de Ciencias Exactas y Naturales, Universidad Nacional de Mar del Plata, Mar del Plata, Argentina
Consejo Nacional de Investigaciones Científicas y Técnicas (CONICET), Buenos Aires, Argentina

M. Cohen
Toxicología Ambiental, Departamento de Ciencias Marinas, Instituto de Investigaciones Marinas y Costeras, Facultad de Ciencias Exactas y Naturales, Universidad Nacional de Mar del Plata, Mar del Plata, Argentina

M.B. Romero & M. Gerpe
Toxicología Ambiental, Departamento de Ciencias Marinas, Instituto de Investigaciones Marinas y Costeras, Facultad de Ciencias Exactas y Naturales, Universidad Nacional de Mar del Plata, Mar del Plata, Argentina
Consejo Nacional de Investigaciones Científicas y Técnicas (CONICET), Buenos Aires, Argentina

F. Garay
INFIQC, Departamento de Físico Química, Facultad de Ciencias Químicas, Universidad Nacional de Córdoba, Pabellón Argentina, Ciudad Universitaria, Córdoba, Argentina

ABSTRACT: Arsenic in natural water is a worldwide concern due to chronic health effects in people exposed through drinking water. The aim was to evaluate the distribution of As(III) and As(V), through ASV, in an Argentine area with low or absent information about these contaminant. The species of inorganic arsenic were quantified in groundwater, surface and drinking water by Anodic Stripping Voltammetry (ASV) using a gold disc electrode. Total As was analyzed after the reduction step, and As(V) was calculated by difference. In some sites, total As values exceed the maximum limit for human consumption (10 μg/L), but the As(III)/As(V) ratio was significantly low. The lowest levels of total As were found in those sites related to possible anthropogenic activities, indicating that the content in the studied water is coming from natural environment. This study constitutes the first report of inorganic As speciation for the southeastern of the Buenos Aires province.

1 INTRODUCTION

Arsenic is a natural contaminant. Inorganic As(III) is the most toxic species, and it is found in both surface and groundwater. As(III) is associated with reducing conditions (aquifers), while As(V) is increased by atmospheric oxygen in surface waters and in shallow groundwaters associated to oxic conditions. Their presence in natural waters is of concern due to the disease caused by chronic exposure through drinking water (HACRE, Regional Chronic Endemic Hydroarsenicism). In Argentina, HACRE is found in a wide geographic distribution. In 2006, the US EPA/WHO established 10 μg L^{-1} as a new maximum level for total As (As_t) in drinking water. The aim was to evaluate the distribution of As(III) and As(V) through Anodic Stripping Voltammetry (ASV) in the SE of the Buenos Aires province (Argentina), an area with low or absent information about this contaminant. The use of this technique is advantageous due to its high sensitivity and the possibility of identifying species of As. The speciation is important in the selection and design of a treatment system (Sorg *et al.*, 2013).

2 METHODS/EXPERIMENTAL

2.1 *Sampling*

Samples were taken from taps, domestic wells and public water supply system, and from stream in the cities of Miramar, Mar del Sur and Otamendi (Buenos Aires province, Argentina). Polypropylene bottles were used, HCl(c) added before and ascorbic acid added *in situ* to posterior sampling to prevent

Table 1. Environmental parameters, As(III) and As(V) concentrations in sampling sites from southeastern Buenos Aires province.

Parameters/Sites	pH	O_2 (mg L^{-1})	Conductivity (mS cm^{-1})	As(III) (µg L^{-1})	As(V) (µg L^{-1})
Miramar Downtown	7.5–8.8	3.5–6.6	1.6–2	0.3–3.8	11.9–34.7
Miramar Close to Stream	7.3–8.5	4.6–6.1	1.5–2	0.9–3.9	7.6–31.1
Miramar Landfill	7.4–8.6	6.3–6.5	1.6–1.7	0.8–1.6	9.2
Miramar Stream	7.5–7.7	8.1–9.9	1.4	1.0–2.4	5.5–18.6
Mar del Sur	8.4–9.4	4.5–6.7	1.8–1.9	1.9–3.8	2.6–30.6
Otamendi	7–8.3	1.2–5.9	0.6–1.3	0.3–3.4	21.5–36.7

oxidation of As(III). Hardness, alkalinity and dissolved oxygen were determined in the laboratory. Temperature, salinity, pH and conductivity were measured *in situ* using a multiparameter analyzer.

2.2 *As(III) and As(V) determination*

Potentiostat POL 150 coupled to MDE 150 polarograph (Radiometer Copenhagen) with the software Trace Master 5 was used. Rotating gold disc (99.99%, area 0.030 cm^2), Ag/AgCl and Pt wire as working, reference and counter electrode, respectively, were employed. The method applied was Square Wave ASV (Bodewig *et al.*, 1982). As(III) is deposited from acid solution onto a rotating gold electrode at a potential of −0.3 V vs. Ag/AgCl and then stripped anodically. After the reduction with Na_2SO_3, As_t was determined as As(III), and As(V) was obtained by difference (Rasul *et al.*, 2002).

3 RESULTS AND DISCUSSION

In all places, concentrations of As(III) were significantly lower ($p < 0.01$) than those of As(V) (Figure 1), whose percentages were 1.5%–16% in relation to As_t. Maximum values were 38.8 µg L^{-1} for As_t, being 3.8 µg L^{-1} of As(III) and 36.7 µg L^{-1} of As(V). Both chemical species showed the lowest mean concentrations in groundwater (close to the landfill) and surface water (stream) of Miramar, while the highest were found in groundwater of Mar del Sur and Otamendi cities. Despite these results, the differences were significant ($p < 0.05$) only for As(V), and not for As(III) ($p > 0.05$). It is noteworthy that the landfill area would not be a source to groundwater. Furthermore, the lower As(III) levels in surface waters than in groundwater would be related to their high oxygen content (Table 1) (Sánchez-Rodas *et al.*, 2005). The concentrations of As_t were higher than the maximum levels for drinking waters (10 µg L^{-1}; US EPA/WHO), with the exception of the sites close to landfill areas.

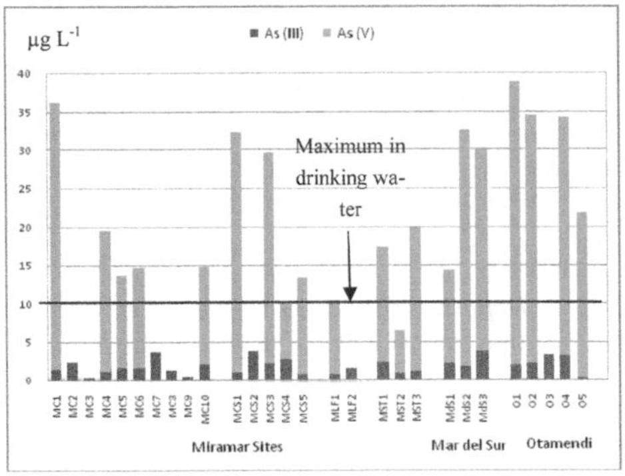

Figure 1. Concentrations (µg L^{-1}) of As(III) and As(V) in sampling sites from the southeastern of Buenos Aires province. MC: Miramar Downtown, MCS: Miramar Close to Stream, MLF: Miramar Landfill, MST: Miramar Stream, MdS: Mar del Sur, O: Otamendi.

Cluster analyses (data not shown) indicated relationships of As(III) with conductivity, pH and dissolved oxygen. Similar results were found by Paoloni *et al.* (2009) in the southern areas of the Pampa plains (Argentina).

4 CONCLUSIONS

This work is the first report for As speciation in the study area. Besides the As values exceeding the maximum limit for human consumption (10 µg L^{-1}), the As(III)/As(V) ratio was significantly low. The obtained result indicated that anthropogenic activities are not As sources. The ASV was useful tool and could be recommendable for the speciation analysis. The information on As speciation would be used in future monitoring and remediation treatments.

ACKNOWLEDGEMENTS

Supports: UNMdP (EXA 640/13), CONICET (PIP 2010/0348).

REFERENCES

Bodewig, F.G., Valenta, P. & Nürnberg, H.W. 1982. Trace determination of As(III) and As(V) in natural waters by differential pulse anodic stripping voltammetry. *Fresenius Journal of Analytical Chemistry* 311: 187–191.

Paoloni, J.D., Sequeira, M.E., Espósito, M.E., Fiorentino, C.E. & Blanco, M.C. 2009. Arsenic in Water Resources of the Southern Pampa Plains, Argentina. *Journal of Environmental and Public Health* 2009: 1–7.

Rasul, S.B., Munir, A.K.M., Hossain, Z.A., Khan, A.H., Alauddin, M. & Hussam, A. 2002. Electrochemical measurement and speciation of inorganic arsenic in groundwater of Bangladesh. *Talanta* 58: 33–43.

Sánchez-Rodas, D., Gómez-Ariza, J.L., Giráldez, I., Velasco, A. & Morales, E. 2005. Arsenic speciation in river and estuarine waters from southwest Spain. *Science of the Total Environment* 345: 207–217.

Sorg, T.J., Chen, A.S.C. & Wang, L. 2013. Arsenic species in drinking water wells in the USA with high arsenic concentrations. *Water Research*. In press.

One Century of the Discovery of Arsenicosis in Latin America (1914–2014) – Litter, Nicolli, Meichtry, Quici, Bundschuh, Bhattacharya & Naidu (Eds)
© 2014 Taylor & Francis Group, London, ISBN 978-1-138-00141-1

Arsenic in groundwater of the southwestern Buenos Aires province, Argentina

D.H. Campaña & A. Airasca
Facultad Regional Bahía Blanca, Universidad Tecnológica Nacional, Provincia de Buenos Aires, Argentina

G. Angeles
Departamento de Geografía, Universidad Nacional del Sur, Provincia de Buenos Aires, Argentina

ABSTRACT: The main objective of this study was to evaluate arsenic in drinking waters of the southwestern part of the Buenos Aires province (Argentina), in order to assess toxicity concerning both human consumption and animal production. More than 25 geographic locations were sampled, most of them twice in a year (including dry and wet seasons). Sampling sites were at Bahia Blanca, Rosales, Pringles, Tornquist, Suarez, Villarino and Tres Arroyos districts and they were GIS referenced. Arsenic concentrations in water samples of shallow wells (up to 40 meters) were measured by atomic absorption spectrometry, and concentrations from 0.015 to 0.267 mg/L were found. It was found that As concentration increases during the dry season.

1 INTRODUCTION

Certain contents of trace metals in the water, without toxicity for the animals, can be accumulated in meat and milk, reaching levels toxic for human consumption (Nriagu, 1994). Drinking water quantity and quality are very important factors for meat and milk bovine production (Fernández Cirelli *et al.*, 2010) and, in some cases, groundwater sources are chemically restricted for livestock consumption (Bavera *et al.*, 1979). Low drinking water quality gives different symptoms in animals and higher mortality index (Perez Carrera, 2009, Sigrist *et al.*, 2010). The SW of the Buenos Aires province (Argentina) presents As and F concentrations exceeding the reference guideline limits indicated by WHO (2004) and CAA (2007).

In this work, more than 25 geographic locations in the SW of the Buenos Aires province were sampled (Bahia Blanca, Rosales, Pringles, Tornquist, Suarez, Villarino and Tres Arroyos districts), in order to assess toxicity in water for both human consumption and animal production.

2 MATERIALS/METHODS

Selected livestock production sites from Bahia Blanca and Coronel Rosales were sampled twice a year (wet and dry weather), mainly from groundwater (5 to 40 m depth) and a few, superficial. The geographic location of the sites were field established using the Global Positioning System (GPS) Garmin Vista HCX, in order to arrange data for GIS, representing spatial distribution of measured concentrations of As and F. The specific layer obtained was overlapped with satellite images from Google Earth server. Mapping with information was developed using POSGAR98 system. SIG model was ARCGis Desktop (version ARCInfo 10).

Quantitative determination of As was done by ICP-AES according to standard method 200.7 (Rev 4.4, EPA, 1997). F was spectrophotometrically analyzed following standard method 340.1 (Rev 4.4, EPA, 1997). Other measured parameters were pH (method 150.1, Rev 4.4, EPA, 1997), and electrical conductivity (method 120.1, Rev 4.4, EPA, 1997).

3 RESULTS AND DISCUSSION

Table 1 shows measured values of pH, EC, As and F at sampling sites at SW of the Buenos Aires province. In the sites indicating two result lines, the first one corresponds to sampling during August 2012, and the second one corresponds to July 2013. The measured values of total As for all samples in this study were similar to those obtained by other researchers at the south of the Buenos Aires province (Esposito *et al.*, 2011). More than 75% sites gave concentrations higher than 0.05 mg/L, and all of them exceeded the reference values (WHO, 2004, CAA, 2007). As water contents at the second

Table 1. pH, electrical conductivity, As and F contents in water samples of SW of the Buenos Aires province.

Site	pH	EC (mS/cm)	As (mg/L)	F^- (mg/L)
Paso Mayor 1	8.10	1.56	0.121	
	8.10	1.45	0.123	3.85
Paso Mayor 2	8.11	2.17	0.138	
	8.20	2.06	0.227	5.80
Paso Mayor 3	7.89	1.93	0.115	
	8.17	1.59	0.134	4.70
Paso Mayor 4	7.97	2.24	0.123	
	8.24	2.27	0.143	0.10
Pehuen-Có 1	7.93	2.62	0.018	
	7.66	3.03	0.029	0.50
Pehuen-Có 2	7.94	2.02	0.022	
	8.16	1.98	0.045	1.80
Cabildo 1	8.26	2.11	0.258	
	8.11	1.99	0.252	4.30
Cabildo 2	8.31	1.15	0.038	
	8.48	1.39	0.059	0.10
Bajo Hondo 1	8.10	2.54	0.166	
	8.15	2.40	0.165	4.40
Bajo Hondo 2	8.18	2.16	0.206	
	8.22	2.08	0.267	11.45
Cabildo 3	8.04	1.20	0.124	
	8.22	1.16	0.118	6.60
Cabildo 4	8.02	1.50	0.114	
	8.19	1.33	0.130	2.50
Camino del Lechero 1	7.79	5.45	0.086	
	7.91	5.40	0.142	1.95
Camino del Lechero 2	8.65	2.10	0.083	
Cabildo 5	7.90	1.40	0.081	
	7.96	1.25	0.142	4.70
Cabildo 6	8.00	1.63	0.135	
	8.19	1.45	0.101	5.10
Cabildo 7	8.09	1.17	0.111	
	8.51	1.09	0.102	4.60
Cabildo 8	7.71	1.01	0.046	
	7.86	0.95	0.069	2.90
Bajo Hondo 3	7.66	2.64	0.143	
	8.05	2.40	0.150	7.60
Bajo Hondo 4	8.03	2.01	0.221	
	8.32	1.78	0.203	7.90
Bajo Hondo 5	7.96	2.09	0.227	
	8.24	1.99	0.218	9.10
Bajo Hondo 6	7.77	2.13	0.203	
	8.17	1.92	0.202	8.20
Bajo Hondo 7	7.96	1.72	0.224	
	8.36	1.71	0.216	7.50
Bahía Blanca	8.03	3.42	0.050	2.00
Lartigau	7.95	1.78	0.034	4.60
Villa Ventana	6.78	0.77	0.050	0.51
Mayor Buratovich	8.10	1.70	0.050	1.90
Coronel Suárez 1	7.98	1.28	0.100	0.98
Coronel Suárez 2	6.99	1.35	0.100	0.90
Tres Arroyos	7.60	2.16	0.016	1.60
	8.00	2.18	0.015	1.10

sampling period (July 2013), corresponding to dry weather (low rain precipitation over the last 6 months before July 2013), were higher than the As contents measured during the wet period.

The values for the recommended maximum limit of As content in drinking water for feedstock range from 0.05 to 0.5 mg/L according several national and international references (Sigrist *et al.*, 2011). Therefore, for the sampling sites involved in this work, average values were almost over the low recommended level (CAA, 2007), and they never reach the maximum values (Law 24051). More than 25% sampling sites showed As contents higher than those indicating a chronic toxicity limit for bovines, 0.15 mg/L (Perez Carrera *et al.*, 2004).

According to the recommended F contents in drinking water quality for human consumption (maximum allowable value: 1.5 mg/L, CAA, 2007), the concentrations measured in all samples were over this value, except those of Sierra de la Ventana region, Coronel Suarez and Villa Ventana. For fluorine in cattle, near 30% of the measured values exceeded 5 mg/L (maximum suggested limit). Concentrations over 15 mg/L, indicated as the chronic toxic limit (Fernández Cirelli *et al.*, 2010) were never found.

4 CONCLUSIONS

Groundwater sources (both for animal and human consumption) contain As levels that would affect a large number of people, most of them living in rural zones, meat and milk production. This information should be useful not only for professionals practicing preventive medicine (in human and in animals), but also for geographical planning and developing.

ACKNOWLEDGEMENTS

This work was supported by Agencia Nacional de la Promoción de la Ciencia y la Tecnología de Argentina, PICTO 2010-0027 (Argentina).

REFERENCES

Bavera G.A., Béguet H.A., Bocco O.A., Rodríguez E.E. y Sánchez J.C. 1979. *Aguas y Aguadas*, Editorial Hemisferio Sur.

CAA, (2007) "*Código Alimentario para Uso de Agua Potable en Argentina*," Cap. XII, Art. 982 (Res. Conj SPRyRS y SAG-PyA N° 68/2007 y N° 196/2007), Buenos Aires.

EPA (1997). *Methods and Guidance for Analysis of Water* Ver 1.0. EPA 821-C-97-001, Abril 1997.

Fernández Cirelli A., Schenone N., Pérez Carrera A., Volpedo A., (2010) *Water quality for the production of*

traditional and non traditional animal species in Argentina, AUGMDOMUS, 1:45–66.
LEY 24051 (1993). *Régimen de Desechos Peligrosos*. Decreto Nacional 831/93, Reglamentación de la Ley 24051.
Nriagu, JO. (1994). *Arsenic in the Environment, Parts I, Cycling and characterization*. New York. John Wiley and Sons Inc. 430 pp.
Perez Carrera, A.; Fernandez Cirelli, A, (2004). *Niveles de arsénico y flúor en agua de bebida animal en establecimientos de producción lechera (Pcia. de Córdoba, Argentina)*. InVet., 6(1): 51–59.
Perez Carrera (2009) AIDIS. *Effect of chronic intake of arsenic-contaminated water on blood oxidative stress indices in cattle-* 1er Congreso de la region centro–Ingeniería Sanitaria y Ambiental – Córdoba – 22–24 septiembre.
Sigrist M., Beldoménico H., Repetti M.R., (2010) *Evaluation of the influence of arsenical livestock drinking waters on total arsenic levels in cow's raw milk from Argentinean dairy farms*. Food Chemistry 121, 487–491.
WHO (World Health Organization) (2004), "*Guidelines for Drinking Water Quality,*" 3rd Ed., Vol. 1, Geneva, Switzerland.

One Century of the Discovery of Arsenicosis in Latin America (1914–2014) – Litter, Nicolli, Meichtry, Quici, Bundschuh, Bhattacharya & Naidu (Eds)
© 2014 Taylor & Francis Group, London, ISBN 978-1-138-00141-1

Relevance of arsenic and other trace elements in natural groundwater in zones of central and west regions of Argentina

M.L. Gomez & F. Donna
Instituto Argentino de Investigaciones de las Zonas Áridas (IADIZA-CCT CONICET), Mendoza, Argentina

Rafael Giordano
Net Log Service, Córdoba, Argentina

ABSTRACT: Arsenic and other trace element naturally occurring in groundwater are an important aspect in many areas of Argentine. Rural and urban activities in NE Mendoza and S Córdoba areas rely exclusively on the use of groundwater that, in many cases, is the only source of drinking water. In our work, we analyzed trace elements (As, F, V, U, Mo, Cr, Sb, Se) and the interaction between F and As in the formation of As-F-complexes in groundwater samples. The study indicates a significant relationship between As and other trace elements in groundwater, corroborating the natural origin of all studied trace elements (volcanic ash in levels and dispersed in the aquifer sediments), especially in "*bucket wells*". The As-F complexes are not present in any of the areas in study, where both elements are originated from aquifer materials.

1 INTRODUCTION

In south of the Córdoba province (C. Moldes, Huinca Renancó and Chaján) and NE of the Mendoza province, many studies have reported the presence of As and F in groundwater and its geochemical conditions, but there are no studies on other trace elements, their concentrations and their interactions in the aquifer. In this preliminary study, one of the main objectives was focused on the interactions between As and F and the potential formation of As-F-complexes. As we mentioned, those elements are present naturally in groundwater of this region (Gomez *et al.*, 2005; Giordano, 2008; Gomez *et al.*, 2009). Groundwater with geothermal characteristics are present in NW and NE of Mendoza (Gomez *et al.*, 2013); therefore, the issue of if As-F complex could have been put in solution and kept in groundwater due to its high stability once formed is of interest. Henke (2009) cited studies where a dominance of $AsO_3F_2^-$ is predicted for natural fluoride-rich waters. However, these predictions have been achieved based on thermodynamical modeling, with the commonly used constant (Wagman *et al.*, 1982), probably overestimating the stability of As-F-complexes. However, its potential formation under natural conditions is still unknown. It is very important to study AsF_6^- complex from a toxicological point of view (Daus *et al.*, 2010). We also analyzed the water chemistry related to As and other trace elements (V, Cr, Se, Mo, Sb, U) in groundwaters of those areas. This work will provide preliminary information to define areas for future specific studies.

In the arid region of NE Mendoza, the outcropping materials are very fine-fine sand and silt, originating mainly from wind and distal fluvial environments. The sediments indicate a composition from intermediate to acid type with important contents of volcanic ash, dispersed in sediments and in discontinuous levels at different depths (Gomez *et al.*, 2013).

In the sub-humid study area of Córdoba, the aquifers are mainly composed of silty sand sediments of aeolian origin, typically loess-like sediments of the Holocene, and part of the study area is situated near igneous-metamorphic basement rocks of the Paleozoic (Gomez *et al.*, 2005). The average chemical composition of the loess sediments closely resembles dacite, and the composition of volcanic glasses is similar to that of rhyolite (Nicolli *et al.*, 1989).

2 METHODS

We selected sixteen wells for groundwater sampling (nine from Córdoba and seven from Mendoza). Groundwater samples were collected considering three windmills and four samples taken from hand-d rilled wells where water is manually drawn with buckets (called "bucket wells"). These wells, of 1 m in diameter, do not reflect natural groundwater conditions due to the evapotranspiration occurred during more than 60 years and low sanitary conditions, but include samples of the water that many rural communities consume (especially indigenous Huarpe communities in NE Mendoza). The samples were collected and filtered (according to laboratory procedure using filters of 0.2 µm cellulose-acetate) during March 2013. Sampling depth ranged between 4 and 230 m. Analysis was performed by IC-ICP-MS at the University of Bayreuth, Germany.

3 RESULTS AND DISCUSSION

The formation of As-F-complexes was not observed in any of the analyzed samples. The result is according to Smedley and Kinniburg (2002) calculations, where fluoroarsenate is a minor species even in samples with the highest concentrations of As and F.

Better correlations are found when samples are analyzed separately, by provinces. In Córdoba samples, slightly good correlation coefficients for As vs. pH (0.69), F vs. V (0.49), F vs. As^{3+} (0.54), Se vs. Cr (0.54), Mo vs. Se (0.83) has been found. The general weak and not so good values may be reflecting the few samples, which would not be sufficient to represent the three different areas in the Córdoba province. Correlations in samples from the Mendoza area show a positive and high correlation between As and F, V, Se, U and between NO_3^- and trace elements (correlation coefficient between 0.42 and 0.98). These relations could be showing the same geochemical conditions that result in highest concentrations. In Mendoza, the "bucket wells" showed the highest concentrations of all trace elements and the lowest concentrations of dissolved oxygen (3.02 to 3.54 mg/L). This condition (low O_2 values indicate the presence of high organic matter contents) might promote the mobilization of As and other trace elements, generating additional increases and may be even responsible for its high correlation (0.63) for NO_3^- vs. As^{3+}, the most toxic species. The NO_3^- content in groundwater from NE Mendoza (1 to 5 mg/L) is related to the organic matter coming from cattle posts and the sanitary conditions of water extraction methods ("bucket wells"). High evaporation rates and poor water renewal rates (low dissolved O_2) could generate chemical conditions favoring the concentration of trace elements, especially As (25 to 923 µg/L). The box plot analyses show the highest dispersion and the median for Mendoza wells, except for F and NO_3^-. In this case, it is likely that the high Ca^{2+} content in groundwater control the availability of F^- in groundwater (Gomez *et al.*, 2013).

4 CONCLUSIONS

Arsenic concentrations and other trace elements in groundwater of the central and west plains of Argentina are conditioned by natural sources from volcanic ash (Gomez *et al.*, 2013), and from the loess-like sediments (Nicolli *et al.*, 1989; Gomez *et al.*, 2009). Loess constituents with different types of minerals, including arsenic, fluoride bearing minerals and other trace elements, like volcanic glass, volcanic lithic fragments and the minerals from the rocks that form the crystalline bedrock of hills of Córdoba are the mainly sources of F^- to the studied aquifers.

In arid zones of Mendoza, the evaporation and sanitary conditions of some water extraction methods can raise the contents of trace elements. The samples taken from "bucket wells" have shown concentrations of trace elements of several orders of magnitude, 2 and almost 3 times in respect to samples from windmills. That point is relevant to the water management, especially in the Mendoza area.

We thus have to conclude that the formation of AsF_6^- does not occur under natural conditions such as those encountered in the study area. Thermodynamical constants are likely overestimating the stability of As-F complexes, and it is unlikely to find them in typical natural groundwaters with neutral pH (slightly acidic-slightly basic) and for typical groundwater temperature (between 19 and 25 °C).

Is important to mention that As and F resulted in much lower concentrations compared with previous analysis of the same wells (Gomez *et al.*, 2005; Giordano, 2008; Gomez *et al.*, 2009; Gomez *et al.*, 2013). Filtration may cause the lower values and this must be considered in future studies, mainly by the potential danger to humans' health, as well as, for As-F complexes studies. It is relevant to study the role of microorganisms and particles of sediments in the concentration of As and other trace elements in groundwater with high input of organic matter and low dissolved oxygen, like in "bucket wells". Human activities may significantly affect the groundwater quality, and local nitrate leaching suggests that human activities and water availability may change groundwater chemistry in the future, increasing nitrate and As^{3+} concentration.

ACKNOWLEDGEMENTS

We want to acknowledge to B. Planer-Friedrich for help with the analysis (trace elements) in the laboratory of Environmental Geochemistry, University of Bayreuth from Germany.

REFERENCES

Besuschio, S., Desanzo, A., Perez, A. & M. Croci. 1980. Epidemiological associations between arsenic and cancer in Argentina. *Biol. Trace Elem. Res.* 2:41–55.

Biagini, R., Rivero, M., Salvador, M. & Cordoba, M. 1978. HACRE. *Arch. Argent. Dermatol.* 48:151–158.

Bocanegra, O. 2002. Informe sobre HACRE en el departamento Lavalle. Mendoza. *Estudio Colaborativo Multicéntrico. Secretaría de Ambiente y Desarrollo Sustentable*. Pp 200.

Daus, B., Weiss, H. & Altenburger, R. 2010. Uptake and toxicity of AsF6 in aquatic organi. *Chemosphere*. 78: 307–312.

Giordano, R. 2008. *Características hidrolitológicas, hidrodinámicas e hidroquímicas del acuífero freático de Huinca Renancó*. unpublished report. Pp. 100.

Gomez, M.L., Blarasin, M. & D. Martínez. 2009. Arsenic and fluoride in a loess aquifer in the central area of Argentina. *Environmental Geology*, 57:143–155.

Gomez, M.L., Aranibar J.N., Wuilloud, R., Rubio, C., Soria, D., Martinez, D., Monasterio, R. & Villagra P. 2013. Hydrogeology of the Gunacache Travesía in the central Monte desert of Mendoza, Arg. Manuscript under revision.

Henke, K. 2009. *Arsenic: Environmental Chemistry, health threats and waste treatment*. John Wiley.

Nicolli, H., Suriano, J.M., Gómez, P., M.A., Ferpozzi, L.H. & Baleani, O. 1989. Groundwater contamination with Arsenic and other trace elements in an area of the La Pampa province of Córdoba, Arg. *Environ. Geol. Water Sci.* 14(1):3–16.

Smedley P. & D. Kinniburgh. 2002. A review of the source, behaviour and distribution of arsenic in natural waters. *Appl. Geochem.* 17:517–568.

One Century of the Discovery of Arsenicosis in Latin America (1914–2014) –
Litter, Nicolli, Meichtry, Quici, Bundschuh, Bhattacharya & Naidu (Eds)

Natural arsenic occurrence in drinking water and assessment of water quality in the southern part of the Poopó lake basin, Bolivian Altiplano

M. Ormachea Muñoz
Instituto de Investigaciones Químicas, Universidad Mayor de San Andrés, La Paz, Bolivia
KTH-International Groundwater Arsenic Research Group, Department of Sustainable Development, Environmental Science and Engineering, KTH Royal Institute of Technology, Stockholm, Sweden

L. Huallpara & E. Blanco Coariti
Instituto de Investigaciones Químicas, Universidad Mayor de San Andrés, La Paz, Bolivia

J.L. García Aróstegui & C. Kohfahl
Instituto Geológico y Minero de España, Madrid, Spain

M.C. Hidalgo Estévez
Departamento de Geología, Universidad de Jaén, Jaén, Spain

P. Bhattacharya
KTH-International Groundwater Arsenic Research Group, Department of Sustainable Development, Environmental Science and Engineering, KTH Royal Institute of Technology, Stockholm, Sweden

ABSTRACT: Drinking water quality and the presence of natural arsenic (As) were studied in a rural, less developed area of the southern part of the Poopó lake basin in the Central Bolivian Altiplano. People in this area use untreated surface- and ground-water directly as drinking water. Water is extracted from excavated wells and from few rivers occasionally present during the rainy season. Forty-one wells and seven different sites along four rivers were sampled as they are common sources for drinking water. The main characteristics of the sampled waters showed a slightly alkaline pH, high electrical conductivity and high salinity where the principal components were sodium, chloride and bicarbonate. Arsenic concentrations reached values up to 623 $\mu g\ L^{-1}$ exceeding the current WHO guideline value (10 $\mu g\ L^{-1}$) in all rivers and in ninety-five percent of the sampled wells. Heavy metals and other trace elements showed relatively low concentrations.

1 INTRODUCTION

The most noteworthy studies related with the presence of high arsenic concentrations in drinking water were carried out in Southeast Asia (India, Bangladesh, China and Taiwan) and in Europe (Spain, Portugal and Hungary) (Bhattacharya *et al.*, 2002; Smedley & Kinniburgh, 2002).

In Latin America, many studies were carried out in Mexico, Nicaragua, Ecuador, Chile, Argentina, Peru, Brazil and Uruguay (Bundschuh *et al.*, 2008), where the most notable problems related with As in water resources are associated with both natural and anthropogenic sources.

In Bolivia, there are few studies dealing with natural As occurrence in drinking water (Van Den Bergh *et al.*, 2010; Ramos *et al.*, 2012; Ormachea *et al.*, 2013).

This research aimed to evaluate the general drinking water quality, with focus on As concentrations, in a rural, less developed area located in the southern part of the Poopó lake basin in the Central Bolivian Altiplano.

2 METHODS

2.1 *Sampling*

Drinking water quality was assessed in 31 communities in a rural area of the southern part of the Poopó lake basin. The number of samples taken in each community depended on community size and accessibility to its drinking water resources. In total, forty-eight samples were collected from seven sites along four rivers and forty-one excavated wells.

Water samples collected from each site were filtered using Sartorius 0.45 μm filters and then divided in two portions of 30 mL bottles, one for anion analyses and another one for cation and trace element analyses. The portion for the cation and trace element analysis was acidified with 14 M HNO_3. Both portions were stored in a refrigerator until further analysis.

pH, Electrical Conductivity (EC), Dissolved Oxygen (DO), temperature and redox potential (Eh) were determined in field using a portable meter, Hanna Instruments HI 9828 multiparameter. Alkalinity was measured *in situ* using a digital Hach titrator and sulfuric acid cartridge, 0.1600 N.

The geographical location of each sampling site was recorded using a hand-held Global Position System, GPSmap 60CSx.

2.2 *Laboratory analyses*

Analyses of major anions and cations were carried out at the Environmental Chemistry Laboratory of the Chemical Research Institute (IIQ) in La Paz, Bolivia. Major anions, chloride (Cl^-), nitrate (NO_3^-), phosphate (PO_4^{3-}) and sulfate (SO_4^{2-}) were determined using an ion chromatography system, Dionex ICS–1100. Major cations, sodium (Na^+), calcium (Ca^{2+}), magnesium (Mg^{2+}) and potassium (K^+) were determined using flame atomic absorption spectrometer, PerkinElmer AAnalyst 200.

The total dissolved content for trace elements including As was determined by Inductively Coupled Plasma-Mass Spectrometry (ICP-MS) analysis at the laboratories of the Scientific Instrumentation Centre of the University of Jaen in Spain. The equipment used for the ICP-MS analysis included a mass spectrometer with a plasma torch ionization source and a quadrupole ion-filter AGILENT SERIE 7500.

3 RESULTS AND DISCUSSION

3.1 *Water quality*

The pH values of the sampled waters were slightly alkaline, ranging from 6.1 to 9.6 with an average of 8.3. Samples showed high EC in a range from 316 to 19670 with an average of 1646 μS/cm. Temperatures ranged from 11 to 26 with an average of 15 °C. Principal components in the samples were sodium and chloride indicating a Na-Cl water type probably due an arid climate and a high evaporation.

3.2 *Arsenic and trace elements*

A statistical summary of selected trace elements in drinking water samples is presented in Table 1.

Table 1. Statistical summary of chemical parameters in the studied wells ($n = 48$).

Parameter	Min	Max	Median	Mean
Al (μg L^{-1})	3.4	44	7	9
As_{Tot} (μg L^{-1})	3.5	623	75	118
Fe (μg L^{-1})	8.3	994	69	124
Li (μg L^{-1})	52	8863	634	1148
Mn (μg L^{-1})	0.9	2787	4	134
V (μg L^{-1})	0.9	39	7	10
Zn (μg L^{-1})	15.6	4034	25	136

Dissolved As concentration in drinking water ranges from 3.5 to 623 μg L^{-1} and averages 118 μg L^{-1} ($n = 48$). Ninety-five percent of the drinking water samples exceeds the WHO guideline (10 μg L^{-1}). Water samples collected from wells located in lower terrains presents the highest levels of dissolved As. Arsenic speciation indicates that the predominant species is As(V).

Correlation analysis did not reveal direct relations between major components nor other trace elements. The distribution of trace elements shows a diversity of concentrations. Zinc concentrations range from 15.6 to 4034 μg L^{-1} (average: 136 μg L^{-1}), Al concentration range from 3.4 to 44 μg L^{-1} (average 9 μg L^{-1}), Li between 52.6 to 8864 μg/L (average 1148 μg L^{-1}) and V ranging from 0.9 to 39 μg L^{-1} with an average of 10 μg L^{-1}.

Among redox sensitive elements, Fe and Mn show wide variability in the ranges of 8.3–994 μg L^{-1} (average 124 μg L^{-1}) and 0.9–2787 μg L^{-1} (average 134 μg L^{-1}), respectively.

4 CONCLUSIONS

The water resources in the area are severely impacted by the presence of high As concentrations and high salinity levels that make the water unsuitable for human consumption. It is assumed that water in contact with alluvial sediments in wells located in lower terrains mobilize As more easily than that in the highest terrains where predominate volcanic and sedimentary rocks. The presence of As in drinking water is also attributed to the oxidation of sulfide minerals such as arsenopyrite and to the dissolution from volcanic rocks as a source of natural contamination.

Further studies in the area will be carried out to develop a conceptual model of the genesis, mobilization and transport of As in the region and potential health impacts on the local population.

ACKNOWLEDGEMENTS

The financial support by the Swedish International Development Cooperation Agency (Sida Contribution: 7500707606) and by the Agencia Española de Cooperación Internacional para el Desarrollo (N° ref. 11-CAP2_1282) is gratefully acknowledged.

REFERENCES

Bhattacharya, P., Frisbie, S.H., Smith, E., Naidu, R., Jacks, G. & Sarkar, B. 2002. Arsenic in the environment: a global perspective. In: Sarkar B (ed) *Handbook of heavy metals in the environment*. Marcell Dekker Inc., New York, 147–215.

Bundschuh, J., Pérez Carrera, A. & Litter, M. 2008. *Distribución del arsénico en las regiones ibérica e iberoamericana. Buenos Aires, Argentina*: Editorial CYTED Desarrollo.

Ormachea Muñoz, M., Wern, H., Fredrick, J., Bhattacharya, P., Sracek, O., Thunvik, R., Quintanilla, J., & Bundschuh, J. 2013. Geogenic arsenic and other trace elements in the shallow hydrogeologic system of Southern Poopó Basin, Bolivian Altiplano, *J. Hazard. Mater.* http://dx.doi.org/10.1016/j.jhazmat.2013.06.078

Ramos Ramos, O.E., Cáceres, L.F., Ormachea Muñoz, M.R., Bhatacharya, P., Quino, I., Quintanilla, J., Sracek, O., Thunvik, R., Bundschuh, J. & García, M.E. 2012. Sources and behavior of arsenic and trace elements in groundwater and surface water in the Poopó Lake Basin, Bolivian Altiplano. *J. Environ. Earth Sci.* 66: 793–807.

Smedley, P.L., & Kinniburgh, D.G. 2002. A review of the source, behavior and distribution of arsenic in natural waters. *Appl. Geochem.* 17: 517–568.

Van Den Bergh, K., Du Laing, G., Montoya, J.C., De Deckere, E. & Tack, F.M.G. 2010. Arsenic in drinking water wells on the Bolivian high plain: Field monitoring and effect of salinity on removal efficiency of iron-oxides-containing filters. *J. Environm. Sci. and Health Part A* 45: 1741–1749.

One Century of the Discovery of Arsenicosis in Latin America (1914–2014) – Litter, Nicolli, Meichtry, Quici, Bundschuh, Bhattacharya & Naidu (Eds)
 ISBN 978-1-138-00141-1

Hydrogeochemical characterization of the presence of arsenic in the Puelche aquifer in the area of Mataderos, Buenos Aires province, Argentina

A.E. Ferral, E. Alaniz & M. Sarmiento Tagle
Carrera de Gestión Ambiental, Universidad Blas Pascal, Córdoba, Argentina

B. Petrusevski
UNESCO-IHE Institute for Water Education, The Netherlands

ABSTRACT: The Chaco-Pampean plain, located in central Argentina, presents arsenic in groundwater related to volcanic glass from loessical sediments. As this region occupies an area of 1 million km^2, approximately, water quality controls are essential. This research gives insight into origin and dynamics of arsenic from the Puelche aquifer in Mataderos, Buenos Aires province, an industrial urban zone. Eighteen wells were monitored and chemically characterized, and arsenic levels, origin and hydrogeological aspects were evaluated. Arsenic concentrations ranged 0.01–0.025 mg L^{-1}. Forty per cent of the samples exceeded the limit set by the Argentine Food Code, 0.01 mg L^{-1}, being this data remarkable because several food industries are settled in the study area. Spatial analysis demonstrated that the south zone presents higher concentration than others. This difference may be related to the natural origin, coming from Pampeano aquifer filtration, where geological, structural and chemical characteristics favor the presence of As.

1 INTRODUCTION

Arsenic in water has become of great interest in the last decade because millions of people are exposed to the ingestion of water with high concentrations of this element in large areas of the world. The guideline value set by the World Health Organization (WHO) for drinking water has been reduced from 0.05 to 0.01 mg L^{-1} for total arsenic and similarly reduced in Argentina by the Argentine Food Code. In our country, mainly in the Chaco-Pampean plain, the availability and quality of groundwater resources are affected by the presence of certain elements (arsenic, vanadium, fluorine, selenium and other trace elements) that come from a natural phenomenon of pollution (volcanic activity), reducing water uses (Nicolli *et al.*, 2012). These features are relevant not only in rural but also in industrialized urban areas, as the Buenos Aires city, where wells are still used as water source for production purposes.

The main objective of this work was to understand the dynamics of the Puelche aquifer system in Mataderos zone, Buenos Aires province, evaluating arsenic concentrations and hydrochemical aspects.

2 EXPERIMENTAL AND METHODS

2.1 *Study area and aquifer explored*

The study area was the Mataderos neighborhood in the southwest of Buenos Aires city. We explored a squared surface of 62.09 km^2 with central geographic coordinates 34°39'38.51"S and 58°28'41.42"O. This is an industrial urban zone. Puelches Arenas are subsurface formations of Pampean, the geological surface unit, made by a sequence of loose quartz sands, fine and medium, with yellow to white shades. They have a fluvial origin and Pliocene age, which take the subsoil of the northeast of the province of Buenos Aires, an area of about 83,000 km^2, most of which are located at the north of the Salado river. They contain the most exploited aquifer of the country, predominantly of safe water for most uses.

The Puelche aquifer is an immense rainwater mass infiltrated into the soil and contained in an underground sand mantle and porous sediments, which lies between two thick layers of clayey and not much permeable sediments. It is found between 15 and 120 meters according to the site of the plain where it is located. Water of Puelche is sodium bicarbonate with a total salinity of less than 1 g L^{-1}. The quality deteriorates towards the basin of the Salado river in the floodplains of the major collectors (Matanza, Reconquista, Luján) and on the coastal plain bordering the Rio de la Plata river. It recharges from the Pampeano aquifer by downward vertical filtering process. The productivity of the Puelche ranges from 30 to 160 m^3 h^{-1} per well and it is used for human consumption, irrigation and industry. It behaves as a semiconfined aquifer.

2.2 *Monitoring assessment*

Two sample campaigns from eighteen wells, representative from the whole study area, were made on dry and wet seasons. The physicochemical variables analyzed were: temperature; pH, conductivity, total suspended solids, hardness, sulfate, alkalinity, total and soluble phosphorus, fluoride, chloride, calcium, magnesium, sodium, potassium, arsenic, cadmium, copper, mercury, lead, chrome and zinc. Sampling, storage, preservation and analysis of water samples were carried out according to APHA (2005). Total arsenic concentration was measured by graphite furnace atomic absorption (CL = 0.001 mg L^{-1}).

3 RESULTS AND DISCUSSION

3.1 *Groundwater hydrochemical characterization*

Figure 1 presents the Piper diagram obtained for eighteen wells in the wet season. This graph provides chemical characterization of groundwater (Piper, 1944). It can be observed that predominant water in the study area can be classified as sodium bicarbonate, calcium bicarbonate and/or magnesium, with low presence of chloride and/or sodium sulfate. Respect to pH values, they ranged from 6.98 to 7.75. Nicolli *et al.* (2012) showed that arsenate mobility in this waters increases with high bicarbonate concentrations and pH greater than 7.5 due to less adsorption of arsenate on the iron and aluminum oxides present in loessic sediments.

3.2 *Arsenic behavior in the study area*

Figure 2 shows the arsenic concentration contour map. Northeast, northwest and south sectors present the highest levels ranging from 0.018 to 0.025 mg L^{-1}. Forty per cent of wells exceeded the threshold set by the Argentine Food Code for this element. Simultaneous analysis of groundwater/piezometric levels/flow direction and arsenic isoconcentration maps show that the highest arsenic concentration found in the northeast area (well 19) corresponds to a depression of the piezometric surface where water flow converges. The same pattern is observed in the south and the northwest.

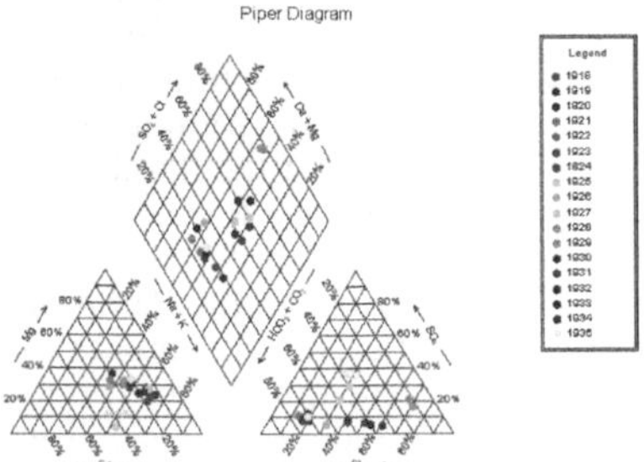

Figure 1. Piper diagram obtained for wet season.

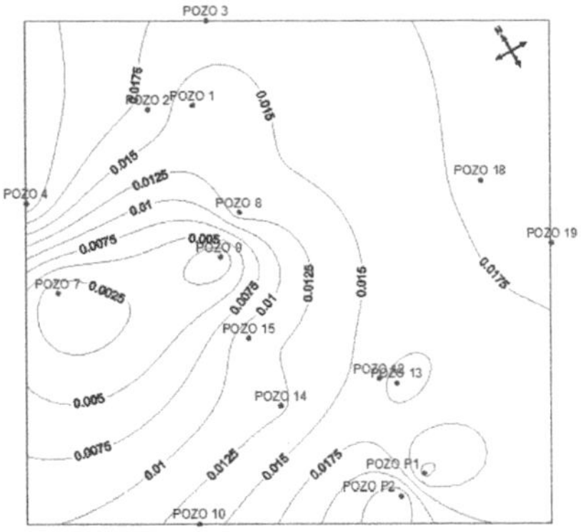

Figure 2. Arsenic isoconcentration curves, in mg L^{-1} units, for the Puelche aquifer. Area equal to 62.09 km^2 with central geographic coordinates 34°39’38.51”S and 58°28’41.42”O.

4 CONCLUSIONS

Arsenic presence in the study area can be attributed to natural origin coming from Pampeano aquifer filtration, where the geological, structural and chemical characteristics favor the occurrence and concentration of this element. Even though concentration values were not alarming high, forty percent of wells analyzed exceed tolerable limits for water consumption according to WHO and Argentine guidelines. Food industries located in the study area must be awarded of these results in order to evaluate their processes in the frame of new water quality considerations.

ACKNOWLEDGEMENTS

This research was funded by PoWER (Partnership for Water Education and Research) and Universidad Blas Pascal, Córdoba, Argentina. The authors want to acknowledge to the team of a joint project, Dr. Nidal Mahmoud, Mr. Omar Zayed and Dr. Qasem Abdel-Jaber from Birzeit University, Palestine. Also to SUDAMFOS company, that was interested in the project considering the relevant issue for the area.

REFERENCES

APHA, AWWA and WEF. 2005. Standard methods for the examination of water and Wastewater, 21st Edition. *American Public Health Association*, Washington D.C.

Nicolli, H.B. Bundschuh, J., Blanco, M.C., Tujchneider, O.C., Panarello, H.O, Dapeña, C. & Rusansky, J.E. 2012. Arsenic and associated trace-elements in groundwater from the Chaco-Pampean plain, Argentina: Results from 100 years of research. *Science of the Total Environment* 429: 36–56.

Piper, A. 1944. A graphic procedure in the geochemical interpretation of water analyses. *Am. Geophys. Union Trans.* 25: 914–923.

One Century of the Discovery of Arsenicosis in Latin America (1914–2014) –
Litter, Nicolli, Meichtry, Quici, Bundschuh, Bhattacharya & Naidu (Eds)
 ISBN 978-1-138-00141-1

Groundwater arsenic in the central-west of the Santa Fe Province, Argentine

M.C. Panigatti, R.M. Boglione, C.A. Griffa & M.C. Schierano
Facultad Regional Rafaela, Universidad Tecnológica Nacional, Rafaela, Santa Fe, Argentina

ABSTRACT: Arsenic is a contaminant in underground aquifers worldwide. The Chaco-Pampean plain, in Argentine, is one of the areas that have been affected by natural inorganic As transport into well waters. In this work, information about the groundwater quality of west Santa Fe province is presented, mainly considering arsenic. Samples were collected from 180 well sites and were analyzed for arsenic, total hardness, nitrite, nitrate, chloride, sulfate, alkalinity, pH, Total Solids and fluoride. Naturally occurring arsenic contamination is common in groundwater in the studied area, and the concentrations are correlated with those of fluoride.

1 INTRODUCTION

Water is essential for life maintenance and reproduction on earth; it is a vital factor for development of biological and geochemical processes. There are many different types of water pollution with varying effects on the environment. Surface or ground water quality is a function of either natural or human influences.

Elevated concentrations of arsenic in well water have been detected in a wide region of northern and central Argentina, including the Santa Fe province. These arsenic levels have resulted from both natural and anthropogenic occurrence.

Arsenic most common sources in the natural environment are volcanic rocks, specifically their weathering products and ash, marine sedimentary rocks, hydrothermal or deposits and associated geothermal waters and fossil fuels, including coals and petroleum (Smedley & Kinniburgh, 2002).

In the last years, there has been an increased amount of research on the occurrence of arsenic in groundwater supplies throughout the world. The most notable cases of regionally elevated arsenic in groundwater include aquifers in Argentina (Smedley *et al.*, 2002; Bundschuh *et al.*, 2004; Bhattacharya *et al.*, 2006).

Arsenic problems occur in groundwater because of a combination of its high toxicity at relatively low concentrations (in the $\mu g\ L^{-1}$ range) and its mobility in water in the pH ranges of many groundwaters and over a wide range of redox conditions (Smedley *et al.*, 2002).

Elevated levels of arsenic are cause of concern because it is associated with a number of adverse health outcomes, including several types of cancer, vascular diseases, dermatological ailments, diabetes, respiratory diseases, cognitive decline, and infant mortality (Mazumder *et al.*, 2005; Rahman *et al.*, 2006; Yang *et al.*, 2003).

Groundwater arsenic concentrations correlate with fluorine, consequently with excessive amounts of one element there is a high concentration of the other, leading to an accumulation of health risk (Pérez & Fernández, 2004; Boglione *et al.*, 2009). Arsenic and fluoride diseases affect between 2 and 8 million people in Argentina (Bundschuh *et al.*, 2012).

The aim of this work is to characterize groundwater in the central-west of Santa Fe province, considering mainly arsenic and fluoride.

2 METHODS

The present study has been carried out since 2008 in the west of Santa Fe Province, Argentina, belonging to Chaco-Pampean plain. The present study involves three departments of Santa Fe (Figure 1).

Groundwater samples were collected from 180 well sites in the zone, using the methodology proposed by APHA (2001). Wells were found at a depth between 12 and 18 m, belonging groundwater to Pampa aquifer.

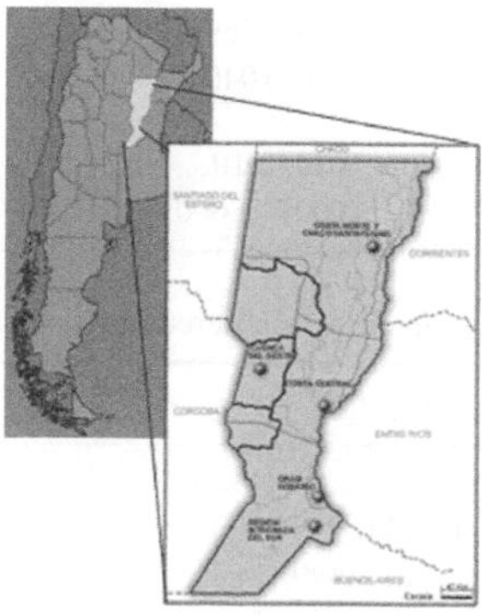

Figure 1. Study area in Santa Fe province map.

During the on-site investigation, each well site was georeferenced using a Nüvi 300 Garmin Global Positioning System (GPS) instrument. Samples were taken from the wells using a peristaltic pump sampling ISCO model 6700. The collected water samples were analyzed for arsenic (As), total hardness ($CaCO_3$), nitrite (NO_2^-), nitrate (NO_3^-), ammonium (NH_4^+) chloride (Cl^-), sulfate (SO_4^{2-}), alkalinity, pH, Total Solids (TS) and fluoride (F^-). The analytical procedures of the analyses are described in APHA (2001).

3 RESULTS AND DISCUSSION

Table 1 gives an overview of the range of values, their mean, maxima, minima and standard deviation of groundwater composition. pH ranges from 6.20 to 8.62, with mainly neutral and alkaline values.

Total solids content varies widely, finding lower concentrations in the center and south of the study area and higher values in the north.

It was found that naturally occurring arsenic in groundwater varied regionally. In shallow aquifers, 95.6% of the studied wells show As at toxic levels, with a maximum of 1.600 mg L^{-1} (Table 2). In 180 sampled wells, the concentrations exceed the Argentine Food Code (2007) and the WHO guideline (2001) limit for safe drinking water (0.01 mg L^{-1}).

Table 1. Results of the analyses of groundwater samples.

	$n = 180$		
Parameter	Range	Mean	Std Dev
pH	6.20–8.62	7.48	0.45
TS (mg L^{-1})	49.0–19213	3688.6	3265.2
SO_4^{-2} (mg L^{-1})	2.8–5947.0	928.9	1066.3
Cl^- (mg L^{-1})	10.5–6084.0	782.7	1028.2
NO_3^- (mg L^{-1})	2.8–681.0	97.7	129.9
NO_2^- (mg L^{-1})	<0.05–0.26	0.13	0.07
NH_4^+ (mg L^{-1})	<0.05–4.40	0.77	1.18
HCO_3^- (mg L^{-1})	25.3–1269.0	737.5	302.4
Total Alkalinity (mg L^{-1})	20.6–1040.3	617.0	234.5
F^- (mg L^{-1})	0.30–3.10	1.16	0.67
As (mg L^{-1})	0.009–1.600	0.120	0.230

Table 2. As concentration percentages at studied wells.

As concentration (mg L^{-1})	%
Lower than 0.010	4.44
Between 0.010 and 0.049	31.11
Between 0.050 and 1.499	63.33
Higher than 1.500	1.11

Analyzing fluoride, a large variability in concentrations was found, ranging from 0.30 to 3.00 mg L^{-1}. Higher values were found in areas with higher concentrations of arsenic.

A great variability, depending on location, depth and economic activity, was determined for the other analytes. In the case of waters in which high levels of nitrate, nitrite and ammonium have been measured, a bacteriological analysis must be carried out.

No correlation was noticed between As and pH ($R^2 = 0.059$) and alkalinity ($R^2 = 0.068$). Moreover, ground water fluoride shows good positively correlation with As concentration ($R^2 = 0.464$, $n = 64$). Further studies are currently in progress to study the arsenic speciation in the same area.

4 CONCLUSIONS

According to the results of the present research, groundwaters in the central-west part of Santa Fe, Argentina, are not suitable for human consumption, since they contain high levels of contaminants such as arsenic, fluoride, nitrate and nitrite.

Arsenic concentration correlates well with fluoride but does not correlate with any other of the parameters examined.

REFERENCES

APHA. 2001. Standard Methods for Examination of Water and Wastewater.

Bhattacharya, P., Claesson, M., Bundschuh, J., Sracek, O., Fagerberg, J. & Jacks, G. 2006. Distribution and mobility of arsenic in the Rio Dulce alluvial aquifers in Santiago del Estero Province, Argentina. *Sci. Total Environ*. 358:97–120.

Boglione, R., Panigatti, M.C., Griffa, C. & Cassina, D. 2009. Estudio de la calidad de las aguas subterráneas de la cuenca oeste de la provincia de Santa Fe. *Libro de resúmenes del XXII Congreso Nacional del Agua*. Vol. 1, pp. 36–37.

Bundschuh, J., Farias, B., Martin, R., Storniolo, A., Bhattacharya, P., Cortes, J., Bonorino, G. & Albouy. R. 2004. Groundwater arsenic in the Chaco-Pampean Plain, Argentina: case study from Robles county, Santiago del Estero Province. *Applied Geochemistry* 19: 231–243.

Bundschuh, J., Litter, M.I., Parvez, F., Roman-Ross, G. & Nicolli, H.B. 2012. One century of arsenic exposure in Latin America: A review of history and occurrence from 14 countries. *Sci. Total Environ.*, 429: 3–36.

Código Alimentario Argentino. 2007. Artículo 982 - (Res Conj. SPRyRS y SAGPyA N° 68/2007 y N° 196/2007).

Mazumder, D.N.G., Steinmaus, C., Bhattacharya, P., von Ehrenstein, O.S., Ghosh, N., Gotway, M., Sil, A., Balmes, J.R., Haque, R., Hira-Smith, M.M. & Smith, A.H. 2005. Bronchiectasis in persons with skin lesions resulting from arsenic in drinking water. *Epidemiology* 16: 760–765.

Pérez Carrera, A. & Fernández Cirelli, A. 2004. Niveles de arsénico y flúor en agua de bebida animal en establecimientos de producción lechera (Pcia. de Córdoba, Argentina). *Investigación Veterinaria.* pp. 51–60.

Rahman, M., Vahter, M., Sohel, N., Yunus, M., Wahed, M.A., Streatfield, P.K., Ekstrom, E.-C. & Persson, L.A. 2006. Arsenic exposure and age- and sex-specific risk for skin lesions: a population- based case-referent study in Bangladesh. *Environ. Health Perspect.* 114: 1847–1852.

Smedley, P.L. & Kinniburg, D.G. 2002. A review of the source, behaviour and distribution of arsenic in natural waters. *Appl. Geochemistry* 17: 517–568.

Smedley, P.L., Nicolli H.B., Macdonald, D.M.J., Barros, A.J. & Tullio, J.O. 2002. Hydrogeochemistry of arsenic and other inorganic constituents in groundwaters from La Pampa, Argentina. *Appl. Geochemistry* 17: 259–84.

WHO. 2001. Guidelines for Drinking-water: Arsenic in Drinking Water. Fact Sheet N° 210. *World Health Organization*, Geneva, Switzerland.

Yang, C.Y., Chang, C.C., Tsai, S.S., Chuang, H.Y., Ho, C.K. & Wu, T.N. 2003. Arsenic in drinking water and adverse pregnancy outcome in an arseniasis—endemic area in northeastern Taiwan. *Environ. Res.* 91: 29–34.

*One Century of the Discovery of Arsenicosis in Latin America (1914–2014) –
Litter, Nicolli, Meichtry, Quici, Bundschuh, Bhattacharya & Naidu (Eds)*
© 2014 Taylor & Francis Group, London, ISBN 978-1-138-00141-1

Arsenic levels in natural and drinking waters from Paracatu, MG, Brazil

E.D. Bidone
Departamento de Geoquímica, Instituto de Química, Universidade Federal Fluminense (UFF), Centro, Niterói, RJ, Brasil

Z.C. Castilhos
Departamento de Geoquímica, Instituto de Química, Universidade Federal Fluminense (UFF), Centro, Niterói, RJ, Brasil
Centro de Tecnologia Mineral (CETEM), Cidade Universitária, Rio de Janeiro, Brasil

M.C.B. Santos, R.S.V. Silva, R.G. Cesar & M. Ferreira
Departamento de Geoquímica, Instituto de Química, Universidade Federal Fluminense (UFF), Centro, Niterói, RJ, Brasil

ABSTRACT: Inorganic Arsenic (As) is a known human carcinogenic (Group I) and the most important human environmental exposure pathway is water ingestion. Gold mining may be an important source of As to the environment. The objective of this study was to evaluate the As levels in natural (superficial and groundwater) and drinking water in Paracatu, where the largest gold mine in Brazil ("Morro do Ouro") is located. The results showed low levels of As in drinking water, tap water and groundwater, which indicates that water ingestion may not be a significant pathway of exposure to local population. In contrast, the highest levels of As, above the legal criterion (10 μg L^{-1}), were found in freshwater of watersheds impacted directly by the present gold mining area or abandoned artisanal gold mining sites.

1 INTRODUCTION

The "Morro do Ouro" gold ore deposit presents Au grade from 0.4 to 0.6 g Au t^{-1} and Arsenic (As) levels > 4,000 mg L^{-1}. High As levels (1,000 mg L^{-1}) in soils and sediments were observed at watersheds located in the same direction of the gold ore body, generally associated with abandoned artisanal gold mines and low energy freshwater environment, and decreasing concentration as the mining site distance increases (Bidone *et al.*, 2014). According to Prohaska & Stingeder (2005), the greatest range and the highest concentrations of As are potentially found in groundwaters. The average values for As in freshwaters range from 0.4 to 80 μg L^{-1}. These values can rise to several hundred μg L^{-1} in streams near industrial and mining areas. Usual groundwater concentrations range from <0.5 to 10 μg L^{-1}. Wells in contaminated areas, e.g. in Bangladesh, can reach 2,500 μg L^{-1}. Given the fact that inorganic As is a known human carcinogenic (Group I) and the most important human environmental exposure pathway is water ingestion (ASTDR 2007), the potential contamination of natural water (superficial and groundwater) and drinking water in Paracatu was investigated. The majority of urban population (97%) is served by treated water, while rural communities consume drinking water from local groundwater sources. Drinking water in urban area is mainly from freshwater source, but, depending on its availability, it is complemented by urban groundwater sources.

2 METHODS

2.1 *Sampling*

Water sampling was performed during dry season (September and October). Under this climatological condition, water quality reflects the highest interaction with soils and rocks, and with pollutant loads, without dilution process and runoff impacts. Twenty eight freshwater (28) samples were collected along the three watersheds, sub-basins of Paracatu River watershed: (i) Córrego Rico watershed—its spring is located inside the open pit mine site and at its medium segment is located at the Paracatu Municipality; (ii) Ribeirão Entre-Ribeiros watershed—it is downstream of tailing dam and potentially receives its effluents; (iii) Rio Escuro watershed—outside the direct influence of gold mining, it is considered as a reference area and it is the main drinking water source for urban population of Paracatu.

Additionally, 29 wells used for urban population consumption after treatment (9), for private human consumption (6), for rural communities consumption (4) and groundwater samples from monitoring wells (10) installed inside of the mining sites (gold mine -2- and lead and zinc mine - 8), spread of all over the municipality were sampled. Thirty-seven (37) drinking water system for urban population were sampled, including 12 watersources (9 groundwater and 3 freshwater), water reservoirs (4) and tap water from 21 houses in Paracatu. These selected houses represent critical areas (end-of-line) in the urban water system distribution.

2.2 *Chemical Analysis*

Physicochemical parameters (pH, electric conductivity-EC, dissolved oxygen-DO, total dissolved solids-TDS, temperature-T and oxy-redox potential-Eh) were measured at field by Multi Sonda (HANNA, Hi 9828). The water samples were keep in plastic jars cleaned, acidified (pH < 2 HNO_3 ultrapure) and refrigerated until analysis. The concentrations of total As were determined by ICP-MS (Model 42 IC-MS, Perkin Elmer MS). The limit of detection of As was 0.5 μg L^{-1}. Precision and accuracy (RCM) was 90% and 95%, respectively. As in groundwater was analyzed by ICP-MS (Agilent, 7500ce) with nebulizer (Micromist and Cross Flow camera). The detection of limit was < 0.1 μg L^{-1}. Precision and accuracy (RCM) was 90% and 95%, respectively.

3 RESULTS AND DISCUSSION

a) Superficial freshwaters. (i) Córrego Rico watershed: As levels resulted from <0.5 μg L^{-1} to 40.10 μg L^{-1} (13.21 ± 12.38 μg L^{-1}; n = 11). At a sampling point, very close to Morro do Ouro, As levels attained 23.60 μg L^{-1}, and where Paracatu city urban area ends, As levels were 40.10 μg L^{-1}. From this point, the arsenic concentration decreased to <10 μg L^{-1} as far as its confluence with Paracatu river. (ii) Ribeirão Entre-Ribeiros watershed: As levels resulted from <0.5 to 29.10 μg L^{-1} (10.75 ± 10.16 μg L^{-1}; n = 11). The highest As levels are associated with abandoned open pits and small scale gold mining sites (SSGM). As level downstream of the tailing dam (7 km) was 5.0 μg L^{-1} (iii) Rio Escuro watershed: As levels resulted from <0.5 μg L^{-1} and 0.80 μg L^{-1} (0.63 ± 0.15 μg L^{-1}; n = 4). As level at the drinking water source that supplies Paracatu population was <0.5 μg L^{-1}. b) Drinking water at the urban system: As levels were lower than the limit of detection (< 0.5 μg L^{-1}) for all samples, including all tap water samples, except for 2 groundwater sources, with low As levels 0.6 μg L^{-1} and 1.8 μg L^{-1}. c) Groundwater samples: As levels in groundwater for rural communities consumption, for private urban human consumption as well as monitoring wells resulted below of limit of detection (<0.1 μg L^{-1}).

4 DISCUSSION

As maximum permitted level in freshwater and groundwater is 10 μg L^{-1} (Resolution n. 357/2005 CONAMA—Conselho Nacional do Meio Ambiente and Ordinance 518/2004 Ministry of Health of Brazil). All of the drinking water (surface and groundwater) samples used to the urban or rural communities' consumption and the tap water samples showed As levels below the quality limit. Actually, almost all samples were below the limit of detection, which is 4 to 10 times lower than the quality limit. On the other hand, 42% of freshwater samples are above 10 μg L^{-1}. These samples are from Corregо Rico and Ribeirão Entre-Ribeiros watershed, under direct influence of gold mining site and/or abandoned artisanal gold mining sites, respectively. Ladeira *et al.* (2002), when investigating arsenic adsorption capacity of soil from Morro do Ouro region, showed the maximum loading capacity between 0.7 and 3.6 mg As / g soil, which results in As concentrations from 700 to 3,600 mg L^{-1}. These values are similar to those measured in soils from Paracatu (Bidone *et al.*, 2014). Therefore, As would be trapped inside the groundwater and transferred to freshwater, mainly at remobilized soils and exposed rocks. In Latosols, as found in Paracatu which are well drained, As may move down a profile with leaching water (as the fine particulate material, complexed with Fe and Al oxides). The organic matter retention at low energy river segments (swamps, small dams and abandoned SSGM open pits) may favor the accumulation of As and, as a consequence, might contribute to the arsenate reduction to the more soluble and toxic chemical form, arsenite. In these sites, physicochemical parameters showed reduction condition (negative or lower levels of Eh, low levels of DO and neutral pH), fact that lead to suppose the stability of the arsenite forms in water. Finally, the equilibrium between water and sterile phyllites (the host rocks of the ore body), with low levels of As, could explain concentrations <0.1 μg L^{-1} in almost all samples of groundwater. These aspects are under study.

5 CONCLUSIONS

As levels were lower than the limit of detection (< 0.5 μg L^{-1}) for all water samples, including tap water, groundwater and surface water that is the main source of the drinking water supply for urban population. Even in rural areas, all drinking water samples showed low levels, even below <0.1 μg L^{-1}. Then, As levels in drinking water in Paracatu are around one order of magnitude lower than the limits established by the Brazilian law. Therefore, the As main human exposure pathway (i.e. oral ingestion of water) might not represent significant risks on human health. As levels above 10 μg L^{-1} were observed in impacted watershed, under direct influence of gold mining site and/or abandoned artisanal gold mining sites. The relationship between As in soils, sediments and natural water are under study.

REFERENCES

ATSDR, US Agency for Toxic Substances and Disease Registry 2007. Toxicological Profile for Arsenic. Atlanta, Georgia.

Bidone, E.D., Castilhos, Z.C., Santos, M.C.B., Silva, R.S.V., Cesar, R.G. & Ferreira, M. 2014. Arsenic in soils and sediments from Paracatu, Brazil. V International Congress of Arsenic in the Environment, Buenos Aires, Argentina, this issue.

Ladeira, A.C.Q., Ciminelli, V.S.T. & Nepomuceno, A.L. 2002. Seleção de solos para imobilização de arsênio. *REM: R. Esc. Minas, Ouro Preto* 55(3): 215–221.

Prohaska, T. & Stingeder, G. 2005. Arsenic and Arsenic Species in Environment and Human Nutrition. In R. Cornelis, H. Crews, J. Caruso & K.G. Heumann (eds), *Handbook of Elemental Speciation II: Species in the Environment, Food, Medicine & Occupational Health*: 69–85. John Wiley & Sons, Ltd.

1.4 Speciation of arsenic and analytical advancements

One Century of the Discovery of Arsenicosis in Latin America (1914–2014) –
Litter, Nicolli, Meichtry, Quici, Bundschuh, Bhattacharya & Naidu (Eds)
© 2014 Taylor & Francis Group, London, ISBN 978-1-138-00141-1

A new method for arsenic analysis in Atmospheric Particulate Matter

M. Gasparon
The University of Queensland, School of Earth Sciences, Brisbane, Australia
INCT-Acqua, Belo Horizonte, Brazil

V.S.T. Ciminelli & G. Cordeiro Silva
INCT-Acqua, Belo Horizonte, Brazil
Departamento de Engenharia Metalúrgica e de Materiais, Escola de Engenharia, Universidade Federal de Minas Gerais, Belo Horizonte, Brazil

V. Dietze
Deutscher Wetterdienst, Zentrum für Medizin-Meteorologische Forschung, Freiburg, Germany

B. Grefalda
School of Earth Sciences, The University of Queensland, Brisbane, Australia

J.C. Ng
EnTox, The University of Queensland, Brisbane, Australia

F. Nogueira
INCT-Acqua, Belo Horizonte, Brazil
Departamento de Engenharia Metalúrgica e de Materiais, Escola de Engenharia, Universidade Federal de Minas Gerais, Belo Horizonte, Brazil

ABSTRACT: Conventional methods for the analysis of arsenic in atmospheric particulate matter provide only bulk concentrations in different particle size fractions. We have developed a method that combines traditional gravimetric and bulk elemental ICP-MS analysis with automated SEM characterization of individual particles. By using this method, it is possible to establish which mineral phases (and non-crystalline matter) and which particle size fractions contribute to the total arsenic content in the atmospheric particulate matter. A less labor-intensive alternative method is currently being trialed using Mineral Liberation Analysis (MLA). Results of the first pilot study carried out in the Paracatu area (Minas Gerais, Brazil), where gold mineralization is associated with arsenopyrite, show that at this specific mine site atmospheric arsenic is primarily associated with organic combustion residues of forest fires, and not with the dispersion of arsenopyrite from the local mine site as previously postulated. This method can be applied in principle to trace the source of any element in atmospheric particulate matter, and can therefore be used to identify and manage activities and sources responsible for the release of potentially toxic elements into the atmosphere.

1 INTRODUCTION

Air pollution caused by the emission of particulate matter (Atmospheric Particulate Matter—APM) into the atmosphere is rapidly becoming one of the main causes of environmental and human health concern in mining areas, where excavation, processing and stockpiling activities have the potential to release large quantities of APM into the surrounding environment. Over the years, mining companies have developed strategies to reduce and control geogenic dust emissions, and to monitor their effectiveness. The monitoring programs typically include the measurement of Total Suspended Particle (TSP) concentrations, and, rarely, concentrations of PM_{10} (particles with diameter smaller than 10 microns), using a range of monitoring active sampler devices and spatial and temporal sampling schedules. In rare cases, the elemental concentrations in the bulk APM samples are measured using a range of methods that include X-ray fluorescence or Inductively Coupled Plasma Mass Spectrometry (ICP-MS). Although these methods are effective in providing information on total APM emissions and its composition, they cannot identify different APM sources, nor provide quantitative information of the elemental concentration in individual sources. This is a serious limitation, because in any environment there are typically different sources that contribute to total APM load (such as agriculture, transport, construction and industrial activities, mining, natural erosion). As a result, mining companies are often forced to establish stringent dust suppression procedures that fail to result in a decrease in total APM load, because APM sources have not been correctly identified.

2 METHODS

2.1 *Sample collection and bulk analysis*

Conventional high-volume active samplers were used to collect TSP and PM_{10} from August 2010 to June 2012. One 24-h TSP sample was collected every six days at an approximate flow rate of 1.0 m^3 min^{-1} from the designated stations. A 24-h PM_{10} sample was also collected at one station with the same sampling flow rate and frequency in accordance to Brazilian National Air Quality Monitoring standards (Anonymous, 1990). Air was pumped through a glass fiber filter, which served as the substrate for particle entrapment (Hayward *et al.*, 2010). Filters were gravimetrically analyzed on site before and after sampling to measure the amount of collected particulate matter. All elemental analyses were carried out at The University of Queensland in Brisbane, Australia (UQ). The procedure for hot plate digestion of the filter material was developed at UQ and based on an optimization of the methods: NIOSH 7303 (Anonymous, 2003) US EPA 200.2 (Martin *et al.*, 1994) and Australian standards (Anonymous, 2007). The USGS W-2 standard was used as calibration standard and a range of rock and atmospheric dust Certified Reference Materials (CRM) were included as quality control unknowns.

Subsamples measuring 1 cm^2 in area were cut from the filter papers using a stainless-steel surgical scalpel. The whole filter was folded in half and six (6) subsamples were taken from each side of the folded filter. These subsamples were weighed and the mass of particles on the subsamples were calculated through ratio, assuming uniform distribution of particles on the surface of the filter. Therefore, each sample consists of six randomly distributed individual subsamples. Subsamples of the same area and amount were also cut from blank (unexposed) filters to be used as filter blanks.

For ICP-MS analysis, samples of ~10 mL solution were transferred to a weighted pre-cleaned ICP-MS tube and diluted with 0.25 ml of internal standards (^{6}Li, ^{61}Ni, ^{103}Rh, ^{115}In, ^{187}Re, ^{209}Bi, ^{235}U). Trace metal concentrations were determined on a Thermo Electron X-7 Series ICP-MS. Calibration based on CRM values and results of the filter blanks were taken into consideration during processing of elemental concentrations for the samples. Detection limits (calculated as three times the standard deviation of replicate instrumental blanks) were always lower than the equivalent to 0.5 ng As per m^3 of air. Typical accuracy and precision for As, based on replicates of CRMs and randomly selected samples, were better than 95%.

2.2 *SEM analyses*

A Sigma-2 passive sampler was used to collect APM for single particle SEM analysis. The Sigma-2 passive sampler consists of a cylinder covered by a cap at the upper end that protects the interior against direct impact of precipitation. Airborne particles freely enter the cylinder through four windows in the cylinder and cap, which are designed to prevent the impact of rain and wind. The interior of the cylinder acts as a stilling chamber, with wind velocity reduced to less than 5% of ambient conditions, resulting in particle settling almost exclusively by gravity (coarse particles dp > 2.5 μm). The particles are collected and fixed on a sampling plate suitable for light microscopy and SEM, with a weather resistant adhesive surface, positioned at the bottom of the cylinder (Schultz, 1993; Fiebig-Wittmaack *et al.*, 2006; Grobéty *et al.*, 2010; Guéguen *et al.*, 2012; Schleicher *et al.*, 2012; VDI, 2013).

SEM images were acquired using an FEI XL 30 Sirion FEG scanning electron microscope (Schottky Field Emitter; 200 eV to 30 KeV; beam current 24 nA). The microscope was equipped with a Secondary Electron (SE) detector, a Centaurus scintillator type Backscattered Electron Detector (BED) and an EDAX Energy Dispersive x-ray Spectrometer (EDS) system equipped with lithium doped silicon detector. The images and data were processed using the AnalySiS Image acquisition program and the EDAX Genesis software version 5.2 with spectrum analysis, elemental mapping and automated particle analysis. The data were further processed manually to eliminate artifacts and to resolve composite particles. Mineral proportions were calculated based on theoretical stoichiometric formulae and considering the mineral phases reported in the literature for the different site lithologies.

3 RESULTS AND DISCUSSION

3.1 *Sampling strategy and particle size distribution*

APM samples were collected at six monitoring stations within the city of Paracatu, around the mine property, and at control sites upwind of the mining area. The monitoring program was established in August 2010 and is still ongoing; however, only the TSP and APM data for the period 9–15 September 2011 are shown here. This period corresponds to the end of the dry season, and to the highest TSP and arsenic content recorded to date. The countryside around the city was affected by major bushfires during this period, with the most significant bushfire recorded on 9 September 2011. The six stations were grouped into two clusters based on their location with respect to the mine pit as a postulated primary emission source. The first cluster is within a three-kilometer radius of the mine pit and includes the stations within the city. The second cluster of monitoring stations is located more than four kilometers upwind of the mine pit. It should be noted that the size of particles produced during mining activities is typically larger than 2.5 μm, with 50% of the particles being typically larger than 10 μm (Figure 1).

The particle size distribution measured in the APM was consistent with a primarily non-mining source, with less than 10% (by number) of the particles larger than 10 μm, and at least 45% of the particles smaller than 2.5 μm (Table 1).

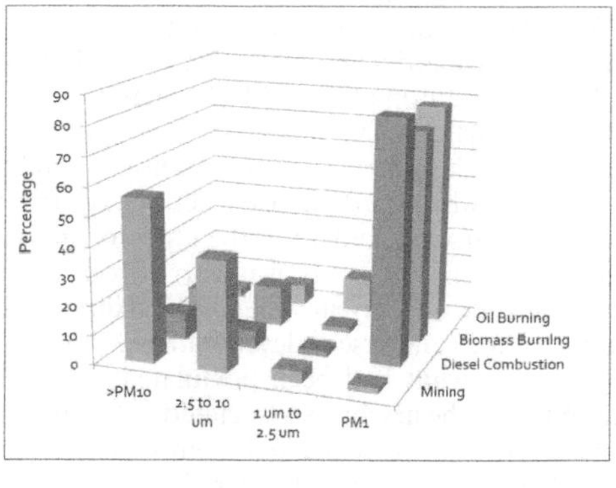

Figure 1. Typical percentage particle size distribution (based on mass concentration calculation) from mining and other sources. Adapted from Anonymous (2000).

Table 1. Distribution of APM particle sizes (percentages of the total particle numbers) at the different monitoring stations for the period 9–15 September 2011. Percentages are calculated from a dataset of over 13,000 individual particle SEM measurements.

Monitoring station	% 1–2.5 μm	% 2.5–10 μm	% 1–10 μm	% >10 μm
1	55.9	39.0	95.0	5.0
2	48.4	43.8	92.2	7.8
3	71.6	25.7	97.2	2.8
4	59.0	35.5	94.5	5.5
5	45.2	50.8	96.0	4.0
6	59.5	36.7	96.2	3.8

3.2 *Bulk ICP-MS analysis*

The most remarkable feature of the As distribution is the almost perfect correlation between As content and TSP (Figure 2). This correlation suggests that the same source controls As distribution, irrespective of the position of the monitoring station relative to the mine pit.

3.3 *Single particle SEM analysis*

Virtually 100% of the As found in the local ore deposits is in arsenopyrite [FeAsS] or, to a lesser extent, scorodite [$FeAsO_4 \cdot 2H_2O$]. Therefore, arsenopyrite (and/or scorodite) should be expected in the APM if the mine were the (main) source of atmospheric arsenic. Table 2 shows the distribution of different mineral classes and carbonaceous matter. The term "carbonaceous" is used here to describe any carbon-based matter, and includes residues of fossil fuel combustion as well as particulate produced during forest fires.

Contrary to expectations, only one arsenopyrite grain (and no scorodite) was found among the >13,000 grains analyzed across the six monitoring stations for the period 9–15 September 2011. Furthermore, the amount of sulfides (other than arsenopyrite) and sulfates in the APM was very low (typically <1%), and was not spatially consistent with dispersion from the mine site. All sites have variable amounts of carbonaceous particles (10–30%) which presumably include combustion particles (including those derived from bushfires), geogenic carbonaceous material (from the local lithologies) and anthropogenic particles (including soot, fertilizers, and other chemicals used in agriculture and farming). Although the EDAX detector is not sufficiently sensitive to measure As at ppm levels, it appears that much of the arsenic is associated with the carbonaceous particles (45% to 84%), irrespective of the location of the monitoring station.

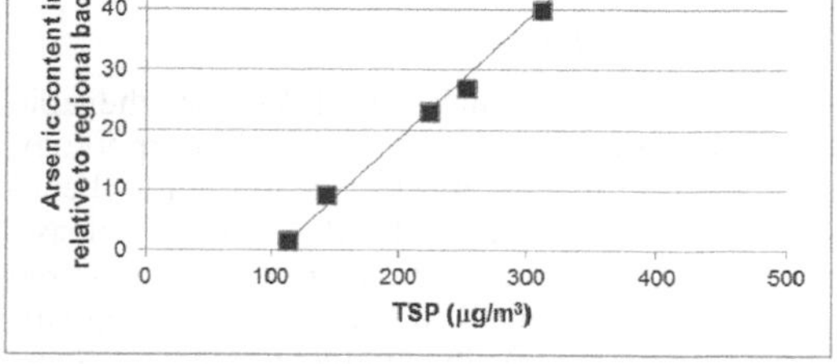

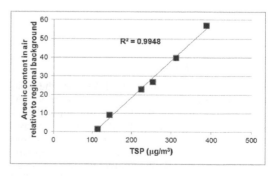

Figure 2. Correlation between TSP content and bulk TSP arsenic content (relative to regional background values) measured by ICP-MS in the six monitoring stations for the period 9–15 September 2011.

Table 2. Percentage distribution of different components in the total APM of the six monitoring stations. "Carbonaceous" refers to particles made primarily of C. "Fe-oxide" includes Fe (oxi)hydroxides. "Sulfide and sulfate" exclude As-bearing sulfides and sulfates. "Not determined" are mineral particles that could not be assigned to any of the other classes, and include mainly aggregates of multiple particles. The % arsenic content was estimated from the EDAX analyses after subtraction of background noise.

Monitoring station	Arsenopyrite %	Carbonaceus %	Silicate %	Carbonate %
1	0.05	22.52	60.42	13.13
2	0.00	10.39	83.55	2.38
3	0.00	16.03	66.97	7.59
4	0.00	21.81	68.08	7.09
5	0.00	29.32	65.64	2.12
6	0.00	19.78	64.77	8.22
% arsenic content in different phases				
1	14	46	4	23
2	0	65	13	8
3	0	63	8	6
4	0	68	7	14
5	0	84	6	8
6	0	45	7	21

Monitoring station	Fe-oxide %	Sulfide and sulfate %	Not determined %
1	2.62	0.79	0.47
2	3.17	0.48	0.03
3	8.73	0.40	0.28
4	3.29	0.40	0.29
5	2.88	0.05	0.00
6	6.24	0.25	0.74
% arsenic content in different phases			
1	2	0	11
2	12	2	0
3	17	0	6
4	3	0	7
5	2	1	0
6	12	0	15

3.4 *MLA analysis*

Mineral Liberation Analysis (MLA) allows the rapid determination of particle size and morphology, combined with mineral identification and chemical analysis. The standard MLA sample consists of mineral grains mounted in epoxy, and analyzed under high vacuum SEM after coating using carbon or other electrically conductive elements (e.g., gold). This procedure, however, cannot be applied to the APM samples, and the standard MLA analysis is not suitable for non-crystalline matter (such as the carbonaceous particles). Our research group is in the process of developing a protocol for sample preparation of APM for MLA analysis. This will significantly reduce analytical time and costs, while at the same time allowing a more accurate definition of mineral phases present in the APM.

4 CONCLUSIONS

In recent years, there has been increasing concern over exposure to arsenic through contaminated atmospheric particulate matter, particularly in mining areas, where excavation, processing and stockpiling activities have the potential to release large quantities of dust into the surrounding environment. Current monitoring programs typically include the gravimetric measurement of particle concentrations, and only in rare cases the elemental composition of the dust. These methods, however, cannot identify different particle sources, and they cannot provide quantitative information of the elemental concentration in individual sources. Synchrotron analyses could theoretically provide this information, however it is unfeasible to collect data for a sufficiently representative number of particles (>1,000 particles per sample).

We have developed a method that combines traditional gravimetric and bulk elemental ICP-MS analysis with automated SEM characterization of individual particles. This method makes it possible to establish which mineral phases (and non-crystalline matter) and which particle size fractions contribute to the total arsenic content in the atmospheric particulate matter. The study was carried out during the driest time of the year, when the maximum dust concentrations and arsenic levels had been recorded. Contrary to what had been postulated, we found that much of the arsenic is associated with carbonaceous particles mostly derived from bushfires, and no evidence was found for the aerial dispersal of As-bearing phases from the mine site.

Mining companies concerned with arsenic dispersal will benefit from this research as it will provide the tools and methodology for assessing the effectiveness of their existing environmental management programs, particularly with regards to dust emission monitoring and control. It will also facilitate the development and implementation of strategies that will ultimately lead to a reduction in environmental management costs, and a better and healthier environment for the workforce and for the general public.

ACKNOWLEDGEMENTS

We wish to thank Kinross Brasil Mineração S.A. for access to the site and assistance with the collection of the samples, and Drs. Sunny Hu and Mario Meier for assistance with the ICP-MS and SEM analysis, respectively. This study is partly funded by the Brazilian "Science Without Borders" program.

REFERENCES

Anonymous, 1990. *Resolução/conama/N.° 003 de 28 de junho de 1990.* In http://www.mma.gov.br/port/conama/res/ res90/res0390.html, accessed 3/10/13.

Anonymous, 2000. *Particulate Matter and Mining.* Sydney: New South Wales Minerals Council Ltd.

Anonymous, 2003. *Elements by ICP: Method 7303.* In http://www.cdc.gov/niosh/docs/2003–154/pdfs/7303.pdf, accessed 3/10/13.

Anonymous, 2007. *Approved methods for the sampling and analysis of air pollutants in New South Wales.* Sydney South: NSW Department of Environment and Conservation (ISBN 978 1 74122 373 6).

Fiebig-Wittmaack, M., Schultz, E., Cordova, A.M. & Pizarro, C. 2006. A microscopic and chemical study of airborne coarse particles with particular reference to sea salt in Chile at 30 ° S. *Atmospheric Environment* 40: 3467–3478.

Grobéty, B., Gieré, R., Dietze, V. & Stille, P. 2010. Airborne particles in the urban environment. *Elements* 6(4): 229–234.

Guéguen, F., Stille, P., Dietze, V. & Gieré, R. 2012. Chemical and isotopic properties and origin of coarse particles collected by passive samplers in industrial, urban, and rural environments. *Atmospheric Environment* 62: 631–645.

Hayward, S.J., Gouin, T. & Wania, F. 2010. Comparison of four active and passive sampling techniques for pesticides in air. *Environmental Science and Technology* 44: 3410–3416.

Martin, T.D., Creed, J.T. & Brockhoff, C.A. 1994. *Method 200.2. Sample preparation procedure for spectrochemical determination of total recoverable elements. Revision 2.8.* In http://water.epa.gov/scitech/methods/cwa/bioindicators/upload/2007_07_10_methods_method_200_2.pdf, accessed 3/10/13.

Schleicher, N., Kramar, U., Dietze, V., Kaminski, U. & Norra, S. 2012. Geochemical characterization of single atmospheric particles from the Eyjafjallajökull volcano eruption event collected at ground-based sampling sites in Germany. *Atmospheric Environment* 48: 113–121.

Schultz, E. 1993. Size-fractionated measurement of coarse black carbon particles in deposition samples. *Atmospheric Environment* 27: 1241–1249.

VDI 2119 2013. Sampling of atmospheric particles >2.5 µm on an acceptor surface using the Sigma-2 passive sampler *VDI Düsseldorf, available at Beuth Verlag 10772 Berlin, Germany.* In http://www.beuth.de/en/technical-rule/vdi-2119/169734792, accessed 4/10/13.

One Century of the Discovery of Arsenicosis in Latin America (1914–2014) –
Litter, Nicolli, Meichtry, Quici, Bundschuh, Bhattacharya & Naidu (Eds)
© 2014 Taylor & Francis Group, London, ISBN 978-1-138-00141-1

Synchrotron-based X-ray spectroscopy and x-ray imaging applied to the study of accumulated arsenic in living systems

C.A. Pérez & E.X. Miqueles
Laboratório Nacional de Luz Síncrotron (LNLS), Campinas, Brazil

Roberto D. Pérez
Universidad Nacional de Córdoba, Córdoba, Argentina

Guillermina A. Bongiovanni
CONICET-Universidad Nacional de Comahue, Neuquén-Cinco Saltos, Argentina

ABSTRACT: Investigation of biological processes as well as determination of chemical composition of tissues and cells require suitable techniques, with high elemental and chemical sensitivity and high spatial resolution. Such requirements can be fulfilled using synchrotron radiation based techniques. In this work, tissues from several organs of rats that were chronically exposed to water containing arsenic were analyzed. SRXRF microprobe measurements have revealed the presence of this metalloid in brain, pancreas and bladder, whereas accumulation happens mainly in kidney, liver, spleen, heart and blood. X-ray fluorescence microtomography (XFCT) was used to explore the 3D distribution of accumulated arsenic in specially prepared tissues. In order to assess also arsenic speciation, arsenic K-edge XANES measurements (XANES/TXRF-XANES) were carried out. By this procedure, it was possible to determine the predominant accumulated species. Main conclusions about the current results as well as future activities will be sketched.

1 INTRODUCTION

It is well established that chronic exposure to arsenic (As) can cause cancerous and non-cancerous health hazards. However, epidemiological and experimental evidence suggests that the development of arsenic-related diseases is not only determined by the dosage of exposure, but also by its chemical form. It is generally accepted that the 3+ methylated arsenic species are more cyto- and genotoxic, and more potent enzyme inhibitors than both their pentavalent counterparts and the inorganic arsenic species. In this regard, it becomes necessary to know the chemical species of accumulated arsenic in target organs in order to do more efficient therapy strategies.

In this work, several tissues from rats that were chronically exposed to drinking water containing arsenic were analyzed. Synchrotron Radiation X-Ray Fluorescence Analysis (SRXRF) microprobe measurements have shown an accumulation of arsenic in the renal cortex. In order to also assess arsenic speciation, arsenic K-edge XANES measurements at grazing incidence geometry (TXRF-XANES) were carried out. Exploratory analyses were also carried out by X-Ray Fluorescence Computed Tomography (XFCT) in order to study the in vivo 3D distribution of arsenic and other elements in small fragments of specific organs.

2 METHODS/EXPERIMENTAL

2.1 *Sample preparation*

Wistar male rats (50 days) received drinking water containing 55 μg/L arsenic (Asa) as sodium arsenite ($NaAsO_2$) for 60 days. These doses of arsenic were well tolerated. After the administration of the last dose, the animals were given a 1-day rest and were sacrificed under light ether anesthesia. Specific organs were removed and freeze drying for μ-XRF and XFCT measurements. For XANES experiments, samples were homogenized in buffer and centrifuged 15 min at 5,000 rpm. Supernatant was then centrifuged 45 min at 12,000 rpm, lyophilized and transported to LNLS, where they were re-suspended in MilliQ water. One μl specimens were pipetted onto Si reflector and Ultralene film for As speciation.

2.2 *Microscopic XRF analysis and XRF tomography (XFCT)*

XRF mapping in sample slices were carried out at DO09B XRF Fluorescence beamline of the Brazilian National Synchrotron Light Laboratory (LNLS, Campinas-SP), according to Pérez et. al. (2006). Briefly, 0.5 mm lyophilized slice from kidney was positioned in the image plane within a three-axis (x, y, z) remote controlled stage. XRF spectra were measured using a standard geometry, and the excitation beam was collimated to 300 μm × 300 μm or was condensed to spot size of around 20 microns in diameter using a glass monocapillary. The step size was 300 or 20 μm/step in both directions according to the spatial resolution used. The counting live-time for each pixel was 10 s. Small fragments from kidney and liver organs as well as tumors from lung and bladder were specially prepared for XRF tomography measurements. XFCT was performed by exciting the samples with a collimated white beam of 100 × 100 μm. In all cases, XRF spectra were recorded with a Si(Li) detector.

2.3 *TXRF-XANES*

Arsenic K-edge XANES measurements in fluorescence mode and grazing incidence geometry were performed at the D09B-XRF beamline from LNLS. During the measurements, the excitation energy was tuned in three steps across the arsenic K-edge at 11,867 eV. Simultaneously, the absorption by an elemental gold foil was recorded in transmission mode. The first inflection point (i.e. the first maximum of the derivative spectrum) of the Au metal foil scan was assumed to be 11,919 eV (Au L3 edge).

3 RESULTS AND DISCUSSION

XRF maps have shown the presence of arsenic in several organs of rat whereas accumulation notably happens on specific tissues. Figure 1 shows an example of its accumulation on renal cortex from rats exposed to 60 days of drinking water containing arsenic (Rubatto Birri et al., 2010).

The TXRF-XANES measurements allowed obtaining characteristic spectra of inorganic accumulated metabolites (see Figure 2) and organic arsenics allowing the identification of abundant arsenic species in tissues that accumulate this contaminant. A high proportion of As(III) and DMA(V) were found in blood and renal supernatant (mitochondrial fraction). The results suggest that these are the predominant species.

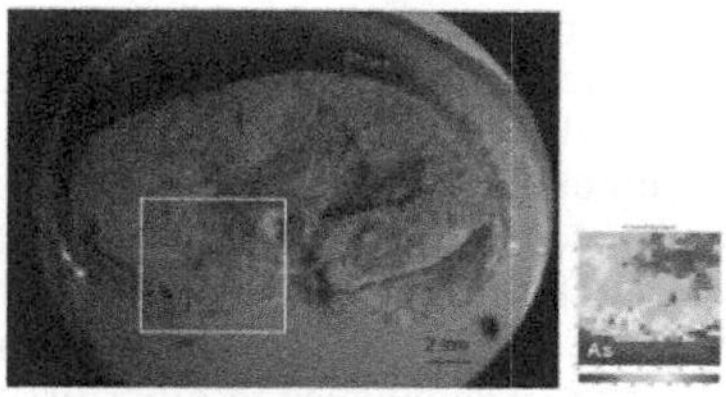

Figure 1. Image of rat's kidney sample (left) and XRF map of arsenic showing its accumulation on the renal cortex (right).

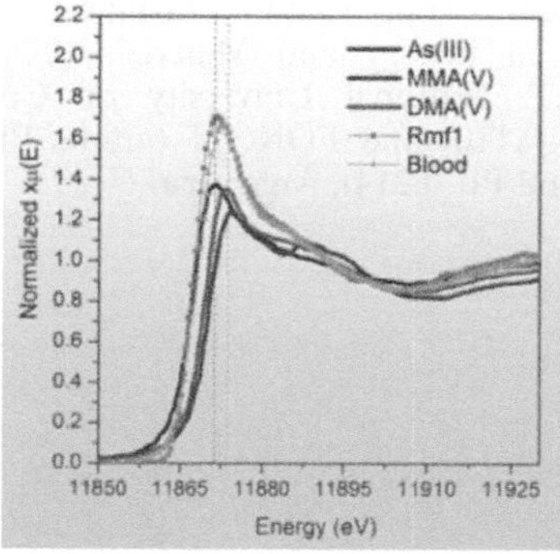

Figure 2. Comparison of (normalized) TXRF-XANES spectra for three samples with organic and inorganic reference compounds.

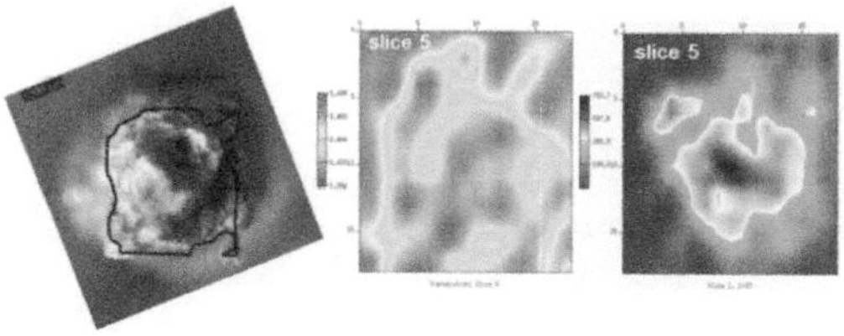

Figure 3. XFCT image on rat kidney exposed to drinking water containing As.

XFCT measurements were performed on fragments of different rat organs revealing the presence and distribution of several elements. Figure 3 shows the reconstruction of the transmission image of one slice of the fragments as well as the elemental distribution of arsenic within this slice (Miqueles et al., 2010).

4 CONCLUSIONS

In this work, we traced the presence, distribution and accumulation of arsenic in several rat tissues by synchrotron-based X-ray fluorescence microscopy techniques. The results for kidney have shown that arsenic is mostly accumulated in renal cortex in discrete structures surrounding the glomerulus.

Using XANES techniques, the main species of arsenic accumulated in rat tissues were determined. As(V) is the predominant species in blood (carrier), whereas As(III) and DMA(V) are the main accumulated species in the renal cortex. The oxidation states of arsenic accumulated in other tissues, but in lesser concentration, are still under study.

ACKNOWLEDGEMENTS

This work was supported by Brazilian Center for Research in Energy and Materials (CNPEM); CONICET, National University of Comahue (grant 04-A110) and FONCyT (grants PICT97-PRH33 and PICT214), Argentina.

REFERENCES

Miqueles, E.X., De Pierro, A.R. 2010. Exact analytic reconstruction in x-ray fluorescence CT and approximate versions, *Phys. Med. Biology*, 55: 1007–1024.

Pérez R.D., Rubio, M. Perez, C.A., Eynard, A.R. Bongiovanni, G.A. 2006. Study of the effects of chronic arsenic poisoning on rat kidney by means of synchrotron microscopic X ray fluorescence analysis. *X Ray Spectrom.*, 35(6): 352–358.

Rubatto Birri, P.N., Perez, R.D., Cremonezzi, D., Perez, C.A., Rubio, M., Bongiovanni, G.A. 2010. Association between As and Cu renal cortex accumulation and physiological and histological alterations after chronic arsenic intake, *Environ. Res*. 110: 417–423.

One Century of the Discovery of Arsenicosis in Latin America (1914–2014) –
Litter, Nicolli, Meichtry, Quici, Bundschuh, Bhattacharya & Naidu (Eds)
 ISBN 978-1-138-00141-1

Arsenic speciation in whole blood samples from Acute Promyelocytic Leukemia (APL) patients after tetra-arsenic tetra-sulfide therapy

B. Chen, W.T. Corns & P.B. Stockwell
P S Analytical Ltd., Orpington, Kent, UK

Y.G. Xu
Xiyuan Hospital, China Academy of Traditional Chinese Medicine, Beijing, China

ABSTRACT: As_4S_4 (realgar) has been used in therapies to treat patients with APL and proven to be successful. The health benefit of using realgar, however, is still up for discussion due to its high arsenic toxicity. In order to understand the mechanism of the therapy, a method for arsenic speciation analysis in blood was developed so that samples collected from the APL patients could be analyzed. A 'one-stop' extraction and clean-up procedure for the speciation of As(III), DMA, MMA and As(V) in whole blood using a C18 cartridge followed by the determination by atomic fluorescence spectrometry is discussed in this presentation.

1 INTRODUCTION

Tetra-arsenic tetra-sulfide (As_4S_4), a traditional Chinese medicine, is effective on acute promyelocytic leukemia with fewer side effects than arsenic trioxide (ATO) (Lu *et al.*, 2002). However, the metabolic pathway of As_4S_4 in human body fluid, especially in the blood, is not yet fully understood. In order to gain insight into the biological functions of As_4S_4 in the human body, speciation of arsenicals in human blood is essential.

The application is challenging and needs to be developed as the extraction of arsenic species from whole blood is not straightforward. Initial studies using spiked arsenic compounds have shown poor recoveries and this problem needs to be analytically investigated. Our initial study will focus on inorganic As metabolites, but it can be extended to other forms of As later in the project. To the best of our knowledge, this the first time that HPLC hydride generation coupled to AFS has been used for this application.

2 EXPERIMENTAL

2.1 *Instrumentation*

Arsenic speciation in the whole blood samples were analyzed using an HPLC-HG-AFS system consisting of an isocratic HPLC pump and a hydride generation atomic fluorescence spectrometer (P S Analytical Millennium Excalibur 10.825). An auto-sampler and an auto-injection valve (P S Analytical 20.400 and N118 A200) were used for sample introduction.

2.2 *Experimental conditions*

Arsenical species are separated isocratically on a Hamilton PRP X-100 anion exchange column (250 × 4.1 mm I.D. 10 μm) with a phosphate buffer based mobile phase (pH 5.9) at 1.0 mL min^{-1}. Four arsenic species (As(III), DMA, MMA and As(V)) were separated and detected. Hydride generation of the eluted species was accomplished by the addition of 10% HCl and 0.7% m/v $NaBH_4$ solutions through a 4-way mixing joint at flow rates of 5 mL min^{-1}. The volatile arsines, along with hydrogen generated are separated from the liquid phase using a gas liquid separator and then atomized in an Argon-Hydrogen flame followed by the detection with an AFS detector. A schematic diagram of the system is shown in Figure 1. The detailed conditions are listed in Table 1.

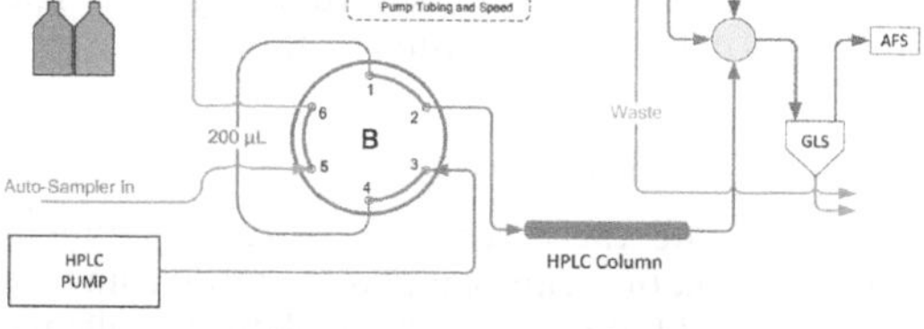

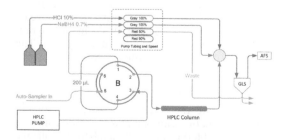

Figure 1. HPLC-HG-AFS speciation system.

Table 1. Analytical conditions for HPLC-HG-AFS.

Parameters	Conditions
Mobile phase	20 mM phosphate buffer, pH 5.9
Mobile phase flow rate	1 mL min^{-1}
HPLC column	PRP-X 100, 250 × 4.1 mm, 10 μm
Carrier solution	10% HCl (v/v)
Reductant	0.7% $NaBH_4$ in 1% NaOH (m/v)
Calibration range	0–10 ng mL^{-1}
Run time	10 min
Detection limit	0.1 ng mL^{-1} As^{III}

3 RESULTS AND DISCUSSION

3.1 *Total arsenic analysis in whole blood samples*

An aliquot of 1 mL whole blood sample was accurately pipetted into a 40 mL glass clean digestion vial followed by the addition of 1 mL concentrated nitric acid and 0.5 mL of hydrogen peroxide. The vial was then loosely capped and heated on a hot-block at 120 °C for 1.5 h until clear digest was obtained. The sample was cooled to the room temperature before it was transferred and diluted to the 10 mL mark in a graded polypropylene vial. An aliquot of the sample digested was pre-reduced by a pre-reductant containing L-cysteine (1%, m/v) and HCl (2%, v/v). Direct hydride generation atomic fluorescence spectrometry was used to quantify the total As concentrations in the whole blood samples. To establish the accuracy of the method a Seronorm quality control sample was prepared with each batch of samples. Whole blood samples were also spiked with As species at different concentrations to check recoveries. The digestion adopted here does not account for non-reducible species such as arsenobetaine.

3.2 *Speciation of As in whole blood samples*

Fresh calibration standards were prepared daily for As(III), DMA, MMA and As(V) by mixing individual species from 10 μg mL^{-1} stock solutions using appropriate dilutions. Mixed arsenical standards were prepared with deionized water in the range of 0–10 ng mL^{-1} for each species. A schematic diagram of the speciation system is shown in Figure 1. The analytical conditions are listed in Table 1.

3.3 *Sample preparation*

To investigate the arsenic metabolite in human body, plasma or serum samples are more common than whole blood due to the analytical challenge associated with this difficult sample. Proteins in the whole blood tend to degrade the LC column quickly which makes it very difficult to obtain repeatable results. Therefore, a cleanup procedure was developed in aid to tackle this very complex matrix. The cleanup procedure is illustrated in Figure 2. An aliquot of the whole blood sample (1 mL) was mixed with 3 mL of acetonitrile in order to precipitate the proteins in the blood as well as extract arsenic into the aqueous phase. The supernatant was then passed through a C18 cartridge with cation exchange function group to remove protein from the sample. After the cleanup, the sample appears colorless. Acetonitrile was evaporated by an argon flow before the sample was re-diluted to the mark with deionized water.

3.4 *Future work—sample analysis*

A Seronorm whole blood Reference Material (RM) was used for the method development and validation. The RM was spiked with interested arsenic metabolite and left to equilibrate overnight before the cleanup procedure. The initial results are shown in Figure 3. Spiked arsenic species were recovered fully from the whole blood samples after

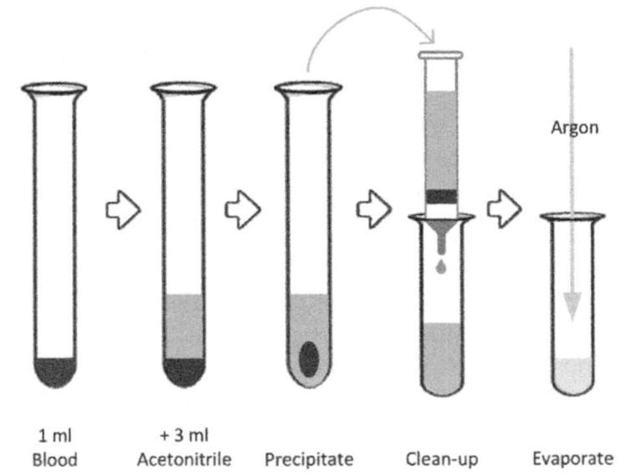

Figure 2. Clean-up procedure for whole blood samples.

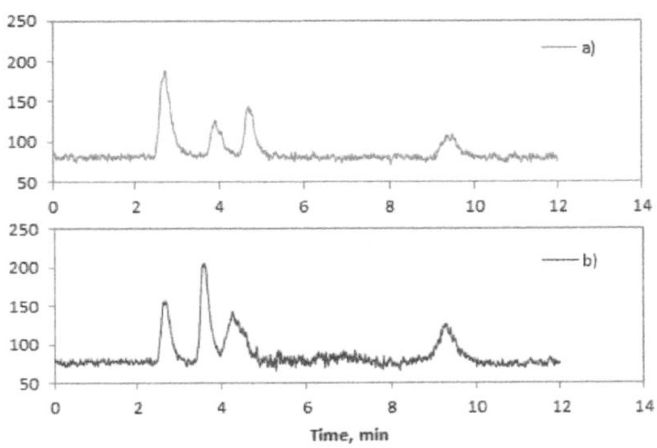

Figure 3. Chromatograms of a) 2 ng mL^{-1} mix arsenical standards; b) whole blood samples after cleanup containing 2 ng mL^{-1} spiked arsenic mix standards.

the cleanup procedure. Arsenic did not appear to be lost on the column which shows a very promising effect of the extraction and cleanup procedure. The whole blood samples collected from APL patients are provided by our collaborator at Xiyuan Hospital in Beijing after each stage of the therapy. The project is currently is on-going and the developed analytical procedure will be used to generate the valuable data for understanding the biological functionality of As_4S_4.

4 CONCLUSIONS

An arsenic speciation method in whole blood samples based on HPLC-HG-AFS was developed. A cleanup procedure effectively removes the proteins in the sample after extraction which facilitates the speciation analysis. The project is still ongoing and the procedure developed will be used for the samples collected from APL patients after each stage of As_4S_4 therapy.

REFERENCE

Lu, D., Qiu, J.Y., Jiang, B., Wang, Q., Liu, K.Y., Liu, Y.R. & Chen, S.S. 2002. Tetra-arsenic tetra-sulfide for the treatment of acute promyelocytic leukemia: a pilot report. *Blood* 99(9): 3136–3143.

One Century of the Discovery of Arsenicosis in Latin America (1914–2014) – Litter, Nicolli, Meichtry, Quici, Bundschuh, Bhattacharya & Naidu (Eds)
© 2014 Taylor & Francis Group, London, ISBN 978-1-138-00141-1

Rapid and nondestructive measurement of labile As and metals in DGT by using Field Portable-XRF

Z. Chen
Research Center for Eco-Environmental Sciences, Chinese Academy of Sciences, Beijing, China
Lancaster Environment Centre, Lancaster University, Lancaster, UK

P.N. Williams
The University of Nottingham, Malaysia Campus, Semenyih, Malaysia
Lancaster Environment Centre, Lancaster University, Lancaster, UK

H. Zhang
Lancaster Environment Centre, Lancaster University, Lancaster, UK

ABSTRACT: This study evaluated the capability of FP-XRF (Field-Portable X-ray Fluorescence) to swiftly generate information of elemental amounts in DGT (Diffusive Gradients in Thin-film gels) devices. Biologically available metal ions in environmental samples passively preconcentrate in the thin films of DGT devices, providing an ideal and uniform matrix for XRF nondestructive detection. Strong correlation coefficients ($r > 0.992$ for Mn, Cu, Zn, Pb and As) between the concentration data from ICP-MS and FP-XRF analysis were obtained for all elements during calibration. When Pb and As co-existed in the solution trials, As did not interfere with Pb detection when using Chelex-DGT. However, there was a significant enhancement of the Pb reading attributed to As when ferrihydrite binding gels were tested, consistent with Fe-oxyhydroxide surfaces absorbing large quantities of As. This study provides a new and simple diagnostic tool for on-site environmental monitoring of labile metals/metalloids.

1 INTRODUCTION

Environmental monitoring needs robust, in situ analytical techniques that return accurate contaminant data quickly and without intensive laboratory work. Techniques such as Diffusive Gradients in Thin-film gels (DGT) technology offer an alternative approach by providing in situ passive sampling of analytes in a quantitatively well-defined manner. Typically, after field deployment, the devices are transported to a laboratory where metals accumulated in the binding gel are eluted and analyzed. The elution and analysis of elements often takes several days. The time incurred in measuring the samples restricts the application of DGT technology in monitoring environmental accidents, which require immediate data feedback.

Field portable-XRF (FP-XRF) provides viable and effective analytical approaches to meet on-site analysis needs for many types of environmental samples. However, the detection limits of FP-XRF are relatively high, varying from several to tens of mg L^{-1}. Thus, it is difficult to directly apply FP-XRF to environmental aqueous samples, in which metal concentrations are typically in the µg L^{-1} to mg L^{-1} range. Sample matrix effects and morphology associated with sorbing the elements in the aquatic phase to a solid phase are considered to be key factors that affect XRF responses. To overcome these limitations, field-collected solid samples are recommended to be dehydrated, sieved and carefully homogenized before FP-XRF analysis. This study investigates whether the DGT technique concentrates As and other metals for FP-XRF on-site measurement of environmental aquatic samples.

Modified DGT devices using well established binding agents of Chelex resin or precipitated ferrihydrite were used to accumulate the metalloid

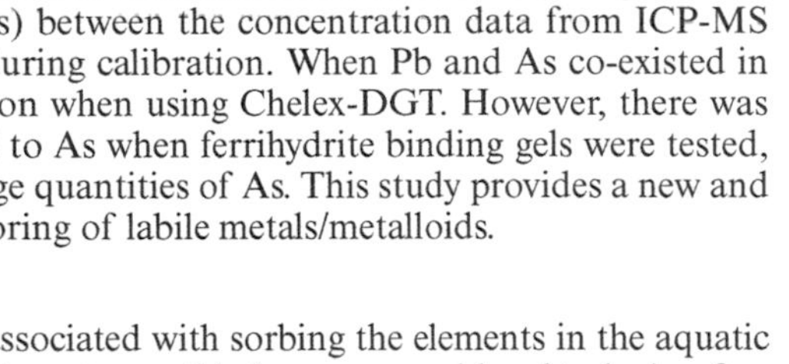

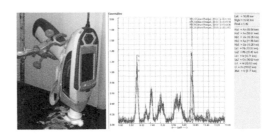

Figure 1. Schematic of the application of field-portable XRF on a modified DGT device and the XRF spectrogram.

arsenic (As), or the metals manganese (Mn), copper (Cu), zinc (Zn), and lead (Pb) for detection by FP-XRF. The signal data from FP-XRF were calibrated with those generated by ICP-MS.

2 METHODS

2.1 *Preparation of the DGT device*

DGT devices for XRF analysis comprised a plastic piston with an exposure window of 3.14 cm^2, a binding layer and an ultra thin diffusion layer. Compared to standard DGT device specifications, the samplers for XRF analysis (DGT-XRF) did not include a diffusive gel. As XRF signals exponentially attenuate with the distance between the detector and samples adsorbed in binding layers, an ultra-thin material diffusive layer (0.014 cm filter membrane plus) was used. Details for preparing those gels and assembling DGT devices have been described previously.

2.2 *FP-XRF set-up and analysis*

All XRF analyses were performed with a Niton® XRF Analyzer (Model XL3t Series) based on standard methods (USEPA Method 6200). Measurements were carried out using a 60 second irradiation, and performed at random locations on the surface of the DGT exposure window and the spot size was maintained at 8 mm.

2.3 *Empirical calibration*

DGT-XRF devices were immersed in 2 L As (as $NaAsO_3$) solution of 50, 250, 1000 µg L^{-1} for 4, 20, 48, and 96 h. All the solutions contained 10 mM $NaNO_3$ as a supporting matrix and were stirred during the experiments. After FP-XRF analysis, the DGT devices were opened to take out the binding gels. Elements in the binding gels were released into solution by incubating in 1 M HNO_3 overnight and measured by ICP-MS (Thermo Elemental X7).

3 RESULTS AND DISCUSSION

3.1 *Calibration of XRF analysis*

Five elements, including Mn, Cu, Zn, Pb, and As, were calibrated. The emission lines (Kα1) of Mn (5.90 keV), Cu (8.03 keV), Zn (8.639 keV), and As (10.54 keV) were used for integrating the peak area. Since the emission line of As Kα1 interferes with the Pb Lα1 (10.55 keV), both of the emission lines of Lα1 and Lβ1 (12.614 kev) were used for the quantitative analysis of Pb. Only the calibration curve for As is shown in Figure 2. The r values

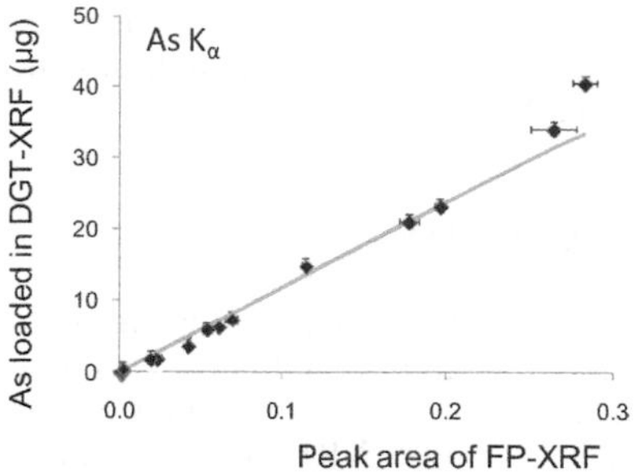

Figure 2. The loading in DGT-XRF versus integrated peak areas of As Kα.

for Mn, Cu, Zn, Pb (Lα), Pb(Lβ) and As were 0.992, 0.998, 0.993, 0.992, 0.994 and 0.993, respectively. The functions of fitted curves for As were $M_{As} = 118.3 \times A_{As} - 0.07$, where $M_{element}$ is the mass loading in DGT-XRF devices, and $A_{element}$ is the peak area of the element's X-ray spectrum.

3.2 *The factors that influence As detection*

The As Kα1 and Pb Lα1 X-ray spectra overlap at 10.6 keV when both elements are present in samples. Consequently, it was important to test whether As and Pb would be co-bound in Chelex and ferrihydrite gel. DGT devices equipped with either Chelex or ferrihydrite binding layers were deployed in a solution of 1 mg L^{-1} Pb and As(V). Pb but not As was found in the Chelex gel. The ferrihydrite gel, however, bound both As and Pb. The amount of Pb was much higher in the Chelex gel than in the ferrihydrite gel. Chelex 100 is able to selectively bind positively charged ions via ion exchange. Its affinity to anions like arsenate is weak. However, its affinity increases in high Fe conditions and Chelex-bound Fe can scavenge As from solution. The implication is that the technique can only be used to measure As in contaminated systems where the Pb concentration is insignificant compared to the concentration of As.

Other anionic species such as phosphate may compete with As(V) on the adsorption sites of ferrihydrite gels and reduce their effective adsorption capacity for As(V). Ferrihydrite DGT-XRF was deployed in solutions of 1 mg L^{-1} As with and without 5 mg L^{-1} PO_4^{3-}. For 3 hour deployments the As accumulated in DGT-XRF devices used in solutions with and without PO_4^{3-} was not significantly different, but the presence of PO_4^{3-} significantly decreased the accumulated As for 21 h deployments (Fig. 2). The result indicates that the influence of competing anions could be significant for longer deployment times. The deployment time of DGT-XRF devices should be carefully controlled to minimize the competition between anions when using FP-XRF as an in situ means of measuring As.

4 CONCLUSIONS

In this study, by combining the techniques of DGT and FP-XRF, quantitative time-integrated concentrations of metals/metalloids were obtained, with field portable XRF providing analysis in a time-scale of minutes. This technology for labile metals/metalloid detection will facilitate the application of DGT as an in situ sensor for monitoring contaminated waters and wetlands.

ACKNOWLEDGEMENTS

This work was supported by the Ministry of Science and Technology, China (2009AA06Z402) and the UK-China Science Bridge project funded by the Research Councils UK (EP/G042683/1).

REFERENCE

Chen, Z., Williams, P.N. & Zhang, H. 2013. Rapid and nondestructive measurement of labile Mn, Cu, Zn, Pb and As in DGT by using Field Portable-XRF. *Environmental Science: Processes & Impacts* 15: 1768–1774.

One Century of the Discovery of Arsenicosis in Latin America (1914–2014) – Litter, Nicolli, Meichtry, Quici, Bundschuh, Bhattacharya & Naidu (Eds)
© 2014 Taylor & Francis Group, London, ISBN 978-1-138-00141-1

Comparison between analytical results of ICP techniques for the determination of arsenic in water matrices

R. Crubellati, L. González, H. Achear & D. Cirello
Servicio Geológico Minero Argentino, Instituto de Tecnología Minera, Argentina

ABSTRACT: In this paper, a comparison between two different analytical methods is described for arsenic determination. Water samples were directly determined for arsenic using Inductively Coupled Plasma Optical Emission Spectrometry (ICP-OES) and Inductively Coupled Plasma Mass Spectrometry (ICP-MS). Data from the study demonstrate differences in the accuracy of the analytical results depending on the As level concentration. Uncertainty plays an important role in this study.

1 INTRODUCTION

Arsenic is a toxic element that is regularly required as a determinant in a suite of elements for many laboratories.

Eight methods are currently approved in 40 CFR 141.23 for the analysis of arsenic in drinking water. Only two of these methods (Inductively Coupled Plasma-Atomic Emission Spectrometry (ICPAES) (Standard Methods 3120 B) and Inductively Coupled Plasma-Mass Spectrometry (ICP-MS) (Standard Methods 3125) are multi-element or multi-analyte, meaning other analytes besides arsenic can be measured during the analysis.

The recommended guideline value of the World Health Organization for arsenic in drinking water is 0,01 mg/L (WHO, 2011). However, in many places, a new value of 0,005 mg/L has been adopted. For this reason, all chemical laboratories have to work in the future trying to adapt methods for reaching this new value.

In this paper, we made a comparison between two ICP methods regarding the detection limit of each technique and studying the different results in the range where both methods are acceptable.

2 METHODS

For the experiments, a Perkin Elmer Optima 5300 DV Inductively Coupled Plasma Optical Emission Spectrometer, equipped with concentric nebulizer and cyclonic spray chamber for use with aqueous matrices (ICP-OES) and a Perkin Elmer ICP Mass Spectrometer Elan DRCe were used.

For ICP-OES, the 188.98 nm wavelength for As was used, and the best conditions to lower the detection limit were studied.

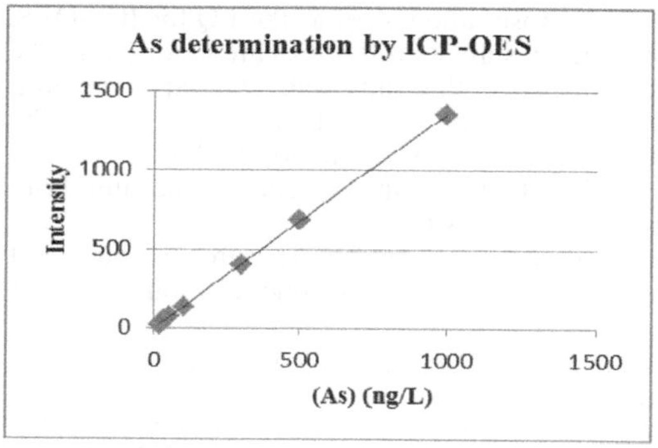

Figure 1. Calibration curve for As by ICP-OES.

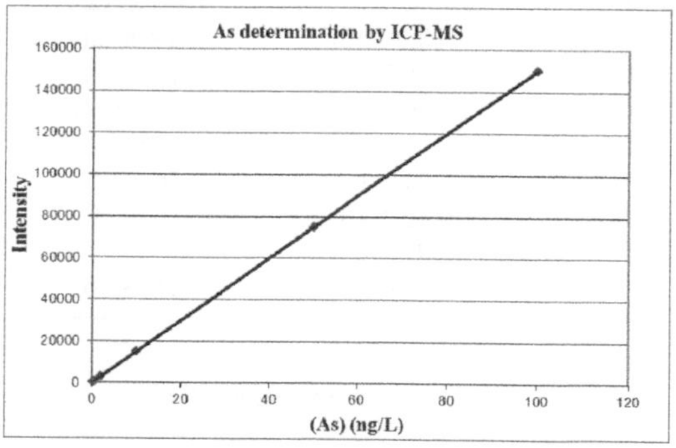

Figure 2. Calibration curve for As by ICP-MS.

LD (Detection Limit) and LQ (Quantification Limit) for As were 20 and 6.0 ng/L, respectively.

For ICP-MS, the determination was carried out at mass 75.

LD (Detection Limit) and LQ (Quantification Limit) for As were 0.05 and 0.15 ng/L, respectively.

Both methods have a range (6–100 ng/L) where it is possible to use any of both methods interchangeably.

3 RESULTS AND DISCUSSION

A study through the participation of the laboratory in proficiency tests of Environmental Canada was attempted. The idea was the evaluation and comparison of analytical results produced by ICP-OES and ICP-MS, taking into account that, in many samples, it was possible the determination indistinctly by one or by the other method.

Ten water samples were distributed between approximately 30 laboratories all over the world to determine about 30 analytes. In the case of As, concentration values were between 1 and 160 ng/L.

The samples for arsenic were analyzed using both ICP methods.

For those samples below the LQ for ICP-OES, only the results with ICP-MS are reported. In the other cases, both results were reported, taking into account the future accreditation of both tests by OAA (Argentine Accreditation Body).

Table 1 shows the assigned value and both results (in ng As/L).

Comparing the results (Figures 3 and 4), it can be seen that below concentration values of 40 ng/L, ICP-MS results are more accurate. For values between 40 and 80 ng/L, both results are similar in accuracy, and above concentration values of 80 ng/L, it is better to report results from ICP-OES.

Table 1. Assigned value and ICP-OES and ICP-MS results.

Assigned value	ICP-OES result	ICP-MS result
9,9	<6	1,1 ± 0,1
6,19	<6	6,1 ± 0,2
8,69	8,9 ± 1,6	8,4 ± 0,3
15,4	16,7 ± 2,0	15,1 ± 0,6
19,8	20,6 ± 2,2	19,1 ± 0,7
25,4	24,0 ± 2,7	25,0 ± 1,0
27,0	29 ± 3	26,0 ± 1,0
40,9	42,2 ± 3	40,0 ± 1,6
81	77 ± 4	73 ± 3
162	163 ± 7	155 ± 10

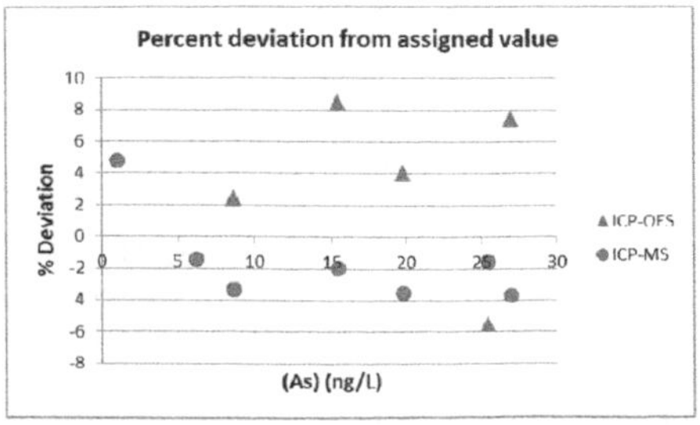

Figure 3. Percent deviations for low As concentrations.

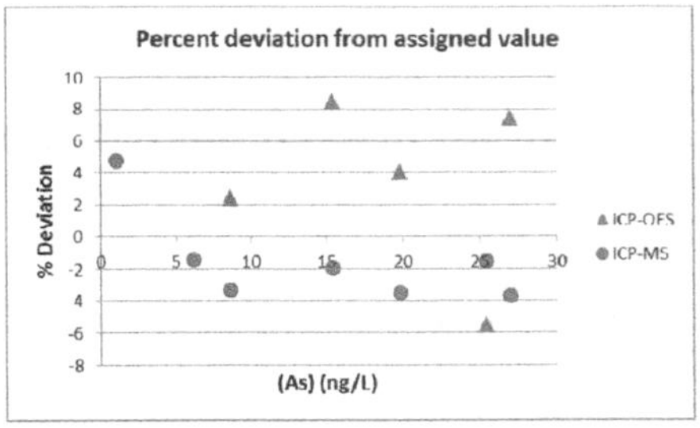

Figure 4. Percent deviations for high As concentrations.

4 CONCLUSIONS

In analytical chemistry, it is not correct to consider only the analytical result of a measurement, but the associated uncertainty must be also considered.

In ICP-MS, as it is shown in Table I, uncertainty is lower than the equivalent one in ICP-OES for values below 40 ng/L. For this reason, in that range, results are more reliable.

In contrast, for results above 80 ng/L, uncertainty in ICP-OES is lower than the equivalent one in ICP-MS. This is largely due to the dilution of samples made prior to the As determination analysis by ICP-MS.

In general, all study results in the laboratory were satisfactory (Z score). Nevertheless, it was possible to evaluate bias and uncertainty in PT to establish a relation between these parameters.

From this evaluation, it was concluded that selection of ICP-OES or ICP-MS depends on the As concentration values. Only in the range 40–80 ng/L, it is possible to use indistinctly any of both techniques.

REFERENCES

Code Federal Regulations (CFR) 141.23 http://ecfr.gpoaccess.gov.

Standard Methods 3120 B, "Inductively Coupled Plasma (ICP) Method" Standard Methods for the Examination of Water and Wastewater, 22th ed., American Public Health Association, 2011.

Standard Methods 3125, "Metals by Inductively Coupled Plasma/Mass Spectrometry, 22th ed., American Public Health Association, 2011.

World Health Organization. Guidelines for drinking-water quality. Fourth edition 2011.

One Century of the Discovery of Arsenicosis in Latin America (1914–2014) –
Litter, Nicolli, Meichtry, Quici, Bundschuh, Bhattacharya & Naidu (Eds)
 ISBN 978-1-138-00141-1

Determination of As(III) and As(V) in bacterial culture by Electro-Thermal Atomic Absorption Spectrometry (ETAAS) after separation with an anion exchange mini-column

D. Maizel
Planta Piloto de Procesos Industriales Microbiológicos (PROIMI-CONICET), Tucumán, Argentina

P. Balverdi
Laboratorio de Análisis de Trazas (LABTRA), Facultad de Bioq., Química y Farmacia, UNT, Tucumán, Argentina

C. Balverdi & A.M. Sales
LABTRA, Facultad de Bioq., Química y Farmacia, UNT, Tucumán, Argentina

M.A. Ferrero
PROIMI-CONICET, Tucumán, Argentina

ABSTRACT: A simple approach is described for the separation and determination of inorganic arsenic species using solid phase extraction and Electrothermal Atomic Absorption Spectrometry (ETAAS). A Dowex 2 × 10 anion exchange mini-column was used to separate As(III) and As(V) from a minimal culture medium, LB_{25} clean medium and LB_{25} after bacterial growth. The chemical (pH, type and concentration of eluent) and physical (flow rate of sample and eluent) parameters affecting the separation were studied. Under optimized conditions, As(V) showed a strong affinity for the mini-column, while As(III) was collected in the effluent. As(V) was recovered by elution with 0.8 M hydrochloric acid. The influence of different culture media on the separation of As(III) and As(V) was also evaluated. Inorganic As species were detected in the supernatants of free-cells from a growth culture of bacteria isolated from drinking water wells of the Tucumán province.

1 INTRODUCTION

Arsenic is a highly toxic element that supports a surprising range of microbial transformations. Specialist bacteria able to obtain energy for growth through redox transformations of arsenic have been evidenced in many environments (Silver *et al.*, 2005). Arsenic transforming bacteria were isolated by enrichment procedures in both minimal and rich growth cultures from drinking water wells in the northeast of the Tucumán province. A variety of separation and detection techniques have been reported for As speciation analysis (Maity *et al.*, 2004; Smichowski *et al.*, 2002; Yalcin *et al.*, 1998). The aim of this study was to develop a simple method based on the use of a mini-column filled with an anion-exchange resin to separate As(III) and As(V). After separation, the As species were quantified by ETAAS. The simplicity and low cost of the method make it suitable for inorganic arsenic speciation analysis in growth medium with bacteria isolated from northeast water wells of the Tucumán province.

2 EXPERIMENTAL

2.1 *Culture media, and bacterial growth*

Four different matrices were tested: Hoeft minimal medium amended with As(III) and As(V), clean Luria-Bertani medium at 25% of concentration (LB25) amended with As(III) and As(V), LB25 amended with As(III) inoculated with bacterial strain *Stenotrophomonas sp.* AE038–8 and LB25 amended with As(III) inoculated with *Pseudomonas* sp. AE038–5, both previously isolated from an As-contaminated water well at Los Pereyra, Tucumán. Inoculum was prepared overnight in LB25. Flasks containing 50 mL of LB25 were inoculated at 10% of the total volume and were incubated 24 h at 30 °C on a rotatory shaker at 150 rpm. After the incubation period, 1 mL samples were taken and centrifuged at 10.000 rpm for 10 min. The supernatant was later used for As speciation analysis.

2.2 *Column packing and conditioning*

The Dowex 2 × 10 resin, Cl^- form, was packed into glass mini-column (10 cm × 3 mm i.d). Before running the sample, the resin was converted into the acetate form by passing 3 mL of 1 M NaOH (1 mL min^{-1} flow rate), followed by 5 mL of 4 M acetic acid (1 mL min^{-1} flow rate). Finally, the mini-column was washed with 3 mL of deionized water (3 mL min^{-1} flow rate).

2.3 *Analytical separation*

Aqueous solutions, minimal and LB_{25} growth medium (1+99), at pH 7 containing As(III) and As(V), and supernatants of LB_{25} with bacterial growth were passed through the column at a flow rate of 5 mL min^{-1} using a peristaltic pump. As(V) was re-tained in the column while As(III) was collected in the effluent. A 0.8 M HCl solution (1 mL min^{-1} flow rate) was used to elute As(V) from the column. After each run, the column was washed with 3 mL of 1 M HCl and then with 3 mL of deionized water.

2.4 *Arsenic determination*

As concentration was determined in the two fractions collected by ETAAS. A Perkin Elmer atomic absorption spectrometer, AAnalyst 100, with graphite furnace HG 800 was used, equipped with a deuterium lamp background corrector and autosampler AS70. Pyrolytically coated graphite tubes with L`vov platforms and hollow cathode lamp were employed. The quantification by direct calibration against aqueous standards is accomplished using peak area measurements determined at 193.759 nm. The linearity of the calibration curve was up to 150 µg L^{-1}. A mixed Pd and $Mg(NO_3)_2$ matrix modifier solution was used. The volume injected was 20 µL of 5% (v/v) nitric acid blank, calibration standards and sample solutions and 5 µL of modifier working solution. The samples of culture media were diluted 1 + 99 to carry out the arsenic determinations without interferences. An optimized program of temperatures was applied. The Detection Limit (DL) of the methodology calculated was 3 µg L^{-1} and the quantification limit was 10 µg L^{-1}. Measurements were conducted by triplicate.

3 RESULTS AND DISCUSSION

The technique was applied to aqueous solutions and different growth media culture supplemented with As to evaluate the column efficiency and the possible interferences produced by the culture media matrix. Under optimized conditions, no interferences were observed. The optimal pH for separation was neutral. Recovery test was performed. The data ranged between 87 and 113% for water samples (data not shown). Lower recovery values (±80%) were obtained for minimal culture media supplemented with As, while LB_{25} medium gave similar data like aqueous solutions. The detection of As(V) in culture media without cells, supplemented with As(III) was negligible. Bacterial cultures with As(III) were performed in LB_{25} and inorganic species of As were detected in the supernatants of the free-cells by the technique previously described. Table 1 shows the average of triplicate values, with a Relative Standard Deviation (RSD) of 8.6%. The results show a minimal oxidation of As(III) by the *Stenotrophomonas* sp. AE038–8 and *Pseudomonas* sp. AE038–5 strains. However, the sum of the arsenical species was very close to the total arsenic recovered from the culture media.

Table 1. Determination of total As, As(III) and As(V) in free-cells supernatants and clean growth media. Concentrations in mg L^{-1}.

Sample	As total	As(III)	As(V)
LB_{25}	4,097	3,855	0,140
Hoeft	4,589	4, 418	0,166
Stenotrophomonas sp. AE038–8	3,700	3,051	0,333
Pseudomonas sp. AE038–5	4,031	3,965	0,199

4 CONCLUSIONS

The procedure proposed, using an anion exchange mini-column for the separation of the As inorganic species and further quantification by ETAAS, provides an adequate method for speciation studies in bacterial culture. There are not interferences caused by the growth media matrix.

ACKNOWLEDGEMENTS

BID-PICT-2008-0312 Project (Ministry of Science and Technology, Argentina).

REFERENCES

Maity, S., Chakravarty, P., Thakur, K., Gupta, S. & Bhattacharjee, R. 2004. Evaluation and standardisation of a simple HG-AAS method for rapid speciation of As(III) and As(V) in some contaminated groundwater samples of West Bengal, India. *Chemosphere*. 54: 1199–1206.

Silver, S. & Phung, L.T. 2005. Genes and enzymes involved in bacterial oxidation and reduction of inorganic arsenic. *Appl. Environm. Microbiol.* 71: 599–608.

Smichowski, P., Valiente, L. & Ledesma, A. 2002. Simple Method for the selective determination of As(III) and As(V) by ETAAS after separation with anion exchange mini-column. *Atomic Spect.* 23(3): 92–97.

Yalcin, S. & Chris Le, X. 1998. Low pressure chromatographic separation of inorganic arsenic species using solid phase extraction cartridges. *Talanta* 47: 787–796.

One Century of the Discovery of Arsenicosis in Latin America (1914–2014) – Litter, Nicolli, Meichtry, Quici, Bundschuh, Bhattacharya & Naidu (Eds)
© 2014 Taylor & Francis Group, London, ISBN 978-1-138-00141-1

Arsenic speciation in seaweeds using liquid chromatography hydride generation atomic fluorescence spectrometry (HPLC-HG-AFS)

L. Morrison
Earth and Ocean Sciences, School of Natural Sciences and Ryan Institute, National University of Ireland, Galway, Ireland

B. Chen & W.T. Corns
P S Analytical Ltd., Kent, UK

ABSTRACT: The content of total and inorganic arsenic in three marine macroalgae species from a pristine location of the North Eastern Atlantic Ocean (Galway Bay, Ireland) was determined using Hydride Generation-Atomic Fluorescence Spectrometry (HG-AFS). Total arsenic concentrations ranged from 27.1 to 83.2 µg g^{-1} (dry weight). Mean inorganic arsenic concentrations ranged from 0.03–3.64 µg g^{-1} (dry weight). *Laminaria digitata* had the highest total and inorganic arsenic content. The data provides baseline concentrations for arsenic compounds in commercially important seaweeds.

1 INTRODUCTION

Inorganic arsenic is a human carcinogen and its presence in the marine environment can be of natural (geogenic) or anthropogenic origin. Seaweeds accumulate metals to levels several orders of magnitude higher than those present in the surrounding seawater and have been widely used as biomonitors of water quality, avoiding the logistical difficulties associated with representative and comparative sampling of seawater and sediments (Morrison *et al.*, 2008). Furthermore, they integrate short-term fluctuation in metal concentrations and reflect bioavailable concentrations in the water. Arsenic exists in both inorganic (i-As) or organic forms (o-As) and toxicity depends on the chemical form (Villaescusa and Bolloinger, 2008).

Although elevated concentrations of arsenic have been reported in groundwater that affects an estimated 226 million people from 105 countries (Murcott, 2012), marine organisms exhibit much higher concentrations of arsenic. A recent European study has highlighted that algae, fish and shellfish remain major sources of arsenic exposure for humans (Sirot *et al.*, 2009). Consequently, the speciation analysis of arsenic in marine biota is important in terms of public health and ecotoxicology. The determination of inorganic and organometallic species of arsenic in marine tissues requires sensitive and selective techniques, involving pre-treatment steps and advanced analytical instrumentation (Bose *et al.*, 2011). The present study investigated arsenic speciation in three commercially important species of brown seaweed.

2 METHODS/EXPERIMENTAL

2.1 *Sampling*

Samples of *Ascophyllum nodosum* Linnaeus, Le Jolis (Fucales, Heterokontophyta), *Fucus vesiculosus* Linnaeus (Fucales, Heterokontophyta), and *Laminaria digitata* (Hudson) J.V. Lamouroux, (Laminariales, Heterokontophyta) were collected in Galway Bay, Ireland (Figure 1). In addition, metal concentrations were determined in different-aged tissue regions of *A. nodosum* up to a maximum age of ten years.

2.2 *Analysis*

Seaweed tissue was freeze-dried and homogenized in an agate ball mill prior to analysis by Hydride Generation-Atomic Fluorescence Spectrometry (HG-AFS). Total arsenic was determined after a hot plate digestion with nitric acid and hydrogen

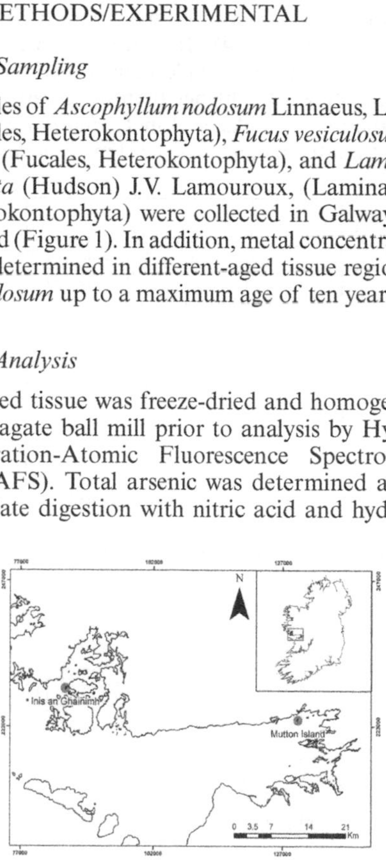

Figure 1. Map of sampling sites in Galway Bay, Ireland (inset).

peroxide. Inorganic arsenic was extracted using 0.28 M nitric acid at 90 °C. Data analysis and graphing were performed using the statistical package R (http://www.R-project) and SigmaPlot® 12.

3 RESULTS AND DISCUSSION

3.1 *Total arsenic*

Arsenic concentrations (Figure 2) ranged 27.1–53.4 μg g^{-1} in *A. nodosum*, 28.9–53.6 μg g^{-1} in *F. vesiculosus* and 54.3–83.2 μg g^{-1} in *L. digitata*. Tukey's HSD showed significant differences between the mean As level in *L. digitata* vs. *A. nodosum* and *F. vesiculosus* vs. *L. digitata*.

There was evidence of significant differences between the average concentrations of As in Years 5 & 6 versus Years 3 & 4 in different-aged tissue of *A. nodosum* (Figure 3).There was no evidence of a significant difference in As levels between the remaining years.

3.2 *Inorganic arsenic*

Inorganic As concentrations were highest in *L. digitata* (Figure 4). In terms of percentage composition, mean % iAs level in *A. nodosum* was approximately 0.1% higher than in *F. vesiculosus*, while the mean % iAs level in *L. digitata* was 5% higher than the mean % iAs level in *A. nodosum*.

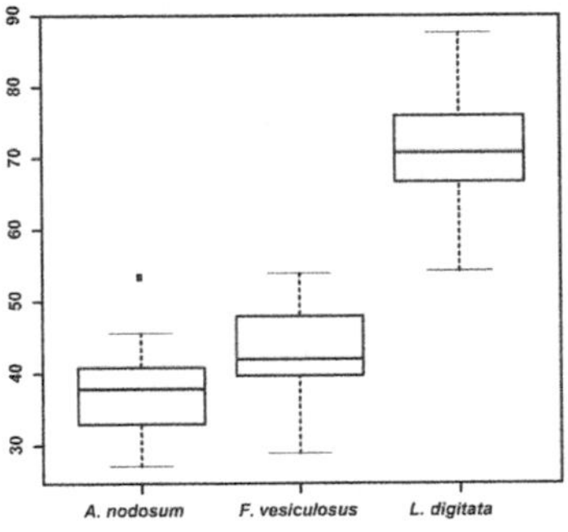

Figure 2. Total arsenic concentration (μg g^{-1}) in seaweed.

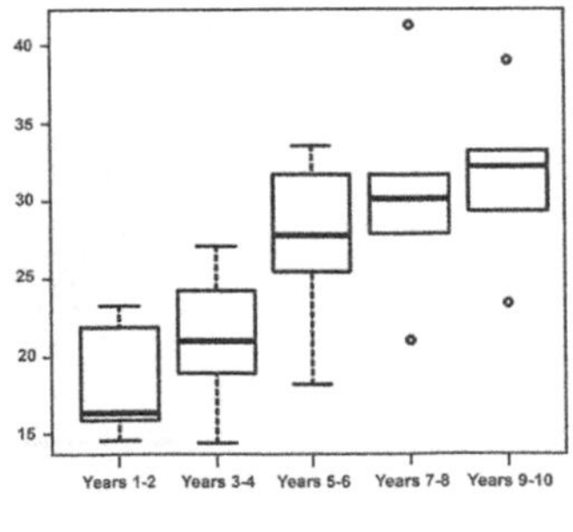

Figure 3. Total arsenic concentration (μg g^{-1}) in different-aged tissue regions of *A. nodosum*.

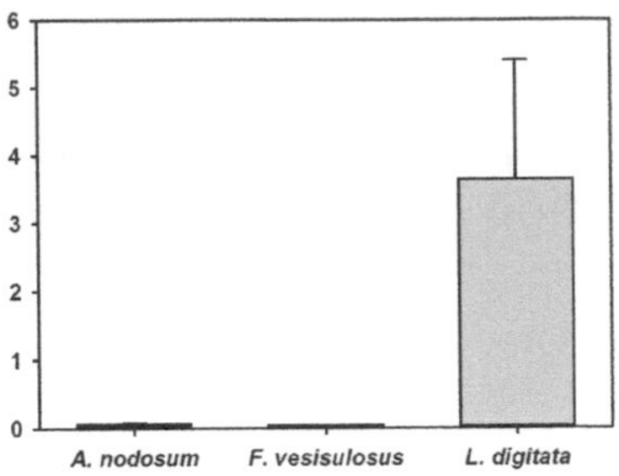

Figure 4. Inorganic arsenic concentration (μg g^{-1}) in seaweed.

4 CONCLUSIONS AND FUTURE WORK

The data from this study represent baseline levels of As in commercially important seaweeds and highlight the variability in concentrations of both total and inorganic As among brown seaweeds. On-going and future work will focus on full arsenic speciation in selected samples using HPLC-HG-AFS.

ACKNOWLEDGEMENTS

The study was funded by an Innovation Voucher (IV 2012 0026T) from Enterprise Ireland. This work includes Ordnance Survey Ireland data reproduced under OSi Licence number NUIG200803. Unauthorized reproduction infringes Ordnance Survey Ireland and Government of Ireland copyright. © Ordnance Survey Ireland, 2012. Funding based on research grant-aided by the Department of Communications, Energy and Natural Resources under the National Geoscience Programme 2007–2013.

REFERENCES

Bose, U., Rahman, M. & Alamgir, M. 2011. Arsenic toxicity and speciation analysis in ground water samples: a review of some techniques. *International Journal of Chemical Technology* 3: 14–25.

Morrison, L., Baumann, H.A. & Stengel, D.B. 2008. An assessment of metal contamination along the Irish coast using the seaweed *Ascophyllum nodosum* (Fucales, Phaeophyceae). *Environmental Pollution* 152: 293–303.

Murcott, S. 2012. *Arsenic Contamination in the World: An International Sourcebook*. IWA Publishing, London, pp. 282.

Sirot, V., Guérin, T., Volatier, J.-L. & Leblanc, J.-C. 2009. Dietary exposure and biomarkers of arsenic in consumers of fish and shellfish from France. *Science of the Total Environment* 407: 1875–1885.

Villaescusa, I. & Bollinger, J.-C. 2008. Arsenic in drinking water: sources, occurrence and health effects (a review). *Reviews in Environment Science and Biotechnology* 7: 307–323.

One Century of the Discovery of Arsenicosis in Latin America (1914–2014) – Litter, Nicolli, Meichtry, Quici, Bundschuh, Bhattacharya & Naidu (Eds)
© 2014 Taylor & Francis Group, London, ISBN 978-1-138-00141-1

Analysis of mercury and arsenic in drinking water in Bogotá DC (Colombia) in 2010 and 2011

N. Patiño Reyes
Laboratorio de Salud Pública, Secretaria Distrital de Salud de Bogotá, Universidad Nacional de Colombia, Colombia

E. Duarte Portocarrero
Laboratorio de Salud Pública, Secretaria Distrital de Salud de Bogotá, Colombia

ABSTRACT: In Bogota, Colombia, the District Department of Health ensures potable water consumption, and performs special tests for arsenic and mercury in rural aqueducts in the periphery of the city. These elements are important chemical parameters to monitor the quality of drinking water in the water supply network of the city. We analyzed 319 samples during the last half of 2010 and the first half of 2011 as a requirement of the Regulation 1575 of 2007. In the 99.38% of the samples, arsenic was no detectable, and the remaining 0.62% was below the reference value (0.01 mg L^{-1}). Regarding to mercury, out of 319 water samples that were analyzed, in 274 (85%) the element was not detectable, and in 45 (15%) results were close to the reference value. In conclusion, the water samples analyzed for mercury and arsenic in 2010 and 2011 yielded no detectable levels that may affect the health of the population in Bogotá.

1 INTRODUCTION

Metals like mercury and arsenic are considered chemical contaminants that threaten the health of the population, if the contaminated water is used for drinking, cooking food or irrigate crops. In early 2010, after an alert caused by a possible mercury contamination in drinking water in the localities of Ciudad Bolivar, Sumpaz and Usme, the District Department of Health of Bogotá, as the government agency responsible for protecting the public health in Bogotá, conducted a study under the parameters established in the 1575 Regulation of 2007 of the Republic of Colombia. This study was conducted by the Public Health Laboratory, and 319 samples of drinking water in the possibly affected localities were analyzed.

2 METHODS/EXPERIMENTAL

We analyzed 319 samples during the last half of 2010 and the first half of 2011, distributed as follows: 235 samples were taken during the last half of 2010 and 84 samples in the first half of 2011.

The arsenic analysis was performed using an Atomic Absorption Spectrophotometer ICE 3000, and the mercury determinations were obtained using a Direct Mercury Analyzer (DMA 80).

The calibration curve for As was made with 7 points ranging from 0.2 to 5.5 mg L^{-1}, obtaining a correlation coefficient of 0.9977. Regarding to mercury, the correlation coefficient was 0.998.

3 RESULTS AND DISCUSSION

Regarding arsenic, it was found that among the 319 water samples analyzed, in 317 (99.38%) As was not detectable and in 2 (0.62%) it was detectable. Out of the 235 samples tested in 2010 and 84 in 2011, none of them yielded values for As.

On the other hand, out of 235 water samples analyzed for mercury in 2010, in 190 (80.8%) Hg was not detectable and 45 samples (19.2%) showed

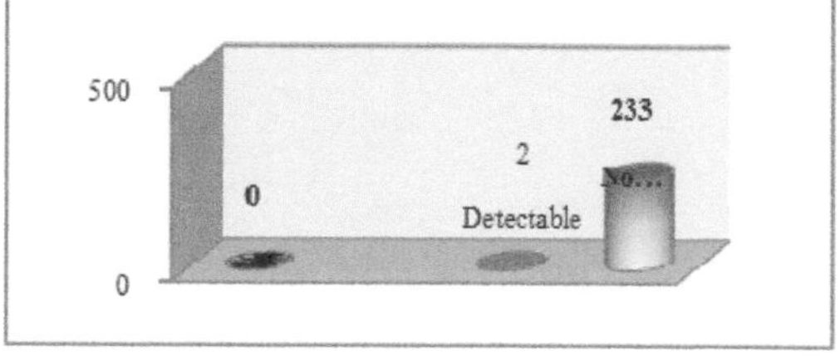

Figure 1. Levels detected for water samples analyzed for arsenic in Bogota-DC (2010).

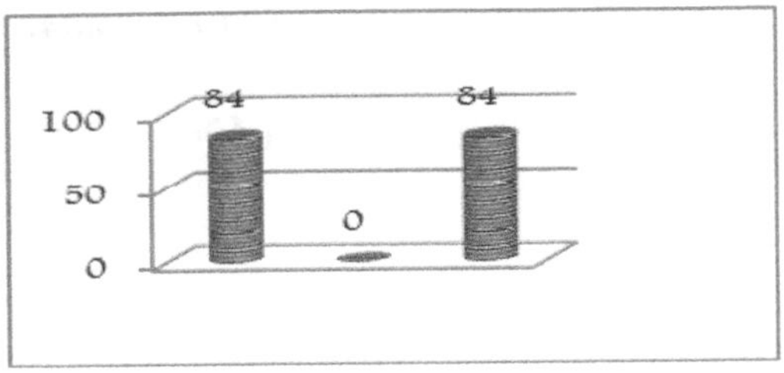

Figure 2. Negative arsenic in water samples in Bogota (2011).

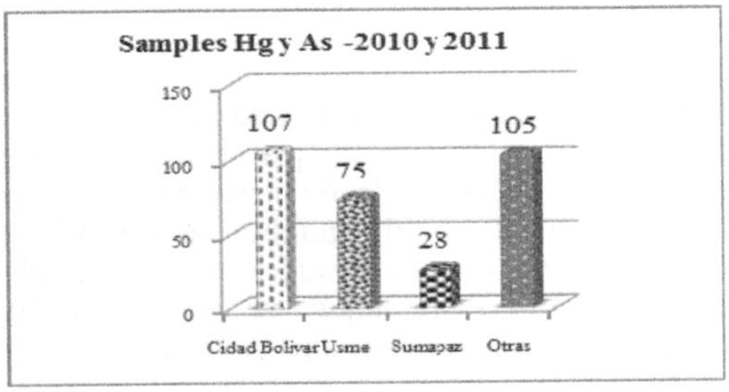

Figure 3. Water test for mercury and arsenic by locality—Bogotá, DC.

values below or close to the norm. In 2011, 84 water samples were analyzed for mercury; in 70 of these samples (91.67%) Hg was no detectable and only 7 (8.33%) yielded values, but below the reference value established by the norm.

4 CONCLUSIONS

The water samples analyzed for mercury and arsenic in the Public Health Laboratory in 2010 and 2011 yielded no detectable levels that may affect the health of the population in Bogotá D.C. The mercury values found ranged from 0.001 to 0.0017 mg L^{-1}. Regarding to arsenic, 0.6% of the samples yielded values below the standard. Routine monitoring of heavy metals in the aqueducts of Bogotá, ensures a good quality water for consumption, preventing diseases that could be caused by the water, considered the most important from the point of view of Public Health.

ACKNOWLEDGEMENTS

To the Laboratory of Public Health, District Department of Health (Bogota), and to all the environmental engineers from the localities of Bogota, who took the samples that were used in the present study.

REFERENCES

Base de Datos SISLAB, 2010–2011.
Decreto de Aguas para Colombia 1575 (2007).
http://www.lenntech.es/faq-contaminantes-del-agua.htm#ixzz2hdJ7BcHL.
Manual del Usuario DMA 80 Milestone.
Manual del usuario ICE 3000 Thermo.
www.greenfacts.org/es/arsenico.
www.who.int/mediacentre/factsheets/fs372.

One Century of the Discovery of Arsenicosis in Latin America (1914–2014) – Litter, Nicolli, Meichtry, Quici, Bundschuh, Bhattacharya & Naidu (Eds)
© 2014 Taylor & Francis Group, London, ISBN 978-1-138-00141-1

Automated unit for separation of arsenic with iron doped calcium alginate

Z. Szűcs
Institute for Nuclear Research of the Hungarian Academy of Sciences, Debrecen, Hungary

B. Kovács
University Debrecen, Debrecen, Hungary

Z. Oláh & R. Dóczi
Technical University, Budapest, Hungary

M. Maiti
Rorkee University, Rorkee, India

S. Lahiri
Saha Institute of Nuclear Physics, Kolkata, India

ABSTRACT: A prototype of an instrument was planned to develop the separation of arsenic from drinking water. A highly iron doped calcium alginate was the chemical agent used to retain the arsenical species (As(III) and As(V)) on a gel column. The objective was to increase the capability of production of the equipment to reach a concentration of 10 μg L^{-1} arsenic in 1 m^3 of purified water with a cost less than 4 USD. During the developing, ^{77}As and ^{75}Se radioisotopes traced the separation. The PLC automated system unit can communicate even through satellite, and the alginate waste satisfies the "green chemistry" concept.

1 INTRODUCTION

The high concentration of toxic and carcinogen arsenic in drinking water is still one of the hottest tasks in Hungary as well as in India. Several technologies are developed for decreasing the concentration level to 10 mg m^{-3} (EU normative). However, they are expensive or create a new danger for environment as wastes (due to Fe precipitation). On the basis of scientific reports of Banerjee *et al.* (2007a), a highly iron doped calcium alginate was chosen for the separation of arsenic from drinking water due to its high efficiency to retain the arsenic species (As(III) and As(V)) anions). The alginate gel contains ~3.8% of ferric ion. The industrial preparation of the modified alginate gel is cost-friendly. The amount of the dried waste is only 2–3% of the mass of the working gel and it does not require any additional treatment. Therefore, the alginate gel is a classical "green chemical". Banerjee *et al.* (2007b) investigated the speciation dependence of arsenic removal studying the reaction time, pH, arsenic concentration and desorption behavior. Their results suggest using the modified alginate in an automated, mobile equipment to be developed. The radioactive tracers, ^{77}As and ^{75}Se are proposed to study the developed system and to trace arsenic and selenium. The dramatic low level of selenium in drinking water in Hungary required checking the concentration of selenium, because of the high affinity of the alginate gel for selenium. Therefore, the planned apparatus contains a selenite adjuster after the elimination of arsenic.

This paper discusses the development of a semi-industrial scale equipment using the above mentioned chemicals for systematic evaluation of parameters to build up an industrial size apparatus producing As-cleaned, cheap drinking water.

2 METHODS/EXPERIMENTAL

3.1 *Chemical part*

The preparation of the highly iron doped calcium alginate was published by Banerjee *et al.* (2007a). Before the design of the equipment, the whole chemical process was simulated and checked at laboratory scale.

3.2 *Design of the equipment*

The equipment operates in batch and semi-batch mode depending on the amount of the ordered cleaned water. The system was built up with three main parts. The first part adjusts the pH of water up to 3, where the most effective retention of seleniate on the gel column takes place. The second part is the cascade of the alginate filled column, where the number of the cascade depends on the concentration of the arsenate and the amount of inlet water. The third part readjusts pH to 7 and adds the water soluble selenite. The level of arsenite in drinking water is lower in 1–2 magnitude than that of arsenate. Therefore, the removing of arsenate can fulfill usually the required 10 mg m^{-3} limit. However, after the pH-readjusting, an additional alginate gel column unit can be connected to decrease the concentration of arsenite.

3.3 *Automation and communication*

The equipment was automated by PLC produced by VACO Company. The system adjusts automatically the pH of the inlet water, and pumps the water through the adequate number of the column unit filled with modified alginate gel. The flow meters, the liquid level sensors and pH-meters generate the remote signals for the system. A one-touch screen helps to check the running parameters, and the system is also able to communicate through the satellite. In that case, a grid of the water treating units can be created covering the full country and can continually updated.

3 RESULTS AND DISCUSSION

3.1 *Chemical processes*

The highly iron doped calcium alginate shows much more efficiency to remove of arsenate from the water than what was claimed by a US patent (Min & Hering, 1998). The alginate gel with the 3.8% of iron content successfully removed arsenic even from 20 g m^{-3} concentration. The necessary contact time with the gel is also reduced from 12 h to 30 min by this modified alginate. The trace radioactive isotopes, ^{77}As and ^{75}Se, help to monitoring the efficiency of the system in situ and online. The saturated gel contains arsenic at mg kg^{-1} level, checked by ICP-MS. This concentration of solid, dried alginate is far from the toxic level stated for solid wastes, even for agricultural soils.

3.2 *Design of apparatus*

The first unit, the pH-adjuster, contains a barrel of 100 dm^3, able to feed 5 gel columns for arsenate removal. The buffer barrel is connected with the acidifying vessel remote by the liquid level controller, the pH-meter and the electrical stirrer. The pH-adjusted water is pushed through the gel column by the membrane pump, controlled by the flow meter, keeping the injected water on the column for 30 min as the critical time for reaching the best arsenic removal. The vessel collects the cleaned water after each column unit. The third part of the equipment is similar to the first one, but here the pH is readjusted up to 7 and it contains an additional vessel to add seleniate to water. The equipment was constructed in a modular form for easy transport and installation. The connection of

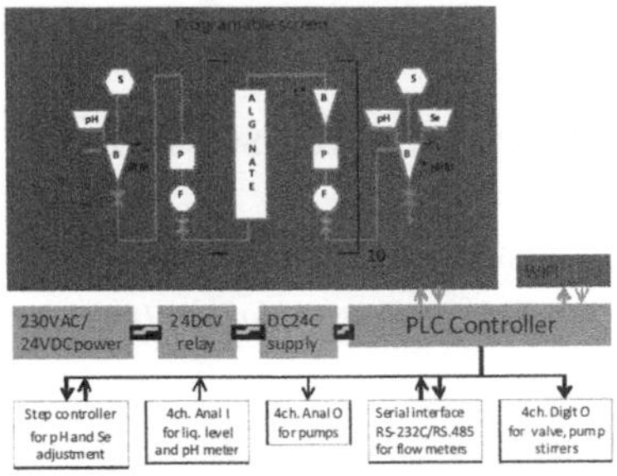

Figure 1. Block scheme of automation and communication of the arsenic remover equipment, where "pH" is the pH adjuster, "S" is the stirrer, "B" is a buffer barrel, "P" is the pump, "F" is the flow meter, "pH M" is the pH meter, "L" is the liquid level sensor and "Se" is the seleniate adjuster.

pipes and columns are flexible and have quick release clamps, as well as all electrical power, and remote cables have quick coupler for quick and easy maintenance.

3.3 *Remote, automation and communication*

Figure 1 shows the remote and automated system with extension of communication. The remote and automation is based on PLC-ETHERNET Programmable Fieldbus Controller. The system is able to communicate through the Internet.

4 CONCLUSIONS

The automated equipment at semi-industrial scale is planned to collect the necessary data for the industrial size instrument. The modified calcium alginate is the chemical agent, and it is easy to be produced and an environmental friendly compound, fulfilling the requirements of "green chemicals". The apparatus autooperates, communicate even through satellite, produce arsenic cleaned water at a cost of 4 USD/m^3.

ACKNOWLEDGEMENTS

The authors thank the Hungarian-Indian bilateral cooperation for the financial support. The contract number is TéT-13-DST-1-2013-0015.

REFERENCES

Banerjee, A., Nayak, D. & Lahiri, S. 2007a. A new method of synthesis of iron doped calcium alginate beads and determination of iron content by radiometric method. *Biochemical Engineering* 33:260–262.

Banerjee, A., Nayak, D. & Lahiri. S. 2007b. Speciation-dependent studies on removal of arsenic by iron-doped calcium alginate beads, *Applied Radiation and Isotopes* 65:769–775.

Min, J.H. & Hering, J.G. 1998. Arsenate sorption by Fe(III)-doped alginate gel. *Water Research* 32(5): 1544–1552; US Patent 6,203,709 B1. 20.03.2001.

One Century of the Discovery of Arsenicosis in Latin America (1914–2014) – Litter, Nicolli, Meichtry, Quici, Bundschuh, Bhattacharya & Naidu (Eds)
© 2014 Taylor & Francis Group, London, ISBN 978-1-138-00141-1

Interlaboratory comparisons for arsenic measurements

A. Hernández & P. Álvarez
Instituto Nacional de Tecnología Industrial (INTI), San Martín, Buenos Aires, Argentina

ABSTRACT: Since 1998, the National Institute of Industrial Technology of Argentina (INTI) is a proficiency test provider, particularly for arsenic and heavy metals in water. Until 2012, there have been eight rounds, and the mean number of participating laboratories was 23. The arsenic measurement was carried out in a synthetic sample matrix (ranged from 31 to 95 μg L^{-1}). The GFAAS method was the main method used for analyzing arsenic in this type of samples, but the Ag-DDTC method was also significantly used. The relative reproducibility standard deviation was about 12%, and this result is in agreement with bibliographic data. The quality criterion used for the proficiency assessment was the z-score. 80% of the participating laboratories obtained satisfactory results.

1 INTRODUCTION

Participation in interlaboratory tests offers a testing laboratory an opportunity to obtain an objective picture of its proficiency.

Interlaboratory comparisons allow determining the competence of individual laboratories to perform specific tests or measurements and monitoring the performance of laboratories over time.

Since 1998, the National Institute of Industrial Technology of Argentina (INTI) is a proficiency test provider in the field of water, in particular for arsenic and heavy metals in water.

This paper shows the participating laboratories, the analytical methods used and an assessment of the laboratory performance along eight interlaboratory comparisons.

2 TECHNICAL REQUIREMENTS—PROFICIENCY TESTING "TRAZAS"

Technical specifications of Proficiency Testing (PT) named "TRAZAS" are:

1. synthetic sample matrix.
2. the participating laboratory may choose the analytical method to be used.
3. two ways are used to obtain the assigned value to the samples: a) by preparing the samples from substances having a precisely known composition, b) by using the results of reference laboratories.
4. the z-score is used to evaluate the laboratory proficiency assessment.
5. the identities of the participating laboratories are treated as strictly confidential, using a randomly coded number for each laboratory.
6. The arsenic concentrations ranged from 31 to 95 μg L^{-1}.

3 RESULTS

3.1 *Participants*

Only laboratories that had the requirements and equipment for the tests to be performed could take part in this interlaboratory test.

It is desirable to ensure that the number of laboratories participating in the interlaboratory tests is sufficiently large and never less than twelve, if statistical parameters have to be derived from the data of the participants.

3.2 *Analytical methods*

The participating laboratory should use the method normally used by the laboratory for analyzing this type of sample.

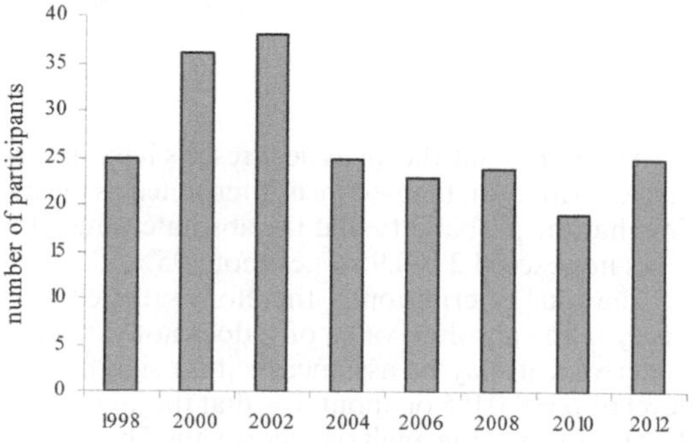

Figure 1. Number of participants in each intercomparison exercise.

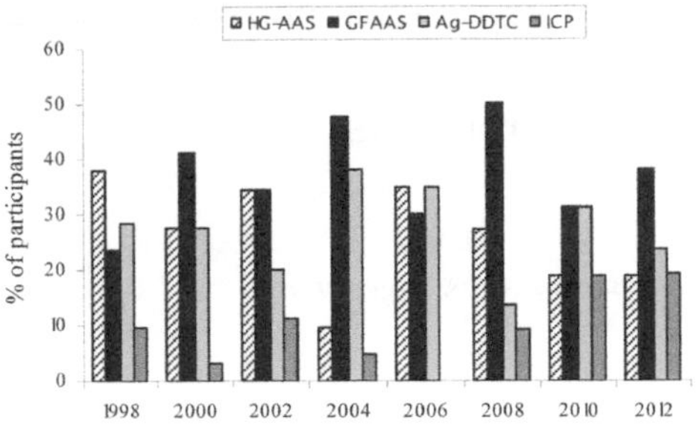

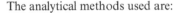

Figure 2. Percent distribution of analytical methods used in each interlaboratory test.

The analytical methods used are:

a. Graphite Furnace Atomic Absorption Spectrometry (GFAAS).
b. Hydride Generation Atomic Absorption Spectrometry (HG-AAS).
c. Silver diethyldithiocarbamate spectrophotometry (AgDDTC)
d. Inductively Coupled Plasma emission spectrometry (ICP).

3.3 *Standard deviation for the proficiency assessment*

The standard deviation for the proficiency assessment ($\tilde{\sigma}$) serves to calculate the quality limits for the analytical results.

The standard deviation for the proficiency assessment was determined from the analytical results of the test participants using statistical methods.

3.4 *Quality of laboratory results*

The use of the z-score is the recommended way of measuring the deviations of the individual laboratory analytical results (y) from the assigned value (x_A).

With the standard deviation for the proficiency assessment ($\tilde{\sigma}$), a z-score is calculated as follows:

$$z = (y - x_A)/\tilde{\sigma}$$

Assuming that the analytical results have a normal distribution, the z-score is interpreted as meaning that the probability that the absolute value of z does not exceed 2 is 0.954, i.e. about 95%.

The quality criterion is therefore satisfied precisely if the absolute value of z does not exceed 2; otherwise, it may be assumed, with a significance level of $\alpha = 0{,}0455$ or about 5%, that the laboratory has performed the analysis incorrectly. Therefore:

- $|z| \leq 2$ satisfactory,
- $2 < |z| < 3$ questionable\
- $|z| \geq 3$ unsatisfactory.

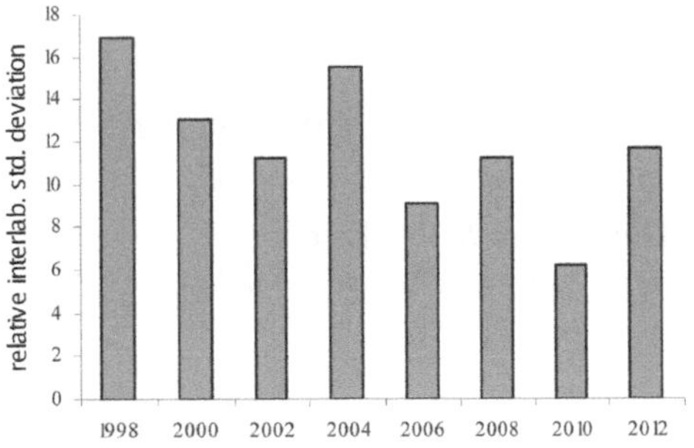

Figure 3. Relative interlaboratory standard deviation in each intercomparison exercise.

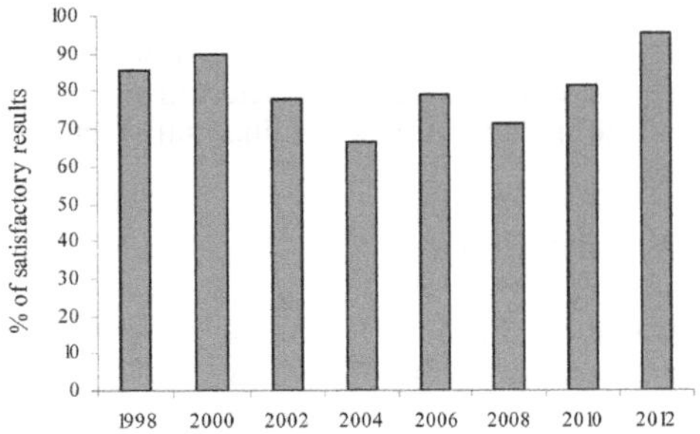

Figure 4. Percent distribution of satisfactory results in each intercomparison exercise.

4 CONCLUSIONS

The number of participating laboratories has increased since 1998 to 2002 (rounds 1 to 3) and since 2004 to 2012 (rounds 4 to 8); this number is about 23 laboratories.

The GFAAS method was the main method used for analyzing arsenic in this type of sample but the Ag-DDTC method was also significantly used.

The use of the HGAAS method was relatively low because the sample has a simple matrix (without interferences).

The relative reproducibility standard deviation decreased from 16% (round 1) to about 12%. These results are in agreement with those reported in the bibliography.

The percentage of laboratories with satisfactory results was better than 70%, with a mean value of 80% and 95% as the maximum.

It can be concluded that the laboratories have shown a quite good performance.

REFERENCES

ISO/IEC 17043:2010—Conformity assessment—General Requirements for proficiency testing.

ISO/TS 20612:2007—Water quality—Interlaboratory comparisons for proficiency testing of analytical chemistry laboratories.

One Century of the Discovery of Arsenicosis in Latin America (1914–2014) – Litter, Nicolli, Meichtry, Quici, Bundschuh, Bhattacharya & Naidu (Eds)
© 2014 Taylor & Francis Group, London, ISBN 978-1-138-00141-1

Ion pair formation with ionic liquids for highly efficient microextraction of arsenic

A. Castro Grijalba & R.G. Wuilloud
Laboratory of Analytical Chemistry for Research and Development, Mendoza, Argentina

L.B. Escudero
Chemistry Institute of San Luis, San Luis, Argentina

ABSTRACT: Several phosphonium-ionic liquids (ILs) were studied to assay their application as ion-pairing agents. Basically, a selective microextraction procedure is proposed based on the formation of As(V)-molibdate complex followed by ion pair reaction with ILs. The ion pair was extracted into a few μL of solvent and As was determined by direct injection in electrothermal atomic absorption spectrometry (ETAAS). A 100% recovery was obtained with several ILs and a preconcentration factor of 62.5 has been achieved. The detection limit was 0.001 μg L^{-1}. The use of organic solvents was significantly diminished, turning this analytical method into an environmentally-friendly alternative fitting the concept of green chemistry.

1 INTRODUCTION

Arsenic (As) is considered one of the most toxic trace elements for humans. For this reason, many analytical methods have been developed for its extraction and determination in several types of samples, such as drinking and natural water (Hung *et al.*, 2004). The use of Ionic Liquids (ILs) for developing extraction procedures for trace As determination is an efficient alternative to replace volatile organic solvents that cause atmospheric pollution (Chen *et al.*, 2008), leading to environmentally-friendly analytical procedures. In this work, different phosphonium ILs were tested for ion-pairing reaction with arsenomolybdate anionic complex $(AsMo_{12}O_{40})^{3-}$ formed between As(V) and molybdate anion. The resulting ion pair was then extracted in a minimum volume of a regular organic solvent followed by direct injection into electrothermal atomic absorption spectrometry (ETAAS) for As determination.

2 EXPERIMENTAL

2.1 *Studied variables*

The ILs evaluated were: trihexyltetradecilphosphonium chloride, trihexyltetradecilphosphonium decanoate, trihexyltetradecilphosphonium dicyanamide and tributylmethylphosphonium methyl sulfate. A conventional ion-pairing agent named cethyltrimethylamonium bromide was used for comparative purposes. The variables studied in the developed microextraction procedure were: volume of sample, acidy and temperature for complex formation, type of solvent (trichloromethane, tetrachloromethane, dichloromethane, trichloroetene and tetrachloroethene), complex formation kinetics, IL concentration, and the effect of ultrasound for ion-pair formation.

2.2 *Preconcentration and microextraction*

To evaluate and optimize conditions for the highest extraction performance of the proposed microextraction procedure, a volume of 5 mL of bidistilled water was chosen. The concentration of As(V) was 1.5 μg L^{-1}. The acidity was varied between 0.01 and 2.0 mol L^{-1} of HCl. The concentration of IL was ranged between 1.5×10^{-4} and 3.0×10^{-3} mol L^{-1} and the temperature of complex formation was in the interval between 20 and 95 °C. For the development of the microextraction procedure, a 1.5 μg L^{-1} As(V) standard solution was placed in a centrifugation tube with 50 μL of 10 mol L^{-1} of HCl, and 500 μL of a 0.12 mol L^{-1} ammonium molibdate solution. Then, 20 μL of 3.4×10^{-4} mol L^{-1} of phosphonium-IL (prepared in toluene) was added. The resulting system was stirred with a vortex for 1 minute. Finally, 80 μL of the extraction solvent (tetrachloromethane) were added to the sample solution. With this volume, it is possible to get a preconcentration factor of 62.5. After 3 min of shaking time, a centrifugation step at 1200 rpm for 5 min was carried out for the separation of final phases. The upper aqueous phase was manually removed with a transfer pipette and the organic phase was directly injected in the graphite furnace of ETAAS instrument.

3 RESULTS AND DISCUSSION

3.1 *Acidity level and IL concentration*

The complex formation between arsenate ion and molybdate is favored under acidic conditions. However, when HCl acid concentration is increased, the protonated form of arsenate ion is the main species and the complex is not formed. Afterwards, the ionic pair will not be formed. This was evidenced by the results obtained for the extractions of As(V) since the optimum level of acidity was obtained at 0.1 M HCl for all the evaluated ILs. In this assay, the IL concentration was constant at 3.4×10^{-4} mol L^{-1}. The results revealed that 0.1 M HCl led to the highest analyte recovery (91%). When the concentration of HCl was increased up to 2 M, the extraction efficiency was lower for all the ILs evaluated and with cethyltrimethylamonium bromide. The extraction efficiencies obtained with phosphonium-ILs were between 53% and 91% at 0.1 mol L^{-1} of HCl. In contrast, the conventional ion-pairing agent cethyltrimethylamonium bromide achieved barely an extraction efficiency of 18% (Figure 1). Then, a 0.1 M HCl concentration was chosen for subsequent experiments.

The optimum IL concentration for the preconcentration and microextraction was 3.4×10^{-4} M. It was observed that extraction efficiencies were decreased with the increase of IL concentration. This can be explained in function of the solubility of the ILs. When its concentration increases, its solubility decreases in the solution, making difficult the ion-pair formation and therefore, the subsequent extraction.

3.2 *Temperature for complex formation and choice of the extraction solvent*

Regarding As extraction, it was found a different behavior for each IL and solvent studied in this work. For trihexyltetradecilphosphonium chloride, the best solvent was tetrachloromethane, for trihexyltetradecilphosphonium decanoate the trichloromethane, for trihexyltetradecilphosphonium dicyanamide the tetrachloroethene and for tributylmethylphosphonium methyl sulfate the trichloromethane. The extraction efficiencies with these solvents were 100, 53, 100 and 78%, respectively. There is no apparent explanation of this behavior based on the polarity of the solvents or the polarity of the ILs. Further studies on additional physic-chemical properties of each IL are required in order to explain this behavior.

On the other hand, As extraction was modified with temperature of the solutions when this was changed during the first step of the microextraction: the formation of the complex.

3.3 *Topics for future evaluation*

In subsequent stages of this work, the following variables will be evaluated: time of the complex

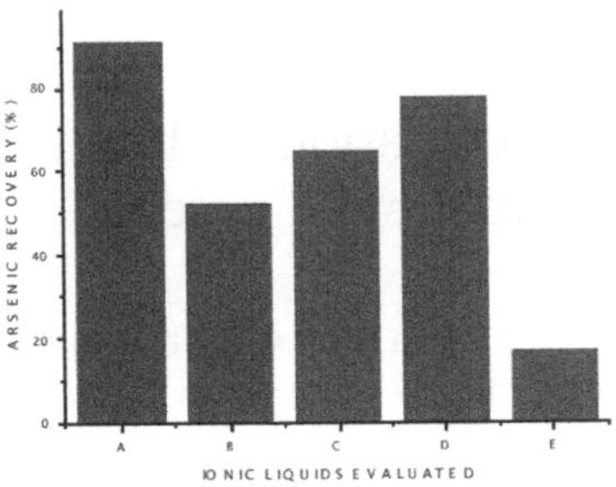

Figure 1. Recovery of Arsenic at 0.1 M HCl and 3.4×10^{-4} M IL. A. trihexyltetradecilphosphonium chloride, B. trihexyltetradecilphosphonium decanoate, C. trihexyltetradecilphosphoniumdicyanamide, D. tributylmethylphosphonium methyl sulfate and E. cethyltrimethylamonium bromide.

formation, effects of the ultrasound in the ionic-pair formation, the effect of ionic strength and the volume of the sample. Once the analytical methodology has been optimized, it will be applied to real samples of water of different sources. In this case, all of the As present in the sample must be as As(V), so a pre-oxidation step will be needed. The arsenomolibdate complex only is formed between the arsenate ion and the molybdenum not with the arsenite species.

4 CONCLUSIONS

In this work, it was demonstrated the possibility of developing a microextraction and preconcentration procedure for sensitive As trace determination based on the formation of a complex between As(V) and molybdate ion followed by ion pairing reaction with phosphonium-ILs as ion-pairing agents. The use of organic solvents was significantly diminished, turning this methodology into an environmentally-friendly alternative fitting the concept of green chemistry.

ACKNOWLEDGEMENTS

This work was supported by Consejo Nacional de Investigaciones Científicas y Técnicas (CONICET), Agencia Nacional de Promoción Científica y Tecnológica (FONCYT)(PICT-BID) and Universidad Nacional de Cuyo (Argentina).

REFERENCES

Chen, Y., Guo, Z., Wang, X. & Qiu, C. 2008. Sample Preparation. *Journal of Chromatgraphy A*. 1184: 191–219.

Hung, D.Q., Nekrassova, O. & Compton, RG. 2004. Analytical methods for inorganic arsenic in water: A review. *Talanta*. 64: 269–277.

Monasterio, R. & Wuilloud, R. 2010. Ionic liquid as ion-pairing reagent for liquid–liquid microextraction and preconcentration of arsenic species in natural waters followed by ETAAS. *Journal of Analytical Atomic Spectrometry* 25: 1485–1490.

One Century of the Discovery of Arsenicosis in Latin America (1914–2014) – Litter, Nicolli, Meichtry, Quici, Bundschuh, Bhattacharya & Naidu (Eds)
© 2014 Taylor & Francis Group, London, ISBN 978-1-138-00141-1

Synthesis, characterization and applications of novel arsenic (III) and (V) complexes

R. Barrera, S. Cano, C. España, J. Rojas & E. Muñoz
Departamento de Química, Facultad de Ciencias, Universidad Nacional de Colombia, Bogotá, Colombia

ABSTRACT: A work, divided in two steps, was developed to quantify arsenic in soil samples. The first step was the synthesis of As(III) and As(V) coordination complexes to, eventually, use them as sensitive complexes in analytical voltammetry protocols. The prepared arsenic compounds were characterized by means of IR, elemental analysis, X-ray diffraction, when suitable, and voltammetry. The spectroscopic analyses confirm the structures of the complexes. In the second step, some studies related to the limit of detection (sample digestion) by voltammetry and to the quantification of arsenic (speciation) in soil samples are also discussed.

1 INTRODUCTION

In the framework of our studies, dealing with environmentally harmful metals and metalloids, we present here some preliminary results about the synthesis of arsenic complexes, their characterization and possible uses in analytical field to improve the protocol figures of merit to test soil samples from different sources. A key point in this study is the digestion of the samples, which make them suitable for electrochemical analysis.

2 METHODS/EXPERIMENTAL

2.1 *Synthesis*

The arsenic (V) complexes were prepared by allowing to react in water Na_2HAsO_4 with MCl_2 (M = Hg, Co, Cr, Ni, Cd) in stoichiometric proportion. The instantaneous precipitate is agitated for 30 minutes and then filtered. Further recrystallization from organic solvents (alcohol, chloroform and dichloromethane) yields the pure compounds. On the other hand, arsenic (III) complexes were obtained by reaction between $AsCl_3$ and dithiocarbamates in alcohol, followed by recrystallization from chloroform/hexane.

2.2 *Characterization*

The obtained complexes were characterized by infrared spectroscopy, elemental analyses and X-ray diffraction when crystals were suitable.

To extract As associated with those complexes, the clay reference material and the soil sample, several extraction media were used as reagents, using ultrasonic extraction (Ultrasonik 19H, 44 kHz) and Microwave Digestion System (MDS, Mars 5 CEM) procedures. About 0.1 g of solid was used. After each step, the suspension was centrifuged and filtered, and the supernatant was separated and stored at 4 °C until analysis.

The electrochemical characterization was made by cyclic voltammetry and anodic square wave voltammetry using an Au, Pt-Ti and Ag/AgCl electrodes as working, auxiliary and reference respectively. The clay reference material was used to test the quality of the analytical procedure and as a starting point for the determination of As in soil samples.

On the other hand, and to evaluate the quality of the electrochemical method, the total As was determined by Inductively Coupled Plasma mass Spectrometry (ICP-MS) (NexION 300D Perkin Elmer). This equipment was equipped with a cell technology to remove polyatomic interferences and elemental isobars, which employs Kinetic Energy Discrimination (KED). KED requires the optimization of the cell pressure for each analyte ion; this is determined by optimizing the flow rate of Helium using gas collision. The system includes a sample introduction (S10 autosampler).

3 RESULTS AND DISCUSSION

The FT-IR characterization as well as the elemental analyses and the X-ray diffraction, when possible, confirmed the structure of the synthesized complexes.

The qualitative (cyclic) and quantitative (anodic stripping square wave) voltammetry of the synthetized coordination complexes, the clay reference material and a natural soil sample were evaluated to establish an all purposes protocol of analysis of As in environmental samples. The possibility of use of the coordination compounds to improve the figures of merit of the standard method of quantification of As using anodic stripping voltammetry was explored as a first step to develop a robust, fast and low cost method to quantify As in different types of samples.

4 CONCLUSIONS

Several complexes bearing As(III) and As(V) were obtained and characterized by means of infrared, elemental analyses and X-ray diffraction, when possible, and voltammetry. The cyclic and the anodic stripping square wave voltammograms between the synthetized, the clay reference material and the natural soil were compared.

ACKNOWLEDGEMENTS

The authors are thankful to Servicio Geológico Colombiano for support for equipment and analysis.

REFERENCES

Levason, W. & Reid, G. 2003. Arsenic, antimony and bismuth. *Compr. Coord. Chem.* II, 3: 465–544.

Pumure, I., Renton, J.J. & Smart, R.B. 2010. Ultrasonic extraction of arsenic and selenium from rocks associated with mountaintop removal/valley fills coal mining: Estimation of bioaccessible concentrations. *Chemosphere* 78: 1295–1300.

Shyam, K. & Narendra, K.K. 1980. Tris-*N*(ethyl, *m*-tolyl) dithiocarbamato complexes of arsenic (III), antimony (III) and bismuth (III). *Termochim. Acta* 41(1): 19–25.

Tsipis, C.A. & Manoussakis, G.E. 1976. Syntheses and spectral study of new iodobis(dialkydithiocarbamate) complexes of arsenic, antimony and bismuth. *Inorg. Chim. Acta* 18: 35–45.

One Century of the Discovery of Arsenicosis in Latin America (1914–2014) – Litter, Nicolli, Meichtry, Quici, Bundschuh, Bhattacharya & Naidu (Eds)
 ISBN 978-1-138-00141-1

Preliminary assessment of the content of total arsenic in waters of the Baratá River (Santander-Colombia) by hydride generation/atomic absorption spectrometry

C.F. Benjumea-Flórez & C.E. Rozo-Correa
Santo Tomas University, Bucaramanga, Colombia

ABSTRACT: Arsenic is considered a highly contaminant metalloid of great care. His presence in hydric sources is undetectable by organoleptic and visual methods. Baratá river waters are used as dumping for wastewaters that come from both formal and informal mining in the area and surrounding villages. The samples were analyzed by hydride generation/atomic absorption spectrometry following the method EPA7062 and IDEAM protocol determination of arsenic in water and sediment by hydride generation/atomic absorption spectrometry.

1 INTRODUCTION

Arsenic is a metalloid that is found naturally on the planet, is a component of over 245 minerals and is considered as a heavy metal for its high density and toxicity. The arsenic distribution in the environment is directly influenced by a combination of natural processes (biological activity, weathering, volmarbles emissions) as well as anthropogenic processes (mining, pesticides, herbicides, preservatives, fossil fuels) (Shukla *et al.*, 2011). Today, arsenic is classified by the International Agency for Research on Cancer in Group 1: carcinogenic agent to humans, creating a special importance in determining their concentrations in water, air and soil, and its effects.

Mining is a human activity that could produce pollution by arsenic in the environment (Sambu & Wilson, 2008). In Colombia, the high mountains of Santander comprise the municipalities of Vetas, California and Surata, all rich in minerals such as gold, silver, iron, among others. The exploitation of these minerals brings waste and leachate consequences, that ends up in water sources causing contamination.

2 METHODS/EXPERIMENTAL

The preliminary study of the content of total arsenic from the Baratá River was carried out through 3 stages: 1) sampling and sample collection, 2) digestion of the problem samples and 3) analysis by hydride generation/atomic absorption spectrometry following the method EPA 7062 and IDEAM protocol for the determination of arsenic in water and sediment by hydride generation/atomic absorption spectrometry (EPA, 1996a). The Institute of Hydrology, Meteorology and Environmental Studies (IDEAM) is a government agency of the Colombian Ministry of the Environment.

2.1 *Sampling and sample collection*

The sampling was taken in three points along the Baratá river: point 1) Tona river (which flows into the Baratá river), point 2) plant of Bosconia which is also located in Baratá River (aqueduct of Bucaramanga) and point 3) Baratá River before joining the Tona river. 500 mL of each sample were taken and were acidified with 65% nitric acid to pH 2 for its preservation.

2.2 *Digestion of the samples*

The 3050B Protocol of the Environmental Protection Agency (EPA, 1996b) was followed for the sample digestion.

2.3 *Analysis by hydride generation/atomic absorption spectrometry*

The analysis of arsenic by AAS-GH is preceded by a series of procedures before the passage of the sample through the equipment. These procedures were: 1) preparation of precursor reagents, i.e. 1% m/v $NaBH_4$ (sodium borohydride) and 10% v/v HCl (hydrochloric acid), 2) preparation of the pre-reduction agent 10% m/v KI (potassium iodide), 3) generation of As^{3+} as in the problem samples as a standard by reaction for 1 hour with the reagents. Finally, the EPA Method 7062 and IDEAM protocol determination of arsenic in water and sediment by hydride generation/atomic absorption spectrometry was followed (EPA 1996a).

3 RESULTS AND DISCUSSION

Arsenic in surface waters of the areas chosen for the study was detected. The values obtained at point 3 represent the highest concentration of arsenic. The lowest concentration of arsenic was found at point 1, while, at point 2, where the water is taken to supply the population of the city of Bucaramanga, an intermediate value of arsenic concentration between those of points 1 and 3 was found. Table 1 presents the results of the arsenic concentration at each point.

The presence of arsenic in Suratá River could be probably attributed to mining and its derivatives, which discharge their wastes into the Suratá River. This is reflected in the results at point 3, which is close to mining. Results at point 2 are probably due to natural geochemical processes. Finally, at point 1, a considerable reduction of arsenic is estimated by the observed increase of water flow (Suratá river + Tona river).

Table 1. Arsenic concentration at each point.

Sample	Absorbance	Concentration ($\mu g\ L^{-1}$)	Colombian regulation ($\mu g\ L^{-1}$)
Point 1	0.040	2.045	
Point 2	0.054	2.841	10
Point 3	0.085	4.602	

4 CONCLUSIONS

As a preliminary study, the research showed the presence of arsenic in surface waters of Suratá River. According to the 1594 decree of the Ministry of Social Protection, Environment, Housing and Territorial Development of Colombia, arsenic values detected in water of Suratá River are below the standards, indicating good water quality for domestic use. However, it is important to have a regular monitoring of this metalloid because a possible contamination can bring affectations to the community of neighboring towns and of the Bucaramanga city.

ACKNOWLEDGEMENTS

Special thanks to Santo Tomás University for permission to use the environmental chemistry laboratory for analyses of this project.

REFERENCES

Environmental Protection Agency. 1996a. Antimony and arsenic (atomic absorption, borohydride reduction).

Environmental Protection Agency. 1996b. Method 3050b, acid digestion of sediments, sludges, and soils.

Sambu, S. & Wilson, R. 2008. Arsenic in food and water a brief history. *Toxicology and Industrial Health* 2: 217–226.

Shukla, E.A., Matsue, N., Henmi, T. & Johan, E. 2011. Arse-nate Adsorption Mechanism on Nano-ball Allophane by Langmuir Adsorption Equation. *Environmental Research, Engineering and Management* 58(4): 5–9.

One Century of the Discovery of Arsenicosis in Latin America (1914–2014) – Litter, Nicolli, Meichtry, Quici, Bundschuh, Bhattacharya & Naidu (Eds)
© 2014 Taylor & Francis Group, London, ISBN 978-1-138-00141-1

Identification of diagenetic calcium arsenates using synchrotron-based micro X-ray diffraction

F. Castillo & N. Martínez-Villegas
IPICyT, Instituto Potosino de Investigación Científica y Tecnológica, División de Geociencias Aplicadas, San Luis Potosí, SLP, México

M. Avalos-Borja
Centro de Nanociencias y Nanotecnología-UNAM, Ensenada, B.C., México
On leave at IPICyT, Instituto Potosino de Investigación Científica y Tecnológica, División de Materiales Avanzados, San Luis Potosí, SLP, México

M. Villalobos
Departamento de Geología, Universidad Nacional Autónoma de México, Ciudad Universitaria, México, D.F., México

H.E. Jamieson
Department of Geological Engineering, Queen's University, Kingston, ON, Canada

ABSTRACT: In this work, we identify diagenetic calcium arsenates in sediment samples using synchrotron-based micro X-ray diffraction analyses. Our results revealed a characteristic (sub)set of diffraction peaks that do not matched any calcium arsenates reported in X-ray diffraction databases and therefore suggests the presence of a unique diagenetic calcium arsenate that has not been identified previously.

1 INTRODUCTION

The diagenetic precipitation of calcium arsenates has been proposed as a process controlling the mobility of arsenic in a shallow calcareous aquifer grossly polluted with arsenic (Martínez-Villegas *et al.*, 2013). Identification of these calcium arsenates by X-ray analysis has, however, proved to be very difficult to carry out because the precipitates of interest are on the microscale and immerse in a matrix of calcite, gypsum, and quartz comprising nearly 100% of the samples. Alternative analytical X-ray techniques include synchrotron-based micro X-ray diffraction analyses. In this work, we identify diagenetic calcium arsenates in sediment samples using synchrotron-based micro X-ray diffraction analyses at Brookhaven National Laboratory.

2 METHODS/EXPERIMENTAL

Liftable thin sections of arsenic containing sediment samples were prepared to identify selected targets of diagenetic calcium arsenates by synchrotron microanalysis using petrographic techniques (Walker *et al.*, 2009). Sections were prepared on pure silica glass slides, removed from the glass slide by soaking in acetone, and lifted on kapton tape to be analyzed at Beamline X26 A at the National Synchrotron Light Source at Brookhaven National Laboratory, New York (Figure 1). Beamline X26 A has proved to be suitable for obtaining high-resolution microdiffraction data on very small (5 μm) crystals. Micro X-ray diffraction was completed in transmission mode using a Rayonix SX-165 CCD Image Plate area detector. The incident X-ray beam was tuned to a wavelength of 0.7093 Å, and the distance from the sample detector was 247 mm. Calibration of the detector was done using the SRM674a diffraction standard α-Al_2O_3 and Ag-Behenate. Calibration, distortions, and conversions of two dimensional figures to one dimensional depiction of the same data were done using Fit2D™ software (Hammersley, 1998). One dimensional X-ray pattern was matched with those compiled in the PDF4 + 2012 database.

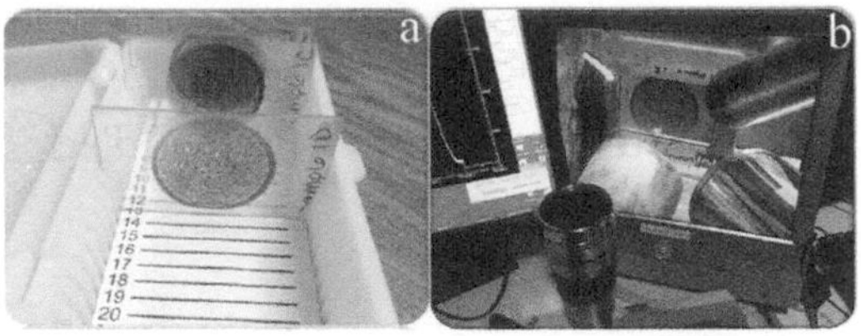

Figure 1. a) Liftable thin section prepared on a pure silica glass slide. b) View of a lifted thin section on kapton tape to be analyzed by micro x-ray diffraction.

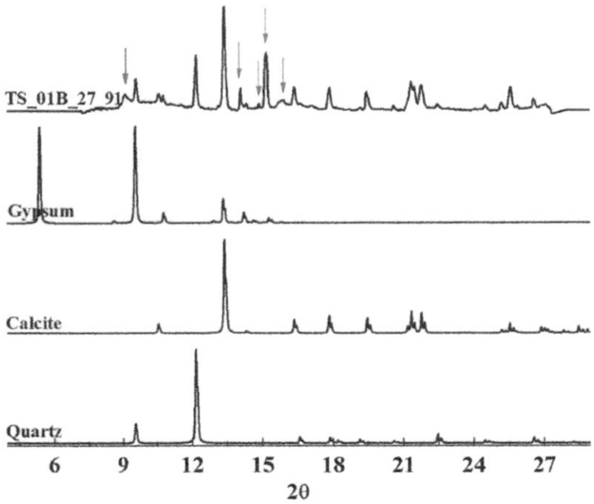

Figure 2. Matching of micro X-ray diffraction data of a calcium arsenate target with gypsum, calcite, and quartz and residual peaks indicated by red arrows.

3 RESULTS AND DISCUSSION

X-ray diffraction data using synchrotron-based microanalyses consistently showed diffraction patters containing the characteristic peaks of gypsum, calcite and quartz plus other additional peaks that showed up at 2θ positions of 9.09, 9.88, 10.12, 11.12, 11.65, 12.64, 14.12, 14.87, 15.17, 15.84, 18.26, 22.52, 22.87 and 24.10 (Figure 2). These additional or residual peaks could not reasonably be explained by any mineral in the PDF4+ 2012 database, including those minerals stoichiometrically similar to diagenetic calcium arsenates studied in this work such as vladimirite, sainfeldite, ferrarisite and guerinite. Residual peaks were therefore explained as a plausible peak (sub)set of the fingerprint diffraction pattern of a new class of diagenetic calcium arsenates which may be comprised not only by the residual peaks but by overlapping peaks with calcite, gypsum, and quartz. These findings are in agreement with previous results (Martínez-Villegas *et al,* 2013) that showed that the diagenetic calcium arsenates studied have the same stoichiometry of guerinite and ferrarisite but a different solubility product.

CONCLUSIONS

Synchrotron-based micro X-ray diffraction analyses on diagenetic calcium arsenates revealed a characteristic (sub)set of diffraction peaks that suggests the presence of a unique diagenetic calcium arsenate not previously reported. Most importantly, our results evidence the lack of fundamental knowledge on arsenic-containing minerals widely spread in the environment after (presumable) stabilization of arsenic in aqueous effluents and smelter gases in metallurgical industries.

ACKNOWLEDGEMENTS

This study was supported by grant numbers CB-2012-183025 and IPICYT S-2694 funded by CONACyT and Curso-Taller de Calidad del Agua y Modelación Hidrogeoquímica, respectively. Synchrotron-based X-ray diffractions were collected at beamline X-26 A, National Synchrotron Light Source (NSLS), Brookhaven National Laboratory. X26 A is supported by the Department of Energy (DOE)—Geosciences (DE-FG02-92ER14244 to The University of Chicago—CARS). Use of the NSLS was supported by DOE under Contract No. DE-AC02-98CH10886. F. Castillo thanks CONACyT for postdoctoral fellowship. Special thanks are due to Ana Iris Peña Maldonado and Araceli Patron from LINAN-IPICyT, Tyler Nash from Queen´s University, and Sue Wirick from University of Chicago.

REFERENCES

Hammersley, A. 1998. Fit2D V10.3 Reference Manual V4.0, European Synchrotron Research Facility, Paper ESRF98-HA01T. Available at: http:www.esrf.fr/computing/expg/subgroups/dataanalysis/FIT2D/

Martínez-Villegas, N., Briones-Gallardo, R., Ramos-Leal, J.A., Avalos-Borja, M., Castañón-Sandoval, A.D., Razo-Flores, E. & Villalobos, M. 2013. Arsenic mobility controlled by solid calcium arsenates: A case study in Mexico showcasing a potentially widespread environmental problem. *Environ. Pollut.* 176: 114–122.

Walker, S.R., Parsons, M.B., Jamieson, H.E. & Lanzirotti, A. 2009. Arsenic mineralogy of near surface tailings and soils: influences on arsenic mobility and bioaccesibility in the Nova Scotia gold mining districs. *The Canadian Mineralogist* 47: 533–556.

One Century of the Discovery of Arsenicosis in Latin America (1914–2014) –
Litter, Nicolli, Meichtry, Quici, Bundschuh, Bhattacharya & Naidu (Eds)
© 2014 Taylor & Francis Group, London, ISBN 978-1-138-00141-1

Arsenic speciation analysis of water in Argentina

N.I. Ward, G. Lord, H. Farnfield, J. O'Reilly & W. Al Rawahi
ICP-MS Facility, Department of Chemistry, University of Surrey, Guildford, UK

M.J. Watts & A.L. Marcilla
British Geological Survey, Keyworth, Nottingham, UK
Patagonia BBS, General Roca, Río Negro, Argentina

ABSTRACT: The arsenic speciation analysis of water (surface and ground) was undertaken in thirteen provinces of Argentina. A field-based method utilizing solid phase extraction cartridges enabled the selective retention of iAs^{III}, iAs^{V}, MA^{V} and DMA^{V}. All fractions and a 'total' arsenic acidified sample were analyzed by ICP-MS using a collision cell. Total arsenic levels varied from <0.2 to 6754 μg L^{-1}, with the highest levels in Neuquén (surface) and in Santiago del Estero and La Pampa (ground). The lowest total As levels were found in Misiones and Chubut (surface; <0.2 μg L^{-1}). Speciation levels varied depending on water type, pH and redox potential. In San Juan and Neuquén, the predominant species is iAs^{III} (surface) whereas in La Pampa (ground) there is variation: Eduardo Castex ($iAs^{III} > iAs^{V}$), Ingeniero Luiggi ($iAs^{III} \sim iAs^{V}$), San Martin ($iAs^{III} < iAs^{V}$). In Buenos Aires and Santa Fe (ground), the main As species is iAs^{V}.

1 INTRODUCTION

In natural waters, As is prevalent as the oxyanions trivalent arsenite (H_3AsO_3) or pentavalent arsenate (H_3AsO_4). The type of arsenic species present strongly influences toxicity, and the inorganic arsenic species typically have the greatest toxicity. Inorganic arsenic compounds can be methylated in biological systems as part of the detoxification process producing monomethylarsonic acid (MA^{V}) and dimethylarsinic acid (DMA^{V}). The mobility of arsenic between different environmental compartments is strongly influenced by pH and redox potential (Eh).

Argentina has one of the largest arsenic affected areas in the world. It is estimated that an area of approximately 106 km^2 of the Chaco-Pampean plain has groundwater with elevated arsenic levels (Smedley *et al.*, 2002). Previous research has been conducted in many of the provinces located in the Chaco-Pampean plain and arsenic levels ranging from below 4 to 14969 μg L^{-1} AsT (total As) have been reported (O'Reilly, 2010; Farnfield, 2013; Farnfield *et al.*, 2012). Most studies in Argentina have reported only total arsenic levels, with very few authors reporting the direct measurement of arsenic species in water.

2 METHODOLOGY

2.1 *Sampling sites*

Between 2005 and 2013, surface and groundwater samples were collected from thirteen provinces of Argentina.

2.2 *Sample collection*

Surface and groundwater samples were collected in 20 ml bottles or 15 mL centrifuge tubes. Surface waters were sampled at a depth of 15 cm.

Samples for dissolved total arsenic determination were filtered (0.45 μm) and acidified to 1% (v/v) HNO_3 acid. Arsenic speciation was carried out in-field using two Solid Phase Extraction (SPE) cartridges; Strong Cation Exchange (SCX) and Strong Anion Exchange (SAX) (Watts *et al.*, 2010). All water samples (total and speciation) were analyzed by Inductively Coupled Plasma Mass Spectrometry (ICP-MS). Validation of the technique was performed using a reference water: TMDA-54.4 mean of 44.5 ± 2.7 μg L^{-1} As_T (certified value 43.6 μg L^{-1} As).

3 RESULTS AND DISCUSSION

3.1 *Total arsenic levels in waters (As_T)*

Table 1 lists the total arsenic levels (μg L^{-1} As_T) in ground- and surface waters from the various provinces of Argentina. Total arsenic levels varied from <0.2 to 6754 μg L^{-1}, with the highest levels in Neuquén (surface) and in Santiago del Estero and La Pampa (ground). The lowest total As levels for surface water were found in Misiones and Chubut (<0.2 μg L^{-1}) associated with major rivers, whilst for groundwater in Neuquén, Chubut and Río Negro (<0.5 μg L^{-1}) the wells were associated with residential properties and farms.

3.2 *Inorganic arsenic speciation levels in waters*

Tables 2 and 3 report the inorganic arsenic speciation analysis of ground—and surface waters. Data is only provided for the two inorganic species that, for most samples contributed to >98% of the sum total of the species concentration.

Arsenic speciation levels varied depending on the type of water, pH and redox potential. There is a significant change in the arsenic species distribution between different wells in a province. For example, in La Pampa, for groundwater from Eduardo Castex, the range of arsenite (iAs^{III}) levels was 5 to 1332 μg L^{-1} in contrast to arsenate (iAs^{V}), which was 0.09 to 592 μg L^{-1}. In northern La Pampa, at Ingeniero Luiggi, the corresponding levels were 32 to 242 μg L^{-1} iAs^{III} and 30 to 277 iAs^{V}. Only low levels of methylated arsenic were found in these waters (typically <5% MA^{V} and <1% DMA^{V}).

Methylated arsenic species were primarily found in the volcanic and geothermal waters of Copahue, Neuquén (Farnfield *et al.,* 2012). The Agrio river (above Lago Caviahue) and at Salto de Agrio had 9.4 to 15.1% MA^{V} and 2.8 to 5.2% DMA^{V} levels.

In terms of the distribution between the inorganic arsenic species, for surface waters, arsenite (iAs^{III}) was the predominant form in San Juan, Neuquén, Cordoba, Santiago del Estero and Tucuman. This suggests that geochemical factors also have a major influence on the distribution of As species.

Table 1. Total As levels in waters of Argentina (μg L^{-1} As_T).

Province	Groundwater		Surface water	
	n	(As_T)	n	(As_T)
San Juan	28	9.2–234	33	77.2–357
Neuquén	8	<0.2–28	34	2.3–3783
Chubut	5	<0.2–3	16	<0.2–6.5
Río Negro	21	0.5–12	48	1.3–17
La Pampa	238	2.7–1326	–	–
Buenos Aires	38	28–191	–	–
Cordoba	–	–	22	60–85
Santa Fe	100	11–983	–	–
Santiago del Estero	17	21–6754	8	235–675
Tucuman	22	85–675	13	93–287
Misiones	6	–	3	<0.2
Jujuy	8	186–1409	4	188–348
Catamarca	17	74–586	17	155–482

n—number of samples.

Table 2. Inorganic arsenic species levels in groundwaters of Argentina (μg L^{-1}).

Province	Ground water		
	n	iAs^{III}	iAs^{V}
San Juan	14	23–346	0.04–76
Neuquén	5	1.2–22	0.04–11
Chubut	2	0.03–1.2	0.08–5.8
Río Negro	14	0.1–4.3	0.6–11.2
La Pampa	68	5–1332	0.09–592
Buenos Aires	12	6.5–145	11.8–155
Cordoba	–	–	–
Santa Fe	100	21.9–197	4.6–369
Santiago del Estero	11	13–5877	8–1654
Tucuman	6	35–409	18–209
Misiones	–	–	–
Jujuy	2	67–789	83–687
Catamarca	8	23–398	32–209

n—number of samples.

Table 3. Inorganic arsenic species levels in surface waters of Argentina (μg L^{-1}).

Province	Surface water		
	n	iAs^{III}	iAs^{V}
San Juan	25	4–138	<0.02–22
Neuquén	18	1.8–226	0.3–20
Chubut	8	1.4–3.2	1.6–6.2
Río Negro	29	0.9–4.8	1.2–15.4
La Pampa	–	–	–
Buenos Aires	–	–	–
Cordoba	14	45–71	18–23
Santa Fe	–	–	–
Santiago del Estero	6	178–398	25–167
Tucuman	7	45–198	32–124
Misiones	2	<0.2	<0.2
Jujuy	2	45–78	67–102
Catamarca	7	102–287	34–292

n—number of samples.

4 CONCLUSIONS

This study provides for the first time an extensive database of arsenic speciation in ground—and surface waters of thirteen provinces of Argentina. The predominant As species are inorganic with a variation in the distribution of arsenite (iAs^{III}) and arsenate (iAs^{V}) depending on the type of water (Farnfield, 2013).

REFERENCES

Farnfield, H.R. 2013. Arsenic (Total and Speciation) in Water from Argentina and its Impact on Human Health, PhD Thesis; University of Surrey.

Farnfield, H.R., Marcilla, A.L. & Ward, N.I. 2012. Arsenic speciation and trace element analysis of the volcanic río Agrio and the geothermal waters of Copahue, Argentina. *Sci. Total Environ.,* 433: 371–378.

O'Reilly, J. 2010. Arsenic Speciation in Environmental and Biological Samples from Argentina: Relationship between Natural and Anthropogenic Levels and Human Health Status. PhD Thesis; University of Surrey.

Smedley, P.L., Nicolli, H.B., Macdonald, D.M.J., Barros, A.J. & Tullio, J.O. 2002. Hydrogeochemistry of arsenic and other inorganic constituents in groundwaters from La Pampa, Argentina. *Appl. Geochem* 17: 259–284.

Watts, M.J., O'Reilly, J., Marcilla, A., Shaw, R.A. & Ward, N.I. 2010. Field based speciation of arsenic in UK and Argentinean water samples. *Environ. Geochem.. Health.*, 32: 479–490.

One Century of the Discovery of Arsenicosis in Latin America (1914–2014) – Litter, Nicolli, Meichtry, Quici, Bundschuh, Bhattacharya & Naidu (Eds)
© 2014 Taylor & Francis Group, London, ISBN 978-1-138-00141-1

Arsenic determination by neutron activation analysis—application to reference materials certification and other sample analysis

S.M. Resnizky, R. Invernizzi & R. Plá
Nuclear Chemistry Department, National Atomic Energy Commission of Argentina, Buenos Aires, Argentina

ABSTRACT: Neutron Activation Analysis (NAA) is an analytical technique with adequate sensitivity and selectivity for determination of trace elements. Instrumental Neutron Activation Analysis (INAA) and Radiochemical Neutron Activation Analysis (RNAA) are used in the Laboratory of Nuclear Analytical Techniques of the National Atomic Energy Commission of Argentina (CNEA) for arsenic determination in biological and other environmental samples. In this paper, results for As concentration in water, hair, oil and for the participation of the laboratory in campaigns of certification of reference materials are presented.

1 INTRODUCTION

Arsenic is a toxic element. Thus, the search for adequate analytical techniques for its determination in biological and environmental matrixes is relevant. Neutron Activation Analysis (NAA) is a multielemental analytical technique suitable for trace analysis, with good accuracy, low detection limits, free of blank and low matrix effects. NAA has been extensively used for As determination in different matrices and it is one of the analytical techniques selected for certification of reference materials (RM).

2 METHODS

2.1 *Generalities*

Neutron Activation Analysis is based on the nuclear transformations occurring in the matter due to neutron interaction, with the production of radionuclides. The technique comprises irradiation of samples with neutrons from a nuclear reactor, and measurement of gamma radiation emitted by radionuclides formed by nuclear reactions.

The laboratory of Nuclear Analytical Techniques (NAT) of the National Atomic Energy Commission of Argentina (CNEA), uses Instrumental Neutron Activation Analysis (INAA) and Radiochemical Neutron Activation Analysis (RNAA) for arsenic determination in different biological, environmental and other matrixes. RNAA has a radiochemical separation step between the end of irradiation and the measurement of the samples.

2.2 *Instrumental Neutron Activation Analysis (INAA)*

The irradiation is performed in the RA-3 reactor, at Ezeiza Atomic Center, 8 Mw nominal power, 5×10^{13} cm^{-2} s^{-1} nominal average thermal flux. Measurements are performed using HPGe detectors (30% efficiency, resolution 1.8 keV at 1332 keV photopeak ^{60}Co), with ORTEC associated electronics and coupled to a 919 E buffer analyzer module. Gamma Vission software is used for spectra acquisition. For calculation of concentrations, a software developed at the laboratory is used.

For As determination, the samples and standards are usually irradiated for 5 hours with a decay time previous of the measurement of about 7 days. The radionuclide ^{76}As ($t_{1/2}$ 26.4 hours) is used for quantification through the measurement of its 559.1 keV gamma-ray.

INAA is used for multielemental determination, including As in different matrixes.

2.3 *Radiochemical Neutron Activation Analysis (RNAA)*

Radiochemical Neutron Activation Analysis is used when it is necessary to have detection limits lower than those obtained with INAA. RNAA was applied to those samples with low As concentration or when the major elements present in the matrix interfere with the gamma ray peak of ^{76}As.

The radiochemical separation developed in the laboratory is based on HNO_3—H_2SO_4 acid digestion of the irradiated samples and arsenic coprecipitation together with HgS; the precipitate is filtered and mounted for measurement.

Measurements and calculation steps are the same as the ones previously described.

3 RESULTS AND DISCUSSION

The laboratory has used NAA for more than 30 years, for multielemental determination in a variety of matrixes such as soils, sediments, rocks, ceramics, foodstuffs, animal tissues, plants, etc., in research projects and for services.

In this work, arsenic results for certification of biological RM and for water, hair and oil samples are presented. Results of the analysis of other RM for analytical Quality Control (QC) purposes are also included.

3.1 *Participation in campaigns of certification of reference materials*

For RM certification, it is necessary to use analytical techniques based on independent physical principles. NAA is one of the analytical techniques usually selected for certification campaigns. Some campaigns with the participation of the laboratory are described below and As results are shown in Table 1.

- *Italian National Program for Research in Antarctica*: The aim of this project was to prepare adequate certified reference materials for environmental monitoring in Antarctica. Within this program, the NAT laboratory participated in the certification of As in MURST-ISS-A2, Antarctic Krill using RNAA and IRMM 813 CRM Adamussium colbecki, with INAA.
- *Certification of RMs on tobacco leaves at Poland Institute of Nuclear Chemistry and Technology (INCT)*: The laboratory participated in the certification of As by RNAA of two materials, Oriental Basma Tobacco Leaves (INCT-OBTL-5) and Polish Virginia Tobacco Leaves (INCT-PVLT-6).
- *Certification of a mollusk reference material at the Chilean Nuclear Energy Commission (CCHEN)*: The CCHEN prepared the MR-CCHEN-002—Almeja (*Venus antiqua*), mainly for analytical quality control of the Chilean fish industry. The laboratory participated in the certification of As by RNAA.

All the provided As results were accepted for the RM certifications showing the good performance of the NAT laboratory.

Table 1. As results in μg g^{-1} for RM certification.

RM	This work Conc.	unc. (*)	Cert. values
MURST-ISS-A2	4.61	0.11	5.03 ± 0.42
IRMM 813 CRM	11.01	0.56	12.5 ± 2.2
INCT-OBTL-5	0.755	0.020	0.668 ± 0.086
INCT-PVLT-6	0.136	0.013	0.138 ± 0.010
MR-CCHEN-002	6.01	0.30	6.05. (5.88–6.22) 95% CI

(*) Combined standard uncertainty for measurements and for quantification standard (68%).

Table 2. As results in μg g^{-1} for water samples.

RM for QC	This work μg g^{-1} Conc.	unc. (*)	Cert. Value μg g^{-1}
NIST, SMR 1643d	0.0567	0.0015	0.05602 ± 0.00073

(*) Combined standard uncertainty for measurements and for quantification standard (68%).

Table 3. As results in μg g^{-1} for hair samples.

RM for QC	This work μg g^{-1} Conc.	unc. (*)	Cert. value μg g^{-1}
IAEA 336	0.649	0.055	0.63 (0.55–0.71) 95% CI

(*) Combined standard uncertainty for measurements and for quantification standard (68%).

Table 4. As results in μg g^{-1} for samples of oils.

RM for QC	This work μg g^{-1} Conc.	unc. (*)	Cert. value μg g^{-1}
NRCC Tort-1	26.7	0.3	24.6 ± 2.2

(*) Combined standard uncertainty for measurements and for quantification standard (68%).

3.2 *RNAA for As determination in water, hair and oil samples*

The analyzed water samples were part of a research project regarding characteristics of ceramic filters. The As concentration in the filtered water was very low varying from 0.02 to 0.13 μg g^{-1}, and a radiochemical separation was needed.

Hair samples related to a forensic report were analyzed by RNAA for As determination, finding values between 0.10 and 0.23 μg g–1.

Samples of oil and by-products were analyzed for As and other heavy metals, in relation to groundwater contamination due to oil extraction activities. The determined As values varied from 0.015 to 0.21 μg g^{-1}.

4 CONCLUSIONS

The results presented show the suitability of Neutron Activation Analysis for As determination in a wide variety of matrices.

The results obtained using both INAA and RNAA show good agreement with the certified values of the reference materials used for analytical quality control.

REFERENCES

Bertini, L.M., Resnizky, S.M., Plá, R.R., Gómez, C.D. & Cohen, I.M. 1987. Determinación de arsénico, selenio y antimonio en productos alimenticios mediante el análisis por activación neutrónica. In *V Congreso Argentino de Ciencia y Tecnología de Alimentos. Salta, Argentina, 1987.*

Ciardullo, S., Held, A., Amato, M., Emmons, H. & Caroli, S. 2005. Homogeneity and stability study of the candidate reference material *Adamussium colbecki* for trace elements. *J. Environ. Monit.* 7: 1295–1298.

Carli, S., Senoforte, O., Caimi, S., Pauwels, J. & Kramer, G.N. 1998. A pilot study for the preparation of a new Reference Material based on Antarctic krill. *Fresenius J. Anal. Chem.* 360: 410–414.

Resnizky, S. & Lantos, E. 1999. Determinación de arsénico y mercurio en muestras de petróleos y derivados aplicando análisis por activación neutrónica. In *Actas XXVI Reunión Anual de la Asociación Argentina de Tecnología Nuclear, Bariloche, Argentina, 1999.*

One Century of the Discovery of Arsenicosis in Latin America (1914–2014) – Litter, Nicolli, Meichtry, Quici, Bundschuh, Bhattacharya & Naidu (Eds)
© 2014 Taylor & Francis Group, London, ISBN 978-1-138-00141-1

ARSOlux—arsenic biosensor based on bioreporter bacteria

S. Hahn-Tomer, K. Siegfried, A. Chatzinotas, A. Kuppardt & H. Harms
Helmholtz Centre for Environmental Research, UFZ, Department of Environmental Microbiology, Leipzig, Saxony, Germany

ABSTRACT: Continuous water testing specifically focused on detecting unsafe levels of arsenic concentration is essential to individuals living on or below $1 per day in Bangladesh. The ARSOlux test for drinking water appears to be the most sustainable first step in the overall arsenic mitigation based on its environmental friendliness, efficiency, preciseness and economic price. This simple test kit is based on bioreporter bacteria, which can be produced worldwide and offer a sustainable alternative to conventional, chemical test kits currently utilized in the field.

1 OVERVIEW

137 million people worldwide consume contaminated drinking water above the World Health Organization's standard of 10 μg arsenic/L (The Associated Press 2007). The country worst affected globally is Bangladesh where around 50 million people consume arsenic concentrations above the recommended standard on a daily basis (Ravenscroft et al. 2009). Continuous intake of high arsenic levels cannot only lead to severe health problems but also pose negative repercussions concerning social and economic aspects.

No sustainable solution approach has yet been found for the arsenic problem. Educational programs need to be established, awareness raised and an overall social mobilization targeted as started but not completed in the Bangladesh Arsenic Mitigation Water Supply Project framework. The arsenic testing method ARSOlux or its follow-up lateral flow test are economically feasible, easily applied and offer an innovative method to test drinking water.

2 ARSENIC DISTRIBUTION

Arsenic is a ubiquitous toxic metalloid causing serious health problems where it contaminates drinking water resources with concentrations higher than the permissible levels of 10 μg As/L. Arsenic is identified as a carcinogen, which can cause death when consumed on a regular basis in high amounts.

In Bangladesh, which contains huge surface water resources and rainwater due to the Ganges-Brahamputra-Meghna River System and overall monsoonal climate, arsenic was only detected in the early 1990s (Ravenscroft et al. 2004). Until the 1970s water from the river system was used as drinking water, which resulted in waterborne diseases such as cholera, dysentery, typhoid, diarrhea and hepatitis, which caused millions of deaths. Mostly driven by the United Nations, six to eleven million hand pump tube wells were installed in Bangladesh during the last three decades (Kinniburgh et al. 2003) to solve this problem.

While curing much of the waterborne disease problem, these wells were the trigger to eventually detect that Bangladesh's groundwater in many parts of the country is highly contaminated with arsenic.

Bangladesh and West Bengal (India) are recognized as seriously arsenic polluted areas. The southeast of Bangladesh is most contaminated; where more than 90 percent of water wells may be affected (Smedley & Kinniburgh 2002).

Furthermore, it is important to realize that the arsenic distribution in Bangladesh is considered patchy, which means that in a certain area, two wells, which are only several meters away from each other, can have different arsenic levels. One can show levels above 50 μg/L and another well can show results below 10 μg/L for example.

3 SOCIAL AND ECONOMIC ASPECTS

Arsenicosis is seen as a collective term for different symptoms connected to drinking arsenic contaminated water. The symptoms, which include hardening of the skin as well as swollen limbs and loss of feeling from hands and legs, are very painful and can lead to socio-economic ostracism.

Health education in Bangladesh is often neglected and many citizens in rural areas mistake arsenicosis for leprosy. People do not necessarily connect the intake of arsenic contaminated water with skin lesions, cancer or death and often do not believe it is poisonous. An American study from 2010states that 1 out of 5 deaths in Bangladesh can be attributed to arsenic exposure in drinking water, if the arsenic water crisis is not controlled (HEALS 2010).

While arsenicosis symptoms might not be a priority for many poor families due to the fact that they have to worry about their everyday lives including work and making ends meet, arsenicosis often is the source of a downward economic spiral.

Bangladesh is one of the poorest countries in the world. There is a correlation between the poor and arsenicosis. Citizens close to or below the poverty line have no financial means to test water sources, drill deeper arsenic free wells, purchase filtration systems or access medical treatment. Many suffer from malnutrition, and have a daily focus on obtaining enough food to survive. Frequently their dwellings are located in lower lying areas where shallow wells are more susceptible to arsenic contamination. Loss of productivity has a damaging impact on family income and causes negative consequences for the overall economy.

4 DETECTION WITH BIOREPORTER BACTERIA

Due to seasonal variations and the heterogeneous distribution of the arsenic concentrations in the groundwater (Bhattacharya et al. 2011), cheap and rapid but sensitive analytical devices allowing repeated large scale arsenic surveys are required.

We are currently developing a robust, precise and easy-to-handle test system, named ARSOlux, which is based on non-pathogenic genetically modified *Escherichia coli* K12 bioreporter bacteria. The patented bioreporter strain contains a fusion of the natural resistance mechanism of *E. coli* against arsenite and arsenate and luciferase as the reporter protein. The bioreporter bacteria are freeze dried in ready to use glass vials and stable for several months (Kuppardt et al. 2009). After rehydration with a water sample the bioreporter bacteria emit light if arsenic is present. The concentration dependent light emission is detected by a portable luminometer and permits analysis of arsenic concentrations down to 5 µg/L. Special software stores the test results electronically and can easily transfer them to any computer.

This practice is best suited for professionally organized, nationwide screenings. In comparison to chemical tests the biosensor is free of toxins. It bears no danger for the individual executing the test and the surrounding environment. The sensor is approved by the German Federal Office of Consumer Protection and Food Safety and deemed not harmful to the human health and environment.

The biosensor is operated in a simple procedure and shall be sold at a lower price than available chemical test kits. This is possible due to the fact that production costs can be kept very low based on utilizing bacteria as the main ingredient.

The performance of the arsenic biosensor was tested successfully in two field campaigns in Bangladesh in 2010. During the pilot campaigns a number of 160 wells were monitored in one day with the biosensor compared to only 60 wells which were measured with chemical field test kits (Siegfried et al. 2012). The suitability of the biosensor for high-volume analyses at considerably low material and infrastructural requirements was demonstrated.

REFERENCES

Bhattacharya, P., Hossain, M., Rahman, S.N., Robinson, C., Nath, B., Rahman, M., Islam, M.M., von Brömssen, M., Ahmed, K.M., Jacks, G., Chowdhury, D., Rahman, M., Jakariya, M., Persson, L.-Å. & Vahter, M. 2011: Temporal and seasonal variability of arsenic in drinking water wells in Matlab, southeastern Bangladesh: A preliminary evaluation on the basis of a 4 year study. *J. of Environmental Science and Health* A46 (11): 1177–1184.

HEALS. 2010. Arsenic exposure from drinking water, and all-cause and chronic-disease mortalities in Bangladesh (HEALS): a prospective in cohort study. *The Lancet*; 376: 252–258.

Kinniburgh, D.G., Smedley, P.L., Davies, J., Milne, C.J., Gaus, I., Trafford, J.M., Burden, S., Ihtishamul Huq, S.M., Ahmad, N. & Ahmed, K.M. 2003. The Scale and Causes of the Groundwater Arsenic Problem in Bangladesh. In: Welch A.H., Stollenwerk KG (Eds) *Arsenic in Ground Water: Geochemistry and Occurrence.* Boston: Kluwer Academic Publishers: 211–257.

Kuppardt, A., Chatzinotas, A., Breuer, U., van der Meer, J. & Harms, H. 2009. Optimization of preservation conditions of As (III) bioreporter bacteria. *Appl Microbiol Biotechnol* 82: 785–792.

Ravenscroft, P., Burgess, W.G., Ahmed, K.M., Burren, M. & Perrin, J. 2004. Arsenic in Groundwater of the Bengal Basin, Bangladesh: Distribution, Field Relations, and Hydrogeological Setting. *Hydrogeology Journal* 13, 727–751.

Ravenscroft, P., Brammer, H. & Richards, K. 2009. *Arsenic Pollution: A Global Synthesis.* Oxford: Willey-Blackwell.

Siegfried, K., Endes, C., Bhuiyan, A.F., Kuppardt, A., Mattusch, J., van der Meer, J.R., Chatzinotas, A. & Harms, H. 2012. Field testing of arsenic in groundwater samples of Bangladesh using a test kit based on lyophilized bioreporter bacteria. *Environ. Sci. Technol.* 46, 3281–3287.

Smedley, P.L. & Kinniburgh, D.G. 2002. A Review of the Source, Behavior and Distribution of Arsenic in Natural Waters. *Applied Geochemistry*, 17, 517–568.

The Associated Press. 2007. Arsenic in Drinking Water Seen as Threat. http://www.usatoday.com/news/world/2007-08-30-553404631_x.htm (07.15.2012).

One Century of the Discovery of Arsenicosis in Latin America (1914–2014) –
Litter, Nicolli, Meichtry, Quici, Bundschuh, Bhattacharya & Naidu (Eds)
© 2014 Taylor & Francis Group, London, ISBN 978-1-138-00141-1

Arsenic determination in atmospheric particulate matter from Paracatu—Brazil by ICP-OES

L.I.D. da Silva
Center for Mineral Technology, Ministry of Science, Technology and Innovation, Rio de Janeiro, Brazil

L. do N. Silva
Center for Mineral Technology, Ministry of Science, Technology and Innovation, Rio de Janeiro, Brazil
Polytechnic School, Universidade Federal do Rio de Janeiro, Brazil

M.C. Carneiro & M.I.C. Monteiro
Center for Mineral Technology, Ministry of Science, Technology and Innovation, Rio de Janeiro, Brazil

Z.C. Castilhos
Center for Mineral Technology, Ministry of Science, Technology and Innovation, Rio de Janeiro, Brazil
Environmental Geochemistry Department, Universidade Federal Fluminense, Niteroi, RJ, Brazil

W.Z. de Mello & J.A. Matos
Environmental Geochemistry Department, Universidade Federal Fluminense, Niteroi, RJ, Brazil

ABSTRACT: Concentration of As in atmospheric suspended particulate matter were collected in glass fiber filters sampled in eight air monitoring stations in a 24-h sampling period, once a week, over a period of 14 months (from May 2011 to June 2012). As was extracted by using the adapted US EPA Method IO-3.1 and determined by ICP-OES. The accuracy of the method was proven by preparing and analyzing a matrix spike, using filter blank and sampled filter spiked with As ICP standard solution and MRC NIST 1648a.

1 INTRODUCTION

This study is a part of the project "Assessment of environmental contamination by arsenic and epi-demiological study of environmental exposure as-sociated in human populations of Paracatu-MG". Paracatu is a town and municipality located in the Northwest of the State of Minas Gerais. Its main economic activities are agriculture and mineral exploitation, including gold, lead and zinc. In Pa-racatu, the gold mining is in an open pit mine and it is important to know the levels of arsenic (As) associated to the total atmospheric suspended par-ticulate matter (TSP), as it is an element classified as a human carcinogen (IRIS, 2013) and its main routes of exposure are environmental water and food ingestion and inhalation. The objective of this study was to determine the As levels in TSP and PM10 collected on 112 glass fiber filters from 8 different air monitoring stations, using High-Volume samplers (Hi-Vol samplers) in a 24-h sampling period, from May 2011 to June 2012 (14 months) in the Municipality of Paracatu.

2 EXPERIMENTAL

2.1 *Equipment*

An ICP-OES, model Ultima 2 (Horiba), was used for As analysis. The extraction of As from the filters was performed by using a heated ultrasonic bath (Unique, 25 kHz, 264 W), a vortex mixer (Phoenix AP-56) and a centrifuge (CELM LS-3, 3200 rpm, 30 min).

2.2 *Reagents and standards*

All reagents used were the highest purity available or at least of analytical grade. Water was purified by an Elix 5/Milli-Q Gradient System (Millipore). The calibration solutions were daily prepared by diluting the 1,000 mg L^{-1} stock solution of As (Merck).

2.3 *Sample collection*

The sampling was conducted in eight air monitoring stations in Paracatu (Fig. 1): Alto da Colina (1), Copasa (2), DER (3), União (4), São Domingos (5), Arena (6), the unique station with TPS and PM_{10} sampling, Barragem (7) and Santo Antonio (8).

Figure 1. Sampling area—the eight air monitoring stations.

High-Volume samplers (Hi-Vol), operating at about 2269 m^3 min^{-1} air flow rate, with glass fiber filters (20.3 cm × 25.4 cm), were used to collect total suspended particulate matter (TSP) and PM_{10}, once a week, over a 24-h period, from May 2011 to June 2012. One sample per month, per station, was selected.

2.4 *Sample treatment*

A method based on the US EPA Method IO-3.1 for determination of trace elements in airborne PM was used. A strip of about 2.5 cm × 20.3 cm of each filter (sampled and blank) was cut on an acrylic plate, with the aid of an acrylic ruler and a surgical stainless steel blade. Each obtained strip was cut into ten small fragments which were carefully piled up and placed into the bottom of a polyethylene centrifuge tube, weighed before and after adding the strip fragments. The masses of strips were measured in order to calculate the TSP and PM_{10} mass. An aliquot of 10 mL of acid mixture (5.55% HNO_3 and 16.75% HCl) was added.

2.5 *Extraction conditions*

The centrifuge tube containing the acid mixture and the strip was placed into a heated ultrasonic bath. The temperature of the ultrasonic bath was maintained at 69 ± 3°C for 3 h. After cooling to room temperature, 10 mL of water was added, the tube was tightly closed, and placed into a vortex mixer for 2 min to complete the extraction. The phases in the tube were separated by centrifugation for 30 min at 3,200 rpm, and they were allowed to stand overnight at room temperature. Finally the samples were filtered through a Nylon 0.45 μm pore size membrane.

2.6 *Accuracy of the method*

A 2.5 cm × 20.3 cm blank glass fiber filter strip was spiked with 1 mL of 1 mg L^{-1} As ICP standard solution; another blank strip was spiked with 0.03 g of the Certified Reference Material (CRM) NIST 1648a (Urban Particulate Matter) and a 2.5 cm × 20.3 cm sampled glass fiber filter strip was spiked with 1 mL of 1 mg L^{-1} As ICP standard solution. The spiked strips were dried and followed the same procedure described above.

3 RESULTS AND DISCUSSION

3.1 *Analytical performance*

The detection limit for As was experimentally determined. It was 7.5 μg L^{-1}, corresponding to 0.52 μg g^{-1} and, considering the air sampled volume, the detection limit in ng m^{-3} was 0.64. The As recovery values for the blank filter spiked with As standard solution and the blank filter spiked with CRM NIST 1648a were 98% and 90%, respectively. For the sampled strip spiked with the As standard solution, the As recovery was 102%. It can be observed that all the obtained recovery values are within the limits proposed by the US-EPA (blank filter: 80 to 120%; sampled filter: 75 to 125%).

3.2 *Results*

Results in Table 1 summarize the data obtained for As in this work, in the eight air monitoring stations, during 14 months.

4 CONCLUSIONS

The results obtained in the recovery test showed that the adapted method can be used to determine the As concentration in TSP and PM_{10} collected in glass fiber filters. The correlation among the

Table 1. Arsenic concentration (ng m^{-3}) in eight air monitoring stations from May 2011 to June 2012.

	(1)	(2)	(3)	(4)	(5)	(6)	(6*)	(7)	(8)
May	7.1	4.4	3.3	2.1	0.6	10	3.7	0.6	0.6
Jun	0.6	0.6	0.6	0.6	0.6	17	0.6	0.6	5.4
Jul	19	0.6	6.2	7.6	0.6	16	0.6	0.6	0.6
Aug	14	2.7	4.2	9.8	0.6	14	9.4	0.6	2.5
Sep	13	3.8	3.3	5.6	0.6	18	12	0.6	3.1
Oct	0.6	3.0	0.6	0.6	5.2	5.4	3.5	0.6	0.9
Nov	2.0	0.6	0.6	0.6	0.6	0.6	3.2	0.6	0.6
Dec	4.7	5.7	2.2	3.6	0.7	3.3	1.7	0.6	0.6
Jan	0.7	2.8	1.6	1.8	0.6	0.6	1.7	0.6	0.6
Feb	2.8	0.6	0.6	1.8	3.0	8.8	7.7	0.6	0.6
Mar	1.0	0.6	0.6	2.9	6.6	0.6	4.0	0.6	3.8
Apr	3.6	1.0	0.6	2.8	5.6	0.6	4.5	0.6	2.1
May	0.6	0.6	1.6	3.0	2.9	6.9	7.8	0.6	2.4
Jun	0.6	1.8	0.6	0.6	0.6	0.6	0.6	0.6	0.6

Legend: (1) = Alto da Colina; (2) = Copasa; (3) = DER; (4) = União; (5) = São Domingos; (6) = Arena (TSP); (6*) = Arena (PM_{10}); (7) = Barragem; (8) = Santo Antônio.

obtained data and the meteorological conditions, as well as its environmental and toxicological significance is presented in other works (Zamboni *et al.*, 2014; De Capitani *et al.*, 2014).

REFERENCES

da Silva, L.I., de Souza Sarkis, J.E., Zotin, F.M., Carneiro, M.C., Neto, A.A., da Silva Ados, S., Cardoso, M.J., Monteiro, M.I. 2008. Traffic and catalytic converter-Related atmospheric contamination in the metropolitan region of the city of Rio de Janeiro, Brazil. *Chemosphere* 71: 677–684.

De Capitani, E.M., Jesus, I.M., Faial, K.R.F., Lima, M.O., Faial, K.C.F., Ferreira, A.P., Domingos, L.M. & Castilhos, Z.C. 2014. Human exposure assessment to arsenic and health indicators in Paracatu, Brazil. As2014.

EUROPEAN Comission. Ambient Air Pollution by As, Cd and Ni Compounds: Position Paper. 2001, 55–56.

IRIS- Integrated Risk Information System. Arsenic (2013). Available in: <www.epa.gov/iris/arsenic>. Accessed in June 28th 2013.

Maggs, R. 2000. *A Review of Arsenic in Ambient Air in the UK*. Department of the Environment, Transport and the Regions, Scottish Executive, The National Assembly for Wales. London.

U.S.EPA. Compendium Method IO-3.1: Selection, Preparation and Extraction of Filter Material. Cincinnati, Ohio, 1999.

WORLD Health Organization Regional Office for Europe. Air Quality Guidelines: 2nd. Edition, 2000.

Zamboni de Mello, M., Matos, J.A., Castilhos, Z.C., da Silva, L.I.D., Carneiro, M.C. & Monteiro, M.I. 2014. Arsenic in atmospheric particulate matter at Paractu-Brazil. As2014.

1.5 Arsenic in mining environments

One Century of the Discovery of Arsenicosis in Latin America (1914–2014) – Litter, Nicolli, Meichtry, Quici, Bundschuh, Bhattacharya & Naidu (Eds)
 ISBN 978-1-138-00141-1

Geochemistry of arsenic in a historical mining zone of Mexico

M.A. Armienta
Instituto de Geofísica, Universidad Nacional Autónoma de México, DF, México

ABSTRACT: Results from 20 years of research on the sources, distribution and environmental impact of As in the Mexican mining zone of Zimapán have been analyzed to determine the main environmental geochemical processes controlling its mobility. Arsenic geochemical behavior was assessed based on detailed chemical, isotopic and mineralogical determinations. Oxidation and dissolution of As-bearing minerals release As in mineralized zones of the deep aquifer and in tailings. Arsenic mobilization in tailings is partly controlled by sorption and secondary minerals formation. In soils, tailings, and river sediments, most of the As is present in low-mobility forms. Presence of limestone plays a major role in the geochemical behavior of As by increasing the pH, retain As and promote Fe-oxyhidroxides formation. Laboratory experiments showed that indigenous limestone sorbs As and may be used to clean As polluted water, and to construct geochemical barriers to treat acid mine drainage.

1 INTRODUCTION

Mining residues are one of the main environmental sources of anthropogenic arsenic. Ore extraction and processing have produced million Tons of wastes that, a few decades ago, and in some countries until recently, and even at present, were disposed of without considering their potential environmental impact. Weathering of mine waste rocks and tailings may release potentially toxic elements including arsenic to nearby water bodies and soils, affecting also the biota and impacting human health. Arsenic may be present in wastes as specific arsenic minerals like arsenopyrite and orpiment or incorporated in other minerals like arsenic-rich pyrite; all of this constitutes a potential environmental source of this contaminant. However, complex geochemical processes occur within tailings that may affect the effective release of arsenic and other potentially toxic elements. In addition, in mining zones, arsenic from natural sources may also reach toxic concentration levels.

Mining has been a wealth source in México for almost five centuries. The economy of several locations has relied on the extraction and processing of base and precious metals (Ag, Au) along these years. However, until the last two decades, assessment of the environmental effects of mining activities was almost absent. Zimapán, in Central México, has been one of the most studied historical mining locations since the discovery of high As concentrations in drinking water. In this manuscript, an overview of the geochemical behavior of arsenic in this zone, including sources, environmental fate and proposed remediation alternatives, will be presented.

2 METHODS

2.1 *Sampling and analytical determinations*

Samples were collected from rocks, wastes, sediments, groundwater, and soils, and analyzed for As mainly by atomic absorption spectrometry with hydride generation and ICP-AES. Physico-chemical determinations and analyses of other toxic elements were also carried out. Mineralogical (XRD, SEM-EDS, SEM-WDS and IR) and chemical characterization, including sequential extractions, have also been performed in solid materials. Finally, the suitability of indigenous limestone to remove As from polluted solutions (water and acid mine drainage) was assessed through batch and column experiments.

3 RESULTS AND DISCUSSION

3.1 *Arsenic sources*

Cretaceous limestones and Tertiary andesites are the main rock types outcropping at Zimapán. Mineralization corresponds to high-temperature carbonate-hosted ore deposits and occurs in chimneys, skarns, veins and sills. Arsenic concentration ranges vary among rock formations, but also between rock samples from the same formation in mineralized and non-mineralized zones. Besides abundant arsenopyrite, other arsenic–bearing minerals have been identified in the ore deposits such as tennantite, geocronite, lollingite, scorodite and adamite. Arsenic occurs mainly as As(V) in groundwater. Mineralogical analyses, physico-chemical parameters and inverse geochemical modeling showed that As is released from arsenopyrite oxidation and, in some locations, also by scorodite dissolution, as:

$$FeAsS + 3.25O_2 + 1.5H_2O \Leftrightarrow Fe^{2+} + SO_4^{2-} + HAsO_4^{2-} + 2H^+ \quad (1)$$

and

$$FeAsO_4 \cdot 2H_2O + H_2O \Leftrightarrow Fe(OH)_3 + H_2AsO_4^- \quad (2)$$

These processes occur mainly in dikes and intrusive bodies. Lowering of the water table in the dry season facilitates arsenopyrite oxidation and rising during the rainy season mobilizes As through fractures within the deep limestone aquifer (Armienta *et al.*, 2001; Rodríguez *et al.*, 2004).

Water from As-rich wells (up to 1.1 mg/L) was contaminated by the natural processes described above, and since As presence was firstly not detected, and later due to the lack of safe water alternatives, As-polluted water was consumed by the population for about two decades.

Tailings from selective flotation accumulated for more than six decades on the town outskirts (Figure 1) contain high As concentrations (up to 39500 mg/kg in one of the impoundments).

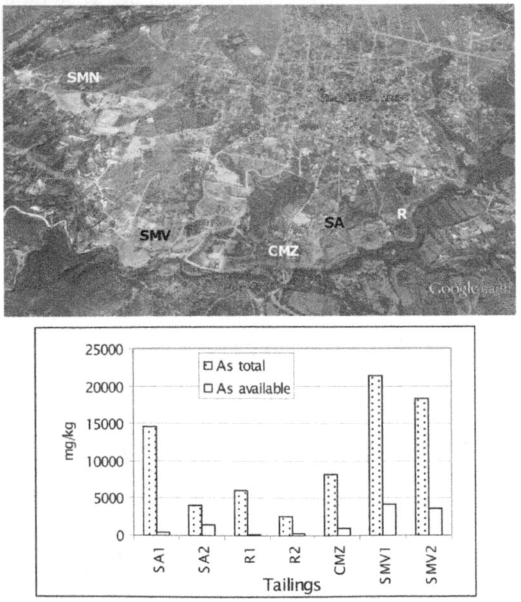

Figure 1. Location of tailings, and As concentrations: total and available (water soluble + carbonate + exchangeable fractions).

Hydrogeochemical characteristics of nearby shallow wells and mineralogical SEM-EDS determinations of tailing samples revealed sulfide oxidation within tailings as the As source to nearby shallow wells containing up to 0.6 mg/L. Dwellers use these waters mainly for house cleaning, and in some sites, for irrigation. Although sulfide oxidation increases sulfate content in natural and tailings-polluted wells, diverse isotopic values of sulfur ($\delta^{34}S$-SO_4) were measured in samples from these wells (from −0.65 to −2.59‰) with respect to natural polluted deep wells (from −9.86 to −12.14‰) reflecting distinct processes and/or sources. Besides, natural As-polluted wells are located upstream the tailing deposits and tap a regional aquifer as revealed by $\delta^{18}O$ and δD values. Samples influenced by tailings are Ca–SO_4–HCO_3 type with higher values of δD, $\delta^{18}O$, and $\delta^{34}S$-SO_4 than natural polluted samples which are Ca–Na–HCO_3 type and more depleted in the analyzed isotopes (Armienta *et al.*, 1997; Romero *et al.*, 2006; Sracek *et al.*, 2010).

3.2 *Arsenic environmental availability from tailings*

However, not all the arsenic is present in available form within tailings. Geochemical speciation applying various sequential fractionation schemes was carried out in samples from tailing deposits differing in oxidation degrees. Results showed that most of the arsenic is present in immobile and low mobility fractions, such as residual and Fe and Al oxyhydroxides (Méndez and Armienta, 2003). However, since the total concentration reaches thousands of mg/kg, arsenic in more available forms (carbonates, exchangeable, soluble) may reach high contents (Figure 1).

Artificial leachates were also obtained from tailings to determine concentrations of soluble As and metals. Arsenic in oxidized tailings varied from 0.41 to 48.7 mg/L while in unoxidized samples the range was much lower (from 0.48 to 3.83 mg/L). Surface charge determined by zeta potential showed a positive charge in oxidized tailings. Mineralogy and pH indicated that arsenopyrite oxidizes releasing As that is further retained through sorption on ferric oxyhydroxides (as observed by SEM) and clays, and also by the formation of secondary As-bearing minerals like beudantite and jarosite (Romero *et al.*, 2006).

Depth profiles were also collected from a tailing deposit located about 9 km northeast of Zimapán town. Detailed mineralogical and physicochemical characteristics were determined in layers showing distinct visible features. Total and soluble concentrations of As, Fe, Sb, Tl, Zn, Cu, Cd, Ni and Cr were analyzed in each layer. The lowest pH was 2.2; however, most layers had near neutral pH values but were potential acid generators. Arsenic was identified in arsenopyrite, As-Pb sulfosalts and tetrahedrite-tennantite; silicates, carbonates and secondary minerals were also recognized by XRD. Concentration trends of As with depth were not observed; however, lowest pH samples had higher soluble As concentrations. Presence of calcite was relevant for As mobility since samples containing 5% or more $CaCO_3$ did not present As in soluble form. X-ray maps showed Fe oxyhydroxides with As. While a hard-pan layer was not observed, higher pH and calcite promote As retention in the solid matrix (Armienta *et al.*, 2012).

3.3 *Arsenic behavior in soils and sediments*

Arsenic concentrations in soils collected in the Zimapán basin had a decreasing trend with distance from pollution sources (tailings and slag heaps) ranging from 14700 to 4 mg/kg. Acid mine drainage, weathering, and eolian transport have polluted nearby soils. Depth profiles revealed also a concentration decrease from the surface. Similarly to tailings, sequential fractionation showed most of the arsenic associated with the Fe and Mn oxyhydroxides fraction (Ongley *et al.*, 2007). Fine sediments (0.063 mm) from the Tolimán River (a non-perennial river), which is the only superficial water body in the Zimapán area, were collected upstream, in front, and downstream tailings settled close to its shore. Clear differences were determined in total As contents depending on the location of samples respect to tailings impoundments. While samples upstream contained up to 126 mg/kg; samples in front or downstream tailings had up to 11,810 mg/kg As. Application of the Geoaccumulation index (Igeo)

$$Igeo = \log 2C/1.5B \quad (3)$$

where C corresponds to the sample concentration and B to the background, classified tailings-influenced sediments as strongly polluted. Calcite, quartz and clays were present in all samples, while pyrite and secondary minerals as jarosite and gypsum were identified only in sediments influenced by oxidized tailings. Most of the arsenic was mainly associated with the residual fraction in the upstream non-polluted zone. In sediments impacted by tailings, arsenic was mostly in the residual and Fe oxyhydroxides fractions in the dry season, and in this later one in the rainy season (Espinosa *et al.*, 2009). Results indicate that acid mine drainage releases arsenic but limestone neutralizes acidity and reacts with iron, producing iron oxyhidroxides that retain an important proportion of the As. However, changes in the physico-chemical conditions, mainly decreasing of redox potential and pH may mobilize As from this fraction.

3.4 *Arsenic removal by indigenous limestone*

Arsenic removal capacity of limestones from Zimapán was determined by batch and column experiments. Batch tests were conducted with artificial polluted solution and with native water (0.6 mg/L and 0.5 mg/L As, respectively) added to rock particles collected from various formations outcropping at Zimapán. Samples from the Soyatal formation had the highest As removal efficiency. Experiments allowed determining the rock-water ratio, particle size, reaction and shaking times to obtain the highest decrease of As content (Ongley *et al.*, 2001). Chemical analyses of Soyatal rocks reported 50.89% CaO and mineralogy showed calcite predominance. A point of zero charge between 9.3 and 9.5 was determined by titration curves with HCl and NaOH at various ionic strengths. Sorption experiments with a 20:10 water:solid ratio carried out with a solution containing 4.2 mg/L As showed about 80% removal in the range pH 7.2 to 9.8 and a maximum retention of 98% at pH 11. Arsenic adsorption on calcite or coprecipitation of complex Ca arsenates are able to remove As from the solution (Romero *et al.*, 2004). Column experiments were also developed to determine the best relationship between particle size and removal efficiency. Particles from 0.84 to 1 mm allowed a flow of 0.060 L/min and decreased As concentration from 0.35 to 0.15 mg/L (Armienta and Micete, 2009).

Various rock types from Zimapán have also been used to evaluate their potential as permeable barriers to treat acid mine drainage flowing from Zimapán tailings. In the first step, kinetic experiments were performed with a Fe-rich solution. Concentration of H^+ with time followed:

$$[H^+] = [H^+_0]e^{(-kt)} \quad (4)$$

where $[H^+_0]$ = initial molar concentration, k = decaying kinetic constant (h^{-1}) and t = time (h). The kinetic constant value was used as a criterion to select the best rock type for barrier construction. Experiments with the above Fe-rich solution showed the formation of an iron coat over rock particles that were identified as schwertmannite from XRD and FTIR-ATR spectra. Batch experiments were carried out with synthetic leachates (produced from tailings), and the selected rock showed more than 99% As removal. Iron was precipitated from the solution as ferrihydrite, schwertmannite, and chamosite. Images obtained by SEM-EDS showed the presence of As and Zn on iron phases. In addition, modeling with Visual MINTEQ indicated also the possible formation of arsenates (Labastida *et al.*, 2013).

4 CONCLUSIONS

Detailed geochemical and mineralogical studies carried out at Zimapán allowed to determine the main geochemical processes involved in the release and mobility of As in the environment. Oxidation and dissolution of As-bearing primary minerals liberates As which thereof is partly retained in the deposits through sorption and formation of secondary minerals. Tailings have also polluted soils and river sediments. Most of the As occurs in low-mobility fractions in tailings, soils and sediments. However, total As concentrations and the possibility of changes in the physico-chemical conditions result on an As environmental threat. Abundance of limestones has a main role on As mobilization. Limestone increases the pH of tailings and promotes formation of iron oxyhidroxides that retain As. Besides, limestone also immobilizes As likely by sorption and precipitation of arsenates. Removal and kinetic experiments showed the feasibility of using indigenous limestone to obtain As-safe potable water and to be used to treat acid mine drainage in permeable barriers.

ACKNOWLEDGEMENTS

The authors acknowledge Ramiro Rodríguez, Guadalupe Villaseñor, Olivia Cruz, Nora Ceniceros, Alejandra Aguayo, Lois K. Ongley, Helen Mango, Alison Lathrop, Ondra Sracek, E. Flores-Valverde, and V. Múgica, who have been part of the scientific team studying the arsenic problem at Zimapán. Numerous Mexican and American students are also acknowledged for their important participation in this research.

REFERENCES

Armienta, M.A., Micete, S. &, Flores-Valverde, E. 2009. Feasibility of arsenic removal from contaminated water using indigenous limestone. In J. Bundschuh, M.A. Armienta, P. Birkle, P. Bhattacharya, J. Matschullat and A.B. Mukherjee (eds.), *Natural Arsenic in Groundwaters of Latin America,, Arsenic in the Environment Vol. 1*, 505–510, Boca Raton: CRC Press.

Armienta, M.A., Rodríguez, R., Aguayo, A., Ceniceros, N. & Villaseñor, G. 1997. Arsenic Contamination of Groundwater at Zimapán, México. *Hydrogeology Journal* 5(2): 39–46.

Armienta, M.A., Villaseñor, G., Cruz, O., N. Ceniceros, N., Aguayo, A. & Morton, O. 2012. Geochemical processes and mobilization of toxic metals and metalloids in an As-rich base metal waste pile in Zimapán, Central Mexico, *Applied Geochemistry* 27: 2225–2237.

Armienta, M.A., Villaseñor, G., Rodriguez, R., Ongley, L.K. & Mango, H. 2001. The role of arsenic-bearing rocks in groundwater pollution at Zimapán Valley, México. *Environmental Geology* 40: 571–581, 2001.

Espinosa, E., Armienta, M.A., Cruz, O., Aguayo, A. & Ceniceros, N. 2009. Geochemical distribution of arsenic, cadmium, lead and zinc in river sediments affected by tailings in Zimapán, a historical polymetalic mining zone of México. *Environmental Geology* 58: 1467–1477.

Labastida, I., Armienta, M.A., Lara-Castro, R.H., Aguayo, A., Cruz, O. & Ceniceros, N. 2013. Treatment of mining acidic leachates with indigenous limestone, Zimapan Mexico, *Journal of Hazardous Materials* (In Press). http://dx.doi.org/10.1016/j.jhazmat. 2012.07.006.

Méndez M. & Armienta M.A. 2003. Arsenic Phase Distribution in Zimapán Mine Tailings, Mexico, *Geofísica Internacional* 42: 131–140.

Ongley, L.K., Armienta, M.A., Heggeman, K., Lathrop, A., Mango, H., Miller, W. & Pickelner, S. 2001. Arsenic Removal from contaminated water by the Soyatal Formation, Zimapán Mining District, Mexico-a potential low-cost low-tech remediation system. Geochemistry, Exploration, Environment, Analysis 1: 23–31.

Ongley, L.K., Sherman L., Armienta A., Concilio A. & Ferguson-Salinas C. 2007. Arsenic in the soils of Zimapán, Mexico. *Environmental Pollution* 145: 793–799.

Rodriguez, R., Ramos, J.A. & Armienta, M.A. 2004. Groundwater arsenic variations: The role of local geology and rainfall, *Applied Geochemistry* 19 (2): 245–250.

Romero, F.M., Armienta, M.A. & Carrillo-Chavez, A. 2004. Arsenic Sorption by Carbonate-Rich Aquifer Material, a Control on Arsenic Mobility at Zimapán, Mexico. Archives of Environmental Contamination and Toxicology 47: 1–13.

Romero F.M., Armienta, M.A., Villaseñor, G. & González, J.L. 2006. Mineralogical constraints on the mobility of arsenic in tailings from Zimapán, Hidalgo, Mexico. *International Journal of Environment and Pollution* 26: 23–40.

Sracek, O., Armienta, M.A., Rodríguez, R. & Villaseñor, G. 2010. Discrimination between diffuse and point sources of arsenic at Zimapán, Hidalgo state, Mexico. *Journal of Environmental Monitoring* 12: 329–337.

One Century of the Discovery of Arsenicosis in Latin America (1914–2014) – Litter, Nicolli, Meichtry, Quici, Bundschuh, Bhattacharya & Naidu (Eds)

ISBN 978-1-138-00141-1

Speciation and bioavailability of arsenic in managing health risks for mine site rehabilitation

B.N. Noller
Centre for Mined Land Rehabilitation, The University of Queensland, Brisbane, Queensland, Australia

V. Diacomanolis
National Research Centre for Environmental Toxicology (Entox), The University of Queensland, Brisbane, Queensland, Australia

J.C. Ng
National Research Centre for Environmental Toxicology (Entox), The University of Queensland, Brisbane, Queensland, Australia
CRC for Contamination Assessment and Remediation of the Environment, Mawson Lakes, Adelaide, South Australia, Australia

ABSTRACT: This paper provides an overview of the rationale of using bioaccessibility to predict arsenic bioavailability for risk assessment of rehabilitated mined land and its application to develop site-specific guidelines. The physiologically based extract test (PBET) *in-vitro* bioaccessibility model was used to assess of arsenic bioaccessibility and validated for its potential to predict bioavailability of mine wastes derived from rat *in-vivo* experiments. The alternative *in-vitro* approach, can replace animal *in-vivo* experiments, which simulate human uptake of arsenic. Comparison with bioaccessibility to predict bioavailability of arsenic then enables the development of a risk-based guideline approach for mine wastes.

1 INTRODUCTION

Mining, smelting of non-ferrous metals and burning of fossil fuels are the major sources for anthropogenic arsenic pollution (IPCS, 2001). Arsenic is associated with base metal mineralogy such as in sulfides and most commonly as arsenopyrite. Arsenite (As^{III}) is the dominant form under reducing condition while arsenate (As^{V}) is the stable oxidized form in the environment. Arsenite is more soluble and mobile than arsenate and the more toxic form (ATSDR, 2007).

This paper aims to: (i) review the significance of speciation and bioavailability of arsenic in managing health risks for mine site rehabilitation and show how mine site rehabilitation with arsenic can be better managed.

2 HUMAN HEALTH RISK ASSESSMENT AND METHODOLOGY

In mineral rich countries such as Australia, mine sites are potential sources of arsenic. If not properly managed, mine activities and dispersion of arsenic may cause adverse effects on the environment and human health. Limiting the effects of mineral processing activities on nearby communities is vital for sustainable development in mining and mineral processing to avoid environment and human health impacts from arsenic (ICMM, 2007). The International Council on Mining and Minerals (ICMM, 2003) developed ten principles for sustainable development in the mining and minerals industry. These principles have been adopted in Australia to promote rigorous mine closure programs before mining completion and identify quantitative indicators of rehabilitation success. In particular, Principle 4 of the ICMM's ten principles is about, 'implementing risk management strategies based on valid data and sound science'(MCA, 2005).

To minimize health risks of arsenic to communities from mining activities, good prediction and planning is required, together with well-designed monitoring to detect adverse trends. This can be achieved by following the Australian enHealth's risk assessment approach (Figure 1), (enHealth, 2012) and relevant supporting guidelines, standards and procedures.

The National Environment Protection Measures (NEPMs) give guidelines for soil contamination from arsenic in Australia (NEPC, 2013) (Table 1). If the Health-based Investigation Level

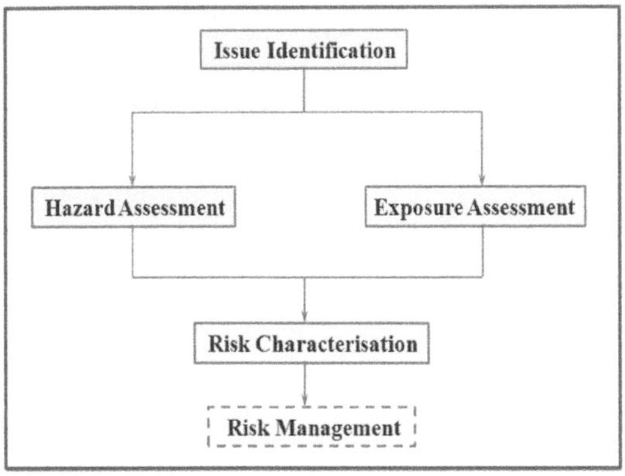

Figure 1. Framework for human health risk assessment (enHealth 2012).

Table 1. Soil contamination health investigation levels in Australia (NEPC, 2013).

Soil contamination health investigation levels			
Residential A (Low-density with garden)	Residential B (High-density minor garden)	Residential C (Open-space recreational)	Residential D (Commercial/ industrial)
Arsenic (mg/kg)			
100	500	300	3000

(HIL) is exceeded, a further Tier II risk assessment of the site needs to be undertaken. In the absence of site specific data for arsenic, the NEPMs may not provide accurate close out criteria for mined land (Ng *et al.*, 2003). Information about the bioavailability of arsenic in contaminated soil can provide this key because, in the absence of site-specific data, bioavailability is usually assumed to be 100% for risk assessment purposes. Commonly, elevated arsenic concentrations in contaminated soils that exceed the HIL values could have adverse health implications.

3 BIOAVAILABILITY AND BIOACCESSIBILITY ASSESSMENT

In-vivo bioavailability of arsenic from contaminated land has been assessed using small mammals including rodents, meadow voles, dogs, pigs, monkeys and more typically guinea pigs or rabbits (NRC, 2003). The rat model has been utilized to estimate arsenic bioavailability and can be applied to provide a risk assessment in calculating the potential exposure route via soil ingestion (Diacomanolis *et al.*, 2007).

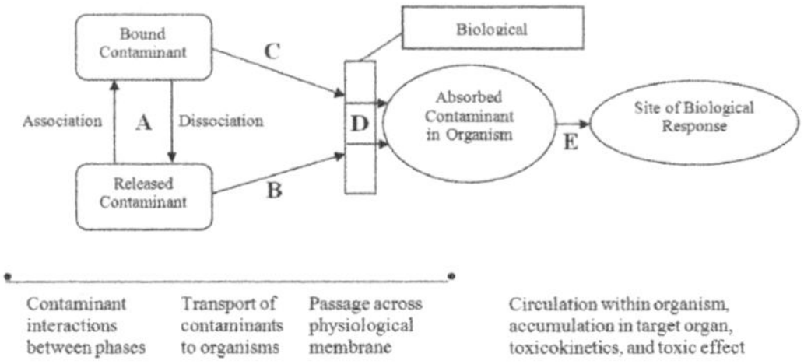

Figure 2. Bioavailability processes in soil or sediment (NRC, 2003).

Table 2. pH setting of PBET gastro-intestinal simulations (Bruce *et al.*, 2007).

	Stomach			Intestinal
Phases	Fasted-solution 1	Mean-solution 2	Fed-solution 3	Small intestine
pH	1.3	2.5	4.0	7.0

Figure 2 shows the processes of bioavailability and includes release of a solid-bound contaminant and subsequent transport, direct contact of a bound contaminant, uptake by passage through a membrane, and incorporation into a living system as illustrated by A, B, C and D, while E denotes the process on which a substance reaches a target site to exert its biological effect (NRC, 2003).

A review of currently available bioaccessibility methods and limitations give general guidance in the NEPM for determining their use and application in contaminant site assessment of arsenic to give prediction of bioavailability (Ng *et al.*, 2013).

Absolute bioavailability is measured via animal uptake but is expensive and time consuming. A more practical approach uses *in-vitro* PBET (physiologically based extraction test) to determine the bioaccessibility of individual soils (Bruce *et al.*, 2007; Ruby *et al.*, 1996).

Comparison of *in-vivo* bioavailability measurement using animal (rat) uptake provides a means to validate the *in-vitro* bioaccessibility measurement. A risk assessment may therefore be conducted on an arsenic survey of surface soils representing different categories of mine wastes by employing the following approaches: (i) *in-vivo* bioavailability measurement of composite wastes using rats (Diacomanolis *et al.*, 2007); and (ii) the *in-vitro* PBET (physiologically based extraction test; Table 2) determination of bioaccessibility of individual soils (Bruce *et al.*, 2007; Ruby *et al.*, 2006).

4 APPLICATION TO MINE SITE REHABILITATION

Bioaccessibility was determined using PBET in 12 randomly selected mine waste soil samples with arsenic concentrations ranging from 102 to 3130 mg/kg. The methods for bioavailability experiments using Sprague-Dawley rats of 7–8 weeks of age and fasted for 5 days prior to experiments in accordance with animal ethics approval (Queensland Health AEC No. 07P05) that included dosage selection, control and treatment dose preparation, mine waste preparation for dosing, protocols for intravenous and oral positive controls and treatment groups (Diacomanolis, 2013).

The dosage for the animal studies was selected based on doses being below arsenic LD50 in order to avoid acute toxicity or mortality. Preparation of 0.5 mg/kg As solution from sodium arsenite and sodium arsenate for As control groups was based on using 0.1 mL As solutions administered using injection while anaesthetized (1:1 mixture of CO_2:O_2 in an open chamber). The injection was administered into the tail vein, after ethanol sterilization, using a 30-gauge (G) needle attached to a 0.5 mL syringe (designed for Insulin injection, BD Micro-Fine™ + 0.5 mL). For mine waste samples dosing, the dosing slurry (mine waste material suspended in 3 mL of Milli-Q water) was administered to the anaesthetized animal via an 8 F rubber feeding tube passed in the mouth and down the esophagus into the stomach. This tube was attached to a 5 mL syringe. Because of the wide range of As concentrations in the different mine waste materials, the rat oral dosing protocol was based on 1 g mine waste for every 200 g of rat body weight. The dosage was then calculated individually, based on 1 g for each mine waste used. The negative control group was administered with an equivalent volume of deionized water. The oral route was selected for the interaction studies, as it is a more realistic scenario for metalloid transfer from mine sites in the real environment. As^V was selected for the interaction experiments, as it is the dominant species found in the oxygenated environment of mine sites (Diacomanolis, 2013).

After dosing, rats were transferred to and kept in individual metabolic cages for 10 days. Rats were bled at the following time points after intravenous injections of control solutions at 0 (pre-dose), 15, 30 and 60 min, followed by 3, 6, 12, 24, 48, 72, 96, 120, 144, 192, 216, and 240 h post-dose. Blood sampling after oral dosing with mine waste were undertaken at 0 (pre-dose), 6, 24, 48, 48, 72, 96, 120, 144, 192, 216, and 240 h post-dose until necropsy 10 days after dosing when, the final blood was collected. Decreasing the blood sample size (one drop ~10 μL) and reagent volume (1 mL) and keeping to the same proportionally-reduced amounts of the reagents enabled the collection of blood samples ethically and at sufficient frequency to collect an appropriate number of samples in order to establish the curve profile for the area under the curve. An alkaline reagent method was used to digest the blood samples taken from rats (Diacomanolis, 2013). Analysis of blood, tissue and urine samples was undertaken using an ICP-MS method (Diacomanolis, 2013).

Absolute bioavailabilities obtained from the blood and urine for arsenic were correlated using linear regression analysis to determine if the urine samples could be used as a surrogate for blood in the determination of absolute bioavailability (ABA). The regression equation showed the following fit that can be used to estimate arsenic bioavailability from both blood and urine:

$$ABA_{Blood} = 0.17ABA_{Urine} + 1.20\ (r^2 = 0.86; p < 0.01)$$

Arsenic bioaccessibility (BAc) in average stomach and intestinal phases in the mine waste samples (Figure 3) ranged 8–36% in the stomach phase and 1–16% in the intestinal phase, with the overall average from both phases of less than 30% (1–27%). No statistical difference ($p > 0.05$) was observed of stomach pH on mean As using ANOVA, suggesting that As solubiity is independent of fasting time, excepting for mine waste sample MIS-SS27 shown in Figure 3. The *t*-test indicated no statistically significant difference between average stomach and intestinal bioaccessibility ($p > 0.05$).

Correlation of bioaccessibility and in-vivo absolute bioavailability of arsenic is given in Figure 4. The linear regressions performed with As in-vivo ABA values for rat blood and all BAc results for the different time intervals and pH states provided information as to which BAc values reflect BA at anyone pH as the best prediction. The statistically significant regression ($p < 0.05$), with the highest goodness of fit value when compared to rats blood, was the pH 7.0 intestinal phase. Hence, the linear equation in Figure 4 can be used to predict

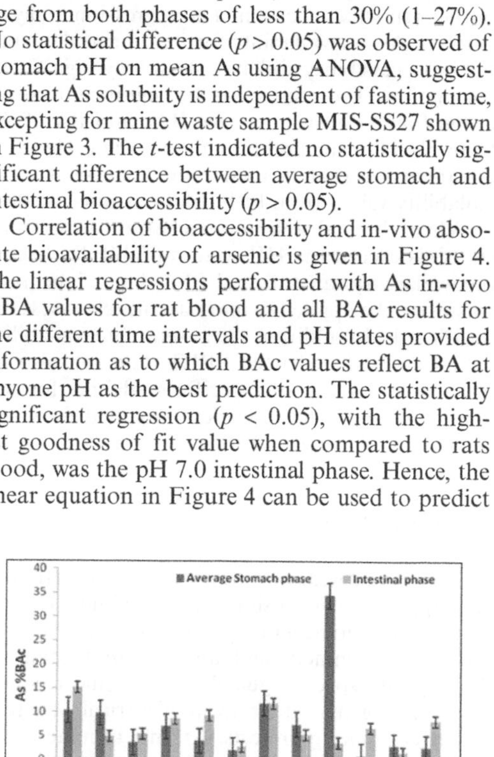

Figure 3. Arsenic bioaccessibility using PBET for average stomach and intestine phases.

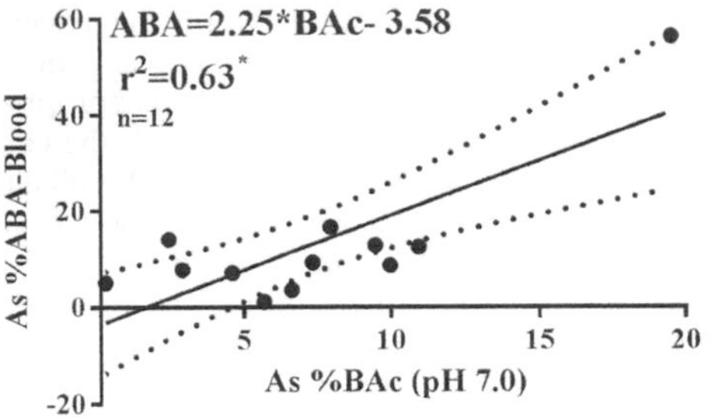

Figure 4. Correlation between bioaccessibility and bioavailability estimates from blood data for mine waste samples. Line of best fit (—) and 95% confidence intervals (----);* statistically significant at $p < 0.05$.

ABA of As for blood from the PBET data. The omission of the sample with the highest level of the As was also tried to see if it would allow more useful correlation analysis; the current figure gave the best correlation. In addition, until a larger collection of mine waste samples are tested whether the higher single point should be considered as an outlier remains uncertain. The regression analysis between As bioavailability from rat urine data and As bioaccessibility did not show any significant correlation. There also was no significant correlation found between average BAc and BA of As from blood, urine and liver tissues data. Hence, a linear equation can be used to predict arsenic bioavailability values from PBET data of mine wastes.

When elevated arsenic concentrations in mine site soils exceed the HIL values (Table 1), arsenic bioavailability values from PBET data of the mine wastes can be applied to more accurately assess the health risk. In addition the arsenic bioavailability values as predicted from PBET data can also be utilized to develop site specific guidelines for soil contamination as has been demonstrated elsewhere (Diacomanolis *et al.*, 2010).

5 CONCLUSIONS

This paper provides an overview of the rationale and application of using bioaccessibility to predict arsenic bioavailability for risk assessment of rehabilitated mined land and its application to develop site specific guidelines. Arsenic residues from mining may arise from historical activities or current mining projects. Arsenic in mine waste can be managed by adopting the procedures within the risk assessment framework. The PBET in-vitro bioaccessibility model (Ruby *et al.*, 1996; Bruce *et al.*, 2007) was used for the assessment of arsenic bioaccessibility and validated for its potential to predict ABA of mine wastes derived from rat *in-vivo* experiments. The alternative *in-vitro* approach, can replace animal *in-vivo* experiments, which simulates human uptake of arsenic.

REFERENCES

ATSDR. 2007. *Toxicological Profile for Arsenic. Atlanta*, U.S. Department of Health and Human Services.

Bruce, S., Noller, B.N., Matanitobua, V. & Ng, J.C. 2007. In-vitro physiologically-based extraction test (PBET) and bioaccessibility of arsenic and lead from various mine waste materials. *Journal of Toxicology and Environmental Health,* Part A. 70: 1700–1711.

Diacomanolis, V. 2013. Interaction of prioritised metals and metalloids in mine wastes with reference to risk assessment PhD Thesis, The University of Queensland, Australia.

Diacomanolis, V., Ng, J., Haymont, R. & Noller, B.N. 2010. Development of site specific guidelines for future land use at the Woodcutters lead zinc mine. *Proceedings of 19th World Congress of Soil Science, Soil Solutions for a Changing World* 1–6 August 2010, Brisbane, Australia. pp. 3.

Diacomanolis, V., Ng, J.C. & Noller, B.N. 2007. Development of mine site close-out criteria for arsenic and lead using a health risk approach" *Proceedings of the Second International Seminar on Mine Closure* 16–19 Santiago, Chile A. Fourie, M. Tibbert & J. Wiertz (eds). pp. 191–198.

enHealth. 2012. *Environmental health risk assessment. Guidelines for assessing human health risks from environmental hazards*. The Environmental Health Committee (enHealth). Australian Government. Canberra.

ICMM. 2003. 10 principles for sustainable development. http://www.icmm.com/our-work/sustainable-development-framework/10-principles. Accessed on 21 January 2010.

ICMM. 2007. Metals environmental risk assessment guidance. International Council of Mining & Metals.

IPCS. 2001. *Environmental health criteria 224: arsenic and arsenic compounds*. Geneva, WHO.

MCA. 2005. *Enduring Value: The Australian Minerals Industry Framework for Sustainable Development.* Minerals Council of Australia.

NEPC. 2013. *National Environmental Protection (Assessment of Site Contamination) Measures*, National Environment Protection Council, Adelaide.

Ng, J.C, Juhasz, A.L., Smith, E. & Naidu, R. 2013. Assessing the bioavailability and bioaccessibility of metals and metalloids.: *Environ Sci. Pollut. Res.*-In press.

Ng, J.C., Noller, B.N. Bruce, S. & Moore. M.R. 2003. Bioavailability of metals and arsenic contaminated sites from cattle dips, mined land and naturally occurring mineralisation origins. Fifth National Workshop on the Assessment of Site Contamination.. A. Langley, M. Gilbey & B. Kennedy (eds). Adelaide, EHPC/enHealth. Adelaide. Pp. 163–181.

NRC. 2003. *Bioavailability of Contaminants in Soils and Sediments: Processes, Tools, and Applications.* Committee on Bioavailability of Contaminants in Soils and Sediments, National Research Council.

Ruby, M.V., Davis, A., Schoof, R., Eberle, S. & Sellstone, C.M. (1996). Estimation of lead and arsenic bioavailability using a physiologically based extraction test. *Environmental Science & Technology* 30(2): 422–430.

One Century of the Discovery of Arsenicosis in Latin America (1914–2014) – Litter, Nicolli, Meichtry, Quici, Bundschuh, Bhattacharya & Naidu (Eds)
© 2014 Taylor & Francis Group, London, ISBN 978-1-138-00141-1

Analysis of factors affecting arsenic speciation in neutral contaminated mine drainage

P. Bauda, T. Risser, S. Devin, B. Sohm, M.A. Dollard, P. Rousselle & C. Pagnout
LIEC, CNRS-UMR 7360, Université de Lorraine, Metz, France

C. Casiot
HSM, CNRS-UMR 5569, Université de Montpellier, Montpellier, France

ABSTRACT: Chemical, physical, climatic and biological variables coming from waters and sediments sampled between 2010 and 2012 in five stations located upstream and downstream the Gabe-Gottes arsenic mine (France) neutral drainage were analyzed together. Principal Component Analysis and Partial Least Square analysis show that arsenite concentration in water correlates with bacterial abundance and arsenic detoxification genes in sediments. Multivariate regression trees reveal that bacterial and archaeal community structures in sediments are respectively explained by sediment iron and arsenic concentrations.

1 INTRODUCTION

In arsenic contaminated aqueous systems, sediments can act as arsenic source or sink, giving rise to strong seasonal variations of arsenic concentrations in surface waters. These variations have been attributed to hydrology, pH and/or bacterial activity. In France, mining of silver-bearing mineralization in hercynian Vosges Mountains (Gabe-Gottes, N 48°12'863, E 007°09'569, Sainte Marie aux Mines, France) have generated punctual high arsenic concentrations in river system downstream. To evaluate the role of bacterial activity in seasonal and space variations of arsenic concentrations and redox status in water phase, water and sediment samples were collected. Physical, chemical and biological variables were statistically analyzed.

2 METHODS/EXPERIMENTAL

2.1 *Sampling and analysis*

Sampling was realized on five stations, from a 2 km long part of a creek upstream and downstream mining area, (December 2010, March 2011, June 2011, September 2011). Sediment samples (2 kg mean samples) were rapidly frozen, water samples (2 L) and sediment interstitial water were filtered through 0.45 μm after centrifugation at 10000 g for 15 min. Sediments were stored at −80 °C for DNA extraction. Water pH and conductivity were measured on site. Mineralization of sediments was performed with the French standard AFNOR NFX 31-151. Organic content was determined from mass loss after 4 hours at 550 °C. Metal and metalloid concentrations were determined after acidification by atomic absorption spectrophotometry, and concentrations of ions by ion chromatography. The total phosphorus, total nitrogen (TKN), chemical oxygen demand, biological oxygen demand of wastewater during decomposition occurring over a 5-day period, and total suspended solids were determined according to French standards (AFNOR, 1994). Arsenic speciation in the water phases was determined by HPLC ICP-MS. Precipitations and flow rate data were considered to describe hydrologic conditions.

2.2 *Microbial community analysis*

PowerMax Soil DNA isolation Kit (MoBio) was used for sediment DNA extraction. Bacterial and archaeal 16S RNA genes were amplified for DGGE. Primers were 341 fGC2 (5′CGCCCGC CGCGCGCGGCGGGCGGGGCGGGGGCA CGGGGGGCCTACGGGAGGCAGCAG 3′) and 907r (5′CCGTCAATTCMTTTGAGTTT3′) for bacterial communities, 934f (5′CCGCCGC GCGGCGGGCGGGGCGGGGGCACGGG GGAATTGGCGGGGGAGCAC 3′) and 1386r (5′GCGGTGTGTGCAAG GAGC 3′) for archaea. PCR program for bacteria was 95 °C 5 min, 20 cycles with 95 °C 30 s, 65 °C 30 s with a 0.5 °C decrement/cycle, 72 °C 45 s, 10 cycles 95 °C 30 s, 55 °C 30 s, 72 °C 45 s, last cycle 72 °C for 7 min.

For archaea, 95 °C 5 min, 14 cycles of 95 °C 30 s, 65 °C 1 min (with a 0.5 °C decrement/cycle), 72 °C 30 s, 20 cycles with 95 °C 30 s, 58 °C 1 min, 72 °C 30 s, last cycle 72 °C for 7 min. 45 to 60% urea was used for archaeal DGGE with 7.5% acrylamide, 40 to 60% urea gradient was used for bacteria with 7% acrylamide. Migration was performed at 60 °C under 100 volts for 16 h. PCR for *arsB*, *ACR3* and *aioA* genes were realized with primers described in Achour *et al.* (2007), and in Heinrich-Salmeron *et al.* (2011).

3 RESULTS AND DISCUSSION

3.1 *Partial Least Square analysis*

We constituted a database informing 69 physical, chemical and biological variables concerning a neutral arsenic mine drainage. PCA were realized on data concerning water and sediment solid phases to identify and eliminate redundant variables. Remaining variables were used in a partial list square with Tanagra software. Arsenic speciation in water was chosen as explained variable. 4 axes explaining 84% of variance were retained. Following the general consensus on PLS interpretation, we focus on variable with VIP values above 1 over the 4 axes. Correlations between those important variables were evaluated through similar trends in model effect loadings. Results indicate that the most important variables to explain arsenic speciation in water are in the water phase: nitrogen sources, phosphates, sulfates, pH, conductivity, hydrology parameters, and in sediment bacterial abundance and arsenic oxidation (*aioA*) and detoxification genes (*ACR3*) abundance. Factorial plane resulting from the PLS analysis shows a strong correlation between arsenite concentration in water and *aioA* and *ACR3* bacterial genes. Correlation is also observed between arsenate concentration in the water column and bacterial abundance in the sediment. Results suggest the implication of sediment bacteria in arsenic speciation in water.

3.2 *Environmental factors explaining the structuration of bacterial and archaeal communities*

Multivariate regression trees were performed to compare microbial fingerprint with environmental factors. Operational taxonomic units obtained by DGGE analysis performed on sediment DNA were exported to R software (Development Core Team, 2011) and compared with the following environmental data: pH, conductivity, arsenite, arsenate, nitrogen sources, sulfates, phosphates, abundance of functional genes, As, Fe and Mn concentrations in sediments. Figure 1 shows that the main factor explaining bacterial community structure is iron concentration in sediment.

The second factor is arsenic concentration in sediment. Iron is probably the main potential electron acceptor for bacteria in the absence of oxygen. Results in Figure 2 indicate that arsenic concentration in sediments is the main factor explaining archaeal community structure in the sediment. Functional roles of archaea are not well understood yet, we have previously shown that thaumarchaeota are well represented in this sediment (Halter *et al.*, 2011). We can say now that arsenic select adapted species among archaeal community.

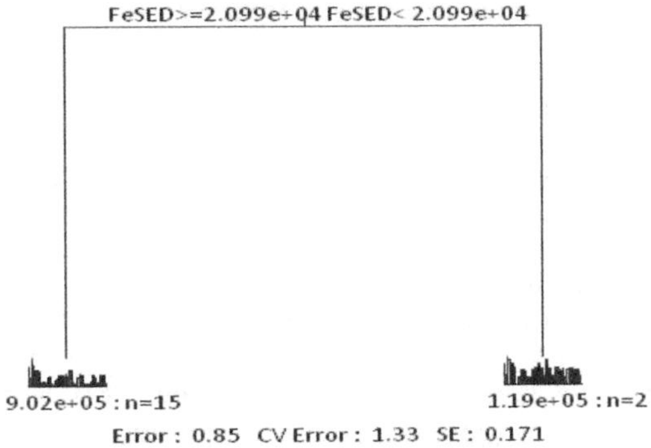

Figure 1. Results of the multivariate regression tree analysis performed on bacteria.

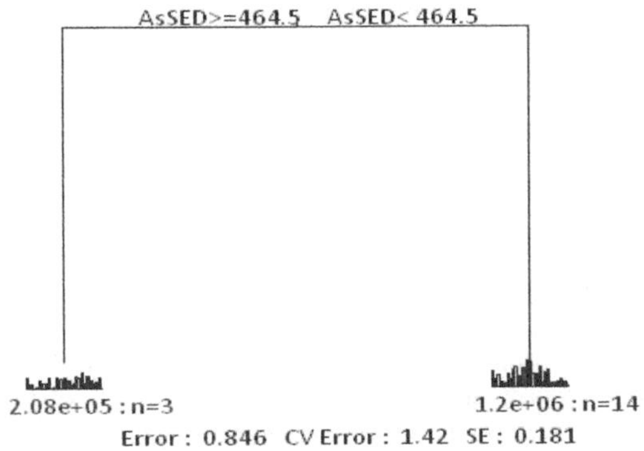

Figure 2. Results of the multivariate regression tree analysis performed on archaea.

4 CONCLUSIONS

Our results indicate that strong variations in arsenic speciation in water from Gabe-Gottes mine drainage have been observed. These variations are correlated with microbial abundance of adapted microbial communities in sediments. Indeed, arsenic detoxification systems influence arsenic speciation. Moreover, both bacterial and archaeal community structures of the sediment are influenced by arsenic concentration in the sediment. Iron concentration in the sediment is the major factor structuring bacterial community in Gabe-Gottes mine drainage sediment.

REFERENCES

Achour, A., Bauda, P. & Billard, P. 2007. Diversity of arsenite transporter genes from arsenic resistant soil bacteria. *Res. Microbiol.* 158(2): 128–137.

Development Core Team. 2011. A language and environment for statistical computing. R Foundation for Statistical Computing, Vienna, Austria. ISBN 3-900051-07-0, URL http://www.R-project.org/

Halter, D., Cordi, C., Gribaldo, S., Gallien, S., Goulhen-Chollet, F., Heinrich-Salmeron, A., Carapito, C., Pagnout, C., Montaut, D., Seby, F., Van Dorsselaer, A., Schaeffer, C., Bertin, P.N., Bauda, P. & Arsène-Ploetze, F. 2011. Taxonomic and functional diversity in mildly arsenic contaminated sediments. *Res. Microbiol.* 162(9): 877–887.

Heinrich-Salmeron, A., Cordi, A. Brochier-Armanet, C., Halter, D., Pagnout, C., Abbaszadeh-fard, E., Montaut, D., Seby, F., Bertin, P.N., Bauda, P. & Arsène-Ploetze, F. 2011. Unsuspected diversity of arsenite-oxidizing bacteria as revealed by widespread distribution of the aoxB gene in prokaryotes. *Appl. Environ. Microbiol.* 77(13): 4685–4692.

One Century of the Discovery of Arsenicosis in Latin America (1914–2014) – Litter, Nicolli, Meichtry, Quici, Bundschuh, Bhattacharya & Naidu (Eds)
© 2014 Taylor & Francis Group, London, ISBN 978-1-138-00141-1

Arsenic extraction from mining tailings: A study case

I. Cano-Rodríguez, A.F. Aguilera-Alvarado, Z. Gamiño-Arroyo, B.E. Rubio-Campos, A. de la Peña-Torres, B. Huerta-Rosas & M.P. Gutiérrez-Valtierra
Departamento de Ingeniería Química, División de Ciencias Naturales y Exactas, Campus Guanajuato, Universidad de Guanajuato, Guanajuato, México

S.H. Soriano-Pérez
Departamento de Ciencias Químicas, Universidad Autónoma de San Luis Potosí, SLP, México

ABSTRACT: Mine tailings represent a generating and rich source of potentially toxic elements. In this work a study was conducted on arsenic extraction in batch systems of samples taken from mine tailings Monte de San Nicolas in the Mining District of Guanajuato, Mexico, using solutions of carbonates and bicarbonates, ammonium acetate, ammonium oxalate, and hydrogen peroxide. Total arsenic, 56 mg/kg, exceeded the value to consider these mining tailings for urban or agricultural use. Arsenic extracted fraction with carbonate solutions as a function of time obtained a maximum concentration of 11%, and released arsenic associated with the oxyhydrate compounds. The release of arsenic associated to sulfide and to carbonate compounds was obtained using ammonium oxalate solution and hydrogen peroxide, respectively; moreover, the greatest amount of released arsenic by these solutions was 80–90%. Therefore, it is possible that arsenic fractions in these mining tailings are mainly associated to sulfide, carbonate, and oxyhydrate compounds.

1 INTRODUCTION

Waste from mines represent a generating source of arsenic and other Potentially Toxic Elements (PTE) (Anawar *et al.*, 2003). Extraction processes generate the release of pollutants, understood as the dissolved mass of a pollutant per dry kilogram of waste (Medel *et al.*, 2008). In the case of mining tailings these pollutants are mainly arsenic, cadmium, lead, nickel, chromium and zinc among others.

Deposits of tailings from the mine of the Monte de San Nicolas of the Mining District of Guanajuato, Mexico, have existed for approximately 70 years of abandonment and presented a varied distribution temporal and spatial concentration of arsenic, cadmium, manganese, and zinc. The concentrations were higher than detected and reported in preliminary work (Rubio-Campos 2010). To explain the extraction and consequent mobility of arsenic from these tailings into aqueous systems, proposes a scenario where the generation of acid drainage and their neutralization by carbonates were carried out when the tailings were active, as previously reported by Ramos-Arroyo & Siebe (2007); simultaneous to it, came the release of arsenic associated to sulfides, among others, and its subsequent adsorption on oxyhydrates of iron or manganese, contained in the same tailings. Therefore, the objective of this work is to determine the effect of different solutions on arsenic extraction from passive mining waste and its functional group association.

2 EXPERIMENTAL

2.1 *Arsenic extraction from tailings in batch systems by solutions*

Samples were collected in three sites at different depths (5, 20 and 40 cm) from the top of tailings of the mine of Monte de San Nicolas, Guanajuato, Mexico. These samples were processed properly according to Mexican Normativity, NOM-141-SEMARNAT-2003, in order to get a compound and representative sample (DOF, 2004).

To assess the effect of the solutions on arsenic extraction from tailings, 1.6 g of a composed sample previously dried and sieved, were in contact with 40 mL of a solution of 0.5 M sodium carbonate, pH 10.5, sodium bicarbonate, pH 8.1, ammonium acetate 1 M, pH 4.5, ammonium oxalate 0.2 M, pH 3, and hydrogen peroxide (30%)

respectively, in continuous agitation as a function of time. This experiment was conducted in triplicate, and deionized water and acidic water were also used as controls.

Extraction processes were stopped by centrifugation and the concentration of arsenic in solution was determined by Spectroscopy of Atomic Absorption Perkin Elmer, Analyst 100, coupled to a hydride generator (AA-HG).

3 RESULTS AND DISCUSSION

3.1 *Arsenic extraction from tailings in batch systems by solutions*

The concentration of arsenic extracted by sodium carbonate solutions as a function of time is presented in Figure 1.

Arsenic content extracted from tailings using carbonate solutions increased depending on the contact time, getting a maximum concentration of approximately 11% of the total arsenic (56 mg/kg), determined according to US EPA Method 3051 (1994).

Using solutions of ammonium oxalate and hydrogen peroxide, a greater amount of arsenic fraction was obtained (80–90%) indicating that these solutions allow the release of arsenic that is associated to iron oxides and sulfides, respectively; with ammonium acetate solution, the amount of arsenic extracted is considerable; thus, arsenic in these tailings may also be associated to carbonate compounds (Table 1).

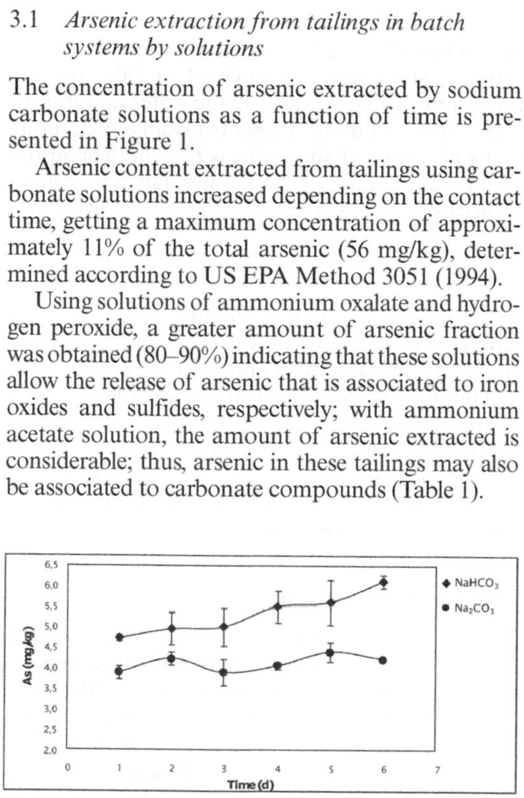

Figure 1. Arsenic from mining tailings extracted by carbonate solutions.

Table 1. Arsenic extracted by solutions.

Solution	%As	Metals associated to
Deionized water	–	Water
Water pH 4	–	Sulfides compounds
Ammonium acetate 1 M	60	Carbonates compounds
Ammonium oxalate 0.2 M	80	Oxides compounds
Hydrogen peroxide 30%	90	Sulfides compounds

4 CONCLUSIONS

The content of total arsenic exceeded the value to consider the study mine tailings for urban or agricultural use. Extraction of arsenic from tailings of the mine of the Monte de San Nicolas was attained 11% by the action of carbonate solutions. This might be explained by its facile and quick desorption as well as its mobility.

ACKNOWLEDGEMENTS

Authors acknowledge to FOMIX-CONACYT 2013 and SEP-PROMEP 2013 *RED DE COLABORACIÓN DE CUERPOS ACADÉMICOS,* projects, for financial support and student scholarships.

REFERENCES

Anawar, H.M., Akai, J., Komaki, K., Teraro, H., Yosioka, T., Ishizuka, T., Safiullah, S. & Kikuo, K. 2003, Geochemical occurrence of arsenic in groundwater of Bangladesh: sources and mobilization processes, *J. Geochem. Explor.*, 77:109–131.

DOF, 2004, Norma Oficial Mexicana NOM-141-SEMARNAT-2003, Establece el procedimiento para caracterizar los jales, así como las especificaciones y criterios para la caracterización y preparación del sitio, proyecto, construcción y operación y postoperación de presas de jales, Secretaría de Medio Ambiente y Recursos Naturales, México, D.F., México.

Medel, A., Ramos, S., Avelar, F.J., Godínez, L.A. & Rodríguez, F. 2008, Caracterización de Jales Mineros y evaluación de su peligrosidad con base en su potencial de lixiviación. Revistas Científicas de América Latina y el Caribe, España y Portugal, 35: 33–35.

Ramos-Arroyo, Y.R. & Siebe, C. 2007, Weathering of sulfide minerals and trace element speciation in tailings of various ages in the Guanajuato mining district, Mexico, Catena, 71(3):497–506.

Rubio-Campos, B.E. 2010, Tesis Doctoral, Departamento de Ingeniería Química, Universidad de Guanajuato, México.

US EPA Method 3051. 1994, Microwave Assisted Acid Digestion of Sediments, Sludges, Soils and Oils.

One Century of the Discovery of Arsenicosis in Latin America (1914–2014) – Litter, Nicolli, Meichtry, Quici, Bundschuh, Bhattacharya & Naidu (Eds)
© 2014 Taylor & Francis Group, London, ISBN 978-1-138-00141-1

Arsenic mobility at alkaline conditions in a gold mine tailings dam

L.C.G. Bissacot
Yamana Gold Corporation, São Paulo, Brazil

V.S.T. Ciminelli
Universidade Federal de Minas Gerais, Minas Gerais, Brazil

ABSTRACT: A mineralogical characterization of several samples from gold mine tailings was performed by X-ray diffraction with Rietveld refinement, indicating a significant presence of sulfides (pyrite, arsenopyrite and pyrrhotite) and carbonates (calcite and dolomite). The results of the water monitoring along the years indicated that the water quality of this reservoir of tailings presents alkaline conditions and some soluble elements, mainly arsenic and sulfate. The objective of this work is to comprehend the mechanisms responsible for the release of arsenic and sulfates in the water of the tailings dam under these alkaline conditions in order to identify the best management practice to prevent arsenic mobility along the operation and closure stages.

1 INTRODUCTION

In the semiarid climate of Brazil, a gold mine is operating since 1984. In 1988, the underground mining of sulfide ores started and, since then, the tailings generated from processing this material (CIP circuit) are being deposited in a tailings dam. A preliminary mineralogical characterization of the tailings was performed by X-ray diffraction indicating a significant presence of sulfides and carbonates. The monitoring results along the years indicated that the tailings dam pond presents alkaline conditions and an increase of As and sulfate concentrations with depth inside the dam lined area. Different from some other trace elements, As can be mobile under neutral to alkaline conditions, depending on the redox conditions of the system (Craw *et al.*, 2003). Since the operation is reaching its closure phase, it is important to understand the mobility of these elements, especially As, in order to propose mitigation actions.

2 METHODS/EXPERIMENTAL

2.1 *Tailings sampling*

Tailings samples were selected for further characterization. The definition of the tailings sampling locations was carried out in joint with the Geology Department. Twenty-one equidistant points every 150 meters were distributed along the tailings area and sampled at different depths. Two fresh CIL tailings samples and one composite sample of precipitated salts were also collected. For the deposited tailings, a cylindrical soil auger with 4.5 meters diameter and 39 cm length was used. The maximum depth achieved is 6 meters. A total of 55 samples of tailings were collected at different depths in the 21 points, totalizing 58 samples sent to the laboratory for further analyses.

2.2 *Tailings characterization*

Several analyzes were performed for understanding the short and long term behavior of these tailings and, especially, the arsenic mobility. The Modified Acid-Base Accounting (MABA) was performed in order to understand the potential of these samples for generating acidy. The Lawrence method (MEND, 1991) was selected for these analyses. The Net Acid Generating test (NAG) was performed as an additional test, according to the procedure described by Stewart (2006) in order to confirm the MABA results. The chemical composition of each sample was conducted by digesting 0.500 g in *aqua regia* at 95 °C for one hour. The extract was then diluted to 10.0 mL and analyzed for metals by combination of OES and MS using Optima 7300 DV for ICP-OES and Elan 9000 for ICP-MS. These results were complemented by mineralogical analyses by X-ray diffraction with Rietveld refinement and kinetic tests in leach columns according to the AMIRA (2002) procedure.

3 RESULTS AND DISCUSSION

3.1 *Acid rock drainage evaluation*

The results of the MABA tests in the tailings indicated an average sulfur concentration of 1.37% while the average sulfide sulfur was 1.10%. This can be related to the partial sulfur content that was oxidized into sulfates. The composite sample collected of the precipitates presented 0.5% of total sulfur on its composition and 0.01% of sulfide sulfur. The average concentration of sulfur can be considered reasonably high, but the average Neutralization Potential (NP) was also elevated, with an average of 106 kg $CaCO_3$ eq/ton. In addition to these results, the NAG tests also indicated an elevated NAGpH for all the tailings tested. According to the NAG test procedure, a sample with a NAGpH greater than 4.5 can be considered as non-acid forming.

The above diagram combines the results of the MABA and NAG tests. The X-axis represents the MABA results as the difference of AP and NP. The Y-axis represents the NAGpH. It is possible to observe that all samples were classified in the non-acid forming area of the diagram.

3.2 *Chemical and mineralogical composition*

The mineralogical characterization by X-ray diffraction with Rietveld was carried out in all the samples. In terms of environmental relevance, the sulfides and carbonates are the most important. The results presented a significant presence of pyrite (1%), arsenopyrite (0.5%) and pyrrhotite (0.7%). Regarding the carbonates, the presence of dolomite (3.7%), calcite (6.1%) and calcite/magnesian (2.9%) was identified. Elevated grades of arsenic, with an average of 3860 mg/kg, were detected in all the 58 samples analyzed, while arsenopyrite was detected only in 34 samples. This can be explained by the limitations of the Rietveld method to identify the arsenopyrite phase under at levels below 1%.

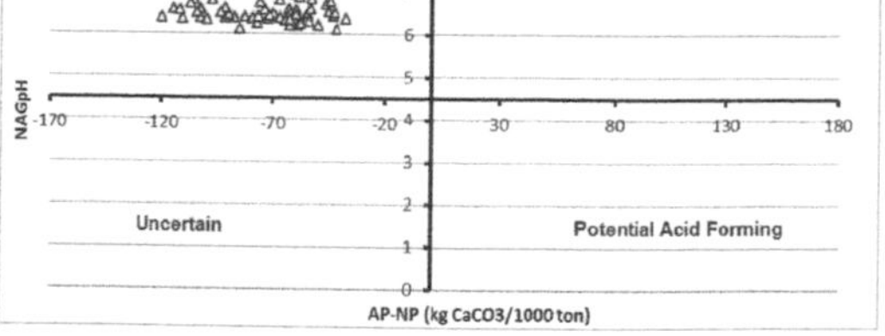

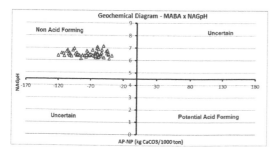

Figure 1. MABA × NAG diagram.

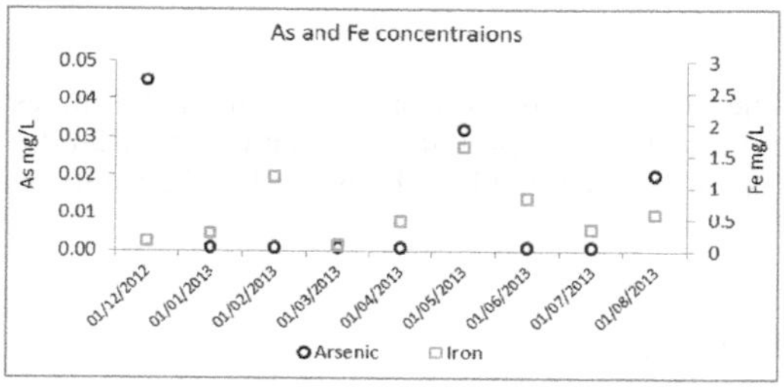

Figure 2. Arsenic and iron leaching after 9 months.

3.3 *Kinetic tests*

The kinetic tests were performed in PVC columns according to the AMIRA 2002 procedure. The results of As and Fe are presented below.

After 9 months of water addition, it is possible to observe that the arsenic is leached from below detection limit to 0.045 mg L^{-1}. The iron concentrations varied from 0.09 to 1.6 mg L^{-1}. The concentrations of arsenic leached in the columns are significantly lower than the concentrations found deep in the tailings reservoir (> 1 mg L^{-1}). It is possible to observe that when iron is present at its highest concentration of 1.6 mg L^{-1}, As concentration was also high. Concentrations of iron of this order of magnitude are mainly in Fe^{2+} form since the Fe^{3+} would precipitate quantitatively at pH 7–8.5. The low concentrations of As in the column can be related to the As co-precipitation with Fe^{3+}, that are stable under the pH-Eh conditions of the other months in the columns. These conditions are likely to be different from those conditions in the tailings pond water, where the redox conditions vary from oxidizing to reducing ones and concentrations of Fe of 5 mg L^{-1} are found at neutral pH. The results highlight the need to control the redox conditions to prevent arsenic mobilization, and have important implications on environmental management in the industrial unit.

4 CONCLUSION

The mineralogical composition is consistent with the MABA results showing alkaline conditions. Arsenic mobility in the tailings water is mostly related to the presence and oxidation of arsenopyrite combined with the redox conditions variation along the tailings deposit (oxidizing to reducing). It is recommended that tailings with potential alkaline conditions and disposed under oxidizing conditions have the redox controlled in order to avoid the long-term arsenic mobility. Iron can be added to the system in order to adsorb and decrease the dissolved arsenic concentrations, but this action should be performed only when it is possible to maintain controlled redox conditions.

ACKNOWLEDGEMENTS

The authors thank the site personnel for samples collection and preparation, and Dr. Claudia Caldeira and Mr. Mark Logsdon for the technical discussions.

REFERENCES

AMIRA International. Ian Wark Research Institute and Environmental Geochemistry *International Pty Ltd ARD Test Handbook*. Australia. 41 p., 2002.

Craw, D., Falconer, D. & Youngson, J.H. 2003. Environmental arsenopyrite stability and dissolution: theory, experiment and field observations. *Chem. Geol.* 199: 71–82.

Mend, 1991. Coastech Research, Acid Rock Drainage Prediction Manual, MEND Project Report 1.16.1b, Ottawa, Ontario.

Stewart, W.A., Miller, S.D. & Smart, R. 2003. Evaluation of the Net Acid Generation (NAG) Test for Assessing the Acid Generating Capacity of Sulfide Minerals. *6th ICARD*. Cairns.

One Century of the Discovery of Arsenicosis in Latin America (1914–2014) –
Litter, Nicolli, Meichtry, Quici, Bundschuh, Bhattacharya & Naidu (Eds)
© 2014 Taylor & Francis Group, London, ISBN 978-1-138-00141-1

Sequential extraction to assess arsenic mobility in sediments affected by acid mine drainage

R.W. Veloso & J.W.V. de Mello
Department of Soils, Federal University of Viçosa, Viçosa-MG, Brazil

S. Glasauer
School of Environmental Sciences, University of Guelph, Guelph, ON, Canada

ABSTRACT: Acid mine drainage is an important environmental problem that may compromise the water quality. Heavy metals and metalloids such as arsenic can be mobilized in acidic water. Soluble arsenic is easily removed from water by sediments, but the stability of arsenic in sediments can be affected by several factors, mainly pH, redox potential (Eh) and Electrical Conductivity (EC). However, these factors can be altered by discharge of acid drainage in water sources. In this study, the stability of As in sediments from creeks surrounding a mining area was evaluated using the sequential extraction. The values of pH were no acidic, but values of EC and Eh suggest the present of salt that could be products of acid drainage. The very poorly crystalline Fe and Al (hydr)oxides fraction may be considered the main responsible for arsenic removal from water.

1 INTRODUCTION

Acid Mine Drainage (AMD) may influence the water quality, as result of oxidation of sulfides. Sediments can remove soluble As from water sources and, therefore, they are important reservoirs of arsenic, but the stability of this element in sediments can be affected by several environmental factors, such as pH, redox potential (Eh) and Electrical Conductivity (EC). Thus, discharges of acid drainage may increase mobilization of heavy metals and metalloids in water.

Although the total content in sediment is important for the evaluation of arsenic contamination, it does not explain its potential mobility in water. Arsenic distribution among the different phases in sediments, as evaluated by sequential extraction procedures, can provide information to predict the potential As mobility. The aim of this work was to evaluate the mobility of arsenic in sediments of creeks on AMD influence.

2 METHODS/EXPERIMENTAL

The study area is a gold deposit located at northwest of Minas Gerais state, Brazil. The location is part of the Bambuí Group, constituted by fine-grained sedimentary rocks. In such rocks, arsenic is associated to quartz boudins in phyllites that contain traces of pyrite, arsenopyrite, pyrrhotite and gold.

Samples of water and sediments were collected in two creeks, labeled as RC and DC, surrounding the mining area, during the dry and wet season, in August of 2012 and February of 2013, respectively. Samples were collected from two points along the creeks. The first points (RC.1 and DC.1) were located close to the mining area for both creeks. The second ones (RC.2 and DC.2) were located downstream. Between these points, there were dikes with limestone drains for both, RC and DC creeks. The drains are supposed to neutralize eventual acid water discharges.

Electrical Conductivity (EC), pH and Eh were measured on site using a portable multi-probe (Hanna model HI 9828). Around 100 mL of water samples were filtered through 0.45-μm filters. The arsenic concentration in water samples was determined by hydride generation atomic fluorescence spectrometry (HG-AFS, PSA 10.055 Millennium Excalibur).

Sequential extraction was performed according to Huang & Kretzschmar (2010) procedure with alterations: (1) arsenic bound to organic matter was extracted by H_2O_2 according to Tessier *et al.* (1979); (2) arsenic associated to sulfides was suppressed; (3) arsenic linked to crystalline Fe and Al (hydr)oxides was extracted by oxalate-ascorbic acid according to Wenzel (2001); and (4) residual As was extracted by acid digestion with HCl, HNO_3 and HF. As concentrations were measured using Graphite Furnace Coupled Atomic Absorption Spectrometry (GFAAS, Perkin-Elmer AAnalyst 800).

3 RESULTS AND DISCUSSION

Values of pH in water ranged from 6.5 to 7.1 for RC creek and from 5.5 to 6.9 for DC. The RC creek showed higher values of EC, 1026 and 694 μS cm^{-1} upstream (RC.1) during dry and wet season, respectively. DC creek showed values ranging from 33 to 195 μS cm^{-1} up stream (DC.1). Those values for RC.1 were higher than the average EC in freshwater rivers (from 200 to 400 μS cm^{-1}). As the pH values do not indicate evidence of acid drainage, the excess of salts in CR.1 can be due to sulfates raised from sulfides oxidation.

In general, redox potential values were low, ranging from −2.7 to 142 mV, indicating slightly reducing conditions. During the dry season, Eh values decreased downstream, but the opposite occurred during wet season.

Soluble arsenic concentrations were higher in samples collected closer to mining areas, except for DC.1 during the dry season. Nevertheless, the difference between upstream and downstream was small for this creek. On the contrary, during the wet season, DC creek presented the highest As concentration, 152 mg L^{-1} at the same point (DC.1), but it decreased more than 10 times downstream. It is noteworthy that such concentration is more than 15 times the standard value for drinking water. This result indicates efficient As removal by sediments and limestone drain. The most efficient As removal from water occurred in wet season.

This efficient As removal by sediments can explain the high arsenic content values for downstream sediment in RC, especially in the dry season, while the arsenic removal by limestone drain in the DC was responsible for the lowest arsenic content in sediments downstream mostly for wet season (Table 1).

Arsenic in upstream samples during dry season was mainly associated with Very Poorly Crystalline (Hydr)oxides (VPCH). This result is relevant

Table 1. As content in sediments* and water for dry and wet season.

	SEASON			
	DRY		WET	
Sample	Sediment mg kg$^{-1\#}$	Water μg/L$^{-1\&}$	Sediment mg kg^{-1}	Water mg L^{-1}
RC.1	2115	36	981	27
RC.2	10279	10	5741	1
DC.1	5214	2.3	8973	152
DC.2	2043	19	588	14

* Total As content in sediment.
\# Detection Limit, DL = 6.8 mg kg^{-1}.
& Detection Limit, DL = 0.30 μg L^{-1}.

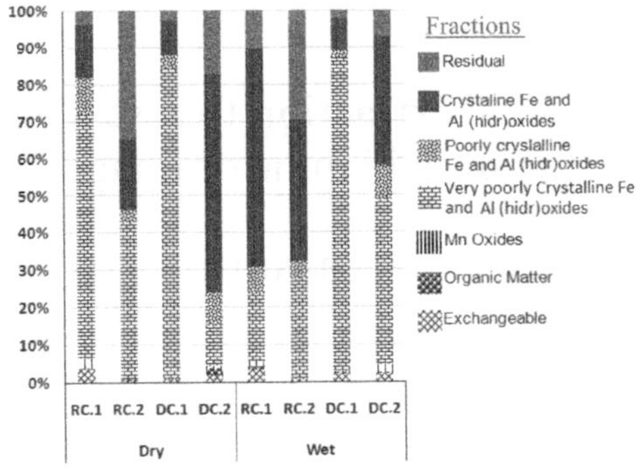

Figure 1. As percentage for 7 fractions in sediments in dry and wet season.

because this fraction can possibly mobilize by acid drainage discharge. The percentage of As associated to VPCH decreased downstream during the dry season for RC samples but, based on As content, there was an increase downstream.

The content of As associated to VPCH increased from dry to wet season for DC upstream samples, from 4321 to 7575 mg kg^{-1} and from 224 to 260 mg kg^{-1} downstream. The opposite was observed for RC creek. This difference may be explained by differences in insoluble arsenic concentrations and water flow in different seasons and creeks. During the wet season, the water flow increases, but it is much higher in RC than in DC, which causes the dragging of Fe (hydr)oxides precipitated. On the other hand, soluble As is much higher in DC than RC in the wet season. Therefore, the co-precipitation of arsenic with VPCH is more likely in DC than RC. Thus, this fraction may be considered the main responsible for arsenic removal from water.

4 CONCLUSIONS

The pH values did not show substantial acidity in water of both creeks. However, the values of EC and Eh, in RC, suggested the presence of salts that could be products of acid drainage. The very poorly crystalline Fe and Al (hydr)oxides fraction may be considered the main responsible for arsenic removal from water for both creeks. The high arsenic content in VPCH fraction was linked directly with soluble arsenic, and with the chance of water flow that may cause the drag of Fe and Al (hidr)oxides precipitates.

ACKNOWLEDGEMENTS

To CNPq, CAPES and FAPEMIG Brazilian agencies.

REFERENCES

Huang, J. H. & Kretzschmar, R. 2010. Sequential extraction method for speciation of arsenate and arsenite in mineral soils. *Anal. Chem.* 82: 5534–5540.

Keon, N.E., Swartz, C.H., Brabander D.J., Harvey C. & Hemond H.F. 2001. Validation of an arsenic sequential extraction method for evaluating mobility in sediments. *Environ. Sci. Technol.* 35: 2778–2784.

Tessier, A., Campbell, P.G.C. & Bisson, M. 1979. Sequential extraction procedure for the speciation of particulate trace metals. *Analytical Chemistry* 51: 844–851.

Wenzel, W.W., Kirchbaumer, N., Prohaska, T., Stingeder, G., Lombi, E. & Adriano, D.C. 2001. Arsenic fractionation in soils using an improved sequential extraction procedure. *Anal. Chim. Acta* 436: 1–15.

One Century of the Discovery of Arsenicosis in Latin America (1914–2014) – Litter, Nicolli, Meichtry, Quici, Bundschuh, Bhattacharya & Naidu (Eds)
 ISBN 978-1-138-00141-1

Arsenic solid speciation in tailings of the abandoned Pan de Azúcar mine, Northwestern Argentina

J. Murray & A. Kirschbaum
Museo de Ciencias Naturales, Instituto de Bio y Geociencias del NOA, Universidad Nacional de Salta, Argentina

M.G. García & L. Borgnino
Centro de Investigaciones en Ciencias de la Tierra (CICTERRA), CONICET and FCEFyN Universidad Nacional de Córdoba, Córdoba, Argentina

E. Mendez Guimaraes
Laboratório de Raios X, Universidade de Brasilia (UnB), Brasil

ABSTRACT: In the Argentina Puna region, there are several metal sulfide mines that ceased their activity twenty years ago without a proper closure plan. The exposure of the waste rocks and tailing impoundments to weathering led to sulfide oxidation and generation of highly acid solutions rich in metal(oid)s, that react with the unaltered waste rocks, partitioning into different solid phases. The As speciation along a 82 cm depth oxidation profile described in the Pan de Azucar mine tailings was studied by standardized sequential extraction procedures. Results show that the highest As concentration remain in the bottom unaltered layer mainly associated with primary and secondary sulfides. In the upper layers, the total As concentrations are ~40 to 70% lower than that of the bottom layer, and most As is associated with amorphous and crystalline oxides. The most bioavailable forms of As (exchangeable and soluble) account for less than 1.5% of the total As concentration.

1 INTRODUCTION

Exposure of metal sulfide mine wastes to the atmosphere, the hydrosphere and the action of microorganisms generates drainage that may be acid and rich in dissolved metal(oid)s and sulfate (Jamieson, 2011; Nordstrom and Alpers, 1999). During weathering, the physical and chemical characteristics of the sediments change over long periods of time. The main process involves the transformation of the original parent sulfide into secondary minerals such as sulfates, oxides, oxyhydroxy-sulfates, etc., and thus releasing potentially toxic elements to water (Lottermoser, 2010). Many of these minerals are fine-grained and have a high capacity for adsorption of potentially toxic metal(oid)s, and therefore can limit the aqueous concentration.

Arsenic is one of the most hazardous contaminants associated with Acid Mine Drainage (AMD). In solution, arsenic occurs as the As(V) species $H_2AsO_4^-$ and $HAsO_4^{2-}$ under acidic to neutral conditions, and as the uncharged As(III) species $H_3AsO_3^0$ that forms under reducing conditions and pH values <8. These species may be partitioned between different phases present in the sediments, such as organic matter, Fe/Al/Mn oxyhydroxides, phyllosilicate minerals, carbonates and sulfides, by different mechanisms (ion exchange, adsorption, precipitation or co-precipitation).

The abandoned Pan de Azúcar district, located in the Argentina Puna (22°32′ S, 66°01′ W–22°38′ S, 66°08′ W; 3700 m a.s.l) was active until 1990 (Figure 1). Pb, Ag, Zn and Sb were extracted from sphalerite and galena ores present in a mineralized hydrothermal vein developed in dacitic rocks and dacitic breccia. Some minoritary arsenopyrite, bournonite and chalcopyrite are also present. Gangue minerals consist of quartz, pyrite and marcasite, as well as some scarce calcite.

The solid mine wastes were stored in piles and tailings impoundments near the mine and abandoned without a proper closure plan. Nowadays, the area shows evidences of advanced acid mine drainage generation and metal release (Kirschbaum *et al.*, 2012 a,b).

The As speciation and mineralogical composition of mine wastes accumulated in a 82 cm depth oxidation profile in the Pan de Azucar tailings was studied in order to identify the As-bearing minerals along the profile and to evaluate its mobility and geochemical behavior during weathering.

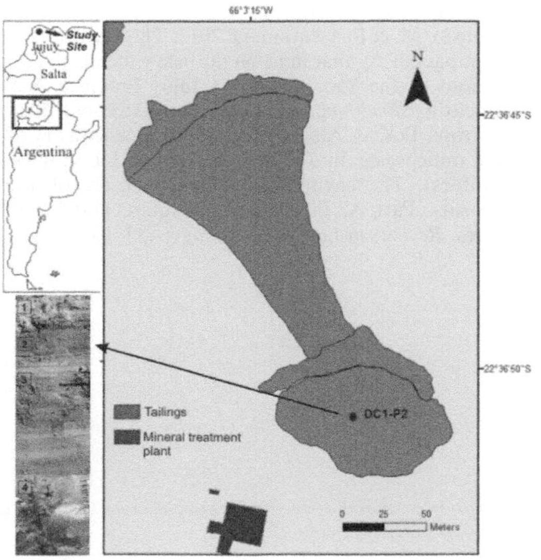

Figure 1. Location of the study area. DC1-P2: profile sampling point.

2 METHODS/EXPERIMENTAL

2.1 *Sampling and mineralogical characterization*

Samples were collected from four oxidation levels described in a 82 cm depth profile (DC1-P2) developed in the tailing impoundment of Pan de Azucar mine (Figure 1). After collection, samples were air-dried and stored in plastic zip-lock bags until analysis. The mineralogy was determined by XR Difractometry and chalcographic microscopy.

2.2 *Bulk chemical composition and sequential extraction*

The bulk chemical composition was determined by ICP/OES after lithium metaborate/tetraborate fusion.

Besides, the 7-steps sequential extraction procedure proposed by Dold (2003) was followed in order to identify the main As-bearing phases. After each extraction, samples were centrifuged at 4000 rpm for 15 min. The obtained supernatants were filtered using 0.22 μm cellulose membranes and the chemical composition was determined by ICP-OES.

3 RESULTS AND DISCUSSION

3.1 *Mineralogical and chemical composition*

The highest total As concentration was determined in the bottom unaltered layer while in the upper layers, the total As concentrations are ~40 to 70% lower.

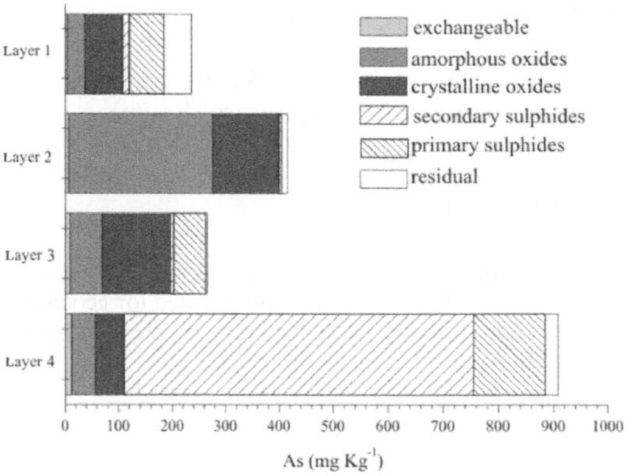

Figure 2. Distribution of As-bearing phases along the studied profile.

The bottom layer (37–82 cm) is composed of pyrite, marcasite, sphalerite, galene, arsenopyrite, quartz and albite. Kaolinite and illite are also present as well as some gypsum eflorescences. In the upper layers (0–37 cm), gypsum, jarosite, alunite, and basaluminite were detected. Primary minerals are quartz, albite, and illite. Amorphous Fe-(hydr)oxides are abundant between 5–37 cm depth.

3.2 *Sequential extractions*

The proportion of different As-bearing phases along the studied profile is shown in Figure 2. In the upper levels, As is mainly associated with amorphous and crystalline Fe oxides, while in the base of the profile, where the rock wastes remain almost unaltered, As is mostly associated with primary and secondary sulfide minerals and organic matter.

4 CONCLUSIONS

Arsenic partitioning was consistent with the mineralogy of the rock wastes. The highest As concentration was determined in the bottom unaltered layer, where As is likely included into the lattice of primary and secondary sulfides. In the upper layers, the most relevant association of arsenic occurs with iron oxyhydroxides. The most bioavailable forms of As (i.e., exchangeable and soluble) account for less than 1.5% of the total As concentration. The release of adsorbed As from the upper layers will be favored under alkaline conditions while structural As in the bottom layer will be released under oxidizing conditions.

ACKNOWLEDGEMENTS

To CONICET, UNC, UNSa (Argentina) and UnB-IG Institute, XR laboratory (Brazil).

REFERENCES

Dold, B. 2003. Speciation of the most soluble phases in a sequential extraction procedure adapted for geochemical studies of copper sulfide mine waste. *Journal of Geochemical Exploration* 80: 55–68.

Kirschbaum, A., Murray, J., López, E., Equiza, A., Arnosio, M. & Boaventura, G. 2012. The environmental impact of Aguilar mine on the heavy metal concentrations of the Yacoraite River, Jujuy Province, NW Argentina. *Environmental Earth Science* 65: 493–504.

Nordstrom, D.K. & Alpers, C.N. 1999. Geochemistry of acid mine waste. In: GS. Plumlee and M.J. Londgson (Editors), The environmental geochemistry of ore deposits. Part A: Processes, techniques, and health issues. *Reviews in Economic Geology*: 133–160.

One Century of the Discovery of Arsenicosis in Latin America (1914–2014) – Litter, Nicolli, Meichtry, Quici, Bundschuh, Bhattacharya & Naidu (Eds)
 ISBN 978-1-138-00141-1

Arsenic spatial distribution and transport at the Allier River (France): The case of an upstream mining impacted watershed

A. Courtin-Nomade, S. Ghorbel, O. Rakotoarisoa & H. Bril
University of Limoges, GRESE, E.A. 4330, Limoges Cedex, France

C. Grosbois
University F. Rabelais of Tours, EA 6293 GéHCo., Tours, France

ABSTRACT: Spatial distribution and transport of As and associated metals were studied in tailings and sediments of a tributary of the Allier River. Because dams are emplaced on this tributary, downstream three mining area (Ag-Pb), this study aims to evaluate if As and Pb may be transported and stored in the sediments of the lake reservoir submitted to occasional flush event.

1 INTRODUCTION

The Allier River is the main tributary of the largest basin of France (the Loire basin). The Allier River is affected by metallic contamination upstream due to the numerous former mining districts and subsequent metallurgical activities in addition to several dams that induce a non-linearity of the stream. The aims of this study are to identify the sources of polymetallic contamination, especially arsenic, and understand the spatial distribution and transport of these elements all along the stream. For this purpose, waste materials from former mining areas (source) and bed sediments (vector and repository) were studied. The objective is to evaluate the released metal content from mining site and metal transport (particulate or dissolved fraction) towards the Allier River and downstream locations.

2 MATERIALS AND EXPERIMENTAL METHODS

2.1 *Site description and materials*

The following results are from the study of the Sioule River watershed, an upstream tributary of the Allier River (Figure 1) where bed sediments were collected. Two hydroelectrical dams are set up along the Sioule River. Because they are located downstream mining areas (extraction of Ag-Pb, Pontgibaud district—PGT), it is of great concern to evaluate how these sediments in the lake reservoir may store metals and how they may also be a source of metallic contamination in case of occasional flush event. The studied waste materials (mainly tailings) originate from the three main mining areas (I, II and III). Each mining area contains several thousand km^3 of material.

2.2 *Experimental methods*

Chemical composition was obtained by ICP after an acid digestion protocol (lithium metaborate/tetraborate fusion, nitric acid digestion plus digestion in Aqua Regia, ACME Laboratories). Mineralogical content was determined using conventional analytical techniques such as X-ray diffraction (XRD), Scanning Electron Microscopy (SEM), Electron Probe Micro-Analysis (EPMA) and micro-Raman spectroscopy (µRS). Leaching experiment results on tailings from the site (I) are presented here to illustrate the impact of former mining activities on the sediments quality of the Sioule River sediments. Three types of waste were chosen with various granulometry (6R: clay; 8R: hardpan—clay and sand mixed; 10R: sand). Stability of the waste materials was evaluated by leaching experiment lasting between 1 day to 1 year. The one-day test was performed according to the normalized protocol EN 12457-2 (solid/solution ratio of 10 l/kg). For longer duration experiments, four different pH leaching conditions were studied:

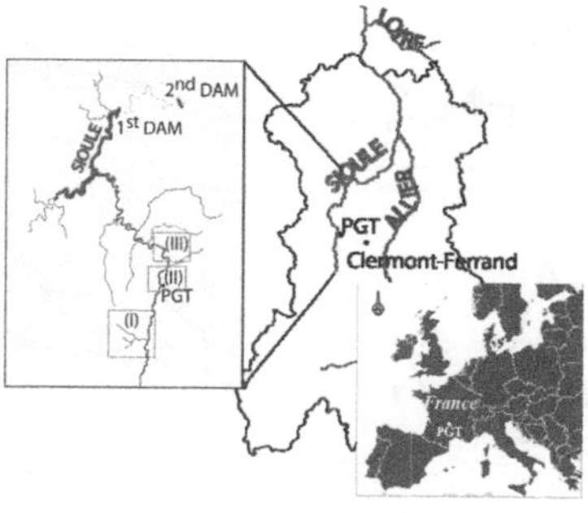

Figure 1. General map of part of the French Massif Central and detailed location of the studied area (PGT) showing the three mining areas along the Sioule River.

2, 5.6, 8 and 12. Only results at pH 5 and 8 will be presented here corresponding to the current pH values of the Sioule River (6.7 < pH < 7.6).

3 RESULTS AND DISCUSSION

3.1 *Mining wastes: Chemistry and mineralogy*

Tailings are mainly silty to sandy materials. Chemical content of mining materials is given in Table 1.

Mineralogy of the waste from the site (I) is dominated by sulfates ($PbSO_4$ or $PbFe_6(SO_4)_4.OH_{12}$). Other secondary phases are identified as oxides (PbO and Fe oxyhydroxides) and phosphates (*e.g.*, $PbFe_3^{3+}(PO_4)_2(OH,H_2O)_6$). Arsenic is preferentially associated to plumbojarosite and to a lesser extent to iron oxides (*e.g.*, goethite).

3.2 *Mining wastes stability over time*

Leaching experiment results from the mining area (I) are presented in Figure 2.

Whatever the pH, the highest concentrations of As (~0.2%) are obtained at the beginning of the experiment (1 day) for the finest samples with no clear relationship between As and Fe. On the contrary, concentrations of S_{tot} and Pb remain high during all the experiments (respectively up to 50% at pH 8 and up to 10% at pH 5.6 of the total content) whatever the granulometry, roughly showing similar trends.

Table 1. Bulk chemical composition (min-max.) of the tailings determined by ICP-MS.

	(I)	(II)	(III)
Pb (%)	1.2–6.6	1.2–3.6	1.1–3.4
As (mg/kg)	443– 8080	600–1335	933–9585
Fe (%)	0.7–8.3	1.0–2.3	1.3–7.1
S_{tot} (%)	0.2–2.3	0.4–1.2	0.4–3.2

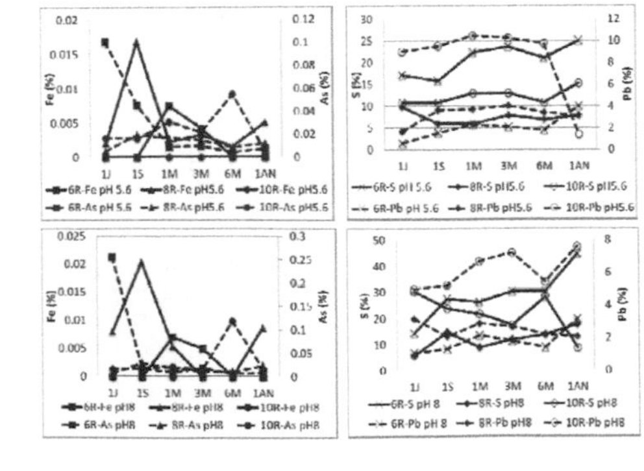

Figure 2. Amount in (%) of As, Fe, Pb and S_{tot} released during a year at pH 5.6 and 8 for three samples presenting different granulometry (from the coarsest to the finest 10R > 8R > 6R).

Table 2. Bulk chemical composition of sediments determined by ICP-MS along the Sioule River.

	As mg/kg	Pb	Fe %	S_{tot}
(I) upstream	54	21	2.7	<0.02
(I) downstream	104	342	2.3	0.04
(II) upstream	106	324	2.3	0.03
(II) downstream	112	1349	1.6	0.04
(III) upstream	96	300	1.9	<0.02
(III) downstream	137	1510	1.6	0.09

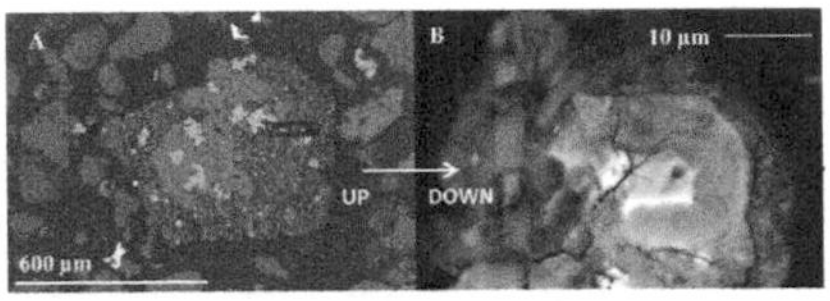

Figure 3. SEM photographs of sediments upstream (A-before site I) and downstream (B-after site III) the Sioule River.

3.3 *Sediments: Chemistry, mineralogy and quality*

Sediments were collected before and after the waste dump of each mining area and their chemical compositions are reported in Table 2.

Sediments quality is directly affected by the proximity of the waste dumps (Figure 3) as suggested by detritic As and Pb-bearing phases but authigenic phases were also identified (*e.g.*, framboidal pyrite). The mineralogy of As and Pb-rich phases is close to the one reported for the tailings plus silicates and clay minerals containing no (or little) As or Pb.

4 CONCLUSIONS

Tailing materials constitute a non-negligible part of the Sioule River sediments. Leaching experiments forecast the stability of these materials and show a higher mobility of S_{tot} and Pb (solid and dissolved transport) comparing to As and Fe (re-precipitation). Solid transport thus mainly explain the As enrichment of sediments downstream the mining site.

ACKNOWLEDGEMENTS

Financial support is provided by the AELB and the FEDER-EPL.

REFERENCE

EN 12457-2, 2002. Characterisation of Waste–Leaching–Compliance Test for Leaching of Granular Waste Materials and Sludges. Part 2. One Stage Batch Test as a Liquid to Solid Ratio of 10 for Materials with Particle Size Below 4 mm (Without or With Size Reduction). AFNOR Norm.

One Century of the Discovery of Arsenicosis in Latin America (1914–2014) – Litter, Nicolli, Meichtry, Quici, Bundschuh, Bhattacharya & Naidu (Eds)
© 2014 Taylor & Francis Group, London, ISBN 978-1-138-00141-1

Geochemical processes controlling mobilization of arsenic and Trace Elements in shallow aquifers in mining regions, Bolivian Altiplano

O.E. Ramos Ramos, I. Quino & J. Quintanilla
Instituto de Investigaciones Químicas (IIQ), Universidad Mayor de San Andrés, La Paz, Bolivia

P. Bhattacharya & J. Bundschuh
KTH-International Groundwater Arsenic Research Group, Department of Sustainable Development, Environmental Science and Engineering, KTH Royal Institute of Technology, Stockholm, Sweden

ABSTRACT: A geochemical approach was applied to understand the factors controlling the mobilization of As and trace elements (TEs) in these mining areas. A total of 30 samples (wells and geothermal water) were collected during the rainy season (2009). As, Cd and Mn concentrations exceed WHO drinking guidelines in some groundwater samples, but Cu, Ni, Pb and Zn do not. Factor analysis of the groundwater chemical data suggests the following geochemical processes: i) plagioclase weathering, ii) dissolution of gypsum and halite, iii) trace element mobilization at acidic pH, iv) sulfide oxidation, and v) release of As following competition with phosphate and bicarbonate for adsorption sites. The As and TEs mobilized in these regions could affect the local water sources, which is a prevalent concern with respect to water resource management in this semi-arid Altiplano region.

1 INTRODUCTION

In the Oruro region located in the Bolivian Altiplano, mineral resources have been exploited since colonial times to present. This has caused environmental impacts such as Acid Rock Drainage (ARD) and Acid Mine Drainage (AMD), which affect the local surface water and groundwater. The objectives of this study are to: i) investigate the extent of As and Trace Elements (TEs) contamination in groundwater in Antequera and Poopó sub-basins, ii) understand the factors and/or processes that cause mobilization of As and TEs in groundwater and iii) identify natural and anthropogenic influences.

2 METHODS

A total of 30 samples were collected during the rainy season (February 2009). Groundwater and thermal water samples were collected from the Antequera-Pazña (GW-1–18) and Poopó (GW-19–30) sub-basins from hand dug wells up to a maximum depth of 20 m and surging sources of thermal water (Figure 1).

Measurements of pH, Eh, EC, alkalinity and TDS were carried out in situ using a Hach portable meter (sensION1) during the sampling fieldwork. TEs and As concentrations in water samples were determined using ICP-OES at Stockholm University, Sweden. The precision and accuracy of all analyses were tested after every 10 samples. Certified standard (NIST 1643e) and synthetic multi-element chemical standards were used and followed the internal quality control of the laboratories.

3 RESULTS AND DISCUSSION

3.1 *Groundwater chemistry*

The field measurements show a pH range for the Antequera sites of 6.6–9.5 (median 8.0) and for the Poopó sites of 7.2–9.7 (median 8.4) in the groundwater samples. The Eh values in groundwater ranged from −86.6 to 203.7 mV (median 64.5 mV).

3.2 *Arsenic and trace elements in groundwater*

In Antequera sub-basin, 89% of the samples exceeded the safe drinking water guideline value (WHO, 2011) and the Bolivian limits (NB-512, 2004) and reached a maximum concentration of 364 μg/L As (GW-13). In the Poopó sub-basin, all the samples exceed the WHO and the Bolivian limits, reaching high concentration of 104.4 μg/L As (GW-19). Groundwater with high As concentrations were found in the plains of the Antequera sub-basin (GW1–4), far away from mining sites. This suggests that high As concentrations are a natural phenomenon.

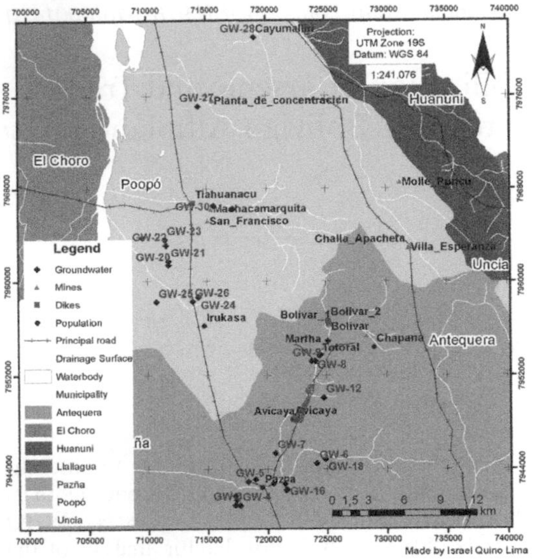

Figure 1. Location of sampling points in study area (dark grey zone) Antequera-Pazña and (light grey zone) Poopó sub-basins.

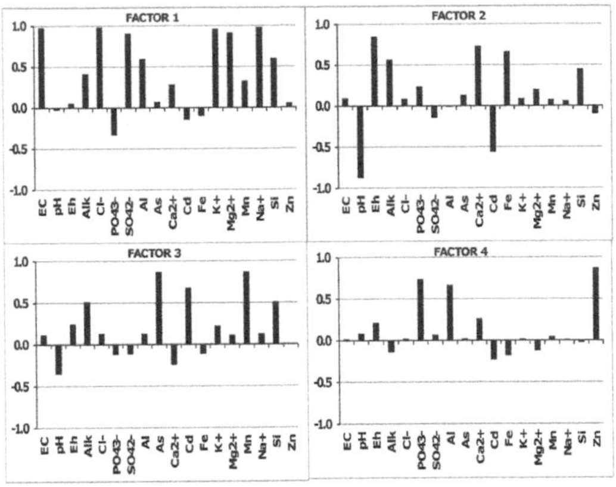

Figure 2. Factor loadings for different geochemical parameters in Poopó groundwater. Loading values of ≥±0.5 represent a significant contribution of the corresponding parameter.

3.3 *Factors responsible for geochemical evolution of groundwater*

Five factors explained 86.7% of the total variance in the geochemical data set from Antequera groundwater samples. Factor 1 shows a positive loading for EC, alkalinity, Cl^-, K^+, Na^+ and Si, which i) account for most of the TDS, representing the dissolution of evaporites, and ii) suggests the weathering of plagioclase minerals as well as carbonates. Factor 2 has a positive loading for Eh, Al, and Fe and a negative loading with respect to pH. The negative correlation of Al and Fe with respect to pH reflects the pH dependency of mineral solubility. Factor 3 has a positive loading for SO_4^{2-}, Ca^{2+}, Mg^{2+} and Mn, and may represent gypsum dissolution, which releases Ca^{2+} and SO_4^{2-} (leaching of salty sediments of Quaternary lakes deposited in drier climate between 8–1.5 kyr BP), and also oxidation of sulfides. Factor 4 has positive loadings for Cd and Zn, which indicates the mobilization of TEs from Hydrous Ferric Oxide (HFO) surfaces. Factor 5 has positive loadings for PO_4^{3-} and As, which is probably linked to competing adsorption of As and PO_4^{3-}, where high PO_4^{3-} concentrations in groundwater effectively de-sorb As(V), thereby increasing dissolved As concentrations.

In Poopó groundwater samples, four factors explained 83.9% of the total variance (Figure 2). Factor 1 shows a positive loading for EC, Cl^-, SO_4^{2-}, Al, K^+, Mg^{2+}, Na^+ and Si, which again represents TDS and the dissolution of primary paleo-evaporite deposits, and suggests the weathering of plagioclase minerals. Factor 2 has positive loadings for Eh, alkalinity, Ca^{2+}, and Fe and negative loadings for pH and Cd. This may represent several geochemical processes: i) the dissolution of calcite minerals and the increase of Fe solubility at low pH, ii) the precipitation of Cd as otavite ($CdCO_3$, e.g., Rötting *et al.*, 2006), and iii) the negative thermodynamic correlation between pH and Eh. Factor 3 has positive loadings for alkalinity, As, Cd, Mn and Si, which suggests that dissolved HCO_3^- and Si may act as a competitor for As at the adsorption sites of Mn and Fe hydroxides, thereby increasing dissolved As in groundwater. Factor 4 has positive loadings for PO_4^{3-}, Al and Zn, which indicates dissolution of gibbsite and mobilization of TEs.

4 CONCLUSIONS

There are four processes that could cause the mobilization of As and TEs in the studied areas: i) desorption from HFO surfaces in samples that are associated with oxidizing conditions, which also implies the mobilization of other oxyanions such as phosphates and bicarbonates, ii) the reductive dissolution of hydrous Mn and Fe oxides and the release of As, which could explain the high As concentrations in sub-surface waters, iii) increased TEs concentrations at acidic pH, and iv) the oxidation of sulfide minerals in oxidizing zones. The high As concentration in groundwater in the alluvial fans of the Antequera-Pazña sub-basin suggests that the characteristics of the sediments are significant in some local areas. The buffer capacity of the surface water in the Poopó sub-basin is higher than for the Antequera sub-basin and seems to be controlling the mobilization of TEs.

ACKNOWLEDGEMENTS

This research was funded through the Swedish International Development Cooperation Agency in Bolivia (Sida Contribution: 7500707606) and CAMINAR project (INCO-CT-2006-032539) funded by the International Cooperation.

REFERENCES

NB-512. 2004. Reglamento Nacional para el Control de la Calidad del Agua para Consumo Humano. Min. Servicios y Obras Públicas, Vice-Min. Servicios Básicos, 59.

Rötting, T.S., Cama, J., Ayora, C., Cortina, J.-L. & De Pablo, J. 2006. Use of caustic magnesia to remove cadmium, nickel, and cobalt from water in passive treatment systems: column experiments. *Environ. Sci. Technol.* 40 (20): 6438–6443.

WHO, 2011. Guideline for drinking water quality, fourth ed. World Health Organization; Singapore. Geneva.

One Century of the Discovery of Arsenicosis in Latin America (1914–2014) – Litter, Nicolli, Meichtry, Quici, Bundschuh, Bhattacharya & Naidu (Eds)
© 2014 Taylor & Francis Group, London, ISBN 978-1-138-00141-1

Arsenic availability in a contaminated area of Vetas-California gold mining district—Santander, Colombia

D.L. Alonso & E. Castillo
Universidad Nacional de Colombia, Bogotá, Colombia

ABSTRACT: The main purpose of this research is to assess the arsenic (As) availability in stream sediments and sludge from a contaminated area close to the Vetas-California gold mining district, Colombia. Availability was determined by using three single extraction procedures. Mobility (MF) and contamination factor (CF) were also calculated. The results indicate arsenic is not readily available due to its low mobility (1.1–1.2), but, through the CF (5.3–245.3) it was demonstrated that there is a very high arsenic contamination downstream from the mining area, with values between 9.1–419.4 mg kg^{-1} in sediments and 8.2–52.0 μg L^{-1} in water of the Suratá River. Furthermore, an ultrasound extraction method was optimized, which achieved arsenic recovery close to 100%.

1 INTRODUCTION

Since the pre-Columbian Era, artisanal and small-scale mining has been carried out in the area formed by the municipalities of California and Vetas, located in the Santander Department (Colombia). The gold mining through the use of mercury and cyanide generates waste that is dumped into the tributaries of the Suratá River. This river provides 40% of the tributary of the aqueduct in the city of Bucaramanga, which provides drinking water to more than one million inhabitants. Although the gold ores in this area are associated with arsenopyrite (FeAsS) and As is released into the environment as a byproduct of the mining processes (Angel & Fierro, 2012), no action has been taken until now to assess its presence and impact. Therefore, we have begun to research the environmental impact of As released by mining activity on the water quality of the Suratá River. In order to assess the As contamination in areas close to Vetas-California gold mining district, water and sediment from the upper basin of the Suratá River and sludge from mine tailings of adjacent mines to tributaries of this river were analyzed.

Conventional heating procedures for As extraction from the solid phase can produce low recoveries due to As volatility. Therefore, ultrasonic extraction (UE) procedures are emerging as an alternative to develop faster methods, in order to enhance desorption of As (Pumure *et al.*, 2010). In this research, UE method was optimized and used in the assessment of As availability in sediments and sludge and also to calculate the contamination factor (CF).

2 MATERIALS AND METHODS

2.1 *Site description and sampling*

The Vetas-California gold mining district is located in NE of Bucaramanga, Colombia, in the center of Santander Massif (72°53′52″W, 7°21′49″N). Eight samples of water, four stream sediments (SD-1 to SD-4) and two sludges (SL-1 and SL-2) were collected along the upper basin of the Suratá River, in areas affected by acid mine drainage (AMD) as well from unaffected areas, to determine the CF. Total As was determined by ICP-MS (Varian 820MS) in waters and by HG-AAS (Perkin Elmer AAnalyst 300) in solid samples.

2.2 *UE optimization*

UE (Ultrasonik 19H, 44 kHz) was optimized in order to improve recovery of As in sediments. Using a 2^2 factorial design, 8 mL of aqua regia and 50 minutes of sonication were determined as optimal conditions when a 1.0 g of sample is used. The recovery was obtained by analysis of two standard reference materials ("Loamy Clay Soil" CRM 052-050, RTC Corp. and "Port Sediment" MAT-SD-0105, MAT Control Labs).

2.3 *Single-step extraction procedure*

For the determination of the As availability in reference materials, sediments and sludge samples, the single-step extractions were performed (Table 1). After each step the suspension was centrifuged and filtered, the supernatant was separated and stored at 4°C until HG-AAS analysis.

Table 1. Single-step extraction procedures (sample: 1.0 g).

Fraction	Chemical agents	Shaking time
Easily soluble	6 ml HPW*	24 h
Acid soluble	25 mL HCl 1.0 M	20 min
NaOH extractable	30 mL NaOH 0.1 M	17 h

* High purity water (at pH = 5.0 ± 0.2 with HOAc 0.5 M).

The mobility factor (MF) was calculated by the ratio between the total As concentration extracted by UE (C_T) with respect to its fraction acid soluble (partial concentration, C_P), according to Equation 1. If MF < 3, then the mobility is low; 3 ≤ MF ≤ 6, moderate and MF > 6, high mobility. The CF was calculated according to the procedure described by Díaz *et al.* (2011).

$$MF = C_T/(C_T - C_P) \quad (1)$$

3 RESULTS AND DISCUSSION

3.1 *ISO-11466 vs. UE*

Table 2 reports the As content in the samples analyzed by the optimized UE method and standard method of ISO Norm 11466. Avoiding heating over 100 °C, according to ISO-11466, loss of small amounts of As by volatilization is prevented, and values close to the total content of As are obtained. The As recovery by UE ranged 95–118%.

3.2 *Availability of arsenic*

Percentages of As for reference materials and samples analyzed by the three extractants used (Figure 1) were evaluated by comparing the value of As obtained by UE (100% As). Overall As availability in soil reference, sediment SD-2 and sludge

Table 2. Comparison between ISO-11466 vs. UE (n = 5).

	As content (mg kg^{-1}, mean ± SD)	
Sample	ISO-11466	UE
CRM-052-050*	28.7 ± 1.3	37.9 ± 1.1
MAT-SD-0105**	17.1 ± 1.5	22.0 ± 1.1
SD-1	8.4 ± 0.5	9.1 ± 0.2
SD-2	398.9 ± 11.1	419.4 ± 5.6
SD-3	1.1 ± 0.1	1.7 ± 0.1
SD-4	158.8 ± 5.6	167.1 ± 2.9
SL-1	25.6 ± 1.4	30.2 ± 1.0
SL-2	22.6 ± 1.3	26.3 ± 0.7

* Certified value = 33.4 ± 1.2 mg As kg^{-1}.
** Certified value = 21.4 ± 1.1 mg As kg^{-1}.

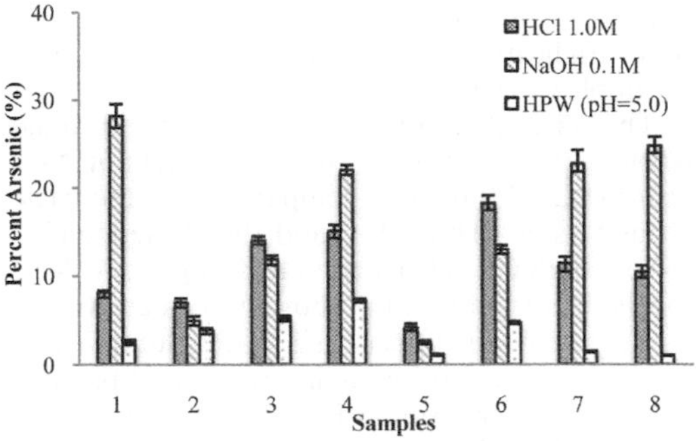

Figure 1. Percent of arsenic with each extractant (n = 5). 1: CRM052-050; 2: MAT-SD-0105; 3: Sediment (SD-1); 4: Sediment (SD-2); 5: Sediment (SD-3); 6: Sediment (SD-4); 7: Sludge (SL-1); 8: Sludge (SL-2).

Table 3. Calculated MF and CF.

	Mobility		Contamination	
Sample	Value	Classification	Value	Classification
SD-1	1.2	Low	5.3	Moderate
SD-2	1.2	Low	245.3	Very high
SD-3*	1.0	Low		
SD-4	1.2	Low	97.7	Very high
SL-1	1.1	Low	17.7	Very high
SL-2	1.1	Low	15.4	Very high

* Unaffected by AMD used to calculate CF.

samples is: easily soluble < acid soluble << NaOH extractable, while in remaining sediments is: easily soluble < NaOH extractable < acid soluble. This indicates that the amount of As easily available (HPW extractable) is minimal compared with the As associated with Ca compounds (acid soluble), iron oxides and partly of organic matter (NaOH extractable).

3.3 *Mobility and contamination factor*

Although the mobility is low we found that the levels of contamination in sediments and sludge is very high (Table 3), showing the impact caused by mining activities and the effect caused on the upper basin of Suratá River. In addition, the contamination is evident when comparing the As content in river water from unaffected areas (0.7–0.8 µg L^{-1}) with waters affected by AMD (8.2–52.0 µg L^{-1}).

4 CONCLUSIONS

The optimized UE method in samples of sediments shows satisfactory results in terms of As recovery

(108 ± 7%) and this can be used to replace extraction with heating and mechanical shaking, in order to simplify analytical procedures.

The As easily available in sediments and sludge is low (1.0–7.2% As), but the high content of As in water (8.2–52.0 μg L^{-1}) compared with the background value (0.8 μg L^{-1}) and the moderate (5.3) to very high CF calculated in the range 15.4–245.3 for sediments and sludge, showed the high As contamination of the upper basin of the Suratá River. The impact caused by mining waste dumped into tributaries of the Suratá River is a potential environmental problem and motive for subsequent investigations.

ACKNOWLEDGEMENTS

We acknowledge the Colombian Geological Survey for the scientific and financial support.

REFERENCES

Angel, A. & Fierro, J. 2012. Análisis y modelamiento del comportamiento de fluidos líquidos de pilas de escombros en minería de oro. Bogotá: Universidad Nacional de Colombia, 5–19. Thesis in Spanish.

Díaz, M., Galindo, M. & Casanueva, M. 2011. Assessment of the metal pollution, potential toxicity and speciation of sediment from Algeciras Bay (South of Spain) using chemometric tools. *Journal of Hazardous Materials* 190: 177–187.

Pumure, I., Renton, J., Smart, R. 2010. Ultrasonic extraction of arsenic and selenium from rocks associated with mountaintop removal/valley fills coal mining: Estimation of bioaccessible concentrations. *Chemosphere* 78: 1295–1300.

One Century of the Discovery of Arsenicosis in Latin America (1914–2014) – Litter, Nicolli, Meichtry, Quici, Bundschuh, Bhattacharya & Naidu (Eds)
 ISBN 978-1-138-00141-1

Groundwater arsenic contamination in Uti Gold Mine Area, Karnataka, India: A potential hazard

M.J. Nandan, B.M.V.S. Sekhar, G.V. Vasavi, P. Harini, T. Balaji & Y. Raghavendra Babu
CSIR-National Geophysical Research Institute, Uppal Road, Hyderabad, India

ABSTRACT: Groundwater contamination with Arsenic (As) in India is of increasing concern due to its severe toxicity. High concentration of arsenic was detected from the bore wells of surrounding villages of Uti gold mine area of Hutti-Muski green stone belt, which is an important gold prospect area in Raichur district of Karnataka. The unscientific dumping of mine tailings, over exploitation of groundwater and excess use of chemical fertilizers by farmers further aggravated the problem in this area. Taking this into concern an attempt has been made to estimate the presence of arsenic in groundwater in Uti gold mine area situated in the Raichur district of Karanataka, India.

1 INTRODUCTION

Arsenic is widely distributed throughout the earth's crust. It is introduced into water through the dissolution of minerals and ores; concentrations in groundwater in some areas are elevated as a result of erosion from surrounding rocks. Occurrence of arsenic in groundwater, exceeding the permissible limits in the Ganga-Brahmaputra basin in India, covering West Bengal, Jharkhand, Bihar, Uttar Pradesh in flood plain of Ganga River and Assam in flood plain of Brahmaputra, is one such large scale groundwater quality disaster, described internationally as the world's biggest natural groundwater calamity to the mankind after Bangladesh. The problem is now evident in the Raichur district of Karnataka, and the people are suffering from arsenic-related skin diseases.

The Precambrian Hutti-Maski Greenstone belt is one of the most important gold deposits, situated in the Raichur district of Karnataka, southern India. The structural setting, mineralization and fluid compositions show many similarities to typical greenstone-hosted gold deposits (Mishra *et al.*, 2005). The Uti gold mine is one of these, and it lies near the northern tip of the schist belt, about 20 km NE of Hutti. The geological set-up of the area indicates arsenic rich sulfide deposits (pyrite, pyrrhotite, and arsenopyrite) and secondary arsenic minerals which may be the primary cause for high arsenic concentrations in the groundwater of surrounding villages (Chakraborti *et al.*, 2002). Taking this into concern, an attempt has been made to estimate the presence of arsenic in groundwater in Uti gold mine area and surrounding villages (Figure 1).

These villages are of significant importance as they are a part of CSIR TECHVIL program covering Raichur TECHVIL of Karnataka.

2 METHODS/EXPERIMENTAL

The study area is located (76.79462^0 N, 16.2714^0 E) in the Raichur district of Southern India, which is traditionally a gold mine area. To estimate the arsenic concentrations in the ground water a network of 27 observation wells (pumping wells, open wells and hand pumps) were established in Palkanmardi, Uti, Wandali, Sunndarkal, Moodalagunda and Madarkal villages, where arsenic problem is persistent.

Water samples were collected in 1,000 mL polyethylene bottles pre-cleaned with double-distilled

Figure 1. Location map of the study area.

water and filtered prior to their analysis in the laboratory. The samples were analyzed by using HACH DR 2800 spectrophotometer and the methodology adopted is silver diethyldithiocarbamate. The test results were obtained by colorimetric measurement at the wavelength of 520 nm.

3 RESULTS AND DISCUSSION

Arsenic concentrations ranging from 0.052–0.145 mg L^{-1} were recorded from 27 bore wells (Figure 2) situated in Uti gold mine area. The highest concentration (0.145 mg L^{-1}) was recorded in Uti village and the lowest concentration (0.052 mg L^{-1}) was recorded in Sunndarkal village. Out of 27 samples, nine recorded arsenic concentrations below the permissible limit. The occurrence of arsenic in the native rocks of Uti gold mine area is believed to be the principal contributor for the increasing arsenic concentrations in the groundwater.

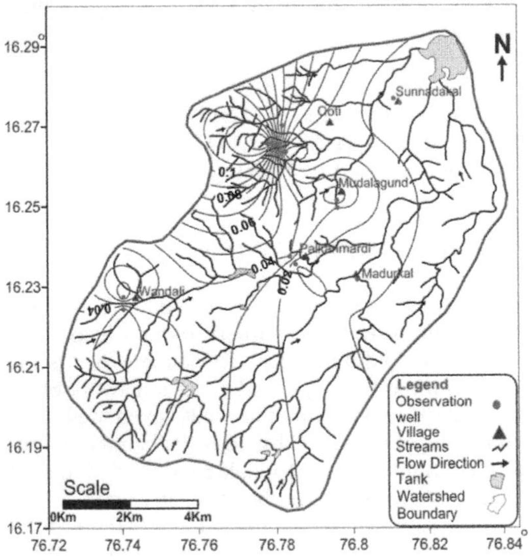

Figure 2. Arsenic concentrations (mg L^{-1}) in the ground water of Uti gold mine area.

4 CONCLUSIONS

The occurrence of elevated concentration of dissolved arsenic in the ground water has put the rural population in and around Uti gold mine area under severe threat. Apart from this, the arsenic groundwater contamination has far-reaching consequences including its ingestion through food chain. The situation is alarming and it will become a future ground water hazard in the state of Karnataka. Preliminary health reports of this area indicates that the people are suffering from severe skin diseases and long term exposure may also lead to different types of cancers. Although the exact sources and mobilization processes of the occurrence of arsenic in groundwater in Uti are yet to be established, the cause is believed to be large scale mining activities in this area.

REFERENCES

Chakraborti, D., Rahman, M.M., Murrill, M., Das, R., Siddayya, Patil, S.G., Sarkar, A., Dadapeer, H.J., Yendigeri, S., Ahmed, R. & Das, K.K. 2002. Environmental arsenic contamination and its health effects in a historic gold mining area of the Mangalur greenstone belt of Northeastern Karnataka, India. *Indian J. Environ. Health.* 44(3): 238–43.

Mishra, B., Pal, N. & Sarbadhikari, A.B. 2005. Fluid inclusion characteristics of the Uti gold deposit, Hutti-Maski greenstone belt, Southern India. *Ore Geology Reviews* 26: 1–16.

One Century of the Discovery of Arsenicosis in Latin America (1914–2014) –
Litter, Nicolli, Meichtry, Quici, Bundschuh, Bhattacharya & Naidu (Eds)
© 2014 Taylor & Francis Group, London, ISBN 978-1-138-00141-1

Diversity and trustworthiness of information sources on environmental contamination and risks of exposure to As in Paracatu, Brazil

F. Mertens, R. Távora & M. Alves
Centro de Desenvolvimento Sustentável, Universidade de Brasília (CDS/UnB), Brazil
Brazil Community of Practice in Ecohealth, Latin America and Caribbean (CoPEH-LAC), Brasília/DF, Brazil

Z.C. Castilhos
Centro de Tecnologia Mineral, Ministério da Ciência, Tecnologia e Inovação (CETEM/MCTI), Rio de Janeiro, Brasil
Brazil Community of Practice in Ecohealth, Latin America and Caribbean (CoPEH-LAC), Brasília/DF, Brazil

ABSTRACT: The largest gold mining activities carried out in Brazil are located in the city of Paracatu, Minas Gerais State. This gold mining may be an important source of arsenic in the environment and might put the population at risk from water consumption and through inhalation. The objective of this study is to analyze the level of awareness of the Paracatu population regarding As issues, as well as to identify the sources of information on the topic and the level of trustworthiness of these sources. Data were collected using face to face interviews, carried out in Amoreira (n = 251) and Paracatuzinho (n = 214), located close and far (few kilometers) away from the gold mining area, respectively. Information sources included Internet, newspapers, television and radio, as well as interpersonal communication. The results have implications regarding the development of communication strategies to promote the diffusion of information on As issues and the implementation of interventions.

1 INTRODUCTION

The largest gold mining activities carried out in Brazil are located in the city of Paracatu, Minas Gerais State. This gold mining may be an important source of Arsenic (As) in the environment and might put the population at risk of exposure from water consumption and through inhalation. In response to these potential threats, an interdisciplinary team of researchers and decision makers teamed up to address As environmental contamination and risk of exposure to the human population of the city. The present study is part of the research project "As environmental contamination and human health risks assessment in Paracatu-MG", which has been coordinated by several Brazilian institutions: CETEM/MCTI, IEC, UFF, UNICAMP/FCM, FIOCRUZ, TECSOMA, CDS/UnB and CoPEH-LAC.

As this interdisciplinary project is aimed at generating a diverse set of results regarding the environmental and health aspects associated to As contamination and exposure, a research component was included in order to guide the development of communication campaigns directed toward the population of Paracatu, in collaboration with local governmental institutions. In order to achieve this goal, it is important to diagnose which information on As issues are available to the population and which are the more efficient channels of diffusion (George *et al.*, 2013). The objective of the present study is to analyze the level of awareness of the Paracatu population regarding As issues, as well as to identify the sources of information on the topic and the level of trustworthiness of these sources.

The expected results have implications regarding the development of communication strategies to promote the diffusion of information (Mertens *et al.*, 2012) on As issues and the implementation of interventions to reduce environmental contamination and risk of exposure.

2 METHODS/EXPERIMENTAL

Data were collected using face to face interviews, carried out in two neighborhoods linked to the Health Family's Assistance Program, Amoreira (n = 251) and Paracatuzinho (n = 214), located close and far (few kilometers) away from the gold mining area, respectively. Specific questionnaires were used to gather information on: a) individuals' knowledge about As environmental sources, toxicity and health effects; b) sources of information on

As issues most accessed by individuals; c) level of trustworthiness of each information source and d) size and composition of personal discussion networks on As issues.

3 RESULTS AND DISCUSSION

In our sample, there is a higher percentage of people aware of the potential health effects associated to environmental exposure to As in the neighborhood of Amoreira (65.7%) than in the one of Paracatuzinho (49.3%). Regarding information sources, overall, 8.4%, 17.9%, 19.9% and 18.3% of the people who were enquired in Amoreira and 6.1%, 11.2%, 15.0%, 18.2% in Paracatuzinho received information on As issues through the Internet, newspapers, television and radio, respectively. The most trusted source of information was the Internet (71.4%) in Amoreira and television (87.5%) in Paracatuzinho.

Furthermore, the mapping of personal discussion networks on environmental and health aspects associated to As issues revealed that a significant percentage of individuals, 38.6% in Amoreira and 17.3% in Paracatuzinho, used to have conversations on these themes with other people, including family members, friends, neighbors, co-workers, members of associations and health agents. Interpersonal relations were considered as the most trusted source of information regarding the issue in the Amoreira district (92.8%), where more people were concerned about the possible health effect associated to As exposure.

4 CONCLUSIONS

Based on the present results, we suggest to consolidate an Internet platform that will present goals, methodology and results of the research and to design television campaigns to communicate the main findings of the project.

The next step will be to further analyze the composition of personal communication network to identify potential opinion leaders on As issues in both neighborhoods. Indeed, the involvement of key individuals in technical workshops to present and debate the research results may have the potential to enhance the diffusion of information regarding As, as a first step toward promoting behavior changes to lower risk of exposure (Valente & Davis, 1999).

These strategies might be carried out in combination and associated with the involvement of local government stakeholders, in order to achieve an increase in local awareness about As environmental contamination and health risks. This approach may also be used as a model for other research projects on mining sites, including abandoned mining sites.

ACKNOWLEDGEMENTS

We thank the UnB′students: Juliana F. de Assis, Lucas V. da Silva, Marcela D. Britto, Marina C.de M. Alves and Sheila L. da Silva for their contribution in collecting field data.

REFERENCES

George C.M., Factor-Litvak, P., Khan, K., Islam, T., Singha, A., Moon-Howard, J. van Geen, A. & Graziano, J.H. 2013 Approaches to Increase As Awareness in Bangladesh an evaluation of an As education program. *Health Education & Behavior* 40: 331–338.

Mertens, F., Saint-Charles, J. & Mergler, D. 2012. Social communication network analysis of the role of participatory research in the adoption of new fish consumption behaviors. *Social Science and Medicine* 75: 643–650.

Valente, T.W. & Davis, R.L. 1999. Accelerating the diffusion of innovations using opinion leaders. *The Annals of the American Academy of Political and Social Science* 566: 55–67.

1.6 Arsenic in soil–plant systems

One Century of the Discovery of Arsenicosis in Latin America (1914–2014) –
Litter, Nicolli, Meichtry, Quici, Bundschuh, Bhattacharya & Naidu (Eds)
 ISBN 978-1-138-00141-1

Arsenic in soils, sediments, and water in an area with high prevalence of chronic kidney disease of unknown etiology

D.L. López
Department of Geological Sciences, Ohio University, Athens, Ohio, USA

A. Ribó, E. Quinteros, R. Mejía, A. López & C. Orantes
Instituto Nacional de Salud, Ministerio de Salud, San Salvador, El Salvador

ABSTRACT: Chronic Kidney Disease of unknown etiology (CKDu) is widespread in Central America and other parts of the world. In El Salvador, the disease seems to be related to agricultural practices and in minor degree to high ambient temperatures. Use of agrochemicals has been proposed to be related to this illness. The environmental phases (surface and groundwater, soils, and sediments) have investigated for nephrotoxic arsenic and heavy metals in an area with high prevalence of the disease (Bajo Lempa). Results show that As is high in soils and groundwater, with agricultural soils having a higher concentration than that of village soils. Possible relationship of this contamination with the prevalence of the illness is suggested as one of possible causes.

1 INTRODUCTION

Central America, Southern Mexico, and other regions of the world have been suffering from chronic kidney disease of unknown etiology (CKDu; e.g. Orantes *et al.*, 2011). Researchers in Sri Lanka have proposed that the origin of the illness is attributed to exposure to agrochemicals and arsenic and heavy metals in soils and groundwater (Jayasumana *et al.*, 2013). In Sri Lanka, the source of arsenic has been found to be fertilizers and pesticides. In Central America, several causes have been suggested for this illness including toxic effects of agrochemicals (e.g. Orantes, 2011) and high ambient temperatures and chronic dehydration (Peraza *et al.*, 2012). In El Salvador (Figure 1), previous research using step-wise multivariate regression found statistically significant correlation between the spatial distribution of the number of sick people per thousand inhabitants and the percent area cultivated with sugar cane, cotton, and beans, and maximum ambient temperature, with sugar cane cultivation as the most significant factor (Vandervort *et al.*, submitted). Ambient temperature has only a minor effect as the regression variable on the incidence of this disease. Spatial distribution of the number of patients per thousand inhabitants for all the stages of the illness, ambient temperature, and the percent of the land cultivated with sugar cane is presented in Figure 2.

In El Salvador, arsenic is abundant in the environment due to the volcanic geology of the region and the presence of several geothermal fields (López *et al.*, 2012). Water contaminated with arsenic has been detected in several regions of the country (Armienta *et al.*, 2008). For that reason,

Figure 1. Location of El Salvador in Central America and study area in the Bajo Lempa region.

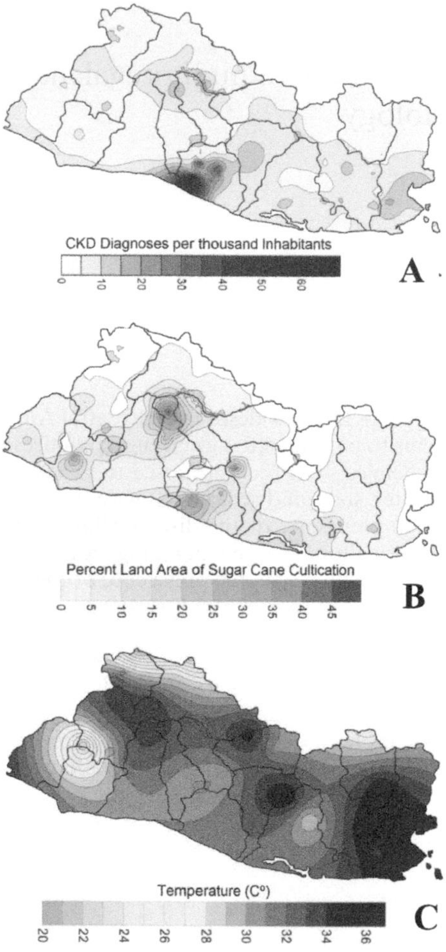

Figure 2. Aerial distribution of A) CKD per thousand inhabitants (2005–2010), B) percentage area cultivated with sugar cane, and C) maximum ambient temperature for El Salvador (after Vandervort *et al.*, submitted).

arsenic can be present in the environment due to natural sources as well as to the use of pesticides and fertilizers.

Arsenic and heavy metals such as cadmium are well recognized nephrotoxic substances that can be present in fertilizers (e.g. Campos, 2002; Chen *et al.*, 2007). For that reason, an environmental study of the concentration of As and heavy metals (Cd, Cu, Cr, Ni, Zn, and Pb) was done in the Bajo Lempa region (Figure 1), which is deeply affected by CKDu. The objective of the study was to verify the presence or not of these contaminants in the different environmental phases, and how their concentration relates to the use of the land.

2 METHODS

Different environmental phases (superficial and ground waters, agricultural and urban soils, and sediments) were sampled for arsenic and heavy metals in the area of Ciudad Romero in Bajo Lempa, Usulután, El Salvador. Sampling was done between October 2012 and March 2013.

2.1 *Water samples*

Water samples were collected from 46 domestic wells located within the Ciudad Romero village and 9 superficial water samples were taken in Lempa River, El Espino stream, and a shrimp farm. Widely accepted protocols for the sampling of surface and groundwaters were used (e.g. USEPA, 1982) within the studied region. Samples were preserved with 5% concentrated nitric acid and transported to the laboratory at around 4 °C.

2.2 *Soil and sediments samples*

For the soil samples, the area of the village where the patients live and the surrounding area where they work were considered. Figure 3 shows the sampling points and the location of the village. For the agricultural soils, composite samples were taken from a depth of 30 cm. For each composite sample, 3 scattered samples were taken in an area of about 30 m^2. For the soils of the village, around 1 kg of superficial soil was collected from an area of 1 m^2. The total number of soil sampling sites was 58 in the village, and 50 in the agricultural land.

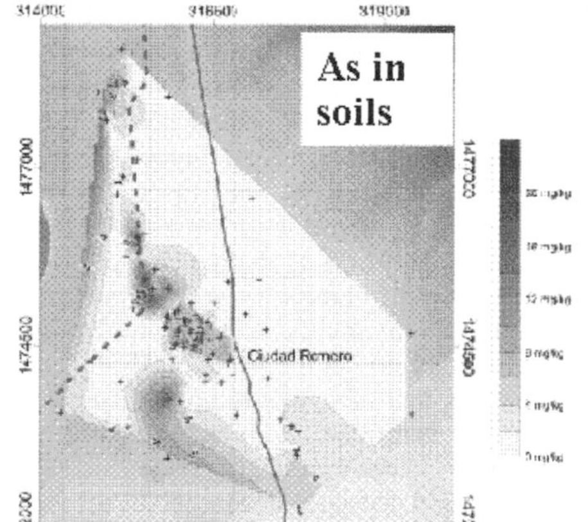

Figure 3. Sampling sites and areal distribution of arsenic in soils of Ciudad Romero and agricultural soils surrounding the village. The gray shaded area represents Ciudad Romero.

Sediments were collected in the same points where superficial water samples were collected. Composite samples were taken from around 10 m^2 to collect around 1 kg of sediment. For the sampling of sediments, a PVC pipe or a VanVeen drag sampler was used.

2.3 *Laboratory work*

The waters, soils, and sediments were analyzed at the National Reference Laboratory of National Institute of Health of the Ministry of Health, El Salvador. Graphite furnace atomic absorption methods were used to analyze the waters with the appropriate standards in an AAnalyst 800 Zeeman equipment. The soils and sediments were first digested using the EPA Method 3050B (USEPA, 1996) to obtain the leachate that was analyzed in a similar way as the waters.

3 RESULTS AND DISCUSSION

3.1 *Water*

For the superficial water samples, only As was detected in all the samples and Cd was detected in the two samples from the shrimp farm (0.016 and 0.009 mg L^{-1}). The USEPA Cd levels for CMC (Criterion Maximum Concentration) and the CCC (Criterion Continuous Concentration) for salty water are 0.040 and 0.0088 mg L^{-1}, respectively (USEPA, 2009). For the groundwaters, only one sample had detectable Cd but a very low level (0.0004 mg L^{-1}), and all the samples presented As contamination. All the other heavy metals were below detection level in the surface waters and only lead was detected in 2 groundwater samples, but the values were well below the EPA standards for drinking water. The As values are presented in Table 1.

Note that the As levels in the groundwater are well above the 0.010 mg L^{-1} USEPA standard for drinking water, with values as high as 0.322 mg L^{-1}.

3.2 *Soil and sediments*

The results for As and heavy metals in soils and sediments are presented in Tables 1 and 2, respectively. Eight samples exceed the Canadian Quality Guidelines for agricultural soils of 12 mg L^{-1} (CCCM, 2002), which suggest that a high fraction of the agricultural soils has too high arsenic contamination and should not be used for agriculture. Figure 3 shows the distribution of As for the agricultural and the village soils. From Table 1 and Figure 3, it is evident that the agricultural soils have a higher As concentration than the village soils, even when both soils should have the same geogenic origin. Statistical comparison of the agricultural and the village soils demonstrates that the agricultural soils have higher As concentration. A parametric Fisher test results in a $p = 10^{-13}$ for the two populations to be equal, and non-parametric Mann-Whtiney test gives $p = 0.0001$.

Table 1. Arsenic concentrations in mg L^{-1} in the surface waters (SF) and groundwater (GW), and in mg/kg for agricultural soils (AS), village soils (VS), and sediments (SE).

	As SW	As GW	As VS	As AS	As SE
Average	0.005	0.068	2.46	3.09	1.23
Standard Dev.	0.003	0.061	1.90	5.37	1.62
Maximum	0.012	0.322	7.10	24.07	5.52
Minimum	0.003	0.008	0.09	0.19	0.43

Table 2. Concentration of heavy metals in sediments, agricultural and village soils in the Ciudad Romero area. Concentrations are in mg kg^{-1}.

	Cu	Cd	Cr	Ni	Pb	Zn
Sediments						
Average	9.85	0.02	2.37	1.48	1.04	20.14
Standard Dev.	5.67	0.01	1.43	0.91	0.57	6.82
Maximum	22.40	0.04	4.99	2.72	2.28	31.46
Minimum	3.70	0.01	0.65	0.37	0.39	8.20
Agricultural soils						
Average	11.08	0.05	5.57	1.42	3.41	21.41
Standard Dev.	7.09	0.03	4.68	0.79	1.91	9.55
Maximum	54.80	0.23	19.97	3.93	8.64	66.69
Minimum	0.40	0.01	0.00	0.01	0.66	1.20
Village soils						
Average	14.01	0.07	20.95	1.16	4.43	63.40
Standard Dev.	4.94	0.07	10.15	0.41	6.06	32.17
Maximum	33.82	0.36	52.72	2.29	47.24	175.89
Minimum	6.68	0.00	1.24	0.64	0.48	1.53
Canadian quality guideline for agricultural soils	64	1.4	63	50	70	200

The results presented in Table 2 show that the contamination of heavy metals in the soils and sediments is below the Canadian Quality Guidelines for agricultural soils (CCCM, 2002). Contamination of heavy metals exists in the soils and sediments of Bajo Lempa, but the concentrations are not a high concern. In comparison, the As contamination in the soils is high with areas not appropriate for agriculture. However, sugar cane and corn are cultivated in the area.

4 CONCLUSIONS

In El Salvador, the prevalence of CKDu seems to be related to agriculture because the illness prevails in men. In Ciudad Romero, 20.7% of the inhabitants had the disease, and in all the Bajo Lempa region 17.9% of the persons are ill (Orantes, 2011), but the prevalence within men is 25.7% and only 11.8% in women. The majority of men work in agriculture. Statistically, the illness seems pretty well correlated to the percentage of area cultivated with several crops (sugar cane, cotton, beans, Vandervort *et al.*, submitted). Pesticides and fertilizers could be introducing to the environment as nephrotoxic chemicals that could combine with the high ambient temperature and other factors to produce the illness. Our work has shown that similarly to Sri Lanka, arsenic is present in anomalous concentrations in soils and groundwater. Arsenic concentrations in soils are so high in some areas that, according to the Canadian Guidelines for agricultural soils, they should not be used for cultivation. This research shows that the agricultural workers and the people living in the area had been exposed to As in soils and in groundwater. Only in the last few years, potable water has been provided to the area. For more than 15 years, since the settlement in the area after 1992, the inhabitants of Bajo Lempa were consuming water high in arsenic.

The origin of arsenic in soils and groundwater can have several sources. One of them could be the obvious geogenic origin due to As present in rocks and geothermal fluids of the region (López *et al.*, 2012). However, the fact that the agricultural soils have statistically significant higher As concentrations than the village soils suggests that As is probably added to the agricultural soils in pesticides or fertilizers. Chemical analysis of the fertilizers and pesticides that are used in the country, as well as in plants, animals, and humans is the obvious next step in this work. This study is only one piece of a comprehensive research that intends to elucidate the factors that could be producing the CKDu in El Salvador and other regions of the world.

ACKNOWLEDGEMENT

We want to thank Lic. Reyna Jovel and the National Reference Laboratory in the Ministry of Health of El Salvador for the chemical analysis of the samples.

REFERENCES

Armienta, M.A., Amat, P.D., Larios, T., & Lopez, D.L. 2008. America Central y Mexico. In J. Bundschuh, A. Pérez Carrera & M.I. Litter (eds*.), Distribucion de Arsenico en las Regiones Iberica e Iberoamericana*, p.p. 187–210. CYTED.

Campos, V. 2002. Arsenic in groundwater affected by phosphate fertilizers at Sao Paulo, Brazil. *Environmental Geology* 42:83–87.

Canadian Council of Ministers of the Environment (CCME), 2002. Canadian sediment quality guidelines for the protection of aquatic life: Summary tables. Updated. In: *Canadian environmental quality guidelines, 1999*. Canadian Council of Ministers of the Environment, Winnipeg. 189 pp.

Chen, W., Chang, A.C. & Wu, L. 2013. Assessing long-term environmental risks of trace elements in phosphate fertilizers. *Ecotoxicology and Environmental Safety* 67(1): 48–58.

Jayasumanam, M.A.C.S., Paranagama, P.A., Amarasinghe, M.D., Wijewardane K.M.R.C, Dahanayake K.S., Fonseca S.I., Rajakaruna, K.D.L.M, Mahamithawa, A.M.P., Samarasinghe, U.D. & Senanayake, V.K., 2013. Possible link of Chronic arsenic toxicity with Chronic Kidney Disease of unknown etiology in Sri Lanka. *Journal of Natural Sciences Research* 3(1): 64–73.

López, D.L., Bundschuh, J., Birkle, P., Armienta, M.A., Cumbal, L., Sracek, O., Cornejo, L. & Ormachea, M. 2012. Arsenic in volcanic geothermal fluids of Latin America. *Science of the Total Environment* 429: 57–75.

O'Donnell, J.K., Tobey, M., Weiner, D.E., Stevens, L.A., Johnson, S., Stringham, P., Cohen, B., & Brooks, D.R. 2011. Prevalence of risk factors for chronic kidney disease in rural Nicaragua. *Nephrology Dialysis Transplantation* 26: 2798–2805.

Orantes, C.M., Herrera, R., Almaguer, M., Brizuela, E.E., Hernández, C.E., Barraye, H., Amaya, J.C., Calero, D.J., Orellana, P., Colindres, R.M., Velázquez, M.E., Núñez S.G., Contreras, V.M. & Castro, B.E. 2011. Chronic Kidney Disease and Associated Risk Factors in the Bajo Lempa Region of El Salvador: Nefrolempa Study, 2009. *MEDICC Review*. 13(4): 14–22.

Peraza, S., Wesseling, C., Aragon, A., Leiva, R., García-Trabanino, R.., Torres, C., Jakobsson, K., Elinder, C.G. & Hogstedt, C. 2012. Decreased Kidney Function Among Agricultural Workers in El Salvador. *American Journal of Kidney Diseases* 54(4): 531–540.

US Environmental Protection Agency, 1982. Handbook for Sampling and Sample Preservation of Water and Wastewater, Environmental Monitoring and Support Laboratory, Office of Research and Development.

US Environmental Protection Agency, 1996. Method 3050B. Acid digestion of sediments, sludges, and soils. http://goo.gl/4GQLV8. Accessed, August 31, 2013.

US Environmental Protection Agency, 2009. National Recommended Water Quality Criteria, Office of Water, Office of Science and Technology, 21 p.

VanDervort, D., López, D.L., Orantes, C.M. & Rodríguez, D.S. (submitted). Comparing the spatial distribution of Chronic Kidney Disease to agricultural land use and ambient temperature in El Salvador. *MEDICC Review*.

One Century of the Discovery of Arsenicosis in Latin America (1914–2014) – Litter, Nicolli, Meichtry, Quici, Bundschuh, Bhattacharya & Naidu (Eds)
 ISBN 978-1-138-00141-1

Arsenic transport and sorption in soils: Miscible displacements and modeling

H.M. Selim
School of Plant, Environmental and Soil Sciences, Louisiana State University, Baton Rouge, LA, USA

ABSTRACT: Miscible displacement experiments were carried out where a pulse of either arsenate (As(V)) or arsenite (As(III)) in a background solution was applied to a uniformly packed soil column where steady flow in soil columns was maintained. Batch experiments were also carried out to quantify As(V) sorption isotherms for each soil. The extent of nonlinearity and kinetics of arsenic retention in different soils was quantified. Breakthrough results from the column experiments and data from kinetic batch experiments were described based on two models based on the assumption of nonlinear kinetic multiple reactions during transport in soils. Two models, the Multireaction and Transport Model (MRTM) and the Second-Order Model (SOM) were successively used to describe As(III) and As(V) effluent results.

1 INTRODUCTION

Arsenic (As) is a toxic trace element and understanding As transport is a prerequisite in assessing As behavior in soils and geological media. Elevated As concentration is a public health concern with the potential to impact wetlands, aquatic environments and the surrounding soils. As adsorption is typically nonlinear and is commonly described using the Langmuir and Freundlich models. However, few studies have investigated the effect of long residence times on As adsorption to soils. Time-dependent reactions exist due to diffusion into inter-particle and intra-particle spaces, sites of different reactivity, or surface precipitation. Several studies have indicated that As adsorption kinetics is quite fast and that longer reaction times have little effect on adsorption.

In this study, the extent of nonlinearity and kinetics of As retention in soils having distinctly different properties was quantified. Miscible displacement experiments were carried out where a pulse of either arsenate As(V) or arsenite As(III) in a background solution was applied to a uniformly packed soil column at a steady flow rate. Olivier loam, Webster loam, and Windsor sandy soils were used in this study. Batch experiments were also carried out to quantify As(V) sorption isotherms for each soil. Breakthrough results from the column experiments and data from kinetic batch experiments were described based on two models based on the assumption of nonlinear kinetic multiple reactions during transport in soils. The two models used were the Multireaction and Transport Model (MRTM) and the Second-Order Model (SOM).

2 EXPERIMENTAL METHODS AND MODELING

Three soils, Olivier loam, Webster loam and Windsor sandy soil, were used in this study. KH_2AsO_4 in 0.01 M KNO_3 was used as background solution for arsenate, and $NaAsO_2$ in 0.01 NaCl as background solution for As(III). Kinetic retention of As(V) was studied using batch methods for a wide range of initial As(V) concentrations (C_o). Miscible displacement methods were used to assess the mobility of As(III) and As(V) in different soils. Acrylic columns (5-cm in length and 6.4-cm i.d.) were packed with soil and slowly water-saturated with a 0.01 M KNO_3 background solution for As(V) columns and 0.01 NaCl for As(III) columns.

Two models were used in this study; a Multireaction and Transport Model (MRTM) and a Second-Order Model (SOM). Basic to both models is that the soil is considered to be composed of several different constituents (e.g. soil minerals, organic matter, iron and aluminum oxides), and that a solute species is likely to react with the various constituents (sites) through different mechanisms. The various reactions mechanisms are illustrated in Figure 1. Here C is the solute concentration (solution phase), S_e is that retained on equilibrium-type sites, S_1 and S_2 are the amount sorbed or retained on kinetic-type sites, and S_s and S_{irr} represent the amount retained irreversibly or partially reversible. MRTM considers reactions as linear nonlinear type reactions of the first-order or n-th order. In contrast, SOM assumes that a fraction of the total sites is kinetic whereas the remaining fractions interact rapidly with the solute in the soil solution.

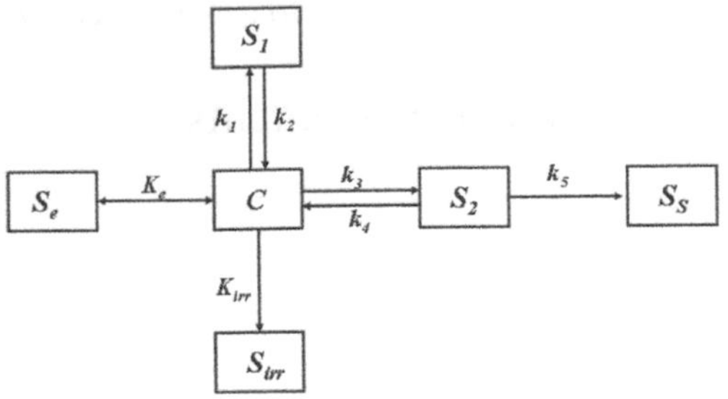

Figure 1. Schematic of multireaction models where C is the solute in the solution phase, Se is the amount retained on equilibrium-type sites, S1 and S1 are the amount retained on kinetic-type sites, Ss is the amount retained irreversibly by consecutive reaction, Sirr is the amount retained irreversibly by concurrent type of reaction, and K_e, k_1–k_5, and k_{irr} are reaction rates.

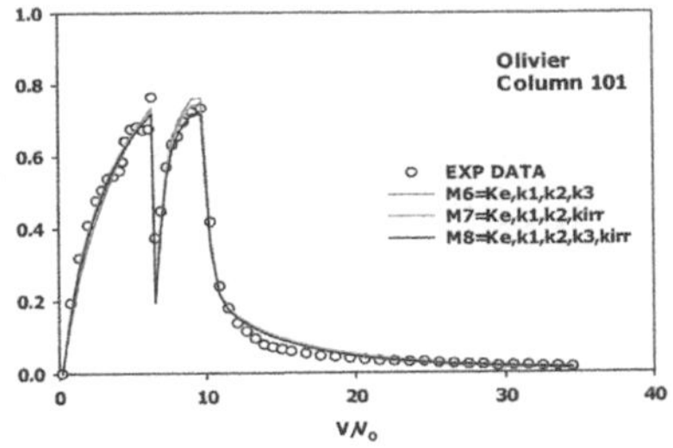

Figure 2. Breakthrough Curves (BTCs) of As(V) from the Olivier soil. Solid curves are MRTM model simulations. M6 through M8 represent different models versions.

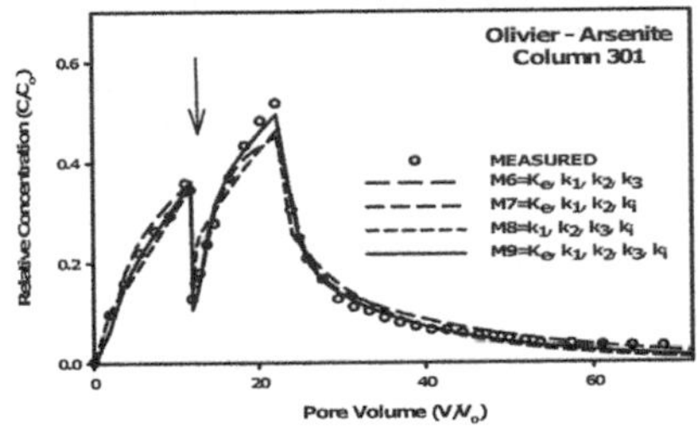

Figure 3. Breakthrough Curves (BTCs) of As(V) from the Olivier soil. Solid curves are SOM model simulations. M6 through M8 represent different models versions.

The controlling mechanisms are considered as a function of the solution concentration of the non-linear type. In a similar manner to MRTM, the second-order approach accounts for two types of sites. The first was of the equilibrium type and the second a kinetically controlled type. In contrast, SOM assumes that sorption is not only dependent on the concentration in solution but also on the amount of vacant sites on the soil matrix surfaces. Sorption will cease when all available sites are occupied as the total amount of available sites (S_{max}) is an intrinsic property of a soil.

3 RESULTS AND DISCUSSION

Typical Breakthrough Curves (BTCs) of effluent concentrations of As(V) and As(V) from Oliver soil columns are shown in Figures 2 and 3. The results are presented as relative concentration (C/C_0) versus pore volume (V/Vo), where Vo is the volume of the entire pore space of each soil column (cm^3). Model simulations clearly illustrate the versatility of the MRTM and SOM models.

In Figure 4 we show results of Olivier and Sharkey soil to illustrate the changes in As(V) concentration in soil solution versus time. The decrease of As soil solution concentrations over time was initially very rapid followed by a slow decline to a residual retention level. While both the multireaction (MRTM) and second order (SOM) models described the experimental data well for all initial concentrations, the MRTM was slightly superior to the SOM and indicated that the use of an irreversible reaction into the model formulations was essential. Moreover, incorporation of the equilibrium sorbed phase into the various model versions for As(V) predictions in these soils should be avoided indicative that kinetic are dominant.

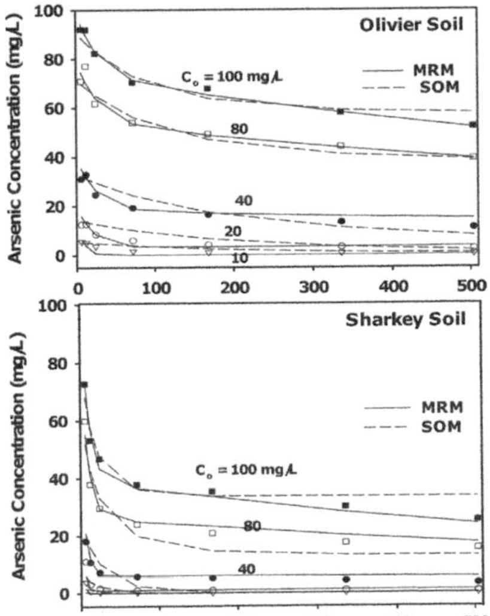

Figure 4. Experimental results of As(V) soil solution decrease concentrations for Olivier and Sharkey soils over time for several initial concentrations (C_0's). The solid and dashed curves were obtained using MRTM and SOM.

REFERENCES

Selim, H.M. & Amacher, M.C. 1997. *Reactivity and Transport of Heavy Metals in Soils*. CRC, Boca Raton, FL (240 p).

Selim, H.M. & Zhang, H. 2007. Arsenic adsorption in soils: second-order and multireaction models. *Soil Science* 72: 144–458.

Zhang, H. & Selim, H.M. 2011. Second-order modeling of arsenite transport in soils. *J. Cont. Hydrol.* 126:121–129.

One Century of the Discovery of Arsenicosis in Latin America (1914–2014) – Litter, Nicolli, Meichtry, Quici, Bundschuh, Bhattacharya & Naidu (Eds)

Natural contents of arsenic in soils of the state of São Paulo, Brazil

C.A. Abreu, R.O. Silva, M. Cantoni & R.M. Coelho
Instituto Agronômico (IAC), Campinas (SP), Brasil

ABSTRACT: This research aims to analyze natural content of As from of 97 soil profiles widely distributed by the State of São Paulo, collected at surface and subsurface horizons. Arsenic background concentration of soil samples ranged from <0.81 to 14.24 mg kg^{-1}. The third quartile was 1.26 mg kg^{-1}, lower than the value of 3.5 mg kg^{-1}, established by the State Regulatory Agency as the quality reference. No significant differences of As concentration were observed between surface and subsurface soil horizons. It was concluded that there should not be one single quality reference value for soil As for the whole State of São Paulo (Brazil). Grouping soils by taxonomic class or by soil attributes such as parent material or CEC can reduce soil variability in regard to As contents and, therefore, facilitate standardization for establishing differential As quality reference values for São Paulo State soils.

1 INTRODUCTION

Arsenic background concentration can be influenced by several physical, chemical and biological soil attributes. Despite the large soil variability, only one arsenic quality reference value (3.5 mg kg^{-1}), a value that defines a clean soil, was set by the State Regulatory Agency for the entire state of São Paulo (CETESB, 2005). However, no detailed information is available to soils of São Paulo State regarding how the As total content variation associates with the soil classification system or how it associates to soil properties. The objectives of this research were, therefore, to evaluate natural contents of As in representative soil samples of the state of São Paulo (Brazil), to evaluate the degree of association between soil As and soil physical, chemical and mineralogical attributes, and to verify adequacy of the current quality reference value for soil As.

2 METHODS

Total arsenic (As) concentration was determined in 97 representative soil profiles of the State of São Paulo, Brazil, sampled at two depths (surface and subsurface), selected from the soils collection of Instituto Agronômico (IAC).

Air-dried soil samples were submitted to the EPA 3051 analytical protocol (USEPA, 2007) and determined using hydride generation-inductively coupled plasma-optical emission spectrometry (HGICP-OES) and prior reduction with KI and 5% ascorbic acid. Validation of the As quantification was obtained using a NIST 2711 certified soil sample in each extraction tray.

Normality of data set was evaluated: As, H^+, Mg^{2+}, sum of bases (S), Cation Exchange Capacity (CEC), Fe and pseudo total Al were transformed in log(x); silt was transformed in square root(x); pH_{H2O} and K^+ were transformed by Johnson's curve or translation system as the choice.

3 RESULTS AND DISCUSSION

3.1 *Soils arsenic concentration*

Mean As concentration in soils ranged from lower than the LQ (0.81 mg kg^{-1}) to 14.24 mg kg^{-1}. The third quartile As concentration values showed similar magnitude for the two soil horizons: 1.26 mg kg^{-1} for surface and 1.42 mg kg^{-1} for subsurface. Those values were lower than the value set by the state regulatory agency, 3.5 mg kg^{-1}, for quality reference.

No significant mean As concentration differences were found between top soil (1.21 mg kg^{-1}) and subsurface soil (1.30 mg kg^{-1}), indicating that As concentration did not vary with soil depth as consequence of pedogenesis.

3.2 *Influence of soil and site attributes on natural As contents*

Autochthonous soil As background concentration was 0.98, 0.92 and 1.46 mg kg^{-1} from soils developed from, respectively, igneous, metamorphic and sedimentary rocks, with the last value (sedimentary) being greater than the remaining two according to Student t test ($p < 0.05$). These results did not agree with the ones reported by the literature, that indicate greater As values in metamorphic rocks, followed by sedimentary and igneous rocks (Zhang, 2006).

In this article, arsenic concentration in autochthonous soils (mg kg^{-1}) was not influenced by bedrock type: gneiss-0.81, sandstone-0.93, basalt-0.97, granite-0.97, diabase-1.02, claystone-1.35.

WRB classification (IUSS, 2006) at the Group level allows grouping soils with similar characteristics. Magnitude of As contents allowed dividing studied soils in two WRB Soil Group clusters: cluster (a) had lower As concentrations (0.81–1.22 mg kg^{-1}) and was formed by Gleysols, Arenosols, Leptosols, Regosols, Podzol, Chernosols, Nitisols and Acrisols; cluster (b) had high As concentrations and was formed by Cambisols (2.82 mg kg^{-1}) and Luvisols (3.20 mg kg^{-1}).

4 CONCLUSIONS

The large range of arsenic (As) concentration in soils of the state of São Paulo and its variation when grouped by soil and site attributes indicates that there should not be one single quality reference value for soil As for the state. Grouping soils by taxonomic class tends to reduce soil variability in regard to As contents and, therefore, standardize establishing differential As quality reference values for soils of the state of São Paulo, Brazil.

ACKNOWLEDGEMENTS

Financial support for this research was provided by FAPESP including a MS-grant to the first author.

REFERENCES

CETESB. Relatório de estabelecimento de valores orientadores para solos e águas subterrâneas no Estado de São Paulo. Companhia de tecnologia de saneamento ambiental, São Paulo, Brazil, edition 2005.

International Union of Soil Sciences. IUSS. (2006). World Reference Base for Soil Resources. World Soil Rresources Report n 103. Rome, WRB. FAO.

United States Environmental Protection Agency. USEPA. 2007. Microwave assisted acid digestion of sediments, sludges, soils and oils: SW-846 method 3051a. Washington, DC, United States Environmental Protection Agency.

Zhang, C. 2006. Using multivariate analyses and GIS to identify pollutants and their spatial patterns in urban soils in Galway, Ireland. *Environmental Pollution* 142(3): 501–511.

One Century of the Discovery of Arsenicosis in Latin America (1914–2014) – Litter, Nicolli, Meichtry, Quici, Bundschuh, Bhattacharya & Naidu (Eds)
© 2014 Taylor & Francis Group, London, ISBN 978-1-138-00141-1

Arsenic adsorption in soils with different mineralogical compositions

I.R. Assis, L.E. Dias, S.C. Silva, J.W.V. Mello & W.A.P. Abrahão
Soil Department, Federal University of Viçosa, Viçosa, MG, Brazil

J. Esper & A. Matos
Kinross Gold Corporation, Paracatu, MG, Brazil

ABSTRACT: The physical, chemical and mineralogical soil characteristics are fundamental to understand the mobility and transfer of arsenic (As) to the food chain. This study aimed at characterizing the As adsorption in soils with different mineralogical compositions. Clay samples were submitted to X-ray diffraction and Thermo Gravimetric Analysis (TGA), allowing, together with data from chemical characterization, to estimate the soil mineralogical composition. The maximum adsorption capacity of arsenic (MCA-AS) was estimated adjusting the data to the Langmuir isotherm. The MCA-AS correlated significantly with the gibbsite and Kaolinite amount, being positive for the first mineral and negative for the second. Hematite and goethite minerals alone did not showed significant correlation; however, the sum of these minerals showed the best correlation with MCA-AS. The soil attribute that best predicts the MCA-AS was the content of iron extracted by dithionite-citrate-bicarbonate.

1 INTRODUCTION

The toxicity of heavy metals and metalloids in the environment has been a constant concern in the environmental and public health. Arsenic (As) is a metalloid with strong potential for toxicity and its overall mean concentration in uncontaminated soils is 5 to 6 mg kg^{-1}, with variations depending on the geological formation. However, the physical, chemical, and mineralogical characteristics of soil regulate its mobility and transfer to the food chain. This study aimed at characterizing the As adsorption in soils with different mineralogical compositions, in order to understand its occurrence and mobility in the environment.

2 MATERIAL AND METHODS

2.1 *Characterization of soil samples*

Samples were collected from 0–20 cm depth of five soils: Cambisols (CX) and Ultisol dystrophic (PVA); one red Oxisol (LV) and two dystrophic red-yellow Oxisol (LVA-V and LVA-M). The samples were characterized chemically, physically and mineralogically. Clay samples were analyzed for X-ray diffraction and thermogravimetric analysis (TGA), allowing, together with the data of the chemical, to estimate the mineralogical composition.

2.2 *Adsorption isotherms*

Solutions were prepared with increasing concentrations of As(V) (0, 6, 12, 18, 30, 42, 54, 66, 84, 102 and 120 mg L^{-1}), using $Na_2HAsO_4.7H_2O$, to determine the maximum adsorption capacity of arsenic (MCA-AS). The solutions were prepared with 10 mM $CaCl_2$ and pH was adjusted to 5.5 (±0.2) with 10 M NaOH or HCl. The MCA-AS and constant related to the binding energy (EL) were estimated with the aid of the Langmuir isotherm, and the data were adjusted for your original equation (hyperbolic).

3 RESULTS AND DISCUSSION

3.1 *Characteristics of soil samples*

The clay is the dominant fraction in soils. Except for the LVA-V with 4 dag kg^{-1} of silt, the other soils had a high proportion of silt, with an average of 17 dag kg^{-1}. In general, all soils are acidic and of low fertility. The ΔpH values were negative, showing that these soils have a predominance of negative charges. There is not a significant variation in the organic matter content, with average of 3.7 dag kg^{-1}. The phosphorus availability in all soils is low, minimizing the effects of competition for the same adsorption sites of this element with As(V).

Kaolinite and gibbsite were identified in all samples. The presence of goethite in the LV sample and the conformation of goethite loss curve for the LVA-M and PVA were not identified, indicating that this mineral is present in less crystalline forms.

3.2 *Arsenic adsorption*

The MCA-AS correlated positively and significantly with the proportion of clay and Fe content extracted by DCB-Fed (Table 1). The importance

Table 1. Correlations between EL and MCA-AS with some soil characteristics.

	MCA-AS	EL
pH	–0,01	–0,38
V%	–0,21	–0,55
OM	–0,31	–0,51
P-rem	–0,81	–0,90*
Clay	0,95*	0,79
Fed	0,98**	0,78
Feo	0,39	–0,17
Fe_2O_3 (Sulfuric acid attack)	0,86	0,73

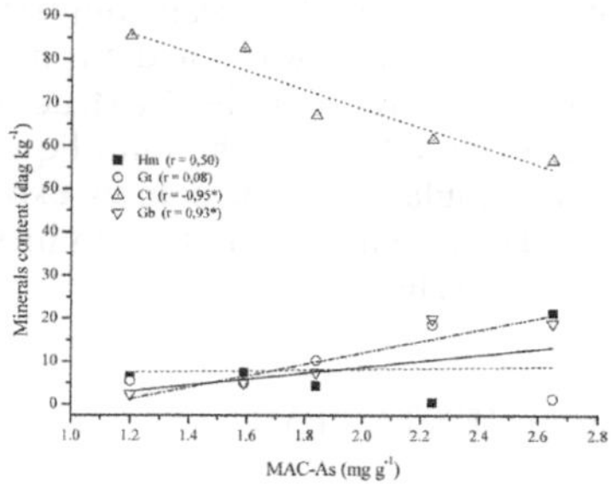

Figure 1. Correlation between MCA-AS and estimated quantities of hematite, goethite, kaolinite and gibbsite.

of the clay is associated with increased reactive surface in comparison with other soil fractions (Curi & Camargo, 1988; Silva, 1999). However, Fed representing the oxides and hydroxides of Fe free crystalline and poorly crystalline, has higher correlation, being able to better predict the adsorption on these soils than the other characteristics. The low value of correlation between the MCA-AS and Fe oxalate—Feo, indicates that the adsorption of these soils is little influenced by poorly crystalline Fe oxides and hydroxides. The EL had significant correlation only with P-rem, and in this case, it was negative. This indicates that, when the value of P-rem is high, there is a low affinity of As with the soil solid phase, with lower EL between them, as observed for phosphate.

The MCA-AS correlated significantly and negatively with the estimated kaolinite and positive with gibbsite present in the clay fraction (Figure 1). Despite there was not a correlation of MCA-AS with hematite and goethite alone, the sum of these correlation was significant ($r = 0.97^{**}$). The sum of Fe and Al oxides (hematite, goethite and gibbsite) also showed a significant correlation with MCA-AS ($r = 0.96^{*}$), but less than the sum of hematite and goethite (Figure 2).

This probably explains the correlation between the amount of kaolinite and MCA-AS, which actually should not be related to this mineral but the sum of others, which have positive charge density at pH values normally found in soils.

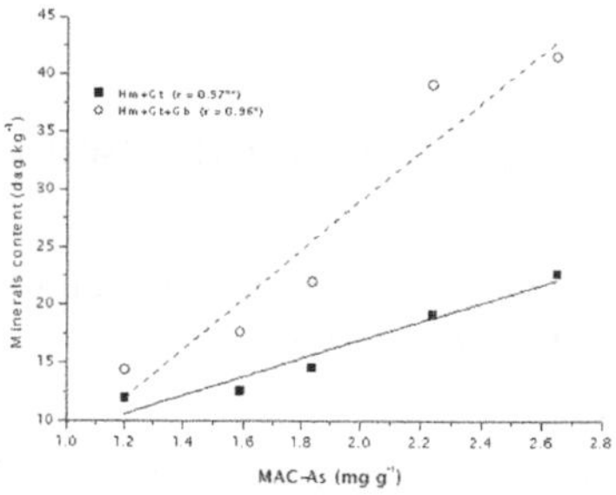

Figure 2. Correlation between MCA-AS and the sum of the estimated hematite + goethite and hematite + goethite + gibbsite.

There was not a significant correlation between MCA-AS and the sum of goethite and gibbsite. This indicates that the adsorption is more related with hematite and gibbsite minerals, different from what occurs for P, where goethite has relevance in the CMAP (Torrent *et al.*, 1994; Borggaard, 1983; Bahia Filho, 1982).

4 CONCLUSIONS

The isolated amount of gibbsite and goethite + hematite sum showed the highest correlations with the maximum adsorption capacity of arsenic in the studied soils.

The iron extracted by dithionite-citrate-bicarbonate was the variable that best described the maximum adsorption of As in soils, suitable for intensely weathered and well-drained soils. The proportion of clay also performed with high potential for such prediction. However, to use these variables for the purpose of prediction, it is necessary to include a greater number of soils.

REFERENCES

Bahia Filho, A.F.C. 1982. Índices de disponibilidade de fósforo em Latossolos do planalto central com diferentes características texturais e mineralógicas. Viçosa, MG, Universidade Federal de Viçosa, 179p. (PhD Thesis).

Borggaard, O.K. 1983. Effect of surface area and mineralogy of iron oxides on their surface charge and anion-adsorption properties. *Clay Clays Miner.* 31: 230–232.

Curi, N. & Camargo, O.A. 1988. Phosphorus adsorption characteristics of Brazilian Oxisols. In *Proceedings of the International Soil Classification Workshop: Classification, Characterization and Utlization of Oxisols.* Part 1. EMBRAPA, SMSS, AID, UPR, Rio de Janeiro, Brasil, pp.56–63.

Silva, J.R.T. 1999. Solos do Acre: caracterização física, química e mineralógica e adsorção de fosfato. Viçosa, MG, Universidade Federal de Viçosa, 117p. (PhD Thesis). Torrent, J., Schwertmann, U.E. & Barrón, V. 1994. Phosphate sorption by natural hematites. *J. Soil Sci.* 45: 45–51.

One Century of the Discovery of Arsenicosis in Latin America (1914–2014) – Litter, Nicolli, Meichtry, Quici, Bundschuh, Bhattacharya & Naidu (Eds)
© 2014 Taylor & Francis Group, London, ISBN 978-1-138-00141-1

Effect of irrigation on arsenic mobility and accumulation in soils

B. Doušová, S. Krejcova, R. Boubinova & M. Lhotka
Institute of Chemical Technology, Prague, Czech Republic

F. Buzek
Czech Geological Survey, Prague, Czech Republic

ABSTRACT: Contrary to the global decline of atmospheric pollutions in all European countries, a significant increase of arsenic in surface water in several Central European regions was noted during the last ten years. The mobilization of arsenic in aerated unsaturated zone is mostly controlled by structural and surface properties of soils and by the stability of ternary OM-Fe-As complexes on the solid-liquid interface. The stability of soil arsenic during a long-term irrigation was simulated via the column experiment using 4 soils from the interest area and As-free natural rainwater. The accumulation of As in mentioned soils was tested in the same manner using As(V) enriched natural rainwater.

1 INTRODUCTION

Acid deposition in Central Europe culminated about 20 years ago. Arsenic in oxidic form was emitted during coal combustion and deposited on soils, in forests, or in surface waters. A high deposition of anthropogenic As did not have significant effect on As concentration in water resources thanks to soil adsorption properties. Despite surface waters sampled after a high deposition period generally had a low As concentration with median values from 0.9 to 1.1 $\mu g\ L^{-1}$ (Vesely & Majer 1996), the same waters sampled recently contained two or three times more As (to 10 $\mu g\ L^{-1}$).

The immobilization of As in soils strongly depends on the presence of Hydrated Ferric Oxides (HFO) forming surface complexes with As oxyanions (Dousova, 2012; Sherman & Randall, 2003). However, oxyanions can be bound via ternary complexes formed by natural organic matter, Fe^{III} and, e.g. As oxyanions. The coexistence of HFO and Dissolved Organic Matter (DOM) can lead to their coprecipitation and/or adsorption of DOM to the HFO surface, forming colloidal particles and their aggregates. The form and structure of Fe-OM colloidal phases can inhibit adsorption of As to mineral phases, enhancing its mobilization and release to water system (Buzek *et al.*, 2013).

The aim of this work was to describe the effect of long-term irrigation to arsenic accumulation and mobility in soils.

2 EXPERIMENTAL PART

2.1 *Sampling area*

The studied area is situated in the central part of the Czech Republic (Central Europe), in the middle part of the Elbe River catchment, a typical agricultural area with unconfined aquifer located in Cretaceous sediments. The total area of 1288 km^2 is located about 30 km east of Prague. Aquents dominate in the Elbe River floodplain, psamments and aquods are developed on higher river terraces. Mollisols, aquents and psamments are developed on the loess. Psamments developed on the marlstones are found on the Elbe River right bank.

2.2 *Soil samples*

Four soils from the catchments were sampled from the margins of cultivated fields. Cores from two 400 cm^2 areas were taken by AMS Auger corer in three depth intervals (0–20 cm, 20–40 cm, 40–60 cm), independently of possible soil horizons. Samples were dried at 50 °C to a constant weight, deprived of stone ballast and organic rests and homogenized. A general characterization of soil samples proved similar pH values of 7.2 to 7.9 and mineralogical composition consisting mainly of muscovite, microcline, kaolinite and quartz, with less participation of illite and anorthite, uniquely albite or calcite. Chemical and surface properties including soil types are summarized in Table 1.

Table 1. Sign and characterization of soil samples.

Sample (soil type)	Horizon (cm)	As (mg g^{-1})	Fe (mg g^{-1})	TOC (%)	S-BET (m^2 g^{-1})
Celakovice	0–20	0.05	25.1	1.8	5.2
(fluvents)	20–40	0.04	19.6	0.6	5.5
	40–60	0.05	21.0	0.6	5.2
Litol	0–20	0.01	25.9	1.2	8.1
(fluvents,	20–40	0.02	26.4	0.6	8.8
psamments)	40–60	0.03	19.5	0.5	8.3
Mlynarice	0–20	0.01	14.5	3.2	1.9
(mollisols,	20–40	<0.01	11.9	2.9	1.9
psamments)	40–60	0.01	6.6	0.4	3.1
Kopanicky	0–20	0.02	19.7	0.7	20.8
brook	20–40	0.02	17.9	0.6	15.5
(lithic groups, psamments, fluvents)	40–70	0.01	20.2	0.4	21.1

2.3 *Column experiments*

Arsenic release and accumulation in soils were studied via column experiments; two columns simulated soil profile of 3 horizons and were put into identical flow glass colonies of 32/300 mm. The volume of soil profile was 225–240 ccm, and horizons were separated with standard sea sand layer (2 mm). One column was leached with As-free natural rainwater (pH ≈ 5.5; $s \approx 57$ µS/cm; As $< 3 \times 10^{-3}$ mg L^{-1}); for the study of As accumulation, the second column was leached with the same natural rainwater enriched with As(V) (pH ≈ 6,1; $s \approx 65$ µS/cm; As ≈ 2.5 mg L^{-1}). The experiments ran at the preset flow of 1.15 mL min^{-1} and 20 °C for 272–327 hours. The leaches were sampled continuously and analyzed on pH, *s* and As, Fe and TOC content.

3 RESULTS AND DISCUSSION

3.1 *Leaching process*

The time-dependent changes in chemical properties of leaches were similar for all profiles and both the (A) and (B) columns; pH varied from 7.0 to 8.2 and the conductivity and Fe/TOC concentrations decreased to balanced value in the ranges dependent on the specific profile. Kopanicky brook with the highest specific surface area (S_{BET}) and lowest content of OM created the most stable soil profile with low concentration changes in the leach, the lowest changes of conductivity and pH.

Litol represented less stable profile with high concentration gradients of Fe and TOC and the highest conductivity decrease. Arsenic flow related to time and mass units (µg/g.day) culminated at different time (Figure 1) depending on the soil

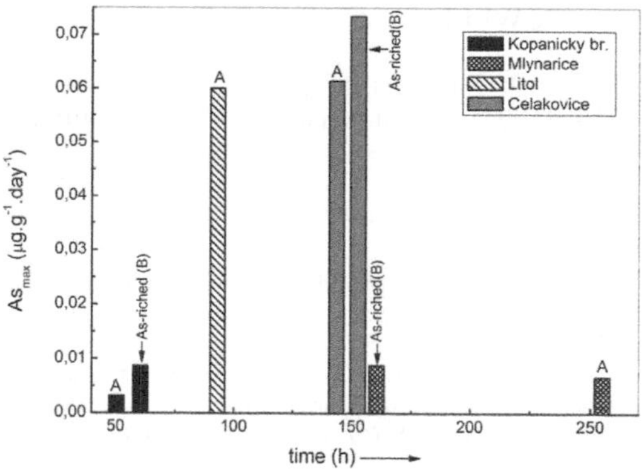

Figure 1. Culmination of As flow in time of leaching.

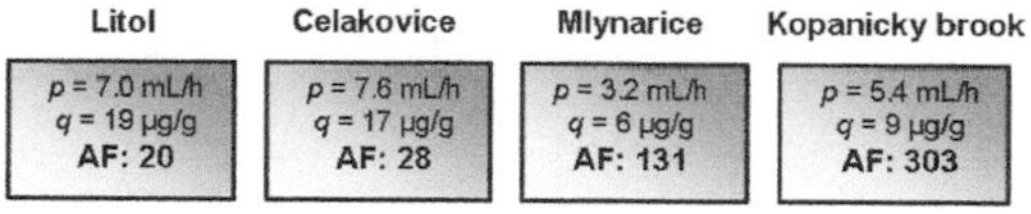

Figure 2. Accumulation properties of soils (p-average water flow rate, q-average adsorption capacity, AF-accumulation factor).

properties. (B) columns illustrated higher amount of released As, mostly at the same time.

3.2 *Arsenic accumulation in soils*

An accumulation ability of soils to incoming As(V), characterized via adsorption capacities and accumulation factors, was estimated according to the data from (B) column experiments. The accumulation factors AF (calculated as As(input)/As(output), at m, t = const), related mostly to the hydraulic conductivity of soil profiles (Figure 2), presented as average water flow rate, and partially to the S_{BET} (Table 1).

The stability of soil arsenic was mostly affected by structural and surface properties of soils. Providing the common presence of HFO particles in soils, chemical composition and soil type denoted rather minor effect. A low hydraulic conductivity and high accumulation factor (Kopanicky brook, Mlynarice) indicated significantly lower As mobility regardless different soil types and general composition (compare Figures 1 and 2). Highly conductive soils (Celakovice, Litol) released multiple higher amount of As (apart from initial As concentrations). The rate of arsenic release related to the specific surface area (Fig. 1). The fractionation of As (Wenzel *et al.*, 2001) during a long-term irrigation process indicated a partial transformation of As binding to more available forms along the soil profile.

ACKNOWLEDGEMENTS

This work was part of projects 13-24155S and P210/10/0938 (Grant Agency of Czech Republic).

REFERENCES

Buzek, F. et al. 2013. Mobilization of arsenic from acid deposition after twenty years. *Appl. Geochem.* 33: 281–293.

Dousova, B. et al. 2012. Adsorption behaviour of arsenic relating to different natural solids: Soils, stream sediments and peats. *Sci. Tot. Environ.* 433: 456–461.

Sherman, D.M. & Randall, S.R. 2003. Surface complexation of arsenic(V) to iron(III) (hydr)oxides: Structural mechanism from ab initio molecular geometries and EXAFS Spectroscopy. *Geochim. Cosmochim. Acta* 67(22): 4223–4230.

Veselý, J. & Majer, V. 1996. The effect of pH and atmospheric deposition on concentrations of trace elements in acidified surface waters: A statistical approach. *Water, Air, Soil Poll.* 88: 227–246.

Wenzel, W. et al. 2001. Arsenic fractionation in soils using an improved sequential extraction procedure. *Analytica Chim. Acta* 436: 309–323.

One Century of the Discovery of Arsenicosis in Latin America (1914–2014) – Litter, Nicolli, Meichtry, Quici, Bundschuh, Bhattacharya & Naidu (Eds)

Arsenic phytotoxicity in Australian soils

D.T. Lamb, M. Kader, M. Megharaj & R. Naidu
Centre for Environmental Risk Assessment and Remediation, University of South Australia, Mawson Lakes, SA, Australia
Cooperative Research Centre for Contamination Assessment and Remediation of the Environment (CRC CARE), University of South Australia, Mawson Lakes, SA, Australia

ABSTRACT: Arsenic phytotoxicity was studied using two tests in a range of soils. The effective concentrations causing a 50% reduction in growth (EC_{50}) for the 4 week growth study was 13 to 235 mg/kg compared to 42 to 452 mg/kg using root elongation. Phytotoxicity thresholds for both tests were strongly correlated to each other. The EC_{50} values were related strongly to soil pH and the Freundlich partitioning (K_f) constants. The EC_{50} values were most consistently related to K_f values.

1 INTRODUCTION

Arsenic (As) is a naturally occurring metalloid in soil, which is toxic to human and ecological receptors at excessive levels. Elevated levels of As in soil result from numerous anthropogenic activities, including cattle dips, pesticide application, smelting of metal ore bodies, animal husbandry, wood preservatives and defoliants (Song *et al.*, 2006). Understanding of As phytotoxicity in soils is complicated by numerous natural factors, including pH, phosphorus, and mineralogical composition, but also by varied methodological approaches used in a study.

In this study, As phytotoxicity was studied in a range of soils using both greenhouse (4 weeks) and root elongation (4 days) studies using the test species *Cucumis sativus* (cucumber). The relationship between toxic thresholds and soil properties is reported.

2 METHODS/EXPERIMENTAL

2.1 *Soils*

Seven uncontaminated soils were sampled with contrasting soil properties from Queensland, South Australia and Victoria, Australia. The soil types include Ferrosols, Vertosols, Kurosols, Calcarosols, Dermosols and Tenosols. Soils were taken from the top 0.2 m with stainless steel trowels, air-dried and sieved through 4 mm sieves. All soils were spiked with As (Na_2HAsO_4) from 0 to 2,000 mg kg^{-1}, depending on the soil properties. Soils were spiked by spraying As solutions to a thin layer of soil and mixing. Soils were incubated for 4 months prior to experimentation. The soils were incubated at water contents of approximately 20–30% (w/w) at ~ 22 °C prior to the plant bioassays. Soil properties were determined using standard procedures (Lamb *et al.*, 2009).

2.2 *Plant growth*

Cucumber (*Cucumis sativa L.*) seeds were sown to plant pots. Each pot was lined with fine nylon mesh and 300 g of air-dry soil was added. Each pot was placed within a collecting tray to collect any excess leachate generated. High purity water was added (18.2 μΩ cm) for watering. Plants were watered by weight to avoid drainage as much as possible and to reach approximately 70% of field capacity throughout the study. Sufficient seed was added to ensure germination. Plant numbers were later reduced to 4 plants per pot. Each soil treatment was replicated 3 times. Plants were grown for 4 weeks from sowing under greenhouse conditions (~16–25 °C).

2.3 *Root elongation*

To 100 g of air dry soil, water was added to reach 70% field capacity. Seeds of *C. sativus* were treated with 10% NaOCl solution and rinsed thoroughly before sowing (6 roots per pot, $n = 3$). After 4 d, plant roots were carefully separated and roots washed for measurements. Root lengths were analyzed with WinRHIZO Arabidopsis software after scanning.

2.4 *Arsenic sorption*

Sorption of As to all soils was studied using a common batch reaction methodology (Smith *et al.* 1999). In brief, 1 g of soil was reacted with 20 mL of solutions containing increasing arsenate concentrations. A 0.03 M $NaNO_3$ background electrolyte was used. Arsenate was reacted with soils for 24 h, then filtered (0.45 μm) and As analyzed by inductively coupled plasma mass spectrometry (Agilent 7500 c). Sorption curves were described with the Freundlich function.

3 RESULTS AND DISCUSSION

3.1 *Soils*

Soil pH_{Ca} in this study ranged from highly acid (4.45) to alkaline (7.79). The clay content of the soils ranged from 6.7 to 20% and Loss on Ignition (%) was 3.89 to 18.3.

3.2 *Arsenic phytotoxicity*

Substantial variability in EC_{50} values was observed between soils (Figure 1). Phytotoxicity thresholds for the 4 d root elongation study ranged from 42 to 452 mg kg^{-1}. In contrast, the EC_{50} data for the 4 week pot study ranged from 13 to 235 mg kg^{-1}. The highest EC_{50} value in both cases was found with the Ferrosol soil (Tarrington) due to its low pH_{Ca} and high iron oxy(hydr)oxide content. The EC_{50} values in the three alkaline soils in the 4 week data set were very low and, in fact, one was below the Australian ecological trigger guideline value of 20 mg kg^{-1}. Despite the root elongation data set yielding higher EC_{50} data for each soil, the EC_{50} values were correlated with root elongation data ($R^2 = 0.82$, slope = 1.8).

3.3 *Phytotoxicity thresholds and soil properties*

Arsenic phytotoxicity was related only to soil pH_{Ca} in this dataset (Figure 2). Other soil parameters showed trends but overall a poor relationship. As soil pH_{Ca} increased, the EC_{50} values were dramatically reduced, indicating substantially higher bioavailable fraction in these alkaline soils. The relationship with pH_{Ca} was stronger for the 4 week pot study than root elongation data. The reduced partitioning of As in these soils was also reflected in the sorption data (K_f ranged 5.80 to 243 L kg^{-1}). Partition constants were strongly related to both shoot and root elongation EC_{50} data, and were the variable best able to explain the phytotoxicity data.

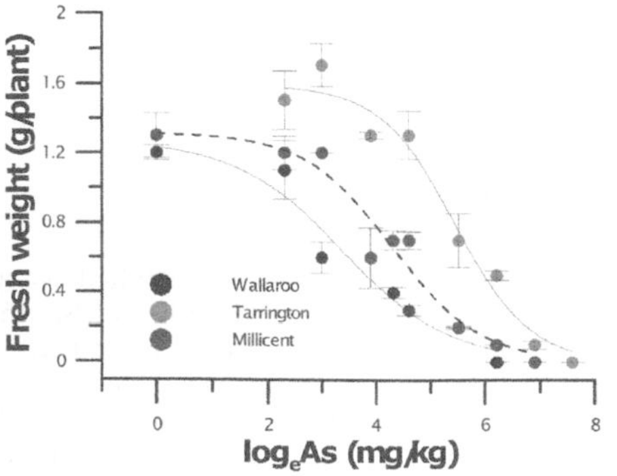

Figure 1. Fresh weight of *C. sativus* in response to increasing As in soils (mean ± standard error, $n = 3$).

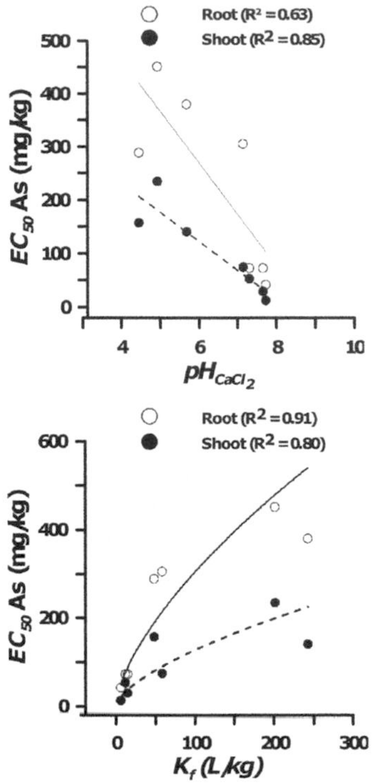

Figure 2. Relationship between soil pH_{Ca}, Freundlich partition constants (K_f) and EC_{50} values for root elongation data and shoot weight.

4 CONCLUSIONS

The EC_{50} values were related strongly to soil pH and the Freundlich Partitioning (K_f) constants. The EC_{50} values were most consistently related to K_f values.

REFERENCES

Lamb, D., Ming, H., Megharaj, M. & Naidu, R. 2009. Heavy metal (Cu, Zn, Cd and Pb) partitioning and bioaccessibility in uncontaminated and long-term contaminated soils. *Journal of Hazardous Materials*. 171: 1150–1158.

Smith, E., Naidu, R. & Alston, A.M. 1999. Chemistry of Arsenic in Soils: I. Sorption of Arsenate and Arsenite by Four Australian Soils. *J. Environ. Qual.* 28: 1719–1726.

Song, J., Zhao, F.-J., McGrath, S.P. & Luo, Y.-M. 2006. Influence of soil properties and aging on arsenic phytotoxicity. *Environmental Toxicology and Chemistry*. 25: 1663–1670.

One Century of the Discovery of Arsenicosis in Latin America (1914–2014) – Litter, Nicolli, Meichtry, Quici, Bundschuh, Bhattacharya & Naidu (Eds)

Arsenic in groundwater, soils and crops of east side area of Ossa mountain, Greece

C.D. Tsadilas, M. Tziouvalekas, E. Evangelou & A. Tsitouras
Institute of Soil Mapping and Classification, Larissa, Greece

ABSTRACT: A survey was conducted in the east side of mountain Ossa, East Greece in waters, soils and plants to estimate As level. Some waters originating from springs on metamorphic basic ophiolitic rocks contain As exceeding the maximum permitted limit, amounting up to 37.20 μg L^{-1}. Soils irrigated with these waters have high As concentration exceeding the usual background values reported worldwide for similar soil types. Vegetables grown on such soils were found to have high As concentration similar to those reported for contaminated with As areas. Further investigation is needed for reducing As involvement in human food chain.

1 INTRODUCTION

Natural Arsenic (As) contamination of groundwater targeted on drinking, irrigation and other common uses has become a serious problem in many regions of the world. Irrigation of plants with As contaminated water may increase soil As concentration and As uptake by the plants. Drinking water, inhaling and eating of plant foods rich in As for a long period may cause serious health problems to humans. Recently, it has become known that in some regions of Greece high As concentration was recorded in groundwater used for irrigation and drinking (Kelepertzis *et al.*, 2006). Due to increasing concern on As contamination of groundwater used in several ways by the people, a survey was conducted in the east side of Ossa mountain, East Thessaly, to investigate the As concentration in ground waters, soils, and irrigated plants.

2 METHODS

2.1 *Description of the study area*

The area studied covers the east lower and mid part of the mountain Ossa, starting from the end of Tempi valley near the village Omolio and extending parallel to coast of the Aegean Sea up to the place "Rakopotamos" (39°40′ and 39°50′ N and 22°45′ and 22°50′ E, Figure 1). The lower part of the area is plain but the mid and higher parts are hilly and mountainous. In the lower part, there is intense agricultural activity including cultivation of corn, kiwifruit, olive trees, and chestnut trees. In the upper parts, the area is covered by forests. The geology of the area includes mainly rocks belonging to Pelagonian zone and the Eohellenic tectonic nape (IGME, 1987). The Pelagonian zone consists of karstic marbles, while the Eohellenic tectonic nape consists of serpentines metamorphic basic ophiolitic rocks and metamorphic rocks of sedimentary origin (mica-chlorite schists and quartz-chlorite-mica schists. The Quaternary sediments of the studied area consist of unconsolidated materials and alluvial sediments.

2.2 *Water, soil, and plant sampling and analyses*

From March to September 2013, water samples were selected coming from natural springs, small rivers, wells and the drinking water network in the eastern side area of Ossa mountain, Greece. From some of these places, surface soils cultivated mostly

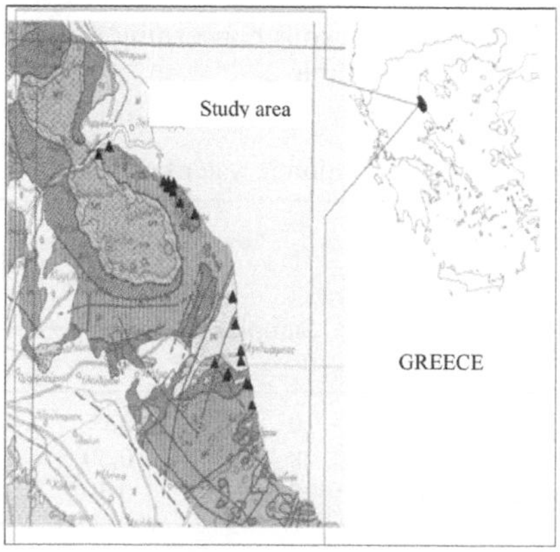

Figure 1. The geology of the area studied. (IGME. 1987. Karitsa sheet Geological Map 1:50.000).

with vegetables as well as vegetables (parsley, lettuce, green pepper, eggplant, cucumber, potato, tomato, and squash) grown in the same places were also selected accordingly. All these samples were analyzed for As by using atomic absorption supplied with a mercury hydride system. Soil samples were air dried, ground to pass a 2 mm sieve and digested with strong acids (65% HNO_3 and 37% HCl plus 30% H_2O_2, Arain *et al.*, 2009). Plant samples were dried at a furnace at 70 °C for 72 hours, dry ashed at 500 °C and the ash was diluted with HNO_3 acid.

3 RESULTS AND DISCUSSION

3.1 *Arsenic concentration in water*

Arsenic concentration in water samples ranged widely depending on their origin (Table 1).

Waters coming from springs were found to have the highest As concentration, amounting up to 37.20 μg L^{-1}, quite higher than the maximum permitted level by the WHO (10 μg L^{-1}). Although it was not possible to safely correlate the As concentration to the geology, it was recorded that the maximum As concentrations were found in areas with preuppercretaceus tectonic nape consisting of metamorphic basic ophiolitic rocks. A more detailed investigation is needed to correlate water As with geology. Water from wells had the lower As concentration ranging from 0.30 to 9.80 μg As L^{-1}, being suitable for any use. However, water from drinking networks had in some cases—obviously where they come from springs with high As concentration—exceeding the maximum permissible level set up by WHO. The important issue is that some people are not adequately aware about the health risk due to As and use this water without stint.

3.2 *Arsenic concentration in soils*

The situation of the soils concerning As contamination is shown in Table 2.

Table 1. As concentration in water samples.

Origin	Samples no	As concentration, μg L^{-1} Min	Max
Springs	10	nd*	37.20
Wells	8	0.30	9.80
Drinking network	17	0.30	36.49

*non detectable.

Table 2. Effect of As contaminated water on soil As concentration (mg/kg soil).

	Irrigation with As contaminated water ($n = 7$)	Non irrigated soils ($n = 12$)
Minimum	26.70	7.00
Maximum	137.70	28.80
Average	54.12	10.82
Median	26.70	8.2

Table 3. As concentration of vegetables irrigated with As contaminated water.

Plant name	As, μg/kg dw
Parsley	2750
Lettuce	350
Green pepper	220
Eggplant	
skin	81
shark	182
Cucumber	
skin	485
shark	640
Potato	
skin	194
shark	nd
Tomato	nd
Squash	
skin	nd
shark	nd

The values of As concentration of non-irrigated soils are in agreement to those reported for similar soil types on the worldwide scale by Kabata Pendias & Mukherjee (2007). From the data of Table 2, it is clear that irrigation with water high in As, elevates As concentration of soils.

3.3 *Arsenic concentration in plants*

Arsenic concentration of vegetable plants studied irrigated with As contaminated water are presented in Table 3.

The highest As concentration was found in parsley (2750 μg/kg dw) and the lowest in tomato and squash (no detectable). In some cases, fruit skin contained different As amount than flesh (eggplant, cucumber, and potato). Cucumber and lettuce contained appreciable amounts of As (640 and 350 μg/kg dw, respectively). These As values are similar to those reported by Alam *et al.* (2003). Vegetables in these areas are irrigated by the same water used for drinking which, as pre-mentioned, when they originated from springs have high As concentration and may contaminate soils and the plants grown on them.

4 CONCLUSION

In the area studied, some waters used by the people contain high amounts of As that may contaminate soils and plants, entering the food chain and threating human health. Further investigation is needed on As origination and soil As contamination to take measures for excluding it from the food chain.

ACKNOWLEDGEMENTS

Thanks are expressed to all the laboratory staff of the Institute of Soil Mapping and Classification that carried out the analyses.

REFERENCES

Alam, M.G.M., Snow, E.T. & Tanaka, A. 2003. *Arsenic and heavy metal contamination of vegetables grown in Samta village, Bangladesh*. The Sci. of the Total Envir. 308: 83–96.

Arain, M.B., Kazi, T.G., Baig, J.A. Jamali, M.K. Afridi, H.I. Ahah, N.A.Q. & Sarfraz, R.A. 2009. Determination of arsenic levels in lake water, sediment, and foodstuff from selected area of Sindh, Pakistan: Estimation of daily intake. *Food and Chem. Tox.* 47: 242–248.

Kabata-Pendias, A. & Mukherjee, A.B. 2007. *Trace elements from soil to human.* Springer-Verlag, Berlin.

Kelepertzis, A., Alexakis, D. & Skordas, K. 2006. Arsenic, antimony, and other toxic elements in the drinking water of Eastern Thessaly in Greece and its possible effects on human health. *Environ. Geol.* 50: 76–84.

One Century of the Discovery of Arsenicosis in Latin America (1914–2014) –
Litter, Nicolli, Meichtry, Quici, Bundschuh, Bhattacharya & Naidu (Eds)
© 2014 Taylor & Francis Group, London, ISBN 978-1-138-00141-1

Arsenic distribution in *Citrullus lanatus* plant irrigated with contaminated water

G. Pineda-Chacón, M.T. Alarcón-Herrera & M.R. Delgado-Caballero
Centro de Investigación en Materiales Avanzados (CIMAV), Chihuahua, México

L. Cortés-Palacios & A. Melgoza-Castillo
Facultad de Zootecnia y Ecología, Universidad Autónoma de Chihuahua, Chihuahua, México

ABSTRACT: The use of groundwater with high arsenic (As) content in agricultural irrigation poses a risk due to the possible bioaccumulation of this metalloid through the trophic chains. As accumulated in soil could cause a long-term decline in crop yields. The effects and distribution of As in watermelon plants (*Citrullus lanatus*) irrigated with As^{III} in a soil mesocosm were evaluated. The As distribution in plant was: root > leaf > stem. In a period of 35 weeks with 1.4 mg L^{-1} of As concentration in irrigation water, *Citrullus lanatus* showed biomass inhibition in root.

1 INTRODUCTION

Groundwater with high concentrations of As is used for agricultural irrigation purposes. This practice has resulted in the accumulation of As in soils to levels that can alter their physicochemical properties, decreasing fertility and/or reducing crop yields (Garg & Singla, 2011, Rahman & Hasan, 2007). Natural sensitivity or tolerance of plants to accumulate As is directly related to the plant species and plant genotypes (Garg & Singla, 2011).

The aim of this study was to determine the distribution of As in plants of watermelon (*Citrullus lanatus*) irrigated with known concentrations of sodium arsenite and to evaluate the effects of this metalloid on the plant biomass development.

2 METHODS/EXPERIMENTAL

2.1 *Experimental development*

Twenty watermelon plants germinated in a greenhouse. Planting was carried out in plastic pots (0.023 m^3). Treatments consisted of solutions prepared with sodium metaarsenite 34 Baker® ($NaAsO_2$) at three concentrations: 0.7, 1.4 and 2.8 mg As L^{-1} (treatments B, C and D respectively) and a control of well water (treatment A). Plants were irrigated three times a week for 35 weeks.

2.2 *Analytical determinations*

Analyses were performed by duplicate. Plant and soil samples were digested in a microwave oven (Marx 5 CEM) according to methods 3050 B and 3051 of the Environmental Protection Agency of the United States. The analytical determination of total As was performed by ICP-OES (Perkin Elmer 8300).

3 RESULTS AND DISCUSSION

3.1 *Accumulation of As in soil*

The As accumulation in soil reached 14.7 ± 0.5 mg kg^{-1} for treatment D (Figure 1). A linear regression analysis determined that the As present in soil was attributed at least 90% to As added via the irrigation water. As in soil can be transformed by microorganisms, however it cannot be removed. Therefore, when added in the irrigation water, this metalloid accumulates in soil (Roychowdhury *et al.*, 2004).

3.2 *As distribution in Citrullus lanatus*

The root showed the highest As accumulation, followed by leaves and finally by stems (Figure 1). For As concentrations between 0.7 and 1.4 mg L^{-1} in the irrigation water, the watermelon plant was able to accumulate in the root until 21 ± 6.4 mg kg^{-1}. When plants were exposed to the highest As concentration (treatment D), the root As accumulation capacity decreased. This can be attributed to saturation and/or to detoxification mechanisms from the plant.

According to Dunnet and Tukey comparative statistical tests ($\alpha < 0.05$), As concentrations in the root and the stems are only significantly higher than those of control subjects for treatments C and D. The leaves were more sensitive to the presence of As, possibly due to metabolic processes inside them. In the case of *Citrullus lanatus*, once As is absorbed by the root, the plant is able to mobilize this metalloid toward the leaves, provoking a significantly

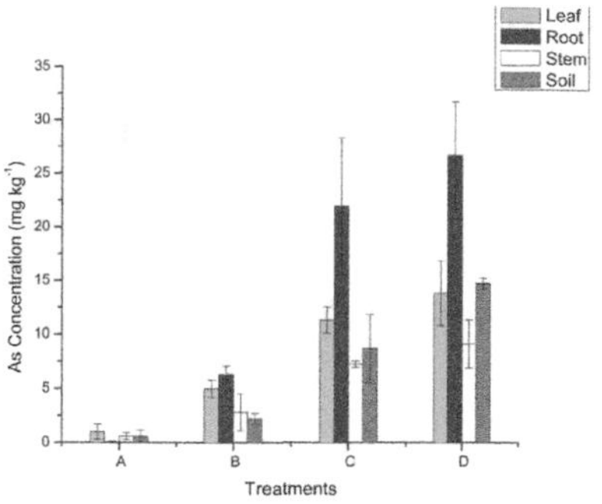

Figure 1. As distribution in the soil mesocosm. Treatments according to As concentration in irrigation water: A (control), B (0.7 mg As L^{-1}), C (1.4 mg As L^{-1}) and D (2.8 mg As L^{-1}). Bars represent the mean values ± SD (n = 5).

greater accumulation there with respect to the control subjects even those which were exposed to the less concentrated treatment (B). According to Pearson's coefficient (0.897) and the proportion of variance obtained, 80% of the As present in the leaf can be attributed to As added in treatments.

3.3 *Biomass development*

Biomass developed by C treatment plants was smaller compared to the other treatments including the control. According to analysis of variance with Bonferroni and Tukey comparisons ($\alpha < 0.05$) compared to treatment A, only the root biomass of treatment C was statistically lower than the control. However, even when the differences are not statistically significant, there is evidence of a trend: As doses higher than treatment B inhibit biomass development compared to the control.

3.4 *Bioaccumulation and translocation of As*

According to the bioaccumulation factor (BAF), when the watermelon plant was exposed to treatment B, it was able to accumulate a concentration equivalent to almost twice the As concentration in the soil. This accumulation capacity decreased with increasing As in the treatments. Higher translocation of arsenic was detected in leaves of the watermelon plant that was exposed to treatment with lower concentration (B). The stem showed approximately half of translocation compared to leaf (Table 1).

Citrullus lanatus is classified as accumulator plant because its BAF > 1. When the concentration of As increased in the treatments, BAF and TF decreased. This fact could be attributable to the detoxification mechanism, including the ability of the plant to return to the soil a portion of the absorbed contaminant and the accumulation reduction of this metalloid in the plant (Zhao *et al.*, 2008). Furthermore, the stress induced by As in plants triggers a series of reactions

Table 1. Bioaccumulation and translocation factors.

Treatment	*BAF*	TF_{leaf}	TF_{stem}
A	1.39	–	–
B	2.13	0.79	0.45
C	1.56	0.52	0.33
D	1.12	0.52	0.34

BAF: Bioaccumulation factor (plant/soil), TF: translocation factor (leaf or stem/root).

that disrupt the cellular transport systems, inhibiting the As translocation (Finnegan & Chen, 2012).

For As concentrations of 1.4 mg As L^{-1} in irrigation water, plant biomass was inhibited, indicating its tolerance limit. The root was the most affected section. As concentration in groundwater used for irrigation could be below those levels employed in this study, but over time, even with lower concentrations, As accumulation in soil can easily exceed the values presented in this study.

4 CONCLUSIONS

As accumulation in soil mesocosms was presented according to the following: root > leaf > stem. *Citrullus lanatus* is an As accumulator plant. Concentrations above 0.7 mg As L^{-1} in irrigation water have adverse effects on the development of the plant. Higher concentrations (1.4 mg As L^{-1}) inhibit the development of the plant. The use of irrigation water contaminated by As causes short-term accumulation of this metalloid in the soil-plant system, causing adverse effects to the crops and human health.

ACKNOWLEDGEMENTS

Thanks to Alejandro Benavides Montoya of Water Quality Laboratory (CIMAV), Mirna E. Cedillo and staff of Environmental Parameters Laboratory, Zootechnics and Ecology Faculty—UACH.

REFERENCES

Finnegan, P.M. & Chen, W. 2012. Arsenic toxicity: the effects on plant metabolism. *Frontiers in Physiology 3: 182.*

Garg, N. & Singla, P. 2011. Arsenic toxicity in crop plants: physiological effects and tolerance mechanisms. *Environmental Chemistry Letters 9*: 303–321.

Rahman, I.M.M. & Hasan, M.T. 2007. Arsenic Incorporation into Garden Vegetables Irrigated with Contaminated Water. *Journal of Applied Sciences and Environmental Management 11(4):* 105–112.

Roychowdhury, T., Tokunaga, H., Uchino, T. & Ando, M. 2004. Effect of arsenic-contaminated irrigation water on agricultural land soil and plants in West Bengal, India. *Chemosphere 58:* 799–810.

Zhao, F.J., Ma, J.F., Meharg, A.A. & McGrath, S.P. 2009. Arsenic uptake and metabolism in plants. *New Phytologist, 181:* 777–794.

One Century of the Discovery of Arsenicosis in Latin America (1914–2014) –
Litter, Nicolli, Meichtry, Quici, Bundschuh, Bhattacharya & Naidu (Eds)
© 2014 Taylor & Francis Group, London, ISBN 978-1-138-00141-1

Arsenic species and antioxidative responses to arsenic in the silver back fern, *Pityrogramma calomelanos* (L.) Link

N.V. Campos, S.A. Silva, I.B. Viana, J.W.V. Mello, M.E. Loureiro & A.A. Azevedo
Universidade Federal de Viçosa, Viçosa, MG, Brazil

B.L. Batista
Universidade de São Paulo, Ribeirão Preto, SP, Brazil
Universidade Federal do ABC, Santo André, SP, Brazil

F. Barbosa Jr.
Universidade de São Paulo, Ribeirão Preto, SP, Brazil

ABSTRACT: This survey aimed at investigating the arsenic (As) speciation and the responses of the enzymatic antioxidant system in *Pityrogramma calomelanos* under As stress. Ferns at 4–5 frond stage were grown in half-strength Hoagland's solution containing 0 or 1 mM of As. After 21 days, the As concentrations in pinnules, raquis + stipe, and roots were around of 3108, 275 and 283 mg As kg^{-1}, respectively. As was present mainly as arsenite (pinnules, 75%) and arsenate (raquis + stipe 74%, roots 92%). Exposure to As increases the activities of APX and POX in pinnules, and SOD and CAT in roots. Induction of antioxidant enzymes is essential to quench reactive oxygen species to a low level, thus, enhancing continuous As accumulation. Reduction of As is also important to increase the translocation and sequestration in shoot. These results help to understand *P. calomelanos* As hyperaccumulation.

1 INTRODUCTION

Arsenic (As) is a metalloid ubiquitously distributed in the environment. The presence of As in drinking water is a serious health problem in many countries due to its genotoxic and mutagenic effects. Areas with high As concentrations derive from natural sources, such as erosion and leaching from geological formations, and from anthropogenic sources, such as mining activities and metal processing (Nriagu, 1994). The cleanup of contaminated water and soils requires the use of appropriate remediation techniques to prevent impacts on the ecosystems.

As-hyperaccumulating ferns provide the possibility of developing cost-effective and ecofriendly As phytoremediation programs. *Pteris vittata* L. (brake fern) was the first As hyperaccumulating fern described and has been widely studied. *Pityrogramma calomelanos* L. (silver back fern) is a less known hyperaccumulating fern that grows naturally in As-contaminated sites. *P. calomelanos* can accumulate up to 8350 mg As kg^{-1} dry weight (DW) (Francesconi *et al.*, 2002) without show symptoms. However, the mechanisms responsible for As hyperaccumulation and detoxification remain unknown.

This study aims at investigating the accumulation of As and its species, and the responses of the enzymatic antioxidant defense system of *P. calomelanos* to high As concentration in hydroponic culture.

2 MATERIAL AND METHODS

2.1 *Experimental conditions and As determination*

Ferns at 4–5 frond stage, obtained from spores, were transferred to a hydroponic system with half-strength Hoagland's solution and continuous aeration. After 40 days of acclimation, 1 mM of As was added in the nutrient solution of half of the plants, supplied as sodium arsenate ($Na_2HAsO_4{\cdot}7H_2O$). The remaining ferns were maintained in Hoagland's solution without As (control treatment). The nutrient solution was renewed every four days and the pH adjusted daily to 5.5. The ferns were harvested after 21 days of exposure and separated into roots, pinnae, and stipe + rachis. Part of this material was oven-dried at 60 °C and then powdered with a knife mill to determine the total As and its species. To evaluate total As, milled samples (0.1 g) were extracted with nitro-perchloric solution (3:1) in a digestion block following procedure described by Tedesco *et al.* (1995). The samples were analyzed by ICP-AES. Arsenic species were evaluated according to Batista *et al.* (2011). Samples (0.2 g) were extracted in 2% (v/v) nitric acid and analyzed in HPLC-ICP-MS.

2.2 *Enzyme extraction and assays and statistical analysis*

To assess enzyme activities, pinna and root fresh weight samples were ground in liquid nitrogen and homogenized in potassium phosphate buffer according to Peixoto *et al.* (1999). The homogenates were centrifuged at 12,000 *g* for 15 min at 4 °C and the supernatant was used as the source of crude enzyme. Activities of ascorbate peroxidase (APX), catalase (CAT), glutathione reductase (GR), peroxidase (POX), and superoxide dismutase (SOD) were determined according to Peixoto *et al.* (1999) and references therein, and expressed per mg of protein. For protein assay, the Bradford method was used.

Statistical analysis was performed using SISVAR program. Student's test was employed for comparison of the means at the 0.05 level of significance.

3 RESULTS AND DISCUSSION

After 21 days of exposure, the pinnules of *P. calomelanos* accumulated up to 3100 mg As kg^{-1} (Figure 1). The As concentrations in the raquis + stipe, and roots were about 275 and 283 mg As kg^{-1}, respectively. The higher As concentrations observed in the pinnules, much higher than those in the roots, imply that a considerable amount of As was translocated from roots to fronds. Arsenic in the pinnules was present mainly as arsenite (75%), with the remainder being arsenate (Table 1). In contrast, in raquis + stipe and roots, arsenate was the predominant form (74 and 92%)

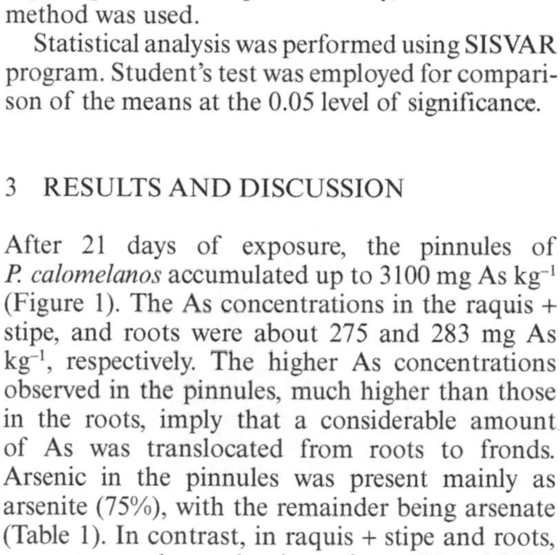

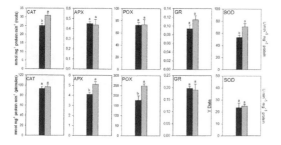

Figure 1. Activities of CAT, APX, POX, GR and SOD in the roots and pinnules of *P. calomelanos* exposed to 0 mM As (black bars) and 1 mM As (gray bars) for 21 days.

Table 1. Concentration of total arsenic and its species in plants of *Pityrogramma calomelanos* under As exposure.

	Concentration (mg kg^{-1})		
As species	Pinnule	Raquis + stipe	Root
Arsenate	258.33	96.26	139.31
Arsenite	761.00	27.48	6.95
MMA	–	–	–
DMA	–	6.24	4.70
Some*	1019.33	129.98	150.96
As Total	3108.28	275.01	283.46
Recovery**	32%	48%	53%

* Sum of arsenic species [arsenate, arsenite, monomethylarsinate (MMA) and dimethylarsinate (DMA)].
** Recovery of As by HPLC-ICP-MS method [100*(Sum of As species/As total)].

and dimethylarsinate was detected as traces. This means that a high percentage of As was reduced to arsenite prior to be accumulated in pinnules.

The reduction of arsenate to arsenite may contribute to the synthesis of enzymatic antioxidants such as SOD and CAT (Meharg & Hartley-Whitaker, 2002). Activities of the antioxidant enzymes were related to plant tissues, with higher activities being in the pinnules than in roots (Figure 1).

Superoxide radicals induced by As stress in *P. calomelanos* are converted to H_2O_2 by the action of SOD. In roots, SOD activity was increased by As treatment. Comparison of the activities of the H_2O_2-scavenging enzymes showed that the lower response of CAT activity to As in pinnules may be compensated by the increased activities of APX and POX, and that the inadequate response of APX and POX activity to As in roots was compensated by the increased activity of CAT. It showed that these enzymes were functioning concurrently to remove H_2O_2 in the different parts of this plant. Similar results were reported by Cao *et al.* (2004) for *Pteris vittata*. GR activity does not change in both organs. The differences in enzyme activities between pinnules and roots may be related to the partition of As species, which lead to particular effects on plant metabolism.

4 CONCLUSIONS

Induction of antioxidant enzymes is essential to quench reactive oxygen species to a low level, thus, enhancing continuous accumulation of As. The arsenic reduction seems to be also important to increase As translocation and sequestration in shoots. These results help us to understand why *P. calomelanos* is a great As hyperaccumulating fern.

ACKNOWLEDGEMENTS

To FAPEMIG (Foundation for Research Support of Minas Gerais) and to CNPq (National Council for Scientific and Technological Development).

REFERENCES

Batista, B.L., Souza, J.M.O., De Souza, S.S. &, Barbosa Jr, F. 2011. Speciation of arsenic in rice and estimation of daily intake of different arsenic species by Brazilians through rice consumption. *J. Hazard. Mat.,* 191: 342–348.

Cao, X., Ma, L.Q. & Tu, C. 2004. Antioxidative responses to arsenic in the arsenic-hyperaccumulator Chinese brake fern (*Pteris vittata* L.). *Environ. Pollut.* 128: 317–325.

Francesconi, K., Visoottiviseth, P., Sridokchan, W. & Goessler, W. 2002. Arsenic species in an arsenic hyperaccumulating fern, *Pityrogramma calomelanos*: a potential phytoremediator of arsenic contaminated soils. *Sci. Total Environ.* 284: 27–35.

Meharg, A.A. & Hartley-Whitaker, J. 2002. Arsenic uptake and metabolism in arsenic resistant and nonresistant plant species. *New Phytol.,* 154: 29–43.

Nriagu, J.O. 1994. *Arsenic in the Environment: Part 1 Cycling and Characterization.* Wiley, New York.

Peixoto, P.H.P., Cambraia, J., Sant'Anna, R., Mosquim, P.R. & Moreira, M.A. 1999. Aluminium effects on lipid peroxidation and the actives of enzymes of oxidative metabolism in sorghum. *Ver. Bras. Fisiol. Veg.* 11:137–143.

Tedesco, M.J., Gianello, C., Bissani, C.A., Bohnen, H. & Volkweiss, S.J. 1995. *Análise de solo, plantas e outros materiais.* Universidade Federal do Rio Grande do Sul, Porto Alegre.

One Century of the Discovery of Arsenicosis in Latin America (1914–2014) – Litter, Nicolli, Meichtry, Quici, Bundschuh, Bhattacharya & Naidu (Eds)
© 2014 Taylor & Francis Group, London, ISBN 978-1-138-00141-1

Arsenic efflux from *Microcystis aeruginosa* under different phosphate regimes

C. Yan & Z. Wang
Key Laboratory of Urban Environment and Health, Institute of Urban Environment, Chinese Academy of Sciences, Xiamen, China

ABSTRACT: In this study, we investigated the arsenic efflux and speciation in *M. aeruginosa* pre-exposed to 10 µM arsenate or arsenite for 24 hours during a short- (12 h) and long-period (13 d) depuration under phosphate–enriched (+P) and depleted (–P) conditions. Arsenate was the predominant species in algal cells during the whole depuration period. During the short-period depuration, arsenic efflux from *M. aeruginosa* was rapid and only arsenate was found in solutions. However, during the long-period depuration, arsenate and dimethylarsinic acid (DMA) were the two dominant arsenic species in solutions under –P condition, but arsenate was the only species under +P condition. The experimental results also suggested that phosphorus has obvious effects on accelerating arsenic efflux and promoting arsenite bio-oxidation in *M. aeruginosa*, and that phosphorus-depletion could reduce arsenic efflux from algal cells, as well as accelerate arsenic reduction and methylation.

1 INTRODUCTION

Microcystis aeruginosa is one of the most common and widespread bloom-forming cyanobacteria in aquatic environments and often causes serious water quality problems. Here we report a series of algal culture experiments designed to demonstrate arsenic (As) efflux from *M. aeruginosa* in freshwater under different phosphate conditions. Changes of arsenic speciation in both alga and culture medium were measured. This is the first report of the effects of phosphate on As efflux from *M. aeruginosa*. This information would provide for a better interpretation of arsenic biogeochemistry in the environment.

2 METHODS/EXPERIMENTAL

2.1 *Arsenic efflux experiments*

Before As efflux experiments, *M. aeruginosa* was exposed to 10 µM arsenate or arsenite separately in 300 mL nutrient solution for 24 h. For each treatment, 30 mL aliquots of algal suspension were centrifuged into a pellet; then, all of them were washed twice with sterile Milli-Q water and incubated in ice-cold phosphate buffer for 10 min. Two of the pellets were prepared to detect intracellular arsenic speciation, and the remainders were respectively resuspended at a concentration of $2–5 \times 10^5$ cells mL^{-1} in 150 mL of BG11 mediums with or without 150 mM phosphate for depuration.

To analyze the characterization of As efflux, the algal depuration phase of arsenic was set up for short- and long-period term with 12 h and 13 d, respectively. During the algal depuration period, 10 mL aliquots of algal suspension were taken from the solutions periodically. The algae were subsequently centrifuged to create an algal pellet, and freeze-dried at a vacuum freeze dryer for arsenic speciation analysis. The supernatant was filtered through 0.45 µm cellulose acetate syringe filters and stored at –20 °C for the analysis of total arsenic and arsenic species.

2.2 *Sample analysis*

Arsenic species in nutrient solutions and algae extracts were determined using HPLC-ICP-MS, as described previously (Wang *et al.*, 2013). Total arsenic concentrations in nutrient solutions were determined by ICP-MS operating in the helium gas mode to remove possible interference of ArCl on m/z 75.

3 RESULTS AND DISCUSSION

3.1 *Arsenic efflux dynamics under ±P conditions*

Figure 1 shows that arsenic elimination phase can be separated into "fast" and "slow" phases. In the fast phase (12 h), intracellular arsenic concentration was linear in relation to depuration time. However, in the slow phase (13 d), intracellular arsenic

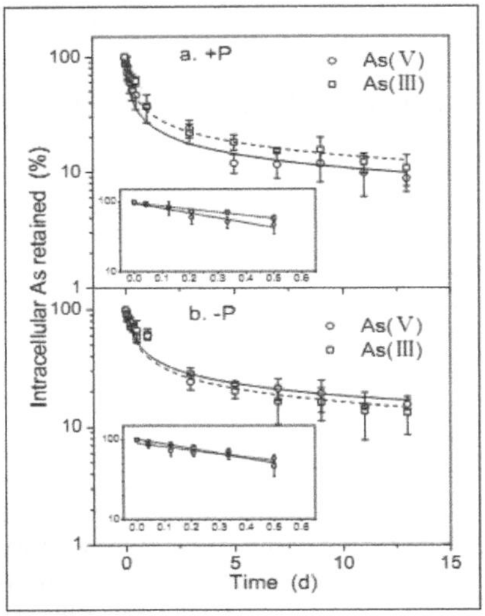

Figure 1. Proportional loss of arsenic over time from *M. aeruginosa* following 24 hours of arsenate or arsenite exposure under different phosphate regimes.

concentration exhibited a decreasing curve pattern in relation to depuration time. In the fast phase, intracellular arsenic concentration decreased rapidly (under +P condition, arsenate: 69% and arsenite: 53%; under –P condition, arsenate: 60% and arsenite: 69%). It appears that significant impacts on arsenic fast efflux occurred in both of +P and –P conditions after arsenate or arsenite pre-exposures. However, in the slow phase, intracellular arsenic concentrations in both arsenate and arsenite treatments decreased by approximately 3.32% and 4.09% per day under +P and –P conditions, respectively. As expected, different P conditions had significant effects on As efflux rate. This type of profile, which reflected biokinetic properties, has been reported by Croteau *et al.* (2004) and Khan *et al.* (2012). At this stage, it can be concluded that As efflux rate depended not only on the species supplied to pre-exposure but also on P conditions.

3.2 *Changes in cellular arsenic fractionation*

During the whole depuration period, arsenate was the predominant species in cells, while arsenite only accounted for not more than 45% of intracellular arsenic (Figure 2). No methylated arsenic species were detected in the algae. For the short-period depuration, the percentages of arsenite in algae under +P condition were two times more than that under –P condition during the depuration time. As for the long-period depuration, the percentages

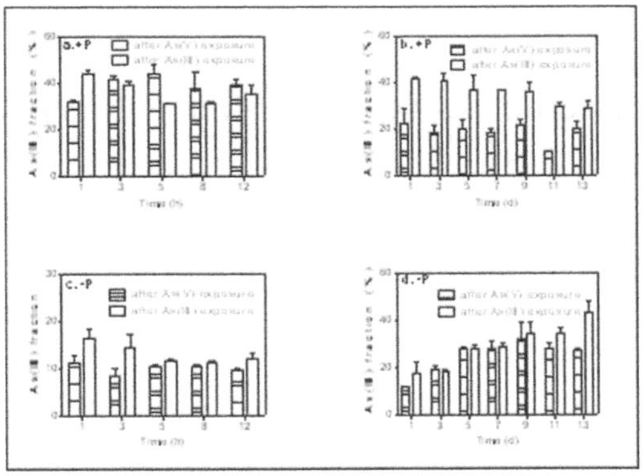

Figure 2. Fraction of arsenite in *M. aeruginosa* over short- (12 h) and long- (13 d) period depuration after 24 h of 10 μM arsenate and arsenite individual pre-exposures.

of arsenite in algae exposed to arsenite decreased gradually with depuration time under +P condition, but it tended to show a general increase with time under –P condition (Figure 2).

3.3 *Changes in total arsenic and its species in mediums*

For arsenite pre-exposure total arsenic concentrations in +P media were approximately 1.5-fold higher compared to arsenate pre-exposure (data not shown). In contrast, for arsenite pre-exposure total arsenic concentrations in –P mediums were lower compared to arsenate pre-exposure. Additionally, it should be noted that only arsenate was detected in mediums under +P conditions during the long-period depuration experiments, suggesting that arsenate was extruded mainly by algae under +P conditions. Similarly, only arsenate was detected in –P mediums during the short-period depuration.

Arsenite and DMA were also detected after 12 h and 3 d of depuration, respectively. This indicated that significant changes of arsenic species in –P mediums occurred over the 13 days of depuration period. Arsenite concentrations fluctuated and accounted for less than 14% and 16% of total arsenic in –P mediums for arsenate and arsenite pre-exposure (data not shown), respectively. DMA concentrations in –P mediums maintained a relatively stable proportion at an average of 49 ± 5% and 40 ± 3% for arsenate and arsenite pre-exposure (data not shown), respectively. This suggested that *M. aeruginosa* had the ability to methylate arsenic and in turn could release it to ambient mediums.

4 CONCLUSIONS

In general, our data indicated that arsenate and arsenite were two major arsenic species in *M. aeruginosa* cells during the short- and long-period depuration under +P or −P condition. Phosphorus has obvious effects on accelerating arsenic efflux and promoting arsenite bio-oxidation of *M. aeruginosa* after arsenate and arsenite pre-exposure. All these findings would be beneficial to understand arsenic biogeochemistry and its potential environmental hazards at different trophic (phosphorus) levels.

ACKNOWLEDGEMENTS

This work was supported by the National Nature Science Foundation of China (21277136) and International Science & Technology Cooperation Program of China (2011DFB91710).

REFERENCES

Croteau, M.N., Luoma, S.N., Lopez, C.B. & Topping, B.R. 2004. Stable metal isotopes reveal copper accumulation and loss dynamics in the freshwater bivalve corbicula. *Environ Sci Technol* 38, 5002–5009.

Khan, F.R., Misra, S.K., Garcia-Alonso, J., Smith, B.D., Strekopytov, S., Rainbow, P.S. Luoma, S.N. & Valsami-Jones, E. 2012. Bioaccumulation dynamics and modeling in an estuarine invertebrate following aqueous exposure to nanosized and dissolved silver. *Environ Sci Technol* 46, 7621–7628.

Wang, Z.H., Luo, Z.X., Yan, C.Z., 2013. Accumulation, transformation, and release of inorganic arsenic by the freshwater cyanobacterium *Microcystis aeruginosa*. *Environ Sci Pollut Res* 20, 7286–7295.

One Century of the Discovery of Arsenicosis in Latin America (1914–2014) – Litter, Nicolli, Meichtry, Quici, Bundschuh, Bhattacharya & Naidu (Eds)
© 2014 Taylor & Francis Group, London, ISBN 978-1-138-00141-1

Arsenic speciation in submerged and terrestrial soil-plant systems

M. Greger & T. Landberg
Department of Ecology, Environment and Plant Sciences, Stockholm University, Stockholm, Sweden

R. Herbert
Department of Earth Sciences, Uppsala University, Uppsala, Sweden

I. Persson
Department of Chemistry, Swedish University of Agricultural Sciences, Uppsala, Sweden

ABSTRACT: In a series of cultivation experiments in unsaturated and water-saturated soil/sediment, As speciation was determined in the terrestrial plant lettuce and the emergent plant common cottongrass. Arsenic species were analyzed in soils and plants using XANES/EXAFS and HPLC-AAS. The transformation process between As species was much faster in plant tissue than in soil. When arsenite was present in soil, plant roots contained more arsenate than arsenite. Lettuce growing in terrestrial soil with no arsenite contained more arsenate in shoots than arsenite, in contrast to common cottongrass. Differences in the arsenate/arsenite-ratio of plant tissue indicated both plant species and environmental differences.

1 INTRODUCTION

Arsenic (As) is a toxic element that exists in various chemical species in the environment, where the inorganic species are the most toxic. Arsenite has oxidation state +III and exists in the form of ortho-arsenite AsO_3^{3-} and meta-arsenite AsO_2^{-}, while arsenate has oxidation state +V and exists as AsO_4^{3-}. In terrestrial soils, the primary form is arsenate, while, in submerged soils, arsenite is also present (Sadiq, 1997). Both oxygen and plant processes may change the speciation of As in soil and plants. Plants take up both arsenite and arsenate and they can be transformed and immobilized, e.g. arsenite may be bound to sulfur in phytochelatins in plants (Smith *et al.*, 2008; Moreno-Jiménez *et al.*, 2012). The aim of this work was to determine the relative changes in As speciation in a terrestrial and a submerged soil-plant system.

2 MATERIALS AND METHODS

In order to study arsenic uptake and speciation in terrestrial and emergent plants from arsenic-rich soils, the terrestrial plant lettuce (*Lactuca sativa*) and the emergent plant common cottongrass (*Eriophorum angustifolium*) were cultivated for 7 days in unsaturated and water-saturated (de-oxygenated) soils and water-saturated sediment. In addition, the unsaturated terrestrial soil was treated with nitrogen gas (14 weeks) in order to decrease the oxygen content and induce a redox transformation from arsenate to arsenite. The soil and sediment were derived from alum shale and from mine tailings, respectively, containing As up to 470 mg kg^{-1}. After 7 days cultivation, the experiment was terminated and the different As species were analyzed in soils (rhizosphere and bulk) and plants (roots and shoots) by XANES/EXAFS and HPLC-AAS.

3 RESULTS AND DISCUSSION

EXAFS measurements of alum soil and $HAsO_4^{2-}$ (aq) standard solutions both indicated As···O scattering lengths of 1.68 Å and four nearest neighbors (Figure 1), confirming that arsenic was present in the soil as inorganic arsenate, most likely as an adsorbed complex. The extractable As, which was loosely bound to the colloids and thus available to plants, was in the range of about 1/100 of total As in soil. When treated with nitrogen in order to remove oxygen, the arsenite content increased and arsenate decreased by about 25%; however, this low level of arsenite could not be resolved in the EXAFS spectra.

For cottongrass grown in alum shale soils, arsenic is present in roots and shoots in multiple oxidation states and coordinates with multiple ligands. Model fits to the k^3-weighted EXAFS spectra of cottongrass roots indicate As···O scattering lengths of ca. 1.78 Å with 3.5 nearest neighbors (Figure 1), corresponding to oxygen bonding with both As(III) and As(V). Furthermore, a scattering length of ca. 2.2 Å is resolved from model fits of the k^3-weighted EXAFS spectra, agreeing with the bond length for As(III)···S scattering.

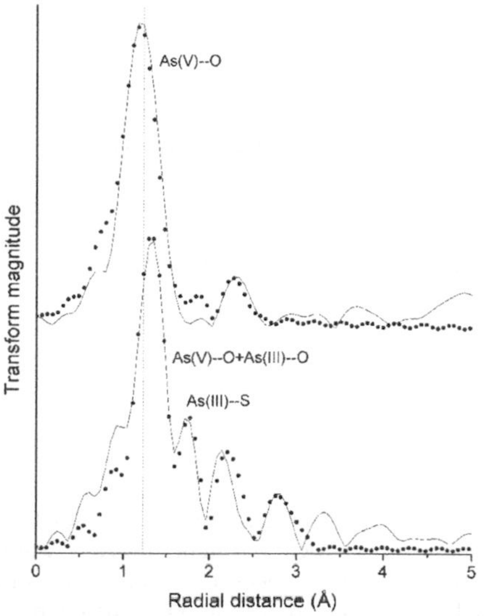

Figure 1. Fourier transformed EXAFS spectra of alum shale (upper) and cottongrass root (lower). Transform data shown as dots, shell fits as lines. Peak position for $HAsO_4^{2-}$(aq) standard shown as vertical dashed line.

Our work shows that transformation in the soil from arsenate to arsenite is slow when the reducing capacity of the soil is low, as in this case, and that the loosely bound fraction of As is low. It is likely that most of the oxygen-bound As(V) is firmly bound to metal oxide surfaces in the soil.

Cottongrass absorbed As and translocated about 10% to the shoot and, independent of arsenate or arsenite concentrations in soil, both arsenate and arsenite were found in the plant tissue (Table 1). When arsenite was present in the soil, plant roots contained significantly more arsenate than arsenite. Lettuce growing in terrestrial soil with no arsenite contained proportionally more arsenite than arsenate in the plant tissue, compared to common cottongrass (Table 1).

In the plant tissue, in contrast to the soil, the transformation process between As species is fast. After a growth period of seven days in soil containing solely arsenate, plants contained both arsenite and arsenate. With arsenite present in the sediment, proportionally more arsenate was present in plant roots compared with only arsenate in the soil. Arsenite is therefore transformed further to arsenate in the plant roots or bound to S in e.g. phytochelatins. The presence of As-S species is confirmed by other studies that indicate that As(III) is often bound with reduced sulfur in phytochelatin complexes (Smith *et al.*, 2008; Moreno-Jiménez *et al.*, 2012). The differences in the arsenate/arsenite-ratio of plant tissue in lettuce and cottongrass indicate plant species differences; lettuce has a higher proportion of arsenite, especially in shoots, relative

Table 1. Distribution of arsenite (As(III)) and arsenate (As(V)) in cottongrass (grown on unsaturated and water-saturated soil) and in lettuce (grown in terrestial soil) in percentage of total arsenic concentration.

	Root (%)		Shoot (%)	
	As(III)	As(V)	As(III)	As(V)
Cottongrass				
Unsat. soil	44.4 ± 0.4	44.8 ± 1.0	5.6 ± 0.3	5.0 ± 0.1
Saturated	40.0 ± 1.1	49.6 ± 0.6	5.2 ± 0.4	5.2 ± 0.4
Lettuce				
Terrest. soil	53.1 ± 2.2	21.2 ± 0.9	19.0 ± 0.3	6.7 ± 0.1

to cottongrass. The transformation of arsenic in the plants might also depend on the environment to which the plant species have adapted.

4 CONCLUSIONS

This work shows that both the plant species and environmental factors influence the As-species found in plants. In soil, the oxygen level and presence of electron donors are important for the rate of As redox transformation; this work demonstrated a very slow reduction from As(V) to As(III) in de-oxygenated soils. The redox transformation is much faster in plants than in the soil. However, the results show that the process is plant-species specific, as both the transformation and translocation differs among plant species. Detailed knowledge about the differences may be useful to minimize toxic levels of arsenic in plants, especially in the edible parts of plants.

ACKNOWLEDGEMENTS

This work was funded by the Swedish Research Council.

REFERENCES

Anawar, H.M., Garcia-Sanchez, A. & Santa Regina, I. 2008. Evaluation of various chemical extraction methods to estimate plant-available arsenic in mine soils. *Chemosphere* 70: 1459–1467.

Mir, K.A., Rutter, A., Koch, I., Smith, P., Reimer, K.J. & Poland, J.S. 2007. Extraction and speciation of arsenic in plants grown on arsenic contaminated soils. *Talanta* 72: 1507–1518.

Moreno-Jiménez, E., Esteban, E. & Peñalosa, J.M. 2012. The fate of arsenic in soil-plant systems. *Rev. Env. Contam. Toxicol.* 215: 1–37.

Sadiq, M. 1997. Arsenic chemistry in soils: an overview of thermodynamic predictions and field observations. *Water, Air and Soil Pollut.* 93: 117–136.

Smith, P.G., Koch, I. & Reimer, K.J. 2008. Uptake, transport and transformation of arsenate in radishes (*Raphanus sativus*). *Sci. Total Environ.* 390: 188–197.

*One Century of the Discovery of Arsenicosis in Latin America (1914–2014) –
Litter, Nicolli, Meichtry, Quici, Bundschuh, Bhattacharya & Naidu (Eds)*
 ISBN 978-1-138-00141-1

Arsenic and trace element concentration in surface water and agricultural soils in the southeast of Córdoba province, Argentina

A. Fernández Cirelli & A. Pérez Carrera
Instituto de Investigaciones en Producción Animal (INPA), CONICET, Facultad de Ciencias Veterinarias, Universidad de Buenos Aires, Buenos Aires, Argentina

M. Billib & P.W. Boochs
Institute of Water Resources Management, Leibniz University of Hannover, Hannover, Germany

ABSTRACT: Arsenic contaminated soils, sediments and water supplies are the major source for the food chain. The main objective of this work is to analyze the arsenic and trace elements content in surface water and agricultural soils in the southeast of Córdoba province, Argentina. Arsenic concentrations in water samples from the Ctalamochita river were between <10 and 18 μg/L, bellow the allowed limit in the Córdoba province. In soils, total As, V and Sr concentration showed no significant differences between the levels found at different depths (30, 60 and 100 cm); values obtained ranged from 3.4 to 9.3 mg/kg, 50.6 to 65.5 mg/kg and 85 to 201 mg/kg, respectively. According to these results, As and V concentration in the studied soils was below the guidelines values considered in Argentina and within the range reported in non contaminated soils in various countries.

1 INTRODUCTION

Arsenic (As) is a toxic pollutant present in the atmosphere as well as in the aquatic and terrestrial environment. Arsenic contaminated soils, sediments and water supplies are the major source of food chain contamination. The Chaco-Pampean plain in Argentina comprises one of the largest areas in the world with high levels of As in groundwater (one million km^2). The province of Cordoba, located in the center of the country, is an important agricultural region highly affected by this contaminant of natural occurrence. In the southeast region of this province, studies have been conducted on water quality for agricultural activities, spatial and temporal variation of As content in both shallow and deep aquifers and biotransference of As from water and forage to the food chain (Pérez Carrera & Fernández Cirelli, 2004, 2005, 2007).

The main objective of this work is to analyze the As and trace elements content in surface water and agricultural soils from our working area, the southeast of Córdoba province.

2 METHODS

The study area is located in the southeast of Cordoba province, Argentina, between 62° 33′ and 62° 57′ west longitude and 32° 12′ and 32° 50′, south latitude.

Surface water samples were collected in seven stations from the main water body in the area (Ctalamochita river). In all cases, the temperature, pH and the electrical conductivity of the water samples were measured *in situ*.

For trace elements, analysis samples were collected in 100 mL plastic bottles previously rinsed with HNO_3 10% and deionized water. For their conservation, samples were acidified (HNO_3 0.2%) and cooled before transportation. Once in the laboratory, the samples were filtered through a cellulose acetate membrane with pore size of 0.45 microns.

Soil samples were collected at 30, 60 and 100 cm depth in agricultural farms and characterized taking into account the physicochemical characteristics (pH, conductivity, CEC) as well as organic matter, clay, silt and sand content. Samples were air dried, sieved and digested by standard methods prior to the quantification of trace elements by inductively coupled plasma-optical emission spectroscopy (ICP-OES).

3 RESULTS AND DISCUSSION

Ctalamochita river constitutes the main drinking water source for the more populated towns in the area (Bell Ville, Morrison, Monte Leña). As level in the water samples were between <10 and 18 μg/L, bellow the allowed limit in the Córdoba province. On the other hand, total As concentration in shallow

Table 1. Trace elements content in the studied soils (mg/kg).

Trace elements content in soils				
	Minimum	Maximum	Average	SD*
As	3.4	9.3	5.6	1.9
V	50.6	65.5	58.5	3.6
Sr	85	201	117	28.8

groundwater ranges between 70 and 4500 μg/L (Pérez Carrera and Fernández Cirelli, 2004).

All soil samples, collected from alfalfa fields, were characterized as molisols, with aptitude for agriculture activities. They showed similar physicochemical characteristics as well as clay, silt and sand contents.

Arsenic and related elements (V and Sr), toxic trace elements, coming from anthropogenic activities (Pb, Cd and Cr), and micronutrients (Cu and Zn) concentrations were determined in order to analyze the risk for agriculture.

In the studied farms, the total As, V and Sr soil concentration in the surface layer (0–100 cm depth), as well than those of the other studied elements, showed no significant differences between the levels found at different depths (30, 60 and 100 cm), $p < 0.05$.

Since levels of Pb, Cd and Cr were below the allowed limits for agricultural soil in Argentina (24051 Law, 1993), attention was focused on As and related elements considering its natural origin. Arsenic, V and Sr soil concentrations are shown in Table 1.

According to these results, As and V concentrations in the studied soils were below the guideline values considered in Argentina for agricultural soils (As: 20 mg/kg, V: 200 mg/kg, Law 24051 (1993)) and within the range of As and V concentrations in non contaminated soils reported in various countries. Regarding the Sr concentrations, there are no regulations since no harmful health effects have been reported until now.

The plots in Figures 1 and 2 display that the concentrations of As are connected with the concentrations of Sr and V. Arsenic is correlated to these elements in a positive linear way. The coefficients of determination show a strong correlation both between Sr and As ($R^2 = 0.84$) and between V and As ($R^2 = 0.86$).

4 CONCLUSIONS

The obtained results indicate that the risk of transference of trace elements to agricultural products is very low. Significant positive correlations found for As, V and Sr suggest a similar origin for these elements in soils and in groundwater. Low As levels in soils are in accordance with the low concentrations determined in surface water.

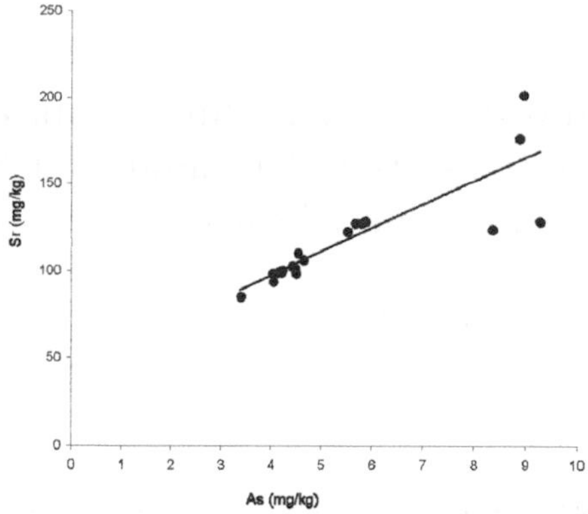

Figure 1. Relationship between arsenic and strontium in the studied soils.

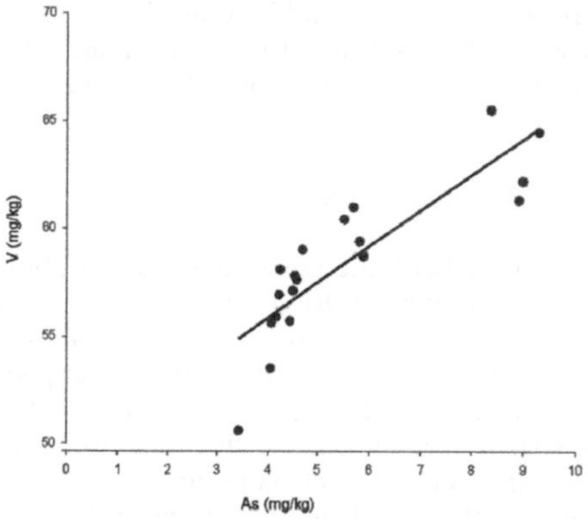

Figure 2. Relationship between arsenic and vanadium in the studied soils.

ACKNOWLEDGEMENTS

The authors are grateful to the University of Buenos Aires, CONICET, MINCyT—BMBF (Bilateral Project Argentina-Alemania, AL/11/07) for financial support.

REFERENCES

Pérez Carrera, A. & Fernández Cirelli, A. 2004. Niveles de arsénico y flúor en agua de bebida animal en establecimientos de producción lechera (Pcia. de Córdoba, Argentina). *Revista de Investigación Veterinaria (INVET)*, 6:51–59.

Pérez Carrera, A. & Fernández Cirelli, A. 2005. Arsenic concentration in water and bovine milk in Cordoba, Argentina. Preliminary results. *Journal of Dairy Research*, 72: 122–124.

Pérez Carrera, A. & Fernández Cirelli, A. 2007. Problemática del arsénico en la llanura sudeste de la provincia de Córdoba. Biotransferencia a leche bovina. *Revista Investigación Veterinaria (INVET)*, 9:123–135.

Law 24051. 1993. Regime for hazardous wastes. Argentine.

One Century of the Discovery of Arsenicosis in Latin America (1914–2014) – Litter, Nicolli, Meichtry, Quici, Bundschuh, Bhattacharya & Naidu (Eds)
 ISBN 978-1-138-00141-1

Soil water management systems decreased arsenic dissolution from the soils and accumulated As on the root iron plaque in As-contaminated soils

T.-H. Huang & Z.-S. Chen
Department of Agricultural Chemistry, National Taiwan University, Taipei, Taiwan

ABSTRACT: Soil water management systems were developed to mitigate arsenic levels in rice. Pot experiments were conducted to evaluate the effects of water management systems on reducing As dissolution from three As-contaminated soils. The ratio of As/Fe in the soil pore water revealed the liability of soil As and Fe. The conventional treatment significantly reduced dissolution of As and Fe in the soil pore water under paddy (Gd-pd) and upland (Gd-up) soils, but no significant effects in alluvial soil (Er) were observed, due to well drainage and soil active Fe content. Water saturated treatment mitigated As dissolution from both Er and Gd soils and accumulation on the iron plaque of rice root.

1 INTRODUCTION

In the paddy field, the uptake and translocation of As is efficient in rice (*Oryza sativa* L.). Several strategies are developed to mitigate the rice contamination. Drainage or aerobic treatments were applied in the soil water management to reduce As dissolution from soil. While physical and chemical properties of soils are different, distinct management systems were adapted to the soil. This study discusses how water managements act on the As and Fe dissolution from soils and precipitation on the iron plaques of root.

2 MATERIALS AND METHODS

2.1 *Pot experiment*

The pot experiment was conducted with 3 arsenic levels in soils and 5 soil water management systems. Rice (*Oryza sativa* L. cv Tainan 11) was grown in the phytotron. The two volcanic soils were sampled from paddy soil (Gd-pd) and upland soil (Gd-up). The alluvial soil (Er) was collected from a representative agricultural area in Taiwan and spiked with 35 mg/kg As.

Five soil water management systems listed as follows: (#1) conventional treatment, drainage for one week at the rice maximum tiller number stage and keep water head at 3–5 cm depth in the resting cultivation; (#2) aerobic before heading treatment (A/F), keep drainage condition for 6 weeks before "heading", and then keep water head at 3–5 cm depth in the resting cultivation; (#3) aerobic after heading treatment (F/A), keep water head at 3–5 cm depth before "heading" and then drainage condition for 6 weeks in the resting cultivation; (#4) water saturated treatment, maintain the soil pore saturated with water and no water head at the soil surface; and (#5) flooding treatment, keep the water head at 3–5 cm depth during the whole cultivation period.

2.2 *Chemical analysis*

Chemical sequential extraction of As in soil was conducted (Wenzel *et al.*, 2001). The amorphous and free iron in the soil was extracted by ammonium oxalate solution (pH 4) and DCB buffer solution. Soil pore water from each treatment was collected during the growing period of rice, and As and Fe concentration were measured within 24 hours.

After harvest, the rice root tissue from the saturated and flooding treatment was kept for subsequent study. Iron plaque extraction process was followed according to Taylor and Crower (1983) at room temperature. Arsenic was determined with ICP/MS or HGAAS, and Fe was determined with ICP/OES. ANOVA and student t-test were conducted with SAS 8.0, and the significant level was set at $p < 0.05$.

3 RESULTS AND DISCUSSION

3.1 *Fractionation of As and Fe in soils*

The results of sequential extraction of As in soils indicate that 60% (Gd-pd) and 40% (Gd-up) of arsenic was distributed in the amorphous material of the soil. The amorphous Fe content was 30% and 20% of the soil iron of the Gd-pd and Gd-up soil, respectively (Table 1). The relative abundance of amorphous Fe in soils corresponded to the result of arsenic sequential extraction.

Table 1. The basic physical and chemical properties of the studied soils (n = 2, mean ± std).

Soil characteristics	Er soil	Gd-pd soil	Gd-up soil
pH	7.7	5.8	5.7
OC (%) †	1.3 ± 0.1	3.6 ± 0.1	2.0 ± 0.1
soil texture	silt loam	silty clay	silty clay
amor. Fe (g/kg)	4	18 ± 1	15.8
free Fe (g/kg)	8.4	27 ± 1	45 ± 3
total Fe (g/kg)‡	35 ± 1	56 ± 2	75 ± 3
amor. Mn (mg/kg)	205 ± 5	100 ± 1	393 ± 3
free Mn (mg/kg)	222 ± 3	160 ± 1	576 ± 28
total Mn (mg/kg)‡	425 ± 10	265 ± 9.5	719 ± 13
amor. Al (g/kg)	0.5	1.8	2.6
free Al (g/kg)	1.4	3.6	5.7 ± 0.2
Total As (mg/kg)	35 ± 1	379 ± 1	531 ± 32

† OC: organic carbon; ‡ aqua regia extraction.

3.1 *Dissolution of Fe and As in the liquid phase*

During the vegetative period, a linear relationship between arsenic and iron concentration explained reduction and mobilization of iron and arsenic in the anaerobic soil condition (Figure 1). In the soil pore water sample, the As/Fe molar ratio in Er and Gd-up soil was nearly equivalent, 0.0111 and 0.0104, respectively, while that of Gd-pd soil was calculated as 0.0033. The reductive dissolution elevated Fe to a concentration higher than that of As in the Gd-pd soil, where a lower ratio was observed.

Compared with the flooding treatment, a significant change of redox potential was observed in the conventional treatment of Gd soils, rather than in the Er soil. The characteristics of poor drainage in Er soil result in higher As and Fe concentrations at the heading period in the conventional treatment. The amorphous material composed of ferric hydroxide is highly reactive with arsenic ions during the drainage period and the soil aerobic condition.

3.2 *Precipitation of As and Fe on the iron plaque*

In the saturated and flooding treatments, the linear relationship between As and Fe was significant, and the molar As/Fe ratio was 0.0134, which was similar to the ratio found in the soil pore water. On the iron plaque, arsenic and iron played the role of both sink and source (Huang *et al.*, 2012).

The concentration of As and Fe on the root iron plaque in the flooding treatment was significant higher than those of saturated treatment among the three soils (Figure 2). Without the standing water on the paddy soil, oxygen was able to diffuse into the bulk soil and rhizosphere. Saturated treatment mitigated the reductive dissolution of As and Fe and decreased the precipitation on the root.

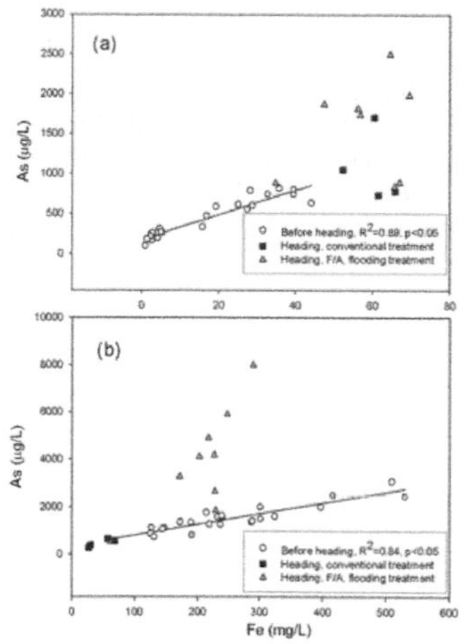

Figure 1. Relationship between arsenic and iron content in the conventional treatment of soils. (a) Er soil; (b) Gd-pd soil.

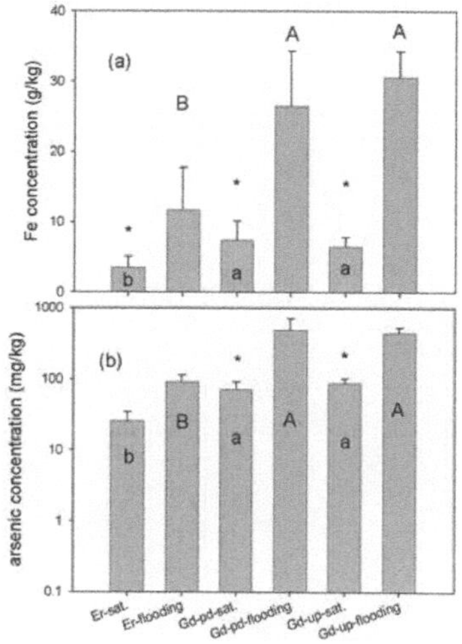

Figure 2. As and Fe concentrations on the root iron plaque in Er, Gd-pd and Gd-up soil. (a) Fe; (b) As. n = 4. mean ± std. * indicates the significant difference between the saturated and flooding treatment.

4 CONCLUSIONS

Soil water managements acted differently on the different soils. The Gd soils were rich in organic carbon and Fe hydroxide, the well drainage leads to reduction of As and Fe concentrations of the soil pore water in the conventional treatment. Arsenic and Fe reprecipitated on the iron plaque as dissolved from the soil. In both Er and Gd soils, water saturated treatment mitigated As dissolution from soil and accumulated on the iron plaque.

REFERENCES

Huang, H., Zhu, Y.G., Chen, Z., Yin, X.X. & Sun, G.X. 2012. Arsenic Mobilization and Speciation during Iron Plaque Decomposition in a Paddy Soil. *Journal of Soils and Sediments* 12:402–410.

Taylor, G.J. & Crowder, A.A. 1983. Use of the dcb technique for extraction of hydrous iron-oxides from roots of wetland plants. *American Journal of Botany* 70:1254–1257.

Wenzel, W.W., Kirchbaumer, N., Prohaska, T., Stingeder, G., Lombi, E. & Adriano, D.C. 2001. Arsenic fractionation in soils using an improved sequential extraction procedure. *Analytica Chimica Acta* 436, 309–323.

One Century of the Discovery of Arsenicosis in Latin America (1914–2014) – Litter, Nicolli, Meichtry, Quici, Bundschuh, Bhattacharya & Naidu (Eds)
 ISBN 978-1-138-00141-1

Oxidative stress in soybean plant subjected to high arsenic concentrations in the soil

C. Bustingorri, R.S. Lavado & K. Balestrasse
Instituto de Investigaciones en Biociencias Agrícolas y Ambientales (INBA, CONICET/UBA), Facultad de Agronomía, Universidad de Buenos Aires, Buenos Aires, Argentina

ABSTRACT: Soybean plants exposed to elevate Arsenic (As) levels show toxicity symptoms. Experiments in pots were performed using soils without or enriched with Phosphorous (P). Results here obtained revealed that As pretreatment causes oxidative stress in plants grown in soils without P, as revealed by an increase in TBARS levels and alterations in GSH content as well as Catalase (CAT) and Peroxidase (GPOX) activities.

1 INTRODUCTION

The use of Arsenic (As)-rich groundwater for irrigation could increase soil concentrations of As, and these soils may negatively affect crop production and food safety. This phenomenon was documented in several countries (Dahal *et al.*, 2008).

Plants exposed to elevated As levels suffer oxidative stress and show toxicity symptoms (Abedin *et al.*, 2002; Pigna *et al.*, 2008). Under normal growth conditions, plants maintain an equilibrium between production and scavenging of reactive oxygen species (ROS), avoiding the damages caused by their accumulation. This equilibrium may be perturbed when plants are subjected to stresses, such as the accumulation of toxic elements. To protect themselves against these toxic oxygen intermediates, plant cells use antioxidant defense systems, including enzymatic and non-enzymatic antioxidants (Balestrasse *et al.*, 2001).

It is known that phosphates have a significant role in As dynamics in soils (Pigna *et al.*, 2008). Due to the similarity between these anions, they compete for the same soil adsorption sites. Phosphate is preferentially adsorbed as compared with As, leading to As an increased solubility when soils received phosphorous fertilizing. Pigna *et al.* (2008) found higher As accumulation in cultivated plants in the presence of Phosphorus (P).

Our objective was to study the oxidative stress of soybean plants subjected to different concentrations of As and the effect of P.

2 METHODS

Experiments in pots (5 L) where soybean was grown were established in a greenhouse. There were two soils: SOIL +P (P > 40 mg kg^{-1}) and SOIL −P (P < 20 mg kg^{-1}). Different concentrations of sodium arsenate were added. To reduce the overestimation of soluble As toxicity, the interaction between the arsenate and soil components was allowed.

There were 4 treatments in each soil: 3 As levels (low: 10 mg As kg^{-1}, medium: 50 mg As kg^{-1} and high: 100 mg As kg^{-1}), and the control treatment. A completely randomized design with three replicates per treatment was established. Pots were irrigated with deionized water maintaining the soil near field capacity.

Pregerminated soybeans seeds (cv Nidera 4613) were sown and thinned to one plant per pot. Each pot received a complete fertilizer. Sixty days after sowing, all plants were rinsed with distilled water and samples of leaves and roots were collected. Then, arsenic was determined by Instrumental Neutron Activation Analysis (INAA) (Alfassi, 1994). Chlorophyll, Thiobarbituric Acid Reactive Substances (TBARS), glutation (GSH), Catalase (CAT) and Peroxidase (GPOX) activities were measured (Balestrasse *et al.*, 2001). Total As was determined on soil samples at seeding.

The results obtained were evaluated using an ANOVA test. When significant differences were found, LSD was applied.

3 RESULTS AND DISCUSSION

All treatments with As showed increased oxidative damage as function of concentration (not shown). Chlorophyll and GSH content decreased in leaves treated with different concentrations of As in the absence of P (−P) compared with those from soils with P. Regarding membrane damage (TBARS), a significant increase was observed in function of the concentration of As in –P soils. These results are consistent with increased enzyme activities of SOD and CAT.

TBARS content and CAT, GPOX activities showed no significant difference in roots treated with different As concentrations in the absence or presence of P. However, GSH content and SOD activity in –P significantly increased compared to +P soils.

These results were consistent with an increase in enzymatic antioxidants defense (SOD, CAT, GPOX) and non-enzymatic (GSH). The effect of the P/As combination was more significant in leaves than in roots of soybean plants.

These results agree with the hypothesis that phosphate competes with arsenate in their absorption in the root and distribution in the whole plant. This is due to the fact that both molecules are chemically similar and have the same sites in the soil as well as in the transport systems. In this way, P could be used to reduce As toxicity in the plants.

Table 1. TBARS (mmol g^{-1} FW) and antioxidants defenses in leaves and roots (GHS—μmol g^{-1} FW; GPOX and SOD—U mg^{-1} protein; and CAT—μmol mg^{-1} protein).

	Control +P	High As +P	Control −P	High As −P
Leaves				
TBARS	49.30	136.84	42.25	173.13
GHS	330.51	85.89	350.00	52.33
GPOX	0.49	1.16	0.45	1.27
SOD	1.72	4.27	2.26	4.96
CAT	32.58	184.91	33.29	193.15
Roots				
TBARS	42.70	133.89	33.10	136.12
GHS	157.35	28.16	170.69	20.69
GPOX	0.49	1.23	0.55	1.10
SOD	0.19	0.51	0.26	0.59
CAT	26.92	57.74	28.22	61.47

4 CONCLUSIONS

These results show that arsenic causes oxidative damage in soybean plants, as observed by the increase of membrane damage and the alterations in the non enzymatic and enzymatic defense system. On the other hand, this oxidative insult was reduced in plants grown in the presence of phosphorous.

ACKNOWLEDGEMENTS

This research was funded by Universidad de Buenos Aires, UBACyT 068/11 project (2011–2014).

REFERENCES

Abedin, M.J., Cottep-Howells, J. & Meharg, A.A., 2002. Arsenic uptake and accumulation in rice (*Oryza Sativa L.*) irrigated with contaminated water. *Plant and Soil* 240: 311–319.

Alfassi, Z.B. 1994. *Chemical Analysis by Nuclear Methods*. John Wiley and Sons, Inc.

APHA, 1993. Standard Methods for the Examination of Water and Wastes. American Public Health Association, Washington DC.

Balestrasse, K.B., Gardey, L. Gallego, S.M. & Tomaro, M.L. 2001. Response of antioxidant defence system in soybean nodules and roots subjected to Cd stress. *Australian Journal of Plant Physiology* 28:497–504.

Dahal, B.M., Fuerhacker, M., Mentler, A., Karki, K.B., Shrestha, R.R. & Blum, W.E.H. 2008. Arsenic contamination of soils and agricultural plants through irrigation water in Nepal. *Environmental Pollution* 155: 157–163.

Pigna, M., Cozzolino, V., Violante, A. & Meharg, A. 2008. Influence of Phosphate on the Arsenic Uptake by Wheat (*Triticum durum L.*) Irrigated with Arsenic Solutions at Three Different Concentrations. *Water, Air, & Soil Pollution* 197: 371–330.

One Century of the Discovery of Arsenicosis in Latin America (1914–2014) – Litter, Nicolli, Meichtry, Quici, Bundschuh, Bhattacharya & Naidu (Eds)
© 2014 Taylor & Francis Group, London, ISBN 978-1-138-00141-1

Adsorption and movement of arsenic in soil

F.V. Molina & A. Montenegro
INQUIMAE, Facultad de Ciencias Exactas y Naturales, Universidad de Buenos Aires, Argentina

C. Bustingorri & R.S. Lavado
INBA, Facultad de Agronomía, Universidad de Buenos Aires, Argentina

ABSTRACT: The effects of both organic matter and available phosphorous on arsenic sorption by typical Argiudol soils was studied. The adsorption of arsenic was investigated by batch and column leaching experiments. The batch experiments showed a maximum adsorption capacity of about 20 mmol kg^{-1} of dry soil. The column experiments revealed As retention by a part of the soil, varying according to the level of organic matter, available phosphorus and As concentration. This behavior could be important from the contamination point of view.

1 INTRODUCTION

A wide variety of methodologies are used to evaluate the mobility and retention of contaminants. Among them, adsorption isotherm determination and column leaching experiments are important examples (Cappuyns & Swennen, 2007). The adsorption isotherms are used to quantify the distribution of an element (adsorbate) between the liquid phase and the solid phase (adsorbent) (Kinniburgh, 1986). These isotherms have some problems, including that the parameters of the model are only valid under certain conditions, they cannot be used as predictive models and do not allow us to understand the ionic adsorption process. However, although other equations and other approaches have been developed, the equilibrium adsorption equations remain the most widely used due to ability for responding adequately to a wide variety of adsorption data, its simplicity and the ease with which their parameters are estimated.

In column leaching dynamic assays, a solution of known concentration leaches through the adsorbent and was collected afterward. Because the outflow is constant, the leachate is less influenced by substances that have percolated. In addition, the renewal of percolate solution has a better similarity with the movement of water in soil, particularly when the elements under study were carried by water (i.e. irrigation). As and P compete for the same adsorption sites in soils. Some authors noted increased availability of arsenates with higher amounts of P (i.e. from fertilizer) (Signes-Pastor *et al.*, 2007). The dynamics of As in soils where contamination is affected by agricultural activity has not been well studied to date. The objectives are to determine the adsorption capacity of As in agricultural soils of different characteristics and to study the changes in the As retention versus dynamic conditions in column leaching experiments.

2 METHODS/EXPERIMENTAL

2.1 *Materials*

All the aqueous solutions were prepared in Milli-Q water. A 250 ppm As(V) stock solution was prepared using sodium arsenate ($Na_2HAsO_4.7H_2O$) (Anedra Chemical). Other concentrations were obtained by diluting the stock solution. The adsorption experiments were performed at pH 6–7.

2.2 *Methods*

Typical Argiudol soils, presenting contrasting features of Organic Matter (OM), Calcium Carbonate ($CaCO_3$) and P contents, were employed. Soils with high and low OM contents (+OM and –OM, respectively), and with high P and carbonate contents (+P and +Ca, respectively) were studied. As adsorption experiments, under constant temperature, humidity and pH were conducted. The time required to reach the apparent equilibrium adsorption was first determined. The adsorption isotherms of As(V) were measured in batch experiments where 0.5 g of soil was mixed with 10 mL solution of varying As concentration (5–250 mg L^{-1}) in Falcon tubes. The As analysis in the supernatant was performed by Inductively-Coupled Plasma (ICP) emission spectrometry and the amount adsorbed was obtained by difference.

Leaching assays of As were carried out in soil columns. Acrylic columns of 4.7 cm internal diameter and 15 cm length were employed. The column was irrigated with solutions enriched with two concentrations of As(V). The amount of solution added was quantified by the Pore Volume (PV). The leach solution was collected every week and, after finishing the assay, the column contents were divided into five layers: 0–2,5; 2,5–5; 5–7,5; 7,5–10; 10–13 cm. As was quantified in each layer. Neutron Activation Analysis (NAA) was used for the determination of As.

3 RESULTS AND DISCUSSION

3.1 *Adsorption isotherm of arsenic*

The arsenic adsorption on a soil reaches the equilibrium in about 2 days. The adsorption isotherm is shown in Figure 1; it shows a maximum adsorption capacity of 1500 mg kg^{-1} or 20 mmol kg^{-1} of soil, which is a relatively large value (Wang & Mulligan, 2008). The data are well described by the Langmuir isotherm.

The amount adsorbed of a given species depends not only on the soil composition but also on the chemical species involved and their concentration in solution.

3.2 *Leaching assays of As in soil columns*

The As accumulation in the soils under study increased over control (1.1 mg As) ($p < 0.05$). Figure 2 shows that the concentration is 72% higher

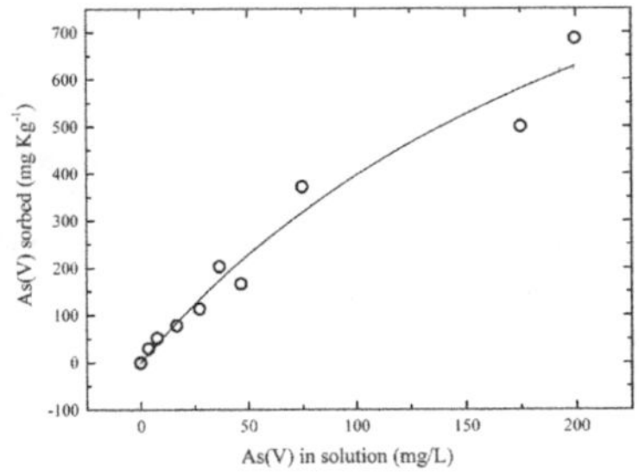

Figure 1. Adsorption isotherm of arsenic in high P soil. The symbols are experimental points and the line is the fit to the Langmuir isotherm.

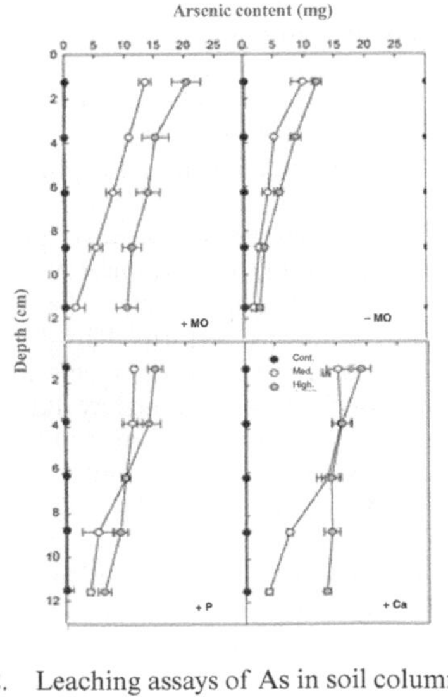

Figure 2. Leaching assays of As in soil columns.

for the high dose compared to the low one. Accumulation differences had the following order: +OM >+ Ca >+ P >– OM. In the first leached fractions the presence of As was less than 0.25 mg L^{-1}, representing between 2–5% of As applied and gradually increased to reach 20% of As applied to the columns at the end of the experiment. The concentration of As in the leachate had a similar dynamic for all soils under study. Results showed that the movement of As in the column depended on its concentration. This behavior could be important from the contamination point of view.

4 CONCLUSIONS

Batch experiments are not fully representative of the natural conditions because they offer large specific surface, and therefore maximum adsorption ability. Instead, column experiments better reflect field conditions, because the flow rate can be controlled and thereby more realistic distribution coefficients can be obtained. In column assays, the As retention by a part of soil was observed, varying according to the level of organic matter, available P and As concentration in percolating water.

ACKNOWLEDGEMENTS

F.V.M. and R.S.L. are researchers of the Consejo Nacional de Investigaciones Científicas y Técnicas (CONICET) of Argentina. Financial support for this work from the Universidad de Buenos Aires, CONICET and the Agencia Nacional de Promoción Científica y Tecnológica (Argentina) is gratefully acknowledged.

REFERENCES

Cappuyns, V. & Swennen, R. 2008. "Acid extractable" metal concentrations in solid matrices: A comparison and evaluation of operationally defined extraction procedures and leaching tests. *Talanta* 75: 1338–1347.

Kinniburgh, D.G. 1986. General purpose adsorption isotherms. *Environmental Science & Technology* 20: 895–904.

Signes-Pastor, A., Burló, F., Mitra, K. & Carbonell-Barrachina, A.A. 2007. Arsenic biogeochemistry as affected by P fertilizer addition, redox potential and pH in a west Bengal (India) soil. *Geoderma* 137: 504–510.

Wang, S. & Mulligan, C.N. 2008. Speciation and surface structure of inorganic arsenic in solid phases: A review. *Environment International* 34: 867–879.

One Century of the Discovery of Arsenicosis in Latin America (1914–2014) – Litter, Nicolli, Meichtry, Quici, Bundschuh, Bhattacharya & Naidu (Eds)
© 2014 Taylor & Francis Group, London, ISBN 978-1-138-00141-1

Effect of arbuscolar mycorrhizae and phosphorous on arsenic availability in a contaminated soil

I. Cattani, M.C. Fontanella, P. Lodigiani & G.M. Beone
Institute of Agricultural and Environmental Chemistry, Università Cattolica del Sacro Cuore, Piacenza, Italy

C. Gonnelli
Institute of Agricultural and Environmental Chemistry, Università Cattolica del Sacro Cuore, Piacenza, Italy
Dipartimento di Biologia, Università degli Studi di Firenze, Firenze, Italy

ABSTRACT: In this experiment, *Zea mays* was cultivated in rhizobox systems filled with soil collected from Scarlino (Italy), containing 250 mg kg^{-1} of As. The influence of an arbuscular mycorrhizal (AM) fungus (*Glomus intraradices*) and phosphorus (P) on the rhizosphere was monitored by Diffusion Gradient in Thin Films (DGT), to estimate the lability of arsenic in the plant. Arsenic availability was generally low and moderately increased close to the maize root, with or without mycorrhiza infection. The relationship between the mass of As accumulated by DGT and that contained in plants showed a good correlation for both shoots and roots for soils having the same P level.

1 INTRODUCTION

Since 7th–6th centuries BC, southern Tuscany (Italy) has been characterized by intensive mining and smelting activity, resulting in a huge release of As to the surrounding soil, where plant cultivation is diffused. It has been demonstrated that phosphate and arbuscular mycorrhizal (AM) fungi are able to enhance the tolerance of plants to soil contamination by As, generally decreasing its concentration in the plant tissues. In this work, the maize plant was chosen to assess the As availability and plant accumulation and translocation in relation to mycorrhizal inoculation and phosphorus application, as it is one of the most representative crop from southern Tuscany.

2 MATERIALS AND METHODS

The used soil was collected from a field at Scarlino (42°55'45"N 10°48'31"E). The analyzed properties were: pH (1:2.5 soil to water), 8.22; organic matter 15.4 g kg^{-1}; total $CaCO_3$, 35.1 g kg^{-1}; total P, 412 mg kg^{-1}; extractable P (Olsen), 4.74 mg kg^{-1}; total As, 194.2 mg kg^{-1} and it was used to fill the rhizobox system as described by Fitz *et al.* (2003). Eight maize (*Zea mays* L.) seeds were placed in each rhizobox, each one with a propagule of the endomychorrhizal fungus *Glomus intraradices* DAOM197198 containing 100 spore/cm^3 (Italpollina, Rivoli Veronese, Italy). Phosphorus was added at a rate corresponding to 100 kg ha^{-1}. Plant cultivation was conducted for a month in a phytotron (14-h photoperiod, relative humidity at 80% and 28 °C/20 °C day/night temperatures). The trials were: i) AM–/P– (not-inoculated and not P-fertilized); ii) AM+/P– (inoculated and not P-fertilized); iii) AM–/P+ (not-inoculated and P-fertilized); iv) AM+/P+ (inoculated and P-fertilized), each one in three replicates. At the end of the experiment, harvested plants were divided into roots and shoots, weighed, air-dried and immediately analyzed. Rhizosphere was separated from the bulk soil using the slicing device of Fitz *et al.* (2003). Ground plant samples (approximately 0.3 g dried weight) were analyzed by ICP-MS (Agilent, 7500ce), following nitric acid-hydrogen peroxide assisted digestion. The used DGT devices consisted of a ferrrihydrite resin (used as a binding agent for As) placed at the back of a well-defined diffusion layer of ion-permeable hydrogel, with a nylon membrane. The As_{DGT} values were calculated using the equation provided by Zhang *et al.* (2001). The deployment duration was 24 h under a controlled humidity. After retrieval, the resin-gel was eluted by HNO_3 acid and analyzed by ICP-MS.

3 RESULTS AND DISCUSSION

Our data on fraction of As bound to the DGT resin (soil labile fraction, Figure 1) suggested that its concentration (from 2 to 6 μg L^{-1}) was very low, if compared to total concentration in bulk soil.

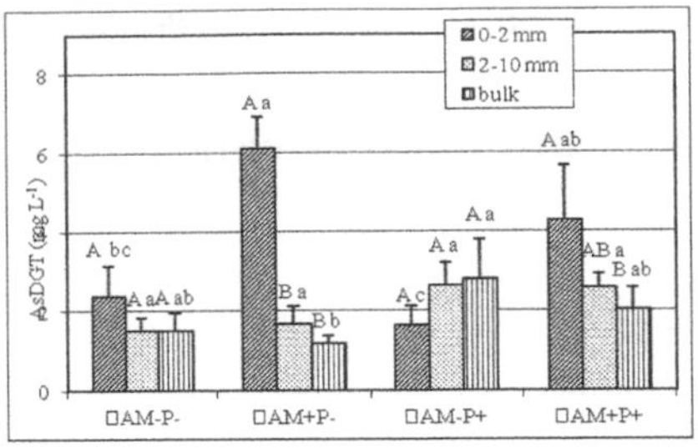

Figure 1. Concentrations of labile As in soil solution after 24 hours DGT deployment in the soil layers from the rhizobox trials: AM−/P− (not-inoculated and no P-fertilized), AM+/P− (inoculated and no P-fertilized), AM−/P+ (not-inoculated and P-fertilized) and AM+/P+ (inoculated and P-fertilized). Mean of three replicates. Significant differences among the soil layers (Tukey HSD test, at least $p < 0.05$) appear with different capital letters if intra-treatment and with different lower case letters if inter-treatments (comparison of the same layer).

No differences were observed for As_{DGT} concentration of the different layers in the control rhizoboxes, whereas the rhizosphere layer showed a higher concentration of labile As in respect to the others, in presence of mycorrhiza infection, with or without P application ($p < 0.05$). Since the measured As_{DGT} concentration is a balance of the bioavailable As and As taken up by plant root, the increase in As_{DGT} concentration in the rhizosphere generated by AM could be due to any possible effect of fungus inoculation on both arsenic mobilization in that layer and plant arsenic uptake. Wang *et al.* (2008), in a study on the influence of *G. mosseae* and *A. morrowiae* inoculation on arsenic uptake and translocation by maize, indicated that mycorrhiza exerted little effects on As translocation within plants, but may have influenced root As efflux and its deposition in the hyphae. Hence, mychorrhizae may significantly increase the available As close to the root and contemporaneously reduce its uptake rate.

On the other hand, comparing the same layer subjected to different treatments, in the two cases of phosphorus addition, it was probable that the fertilizer improved the As mobilization and contemporaneously plant uptake and, consequently, the balance of rhizosphere As consisted in a decrease of it. In general, it has been demonstrated that, when P is deficient (as well as in our soil), plants release much more root exudates, which could play a role in the As assimilation and mobilization in the rhizophere; however, it is also well-proven that phosphorus addition increases extractable fraction of As (Fitz *et al.*, 2002).

As mentioned by Cattani *et al.* (2009), several studies demonstrated that DGT measured concentrations of some trace element can be correlated to their uptake by some plants in a wide range of soils.

The relationship between As_{DGT} (measured after 24 h-deployment) and the concentration of

Table 1. Relationships between mass of labile As bound by the DGT resin after 24 hour-deployment and total As taken up by plants per rhizobox.

		Shoots	Roots
		As_{DGT} (ng)	As_{DGT} (ng)
Plant As (ng)	P+	$y = 0.91x - 1.69$ $R^2 = 0.50$	$y = 0.06x - 31.60$ $R^2 = 0.52$
	P−	$y = 0.43x - 4.12$ $R^2 = 0.91$	$y = 0.01x + 0.34$ $R^2 = 0.88$

As measured in plants (accumulated amount/n. of cultivation days) in our experiment is shown in Table 1. As translocation of arsenic is very limited in corn, we tested separately root and leaves and we divided the results according to the phosphorus addition, which could modify As uptake and transport to shoot. Correlation between DGT and plant accumulation was similar vs. both shoots and roots under the same P level. This relationship was stronger when the soil was not fertilized, and this highlighted the effect of P on As assimilation by the plant.

4 CONCLUSION

In maize plant, accumulation of As is mainly regulated at root level, where the concentration is higher. The labile fraction of arsenic in the examined soil is low and it may change if AM or phosphorus are applied, as suggested by DGT analysis. AM and fertilizer modified differently As mobilization and plant uptake, but the effect of P was much more evident.

REFERENCES

Cattani, I., Capri, E., Boccelli, R. & Del Re, A.A.M. 2009. Assessment of arsenic availability to roots in contaminated Tuscany soils by a diffusion gradient in thin films (DGT) method and uptake by *Pteris vittata* and *Agrostis capillaris*. *Eur. J. Soil Sci.* 60: 539–548.

Fitz, W.J. & Wenzel, W.W. 2002. Arsenic transformations in the soil-rhizosphere-plant system: fundamentals and potential application to phytoremediation. *J. Biotech.* 99, 259–278.

Fitz, W.J., Wenzel, W.W., Wieshammer, G. & Istenic B. 2003. Microtome sectioning causes artifacts in rhizobox experiment. *Plant Soil* 256: 455–462.

Wang, Z.-H., Zhang, J.-L., Christie, P. & Li, X.-L., 2008. Influence of inoculation with *Glomus mosseae* or *Acaulospora morrowiae* on arsenic uptake and translocation by maize. *Plant Soil* 311: 235–244.

Zhang, H., Zhao, F.J., Sun, B., Davison, W. & McGrath, S.P. 2001. A new method to measure effective soil solution concentration predicts copper availability to plants. *Environ. Sci. Technol.* 35: 2602–2607.

One Century of the Discovery of Arsenicosis in Latin America (1914–2014) – Litter, Nicolli, Meichtry, Quici, Bundschuh, Bhattacharya & Naidu (Eds)
© 2014 Taylor & Francis Group, London, ISBN 978-1-138-00141-1

Speciation of arsenic in ferns from the Iron Quadrangle, Minas Gerais, Brazil

L.M.L.A. Auler, H.E.L. Palmieri & M.A.B.C. Menezes
Centro de Desenvolvimento da Tecnologia Nuclear (CDTN/CNEN), Belo Horizonte, MG, Brazil

J. Ascari
Bolsista de Pós-Doutorado CDTN/CNEN, Belo Horizonte, MG, Brazil

R. Jaćimović, V. Stibilj, Z. Šlejkovec, I. Falnoga & D. Mazej
Department of Environmental Sciences, Jožef Stefan Institute, Ljubljana, Slovenia

V.P. Amaral
Bolsista Iniciação Científica CDTN/CNEN, Belo Horizonte, MG, Brazil

ABSTRACT: Arsenic speciation was determined in native ferns (*Pteris vittata* and *Macrothelypteris torresiana*) growing in an area near Nova Lima City, which has high background levels of arsenic due mainly from mine tailings. The arsenic extraction process for speciation was optimized. A methodology was defined for the speciation by HPLC-Diode Array UV detector.

1 INTRODUCTION

The Iron Quadrangle, located in the Brazilian State of Minas Gerais, is one of the richest and best-known mineral deposit worldwide. A great number of active and inactive gold mines can be found in this region. The gold ore from these mines is rich in arsenic (As) with As/Au ratios ranging from 300 to 3000 (Borba, 2002).

Although the present mining operations may no longer contribute significantly to the contamination of soils and sediments, there are many potential risks for As intoxication caused by the dispersion of old tailings, human occupation of polluted soils, and the consumption of contaminated surface and groundwater. A large number of native fern species and medicinal plants grow in this region.

This study determined the speciation of As in two native ferns (*Pteris vittata* and *Macrothelypteris torresiana*) growing in soil near Nova Lima City which typically has high backgrounds concentrations of soil As due to mine tailings contamination. While *Pteris vittata* is a well known As hyperaccumulator plant with great potential as phytoremediator of As-contaminated soils little is known about the As uptake potential of *Macrothelypteris torresiana.*

The aims of this study were to improve an extraction method for As in plants, to determine the total As concentration by ICP-MS, to use High Performance Chromatography (HPLC) with Diode Array UV detector, to separate the As species and to establish an intercomparison of the results obtained at IJS as part of a International bilateral Cooperation Project.

2 METHODOLOGY

Pteris vittata and *Macrothelypteris torresiana* were analyzed to obtained the total As concentration and to determine As speciation as either organic monomethylarsenium (MMA) and dimethylarsenium (DMA) or inorganic species As(III) and As(V).

Total arsenic was determined by Hydride Generation Atomic Fluorescence Spectrometry (HGAFS) at Jözef Stefan Institute (JSI-Slovenia). For determination of the species, an extraction method was tested and its recovery was evaluated by Inductive Plasma Mass Spectrometry (ICP-MS).

3 RESULTS AND DISCUSSION

After tests, it was verified that one of the extraction methods was adequate for total As determination. Arsenic was determined by ICP-MS and the results were compared with those obtained at JSI.

The methodology consisted of extraction with methanol-water 2:1, extraction with nitric acid and water, and finally analysis by ICP-MS. The concentration results ($mg.kg^{-1}$) of total, soluble and insoluble arsenic in fern samples are shown in Table 1 and the results of the species obtained at JSI by using Hydride Generation Atomic Fluorescence Spectrometry (HGAFS) technique are shown in Table 2.

Sugars and proteins were identified as interferences in the quantification of As(III) species.

Table 1. Total, soluble and insoluble arsenic concentration ($mg\ kg^{-1}$) in fern samples (dry weight) determined by ICP-MS.

Ferns/As concentration	*Macrothelypteris torresiana*	*Pteris vittata*
Soluble As (IJS)	0.22 ± 0.01	1897 ± 68
Soluble As (CDTN)	0.16 ± 0.05	2283 ± 57
Insoluble As (IJS)	3.77 ± 0.03	658 ± 85
Insoluble As (CDTN)	3.96 ± 0.34	592 ± 149

Table 2. Concentration of arsenic species ($mg.kg^{-1}$) in fern samples (dry weight) determined by HGAFS.

Species/ Fern	DMA	MMA	As(III)	As(V)
M.torr.	0.39 ± 0.06	0.06 ± 0.01	4.55 ± 0.16	9.62 ± 0.87
P.vitt.	ND	ND	2170 ± 41	410 ± 59

ND-Not detected.

4 CONCLUSION

The extraction process of arsenic and its species in two fern species was optimized and a methodology was defined for the speciation by HPLC-Diode Array UV detector. Sugars and proteins were identified as interferences in the quantification of As(III) species.

The methodology could be applied to water samples after optimization of other types of chromatography in order to separate the interferences and apply this process to different plant samples.

ACKNOWLEDGEMENTS

We thank CDTN/CNEN and CNPq (Conselho Nacional de Desenvolvimento Científico e Tecnológico) for their financial support.

REFERENCES

Borba, R.P. 2002. Arsênio em ambiente superficial: processos geoquímicos naturais e antropogênicos em uma área de mineração aurífera. 113f. Geoscience Institute, State University of Campinas (PhD thesis).

Deschamps, E. & Matschullat, J. 2007. Arsênio antropogênico e natural. Um estudo em regiões do Quadrilátero Ferrífero. 1a edição, FEAM, 330p.

Stibilj, V., Smrkolj, P., Jaćimović, R. & Osvald, J. 2011. Selenium uptake and distribution in chicory (Chichorium intybus L.) grown in an aeroponic system. *Acta agric. Slov*. 97: 189–196.

Šlejkovec, Z., Elteren, J.T. van, Glass, H.J., Jeran, Z. & Jaćimović, R. 2010. Speciation analysis to unravel the soil-to-plant transfer in highly arsenic-contaminated areas in Cornwall (UK). *Int. J. Environ. Anal. Chem.* 10(90): 784–796.

One Century of the Discovery of Arsenicosis in Latin America (1914–2014) – Litter, Nicolli, Meichtry, Quici, Bundschuh, Bhattacharya & Naidu (Eds)
 ISBN 978-1-138-00141-1

Arsenic removal by *in-situ* formation of iron and sulfur minerals

A.S. Vega, E.A. Leiva, G.E. Pizarro & P.A. Pasten

Departamento de Ingeniería Hidráulica y Ambiental, Pontificia Universidad Católica de Chile, Santiago, Chile

ABSTRACT: The fate of arsenic (As) is determined by the redox cycle of iron and sulfur. Its immobilization is controlled by sorption and co-precipitation with iron and/or sulfur minerals. At changing redox conditions, As is liberated by dissolution of the binding mineral; its enrichment is limited by the new mineral being formed. Although changing redox conditions occur in As-rich environments (like sediments), few studies of removal by minerals formed in the presence of As have been performed. We used ferrihydrite and mackinawite as representative minerals of iron oxic and iron-sulfur anoxic environments, respectively. Batch and column experiments were performed to compare the removal of As by minerals formed in its absence and presence. The removal of As was larger in minerals formed when As is present, except for As(III) at low concentration. In redox gradient column, As was effectively attenuated at oxic and reduced conditions.

1 INTRODUCTION

The fate of As has been widely studied for fully oxic and anoxic systems. In oxic systems, sorption on iron oxy/hydroxides plays a fundamental role, whereas in anoxic systems, the precipitation and/or sorption with sulfide minerals are relevant. Under changing redox conditions, the binding minerals may be dissolved, and As is released to the aqueous phase, until the thermodynamic conditions are favorable and a new mineral is formed. During its formation, the new mineral is able to sorb As.

Despite the importance of sorption by minerals formed in the presence of the sorbate, most removal studies are developed with preformed minerals. Contradictions exist regarding the capacity of mackinawite to remove As. Studies show that preformed mackinawite is a good sorbent of As (26%wt.) (Farquhar *et al.*, 2002), while other experiments, where the mineral was formed in the presence of As, proved to be a weaker sorbent (<0.1%wt.) (Kirk *et al.*, 2010).

This research evaluates As immobilization by ferrihydrite and mackinawite formed when As is present. We approached this through batch experiments, and with diffusion-limited laboratory columns where a redox gradient is formed.

2 METHODS

2.1 *Batch experiments*

Ferrihydrite and mackinawite were synthesized by standardized protocols. Two initial concentrations of As were used: 0.05 and 1 mM. In experiment type A, the preformed mineral was added to a solution with dissolved As; in experiment type B, the chemicals to form the minerals were added to the solution. We used a solid concentration of 1 mM. All experiments were conducted at circumneutral pH (with HEPES buffer) and an electrolyte background of 0.1 M of $NaNO_3$. Anoxic experiments were performed in a nitrogen atmosphere. The mineral synthesized was verified by IR and XRD analyses. Additionally, formation of ferrihydrite and mackinawite in oxic and anoxic conditions, respectively, were verified in the experiments of sorption in the presence of low As concentrations.

Sorbed As concentration was obtained by difference between As added and the measured concentration in the aqueous phase after 24 hours. These were measured in the aqueous phase by TXRF and ICP-OES.

The PHREEQC model was used to calculate the mineral saturation with the wateq.4f database.

2.2 *Redox gradient column*

To elucidate the interaction between diffusion-limited mass transport, the kinetics of chemical reactions of the As-Fe-S system, and As attenuation (representative of sediment column), we used columns filled with glass porous media and controlled boundary redox conditions (oxygen saturated water and 1 mM sulfide, respectively). The columns were filled with 1 mM of Fe(II), 0.1 M of $NaNO_3$, HEPES and 0.1 mM As(III). At the end, the solid was analyzed for elemental composition by SEM-EDS.

3 RESULTS AND DISCUSSION

3.1 *As removal by minerals formed in its absence and presence*

Both experiments with ferrihydrite (Type A and B), attenuated the concentration of As in the aqueous phase (see Figure 1). Experiment type B showed a better removal than type A, probably due to the formation of smaller particles. This is an important result for the treatment of wastewaters with high As concentration. At low As concentration (0.05 mM), the difference is minimal.

Figure 2 shows As removal by mackinawite. For As(V), the behavior was similar to that of ferrihydrite, showing a better removal for experiment type B, but with very low removal at low concentration. As(III) at low coverage showed a lower removal in experiment type B than in experiment type A. At high concentration, it shows the same pattern of ferrihydrite (B is better than A). This suggests a different mechanism of removal of As(III) at high and low concentrations. The lower removal in experiment type B, at low concentrations, could be explained by the formation of thioarsenic species, which do not precipitate at low concentrations (only at high As concentration, the geochemical model shows saturation by As_2S_3). Mackinawite, at experimental conditions, shows to be a good sorbent of As; the percentage of removal was similar to that of ferrihydrite, except for As(V) at low concentrations.

3.2 *As removal in redox gradient column*

Figure 3 shows As concentration in the column sampling ports. At the start port, the column is aerated, and contacted with a solution of sulfide at the end port. Port 1 showed a typical behavior of kinetics removal of As for ferrihydrite formed in its presence, a high initial removal with subsequent release. In the oxidized zone, As is sorbed by minerals of Fe (probably ferrihydrite). In the reduced zone, As is sorbed by minerals of Fe-S (probably mackinawite). The elemental composition of the minerals is confirmed by SEM-EDS showing As in the solid. At intermediate redox potentials, As is mobile.

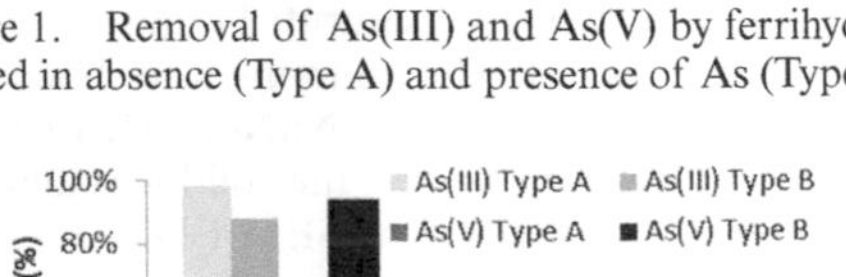

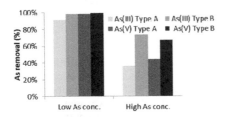

Figure 1. Removal of As(III) and As(V) by ferrihydrite formed in absence (Type A) and presence of As (Type B).

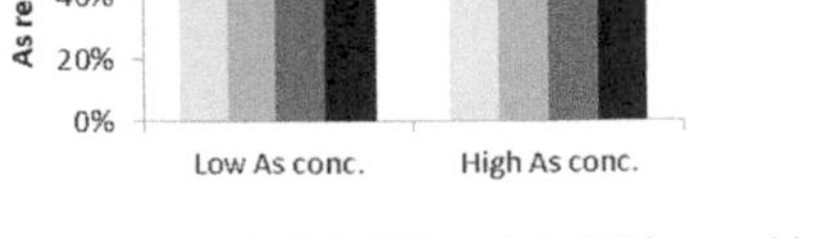

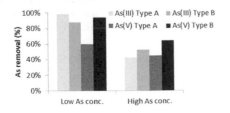

Figure 2. Removal of As(III) and As(V) by mackinawite.

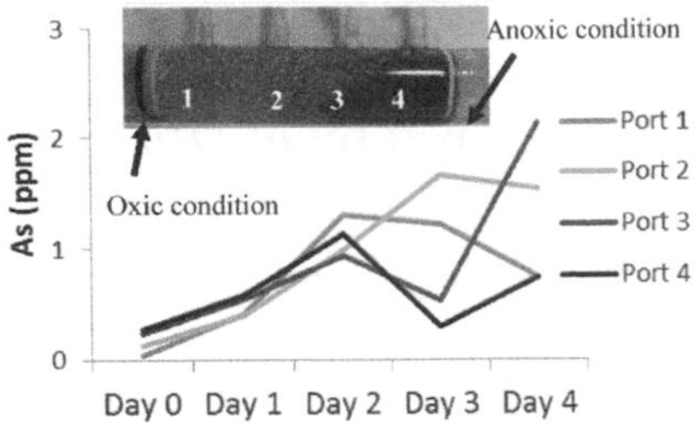

Figure 3. As(III) concentration in redox gradient column. In the image we can see a black precipitate in the anoxic zone that is formed with time (mackinawite) and, in the oxic zone, an orange or brown precipitate that decreases with time (ferrihydrite).

4 CONCLUSIONS

Minerals, formed in absence and presence of As, show different sorption capacity. This finding imply that experimental parameters of removal by preformed minerals could not be apply to natural conditions, where minerals are formed in the presence of As and other contaminants.

As can be attenuated by reactions with Fe and S at oxic and anoxic conditions. We have observed that mackinawite immobilizes As in batch and diffusion-limited transport columns under abiotic conditions, in contrast with Kirk *et al.* (2010) results under biotic conditions. Thus, it is necessary to improve the knowledge of Fe-S-As systems to design innovative cost-effective treatment systems for As and prevent the risk of contamination.

ACKNOWLEDGEMENTS

The authors acknowledge CONICYT/FONDAP 15110020, Scholarship CONICYT.

REFERENCES

Farquhar, M.L., Charnock, J.M., Livens, F.R. & Vaughan, D.J. 2002. Mechanisms of arsenic uptake from aqueous solution by interaction with goethite, lepidocrocite, mackinawite, and pyrite: An X-ray absorption spectroscopy study. *Environmental Science & Technology,* 36(8): 1757–1762.

Kirk, M.F., Roden, E.E., Crossey, L.J., Brearley, A.J. & Spilde, M.N. 2010. Experimental analysis of arsenic precipitation during microbial sulfate and iron reduction in model aquifer sediment reactors. *Geochimica et Cosmochimica Acta,* 74(9): 2538–2555.

One Century of the Discovery of Arsenicosis in Latin America (1914–2014) – Litter, Nicolli, Meichtry, Quici, Bundschuh, Bhattacharya & Naidu (Eds)
 ISBN 978-1-138-00141-1

Arsenic sorption properties in a chronosequence of soils derived from clay parent material

M.L. Bonis & G. Echevarria
Laboratoire Sols et Environnement (LSE), Université de Lorraine (INPL (ENSAIA)/INRA), Vandœuvre-lès-Nancy Cedex, France

Y. Thiry
Agence Nationale Pour la Gestion des Déchets Radioactifs (Andra), Châtenay Malabry, France

M.O. Simonnot
Laboratoire Réactions et Génie des Procédés (LRGP), Université de Lorraine-CNRS, Nancy Cedex, France

ABSTRACT: The aim of this study was to characterize arsenic mobility in soils developed from clayey bedrocks. A soil chronosequence was determined that represents all possible pedogenesis in French north-east climate conditions (from young Technosol to stagnic Luvisol and albic Planosol). Each soil represents a pedogenetic step derived from young Technosol (Callovo-Oxfordian clay). Soil horizons were characterized for their physicochemical properties. Batch experiments were conducted to adsorb arsenic on soil samples and to determine the distribution coefficient–Kd in linearity conditions. Adsorption isotherms were obtained for three surface horizons and seemed to follow Langmuir model. Kd of the horizons were obtained in the linear part of the isotherm: young Technosol Kd: 843.04 L kg^{-1} ($R^2 = 0.9907$), albic Planosol Kd: 649.24 L kg^{-1} ($R^2 = 0.9998$) and vertic Cambisol Kd: 194.63 L kg^{-1} ($R^2 = 0.997$). By Principal Component Analysis (PCA), it was determined that Kd seems to be much higher in younger than in developed soils.

1 INTRODUCTION

Arsenic (As) is an ubiquitous metalloid present in radiferous radioactive wastes. These wastes are planned to be stored in the long life low activity (FAVL) radioactive waste storage facility of the Andra (French nuclear waste agency). At the beginning of the study, FAVL area was not yet located. The site will be built in a clayey environment. An argillite (160 million years)—excavated during the construction of Andra underground research laboratory (France)—has been chosen as an analogue. The excavated rock was spread on the surface before being vegetated. In the long term, argillite will be weathered leading to the formation of soil.

The objective of the study is to determine As mobility in soils in the case of an As discharge which could contaminate the soil cover. Arsenic sorption onto various soils that belong to the soil (initial) chronosequence has been determined as well as distribution coefficient (Kd) evolution in relation to soils evolution. Kd were also measured in each soil horizon.

2 MATERIALS AND METHODS

Eight soils were chosen depending on two criteria: parent geological material (Callovo-Oxfordian argillite or a clayey material with similar properties) and land use. These soils represent all possible pedogenetic evolutions under a temperate oceanic climate. Each soil is a pedogenetic step of development from a young Technosol to stagnic Luvisol and albic Planosol (World Reference Base, FAO-WRB).

Arsenic sorption on soils was studied to obtain Kd (L kg^{-1}) values in domains of linearity of sorption isotherm:

To confirm this, some batch tests were realized with three contrasting surface soil horizons (young Technosol 0–10 cm, vertic Cambisol 0–25 cm and albic Planosol 0–10 cm). The reagent used in this study was an arsenate ($Na_2HAsO_4 \cdot 7H_2O$). Initial concentrations were: 0; 0.25; 1.5; 2.5; 4; 7; 12; 17 and 30 mg As L^{-1}. Three experimental times of 24, 48 and 72 h were considered to determine equilibrium. At the end of the tests, As solution concentration—C_{eq} and As adsorbed quantity—Q_{eq} were obtained:

$$Q_{eq} = (V/M)\,(C_i - C_{eq})$$

Q_{eq}: equilibrium adsorbed quantity (mg kg^{-1}),
C_{eq}: equilibrium concentration in solution (mg L^{-1})
V: solution volume (L)
M: mass of soil (kg).

Batch experiments were carried out for each soil horizon. Equilibrium time was 48 h. As initial concentrations were 0.25, 1.5 and 2.5 mg As L^{-1}. These concentrations were chosen because they were in the linearity domain of the sorption isotherm. At the end, a principal component analysis (PCA) was realized in order to carry out a potential comparison between the degree of pedogenesis and the Kd of As.

3 RESULTS AND DISCUSSION

3.1 *Batch experimental*

Sorption isotherms of the three surface horizons (Technosol, vertic Cambisol and Planosol) followed reasonably a Langmuir model:

$$Q_{eq} = Q_{\max}\frac{K_L C_{eq}}{1 + K_L C_{eq}}$$

$Q_{\max}$: sorption maximum capacity (mg kg^{-1}),
K_L: Langmuir coefficient (L mg^{-1})

For each isotherm, a linearity domain at low equilibrium As concentrations (C_{eq}: 0 to 0.16 mg L^{-1} or Ci: 0 to 2.5 mg L^{-1}) was observed (Figure 1). Kd was then computed for the 3 surface horizons as the slope:

- Young Technosol Kd: 843 L kg^{-1} ($R^2 = 0.9907$),
- Vertic Cambisol Kd: 195 L kg^{-1} ($R^2 = 0.997$),
- Albic Planosol Kd: 649 L kg^{-1} ($R^2 = 0.9998$).

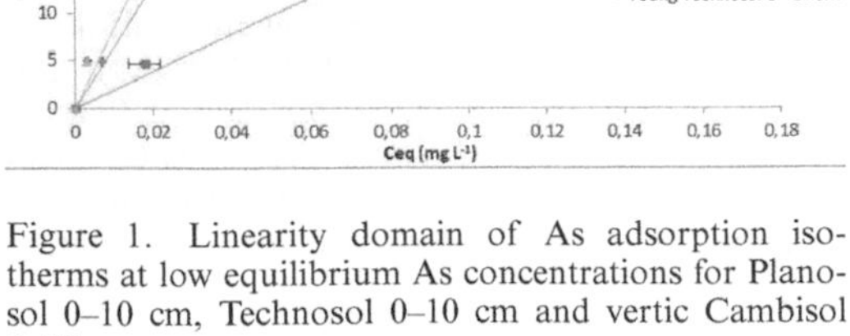

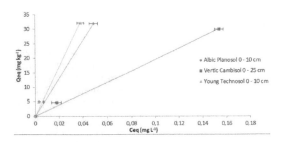

Figure 1. Linearity domain of As adsorption isotherms at low equilibrium As concentrations for Planosol 0–10 cm, Technosol 0–10 cm and vertic Cambisol 0–25 cm.

3.2 *Batch experiments*

Batch experiments were carried out with a single As initial concentration (Ci: 1.5 mg L^{-1}) and a provisional Kd value was determined for each horizon.

The highest Kd were obtained for the following horizons:

- Young Technosol 80–100 cm: 1513.9 L kg^{-1}
- Calcaric Cambisol 120 cm: 6642.3 L kg^{-1}
- Calcaric Cambisol 165 cm: 1654.2 L kg^{-1}.

The lowest Kd were obtained for the calcic Calcisol (from 11.4 L kg^{-1} at 17–45 cm to 28.5 L kg^{-1} at 45–70 cm).

3.3 *Principal Component Analysis (PCA)*

PCA was performed by taking into account the physicochemical properties of the soils; Kd and soil horizons were also positioned on the same plane (Figures 2 and 3). The aim was to track potentially relationships between Kd and soils properties and consequently to pedogenesis.

The degree of pedological evolution is a factor that clearly discriminates the samples in the PCA.

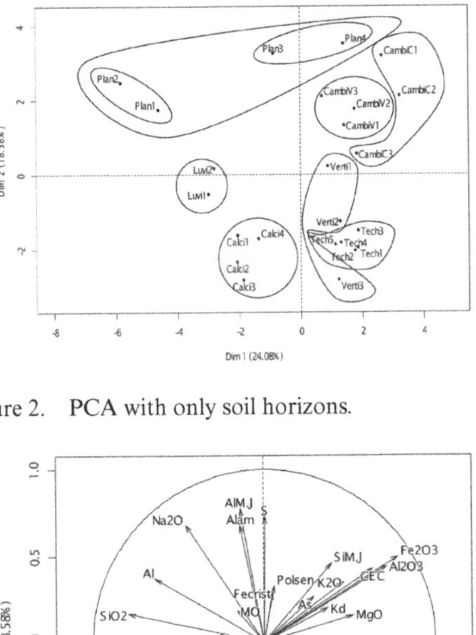

Figure 2. PCA with only soil horizons.

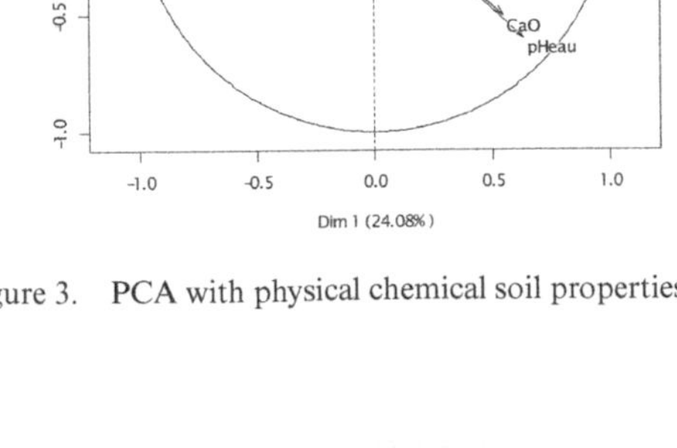

Figure 3. PCA with physical chemical soil properties.

Among key parameters in the discrimination, are pH in water, total Ca and exchangeable Al.

Kd values were higher in younger (Ex: calcaric Cambisol) than in developed soils (Ex: vertic stagnic Calcisol and stagnic Luvisol). They also tended to be higher in B horizons where total Fe (i.e. Fe_2O_3) is maximum.

4 CONCLUSIONS

This work was aimed at characterizing As mobility in soils developed from clayey bedrocks. Therefore, a soil chronosequence was considered that represents all possible pedogenesis in temperate oceanic climate conditions. Each soil sample represents a pedogenetic step initially derived from young Technosol to Luvisols/Planosols. In batch experiments, adsorption isotherm in representative soil horizons followed Langmuir model with a linearity domain in which Kd could be calculated for all soils. Kd values in soils derived from clay parent material are probably linked to high total Fe contents and tend to decrease with the degree of soil evolution.

One Century of the Discovery of Arsenicosis in Latin America (1914–2014) – Litter, Nicolli, Meichtry, Quici, Bundschuh, Bhattacharya & Naidu (Eds)
© 2014 Taylor & Francis Group, London, ISBN 978-1-138-00141-1

Phytoremediation potential of native flora of arsenic-contaminated soils

P.J.C. Favas
University of Trás-os-Montes e Alto Douro, UTAD, School of Life Sciences and the Environment, Quinta de Prados, Vila Real, Portugal

J. Pratas
Faculty of Sciences and Technology, Department of Earth Sciences, University of Coimbra, Coimbra, Portugal

ABSTRACT: The aim of this study is to evaluate the phytotechnological potential of native flora of soils enriched with As in distinct abandoned mining areas of Portugal. Significant concentrations of As in soils suggest that As-contamination is a matter of great concern in the studied mining areas. Arsenic concentrations higher than the toxic level in some species like *Agrostis castellana, Holcus lanatus, Pinus pinaster*, and *Pteridium aquilinum* indicate that their utility for phytoremediation is possible.

1 INTRODUCTION

In recent decades, many studies have been conducted in contaminated mining and industrial areas and in natural metalliferous soils in order to inventory and screen the indigenous species and evaluate their potential for phytoremediation of contaminated soils.

The aim of this study is to evaluate the phytotechnological potential (phytoremediation, phytomining, bioindication, biogeochemical prospecting) of native flora of soils enriched with arsenic in distinct abandoned mining areas of Portugal.

2 METHODS

2.1 *Sampling and sample preparation*

In the old mining areas studied, several line transects were made in mineralized and non-mineralized zones as well as tailings. Soils and plants were collected at 20 m intervals along the line transects (0, 20, 40 m, etc.) in circle of ≅2 m radius. At each location, four random partial soil samples weighing 0.5 kg each were collected from 0 to 20 cm depth and mixed to obtain one composite sample to save time and costs. They were oven-dried at a constant temperature, manually homogenized and quartered. Two equivalent fractions were obtained from each quartered sample. One was used for the determination of pH, and the other for chemical analysis. The samples for chemical analysis were sieved using a 2 mm mesh sieve to collect plant matter and subsequently screened to pass through a 250 μm screen. Samples were also obtained from all species of plants whenever found growing within the 2 m radius of each sampling point. They consist of the aerial parts, taking into consideration similar maturity of the plants and the proportionality of the different types of tissues, or the separation of different types of tissues (leaves and stems) in some species. In the laboratory, the vegetal material was washed thoroughly, first in running water followed by distilled water, and then dried in a glasshouse. When dry, the material was milled into a homogenous powder. The soil and plant samples were acid-digested for elemental analysis.

2.2 *Analytical methods*

Soil pH was determined in water extract (1:2.5 v/v). The determination of total element contents in digested soil and plant samples was performed by Atomic Absorption Spectrophotometry (AAS, Perkin-Elmer, 2380) and hydride generation system (HGS). Data quality control was performed by inserting triplicate samples into each batch. Certified references materials were also used.

3 RESULTS AND DISCUSSION

The studied areas included several abandoned mines (Sarzedas mine, Fragas do Cavalo mine, Tarouca mine, Vale das Gatas mine, Adoria mine, Ervedosa mine, Regoufe mine, and Rio de Frades mine). Results obtained from Sarzedas (Central Portugal) and Vale das Gatas mines (Northern Portugal) are presented.

High maximum values for As (651 mg kg^{-1}) were observed in soils at the Sarzedas mine. High levels of sulfides, in particular pyrite and arsenopyrite that are easily weathered, favor the dissolution of toxic elements allowing a high dispersion and bioavailability.

In the flora of Sarzedas mine area, As was accumulated in aerial tissues of *Pinus pinaster* and *Digitalis purpurea*. Therefore, these species are suited for recognizing the anomaly. High accumulation of As was present in leaves (Figure 1), and it increased in the older tissues. This translocation is a common mechanism in plants to avoid toxicity in young leaves as their metabolic activity is higher (Pratas *et al.*, 2005).

It was concluded that the species and organs best suited as indicators of excessive As and/or with potential for mine restoration in the Sarzedas mine area are by order of importance: old needles of *P. pinaster*, aerial tissues of *C. vulgaris*, *Chamaespartium tridentatum*, leaves of *C. ladanifer*, *Erica umbellata* and *Quercus ilex* subsp. *ballota*.

Very high maximum values for As (5,770 mg kg^{-1}) were observed in soils at the Vale das Gatas mine. In general, the content variations in plant materials were strongly related to the content variations in soils. It has also been verified that in contaminated locations or tailings, the concentration of metals in plant tissues is high due to the high metal concentrations in the soil. *Holcus lanatus*, *Pteridium aquilinum* and *Agrostis castellana* were the main accumulators of As (Figure 2).

The *Pinus pinaster* trees growing on the tailings and contaminated soils of Vale das Gatas mine accumulated higher As contents than those observed in plants representative of the local biogeochemical background (0.2 to 10.8 mg kg^{-1} dry weight, depending on plant organ and organ age).

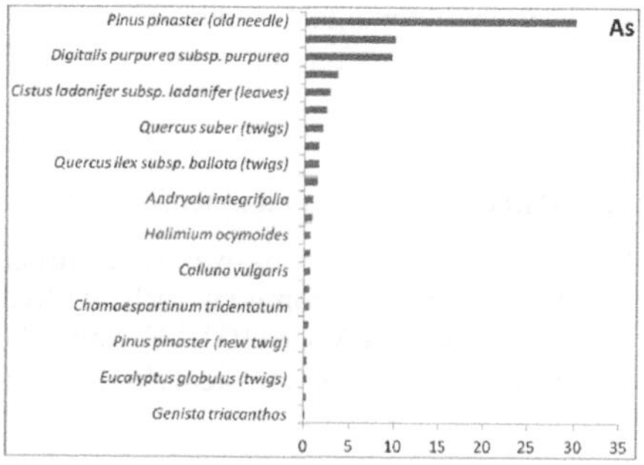

Figure 1. Accumulation of As (mg kg^{-1} dry weight) in plant species of the Sarzedas mining area.

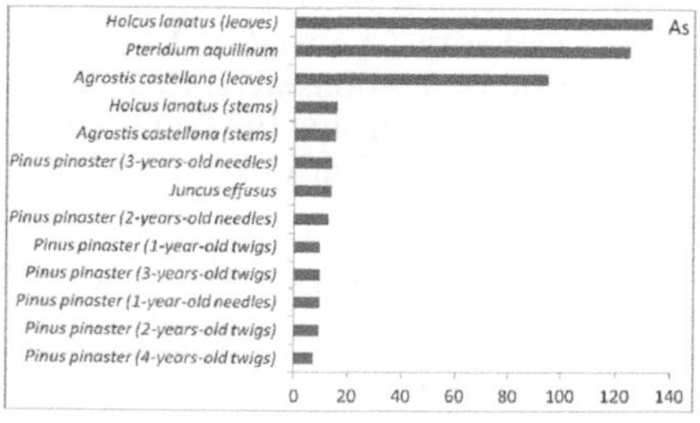

Figure 2. Accumulation of As (mg kg^{-1} dry weight) in plant species of the Vale das Gatas mining area.

These values were also higher than those typically observed in this species.

In the *P. pinaster* samples from tailings and contaminated soil locations, the older needles (2- and 3-years-old) show a tendency to accumulate higher concentrations of As (Favas *et al.*, 2013). This allowed to conclude that the As concentrations in plants depend as much on the plant organ as on its age. The species showed a great variability in the accumulation behavior of As with the age of the organ.

4 CONCLUSIONS

Significant accumulation of As in both soils and native wild flora suggests that As-contamination is a matter of great concern in the studied mining areas. The native flora displayed its ability to withstand high concentrations of metalloid in the soil. However, accumulation patterns in the plants tested differed. Arsenic concentrations higher than the toxic levels (1 to 20 mg kg^{-1}; Alloway, 1995) in some species like *Agrostis castellana*, *Holcus lanatus*, *Pinus pinaster*, *Pteridium aquilinum*, indicate that internal detoxification tolerance mechanisms might also exist; therefore, their utility for phytoremediation is possible. Furthermore, the plants could grow and propagate in substrata with low nutrient conditions which would be a great advantage in the revegetation of mine tailings.

ACKNOWLEDGEMENTS

This study was partially supported by the European Fund for Economic and Regional Development (FEDER) through the Program Operational Factors of Competitiveness (COMPETE) and National Funds through the Portuguese Foundation for Science and Technology (PEST-C/MAR/UI 0284/2011, FCOMP 01 0124 FEDER 022689).

REFERENCES

Alloway, B. 1995. Soil processes and the behaviour of heavy metals. In B.J. Alloway (ed), *Heavy Metals in Soils*, second edition: 11–37. London: Blackie Academic & Professional.

Favas, P.J.C., Pratas, J. & Prasad, M.N.V. 2013. Temporal variation in the arsenic and metal accumulation in the maritime pine tree grown on contaminated soils. *International Journal of Environmental Science and Technology*, 10(4): 809–826.

Pratas, J., Prasad, M.N.V., Freitas, H. & Conde, L. 2005. Plants growing in abandoned mines of Portugal are useful for biogeochemical exploration of arsenic, antimony, tungsten and mine reclamation. *Journal of Geochemical Exploration* 85: 99–107.

One Century of the Discovery of Arsenicosis in Latin America (1914–2014) –
Litter, Nicolli, Meichtry, Quici, Bundschuh, Bhattacharya & Naidu (Eds)
© 2014 Taylor & Francis Group, London, ISBN 978-1-138-00141-1

Arsenic in agricultural Chilean soils: Solvent extraction tests and risk assessment

L. Cornejo
Laboratorio de Investigaciones Medioambientales de Zonas Áridas, LIMZA, Escuela Universitaria de Ingeniería Industrial, Informática y Sistemas, EUIIS, Universidad de Tarapacá, Arica, Chile
Centro de Investigaciones del Hombre en el Desierto, CIHDE, Arica, Chile

J. Acarapi, H. Lienqueo, M.J. Arenas & P. Vilca
Centro de Investigaciones del Hombre en el Desierto, CIHDE, Arica, Chile

ABSTRACT: A Sequential Extraction Procedure (SEP) and a Physiologically Based Extraction Test (PBET) were carried out with the aim of studying the mobility, bio-availability (for plants and human beings) and the spatial variation in agricultural Chilean soils in the valleys of Arica and Parinacota (northern Chile). For this purpose, five samples were collected in two local valleys (Azapa and Camarones). The results showed that arsenic is mainly associated with Fe and Al either amorphous or poorly crystalline phases, and with Fe and Al crystalline phases. In addition, the Chronic Daily arsenic Intake (CDI) of arsenic values obtained from the PBET do not exceed the maximum reference dose for chronic oral exposure (RfD = 0.3 μg kg^{-1} d^{-1}). Therefore, the solvent extraction tests can be used as reliable information and useful for soil arsenic exposure risk assessment.

1 INTRODUCTION

Arsenic in the Chilean region of Arica and Parinacota is of natural origin, as a result of the geothermal activity in the Cordillera of Andes. This element is associated with several soil phases by different binding forces as it happens with other trace elements. The forces and interactions between the substratum (arsenic) and the solid phase restrict the element availability in the natural environment as well as its movement to the water bodies (like rivers, lakes and slopes) and plants and animals. In particular, the movement of inorganic arsenic from soils to the potable water is a source of risk for the human health, even at low concentrations. The focus of this study was to assess the mobility, bioavailability (for plants and human beings) and spatial variation of arsenic in agricultural soils for two valleys, Camarones (study case) and Azapa (control valley). The study was made by a sequential extraction procedure (SEP) and using a physiologically based extraction test (PBET).

2 METHODS/EXPERIMENTAL

2.1 *Sampling*

Ten samples were collected from two valleys located at the Arica and Parinacota regions: Camarones (high levels of environmental arsenic) and Azapa (low levels of environmental arsenic). These locations showed total arsenic ranging between 30 and 700 mg kg^{-1}.

2.2 *Total arsenic in soils and arsenic fractions*

Total arsenic in the fractions was determined by atomic absorption spectrometry with an hydride generator accessory (AAS-HG), Varian 280FS. Experimental conditions: air-acetylene reducing flame, 0.5 nm slit, 193.7 nm.

2.3 *Total arsenic in soils and arsenic fractions*

The total arsenic was determined in the fractions by Atomic Absorption Spectrometry with Hydride Generator accessory (AAS-HG), Varian 280FS. The experimental conditions were: Reducing flame of air-acetylene, 0.5 nm slit, 193.7 nm wavelength and a lamp current of 10 mA.

2.4 *Arsenic fraction available to humans*

The bioavailability of arsenic for plants was considered as the extracted fraction with hydrochloric acid 1 M. The experimental conditions were: 1 g sample, magnetic agitation for 4 h, 25 mL 1 M HCl, filtration by 0.45 μm nylon membrane. The bio-availability

Table 1. Sequence and description of extracting solutions for the used SEP.

Fraction	Extracting solution composition	Extraction conditions	Soil: Solution ratio
1[a)]	$(NH_4)_2SO_4$ 0.05 M	4 h shaking, 20 °C	1:25
2[b)]	$NH_4H_2PO_4$ 0.05 M	16 h shaking, 20 °C	1:25
3[c)]	NH_4^+– oxalate 0.2 M, pH 3.25	4 h shaking, 20 °C	1:25
4[d)]	NH_4^+– oxalate-ascorbic acid 0.2 M, pH 3.25	0.5 h shaking, 96 °C	1:25
5[e)]	HCl/HNO_3/FH, microwave digestion	2 h, 150 °C	1:25

Association of the arsenic fraction with the soil solid phase: a) non-specifically sorbed; b) specifically-sorbed; c) amorphous and poorly-crystalline hydrous oxides of Fe and A; d) well-crystallized hydrous oxides of Fe and Al; e) residual phases.

of arsenic for human beings was determined by physiologically based extraction test (PBET). The experimental conditions were: 1 g of soil, 150 mL of gastric solution (pH 1.8, 1 M HCl, 0.15 M NaCl and 1% porcine pepsin, magnetic agitation for 1 h at 37 °C and centrifugation of the supernatant.

3 RESULTS AND DISCUSSION

3.1 *Arsenic-soil solid phase interaction*

The results of SEP show that arsenic is mainly associated with less mobile fractions (3 and 4): amorphous or poorly crystalline iron and aluminum phases, and Fe and Al crystalline phases (Figures 1 and 2).

3.2 *Bioavailability and risk assessment*

The chronic daily arsenic intake (CDI) determined by PBET does not exceed the maximum reference dose for chronic oral exposure (RfD = 0.3 $\mu g\ kg^{-1}\ d^{-1}$) established by USEPA (Tables 2 and 3).

In all cases, a significant statistic correlation was found between the variables studied, which validated the regression models for later predictions of arsenic in fractions as a function only of the total arsenic (Table 2).

The bioavailability of arsenic in human beings is referred to its solubility in soils by means of the stomachic liquid and its later transmission through the gastrointestinal tract. The bioavailable arsenic

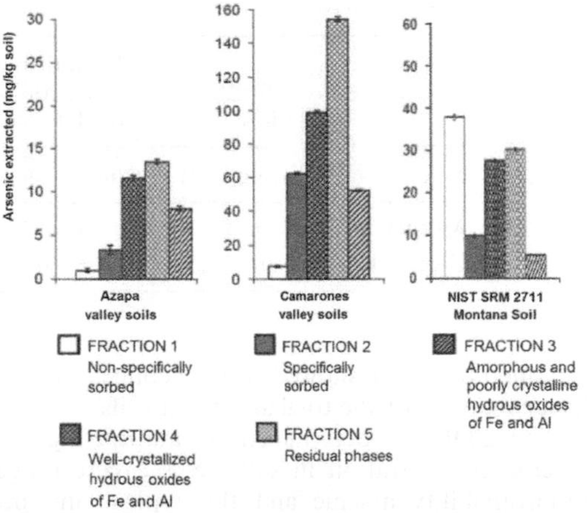

Figure 1. Partitioning of arsenic in soils of Azapa and Camarones valleys and the Standard Reference Material SRM2711 "Montana Soil", U.S. National Institute of Standards and Technology.

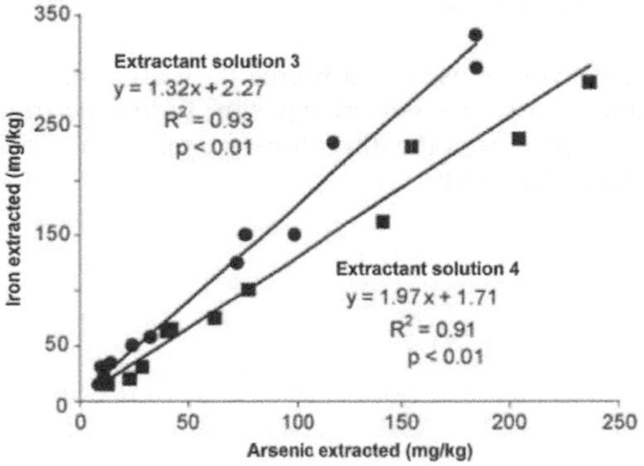

Figure 2. Correlation between extracted arsenic and extracted iron: Fractions 3 and 4.

Table 2. Correlation parameters for relationship between arsenic fractions, bioavailable arsenic and total arsenic in soils of Azapa and Camarones valleys.

Soil arsenic fractionation			
Fractions	Equation	R^2	P
1	$[As]_{F1} = 0.5\ [As]_{Total} + 1.10$	0.963	<0.01
2	$[As]_{F2} = 0.19\ [As]_{Total} + 1.16$	0.908	<0.01
3	$[As]_{F3} = 0.22\ [As]_{Total} + 2.08$	0.951	<0.01
4	$[As]_{F4} = 0.33\ [As]_{Total} + 3.24$	0.940	<0.01
5	$[As]_{F5} = 0.16\ [As]_{Total} + 1.1$	0.918	<0.01
Soil arsenic bioavailability			
Living organism	Equation	R^2	P
Humans	$[As]_{bioavailable,\ humans} = 0.04\ [As]_{Total} + 1.22$	0.953	<0.01

Table 3. Arsenic bioavailability in different extracts and chronic daily arsenic intake (CDI)*.

Sample code	Total arsenic in soil	Bioavailable arsenic in soil for humans (PBET)		Bioavailable arsenic in soil for plants (HCl)		CDI
	(mg kg^{-1})	(mg kg^{-1})	(%)	(mg kg^{-1})	(%)	(µg As kg^{-1} d^{-1})
AZAPA	140.1 ± 1.0	2.9 ± 0.7	2.4 ± 0.7	55 ± 1	42 ± 4	0.037
CAMA-RONES	369.5 ± 1.1	5.2 ± 0.6	2.1 ± 0.1	123 ± 2	36 ± 2	0.080

obtained by PBET showed average concentrations lower than 3% of the total arsenic in soils.

The CDI was determined by multiplying the arsenic concentration in soils by the percentage bio-availability arsenic and the typical amount of soil ingested by a child. This value was not exceeded in any case (Table 3), which proved that the soils studied are not toxic.

4 CONCLUSIONS

SEPs can be used as a source of useful and reliable for risk assessment exposure to arsenic from soil, apart from the direct use of total arsenic (risk overestimation).

ACKNOWLEDGEMENTS

The authors wish to thank the FONDECYT Project N°1120881 (Chile).

REFERENCES

Cornejo, L., Acarapi, J., 2011. Fractionation and bioavailability of arsenic in agricultural soils: Solvent extraction tests and their relevance in risk assessment, *Journal of Environmental Science and Health*, Part A, 46: 1247–1258.

Lihareva, N., 2005. Arsenic solubility, mobility and speciation in the deposits from a copper production waste storage. *Microchem. J.*, 81(2): 177–183.

1.7 Arsenic-biotransformations and microbial interactions

One Century of the Discovery of Arsenicosis in Latin America (1914–2014) – Litter, Nicolli, Meichtry, Quici, Bundschuh, Bhattacharya & Naidu (Eds)
© 2014 Taylor & Francis Group, London, ISBN 978-1-138-00141-1

Arsenite-oxidizing bacterial abundance and diversity in the groundwater of Beimen, a blackfoot disease endemic area of southwestern Taiwan

J.-S. Jean, S. Das & S. Kar
Department of Earth Sciences, National Cheng Kung University, Tainan, Taiwan

ABSTRACT: The Most Probable Number-Polymerase Chain Reaction (MPN-PCR) and Denaturing Gradient Gel Electrophoresis (DGGE) was applied to monitor depth-wise abundance and diversity of aerobic arsenite oxidizers in arsenic-enriched groundwater of Beimen, southwestern Taiwan, with the aim of improving our understanding of the phylogeny of As(III)-oxidizers in As-contaminated aquifers. As-oxidizing bacterial communities in As-enriched groundwater at different depths from Beimen were dominated by *Betaproteobacteria* (61%), *Alphaproteobacteria* (28%) and *Gammaproteobacteria* (11%); 78% As-oxidizers were observed in less As-enriched groundwater of Beimen 2B at 60 m depth while 17% As-oxidizers were observed in high As-enriched groundwater of Beimen 1 at 200 m depth. *Pseudomonas* sp. was found only in groundwater containing high arsenic levels at Beimen 1 and Beimen-Jinhu, while arsenite oxidizers belonging to *Alpha-* and *Betaproteobacteria* were dominated in groundwater containing low arsenic. Dissolved As concentration, Eh and oxygen levels were the important variables among several factors influencing As(III)-oxidizer abundance and diversity.

1 INTRODUCTION

In a dissolved state, As occurs mainly in the form of the inorganic species i.e. arsenite [As(III)] and arsenate [As(V)], and the former species, which is more bioavailable, is usually more toxic than As(V) (Lievremont *et al.*, 2009). Oxidation of As(III) leads to the formation of the less bioavailable and less toxic As(V), which can be either precipitated or adsorbed by metal (hydr)oxides (Silver & Phung, 2005). As(III)-oxidizing bacteria can thus contribute to the natural remediation processes, as observed in different contaminated aquifers (Lievremont *et al.*, 2009). The role of arsenite oxidizers in natural attenuation of arsenic pollution necessitates studies on their abundances and diversity in arsenic-contaminated aquifers.

The incidence of high As in groundwater of the Chianan Plain, southwestern Taiwan, is known for its association with endemic Blackfoot Disease (BFD), a peripheral vascular disease (i.e., gangrene) (Tseng, 1977). Despite considerable research over the past three decades into the occurrence and distribution of arsenic in groundwater, studies investigating the microbiological processes that influence the speciation and mobility of the element in this aquifer are limited. In this study, we used MPN-PCR and DGGE of the *aoxB* gene to monitor the structure, diversity and abundance of As(III)-oxidizers in As-enriched groundwater of Beimen, southwestern Taiwan, at different depths, with the aim of improving our understanding of the phylogeny of As(III)-oxidizers in As-contaminated aquifers.

2 METHODS

2.1 *Groundwater sampling and analysis*

Groundwater samples were collected at different depths (60 m–300 m) from Beimen, an area of Chianan Plain, southwestern Taiwan (Figure 1), which is known for endemic blackfoot disease. Groundwater temperature, pH, $Eh_{S.H.E.}$ (relative to

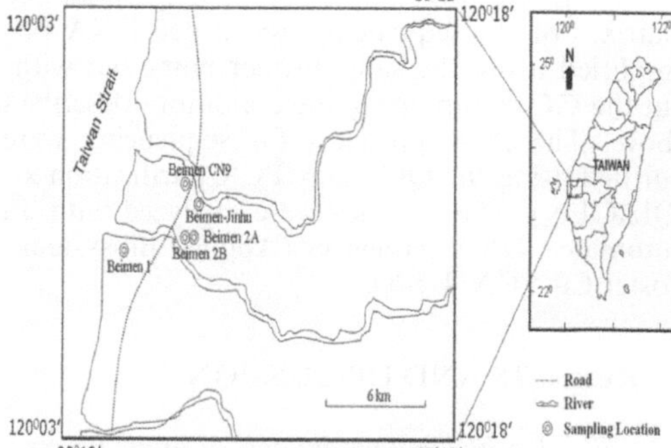

Figure 1. Geographical representation of sampling locations.

standard hydrogen electrode), Dissolved Oxygen (DO), Electrical Conductivity (EC), and Total Dissolved Solids (TDS), were measured at the field sites and total As and As speciation were measured using an Inductively Coupled Plasma Mass Spectrophotometer (ICP-MS; Hewlett–Packard, Yamanashi-Ken, Japan) as described by Kar *et al.* (2010).

2.2 *Genomic DNA extraction from groundwater and PCR*

Genomic DNA from groundwater samples was extracted using the method proposed by Watanabe *et al.* (2000). PCR amplification targeting bacterial 16S rRNA and *aoxB* genes was performed according to Quemeneur *et al.* (2010).

2.3 *MPN PCR and DGGE*

MPN-PCR was performed according to the procedure described in Nesme *et al.* (1995). DGGE analysis of the amplified *aoxB* gene sequences was performed in 8% polyacrylamide gels with a denaturing gradient of 30–70%. 200 ng PCR products were loaded in each well, and electrophoresis was performed at a constant voltage of 60 V for 14 h in 7L 1 × TAE running buffer at 60 °C in the DGGE tank (Biorad, USA). Gels were stained in sterile milli-Q purified water containing ethidium-bromide (0.5 mg L^{-1}) for 10 min, and then distained in sterile milli-Q purified water for 20 min and photographed with UV illumination using a Bio-Rad gel Doc XR system (USA). Replicate DGGE analyses were performed for each sample.

2.4 *Partial aoxB gene sequencing of excised DGGE band*

The bands of interest were excised from the gel using a sterile blade and incubated overnight at 4 °C in 20 µL sterile milli-Q purified water for 24 h to allow DNA diffusion out of the polyacrylamide matrix. For the sequencing, the eluted DNA was amplified using the same primer pairs, but without the GC clamp, using the conditions described above. The PCR products for sequencing were purified using the QIAquick PCR purification kit (QIAGEN). The samples were analyzed with an automated DNA sequencer (Applied Biosystems, Foster City, CA, USA).

3 RESULTS AND DISCUSSION

3.1 *Physicochemical parameters of groundwater at different depths*

The arsenic concentrations of the groundwater samples were much higher than both the World Health Organization (WHO) standard and the Taiwan Drinking Water Standard (TDWS) of 10 µg L^{-1}. The total As concentration in groundwater varied from 230 to 644 µg L^{-1}, with the lowest value in Beimen 2B at a depth of 60 m and the highest value in Beimen 1 at a depth of 200 m. Groundwater was predominantly near neutral to mildly alkaline (pH value 7.1–7.8). The fraction of As(III) to total As in the groundwater increased along with the depth, and was significantly correlated with the decrease in $Eh_{S.H.E.}$ ($r = -0.989$, $p < 0.001$) and DO ($r = -0.972$, $p < 0.001$). The groundwater temperature at different depths varied from 24.8 to 28.7 °C with the lowest in Beimen 2B at 60 m and the highest in Beimen CN9 at 100 m. The $Eh_{S.H.E.}$ values ranged from 61 to 89 mV, with the highest value in Beimen 2B at 60 m and the lowest value in Beimen 1 at 200 m. Dissolved oxygen ranged from 1.2 to 1.8 mg L^{-1}, with the highest value in Beimen 2B at 60 m and the lowest value in Beimen 1 at 200 m. Electrical conductivity and TDS varied wildly, from 1420 to 67900 µS cm^{-1} and 682 to >2000 mg L^{-1}, respectively, with the highest values observed in Beimen 2B at 60 m and the lowest in Beimen-Jinhu at 300 m.

3.2 *Abundance and diversity of As(III)-oxidizing bacterial population in groundwater*

The copy numbers of *aoxB* genes retrieved from groundwater of Beimen at different depths ranged from 2.8×10^2 to 2.6×10^4. Their abundances, i.e., ratio of *aoxB* gene copies relative to universal bacterial 16S rRNA gene copies, ranged from 0.04 to 0.22. The lowest ratio was observed in the most As-enriched and comparatively more reduced groundwater of Beimen 1 at a depth of 200 m, while the highest ratio was observed in the less As-enriched and comparatively less reduced groundwater of Beimen 2B at a depth of 60 m. The decrease in the abundances of *aoxB* genes found may be due to the comparatively high As(III) concentration and the comparatively more reduced conditions prevailing in the deep groundwater.

Different DGGE profiles of As(III)-oxidizing bacterial populations consisting of bands with various intensities and positions were obtained from As-enriched groundwater taken from Beimen at different depths. The most complex DGGE profile (14 bands) was obtained for the less As-enriched (230 µg L^{-1}) and comparatively less reduced groundwater from Beimen 2B at a depth of 60 m, whereas highly As-enriched (644 µg L^{-1}) and comparatively more reduced groundwater from Beimen 1 at a depth of 200 m exhibited the simplest DGGE profile (3 bands) and thus lowest *aoxB* gene richness.

The majority of *aoxB* gene sequences retrieved from the Beimen groundwater were affiliated with

Betaproteobacteria (61%), whereas only 28 and 11% sequences were affiliated with *Alphaproteobacteria* and *Gammaproteobacteria*, respectively. Most of the sequences were closely related to the *aoxB* gene sequences retrieved from various geographical areas around the world, indicating that As(III)-oxidizing bacteria are ubiquitous in As-polluted environments. The aoxB gene sequence more similar to *Pseudomonas* sp. 72 *aoxB* sequence (99% identity), was present in the highly As-enriched groundwater of Beimen 1 and Beimen-Jinhu, but was not detected in less As-enriched groundwater, suggesting the resistance of this bacterium to high As concentrations. *Hydrogenophaga defluvii aoxB* sequence (81 to 86% identity), were observed in the less As-enriched groundwater at Beimen 2B. *Acidovorax* sp. GW2 (78% identity), *Leptothrix* sp. S1-1 (82% identity), *Leptothrix* sp. S1-1 (81% identity) and *Acidovorax* sp. 75 (90% identity) *aoxB* sequences, respectively, were located in the comparatively less As-enriched groundwater of Beimen 2B, Beimen CN9 and Beimen 2A, but not in the highly As enriched groundwater of Beimen 1 and Beimen-Jinhu. *Alphaproteobacteria*, were only located in the less As-enriched and comparatively less reduced groundwater of Beimen 2B. We expected that the high salinity in groundwater (depth 60 m) from Beimen 2B may hinder the growth of As(III)-oxidizing bacterial communities. However, surprisingly, we found that 78% of the As(III)-oxidizers communities consisting of the members of *Betaproteobacteria* and *Alphaproteobacteria* were dominant in the groundwater of Beimen 2B, indicating their resistance to high salinity. As(III)-oxidizers belonging to *Betaproteobacteria* and *Alphaproteobacteria* were reported in the hyper-saline Mono Lake water of California (Oremland *et al.*, 2004). Interestingly, *Thiomonas* sp. *aoxB* sequence (72% identity), were present in the groundwater of Beimen CN9, Beimen 1, Beimen 2A and Beimen-Jinhu, but not in the less As-enriched groundwater of Beimen 2B. Members of the *Thiomonas* genus are ubiquitous in extreme environments contaminated by As, and have a long-term role in the natural remediation process occurring in the As-contaminated sites (Marchal *et al.*, 2011). More importantly, such remediation processes have an important selective advantage by allowing the strain to survive in harsh conditions (Marchal *et al.*, 2011).

3.3 *Impacts of environmental factors on As(III)-oxidizing bacterial abundance*

In order to obtain a better understanding of As(III)-oxidizing bacterial abundance and diversity, and thus As transformation and its mobility, it is of great importance to elucidate the impacts of various environmental factors on the abundance and diversity of this functional group. The abundance of As(III)-oxidizing bacteria in groundwater at different depths from Beimen was concomitant with changes in several of the physico-chemical parameters studied. Even though all the environmental factors significantly influenced As(III)-oxidizing bacterial abundance, $Eh_{S.H.E.}$ ($R^2 = 0.989$, $p < 0.001$) and DO ($R^2 = 0.958$, $p < 0.001$) exerted the greatest impact. However, it was not surprising that As(III)-oxidizing bacteria, which are mostly aerobic, prefer to grow in oxic and/or less reduced aquifers.

4 CONCLUSIONS

This study examined the abundance and diversity of As(III)-oxidizers in the As-enriched groundwater of Beimen, southwestern Taiwan, with the dissolved As concentration, Eh and oxygen levels being important variables among several other factors influencing As(III)-oxidizer abundance and diversity. As-oxidizing bacterial communities in As-enriched groundwater obtained at different depths from Beimen were dominated by *Betaproteobacteria* (61%), *Alphaproteobacteria* (28%) and *Gammaproteobacteria* (11%), with 78% As-oxidizers observed in the less As-enriched groundwater of Beimen 2B at 60 m, while 17% As-oxidizers were observed in high As-enriched groundwater of Beimen 1 at 200 m. *Alphaproteobacteria* comprising members of *Ancylobacter* sp., *Aminobacter* sp., *Rhizobium* sp. were found in the less As-enriched groundwater of Beimen 2B, whereas *Pseudomonas* sp. (*Gammaproteobacteria*) was found in the high As-enriched groundwater of Beimen 1 and Beimen-Jinhu. The study of As(III)-oxidizers abundance and diversity carried out in this work can not only help to predict As transformation and mobility in As-enriched aquifers, but also help to develop bioprocesses for the treatment of As-contaminated water.

ACKNOWLEDGEMENTS

This work was supported by the National Science Council of Taiwan (Project Grant No. NSC 100-2116-M-006-009).

REFERENCES

Jarvis, B., Wilrich, C. & Wilrich, P.T. 2010. Reconsideration of the derivation of Most Probable Numbers, their standard deviations, confidence bounds and rarity values. *J. Appl. Microbiol.* 109: 1660–1667.

Kar, S., Maity, J.P., Jean, J.-S., Liu, C.-C., Nath, B., Yang, H.-J. & Bundschuh, J. 2010. Arsenic-enriched aquifers: Occurrences and mobilization of arsenic in groundwater of Ganges Delta Plain, Barasat, West Bengal, India. *Appl. Geochem.* 25: 1805–1814.

Lievremont, D., Bertin, P.N. & Lett, M.C. 2009. Arsenic in contaminated waters: biogeochemical cycle, microbial metabolism and biotreatment processes. *Biochimie* 91: 1229–1237.

Marchal, M., Briandet, R., Halter, D., Koechler, S., DuBow, M.S., Lett, M.C. & Bertin, P.N. 2011. Subinhibitory arsenite concentrations lead to population dispersal in *Thiomonas* sp. *PLoS ONE* 6(8): 1–8.

Nesme, X., Picard, C. & Simonet, P. 1995. Specific DNA sequences for detection of soil bacteria. In: J.T. Trevors & J.D. Van Elsas (eds.), *Nucleic acids in the environment. Methods and applications*, pp 111–139. Springer- Verlag, Berlin Heidelberg New York.

Oremland, R., Stolz, J.F. & Hollibaugh, J.T. 2004. The microbial arsenic cycle in Mono Lake, California. *FEMS Microb. Ecol.* 48: 15–27.

Quemeneur, M., Cebron, A., Billard, P., Battaglia-Brunet, F., Garrido, F., Leyval, C. & Joulian, C. 2010. Population structure and abundance of arsenite-oxidizing bacteria along an arsenic pollution gradient in waters of the upper Isle river basin, France. *Appl. Environ. Microbiol.* 76: 4566–4570.

Silver, S. & Phung, L.T. 2005. Gene and enzymes involved in bacterial oxidation and reduction of inorganic arsenic. *Appl. Environ. Microb.* 71: 599–608.

Tseng, W.-P. 1977. Effects and dose-response relationships of skin cancer and blackfoot disease with arsenic. *Environ. Health Perspect.* 19: 109–119.

Watanabe, K., Watanabe, K., Kodama, Y., Syutsubo, K. & Harayma, S. 2000. Molecular characterization of bacterial populations in petroleum-contaminated groundwater discharged from underground crude oil storage cavities. *Appl. Environ. Microbiol.* 66: 4803–4809.

*One Century of the Discovery of Arsenicosis in Latin America (1914–2014) –
Litter, Nicolli, Meichtry, Quici, Bundschuh, Bhattacharya & Naidu (Eds)*
 ISBN 978-1-138-00141-1

Arsenic biotransformations: The cycle of arsenic methylation and demethylation

B.P. Rosen, K. Marapakala, A.A. Ajees & M. Yoshinaga
Herbert Wertheim College of Medicine, Florida International University, Miami, FL, USA

ABSTRACT: Arsenic is the most prevalent environmental toxic element and causes health problems throughout the world. Here we will describe the connection between arsenic geochemistry and biology, with emphasis on redox and methylation cycles, as well as other arsenic biotransformations. The arsenic methylation cycle remodels the environment, with some members of microbial communities converting inorganic arsenic into organic species, and others breaking down organic arsenicals such as the arsenical herbicide monosodium methylarsenate (MSMA) into inorganic arsenic. The microbes, genes and proteins that carry out these pathways were identified. The enzymes were characterized, and their crystal structure solved. A new enzymatic mechanism for arsenic methylation by human AS3MT that accounts for the production of carcinogenic species is proposed.

1 INTRODUCTION

Arsenic methylation is catalyzed by the enzyme AS3MT (or ArsM) using S-adenosylmethionine (SAM). In microbes ArsM has been demonstrated to detoxify inorganic arsenic. We cloned *arsM* from the acidothermophilic red alga *Cyanidioschyzon merolae* (Figure 1) and showed it confers arsenic resistance in *E. coli* (Qin *et al.* 2009). Purified ArsM is most active at 50–70 °C and produces volatile TMAs(III) (Marapakala *et al.* 2012). In human liver AS3MT methylates arsenic, presumably for detoxification, but the nature of the products is controversial. Human AS3MT was originally proposed to form both pentavalent and trivalent methylated species, with relative nontoxic and noncarcinogenic DMAs(V) as primary product, thus detoxifying inorganic arsenic. An alternate proposal is that substrates are GSH conjugates, $As(GS)_3$ and $MAs(GS)_2$, and the products are carcinogenic trivalent species MAs(III) and DMAs(III). The products of the former are less carcinogenic than inorganic As(III), while the products of the latter are more carcinogenic. To differentiate between these, we analyzed ArsM. All As(III) SAM methyltranferases have four conserved cysteines. Mutants in any of the four cannot methylate As(III). In contrast, a mutant retaining Cys174 and Cys224 still methylates MAs(III). Thus, all four are required for the first methylation but only two for the second round of methylation. $As(GS)_3$ and $MAs(GS)_2$ bind to ArsM 100-fold faster than the free metalloids, supporting the hypothesis that glutathionylated arsenicals are preferred substrates. ArsM was crystallized, and the structure solved with a variety of ligands including SAM, SAH, As(III) and PhAs(III). We propose a new catalytic model in which transient pentavalent intermediates are reduced by a disulfide bond cascade. We propose that the first product of As(III) SAM methyltransferases, MAs(III), has higher affinity for two of the conserved cysteines than the initial substrate, As(III), and remains enzyme-bound. This facilitates the second round of methylation to DMAs(III), which has lower affinity for the binding site and dissociates faster than the third round of methylation occurs. In this reaction

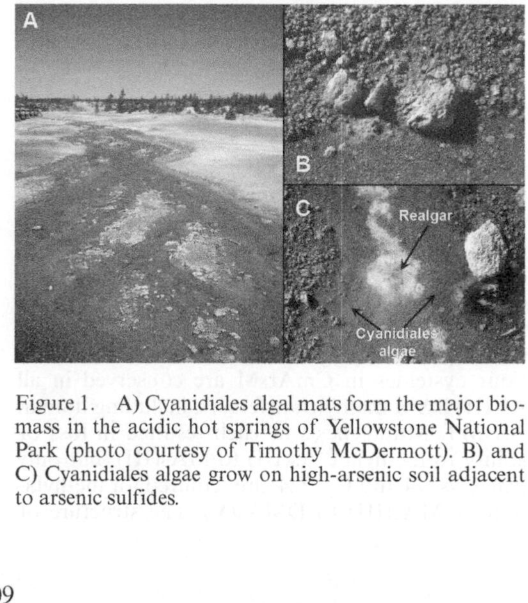

Figure 1. A) Cyanidiales algal mats form the major biomass in the acidic hot springs of Yellowstone National Park (photo courtesy of Timothy McDermott). B) and C) Cyanidiales algae grow on high-arsenic soil adjacent to arsenic sulfides.

scheme, the products are the carcinogenic species MAs(III) and DMAs(III). These results signify that arsenic methylation transforms inorganic arsenic into more carcinogenic organic species.

Methylated arsenicals are produced by many organisms and are also used as pesticides and herbicides for weed control, especially for cotton, ornamental plants, lawns and golf courses. Annually, 1.4 M kg of MSMA is applied commercially in the USA. Much of the MSMA is demethylated to more toxic As(III). Recently, we showed that demethylation of MSMA is the result of the activity of a microbial community isolated from golf course soil. We identified several bacterial species capable of reducing MSMA to MAs(III) and several that demethylate MAs(III) to As(III). In mixed culture, the two types of bacteria demethylated MSMA to As(III). This is a novel pathway for demethylation of an organic arsenical utilizing a communal sequential reduction and demethylation reactions catalyzed by different soil microorganisms.

2 METHODS

2.1 *Methylation*

Purification and crystallization of ArsM from *C. merolae* has been described (Marapakala *et al.* 2010). Enzymatic and structural analysis were accomplished as described (Ajees *et al.* 2012; Marapakala *et al.* 2012).

2.2 *Demethylation*

Organisms in microbial communities that demethylate MSMA were identified as described (Yoshinaga *et al.* 2011).

3 RESULTS AND DISCUSSION

3.1 *Arsenic methylation*

The gene for CmArsM was cloned from total DNA from *C. merolae* (Qin *et al.* 2009), and the enzyme purified from expression in *E. coli* (Marapakala *et al.* 2010). The purified enzyme methylated As(III) to MAs(III) at short times, demonstrating that the methylation reaction produces trivalent products rather than pentavalent arsenicals. At later times, DMAs(V) is formed, but this is quite likely due to oxidation of DMAs(III). Four cysteines in CmArsM are conserved in all As(III) SAM methyltransferases, including human AS3MT. Mutagenesis of each resulted in loss of ability to methylate As(III) to MAs(III), but only two, Cys174 and Cys224, are required for methylation of MAs(III) to DMAs(V). The structure of CmArsM was determined (Figure 2). It has three domains. The N-terminal domain has the SAM binding site. The middle domain has the As(III) binding site. The function of the C-terminal domain is not known but is conserved in arsenite methyltransferases.

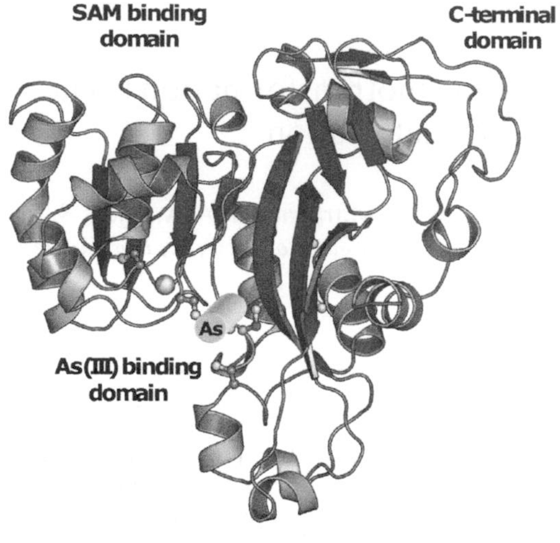

Figure 2. Ribbon representation of CmArsM (PDB ID: 3P7E). The position of the three domains are indicated. The cysteine residues forming the arsenic binding site are shown in ball and stick. The bound arsenic atom is identified as a cylinder.

3.2 *Methylarsenic demethylation*

Arsenic is one of the most wide-spread environmental carcinogens and has created devastating human health problems world-wide; yet little is known about mechanisms of biotransformation in contaminated regions. Methylarsonic acid (MAs(V)), extensively utilized as an herbicide, is largely demethylated to more toxic inorganic arsenite, which causes environmental problems. To understand the process of demethylation of methylarsenicals, soil samples commonly used on Florida golf courses were studied. Several soil extracts were found to demethylate MAs(V) to inorganic arsenite [As(III)]. From these extracts, a bacterial isolate was capable of reducing MAs(V) to MAs(III) but not of demethylating to As(III) (Yoshinaga *et al.* 2011). A second bacterial isolate was capable of demethylating MAs(III) to As(III) but not of reducing MAs(V). A mixed culture could carry out the complete process of reduction and demethylation, demonstrating that demethylation of MAs(V) to As(III) is a two step process (Figure 3).

Analysis of the 16S ribosomal DNA sequences of the three organisms identified the MAs(V)-reducing

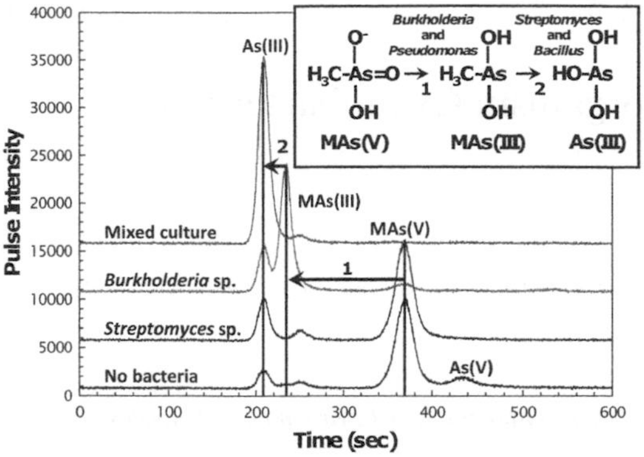

Figure 3. Microbial communities demethylate MSMA to inorganic As(III). Curve 1: control. Curve 2: Streptomyces reduces MSMA (MAs(V)) to MAs(III). Curve 3: *Burkholdaria* alone cannot demethylate MSMA. Curve 4: In mixed culture the two completely demethylate MSMA to As(III).

isolate as belong to *Burkholderia*, and laboratory strain of *Pseudomonas putida* could also reduce. *Streptomyces* and *Bacillus* species could demethylate MAs(III) but not reduce MSMA. We propose that this pathway of carbon-arsenic bond cleavage by sequential reduction and demethylation in a microbial soil community plays a significant role in the arsenic biogeocycle.

4 CONCLUSIONS

Given the widespread occurrence of arsenic in water and soil, microbial transformation into methylated species takes place continuously worldwide. In addition, methylated arsenicals are used as herbicides and for other purposes. Although difficult to quantify, global microbial production of methylarsenicals probably far exceeds anthropogenic introduction. It is no wonder that other microbes have evolved pathways for degradation of the metabolites of other microbes and for herbicide resistance, and characterization of these pathways is essential for understanding environmental cycling of arsenic.

ACKNOWLEDGEMENTS

This work was supported by NIH grant GM55425 to BPR.

REFERENCES

Ajees, A.A., K. Marapakala, C. Packianathan, B. Sankaran & Rosen, B.P. (2012). Structure of an As(III) S-adenosylmethionine methyltransferase: insights into the mechanism of arsenic biotransformation. *Biochemistry* 51(27): 5476–5485.

Marapakala, K., A.A. Ajees, J. Qin, B. Sankaran & Rosen, B.P. (2010). Crystallization and preliminary X-ray crystallographic analysis of the ArsM arsenic(III) S-adenosylmethionine methyltransferase. *Acta Crystallogr. Sect F Struct. Biol. Cryst. Commun.* 66(Pt 9): 1050–1052.

Marapakala, K., J. Qin & Rosen, B.P. (2012). Identification of catalytic residues in the As(III) S-adenosylmethionine methyltransferase. *Biochemistry* 51(5): 944–951.

Qin, J., C.R. Lehr, C. Yuan, X.C. Le, T.R. McDermott & B.P. Rosen (2009). Biotransformation of arsenic by a Yellowstone thermoacidophilic eukaryotic alga. *Proc. Natl. Acad. Sci U S A* 106(13): 5213–5217.

Yoshinaga, M., Y. Cai & Rosen, B.P. (2011). Demethylation of methylarsonic acid by a microbial community. *Environ. Microbiol.* 13(5): 1205–1215.

One Century of the Discovery of Arsenicosis in Latin America (1914–2014) –
Litter, Nicolli, Meichtry, Quici, Bundschuh, Bhattacharya & Naidu (Eds)

Arsenic biotransformation by model organism Cyanobacterium *Synechocystis* sp. PCC6803

Y.-G. Zhu, X.-M. Xue & J. Ye
Key Lab of Urban Environment and Health, Institute of Urban Environment, Chinese Academy of Sciences, Xiamen, China

Si-Yu Zhang & Xi-Xiang Yin
State Key Laboratory of Urban and Regional Ecology, Research Center for Eco-Environmental Sciences, Chinese Academy of Sciences, Beijing, China

ABSTRACT: Arsenic speciation determines its mobility and toxicity. Speciation changes are largely driven by biological actions, such as microbes and plants. This paper is a summary based on our studies using cyanobacterium *Synechocystis* sp. PCC6803 as a model organism for the elucidation of molecular mechanisms of arsenic biotransformation. While it serves as a mini-review, we will also discuss new unpublished results, including the molecular mechanism of arsenosugar biosynthesis.

1 INTRODUCTION

Arsenic transport and transformation in the environment are governed by geochemical as well as biological processes, generating an As biogeochemical cycle (Wang & Mulligan, 2006). Microbiology mediated biogeochemical transformations primarily determine the biogeochemistry and toxicity of arsenic in various natural ecosystems, especially arsenic redox processes which influence the two most abundant inorganic arsenic forms (arsenite As(III) and arsenate As(V)) in the environment (Oremland & Stolz, 2003). Recently, arsenic biomethylation and volatilization, generating the less toxic arsenic species, such as methylarsenate (MMA(V)), dimethylarsenate (DMA(V), and trimethylarsine oxide (TMAO), have also been identified in many microorganisms, and contributed to arsenic biogeochemical cycle significantly (Zhang *et al.*, 2013; Yin *et al.*, 2011).

Due to their ability to adapt rapidly to environmental changes and rapid growth rates, cyanobacteria are often key players in toxic algal blooming in various aquatic environments (Bianchi *et al.*, 2000; Dokulil & Teubner, 2000). However, few studies have been focused on the role of the freshwater and soil algae in arsenic biogeochemistry, which might make great contribution to arsenic biogeochemical cycle, not to mention the biochemical and molecular mechanisms.

2 METHODS/EXPERIMENTAL

2.1 *Alga culture and arsenic species analysis*

Axenic cultures of *Synechocystis* sp. strain PCC6803 wild type and mutant strains were grown in 250 mL Erlenmeyer flasks containing 100 mL BG-11 medium at 30 °C under white light illumination (50 μmol photons $m^{-2}\ s^{-1}$). Cells were grown to logarithmic phase, and preculture cells were diluted with fresh sterile medium to give an OD730 of 0.1. Arsenic speciation in cells and mediums was determined by HPLC- ICP-MS.

2.2 *Deletion of arsM from Synechocystis sp. PCC6803 and complementation of $^{\Delta}$arsM*

The gene *arsM* encoding an arsenite methyltransferase from *Synechocystis* was deleted as follows: approximately 1 kb of arsM was amplified according to Yin (Yin *et al.*, 2011). The plasmid p18T-arsM then was inactivated by inserting a kanamycin gene. A complete arsM containing upstream and downstream sequence and a chloramphenicol gene were inserted into a plasmid pKW1188. Two plasmids were transformed into *Synechocystis* sp. strain PCC6803 wild type and $^{\Delta}$arsM successively in light of previous method (Gao *et al.*, 2007). In order to test whether *Synechocystis* mutant strains were completely segregated, genomic DNA from the mutants and WT as templates were amplified with diagnostic primers.

2.3 *Purification of ArsM and in vitro assay*

The *arsM* was inserted into vector pET22b to form pET22b-*arsM* plasmid. ArsM was expressed in *E. coli* strain Rosetta, purified by Ni(II)-NTA chromatography according to the manufacturer's instruction. Fractions containing purified ArsM were pooled and concentrated by using a 10-kDa cutoff Amicon Ultrafilter (Millipore). As(III) methylation with purified ArsM were performed in a buffer consisting of 50 mM K_2HPO_2, pH 7.4, containing 8 mM GSH, 0.3 mM SAM, 10 μM As(III), and either 5 μM ArsM at 37 °C for 12 h (Yin *et al.*, 2011).

3 RESULTS AND DISCUSSION

3.1 *Arsenic oxidation in Synechocystis sp. PCC6803*

Our results demonstrated that in the presence of cyanobacteria, arsenic redox dynamics in aquatic environment was influenced by phosphate levels. As(III) oxidation by *Synechocystis* sp. PCC6803 appeared to be more effective with increased phosphate levels, indicating that phosphate concentrations should be taken into consideration when predicting the effect of biological As(III) oxidation. Arsenic redox changes by *Synechocystis* sp. PCC6803 under phosphate-limited conditions is a dynamic cyclic process that includes: surface As(III) oxidation (either in the periplasm or near the outer membrane), As(V) uptake, intracellular As(V) reduction and As(III) efflux. These results indicate the biogeochemical coupling of arsenic and phosphorus in aquatic and wetland environments.

3.2 *Arsenic reduction in Synechocystis sp. PCC 6803*

Our results showed that *Synechocystis* sp. strain PCC6803 had the ability of reducing As(V). Arsenic species in cells was determined after exposed for 2 weeks with different levels of As(V) using HPLC-ICP-MS. As(III) and As(V) were detected only as species when exposed to low As(V) concentrations (10–50 μM). It was found that As(V) was the predominant species when exposed to As(V), accounting for 81–84% of the total arsenic. As(III) is the minor species, accounting for 16–19% of the total (Yin *et al.*, 2012).

3.3 *Arsenic methylation in Synechocystis sp. PCC6803*

Purified ArsM was shown to methylate As(III) *in vitro* into DMA and TMAO (Yin *et al.*, 2011). We treated cyanobacteria *Synechocystis* sp. strain PCC6803 wild type, $^{\Delta}$*arsM* and $^{\Delta}$*arsM::arsM* with As(V). Methylated arsenic were found in wild type and $^{\Delta}$*arsM::arsM*, not detected in $^{\Delta}$*arsM*. The results showed that *Synechocystis* sp. PCC6803 had an ability to methylate arsenic and ArsM played an important role in arsenic methylation.

4 PERSPECTIVES AND CONCLUSIONS

Metabolic processes of incorporated As(V) in axenic cultures of *Synechocystis* sp. PCC6803 were examined. Analyses of arsenic compounds in cyanobacterial extracts and culture medium showed that *Synechocystis* sp. PCC6803 have the ability to oxidize As(III) into As(V), reduce As(V) to As(III), methylate As(III) to TMAO. In addition, *Synechocystis* sp. PCC6803 treated with As(V) for three weeks was extracted with chloroform and methanol. Oxo-arsenosugar-glycerol, oxo-arsenosugar-PO4, oxo-arsenosugar-SO3, and arsenosugar phospholipids were identified. These findings suggest that arsenosugars as well as methylated arsenic possibly are the intermediate of arsenosugar phospholipids. We postulate (Figure 1) that arsenosugar phospholipids are biosynthesized by the mechanisms outlined initially Edmonds (Edmonds & Francesconi, 1987) for the methylation of inorganic arsenic by microorganisms, and involving sequential reduction and methylation by As(III) S-Adenosylmethionine methyltransferase, then DMA undergoes glycosylation to format arsenosugars. Finally, oxo-arsenosugar-PO4 is transferred into arsenosugar phospholipids by adding different phospholipids

So far, the known genes related to arsenic metabolism in *Synechocystis* sp. PCC6803 are found in three locations on the chromosome (Ye *et al.*, 2012). Three genes, *acr3*, *arsH*, and *arsC*, form the first ars operon (Figure 2). The gene *acr3* encodes an As(III) efflux pump, and *arsC* encodes an arsenate reductase. *ArsC* reduces As(V) to As(III), which can be excluded from the cell by *Acr3*, making the organism resistant to both As(V) and As(III). The function of gene product of *arsH* has not been identified. However, it has been found to have quinone reductase activity (Hervas *et al.*, 2012). This three-gene operon is regulated by arsR

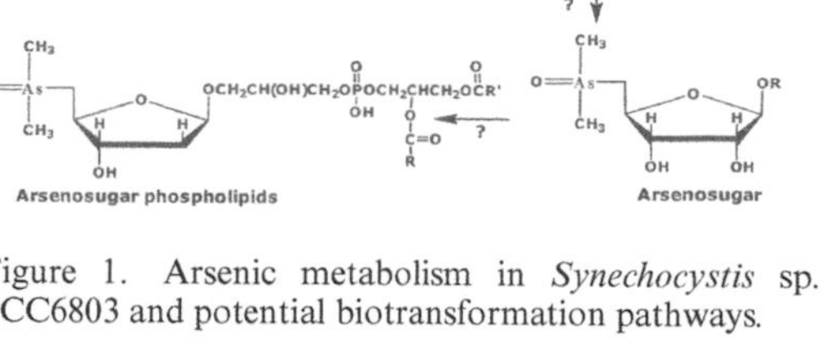

Figure 1. Arsenic metabolism in *Synechocystis* sp. PCC6803 and potential biotransformation pathways.

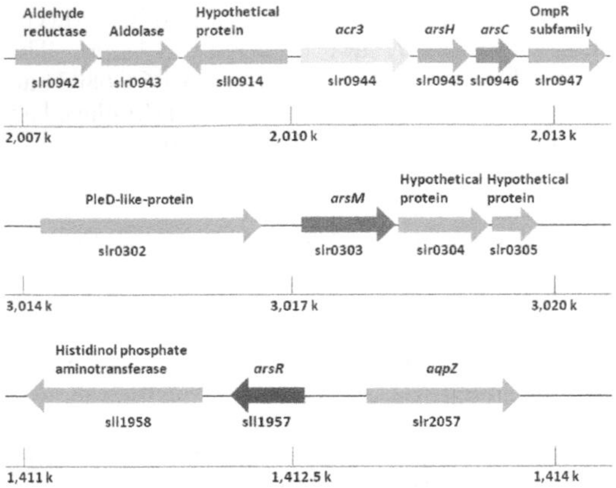

Figure 2. The ars genes of *Synechocystis* sp. PCC6803 (Ye *et al.*, 2012).

(Lopez-Maury *et al.*, 2003), which is located far away. The gene arsM is not near either the three-gene operon or arsR. It may not be regulated by ArsR. Instead, it could be constitutively expressed (Ye *et al.*, 2012).

There could be other genes related arsenic metabolism remaining to be identified in *Synechocystis* sp. PCC6803. For example, genes involved in arsenosugar or arsenolipid synthesis and metabolism have yet to be found. There is evidence that with this cyanobacterium, As(III) oxidation became more effective with increased phosphate concentrations in the medium. However, there is no known arsenite oxidase identified. Thus, it requires further studies to fully understand every aspect of arsenic metabolism in this model organism.

ACKNOWLEDGEMENTS

Our research is supported by the National Natural Science foundation of China (31070101) and the Ministry of Science and Technology of China (2011DFB91710).

REFERENCES

Bianchi, T.S., E. Engelhaupt, P. Westman, T. Andren, C. Rolff, and R. Elmgren. 2000. Cyanobacterial blooms in the Baltic Sea: natural or human-induced? *Limnology and Oceanography* 45 (3):716–726.

Dokulil, M.T., and K. Teubner. 2000. Cyanobacterial dominance in lakes. *Hydrobiologia* 438 (1–3):1–12.

Edmonds, J. & Francesconi, K. 1987. Transformations of arsenic in the marine environment. *Experientia* 43 (5): 553–557.

Gao, H., Tang, Q. & Xu, X. 2007. Construction of copper-induced gene expression platform in Synechocystis sp. PCC6803. *Acta Hydrobiologica Sinica* 31 (2): 244.

Hervas, M., Lopez-Maury, L., Leon, P., Sanchez-Riego, A.M., Florencio, F.J. & Navarro, J.A. 2012. ArsH from the cyanobacterium Synechocystis sp. PCC 6803 is an efficient NADPH-dependent quinone reductase. *Biochemistry* 51 (6): 1178–1187.

Lopez-Maury, L., Florencio, F.J. & Reyes, J.C. 2003. Arsenic sensing and resistance system in the cyanobacterium Synechocystis sp. strain PCC 6803. *J. Bacteriol.* 185 (18): 5363–5371.

Oremland, R.S. & Stolz, J.F. 2003. The ecology of arsenic. *Science* 300 (5621): 939–944.

Wang, S. &. Mulligan, C.N. 2006. Occurrence of arsenic contamination in Canada: sources, behavior and distribution. *Sci. Total Environ.* 366 (2–3): 701–721.

Ye, J., Rensing, C., Rosen, B.P. & Zhu, Y.G. 2012. Arsenic biomethylation by photosynthetic organisms. *Trends in Plant Science* 17 (3): 155–162.

Yin, X.-X., Wang, L.H., Bai, R., Huang, H. & Sun, G.-X. 2012. Accumulation and Transformation of Arsenic in the Blue-Green Alga Synechocysis sp. PCC6803. *Water, Air, & Soil Pollution* 223 (3): 1183–1190.

Yin, X.X., Chen, J., Qin, J., Sun, G.X., Rosen, B.P. & Zhu, Y.G. 2011. Biotransformation and Volatilization of Arsenic by Three Photosynthetic Cyanobacteria. *Plant Physiol.* 156 (3): 1631–1638.

Zhang, S.-Y., Sun, G.-X., Yin, X.-X., Rensing, C. & Zhu, Y.-G. 2013. Biomethylation and volatilization of arsenic by the marine microalgae Ostreococcus tauri. *Chemosphere* 93 (1): 47–53.

One Century of the Discovery of Arsenicosis in Latin America (1914–2014) – Litter, Nicolli, Meichtry, Quici, Bundschuh, Bhattacharya & Naidu (Eds)

Titanium dioxide nanoparticles effects on arsenic accumulation and transformation by freshwater cyanobacterium *Microcystis aeruginosa*

Z. Luo, Z. Wang & C. Yan
Key Laboratory of Urban Environment and Health, Institute of Urban Environment, Chinese Academy of Sciences, Xiamen, China

ABSTRACT: Increased use of engineered nanoparticles raises concerns about their environmental impacts, but the effects of titanium dioxide nanoparticles (nano-TiO_2) on environmental processes of other vital co-contaminants remain largely unknown. In this study, *Microcystis aeruginosa* was exposed to inorganic arsenic (arsenate and arsenite), to study changes in As accumulation and As speciation in the presence of nano-TiO_2. Results showed that nano-TiO_2 improve arsenate accumulation but decrease arsenite accumulation. Also, nano-TiO_2 promote arsenic methylation in algae under both arsenate and arsenite treatments.

1 INTRODUCTION

Nanotechnology is advancing rapidly and could soon become a trillion-dollar industry. Subsequently, release of substantial amount of engineered nanoparticles into the environment is inevitable. Understanding the safety, environmental and human health implications of nanotechnology-based products is of worldwide importance. Titanium dioxide nanoparticles (nano-TiO_2) have drawn considerable attention because of their unique properties and widespread uses in sunscreens, toothpastes, surface coatings, and water treatment. Currently, researches on the risk of nano-TiO_2 are one of the critical topics about the health and safety of nanomaterials, and also are urgently significant issues during the scientific development of nanotechnology in China and other worlds. However, little information, so far, is available for the interaction of nano-TiO_2 with other contaminants in freshwater algae. As a result, it is extremely unclear what are the effects of the interaction on the bioavailability, bioaccumulation and transformation of other contaminants and what are the mechanisms involved.

Arsenic (As) as a major hazardous metalloid is affected by phytoplankton in many aquatic environments. Although As can be found in the environment in several oxidation states, the trivalent (As(III)) and pentavalent oxyanions (As(V)) are more prevalent than the organic forms in freshwaters, and also more toxic. The As(III) form is about 60 times more toxic and mobile than As(V), and conversely, As(V) is about 70 times more toxic than methylated species such as Monomethylarsonic Acid (MMA) and Dimethylarsinic Acid (DMA). MMA(V) and DMA(V) are only slightly toxic. In aquatic systems, the dominant inorganic As is incorporated into microorganisms such as phytoplankton, and converted to methylarsenicals and/or high order organic As such as arsenosugars (AsS). Thus, algae play an important role in As bioaccumulation and biotransformation in the aquatic environment, because they show very high As bioaccumulation from surrounding water and thereby determine the amount of As available to higher organisms and their subsequent transformation.

As the most common toxic cyanobacterium in eutrophic freshwater, *Microcysis aeruginosa* can form harmful algal blooms causing animal poisoning and presenting risks to human health. Therefore, researches on the uptake, bioaccumulation and transformation of arsenic pose vital theoretic roles and practical values. The main objective of this work was to determine the changes of accumulated arsenic and arsenic speciation impacted by nano-TiO_2. The proposed project will provide insights into the impacts of engineered nano-TiO_2 on the behavior of other contaminants in algae and on the environmental risk of engineered nanomaterials.

2 METHODS

2.1 *Preparation and treatments*

Nano-TiO_2 (anatase TiO_2) with claimed particle size of less than 25 nm and purity > 99.7% was used as purchased from Sigma-Aldrich Co. A nano-TiO_2 stock solution (1 g L^{-1}) was prepared by suspending the nanoparticles in an Erlenmeyer flask using ultrapure water, and sonicating the suspension for 30 min. The size distribution of nano-TiO_2 particles in the stock solution was determined using a Laser Particle Size Analyzer. The average hydrodynamic size of TiO_2 nanoparticles was 193 ± 10 nm. Then, a diluted 100 μg L^{-1} nano-TiO_2 suspension was prepared from the stock solution.

In this study, the toxic dominant algae *Microcystis aeruginosa* was exposed to inorganic arsenic (arsenate and arsenite) for 15 days in BG11 culture medium. The arsenite concentrations were 0, 5, 10, 20, 50 μM, while the arsenate concentrations were 0, 5, 10, 20, 50, 100 μM. The control was the similarly spiked arsenic solution without nano-TiO_2.

2.2 *Algae arsenic determination*

For the estimation of the accumulated intracellular total As, the samples of algae were prepared and analyzed according to the method of Yin *et al.* (2012). About 0.02 g oven-dried algae samples were digested with 5 mL concentrated HNO_3 (65%, guaranteed reagent, Fairfield, OH, USA) overnight and heated in a microwave accelerated reaction system (CEM Microwave Technology Ltd, Matthews, NC, USA). Total As (TAs) in the medium and algae were measured by ICP-MS. Approximately 0.02 g freeze-dried algae samples were placed in 15 mL centrifuge tubes and treated with 5 ml of 1% HNO_3 overnight. Then, they were digested in a microwave-accelerated reaction system. The digests were filtered through 0.45 μm cellulose acetate filters. The algae extracts as well as the solutions were analyzed by HPLC-ICP-MS (Agilent LC1100 series and Agilent ICP-MS 7500a; Agilent Technologies) for arsenic speciation with anion exchange columns.

3 RESULTS AND DISCUSSION

3.1 *Arsenic accumulation*

Figure 1 showed arsenic accumulation in algae impacted by nano-TiO_2. Total arsenic (TAs) in the presence of nano-TiO_2 was higher than in its absence in arsenate treatments. However, TAs in the presence of nano-TiO_2 was lower than in the absence in arsenite treatments. This indicates that nano-TiO_2 can improve arsenate accumulation but not arsenite accumulation in algae. The reasons were possibly related to changes of properties after the interaction of the chemical and algal species.

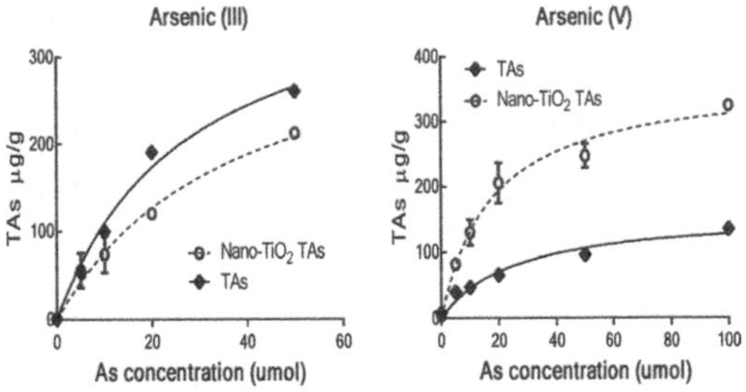

Figure 1. Arsenic accumulation in algae impacted by nano-TiO_2.

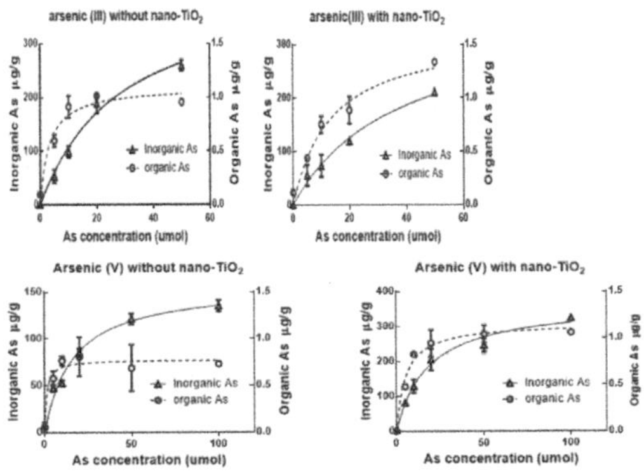

Figure 2. Arsenic speciation in algae impacted by nano-TiO_2.

3.2 *Arsenic speciation*

Figure 2 shows the arsenic speciation in algae impacted by nano-TiO_2. Results indicate that both arsenate and arsenite increased the concentrations of their organic amount in algae under the nano-TiO_2 treatments compared with results in the absence of the nanoparticles. Therefore, nano-TiO_2 promoted inorganic arsenic methylation in algae, indicating that arsenic toxicity to algae and other organisms is reduced. However, it is necessary to investigate the potential mechanism involved in further studies.

4 CONCLUSION

Nano-TiO_2 improves arsenate accumulation but decreases arsenite accumulation. In addition, the nanoparticles promote arsenic methylation in algae under both arsenate and arsenite treatments. Further studies are needed on arsenic changes in algae impacted by nano-TiO_2.

ACKNOWLEDGEMENTS

This work was supported by the National Nature Science Foundation of China (41271484).

REFERENCES

Duester, L., van der Geest, H.G., Moelleken, S., Hirner, A.V. & Kueppers, K. 2011. Comparative phytotoxicity of methylated and inorganic arsenic- and antimony species to Lemna minor, Wolffia arrhiza and Selenastrum capricornutum. *Microchemical Journal* 97: 30–37.

Duncan, E., Foster, S. & Maher, W. 2010. Uptake and metabolism of arsenate, methylarsonate and arsenobetaine by axenic cultures of the phytoplankton Dunaliella tertiolecta. *Bot. Mar*. 53: 377–386.

Luo, Z., Wang, Z., Li, Q., Pan, Q., Yan, C. & Liu, F. 2011. Spatial distribution, electron microscopy analysis of titanium and its correlation to heavy metals: occurrence and sources of titanium nanomaterials in surface sediments from Xiamen Bay, China. *J Environ. Monit.* 13(4): 1046–1052.

Wang, Z., Luo, Z. & Yan, C. 2013. Accumulation, transformation, and release of inorganic arsenic by the freshwater cyanobacterium Microcystis aeruginosa. *Environ. Sci. Pollut. Res.*, DOI: 10.1007/s11356-013-1741-7.

Yin, X.-X., Wang, L.H., Bai, R., Huang, H. & Sun, G.-X. 2012. Accumulation and Transformation of Arsenic in the Blue-Green Alga *Synechocystis* sp. PCC6803. *Water, Air, & Soil Pollution* 223, 1183–1190.

One Century of the Discovery of Arsenicosis in Latin America (1914–2014) – Litter, Nicolli, Meichtry, Quici, Bundschuh, Bhattacharya & Naidu (Eds)
© 2014 Taylor & Francis Group, London, ISBN 978-1-138-00141-1

Characterization of bacterial biofilms formed on As(V)-containing water

S.E. Rastelli
Centro de Investigación y Desarrollo de Pinturas (CIDEPINT, CICPBA—CCT La Plata CONICET), La Plata, Argentina
Facultad de Ciencias Naturales y Museo, Universidad Nacional de La Plata, Argentina

B.M. Rosales
Centro de Investigación y Desarrollo de Pinturas (CIDEPINT, CICPBA—CCT La Plata CONICET), La Plata, Argentina

M.R. Viera
Centro de Investigación y Desarrollo de Pinturas (CIDEPINT, CICPBA—CCT La Plata CONICET), La Plata, Argentina
Facultad de Ciencias Exactas, Universidad Nacional de La Plata, Argentina

ABSTRACT: Microorganisms, particularly those associated to surface (sessile microorganisms), play a role in the process of arsenic mobilization. Several studies showed the existence of specific genes associated to arsenic resistance in microorganisms exposed to arsenic environments. The aim of this work was to study the tolerance to As(V) of sessile bacteria grown on materials for drinking water distribution in the presence of As(V). Bacterial counts on Fe and Zn were higher than those obtained on Cu and polypropylene and they were higher in the presence of As. Culturable As-tolerant bacteria able to grow in the presence of high As(V) concentration (up to 1000 mg L^{-1}) were obtained from all the biofilms except Cu-biofilms, which grew in the presence of up to 300 mg L^{-1}. Seventeen different species were identified: 64% Bacilli, 12% α-Proteobacteria and Actinobacteria, 6% β-Proteobacteria and γ-Proteobacteria. The arsC gene was detected in several sessile communities and bacterial isolates.

1 INTRODUCTION

Arsenic is widely distributed in the environment as the result of natural geochemical phenomena and anthropogenic activities. Arsenic can exist in different oxidation states, being As(V) and As(III) the most common in nature. Both As species exhibit variation in solubility, mobility, bioavailability, and toxicity (Macur *et al.*, 2004). Microorganisms, particularly those associated to surface (sessile microorganisms) play a role in the process of arsenic mobilization (Oremland & Stolz, 2005). Several studies have shown the existence of specific genes associated to arsenic resistance in microorganisms exposed to arsenic environments (Macy *et al.*, 2000). The aim of this work was to study the tolerance to As(V) of sessile bacteria grown on materials for distribution of drinking water in the presence of As(V). These microorganisms may play a role in a bioremediation process.

2 METHODS/EXPERIMENTAL

2.1 *Experimental setup*

Two laboratory simulated water distribution circuits consisting each in a 50 L polyethylene storage tank and a closed loop of polypropylene tubes (inner diameter: 2.32 cm; length: 200 cm) with a removable 20 cm acrylic cell were used. La Plata City drinking water was pumped from the tank through the loop at a laminar flux with 30/60 minutes work/stop periods along the day and no flow at night to simulate domestic network operating cycles. To study the settlement of bacteria on different water distribution network materials, coupons of 1 cm × 1 cm × 0.02 cm of commercial low carbon steel (Fe), zinc (Zn), copper alloy (Cu) and polypropylene (PP) were placed in the acrylic cell. To study the influence of arsenic, 5 mg L^{-1} As(V) were added in one of the circuits.

2.2 *Enumeration of total and As-tolerant sessile bacteria*

After 45–60 days of water recirculation, 4 coupons of each material were withdrawn from each circuit, and replaced by new coupons (this procedure was repeated 7 times). Biofilms were scrapped and poured in 1 mL physiological solution. Enumeration of heterotrophic sessile bacteria was made by plate count on nutrient agar. Arsenic-resistant sessile bacteria was evaluated by culturing in nutritive broth with 50 to 1000 mg L^{-1} As(V).

2.3 *Identification of As resistance bacteria and detection of ars gene*

Isolated colonies formed on the nutrient agar plate were picked, inoculated on a new plate and DNA was extracted by boiling, amplified by PCR of the 16S rRNA using the primers 27F and 1492R and sequenced by MACROGEN (Korea). Sequence data were compared for initial identification with the closest relatives represented by the retrieved sequences obtained by homology searches using the Blast algorithm at the NCBI (http://www.ncbi.nlm.nih.gov / blast/). The presence of the arsC gene in the isolated bacteria and in the whole sessile communities was analyzed by PCR using four set of primers: arsC-gram(+) 16F and 317R; arsC-Pseud. 6F and 365R (Macur *et al.* 2004); amlt 42F and 376R; smrc 42F and 376R (Drewniak *et al.*, 2008).

3 RESULTS AND DISCUSSION

3.1 *Enumeration of sessile bacteria*

Mean total heterotrophic sessile bacteria counting showed differences amongst the average counting obtained on the four tested materials but also an influence of the presence of As(V) (Figure 1). In general, bacterial counts on corrosion susceptible materials (Fe and Zn) were higher than those obtained on materials less susceptible to bacterial attack (Cu and PP) and they were higher in the presence of As (except for the case of copper), suggesting that the presence of the toxic in the liquid medium favored the election of the sessile form of living by the microorganisms. The statistical analysis indicated that there were significant differences between Zn vs. Cu ($p = 0.01$) and Zn vs. PP ($p < 0.01$) in the presence and in the absence of As(V).

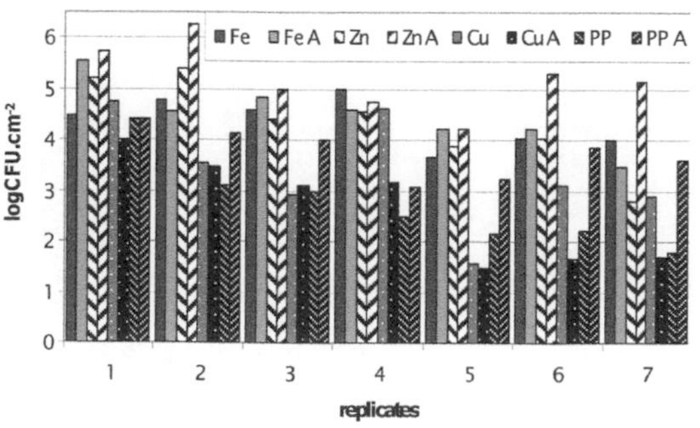

Figure 1. Enumeration of total heterotrophic sessile bacteria obtained from all materials assayed from both circuits. A: indicates biofilm from the As-containing circuit.

Culturable As-tolerant bacteria were obtained from all the biofilms except Cu-biofilms cultured in the highest As(V) concentrations. In general, the number of arsenic-tolerant bacteria obtained by the dilution to extinction method were higher in biofilms from the As-containing circuit and diminished as the As(V) concentration increased (Table 1).

3.2 *Identification of bacteria*

A total of 38 colonies were classified according to their similarity to sequences in the GenBank database, resulting in 17 different species (with similarity higher than 94%). 64% of them belonged to the Class Bacilli: 10 different species of *Bacillus* found on all biofilms and *Paenibacillus tylopili*OL-8 developed on Fe(A). 12% to Class α-Proteobacteria: *Brevundimonas* sp. OS16 found on all biofilm samples and *Sphingomonas* sp. XJ3 found on Fe and Cu biofilms. 12% to the Class Actinobacteria: *Janibacter* sp. TS20 found only on PP(A) and *Kokuria* sp. TS13 found on Cu(A) sample. 6% to the β-Proteobacteria: *Delftia* sp. TS33 detected on PP(A) and Zn(A) biofilms; and 6% to the γ-Proteobacteria: *Acinetobacter* sp. CNE4 found on Zn biofilm. A higher diversity of species was found in those biofilms developed in the As-containing circuit.

Table 1. Enumeration of Arsenic tolerant bacteria in the biofilms formed on the different substrata from both circuits, cultured at different As(V) concentrations.

	As (V) concentration (mg L^{-1})					
Sample	50	100	200	300	500	1000
Zn (A)	10^5–10^6	10^5–10^6	10^5–10^6	10^5–10^6	10^5–10^6	10^4–10^5
Zn	10^5–10^6	10^5–10^6	10^5–10^6	10^5–10^6	10^4–10^5	10^3–10^4
Fe (A)	10^5–10^6	10^5–10^6	10^5–10^6	10^5–10^6	10^5–10^6	10^5–10^6
Fe	10^5–10^6	10^5–10^6	10^5–10^6	10^5–10^6	10^3–10^4	10^4–10^5
Cu (A)	10^4–10^5	10^5–10^6	10^5–10^6	10^5–10^6	ng	ng
Cu	10^4–10^5	10^5–10^6	10^5–10^6	10^5–10^6	ng	ng
PP (A)	10^4–10^5	10^5–10^6	10^5–10^6	10^5–10^6	10^2–10^3	10^2–10^3
PP	10^4–10^5	10^5–10^6	10^5–10^6	10^5–10^6	nd	nd

A: indicates biofilm from the As-containing circuit; ng: no grow detected; nd: no data.

The presence of a genetic component associated to arsenic resistance was confirmed in the biofilms formed on Cu and PP in the As-containing circuit and in the biofilms on Fe and Zn from the other circuit. Besides, the presence of genes associated to arsenic resistance was also detected in several bacterial isolates: *Bacillus licheniformis* SeaHAs1 W, *Delftia* sp. TS33, *Bacillus megaterium* MBFF6 and *Paenibacillus tylopili* OL-8.

4 CONCLUSIONS

Zn and Fe coupons were the most susceptible materials for bacterial colonization. In general, bacterial numbers were higher in the samples from the As-containing circuit. Biofilm bacteria from both circuit were able to grow in the presence of high As(V) concentration (up to 1000 mg L^{-1}).

The arsC gene was detected in several sessile communities and bacterial isolates.

ACKNOWLEDGEMENTS

This work was financially supported by the governmental ANPCYT (PICT 38380) of Argentina.

REFERENCES

Drewniak, L., Styczek, A., Majder-Lopatka, M. & Sklodowska, A. 2008. Bacteria, hypertolerant to arsenic in the rocks of an ancient gold mine, and their potential role in dissemination of arsenic pollution. *Environ. Pollut.* 156: 1069–1074.

Macur, R., Jackson, C., Botero, L., Mc Dermott, T. & Inskeep, W. 2004. Bacterial populations associated with the oxidation and reduction of arsenic in an unsaturated soil. *Environ. Sci. & Technol.* 38(1):104–111.

Macy, J., Santini, J., Pauling, B., O'Neill, A. & Sly, L. 2000. Two new arsenate/sulfate-reducing bacteria: mechanisms of arsenate reduction. *Archiv. Microbiology*, 173(1): 49–57.

Oremland, R. & Stolz, J. 2005. Arsenic, microbes, and contaminated aquifers. *Trends in Microbiology* 13(2): 45–49.

One Century of the Discovery of Arsenicosis in Latin America (1914–2014) – Litter, Nicolli, Meichtry, Quici, Bundschuh, Bhattacharya & Naidu (Eds)
 ISBN 978-1-138-00141-1

Global genomic expression of active bacteria populations that transform arsenic in sediments of Salar de Ascotán, Chile

M.A. Ferrero & J.A. Lara
Planta Piloto de Procesos Industriales Microbiológicos (PROIMI-CONICET), Tucumán, Argentina

L. Escudero
Centro de Investigación Científica y Tecnológica para la Minería (CICITEM), Antofagasta, Chile

G. Chong
Departamento de Ciencias Geológicas, Universidad Católica del Norte, Antofagasta, Chile

C. Demergasso
Centro de Biotecnología, Universidad Católica del Norte, Antofagasta, Chile
CICITEM, Antofagasta, Chile

ABSTRACT: Arsenic mobility and bioavailability in sediment can be controlled by microbial mediated transformations between the most prevalent types of dissolved arsenic: arsenite and arsenate. Specialist bacteria able to obtain energy for growth through redox transformations of arsenic have been evidenced in Salar de Ascotán, Antofagasta region, Chile. The global genomic expression has been analyzed by RAP-PCR using arbitrary primers, to detect microorganisms with active metabolism in Laguna Turquesa. Several protein sequences revealed that many genes related to arsenic resistance have been expressed in this environment: *arsM* gene, responsible for the removal of arsenic as the volatile arsines: *MraW* protein, homolog also with *ArsM* protein; acetyl-CoA acetyltransferase, which present similarity with a novel arsenate resistance gene, *arsN*. Our results suggest that novel bacteria with important biotechnological properties could be exploited in this unique and stressed saline environment.

1 INTRODUCTION

The Region of Antofagasta, Northern Chile, is a barren, arid to hyperarid region where arsenic is widely present in water, rocks and soil. It is generally assumed that arsenic is produced by volcanic activity and abundant arsenic containing minerals. Dissimilatory Arsenate Reducing (DsAR) prokaryotes able to respire using arsenate as the electron acceptor have been described and cultured from the Salar de Ascotán (Demergasso *et al.*, 2007). The isolates can reduce arsenate and sulfate causing the precipitation of arsenic sulfides. In this study, water from one salt lake (Laguna Turquesa) and one spring or "vertiente" (V10) of Salar de Ascotán was collected in order to investigate the global expression of genes in this hypersaline environment and to identify those involved in arsenic transformations.

2 METHODS

2.1 *Sampling and RNA extraction*

Salar de Ascotán was visited in June 2006 (winter) during an intensive sampling expedition (Lara *et al.*, 2012). To extract RNA, cells from Laguna Turquesa (LT) and Vertiente 10 (V10) were collected by filtering 1 L of water through a 0.2 μm pore size membrane and samples were stored at −80 °C. RNA was extracted from the filters and samples were treated with DNAse for 1 hour at 37 °C and then stored at 80 °C.

2.2 *RAP-PCR amplification, cloning and sequencing*

RAP-PCR experiments were performed, following previously described methods (Paulino *et al.*, 2002) with modifications. OPJ-10 was the arbitrary primer tested in this experiment. After amplifications, the gel was run for 5 h at 1500 V on a sequencer Sequi-Gen®GT Nucleic Acid Electrophoresis Cell. After electrophoresis, the gel was revealed and analyzed in a Cyclone® Plus Storage Phosphor System. The differentially expressed RAP-PCR bands were excised from the gel and the DNA was eluted. All bands obtained from differential display gel were cloned in pGEM®-T Easy Vector System cloning kit and screened for white colonies. Inserts were amplified by using M13 forward and reverse primers. All clones with different sizes of inserts were selected for sequencing with M13 primers.

3 RESULTS AND DISCUSSION

A total of 18 bands were excised from the gel and reamplified using primer random OPJ10, according to previously RAP-PCR program. Bands retrieved of natural and differentially expressed genes in Laguna Turquesa and Vertiente 10 were sequenced. From the 83 sequences retrieved, 11 belonged to 16S rRNA genes (Table S2). The genes retrieved by RAP-PCR belong to *Archaea, Bacteroidetes, Firmicutes, Proteobacteria, Verrucomicrobia, Spirochaetes* and *Actinobacteria* according with BLAST analysis. *Firmicutes, Proteobacteria, Bacteroidetes, Spirochaetes* and *Actinobacteria* Phyla were also evidenced as bacterial community members by DGGE analysis of 16S rRNA fragments (Lara *et al.*, 2012). The sequences retrieved were related to genes involved in the functional categories of sugar metabolism, methanogenesis, aminoacid and nucleotide metabolism, energy production and conversion, metabolism and

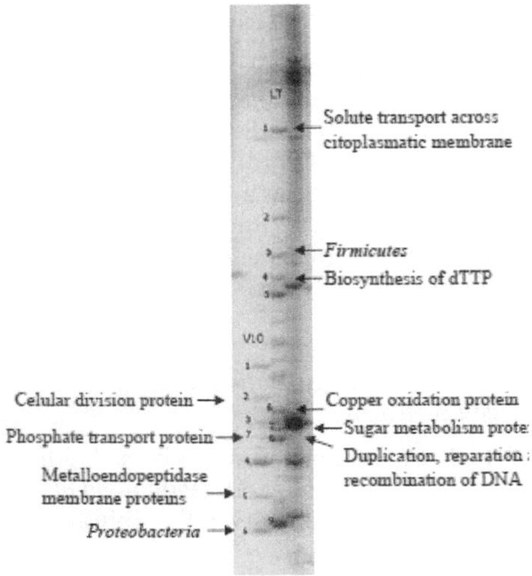

Figure 1. Differential display gel: (1) Vertiente 10 (V10). (2) Laguna Turquesa (LT). cDNA product of RAP-PCR from each street were obtained with OPJ-10 primers.

metabolism of other aminoacids, lipid transport and metabolism, signal transduction mechanisms, nucleotide transport, cell wall/membrane biogenesis, cell cycle control and multifunctional. Six other sequences without related function were isolated. Two sequences related to genes encoding for S-adenosylmethionine (SAM)-dependent methyltransferase and to acetyl-CoA acetyltransferase homologues were retrieved from the arsenic bearing environment.

S-adenosylmethionine (SAM)-dependent methyltransferase and acetyl-CoA acetyltransferase have been reported to be genes conferring arsenic resistance (Qin *et al.*, 2006; Chauhan *et al.*, 2009), and they were termed arsM and arsN, respectively.

4 CONCLUSIONS

The analysis of global genomic expression in sediment samples of Salar de Ascotán allowed to detect microorganisms with active metabolism related to arsenic transformation in the communities present in the Laguna Turquesa sediments. Our results suggest that novel bacteria with important biotechnological properties could be exploited in this unique and stressed saline environment.

ACKNOWLEDGEMENTS

BBVA Foundation project BIOARSENICO, FONDECYT (Project 1100795) from the Science and Technology Chilean Commission (CONICYT) and CONICET, from Ministry of Science and Technology (Argentina).

REFERENCES

Chauhan, N.S., Ranjan, R., Purohit, H.J., Kalia, V.C. & Sharma, R. 2009. Identification of genes conferring arsenic resistance to Escherichia coli from an effluent treatment plant sludge metagenomic library. FEMS *Microbiol. Ecology* 67: 130–139.

Demergasso, C., Chong, G., Escudero, G.L., Pueyo, J.J. & Pedrós-Alió, C. 2007. Microbial precipitation of arsenic sulfides in Andean salt flats. *Geomicrobiology Journal* 24: 111–123.

Lara, J., Escudero G.L., Ferrero, M., Chong Díaz, G., Pedrós-Alió, C. & Demergasso, C. 2012. Enrichment of arsenic transforming and resistant heterotrophic bacteria from sediments of two salt lakes in Northern Chile. *Extremophiles* 16: 523–538.

Paulino, L.C., de Mello, M.P. & Ottoboni, L.M. 2002. Differential gene expression in response to copper in *Acidithiobacillus ferrooxidans* analyzed by RNA arbitrarily primed polymerase chain reaction. *Electrophoresis* 23 (4): 520–527.

Qin, J., Rosen, B.P., Zhang, Y., Wang, G., Franke, S. & Rensing, C. 2006. Arsenic detoxification and evolution of trimethylarsine gas by a microbial arsenite S-adenosylmethionine methyltransferase. *PNAS-USA* 103: 2075–2080.

One Century of the Discovery of Arsenicosis in Latin America (1914–2014) – Litter, Nicolli, Meichtry, Quici, Bundschuh, Bhattacharya & Naidu (Eds)

Identification of arsenite S-adenosylmethionine methyltransferase in anaerobic archaea

G.-X. Sun, P.-P. Wang, P. Bao & Y.-G. Zhu
State Key Laboratory of Urban and Regional Ecology, Research Center for Eco-Environmental Sciences, Chinese Academy of Sciences, Beijing, China

ABSTRACT: Arsenic methylation is an important process frequently occurring in anaerobic environments by microorganisms such as methanoarchaea. However, little is known regarding the enzymatic mechanism of As methylation by archaea. In our study, a new As methyltransferase genes, *MaarsM*, was cloned from methanoarchaea *Methanosarcina acetivora* C2A. Heterologous expression of *MaarsM* conferred As resistance and the ability of methylating As to an As-sensitive strain of *E. coli*. Purified enzymes MaArsM catalyzed the formation of methylated As from arsenite. Results showed that MaArsM exhibited high capacity of As(III) methylation both *in vivo* and *in vitro*. These results provided the characteristics of As methyltransferases from archaeal As methyltransferase. Due to its ubiquity in sediments and wetlands, methylation and volatilization mediated by archaea are proposed to be the important component in As biogeochemical cycling of anaerobic environments.

1 INTRODUCTION

Arsenic is the most common toxic substance in the environment. Naturally anaerobic environments, such as sediments, paddy soils and wetlands, typically play major roles in the transformation, mobilization and biogeochemical cycling of arsenic (As) originating from natural and anthropogenic sources. Anaerobic microorganisms existent in these ecosystems are major contributors for As transformation including methylation. Arsenic methylation and volatilization have been reported in many wetland systems like paddy soils (Huang *et al.*, 2012; Jia *et al.*, 2012), and *arsM* is the key functional gene responsible for microbial As methylation. However, studies aimed at identifying As methylation genes or enzymes have focused almost entirely on eukaryotic or aerobic microorganisms (Wang *et al.*, 2013). Few evidences for As methylation mechanism has been presented in anaerobic microorganisms so far (Qin *et al.*, 2006).

Methanogenic archaea as typical anaerobic prokaryotes are widely distributed in anaerobic ecosystems. Arsenic methylation was proposed to be an inherent feature of methanoarchaea (Meyer *et al.*, 2008). Despite a methylcobalamin dependent non-enzymatic methylation of As has been reported in methanoarchaea, the enzymatic mechanism for As methylation in archaea has remained unclear.

Methanosarcina acetivora C2A was chosen as model of anaerobic archaeal organisms. Our main objectives were to provide genetic evidence for As methylation, (2) provide enzymatic evidence for As methylation, and (3) identify the dominant As methylation mechanism.

2 METHODS/EXPERIMENTAL

2.1 *Culture of strain*

M. acetivorans C2A obtained from American Type Culture Collection 35395 was grown in ATCC Medium 1355 at 37 °C in the dark. The headspace of the serum bottle containing the medium was filled with a filtered gas mixture (80% H_2–20% CO_2) before inoculation.

2.2 *Purification of ArsM enzymes*

Purification of ArsM enzymes was conducted by expression of *arsM* in *E. coli* BL21 and separation through a Ni(II)-NTA column. Arsenic speciation was analyzed by HPLC-ICP-MS (7500a; Agilent Technologies) as described (Zhu *et al.*, 2008)

3 RESULTS AND DISCUSSION

3.1 *Arsenic methylation by methanoarchaea*

Results showed that archaea *M. acetivora* C2A had the capacity of methylating As, as monomethylarsenate [MMA(V)] was detected in the medium added with As(III); in contrast, no dimethylarsenate [DMA(V)] was detected (Figure 1, curve 3). The percentage of MMA(V) was about 8% of total

As in the medium spiked with 10 μM As(III), and 14% in that with 5 μM As(III). There was no methylated As detected in the negative controls (no bacterial inoculation) (Figure 1, curve 2).

3.2 *Subsection 3.2*

E. coli AW3110 expressing either *MaarsM* gene obtained the capacity of methylating As because of the formation of methylated As in their media (data not shown), indicating that the ArsM enzyme encoded by *arsM* gene indeed catalyzed the reaction of As(III) methylation. To further elucidate the mechanism of As methylation by archaeal enzyme, MaArsM protein was purified from recombinant *E. coli* cytosols and assayed for As(III) methyltransferase activities *in vitro*.

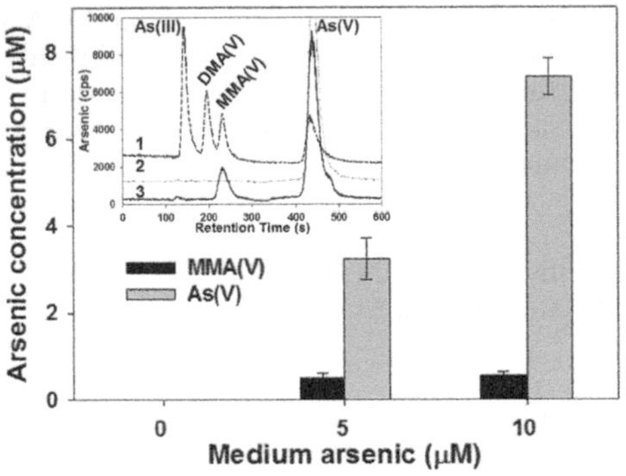

Figure 1. Arsenic transformations by *M. acetivora* C2A exposed to 5 and 10 μM arsenite for 14 days. Curve 1: Standard. Curve 2: Control (without bacteria). Curve 3: Strains. The error bars indicate the standard errors of three triplicates.

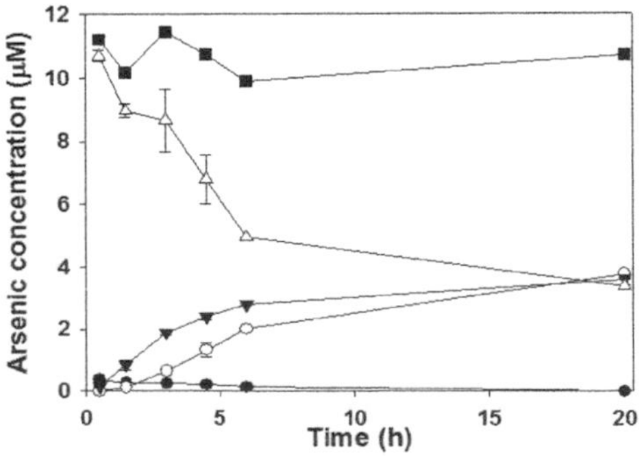

Figure 2. Time courses of As(III) methylation by purified ArsM enzyme. Each assay contained 10 μM As(III), 0.5 mM AdoMet, 8 mM GSH, and 10 μM purified MaArsM in 500 μL MOPS buffer (pH 7.4). The reactions were kept at 37 °C. Arsenic speciation of the samples were analyzed at the indicated times by HPLC-ICP-MS. The error bars indicate the standard errors of three triplicates.

The concentrations of DMA(V) and MMA(V) kept on increasing during the reaction, concomitant with the decrease of inorganic As (Figure 2). In addition, the level of trimethylarsine oxide (TMAO) was maximum at the beginning of reaction (0.5 h) and then decreased with time. At 20 h, DMA(V) and MMA(V) were the dominant organic species and TMAO disappeared (Figure 2).

4 CONCLUSIONS

In summary, the identification and characterization of As methylation genes in archaea demonstrated the roles of As methylation and cycling in anaerobic environments. Given the high As methylation capacity of archaea, this microorganism might play a key role in some anaerobic environments, especially in the extreme ones according to its prevalence in these ecosystems.

ACKNOWLEDGEMENTS

This project was financially supported by the Major Program of the Natural Science Foundation of China (No. 41090284) and the State Key Program of Natural Science Foundation of China (No. 41330853).

REFERENCES

Huang, H., Jia, Y., Sun, G.X. & Zhu, Y.G. 2012. Arsenic speciation and volatilization from flooded paddy soils amended with different organic matters. *Environmental Science and Technology* 46: 2163–2168.

Jia, Y., Huang, H., Sun, G.X., Zhao, F.J. & Zhu, Y.G. 2012. Pathways and relative contributions to arsenic volatilization from rice plants and paddy soil. *Environmental Science and Technology* 46: 8090–8096.

Meyer, J., Michalke, K., Kouril, T. & Hensel, R. 2008. Volatilisation of metals and metalloids: An inherent feature of methanoarchaea? *Systematic and applied microbiology* 31: 81–87.

Qin, J., Rosen, B.P., Zhang, Y., Wang, G., Franke, S. & Rensing, C. 2006. Arsenic detoxification and evolution of trimethylarsine gas by a microbial arsenite S-adenosylmethionine methyltransferase. *Proceedings of the National Academy of Sciences of the United States of America* 103: 2075–2080.

Wang, P.P., Sun, G.X., Jia, Y., Meharg, A.A. & Zhu, Y.G. (2013) Completing arsenic biogeochemical cycle: Microbial volatilization of arsines in environment. *Journal of Environmental Sciences* DOI: 10.1016/S1001-0742(13)60432-5.

Zhu, Y.G., Sun, G.X., Lei, M., Teng, M., Liu, Y.X., Chen, N.C., Wang, L.H., Carey, A.M., Deacon, C., Raab, A., Meharg, A.A. & Williams, P. 2008. High percentage inorganic arsenic content of mining impacted and nonimpacted chinese rice. *Environmental Science and Technology* 42: 5008–5013.

One Century of the Discovery of Arsenicosis in Latin America (1914–2014) – Litter, Nicolli, Meichtry, Quici, Bundschuh, Bhattacharya & Naidu (Eds)
© 2014 Taylor & Francis Group, London, ISBN 978-1-138-00141-1

SensAr: An arsenic biosensor for drinking water

A.D. Nadra
Departamento de Química Biológica, IQUIBICEN-CONICET, Facultad de Cs. Exactas y Naturales, Universidad de Buenos Aires, Argentina

B. Basanta, H. Bonomi, N. Carlotto, M. Giménez, A. Grande, N. Nieto Moreno, F. Barone, F. Dorr, L. Marasco, S. Mildiner, I. Patop, S. Sosa, L. Vattino & F. Vignale
Facultad de Cs. Exactas y Naturales, Universidad de Buenos Aires, Argentina

ABSTRACT: Based on synthetic biology tools, a biosensor was designed to be specific for arsenic detection in water. It was designed to be cheap and easy to use (image based instructions will be enough). The device has a modular design which enables us to detect other water pollutants, with minor genetic modifications. In the presence of arsenic, the system develops a red color in a dose dependent manner that can be evaluated by naked eye and compared to an internal standard. Laboratory proof of concept worked nicely and the dynamic range of the assay (between 0 and 500 ppb) covers the critical arsenite concentrations according the World Health Organization. The current effort is focused on building a prototype that could be delivered for domestic use.

1 INTRODUCTION

Limited access to potable water is a serious problem that tends to increase with the years. Pollution that turns potable water into non-potable water can vary from a single toxic (e.g. arsenic) to a wide and complex mixture of different types and quantities of harmful substances, as can be found in diverse hydrographic watersheds (e.g. Salí Dulce, Matanza-Riachuelo, among others in Argentina). Even if it was not possible to potabilize polluted water, information about levels of pollutants should be easily used to modify consumption patterns and search for alternative water sources. Nowadays, water pollutant measurement is limited by the difficulties in sample processing and the elevated costs of consumables, skilled staff and/or mobility.

Our aim is to develop a cheap and easy way to warn consumers about possible pollutants present in drinking water.

2 SECTION 2 (METHODS/EXPERIMENTAL)

Based on standard parts from the Registry of Standard Biological Parts delivered as part of the iGEM student competition, we designed a genetic circuit that is able to report arsenic presence in health relevant concentrations by the naked eye based on color. The system works in *E. coli* as chassis. Detection is based on accumulation of mRFP (Monomer Red Fluorescent Protein), which is amplified by LuxI from *Vibrio fischeri* and expressed under the LuxR promoter. To have a robust signal, we designed a pulse of color generation. As we need a delay to allow color production in our incoherent feed-forward system, we decided to introduce an intermediary protein to produce the repression. Thus, the expression of the repressor C2P22 under the regulation of the pLL promoter is induced by the transcription activator PSP3, which is under the regulation of the pLux promoter. According to the mathematical model results, this design was feasible.

3 RESULTS AND DISCUSSION

Our designed system is able to produce a colored output proportional to arsenite concentration between 0 and 500 μg/L. Furthermore, we can discern among the critical arsenite concentrations according to the World Health Organization.

mRFP production responds efficiently under inducible conditions, both over time and different arsenite concentrations. The mRFP stability is acceptable for the aim of our project. However, the visibility to the naked eye may not be sufficient at very low arsenite concentrations.

The physical design of our device consists of a plastic support in which we will deliver the bacteria

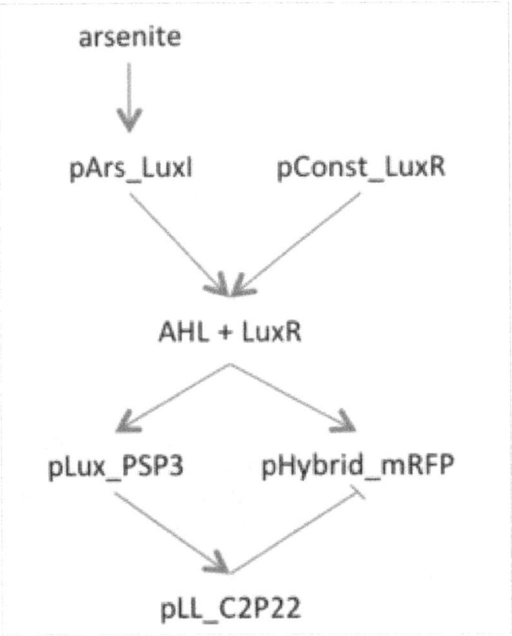

Figure 1. Scheme of the synthetic genetic circuit introduced in *E. coli.*

lyophilized. Bacteria wells are sealed to avoid accidental release. The usage of the device consists of:

I. Collecting a sample of groundwater.
II. Adding the sample to the testing well and distilled water (provided with the device) to the standard pattern wells at the same time. As distilled water is added, arsenite in the standard wells dissolves and gets in contact with bacteria. At the same time, bacteria are recovered from the lyophilized state by the added water,
III. Incubation for some hours, until color differences are appreciable.
IV. After usage, bacteria die as they enter in contact with a bleach solution that is contained in a capsule which is dissolved some hours after entering in contact with water.

Finally, the user will be able to know if the water tested is drinkable, around the limit of arsenite allowed by the World Health Organization or above it, by comparison with the included standard pattern.

4 CONCLUSION

We have designed a prototype for arsenic detection in water. Laboratory proof of concept worked nicely. A physical prototype was designed and it is in the process to be evaluated.

ACKNOWLEDGEMENTS

To Romina Mathieu, Luciana Feo Mourelle and Adrián Teijeiro for prototype design. To Facultad de Ciencias Exactas y Naturales for hosting the project. To Bolland y Cía. and Ministerio de Ciencia, Tecnología e Innovación Productiva (Argentina) for financial support. ADN is staff scientist from CONICET, AG, HB and NNM hold fellowships from CONICET.

REFERENCES

Alon, U. 2007. *An introduction to systems biology: design principles of biological circuits.* Chapman and Hall CRC, pp. 57–64.

Basu, S., Mehreja, R., Thiberge, S., Chen, M.-T. & Weiss, R. 2004. Spatiotemporal control of gene expression with pulse-generating networks. *PNAS*, 101(17):6355–6360.

iGEM competition: http://igem.org.

Registry of Standard Biological Parts: http://parts.igem.org/.

One Century of the Discovery of Arsenicosis in Latin America (1914–2014) – Litter, Nicolli, Meichtry, Quici, Bundschuh, Bhattacharya & Naidu (Eds)
© 2014 Taylor & Francis Group, London, ISBN 978-1-138-00141-1

Formation and use of monothioarsenate during dissimilatory arsenate reduction by the chemoautotrophic arsenate respirer strain *MLMS-1*

B. Planer-Friedrich, C. Haertig & R. Lohmayer
Environmental Geochemistry Group, University of Bayreuth, Germany

E. Suess
Institute of Biogeochemistry and Pollution Dynamics, ETH Zürich, Switzerland

S. Hoeft McCann & R. Oremland
United States Geological Survey, Menlo Park, California, USA

ABSTRACT: Monothioarsenite ($As^{III}SO_2^{3-}$) was previously reported as transient species upon reduction of arsenate to arsenite in the sulfidic, highly alkaline waters of Mono Lake by the chemoautotrophic strain *MLMS-1*. The formation and decomposition of monothioarsenite was attributed exclusively to chemical reactions, not the metabolism of *MLMS-1*. Using IC-ICP-MS and XAS analysis, we were able to correctly identify said species as monothioarsenate ($As^{V}SO_3^{3-}$). We show that monothioarsenate forms indeed abiotically, namely from arsenite and polysulfides (S_n^{2-}). Polysulfides form as intermediates from the microbial oxidation of sulfide to sulfate. The decomposition, however, is not abiotic, since monothioarsenate is kinetically stable in sulfidic solutions. We were able to show that *MLMS-1* actively uses monothioarsenate for growth, producing arsenite. These findings reinforce that studies on microbial reactions of arsenic in sulfidic environments need to consider thioarsenic (trans)formation.

1 INTRODUCTION

Mono Lake, a terminal, highly alkaline (pH 9.8) and saline lake in California, USA, contains high concentrations of arsenic (approximately 200 μM). In laboratory experiments, arsenate added to anoxic sulfidic bottom water samples was found to be quickly reduced to arsenite (Hoeft *et al.*, 2004), even though abiotic arsenate reduction only occurs at low pH (<4) and is kinetically unfavorable at alkaline conditions (Rochette *et al.*, 2000). Strain *MLMS-1*, a gram-negative, motile curved rod has previously been isolated from Mono Lake (Hoeft *et al.*, 2004) and was found to grow chemoautotrophically by oxidizing sulfide to sulfate while reducing arsenate to arsenite. In these laboratory experiments, transient accumulation of arsenic-sulfur complexes was observed upon reduction of arsenate under sulfidic conditions. The dominant species was identified as "monothioarsenite" ($As^{III}SO_2^{3-}$). The fate of "thioarsenites" was proposed to merely reflect favorable chemical conditions, first their formation, and then their transformation, due to their general instability. Direct production or consumption by the metabolism of strain *MLMS-1* was excluded.

"Thioarsenites" ($As^{III}S_nO_{3-n}^{3-}$; n = 1–3) have previously been reported to dominate arsenic speciation in the sulfidic bottom water of natural Mono Lake water samples (Hollibaugh *et al.*, 2005). Later studies have, however, shown that the applied technique (IC-ICP-MS) does not yield thioarsenites but thioarsenates ($As^{V}S_nO_{4-n}^{3-}$; n = 1–4) (Planer-Friedrich *et al.*, 2010; Suess *et al.*, 2009), which also dominate in Mono Lake waters (Fisher *et al.*, 2008). Regarding the importance of thioarsenites, only trithioarsenite could be identified by X-ray Absorption Spectroscopy (XAS), and it was found to be a necessary precursor for the formation of di-, tri- and tetrathioarsenate from arsenite and sulfide solutions (Planer-Friedrich *et al.*, 2010).

The aim of the current study was to repeat the previous experiment of Hoeft *et al.* (2004) with unambiguous identification of the formed thioarsenic species by IC-ICP-MS and XAS and clarification of their (trans)formation as a purely chemical reaction or part of the microbial metabolism.

2 METHODS

Strain *MLMS-1* was received from the Oremland group, USGS, Menlo Park. Abiotic and biotic incubations were done in triplicate at 28 °C in anaerobic septum bottles supplied with artificial Mono Lake medium (Hoeft *et al.*, 2004) plus 10 mM arsenate or synthesized monothioarsenate as electron acceptor and 5 mM sulfide as electron donor. Cell growth and species transformation were followed over more than 6 weeks. Cell counting was done after staining with SybrGold with a Neubauer Chamber on a Zeiss Axioplan. Aqueous samples were taken in a glovebox (95% N_2/5% H_2) to avoid oxidation. Sulfide was determined immediately by the methylene blue method. Polysulfides were derivatized with methyl trifluoromethane-sulfonate and stored frozen until analysis by HPLC-UV with separation on a C18 column (Waters-Spherisorb, ODS2, 5 μm, 250 × 4.6 mm) and methanol gradient elution according to Rizkov *et al.* (2004). The detection was performed at a wavelength of 230 nm. Samples for arsenic speciation were flash-frozen until analysis by IC-ICP-MS using 20–100 mM NaOH as eluent and a AS16 column (Dionex) according to a previously published method (Planer-Friedrich *et al.*, 2007). XAS identification of (thio)arsenic species was conducted at the ESRF, France, and followed previously published routines (Suess *et al.*, 2009).

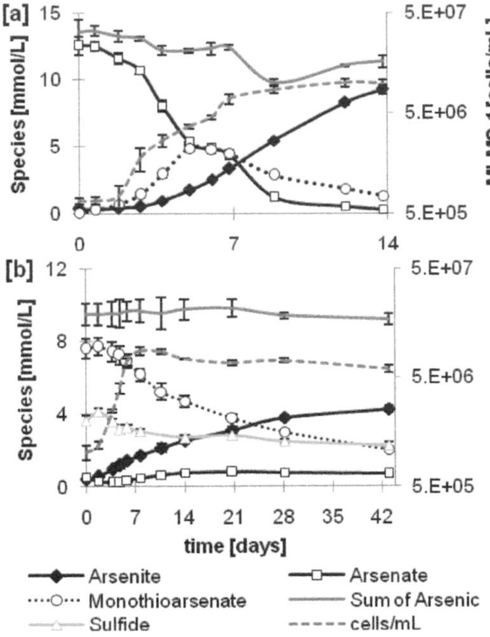

Figure 1. Reduction of arsenate (a) and monothioarsenate (b) in sulfidic solutions abiotically and in the presence of *MLMS-1*.

3 RESULTS AND DISCUSSION

Repeating the experiments of Hoeft *et al.* (2004), we confirmed that *MLMS-1* reduces arsenate to arsenite completely within 14 days. An As-S-species forms as transient species with a maximum concentration after 6 days (Figure 1a). The production of this As-S-species concurs with a steep cell number increase.

XAS results (not shown) indicated that the As-S-species was not trivalent. Analysis by IC-ICP-MS revealed it to be monothioarsenate instead of the previously assumed monothioarsenite. No other thioarsenic species were determined. No arsenate reduction and thioarsenate formation was observed in abiotic controls.

Investigating the formation of monothioarsenate, we observed significant amounts of polysulfides in the biotic incubations. Separate abiotic experiments indicate that the addition of polysulfide-bound elemental sulfur to an arsenite solution leads to direct formation of almost exclusively monothioarsenate, while arsenite-sulfide solutions yield di-, tri- and tetrathioarsenate via trithioarsenite (Planer-Friedrich *et al.*, 2010). Due to the quick abiotic reaction, it was impossible to distinguish whether monothioarsenate formation could also be microbially triggered.

Experiments to determine the influence of MLMS-1 on monothioarsenate decomposition (Figure 1b) showed that abiotically monothioarsenate remained stable over the whole duration of the experiment. In biotic experiments, however, we observed cell growth parallel to rapid transformation of monothioarsenate to arsenite. The decomposition of monothioarsenate observed during arsenate reduction (Figure 1a) is similarly assumed to be microbial.

4 CONCLUSIONS

Our study has shown that thioarsenic species do not just form as transient species upon favorable chemical conditions in As-S-solutions (As^{III} + S^{-II} = $As^{III}S_nO_{3-n}^{\ 3-}$), but that they induced redox reactions themselves (As^{III} + S^0 = $As^{V}S_nO_{4-n}^{\ 3-}$) or can be used directly for microbial growth. Their correct identification and quantitative determination is thus mandatory in all As redox studies in sulfidic environments.

ACKNOWLEDGEMENTS

We acknowledge generous funding by the German Research Foundation within the Emmy Noether program (Grant No. PL 302/3-1) and collaboration with the groups of Andreas Scheinost, ESRF, France, and Tim Hollibaugh, Georgia University, Athens, USA.

REFERENCES

Fisher, J.C., Wallschlaeger, D., Planer-Friedrich, B. & Hollibaugh, J.T. 2008. A new role for sulfur in arsenic cycling. *Environ. Sci. Technol.* 42: 81–85.

Hoeft, S.E., Kulp, T.R., Stolz, J.F., Hollibaugh, J.T. & Oremland, R.S. 2004. Dissimilatory arsenate reduction with sulfide as electron donor: Experiments with mono lake water and isolation of strain MLMS-1, a chemoautotrophic arsenate respirer. *Appl. Environ. Microbiol.* 70(5): 2741–2747.

Hollibaugh, J.T., Carini, S., Gürleyük, H., Jellison, R., Joye, S.B., LeCleir, G., Meile, C., Vasquez, L. & Wallschläger, D. 2005. Arsenic speciation in Mono Lake, California: Response to seasonal stratification and anoxia. *Geochim. Cosmochim. Ac.* 69(8): 1925–1937.

Planer-Friedrich, B., London, J., McCleskey, R.B., Nordstrom, D.K. & Wallschläger, D. 2007. Thioarsenates in geothermal waters of Yellowstone National Park: determination, preservation, and geochemical role. *Environ. Sci. Technol.* 41(15): 5245–5251.

Planer-Friedrich, B., London, J., McCleskey, R.B., Nordstrom, D.K. & Wallschläger, D. 2007. Thioarsenates in geothermal waters of Yellowstone National Park: determination, preservation, and geochemical role. *Environ. Sci. Technol.* 41(15): 5245–5251.

Rizkov, D., Lev, O., Gun, J., Anisimov, B. & Kuselman, I. 2004. Development of in-house reference materials for determination of inorganic polysulfides in water. *Accredit. Qual. Assur.* 9(7): 399–403.

Rochette, E., Bostick, B., Li, G. & Fendorf, S, 2000. Kinetics of arsenate reduction by dissolved sulfide. *Environ. Sci. Technol.* 34: 4714–4720.

Suess, E., Scheinost, A., Bostick, B.C., Merkel, B.J., Wallschaeger, D., Planer-Friedrich, B. 2009. Discrimination of Thioarsenites and Thioarsenates by X-ray Absorption Spectroscopy. *Anal. Chem.* 81: 8318–8326.

One Century of the Discovery of Arsenicosis in Latin America (1914–2014) – Litter, Nicolli, Meichtry, Quici, Bundschuh, Bhattacharya & Naidu (Eds)
© 2014 Taylor & Francis Group, London, ISBN 978-1-138-00141-1

Biotransformation of arsenic oxyanions by cyanobacteria from mining areas

M.W. Franco, F.A. Guedes & F.A.R. Barbosa
Laboratório de Limnologia Ecotoxicologia e Ecologia Aquática, Universidade Federal de Minas Gerias, Brazil

B.L. Batista
Faculdade de Ciências Farmacêuticas de Ribeirão Preto, Universidade de São Paulo, Brazil
Centro de Ciências Naturais e Humanas, Universidade Federal do ABC, Brazil

E.S. Paula & F. Barbosa Jr.
Faculdade de Ciências Farmacêuticas de Ribeirão Preto, Universidade de São Paulo, Brazil

S.M.S. Magalhães
Faculdade de Farmácia, Universidade Federal de Minas Gerais, Brazil

ABSTRACT: Ecological studies have evaluated the potential of cyanobacteria species to bioaccumulate and biotransform arsenic (As) in less toxic organoarsenic compounds in order to investigate the role of these organisms in the biogeochemical cycle of this metalloid. Cyanobacteria strains were isolated from a contaminated area with several heavy metals and As near a gold processing industry. When exposed to As(III) and As(V), Synechococcus nidulans transformed inorganic arsenic species in organic compounds as a detoxification mechanism. This work shows that cyanobacteria are tolerant to high levels of arsenic and participate in the biogeochemistry of this element.

1 INTRODUCTION

Although mining activity is one of the pillars of Brazilian economy, particularly gold mining activity moves various heavy metals and Arsenic (As) from their natural deposits, causing impacts to aquatic environments. To investigate the richness of phytoplankton species in contaminated environments and their mechanisms of As tolerance is important for understanding the bioavailability of this element for the subsequent trophic chain levels. Furthermore, cyanobacteria species are able to bio-transform arsenic into less toxic organic arsenic compounds thus participating in the biogeochemical cycle of this metalloid. This study was conducted in two stages, the first aiming to collect and isolate cyanobacteria strains from an arsenic contaminated environment and the second one aiming to evaluate the potential of *Synecochocus nidulans* to bio-transform arsenic.

2 METHODS

2.1 *Obtaining the cyanobacteria strains*

Water samples were collected from a stream (19° 58′74.8″ S; 43°49′25.9″W), Brazil, where contamination by As and Mn at levels above those recommended by Brazilian legislation for water use in domestic activities has been verified. 1 mL of sample was added to 20 mL of BG-11 culture medium and incubated at controlled conditions of light and temperature (21 ± 1°C). The colonies were successively replicated until mono-specific algal cultures were obtained. Taxonomic identification was based on morphological characteristics. The obtained strains have been maintained in the algae culture bank of LIMNEA/UFMG.

2.2 *Exposure of cyanobacteria to arsenic*

Culture medium (BG-11) was prepared and added with 6 mg L^{-1} As(III) or 400 mg L^{-1} As(V), inoculated with cyanobacteria in triplicate. The control consisted of medium without inoculation. Total arsenic determinations were performed by Inductively Coupled Plasma Optical Emission Spectrometry (ICP-OES) in the first day (initial) in supernatant and after 30 days (end point) in supernatant and cyanobacteria biomass previously mineralized by microwave acid digestion using HNO_3 (3 mL) and hydrogen peroxide (1 mL). Speciation analyses were performed in supernatant and biomass after 30 days of incubation by

HPLC-ICP-MS according to Batista *et al.* (2011). The samples were extracted with HNO_3 + methanol 2% v/v each followed by overnight rotational agitation and filtration.

3 RESULTS AND DISCUSSION

3.1 *Cyanobacteria isolates*

Microalgae and cyanobacteria species were obtained: *Phormidium* cf *tergestinum, P. ambigum, P. inundatum, P. autumnale, Pseudanabaena minima, P. limnetica, Scytonema* sp., *Chlorella vulgaris, Anabaena* sp. *Nostoc* sp., *Synechococcus nidulans.* The last one was chosen due to the rapid growth characteristics, easy homogenization, and not possessing the gens for producing toxins, desirable characteristics for handling in the laboratory.

3.2 *Arsenic speciation analysis*

The total As concentrations in *S. nidulans* biomass exposed to As(III) and As(V) were 61.6 and 60 $\mu g\ g^{-1}$ respectively; arsenic in the culture medium remained constant (paired t test) over a period of 30 days (Table 1). Although bioaccumulation has been observed, the ratio between the mean dry biomass at the end of tests (0.4 g L^{-1}) and the concentration of As was not significant due to the high concentrations of As used relatively to the average biomass in the culture.

Mechanisms of biotransformation in order to As detoxification were observed in *S. nidulans* (Figure 1).

Possibly, the mechanisms of methylation and excretion are more intensely activated upon exposure to As(III) which led to the production of higher proportions of MMA and DMA intracellularly (Figure 1A) compared to tests with As(V) (Figure 1B). Both species were also observed in the As(III) test supernatant to a greater extent

Table 1. Total arsenic concentrations in tests.

Tests	$[As]_{nominal}$ (mg L^{-1})	$[As]_{initial}$ (mg L^{-1})	$[As]_{final}$ (mg L^{-1})	$[As]_{biomass}$ ($\mu g\ g^{-1}$)
As(III) + C	6	5.8 ± 0.04	5.7 ± 0.014	61.61 ± 8.43
As(III)Ø	6	6.0 ± 0.01	5.8 ± 0.02	–
As(V) + C	400	416.3 ± 4.75	421.4 ± 19.9	60.11 ± 1.37
As(V)Ø	400	404.8 ± 1.42	402.7 ± 4.19	–

C: inoculation with cyanobacteria; Ø: control.

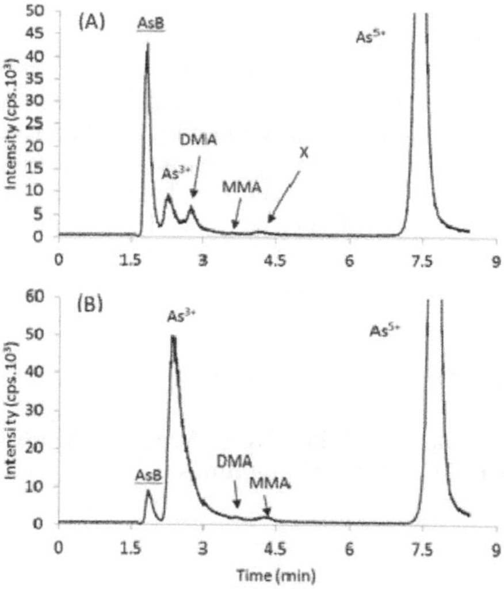

Figure 1. Chromatogram obtained in ICP-MS for the biomass of *S. nidulans* exposed to (A) As(III) and (B) As(V). DMA: dimethylarsinic acid, MMA: monomethylarsonic acid; AsB: possible arsenobetaine X: arsenic species not identified.

compared to As(V) test supernatant (data not shown).

Similarly, As biotransformation was found in the biomass of *Synechocystis* sp. after exposure to 7.4 mg L^{-1} As(III) (Yin *et al.*, 2011). In As(III) speciation control only As(V) was found. Oxidation of As(III) should be considered by Fe(III) present in culture medium (Mccleskey *et al.*, 2004). Under exposure to As(III), another two species found were not identified: peaks corresponding to species X and AsB. The latter has the same retention time as arsenobetaine; however, more specific techniques are necessary to identify it. A complex of the type $As(GS)_3$—Glutathione could be the case for the X unidentified fraction, although a confirmation is necessary. The extraction protocol was used for the speciation of As allowed in biomass, measuring only 5% of the total content of intracellular arsenic. It is probable that a fraction of As may have remained within the thylakoid membranes. This possibility should be considered in *S. nidulans* due to its dense organization of thylakoids, filling approximately two-thirds of the volume of the cell (Smarda *et al.*, 2002).

4 CONCLUSIONS

The *S. nidulans* cyanobacteria, collected in aquatic environment contaminated with As, participates in the biogeochemistry of this metalloid through the biotransformation of inorganic forms to organic As compounds, possible as a detoxification mechanism. It is likely that cyanobacteria cells prevent further uptake after accumulating a certain level of As for some time. As demonstrated, the same levels of accumulation occurred despite the distinct concentrations of As oxyanions. Investigations on biochemical changes at membrane level could help to clarify this mechanism.

ACKNOWLEDGEMENTS

We thank to CNPq/INCT Acqua, CAPES and FAPEMIG for providing funds.

REFERENCES

Batista B.L., Souza, J.M.O, De Souza, S.S. & Barbosa Jr., F. 2011. Speciation of arsenic in rice and estimation of daily intake of different arsenic species by Brazilians through rice consumption. *J. Hazard. Mat.* 191: 342–348.

Mccleskey, R.B., Nordstrom, D.K. & Maest, A.S. 2004. Preservation of water samples for arsenic (III/V) determinations: an evaluation of the literature and new analytical results. *Appl. Geochem.* 19: 995–1009.

Šmarda, J., Šmajs, D., Komrska, J. & Krzyžánek, V. 2002. S-layers on cell walls of cyanobacteria. *Micron.* 33: 257–277. Yin, X.-X., Chen, J., Qin, J., Sun, G.-X., Rosen, B.P. & Zhu, Y.-G. 2011. Biotransformation and volatilization of arsenic by three photosynthetic cyanobacteria. *Plant Physiol.* 156, 1631–1638.

One Century of the Discovery of Arsenicosis in Latin America (1914–2014) –
Litter, Nicolli, Meichtry, Quici, Bundschuh, Bhattacharya & Naidu (Eds)
© 2014 Taylor & Francis Group, London, ISBN 978-1-138-00141-1

Mobilization of arsenic on nano-TiO_2 in soil columns with sulfate-reducing bacteria

C. Jing & S. Liu
Research Center for Eco-Environmental Sciences, Chinese Academy of Sciences, Beijing, China

ABSTRACT: Arsenic remediation in contaminated water using nanoparticles is promising; however, the fate and transport of As associated with nano-adsorbents in natural environment is poorly understood. To investigate the redox transformation and mobility of As adsorbed on nano-TiO_2 in the subsurface, we inoculated a Sulfate-Reducing Bacterium (SRB), *Desulfovibrio vulgaris* DP4, in a soil column and added the As(V)-TiO_2 suspension in the influent groundwater matrix. As(III) accounted for 21 ± 7% of the total dissolved As, whereas no As(III) was detected in the abiotic control. Our thermodynamic pe-pH calculation and XANES analysis demonstrated that biogenic magnetite as a secondary iron mineral was responsible for As retention in the soil column, whereas the As-S precipitate was not detected. While SRB restrained the release of dissolved As, it facilitated the transport of particulate As with biogenic magnetite.

1 INTRODUCTION

Arsenic (As) is a redox active and toxic element that poses considerable human health risks. Recently, adsorption on nano-TiO_2 has been demonstrated to be a promising technique for removing As from groundwater and industrial wastewater (Luo *et al.*, 2010). However, the fate of As, once it is removed from the water, is not well known. The pressing need to predict the fate and transport of As adsorbed onto nano-TiO_2 during environmental redox transition has motivated this study.

The shift in the redox environment from oxidizing to reducing, as mediated by microbial activities, plays an important role in As biogeochemical cycling. Sulfate-reducing bacteria (SRB) are ubiquitous in the anoxic subsurface (Luo *et al.*, 2013), and biogenic sulfide is an efficient reductant for As(V) and Fe(III). Consequently, the reduced As(III) may form As-sulfide precipitates such as realgar and orpiment, which regulates the dissolved As levels in the aquifer. Moreover, the formation of Fe(II)-sulfide minerals provides an additional sink for As immobilization.

However, to the best of our knowledge, no publications report on the effects of SRB on the transport and potential risks of nanosized-TiO_2-containing As in the natural environment.

The objectives of this research are to study the redox reactions and release of adsorbed As on nano-TiO_2 in the presence of SRB in soil columns. The soil column and effluent were monitored for As- and Fe-speciation and TiO_2-transport using complimentary analytical techniques, including XANES and μ-XRF, and were analyzed by thermodynamic calculations. This study provides the first evidence regarding the mobilization of adsorbed As on nano-TiO_2 with SRB in natural soil under anoxic conditions.

2 METHODS/EXPERIMENTAL

2.1 *As(V) loading on nano-TiO_2*

The adsorption of As(V) on TiO_2 was conducted in a glovebox (100% N_2). The As(V) solution was mixed with the preloaded P-TiO_2 solids for 24 h at pH 8.0. The final load of As(V) on TiO_2 was approximately 390 mg g^{-1} by mass balance calculation. The soluble As concentrations in the suspension were 2.5 mg L^{-1}.

2.2 *Column experiment*

Soil samples were collected at 1 m depth at a site in Shanxi, China. *Desulfovibrio vulgaris* DP4 was isolated from the soil. The TiO_2-As(V) suspension was delivered to the column in an upward flow using a peristaltic pump at three pore volumes (PV) per day (11 μL min^{-1}). Sulfate ($MgSO_4$, 5.5 mM) was added as an electron acceptor; the electron donor was lactate (Na-lactate, 10 mM). Two abiotic columns, without the addition of SRB, were also set up for comparison. The influent stream for one abiotic column was the TiO_2-As(V) suspension, which was used as a control. Synthetic groundwater was used as background in the other column.

2.3 *μ-XRF and XANES analysis*

The solids in the bottom (influent front) and upper (end) layers of the column were collected and analyzed by μ-XRF at beamline 15U at Shanghai Synchrotron Radiation Facility (SSRF), China. The As and Fe K-edge XANES spectra were collected at beamline 01C1 at the National Synchrotron Radiation Research Center (NSRRC), Taiwan. Spectra were acquired at cryogenic temperatures (77 K) using a cryostat to prevent the beam-induced oxidation of As(III) and Fe(II).

3 RESULTS AND DISCUSSION

3.1 *As release and reduction in soil columns*

The release of As can be divided into a lag phase (<40 PV, 13 d) and an acceleration phase (>40 PV). In contrast, only 11 μg L^{-1} As(V), on average, were detected in the effluent of the background column, suggesting that As in the soil (8.6 mg kg^{-1}) contributed little to the overall As release. Even the presence of SRB resulted in only an 8 μg L^{-1} As release from the soil to the groundwater matrix.

Dissolved As(III) in the effluent was observed only from the biotic column after 3 PV; its concentration was increased to 178 μg L^{-1} after 120 PV in the acceleration phase. As(III) concentrations were 21 ± 7% of the total dissolved As in the effluent, indicating that SRB induced the As(V) reduction in column. In agreement with the observed As speciation, redox potential measurements superimposed on the pe-pH diagram suggest that As(III) was the stable As species in the reducing environment ($-1.6 < pe < -3.6$) in biotic column, whereas As(V) predominated in the abiotic control and background columns.

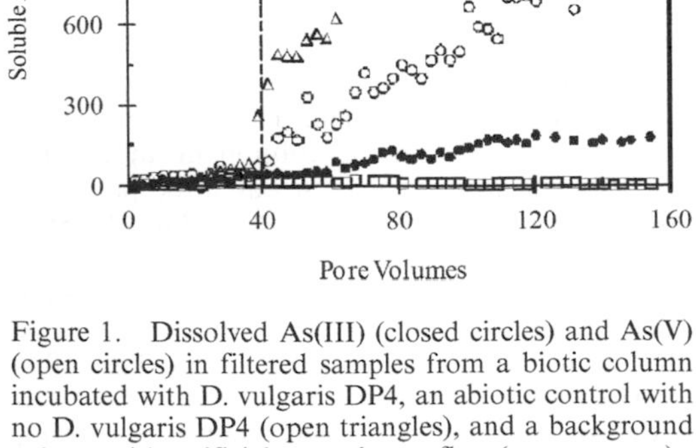

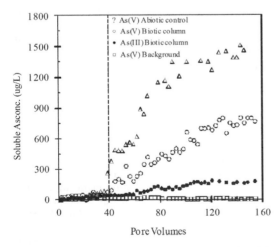

Figure 1. Dissolved As(III) (closed circles) and As(V) (open circles) in filtered samples from a biotic column incubated with D. vulgaris DP4, an abiotic control with no D. vulgaris DP4 (open triangles), and a background column with artificial groundwater flow (open squares).

3.2 *Sulfur and iron behavior in soil column*

Effluent SO_4^{2-} in control and background columns ranged from 5.2 to 5.5 mM. In contrast, the average effluent SO_4^{2-} was 3.6 mM in the biotic column. The loss of aqueous SO_4^{2-} (1.9 mM, 34%) can be attributed to the SO_4^{2-} reduction by SRB. As a result, sulfide levels increased to 61 mg L^{-1} (1.9 mM) at the end of the column experiment, accounting for 34% of the total SO_4^{2-} input.

Coincident with the increase of sulfide, an increase in Fe(II) up to 180 μg L^{-1} after 70 PV (24 d) was observed in biotic column. Sulfide and Fe(II) were not detected in the abiotic control and background columns. The total Fe in the effluent increased from approximately 3.2 to 13 mg L^{-1} in the inoculated column, while remaining stable (3.2 mg L^{-1} on average) in the abiotic control and background columns.

3.3 *TiO_2 transport in soil column*

The mass balance of TiO_2 showed that over 99.9% TiO_2 was retained in the biotic column as evidenced by the negligible amount of Ti in the effluent (3–450 μg L^{-1}) and the considerable amounts in the solids (1,943–2,634 mg kg^{-1}). In addition, no significant correlation was found between the total As and Ti in the effluent, indicating that the particulate As release might not result from the transport of nano-TiO_2.

3.4 *Analysis with synchrotron techniques*

The μ-XRF images showed that the majority of adsorbed As in the column was still associated with TiO_2, which was mainly retained in the influent front of the soil column during the transport process.

The XANES analysis implied that As(V) was the primary As species in the column as evidenced by its peak position at 11,874 eV. As-S precipitates, such as realgar and orpiment, were not detected by XANES, in line with the pe-pH diagram predication.

In agreement to the pe-pH diagram, the Fe k-edge XANES analysis indicates the presence of magnetite in the column.

4 CONCLUSIONS

Our results indicate that SRB inhibit the soluble As release due to its re-adsorption on biogenic magnetite. Moreover, the SRB facilitate the migration of secondary Fe-minerals associated with As. The transport of these minerals might be due to their submicron or nanometer particle size, and therefore, lead to the As mobility and contamination in the field.

ACKNOWLEDGEMENTS

We acknowledge the beamline SSRF 15U and NSRRC 01C1.

REFERENCES

Luo, T., Cui, J., Hu, S., Huang, Y. & Jing, C. 2010. Arsenic removal and recovery from copper smelting wastewater using TiO_2. *Environ. Sci. Technol.* 44: 9094–9098.

Luo, T., Tian, X., Guo, Z., Zhuang, G. & Jing, C. 2013. Fate of arsenate adsorbed on nano-TiO_2 in the presence of sulfate reducing bacteria. *Environ. Sci. Technol.* 47: 10939–10946.

One Century of the Discovery of Arsenicosis in Latin America (1914–2014) – Litter, Nicolli, Meichtry, Quici, Bundschuh, Bhattacharya & Naidu (Eds)
© 2014 Taylor & Francis Group, London, ISBN 978-1-138-00141-1

Influence of organic matters on the bacterial As(III) oxidation in polluted soils

T. Lescure
BRGM, ISTO, UMR 7327, Orléans, France
LIEC UMR 7360, University of Lorraine, Bridoux Campus, Metz, France

F. Battaglia-Brunet & C. Joulian
BRGM, ISTO, UMR 7327, Orléans, France

P. Bauda
LIEC UMR 7360, University of Lorraine, Bridoux Campus, Metz, France

C. Hénault
INRA, Research Center of Orléans, Soil Science Unit, Orléans, France

ABSTRACT: The global bacterial As(III)-oxidizing activity tends to decrease As toxicity in soils and its transfer toward underlying aquifers. The influence of different types of organic matter on arsenic speciation by pure strains and soils microflora was determined. Experimental results show that the presence of organic matter has a negative effect on the specific rate of As(III) oxidation by pure strains presenting different metabolisms. In parallel, experiments are performed with various soils from different types of polluted sites: former mining and industrial sites, forest and agricultural soils impacted by mining activities. The effect of yeast extract on the As(III)-oxidizing activity of the soil microflora is evaluated while considering the potential influence of soil parameters. Results suggest a variable influence of yeast extract on As(III) oxidation kinetics according to the range of yeast extract concentration. As a whole, experimental data strongly suggest a negative effect of organic matter on bacterial As(III) oxidation.

1 INTRODUCTION

Microorganisms play a major role in the behavior of metals and metalloids in soils. Arsenic is widely found in soils, released from both natural and anthropogenic sources. Natural emissions are principally due to igneous activity, whereas human emissions arising primarily from smelting of metals, combustion of fuels and use of pesticides. Bacteria are known to govern As speciation through oxidation and reduction processes. Many bacteria able to oxidize or/and reduce As were found in soils of diverse origins (Macur *et al.*, 2004; Majumder *et al.*, 2013). The global bacterial As(III)-oxidizing activity tends thus to decrease As toxicity in soils and transfer of this pollutant toward underlying aquifers.

A current BRGM project aims to quantify the influence of organic matter on arsenic speciation determined by soils microflora, and the consequences of this biogeochemical phenomenon on arsenic mobility. Experiments are performed with various soil samples from different types of polluted sites: former mining and industrial sites, forest and agricultural soils impacted by mining activities. The effect of organic matter on the As(III)-oxidizing activity of the soil microflora is evaluated while considering other soil parameters impacting bacterial activities. In parallel, the effect of organic substances on the specific As(III)-oxidizing rate by pure bacterial strains is evaluated.

2 METHODS

2.1 *As(III)-oxidizing pure strains*

Microorganisms used in the present study were a mixotrophic strain (*Thiomonas delicata arsenivorans* strain T. ars) and a heterotrophic strain (*Herminiimonas arsenicoxydans*), which is a well characterized heterotrophic bacteria. Among the possible substrates, succinate and acetate were tested as simple organic materials, and yeast extract as complex organic matter at different concentrations.

The experiments were performed in batch at a relatively high concentration of arsenic (75 mg L^{-1} or 1 mM), and at low levels of As(III), more compatible with concentrations potentially found in the pore water of the soils (2 mg L^{-1} As(III)). Bacterial growth was followed by microscopic observations and measurements of optical density at 620 nm.

2.2 *Polluted soils*

Experiments of As(III) oxidation by global soil microbial communities were carried out in the same conditions as the assays with pure strains. A synthetic medium was inoculated with raw soil and amended with different concentrations of organic substrates. As(III) oxidation kinetics was followed during one week. Soils from different French sites were used: two former mining areas (Salsigne, Aude and a Cheni, Limousin), two impacted industrial sites (Auzon, Haute-Loire, Saint-Laurent Le Minier, Gard). The soils were characterized in a physico-chemical point of view: pH, intrinsic soil organic matter, nitrogen and phosphorus, As concentration and speciation, determination of particle size. As(III) oxidation rate constants were calculated and a statistical method (Principal Components Analysis) was used in order to determine which factors influence arsenite oxidation by microorganisms in those soils.

3 RESULTS AND DISCUSSION

3.1 *As(III) oxidation by bacterial pure strains*

The overall and specific As(III)-oxidation rates were calculated from the kinetics obtained with both pure strains. Figure 1 shows specific As(III)-oxidation rate by *Thiomonas delicata* (strain T. ars) in presence of yeast extract or succinate.

In the presence of yeast extract, the specific rate decreased significantly with the increasing concentration of substrate. Indeed, the specific activity decreased by 38% when yeast extract concentration varies from 0.05 g L^{-1} to 0.2 g L^{-1}, and by 70% at 1 g L^{-1}. In presence of succinate, a simple substrate, the specific oxidation rate was not significantly different between 0.05 g L^{-1} and 0.2 g L^{-1}. However, it decreased dramatically with 1 g L^{-1} succinate (almost 85% less). Comparable results were obtained with *Herminiimonas arsenicoxydans* on yeast extract and acetate. These results confirm an inhibitory effect of organic matter on the oxidation of As(III).

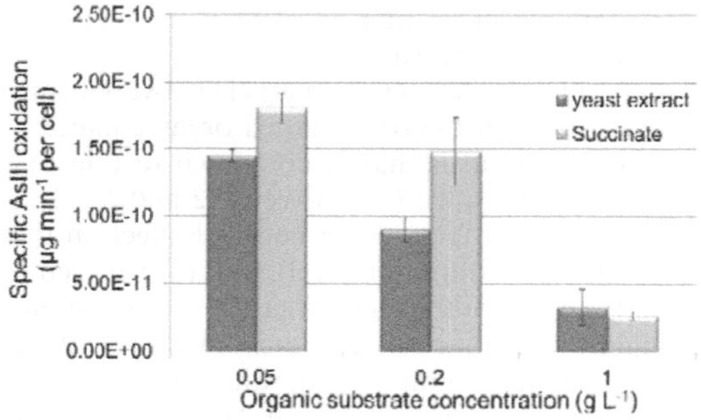

Figure 1. Specific As(III)-oxidation rate at 2 mg L^{-1} initial As(III) by strain T. ars in presence of 0.05, 0.2 and 1 g L^{-1} of organic substrates.

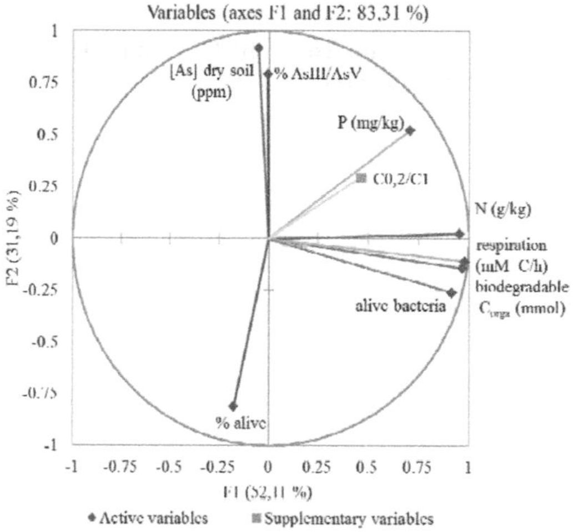

Figure 2. Example of result obtained by principal components analysis with $C_{0.2}/C_1$, the ratio of rate constants for oxidation of 0.2 g L^{-1} and 1 g L^{-1} yeast extract. The ratio was treated as a supplementary variable.

3.2 *Influence of soil parameters on As(III) oxidation by soil communities analyzed by Principal Component Analysis (PCA)*

The bacterial oxidation of arsenic in batch was treated as a 1st order reaction. A calculation of the rate constants was made for each soil incubated in the presence of four concentrations (0, 0.2, 1 and 10 g L^{-1}) of yeast extract. Figure 2 shows an example of the Principal Components Analysis of all data from 9 soils. Here, the PCA shows the correlation of the ratio of rates constants obtained at 0.2 and 1 g L^{-1} yeast extract with soil parameters. When the ratio is high, the increasing concentration of yeast extract exerted a negative effect on As(III) oxidation. Conversely, when the ratio is low, there was a positive effect of increasing concentration of yeast extract.

Globally, results suggest that: (1) between 0 and 0.2 g L^{-1} of yeast extract (added organic matter), the rate constant is mainly correlated to total As content of the soils; (2) between 0.2 and 1 g L^{-1}, soil organic matter exerts a negative effect on the rate constant (inhibition level), and (3) between 1 and 10 g L^{-1}, both As and organic matter content stimulate oxidation. These phenomena might be related to the stimulation of different microbial communities from the soils, i.e. "autotrophic" vs. "heterotrophic", according to the yeast extract concentration added in the culture medium.

ACKNOWLEDGEMENTS

This work is cofunded by BRGM and ADEME (convention TEZ 11-16).

REFERENCES

Macur, R.E., Jackson, C.R., Botero, L.M., McDermott, T.R. & Inskeep, W.P. 2004. Bacterial populations associated with the oxidation and reduction of arsenic in an unsaturated soil. *Environmental Science and Technology* 38: 104–111.

Majumder, A., Bhattacharyya, K., Bhattacharyya, S. & Kole, S.C. 2013. Arsenic-tolerant, arsenite-oxidising bacterial strains in the contaminated soils of West Bengal, India. *Science of the Total Environment* 463–464: 1006–1014.

One Century of the Discovery of Arsenicosis in Latin America (1914–2014) – Litter, Nicolli, Meichtry, Quici, Bundschuh, Bhattacharya & Naidu (Eds)
© 2014 Taylor & Francis Group, London, ISBN 978-1-138-00141-1

Volatile Organic Compounds (VOCs) and siderophores in arsenite-tolerant bacteria isolated from grapevine plants

M.I. Funes Pinter, M.V. Salomón, R. Bottini & P. Piccoli
Instituto de Biología Agrícola de Mendoza (IBAM), CONICET, Mendoza, Argentina

ABSTRACT: This study tested the arsenite tolerance at different pH values of nine bacterial strains isolated from grapevine roots and soils. Since *Micrococcus luteus* was the most arsenite tolerant bacteria (40 mM $NaAsO_2$ at pH 9), it was selected to analyze siderophore and VOCs production. It produced siderophores even in the presence of arsenite, but its production is significantly affected. Also, *M. luteus* increases and modifies the VOCs profile (aliphatic and ketones) in the presence of arsenite. The VOCs produced after arsenite influence were 1,3-hexadien-3-yne, 2-heptanone, 2-nonanone and 2-decanone, which was reported to have Plant Growth Promoting Rhizobacteria (PGPR) activity. The selected strain is a potential candidate to be used for rhizoremediation in vineyards.

1 INTRODUCTION

Arsenic is widely spread over the world and constitutes a global health problem. Many studies report arsenic tolerant bacteria as a potential solution to remediate or minimize the effects of high arsenic concentration in soils. In addition, some Plant Growth Promoting Rhizobacteria (PGPR) can enhance the tolerance of plants to heavy metals (Dary *et al.*, 2010). The objective of this study was to evaluate the arsenite tolerance of bacteria isolated from root and rhizosphere of grapevine plants, and the effects of arsenite on some aspects of bacteria metabolism like siderophore and production of Volatile Organic Compounds (VOCs). This study attempts to select bacteria to be used in rhizoremediation in grapevine soils with high arsenite concentrations.

2 METHODS/EXPERIMENTAL

2.1 *Screening of arsenite tolerant bacteria*

Nine strains previously isolated and characterized from grapevine (Salomon *et al.*, 2013), were tested in order to select tolerant bacteria. Each bacterium was grown to exponential phase in 20 mL of Luria-Broth (LB) medium at 28° C and pH 7. Then, aliquots of 5 μL (10^8 CFU mL^{-1}) of each culture, were plated in Petri dishes containing LB agar supplemented with different concentrations of $NaAsO_2$ (10, 20, 30 and 40 mM) and different pH (5, 7 and 9). As control, LB agar without $NaAsO_2$ was used. After 5 days of incubation at 30 °C, bacterial growth was evaluated. The experiment was done by triplicate. The maximal tolerable concentration (MTC), defined as the maximal concentration of an element not affecting the bacterial growth, is used to evaluate the tolerance (Dary *et al.*, 2010). The most tolerant strain was selected for further assays.

2.2 *Effects of arsenite and pH on production of siderophores by bacteria*

Production of siderophores was evaluated using the Chrome Azurol S-agar (CAS-AGAR) protocol according to Milagres *et al.* (1999). The plates were prepared containing a half of blue CAS-agar and other half of LB-agar supplemented with/out $NaAsO_2$ (10 mM) at pH 7 and 9. The half plate containing LB medium was inoculated with the selected bacterium. After 15 days of incubation at 30 °C, a change of color (from blue to orange) on the half of the CAS-agar plate was observed and the halo diameter was measured. Plates without arsenite were used as control.

2.3 *VOCs production by arsenite tolerant bacteria*

VOCs production was evaluated to assess effects of arsenite on the bacterium metabolism. The compounds were measured by gas chromatograph-electron impact mass spectrometer (GC-EIMS, Clarus 500, Perkin Elmer, Shelton, CT, USA) using the SPME method (Farag *et al.*, 2010) with modifications. The bacterium was grown in vials containing 5 ml of Murashide & Skoog medium supplemented with 5 mM $NaAsO_2$ at pH 7. After 48 h at 28 °C, samples were preheated at 50 °C in a heating block for 20 min and, then, 0.3 g of NaCl and 100 ng of 4-metil-2-pentanone as internal standard were added.

3 RESULTS AND DISCUSSION

3.1 *Screening of arsenite tolerant bacteria*

Table 1 shows the bacteria tolerance to arsenite at five different concentrations and different pH values. At pH 5, no bacteria were grown (data not shown). At pH 7, *M. luteus* and *B. licheniformis* were able to growth at 30 and 20 mM $NaAsO_2$, respectively, showing the highest tolerance to the metalloid at pH 9 (40 and 30 mM $NaAsO_2$, respectively).

3.2 *Production of siderophores*

Since *M. luteus* showed the highest arsenite-tolerance, it was selected to test the production of siderophores. The strain was able to produce siderophores with and without $NaAsO_2$ at 10 mM and at different pH values, but, in the presence of the metalloid, the halo diameter was significantly smaller. In addition, pH affected those treatments with arsenite; at pH 9, the halo size was the smallest (14 ± 0.58 mm), while, at pH 7, was 21.67 ± 0.67 mm.

Table 1. Screening of arsenite tolerant bacteria.

		Arsenite concentration (mM)					
Strain	pH	0	5	10	20	30	40
Arthrobacter parietes	7	+++	–	–	–	–	–
	9	+++	–	–	–	–	–
Bacillus licheniformis	7	+++	+++	+++	+++	–	–
	9	+++	+++	+++	+++	++	–
Brachybaterium faecium	7	+++	+++	–	–	–	–
	9	+++	+++	+++	–	–	–
Kocuria eryhtromyxa	7	+++	–	–	–	–	–
	9	+++	+	–	–	–	–
Microbacterium imperiale	7	+++	+++	–	–	–	–
	9	+++	+++	++	–	–	–
Micrococcus luteus	7	+++	+++	+++	+++	+++	–
	9	+++	+++	+++	+++	+++	++
Planococcus sp.	7	+++	+++	–	–	–	–
	9	+++	+++	–	–	–	–
Pseudomonas fluorescens	7	+++	+++	–	–	–	–
	9	+++	+++	–	–	–	–
Terribacillus saccharophilus	7	+++	+++	–	–	–	–
	9	+++	+++	–	–	–	–

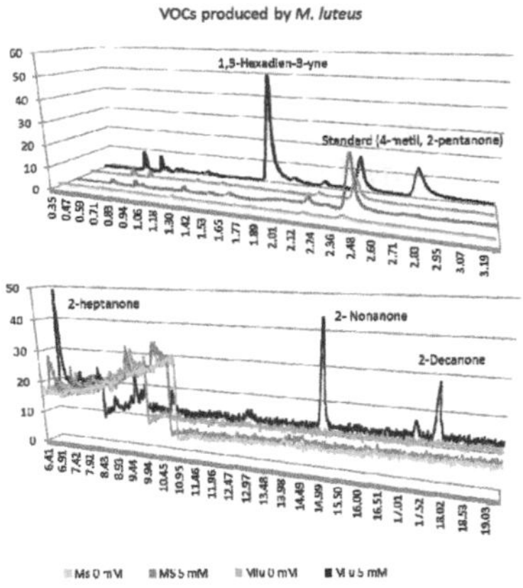

Figure 1. Volatile organic compounds produced by *Micrococcus luteus* growing in arsenite conditions.

On the other hand, the halo diameter was greater in those treatments without arsenite, and there was no difference between pH 7 and 9 (28 ± 1.15 and 29.67 ± 0.33 mm respectively).

3.3 *VOCs production by arsenite tolerant bacteria*

Figure 1 shows four VOCs produced by *M. luteus*: 1,3-hexadien-3-yne, 2-heptanone, 2-nonanone, and 2-decanone. The 1,3-hexadien-3-yne was produced by the strain with/without $NaAsO_2$, but, at higher metalloid concentration, the production of VOCs was highly augmented. The ketones were only detected in cultures supplemented with $NaAsO_2$, indicating that arsenite induced *de novo* synthesis of this compounds. Some studies reported that the detected compounds have plant growth promotion activity, especially by acting as nematicides.

4 CONCLUSIONS

M. luteus was the most arsenite tolerant bacterium. The capacity of production of siderophores was significantly affected by the presence of arsenite, especially at pH 9. The strain increased and modified the profile of VOCs of the aliphatic and ketone compounds in the presence of arsenite. *Micrococcus luteus* is a potential candidate to be used in rhizoremediation.

ACKNOWLEDGEMENTS

The authors are grateful to L. Bolcatto for technical assistance in the GC-MS determination.

REFERENCES

Dary, M., Chamber, M.A., Palomares Diaz, A.J. & Pajuelo Dominguez, E. 2010. "in Situ" Phytostabilisation of Heavy Metal Polluted Soils Using Lupinus Luteus Inoculated With Metal Resistant Plant-Growth Promoting Rhizobacteria. *Journal of Hazardous Materials* 177: 323–330.

Farag, M.A., Ryu, C., Sumner, L.W. & Paré P.W. 2006. GC-MS SPME profiling of rhizobacterial emissions reveals prospective inducers of growth promotion and induced systemic resistance in plants. *Phytochemistry* 67: 2262–2268.

Milagres, A.M.F., Machuca, A. & Napoleaõ, D. 1999. Detection of siderophore production from several fungi and bacteria by a modification of chrome azurol S (CAS) agar plate assay. *Journal of Microbiological Methods* 37: 1–6.

Salomon, M.V., Bottini, R., De Souza Pilho, G.A., Cohen, A., Moreno, D., Gil, M. & Piccoli, P. 2013. Bacteria isolated from roots and rhizosphere of *Vitis vinifera* retarded water losses via production of abscisic acid and induced synthesis of defense-related terpenes in *in vitro* cultured grapevine. *Physiologia Plantarum* DOI: 10.1111/ppl.12117.

1.8 Arsenic bioavailability and ecotoxicology

One Century of the Discovery of Arsenicosis in Latin America (1914–2014) – Litter, Nicolli, Meichtry, Quici, Bundschuh, Bhattacharya & Naidu (Eds)
ISBN 978-1-138-00141-1

Recent studies of arsenic mineralization in Iran and its effect on water and human health

D.K. Nordstrom
US Geological Survey, Boulder, CO, USA

R. Sharifi
Shiraz University, Shiraz, Iran

ABSTRACT: Few studies have been done on As mineralization and its effect on groundwater and human health in Iran yet considerable potential exists for high As groundwaters because of the occurrence of several areas of Au mineralization and the known relationship of hydrothermal As with Au. Recent studies in northwestern Iran have now shown that not only is there high As mineralization as expected for Iranian tectonics but high As in travertine springs and nearby groundwaters have been reported for the Takab-Qorveh area and human health problems have been documented for Bijar, Kurdistan.

1 INTRODUCTION

Very limited information exists about the source, distribution, occurrence, and human-health effects of arsenic-contaminated surface waters and groundwaters in Iran (Ravenscroft *et al.*, 2009). Yet the geological conditions would suggest a strong probability of human health risks from As because of the occurrence of several orogenic Au deposits in Iran and the strong relationship of arsenic with Au mineralization in particular tectonic settings (Nordstrom, 2012). Furthermore, numerous thermal springs also occur in Iran which may contain As but have not been analyzed adequately and may affect groundwater quality. Arsenic poisoning is prevalent in the Kurdistan province, identified in 1986 when a woman, in Bijar, lost her legs from gangrene following intense skin lesions (Mosaferi *et al.*, 2008). Pazirandeh *et al.* (1998) and Mosaferi *et al.* (2005) found a positive correlation between As in hair and As dosage in drinking water for residents of Bijar. Mosaferi *et al.* (2003) reported As concentrations in drinking water for several villages to be in the tens to hundreds of µg/L and the highest concentration greater than 1 mg/L. Among 752 participants studied, 49 were suffering from hyperkeratosis and 20 were suffering from hyperpigmentation (Mosaferi *et al.*, 2008). Barati *et al.* (2010) sampled 530 drinking water sources in Bijar and Qorveh and found As concentrations ranging from 42 to 1500 µg/L. They also found that 31% of the 587 people who received a clinical examination for chronic As poisoning showed symptoms such as pigment disorders, keratosis, and prevalence of Mee's lines (lines of discoloration across toenails and fingernails). This paper summarizes recent studies in Iran on As mineralization and points to the need for systematic studies of As in water supplies that could be guided by the geological conditions. Tying the water quality conditions to the geologic framework is essential for understanding the sources and extent of As contamination in water bodies.

2 GEOLOGIC SETTING

2.1 *Tectonics*

The largest mountain chain in Iran, the Zagros orogenic belt, is part of the Alpine-Himalyan orogeny developed from the collision of the African-Arabian plate with the Eurasian (or Iranian) plate. The mountain range extends about 2000 km from eastern Turkey to southern Iran in a NW-SE direction and is composed of three tectonically related parallel zones: (1) the Uromieh-Dokhtar Magmatic Assemblage, (2) the Sanandaj-Sirjan Zone, and (3) the Zagros Fold-Thrust Zone (Alavi, 1994). The first 2 zones host several gold mineral deposits containing arsenic mineralization. Plate convergence and compression occurred during the period of Late Cretaceous through Paleocene and was followed by gold emplacement. During plate convergence, a thick sequence of volcanic and sedimentary rocks of variable composition were folded, faulted, and metamorphosed. Substantial thrust fault motion led to older Tethyan sea sediments and Neo-Tethyan sediments to be emplaced on younger rocks along with obducted ophiolites (Alavi, 1994).

2.2 *Gold deposits*

Gold was reported from the Sanandaj-Sirjan Zone in 1823 and discoveries have continued to the present day, although production is modest (ca. 600 kg yr^{-1}, Aliyari *et al.*, 2012). At least five Au ore deposit types have been identified, orogenic, epithermal, Carlin-type, intrusion-related, and volcanic massive sulfide (Aliyari *et al.*, 2012). Pyrite is a common accessory mineral along with pyrhotite, arsenopyrite, arsenian pyrite, chalcopyrite, and sphalerite. Native Au occurs as well as inclusions and solid solutions in pyrite, arsenian pyrite, and arsenopyrite (Aliyari *et al.*, 2012; Ashrafpour *et al.*, 2012). Host rocks are highly variable and include phyllites, schists, metavolcanic rhyolites and andesites, limestones, granites, and tuffs. Gold deposits are found in northern areas near the Armenian border (Sungun and Sharafabad), central areas (San Gunay and Dali), southern areas (Meiduk and Sarchesmeh), and eastern Iran (Argash and Shadan; Ashrafpour *et al.*, 2012; Nehbandan; Damshenas *et al.*, 2010). Gold deposits with accompanying arsenic mineralization are still being discovered in this same orogenic belt (e.g. Çolakoğlu *et al.*, 2011).

3 ARSENIC AND GOLD IN TAKAB-QORVEH AREA

3.1 *Takab area, northwestern Iran*

The Takab area is a sparsely vegetated, semi-arid, mountainous region north of Takab and northwest of Bijar. It is one of the most important Au regions of Iran and includes an active As mine, Zarshuran, and two Au deposits, Zarshuran and Agdarreh. There is evidence for ancient mining of Au, As, Sb, and other base metals in this area. Takab is also a geothermal field located in the northern part of the Sanandaj-Sirjan Zone and comprising part of the Uromieh-Dokhtar volcanic belt. The Agdarreh and Zarshuran Au deposits are hosted by carbonate-rich sedimentary rocks that may be part of the same mineralization system although they are of vastly different ages. The Agdarreh disseminated Au deposit occurs in a hydrothermally altered Miocene reefal limestone with low Cu but enriched in As, Sb, Hg, Te, Se, Tl, Ba, Zn, Ag, Cd, Bi, and Au. The Zarshuran Au deposit is hosted by Precambrian carbonate and black shale formations intruded by a weakly mineralized granitoid. Orpiment is the most abundant sulfide mineral, the primary As ore mineral, and contains the highest concentrations of Au. About 30 sulfide minerals and sulfosalts have been identified at Zarshuran including pyrite, arsenian pyrite, orpiment, realgar, stibnite, getchelite, cinnabar, and thallium minerals (Modabberi and Moore, 2004). The highest concentrations of Au were detected in microcrystalline orpiment, carbonaceous shale, and silicified shale with pyrite and As minerals (Asadi *et al.*, 2000; Daliran, oral comm., 2005).

3.2 *Zarshuran mine water and downstream effects*

Many years of mining at Zarshuran have produced large volumes of mine wastes that have undergone weathering and erosion. Waste piles contained As up to 40 wt.%, Sb up to 0.5%, and Hg up to 0.013% (Modabberi and Moore, 2004). Mine waters are circumneutral in pH but contain up to 40 mg/L As and up to 1.1 mg/L Sb. These concentrations are among the largest known from a mineral deposit not draining As_2O_3 from smelter emissions. Drainage from the Zarshuran and Agdarreh mines are tributaries to the Sarouq River, which discharges into a reservoir used as a drinking water supply for several cities such as Tabriz, Saqqez, Bonab, Malekan, and many small villages. Although the Sarouq River supplies one third of the annual recharge to the reservoir, it is diluted substantially by clean tributaries before entering it. An As concentration of 25 μg/L was reported by Modabberi and Moore (2004) and no data are available on seasonal variability.

3.3 *Travertines and mineral springs*

Relatively little data is available on As occurrences in carbonate-hosted lithologies and mineral-spring associations. Clarke (1924) mentions an analysis of a travertine with 0.27% As, Bernasconi *et al.* (1980) describes travertine at the Senator mine in western Turkey with high As concentrations from hydrothermal mineralization depositing dussertite, scorodite, and orpiment. Dessau (1968) reported As concentrations in travertine up to 1600 ppm. In the Takab area thermal springs precipitate travertine and form muds that contain Cu, Se, Sb, Pb, and As in the percent range, up to several thousand ppm Hg and Te, and up to 1,000 ppm Au and Ag (Daliran, 2003). Cold and hot springs in the Takab-Qorveh area are mostly artesian, temperatures range from 8 to 45 °C, pH values are 6.2–6.6, and they are mineralized (0.3–3.0 g/L dissolved solids) of Na-Cl-HCO_3 water type with high As concentrations (212–987 μg/L; see Table 1; Keshavari *et al.*, 2011). Arsenic is found predominantly in the reduced trivalent state. The composition of these waters have typical geothermal signatures with their high As, B, and Li concentrations. High As concentrations in travertines and smithsonite mineral deposits in the region were reported by Boni *et al.* (2007).

The original As mineralization was arsenopyrite, with some Ni and Co arsenides, but in a late

Table 1. Selected analyses of travertine springs, NW Iran (from Keshavarzi *et al.*, 2011; concentrations in mg/L).

Parameter	Qorveh	Bijar	Takab
Temperature	16.4	15.1	18.9
pH	6.44	6.53	6.21
Na	173	452	115
K	569	236	26.3
Ca	129	190	413
Mg	101	310	110
Cl	10300	1050	851
HCO_3	3340	4534	1230
SO_4	360	863	644
As	0.279	0.842	0.987
B	221	189	6.3
Li	37.1	11.0	0.30

Table 2. Selected mean values for analyses of groundwaters, NW Iran (from Keshavarzi *et al.*, 2011; concentrations in mg/L).

Parameter	Mean (dry season)	Mean (wet season)
pH	7.02	7.6
Na	129	124
K	8.9	9.5
Ca	81.9	164
Mg	36.5	44.2
Cl	64.7	61.3
HCO_3	316	417
SO_4	146	151
As	0.079	0.112
B	1.48	1.12
Li	0.115	0.090

stage alteration high-carbonate oxidizing solutions altered arsenopyrite and galena to mimetite, $Pb_5(AsO_4)_3Cl$, which is moderately soluble.

3.4 *Groundwaters*

Keshavarzi *et al.* (2011) sampled 38 groundwater wells once during the dry season and again during the wet season. Groundwaters are Na-Ca-HCO_3 type occasionally with elevated Cl and SO_4 concentrations. Saturation indices show saturation to supersaturation with respect to calcite. Arsenic concentrations ranged from 0.4 to 689 μg/L and the As was predominantly oxidized to arsenate. Little effect could be seen from the seasonal data but different amounts of mixing of groundwater with mineral springs were evident from the variable Cl concentrations that were as high as 569 mg/L. There were higher As, B, Cl, Sr, Cs, Br, and Li concentrations in wells closer to travertine springs.

4 CONCLUSIONS

Although few studies have been completed on evaluating As distribution in Iranian groundwaters that may be a source of drinking or irrigation water supplies, substantial evidence from geological and mineralogical data exists to suspect several areas may be detrimentally affected. Recent studies in northwestern Iran confirm areas of considerable As enrichment in travertine and Au-As mineral deposits that provide a source of As to groundwaters. These anomalous As concentrations are related to waning stages of Quaternary volcanic activity, manifested by travertine-forming springs. Future investigations need to locate similarly mineralized areas that are spatially proximal to villages and cities, especially those that use groundwater for domestic use, to determine As concentrations and any occurrences of chronic As symptoms among the populace.

ACKNOWLEDGEMENTS

The senior author acknowledges the support of the National Research Program of the US Geological Survey and his Iranian colleagues for their willingness to collaborate on this paper. We are very grateful to Dr. Farid Moore for his support and advice for this important research.

REFERENCES

Alavi, M. 1994. Tectonics of the Zagros orogenic belt of Iran: new data and interpretations. *Tectonophysics* 229: 211–238.

Aliyari, F., Rastad, E. & Mohajjel, M. 2012. Gold deposits in the Sanandaj-Sirjan Zone: Orogenic gold deposits or intrusion-related gold systems? *Resource Geology* 62(3): 296–315.

Asadi, H.H., Voncken, J.H.L., Kühnel, R.A. & Hale, M. 2000. Petrography, mineralogy and geochemistry of the Zarshuran Carlin-like gold deposit, northwest Iran. *Mineralium Deposita* 35: 656–671.

Ashrafpour, E., Ansdell, K.M. & Alirezaei, S. 2012. Hydrothermal fluid evolution and ore genesis in the Argash epithermal gold prospect, northeastern Iran. *Journal of Asian Earth Sciences* 51: 30–44.

Barati, A.H., Maleki, A. & Alasvand, M. 2010. Multi-trace elements level in drinking water and the prevalence of multi-chronic arsenical poisoning in residents in the west area of Iran. *Science of the Total Environment* 408: 1523–1529.

Bernasconi, A., Glover, N. & Viljoen, R.P. 1980. The geology and geochemistry of the Senator antimony deposit – Turkey. *Mineralium Deposita* 15: 259–274.

Boni, M., Gilg, A., Balassone, G., Schneider, J., Allen, C.R. & Moore, F. 2007. Hypogene Zn carbonate ores in the Angouran deposit, NW Iran. *Mineralium Deposita* 42: 799–820.

Clarke, F.W. 1924. *The Data of Geochemistry*. 5th edition, US Geological Survey Bulletin 770.

Çolakoğlu, A.R., Oruç, M., Arehart, G.B. & Poulson, S. 2011. Geology and isotope geochemistry (C-O-S) of the Diyadian gold deposit, Eastern Turkey: A newly-discovered Carlin-like deposit. *Ore Geology Reviews* 40: 27–40.

Daliran, F. 2003. Discovery of 1.2 kg/t gold and 1.9 kg/t silver in mud precipitates of a cold spring from the Takab geothermal field, NW Iran, In D.G. Eliopoulos, *et al.* (eds) *Mineral Exploration and Sustainable Development*: 461–464.

Damshenas, E., Dahrazma, B., Sadeghian, M. & Askari, A. 2010. Assessment of the arsenic contamination in groundwater in Hired gold mine zone (northwest of Nehbandan – Iran), In J.-S. Jean, J. Bundschuh & Bhattacharya, P. (eds.) *Arsenic in Geosphere and Human Diseases, Arsenic in the Environment – Proceedings*: 378–379. Boca Raton, CRC Press.

Dessau, G. 1968. Il berillo e l'arsenico nei travertine dell'Italia central. *Atti della Societa Toscana de Scienza Naturli di Pisa* 75: 690–711.

Grove, A.T. 1980. Geomorphic evolution of the Sahara and the Nile. In M.A.J. Williams & H. Faure (eds), *The Sahara and the Nile*: 21–35. Rotterdam: Balkema.

Jappelli, R. & Marconi, N. 1997. Recommendations and prejudices in the realm of foundation engineering in Italy: A historical review. In Carlo Viggiani (ed.), *Geotechnical engineering for the preservation of monuments and historical sites; Proc. intern. symp., Napoli, 3–4 October 1996*. Rotterdam: Balkema.

Johnson, H.L. 1965. Artistic development in autistic children. *Child Development* 65(1): 13–16.

Keshavarzi, B., Moore, F., Mosaferi, M. & Rahmani, F. 2011. The source of natural arsenic contamination in groundwater, west of Iran. *Water Quality, Exposure and Health* 3, 135–147.

Modabberi, S. & Moore, F. 2004. Environmental geochemistry of Zarshuran Au-As deposit, NW Iran. *Environmental Geology* 46: 796–807.

Mosaferi, M., Yunesian, M., Dastgin, S., Mesdaghinia, A.R. & Esmailnasab, N. 2008. Prevalence of skin lesions and exposure to arsenic in drinking water in Iran. *Science of the Total Environment* 390: 69–76.

Mosaferi, M., Yunesian, M., Mesdaghinia, A.R., Nasseri, S., Mahvi, A.H. & Nadim, A. 2005. Correlation between arsenic concentration of drinking water and hair. *Iranian Journal of Environmental Health Science and Engineering* 2: 11–23.

Mosaferi, M., Yunesian, M., Mesdaghinia, A.R., Nadim, A., Nasseri, S. & Mahvi, A.H. 2003. Occurrence of arsenic in Kurdistan Province of I.R, Iran, In M.F. Ahmed, M.A. Ali & Z. Adeel (eds), *Fate of arsenic in the environment*, BUET-UNU International Symposium, Bangladesh University of Engineering and Technology, United Nations University, Tokyou, 2003, 1–6.

Nordstrom, D.K. 2012. Arsenic in the geosphere meets the anthroposphere, In J.C. Ng, B.N. Noller, R. Naidu, J. Bundschuh & P. Bhattacharya (eds), *Understanding the Geological and Medical Interface of Arsenic*, 4th International Congress on Arsenic in the Environment, Cairns, Australia, July 22–27, 2012, 15–19.

Pazirandeh, A., Brati, A.H. & Marageh, M.G. 1998. Determination of arsenic in hair using neutron activation. *Applied Radiation Isotopes* 49(7): 753–759.

Polhill, R.M. 1982. *Crotalaria in Africa and Madagascar*. Rotterdam: Balkema.

Ravenscroft, P., Brammer, H. & Richards, K. 2009. *Arsenic Pollution: A Global Synthesis*. Oxford: Wiley-Blackwell.

One Century of the Discovery of Arsenicosis in Latin America (1914–2014) – Litter, Nicolli, Meichtry, Quici, Bundschuh, Bhattacharya & Naidu (Eds)
© 2014 Taylor & Francis Group, London, ISBN 978-1-138-00141-1

Bioaccessibility of arsenic in a gold mine area in Brazil: Why is it so low?

L.R.G. Guilherme & F.B. Ono
Department of Soil Science, Federal University of Lavras (UFLA), Lavras, Minas Gerais, Brazil

M. Cantoni, C.A. de Abreu & A.R. Coscione
Agronomic Institute of Campinas (IAC), Campinas, São Paulo, Brazil

R. Tappero
National Synchrotron Light Source, Brookhaven National Laboratory, Upton, New York, USA

D. Sparks
University of Delaware, Delaware Environmental Institute, Newark, Delaware, USA

ABSTRACT: Research has shown the presence of high concentrations of Arsenic (As) and very low As bioaccessibility in tailings from a gold mining area in Brazil. Assessment of As speciation in the tailings is needed to explain this low bioaccessibility. Synchrotron-based bulk-X-ray Absorption Near Edge Structure (XANES) spectroscopy, μ-XANES and μ-X-Ray Fluorescence (μ-SXRF) spectroscopy were used in this study to assess As speciation in tailing samples collected at four different locations in the mining area. Bulk-XANES spectra indicated that As occurred as As(V). Micro-XANES and μ-SXRF analyses revealed that As occurred also as arsenopyrite and its weathering products, but mostly as poorly crystalline ferric arsenate. This supports the findings of low bioaccessible As. For comparison, As bioaccessibility was performed in soils with high contents of either naturally-occurring or anthropogenically-added As. The naturally occurring high-As soil showed very low As bioaccessibility, similarly to samples collected in the gold mining area.

1 INTRODUCTION

Arsenic (As) is a trace element of great interest in studies concerning environmental contamination due to its high toxicity and consequent risks to humans. Natural sources of As are frequently associated with gold deposits containing sulfide minerals (Drahota & Filippi, 2009, Deschamps *et al.*, 2002). Generally, As exists in these materials in the form of reduced sulfide minerals; however, the gold mining activities enable the weathering of sulfides, releasing As and Fe sulfates and hydroxides, generating secondary compounds and forming more oxidized As forms (Murciego *et al.*, 2011).

The release of As from mine wastes is controlled by precipitation-dissolution and adsorption-desorption reactions involving the secondary compounds generated during the weathering of arsenopyrite (Lengke *et al.*, 2009). In mine tailings, As usually occurs integrated with sulfide minerals (Foster *et al.*, 1998). In soils from mining areas, however, As is mainly associated with amorphous Fe oxyhydroxides (Filippi *et al.*, 2004). In well-aerated environments, As(V) is the predominant form of As, which is mostly adsorbed onto Fe oxyhydroxides (Grossl *et al.*, 1997).

Previous studies have shown high contents of As in tailings from an open pit gold mine in Brazil (Ono *et al.*, 2012). However, despite the high levels of total As found in this mine area, the samples showed very low contents of bioaccessible As. It is well known that the bioavailability, mobility and toxicity of As depends on its speciation (Walker *et al.*, 2005). However, studies concerning the solid-phase chemical speciation of As have not yet been carried out for samples collected at this mining area.

This study aimed at identifying the solid-phase speciation of As in samples from a gold mine area using synchrotron-based spectroscopy analyses such as bulk-X-ray absorption near-edge structure (bulk-XANES), micro-X-ray fluorescence (μ-SXRF) and micro-X-ray absorption near-edge structure (μ-XANES) spectroscopy. For comparison, we also evaluated bioaccessible As in a naturally-occurring high-As soil, as well as in As-spiked soils.

2 EXPERIMENTAL

2.1 *Site 1: Samples from the gold mining area*

Mine samples (hereafter called PNR, B1, Exp.B1 and pond tailings) were collected on an open pit gold mine located in the State of Minas Gerais, Brazil. In a previous study (Ono *et al.*, 2012), five composite samples were collected from each of these five areas at two soil depths, and the total As concentration was determined. In addition, As bioaccessibility was analyzed using the IVG protocol (Rodriguez *et al.*, 1999). These materials were once again analyzed in the present study. We selected the sample with the highest As content from each area at the 0–2 cm depth.

Besides total and bioaccessible As, we also measured non-specifically adsorbed As and specifically adsorbed As in the samples (<250 μm) using the first two steps of the sequential-extraction method developed by Wenzel *et al.* (2001). The first step (0.05 M $(NH_4)_2SO_4$, 4-h shaking, 20 °C) removes the non-specifically bound As fraction (F1), and the second step (0.05 M $(NH_4)_2HPO_4$, 16-h shaking, 20 °C) is used to extract the specifically sorbed As fraction (F2).

The oxidation state of As in the samples was determined by bulk-XANES spectroscopy. The analyses were conducted at the As K-edge (11,867 eV) at the Brazilian Synchrotron Light Laboratory (Campinas, Brazil) on the D08B-XAFS2 beamline. The spectra were recorded at room temperature using a Si (111) double crystal monochromator with an upstream vertical aperture of 0.5 mm and calibrated with Au L3-edge (11,919 eV). The samples were mounted behind Kapton tape in a Teflon holder and analyzed in fluorescence or transmission mode. Three scans were collected for each sample, and then merged. Natural minerals were used as reference standards for As.

Micro-SXRF and μ-XANES spectroscopic analyses in the samples were conducted at the beamline X27A of the National Synchrotron Light Source, Brookhaven National Laboratory, USA, to determine elemental distributions and direct solid-phase speciation of As. The fine particulates were mounted behind Kapton tape in a cardboard slide-holder. The μ-SXRF maps were 3 × 3 mm in size, with a 10 μm step size, and beam energy was set to 12.5 keV. Micro-XANES spectra were then collected for the As K-edge (11,867 eV), at points of interest observed in μ-SXRF maps. Linear-combination fitting (LCF) of the μ-XANES spectra of samples with those of various model compounds was performed using the Athena software to estimate As species in the mine samples. The following materials were included as As standards: arsenopyrite, realgar, sodium arsenite, sodium arsenate, scorodite, and poorly crystalline ferric arsenate.

2.2 *Site 2: Arsenic-rich and As-spiked soils*

Soil samples were collected from the subsoil horizon of a Quartzarenic Neosol (QN), Red Latosol (RL) and Cambisol (CA) located in the cities of São Pedro, Campinas and Iporanga, respectively, all in the state of São Paulo, Brazil. The CA soil has natural As contamination. The RL and QN soils have low As levels. Thus, these soils were spiked with sodium arsenate (simulating anthropogenic As contamination) to achieve a final As concentration of 1,600 and 400 mg kg^{-1}, respectively, and kept at 70% soil water holding capacity. Soils were stored in plastic bags and aged for up to 60 days.

The total As concentration of the samples (<250 μm) was analyzed by ICP-OES after aqua regia digestion. The SBET protocol (Kelley *et al.*, 2002) was used to evaluate the bioaccessibility of As in the soils (fraction <250 μm). This protocol only has the gastric phase because the addition of the intestinal phase does not improve the estimation of bioaccessible As (Rodriguez *et al.*, 1999).

The fractions corresponding to non-specifically adsorbed As and specifically adsorbed As in the samples (<250 μm) were determined using the first two steps (F1 and F2) of the sequential-extraction method developed by Wenzel *et al.* (2001). Arsenic concentration in extracts was determined by ICP-OES.

3 RESULTS AND DISCUSSION

3.1 *Site 1: Samples from a gold mining area*

A previous study (Ono *et al.*, 2012) showed high As concentrations in the mine samples, except for the control area (Table 1). Arsenic bioaccessibility in the same samples was very low (<4.4%). In fact, As fractionation by $(NH_4)_2SO_4$ (easily exchangeable) and $NH_4H_2PO_4$ (inner-sphere surface bound As) confirm that As was mostly in non-labile forms in the mine samples. The proportion of exchangeable and specifically sorbed As represented less than 0.5 and 4.6% of the total As, respectively (Table 1). Therefore, these results confirmed the low As availability found in the samples through a bioaccessibility test, and indicated that the greatest proportion of As in the mine tailings was probably associated with Fe oxyhydroxides and/or the residual phase.

To further confirm the very low availability of As, mine tailings were analyzed by bulk-XANES spectroscopy to determine the predominant oxidation state of As (Figure 1). When comparing the absorption edge position of mine tailings and reference compounds, the XANES spectra indicated that in all mine tailings As occurred mainly in the As(V) oxidation state. This result could have been expected, since the mine tailings were collected at 0–2 cm and were therefore well aerated, except for the pond tailings sample, which was originally

Table 1. Total As content and available As fractions in the mine tailings from a gold mining area.

Samples	Total As[&] mg kg^{-1}	B[&] (%)	F1[§] (%)	F2[§] (%)
Control	37	<QL	<QL	<QL
B1	262	3.1	0.2	2.2
Exp. B1	335	2.0	0.4	2.0
Undisturbed	527	1.6	0.5	4.6
Pond tailings	2666	4.4	0.1	2.1

[&]Modified according to Ono *et al.* (2012). B = bioaccessible. <QL = below the quantification limit (0.6 mg As kg^{-1}).
[§]As fractions associated with (F1) non-specifically sorbed—extracted by $(NH_4)_2SO_4$—and (F2) specifically sorbed—extracted by $NH_4H_2PO_4$—according to Wenzel *et al.* (2001).

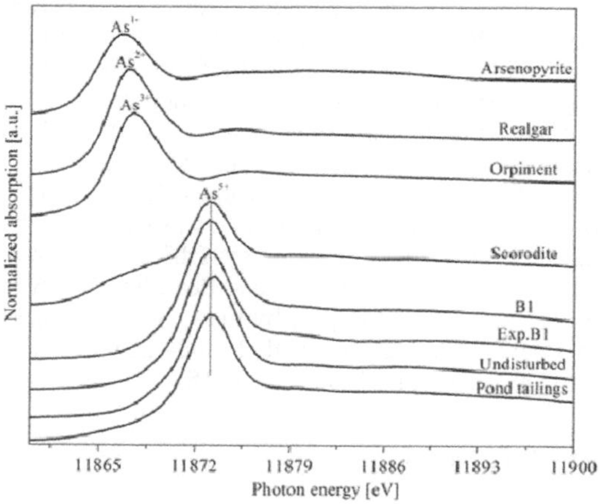

Figure 1. Normalized As K-edge XANES spectra of mine samples. Also shown are XANES spectra of standards for comparison of the edge positions.

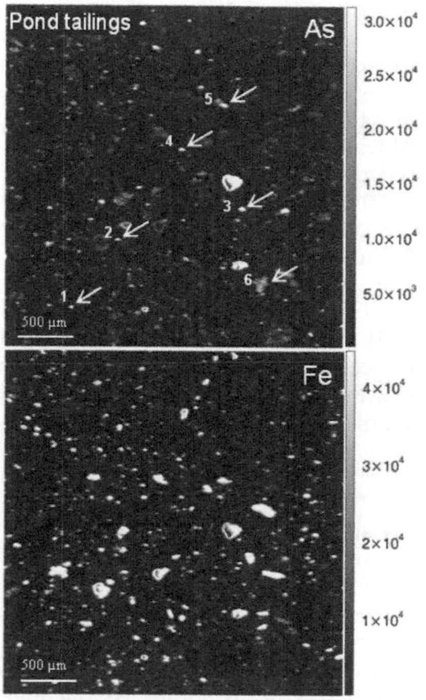

Figure 2. Micro-SXRF maps using an AsFe filter for the Pond tailings sample. The numbers indicate the spots where µ-XANES spectra were collected.

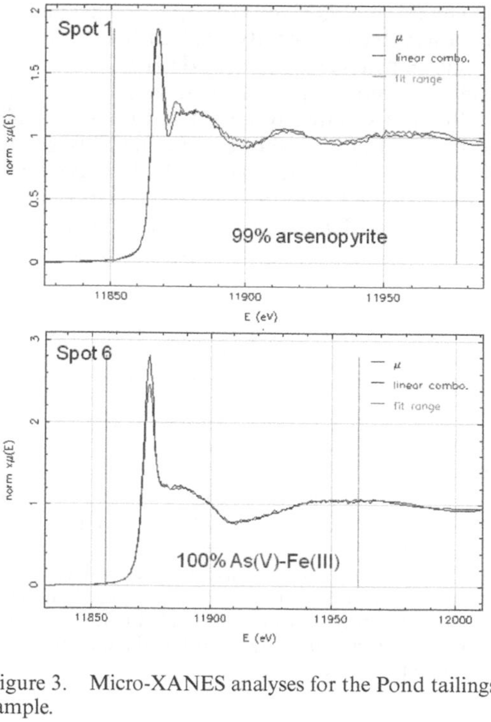

Figure 3. Micro-XANES analyses for the Pond tailings sample.

under water-saturated conditions, even though the sample was not kept like that during the measurements. This condition may have caused the oxidation of lower-valent As.

Micro-SXRF and µ-XANES spectroscopic analyses were used to generate elemental distribution maps and to determine the As species of mine samples (Figures 2 and 3). Micro-SXRF images taken from powder-on-tape mine samples showed evident association of As and Fe (Figure 2). Usually, Fe is the element that best associates with As in soil/sediment/tailing samples, which occurs due to its high concentrations and the strong binding of As with Fe-oxides (Smedley & Kinniburgh, 2002).

The LCF results from µ-XANES spectra of mine tailings showed that As was also present as arsenopyrite and its transformation products, but mostly it was in the form of As(V) as poorly crystalline ferric arsenate (Figure 3). Our results did not indicate the presence of scorodite—which is the most common arsenate mineral found in the weathering environment of As-bearing sulfide deposits (Dove & Rimstidt, 1985)—or the sulfide mineral realgar (As_4S_4) in the samples.

Based on the As species and the proportions found in the present study, we can infer that the As fractions present in these tailings were not readily available and therefore are not considered a great source of As release to the environment, as long as no physicochemical changes occur (e.g. pH, redox potential). In addition, this finding highlights the importance of Fe oxides in controlling the mobility and availability of As in tailings following dissolution of the sulfide minerals of As.

3.2 *Site 2: Arsenic-rich and As-spiked soils*

The CA soil, which has naturally-occurring high contents of As, showed very high total As concentration (4720 mg kg^{-1}) (Table 2). Despite that, its bioaccessible As was very low (Table 2), showing a behavior similar to that of the mine samples evaluated at Site 1 (Table 1). The non-specifically sorbed and specifically sorbed As fractions confirmed that As is mostly in non-labile forms in this soil (Table 2). Thus, these results indicate that the greatest proportion of As in this sample was probably associated with Fe oxyhydroxides and/or the residual phase.

The RL and QN soils (As-spiked soils) had the highest As bioaccessibility (Table 2). In fact, As fractionation by $(NH_4)_2SO_4$ and $NH_4H_2PO_4$ confirmed that As is mostly labile in these samples. The sum of the non-specifically sorbed and the specifically sorbed As fractions showed an availability greater than 50%. The difference of As bioaccessibility between the RL and the QN soils could be attributed to the difference of particle size (texture), the RL being a clayey soil and the QN a sandy soil.

Table 2. Total As content and available As fractions in the soils and mine tailing samples.

Samples	Total As mg kg^{-1}	B (%)	F1[§] (%)	F2[§] (%)
RL	1474	26	13	40
QN	381	80	27	35
CA	4720	1	<QL	7

B = bioaccessible. [§]As fractions associated with (F1) non-specifically sorbed—extracted by $(NH_4)_2SO_4$—and (F2) specifically sorbed—extracted by $NH_4H_2PO_4$—according to Wenzel *et al.* (2001). QL = below the quantification limit (0.3 mg As kg^{-1}).

4 CONCLUSIONS

Arsenic mainly occurred in the pentavalent form in the mine samples. It was also present associated with arsenopyrite and its weathering products, but it was mostly associated with poorly crystalline Fe arsenate. This supports the finding of low bioaccessible As and highlights the importance of Fe oxides in immobilizing As in the environment.

The CA soil, which a had high naturally-occurring As content, showed low availability of As, with a behavior similar to that of the As-carrying soil material found in the gold mine area (Site 1). XAS analysis could explain the low As availability in this soil sample.

ACKNOWLEDGEMENTS

We gratefully acknowledge funding received from CNPq, CAPES, FAPEMIG, and FAPESP (2010/11117-8). We thank the Brazilian Synchrotron Light Laboratory technical, scientific, and administrative staff for assisting with the XAS analysis (project 11781, supported by LNLS/ABTLuS/MCT). Use of the National Synchrotron Light Source, Brookhaven National Laboratory, was supported by the U.S. Department of Energy, Office of Science, Office of Basic Energy Sciences. Beamline X27A at NSLS is supported in part by the U.S. Department of Energy—Geosciences (DE-FG02-92ER14244 to The University of Chicago—CARS).

REFERENCES

Deschamps, E., Ciminelli, V.S.T., Lange, F.T., Matschullat, J., Raue, B. &, Schmidt, H. 2002. Soil and sediment geochemistry of the Iron Quadrangle, Brazil: the case of arsenic. *Journal of Soils and Sediments* 2:216–222.

Dove, P.M. & Rimstidt, J.D. The solubility and stability of scorodite $FeAsO_4.2H_2O$. 1985. *American Mineralogist* 70:838–844.

Drahota, P. & Filippi, M. 2009. Secondary arsenic minerals in the environment: a review. *Environment international* 35:1243–1255.

Filippi, M., Goliáš, V. & Pertold, Z. 2004. Arsenic in contaminated soils and anthropogenic deposits at the Mokrsko, Roudný, and Kašperské Hory gold deposits, Bohemian Massif (CZ). *Environmental Geology* 45:716–730.

Foster, A.L., Brown, Jr. G.E., Tingle, T.N. & Parks, G.A. 1998. Quantitative As speciation in mine tailings using X-ray absorption spectroscopy. *American Mineralogist* 83:553–568.

Grossl, P.R., Eick, M., Sparks, D.L., Goldberg, S. & Ainsworth, C.C. 1997. Arsenate and chromate retention mechanisms on goethite. 2. Kinetic evaluation using a pressure-jump relaxation technique. *Environmental Science and Technology* 31:321–326.

Kelley, M.E., Brauning, S.E., Schoof, R.A. & Ruby, M.V. 2002. *Assessing Oral Bioavailability of Metals in Soil*. Battelle Press: Columbus, Ohio.

Lengke, M.F., Sanpawanitchakit, C. & Tempel, R.N. 2009. The oxidation and dissolution of arsenic-bearing sulfides. *The Canadian Mineralogist* 47:593–613.

Murciego, A., Alvarez-Ayuso, E., Pellitero, E., Rodríguez, M., García-Sánchez, A., Tamayo, A., Rubio, J., Rubio, F. & Rubin, J. 2011. Study of arsenopyrite weathering products in mine wastes from abandoned tungsten and tin exploitations. *Journal of Hazardous Materials* 186:590–601.

Ono, F.B., Guilherme, L.R.G., Penido, E.S., Carvalho, G.S., Hale, B., Toujaguez, R. & Bundschuh, J. 2012. Arsenic bioaccessibility in a gold mining area: a health risk assessment for children. *Environmental Geochemistry and Health* 34:457–465.

Rodriguez, R.R., Basta, N.T., Casteel, S.W. & Pace, L.W. 1999. An in vitro gastrointestinal method to estimate bioavailable arsenic in contaminated soils and solid media. *Environmental Science and Technology* 33(4): 642–649.

Smedley, P.L. & Kinniburgh, D.G. A review of the source, behaviour and distribution of arsenic in natural waters. 2002. *Applied Geochemistry* 17:517–568.

Walker, S.R., Jamieson, H.E., Lanzirotti, A., Andrade, C.F. & Hall, G.E.M. 2005. The speciation of As in iron oxides in mine wastes from the Giant gold mine, N.W.T.: application of synchrotron micro-XRD and micro-XANES at the grain scale. *The Canadian Mineralogist* 43:1205–1224.

Wenzel, W.W., Kirchbaumer, N., Prohaska, T., Stingeder, G., Lombi, E. & Adriano, D.C. 2001. Arsenic fractionation in soils using an improved sequential extraction procedure. *Analytica Chimica Acta* 436:309–323.

One Century of the Discovery of Arsenicosis in Latin America (1914–2014) – Litter, Nicolli, Meichtry, Quici, Bundschuh, Bhattacharya & Naidu (Eds)
© 2014 Taylor & Francis Group, London, ISBN 978-1-138-00141-1

Arsenic chemical form by XANES, bioaccessibility measurement and prediction in environmental samples for human health risk assessment purposes

R. Taga
Centre for Mined Land Rehabilitation, The University of Queensland, Brisbane, Queensland, Australia
National Research Centre for Environmental Toxicology (Entox), The University of Queensland, Brisbane, Queensland, Australia

B.N. Noller
Centre for Mined Land Rehabilitation, The University of Queensland, Brisbane, Queensland, Australia

J.C. Ng
National Research Centre for Environmental Toxicology (Entox), The University of Queensland, Brisbane, Queensland, Australia
CRC for Contamination Assessment and Remediation of the Environment, Mawson Lakes, Adelaide, South Australia, Australia

J. Aitken
School of Chemistry, The University of Sydney, New South Wales, Australia

H.H. Harris
School of Chemistry and Physics, Adelaide University, Adelaide, South Australia, Australia

ABSTRACT: The Physiologically Based Extract Test (PBET) was used to measure *in-vitro* arsenic bioaccessibility in mine wastes from mine sites in Northern Australia, and X-ray Absorption Near Edge Spectroscopy (XANES) was used to obtain arsenic speciation data directly in the solid phase of mine waste samples. Because Arsenate (As(V)), the oxidized form of arsenic, is more likely to be found in mine waste materials, the XANES results from fitting with As model compounds support the finding that measured bioaccessibility was in good agreement with the predicted values and show a strong association with the presence of ferric arsenate and As sulfides. The results indicate that soil intake adjusted for bioaccessibility is less than the default and conservative assumption of 100% bioavailability often employed for the risk assessment calculation for health risk estimate of local residents exposed to the mine waste.

1 INTRODUCTION

Arsenic (As) in mining waste is characteristically associated with base metal sulfide mineralogy. Arsenate (As(V)) is the stable oxidized form in the environment. It has been demonstrated that in-vitro bioaccessibility (BAc) of metals, the surrogate of bioavailability, can be predicted by summing the contributions of chemical species that exist in environmental samples from XANES fitting proportionally with the measured bioaccessibility of the respective pure metal compounds (MacLean *et al.*, 2011; Rasmussen *et al.*, 2011). The application of this technique for estimating arsenic bioaccessibility is demonstrated here. This paper aims to: (i) measure the bioaccessibility of pure arsenic compounds and rank these against sodium arsenite as 100%; and (ii) show the regression of measured bioaccessibility vs. predicted bioaccessibility based on proportions of arsenic compounds estimated from X-ray Absorption Near Edge Spectroscopy (XANES) Linear Combination Fitting (LCF).

2 EXPERIMENTAL

Sixty-one mine wastes were collected from varying locations and origins in Northern Australia. Sample types ranged from waste rock, tailings, ore and concentrate dusts, mine community soils, carpet dusts, fall out and haul road soils. Mine waste samples initially sieved to <250 μm for bioaccessibility measurement (Ng *et al.*, 2013) were ground to <20 μm for X-ray analysis. Total As concentrations in the <250 μm fractions were determined using ICP-MS on *aqua regia* digests of samples by Australian Laboratory Services Pty Ltd, Brisbane.

Bioaccessibility of arsenic was measured in model compounds used for XANES speciation and mine waste samples using the *in-vitro* Physiologically Based Extraction Test (PBET) (Ruby *et al.*, 1993).

Arsenic K-edge X-ray absorption spectra were recorded at the Australian National Beamline Facility (ANBF, Beamline 20B at the Photon Factory, KEK, Tsukuba, Japan). The spectra were recorded at room temperature in fluorescence mode, using a thirty six-element germanium-array detector. BL–20B was equipped with a channel-cut Si (111) monochromator which was detuned 50% to reject harmonics. The scan parameters were: (i) the pre-edge section, 11.64–11.84 keV (10-eV steps); (ii) the X-ray absorption near-edge structure (XANES) section, 11.84–11.92 (0.25-eV steps); and (iii) the post-edge section, 11.92–12.10 keV (7 Å, 0.2 Å steps in the k-space). Sodium arsenate was used as an internal standard for energy calibration, with the first peak of the first derivative assumed to be 11,873.6 eV. Solid samples (ground to <20 μm) were pressed into aluminum spacers between two 63.5-lm Kapton tape windows (window size, 2*10 mm). Model compounds were: aluminum arsenate, arsenic (III) oxide, arsenic (V) oxide, arsenic (III) sulfide, arsenic (V) sulfide, calcium arsenite, calcium arsenate, arsenopyrite, ferric arsenate, sodium arsenate, sodium arsenite, scorodite, α-arsenic and arsenic tri-iodide. Data analysis, including calibration, averaging and background subtraction of all spectra and Principal Component Analysis (PCA), target and linear regression analyses of XANES spectra were performed using the EXAFSPAK software package (George & Pickering, 2000). The XANES analysis consisted of fitting linear combination of model spectra to sample spectra using the program DATFIT (George & Pickering, 2000). The precision of this fit procedure was determined to be ~10% based on analyses of control model compound mixtures.

Bioaccessible As was divided by total As to calculate the percentage of As bioaccessibility (As BAc/As Total * 100). Presentation of As bioaccessibility results relative to the soluble As reference salt (sodium arsenite) permitted comparison of *in-vitro* relative bioaccessibility using linear regression (MacLean *et al.* 2011; Rasmussen *et al.* 2011). Statistical analysis was conducted with SPSS 21.

3 RESULTS AND DISCUSSION

Table 1 shows that total As concentrations are skewed when mean and median are compared.

Table 1. Comparison of total concentration and measured % bioaccessibility data sets using PBET for arsenic in mine waste samples.

	Arsenic ($n = 61$)	
Statistical feature	Total (mg kg^{-1})	% BAc
Mean	2027	18.2
SD	6028	14.5
Minimum	13	0.7
25th percentile	170	6.3
Median	298	14.3
75th percentile	615	28.5
95th percentile	10000	40.0
Maximum	32000	56.0

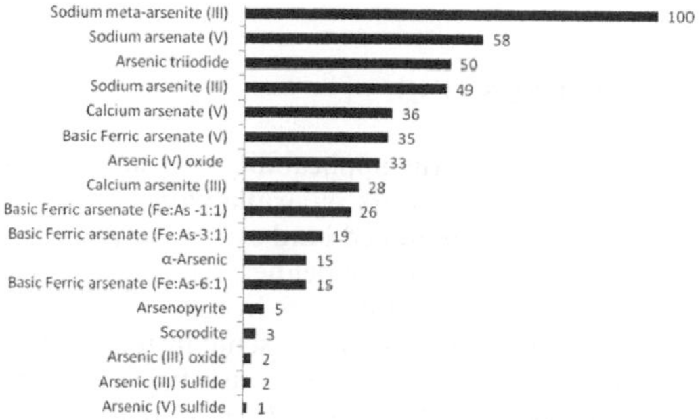

Figure 1. Relative bioaccessibility of As model compounds for As speciation in mine waste samples from XANES analysis.

Below median concentrations are relatively low with low As bioaccessibility compared with the 75th percentile concentration which is associated with processed mine wastes and higher As bioaccessibility. The predicted As bioaccessibility in dust samples was estimated using percentages of each As species from XAS speciation weighted by their bioaccessibility relative to sodium arsenite (RBA) (Figure 1).

Figure 2 shows these estimates plotted against PBET measured As bioaccessibility of mine site samples. The strong correlation ($r^2 = 0.722$; $n = 61$; $p < 0.0001$) between estimated BAc from XANES speciation and measured BAc using PBET indicates the usefulness of XANES spectral modelling for BAc prediction.

When elevated As concentrations in mine site soils exceed the Health Investigation Levels (HILs) (NEPC, 2013), As BAc from PBET measurement of mine wastes can be applied to more accurately assess the health risk. In addition, As BAc data from PBET can also be utilized to develop site-specific guidelines for soil contamination as demonstrated elsewhere (Diacomanolis *et al.*, 2010).

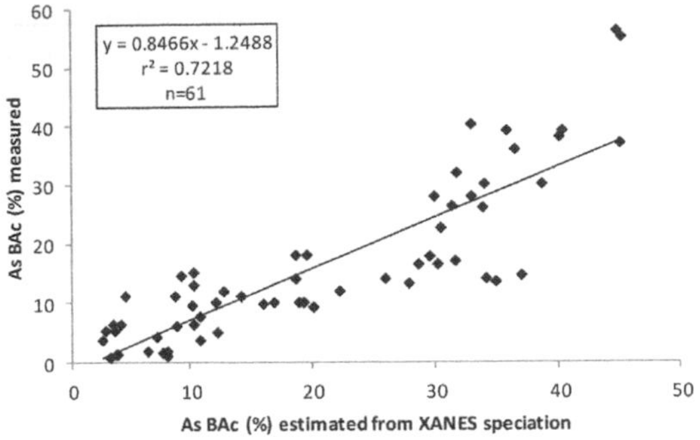

Figure 2. Regression of measured As bioaccessibility (%BAc) and estimated %BAc from XANES spectra fitting of model As compounds (all mine waste samples $n = 61$).

4 CONCLUSIONS

This paper shows the applicability of using bioaccessibility to predict As bioavailability for health risk assessment of mined land and its application to develop site specific guidelines. Arsenic residues from mining may arise from historical activities or current mining projects. Arsenic in mine waste can be managed by using a risk-based assessment approach. The PBET *in-vitro* BAc model (Ruby *et al.*, 1996; Bruce *et al.*, 2007) was used for the assessment of As BAc and validated for its potential to predict bioavailability of mine wastes derived from rat *in-vivo* experiments. The alternative *in-vitro* approach can replace animal *in-vivo* experiments, which simulate human uptake of As.

ACKNOWLEDGEMENTS

This work was performed at the BL20B Beamline Facility Photon Factory (PF), KEK Tsukuba-Japan with support from the Australian Synchrotron Research Programme, which is funded by the Commonwealth of Australia under the major National Research Facilities. Entox is a partnership between Queensland Health and the University of Queensland.

REFERENCES

Bruce, S., Noller, B.N., Matanitobua, V. & Ng, J.C. 2007. In-vitro physiologically-based extraction test (PBET) and bioaccessibility of arsenic and lead from various mine waste materials. *Journal of Toxicology and Environmental Health,* Part A. 70: 1700–1711.

Diacomanolis, V., Ng, J., Haymont, R. & Noller, B.N. 2010. Development of site specific guidelines for future land use at the Woodcutters lead zinc mine. *Proceedings of 19th World Congress of Soil Science, Soil Solutions for a Changing World* 1–6 August 2010, Brisbane, Australia. pp. 3.

George, G.N. & Pickering, I.J. 2000. EXAFSPAK: A Suite of Computer Programs for Analysis of X-ray Absorption Spectra. In *Stanford Synchrotron Radiation Laboratory*: Stanford, CA.

MacLean, L.C.W., Beauchemin, S. & Rasmussen, P.E. 2011. Lead Speciation in House Dust from Canadian Urban Homes Using EXAFS, Micro-XRF, and Micro-XRD. *Environmental Science and Technology* 45: 5491–5497.

MacLean, L.C.W., Marro, L., Jones-Otazo, H., Petrovic, S., McDonald, L.T. & Gardner, H.D. 2011. Canadian House Dust Study: Lead Bioaccessibility and Speciation. *Environmental Science and Technology* 45:4959–4965.

NEPC National Environmental Protection (Assessment of Site Contamination) Measures. 2013. *National Environment Protection Council,* Adelaide.

Ng, J.C, Juhasz, A.L., Smith, E., & Naidu, R. 2013. Assessing the bioavailability and bioaccessibility of metals and metalloids. *Environmental Science and Pollution Research,* in press.

Rasmussen, P.E., Beauchemin, S., Chenier, M., Levesque, C., Ruby, M.V., Davis, A., Schoof, R., Eberle, S. & Sellstone, C.M. 1996. Estimation of lead and arsenic bioavailability using a physiologically based extraction test. *Environmental Science and Technology* 30(2): 422–430.

Ruby, M.V., Davis, A., Link, T.E., Schoof, R., Chaney, R.L., Freeman, G.B. & Bergstrom, P. 1993. Development of an in vitro screening test to evaluate the in vivo bioaccessibility of ingested mine-waste lead. *Environmental Science and Technology* 27:2870–2877.

Ruby, M.V., Davis, A., Schoof, R., Eberle, S. & Sellstone, C.M. (1996). Estimation of lead and arsenic bioavailability using a physiologically based extraction test. *Environmental Science and Technology* 30(2): 422–430.

One Century of the Discovery of Arsenicosis in Latin America (1914–2014) – Litter, Nicolli, Meichtry, Quici, Bundschuh, Bhattacharya & Naidu (Eds)
© 2014 Taylor & Francis Group, London, ISBN 978-1-138-00141-1

Arsenic bioavailability regulated by magnetite in copper tailings: As mobilization into pore water and plant uptake

Y.J. Liu, L. Zhao & L. Huang
Centre for Mined Land Rehabilitation (CMLR), Sustainable Minerals Institute (SMI), The University of Queensland, Brisbane, Queensland, Australia

ABSTRACT: The present study aimed at investigating the hypothesis that the solubility and bioavailability of As could be increased by lowering magnetite contents in copper-tailings under direct revegetation. Both new (5–7% magnetite) and old (20–30% magnetite) tailings were amended with 5% sugarcane residues as a basal treatment, in combination with 0, 1 and 5% pine-biochar, in which native grass (*Iseilema Vaginiflorum*) plants were grown for 4 weeks under well-watered conditions. Total As concentration in the pore water of new tailings with low magnetite content (5–7%) was increased by 6–7 folds and As in shoots increased by at least 50%, compared to those in the old tailings, despite that the total As content in the new tailings was only 37% of the old tailings. The results demonstrated the importance of magnetite in Cu-tailings for mobilization of As in the Cu-tailings and potential impacts of ore processing on environmental pollution risks of tailings.

1 INTRODUCTION

Arsenic (As) is commonly present in the copper tailings due to its association with primary minerals in ores, such as pyrite, chalcopyrite, apatite (Wang & Mulligan, 2008). Amorphous and poorly crystalline Fe mineral such as magnetite and hematite are some of the most important adsorbents which can coprecipitate and adsorb As (Drahota & Filippi, 2009). In a Cu mine located in Northwest Queensland, its tailings prior to magnetite recovery contained 20–30% magnetite (Fe_3O_4), which has been reduced to about 5–7% after further recovering magnetite for economic benefit. Arsenic in the minerals may be mobilized into solution phase by microbial oxidation-reduction, which can be enhanced by microbial access to organic carbon (such as organic matter used to amend tailings). As a result, it was hypothesized that the significant lowering magnetite in the Cu-tailings would increase the mobilization of As into solution phase and the availability for plant uptake, in comparison with the tailings of high magnetite content.

2 METHODS/EXPERIMENTAL

2.1 *Materials preparation*

The new (low magnetite) and old (high magnetite) tailings were bulk-collected from Cu-mine tailings storage dam located in a semi-arid region, which were air-dried and stored in a glasshouse. Before potting, the tailings were dried for 6–7 days and passed through a 2 mm sieve. The organic matter Sugarcane Residue (SR) mulch and pine biochar (BC) were also grinded and sieved through 2 mm sieve. The treatments of tailings included amendments by SR and BC on weight basis: 5% SR; 1% BC + 5% SR; 5% BC + 5% SR, respectively. Each treatment had four replicates.

2.2 *Glasshouse experiment*

A native grass species Red Flinders grass (*Iseilema Vaginiflorum*) was used as test plant, which was grown in a glasshouse with temperature ranged from 26 °C to 32 °C between February and May 2013 in Brisbane, Queensland. The seeds were germinated directly into seeding trays with a sandy potting mix soil. After 20 days, uniform seedlings were transplanted into the tailings treatments watered by capillary suction of water from the base in "twin-pot-system". One pore water sampler was inserted horizontally to each of the pots and pore water samples were collected weekly for total As analysis by means of Inductively Coupled Plasma Mass Spectrometry (ICP-MS, Agilent Technologies).

In 4 weeks after transplanting, the shoots and roots were harvested. Plant samples were washed thoroughly with DI water and were dried at 65 °C for 3 days for digestion. The dried shoot and root biomass were accurately weighted (0.1 g for shoots and 0.05 g for roots) and digested with 4 mL HNO_3 using a Milestone microwave digestion system. The aliquot in each digested tube was diluted to

20 mL with Millipore water for As analysis by an ICP-MS.

3 RESULTS AND DISCUSSION

3.1 *Total As concentration in pore-water*

Total As concentration in pore water of the new tailings was much higher ($p \leq 0.05$) than those of the old tailings (Figure 1). The average As concentrations were 688, 644 and 406 µg L^{-1} in the new tailing amended with SR and 1% or 5% BC, respectively, while they were 34, 62 and 60 µg L^{-1} in the old tailing containing high contents of magnetite. There was no significant difference in different biochar treatments, though a general declining trend was observed with increasing BC rate.

Iron rich soils have a high As adsorbing capacity (Wenzel *et al.*, 2002). The Fe-mineral-magnetite in the tailings seemed to have played a major role in the mobilization of As into the pore water. The presence of other forms of Fe was not investigated, but high concentrations of soluble Fe and Mn were observed in the pore water of new tailings. It is possible that metal reducing bacteria reduced Fe(III) into Fe(II) in the well-watered tailings amended SR (Zachara *et al.*, 2002). This As mobilization had led to an increased bioavailability of As for plant uptake. Although there was no significant difference among the BC treatments due to the high variability in 1% BC treatment, there seemed a declining trend of total As concentration in the pore water of the new tailings.

3.2 *Arsenic uptake by plants*

In general, the plants absorbed more As when being grown in the new tailings than in the old tailings (Figure 2). Most of the As in plant seemed to be in roots, due to the presence of Fe plague on the outer surfaces of roots (Figure 2) (Mei *et al.*, 2009). Overall speaking, the grass absorbed more As from

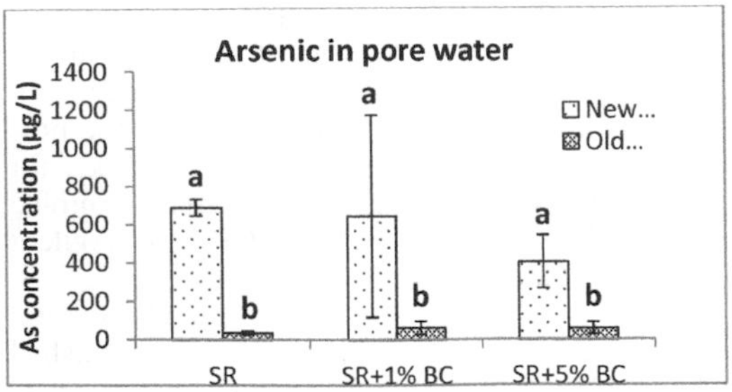

Figure 1. Total As concentration in pore water; "a" and "b" represented statistically significant differences of total As concentration in pore water among different treatments (LSD, $p \leq 0.05$).

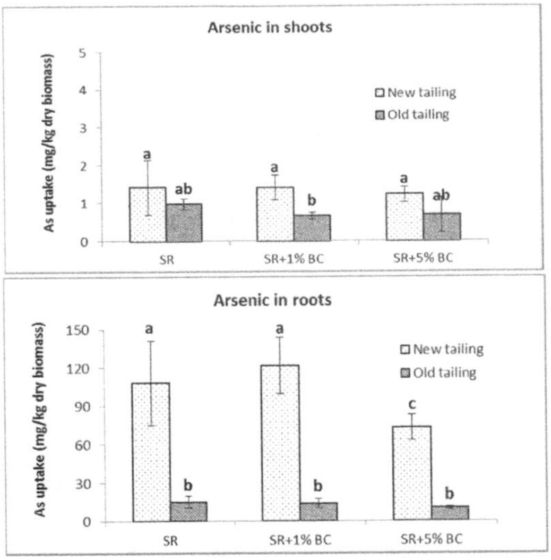

Figure 2. As concentrations in shoots and roots of Red Flinders Grass (*Iseilema Vaginiflorum*). The bars represented standard deviations of corresponding means in the graphs. Different letters represented statistically significant differences of total As concentrations among different treatments (LSD, $p \leq 0.05$).

the new tailings than the old tailings. For different level of BC treatments, As concentration in 5% BC new tailing treatment was significantly lower than those of 0% and 1% treatment ($p \leq 0.05$), but this trend was not obvious in old tailing treatment.

Despite the much reduced total As concentration in the new tailings, much more As was mobilized into the pore water, compared to the old tailings containing high contents of magnetite. Microbial transformation of the As forms would have occurred in the tailings amended with SR, which may have been regulated by the presence of magnetite and other Fe forms in the tailings.

4 CONCLUSION

Arsenic solubility in pore water of Cu-tailings was increased with lowering magnetite content in new Cu-tailings amended with organic matter under the well-watered conditions, leading to increased As uptake by Red Flinders Grass (*Iseilema Vaginiflorum*).

ACKNOWLEDGEMENTS

This research was funded by Mt Isa project. ICP-MS analysis was provided by Ms. Xiaohong Yang, Forensic and Scientific Services Queensland Health.

REFERENCES

Drahota, P. & Filippi, M. 2009. Secondary arsenic minerals in the environment: A review. *Environment International* 35(8): 1243–1255.

Mei, X.Q., Ye, Z.H. & Wong, M.H. 2009. The relationship of root porosity and radial oxygen loss on arsenic tolerance and uptake in rice grains and straw. *Environmental Pollution* 157(8–9): 2550–2557.

Wang, S. & Mulligan, C.N. 2008. Speciation of Inorganic Arsenic in Solid Phases and Surface Complexation Modeling. *Environment International* 34: 867–879.

Wenzel, W.W., Brandstetter, A., Wutte, H., Lombi, E., Prohaska, T., Stingeder, G. & Adriano, D.C. 2002. Arsenic in field-collected soil solutions and extracts of contaminated soils and its implication to soil standards. *Journal of Plant Nutrition and Soil Science* 165(2): 221–228.

Zachara, J.M., Kukkadapu, R.K., Fredrickson, J.K., Gorby, Y.A. & Smith, S.C. 2002. Biomineralization of poorly crystalline Fe(III) oxides by dissimilatory metal reducing bacteria (DMRB). *Geomicrobiology Journal* 19: 179–207.

One Century of the Discovery of Arsenicosis in Latin America (1914–2014) –
Litter, Nicolli, Meichtry, Quici, Bundschuh, Bhattacharya & Naidu (Eds)

Selective sequential extractions for the solid-phase fractionation of arsenic: Do they influence the interpretation of bioaccessibility estimates?

C. Patinha, A.P. Reis & E. Ferreira da Silva
GEOBIOTEC, Universidade de Aveiro, Campus de Santiago, Aveiro, Portugal

A.J. Sousa
CERENA, Technical Superior Institute, Lisbon, Portugal

ABSTRACT: Sequential selective extraction techniques are commonly used to fractionate the solid-phase forms of metals in soils. Many sequential extraction procedures have been developed, and, despite numerous criticisms, they remain very useful. The aim of this study is the comparison of two sequential extraction schemes to study the solid phase distribution of arsenic in soil samples from an abandoned mining area. The results suggest that the leaching solutions used in Scheme II are more selective for the desired target phases.

1 INTRODUCTION

Arsenic (As) is a harmful metalloid that is widely distributed in the earth's crust. Arsenic may be derived from naturally occurring minerals or be introduced anthropogenically by industrial activities as mining and processing of metalliferous ores.

Arsenic interactions with soil constituents are very diverse. The relative predominance of the forms of As in soils depends on the type and abundance of adsorbing components of the soil, the pH, and the redox potential (Buschmann *et al.*, 2006).

In the solid phase, As occur as a complex mixture of compounds that includes discrete mineral phases, coprecipitated and sorbed species associated with soil minerals or organic matter, and dissolved species that may be complexed by a variety of organic and inorganic ligands. Sequential Extraction Procedures (SEP) can be used to determine the above mentioned binding fractions of elements in solid environmental samples. In this process, the sample is submitted to specific reagents of successively stronger dissolving power that, under controlled conditions, will remove metals from the particular phases of concern (Bird *et al.*, 2005). Depending on the fractions of interest, a broad range of chemical extractants can be used and, consequently, numerous SE schemes are available in the literature.

The aim of this study is the comparison of two sequential extraction schemes to study the solid phase distribution of As. The efficiency and suitability of these methods and their corresponding extraction steps for As bound soil phases were carried out using soil samples from an abandoned mining area.

2 MATERIAL AND METHODS

2.1 *Soil samples*

The study area is located in the vicinity of Marrancos, near Braga city, in northwest Portugal. The sulphide—gold mineralization occurs in an N-45°E quartz breccia at a hornfels strip, formed at the contact between Silurian metasediments and Hercynian granite. Following quartz, arsenopyrite and pyrite are the most important minerals of the deposit.

This study was carried out in 10 select topsoil samples from the 144 available. Five of those samples fit in the more disturbed soils of the southern part of the area, and the other five are undisturbed soils from the northern part of the study area.

The size fraction analyzed is the <77 µm fraction of the 144 soil samples. Samples were analyzed by ICP–MS for low to ultra-low determination of 53 elements at ACME Analytical Laboratory (ISO 9002 Accredited Co.), Vancouver, BC—Canada (Reis *et al.*, 2007).

2.2 *Sequential extractions procedures*

The sequential extraction schemes used are summarized in Tables 1 and 2. The leaches from scheme I were analyzed by ICP/MSat Actlabs Analytical

Laboratory (ISO 9002 Accredited Co.), Ontario, Canada. The recoveries were calculated for each extraction with values in the range of 80 to 120% of recovery (Reis *et al.*, 2012). The leaches from scheme II has been analyzed by AAS-HG.

3 RESULTS AND DISCUSSION

3.1 *Arsenic concentration*

Pyrite and arsenopyrite are the major sulphide minerals of the Marrancos deposit, and As occurs in both sulphides. In the topsoil, As concentrations are high, with values ranging from 30 to 19.950 mg kg^{-1} and an average value of 883 mg kg^{-1}. The superficial dispersion pattern of As is strongly controlled by anthropogenic activities related to the mining and exploration of gold (Reis *et al.*, 2012).

Table 1. Extraction conditions for Scheme I (Actlabs method).

Step	Target phase	Extraction agent and conditions
F1	For exchangeable cations adsorbed by clay and elements co-precipitated with carbonates	Sodium acetate Leach pH 5
F2	Mn oxides and amorphous Fe oxides	hydroxylamine leach cold
F3	For amorphous and crystalline Fe oxides and crystalline Mn oxides	hydroxylamine leach hot
F4	Will leach sulphide species and clay minerals	aqua regia

Table 2. Extraction conditions for Scheme II (Patinha *et al.*, 2004).

Step	Target phase	Extraction agent and conditions
F1	Dissolved exchangeable and ions specifically adsorbed by clay	Ammonium acetate (pH 7)
F2	Acid soluble	Ammonium acetate (pH 4.5)
F3	For Mn oxyhidroxides	Hydroxylamine hydrochloride (pH 2)
F4	For amorphous Fe oxides	Oxalic acid/ammonium oxalate (dark)
F5	For to organic matter; sulphide (partially)	H_2O_2 35%
F6	Crystalline Fe oxides	Oxalic acid/ammonium oxalate (under UV radiation)
F7	Residue and sulphide	Aqua Regia

3.2 *Selective extractions*

To attain chemical selectiveness the choice of the type of extractant should be strictly correlated with: (a) the nature of metal; (b) the chemical form of metal; (c) the matrix from which the compounds are to be extracted; and (d) the analytical techniques available in the laboratory for the final determination. The $NH_2OH.HCl$ extraction used in scheme I was found to dissolve completely a pure phase of amorphous iron hydroxide, as expected, so all the As adsorbed on it was released. However, in a mixture of amorphous and crystalline iron oxides, this reagent appeared to be ineffective in releasing As from the solid phases (Wenzel *et al.*, 2001). However, scheme II uses oxalic acid/ammonium oxalate at dark and under UV radiation, which allows discriminating the amorphous from the crystalline forms. Therefore, a sequential protocol was designed to overcome this limitation (Patinha *et al.*, 2004).

The results show that the percentage of As leached in most labile phases (F1 (scheme I) and F1 + F2 (scheme II) is insignificant. The most important solid phase for As are the sulfides. In sample 4, almost 10% of As associated with amorphous Fe oxides corresponds to 900 mg kg^{-1} of As in the topsoil, which is well above the national guideline values for soil quality established in EU countries. This sample was collected closer to a football field. Here, the probability of exposure is significantly enhanced as soil dust ingestion is one of the common pathways of exposure. Most of As adsorbed on the surface and some of As inside the matrix of Fe oxide are likely to be dissolved in the GI system (Rodrigues *et al.*, 2003).

4 CONCLUSIONS

The results suggest that the leaching solutions used in Scheme II are more selective for the desired target phases. In addition, adequate discrimination among amorphous and crystalline iron phases seems to be achieved. The application of this scheme may provide interesting environmental results because enables differentiation of the most mobilizable forms of arsenic, which may become labile forms, and highly insoluble minerals are relegated to the residual phase.

ACKNOWLEDGEMENTS

The authors wish to acknowledge the Foundation for the Science and the Technology (FCT, Portugal) to financially support projects PDCT/ECM/58216/2004 and PEst-C/CTE/UI4035/2011.

REFERENCES

Bird, G., Brewera, P.A., Macklina, M.G., Serbanb, M., Balteanub, D. & Drigab, B. 2005. Heavy metal contamination in the Aries river catchment, western Romania: Implications for development of the Rosia Montana gold deposit. *J Geochem Explor* 86: 26–48.

Buschmann, J., Kappeler, A., Lindauer, U., Kistler, D., Berg, M. & Sigg, L. 2006. Arsenite and Arsenate Binding to Dissolved Humic Acids: Influence of pH, Type of Humic Acid, and Aluminum. *Environ. Sci. Technol.* 40: 6015–6020.

Patinha, C., Ferreira da Silva, E. & Cardoso Fonseca, E. 2004. Mobilisation of arsenic at Talhadas old mining contaminated área (Central Portugal). *Journal of Geochemical Exploration* 84: 167–180.

Reis, A.P., Ferreira da Silva, E., Sousa, A.J., Patinha C., Nogueira, P. & Martins, E. 2007. Assessing the analytical quality of two analytical methods used in the same soil samples. *In* E. Preto (ed.), *Abstract book of the XV geochemistry week—VI geochemistry Iberian congress 235–238. Vila Real, Portugal: UTAD.*

Reis, A.P., Patinha, C., Ferreira da Silva, E. & Sousa, A.J. 2012. Metal fractionation of cadmium, lead and arsenic of geogenic origin in topsoils from the Marrancos gold mineralisation, northern Portugal. *Environ. Geochem. Health* 34: 229–241.

Rodriguez, R.R.,. Basta, N.T, Casteel, S.W., Armstrong F.P. & Ward, D.C. 2003. Chemical Extraction Methods to Assess Bioavailable Arsenic in Soil and Solid Media. *J. Environ. Qual.* 32: 876–884.

Wenzel, W.W., Kirchbaumer, N., Prohaska, T., Stingeder, G., Lombi, E. & Adriano, D.C. 2001. Arsenic fractionation in soils using an improved sequential extraction procedure. *Anal. Chim. Acta* 436: 1–15.

One Century of the Discovery of Arsenicosis in Latin America (1914–2014) – Litter, Nicolli, Meichtry, Quici, Bundschuh, Bhattacharya & Naidu (Eds)
© 2014 Taylor & Francis Group, London, ISBN 978-1-138-00141-1

Lethal and sublethal effects of As(V) on early larval development of *Rhinella arenarum* and utility of AMPHITOX as an ecotoxicological test for As(V) solutions after treatment with zerovalent iron nanoparticles

C.S. Pérez Coll
Instituto de Investigación e Ingeniería Ambiental, UNSAM, San Martín, Argentina

D.C. Pabón Reyes, J.M. Meichtry & M.I. Litter
Gerencia Química, CAC-CNEA, San Martín, Argentina

ABSTRACT: The present study shows the lethal and sublethal effects of As(V) on early larval development of *Rhinella arenarum*, the common South American toad, a widely distributed species in Argentina, and the utility of the AMPHITOX test for monitoring changes in the toxicity of As(V) solutions after treatment with zerovalent iron nanoparticles.

1 INTRODUCTION

Although the adverse impacts of the consumption of As-contaminated water on human health are well documented (Mandal & Suzuki, 2002), the possible impacts on the ecosystems and wildlife of As-contaminated water are rather unknown. Moreover, the decline and extinction of amphibians is a major concern for biodiversity protection worldwide, already alerted since the 1960's (Simms, 1969).

Frogs and toads are more susceptible than most vertebrates to a wide diversity of pollutants at embryonic and larval stages, which make them good indicators of environmental quality. They are widely used in hazard assessment and ecotoxicological studies through standardized bioassays such as the AMPHITOX test (Herkovits & Pérez-Coll, 2003).

On the other hand, one of the innovative treatments for As removal from water is the use of zerovalent iron nanoparticles (nZVI) (Litter *et al.*, 2010). The aims of this work were to evaluate the As(V) toxicity on *Rhinella arenarum* larvae together with the utility of AMPHITOX for monitoring changes in the toxicity of As(V) solutions submitted to treatment with nZVI.

2 MATERIALS AND METHODS

2.1 *Reagents*

The zerovalent iron nanoparticles (nZVI) were provided by NANO IRON (NANOFER 25®) as aqueous suspension. Sodium arsenate dibasic 7-hydrate ($Na_2HAsO_4.7H_2O$, Baker) and all other chemicals were of the highest purity.

2.2 *As removal with nZVI*

To 200 mL of an aqueous solution of 10 mg L^{-1} As(V) (pH 7) in a thermostatted (25 °C) glass cylindrical cell, drops of nZVI suspension were added to reach 745 mg L^{-1} nZVI (1:100 As:Fe molar ratio), and the system was stirred with a paddle stirrer for 5, 15, 30, 45 and 60 min. nZVI were removed by centrifugation, and the supernatant was used in toxicity bioassays (nZVIR). As(V) was measured at the end of the time periods by spectrophotometry (Lenoble *et al.*, 2003).

2.3 *Obtaining* Rhinella arenarum *larvae*

R. arenarum adults weighing approximately 200–250 g were acquired in Lobos (Buenos Aires province, Argentina: 35° 11′ S; 59° 05′ W). Toad care, breeding, embryo acquisition and analysis were conducted according AMPHITOX protocols (Herkovits & Pérez-Coll, 2003). Oocytes were fertilized *in vitro* using fresh sperm suspended in AMPHITOX solution (36.0 mg L^{-1} NaCl, 0.5 mg L^{-1} KCl, 1.0 mg L^{-1} $CaCl_2$ and 2.0 mg L^{-1} $NaHCO_3$). Embryos were kept in AMPHITOX solution and maintained at 20 ± 2 °C, until organisms reached the complete operculum stage, S.25.

2.4 *Toxicity bioassays*

Groups of ten larvae at early S.25 were randomly placed by triplicate in 10 cm-diameter glass Petri dishes containing 40 mL of solution as follows: 1) 10 mg L^{-1} As(V) water solution treated with nZVI for the indicated reaction time (nZVIR 5 to 60 min), 2) 10 mg L^{-1} As(V) solution in AMPHITOX solution, 3) absolute control: AMPHITOX solution, 4) supernatant of a pure aqueous 745 mg L^{-1} nZVI suspension centrifuged after 60 min (nZVIR) and 5) pure aqueous 745 mg L^{-1} nZVI suspension without centrifugation (nZVI). Larvae in water showed no difference with absolute control. Early larvae were continuously exposed for acute (96 h), short-term chronic (168 h) and chronic (336 h) periods. A 10% of difference with the absolute control was set as the significance level. Bioassays were semistatic, and test solutions were entirely replaced every 48 h. Lethal and sublethal effects were evaluated every 24 h. Larvae were fed with balanced fish food.

3 RESULTS AND DISCUSSION

After contact of As solutions with nZVI, the initial As(V) concentration fell from 10 to 5.6, 3.7, 3.6, 3.1 and 2.4 mg L^{-1}, after 5, 15, 30, 45 and 60 min of contact, respectively (nZVIR 5 to 60 min).

Figure 1 shows the survival curves of *Rhinella arenarum* larvae exposed to different experimental conditions. The exposure of *R. arenarum* to 10 mg L^{-1} As(V) resulted in time-dependent lethality. Thus, although there were no lethal effects during acute exposure (up to 72 h), survival was reduced by 20% near short-term chronic exposure and by 40% arriving to the chronic period. The exposure of the larvae to As solutions treated with nZVI for 5, 15 and 30 min resulted in only 13% lethality regardless time of treatment, while the survival to As solutions treated for 45 and 60 min was similar to the absolute control in AMPHITOX solution. The results of the toxicity bioassay reflect the diminution in As(V) concentration in the treated solutions and show the high usefulness of nZVI to reduce the toxicity of solutions containing As(V). The nZVIR solution also showed no significant difference with the absolute control, indicating the safety of nZVI to remove As. Contrarily, the exposure to nanoparticles not removed from the suspension (nZVI in Figure 1) resulted highly toxic for the larvae: 80% lethality at 24 h exposure, and 100% lethality at 336 h were registered. By comparing results of nZVIR and nZVI, the need to remove the nanoparticles after the treatment and the safety of the treated water were shown.

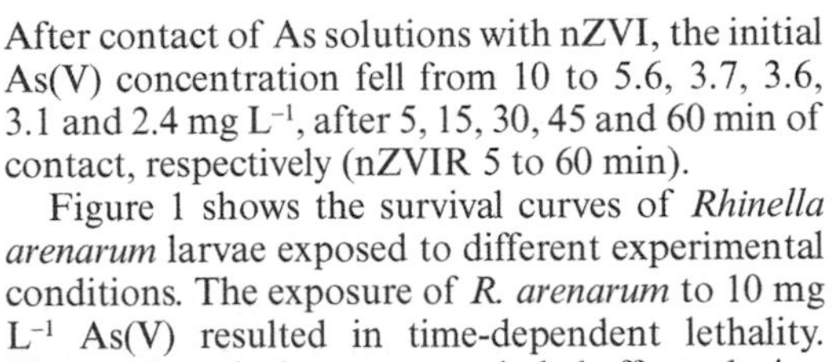

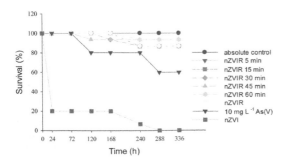

Figure 1. Survival curves of *Rhinella arenarum* larvae exposed to different experimental conditions.

With regards to sublethal effects, larvae exposed to 10 mg L^{-1} As(V) showed edema, alterations in the pattern of pigment, spasmodic contractions, weakness, narcosis and non-feeding behavior. Although this last alteration did not manifest in significant differences neither in the length nor in the dry weight of larvae at 336 h, it would be convenient to extend the exposure beyond 15 days to evaluate eventual differences. On the other hand, sublethal effects were almost absent after nZVI treatment.

From a toxicological point of view, As exposure is associated with diverse adverse effects involving sublethal effects and eventually death of living organisms. These sublethal effects, such as skin lesions including tumors, hyperkeratosis and pigmentation changes (Shannon & Strayer, 1989) were observed here. Also neuropathies expressed as lethargy, seizures, numbness, loss of equilibrium, and general weakness were observed. Edema, an osmoregulation failure, could be related to the effect of As on the renal function. According to Mandal & Suzuki (2002), all these effects were obtained in other experimental models by As exposure.

4 CONCLUSIONS

In this study, lethal and sublethal effects of As on early larval development of *R. arenarum*, and the utility of the AMPHITOX test for monitoring changes in the toxicity of the As(V) solutions by nZVI treatment were provided. It should be emphasized that studies of this type with other vertebrate groups are scarce and that there are only few references about the toxicity of As for amphibians (Moriarty *et al.*, 2013). In addition, this study alerts on the conservation of *Rhinella arenarum*, the common South American toad, widely distributed in the Chaco-Pampean plain of Argentina, one of the largest regions with high As concentrations in groundwater.

ACKNOWLEDGEMENTS

This work was funded by ANPCyT (Argentina) PICT 463. All authors are members of CONICET (Argentina).

REFERENCES

Herkovits, J. & Pérez-Coll, C.S. 2003. AMPHITOX: A customized set of toxicity tests employing amphibian embryos. Symposium on multiple stressor effects in relation to declining amphibian populations, In G.L. Linder, S. Krest, D. Sparling & E.E. Little (eds), *Multiple Stressor Effects in Relation to Declining Amphibian Populations*, ASTM International STP 1443,46–60, USA.

Lenoble, V., Deluchat, V., Serpaud, B. & Bollinger, J.C. 2003. Arsenite oxidation and arsenate determination by the molybdene blue method. *Talanta* 61: 267–276.

Litter, M.I., Morgada, M.E. & Bundschuh, J. 2010. Possible treatments for arsenic removal in Latin American waters for human consumption. *Environ. Pollution* 158: 1105–1118.

Mandal, B.K. & Suzuki, K.T. 2002. Arsenic round the world: a review. *Talanta* 58: 201–235.

Moriarty, M.M., Koch, I. & Reimer, K.J. 2013. Arsenic species and uptake in amphibians (*Rana clamitans* and *Bufo americanus*). *Environ. Sci. Proc. Impacts* 15: 1520–1528.

Shannon, R.L. & Strayer, D.S. 1989. *Hum. Exp. Tox.* 8: 99–104.

Simms, C. 1969. Indications of the decline of breeding amphibians at an isolated pond in marginal land, 1954–1967. *British J. Herpetology* 4: 93–96.

One Century of the Discovery of Arsenicosis in Latin America (1914–2014) – Litter, Nicolli, Meichtry, Quici, Bundschuh, Bhattacharya & Naidu (Eds)
© 2014 Taylor & Francis Group, London, ISBN 978-1-138-00141-1

Potential of *Lemna gibba* for removal of arsenic in nutrient solution

L.V.R.A.B. Pandolpho
Department of Soil, Federal University of Vicosa, Vicosa, Minas Gerais, Brazil

V.L.M. Guarda
Department of Pharmacy, Federal University of Ouro Preto, Ouro Preto, Minas Gerais, Brazil

ABSTRACT: This study aimed at investigating the potential for removal of different concentrations of arsenic (As) in nutrient solution by a macrophyte species, Lemna gibba by peroxidase activity and growth rate. The growth patterns were carried out using the method of Hunt. The determination of As was made by colorimetry comparing with different dosages of sodium arsenate ($Na_2AsHO_4.7H_2O$): 0; 1.0; 2.5; and 5.0 mg L^{-1}. The analysis of peroxidase activity was calculated using the extinction coefficient after measuring the absorbance of purpurogallin production in 1 minute.

1 INTRODUCTION

Many heavy metals and metalloids, such as arsenic (As) have been commonly found as environmental pollutants, being not essential and highly toxic to plants. They are often improperly disposed of in the environment (Lasat, 2002).

Lemna gibba has important characteristics required for bioremediation, such as high growth rates, wide distribution and property of accumulating heavy metals in aquatic environments (Demirrezen *et al.*, 2007). This, this species represent a cost effective alternative for As removal from aquatic environments (Lasat, 2002).

This study aimed at evaluating the potential of *L. gibba* for remediation of aquatic environments artificially contaminated by arsenic.

2 METHODS/EXPERIMENTAL

2.1 *Biological material and plant acclimation*

L. gibba was collected in the Botanical Garden tanks belonging to the Plant Biology Department (DBV) - Federal University of Viçosa (UFV). The plants were disinfected using sodium hypochlorite solution (1%) and rinsed in deionized water. Plants were accommodated in containers with 3 L of Hoagland solution (1/4 ionic strength, pH 6.5) and grown in a greenhouse.

2.2 *Plant exposure to arsenic*

Macrophytes were transferred to pots with nutrient solution containing arsenic in concentrations of 0.0, 1.0, 2.5 and 5.0 mg L^{-1}, in triplicates, for seven days, with controlled temperature and light (25 ± 2 °C, 230 μE s^{-1} m^{-2}), 16 hours photoperiod. pH was measured and adjusted daily to 6.5 with hydrochloric acid (HCl) or sodium hydroxide (NaOH). After 10 days, the plants were washed with 0.1 M HCl solution and dried at 80 °C until constant dry weight.

2.3 *Accumulation of arsenic*

The dried plants were subjected to wet digestion with 3 mL of a 2:1 mixture of nitric and perchloric acid at 220 °C. Mineralized samples were analyzed by atomic absorption spectrometry, on Water Quality Laboratory (LaQua), School of Pharmacy, Federal University of Ouro Preto (UFOP).

2.4 *Effect of arsenic on the relative growth rate*

The Relative Growth Rates (RGR) of plants were calculated using the equation proposed by Hunt (1978) (Figure 1).

2.5 *Extraction and determination of peroxidase activity (POXs, EC 1.11.1.7)*

Plant samples were soaked in liquid nitrogen and received 2 mL of extraction buffer consisting of 0.1 M potassium phosphate, 0.1 mM EDTA, 1 mM phenylmethylsulfonyl fluoride (PMSF) and 1% polyvinyl polypyrrolidone (PVPP) (w/v), pH 6.8. The homogenate was centrifuged at 12,000 × g for 15 min at 4 °C and the supernatant was used as the crude enzyme extract for assay The peroxidase

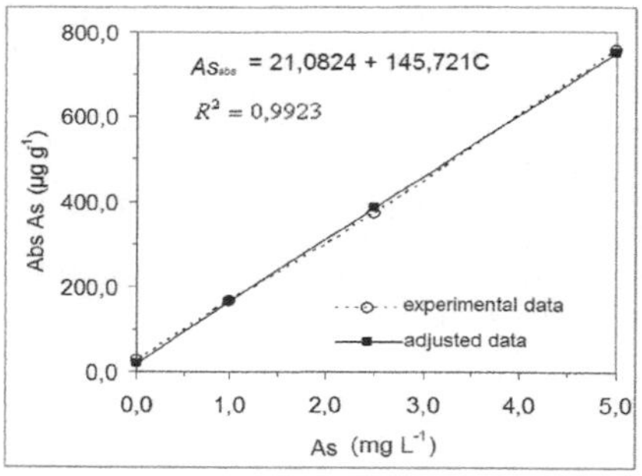

Figure 1. Arsenic absorption as a function of As concentration.

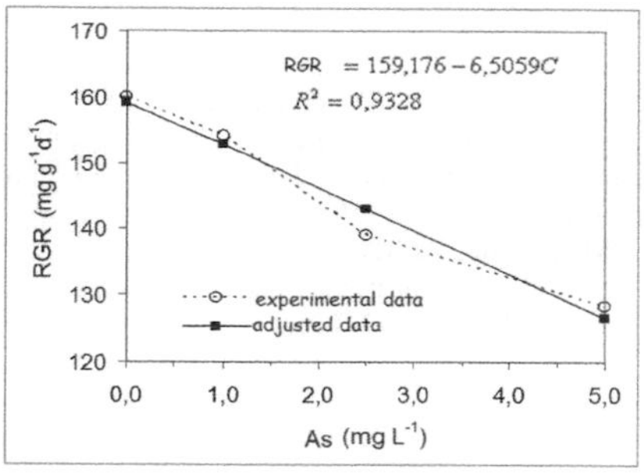

Figure 2. Relative Growth Rate (RGR) of plants.

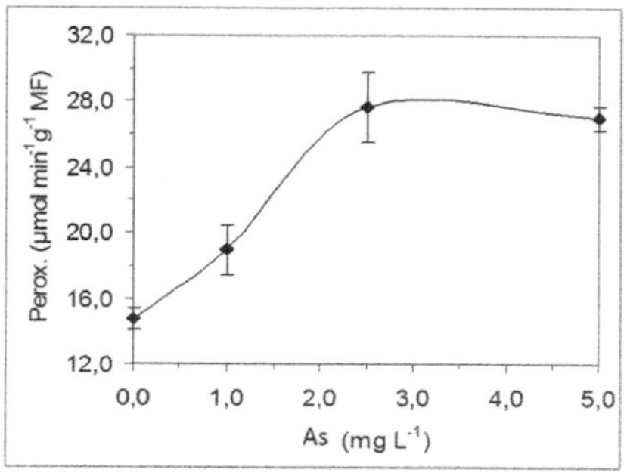

Figure 3. Peroxidase activity.

activity was determined by adding 0.1 mL of crude enzyme extract to 4.9 mL of reaction medium containing 25 mM potassium phosphate buffer, 20 mM pyrogallol and 20 mM hydrogen peroxide, pH 6.8. Purpurogallin production was determined by absorbance (420 nm) after the first minute of reaction at 25 °C. The enzyme activity was calculated using the molar extinction coefficient of 36 M^{-1} cm^{-1} and expressed in mmoles of H_2O_2 min^{-1} g^{-1} Wet Weight (WW).

3 RESULTS AND DISCUSSION

3.1 *The accumulation*

The higher the As concentration in solution resulted in the higher metal concentration found in *L. gibba* tissues (Figure 3). For treatments with 1.0, 2.5 and 5.0 mg L^{-1}, these plants showed an absorbance of As 166.9, 372.4 and 756.2 µg g^{-1} Dry Weights (DW), respectively.

The plants were able to remove arsenic from the medium, but the removal efficiency decreased with the increasing As concentration. The removal efficiencies obtained were 12.4, 9.9 and 9.7%, for the respective treatments. Soares *et al.* (2005) studied the potential of *L. gibba* to remove the pollutant from aqueous solution, and they verified that this species removed 4.41% of the arsenic present in the medium (0.5 mg L^{-1}). It was also observed that the concentration of As in solution increased, and it reduced the Relative Growth Rates of plants (RGR). Despite the potential of As absorption presented by *L. gibba* increased, a decrease on the size of the leaves of plants exposed to solutions at all As concentrations was observed (Figures 1 and 2).

3.2 *Peroxidase activity*

This system is important in the prevention of oxidative stress by As in plants, since the activity of one or more of these enzymes generally increases with increasing stress by As (Li *et al.*, 2006). As shown in Figure 3, the pyrogallol peroxidase activity increased with the increasing of As concentration, stabilizing at 2.5 mg L^{-1}. The trend of Figure 3 clearly follows a Michaelis-Menten kinetics. The average peroxidase activity subjected to 2.5 and 5.0 mg L^{-1} As was statistically similar, according with a Tukey test at 5% probability ($p < 0.05$).

4 CONCLUSION

Increasing As concentration in the solution resulted in an increased accumulation of this metalloid by *L. gibba*. However, it was not possible to verify the potential of As absorption at concentrations above 5.0 mg L^{-1}. Further studies with higher quantities of As are necessary to determine the maximum tolerable amount accumulated by this plants.

As accumulation by these plants decreases with the As concentration in solution. However, it was proved that they possess a high potential to be used in aquatic environments contaminated with medium to low As concentrations.

ACKNOWLEDGEMENT

To Brazilian financial support provided by CAPES, FAPEMIG and CNPq.

REFERENCES

Demirezen, D., Aksoy, A. & Uruc, K. 2007. Effect of population density on growth, biomass and nickel accumulation capacity of *Lemna gibba* (Lemnaceae). *Chemosphere* 66: 553–557.

Lasat, M.M. 2002. Phytoextration of toxic metals: A review of biological mechanisms. *Journal of Environmental Quality* 31: 109–120.

Li, M., Hu, C.W., Zhu, Q., Chen, L., Kong, Z.M. & Liu, Z.L. 2006. Copper and zinc induction of lipid peroxidation and effects on antioxidant enzyme activities in the microalga *Pavlova viridis* (Prymnesiophyceae), *Chemosphere* 62: 565–572.

Soares, C.R.F.S., Siqueira, J.O., Carvalho, J.G. & Moreira, F. M.S. 2005. Fitotoxidez de arsênio para *Eucalyptus maculata* e *E. urophylla* em solução nutritiva. *R. Árvore*. 29: 175–183.

One Century of the Discovery of Arsenicosis in Latin America (1914–2014) – Litter, Nicolli, Meichtry, Quici, Bundschuh, Bhattacharya & Naidu (Eds)
 ISBN 978-1-138-00141-1

Arsenic toxicity in the freshwater microalga *Euglena gracilis*

P. Nannavecchia
Facultad de Ciencias Exactas y Naturales, Universidad de Buenos Aires, Argentina

V. Conforti & Á. Juárez
Facultad de Ciencias Exactas y Naturales, Universidad de Buenos Aires, Argentina
IBBEA-CONICET, Buenos Aires, Argentina

M. del C. Ríos de Molina
Facultad de Ciencias Exactas y Naturales, Universidad de Buenos Aires, Argentina
IQUIBICEN-CONICET, Buenos Aires, Argentina

ABSTRACT: Metals and metalloids may be toxic to algae and can generate reactive oxygen species, leading to oxidative stress, which impairs antioxidant defenses and produces oxidation of essential molecules. As a consequence of the development of water-dependent activities such as mining, industry, and agriculture, arsenic levels in aquatic ecosystems have been increasing. The microalga *Euglena gracilis* is one of the few planktonic microorganisms that grow in highly contaminated environments with arsenic, such as in mining tailings. The toxicity of sodium arsenite in two strains of the freshwater microalga *E. gracilis* was studied by comparing culture growth and by measuring some indicators of the antioxidant response and oxidative damage. We show evidence for the two *E. gracilis* strains that the toxicity response to sodium arsenite involves oxidative stress processes.

1 INTRODUCTION

Over the years, the development of mining has led to the deterioration of water quality and biodiversity in many aquatic ecosystems. Acidic effluents generated in mines contain metals such as arsenic. Through the discharge of mining waste, this metalloid originates from the erosion of arsenical pyrite and accumulates in high concentrations, up to 350 mg L^{-1} in natural waters (Morin & Calas, 2006). The acidity and high concentration of arsenic in affected environments generate conditions that are toxic to the biota. However, some microorganisms such as bacteria, archaea, fungi, and protists are able to live in extreme conditions (Zettler *et al.*, 2002). In particular, microalgae of the genus *Euglena* are among the few eukaryotes present in these systems (Miot *et al.*, 2009). Detoxifications mechanisms in *Euglena gracilis* species under extreme acid mine drainage conditions were studied by Miot *et al.* (2008), who proposed that in these cells the arsenic binds in the form of arsenic-trisglutathione complexes ($As\text{-}(GS)_3$), or arsenic-phytochelatin complexes (As-PC) via glutation transferase. These complexes can be exported out the cell.

Heavy metals and metalloids may induce the generation of Reactive Oxygen Species (ROS). This increase triggers oxidative damage and activates enzymatic antioxidant responses (such as superoxide dismutase, catalase, etc.) and non enzymatic antioxidant responses (as reduced glutathione). If these antioxidant defenses are overcome, oxidative stress occurs, which causes damage to essential macromolecules in maintaining structure and cell physiology (lipids, proteins and nucleic acids). In this context, the toxicity of sodium arsenite in the freshwater microalga *Euglena gracilis* was studied by comparing culture growth and by measuring some indicators of the antioxidant response and oxidative damage.

2 METHODS/EXPERIMENTAL

2.1 *Organism and culture*

Euglena gracilis MAT strain was isolated from Matanza River, a highly polluted water body from Buenos Aires, Argentina, and the UTEX 753 strain was acquired from Culture Collection of Algae of the Texas University, USA. Axenic stock cultures of both strains were grown in a mineral medium (Cramer & Myers) and in an organic medium (EGM), at 24 ± 1°C, 12L: 12D photoperiod with cool-white fluorescent at 150 µE m^{-2} s^{-1}.

2.2 *Toxicity bioassays*

Experiments were performed in 250 mL of culture medium, inoculated with 5×10^4 cell mL^{-1} of *E. gracilis*. $NaAsO_2$ was added from a stock solution to a total concentration of 100, 200, 400 and 600 mg L^{-1} for the mineral medium, and 50, 100, 200 and 500 mg L^{-1} for the organic medium.

Control (without arsenite) and treated cultures were maintained at 26 ± 1°C with cool-white fluorescent continuous light at 150 μE m^{-2} s^{-1} during 96 hours. Cellular density was determined using a Neubauer chamber. Results are expressed as cells/ml. Inhibitory Concentration 50 (IC50%) was estimated using statistical program Linear Interpolation (ICp).

Indicators of antioxidant response and oxidative damage were determined: reduced glutathione levels (GSH), activity of the antioxidant enzyme superoxide dismutase (SOD) and detoxifying enzyme (GST), and the lipid peroxidation levels (TBARS). Mean and standard deviations were obtained from the triplicates of each concentration. The one-way analysis of variance was used to compare the results obtained in each bioassay, then Dunnett's Multiple Comparison test was used, with significance level $\alpha = 0.05$. For this analysis, the GraphPad Prism 5 was used.

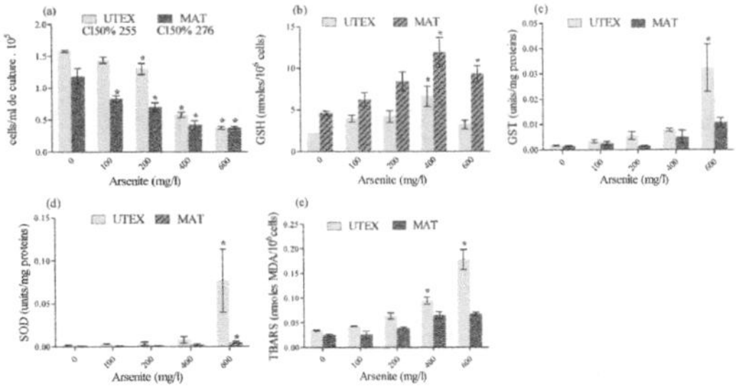

Figure 1. (a) Growth, (b) GSH levels, (c) GST activity, (d) SOD activity and (e) TBARS levels, in the MAT and UTEX strains of *E. gracilis* in mineral medium. Data are means of three replicas with standard deviation. Asterisk (*) denotes significant differences regarding the control.

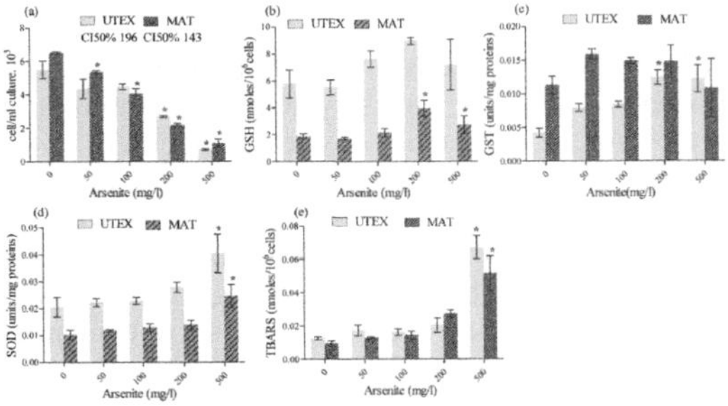

Figure 2. (a) Growth, (b) GSH levels, (c) GST activity, (d) SOD activity and (e) TBARS levels, in the MAT and UTEX strains of *E. gracilis* in organic medium. Data are means of three replicas with standard deviation. Asterisk (*) denotes significant differences regarding the control.

3 RESULTS AND DISCUSSION

Increased $NaAsO_2$ concentrations caused in both strains and both culture media growth inhibition, increased lipid damage, and an increase in detoxifying and antioxidant defenses.

In mineral medium, IC50% values for *E. gracilis* strains (Figure 1a) was 276 mg L^{-1} for the MAT strain and 255 mg L^{-1} for the UTEX strain. Toxic tolerance of the UTEX strain could be related to its detoxifying and antioxidant response, resulting in GST and SOD activity levels (Figure 1c,e), which were higher than in the MAT strain. In both strains, these parameters increased, but in the UTEX strain these changes occurred in $NaAsO_2$ concentrations that were lower than in the MAT strain. However, damage to lipids was greater in the UTEX strain (Figure 1d), which could be related to the lower levels of GSH detected (Figure 1b). These levels could be explained by their participation in conjugation reactions catalyzed by GST, enzyme that showed greater activity in the UTEX strain (Figure 1c).

In organic medium, IC50% values for *E. gracilis* strains (Figure 1a) was 143 mg L^{-1} for the MAT strain and 196 mg L^{-1} for the UTEX strain. In the UTEX strain, there was an increased GST activity from 200 mg L^{-1} (Figure 2c), without alterations in GSH levels (Figure 2b). MAT showed an increase of GSH levels in presence of 200 and 500 mg L^{-1} respectively (Figure 2b), without alterations in GST activity (Figure 2c). In both strains, increase SOD activity and TBARS levels was observed at 500 mg L^{-1} (Figure 2c,d).

4 CONCLUSIONS

The $NaAsO_2$ toxicity of *E. gracilis* is dose dependent. Despite the increase in enzymatic and non enzymatic antioxidant defenses measured, these responses were not sufficient to prevent lipid damage.

We show evidence that the toxicity of this metalloid involves oxidative stress processes. Even though the arsenic sensitivity of both strains in both culture media was not significantly different, the stress response of each strain differed. The results show that the antioxidant response prevails in the MAT strain, while in the UTEX strain the detoxifying response prevails.

ACKNOWLEDGEMENTS

This investigation was supported by grants to VC, UBACYT 01/W290 and CONICET-PIP 283. The authors are grateful to Dr. Cristian Solari for the English text revision.

REFERENCES

Morin, G. & Calas, G. 2006. Arsenic in soils, mines tailings and former industrial sites. *Elements* 2: 97–101.

Miot, J., Morin, G., Skouri-Panet, F., Férard, C., Aubry, E., Briand, J., Wang, Y., Ona-Nguema, G., Guyot, F. & Brown, G. 2008. XAS Study of arsenic coordination in *Euglena gracilis* exposed to arsenite. *Environ. Sci. Technol.* 42: 5342–5347.

Miot, J., Morin, G., Skouri-Panet, F., Férard, C., Poitevin, A., Aubry, E., Ona-Nguema, G., Juillot, F., Guyot, F. & Brown, G. JR. 2009. Speciation of Arsenic in *Euglena gracilis* Cells Exposed to As(V). *Environ. Sci. Technol.* 43: 3315–3321.

Zettler, L.A, Gómez, F., Zettler, E., Keenan, B.G, Amils, R. & Sogin, M.L. 2002. Eukaryotic diversity in Spain's River of Fire. *Nat-Lond.* 417(6885): 137.

One Century of the Discovery of Arsenicosis in Latin America (1914–2014) –
Litter, Nicolli, Meichtry, Quici, Bundschuh, Bhattacharya & Naidu (Eds)
 ISBN 978-1-138-00141-1

Arsenic contamination, size, and complexity: A volvocine green algae case study

J.P. Basualdo, V.J. Galzenati, C.A. Solari & V. Conforti
Laboratorio de Biodiversidad, Ultraestructura y Ecofisiología de Microalgas, Instituto de Biodiversidad y Biología Experimental Aplicada, CONICET, Universidad de Buenos Aires, Buenos Aires, Argentina

ABSTRACT: Evolution occurs not only through mutational changes, but also during evolutionary transitions—when groups become a new higher-level individual. The unicellular-multicellular transition was one of these important events in life. The volvocine green algae are an ideal model system to study this transition since they range from unicellular (e.g., *Chlamydomonas*), to undifferentiated colonies (e.g., *Eudorina*), to multicellular forms with complete germ-soma differentiation (e.g., *Volvox*). How does the evolution of traits that are necessary for multicellularity such as an extra-cellular matrix and germ-soma separation, alter the response to harmful chemicals? To test this, Volvocales of different size and complexity, but of similar cellular biology and development, were grown at different arsenic concentrations to measure their ecotoxicological response. We found evidence that Volvocales of larger size and complexity are more sensitive to arsenic contamination than their simpler and smaller counterparts.

1 INTRODUCTION

The volvocine green algae are an ideal model system to study the unicellular-multicellular transition since they range from unicellular (e.g., *Chlamydomonas*), to 8–64 celled colonies composed of undifferentiated cells (e.g., *Gonium*, *Eudorina*) to 500–50000 celled multicellular species with complete germ-soma differentiation (e.g., *Volvox*, Figure 1, Kirk, 1998; Solari *et al.*, 2006). These species have biflagellated motile cells, worldwide distribution, and coexist in the euphotic zone of freshwater ponds and lakes.

Arsenic is well known as a ubiquitous metalloid widely distributed in marine, freshwater and soil environments. Different studies have been carried out with microalgae and macroalgae (e.g., Levy *et al.*, 2005), but none has evaluated the effect of arsenic on organisms of different size and complexity, but with similar cellular biology and development.

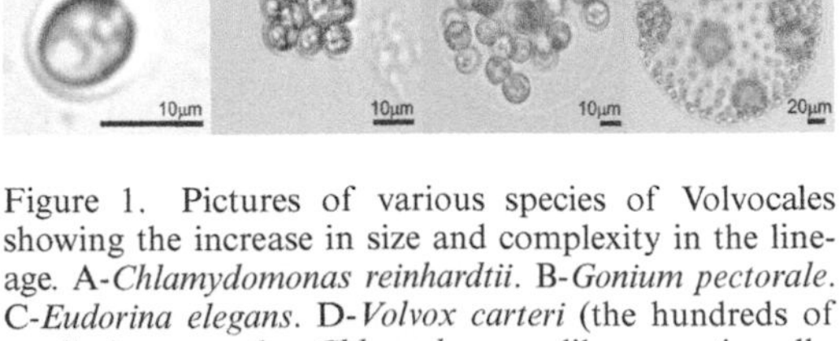

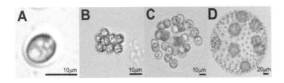

Figure 1. Pictures of various species of Volvocales showing the increase in size and complexity in the lineage. A-*Chlamydomonas reinhardtii*. B-*Gonium pectorale*. C-*Eudorina elegans*. D-*Volvox carteri* (the hundreds of small dots are the *Chlamydomonas*-like somatic cells; large cells are the germ cells).

To investigate the relationship between contamination, size, and complexity, we calculated the inhibitory concentration 50% (IC_{50}) of arsenic for four volvocine species of different size and complexity.

2 METHODS

We performed 96 hours experiments with four volvocine species that reflect the differences in size and complexity in the group but have the same "palintomic" developmental mode, unicellular *Chlamydomonas reinhardtii* (UTEX89, Figure 1A), 1–16 cells colonial *Gonium pectorale* (UTEXLB826, Figure 1B), 8–64 cells colonial *Eudorina elegans* (UTEX1201, Figure 1C), and 500–4000 cells multicellular *Volvox carteri* (EVE strain, kindly provided by D.L. Kirk, Figure 1E). Axenic populations for inoculation were kept at exponential phase in standard *Volvox* medium (SVM, Kirk & Kirk, 1983), and illuminated by homogeneous cool white light (~10,000 lux) in a daily cycle of 16 hours light (at 25 °C) and 8 hours in the dark (at 23 °C).

The algae were inoculated to an initial concentration of 2×10^4 cells/mL with sterile serological pipettes in 300 mL SVM sterilized Erlenmeyer flasks (e.g., *Chlamydomonas* experiments started at 2×10^4 cells/mL, *Volvox* experiments started at 20 colonies/mL if colonies had on average 1000 cells/colony). All experiments were continuously illuminated for the 96 hours. Concentrations of arsenic ($NaAsO_2$)

in the range 0–400 mg L^{-1} were used in duplicate replicates.

Digital images were taken under high magnification and imaging processing software (ImageJ, NIH, USA) was used to measure cell sizes, and cell counting for multicellular individuals (10 individuals). Automatic population counts and size distributions were performed with the Invitrogen Automated Cell Counter for *C. reinhardtii*, Neubauer chamber was used for *G. pectorale* and *E. elegans*, and large *Volvox* colonies were counted with Bogorov chambers. Chlorophyll a and b were estimated using spectrophotometric determination with 80% acetone extraction following Lichtenthaler (1987).

The half maximal inhibitory concentration (IC_{50}) for cells/ml and chlorophyll a+b/mL was calculated using linear interpolation (Norberg-King 1993). Error bars are standard deviations.

3 RESULTS AND DISCUSSION

Figure 2 shows the dose-dependent response of the population growth for each species. It is clear that *C. reinhardtii* cells are less affected to arsenic than the multicellular species.

When plotting the IC_{50} for cells/mL vs. cell number, we can observe an increase in arsenic sensitivity as size increases (Figure 3). For all the species, the calculated chlorophyll a+b/mL IC_{50} values were higher than the ones for cells/mL, but not significantly different. *E. elegans* (Figure 1C) and *V. carteri* (Figure 1D), the two species with an extra-cellular matrix structure that forms a hollow sphere, were significantly more sensitive to arsenic than unicellular *C. reinhardtii*. Our data points were not sufficient to get a significant allometric linear regression ($r^2 = 0.47$; slope $p = 0.15$).

Colony cell number decreased in *V. carteri* colonies treated with arsenic (800 vs. 500–600 cells), but we found no change in *E. elegans* (Figure 4).

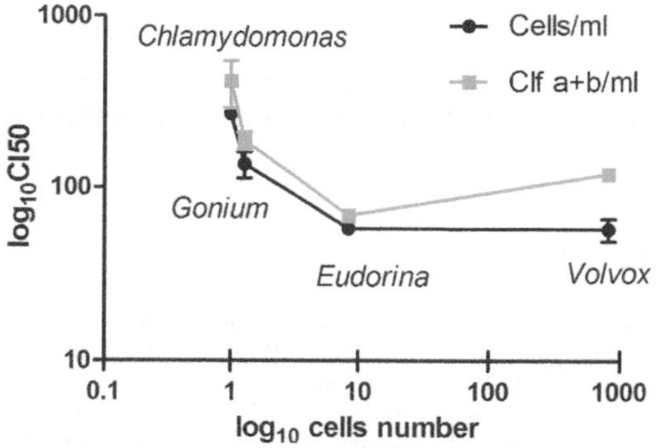

Figure 3. Half inhibitory concentration (IC_{50}) estimated with cells counts and chlorophyll a+b concentration for *C. reinhardtii*, *G. pectorale*, *E. elegans* and *V. carteri*.

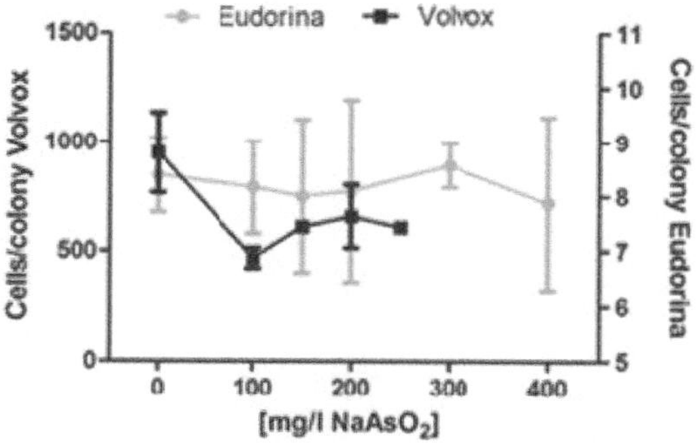

Figure 4. Number of cells/colony as a function of arsenic concentration.

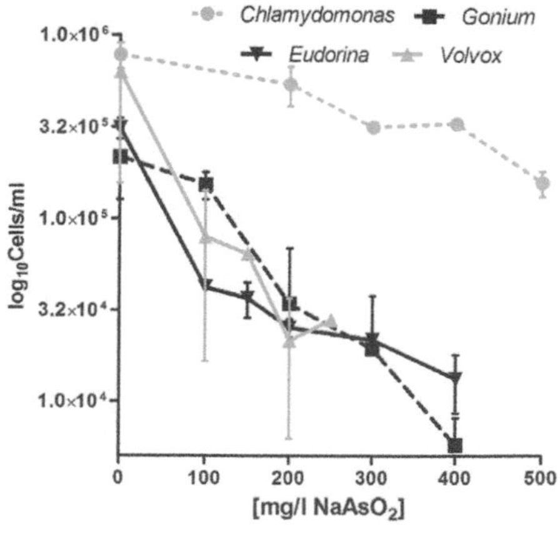

Figure 2. Population size after 96 h for different As concentrations. *C. reinhardtii* (●), *G. pectorale* (■), *E. elegans* (▼) and *V. carteri* (▲).

4 CONCLUSIONS

In the Volvocales we found evidence that larger multicellular species are more sensitive to arsenic contamination than unicellular ones. Species with structural complexity were the most affected in our study. We found no evidence that germ-soma differentiation in *Volvox* benefits or impairs the response. In the future, we plan to perform detail measurements of the stress response and study more species to derive the allometric relationships.

ACKNOWLEDGEMENTS

This work was supported by CONICET PIP 283, MINCYT PICT 1435 and Universidad de Buenos Aires grants.

REFERENCES

Kirk, D.L. 1998. Volvox: *Molecular-genetic origins of multicellularity and cellular differentiation*. Cambridge University Press, Cambridge.

Kirk, D.L. & Kirk, M.M. 1983. Protein synthetic patterns during the asexual life cycle of Volvox carteri. *Developmental Biology* 96: 493–506.

Levy J.L., Stauber J.L., Adams M.S., Maher W.A., Kirby J.K., Jolley D.F. 2005. Toxicity, biotransformation, and mode of action of arsenic in two freshwater microalgae (Chlorella sp. and Monoraphidium arcuatum). *Environmental Toxicology and Chemistry* 24(10):2630–9.

Lichtenthaler H.K. 1987. Chlorophylls and carotenoids: pigments of photosynthetic biomembranes. *Methods in Enzymology* 148: 350–382.

Norberg-King, T.J. 1993. *A linear interpolation method for sublethal toxicity: The inhibition concentration (ICp) approach.* Version 2.0. National Effluent Toxicity Assessment Center Technical Report 03-93, Environmental Research Laboratory, Duluth, MN 55804. June 1993.

Solari C.A., Kessler J.O., & Michod R.E. 2006. A Hydrodynamics Approach to the Evolution of Multicellularity: Flagellar motility and the evolution of germ-soma differentiation in volvocalean green algae. *The American Naturalist* 167:537–554.

One Century of the Discovery of Arsenicosis in Latin America (1914–2014) –
Litter, Nicolli, Meichtry, Quici, Bundschuh, Bhattacharya & Naidu (Eds)
 ISBN 978-1-138-00141-1

The *in-vitro* and *in-vivo* influence of arsenic on arbuscular mycorrhizal fungi

F.N. Spagnoletti
Cátedra de Microbiología Agrícola, Facultad de Agronomía, UBA, Buenos Aires, Argentina
Instituto de Investigaciones en Biociencias Agrícolas y Ambientales—INBA, (CONICET-UBA), Ciudad Autónoma de Buenos Aires, Argentina

N.E. Tobar
Instituto de Investigaciones en Biociencias Agrícolas y Ambientales—INBA, (CONICET-UBA), Ciudad Autónoma de Buenos Aires, Argentina

V.M. Chiocchio
Cátedra de Microbiología Agrícola, Facultad de Agronomía, UBA, Buenos Aires, Argentina
Instituto de Investigaciones en Biociencias Agrícolas y Ambientales—INBA, (CONICET-UBA), Ciudad Autónoma de Buenos Aires, Argentina

R.S. Lavado
Instituto de Investigaciones en Biociencias Agrícolas y Ambientales—INBA, (CONICET-UBA), Ciudad Autónoma de Buenos Aires, Argentina

ABSTRACT: Due to the potential occurrence of arsenic (As) in agricultural soils and the ecological significance of mycorrhizal fungi, it is proposed to assess the effect of this metalloid on these fungi both *in vitro* and *in vivo*. To this end, germination of spores, hyphal length and the percentage of radical colonization were evaluated. A decrease in the first two parameters when the levels of As were increased in the media was recorded. However, the colonization percentages do not reflect the toxicity generated by As.

1 INTRODUCTION

The arbuscular mycorrhizas are a symbiotic relationship obliged between roots and fungi belonging to the *Phylum Glomeromycota* (Wilcox, 1990). Their life cycle begins with the germination of the spores, which is one of its most important stages of the cycle because the beginning of the radical infection depends on it.

Accumulation of toxic elements in the soil can affect the fungi which form arbuscular mycorrhizal (Ortega-Larrocea *et al.*, 2007). As a result, a decrease of the abundance of spores and even complete inhibition of their germination take place. Arsenic is a toxic element that poses a danger to microorganisms, plants, animals and human beings.

The aim of this study is to evaluate the effect *in vitro* and *in vivo* of As on the pre-symbiotic and symbiotic stages of the fungi that form arbuscular mycorrhizas.

2 METHODS

2.1 *In-vitro test*

A test was conducted where the influence of As on the spore's germination and on the germ's tube length was evaluated. Agar-water 1% was used as a medium and As was added at the doses of 0, 0.5, 1, 5, 10 and 25 mg As/L. *Rhizophagus intraradices* spores, obtained from monosporic growth from a soil located in the FAUBA (Facultad de Agronomía-UBA), where previously sterilized with a solution of Chloramine T (2% *m*/*v*) with the addition of 200 μg/mL of streptomycin and traces of Tween 80. Then, they were washed three times with sterile distilled water (Mosse, 1962). Each treatment was replicated twenty times and five spores were sowed in each replication. It was incubated at 26 °C and in the darkness. The spore germination and the hyphal length where observed on day 23 using an optical microscope and an ocular micrometer.

2.2 *In-vivo test*

A test was conducted in a greenhouse using soybean plants as a host. These plants were sown in a substrate previously sterilized, composed of a mixture of soil and sand (7:3) in pots with a capacity of 1 L. The treatments were inoculated and non-inoculated plants with *R. intraradices* and periodically watered with distilled water enriched with sodium arsenate at doses of 0, 0.5, 1, 10 and 25 mg As/L. Each treatment consisted of five pots for treatment. The experiment concluded when plants reached the phonological R4 stage.

Roots were stained according to the methodology described by Phillips and Hayman (1970). Subsequently, the percentage of mycorrhization was determined according to Mc Gonigle *et al.* (1900).

3 RESULTS AND DISCUSSION

3.1 *In-vitro test*

After incubation, no significant differences in the germination percentage between the 0.5, 1 and 5 mg As/L doses and the control were observed.

When the As dose reach 10 mg As/L, germination decreased differing from the control but not from the other doses. The dose of 25 mg As/L showed the lowest percentage and differed significantly from all the other treatments (Figure 1). On day 23, the hyphal length, significantly decreased when the As levels increased above 10 mg As/L. This dose and 25 mg As/L differed from the rest of the doses and also among them, showing the lowest values of hyphal length (Figure 2).

These results would indicate that As has a toxic effect over the pre-symbiotic stage of the mycorrhizal colonization, causing a lower percentage of spore germination and lower values of hyphal length, thereby causing a low probability that these fungi can colonize possible host plants.

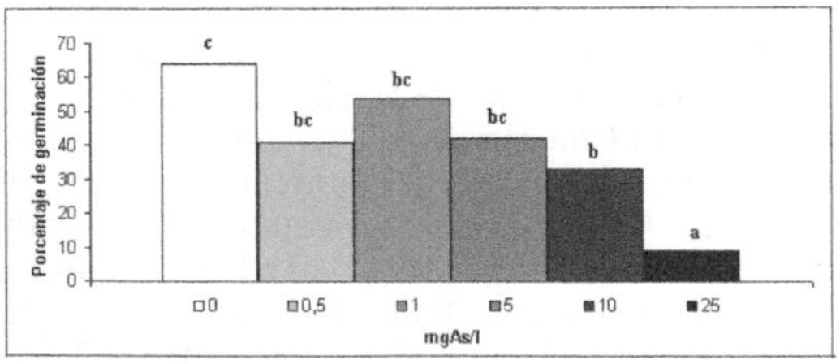

Figure 1. Effect of As on spore's germination of *R. intraradices*. Different letters indicate significant differences determined by Tukey's test ($p < 0.05$).

3.2 *In-vivo test*

Table 1 shows the percentages of colonization when the soybean plants reached the R4 stage. They were higher than 64% on inoculated plants, not showing significant differences between the As doses. Colonization was not observed in those plants that were not inoculated. These results suggest that the inoculum was active even in soils that were exposed to high levels of As, as opposed to the *in vitro* test, which recorded decreasing in the percentages of germination and hyphal length.

These results, apparently contradictory, could be due to the fact that the levels of As in the soil were increasing with successive irrigations. Therefore, it is possible that fungus colonization was established before As accumulated in the soil. Evidently, the fungi on that stage tolerate the toxic effect of As.

Additionally, the As in the *in vitro* test is in available form, whereas *in vivo* test, the As solubility could be affected by soil factors. Among them, the content of Fe and Al, the concentration of phosphate and pH (Sakata, 1987; Zarei *et al.*, 2010).

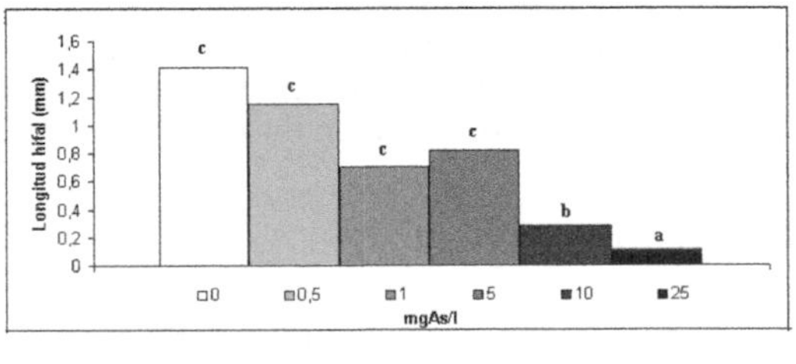

Figure 2. Effect of As on the hyphal length of *R. intraradices*. Different letters indicate significant differences determined by Tukey's test ($p < 0.05$).

Table 1. Colonization by *R. intraradices* in soybean roots.

Concentration of As in irrigation water (mg As/L)	% colonization
0	77.6 ± 6.0
0.5	69.9 ± 24.6
1	76.3 ± 11.4
5	69.3 ± 6.3
10	70.0 ± 9.6
25	64.1 ± 6.9

4 CONCLUSIONS

The effect of As on a species of fungi that forms arbuscular mycorrhizal was studied. A decrease of germinated spores and of the hyphal length was observed when the concentration of As was incremented in the soil media.

The high levels of radical colonization do not imply a high tolerance of these fungi to As, but this must be confirmed by *in vitro* tests such as the ones presented in this research or by stains that account for the vitality of the infection.

ACKNOWLEDGEMENTS

This work was funded by Universidad de Buenos Aires, UBACyT 2011-2014 068/2011 Project and by Consejo Nacional de Investigaciones Científicas y Técnicas (CONICET, Argentina), Project PIP 0148.

REFERENCES

Mc Gonigle, T.P., Miller, M.H., Evans, D.G., Fairchild, G.S. & Swan. J.A. 1990. A new method which gives an objective measure of colonization of roots by vesicular-arbuscular mycorrhizal fungi. *New Phytologist* 115:495–501.

Mosse, B. 1962. The establishment of vesicular arbuscular mycorrhizae under aseptic conditions. *Journal of General Microbiology* 27:509–520.

Ortega-Larrocea, M.P., Siebe, C., Estrada, A. & Webster. R. 2007. Mycorrhizal inoculum potential of arbuscular mycorrhizal fungi in soils irrigated with wastewater for various lengths of time, as affected by heavy metals and available P. *Applied Soil Ecology* 37:129–138.

Phillips, J.M. & D.S. Hayman. 1970. Improved procedures for clearing roots and staining parasitic and vesicular-arbuscular mycorrhizal fungi for rapid assessment of infection. *Transactions of the British Mycological Society* 55:158–161.

Sakata, M. 1987. Relationship between adsorption of arsenic (III) and boron by soil and soil properties. *Environmental Science & Technology*. 21: 1126–1130.

Wilcox, H.G. 1990. Mycorrhizae. In: *Plant Roots. The Hidden Half*. Y. Waisel, A. Eshel & U. Kafkafi (eds.), New York. U.S.A., p. 948.

Zarei, M., Hempel, S., Wubet, T., Schäfer, T., Savaghebi, G., Jouzani, G., Nekouei M.K. & Buscot, F. 2010. Molecular diversity of arbuscular mycorrhizal fungi in relation to soil chemical properties and heavy metal contamination. *Environmental Pollution* 158:2757–2765.

One Century of the Discovery of Arsenicosis in Latin America (1914–2014) – Litter, Nicolli, Meichtry, Quici, Bundschuh, Bhattacharya & Naidu (Eds)

Does cadmium influence arsenic phytotoxicity?

M. Kader, D.T. Lamb, M. Megharaj & R. Naidu
Centre for Environmental Risk Assessment and Remediation, University of South Australia, Mawson Lakes, SA, Australia
Cooperative Research Centre for Contamination Assessment and Remediation of the Environment (CRC CARE), University of South Australia, Mawson Lakes, SA, Australia

ABSTRACT: Arsenic (As) and Cadmium (Cd) phytotoxicity was studied individually and as a mixture in solution as well as in soil. The effective concentrations causing a 50% reduction in growth for root elongation to pot study was significantly varied though having significant correlation. The interaction of As and Cd was antagonistic in solution. In addition, As toxicity was significantly negatively affected in soil. The reduced binding ability of acidic soils for Cd may have impacted As phytotoxicity as the binding constant is 3–4 times less in alkaline soil.

1 INTRODUCTION

Metal(loid)s such as As and Cd are matters of serious concerns due to their high toxicity and pervasiveness in urban and agricultural land. Arsenic and Cd are wide spread in the environment as result of various anthropogenic and natural processes e.g. smelting and mining activities, sewage sludge application and various industrial activities. Soil contamination typically involves several contaminants concurrently. As a result, phytotoxicity may increase (synergistic), decrease (antagonistic) or even show no change (additive) in complex soil media. However, ecological risk assessment is mainly focused on the exposure of individual chemicals on the basis of acute toxicity data (Groten *et al.*, 2001) and environmental authorities set the threshold limit value on the basis of single contaminant systems. Approximately 95% of toxicology research conducted is based on single toxicant systems (Yang, 1994). The ability to predict how contaminants will interact with other co-contaminants to affect the health of ecological receptors in soil remains poorly understood.

The aim of this study was to investigate the interaction of binary mixtures of As and Cd on different dose level to *Cucumis sativus* L. (cucumber) in solution as well as soil.

2 METHODS/EXPERIMENTAL

2.1 *Soils*

Three different uncontaminated surface soils (0–20 cm) were collected from South Australia and Victoria, Australia. These include a Ferrosol, Kurosol, and a Tenosol. Soils were air-dried and sieved (4 mm). Individually, all soils were spiked with As(Na_2HAsO_4) and Cd ($Cd(NO_3)_2$) at 8 different concentrations ranging from 0 to 2000 mg/kg and 0 to 500 mg/kg, respectively, depending on the soil properties. For combinations, equitoxic mixtures were used based on EC_{50} values of each soil. For example, EC_{50} of As: EC_{50} of Cd, 1/2 EC_{50} of As: 1/2 EC_{50} of Cd. In addition, soils were spiked at Australian Ecological Investigation Levels for both contaminants. Soils were spiked by spraying with appropriate solutions and mixed thoroughly. Separate solution of As and Cd were used for mixture spiked (i.e. As + Cd). Soils were incubated at 60–70% field capacity for 2 months prior to experimentation. Soil properties were analysis using standard procedures (Lamb *et al.*, 2009).

2.2 *Plant growth*

The pot study was carried out in greenhouse conditions (~16–25 °C) for 4 weeks ($n = 3$). Cucumber (*Cucumis sativa* L.) seeds (8–10) were sown to pots containing 300 g air-dried soil. Water content of ~70% of field capacity was maintained throughout the period by adding high purity water by weight. Five days post-germination plant numbers were thinned to 4 plants per pot.

2.3 *Root elongation*

In the solution study, seedlings were exposed to 10 different concentrations of As and Cd separately and in combination (0 to 64 mg/L; $n = 3$). Equiconcentration was used in combination e.g. 1 mg/L As: 1 mg/L Cd, 2 mg/L As: 2 mg/L Cd. Seeds were sown in a 250 mL polypropylene pot containing glass beads to maintain a fixed volume of solution (7 mL). In all treatments, seeds were incubated for

48 hours in 25 ± 2 °C chambers for germination and a further 48 hours for root elongation.

In the soil study, seeds were grown in 100 g of air dried soil (5 roots per pot, $n = 4$). Water was added to maintain 70% field capacity. After 4 days of sowing seeds, roots were carefully separated and washed for measurements. Root lengths were analyzed with WinRHIZO Arabidopsis software after scanning.

2.4 *Statistical analysis*

EC_{50} calculation was performed by a sigmoidal dose—response model using R program and Grapher 9 (Surtek, Inc. USA). The following logistic formula was used for data fitting:

$$y = \frac{a}{1 + e^{b(x-c)}}$$

where y is the tested parameter, a is the uninhibited value of y, b is a slope factor, x is the logarithm value of concentration and c is the logarithm of the EC_{50} value, respectively. Interactions were analyzed in the solution study by both the additive and the multiplicative model.

3 RESULTS AND DISCUSSION

3.1 *As and Cd phytotoxicity in solution*

The EC_{50} value of A sand Cd was 0.415 and 2.07 mg/L respectively. The interaction of As and Cd was antagonistic at low concentrations but additive in the multiplicative model at higher dose levels.

3.2 *As phytotoxicity in soil*

EC_{50} values were substantially varied between soils. Arsenic phytotoxicity thresholds (EC_{50}) for the 4 d root elongation study ranged from 42 to 452 mg/kg whereas it ranged from 13 to 235 mg/kg in the pot study. EC_{50} values were highest in the Ferrosol (Tarrington) due to its low pH_{Ca} and high iron oxy(hydr)oxide content. Though the root elongation data yielded higher EC_{50} values, it was correlated with the pot study ($R^2 = 0.82$, slope = 1.8).

3.3 *As and Cd phytotoxicity in soil*

The shape of the dose response curve expressed as Toxicity Units (TU) was different between the As and As + Cd treatments, and to a lesser extent, soil type. The data indicated that in the acidic soils there was an impact in the As + Cd soils at low TU levels, i.e. an initial plateau followed by a large reduction in root growth was not observed. In the alkaline Tenosol, an initial plateau was observed followed by a rapid reduction in root growth for both treatments. Interestingly, for As alone, the typical dose-response

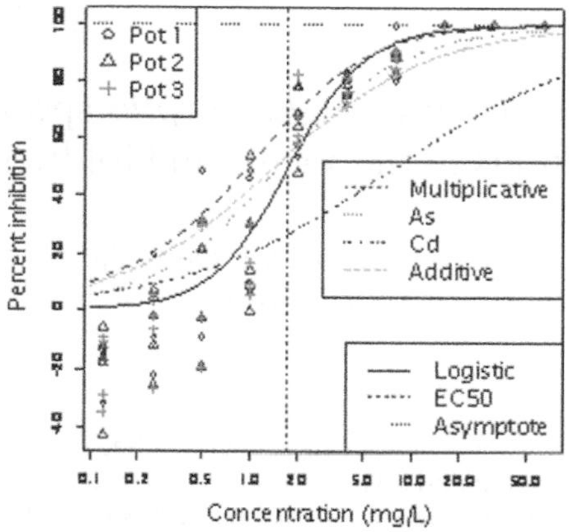

Figure 1. Interactive response of As in the presence of Cd at different dose levels.

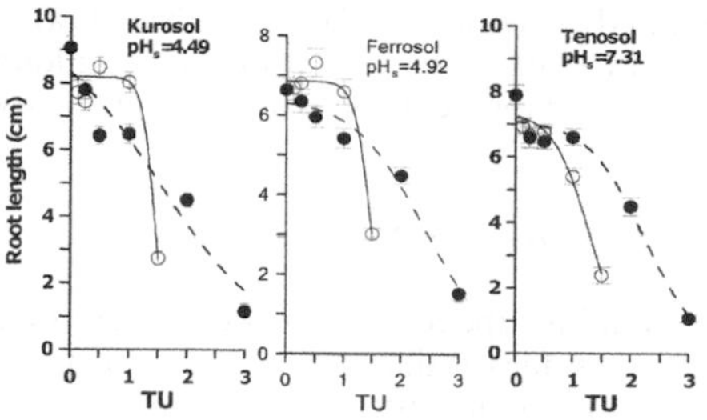

Figure 2. Cucumber root elongation response to As (open symbol) and As + Cd (closed symbol) in three different soils (Kurosol, Ferrosol, Tenosol). Toxicity Unit (TU) approach was used, where TU = 1 = EC_{50}.

shape was observed in all soils. The Cd Freundlich constants (K_f L/kg) for the 3 soils were 37.5, 80.1 and 205 for the Kurosol, Ferrosol and Tenosol, respectively. The reduced ability of the acidic soils to bind Cd may in part explain the impact of the added Cd to root elongation at low concentrations, as the binding constant in the alkaline soil is 3–4 times higher than the two acidic soils.

REFERENCES

Groten, J.P., Feron, V.J. & Suhnel, J. 2001. Toxicology of simple and complex mixtures. *Trends in pharmacological sciences*, 22: 316–322.

Lamb, D., Ming, H., Megharaj, M. & Naidu, R. 2009. Heavy metal (Cu, Zn, Cd and Pb) partitioning and bioaccessibility in uncontaminated and long-term contaminated soils. *Journal of Hazardous Materials*. 171:1150–1158.

Yang, R. 1994. Toxicology of chemical mixtures derived from hazardous waste sites or application of pesticides and fertilizers. *Toxicology of Chemical Mixtures* (Yang, R., ed). New York: Academic Press, 99, 117.

One Century of the Discovery of Arsenicosis in Latin America (1914–2014) – Litter, Nicolli, Meichtry, Quici, Bundschuh, Bhattacharya & Naidu (Eds)
 ISBN 978-1-138-00141-1

Performance of *Eucalyptus* species on capped arsenic-rich gold mine tailings in the Victorian Goldfields, Australia

J.T. Sanchez-Palacios, A.I. Doronila, A.J.M. Baker & I.E. Woodrow
School of Botany, The University of Melbourne, Australia

ABSTRACT: An experimental tailings research facility was constructed to allow long-term trials to be conducted and investigate the performance and variation in growth and arsenic foliar content in *Eucalyptus* spp. A cover of slurried oxide waste-residue was poured on to consolidated arsenic-rich sulfidic gold tailings (≈2000 mg As kg^{-1}) and capped with local topsoil. This study focuses on the growth responses of candidate Eucalyptus species in relation to arsenic-rich substrate. Three provenances of *Eucalyptus cladocalyx* grew the fastest and, on average, produced the largest stem volumes. The local provenance *E. goniocalyx* was the poorest. Among the others, *Corymbia maculata* ranked second, *E. camaldulensis* ranked third, and *E. tricarpa* ranked after these. Owing to its ability to grow under arsenic-rich conditions, more detailed testing of *E. cladocalyx* involving long-term monitoring of growth and foliar arsenic content is required to improve the selection of suitable species for use in arsenical mineral waste rehabilitation.

1 INTRODUCTION

Gold mining operations generate large volumes of fine residues (tailings), with storage facilities often being the largest area on mine sites requiring rehabilitation. Traditional covers required at least a 300 mm thick compacted clay layer over the tailings to seal off the toxic wastes, plus a greater thickness of protective neutral cover material. There is a need to mitigate high levels of As contamination in terrestrial ecosystems in order to reduce the threat to humans and the environment. King *et al.* (2008) examined four *Eucalyptus* species and demonstrated that *E. cladocalyx* may have potential for As phytostabilization. However, there has been no study examining the performance. The aim of this work was to examine variation in growth and foliar As content in 16 taxa of eucalypts comprising five *Eucalyptus* spp. with various provenances including *E. cladocalyx*, and to provide a more comprehensive basis for further comparison and provenance selection.

2 METHODS/EXPERIMENTAL

A field trial was conducted on a Tailings Experimental Research Facility (TERF) dam located at Stawell Gold Mine (37°03′59 S latitude, 142°48′15 E longitude and altitude 203 m) in Western Victoria, Australia. The final TERF has ~3 m high walls, forming an experimental cells of 75 × 155 m. Seedlings of five species of eucalypts—*E. cladocalyx* (sugar gum), *E. camaldulensis* (river red gum), *Corymbia maculata* (spotted gum), *E. tricarpa* (red ironbark), and *E. goniocalyx* (long-leaved box)—were transplanted into the trail plots in late spring (September) 2007. A Latin Rectangle method was used to provide randomization. The performance of plants was measured during the spring season (September to November) and from 2007 to 2010. Estimates of relative stem height SH and diameter growth SD were made using the equation $d\log_e(SH/SD)/dt$, with a linear model fitted. Foliar material was collected in 2010 using the method from King *et al.* (2008). Briefly, tissue was washed to remove traces of As from any surface deposits, and submerged in As-free phosphate solution to further remove As from the leaf stomata pores. After washing, all plant tissues were oven-dried before acid digestion. Statistical tests were performed using Sigma plot. Variables were transformed using their square-root values to satisfy normality. A General Linear *r* Model approach and the mean slope of relative stem volume growth rate were used to assess differences in stem volume increase with time.

3 RESULTS AND DISCUSSION

3.1 *Performance*

Appraisal of performance using the stem volume growth rate showed that *E. cladocalyx* provenance Wirrabara from SA produced the fastest-growing trees (Figure 1). *E. cladocalyx* provenance Bordertown from SA was another

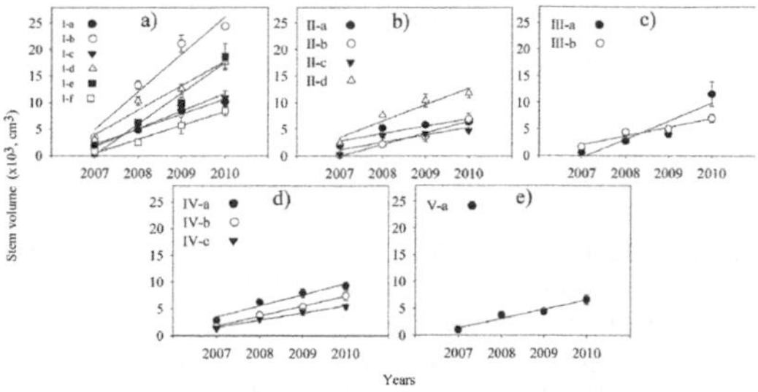

Figure 1. Relationship between stem volume and time for the *Eucalyptus* spp. provenances; *E. cladocalyx* provenances (a): Flinders (I-a), Wirrabara (I-b), Kersbrook (I-c), Bundaleer (I-d), Bordertown (I-e) and Manjimup (I-f); *E. camaldulensis* provenances (b): Melrose (II-a), Lake Albacutya (II-b), Manjimup growth (II-c) and Manjimup salt-tolerant (II-d); *C. maculata* provenances (c): Barclays-Deniliquin (III-a) and Manjimup growth (III-b); *E. tricarpa* provenances (d): Martin's Creek (IV-a), Tucker Box (IV-b) and Manjimup growth (IV-c); and *E. goniocalyx* provenance (d) Stawell (V-a). Data points are means ±se. A linear regression was applied for each provenance from 2007 to 2010. One-way ANOVA was performed between provenances using the slope values to compare the stem volume growth rate, and significant differences were detected ($F = 10.02, p < 0.001$).

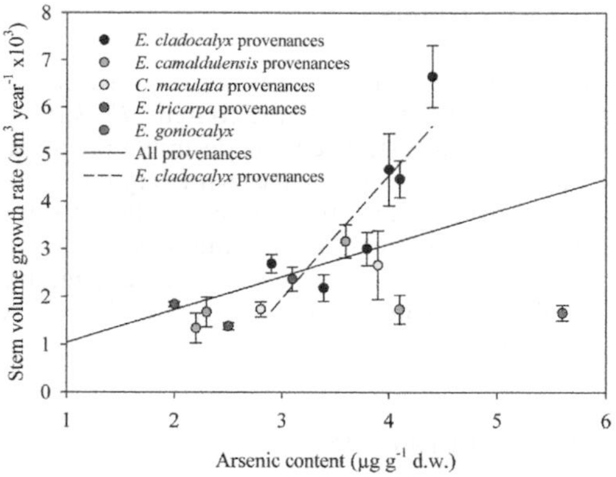

Figure 2. Relationship between stem volume growth rate and foliar arsenic content of *Eucalyptus* spp. provenances. Regressions were performed using the mean foliar arsenic content in 2011 and the mean stem volume growth rate for each provenance. Across all provenances a weak positive relationship (solid line; $r^2 = 0.20$; $F = 5.51$, $p < 0.05$) was found. A moderate positive relationship was found for the *E. cladocalyx* provenances (dotted line; $r^2 = 0.71$; $F = 9.87, p < 0.05$).

strong performer, ranking second. Another strong performer was *E. cladocalyx* provenance Bundaleer from SA, ranking third in growth rate. These three provenances were delimited as the top performing group. Interestingly, the best performing trees did not come from local Victorian provenances; five of the top six provenances are from SA. One factor influencing performance may have been root development, with the SA provenances possibly being more deep-rooted than their Victorian counterparts. Deep rooting may be an advantage on tailings, given that they are very wet at depth and tend to dry out in the upper layers. On the other hand, *E. camaldulensis* provenance Manjimup, CSO-Growth from WA was the poorest performer in stem volume growth rate. It is noteworthy that significant insect damage was observed in *E. camaldulensis* and to a greater degree in *E. goniocalyx*, with the majority of the damage done at the top part of the tree (data not shown). These results show the potential importance of the selection of *Eucalyptus* provenances, and that the above-mentioned three *E. cladocalyx* provenances are the best option for growth under those environmental and edaphic conditions.

3.2 *Relationship between growth and As content*

At age four years in 2010, the average foliar arsenic content for *E. goniocalyx* was 5.6 ± 0.3 µg As g^{-1} d.w., the highest of all species. *Eucalyptus cladocalyx* was lower at 3.8 ± 0.2 µg As g^{-1} d.w., followed by *C. maculata* at 3.4 ± 0.6 µg As g^{-1} d.w., *E. camaldulensis* at 3.1 ± 0.5 µg As g^{-1} d.w., and *E. tricarpa* at 2.5 ± 0.3 µg As g^{-1} d.w., which was half the concentration for *E. goniocalyx*. The relationship between stem growth rate and foliar arsenic accumulation for provenances was investigated (Figure 2). A regression was performed including all provenances using the mean foliar arsenic concentrations in 2010 and the mean stem volume growth rate. A weak positive relationship was detected ($r^2 = 0.2048$; $F = 5.51, p < 0.05$; Figure 2). A regression at the level of individuals between the foliar arsenic concentrations and the stem volume growth rates showed a weaker but more significant positive relationship ($r^2 = 0.0430$; $F = 13.93$, $p < 0.001$). There was a particularly good correlation for the six *E. cladocalyx* provenances, which is important given that this species accounted for the largest number of provenances in the trial ($r^2 = 0.71$; $F = 9.87, p < 0.05$). This relationship could relate to the As tolerance mechanism such that fast-growing provenances are more tolerant than slow-growing provenances to As. These results are in line with previous finding by King *et al.* (2008) involving *Eucalyptus* spp. under similar tailings conditions. Perhaps the most concerning aspect is the potential impact on herbivore performance and As transfer into the food chain.

4 CONCLUSIONS

Species and provenances differed significantly in their growth and foliar arsenic content. *E. cladocalyx* grew the fastest. The most desirable phytostabilization system involves trees with very low foliar As levels, to minimize the risk of transfer into the food chain. It is worth considering selective breeding of desirable genotypes since it can be suggested from the results here that foliar As uptake can be influenced at root level.

ACKNOWLEDGEMENTS

Thanks to the University of Melbourne, the Mexican Council for Science and Technology (CONACYT), and to Stawell Gold Mine Victoria for its continued support of this research.

REFERENCE

King, D.J., Doronila, A.I., Feenstra, C., Baker, A.J.M. and Woodrow, I.E. 2008. Phytostabilisation of arsenical gold mine tailings using four *Eucalyptus* species: growth, arsenic uptake and availability after five years, *The Science of the Total Environment*, 406:pp. 35–42.

One Century of the Discovery of Arsenicosis in Latin America (1914–2014) –
Litter, Nicolli, Meichtry, Quici, Bundschuh, Bhattacharya & Naidu (Eds)
© 2014 Taylor & Francis Group, London, ISBN 978-1-138-00141-1

Microbenthic communities and their role in arsenic fate in fluvial systems

H. Guasch
Universitat de Girona, Cataluña, Spain

M.C. Rodriguez Castro & A.D.N. Giorgi
Universidad Nacional de Luján, Buenos Aires, Argentina
CONICET, Argentina

ABSTRACT: The aim of this study was to evaluate the effect of phosphate and arsenate in natural microbenthic communities in fluvial systems. Communities were grown in artificial substrata placed in seven streams with different As/P ratios. After colonization, photosynthetic parameters were measured. Algal biomass was low in streams with elevated As, fact that suggests that As limits growth. Photosynthetic efficiency was lower in streams where As/P ratio was high. This suggests that, beyond the concentration of As in the fluvial system, the presence of P can lessen the impact of As in the growth of algae.

1 INTRODUCTION

In Argentina, the increasing concern of arsenic (As) toxic effects on humans due to intake of As-contaminated waters leads us to pay attention to its effects in the environment and in the role of aquatic organisms in its biogeochemical cycle and environmental factors that influence it.

Due to its characteristics, As is considered to be a highly toxic element. Occurrence, distribution, speciation and biotransformation of As in aquatic environments have been studied thoroughly in the last two decades. However, most of these studies were performed in marine environments and little is known about these processes in freshwater environments, particularly in lotic systems (Rahman *et al.*, 2012).

In the Pampean region of Argentina, naturally occurring high levels of As are found in water systems. In a recent study, Rosso *et al.* (2011) studied As concentration in 39 streams from the southern part of the Pampean region and found high average As concentrations (114 $\mu g\ L^{-1}$).

Chemical similarity between As(V), toxic metalloid, and phosphate (PO_4), nutrient that is fundamental to life, has important consequences to organisms (Blanck & Wängberg, 1991). As(V) enters the cell competing for PO_4 transporters and replaces PO_4 in metabolic pathways (e.g. formation of ADP-arsenate instead of ATP), causing toxicity (Krogmann *et al.*, 1959).

The presence of PO_4, both outside and inside the cell, interferes with As(V) at different levels. PO_4 availability has well known effects on algae. In PO_4 limiting conditions, growth rate of algae is reduced, but there is an increase in the efficiency of nutrient uptake. Starved algae have a more efficient uptake of PO_4 due to several mechanisms such as increasing high affinity PO_4 transporters in their plasmatic membrane. If these cells are exposed to As(V), intake is also increased, leading to a higher bioaccumulation and toxicity compared with not starved cells (Wang *et al.*, 2013).

Inside the cell, PO_4 chemical species may compete with As(V) species, affecting the toxicity of As. Wang *et al.* (2013) support the theory that As toxicity depends on PO_4 levels and, more specifically, on the As/P ratio inside the cell. However, the study also shows that different algal differ in their tolerance to As, a fact attributed to its different regulatory capacity depending on the concentration of the entry and exclusion rates.

Microbenthic communities are complex microbial communities composed by algae, bacteria and fungi attached to substrata. These communities integrate the physiological variety of the organisms of which they are formed (Sabater *et al.*, 2007) and are able to reflect the effects of As and PO_4 with high ecological realism. The aim of this study was to evaluate the effect of PO_4 and As(V) in natural microbenthic communities in a real field situation.

2 METHODS

During June to September of 2013, a study was performed with the aim to evaluate the effect of phosphate and arsenate in natural communities in a real field situation, and the impact of this interaction on the ecology of the system.

Seven sections of different streams with similar ecologic characteristics were chosen. Water samples were taken and total arsenic, phosphate (as Soluble Reactive Phosphorus, SRP) concentrations were determined, beside physical and chemical parameters (pH, temperature, conductivity, dissolved oxygen).

In each stream section, concrete blocks with 1.4 cm^2 etched glasses were placed underwater and parallel to the current. These glasses acted as substrata for colonization of microbenthic communities. Colonization lasted 30 days. After colonization, substrata were gently separated from the block and taken to the laboratory were fluorescence parameters were measured. An estimation of algal biomass (minimum fluorescence), photosynthetic efficiency (Yeff) and photosynthetic capacity (Ymax) were measured by means of a pulse-modulated fluorometer (PAM) (Walz, Germany).

3 RESULTS AND DISCUSSION

The results show that superficial freshwaters have different arsenic concentrations. In some cases, As levels are higher than 75 μg L^{-1}, upper limit concentration stated by EPA as EC50 to freshwater algae. The selected streams differ not only in As concentration but also in their eutrophic level.

The ratio between As and P was separated in high (>2), medium (= 1) or low (<1) values. Of the seven streams sampled, only El Pescado had a high As/P ratio, four had a medium ratio and other two, El Moro and Las Flores had a low ratio (Table 1).

Las Mostazas and Indio Rico had high As and SRP levels. However, algal biomass was low, fact that suggests that As limits growth. In Las Mostazas, the photosynthetic efficiency is elevated, suggesting adaptation of the community. Furthermore, photosynthetic efficiency is lower when As/P ratio is high and increased when it is low. This suggests that, beyond the concentration of As in the fluvial environment, the presence of P can lessen the impact of As in the growth of algae.

Table 1. Arsenic and Soluble Reactive Phosphorus (SRP) concentrations and As/P ratio of the seven stream sections sampled.

Stream	As μg L^{-1}	SRP μg L^{-1}	As/P
1	30.28	6.45	4.69
2	26.8	29.4	0.91
3	46.0	47.2	0.97
4	125.3	93.9	1.33
5	113.4	90.5	1.25
6	52.4	300.0	0.17
7	43.2	80.3	0.54

Table 2. Average and standard error (between brackets) of fluorescence parameters of algal community after four weeks of colonization on artificial substrates ($n = 5$). Algal biomass as minimum fluorescence (Fo); Photosynthetic efficiency (Yeff) and photosynthetic capacity (Ymax). *Bdl* stands for "below detection limits."

Stream	Fo	Yeff	Ymax
El Pescado	195.6 (95)	0.272 (0.067)	0.501 (0.070)
Grande	71.0 (18.8)	0.306 (0.038)	0.549 (0.016)
La Carolina	297.4 (90)	0.284 (0.042)	0.650 (0.020)
Indio Rico	31.3 (9.1)	Bdl	Bdl
de las Mostazas	43.7 (12.9)	0.339 (0.057)	0.571 (0.063)
Las Flores	396.6 (93.7)	0.523 (0.035)	0.606 (0.062)
El Moro	167.3 (31.9)	0.251 (0.081)	0.648 (0.012)

4 CONCLUSIONS

Even though the protective effect of P must be demonstrated experimentally, it is interesting to remark that As/P ratio could explain the low development of algal biomass in some environments that, according to their nutrient status, should bear a higher amount of microbenthic organisms; however, growth is inhibited by higher As concentrations.

ACKNOWLEDGEMENTS

To Ministry of Sciences of Spain that supported the staying of H. Guasch in Argentina. To Eng. Agr. Gustavo Giaccio of experimental station of Barrow (INTA) for their help.

REFERENCES

Blanck, H., & Wängberg, S.-A. 1991. Pattern of cotolerance in marine periphyton communities established under arsenate stress. *Aquatic Toxicology* 21: 1–14.

Hellweger, F.L., Farley, K.J., Lall, U. & Di Toro, D.M. 2003. Greedy algae reduce arsenate. *Limnology and Oceanography* 48(6): 2275–2288.

Krogmann, D.W., Jagendorf, A.T., & Avron, M. 1959. Uncouplers of spinach chloroplast photosynthetic phosphorylation. *Plant physiology* 34(3): 272–277.

Rahman, M.A., Hasegawa, H. & Lim, R.P. 2012. Bioaccumulation, biotransformation and trophic transfer of arsenic in the aquatic food chain. *Environmental research* 116: 118–135.

Rosso, J.J., Troncoso, J.J. & Fernandez Cirelli, A. 2011. Geographic distribution of Arsenic and Trace metals i nlotic ecosystems of the Pampa Plain, Argentina. *Bull. environ. contam. toxicol.* 86(1): 129–132.

Sabater, S., Guasch, H., Ricart, M., Romaní, A., Vidal, G., Klünder, C. & Schmitt-Jansen, M. 2007. Monitoring the effect of chemicals on biological communities. The biofilm as an interface. *Analytical and bioanalytical chemistry* 387(4): 1425–1434.

Wang, N.-X., Li, Y., Deng, X.-H., Miao, A.-J., Ji, R., & Yang, L.-Y. 2013. Toxicity and bioaccumulation kinetics of arsenate in two freshwater green algae under different phosphate regimes. *Water Research* 47(7): 2497–2506.

One Century of the Discovery of Arsenicosis in Latin America (1914–2014) – Litter, Nicolli, Meichtry, Quici, Bundschuh, Bhattacharya & Naidu (Eds)

Arsenic biotargets in volcanic environments

P.A. Lamela & G.A. Bongiovanni
IDEPA, Multidisciplinary Institute of Scientific Research and Development from North Patagonia, (CONICET-CCT COMAHUE), National University of Comahue, Neuquén y Cinco Saltos, Argentina

R.D. Pérez
School of Mathematics, Astronomy and Physics, (CONICET-CCT CORDOBA), National University of Cordoba, Cordoba, Argentina

C.L. Vodopivez
Argentinean Antarctic Institute (IAA), Buenos Aires, Argentina

ABSTRACT: In Patagonian Andes (Argentina) we found concentrations of As in superficial water up to 25 times the quality standards of drinking water for humans and 17 times for the life of aquatic species. The Andes continue beyond Tierra del Fuego, under the sea, reappearing as islands and as Antartandes, the great mountain chain of the Antarctic Peninsula. In these regions, we also found As. Here, we report arsenic bioaccumulation as well as elemental composition of Patagonian and Antarctic organisms measured by X-ray Fluorescence Spectrometry (SR-XRF) using synchrotron radiation. In order to determine molecular As-targets, the As-protein association was analyzed by this methodology. The highest As concentration in naturally exposed species was 2.26 $\mu g\ g^{-1}$ of dry weight in mussels with two As-binding proteins. In an ecologic context, As-accumulating organisms may serve as a critical link in the biotransference of arsenic within both, aquatic and terrestrial communities, including humans.

1 INTRODUCTION

Arsenic (As) is an abundant toxic metalloid in the earth crust and it is naturally introduced in aquatic system by leaching from soil/rock (mainly volcanic ones) (Bundschuh *et al.*, 2012). Thus, water is the major source of contamination and since arsenic can be bioaccumulated in microorganisms, plants and animals, being magnified in the food chain, it threatens environmental health, including human health (Arribére *et al.*, 2010; Zhao *et al.*, 2009; Pérez *et al.*, 2006; Rubatto Birri *et al.*, 2010). In the South Andean Range, the Southern Volcanic Zone of the Andes is the Quaternary volcanic area developed on the western margin of the Argentine Patagonia (between Chile and Argentina). Furthermore, the Andes continue beyond Tierra del Fuego, under the sea, reappear as islands and then emerge again as Antartandes, the great mountain chain of the Antarctic Peninsula. Despite the role that bioaccumulating species play in transporting contaminants through food webs, studies in Patagonia or Antarctic Peninsula are almost inexistent.

The aim of this work was to determine arsenic biotargets by X-ray Fluorescence Spectrometry using Synchrotron radiation (SR-XRF) Light Laboratory (LNLS). Furthermore, this technology allowed us to determine the concentration of the elements of the Periodic Table of Elements from P (phosphorus) to Zr (zirconium).

2 METHODS

2.1 *Sample collection and preparation*

Since samples were collected in protected natural areas, the permissions from Provincial Direction of Natural Protected Areas of Neuquén, Ministry of Environment and Sustainable Development Control of the Province of Chubut, Ministry for Territorial Development of the Province of Neuquén, Council of Ecology and Environment of the Province of Rio Negro, National Park Administration, and Permission for sampling in Antarctic Specially Protected Areas were obtained.

All biological sampled specimens were rinsed *in situ*, cooled in ice and carried to the laboratory immediately. Then, they were extensively washed. The vegetable samples were dried in an oven at 37 °C, and then powdered in liquid nitrogen using a porcelain mortar. The animal samples were freeze-dried, powdered and sieved (1-mm mesh).

2.2 *Digestion of samples*

Samples were digested with *aqua-regia* solution (3:1 HCl:HNO_3; all acids used were Fisher trace-metal grade). The tubes were heated on a Dry Block heater until the solution was evaporated. In order to assess element recovery, two additional vegetable samples were digested containing 3 μg g^{-1} Yttrium (TraceCERT® Yttrium Standard for ICP from Sigma-Aldrich Co.) as internal standard.

2.3 *SDS-PAGE*

The protein from As-accumulating species were separated on 10% SDS-PAGE, transferred onto nitrocellulose membrane and protein lanes were visualized by Ponceau S staining.

2.4 *Multielemental analysis by X-ray fluorescence analysis using synchrotron radiation*

Chemical analysis of samples was carried out by X-ray Fluorescence Spectrometry in grazing incidence geometry (SR-TXRF) using a white beam into the DO09Bbeam line at the LNLS (Brazilian Synchrotron Laboratory). Fluorescent intensity was normalized by using 10 mg L^{-1} Y as internal standard. The certified water standard (NIST 1640) was used as reference material.

The Fe, Cu, Zn and As-binding proteins were determined by X-ray Fluorescence Spectrometry in conventional incidence geometry (SR-μXRF), in the DO09B beamline from LNLS exciting nitrocellulose membrane along protein lines with a collimated white beam (X-ray in Figure 1).

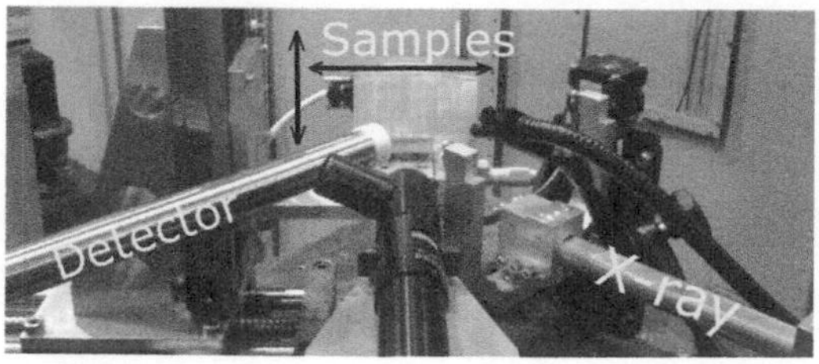

Figure 1. Instrumental setup. Each protein line was positioned (arrows) with a three-axis (x, y, z) remote controlled stage.

The fluorescence spectra were recorded with a Si(Li) detector in air and the elemental concentrations of specimens or proteins were mainly obtained by the area of kα peak count of Si, P, S, Cl, K, Ca, Ti, V, Mn, Fe, Cu, Zn, As, and Br.

3 RESULTS AND DISCUSSION

3.1 *As-accumulating species*

In arsenic rich environments, As can be accumulated in roots, shoots, leaves or grains (Zhao *et al.*, 2009). Although fifteen vegetable species were analyzed, As was only found in one. This sample consisted in leaves from a Pehuén tree (*Araucaria araucana* (Mol.) K. Koch) growing closed (50 cm) to As-contaminated river (value in Table 1). Other similar sample from Pehuén trees at 300 meters from the river did not contain As. Other abundant elements of Andes arc rocks as S, K, Ti, and Fe, were enriched in the analyzed vegetables (not shown).

The metal accumulation capacity of bivalves is widely known. Thus, it was not surprising the presence of As in the analyzed species. In this report, 1.39 μg g^{-1} of dry weight (DW). As were found in mussels from Atlantic costs. In similar context, because As in hair have been used to assess As exposure (McClintock *et al.*, 2012), the absence of As in hair of llama (*Lama glama*), would suggest no contamination in their living zone. Table 1 shows only As-accumulating species.

3.2 *As-binding proteins*

As-binding proteins at 38.7; 46.9; 67.4; 75.6; 85.5 and 95.3 kDa were detected in As-accumulating species by SR-μXRF (not shown).

Table 1. As-accumulating species. Data are expressed as μg of As per g of dry weight (μg g^{-1} DW) and represent the mean of two independent determinations by SR-TXRF.

Patagonian samples	As
Leaves of *Araucaria araucana* (Mol.) K. Koch (Patagonian tree)	0.05
Diplodon chilensis (river clams)	0.22
Mytilus edulis (mussels from Atlantic cost-1)	1.39
Mytilus edulis (mussels from Atlantic cost-2)	2.09
Antarctic samples	
Kidney of *Stercorarius antarcticus lonnbergi* (Antarctic bird)	0.38
Euphausia superba (Antarctic Krill)	0.19

4 CONCLUSIONS

In agreement with other authors, the results by SR-μXRF indicate that some trees and marine ecosystems accumulate arsenic from the environment, and that mussels are an important source of iron and arsenic. Additionally, high As concentration was seen in few proteins.

SR-XRF methodologies are a powerful tools to investigate in LNLS arsenic levels in environment as well as to characterize metalloproteinases, including arsenic acceptors.

ACKNOWLEDGEMENTS

The authors would like to thank LNLS-Campinas, Brazil, CONICET, UNCo and FONCyT, Argentina.

REFERENCES

Arribére, M.A., Campbell, L.M., Rizzo, A.P., Arcagni, M., Revenga, J. & Ribeiro Guevara S., 2010. Trace Elements in Plankton, Benthic Organisms, and Forage Fish of Lake Moreno, Northern Patagonia, Argentina. *Water Air Soil Pollut.* 21: 167–182.

Bundschuh, J., Litter, M.I., Parvez, F., Román-Ross, G., Nicolli, H.B., Jean, J-S., Liu, C-W., López, D., Armienta, M.A., Guilherme, L.R.G, Gomez Cuevas, A., Cornejo, L., Cumbal L. & Toujaguez, R. 2012. One century of arsenic exposure in Latin America: A review of history and occurrence from 14 countries. *Sci. Total Environ.* 429: 2–35.

McClintock, T.R., Chen, Y., Bundschuh, J., Oliver, J.T., Navoni, J., Olmos, V., Villaamil Lepori, E., Ahsan, H. & Parvez, F. 2012. Arsenic exposure in Latin America: Biomarkers, risk assessments and related health effects. Review. *Sci. Total Environ,* 429: 76–91

Pérez, R.D., Rubio, M., Pérez, C.A., Eynard, A.R. & Bongiovanni, G.A. 2006. Study of the effects of chronic arsenic poisoning on rat kidney by means of synchrotron microscopic X ray fluorescence analysis. *X Ray Spectrom,* 35(6): 352–358.

Rubatto Birri, N., Pérez R.D., Cremonezzi, D., Pérez, C.A., Rubio, M. & Bongiovanni, G.A. 2010. Association between Cu and As renal cortex accumulation and physiological and histological alterations after chronic arsenic intake. *Environ. Res.* 110: 417–423

Zhao, F.J., Ma, J.F., Meharg, A.A. & McGrath, S.P. 2009. Review: Arsenic uptake and metabolism in plants. *New Phytologist* 181: 777–794.

One Century of the Discovery of Arsenicosis in Latin America (1914–2014) – Litter, Nicolli, Meichtry, Quici, Bundschuh, Bhattacharya & Naidu (Eds)
© 2014 Taylor & Francis Group, London, ISBN 978-1-138-00141-1

Arsenic-containing lipids in five species of edible algae

S. García-Salgado
School of Civil Engineering, Technical University of Madrid, Madrid, Spain

G. Raber & K.A. Francesconi
Institute of Chemistry, Analytical Chemistry, Graz, Austria

ABSTRACT: The arsenolipids in five types of edible algae were investigated. The arsenolipids were extracted from the dried algae products with methanol/chloroform and partially purified by passage through a small silica column and elution with methanol/aqueous ammonia. The resultant fraction was examined by simultaneous HPLC/ICPMS and HPLC/ESMS, whereby the major arsenolipids were identified as arsenosugar-containing phospholipids and arsenic-containing hydrocarbons. These compounds constituted about 2–4% of the total arsenic originally present in the algae.

1 INTRODUCTION

Toxicological interest in arsenic compounds in foods has focused on distinguishing between inorganic arsenic and, collectively, the organic arsenic species. The organic arsenic species have been dominated by arsenobetaine, the major arsenical in most marine animals, and arsenosugars, which are found in algae. Over the last five years, however, a new group of arsenic species, arsenolipids, has been identified. The biological and toxicological properties of these compounds have still to be investigated.

Arsenic-containing fatty acids or hydrocarbons are found at significant concentrations in fish oils (Rumpler *et al.*, 2008) and in the tissues of fatty fish such as sashimi tuna (Taleshi *et al.*, 2010) and cod (Arroyo-Abad *et al.*, 2010). An arsenosugar phospholipid was first identified in an alga in 1988 (Morita & Shibata, 1988). However, it is only in the last two years that additional arsenosugar phospholipids have been identified (García-Salgado *et al.*, 2012; Raab *et al.*, 2013). It is still not known how widespread these arsenolipids are. We report an investigation into the arsenolipids in edible algae by using HPLC/mass spectrometry.

2 METHODS/EXPERIMENTAL

2.1 *Collection and preparation of algae samples*

Edible algae samples were obtained as dried packaged products from food outlets in Spain. Five types of algae were investigated: Arame (*Eisenia arborea*), sea spaghetti (*Himanthalia elongate*), Kombu (*Laminaria ochroleuca*), Wakame (*Undaria pinnatifida*), and Nori (*Porphyra umbilicalis*).

Extraction of arsenolipids from the algae was performed in triplicate for each of the five samples of algae. A typical extraction procedure was: a portion (1.0 g dry mass) of alga was extracted with a mixture of chloroform/methanol (2+1 v/v; 25 g) in a mechanical shaker overnight. The mixture was centrifuged, and the supernatant was washed with bicarbonate solution (1% m/v; 2 × 20 mL) to remove the water-soluble arsenic compounds. The chloroform layer was separated and evaporated to dryness to yield a residue (crude As-lipid fraction), which was redissolved in a mixture of chloroform/acetone (1+1 v/v, 1 mL). A portion (500 μL) of this solution was applied to a "plug" of silica (conditioned with chloroform/acetone/1+1 containing 1% formic acid), packed into a Pasteur pipette. The silica was washed with the conditioning solvent mixture (5 × 1 mL), then methanol (3 × 1 mL) and finally methanol containing 1% aqueous ammonia (10 × 1 mL). Arsenic was located in the fractions by using graphite furnace atomic absorption spectrometry. Those methanol/ammonia fractions containing most of the total arsenic applied to the column were combined and evaporated to dryness (purified As-lipid fraction). This material was redissolved in methanol (200 μL) and, together with the crude As-lipid fraction, analyzed by HPLC/ICPMS and HPLC/ESMS.

2.2 *HPLC/mass spectrometry*

Separation of the lipid-soluble arsenic species was performed under reversed-phase HPLC conditions by using a Zorbax Eclipse XDB-C8 column (4.6 mm × 150 mm; 5 µm i.d.; Agilent Technologies, Waldbronn, Germany) and a mobile phase comprising acetic acid (10 mM at pH 6.0, adjusted with aqueous ammonia) and methanol under the following gradient elution conditions: 0–25 min, 50%–95% methanol; 25–40 min, 95% methanol. The column effluent was split whereby 10% was directed to the ICPMS and 90% to the ESMS using an Agilent G1968D active splitter and introducing a sheath flow of 0.2 mL min^{-1} (5% methanol and 0.1% formic acid). To prevent deposition of carbon on the interface cones of the ICPMS, an optional gas (1% oxygen in argon) was introduced. Electrospray MS data were obtained by selected ion monitoring (SIM) in positive ion mode at a fragmentor voltage of 150 V.

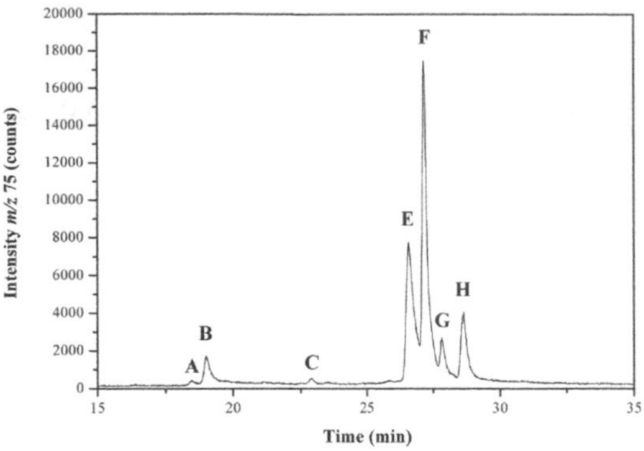

Figure 1. Reversed-phase HPLC/ICPMS chromatogram of lipid-extract from the edible alga Arame. Conditions, see experimental. The major arsenolipid (Peak F) was assigned the structure shown in Figure 2.

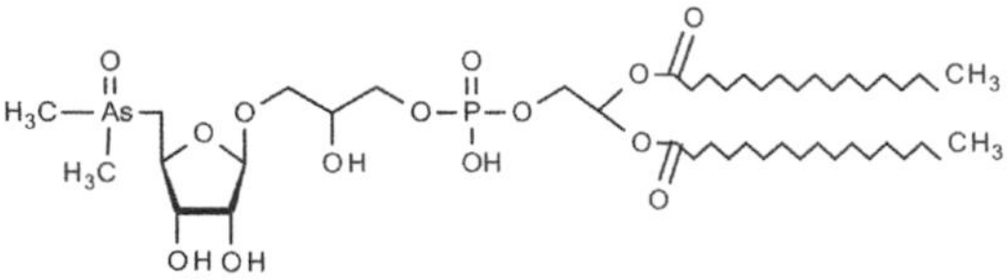

Figure 2. The major arsenolipid in algae.

3 RESULTS AND DISCUSSION

3.1 *Lipid extraction and clean-up procedures for arsenolipids*

Chloroform/methanol was an effective extractant for the arsenolipids. A subsequent back extraction of the organic phase with bicarbonate solution removed possible traces of polar arsenicals, and ensured that the resultant chloroform layer contained only lipid-soluble arsenic. In this way, the lipid content of the algae was found to constitute 2–4% of the total arsenic content of the algae (Table 1).

ICPMS can serve as a robust detector when coupled to HPLC for determining arsenic species. It is also sufficiently sensitive to analyze crude extracts. To obtain molecular information, however, electrospray MS must be used as detector, and this ionization suffers from severe matrix effects. Thus, a clean-up step is essential. We found that passage of the crude lipid extract through a simple silica column was very effective in concentrating the arsenolipids: it removed >90% of the total mass while retaining >74% of the arsenic.

Table 1. Total arsenic content (µg As g^{-1} dry mass) and % lipid arsenic in five types of edible algae. The total arsenic values represent mean ± SD from 3 measurements; the % lipid values represent mean ± SD from 3 extraction experiments.

Sample	Total As (µg g^{-1})	% lipid As
Arame	35 ± 9	4.0 ± 1.0
Sea spaghetti	23 ± 1	3.0 ± 0.5
Kombu	46 ± 2	2.5 ± 0.4
Wakame	33 ± 4	3.7 ± 0.7
Nori	40 ± 3	2.0 ± 0.1

3.2 *Identification of arsenolipids*

The use of HPLC with, simultaneously, both ICPMS and electrospray MS showed the presence of several arsenosugar phospholipids and arsenic-containing hydrocarbons.

The procedure by which structures can be assigned by this method has been previously reported (García-Salgado *et al.*, 2012). An example of an HPLC/ICPMS chromatogram is given in Figure 1. In this way, the major arsenolipid in all five species of edible algae was identified as the arsenosugar phospholipid-containing palmitic acid residues in the glycerol side chain (Figure 2). This arsenolipid was first reported by Morita & Shibata (1988). These authors also predicted the likely presence of other related arsenolipids in algae.

4 CONCLUSIONS

Edible algae contain high levels of arsenic, but only a small part (2–4%) of that is present as arsenolipids, primarily arsenosugar-containing phospholipids. The origin and possible biological role of these compounds are currently unknown.

ACKNOWLEDGEMENTS

We thank the Austrian Science Fund (FWF) project number P23761-N17 for support.

REFERENCES

Arroyo-Abad, U., Mattusch, J., Mothes, S., Moeder, M., Wennrich, R., Elizalde-Gonzalez, M.P. & Matysik, F.M. 2010. Detection of arsenic-containing hydrocarbons in canned cod liver tissue. *Talanta* 82: 38.

García-Salgado, S., Raber, G., Raml, R., Magnes, C. & Francesconi, K.A. 2012. Arsenosugar phospholipids and arsenic hydrocarbons in two species of brown macroalgae. *Environmental Chemistry*, 9: 63–66.

Morita, M. & Shibata, Y. 1988. Isolation and identification of arseno-lipid from a brown alga, *Undaria pinnatifida* (Wakame). *Chemosphere* 17: 1147.

Raab, A., Newcombe, C., Pitton, D., Ebel, R. & Feldmann, J. 2013. Comprehensive Analysis of Lipophilic Arsenic Species in a Brown Alga (*Saccharina latissima*). *Anal Chem.* 85: 2817–2824.

Rumpler, A., Edmonds, J.S., Katsu, M., Jensen, K.B., Goessler, W., Raber, G., Gunnlaugsdottir, H. & Francesconi, KA. 2008 Arsenic-containing long-chain fatty acids in cod liver oil: a result of biosynthetic infidelity. *Angew. Chem. Int. Ed.* 47: 2665.

Taleshi, M.S., Edmonds, J.S., Goessler, W., Ruiz-Chancho, M.J., Raber, G., Jensen, K.B. & Francesconi, KA. 2010. Arsenic-containing lipids are natural constituents of sashimi tuna. *Environmental Science & Technology* 44: 1478–1483.

One Century of the Discovery of Arsenicosis in Latin America (1914–2014) – Litter, Nicolli, Meichtry, Quici, Bundschuh, Bhattacharya & Naidu (Eds)
© 2014 Taylor & Francis Group, London, ISBN 978-1-138-00141-1

Arsenic in common Australian bivalve mollusks

W.A. Maher, F. Krikowa, J. Waring & S. Foster
Ecochemistry Laboratory, Institute for Applied Ecology, University of Canberra, Bruce, Australia

ABSTRACT: Arsenic concentrations and species in tissues of nine species of Australian bivalve mollusks from South-East Australia are presented. Arsenic concentrations were variable: whole tissues (18–109 mg/kg), viscera (18–158 mg/kg), muscle (1–42 mg/kg) and gills/mantle (19–60 mg/kg). Organisms living in sediments such as Tellina deltoidalis have higher arsenic concentrations. The suspension feeding omnivores contained mostly arsenobetaine and traces of dimethyl arsenate, arsenocholine and tetramethylarsonium ion. The deposit feeders contained less arsenobetaine and appreciable amounts of inorganic arsenic and arsenosugars. Arsenic concentrations and species are organism specific and determined by a variety of factors including exposure, diet and physiology.

1 INTRODUCTION

Mollusks are important components of most marine food-webs and as filter feeders can process large quantities of water, algae or suspended sediments and thus play key roles in the cycling of arsenic in marine ecosystems (Maher *et al.*, 2009). Along the South East coast of Australia, there are 15 main species of mollusks that can be found on rocks or in sandy or silty sediments. In this paper, we report the arsenic concentrations in tissues of nine organisms together with the arsenic species present on a whole organism basis.

2 METHODS

2.1 *Collection of organisms*

A wide geographic distribution of uncontaminated locations was sampled across three broad habitat types to encompass deposit and suspension feeding organisms. Specimens were rinsed with sea water and depurated for 48 h in aerated seawater from the location from which they were collected.

2.2 *Arsenic analysis*

Samples were freeze-dried for 48 h and then milled with an IKEA A11 micromill and digested with concentrated nitric acid and microwave heating (Baldwin *et al.*, 1994) and arsenic measured using inductively coupled plasma mass spectroscopy (Maher *et al.*, 2001).

Arsenic species were measured in methanol/water (1:1 v/v) extracts of tissues by high pressure liquid chromatography and inductively coupled plasma mass spectroscopy (Foster *et al.*, 2007, 2008; Kirby *et al.*, 2004).

3 RESULTS AND DISCUSSION

3.1 *Arsenic concentrations in tissues*

Organisms living in sediments *(A. trapezia, S. biradiata* and *T. deletoidalis*) have higher arsenic concentrations than those living in the water column (Figure 1). Hyper-accumulation of arsenic as reported for some polychaetes (Gibbs *et al.*, 1983) was not evident. Generally, viscera and gills/mantle tissues have higher arsenic concentrations than muscle tissues reflecting exposure and processing of arsenic from water, food and sediments.

3.2 *Arsenic species in whole organisms*

The suspension feeding omnivores contained mostly arsenobetaine (AB) (>90%) and traces of Dimethyl Arsenate (DMA), Arsenocholine (AC), tetraarsonium ion (TETRA) and some arsenosugars (AS) (Table 1). The deposit feeders (*T. deltoidalis* and *S. biradiata*) contained less arsenobetaine (55–68%) and appreciable amounts of inorganic arsenic (As(V)) and AS reflecting their processing or incidental ingestion of sediment containing benthic algae. The high proportion of AB in carnivores is a reflection of the assimilation and retention of AB, which is known to be efficiently retained by animals (Francesconi *et al.*, 1989) whereas other arsenic species such as arsenoribosides are degraded to DMA or excreted.

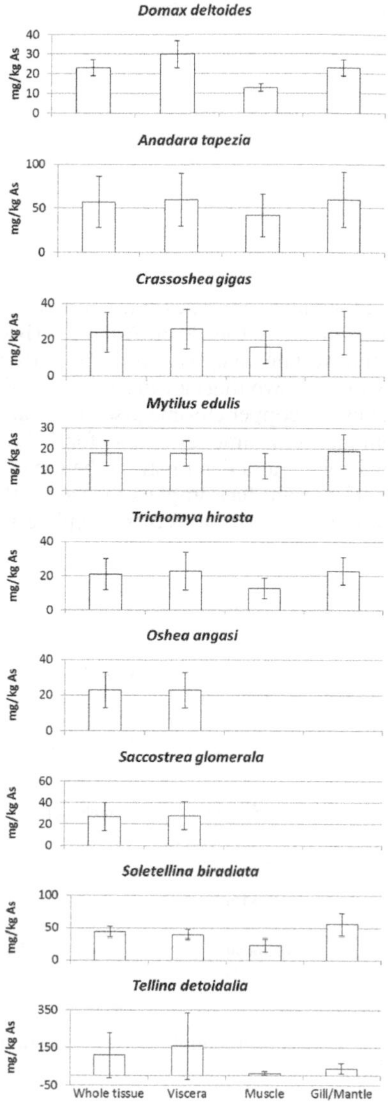

Figure 1. Arsenic concentrations in Australian mollusks.

Table 1. Major arsenic species in Australian mollusks.

Organism	AB %	DMA %	AC %	TETRA %	As(V) %	AS %
D. deltoides	95	5	–	–	–	–
A. trapezia	99	0.5		0.1	–	–
C. gigas	90	5	0.1	–	–	8
M. edulis	93	4	0.2	–	–	2
T. hirusta	96	1	–	–	–	2
O. angasi	70	5	–	–	6	3
S. glomerata	94	2	–	–	1	3
S. biradiata	68	7	–	0.4	8	9
T. deltoidalis	55	5	–	–	5	10

4 CONCLUSIONS

Arsenic concentrations are in the range expected for Australian marine organisms. Organisms living in sediments have higher arsenic concentrations than those living in the water column. Arsenic concentrations and species are organism specific and determined by a variety of factors including exposure, diet and physiology. AB is the major arsenic species (55–96%) with sediment containing organisms with appreciable amounts of inorganic arsenic and arsenosugars.

ACKNOWLEDGEMENTS

JW received an Australian Postgraduate award.

REFERENCES

Baldwin, S., Deaker, M. & Maher, W. 1994. Low volume microwave digestion of marine biological tissues for the measurement of trace elements. *Analyst* 119: 1701–1704.

Foster, S., Maher, W. & Krikowa, F. 2008. Changes in proportions of arsenic species within an *Ecklonia radiata* food chain. *Environmental Chemistry* 5: 176–183.

Foster, S., Maher, W., Krikowa, F. & Apte, S. 2007. A microwave assisted sequential extraction of water and dilute acid soluble arsenic species from marine plant and animal tissues. *Talanta* 71: 537–549.

Francesconi, K.A., Edmonds, J.S. & Stick, R.V., 1989. Accumulation of arsenic in yellow eye mullet (*Aldrichetta forsteri*) following oral administration of organoarsenic compounds and arsenate. *Science of the Total Environment*, 79: 59–67.

Gibbs, P.E., Langston, W.J., Burt, G.R. & Pascoe, P.L.1983. Tharyx marioni (Polychaeta): a remarkable accumulator of arsenic. Journal of the Marine Biological Assocation U. K. 63: 313–325.

Kirby, J., Maher,W., Ellwood, M. & Krikowa, F. 2004. Arsenic species determination in biological tissues by HPLC-ICP-MS and HPLC-HG-ICP-MS. *Australian Journal of Chemistry* **57**: 957–966.

Maher, W., Foster, S. & Krikowa, F. 2009. Arsenic species in Australian temperate marine food chains. *Marine and Freshwater Research* 60: 885–892.

Maher, W., Foster, S., Krikowa,F., Snitch, P., Chapple, G. & Craig, P. 2001. Measurement of trace metals and phosphorus in marine animal and plant tissues by low volume microwave digestion and ICPMS. *Analytical Spectroscopy* 22: 361–370.

One Century of the Discovery of Arsenicosis in Latin America (1914–2014) –
Litter, Nicolli, Meichtry, Quici, Bundschuh, Bhattacharya & Naidu (Eds)
© 2014 Taylor & Francis Group, London, ISBN 978-1-138-00141-1

Arsenic and air pollution in New Zealand

P.K. Davy, T. Ancelet, W.J. Trompetter & A. Markwitz
GNS Science, Lower Hutt, New Zealand

ABSTRACT: Air particulate matter samples have been collected on filter substrates at multiple sites across New Zealand for subsequent determination of elemental concentrations by Ion Beam Analysis (IBA). The resulting time-series data matrices have provided the basis for multivariate analysis and the apportionment of sources by positive matrix factorization. The elemental analysis has shown that elevated concentrations of arsenic occur every winter in many urban areas as a result of the use of copper chrome arsenate treated timber as fuel for domestic space-heating appliances. Here we present multiyear time-series data for daily arsenic concentrations at a number of urban locations throughout New Zealand and demonstrate the use of multivariate techniques to identify the source(s) of the contamination, The results have implications for both acute and chronic health effects associated with exposure to elevated winter concentrations of particulate arsenic.

1 INTRODUCTION

Particulate matter concentrations in New Zealand urban environments have been shown to exceed national and international air quality standards regularly during winter months in urban areas (MfE, 2009). Analysis of particulate matter composition coupled with multivariate techniques have therefore been used to identify the sources of pollution for regulatory control and policy intervention (Davy *et al.*, 2012, Davy *et al.*, 2005, Petersen *et al.*, 2009). Time-series analysis of individual elemental components revealed a distinct seasonal cycle in arsenic concentrations associated with particulate matter and that the annual average arsenic concentrations were likely to exceed health guidelines.

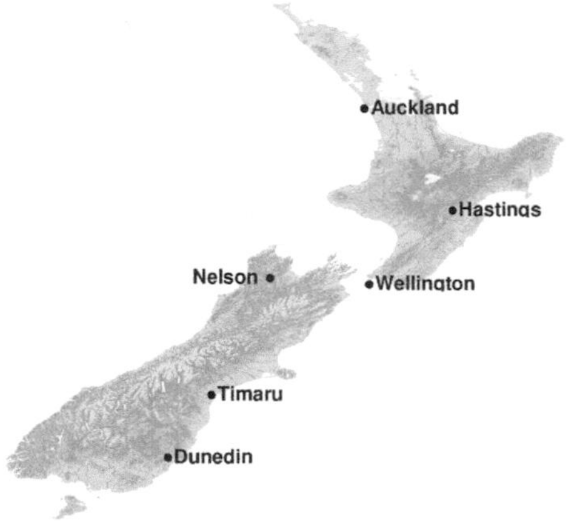

Figure 1. Location of sampling sites throughout New Zealand.

2 EXPERIMENTAL

2.1 *Particulate matter sampling*

Ambient air monitoring for particulate matter was conducted at twelve urban locations (six in Auckland City) across New Zealand between 2006 and 2013, covering a population base of approximately 2 million people (Figure 1). Full descriptions of the sampling setups have been reported (Davy *et al.*, 2012, Davy *et al.*, 2005, Petersen *et al.*, 2009). Briefly, at each urban location, particulate matter filter samples of PM_{10} and $PM_{2.5}$ were collected on a 24-hour, one-day-in-3 basis. In total, more than 10,000 size resolved particulate matter samples have been collected so far across all urban areas.

2.2 *Sample analysis and source identification*

IBA measurements for this study were carried out at the New Zealand IBA Facility operated by GNS Science at Gracefield, Lower Hutt (Trompetter *et al.*, 2005). Black Carbon (BC) concentrations on filters were determined by light reflection using a M43D Digital Smoke Stain Reflectometer. Receptor modeling and apportionment of particulate matter mass was performed using the PMF2 program (Paatero, 1997). Data screening and receptor modeling were performed as reported previously (Ancelet *et al.*, 2012, Davy *et al.*, 2011).

3 RESULTS AND DISCUSSION

Each of the twelve urban locations studied were found to have a number of particulate matter sources, including wood combustion for home heating, motor vehicles, marine aerosol, crustal matter and others depending on the nature of activities impacting on the monitoring site. In each urban location, elevated arsenic concentrations were also observed during winter (June-August) as shown in the composite plot presented in Figure 2.

A time-series plot of wood combustion extracted from the receptor modeling analyses shows a similar temporal pattern as presented in Figure 3.

Arsenic was found to be associated with the wood combustion source profile, indicating that at least some residences were using Copper Chrome Arsenate (CCA)-preserved timber as fuel for domestic heating purposes. Figure 4 presents a typical wood combustion source profile with associated arsenic.

The average annual arsenic concentration data presented in Figure 5 indicate that New Zealand

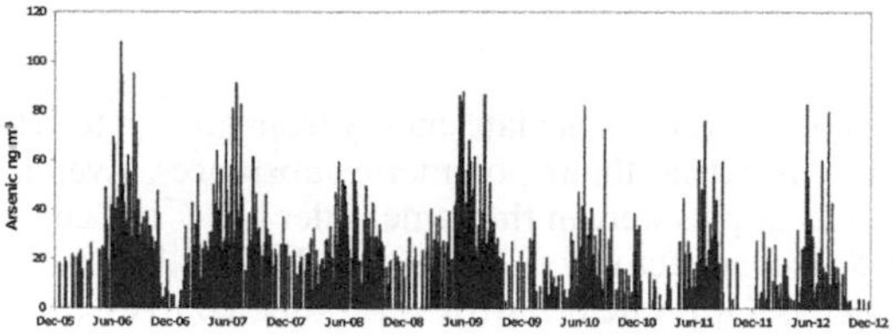

Figure 2. Daily arsenic concentrations at all locations.

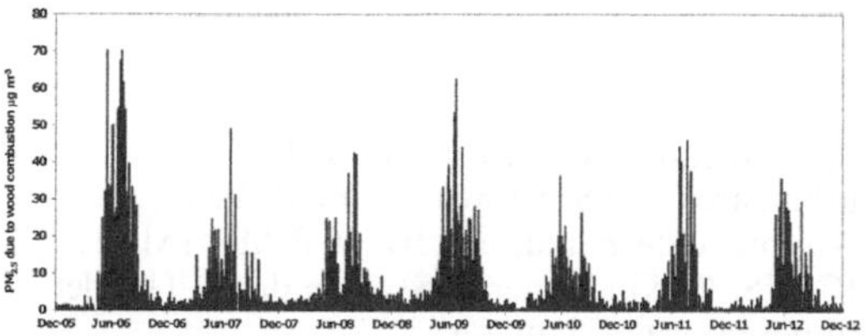

Figure 3. Particulate Matter ($PM_{2.5}$) concentrations associated with wood combustion at all locations.

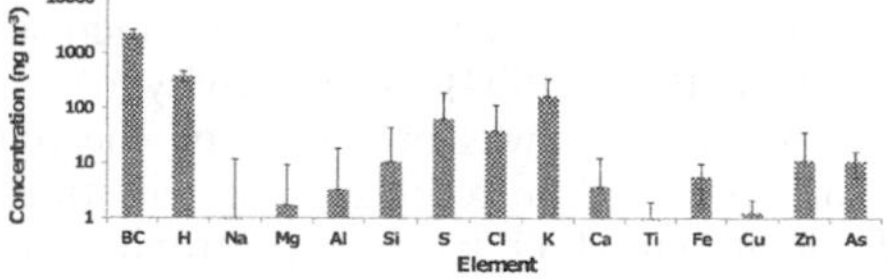

Figure 4. Typical wood combustion source profile showing the association of arsenic.

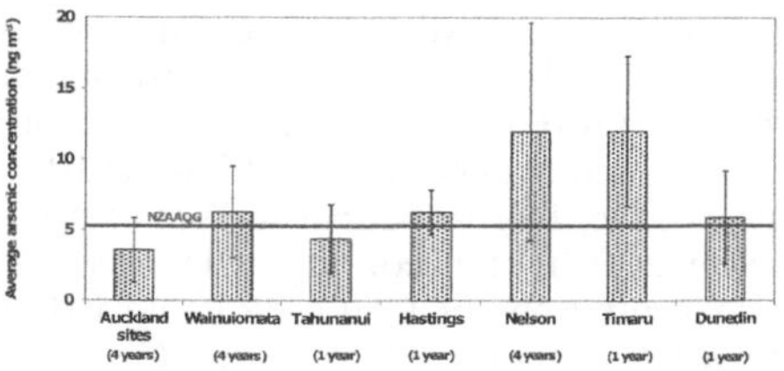

Figure 5. Annual average arsenic concentrations at New Zealand urban locations.

Ambient Air Quality Guideline (NZAAQG) of 5.5 ng m^{-3} as an annual average may be exceeded in some locations and close to it in others.

4 CONCLUSIONS

Elemental speciation and receptor modeling has shown that the combustion of arsenate-treated timber is responsible for elevated arsenic concentrations in a number of New Zealand urban locations resulting in widespread population exposure. The potential health effects associated with the repeated winter exposure to elevated concentrations of arsenic are unknown and warrant further investigation.

ACKNOWLEDGEMENTS

The authors thank the following councils for use of their data: Auckland Council, Hawkes Bay Regional Council, Greater Wellington Regional Council, Otago Regional Council, Nelson City Council and Environment Canterbury.

REFERENCES

Ancelet, T., Davy, P.K., Mitchell, T., Trompetter, W.J., Markwitz, A. & Weatherburn, D.C. 2012. Identification of particulate matter sources on an hourly timescale in a wood burning community. *Environmental Science and Technology* 46: 4767–4774.

Davy, P.K., Ancelet, T., Trompetter, W.J., Markwitz, A. & Weatherburn, D.C. 2012. Composition and source contributions of air particulate matter pollution in a New Zealand suburban town. *Atmospheric Pollution Research* 3: 143–147.

Davy, P.K., Gunchin, G., Markwitz, A., Trompetter, W.J., Barry, B.J., Shagjjamba, D. & Lodysamba, S. 2011. Air Particulate Matter Pollution in Ulaanbaatar, Mongolia: Determination of composition, source contributions and source locations. *Atmospheric Pollution Research* 2: 126–137.

Davy, P.K., Markwitz, A., Trompetter, W.J. & Weatherburn, D.C. 2005. Elemental analysis and source apportionment of ambient particulate matter at Masterton, New Zealand. *Int. J. PIXE* 15: 225–231.

MFE. 2009. Air Quality (Particulate Matter – PM10) Environmental Report Card. . Wellington: Ministry for the Environment.

Paatero, P. 1997. Least squares formulation of robust non-negative factor analysis. *Chemom. Intell. Lab. Syst.* 18: 183–194.

Petersen, J., Davy, P.K., Trompetter, W. & Markwitz, A.A. 2009. Multi-site, multi-year source apportionment study of particulate matter in Auckland. In *19th International Clean Air and Environment Conference, 2009 Perth.* Clean air Society of Australia and New Zealand, 7.

Trompeter, W.J., Markwitz, A. & Davy, P.K. 2005. Air Particulate Research Capability at the New Zealand Ion Beam Analysis Facility Using PIXE and IBA Techniques. *International Journal of PIXE* 15(3–4): 249–255.

One Century of the Discovery of Arsenicosis in Latin America (1914–2014) – Litter, Nicolli, Meichtry, Quici, Bundschuh, Bhattacharya & Naidu (Eds)
© 2014 Taylor & Francis Group, London, ISBN 978-1-138-00141-1

Microbial activities response to As exposure in soil and sediments surrounding a gold mining area at Paracatu—Brazil

E. Sabadini-Santos & E.D. Bidone
Curso de Pós-graduação em Geoquímica Ambiental, Universidade Federal Fluminense (UFF), Brazil

Z.C. Castilhos
Centro de Tecnologia Mineral, Ministério da Ciência, Tecnologia e Inovação (CETEM/MCTI), Brazil

M.A.C. Crapez
Curso de Pós-graduação em Biologia Marinha, Universidade Federal Fluminense (UFF), Brazil

ABSTRACT: Bacterial community exposed to high As concentrations deviate energy from growth to cell maintenance modifying enzymatic activity and increasing its extracellular polymeric substances. Even if the number of bacterial cells presented in soil and sediment samples were in the same order of 10^7 cell cm^{-3}, they declined in soil samples closer to mining area. Dehydrogenase activity (0.6–70.0 µg INT-F g^{-1}) showed the same trend, suggesting inhibition by toxic effect of metals. Opposite behavior was observed to esterase activities (2.2–7.9 µg FDA h^{-1} g^{-1}) that are representative of increasing energy demand by the community under environmental stress. The labile organic matter was mainly carbohydrate (1.4–10.7 mg C g^{-1}) of allochthonous source and its higher concentration in sediment from station 9 explains the rise of esterase activities at that site.

1 INTRODUCTION

Due to their function and ubiquitous presence, bacterial community can act as an environmentally very relevant indicator of pollution. A clear negative influence of metals on microbiological mediated processes and on the structure and diversity of microbial community have been reported worldwide (Beelen & Doelman, 1997; Harrison, 2007). Our objective is to understand the microbial community response to increasing As concentrations, searching for alterations of enzymatic activities and biomass, that should vary according to the development of resistance. This study is part of the environmental and health assessment conducted by Brazilian research institutions under the general coordination of CETEM, at Paracatu municipality, where the largest gold mine in Brazil ("Morro do Ouro") is located.

2 METHODS

2.1 *Sampling and As determination*

The sediments and soils samples were collected during the dry season along the Córrego Rico watershed—its spring is located inside the open pit mine site and at its medium segment is located the Paracatu city: St2 output of the mine area, St1 urban area and St9 downstream of the urban area. Arsenic determinations by ICP-MS (Model 42 IC-MS, Perkin Elmer MS) were done. The detection limit was 0.5 mg kg^{-1}.

2.2 *Microbial analyses*

The total number of bacterial cells (CELL) was quantified by epifluorescent microscopy (Kepner & Pratt, 1994). The activity of dehydrogenase enzymes was measured by the reduction of INT (iodonitrotetrazolium chloride) to INT—formazan (INT-F) in a spectrophotometer (Trevors 1984). Esterase enzymes activities (EST) were analyzed according to Stubberfield & Shaw (1990) based on fluorogenic compounds (fluorescein diacetate—FDA). Concentrations of total biopolymers (carbohydrate, lipids and proteins) were determined by spectrophotometric methods. Carbohydrates (CHO) were quantified according Gerchacov and Hatcher (1972). Lipids (LIP) were analyzed according to Marsh & Wenstein (1966) and proteins (PRT), according to Hartree (1972). The concentrations of CHO, PRT and LIP were expressed in carbon equivalent (mg C. g^{-1}) using the conversion factors 0.40, 0.49 and 0.75 respectively (Fabiano *et al.*, 1995). All determinations described above were done in triplicates of 1 g of sediment samples.

3 RESULTS AND DISCUSSION

Microorganisms are present in sediments in numbers about 10^{10} cells g^{-1} *d.w.* (Meyer-Reil & Köster, 2000). Their biomass are greater than all other benthic organisms and through their organization in multispecies biofilms, bacteria develop physicochemical gradients where organic matter can be metabolized by hydrolytic enzymes (Flemming & Wingender, 2010). Even if the number of CELL presented in soil and sediment samples were in the order of magnitude of 10^7 cell cm^{-3}, they declined smoothly in soil samples closer to mining area (Figure 1). Bacterial community exposed to high As concentrations deviates energy from growth to cell maintenance functions (Beelen & Doelman, 1997; Harrison, 2007).

DHA shows opposite trend to As concentration in soil and sediment (Figure 1). The activity of dehydrogenase enzymes is based on the fact that INT acts specifically as an artificial electron acceptor, when the succinate dehydrogenase complex in the electron transport chain is reoxidized (Stubberfield & Shaw, 1990). It allows the generation of energy (adenosine triphosphate—ATP), which is only made by viable cells. Consequently, DHA activity declines as the number of bacterial cells reduces or if its activity is inhibited by the toxic effect of metals including As (Beelen & Doelman, 1997; Harrison, 2007).

In contaminated environments, bacteria need more energy to survive, and thus they might be used as sensitive indicators of pollution. This assumption is supported by the increasing trend of EST of the community under high concentrations of As (Figure 1). EST act on biopolymers and transform them into low-molecular-weight organic carbon that could be degraded intracellular. Its increasing trend with arsenic concentration reflects the energy demand rising of the community. The EST activity did not decline at St 9 because of the large amount of organic matter.

Biopolymers present higher concentrations at sediments from St 9. PRT and LIP were lower than CHO, ranging from 0.02 to 0.37 and 0.12 to 0.85 mg C g^{-1}, respectively. CHO varied from 1.35 to 10.65 mg C g^{-1} and represent more than 90% of the biopolymeric carbon (BPC), the sum of PRT, LIP and CHO that represented the amount of bioavailable carbon present in samples (Figure 1). It suggests that this organic matter consist of an allochthonous rather than autochthonous origin and has low nutritional value to local biota.

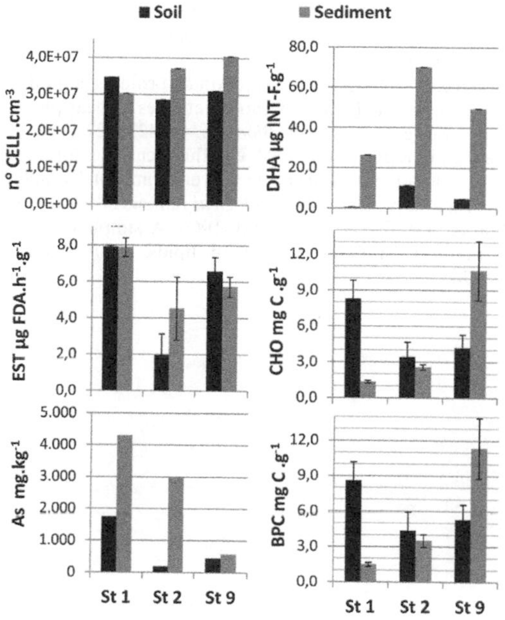

Figure 1. Distribution of bacterial cells, DHA activity, EST activities, As, CHO and CBP concentrations in samples of soil and sediments.

4 CONCLUSIONS

The number of bacterial cells and DHA activity showed a smooth decline tendency in soil samples closer to mining area, suggesting activity inhibition by toxic effect of As. Opposite behavior was observed to EST activities that are representative of increasing energy demand by the community to its maintenance under environmental stress. The labile organic matter was mainly CHO of allochthonous source and its higher concentration in sediment from downstream of the urban area explains the rise of EST activities at that site.

REFERENCES

Beelen, P.V. & Doelman, P. 1997. Significance and application of microbial toxicity tests in assessing ecotoxicological risks of contaminants in soil and sediment. *Chemosphere* 34: 445–499.

Fabiano, M., Povero, P. & Danovaro, R. 1993. Distribution and composition of particulate organic matter in the Ross Sea (Antarctica). *Polar Biol.* 13: 525–533.

Flemming, H.C. & Wingender, J. 2010. The Biofilm Matrix. *Nature Reviews* 8: 623–633

Gerchacov, S.M. & Hatcher, P.G. 1972. Improved technique for analysis of carbohydrates in sediment. *Limnol. Oceanogr.* 17: 938–943.

Harrison, J.J., Ceri, H. & Turner, R.J. 2007. Multimetal resistance and tolerance in microbial biofilms. *Nature* 5: 928–938.
Hartree, E.F. 1972. Determination of proteins: a modification of the Lowry method that gives a linear photometric response. *Anal. Biochem.* 48: 422–427.
Kepner, Jr. & Pratt, J.R. 1994. Use fluorochromes for direct enumerations of total bacteria in environmental samples: past and present. *Microbiological Reviews* 58: 603–615.
Marsh, J.B. & Wenstein, D.B. 1966. A simple charring method for determination of lipids. *J Lipid Res* 7: 574–576.
Meyer-Reil, L.A. & Köster, M. 2000. Eutrophication of marine waters: effects on benthic microbial communities. *Marine Pollution Bulletin* 41: 255–263.
Stubberfield, L.C.F. & Shaw, P.J.A. 1990. Enzymatically hydrolyzable protein and carbohydrate sedimentary pools as indicators of the trophic state of detritus sink systems: A case study in a Mediterranean Coastal Lagoon. *Journal of Microbial Methods* 12: 151–162.
Trevors, J. 1984. Effect of substrate concentration, inorganic nitrogen, 02 concentration, temperature and pH on dehydrogenase activity in soil *Water Research* 77: 285–293.

One Century of the Discovery of Arsenicosis in Latin America (1914–2014) – Litter, Nicolli, Meichtry, Quici, Bundschuh, Bhattacharya & Naidu (Eds)
 ISBN 978-1-138-00141-1

Ecotoxicological assessment of arsenic contaminated soil and freshwater from Paracatu, Minas Gerais, Brazil

S. Egler, R. Zacaron, B. Pinheiro & Z.C. Castilhos
Centre of Mineral Technology, Rio de Janeiro, Brazil

E.D. Bidone
Fluminense Federal University, Niteroi, Brazil

ABSTRACT: Samples of freshwater and soils were submitted to ecotoxicological tests to estimate the arsenic bioavailability and its toxic effects on aquatic and terrestrial biota. Aquatic organism bioassays indicated no acute but chronic toxicity in tested freshwater from Paracatu. Although the As bioaccumulation factor in soil at Corrego Rico resulted lower than in controls, the actual bioavailability of As is high because As levels resulted 30 times higher than in controls. This indicates that As from soils at Corrego Rico is bioavailable, but the earthworms can mitigate As toxicity by biological saturation controls. Despite of this fact, bioassays with earthworms suggested no acute toxic effects.

1 INTRODUCTION

Toxicity of metals to soil organisms has been related to total metal activity in pore water and to bounded-organic matter digested in guts. The two major routes of metal accumulation in aquatic organisms are dissolved uptake and dietary assimilation, whereas excretion played a dominant role in arsenic elimination, maybe as arsenosugars. Evaluation of the degree of contamination of aquatic environments must not take in account only its chemical characterization but it should be complemented with biological assays, which assess potential toxic effects to aquatic biota as a results of synergic, antagonistic, or additive effects of substances (Palma *et al.*, 2010). These methods allow the characterization of water quality in a shorter period and the assessment of multicomponent long-term effects of water contamination. The objective of this work was to evaluate the bioavailability and ecotoxicity of arsenic in waters and soils from Paracatu on aquatic and terrestrial biota.

2 METHODS/EXPERIMENTAL

Six stream surface water and two soil samples were collected from a gold mining area. The surface water were collected from 4 streams: Córrego Rico (points 2 and 9), Santo Antônio (3 and 20) and Neto river (Y) and Santa Isabel river (22). Soils were sampled from point 2 (Córrego Rico area) and 22 (Santa Isabel area). Points 2 and 3 are close to the gold mining and tailings dam, respectively, whereas point 22 is almost 50 km far from gold mining and it is the Paracatu water supply. After collected, the water sample was frozen, and the soil was air-dried and sieved to < 2.0 mm prior to total arsenic measurement by ICP-OES at CETEM.

In order to evaluate the ecotoxicity of surface water, the cladoceran species was chosen as indicated by the Brazilian legislation. Ecotoxicity of soils was assessed by acute assay by using the earthworm species *Eisenia andrei*.

Daphnia similis (Cladocera, Crustacea) were housed in 2-L glass beakers at environmental chamber with constant temperature (20 ± 2°C) and photoperiod (16:8 h light:dark). Organisms were maintained in dilute MS medium and fed daily *Pseudokirchneriella subcapitata* at a rate of $3{,}3 \times 10^6$ cells/mL per organisms. Toxicity tests employed 100% stream water samples and a control (MS medium) group per test. Five neonates (24 h old) were randomly placed into a 25-mL glass exposure chamber containing 20 mL of sample or MS medium. Four replicate exposure chambers were employed per sample or control group. Daphnids were not fed during tests, and all tests were conducted at the same condition of culture. Mortality or immobilization was assessed for individuals in each container after 48-h exposure. The algal growth inhibition test with *P. subcapitata* was performed based on monospecific algal cells growth in test flasks incubated on a shaker (130 rpm) with continuous illumination of

4500 lux at 20 ± 3°C. The inocula came from exponential phase of cell growth in LC Oligo medium at 24°C and continuous lighting. During the tests, the algae are incubated with 100 mL sample and LC Oligo medium as control group. Three replicate exposure chambers were employed per point or control group. Algal biomass was assessed for each replicates after 96 h.

Samples were monitored at the start and conclusion of each test for pH and dissolved oxygen. Percentage of inhibition is: I = (mean control biomass—mean sample biomass)/(mean control biomass*100).

Eisenia andrei (Lumbricidae, Annelida) were kept in plastic boxes in a bedding of cow manure, held under continuous lighting at 20 ± 2°C. The artificial soil used as control soil was prepared by mixing 70% quartz sand, 20% kaolin clay and 10% coconut shell powder. Moisture content of samples and control soils were brought to 55% (w/w) by adding demineralized water. Test soils (500 g dw) were placed in 1 L glass test containers and 10 adult earthworms were added at each of the three replicates. The total weight of earthworms added to each replicate was recorded. The survival earthworms were placed on moist filter paper for 24 h to allow them to void its gut contents. The earthworms were rinsed with distilled water, and then they were killed at – 20°C. After they were freeze-dried for 24 h and then ground to a fine powder with an agate pestle and mortar for metal chemical analysis.

The endpoints were lethality (%L) and weight loss (%WL). The bioaccumulation factor (BAF) is the metal levels ratio between tissue and soil.

3 RESULTS AND DISCUSSION

The water samples were chosen according to their As levels: four samples are higher than the Brazilian legislation limit (10 μg L^{-1}) while two samples are lowest than the limit. Total As levels in artificial soil were lower than the prevention value (15 mg kg^{-1}); sample 22 was slightly higher and soil sample 2 was 10 times higher than industrial (150 mg kg^{-1}) intervention criterion (Table 1).

No water acute toxic effect was observed in daphnids bioassays, but chronic effect was observed in algal bioassays. Fikirdeşici *et al.* (2012) found $LC_{50,\ 24h}$ = 0.509 mg L^{-1} to As concentration for *D. magna* bioassays. Other authors found higher $LC_{50,\ 24h}$ (from 2.6 to 2.9 mg L^{-1}). At the present study, As levels were lower than those LC_{50} and not toxic to the daphnids tested. Algal bioassays may be a good indicator, as all points tested (2, 3, 9, 20, 22 and Y) were significantly different from the control (p< 0.05) and showed% I = 37, 40, 37, 48, 19 and 21 for points studied, respectively. Duester *et al.* (2011), testing As species (III and V) toxicity for *P. subcapitata*, found that both $EC_{50,\ 72h}$ were >100 mg L^{-1}. Although the total As concentrations of surface water (Table 1) were lower, they showed chronic effects on the algal species tested.

Table 1. Total As levels in surface water and soils of Paracatu.

Samples and control	Total As	
	Surface water (μg L^{-1})	Soil (mg kg^{-1})
2	23.6	1752.9
3	5.0	
9	40.1	
20	18.9	
22	0.5	16
Y	22.1	
Artificial soils	–	0.8

Table 2. Earthworms As contamination and bioassays results.

Soil samples	L (%)	WL (%)	As (mg kg^{-1})		
			Earthworms	Soil	BAF
Control	0	16.2	1.6	0.8	2
2	0	18.0	69	1752.9	0.04
22	20	18.2	2.1	16	0.1

L—Letality; WL—Weight Loss.

Earthworms bioassay showed no acute and sub-lethal (weight loss) toxic effect by exposure to soils from Paracatu. The results in Table 2 showed higher BAF values for artificial soil and sample 22 (where soil arsenic is relatively low) than for sample 2, which has As levels two orders of magnitude higher. It has been reported by Watts *et al.* (2008), suggesting that elimination rates of arsenic may be increased at higher concentrations of arsenic. On the other hand, earthworm exposed to high levels of As in soils (Corrego Rico, point 2) showed As levels close to 30 times higher than the control (and Santa Isabel—point 22). This indicates that As from soils at Corrego Rico is bioavailable but the earthworms can mitigate As toxicity by biological saturation controls.

4 CONCLUSIONS

Aquatic organism bioassays indicated no acute but chronic toxicity in tested freshwater from Paracatu. This indicates that strong contamination in the streams studied may occur but continuous inputs of contaminants may not cause acute toxicity. Although As BAF result at Corrego Rico was lower than in control, the actual bioavailability of As is high because the result of earthworms As level was 30 times higher than in control. This indicates that As from soils at Corrego Rico is bioavailable but the earthworms can mitigate As toxicity by biological saturation controls. Despite of this fact, bioassays with earthworms suggested no acute toxic effects.

REFERENCES

Duester, L., van der Geest, H.G., Moelleken, S., Hirner, A.V. & Kueppers, K. 2011. Comparative phytotoxicity of methylated and inorganic arsenic- and antimony species to *L. minor*, *W. arrhiza* and *S. capricornutum*. *Microchem J.* 97: 30–37.

Fikirdeşici, S., Altindağ, A. & Ozdemir, E. 2012. Investigation of acute toxicity of cadmium-arsenic mixtures to *D. magna* with toxic units approach. *Turk. J. Zool.* 36(4): 543–550.

Palma, P., Alvarenga, P., Palma, V., Matos, V., Fernandes, R. Soares, A. & Barbosa, I. 2010. Evaluation of surface water quality using an ecotoxicological approach: a case study of the Alqueva Reservoir (Portugal). *Environ. Sci. Pollut. Res.* 17(3): 703–716.

Section 2: Arsenic in food

2.1 Overview and analytical aspects

One Century of the Discovery of Arsenicosis in Latin America (1914–2014) – Litter, Nicolli, Meichtry, Quici, Bundschuh, Bhattacharya & Naidu (Eds)
© 2014 Taylor & Francis Group, London, ISBN 978-1-138-00141-1

Arsenic occurrence in rice-based foods for infants and children with celiac disease

A.A. Carbonell Barrachina, S. Munera-Picazo, A. Ramírez, F. Burló & C. Castaño-Iglesias
Departamento de Tecnología Agroalimentaria, Universidad Miguel Hernández, Ctra. Beniel, Orihuela, Alicante, Spain

ABSTRACT: Celiac disease is an autoimmune disease that affects the villi of the small intestine causing abdominal pain, gas, diarrhea or bad absorption due to gluten intolerance. Its only treatment consists of a lifelong gluten free diet. The current study demonstrated that products for celiac children with a high percentage of rice contained high concentrations of arsenic (total-As: 256 and inorganic-As: 128 μg kg^{-1}). New legislation is needed to force companies to include key information in the labeling of these products, including for instance rice percentage, geographical origin and cultivar of the used rice.

1 INTRODUCTION

Celiac Disease (CD) is a digestive illness that damages the mucous membrane of the small intestine and interferes with absorption of nutrients from food (ESPGHAN, 2012; NIDDK, 2008). This illness is caused by intolerance to gluten proteins from *Triticeae* cereals group: wheat (gliadine), rye (secaline) and barley (hordeine). The world prevalence of CD is estimated at 1%, although this prevalence may be much higher because a large proportion of cases remain undetected.

Currently, the only effective treatment for CD is lifelong adherence to a gluten-free diet as there is not specific drug treatment. This diet basically consists of eliminating all foods having wheat, rye and barley (Polanco, 2008; Rodrigues & Jenkins, 2006). Rice is especially used in children and celiac foods due to its blandness, material properties, low allergen potential and nutritional value (Meharg *et al.*, 2008). Rice is therefore essential for the manufacturing of commodities for celiac people and reaches high percentages in their formulations.

However, rice tends to accumulate arsenic (As) due to its cultivation under flooded conditions. In rice, As content usually ranges from 100 to 400 μg kg^{-1} (Meharg *et al.*, 2009; Sun *et al.*, 2008). Recently, Burló *et al.* (2012), Carbonell-Barrachina *et al.* (2012) and Hernández-Martínez & Navarro-Blasco (2013) reported that rice-based infant foods contain significant levels of As, and indicated that celiac products may also be contaminated.

Summarizing, celiac foods require special attention with respect to their inorganic As (i-As) content. The three objectives of this study were: (i) to determine the content of total As (t-As) and i-As in the main foods for children aged 1–5 with the CD (control samples without rice will be studied as control products), (ii) to estimate the dietary intake of i-As and to model and predict the health risks in children (1–5 years of age) as a result of their exposure to i-As from rice-based foods, and (iii) to give a series of recommendations about labeling and appropriate selection of rice cultivars and geographical sources to manufacturers, consumers and authorities.

2 METHODS/EXPERIMENTAL

2.1 *Samples*

In this study, samples from five foods groups were analyzed: pasta, bread, breakfast cereals, chocolate wafers and biscuits. All products were purchased in national supermarket chains from the provinces of Alicante and Murcia (Spain) and were analyzed for t-As and i-As contents. The products under analysis were: 5 pasta samples, 3 bread samples, 2 breakfast cereals with chocolate, 2 chocolate wafers, and 4 biscuits, making a total of 16 products (in triplicate).

All products were targeted for children above 1 year of age and suffering from the celiac disease. For each group, there was a control product containing no rice; in most of the cases, this ingredient was replaced by corn. Products within each group had rice percentages ranging from 5 to 95% rice in their formula (Table 1).

2.2 *Total and inorganic arsenic (t-As and i-As)*

All samples were analyzed for t-As by Hydride Generation Atomic Absorption Spectrometry (*HG-AAS*). A 0.5 g (pasta) or 1 g portion (other products) of dried, ground and homogenized

Table 1. Total (t-As) and inorganic arsenic (i-As) concentrations in Spanish foods for children with the celiac disease.

Group	Rice (%)	t-As (μg kg^{-1})	i-As (μg kg^{-1})
Control Samples			
1. Pasta	0	nd†	nd
2. Bread	0	nd	nd
3. Cereals	0	nd	nd
4. Wafers	0	nd	nd
5. Biscuits	0	nd	nd
Rice-based Samples			
1. Pasta	35.0	46.3 de‡	30.9 c
	90.0	256 a	128 a
	93.4	128 c	75.3 b
	95.0	202 b	135 a
2. Bread	30.0	71.6 d	34.1 c
	30.0	62.0 d	34.4 c
3. Cereals	75.0	136 c	124 a
4. Wafers	5.0	nd f	nd e
5. Biscuits	10.0	16.8 e	12.0 d
	15.0	28.1 e	14.8 d
	15.0	31.8 e	35.3 c

† Values followed by different letters, in the same column, were significantly different ($p < 0.05$), according to the Tukey's test.

sample was weighed and digested using the method first described by Muñoz *et al.* (2000). The generated solutions were analyzed using a Unicam Model Solaar 969 atomic absorption spectrometer equipped with a continuous hydride generator Unicam Solaar VP90.

Speciation extraction procedure followed that first described in Zhu *et al.* (2008) and later by Carbonell-Barrachina *et al.* (2012), using microwave heating and 1% HNO_3. For As speciation analysis, a HPLC-HG-AFS (PSA 10.044 Excalibur, PS Analytical) was used, which also contained a Hamilton PRP X-100 anion-exchange column.

2.3 *Statistical analyses*

All data were subjected to analysis of variance (ANOVA) and the Tukey's least-significant difference multi-comparison test to determine significant differences among samples (food type). The statistical analyses were performed using SPPS 14.0 (SPSS Science, Chicago, USA).

3 RESULTS AND DISCUSSION

3.1 *Total and inorganic arsenic in control and rice-based samples*

The first important result of this study is the statement that all control samples (without rice in their composition) did not contain measurable amounts of As, neither i-As nor organic As (o-As). These control products were based on corn flour and potato starch. Consequently, it can be concluded that foods for CD children based on cereals others than rice (basically corn-based) are arsenic-free. These results agreed with the study by Matos-Reyes *et al.* (2010), who reported that As content in corn flour was below the detection limit of their analytical equipment, 0.5 μg/kg.

On the other hand, the highest values of As content in rice-based foods for CD children (Table 1), were found in samples 1B, 1D, 3A and 1C: 256, 202, 136 and 128 μg t-As kg^{-1} and 153, 131, 88.4, and 74.2 μg i-As kg^{-1}, respectively. These high values of As contents seemed to be related to high percentages of rice; rice percentages for these samples were 95, 90, 75, and 93.4%, respectively. Figure 1 shows positive relationships among the rice percentage used in the manufacturing of foods and the contents of t-As ($R^2 = 0.8909$) and i-As ($R^2 = 0.9003$) in it. This positive relationship means that the higher the rice percentage, the higher the t-As and i-As in the sample.

The groups with the highest t-As and i-As were pasta and breakfast cereals, with mean values of 158 ± 15 and 136 ± 6 μg t-As kg^{-1} and 92.2 and 124 μg i-As kg^{-1}, respectively. As can be seen, the breakfast cereals were characterized by a high i-As/t-As ratio, with a value of 0.91.

On the other hand, the chocolate wafers contained no measurable As without any doubt because of the low percentage of rice used in their manufacturing (5%). Bread and biscuits samples presented intermediate contents of both t-As and i-As.

The i-As/t-As ratio presented a mean value of 0.67 ± 0.06, a median of 0.63 and ranged between

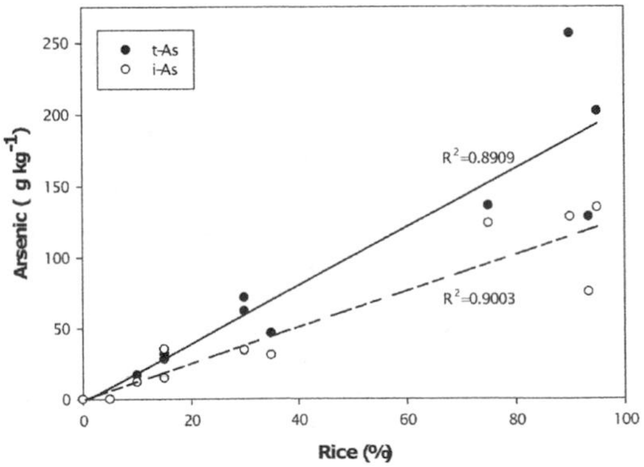

Figure 1. Relationship between total (t-As) and inorganic arsenic (i-As) concentrations and rice percentage in Spanish rice-based foods for children with the celiac disease.

0.48 and 1.11. These values agree quite well with those reported by Carbonell-Barrachina *et al.* (2012) in Spanish infant foods based on rice; the values found by these researchers ranged 0.36–0.89, with a mean of 0.64. In this same studied, it was reported that the i-As/t-As ratio presented mean values of 0.76, 0.55, 0.71 and 0.53 for rice-based infant foods from China, USA, UK and Spain, respectively. Sun *et al.* (2009) found high values of the i-As/t-As ratio, with mean of 0.81, in rice products, such as breakfast cereals, rice crackers and Japanese rice condiments.

Regarding the o-As species, DMA was predominant and represented approximately 80% of the total o-As content. Signes *et al.* (2008) reported that all As present in rice from West Bengal (India) was present as DMA; no measurable MMA was found. Similarly Meharg *et al.* (2008) reported that all o-As was present as DMA in baby rice.

It can be concluded that the content of i-As followed the order: pasta > cereals > bread > biscuits > wafers, and that these contents were positively correlated with the amount of rice in the product formulation.

3.2 *Estimation of arsenic intake in celiac children*

After determining the contents of t-As and i-As in rice-based food for celiac children, it seemed desirable to estimate whether the weekly intake could represent a significant risk for celiac children aged 2 to 5 years old. The weight information as affected by the children age was obtained from the Generalitat Valenciana (2012). In this study, low weight percentiles were used: (i) 10 and 14 kg corresponding to the percentile 5 and 50 for children of 2–3 years, and (ii) 14 and 18 kg corresponding to the percentiles 5 and 50 for children of 4–5 years (WHO, 2013).

A menu containing the foods analyzed in the present study was designed according to the needs of children aged 2 to 5 years.

The estimated dietary exposures to i-As of Spanish children suffering from the CD ranged from 0.60 and 0.84 $\mu g\ kg^{-1}\ d^{-1}$ and from 0.61 and 0.78 $\mu g\ kg^{-1}\ d^{-1}$ in children aged 2–3 and 3–5 years old, respectively. The European Food Safety Authority (EFSA reported) in 2009 that the exposure levels in a normal consumer ranged from 0.13 to 0.56 $\mu g\ kg^{-1}\ d^{-1}$, while the exposure level in consumers highly exposed to i-As ranged from 0.37 to 1.22 $\mu g\ kg^{-1}\ d^{-1}$. Consequently, the experimental levels reported in this study completely agreed with the EFSA ranges, and were typical of a highly exposed to i-As group. In this particular case, the situation is worse because there are no many options for the restricted diet of children with the CD. Besides, at these young ages, it has been reported that the human organism is very susceptible to the toxic effects of i-As (Rahman *et al.*, 2009).

4 CONCLUSIONS

There was a positive correlation between rice and As in rice-based products for celiac children; this means that the greater the percentage of rice used in the formulation of the product, the higher the content of As. Gluten-free products that do not contain rice in their formulations, do not contain As. The highest values found were 256 and 128 $\mu g\ kg^{-1}$ in t-As and i-As, respectively, and corresponded to pasta samples. The daily i-As intake from the studied rice-based products ranged from 0.61 to 0.78 $\mu g\ kg^{-1}$ body weight (bw) and it was within the $BMDL_{01}$ considered safe by the EFSA Panel (0.3–8.0 µg/(kg bw *per* day). Therefore, there is no serious and immediate risk to the health of children between 2 and 5 years suffering from the celiac disease. However, these values could be considered as deserving attention because the exposure of celiac people to i-As from rice-based products will start with the detection of their illness and will last all their life. Besides, the addition of other foods to the diet can increase the intake of this contaminant to higher levels.

REFERENCES

Burló, F., Ramírez-Gandolfo, A., Signes-Pastor, A.J., Haris, P.I. & Carbonell-Barrachina, A.A. 2012. Arsenic Contents in Spanish infant rice, pureed infant foods, and rice. *Journal of Food Science* 71: 15–19.

Carbonell-Barrachina, A.A., Wu, X., Ramírez-Gandolfo, A., Norton, G.J., Burló, F., Deacon, C. & Meharg, A.A. 2012. Inorganic arsenic contents in rice based infant foods from Spain, UK, China and USA. *Environmental Pollution* 163: 77–83.

EFSA (Panel on Contaminants in the Food Chain). 2009. Scientific Opinion on Arsenic in Food. *EFSA Journal* 7: 1351.

ESPGHAN (European Society for Pediatric Gastroenterology, Hepatology, and Nutrition Guidelines). 2012. Guidelines for Diagnosis of Coeliac Disease. *JPGN* 54: 1.

Generalitat Valenciana. 2012. *Cartilla de Salud Infantil.* Valencia (Spain): Conselleria de Sanitat, Direcció General de Salut Pública.

Hernández-Martínez, R. & Navarro-Blasco, I. 2013. Survey of total mercury and arsenic content in infant cereals marketed in Spain and estimated dietary intake. *Food Control* 30: 423–432.

Matos-Reyes, M.N., Cervera, M.L., Campos, R.C. & de la Guardia, M. 2010. Total content of As, Sb, Se, Te and Bi in Spanish vegetables, cereals and pulses and estimation of the contribution of these foods to the Mediterranean daily intake of trace elements. *Food Chemistry* 122: 188–194.

Meharg, A.A., Sun, G., Williams, P.N., Adomako, E., Deacon, C., Zhu, Y.G., Feldmann, J. & Raab, A. 2008. Inorganic arsenic levels in baby rice are of concern. *Environmental Pollution* 152: 746–749.

Meharg, A.A., Williams, P.N., Adomako, E., Lawgali, Y.Y., Deacon, C., Villada, A., Cambell, R.C.J., Sun, G., Zhu, Y.G., Feldmann, J., Raab, A., Zhao, F.J., Islam, R., Hossain, S. & Yanai, J. 2009. Geographical variation in total and inorganic arsenic content of polished (white) rice. *Environmental Science and Technology* 43: 1612–1617.

Muñoz, O., Devesa, V., Suñer, M.A., Vélez, D., Montoro, R., Urieta, I., Macho, M.L. & Jalón, M. 2000. Total and inorganic arsenic in fresh and processed fish products. *Journal of Agricultural and Food Chemistry* 48: 4369–4376.

NIDDK (National Institute of Diabetes and Digestive and Kidney Diseases). 2008. Celiac disease. http://digestive.niddk.nih.gov/ddiseases/pubs/celiac/celiac.pdf [Accessed August 2013].

Polanco, I. 2008. *Libro Blanco de la Enfermedad Celíaca.* Madrid (Spain): Consejería de Sanidad de la Comunidad de Madrid and Editorial ICM.

Rahman, A., Vahter, M., Smith, A.H., Nermell, B., Yunus, M., El Areifeen, S., Persson, L.A. & Ekström, E.C. 2009. Arsenic exposure during pregnancy and size at birth: a prospective cohort study in Bangladesh. *American Journal of Epidemiology* 169: 304–312.

Rodrigues, A.F. & Jenkins, H.R. 2006. Coeliac disease in children. *Current Paediatrics* 16: 317–321.

Signes, A., Mitra, K., Burló, F. & Carbonell-Barrachina. A.A. 2008. Effect of two different rice dehusking procedures on total arsenic concentration in rice. *European Food Research and Technology* 226: 561–567.

Sun, G.X., Williams, P.N., Carey, A.M., Zhu, Y.G., Deacon, C., Raab, A., Feldmann, J., Islam, R.M. & Meharg, A.A. 2008. Inorganic arsenic in rice bran and it products are an order of magnitude higher than in bulk grain. *Environmental Science and Technology* 42: 7542–7546.

Sun, G.X., Williams, P.N., Zhu, Y.G., Deacon, C., Carey, A.M., Raab, A., Feldmann, J. & Meharg, A.A. 2009. Survey of arsenic and its speciation in rice products such as breakfast cereals, rice crackers and Japanese rice condiments. *Environment International* 35: 473–475.

WHO (World Health Organization). 2013. The WHO Child Growth Standards http://www.who.int/childgrowth/standards/en/ [Accessed August 2013].

Zhu, Y.G., Sun, G.X., Lei, M., Teng, M., Liu, Y.X., Chen, C., Wang, L.H., Carey, A.M., Deacon, C., Raab, A., Meharg, A.A. & Williams, P.N. 2008. High percentage inorganic arsenic content of mining impacted and nonimpacted Chinese rice. *Environmental Sciences and Technology* 42: 5008–5013.

One Century of the Discovery of Arsenicosis in Latin America (1914–2014) –
Litter, Nicolli, Meichtry, Quici, Bundschuh, Bhattacharya & Naidu (Eds)
 ISBN 978-1-138-00141-1

Arsenic speciation in edible plants and food products: Environment and health implication

L.Q. Ma
State Key Lab of Pollution Control and Resource Reuse, School of Environment, Nanjing University, China
Soil and Water Science Department, University of Florida, Gainesville, USA

J. Luo & D. Zhao
State Key Lab of Pollution Control and Resource Reuse, School of Environment, Nanjing University, China

ABSTRACT: Arsenic is a known carcinogen and is ubiquitous in the environment. Due to its wide presence in the environment, crop plants take up arsenic and make it into food chain, causing adverse impacts on human health. Besides incidental soil ingestion and intake from drinking water, arsenic from food is a major pathway for human exposure to arsenic. Arsenic accumulations in edible plants and food products are mainly associated with the presence of As in soils and irrigation waters. Elevated arsenic has been reported in rice, vegetables and fish, with both organic and inorganic species being detected. This presentation will cover arsenic toxicity and speciation in plants and food, focusing on its environmental and health implication.

1 INTRODUCTION

Arsenic is a known carcinogen and is ubiquitous in the environment. Both natural and anthropogenic activities have elevated arsenic concentrations in terrestrial and aquatic environment (Figure 1). Arsenic has four oxidation states (–III, 0, III and V) and is present as both organic and inorganic forms. Arsenate (As(V)) is more predominant in aerobic soils (typical agricultural soils) and arsenite (As(III)) is present in anaerobic environment (rice field, wetland, and sediment).

As(V) and As(III) are the common arsenic species in the environment, which changes with Eh and pH (Figure 2). While As(V) is present as oxyanion ($HAsO_4^{2-}$ and $H_2AsO_4^-$), As(III) is mostly as neutral species ($HAsO_2^{o}$).

Hence, both oxidation/reduction and methylation/demethylation are important in controlling arsenic species, toxicity and mobility in the environment. Human can be exposed to arsenic via soil, water, and food.

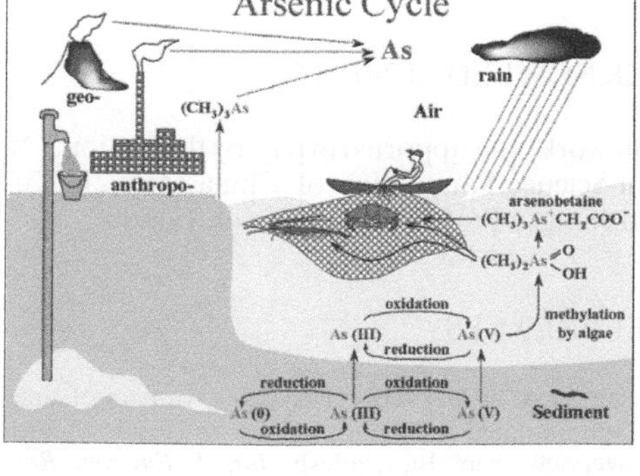

Figure 1. Transformation and cycling of arsenic in the environment.

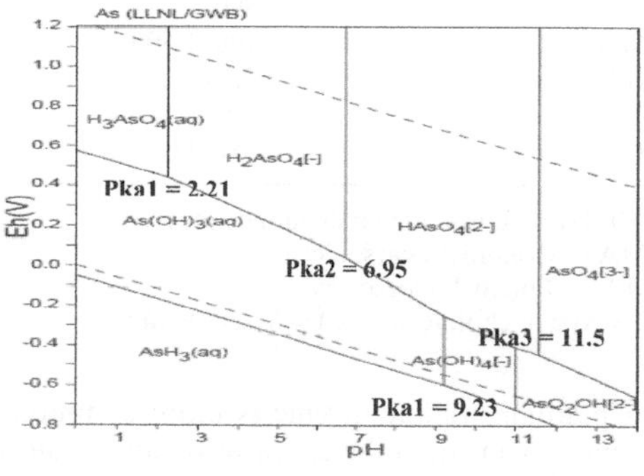

Figure 2. Changes in As(V) and As(III) species under different Eh and pH (Takeno, 2005).

2 ARSENIC SPECIES IN TERRESTRIAL AND AQUATIC ENVIRONMENT

This presentation will discuss various arsenic species in terrestrial and aquatic environment and their associated toxicity. While inorganic arsenic dominates terrestrial environment (soil, groundwater

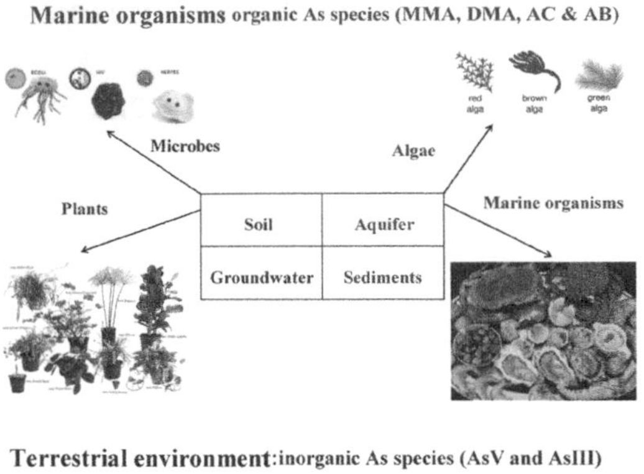

Figure 3. Typical arsenic species present in organisms marine and terrestrial environment.

Table 1. Acute LD_{50} values for arsenic compounds based on oral administration to mice and rats (Hedegaard and Sloth, 2011).

Arsenic species	LD_{50} values ($mg \cdot kg^{-1}$)
AS(III)	15–42
AS(V)	20–800
TETRA	890
MMA	700–1,800
DMA	1,200–2,600
AC	6,500
AB	>10,000

*TETRA = Tetramethylarsonium cation;
MMA = Monomethylarsonic acid;
DMA = Dimethylarsonic acid;
AC = Arsenocholine and AB = Arsenobetaine.

and plants), organic arsenic is more common in marine organisms (algae, seaweed, and seafood) (Figure 3).

In general, organic arsenic is less toxic than inorganic arsenic, with arsenite (As(III)) being more mobile and toxic than arsenate (As(V)) (Table 1).

3 ARSENIC UPTAKE BY FOOD CROPS

As(V) is a P analog and is taken up by biota via P-transporters, inhibiting oxidative phosphorylation and disrupting energy-generation system. As(III) is present as an uncharged species at pH < 9.2 and enters the cell via aquaglycerolporins, binding sulfhydryl groups (SH–) and impairing protein functions.

Plants take up both As(III) and As(V) with different efficiency. Most plants do not take much arsenic, including most crop plants. However, since rice grows in anaerobic environment with As(III) being dominant, rice accumulate arsenic more efficiently than other crops. Hence, arsenic intake through rice consumption has been a global concern. In addition to rice, seaweeds have been shown to contain elevated arsenic (Figure 4).

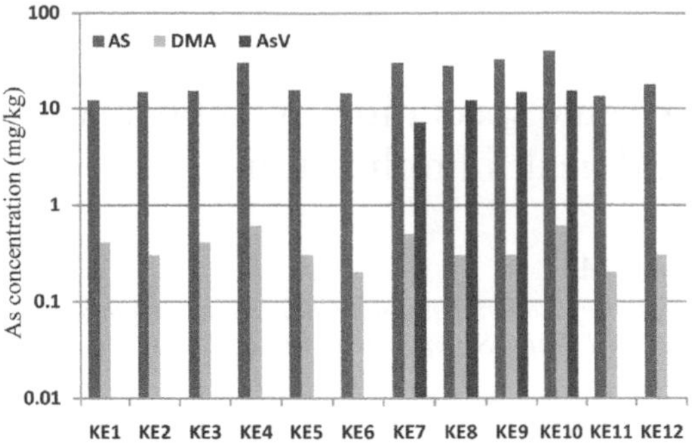

Figure 4. Arsenic speciation in 18 edible seaweeds ($As = 25–83\ mg\ kg^{-1}$) (Nischwitz & Pergantis, 2006).

4 ARSENIC BIOTRANSFORMATION AND TOXICITY

The presentation will also cover biotransformation and toxicity of arsenic by biota. Arsenic oxidation/reduction processes are considered detoxification mechanism for many organisms. While microbial arsenic oxidation and reduction reduce toxicity, it significantly impacts arsenic toxicity in the environment.

Though marine organism contains higher level of arsenic compared to terrestrial organisms, it is mostly present as less toxic organic species including arsenobetaine (AB), arsenocholine (AC) and arsenosugars (AS) (Figure 4). However, some may contain high levels of inorganic arsenic, making it unsafe to consume (Figure 3). Arsenic speciation is important not only for determining arsenic toxicity in the environment but also helps to enhance food safety and minimize human exposure.

ACKNOWLEDGEMENTS

This work was supported in part by the National Natural Science Foundation of China (No. 21277070) and Jiangsu Provincial Innovation Fund.

REFERENCES

Ahsan, D.A. & Del Valls, T.A. 2011. Impact of arsenic contaminated irrigation water in food chain: an overview from Bangladesh. *Int. J. Environ. Res.* 5: 627–638.

Bhattacharya, S., Gupta, K., Debnath, S., Ghosh, U., Chattopadhyay, D. & Mukhopadhyay, A. 2012. Arsenic in the human food chain: the Latin American perspective. *Toxicological Environmental Chemistry* 94: 429–441.

Bundschuh, J., Nath, B., Bhattacharya, P., Liu C-W., Armienta, M.A., Moreno López, M.V., Lopez, D.L., Jean, J.S., Cornejo, L., Lauer Macedo, L.F. & Filho, A.T. 2012. Arsenic in the human food chain: the Latin American perspective. *Science Total Environment* 429: 92–106.

Carbonell-Barrachina, A.A., Signes-Pastor, A.J., Vazquez-Araujo, L., Burl, F. & Sengupta, B. 2009. Presence of arsenic in agricultural products from arsenic-endemic areas and strategies to reduce arsenic intake in rural villages. *Molecular Nutrition Food Research* 53: 531–541.

Hedegaard, R.V. & Sloth, J.J. 2011. Speciation of arsenic and mercury in feed: why and how? *Biotechnol. Agron. Soc. Environ.* 15: 45–51.

Jones, F.T. 2007. A broad view of arsenic. *Poultry Science* 86: 2–14

Liu, Xi.-J., Zhao, Q.-L., Sun, G-X., Williams, P., Lu, X.-J., Cai, J.-Z., & Liu, W.J. 2013. Arsenic speciation in Chinese herbal medicines and human health implication for inorganic arsenic. *Environmental Pollution* 172: 149–154.

Maher, W., Foster, S. & Krikowa, F. 2009. Arsenic species in Australian temperate marine food chains. *Marine Freshwater Research* 60: 885–892.

Nischwitz, V. & Spiros, A.P. 2006. Improved arsenic speciation analysis for extracts of commercially available edible marine algae using HPLC-ES-MS/MS. *J. Agric. Food Chem.* 54: 6507–6519

Rahman, M.A., Hasegawa, H. & Lim, R.P. 2012. Bioaccumulation, biotransformation and trophic transfer of arsenic in the aquatic food chain. *Environmental Research* 116: 118–135.

Roy, P. & Saha, A. 2002. Metabolism and toxicity of arsenic: A human carcinogen. *Current Science* 82: 38–45.

Takeno, N. 2005. Atlas of Eh-pH diagrams: intercomparison of thermodynamic databases. Geological Survey of Japan Open File Report No.419. National Institute of Advanced Industrial Science and Technology.

Uneyama, C., Toda, M., Yamamoto, M. & Morikawa, K. 2007. Arsenic in various foods: cumulative data. *Food Additives Contaminants* 24: 447–534.

Van Hulle, M., Zhang, C., Zhang, X. & Cornelis, R. 2002. Arsenic speciation in Chinese seaweeds using HPLC-ICP-MS and HPLC-ES-MS. *Analyst.* 127: 634–640.

One Century of the Discovery of Arsenicosis in Latin America (1914–2014) – Litter, Nicolli, Meichtry, Quici, Bundschuh, Bhattacharya & Naidu (Eds)
© 2014 Taylor & Francis Group, London, ISBN 978-1-138-00141-1

Total arsenic levels by flow injection hydride generation atomic absorption spectrometry in selected food from Santa Fe, Argentina

N. Hilbe, L. Brusa, H. Beldoménico & M. Sigrist
Programa de Investigación y Análisis de Residuos y Contaminantes Químicos, PRINARC, Facultad de Ingeniería Química, Universidad Nacional del Litoral, Santa Fe, Argentina

ABSTRACT: The total arsenic concentration in selected high consumption food from Santa Fe, Argentina, was determined by Flow Injection Hydride Generation Atomic Absorption Spectrometry (FI-HGAAS) in order to estimate their contribution to the daily dietary intake. A dry ashing procedure was used for the mineralization of the samples. An aqueous standard curve was used to the quantification of arsenic in all matrices. Limits of detection and quantification were 6 and 18 μg kg^{-1}, respectively. The reliability of the method was demonstrated by recovery assays. The average recovery range at a fortification level of 200 μg kg^{-1} was 80–110%. From data, foods that more contributed to arsenic dietary intake were rice and wheat flour.

1 INTRODUCTION

Arsenic exposure by drinking water and food has been associated to adverse effects on human health involving non-cancerous and cancerous diseases. In order to estimate of the total arsenic intake, the contribution of each food group should be considered. Nevertheless, relevant information can be achieved by selecting high consumption food with potentially significant arsenic concentrations. The highest values of arsenic in food have been reported from fish, mainly as organic arsenic compounds, like arsenobetaine, which seems to be non-toxic for humans. Meat and cereals, especially rice, also can be a significant dietary arsenic source. Vegetables and fruit do it lesser extent. Since the dietary arsenic exposure is referred as inorganic arsenic, its contribution to total arsenic can be achieved considering assumptions from the literature data.

In the present study, the measurement of the total arsenic levels of 105 food samples commonly consumed in the central-east region of the Santa Fe province was performed using dry ashing of samples and flow injection hydride generation atomic absorption spectrometry (FI-HGAAS). Subsequently, the contribution of these foods to the total dietary arsenic intake was evaluated and compared with the arsenic contribution by drinking water.

2 EXPERIMENTAL

2.1 *Instrumentation*

A Perkin-Elmer Model 3110 flame atomic absorption spectrometer was used as detector (arsenic hollow cathode lamp set at 193,7 nm wavelength, 11 mA lamp current and 0.7 nm slit width). A Perkin-Elmer FIAS 100 flow injection hydride generation system with a heated quartz tube atomizer was used for hydride generation and coupled to the AAS.

2.2 *Reagents, solutions and samples*

Hydrochloric acid solutions used as carrier solution (12 M) and for dissolution of ashes (4.5 M) were prepared from concentrated HCl (J.T. Baker, USA). Sodium tetrahydroborate solution 0.2% (m/v) used as reductant was prepared daily by dissolving $NaBH_4$ (Merck, Germany) in 0.025% (m/v) sodium hydroxide solution. Ashing aid suspension used for the sample mineralization was prepared by stirring 50 g $Mg(NO_3)_2 \cdot 6H_2O$ Merck (Darmstadt, Germany) in 100 mL of deionized-distilled water. A pre-reducing solution containing 5% potassium iodide (Merck, Germany)—5% ascorbic acid (Merck, Germany) was used to reduce As(V) to As(III) in standard and sample solutions.

Working solutions for external calibration curve and recovery assays were prepared from a 1000 mg L^{-1} As (V) standard (Merck, Germany).

Food samples were purchased from supermarkets and commercial food stores in three cities (Santa Fe, Rafaela, Gálvez) from Santa Fe province (located in the central region of Argentina). All samples, except milk, were milled and homogenized in a food processor and then were subsampled in polyethylene bottles and refrigerated until analyzed shortly.

2.3 *Total arsenic determination by FI-HGAAS*

A dry ashing procedure was employed to mineralize the food samples. A volume of 4 mL of ashing aid suspension was added to 5.0 g ± 0.1 g of homogenized sample in a 50 mL glass beaker. The samples were dried at 100 °C and then placed into a muffle furnace following a heating program of 100 °C/h until 500 °C (12 h). The carbonaceous residue was

Table 1. Analytical figures of merit.

LD[a] µg kg^{-1}	LQ[a] µg kg^{-1}	Mean recovery ± SD %	Precision, RSD[b] %
6	18	99.3 ± 9.4	6.2

[a]Limits of Detection (LD) and Quantification (LQ) were calculated as 3.3 and 10 times Standard Deviation (SD) of the absorbance of 10 reagent blanks divided by the calibration slope.
[b]Relative deviation standard for n = 3 determinations (interday).

wetted with 2 mL of HNO_3 50% (v/v) and a second ashing step was performed. The white ashes were dissolved with 10 mL of 4.5 M HCl. An aliquot of 5 mL was transferred to a 25 mL volumetric flask. To each aliquot, 3.5 mL of concentrated HCl and 2.5 mL of pre-reducing solution were added (45 min of rest). An external calibration curve at concentration levels ranging from 5 to 30 µg L^{-1} of As(III), prepared by reducing of As(V) solutions, was used for quantification of arsenic in the samples.

3 RESULTS AND DISCUSSION

3.1 *Matrix effect and recovery studies*

A study on the combined effect of matrix interferences and analyte recovery was tested by samples fortified prior to the mineralization. The analytical slope of standard addition line on rice samples was compared to the As(III) aqueous standard calibration line. The results obtained showed that the standard addition line slope (0.01203 L µg^{-1}) was statistically comparable with the aqueous standard line (0.01330 L µg^{-1}) so that no effect matrix was observed. The average recovery range obtained for all food at a fortification level of 200 µg kg^{-1} was 80–110%. The use of an As(III) aqueous standard calibration line for arsenic quantification in the food samples allowed a significant simplification of the methodology (replicates $n = 3$; value-$p = 0.01$).

3.2 *Analysis of food by FI-HGAAS*

Data from the analysis of food samples are showed in Table 2. The contribution to arsenic intake was calculated by multiplying the mean arsenic content of each food by its mean consumption in Argentina per person/day as collected in Argentine consumption tables for adults (ENNS, 2007; for women 10–49 years old). Half the LQ was assumed in cases for which the concentration was lower to the limit of quantification. The proportion of inorganic arsenic can be assumed to vary from 50 to 100% of the total arsenic (other than fish) with 70% considered as best, reflecting an overall average. In fish, the relative proportion is much smaller (EFSA, 2009). Other specific factors, such as traditional cooking processes, would be considered for better

Table 2. Arsenic concentration in food (µg g^{-1}, fresh weight).

Sample	n	Range µg kg^{-1}	Mean µg kg^{-1}
Cereals and derivates			
Rice (polished)[a]	27	87–316	180
Wheat flour (leavening)	8	<LQ-73	36
Corn flour	5	ND	ND
Oatmeal	4	<LQ-25	18
Breakfast cereals (bars, flakes)	3	<LQ-70	45
Legumes			
Lentil	5	<LQ-19	<LQ
Soybean	2	<LQ-21	<LQ
Chickpea	2	20–44	32
Lima bean	2	<LQ	<LQ
Vegetables			
Potato	4	<LQ	<LQ
Milk and dairy products			
Cow's whole milk (raw)	10	ND	ND
Cheese[b]	6	<LQ-77	28
Egg	5	ND	ND
Meat			
Beef	5	<LQ-29	<LQ
Fish	17	152–439	330

[a]Varieties: Doble Carolina, Parboil, Largo Fino, Largo Ancho.
[b]Types: Criollo, Chubut, Reggianito, Muzzarella, Pategras. ND: not detected.

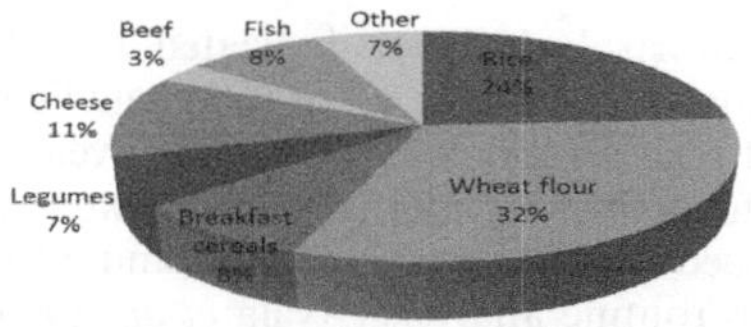

Figure 1. Percent contribution to As intake for each food.

intake estimation. From data, the inorganic arsenic exposure from selected food (estimated intake of 34 µg total As day^{-1}) would be significantly lower than that from drinking water containing 50 µg As L^{-1} (usually as inorganic arsenic) for an consumption of 2 L day^{-1} (100 µg As day^{-1}).

4 CONCLUSIONS

A reliable methodology for the determination of total arsenic in food at trace levels is provided. Data contribute to the nutritional information on the background arsenic levels in the diet from Santa Fe.

REFERENCES

EFSA Panel on Contaminants in the Food Chain (CONTAM) (2009). Scientific Opinion on Arsenic in Food.
Encuesta Nacional de Nutrición y Salud (2007). Plan Federal de Salud. Ministerio de Salud de la Nación Argentina.

*One Century of the Discovery of Arsenicosis in Latin America (1914–2014) –
Litter, Nicolli, Meichtry, Quici, Bundschuh, Bhattacharya & Naidu (Eds)*

Ultrasonic-Assisted Enzymatic Digestion (USAED) for total arsenic determination in cooked food: A preliminary study

E.E. Buchhamer & M.C. Giménez
Departamento de Química Analítica, UNCAus, Pcia, R. Sáenz Peña, Chaco, Argentina

J.A. Navoni & E.C. Villaamil Lepori
Cátedra de Toxicología y Química Legal, FFyB, UBA CABA, Buenos Aires, Argentina

ABSTRACT: The aim of this study was to compare different enzymatic hydrolysis procedures for the extraction and determination of total arsenic in cooked foods by FI-HG-AAS. According to the characteristics of the studied food (rice stew, vegetables and meat cooked with water with arsenic concentration of 220 mg/L), two enzymes (cellulase and pancreatin) were selected and combined with two processes of agitation (thermostatic bath/ultrasound). The reliability of the procedures was tested with certified reference material (Brown rice SYTED105PI0272) and the results were compared with the conventional method for extracting total arsenic Dry Ashing. The recoveries obtained were as follows: 101% for the circulator-pancreatin, 104% for thermostatic bath-cellulase, 99% for ultrasound-cellulase and 85% for ultrasound-pancreatin. The mean comparison test T ($p < 0.05$) showed that the values of total arsenic extracted in the first three procedures were similar, with very good yields, being different and less ultrasound-pancreatin combination.

1 INTRODUCTION

Many efforts have been dedicated to solve the problems of extraction of organometallic species in food-related samples. Although solvent extraction (chloroform or methanol) have worked well, the procedures are long, laborious and difficult to apply in routine analysis (Ayala *et al.*, 1999). The efficacy of enzymes is well known as a procedure to extract biomolecules attached to organometallic species. The results obtained with trypsin showed its effectiveness in arsenic removal in materials with a high protein content (fish and shellfish), as well as cellulase and pancreatin enzymes on vegetal material for which trypsin is not useful (Morán *et al.*, 2003). As far as arsenic extraction is concerned, the enzymatic hydrolysis of biological samples accelerated by ultrasound (known as ultrasonic assisted enzymatic digestion, USAED) is currently the most powerful tool used to extract the total content of As whilst maintaining As-species integrity. It can be used for total elemental determination and elemental speciation (Vale *et al.*, 2008). The aim of this study was to compare different enzymatic hydrolysis procedures for the extraction and quantitative determination of total arsenic in cooked foods by flow injection hydride generation-atomic absorption spectrometry (FI-HG-AAS).

2 METHODS/EXPERIMENTAL

2.1 *Food samples*

Samples were prepared in the laboratory. These samples were cooked in accordance with the cooking in customs of the population in the region central of Chaco, Argentina. The raw products were collected from various businesses using a random collection. The cooking treatments applied to the food were those indicated in the dietary recall questionnaires as the most frequent. Rice stew, vegetables and meat were selected for this study, and were cooked with water with an arsenic concentration of 200 μg/L). The sample was dried and kept in the freezer at 4 °C until analysis.

2.2 *Digestion procedure and measure of total arsenic*

Two enzymes (cellulase and pancreatin) were selected and combined with two agitation processes; the first was assisted by a heated water bath (12 h at 37°C and 100 rpm) and the second by ultrasound (20 min at 400 W of power and 40 kHz of frequency). For the validation of the methods, a certified reference material was analyzed, Brown Rice CYTED105PI0272, and compared with the conventional method for extracting total arsenic Dry Ashing.

Determination of total arsenic in cooked foods was performed using an atomic absorption spectrometer Varian SpectrAA 220 by flow injection hydride generation-atomic absorption spectrometry (FI-GH-AAS). Arsenic concentrations found in the enzymatic extracts were analyzed using one-way ANOVA ($p < 0.05$), using the T test for comparison of means using the statistical software SPSS 17.0.

3 RESULTS AND DISCUSSION

Table 1 showed that the percentage of recovery obtained was 102% for water Bath-Pancreatin (BP), 100% for water Bath-Cellulase (BC), 99% for Ultrasound-Cellulase (UC) and 85% for Ultrasound-Pancreatin (UP), compared with the conventional method for extracting total arsenic Dry Ashing (DA).

Arsenic concentration found in the enzymatic extracts were analyzed and compared using one-way ANOVA ($p < 0.05$), with the T test for comparison of means using the statistical software SPSS 17.0. The first three methods showed no significant differences ($p < 0.05$), but not the combination ultrasound-pancreatin, which showed a lower yield.

Enzymes can be used as a method for extracting soft organometallic species linked to biomolecules. A general drawback is the long incubation time and the high cost of reagents. This led researchers to develop a rapid and simple method consisting of a combination of ultrasound and the use of enzymes. The ultrasound breaks cell walls, facilitating the interaction of the enzymes with their components and, reducing the treatment time to only minutes.

The values obtained allow to conclude that the cellulase enzyme has an improved performance for AST removal in cooked food samples selected for this study. It is also concluded that the ultrasonic-enzyme combination as enzymatic hydrolysis process has the advantage of reducing treatment times and reagent consumption.

Table 1. Concentration of arsenic and % of recovery in the different extraction methods applied in cooked food samples and the MR (Brown rice).

Method	Cooked foods As (µg/g)	Recovery %	Brown rice As (µg/g)	Recovery %
DA	0.757 ± 0.030	–	0.492 ± 0.016	–
BC	0.788 ± 0.048	104%	0.471 ± 0.040	100%
BP	0.763 ± 0.052	101%	0.492 ± 0.001	102%
UC	0.750 ± 0.031	99%	0.501 ± 0.021	100%
UP	0.641 ± 0.045	85%	0.492 ± 0.051	96%

Value of Reference MR Brown rice CYED105PI0272: As 0.508 ± 0.032 as AsT.

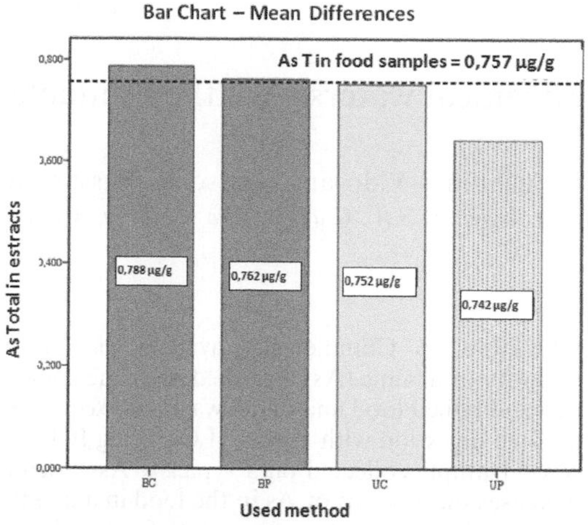

Figure 1. Mean differences—total arsenic in extracts.

4 CONCLUSIONS

For both procedures, the cellulase enzyme showed their effectiveness in the extraction of AsT in the cooked food samples considered, but not for the case of pancreatin enzyme. The use of ultrasonic-assisted enzymatic digestion as enzymatic hydrolysis procedure for extraction of arsenic in cooked food samples has the great advantage of being simple, effective and reduce treatment times and reagent consumption. It was not possible to obtain good results with the pancreatin enzyme and further studies are needed to improve its effectiveness.

ACKNOWLEDGEMENTS

This work was supported with funds from PI 36/00005 UNCAus (Universidad Nacional del Chaco Austral).

REFERENCES

Ayala, J., Garro, O.A. & Giménez, M.C. 1999. Extracción enzimática de arsénico en muestras de origen vegetal. *Acta Reunión de Comunicaciones Científicas y Técnicas* (V): 67–70.

Morán, E., Osicka, R.M., Giménez, M.C. & Garro, O.A. 2003. Utilización de enzimas para la extracción de As en muestras vegetales y su determinación por espectrometría por absorción atómica con generación de hidruro (HG-AAS) E-059. *http:www1.unne.edu.ar/cyt2003/comunicaciones/cyt.htm.*

Vale, G.A., Rial-Otero, R., Mota, A.M., Fonseca, L.P. & Capelo, J.L. 2008. Ultrasonic—assisted enzymatic digestion (USAED) for total elemental determination and speciation: A tutorial. *Talanta*. 75: 872–884.

*One Century of the Discovery of Arsenicosis in Latin America (1914–2014) –
Litter, Nicolli, Meichtry, Quici, Bundschuh, Bhattacharya & Naidu (Eds)*
 ISBN 978-1-138-00141-1

Arsenical waters: Are they suitable for cooking food?

I. Cabanillas-Vidosa, E. Rosso, R. Bassani, M. Corelli & R. Sandrini
JLA Argentina S.A., Gral. Cabrera, Córdoba, Argentina

ABSTRACT: Commercially available rice and soybean were cooked and analyzed to determine the quantity of arsenic (As) absorbed from the cooking water. The evaluation for the total As content in the raw and boiled food and in the water on excess from the cooking process was determined after microwave assisted digestion with HNO_3-H_2O_2, using ICP-MS. The study shows that the cooking process using tap water (sample collected on a typically As-endemic area as General Cabrera City, Córdoba, Argentina) increases the amount of As in the food in a quantity that depends upon cooking conditions: the cooking time combined with the amount of food and the volume of water used. On the other hand, under the experimental conditions of this work, the use of arsenic-free water leads to a decrease in the amount of As present in the food, due to an arsenic-transfer from the sample to the cooking water.

1 INTRODUCTION

In its most recent report (September 6, 2013), The Food and Drug Administration of United States (FDA) notes that "rice is a life-long dietary staple for many people" and does not recommend changes by consumers regarding their consumption of rice and rice products. Scientists from this agency "determined that the amount of detectable arsenic is too low in the rice and rice product samples to cause any immediate or short-term adverse health effects". The assessment developed for these types of communications, focuses on the raw product and generally does not consider the preparation for its consumption. The last point becomes important in regions such as Argentina, with high levels of arsenic in tap water, where the effect actually is not totally clear. Therefore, the main purpose of the present study was to evaluate the effects of traditional cooking process of foods, using waters samples with different levels of arsenic.

2 EXPERIMENTAL

2.1 *Sampling*

For this study, water with three different levels of arsenic was used: groundwater from the rural area of Carnerillo, Córdoba (depth: 25 m) containing high levels of arsenic; tap water from General Cabrera, Córdoba (As-endemic area), and commercially available mineral water from Mendoza, Argentina. This samples were evaluated in the cooking process of foodstuffs and, for this purpose, rice and soybean (randomly purchased on local markets of General Cabrera), foods typically with high and low content of arsenic, respectively, were used.

2.2 *Cooking processes*

The boiling processes were carried out in a gas stove, since this is the most used by the population of Argentina. On the case of rice, approximately 90 g of sample were added to boiling water (420 mL) in an open glass vessel (12 cm in diameter), and the time for the cooking process was 11 minutes. As a result, remaining water of cooking process and boiled rice were obtained. On the other hand and prior to the cooking process, the soybean (90 g) was placed in the glass vessel containing 570 mL of water and left to stand by 12 hours. After that, soaked soybean and water were boiled by 75 min in the same conditions as the rice, and remaining water of cooking process and boiled soybean were collected to perform its analysis.

2.3 *Analytical procedure*

The level of As was determined by ICP-MS, previous digestion of the samples. About 0.5 g of homogenized boiled food (or 2 mL of remaining water of cooking process) were digested with 4 mL of HNO_3 plus 2 mL of H_2O_2 in quartz vessels with a microware digestion system (Anton Paar Multiwave 3000), heated up to 200–210 °C for 30 min. The resulting solutions were diluted to a final volume of 50 mL with 5% HNO_3. The instrumental technique chosen for the arsenic determination was Inductively Coupled Plasma Mass Spectrometry (ICP-MS, Perkin Elmer NexION 300X), and the limit of detection of the method developed in JLA Argentina S.A. are 0.1 μg L^{-1} and 0.01 mg kg^{-1} for aqueous and foodstuff matrices, respectively, using germanium as internal standard.

2.4 *Quality control*

All determinations were developed in duplicate, where the relative difference between replicates was <10%. For sets of every eight digested samples, reagent blank and spiked sample were performed to check possible cross-contamination, interferences and recovery rates (ranged between 80 and 110%).

3 RESULTS AND DISCUSSION

3.1 *Characterization of the original samples*

The concentration of As determined in the raw food and original water samples are listed in Table 1.

As can be seen, the levels of arsenic in both food samples, as well as in the three water samples are very different from each other. These differences allow carrying out an unequivocal study about the effect of cooking process on the As concentrations.

3.2 *As in products of cooking process*

Table 2 summarizes the As concentrations in products of the cooking process.

Table 1. Arsenic concentration in the original samples.

	Arsenic	
Sample	mg kg^{-1}	mg L^{-1}
Raw rice	0.251	
Raw soybean	0.030	
Mineral water		0.001
Tap-water		0.031
Groundwater		0.261

Table 2. As concentration in products of cooking processes.

	Arsenic in products of cooking processes	
Water	Rice	Soybean
Mineral	Boiled: 0.074 mg kg^{-1} Water: 0.037 mg L^{-1}	Boiled: 0.012 mg kg^{-1} Water: 0.009 mg L^{-1}
Tap	Boiled: 0.091 mg kg^{-1} Water: 0.067 mg L^{-1}	Boiled: 0.069 mg kg^{-1} Water: 0.108 mg L^{-1}
Ground	Boiled: 0.241 mg kg^{-1} Water: 0.329 mg L^{-1}	Boiled: 0.335 mg kg^{-1} Water: 0.594 mg L^{-1}

As a general trend, the contamination levels increase on cooked food as a function of the arsenic concentration in the water used. The total quantity of As on the whole system remains about constant during the process, but there is a remarkable increasing of the concentration due to the water evaporation. The weight of the boiled rice was about 290 g, and boiled soybean, 204 g. This indicates that 200 g and 114 g of water cooking were absorbed in the processes by the rice and soybean, respectively.

Mineral water. The arsenic concentration in the boiled food decreases relative to the raw samples but, in the cooking process: is there a true arsenic-transfer from the food to the remaining water? The analysis of the amounts (and not concentrations) of arsenic in the system should answer this question. The samples of 90 g of original raw food have 22.6 μg (rice) and 2.8 μg (soybean) of arsenic. After the cooking processes, the boiled rice (290 g) has 21.5 μg, whereas the boiled soybean (204 g), 2.4 μg. These amounts indicate that, under the present experimental conditions, the 4.9% and 11.1% of the total original amount of arsenic is transferred from the rice and soybean samples to the water, respectively.

Tap water. For the rice case, the 0.031 mg L^{-1} of As on the water leads to the sample absorbs 4.1 μg in the process, which represents an increasing of 18.1% relative to the amount original present in the raw food. It is in agreement with the results of previous studies (Rahman *et al.*, 2006; Bae *et al.*, 2002; Perelló *et al.*, 2008), where the concentration of As in cooked rice was higher than in raw rice, suggesting a chelating effect by rice grains or a concentration of As because of water evaporation during cooking or both. For soybean, the same but stronger trend was observed: the quantity of As absorbed from the water was 11.4 μg, which is an increase of 422.2%.

Groundwater. The original amounts of 22.6 μg and 2.8 μg of arsenic in 90 g of raw food increase to 69.4 and 71.0 μg in the boiled rice and soybean samples, respectively. As expected, the groundwater is inappropriate to be used in the process of cooking food.

4 CONCLUSIONS

Water with different levels of arsenic was evaluated in the cooking process of foodstuff. The use of arsenic-free water leads to a decrease in the amount of As present in the food, due to an arsenic-transfer from the sample to the cooking water. In addition, the use of arsenical tap water (from typical endemic areas) should be avoided, because, in this case, the transfer occurs from the water to the foodstuff.

ACKNOWLEDGEMENTS

The authors would like to express gratitude to Dr. S. Farías (from CNEA) for her valuable comments on the ICP-MS analysis, and to J.M. Leek for providing the facilities to carry out this work.

REFERENCES

Bae, M., Watanabe, C., Inaoka, T., Sekiyama, M., Sudo, N., Bokul, M.H. & Ohtsuka, R. 2002. Arsenic in cooked rice in Bangladesh. *The Lancet* 360: 1839–1840.

Food and Drug Administration of United States (FDA). 2006. http://www.fda.gov/food/foodborneillnesscontaminants/metals/ucm319870.htm

Perelló, G., Martí-Cid, R. & Llobet, J.M. 2008. Effects of various cooking processes on the concentrations of arsenic, cadmium, mercury, and lead in foods. *J. Agric. Food Chem.* 56: 11262–11269.

Rahman, M.A., Hasegawa, H., Rahman, M.A., Rahman, M.M. & Miah, M.A.M. 2006. Influence of cooking method on arsenic retention in cooked rice related to dietary exposure. *Sci. Total Environ.* 370: 51–60.

One Century of the Discovery of Arsenicosis in Latin America (1914–2014) – Litter, Nicolli, Meichtry, Quici, Bundschuh, Bhattacharya & Naidu (Eds)
© 2014 Taylor & Francis Group, London, ISBN 978-1-138-00141-1

Intake estimated of inorganic arsenic and bioaccessible fraction in the diet from Pastos Chicos, Susques, Argentina—preliminary study

D. Choque, A. Quiquinto & G. Bovi Mitre
Grupo INQA, Facultad de Ciencias Agrarias, Universidad Nacional de Jujuy, Argentina

J.A. Navoni & E.C. Villaamil Lepori
Cátedra de Toxicología y Química Legal, Facultad de Farmacia y Bioquímica, Universidad de Buenos Aires, Argentina

ABSTRACT: Surface water and groundwater in Pastos Chicos, Jujuy, are naturally contaminated with arsenic (As). With the aim to estimate the intake of As present in the diet of the population, the concentration of total and inorganic arsenic (TAs and IAs) in drinking water and prepared food (solid and liquid), and the bioaccessible fraction in solid food, was investigated by hydride generation atomic absorption spectrometry. The TAs in water was in a range from 700 and 820 μg L^{-1} and in food from 20 to 1920 μg kg^{-1} f.w. (fresh weight), corresponding to IAs 82 to 115%. The IAs bioaccessible fraction was 63 ± 9% of TAs in solid food. Taking into account drinking water and cooked food, the IAs intake exceeded the Provisional Temporary Weekly Intake (PTWI) in 15 times. High levels of arsenic in cooked food by the use of water with As suggest that they must be considered in the evaluation of exposure.

1 INTRODUCTION

Pastos Chicos at Jujuy Province has a history of natural water pollution by As, according to data reported by De Sastre *et al.*, (1987) and Tschambler *et al.*, (2007). In 2012, new samplings were performed, and the presence of this toxic in different water sources in this place both in ground- and surface water was confirmed. The main source of drinking water for the population contains 820 μg/L of As (Choque *et al.*, 2013). Not only the prepared food but also the agricultural products are cultivated or irrigated using contaminated water.

The aim of this study was to estimate the intake of inorganic arsenic (IAs), taking into account the bioaccessible fraction, for the adult rural population of Pastos Chicos.

2 METHODS/EXPERIMENTAL

2.1 *Sample collection*

The sample collection was performed in the period January–April 2012. Water samples of each family and 79 samples of food ready to consume, corresponding to the four basics daily meals (breakfast, lunch, snack and dinner), were collected from families of the studied area.

2.2 *TAs and IAs quantification*

Total Arsenic (TAs) and Inorganic Arsenic (IAs) water and food content was quantified by Hydride Generation-Atomic Absorption Spectrometry (HG-AAS) following the methods described by Muñoz *et al.*, (2002) (Navoni, 2012).

2.3 *Bioaccesibility assay*

Food ready to consume (N = 13) were analyzed following the *in vitro* static method of gastrointestinal digestion described by Torres Escribano *et al.*, (2011). Then, the soluble bioaccessible fraction was determined by HG-AAS.

2.4 *Exposure estimation*

The As intake estimation was calculated taking into account the amount of As in food and water for each day, the weight of each portion of food, the volume of liquid ingested and the body weight of 60 kg.

The estimated IAs intake were calculated as indicated by EPA (1998) and the results were compared with the Provisional Tolerable Weekly Intake (PTWI) recommended by FAO (WHO/FAO, 1989).

3 RESULTS AND DISCUSSION

3.1 *TAs in water and food*

The concentration of As in drinking water was between 700 and 820 μg L^{-1} (N = 3). Total arsenic content in solid food (N = 47) was in a range between 20 to 1920 μg kg^{-1} f.w. (fresh weight), and in liquid foods (N = 32) between 477 to 1162 μg L^{-1}

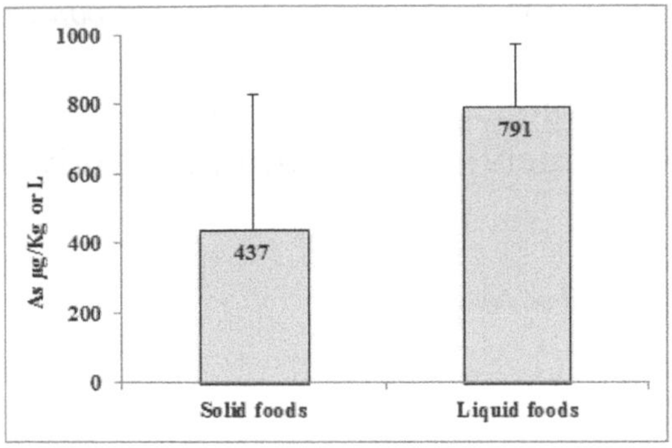

Figure 1. Total Arsenic in solid and liquid food. Results were expressed as mean and standard deviation.

with mean value of 437 ± 388 µg kg^{-1} f.w. and 791 ± 181 µg L^{-1} respectively (Figure 1).

3.2 *IAs in food*

AsI concentration was in a range of 50 to 1300 µg kg^{-1} f.w., which corresponds to 82.2 and 115% of the AsT found (Wong *et al.*, 2013). These results are in agreement with those previously reported in non-marine food (EFSA, 2009).

3.3 *Bioaccesibility*

Bioavailable fraction for gastrointestinal absorption of IAs was between 26 to 107% of the TAs present in the daily food. Most of the samples analyzed (77%) overcame the 50%. In solid food, the IAs bioaccessible fraction of IAs was 63% ± 9%.

3.4 *Intake estimation*

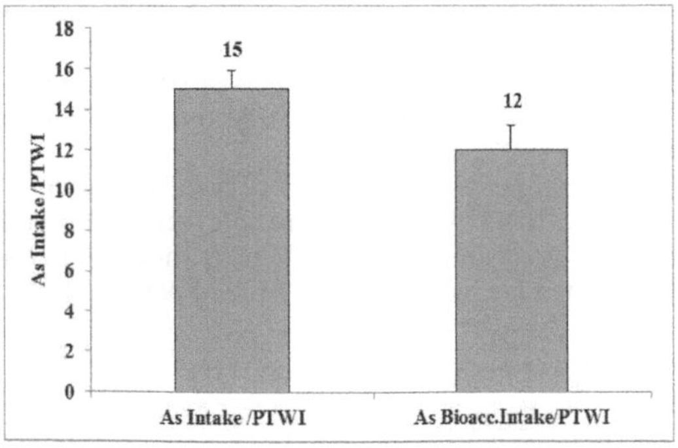

Figure 2. Comparison of estimated intake of AsI and bioaccessible fraction with the Provisional Tolerable Weekly Intake (PTWI) of 15 µg kg^{-1} day^{-1}.

4 CONCLUSIONS

High levels of arsenic in cooked food due to the use of water containing arsenic increase the intake of arsenic and must be considered in the evaluation of exposure.

The exposure was almost exclusively related to inorganic arsenic species.

Most of arsenic in the diet is inorganic and bioaccessible, becoming a risk to human health.

The estimates performed suggest an effective exposure through diet of the population to the toxic arsenic.

ACKNOWLEDGEMENTS

This work was financially supported by project PICTO UNJu 2008/140 "Contaminantes, riesgos y alternativas para el desarrollo sustentable de agricultores localizados entre los valles fértiles andinos de la Quebrada de Humahuaca y la baja Puna Jujeña".

REFERENCES

Choque, D., Navoni, J., Farías, S., Bovi Mitre, G. & Villaamil Lepori, E. 2013. Arsénico y otros elementos en aguas de Pastos Chicos y Susques - Puna Jujeña. In *XVIII Congreso Argentino de Toxicología*, Bs. As. *Acta Tox. Arg.* 21 (sup) 45–46.

EFSA. 2009. *Scientific Opinion on Arsenic in Food*. European Food Safety Authority Journal 7(10): 1351.

Muñoz, O., Díaz, O.P., Leyton, I., Núñez, N., Devesa, V., Súñer, M.A., Vélez, D. & Montoro, R. 2002. Vegetables collected in the cultivated Andean area of the Nothern Chile. Total and inorganic arsenic contents in raw vegetables. *J. Agric. Food Chem.* 50: 642–647.

Navoni, J.A. 2012. Toxicidad del arsénico: *Evaluación de riesgo en poblaciones expuestas crónicamente al arsénico. PhD thesis*. Cátedra de Toxicología y Química Legal. Facultad de Farmacia y Bioquímica. UBA.

Torres Escribano, S. 2011. Bioaccesibilidad de arsénico y mercurio en alimentos con potencial riesgo toxicológico. PhD thesis. Universidad de Valencia.

Tschambler, J., Cabrera, R. & Bovi Mitre, M.G. 2007. Georreferenciamiento del contenido de Arsénico en aguas de la Provincia de Jujuy-Argentina. In *XV Congreso Argentino de Toxicología, 26 al 28 de septiembre, Neuquén*, 48.

WHO. 1989. Evaluation of certain food additives and contaminants. 33rd report of the Joint FAO/WHO expert Committee on Food Additives. WHO Technical Report Series 759 (Geneva, WHO).

Wong, W.K., Chung, S.W., Chan, B., Ho, Y.Y., Xiao, Y. 2013. Dietary exposure to inorganic arsenic of the Hong Kong population: Results of the first Hong Kong Total Diet Study. *Food and Chemical Toxicology* 51: 379–385.

One Century of the Discovery of Arsenicosis in Latin America (1914–2014) – Litter, Nicolli, Meichtry, Quici, Bundschuh, Bhattacharya & Naidu (Eds)
© 2014 Taylor & Francis Group, London, ISBN 978-1-138-00141-1

Determination of arsenic in food from Brazilian dietary

G.C. Silva, V.S.T. Ciminelli, G. Duarte, A.M. Oliveira & T. Paschoali
Departamento de Engenharia Metalúrgica e de Materiais, Escola de Engenharia, Universidade Federal de Minas Gerais, Belo Horizonte, Brazil
INCT-Acqua, Belo Horizonte, Brazil

J.C. Ng
INCT-Acqua, Belo Horizonte, Brazil
EnTox, The University of Queensland, Brisbane, Australia

M. Gasparon
INCT-Acqua, Belo Horizonte, Brazil
School of Earth Sciences, The University of Queensland, Brisbane, Australia

ABSTRACT: A food survey was conducted focusing on most commonly consumed staple dietary items obtained from local markets of three different towns in the State of Minas Gerais, Brazil. The samples were analyzed for total arsenic. The daily intake was estimated based on a Brazilian food consumption statistics and compared with the recommendation of international organizations. The data represent national samplings; the analyses were carried out by laboratories approved in inter-laboratory evaluations. Rice and beans alone contribute to approximately 90% of the total As intake from food. The calculated daily intake (0.399 μg/kg of body weight per day) is below the $BMDL_{0.5}$ of 3.0 μg kg^{-1} of body weight per day.

1 INTRODUCTION

Arsenic contamination in food is internationally recognized as a public health concern (JECFA, 2011; EFSA, 2009). Considerable As levels found in food contribute significantly to As intake in different parts of the world (JECFA, 2011). However, little is known about arsenic levels in Brazilian staple food and its contribution to the daily arsenic intake, especially in gold mining regions, where arsenic naturally occurs.

According to the aforementioned context, the present work aims to analyze food samples purchased from towns in the State of Minas Gerais, Brazil, for total arsenic, and estimate the daily intake based on a Brazilian food consumption statistics.

2 METHODS/EXPERIMENTAL

A food survey was conducted focusing on most commonly consumed staple dietary items (totaling 71 samples) obtained from local markets. Fresh vegetables and other dried food items were transported to UFMG for freeze drying and powdering. Moisture content was determined in the fresh products. Rice, beans and pasta were already in dried form when obtained. Aliquots of about 10–20 g were put into plastic zip-lock bags and hand-carried back to Australia for analysis (Entox, the University of Queensland).

The powdered samples were further ground to finer powder with a high-speed coffee blender to improve homogeneity of the sample matrix. An aliquot of about 0.3 g was accurately weighed for acid digestion using a microwave digestion system in accordance with a standard operation procedure from a NATA registered laboratory before ICP-MS analysis. Arsenic concentrations were adjusted for the moisture content and expressed in fresh weight with the exception of dried items such as rice, pasta and beans.

Certified Reference Materials (CRM: Tomato Leaves NCS-ZC-85006, Parsnip Root Powder CS-PR-2, Dog Fish Muscle DORM-3 and Bovine Liver 1577b were analyzed in the same manner.

3 RESULTS AND DISCUSSION

Food results adjusted for moisture contents are shown as mean ± SD of each food category in Table 1. The mean concentration of each food

Table 1. Range, average and standard deviation (SD) of arsenic concentrations (mg/kg) of food items surveyed expressed in fresh weight unless specified by * (* = dry weight as purchased).

Food Item	Range [As]	Mean [As] ± SD#
Egg	0.011–0.015	0.013 ± 0.002
Chicken	0.010–0.045	0.021 ± 0.015
Meat	0.012–0.030	0.021 ± 0.007
Fish	0.048–0.417	0.233 ± 0.261
Potato	0.009–0.010	0.009 ± 0.001
Carrot	0.005–0.008	0.007 ± 0.001
Tomato	0.003–0.008	0.005 ± 0.002
Garlic	0.008–0.020	0.013 ± 0.005
Cabbage	0.004–0.009	0.006 ± 0.002
Lettuce	0.007–0.031	0.017 ± 0.011
Milk	0.002–0.004	0.003 ± 0.001
Coffee*	0.038–0.076	0.049 ± 0.060
Beans*	0.046–0.223	0.076 ± 0.060
Rice*	0.043–0.150	0.122 ± 0.043
Pasta*	0.042–0.048	0.045 ± 0.002

SD indicates the broad range of concentrations in different brands of each food item.

Table 2. Dietary inorganic arsenic (iAs) intake from the food survey calculated from the average arsenic concentrations.

Food	iAs (%)#	DC (g)	Intake (μg/kg/d)	% of Total
Egg	50	12	0.0011	0.28
Chicken	50	37	0.0056	1.40
Meat	50	63	0.0095	2.38
Fish	10	23	0.0077	1.93
Potato	50	15	0.0010	0.25
Carrot	50	3	0.0002	0.05
Tomato	50	3	0.0001	0.03
Garlic	50	3	0.0003	0.08
Cabbage	50	3	0.0001	0.03
Lettuce	50	3	0.0004	0.10
Milk	50	35	0.0008	0.20
Coffee*	50	30	0.0105	2.63
Beans	50	183	0.0993	24.87
Rice	90	160	0.2510	62.88
Pasta	50	36	0.0116	2.91
Daily total			0.3992	100.00

= arsenic speciation assumption; * assume 30 g of coffee beans in 215 mL of brewed coffee; DC = daily consumption. For small sample size of each food category, the mean value is appropriate for risk calculation, whereas for larger sample size, medium and 75th percentile values are often used.

category was used to calculate the arsenic daily intake. Arsenic concentrations of Certified Reference Materials (CRMs) were in general agreement of those certified values.

Relatively higher total arsenic concentrations were found in fish (0.233 ± 0.261 mg/kg, or μg/g), rice (0.122 ± 0.043 mg/kg), beans (0.076 ± 0.060 mg/kg), coffee (0.049 ± 0.014 mg/kg) and pasta (0.045 ± 0.002 mg/kg). Other items had relatively low arsenic concentrations. It should be noted that the consumption of fish in a typical Brazilian diet is relatively low and that the inorganic arsenic in fish is likely to be lower than the total arsenic concentration.

A recent survey of the consumption pattern of a Brazilian diet (IBGE, 2008/09) was used together with our own dietary elemental concentrations to calculate the arsenic daily intake (μg/kg of body weight (bw)/day) considering an adult of 70 kg (Table 2). It is also important to know the inorganic arsenic (arsenic speciation) component of the total arsenic level. Although arsenic speciation has not been done for this work, assumptions were made based on literature data (JECFA, 2011).

The results have led to As in rice and beans being consumed daily at higher amounts than from the other food items. These two items alone contributed to approximately 90% of the total intake from food. Rice is recognized as a significant source of inorganic arsenic especially in Asian countries and other countries where rice is a staple food (JECFA, 2011). Brazil is considered to be a country where rice is also a staple food item. The World Health Organization (WHO) has recently reviewed the rice arsenic issues and recommends 0.3 mg/kg as the maximum level for arsenic in rice (WHO, 2012). The mean value found for rice in this work (0.12 mg/kg) is similar to the mean value found for rice in the world (0.14 mg/kg) (WHO, 2012).

Despite the variations among the reviews and standards set by various agencies, the Brazilian arsenic exposure from the diet is within the range of available arsenic intake data from many parts of the world (JECFA, 2011).

The Joint FAO/WHO Expert Committee on Food Additives (JEFCA) determined the inorganic arsenic Benchmark Dose Lower Limit for a 0.5% incremental increase in incidence of lung cancer, $BMDL_{0.5}$ (3.0 μg/kg of body weight per day) by using a range of assumptions to estimate exposure from drinking-water and food with differing concentrations of inorganic arsenic (JECFA, 2011). The risk associated with the arsenic exposure from food—0.399 μg/kg bw/d—can be considered low by comparison to the $BMDL_{0.5}$.

4 CONCLUSIONS

Considering the average concentrations of arsenic in the food samples, the calculated daily intake (0.399 μg/kg bw/d) is below the $BMDL_{0.5}$ (3.0 μg/kg bw/d). Rice and beans are the main contributors to the As intake from food (90%).

ACKNOWLEDGEMENTS

Authors are grateful to the Brazilian agencies (CNPq, CAPES and FAPEMIG) for financial support and fellowships. The support from Kinross Brazil Mineração S/A is also acknowledged.

REFERENCES

EFSA Panel on Contaminants in the Food Chain (CONTAM). 2009. Scientific Opinion on Arsenic in Food. *EFSA Journal* 7(10): 1351.

IBGE. 2008/2009. Análise do Consumo Alimentar Pessoal no Brasil (Dietary consumption pattern of a Brazilian population).

JECFA. 2011. Evaluation of certain contaminants in food. The seventy-second report of Joint FAO/WHO Expert Committee on Food Additives. WHO, 115 pp.

WHO. 2012. JOINT FAO/WHO FOOD STANDARDS PROGRAMME CODEX COMMITTEE ON CONTAMINANTS IN FOODS Sixth Session Maastricht, The Netherlands, 26–30 March 2012 PROPOSED DRAFT MAXIMUM LEVELS FOR ARSENIC IN RICE.

2.2 Arsenic in animal-based foods

One Century of the Discovery of Arsenicosis in Latin America (1914–2014) – Litter, Nicolli, Meichtry, Quici, Bundschuh, Bhattacharya & Naidu (Eds)
 ISBN 978-1-138-00141-1

Unusual arsenic speciation in urine of ruminants

W. Goessler & S. Braeuer
Institute of Chemistry, Analytical Chemistry, University of Graz, Austria

ABSTRACT: We determined the arsenic speciation in urine samples of cows of two Austrian farms and other terrestrial animals from the Zoo of Vienna (Tiergarten Schönbrunn Vienna, Austria). Total arsenic concentrations in the urine samples ranged from 5 to 125 µg L^{-1}. The highest arsenic concentrations of the zoo animals were found in the urine of bison (70 ± 20 µg L^{-1}) and water buffalo (77 µg L^{-1}). The urine of cows from one farm showed also quite high arsenic concentrations (40 to 125 µg L^{-1}). The dominant arsenic compounds in the urine of ruminants were dimethylarsinic acid (50–70%), methylarsonic acid (10–15%), and inorganic arsenic (5–25%). Besides these common arsenicals, unknown arsenicals were frequently detected at significant concentrations (up to 25%). Currently, we are trying to identify these compounds with molecular mass spectrometry. Our results indicate differences in the arsenic metabolism in ruminants compared to other mammals.

1 INTRODUCTION

Arsenic speciation analysis has been carried out in a variety of biological systems. Especially, the marine ecosystem has been studied in detail because of the higher arsenic concentrations normally present. With the improvement of the analytical methods, more work has been devoted to the terrestrial ecosystem (Kuehnelt & Goessler, 2003).

For mammals, it is well known that kidneys are the major route for arsenic excretion. The arsenic speciation in urine of humans, monkeys, and rodents has been well studied (Watanabe & Hirano, 2012). Dimethylarsinic acid (DMA) is the dominating arsenic compound in urine of humans after exposure to inorganic arsenic. Additionally, methylarsonic acid (MA) and inorganic arsenic are usually found. Typically, DMA accounts for 60–80%, MA for 10–20% and inorganic arsenic for 10–30%. This ratio is very constant as long as inorganic arsenic is the only route of exposure (Vahter, 2002). When exposed to more complex arsenic compounds such as arsenosugars, the picture gets more complex. After exposure to an arsenosugar, at least twelve arsenic compounds were determined, of which dimethylarsenoylethanol, trimethylarsine oxide, dimethylarsenoylacetate, thio-dimethlyarsenoacetate, thio-dimethylarsenoethanol, and the thio-arsenosugar could be identified, besides dimethylarsinic acid (Raml *et al.*, 2005). After ingestion of arsenolipids, even longer fatty acids were found in urine (Schmeisser *et al.*, 2006).

Although many scientists worked on the arsenic speciation in urine of mammals, almost no data exist on the arsenic speciation in urine of ruminants. In the present work, we determined the arsenic speciation in urine samples of cows and other so far not studied terrestrial animals.

2 EXPERIMENTAL

2.1 *Samples and sample preparation*

Urine samples of 13 different cows from two Austrian farms were obtained by directly collecting them in polypropylene tubes from the urinating cattle. From the Zoo of Vienna (Tiergarten Schönbrunn), we got urine samples from 19 different terrestrial mammal species, of which 12 were ruminants.

All samples were stored at 4 °C. Prior to analysis, the samples were filtered through 0.2 µm Nylon® filters. An aliquot of the samples was also analyzed after addition of H_2O_2 to a final concentration of 10%.

2.2 *Total arsenic*

The samples were diluted 1 + 9 with 10% (v/v) nitric acid in polystyrene tubes and were analyzed directly with ICPMS (Agilent 7500ce, Waldbronn, Germany) at m/z 75 with an optional gas (1% CO_2 in Ar) added to the carrier gas and helium (4 mL/min) as collision gas. Part of the samples was reanalyzed after a microwave assisted digestion with nitric acid.

2.3 *Speciation analysis*

For the determination of the arsenic compounds, an HPLC (Agilent 1200, Waldbronn, Germany) was coupled with an ICPMS (Agilent 7700x,

Waldbronn, Germany) as element-selective detector. The arsenic compounds were chromatographed at two different chromatographic conditions.

The 'anionic' arsenic compounds were separated on a PRP-X100 anion-exchange column (150 × 4.6 mm, 5 μm; Hamilton, Bonaduz, Switzerland) with an aqueous 20 mM ammoniumphosphate buffer at pH 6.0 (adjusted with an aqueous ammonia solution) as mobile phase at 1.0 mL min^{-1}. The column was operated at 40 °C and 20 μL of the urine samples were injected. For the characterization of the unknown compounds, buffers at other pH values were prepared.

The 'cationic' arsenic compounds were separated on a Zorbax 300 SCX cation-exchange column (150 × 4.6 mm, 5 μm; Agilent, Waldbronn, Germany) with a 10 mM pyridine solution at pH 2.3 (adjusted with formic acid) as mobile phase at 1.5 mL min^{-1}. The column was operated at 30 °C and 20 μL of the urine samples were injected. For the characterization of the unknown compounds, buffers with different MeOH (0–15%, v/v) were additionally prepared.

To get a better understanding about the unknown compounds, the influence of pH on the retention of the arsenic compounds was studied on the anion-exchange column from pH 4.0 to 8.0.

The arsenic compounds were identified by co-chromatography with standards. The quantification was done with the individual compounds. The un-known compounds were quantified with dimethyl-arsinic acid on the anion-exchange chromatography and arsenobetaine on the cation-exchange chromatography.

3 RESULTS AND DISCUSSION

3.1 *Total arsenic concentrations*

The total arsenic concentrations in the urine samples ranged from 5 to 125 μg L^{-1} (Figure 1). Highest arsenic concentrations of the zoo animals were found in the urine of bison (81 ± 1 μg L^{-1}) and water buffalo (77 ± 1 μg L^{-1}). High arsenic concentrations were found in urine of cows from a farm located in Southeast Styria (40 to 125 μg L^{-1}). The total arsenic concentrations are rather high compared to arsenic concentrations typically found in urine of unexposed humans (<10 μg L^{-1}). Arsenic might originate from the drinking water, pasture, or feed supplements. Checking the feed supplements did not reveal any high arsenic concentration. Very likely, the animals take up part of the soil during grazing.

3.2 *Anionic arsenic compounds*

The anion-exchange conditions used are commonly employed to separate the most toxic arsenicals in urine. The urinary metabolites found (Table 1) are similar to arsenic profiles published for other terrestrial mammals. DMA accounted for 50–70%, MA for 10–15%, and inorganic arsenic for 5–25%. In Table 1, the cationic arsenic compounds (eluting in the solvent front of the anion-exchange chromatography) are not listed.

Besides the commonly detected arsenicals, we found significant amounts of three unknown compounds. UNK1 + UNK2 eluted right after methylarsonic acid while UNK3 was even longer retained than As(V). A typical anion-exchange chromatogram is shown in Figure 2. For the urine samples of the cattle from Upper Styria, the sum of the unknowns accounted for even 25% of the total arsenic. It is well known that trivalent or thio-arsenicals are readily oxidized in the presence of H_2O_2 (Scheer *et al.*, 2012). We observed that UNK1 decreased only slowly in the presence of H_2O_2 (10%v/v) after several hours at room temperature, while UNK3 increased at the same time.

In order to get some idea about the chemical structures of the unknown compounds, we chromatographed one urine sample at different pH values of the mobile phase.

As demonstrated in Figure 3, changing the pH does not influence the retention times of UNK1 and UNK3. However, the retention time of the third unidentified peak (UNK2) decreases from pH 4.0 to 6.0 At pH 7.0 and 8.0, a third signal was observed at much higher retention times. Whether it can be assigned to UNK2 an answer cannot be given at present.

Additional experiments with a silica based anion-exchange column revealed that the unknowns are much less retained than with the polymer based PRP-X100 column, indicating a strong interaction with the back bone.

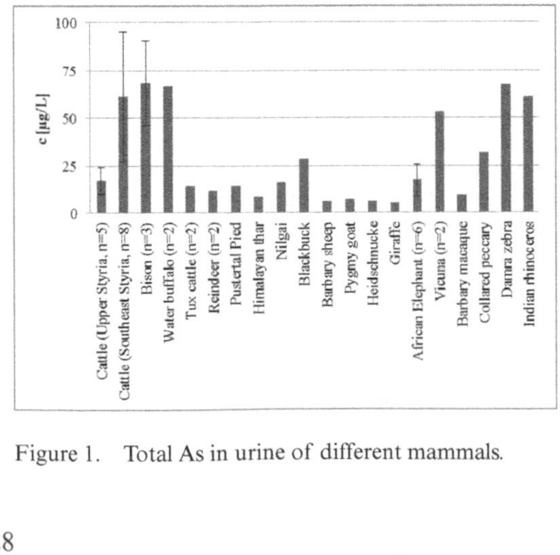

Figure 1. Total As in urine of different mammals.

Table 1. As species in urine of different mammals (*Upper Styria; **South East Styria; unknown compounds = UNK).

	DMA [µg L^{-1}]	MA [µg L^{-1}]	iAs [µg L^{-1}]	UNK1+2 [µg L^{-1}]	UNK3 [µg L^{-1}]
Cattle* (*n* = 5)	7.3	1.1	0.68	2.6	1.9
Cattle** (*n* = 8)	44	6.2	3.5	7.1	1.2
Bison (*n* = 3)	29	5.0	22	3.1	4.2
Water buffalo (*n* = 2)	47	6.8	1.9	2.7	3.5
Tux cattle (*n* = 2)	6.2	1.4	0.21	0.70	2.8
Reindeer (*n* = 2)	5.2	0.35	2.6	0.79	0.63
Pustertal Pied	6.4	1.6	1.0	0.94	3.0
Himalayan thar	4.9	0.36	0.29	<0.1	2.5
Nilgai	6.6	0.73	0.65	1.8	2.1
Blackbuck	10	1.3	6.0	1.9	2.2
Barbary sheep	3.7	0.36	0.56	<0.1	<0.1
Pygmy goat	2.6	0.59	1.2	<0.1	0.23
Heidschnucke	2.5	0.25	2.3	<0.1	<0.1
Giraffe	3.1	0.39	<0.1	<0.1	<0.1
African Elephant (*n* = 6)	4.4	1.5	4.6	0.40	<0.1
Vicuna (*n* = 2)	41	1.0	2.5	0.85	3.8
Collared Peccary	25	2.7	1.0	1.0	<0.1
Damara zebra	51	2.1	1.4	5.2	0.89
Indian rhinoceros	40	5.9	1.5	4.5	1.0

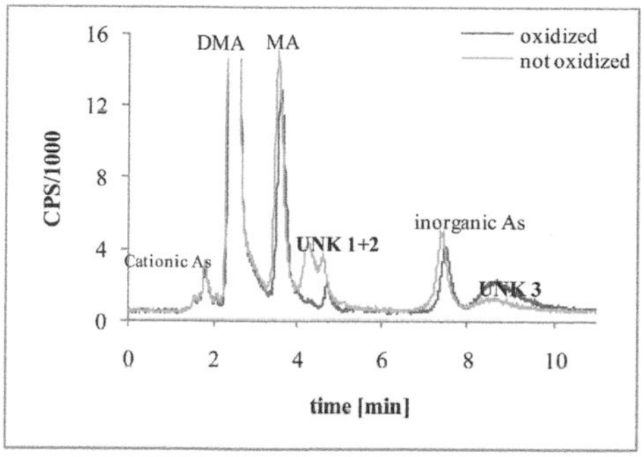

Figure 2. A comparison of anion-exchange chromatograms of a cow's urine not oxidized and oxidized with H_2O_2.

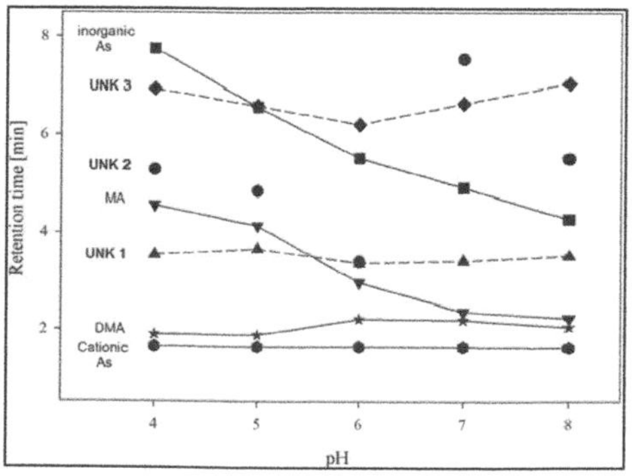

Figure 3. Retention behavior of the detected unknown compounds at different pH values of the mobile phase with the polymer based PRP-X100 column.

3.3 *Cationic arsenic compounds*

Co-chromatography and spiking experiments confirmed the presence of arsenobetaine (AB) and also traces of trimethylarsine oxide (TMAO) in most urine samples of ruminants. Arsenocholine (AC) and the tetramethylarsonium ion (TETRA) were below the detection limit. Instead, we obtained two unknown signals, of which one could be assigned to UNK3 after isolation from the anion-exchange chromatography (Figure 4).

Addition of methanol to the mobile phase did not affect AB or TMAO, but it shifted the unknown peaks to shorter retention times from 9.5 and 17 minutes to about 7 and 12 minutes.

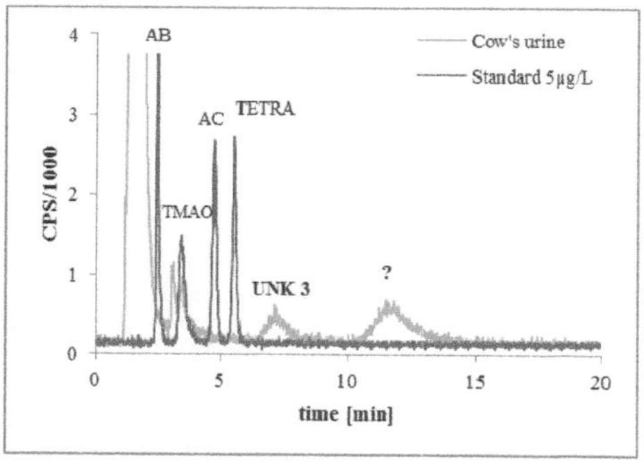

Figure 4. Cation-exchange chromatogram of a urine of cow (not oxidized) and overlaid a chromatogram of a standard solution containing 5 µg L^{-1} of AB, TMAO, AC and TETRA. (10% MeOH v/v were added to the mobile phase).

The behavior of UNK 3 seems quite unusual as it is well retained both with anion—and cation-exchange chromatography.

4 CONCLUSIONS

Although the gut flora of the ruminants is quite different compared to humans, we found quite similar arsenic metabolites in the urine. Additionally, up to 25% of unknown arsenicals were observed, which we are trying to identify at present.

ACKNOWLEDGEMENTS

Special thanks to the Tiergarten Schönbrunn, Vienna, Austria, for the cooperation and collection of urine samples.

REFERENCES

Kuehnelt, D. & Goessler, W. 2003. Organoarsenic compounds in the terrestrial environment. In: Craig, P.J. (ed.), *Organometallic Compounds in the Environment 2nd Edition. Wiley, Chichester,* 223–275.

Raml, R., Goessler, W., Traar, P., Ochi, T. & Francesconi, K.A. 2005. Novel Thioarsenic Metabolites in Human Urine after Ingestion of an Arsenosugar, 2′,3′-Dihydroxypropyl 5-Deoxy-5-dimethylarsinoyl-b-D-riboside. *Chemical Research in Toxicology* 18: 1444–1450.

Schmeisser, E., Rumpler, A., Kollroser, M., Rechberger, G., Goessler, W. & Francesconi, K.A. 2006. Arsenic fatty acids are human urinary metabolites of arsenolipids present in cod liver *Angewandte Chemie International Edition* 45: 150–154.

Vahter, M. 2002. Mechanisms of arsenic biotransformation. *Toxicology* 181–182: 211–217.

Watanabe, T. & Hirano, S. 2012. Metabolism of arsenic and its toxicological relevance. *Archives of Toxicology* 87: 969–979, and references therein.

One Century of the Discovery of Arsenicosis in Latin America (1914–2014) – Litter, Nicolli, Meichtry, Quici, Bundschuh, Bhattacharya & Naidu (Eds)

Effect of arsenic on chicken performance in Bangladesh

M.A.S. Khan, M.A.L. Promige & M.A. Ali
Faculty of Animal Husbandry, Bangladesh Agricultural University, Mymensingh, Bangladesh

ABSTRACT: The study was conducted to investigate the effects of arsenic on the performance of broiler and *desi* chicken, and the transmission into meat. Four levels of arsenic were given in two groups. Body weight gain, feed consumption, FCR, mortality (%), dressing (%) and As level in meat were examined. In trial I, 60 broiler chicks were allocated into four groups and 0.0, 250, 500, 1000 μg L^{-1} As were supplied. In trial II, 58 *desi* chicks were also allocated into four groups. Arsenic contaminated feed was also supplied. Results demonstrated that body weight, feed consumption, feed conversion ratio (and dressing (%)) were decreased with increasing of arsenic concentration. The mortality (%) was also increased. Arsenic was also found in edible meat, internal organs and droppings. It could be noted that the presence of As in poultry meat and organs was below to the harmful level for consumer health (>10 μg L^{-1}).

1 INTRODUCTION

Arsenic is a naturally occurring omnipresent trace element found in both organic and inorganic forms. Bangladesh has been devastatingly suffering from a serious public health problem due to drinking of arsenic contaminated groundwater. Public health hazard by drinking arsenic contaminated groundwater in Bangladesh was first identified in 1993 at the Nawabgonj district (Smith *et al.*, 2000). Almost a number of 57 million people out of 140 million (Mahmood, 2002) in the area of 61 districts out of 64 in Bangladesh were reported to have dangerous levels of inorganic arsenic (>10 μg L^{-1}), which caused the contamination of tube-well water, the main source of water for drinking and cooking in the affected area (DPHE, BGS and MML, 1999; BAMWSP, 2001). In Bangladesh, elaborated data for arsenic are available only on tube-well water; however, information on the presence of arsenic in livestock and livestock products, as well as on human and animal food chain is very scanty. However, evidence suggests that arsenic is not only a problem for humans but it may also accumulate in animal tissues and their products (Calvert and Smith, 1980; Awal, 2007). Thus, human exposure to As contamination may occur through the food chain. Therefore, information on detection of arsenic in livestock and livestock products, and on the treatment and prevention of arsenicosis in man and animal will be expecting new findings. Keeping all above things in mind, this work was undertaken with the following objectives:

i. To evaluate the effect of As levels of contaminated water and feed on chicken performance and meat quality.
ii. To determine the flow of As contamination in the chicken meat from food chain.

2 METHODS/EXPERIMENTAL

The experiment was conducted at Bangladesh Agricultural University Poultry Farm, Mymensingh, Bangladesh. Two independent trials were performed; one with commercial broiler and another with *desi* chickens. In trial I, a total of 60 one day-old commercial broiler chicks (Hubbard classic) were distributed into four treatments having 3 replications in each. In the case of trial II, a total of 56 one day-old unsexed *desi* chicks were distributed into four treatments having two replications in each. The four As concentrations (treatments) were T_0 (0 μg L^{-1}), T_1 (250 μg L^{-1}) T_2 (500 μg L^{-1}) and T_3 (1000 μg L^{-1}). These concentrations of arsenic were supplied to the birds through drinking water. Weekly body weight, feed consumption, weight gain, and feed conversion ratio (F.C.R.) records were recorded. Muscle and organ samples were collected and arsenic in thigh and breast muscle, and in liver were estimated by flow injection hydride generation atomic absorption spectrophotometer (UNICAM model No. 969). The collected and calculated parameters were analyzed as Completely Randomized Design (CRD) with the help of SPSS software. Differences among the treatment groups were separated by Duncan New Multiple Range Test (DMRT).

3 RESULTS AND DISCUSSION

Feeding As to the broiler for short duration (7–21 days) affected the growth response differently.

Table 1. The overall performance of broiler fed different concentration of Arsenic in drinking water.

Treat.	Body wt. (g)	Feed consumption (g)	FCR	Dressing yield (%)	Mortality (%)
T_0	$606.0^{a} \pm 5.0$	$865.0^{a} \pm 21.66$	1.43 ± 0.15	63.00 ± 1.00	$13.33^{c} \pm 5.78$
T_1	$582.67^{b} \pm 6.81$	$763.33^{b} \pm 61.53$	1.53 ± 0.05	60.33 ± 1.52	$16.67^{c} \pm 5.77$
T_2	$554.0^{c} \pm 5.29$	$772.3^{b} \pm 12.01$	1.52 ± 0.04	60.00 ± 1.00	$25.00^{b} \pm 5.00$
T_3	$535.0^{d} \pm 5.0$	$745.0^{b} \pm 18.52$	1.55 ± 0.05	59.50 ± 0.70	$43.33^{a} \pm 49.07$
Level of sig.	**	**	NS	NS	*

Superscripts with similar alphabet in the figures do not differ significantly. **Significant ($p < 0.05$); NS = Non Significant.

Table 2. The overall performance of *desi* chickens with different concentration of arsenic in drinking water.

Treat.	Body wt. (g)	Feed consumption (g)	FCR	Dressing yield (%)	Mortality (%)
T_0	$624.50^{a} \pm 3.54$	$1867^{a} \pm 8.49$	3.11 ± 0.028	$70.50^{a} \pm 0.71$	$0.00^{c} \pm 0.00$
T_1	$543.50^{c} \pm 3.54$	$1667.50^{c} \pm 9.19$	$3.22^{a} \pm 0.00$	$69.00^{ab} \pm 1.41$	$8.00^{b} \pm 2.83$
T_2	$576.00^{b} \pm 5.66$	$1781.50^{b} \pm 17.68$	$3.23^{a} \pm 0.064$	$66.50^{bc} \pm 0.71$	$10.0^{ab} \pm 0.00$
T_3	$528.50^{d} \pm 4.95$	$1637.50^{c} \pm 9.19$	$3.25^{a} \pm 0.007$	$65.50^{c} \pm 0.71$	$14.00^{a} \pm 1.41$
Level of sig.	**	**	*	*	**

However, when the birds were exposed to As for a long period (28 days), growth was reduced significantly. The results disagree with the previous observation of Chapman and Johnson (2002), who reported that the use of 3-nitro-4-hydroxyphenylarsonic acid (*Roxarsone*, abbreviated ROX), as feed additives increased weight gain. Arsenic feeding in drinking water reduced feed consumption even at 250 ppb level. However, the exposure of As for a longer period i.e. up to 28 days, significantly reduced feed consumption. The results were consistent with the report of Chen and Chiou (2001), who observed that the administration of arsenic at a dose of 11.36 mg/d for 1 week significantly depressed feed intake. It is evident that feed conversion ratios were also reduced in broiler fed different concentration of Arsenic. The results were not consistent with those of Morrison *et al.*, (1954) who observed a very slight improvement in feed efficiency when arsenicals were fed in practical diets.

Table 3. Arsenic level in edible meat, organs and manure of Broiler (ppm) (Mean ± SD).

Treatment	Thigh	Breast	Liver	Manure
T_0	$0.15^{bc} \pm 0.0$	0.11 ± 0.0	$0.12^{c} \pm 0.5$	$0.96^{c} \pm 0.0$
T_1	$0.16^{bc} \pm 0.0$	0.17 ± 0.0	$0.14^{bc} \pm 0.0$	$1.17^{bc} \pm 0.1$
T_2	$0.26^{b} \pm 0.1$	0.21 ± 0.0	$0.33^{ab} \pm 0.1$	$1.35^{b} \pm 0.4$
T_3	$0.39^{a} \pm 0.0$	0.22 ± 0.0	$0.42^{a} \pm 0.0$	$2.75^{a} \pm 0.1$
Level of sig.	*	NS	**	**

Feeding arsenic in drinking water at different concentrations significantly reduced body weight of *desi* chickens at all ages even at 250 μg L^{-1} level. The growth depression of arsenic level is dose dependent. It is evident that feed consumption of chickens fed with drinking water containing arsenic significantly reduced even at 250 μg L^{-1} level in *desi* chickens. The F.C.R. of chicken fed different concentration of arsenic solutions reduced As but the differences was non-significant ($p > 0.05$).

4 CONCLUSIONS

It may be concluded from the study that arsenic contamination was found in the edible meat and in organs of control birds. The higher arsenic contamination in water increased arsenic level in edible meat and organs. Arsenic contamination in feed and water reduced performance, meat yield and increased mortality of chickens. The arsenic content in edible meat and organs was much lower than the safety limit level (>10 μg L^{-1}).

REFERENCES

Awal, M.A. 2007. Detection of arsenic in the food chains and animal samples and study of the preventive measure using the cost effective agricultural products based spirulin against arseniasis in man and livestock. *Annual Research Report (2006–2007), USDA-Bangladesh collaborative research. Bangladesh.*

Bangladesh Arsenic Mitigation Water Supply Project (BAMWSP), 2001. Status report of Bangladesh Arsenic Mitigation Water Supply Project, December, 2001.
Calvert, C.C. & Smith, L.W. 1980. Arsenic in tissues of sheep and milk of dairy cows fed arsanilic acid and 3-nitro-4-hydroxyphenylarsonic acid. *J. Anim. Sci*, 51: 414–421.
Chapman, H.D. & Johnson, Z.B. 2002. Use of Antibiotics and Roxarsone in Broiler Chickens in the USA. *Poult. Sci.* 81: 356–364.
Chen. K.L. & Chiou, P.W.S. 2001. Oral treatment of Mule Ducks with arsenicals for inducing fatty liver. *Poult. Sci.* 80: 295–301.
Department of Public Health Engineering (DPHE), British Geological Survey (BGS) and Mott MacDonald Ltd. (MML). 1999. Groundwater studies for Arsenic contamination in Bangladesh, Final Report, Phase-1 January, EAWAG–SANDEC, 1998. SODIS News, No. 3, August.
Mahmood, S.A.I. 2002. Arsenic is the silent killer. The Bangladesh Observer, 9th May, 2002.
Morrison, A.B., Hunsaker W.G. & Aitken J.R. 1954. Influence of environment on the response of chicks to growth stimulants. *Poult. Sci.* 33: 491–494.
Smith, B.P.G., Koch, L. & Reimer, K.J. 2007. An investigation of arsenic compounds in fur and feathers using x-ray absorption spectroscopy speciation and imaging, *Sci. of the total Environment* 390: 198–204.

One Century of the Discovery of Arsenicosis in Latin America (1914–2014) – Litter, Nicolli, Meichtry, Quici, Bundschuh, Bhattacharya & Naidu (Eds)

Total arsenic in fish and water in four different aquatic environments in Argentina

E. Avigliano, N.F. Schenone, A.V. Volpedo & A. Fernández Cirelli
Instituto de Investigaciones en Producción Animal (INPA-CONICET-UBA), Facultad de Ciencias Veterinarias, Universidad de Buenos Aires, Buenos Aires, Argentina

W. Goessler
Institut für Chemie, Analytische Chemie, Graz, Austria

ABSTRACT: Concentrations of total As was determined in water and muscle of the silverside (*Odontesthes bonariensis*) in four different water bodies (De la Plata River, Adela and Barranca Lagoons and Chasicó Lake, Argentina) using ICP-MS. In all cases, total As concentration in water (28.40–367.16 μg L^{-1}) exceeded national and international guidelines for the protection of the aquatic life. High concentrations of As (0.05–1.20 g kg^{-1} dry wt) were found in fish muscle; however, the values were below international recommended limits for human consumption. The implication of As in fish products in South America is becoming an urgent issue due to the need to asses exposure limits to human consumption.

1 INTRODUCTION

Arsenic (As) is related to adverse health effects, not only in fish but also for consumers, depending on the concentration levels and the intake rate. The South American Chaco-Pampean Plain presents As in surface water coming from both natural and anthropic origin (Rosso *et al.*, 2013). The silverside *Odontesthes bonariensis* not only posses an ecological relevance (it could represented the largest fish biomass, especially in low diversity euryhaline lakes, but it also has regional socioeconomic relevance due to commercial and sport fishing, and aquaculture in South America (Baigún & Delfino, 2001). This species is distributed in the low Plata Basin (Avigliano & Volpedo, 2013) in lentic inland water bodies such as ponds and lagoons in the Pampean plain, being the second fishery resource in this region, not only for Argentina but also for Uruguay.

The objective of the present study was to evaluate the arsenic concentration in silverside muscle in four different aquatic environments and to present for the first time a baseline for this particular region and its relationship with human health.

2 METHODS

2.1 *Sample collection and preparation*

During August 2012, a sampling campaign was carried out to obtain fish and water samples from De la Plata River, Barranca Lagoon and Adela Lagoon, and during August 2011, from Chasicó Lake (Figure 1). Water samples were taken in triplicate. The capture method in the lagoons was performed over night with a 3 layer 3 × 3 cm mesh net of 10 m length in two different points. Fish samples (Table 1) were then transported in individual Ziploc® bags in ice to the laboratory facility. All laboratory ware was soaked in 10% HNO_3 for 48 h, and rinsed five times with distilled water, and then five times with ultrapure Milli-Q water prior to use. After dissection, the muscle tissue was weighed and freeze-dried. The water samples were filtered and acidified to 0.2% (v/v) by addition of concentrated nitric acid (Merck® p.a.) and stored at 4°C.

2.2 *Determination of total elements by ICP-MS*

Water samples were analyzed directly by ICP-MS without digestion. The freeze-dried muscle samples were pulverized in a coffee-mill. Samples were

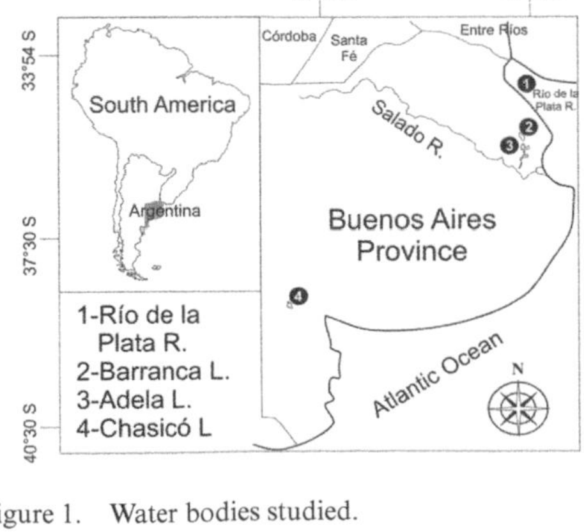

Figure 1. Water bodies studied.

Table 1. Total Length (TL) of the captured fish (cm) (mean ± SD).

Water body	n	Total length
De la Plata River	5	36.1 ± 1.3
Barranca Lagoon	3	22.8 ± 1.8
Adela Lagoon	3	22.0 ± 2.7
Chasicó Lake	4	37.0 ± 4.7

digested in a Milestone ultraCLAVE III microwave digestion system with 5 mL of concentrated nitric acid for total element determination. As concentration was determined with an Agilent 7500 inductively coupled plasma mass spectrometer (ICP-MS) equipped with a Micro Mist nebulizer and a Scott double pass spray chamber. All measurements were performed in triplicate.

The certified reference materials DORM-2 (dogfish muscle, National Research Council of Canada) was used for quality control. Water used was from a Milli-Q Academic water purification system with a specific resistivity of 18.2 MΩ.cm.

3 RESULTS AND DISCUSSION

The concentration of As in water present minimum values in De la Plata River, and maximum values in Chasicó Lake. Adela and Barranca Lagoons present intermediate concentration of As (Table 2). Arsenic showed in muscle the lowest values in fishes of De la Plata River and highest concentrations in fishes of Chasicó Lake. The concentrations of As in muscle of fishes of Adela and Barranca Lagoons present intermediate values and no significant differences were observed between them (Table 2).

According to the Argentine National Guidelines for the Aquatic Biota Protection (NGABP), the As concentration in water in all the studied sites, except in De la Plata River, was over the recommended level (15 μg L^{-1}). When considering the Canadian Guidelines for the Aquatic Biota Protection, the concentration of As in water was over the recommended limit (As 5 μg L^{-1}) in all sites except in De la Plata River.

Arsenic is categorized as a toxic element; it does not play any metabolic function but can be harmful for humans, even at low concentrations, when ingested over a long period of time (Sommers, 1974). The silverside muscle samples showed significant concentrations of As. According to the Argentine Food Code, the observed concentrations were below the recommended maximum limits (As: 1 mg kg^{-1}). The US Environmental Protection Agency has considered an upper limit of 1.3 mg kg^{-1} of wet weight for As (Eisler, 1994). The concentrations for these elements obtained in this study were below the maximum recommended limits.

Because of the uncertainties in the exposure in the key epidemiological studies, the European Food Safety Authority (EFSA, 2010) CONTAM Panel

Table 2. Concentration As in water (μg L^{-1}) and muscle (mg kg^{-1}). Values with non common letter superscript are significantly different ($p < 0.05$) within same row.

	De la plata River	Adela Lagoon	Barranca Lagoon	Chasicó Lake
Water	3.9 ± 0.4^{a}	28.4 ± 1.1^{ab}	42.9 ± 1.8^{ab}	367.1 ± 2.3^{b}
Muscle	0.05 ± 0.02^{a}	0.14 ± 0.04^{b}	0.13 ± 0.04^{b}	1.2 ± 0.4^{c}

identified a lower confidence limit of the benchmark dose (BMDL) of 1% extra risk instead of a single reference point, for inorganic arsenic. The BMDL values for the relevant health endpoints ranged from 0.3 to 8.0 μg kg^{-1} body weight (bw) per day. If we consider these values, a 60 kg person should consume 0.52–14 kg $week^{-1}$ silverside muscle to be exposed to any health issue. Nevertheless, these values are set for inorganic arsenic and no for total arsenic. In this case, the consumption rate could be higher taking into account that only a small percentage of the total arsenic is inorganic.

4 CONCLUSIONS

The concentration of As was below the limits of the Argentine legislation for fish consumption. However, it is highly probable that a fraction of the population exceeds the maximum consumption limits.

ACKNOWLEDGEMENTS

To CONICET, MINCYT, BMWF project WTZ AR 10/2011, University Graz, UBA and ANPCyT.

REFERENCES

Avigliano, E. & Volpedo, A.V. 2013. Use of otolith strontium:calcium ratio as indicator of seasonal displacements of the silverside (*Odontesthes bonariensis*) in a freshwater-marine environment. *Marine and Freshwater Research* 64(8): 746–751.

Baigún, C. & Delfino, R. 2001. Consideraciones y criterios para la evaluación y manejo de pesquerías de pejerrey en lagunas pampásicas. In. F. Grosman (Ed.), *Fundamentos biológicos, económicos y sociales para una correcta gestión del recurso pejerrey*. Buenos Aires: Astyanax. 132–145.

EFSA, 2010. Scientific Opinion on Lead in Food. *The European Food Safety Authority Journal* 8(4): 15–70.

Eisler, E. 1994. A review of arsenic hazards to plants and animals with emphasis on fishery and wildlife resources, In. J.O. Nriagu (Ed.), *Arsenic in the Environment* (Part II). Wiley, New York.

Rosso, J.J., Schenone, N.F., Perez Carrera, A. & Fernández Cirelli, A. 2013. Concentration of arsenic in water, sediments and fish species from naturally contaminated rivers. *Environmental Geochemistry Health* 35: 201–214.

Sommers, E. 1974. The toxic potential of trace metals in foods. A review. *Journal of Food Science* 39: 215–217.

One Century of the Discovery of Arsenicosis in Latin America (1914–2014) – Litter, Nicolli, Meichtry, Quici, Bundschuh, Bhattacharya & Naidu (Eds)
ISBN 978-1-138-00141-1

Arsenic levels in bovine kidney and liver from an arsenic affected area in Argentina

A. Pérez Carrera, C.V. Alvarez Gonçalvez & A. Fernández Cirelli
Instituto de Investigaciones en Producción Animal (INPA), CONICET, Facultad de Ciencias Veterinarias, Universidad de Buenos Aires, Buenos Aires, Argentina

S. Braeuer & W. Goessler
Institute of Chemistry, Analytical Chemistry, University of Graz, Austria

ABSTRACT: The study of arsenic occurrence and its transference to food is fundamental to estimate the risk for the consumers. Arsenic can be transferred from drinking water to cattle tissues and might be dangerous for humans. The aim of this study is to determine arsenic levels in liver and kidney from cattle in the southeast of Córdoba province, an arsenic affected area in Argentina. For that, we quantified the arsenic levels in bovine livers and kidneys. The total arsenic concentration in liver samples ranged from 25 to 150 μg kg^{-1} (dry weight) and in kidneys from 53 to 500 μg kg^{-1} (dry weight). Those levels are not dangerous for human consumption. Still, some samples showed values higher than previously reported.

1 INTRODUCTION

In Argentina, cow meat constitutes a fundamental part of the human diet, which requires high quality standards and safety controls. This has made the quantification of trace metals a new challenge for agriculture companies. In spite of the fact that animals in general reduce the human exposition to trace metals, some elements, such as arsenic (As), have been identified that they can be present in bovine diet in acceptable concentrations but they can be accumulated in tissues in dangerous concentrations for humans (NRC, 2005). Highest As concentrations are registered in skin, hoof, hair, liver and kidney (Salisbury *et al.*, 1991; Lopez-Alonso *et al.*, 2000; 2002; Pérez-Carrera *et al.*, 2010). Arsenic levels in tissues of cattle may depend on the As levels in food and water for the livestock. The aim of this study is to determine As levels in liver and kidney in cattle from an As affected area in Argentina.

2 METHODS

Bovine liver and kidney tissues from the southeast of Córdoba province were collected directly after slaughtering and then the 16 samples were freeze-dried and stored for further analysis. For the determination of total As, the samples were digested with nitric acid. Then they were diluted with ultrapure water (18.2 MΩ * cm) to an acidity of 10% and measured with Inductively Coupled Plasma Mass Spectrometry (Agilent ICPMS 7500ce, Waldbronn, Germany). For As analysis, Dogfish Liver (DOLT-3) Certified Reference Material for Trace Metals, NRC-CNRC (Otawa, Canada), was used. The detection limit for As was 10 μg kg^{-1}. The median values of the corresponding digestion blanks were subtracted from the concentrations.

3 RESULTS AND DISCUSSION

The total As concentrations in bovine liver samples (figure 1) ranged from 25 to 150 μg kg^{-1} (dry weight). Some of the values are higher than previously reported in cattle exposed to As via groundwater used as drinking water by Salysbury *et al.*. (1991), Lopez Alonso *et al.* (2000); Pérez Carrera, 2007, 2010.

Nevertheless, these values were below the maximum limit proposed by Plan CREHA, SENASA, Argentina (1000 μg kg^{-1} dry weight). The variability of the As concentration was high.

The total As concentration in kidney samples was highly variable ranging from 53 to 500 μg/kg (dry weight) (Figure 2). Total As concentrations were higher than previously reported in other As affected regions (Lopez Alonso *et al.*, 2000; Vos *et al.*, 1987). Nevertheless, values were below the maximum allowed limit in Argentina.

This study shows that the As content in bovine kidneys generally is greater than in bovine livers.

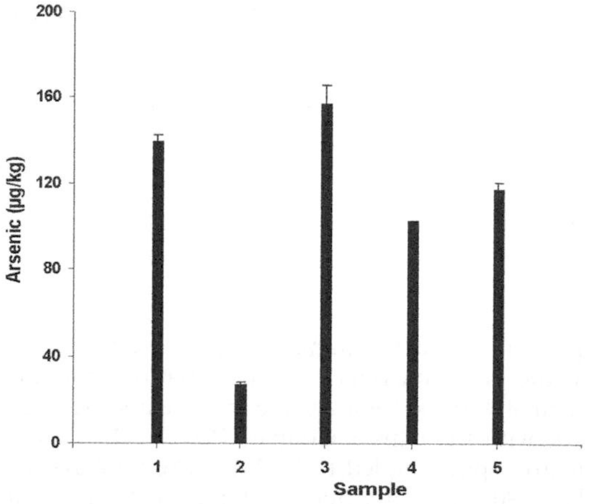

Figure 1. Arsenic levels in cattle liver from Córdoba, Argentina.

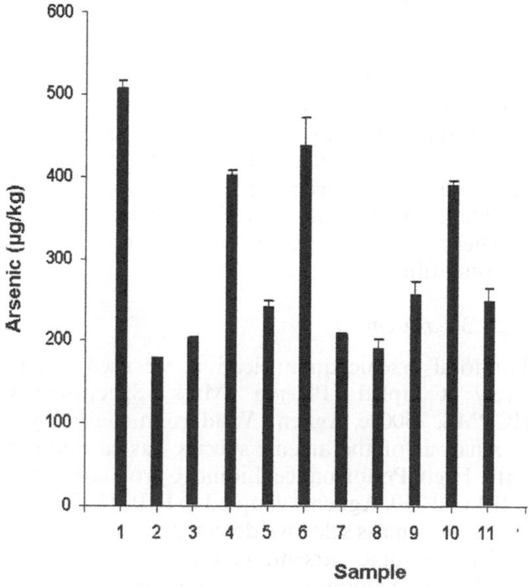

Figure 2. Arsenic levels in cattle kidney from Córdoba, Argentina.

This result is in accordance with the values reported by other authors (Vos *et al.*, 1987; Lopez Alonso *et al.*, 2000; Pérez Carrera *et al.*, 2006; 2010).

4 CONCLUSIONS

The As concentrations, in the analyzed samples, were below the limits considered in Argentina, and therefore are not dangerous for human consumption. However, many of the samples had higher As concentrations than values from previous reports. Generally, the levels in kidneys were greater than in livers. This may suggest that most of ingested As is excreted through the urine. It is important to note that more information is needed to understand the pathway of As in our food chain.

ACKNOWLEDGEMENTS

The authors are grateful to the University of Buenos Aires, CONICET, MINCyT—BMWF (Bilateral Project Argentina-Austria, AU/12/15) and OEAD WTZ (Österreich Argentinien, Projekt 08/2013 2013–2015) for financial support.

REFERENCES

López Alonso, M., Benedto, J., Miranda, M., Castillo, C., Hernandez, J. & Shore, R. 2000. Toxic and trace elements in liver, kidney and meat from cattle slaughtered in Galicia (NW Spain). *Food Additives and Contaminants*, 17: 447–457.

López Alonso, M., Benedto, J., Miranda, M., Castillo, C., Hernandez, J. & Shore, R. 2002. Cattle as biomonitors of soil arsenic, copper and zinc concentrations in Galicia (NW Spain). *Environmental Contamination and Toxicology*, 43: 103–108.

National Research Council. 2005. Mineral tolerance of animals. 2nd revised edition. *Committee on Minerals and Toxic Substances in Diets and Water for Animals*. The National Academy Press, Washington D.C., 495 pp.

Pérez Carrera, A. & Fernández Cirelli, A. 2004. Niveles de arsénico y flúor en agua de bebida animal en establecimientos de producción lechera (Pcia. de Córdoba, Argentina). *Revista de Investigación Veterinaria (INVET)*, 6: 51–59.

Pérez Carrera, A. & Fernández Cirelli, A. 2007. Problemática del arsénico en la llanura sudeste de la provincia de Córdoba. Biotransferencia a leche bovina. *Revista Investigación Veterinaria (INVET),* 9: 123–135.

Pérez Carrera, A., Pérez Gardiner, M.L. & Fernández Cirelli, A. 2010. Presencia de arsénico en tejidos de origen bovino en el sudoeste de la provincia de Córdoba, Argentina. *Revista Investigación Veterinaria (INVET)*, 12: 59–67.

Plan Nacional de Control de Residuos e Higiene en Alimentos (Plan CREHA). 2012. Plan Anual de toxicinas y residuos en alimentos de origen animal. SENASA. Min. de Agricultura Ganadería, Pesca y Alimentación. Argentina.

Salisbury, C.D.C., Salisbury, W. Chan, P.W. & Saschenbrecke, P. 1991. Multielement concentrations in liver and kidney tissues from five species of Canadian slaughter animals. *Journal of the Association of Official Analytical Chemists*, 74:587–591.

Vos, G., Hovens, J. & Van Delft, W. 1987. Arsenic, cadmium, lead and mercury in meat, liver and kidneys of cattle slaughtered in the Netherlands during 1980–1985. *Food Additives and Contaminants*, 4:73–88.

One Century of the Discovery of Arsenicosis in Latin America (1914–2014) –
Litter, Nicolli, Meichtry, Quici, Bundschuh, Bhattacharya & Naidu (Eds)

Arsenic and its compounds in tissue samples from Austrian cattle

S. Braeuer & W. Goessler
Institute of Chemistry, Analytical Chemistry, University of Graz, Austria

ABSTRACT: Arsenic has been investigated in samples of different mammalian species, but there is only little information about arsenic in ruminants, like cattle, and there is a lack of information about the arsenic speciation in bovine tissues. For this reason, we analyzed the total arsenic and the arsenic compounds of liver and kidney samples from Austrian cattle with Inductively Coupled Plasma Mass Spectrometry (ICPMS) and high performance liquid chromatography coupled to ICPMS. The total arsenic levels in kidneys ranged from 100 to about 1400 (median 290) μg/kg dry mass and in livers from about 20 to 170 (median 50) μg/kg dry mass. Arsenic concentrations in livers and kidneys taken from the same animal were highly correlated. First results show that the major arsenic compounds in liver and kidney were methylarsonic acid, dimethylarsinic acid and inorganic arsenic. Additionally, we found some cationic arsenic compounds and two unknown arsenicals in the tissues samples.

1 INTRODUCTION

The fate of arsenic in the bodies of mammals has been a topic of interest for a long time. There are several studies dealing with the total arsenic concentrations in samples of terrestrial animals (especially rodents and monkeys) and humans. Often studied organs are liver and kidney. Still there is only little information about the arsenic speciation in those tissues. It has been suggested that the liver plays an important role in the methylation of arsenic (reviews by Vahter, 1999 and Watanabe and Hirano, 2012).

Arsenic levels in bovine tissues have been reported several times, for example by Yabe et al., 2012 and Roggeman et al., 2014, but there is a lack of data about the different arsenic compounds in bovine livers or kidneys.

We investigated the total arsenic and the arsenic speciation in tissue samples from Austrian cattle to get a better understanding of the arsenic metabolism of ruminants.

2 EXPERIMENTAL

2.1 *Sample preparation*

Matching liver and kidney samples from 10 Austrian cattle were collected at the slaughterhouse in Graz, Austria with detailed information about their origin and age. They were cut into small pieces with a stainless steel knife and then freeze-dried.

For total element analysis, an aliquot (~0.5 g) of the freeze dried material was suspended with nitric acid (subboiled) and then digested in a microwave heated autoclave up to 250 °C. Afterwards, the samples were diluted with ultrapure water (18.2 MΩ * cm) to a final acidity of 10%.

For speciation analysis, the freeze-dried tissue samples were extracted with 20 mM trifluoracetic acid and 1% hydrogen peroxide, sonicated for 15 minutes and centrifuged for 20 minutes at 3300 * g.

The supernatant was filtered through 0.2 μm Nylon® filters.

2.2 *Measurement*

For total arsenic quantification, we used Inductively Coupled Plasma Mass Spectrometry (ICPMS, 7500ce, Agilent, Waldbronn, Germany).

Analysis of the arsenic species was carried out with High Performance Liquid Chromatography (HPLC, 1260, Agilent) coupled to ICPMS (7500ce, Agilent) as mass selective detector.

The "anionic" arsenic compounds were separated with an anion-exchange column (PRP-X100, Hamilton, Bonaduz, Switzerland) and 20 mM ammonium phosphate, pH 6.0, as mobile phase.

Analysis of total arsenic as well as speciation analysis was performed in collision gas mode with helium as collision gas (4 mL/min). Furthermore, CO_2 (1% in Argon) was used as an optional gas which was added to the carrier gas of the ICPMS to enhance the arsenic signal.

3 RESULTS AND DISCUSSION

3.1 *Total arsenic concentrations*

From all analyzed samples, arsenic was generally higher in kidneys. In 9 of 10 livers, we found 23 to 75 μg As/kg Dry Mass (dm) and, in the corresponding kidneys, there were 106 to 480 μg As/kg dm. In one liver and kidney pair, we determined exceptional high concentrations of 168 and 1387 μg As/kg dm. The arsenic concentration of liver and kidney from the same animal are highly correlated as shown in Figure 1.

3.2 *Arsenic speciation*

We were able to extract between 20 and 65% of the total arsenic from the tissue samples.

First results show that, according to the retention on the anion-exchange column, methylarsonic acid (MA) is a major compound in the investigated livers (30–50% of the extractable arsenic) and even more in kidneys (45–65%). There are also significant amounts of dimethylarsinic acid (DMA) and inorganic arsenic. DMA is more abundant in livers (15–30%) than in kidneys (10–15%). There is 5–30% inorganic As in livers as well as in kidneys, but generally inorganic As tends to be higher in livers. Cationic arsenic species could only be detected in small amounts in kidneys (5–10%). In addition to the common arsenicals, we found one unknown peak with a retention time of about 4.6 minutes in 6 kidney and 2 liver samples (4–9%). We were able to detect another broad peak at around 8.6 minutes in 3 of the kidney samples (around 3%).

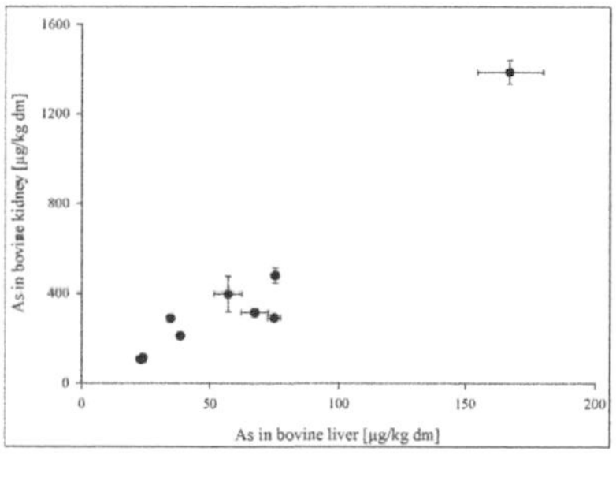

Figure 1. Correlation of As levels in livers and kidneys of Austrian cattle.

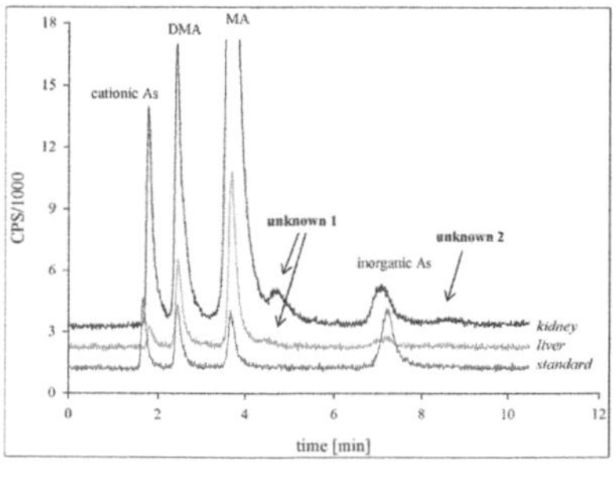

Figure 2. Anion-exchange chromatograms of bovine liver (offset: +1000 CPS), kidney (offset: +2000 CPS) and a standard solution with 1 μg/L of AB, DMA, MA, As (III) and As (V), oxidized with H_2O_2.

The high abundance of MA stands in high contrast to typical ratios found in cattle urine, where DMA is the major arsenic species (50–70%) and MA only accounts for around 10–15% of the total arsenic. Chromatograms of liver and kidney are shown in Figure 2.

4 CONCLUSIONS

Arsenic levels in Austrian cattle are higher in kidneys than in livers. The major arsenic species in livers and kidneys is MA (accounting for 30–65% of the extractable arsenic). Additionally, we found significant amounts of DMA and inorganic arsenic. The ratio of DMA and inorganic arsenic versus MA is really unusual compared to the arsenic speciation typically found in urine.

As the samples were obtained from the slaughterhouse we do not have information about the nutrition (water, feed) of the studied animals. Therefore, we can only speculate about the intake of the arsenic. The origin of arsenic might be drinking water, pasture, or feed supplements. None of these sources is known to contain high amounts of MA. Therefore, this compound could be formed by methylation of inorganic arsenic or by demethylation of more complex arsenic compounds. At present, we are looking for the source of the MA.

Besides this unusual arsenic profile in the livers and kidneys, we were able to detect two unknown arsenic compounds which did not co-chromatograph with the standards available to us. We are currently working on the identification of these unknown arsenicals with molecular mass spectrometry.

ACKNOWLEDGEMENTS

The authors want to thank OEAD WTZ (Österreich Argentinien, Projekt 08/2013 2013-2015) for financial support.

REFERENCES

Roggeman, S. et al. 2014. Accumulation and detoxification of metals and arsenic in tissues of cattle (Bos taurus), and the risks for human consumption. *Science of the Total Environment* 466–467: 175–184.

Vahter, M. 1999. Methylation of inorganic arsenic in different mammalian species and population groups. *Science Progress* 82(1): 68–88 and references therein.

Watanabe, T. and Hirano, S. 2012. Metabolism of arsenic and its toxicological relevance. *Archives of Toxicology* 87: 969–979 and references therein.

Yabe, J. et al. 2012. Accumulation of metals in the liver and kidneys of cattle from agricultural areas in Lusaka, Zambia. *Journal of Veterinary Medical Science* 74(10): 1345–1347.

2.3 Arsenic in rice and other crops

*One Century of the Discovery of Arsenicosis in Latin America (1914–2014) –
Litter, Nicolli, Meichtry, Quici, Bundschuh, Bhattacharya & Naidu (Eds)*
 ISBN 978-1-138-00141-1

Effects of water management and soil amendments on arsenic and cadmium uptake by rice plant

T. Honma, H. Ohba & A. Kaneko (Kadokura)
Niigata Agricultural Research Institute, Nagaoka, Japan

T. Makino & K. Nakamura
National Institute for Agro-Environmental Sciences, Tsukuba, Japan

ABSTRACT: Arsenic and cadmium contents in rice grains are of public concern for human health. We carried out field experiments to investigate the effects of water management and soil amendments applied into paddy on arsenic and cadmium uptake by rice. We also investigate the effects of those factors on arsenic speciation in rice grain. Prolonged ponding water management (3 weeks before and after heading) led to increase the arsenic concentration in soil solution and in rice grain. Ponding water release management which is non-irrigation to harvest after midseason drainage led to increase cadmium concentration in soil solution and in rice grain. Arsenic uptake by rice decreased with application of short-range-order iron hydroxide. On the other hand, cadmium uptake decreased with converter furnace slag. Additionally, the main arsenic species in rice grain was As(III), and there was no difference in the ratio of several species among water management and soil amendments.

1 INTRODUCTION

Inorganic arsenic (As) causes cancer of lung, urinary tract and skin. Rice is a major source of dietary inorganic As in Asian populations. Cadmium (Cd) causes serious damages to the human health such as the Itai-itai disease. This illness has been generally recognized since the 1950's in Japan, where people eat rice grain grown in paddy fields polluted by high levels of Cd coming from current or abandoned mines.

For reduction of Cd concentration in rice grain, a ponding management during the period from 3 weeks before heading to 3 weeks after heading has been practically adopted in paddy field in Japan. Although ponding of paddy fields effectively reduces grain levels of Cd (Kyuma, 2004), anaerobic conditions in paddy soil lead to As mobilization, which can consequently increase the uptake of As by rice (Koyama, 1975). Thus, it is well known that there is a compromise between the As and the Cd uptake by rice (Arao *et al.*, 2009). On the other hand, application of soil amendments like the incorporation of iron-related materials into paddy field has been expected as an available technique for reducing As and Cd uptake by rice.

The main objective of the present study was to investigate the effects of the water management and application of amendments in paddy soils on As and Cd uptake by rice and on As speciation in rice grains.

2 METHODS

2.1 *Experimental field*

Field experiments were performed in 2012 at a paddy field which was classified as a Sandy mesic Typic Hydroaquents by US Soil Taxonomy. A soil in plowed layer contained 0.76% total C, 0.06% total N, 4.3 mg kg^{-1} total As, 0.48 mg kg^{-1} total Cd, 45% coarse sand, 35% fine sand and 7% clay and it had a pH of 5.5.

2.2 *Cultivation experiment*

Seedlings of rice (*Oryza sativa* L. cv. Koshihikari) were transplanted at 13 May 2012. The rice was grown under flooded conditions for a month from transplanting, followed by mid-season drainage on 27 June 2012 and harvested on 19 September 2012.

Two experimental plots in this field were set to investigate the effect of water managements on As and Cd uptake by rice. The first was ponding water release management, non-irrigation to harvest after mid-season drainage to reduce the As uptake. The second was prolonged ponding management ponding for 6 weeks from 3 weeks before heading to 3 weeks after heading to reduce the uptake of Cd.

Additionally, we investigated the effects of two iron-related materials, Converter Furnace Slag (C.F.S.) and short-range-order iron hydroxide

(I.Othe.) on the As and Cd uptake by rice plant under above water managements. These soil amendments were applied into paddy soil at a rate of 5000 kg ha^{-1} with basal fertilizer. All experiments were performed by triplicate.

Soil redox potential (Eh) was measured at a depth of 15 cm using a platinum electrode. A soil moisture sampler (DIK301B, Daiki Rika Kogyo Co., Saitama, Japan) was buried at the same depth for collecting soil solution. The soil solution was sampled from mid-June to early September at intervals of 1–2 weeks, diluted with 10% HNO_3 at a ratio of 9:1 immediately after collection and filtered through a sterilized 0.45 μm filter. As speciation in rice grain, As(III), As(V) and dimethylarsinic acid (DMA) were determined by HPLC/ICP-MS.

3 RESULTS AND DISCUSSION

3.1 *Soil solution*

In prolonged ponding management, As concentration in soil solution decreased with an increase in Eh because of the mid-season drainage. After that, As concentration in soil solution increased gradually with a decrease in Eh because of re-ponding. After the mid-season drainage, As concentration in soil solution applied I.O. kept lower than that of control plot until 11 September, whereas As concentration in soil solution applied C.F.S. was higher than that of the control plot. As concentration in soil solution in all plots were below the detection limit at harvest because of drainage (Figure 1).

On the other hand, in ponding water release management, Cd concentration in soil solution increased gradually toward harvest with an increase in Eh. Cd concentration in soil solution applied C.F.S. and I.O. kept lower than that of the control plot. Cd concentration in soil solution was very low compared with As concentration (Figure 2). The soil pH applied C.F.S. was higher than that of the control plot during the growing period. On the other hand, soil pH applied I.O. was the same as the control plot since August, though higher from transplanting to 25 July.

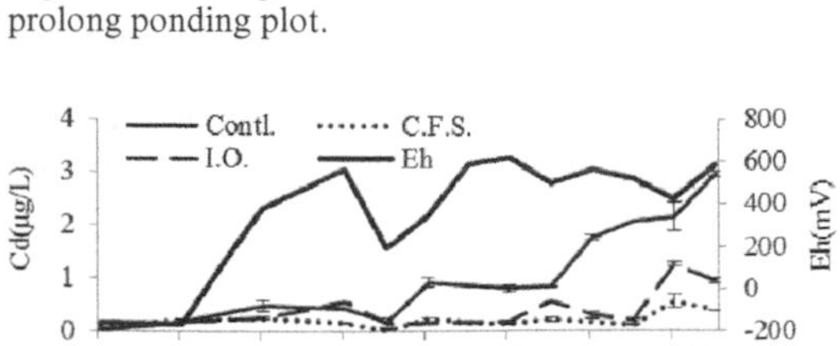

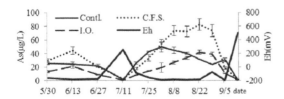

Figure 1. Changes of soil Eh and As concentration in prolong ponding plot.

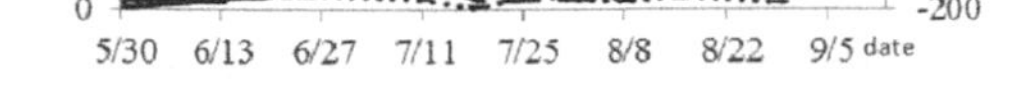

Figure 2. Changes of soil Eh and Cd concentration in ponding water release plot.

3.2 *As and Cd concentration in rice straw and grain*

In prolonged ponding water management, As concentration in rice straw applied I.O. was lower than that from the control plot statistically. Additionally, As concentration applied C.F.S. was lower too, though As concentration in soil solution was higher than that from the control plot. Furthermore, As concentration in rice straw in ponding water release management plot was about ten times lower than that of prolong ponding water management. As concentration in rice grain applied I.O. was lower than that from the control plot. No difference was observed between As concentration applied C.F.S. and control. Cd concentration in rice straw and grain applied C.F.S. was lower than that from the control plot statistically. However, Cd concentration applied I.O. was the same as control plot (Table 1).

3.3 *As speciation in rice grain*

The main arsenic species in rice grain was As(III) in all the treatment plots without any difference in the ratio of several species among water management and soil amendments. The ratio of inorganic As content were above 90% in all experimental plots.

Table 1. As and Cd concentration in rice shoots and grains.

Plot	Amendments	As Shoots	As Grains	Cd Shoots	Cd Grains
Prolong ponding	Contl.	8.83 a	0.39 a	1.01 a	0.07 a
	C.F.S.	6.24 b	0.34 ab	0.13 b	0.02 b
	I.O.	4.48 bc	0.29 bc	0.70 a	0.03 b
Ponding water release	Contl.	0.85 a	0.11 a	3.64 a	0.44 a
	C.F.S.	0.52 b	0.10 a	0.81 b	0.13 b
	I.O.	0.37 b	0.09 a	2.35 a	0.31 a

C.F.S.: Converter Furnace Slag.
I.O.: Short-range-order Iron hydroxide.
Data are mean values (n = 3).
Results followed by the same letter are not significantly different at the 5% level.

4 CONCLUSIONS

The water management affected the As and Cd uptake by rice plant. Application of I.O. was effective to decrease As uptake under prolong ponding water management. In the same way, C.F.S. was effective to decrease Cd uptake under ponding water release management. Additionally, there was no difference in the ratio of several As species in rice grains among water management and soil amendments.

ACKNOWLEDGEMENTS

This work was supported by a grant from the Ministry of Agriculture, Forestry and Fisheries of Japan (Research project for ensuring food safety from farm to table AC-1122).

REFERENCES

Arao, T., Kawasaki, A., Baba, K., Mori, S. & Matsumoto, S. 2009. Effects of water management on cadmium and arsenic accumulation and dimethylarsinic acid concentrations in Japanese rice. *Environ. Sci. Technol.*, 43 (24): 9361–9367.

Koyama, T. & Shibuya, M. 1976. Studies on the relation between arsenic in soils and paddy rice growth (2) – soluble As with special reference to 1 N soluble As. *Jpn. J. Soil Sci. Manure*, 47(3): 93–98 (in Japanese).

Kyuma, K. 2004. *Paddy Soil Science*. Kyoto, Kyoto Univ. Press.

One Century of the Discovery of Arsenicosis in Latin America (1914–2014) – Litter, Nicolli, Meichtry, Quici, Bundschuh, Bhattacharya & Naidu (Eds)
© 2014 Taylor & Francis Group, London, ISBN 978-1-138-00141-1

Effect of irrigation management and incorporation of plant residues on arsenic species and straighthead in rice

J. Panozzo, C. Quintero, R. Befani, M. Zamero, E. Díaz & G. Boschetti
Facultad de Ciencias Agropecuarias, Universidad Nacional de Entre Ríos, Paraná, ER, Argentina

S.S. Farías, J.A. Londonio, Y. Morisio, P. Smichowski & R.E. Servant
Departamento de Química Analítica, Gerencia Química, Comisión Nacional de Energía Atómica, Argentina

ABSTRACT: The biogeochemical reactions that occur under different rice growing conditions affect arsenic speciation, availability and uptake. Using two irrigation systems, continuous and interrupted, both in combination with the incorporation of vegetable residues, As concentration (speciation) in grain and straw, grain yield and unfilled kernels were measured. The occurrence of straighthead, a physiological disorder associated to the presence of As, which cause empty and deformed grains, was also determined. Continuous irrigation in the presence of vegetable residues resulted in high availability of DMA, favoring straighthead and resulting in yield losses. Simple interruption of irrigation during the main growth period resulted in higher yield, lower concentrations of total As and minimization of the impact of "straighthead" in grain.

1 INTRODUCTION

The understanding of biogeochemical cycling of arsenic (As) in soils, and As uptake and metabolism by rice plants has increased exponentially in recent years (Meharg & Zhao, 2012). Inorganic As (As-i) and its methylated organic forms, such as mono-methylarsonate (MMA) and dimethylarsinate (DMA) have all been found in rice grain (Zhao *et al.*, 2013). Rice grown in Asia and India has a higher proportion of As-i compared to rice grown in the U.S. and Europe, which has greater levels of organic As (Zavala *et al.*, 2008). In Entre Rios (Argentina), particularly high total levels of As have been detected in rice grains, in which the As is mainly in organic forms (Quintero *et al.*, 2011). There is considerable evidence that the rice plant is unable to methylate As-i (Lomax *et al.*, 2012). DMA is translocated at much higher speeds than As-i towards the grain and may be a major factor in the induction of "straighthead", since recent evidence indicates that the methylated forms are particularly toxic (Zheng *et al.*, 2013).

The "straighthead" appears severely, on some occasions, under permanent flooding if crop residues are abundant (Panozzo *et al.*, 2011). The aim of this study was to evaluate the concentration of different species of As in rice plants and its relationship to the occurrence of "straighthead".

2 MATERIALS AND METHODS

In the growing season (October/March) 2010/11, a trial was conducted in a commercial rice field, located in the south of Corrientes, Argentina. Two factors were tested for influencing As uptake by rice, water management and incorporation of plant residues. The experiment was carried out in 10 × 4.40 m parcels with 3 replicates in a randomized complete block design. The treatments were: continuous irrigation, with plant debris (Rc-Cr); continuous irrigation, without plant debris (Rc-Sr); irrigation interrupted (drainage)-with plant debris (Ri-Cr); irrigation interrupted, without plant debris (Ri-Sr). Water management consisted of a constant flooding treatment from 4 leaves until maturity (Rc) and a similar one but with a drainage 15 days before the panicle differentiation phase, then turned to be flooded to maturity (Ri). As in water was around 10 $\mu g\ L^{-1}$. Plant debris (naturally dry grass) was incorporated at a rate of 24 t/ha before seeding (Cr) and no waste treatment (Sr) was included as a control (Panozzo *et al.*, 2011). Redox potential in the soil was determined every 7 days with a redox electrode at 10 cm depth. Biomass was evaluated (grain and straw) at maturity and yield characteristics assessed: panicles/m^2, grains/panicle, 1000 grain weight, percentage of unfilled grains and empty hulls deformed. Yield was corrected to 14% moisture content.

Whole grain samples (with husk) and rice straw were digested using 0.28 M nitric acid (Huang *et al.*, 2010). As speciation was performed in extracts by means of high resolution liquid chromatography coupled to atomic fluorescence spectrometry (HPLC-AFS). Matrix certified reference material (NIST SRM 1568a "rice flour") was used to evaluate bias of our results (Farías *et al.*, 2012).

Data was analyzed by ANOVA and Fisher test.

3 RESULTS AND DISCUSSION

Good yields were obtained for intermittent irrigation treatments (10935 and 10901 kg/ha) for Ri-Sr and Ri-Cr, respectively. Permanent flooding with the addition of plant debris significantly adversely affected the crop and show lowest redox potential. The number of panicles, number of flowers per panicle and filled grain weight was not affected by treatments. However, there was a significant effect on the number of empty grains with the presence of deformed glumellae characteristic of "straighthead". Grain yield was significantly lower for Rc-Cr (6280 kg/ha), caused by a decrease of the reproductive destinations (48.3% sterility). Conversely, the yield was greater for both systems Ri, with the lowest percentage of empty grains (6.8%). As speciation in both straw and grain varied significantly with treatments. Although there were similar levels of total As, As-i were higher in straw, while organic species were more dominant in grain (Tables 1 and 2). Water management had a significant effect on the proportion of DMA in the straw (Table 2).

The highest concentrations of DMA corresponded to the highest percentages of vain grains, deformed grains and lower yields. Our results agree with recent literature indicating that anaerobic conditions stimulate microbial production of DMA in the soil (Jia *et al.*, 2013). DMA is absorbed by rice and translocated to the grain, causing deformation and sterility of flowers (straighthead). Thus, in the study area, the combination of soil and microbial flora results in highly reducing conditions that may produce elevated levels of methylated As species, toxic to the rice, that could reduce yield in susceptible cultivars.

Table 1. Variation of As species and total As in paddy rice grains with treatment.

	As(V) (ng/g)	As(III) (ng/g)	DMA (ng/g)	MMA (ng/g)	As Total (ng/g)
Rc-Cr	281	36	2799 B*	67	3182 B*
Rc-Sr	375	26	2654 B	22	3077 B
Ri-Cr	148	5	2158 AB	47	2358 AB
Ri-Sr	355	24	1390 A	32	1800 A

Table 2. Variation of As species and total As in paddy rice straw with treatment.

	As(V) (ng/g)	As(III) (ng/g)	DMA (ng/g)	MMA (ng/g)	As Total (ng/g)
Rc-Cr	1500	1,3	664 A*	120 AB*	2286 A*
Rc-Sr	2229	2,0	1346 B	187 B	3764 B
Ri-Cr	1917	2,3	412 A	103 A	2434 A
Ri-Sr	2178	0,7	530 A	133 AB	2841 AB

*Different letters indicate significant differences (Fisher $p \leq 0{,}05$).

In similar manner to that reported by Hua *et al.* (2013), intermittent irrigation reduced soil anoxia prior to reaching reproductive status, decreased the concentration of DMA and resulted in no observable symptoms of "straighthead".

4 CONCLUSIONS

Continuous irrigation of rice with crop residues generated low redox potentials, high concentration of DMA in the grain and low grain yield. The interruption of irrigation in the vegetative state allowed sufficient oxygenation of the soil to reduce the levels of available DMA and avoid toxicity, even when abundant vegetable waste was incorporated.

Speciation of As in soils and plants can play an important role in the understanding of physiological disease such as straighthead, including techniques for its management and selection of resistant varieties.

ACKNOWLEDGMENTS

This work was funded by project PICT-2011, N° 1955. ANPCYT. "Presencia de arsénico y metales traza tóxicos en arroces argentinos. Un desafío multidisciplinario para su minimización".

REFERENCES

Farías, S., Befani, R., Temporetti, C., Morisio, Y. Londonio, J., Quintero, C. & Smichowski, P. 2012. Speciation analysis of arsenic in white rice (Oryza Sativa L. CV Camba) by HPLC-HG-AFS. In *Proceedings of XII Rio Symposium on Atomic Spectroscopy*. F. do Iguaçu. Brazil.

Hua, B., Yan, W. & Yang, J. 2013. Response of rice genotype to straighthead disease as influenced by arsenic level and water management practices in soil. *Sci. Tot. Environ.* 442: 432–436.

Huang, J.-H., Fecher, P. & Ilgen, G. 2010. Quantitative chemical extraction for arsenic speciation in rice grains. *J. Anal. At. Spectrom.* 25: 800–802.

Jia, Y., Huang, H., Zhong, M., Wang, F., Zhang, L. & Zhu, Y. 2013. Microbial Arsenic Methylation in Soil and Rice Rhizosphere. *Environ. Sci. Technol.* 47: 3141–3148.

Lomax, C., Liu, W., Wu, L., Xue, K., Xiong, J., Zhou, J., Mcgrath, S.P., Meharg, A.A., Miller, A.J. & Zhao, F.J. 2012. Methylated arsenic species in plants originate from soil microorganisms. *New Phytol.* 193: 665–672.

Meharg, A. & Zhao, F.J. 2012. *Arsenic & Rice.* Springer: Dordrecht.

Panozzo, A.J., Quintero, C.E., Arévalo, E.S., Zamero, M.A. & Befani, R. 2011. Efecto de manejo de riego e incorporación de restos vegetales sobre vaneo fisiológico "straighthead" en arroz. *VII Congreso Brasileiro de Arroz irrigado.* Anais 2:40–43. Camboriú, Brazil.

Quintero, C.E., Farias, S., Befani, R., Temporetti, C., Londonio, A., Morisio, Y., Díaz, E., Smichowski, P. & Servant, R. 2011. Concentración, especies y origen del arsénico en arroz cv Cambá cultivado en Entre Ríos. P. 157–163. *Volumen XX. Resultados Experimentales. Jornada Técnica Nacional Cultivo Arroz*. Concordia.

Zavala, Y.J., Gerads, R., Gürleyük, H. & Duxbury, J.M. 2008. Arsenic in rice. II. Arsenic speciation in USA grain and implications for human health. *Environ. Sci. Technol.* 42: 3861–3866.

Zhao, F.J., Zhu, Y.G. & Meharg, A. 2013. Methylated Arsenic Species in Rice: Geographical Variation, Origin, and Uptake Mechanism. Environ. *Sci. Technol.* 47: 3957–3966.

Zheng, M.Z., Li, G., Sun, G.X., Shim, H. & Cai, C. 2013. Differential toxicity and accumulation of inorganic and methylated arsenic in rice. *Plant Soil.* 365: 227–238.

One Century of the Discovery of Arsenicosis in Latin America (1914–2014) – Litter, Nicolli, Meichtry, Quici, Bundschuh, Bhattacharya & Naidu (Eds)
© 2014 Taylor & Francis Group, London, ISBN 978-1-138-00141-1

Concentration and origin of arsenic species in rice *cv Cambá* grown in Entre Ríos (Argentina)

C. Quintero, R. Befani, C. Temporetti & E. Díaz
Facultad de Ciencias Agropecuarias, UNER, Entre Rios, Argentina

S.S. Farías, J.A. Londonio, Y. Morisio, P. Smichowski & R.E. Servant
Departamento de Química Analítica, Gerencia Química, Comisión Nacional de Energía Atómica, Argentina

ABSTRACT: Few works on arsenic (As) species levels in rice grains have been reported in South America. This work was performed to determine inorganic/total arsenic in *Cambá* cultivar and its possible origin. Samples of rice, soil and water in 20 rice paddy of Entre Ríos, Argentina, were collected, and total As and its species were determined. Irrigation water showed low values of As (5–38 μg L^{-1}), but medium levels of the metalloid found in soils may be As source (2,1–3,7 mg.kg^{-1}). Rice grains had low inorganic As concentrations (0.023–0.140) μg kg^{-1}. Dimethylarsinic acid (DMA) was the main species detected in the studied rice grains.

1 INTRODUCTION

Arsenic (As) is a naturally occurring and ubiquitous element, toxic to humans, that may come from water or food. Arsenic chemistry and cultivation form cause the rice to absorb it in larger quantities than other cereals. In some circumstances, along with diets with medium to high consumption of rice, daily intakes involve undesirable levels, above those recommended by the World Health Organization.

The presence of As in rice has been subject of extensive study in the last decade. However, there has been surprisingly little research on Latin American rice. A survey of the contents of As in 30 samples of rice from Entre Rios, Argentina, made by Quintero *et al.* (2010) reported mean values of total As (As_{tot}) of 340 μg kg^{-1}, higher than those reported by William *et al.* (2005). The study showed significant differences between producing areas and the rice varieties used. The northern part of Entre Rios presented values of As_{tot}: 500 μg.kg^{-1} and one of the more sowed varieties planted (*Cambá*) exceeded this value on average. 25% of the samples showed values of 460–960 μg kg^{-1}. It is unknown what proportion of As_{tot} in the rice grain exists as inorganic form and would therefore produce the effect on the population consuming rice. The aim of this study was to evaluate total As and its chemical species content in rice grains of *Cambá* variety and their possible sources or origins.

2 MATERIALS AND METHODS

Soil samples, water for irrigation and rice grains were collected during the campaign 2009/10. They belonged to different areas of the Province of Entre Rios, totalizing 20 sampling sites. The cultivar selected in all cases was CAMBA (INTA-ROARROZ, Argentina). The analysis of total As in soils was per-formed by means of a leaching methodology as indicated by EPA 3050B standard, digesting the sample sequentially with 1:1 nitric acid, concentrated nitric acid followed by a 30% hydrogen peroxide digestion step. Samples of rice grains were digested using nitric acid 0.28 M (Huang *et al.*, 2010) for analysis of As speciation.

Total As was determined by inductively coupled plasma optical emission spectroscopy (ICP-OES), and analysis of As speciation was performed by high resolution liquid chromatography coupled to hydride generation atomic fluorescence spectrometry (HPLC-HG-AFS). An anionic exchange column was employed to separate As species.

For water samples, total As and soluble As were evaluated by ICP-OES. Prior to total As determination, samples were digested with concentrated nitric acid, in a hot plate at 90 °C.

All samples were analyzed in the Department of Analytical Chemistry, Chemical Management, National Atomic Energy Commission (CNEA).

3 RESULTS AND DISCUSSION

3.1 *As in water*

Arsenic levels detected in studied waters, used to irrigate rice are consistent with previous reports (Díaz *et al.*, 2008). Analyzed groundwater showed higher amounts of As, with a mean of 25 μg L^{-1}, while surface waters had a mean of 10 μg L^{-1}. Found values can be considered low and suitable even for human consumption in many cases, and very low when compared with waters of paddy rice areas with pollution problems (like Bangladesh), where As levels are ten to one hundred times higher than the levels found in Entre Rios, Argentina (Hossain *et al.*, 2008).

3.2 *As in soil*

The concentration of As in unpolluted soils depends on sediments that originated them and varies widely (from 0.1 to 40 mg kg^{-1}); nevertheless, the average is around 3–4 mg kg^{-1} (Bundschuh *et al.*, 2008). The fluvial sediments that sustain the production at La Paz area, Entre Ríos, exhibit the lowest As values (1.6 ± 0.2) mg kg^{-1}. The Vertisols of the central-South showed values of (2.9 ± 0.6) mg kg^{-1}, while in the Argiacuoles or wetlands area in the north, values of As in soil averaged (4.1 ± 1.2) mg kg^{-1}. In this area, in previous studies, the highest values of As in grain found in Entre Rios were reported (Quintero *et al.*, 2010).

3.3 *As in grain*

Total As values in grains (expressed as sum of species: As(III) and dimethylarsinic acid (DMA)) showed a mean 325 μg kg^{-1} (Table 1), similar to previously reported average of 340 μg kg^{-1} (Quintero *et al.*, 2010). These levels are in the range of the highest values informed by Williams *et al.* (2005) who reported values of total As in the United States of 260 μg kg^{-1}, followed by the levels found in Europe (180 μg kg^{-1}), Bangladesh (130 μg kg^{-1}) and India (50 μg kg^{-1}).

In this data set, it does not appear differentiation by zones, since high values of As are presented both in the north and in the center-south zones of Entre Rios.

It is very important to emphasize that the proportion of inorganic (As(III)) detected in this study was very low, even lower than the averages quoted by William *et al.* (2005) for the United States (42%), Europe (64%), Bangladesh (80%) and India (81%). In this report, all values of As found in rice grains were below the limit of 150 μg/kg of inorganic As / kg, proposed by China authorities. This would mean that while there are high levels of total As in grains, their toxicity for human would be much lower, because As is mostly present as organic species.

Table 1. Values of As in husked and polished rice grains.

Zone	Source location	As(III) (μg kg^{-1})	DMA (μg kg^{-1})
Center	San Salvador	28	147
Center	San Salvador	74	93
Center	Villa Clara	67	93
Center	Villa Clara	73	393
Center	San Salvador	83	748
Center	Villa Clara	74	163
South	Sajaroff	46	238
South	Sajaroff	34	372
La Paz	La Paz	63	82
La Paz	La Paz	140	118
North	Federal	77	85
North	Federal	27	270
North	Federal	59	145
North	San Jaime	123	39
North	Los Conq.	50	48
North	Los Conq.	23	185
North	Feliciano	34	311
North	Los Conq.	96	615
North	Los Conq.	55	957
North	Los Conq.	37	135

4 CONCLUSIONS

Total As levels in polished rice grain, *Camba* variety, grown in Entre Rios, are particularly high. Fortunately, toxicity is reduced as the proportion of inorganic As is very low and the values are below the limit proposed for human consumption.

Water used to irrigate rice contains low concentrations of As and seems not to be the main source of As in rice. Since no agrochemicals with As are used and there is no other contamination with As, it is assumed that the main contribution comes from soil.

In the near future, it would be required more research regarding: varieties, availability of As in soils and optimization of analytical techniques for the determination of total As and its species. The processing of a greater number of samples would allow more reliable evidence, based on accurate statistics.

ACKNOWLEDGEMENTS

This work was supported by the Project PICT-2011, N° 1955, ANPCYT: "Presencia de arsénico y metales traza tóxicos en arroces argentinos. Un desafío multidisciplinario para su minimización", and PROARROZ Foundation.

REFERENCES

Bundschuh, J., Pérez Carrera, A. & Litter, M.I. 2008. IBEROARSEN. *Distribución del Arsénico en las regiones Ibérica e Iberoamericana*. CYTED. Argentina.

Díaz, E.L., Lenzi, L.L. & Perusset, A. 2008. *Evaluación de residuos de plaguicidas en suelos y aguas cultivados con arroz en Entre Ríos. Resultados Experimentales 2007–2008*. Volumen XVII: 153–160. Concordia.

Hossain, M., Islam, M.R., Jahiruddin, M., Abedin, A., Islam, S. & Meharg, A.A. 2008. Effects of arsenic-contaminated irrigation water on growth, yield, and nutrient concentration in rice. *Commun. Soil Sci. Plant Anal.* 39: 302–313.

Huang, J.-H.; Fecher, P. & Ilgen, G. 2010. Quantitative chemical extraction for arsenic speciation in rice grains. *J. Anal. At. Spectrom.* 25: 800–802.

Quintero, C., Duarte, O., Díaz, E. & Boschetti, G. 2010. Evaluación de la concentración de arsénico en arroz. In *XIX Jornada Técnica Nacional Cultivo Arroz. Concordia*. 129–134.

Williams, P.N., Price, S.H., Raab, A., Hossain, S.A., Feldmann, J. & Meharg, A.A. 2005. Variation in Arsenic Speciation and Concentration in Paddy Rice Related to Dietary Exposure. *Environ. Sci. Technol.* 39: 5531–5540.

One Century of the Discovery of Arsenicosis in Latin America (1914–2014) –
Litter, Nicolli, Meichtry, Quici, Bundschuh, Bhattacharya & Naidu (Eds)
 ISBN 978-1-138-00141-1

Analysis of arsenic species in rice by HPLC-HG-AFS

S.S. Farías, J.A. Londonio, Y. Morisio, P. Smichowski & R.E. Servant
Departamento de Química Analítica, Gerencia Química, Comisión Nacional de Energía Atómica, Argentina

R. Befani, C. Quintero, C. Temporetti & E. Díaz
Facultad de Ciencias Agropecuarias, UNER, Entre Ríos, Argentina

A.B. Livore & E.M. Soro
EEA INTA, Concepción del Uruguay, Entre Ríos, Argentina

ABSTRACT: The distribution of As and its species in rice is mandatory to determine the uptake, transformation and potential risk posed by As contaminated rice. In this context, a study was carried out to develop, optimize and validate an accurate and reliable to quantify the concentration of four As species, namely, As(III), DMA, As(V) and MMA as well as to characterize statistically the distribution of As species in samples of domestically grown white rice marketed for direct food use in Argentina. Chromatographic separation coupled to atomic fluorescence detection resulted good enough to separate As species and quantify them properly and rapidly at ultratrace levels (0.020–0.025 µg .g^{-1}). Precision (expressed as intermediate precision) ranged between 2 and 10% and trueness (expressed as bias) ranged from 4 to 9% for the studied species.

1 INTRODUCTION

Rice is a staple food and the only source of carbohydrates for many populations throughout the world. The consumption of rice contaminated with arsenic (As) is an important issue for being an exposure route for humans. Studies related to the presence of As and its species in rice are of prime importance in countries where, for rice cultivation, water contaminated with As are employed for irrigation, or the presence of As in soils is suspected. The distribution of As and its species in the rice plant are hence necessary to determine the uptake, transformation and potential risk posed by As contaminated rice.

The arsenic chemistry is responsible for the elevated concentration of the metalloid found in rice grains. Rice absorbs more As than other gramineous because anaerobiosis generated by flooding reduces arsenate to arsenite, which is far more mobile in solution at typical pH values (H_3AsO_3, pK_a 9.2). Due to its small size and since it has no charge, As is easily absorbed by rice roots. Fortunately, over 95% of the As remains in the roots and only 1% reaches the grain (Rahman *et al.*, 2007).

In this context, a study was carried out to develop, optimize and validate an accurate and reliable method to quantify the concentration of four As species, namely, As(III), dimethyl arsenic acid (DMA), As(V) and monomethylarsonic acid (MMA) as well as to statistically characterize the distribution of As species in samples of domestically grown white rice marketed as a suitable human food. Extraction of species was performed using 0.28 M nitric acid as described by Huang *et al.* (2010), a simple method that recovered not only As quantitatively from rice grains but also preserved As speciation completely. For speciation analysis a powerful separation technique such as HPLC was coupled with a very sensitive analytical technique such as AFS, previous the HG of the different species. Studies comprehend optimization of working parameters, kinetic evaluation of the As reduction reaction, optimization, and validation of the analytical method including the estimation of uncertainty of results.

2 EXPERIMENTAL

2.1 *Preparation of samples*

Sixty rice samples were collected in rice fields located in the Entre Ríos province, Argentina. Once in the laboratory, samples were dried at 103 ± 2°C and digested with 0.28 M HNO_3 at 95 ± 2°C in a heating plate provided with a sand bath during 90 minutes. Once they reached room temperature, they were filtered and diluted to volume in a volumetric flask.

2.2 *Analytical determinations*

Aliquots of 100 μL of sample were injected through a Rheodyne 7725i valve coupled to a Jasco PU-2089s pump. Separation of As species was performed by means of a heated anion exchange chromatographic column Hamilton PRP-X100 (250 mm, 4.6 mm i.d., p.s. 10 μm) coupled to an atomic fluorescence spectrometer PSA 10.055 Millenium Scalibur. Standards were prepared using Certipur Merck solutions for As(III) and As(V) and analytical grade acids for MMA and DMA. Stock solutions of inorganic and organic species were prepared separately. Working standards were prepared by proper dilutions of mother solutions in the used buffer and were injected from lowest to highest to obtain a calibration curve (daily). NIST SRM 1568a "Rice Flour" was used to evaluate test bias. Analytical data was processed by the spectrometer software (SAMS V 2.02), and statistic evaluations were performed by using ORIGIN 6.0.

2.3 *Method optimization and method validation*

Operating parameters optimization was performed one at a time for a better understanding of phenomena related to each of them and by using a Doehlert type statistical experiment design (Ferré *et al.*, 2002). A full validation protocol that included linearity, limits of quantification (LQ) (0.020 μg g^{-1} As(III)–0.025 μg g^{-1} the other three species), precision (intermediate precision) (2% As(III) and DMA up to 10% for the other two species), trueness (bias) (4% As(III) – up to 9% for other three species) and uncertainty (ranging from 9.5 for As(III) to 19% for As(V) (Magnusson *et al.*, 2012), was carried out.

3 RESULTS AND DISCUSSION

The validated method was used to analyze samples coming from different commercial production fields located in the Entre Ríos province, Argentina, in the frame of a research project devoted to study and perform crop strategies and genetic improvement of a rice cultivar (*cv Cambá*). Results obtained demonstrated that the digestion method and HPLC-HG-AFS technique are suitable for this purpose. Approximately 86% of studied samples contained less than 40% of As(III), the most toxic species and almost 93% of samples had more than 40% of DMA, an almost harmless As species. Contents of As(V) were much lower than that of As(III). MMA was almost ever below LQ for practically all samples.

It is important to mention that the mean content of i-As in the 60 studied field rice grains was much smaller (approximately 26%) than the mean values cited by William *et al.* 2005 for USA (42%), Europe (64%), Bangladesh (80%) and India (81%).

Speciation analysis results showed that there was a good lineal correlation between DMA and Total As present in the digested rice grains ($r^2 = 0.9768$, As(III) type rice, Zavala *et al.*, 2008). This finding highlighted that when As content in the grain increased, the content of DMA, a less toxic species, also increased. These results do not match neither those published by Huang *et al.* (2010), nor some of the results published by Zavala *et al.* (2008) for commercial rice from Arkansas, Texas and California (USA) (although few results were coincident with our results DMA > As(III)).

4 CONCLUSIONS

HPLC separation coupled to AFS detection allowed the separation and quantification of As species properly and rapidly at ultratrace levels, as depicted in Figure 1. LQ are compatible with the contents of As species present in the studied rice cultivar. Precision and trueness obtained with this method allowed to obtain levels of uncertainty consistent with the requirements for the analysis of residues (FAO-OMS/ Joint FAO/WHO, 2012). Developed methodology allowed us to perform speciation analysis of As(III), DMA, MMA and As(V) in about 10 minutes, with a sample throughput of about 30 samples a day including quality control samples. DMA was found to be the prevalent species in cv *Cambá* rice, indicating that it might be less toxic than rice varieties cultured in Europe and Asia.

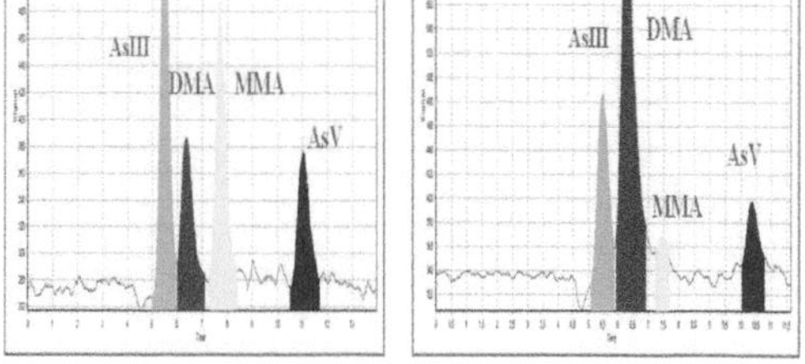

Figure 1. Chromatograms showing differences between species contents at ultratrace levels, both for standards and real samples. DMA was found to be the prevalent species in *cv Cambá*.

ACKNOWLEDGEMENTS

This work was funded by project PICT-2011, No. 1955, ANPCYT (Argentina).

REFERENCES

Ferré, J. & Rius, F.X. 2002. Introducción al diseño estadístico de experimentos. *Técnicas de Laboratorio* 274: 648–653.

Huang, J.-H., Fecher, P. & Ilgen, G. 2010. Quantitative chemical extraction for arsenic speciation in rice grains. J. Anal. At. Spectrom. 25: 800–802.

Joint FAO/WHO Food Standards Programme. Codex Comittee on Contaminants in Foods. Sixth Session. Maastricht, The Netherlands. Proposed draft maximum levels for arsenic in rice. January 2012.

Magnusson, B., Näykki T., Håvard, H. & Krysell, M. 2012. *Handbook for Calculation of Measurement Uncertainty in Environmental Laboratories.*

Rahman, M.A., Hasegawa, H., Rahman, M.M., Rahman, M.A. & Miah, M.A. 2007. Accumulation of arsenic in tissues of rice plant (*Oryza sativa* L.) and its distribution in fractions of rice grain. *Chemosphere* 69: 942–948

Williams, P.N., Price, S.H., Raab, A., Hossain, S.A., Feldmann, J. & Meharg, A.A. 2005. Variation in arsenic Speciation and concentration in paddy related to dietary exposure. *Environ. Sci. Technol.* 39: 5531–5540.

Zavala, Y.J., Gerads, R., Gorleyok, H. & Duxbury, J.M. 2008. Arsenic in rice: II Arsenic speciation in USA grain and implications for human health. *Environ. Sci. Technol.* 42(10): 3861–3866.

One Century of the Discovery of Arsenicosis in Latin America (1914–2014) – Litter, Nicolli, Meichtry, Quici, Bundschuh, Bhattacharya & Naidu (Eds)

ISBN 978-1-138-00141-1

Processing of raw rice grains (*Oryza sativa* L.) influences the concentration of arsenic species in Brazilian cultivars

B.L. Batista
Faculdade de Ciências Farmacêuticas de Ribeirão Preto, Universidade de São Paulo, São Paulo, Brazil
Centro de Ciências Naturais e Humanas, Universidade Federal do ABC, Brazil

E.S. Paula, J.M.O. Souza, G.R.M. Barcelos & A.C.C. Paulelli
Faculdade de Ciências Farmacêuticas de Ribeirão Preto, Universidade de São Paulo, São Paulo, Brazil

G. Rodrigues
Centro de Ciências Naturais e Humanas, Universidade Federal do ABC, Brazil

V.S. Santos
Faculdade de Ciências Farmacêuticas de Ribeirão Preto, Universidade de São Paulo, São Paulo, Brazil

P.C. Neves
Centro Nacional de Pesquisa em Arroz e Feijão, EMBRAPA, Brazil

C.V. Barião
Faculdade de Ciências Farmacêuticas de Ribeirão Preto, Universidade de São Paulo, São Paulo, Brazil
Centro Universitário Barão de Mauá, Brazil

F. Barbosa Jr.
Faculdade de Ciências Farmacêuticas de Ribeirão Preto, Universidade de São Paulo, São Paulo, Brazil

ABSTRACT: Rice is an important component of the world population and Brazilian's basic diet. The plant may accumulate considerable amounts of As in the grains, which is presented in different forms (As^{3+}, As^{5+}, DMA and MMA) with different toxicities. For the present study, grains were collected from four varieties (flooded cultivation). First, the grains were polished during: i) zero second (s), non-polished (brown rice); ii) 30 s or; iii) 60 s: (white grains). Then, the grains were analyzed by HPLC-ICP-MS. The grains that were polished presented reduced As^{3+} levels in comparison with the brown grains for all cultivars. Regarding As^{5+} and organic As, we detected different results between cultivars. These results show a manner to decrease the As content in rice and reduce the toxicological/nutritional risk of consumption.

1 INTRODUCTION

Rice (*Oryza sativa* L.) is the second global most produced cereal, and an important part of the world population's basic diet. Numerous countries in the world (Asia and Africa) are highly dependent on this grain as a resource of foreign exchange earnings and government income. Regarding the American continent, Brazil is the most important producer, followed by the United States. In Brazil, the consumption of rice achieves 74–76 kg/inhabitant/year (rice with husk) (Batista *et al.*, 2011).

This plant has a special mechanism in the roots that improves the absorption of As, especially arsenite (As^{3+}), when grown in soil at flooded conditions. Several studies showed ways to reduce the concentration of As in the grains. Methods such as cultivation at non-flooded soil, cooking and rinse washing are some important procedures to reduce As in the rice grains (Raab *et al.*, 2009).

In the present study, we aimed to evaluate the concentration of As species (As^{3+}, dimethyl arsenic (DMA), monomethylarsenic (MMA) and arsenate (As^{5+}) in rice grains of 4 Brazilian cultivars processed in 3 different times of polishing (zero, 30 or 60 seconds).

2 MATERIAL AND METHODS

2.1 *Samples and rice sampling*

Four samples of rice varieties cultivated at flooded conditions were supplied by EMBRAPA (Brazil). These samples were peeled and: a-) packaged (brown rice); b-) polished by 30 s; 3-) polished by 60 s. The grains were polished by Suzuki (model MT-10) equipped with cylinder number 1. After quartering, fifty grams of samples were individually milled, homogenized and sieved (0.25 mm) with an electric ultra-centrifugal mill (Retsch ZM200) at 6000 rpm. Then, they were placed into 50 mL conic tubes until analysis.

2.2 *Total As determination*

Grains (250 mg) were accurately weighed in a PFA digestion vessel and then 5 mL of nitric acid 20% v/v + 2 mL of 30% (v/v) H_2O_2 were added. Decomposition was carried out following the program: 160 °C for 4.5 min; 160 °C for 0.5 min; 230 °C for 20 min and 0 °C for 20 min (Milestone Ethos D, Italy). Then, the volume was made up to 50 mL and analyzed by ICP-MS (Elan DRCII, Perkin Elmer, Sciex, Norwalk, CT, USA).

2.3 *Arsenic speciation*

0.2 g of milled sample (in triplicate) was weighed into a 50 mL tube followed by the addition of 10 mL of 2% (v/v) HNO_3. The tube was then closed and heated in a water bath from 25 to 95 °C for 0.75 h and for 1.5 h at 95 °C. Then, samples were cooled at room temperature, filtered by a 0.20 µm cellulose filter and injected into the HPLC-ICP-MS system for chemical speciation (HPLC series 200 Perkin Elmer hyphenated to the ICP-MS). Certified reference material SRM NIST 1568a and blanks were run with each extract batch (Batista *et al.*, 2011).

2.4 *Statistics*

Statistical were performed by SigmaStat software (SigmaStat Version 3.5). The p-value was set at $p < 0.05$.

3 RESULTS AND DISCUSSION

Figure 1 shows the concentration of the different As species after polishing (zero or brown, 30 s or 60 s). After being received in the industry, the rice

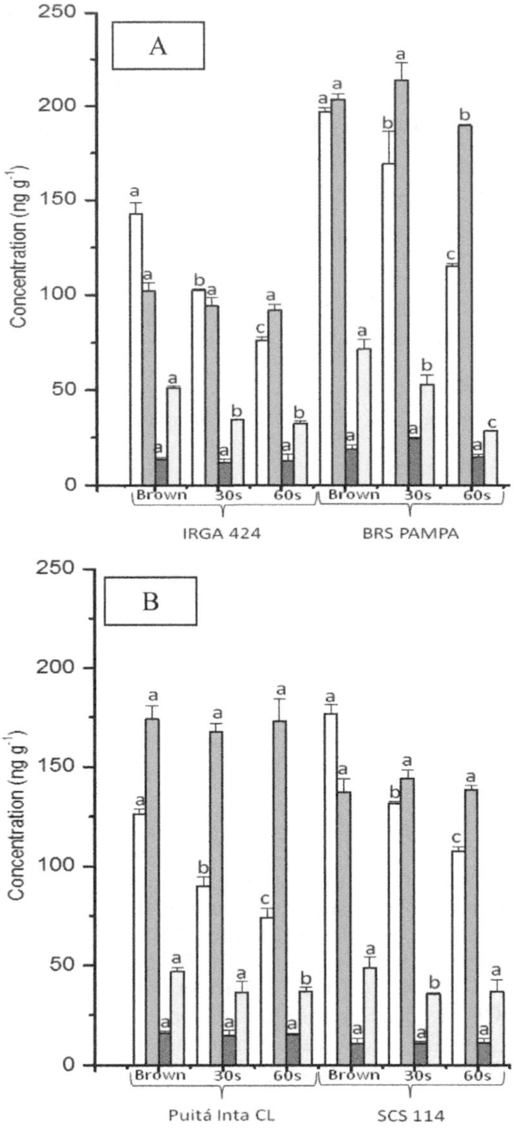

Figure 1. Concentration of As^{3+} (white), DMA (gray), MMA (dark gray) and As^{5+} (yellow) in four different Brazilian rice cultivars. ^{a,b,c}: means with the same letter do not differ statistically ($p > 0.05$) and with different letters are statistically different ($p < 0.05$); A: IRGA 424 and BRS PAMPA; B: Puitá Inta CL and SCS 114.

undergoes a drying process (when the humidity is higher than 13.5%), stripping, polishing, separation (broken grain and husk) and finally, storage and marketing. The mechanical action exerted on the grains, particularly during the polishing can damage the grains. Otherwise, this process removes the bran and increases the expiration time of the cereal. Here we focused on times (30 and 60 s) to

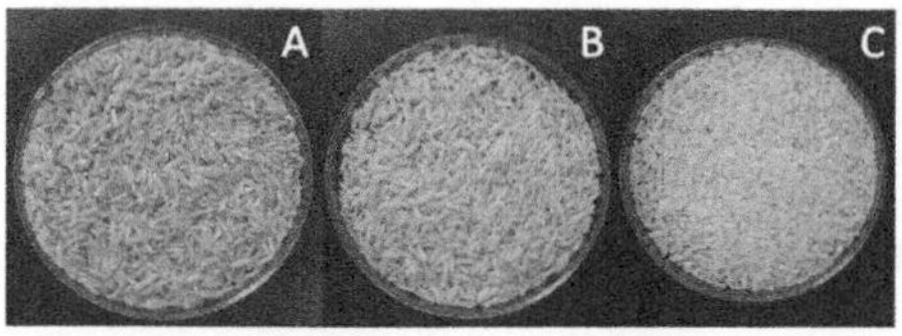

Figure 2. Cultivar IRGA 424 in different degrees of polishing. A: brown; B: 30 s; C: 60 s.

warranty the integrity (Figure 2) of the grains and to increase the removal of external layer.

According to Meharg *et al.* (2008), the inorganic As is mainly present in the pericarp to aleurone layers (brown rice). Our findings agree with theirs. Furthermore, the inorganic:organic As ratio decreases from 0.90 to 0.60, 1.60 to 0.96, 1.77 to 1.04 and, 1.21 to 0.70 for Puitá Inta CL, SCS 114, IRGA 424 and BRS Pampa, respectively. IRGA 424 presented the highest reduction of As^{3+} (46%, $p < 0.05$) after a polishing of 60 s. The other cultivars decrease around 41% ($p > 0.05$).

As shown in Figure 1, inorganic As (sum of As^{3+} and As^{5+}) is the predominant form in all cultivars. In BRS Pampa, we found the highest concentrations of organic As (MMA + DMA) and a constant decrease of As^{5+}. On the other hand, IRGA 424 presented the lowest levels of these forms and, interestingly, the highest decrease of As^{3+} and of As^{5+} after 30 s.

4 CONCLUSIONS

Our findings focused on cultivars planted in traditional rice farms from Brazil. These cultivars, after harvesting, can be polished during 60 s to reduce the toxicological/nutritional risk of rice consumption, without reduction of the grains quality.

ACKNOWLEDGEMENTS

To Fundação de Amparo à Pesquisa do Estado de São Paulo (FAPESP) and to Conselho Nacional de Desenvolvimento Científico e Tecnológico (CNPq).

REFERENCES

Batista, B.L., Souza, J.M.O., Souza & S.S., Barbosa, F. 2011. Speciation of arsenic in rice and estimation of daily intake of different arsenic species by Brazilians through rice consumption. *Journal of Hazardous Materials* 191: 342–348.

Meharg, A.A., Lombi, E., Williams, P.N. Scheckel, K.G., Feldmann, J., Raab, A., Zhu, Y. & Islam, R. 2008. Speciation and Localization of Arsenic in White and Brown Rice Grains. *Environmental Science and Technology* 42: 1051–1057.

Raab, A., Baskaran, C., Feldmann, J. & Meharg, A.A. 2009. Cooking rice in a high water to rice ratio reduces inorganic arsenic content. *Journal of Environmental Monitoring* 11: 41–44.

One Century of the Discovery of Arsenicosis in Latin America (1914–2014) – Litter, Nicolli, Meichtry, Quici, Bundschuh, Bhattacharya & Naidu (Eds)
 ISBN 978-1-138-00141-1

Investigation of arsenic content in polished white rice (*Oryza sativa* L.) in Ghaemshahr city (Vahdat Center, North of Iran): Its weekly intake

H. Boudaghi Malidareh
School of Public Health, Tehran University of Medical Sciences, Tehran, Iran

A.H. Mahvi, M. Yunesian, M. Alimohammadi, Sh. Nazmara & S.S. Hosseini
Center for Solid Waste Research, Institute for Environmental Research, Tehran University of Medical Sciences, Tehran, Iran

P. Boudaghi Malidareh
Antibiotic Sazi IRAN Co., (ASICO), Mazandaran, Iran
Islamic Azad University, Ghaemshahr, Iran

Z. Boudaghi Malidareh
Mazandaran Province Remedy & Social Security, Iran

S. Namavar
National Iranian Oil Products Distribution Company (NIOPDC), Yasuj, Kohgiluyeh Boyer Ahmad, Iran

ABSTRACT: Rice is the dominant agricultural product in Ghaemshahr city. The aims of this study were the investigation of arsenic content in polished white rice (*Oryza sativa* L.) in Ghaemshahr city (Vahdat Center, North of Iran) and the analysis of its weekly intake. 20 polished white rice samples were collected at harvesting of rice in field and, in the laboratory, grains of rice were milled and extracted by acid digestion. The As content in the extracts was measured by ICP-OES. The results indicated that the value of As concentration in these samples was in the range 0.005–0.051 mg kg^{-1} on dry wt. Since in Iran there is no standard for As content in rice, Chinese standards (0.2 mg/kg) were used. According to the results, the weekly As intake from rice was down the maximum weekly intake recommended by WHO/FAO.

1 INTRODUCTION

Rice is the dominant agricultural product in Ghaemshahr city (Vahdat Center, North of Iran). The quality of rice is very important for human health (Zhao *et al.*, 2009). Chemical fertilizers containing toxic metals and metalloids such as arsenic are considered one of the important sources of pollution (Rui *et al.*, 2008). According to studies conducted, it is feasible to estimate the daily intake of heavy metals through rice in South Korea (Jung *et al.*, 2005). Arsenic is usually present in subsoil and it is carcinogen in humans (Tuli *et al.*, 2010). The aims of this study were the investigation of arsenic content in polished white rice (*Oryza sativa* L.) in Ghaemshahr city during the agricultural year of 2010 and the analysis of its weekly intake.

2 MATERIALS AND METHODS

2.1 *Site study*

Ghaemshahr is one of the active zones of agriculture (rice) in Mazandaran province (northern Iran). The area of rice cultivation for 2009–2010 in this city is 15,650 ha. Vahdat Center is one of the most important areas producing different types of high-yield rice.

2.2 *Sampling strategy*

According to the geographical conditions, rice type, level under harvest, type & size of chemical fertilizer consumed, and by consulting the experts of the Agriculture Office in Ghaemshahr, 20 polished white rice samples were collected from rice in field. In the laboratory, grains of rice were milled to determine the As content. 0.5 g of each sample were refluxed in 10 mL of conc. nitric-sulfuric-perchloric acid mixture (4/1/1, v/v/v) for 1 h. Formic acid (90%) was then added dropwise until the red-brown gas formed disappeared. Afterwards, deionized water was added to bring the digest to 25 mL. The resulting solution was reacted with an aqueous solution of 1% $NaBH_4$ and 1% NaOH (Lin *et al.*, 2004). Rice was extracted and measured by ICP-OES. Statistical analysis comparison of the heavy metal content was done through SPSS software v.11.5.

Table 1. Arsenic contents in polished white rice in Ghaemshahr (Vahdat center).

(mg kg^{-1} on dry wt.)

No	As	No	As
1	0.025	11	0.005
2	0.020	12	0.005
3	0.022	13	0.005
4	0.018	14	<0.005
5	0.016	15	0.032
6	0.017	16	0.051
7	0.016	17	0.017
8	0.017	18	0.016
9	0.021	19	0.019
* Mean	0.0166	Std.	0.0114
Min.	0.0025	Max.	0.0510

Note: the one-half the detection limit was used for non-measurable quantities. Wavelength: 189.042 nm.

Table 2. Intake of As via rice (weekly dietary intake of As by eating rice, μg kg^{-1} body weight/week).

Item	Average	Range
Daily Rice Consumption (g day^{-1})	165	158–178
Content (μg g^{-1})	0.0166	0.0025–0.051
*Weekly intake	0.319	0.046–1.059
(μg kg^{-1} body weight/week)	15	–

*Provincial tolerable weekly intake—iAs.

3 RESULTS

The results of As contents in 20 samples of polished white rice from various areas in Ghaemshahr were in the range 0.005–0.051 mg kg^{-1} on dry wt. (Table 1).

4 CONCLUSIONS

As contents in 20 samples of raw rice from various areas in Ghaemshahr (North of Iran) indicated that the value of As concentration in rice was in the range 0.005–0.051 mg kg^{-1} on dry wt. (Table 1). Since in Iran there is no standard for As content in rice, Chinese standards were used. In China, the established Maximum Level for inorganic arsenic in rice and rice-based products is 0.2 mg/kg. For comparison, arsenic levels in Brazilian rice samples are reported to vary from 58.8 to 216.9 ng g^{-1} (0.0588 to 0.2169 mg kg^{-1} (Batista *et al.*, 2010). Arsenic concentration in rice in the western part of Kocani Field (Macedonia) was 0.53 μg g^{-1} (Rogan *et al.*, 2009). Distribution of Pb, Cd and As in roots of rice plant were found higher than in any other parts of the plant, in the order: root >> shoot > husk > whole grain (Lei *et al.*, 2011).

The Joint FAO/WHO Expert Committee on Food Additives (JECFA) has set PTWI (Provisional Tolerable Weekly Intake) for iAs equal to 15 μg kg^{-1} of body weight (FAO, 2010). According to the published papers, average daily consumption of rice in Asian countries is 165 g/person each day (ranging from 158–178 g/person-day (Nogawa & Ishizaki, 1979; Rivai *et al.*, 1990). According to the results of this study (Table 2), the weekly intake of As from rice was 0.319 (0.046–1.059) μg kg^{-1} body weight/week, that was lower than that of total dietary As intake; weekly As intake from rice was down the maximum weekly intake recommended by WHO/FAO. Although the amount of arsenic in rice is less than the standard level, chemical fertilizers are being used more and more by farmers with the consequent risk of As uptake by the plants.

Since people use various rice brands, it is expected that health authorities study the annual level of heavy metals and arsenic either in exported or imported rice.

ACKNOWLEDGEMENTS

The authors would like to acknowledge the support of this research provided by the School of Health at the Tehran University of Medical Sciences.

REFERENCES

Batista, B.L., De Oliveira Souza, V.C., Da Silva, F.G. & Barbosa, J.F. 2010. Survey of 13 trace elements of toxic and nutritional significance in rice from Brazil and exposure assessment. *Food Additives & Contaminants: Part B: Surveillance* 3(4): 253–262.

FAO, 2010. *Evaluation of certain contaminants in food. Food and Agriculture Organization of the United Nations, World Health Organization. FAO, Evaluation of certain food additives and contaminants.* Food and Agriculture Organization of the United Nations, World Health Organization.

Jung, M.C., Yun, S.-T., Lee, J.-S. & Lee, J.-U. 2005. Baseline study on essential and trace elements in polished rice from South Korea. *Environ. Geochem. Health* 27(5–6): 455–464.

Lei, M., Tie, B., Williams, P.N., Zheng, Y. & Huang, Y. 2011. Arsenic, cadmium, and lead pollution and uptake by rice (*Oryza sativa* L.) grown in greenhouse. *Journal of Soils and Sediments* 11(1): 115–123.

Lin, H.T., Wong, S.S & Li, G.C. 2004. Heavy metal content of rice and shellfish in Taiwan. *J. Food Drug Anal.* 12: 167–174.

Nogawa, K. & Ishizaki, A. 1979. A comparison between cadmium in rice and renal effects among inhabitants of the Jinzu River basin. *Environmental Research* 18(2): 410–420.

Rivai, I.F., Koyama, H. & Suzuki, S. 1990. Cadmium content in rice and its daily intake in various countries. *Bulletin of Environmental Contamination and Toxicology* 44(6): 910–916.
Rogan, N., Serafimovski, T., Dolenec, M., Tasev, G. & Dolenec, T. 2009. Heavy metal contamination of paddy soils and rice (*Oryza sativa* L.) from Kocani Field (Macedonia). *Environmental Geochemistry and Health* 31(4): 439–451.
Rui, Y.K., Shen, J.B. & Zhang, F.S. 2008. Application of ICP-MS to determination of heavy metal content of heavy metals in two kinds of N fertilizer. *Guang Pu Xue Yu Guang Pu Fen Xi* 28(10): 2425–2427.
Tuli, R., Chakraborty, D., Trivedi, P.K. & Tripathi, R.D. 2010. Recent advances in arsenic accumulation and metabolism in rice. *Molecular Breeding* 26(2): 307–23.
Zhao, K., Zhang, W., Zhou, L., Liu, X., Xu, J. & Huang, P. 2009. Modeling transfer of heavy metals in soil-rice system and their risk assessment in paddy fields. *Environmental Earth Sciences* 59(3): 519–27.

One Century of the Discovery of Arsenicosis in Latin America (1914–2014) – Litter, Nicolli, Meichtry, Quici, Bundschuh, Bhattacharya & Naidu (Eds)
 ISBN 978-1-138-00141-1

Accumulation and arsenic speciation in maize crop (*Zea mays*) in San Luis Potosí, México

L. Hinojosa-Reyes, J.M. Rosas-Castor, J.L. Guzmán-Mar, A. Hernández-Ramírez, E. Ruiz-Ruiz & J.M. Alfaro-Barbosa

Facultad de Ciencias Químicas, Universidad Autónoma de Nuevo León, Nuevo León, México

ABSTRACT: The presence of arsenic in agricultural products is a matter of concern due to potential adverse human health effects even at low concentrations. In this study, total arsenic and arsenic speciation was determined in maize crops and agricultural soils from a mining area of San Luis Potosi, Mexico, and the contribution of the rhizosphere conditions to As accumulation and translocation was evaluated. Elevated levels of Fe and Mn were more highly correlated with As translocation in maize than most of the other evaluated parameters, while Ca was highly correlated with the total content of As in soil. While As(III) was the major species detected in maize stem, As(V) was the main species present in maize leaf and root.

1 INTRODUCTION

Arsenic (As) is a highly toxic element ubiquitous in the nature that is a potential environmental threat to human and animal health. As occurs in the natural environment as either inorganic (As(III), As(V)) or organic species (MMA and DMA). Chronic exposure to inorganic As has been associated with skin lesions and cancer (Cubadda *et al.*, 2010). Humans may be exposed to As through many different pathways, but direct water intake and As transfer through the crops-soil-water system are the two major dietary pathways (Khan *et al.*, 2009). Plants and vegetables can access As present in agricultural soil and irrigation water through their root systems resulting in food chain contamination. However, the phytoavailability of As in agricultural soils to crops can vary dramatically from one location to another depending on the soil conditions. Maize (*Zea mays* L.), is the most commonly grown cereal crop in the world (883 million tons per year in 2011, FAOSTAT 2013), and is also an important animal feed and a staple food for humans (grain maize). In this study, the effect of soil conditions on As accumulation by maize crops, as well as, the species distribution within the soil-water-maize system were evaluated in a mining area of San Luis Potosi, Mexico.

2 METHODS/EXPERIMENTAL

2.1 *Sampling agricultural soil and maize*

Samples of soil, irrigation water and maize were collected from three agricultural zones in San Luis Potosi, Mexico (Matehuala zone A, MTA; Matehuala zone B, MTB; and Villa de Ramos, VR). Six samples of soil and maize crop were collected per sampling site. Soil samples were dried at room temperature and then sieved to obtain the clay fraction (<63 µm). The maize plants were cut (root, stem, leaf and grain), dried at 70 °C for 3 days, and sieved through a mesh of 250 µm. Before drying, corn roots were washed in 5 mM $CaCl_2$. Water samples were filtering through a 0.45 µm cellulose filter and, then, EDTA and acetic acid were added as preservatives. The samples were hermetically sealed and stored at 4 °C in the absence of light.

2.2 *Soil and irrigation water characterization*

The pH (pH_{WE}; potentiometric method, Thermo scientific Orion 3 star), total organic carbon (TOC_{WE}; Shimadzu TOC-VCSH), chloride (Cl^-_{WE}; potentiometric method, Thermo Orion 720 A+), phosphate ($PO_4^-{}_{WE}$; spectrophotometric method, NMX-AA-029-SCFI-2001), and conductivity (potentiometric method, Thermo scientific Orion 3 star) were determined from the water-extract soil, whereas the content of calcium (Ca_T; Atomic Absorption Spectrometry), phosphate ($PO_4^-{}_T$; spectrophotometric method, NMX-AA-029-SCFI-2001) iron, and manganese (Mn_T and Fe_T, respectively; NMX-AA-051-SCFI-2001) from the digests. All samples were analyzed in triplicate.

2.3 *Determination of total and species arsenic*

Total As (As_T) was determined in maize crop (root, stem, leaf and grain), agricultural soil and water samples. Solid samples were processed according to the slightly modified microwave digestion

3052 method (EPA). After digestion, samples were diluted and analyzed by HG-AFS (Rayleigh AF-640). The determination of As species (As(III), As(V), MMA, and DMA) was performed by IC-HPLC-HG-AFS (Hamilton PRP-X100 column, 250 × 4.6 mm); mobile phase: phosphate buffer in gradient elution mode from 5 to 60 mM; and flow rate: 1 mL min^{-1}. The extraction of As species from maize plant was carried out using 0.15 N HNO_3 at 80 °C during 120 min (Zheng *et al.*, 2008).

3 RESULTS AND DISCUSSION

3.1 *Rhizosphere conditions effect in arsenic accumulation by maize*

The correlation between rhizosphere conditions and accumulated As in corn root was determined using principal component analysis (PCA; IBM SPSS Statistics software), where three main components (criterion: eigenvalues greater than 1) accounted for 85% of the total variability. Ca_T concentrations were positively correlated with As_T, which may be due to the favorable adsorption of As by organic matter in the presence of Ca_T (Wang, 2006). A negative correlation was found between pH_{WE} and As in root. It is known that low P concentration in soil induces organic acid exudation by corn plant. Organic acids tend to increase the nutrient phytoavailability, and therefore, As solubility in the soil (Marwa *et al.*, 2012). The Fe_T and Mn_T content presented a negative correlation with the translocation of As from root to stem. Mallick *et al.* (2011) reported that the negative correlation between As mobility to plant aerial parts and the concentrations of Fe and Mn in root can be attributed to As adsorption on (hydr)oxides of these metals.

3.2 *Arsenic speciation*

As_T concentration in the maize plant parts (root, stem, leaves and grain) was determined for each sampling point (Table 2). The concentration of As in the maize crop was root >> stem > leaf > grain.

Table 2. Total arsenic concentration in agricultural soil and maize crop (minimum—maximum value), LOD, 0.05 mg kg^{-1}.

Sampling zone	Total As (mg kg^{-1}), $n = 6$			
	Soil	Root	Stem	Leaf
MTA	4.1–10.1	21.6–54.2	0.7–2.9	0.9–2.1
MTB	24.2–43.7	2.1–7.6	0.24–3.1	0.1–1.8
VR	4.4–10.3	0.3–1.5	<LOD	<LOD-0.1

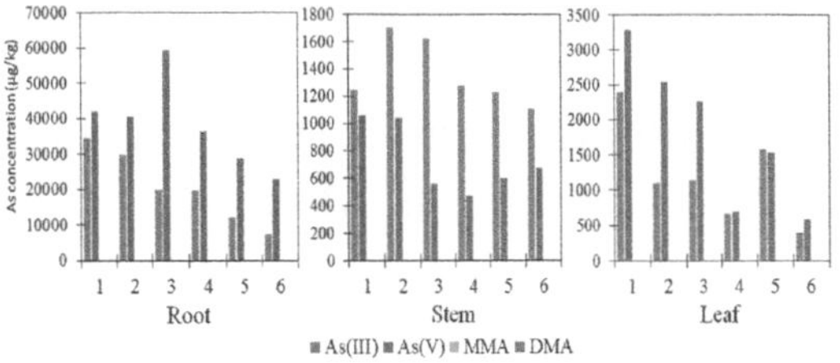

Figure 1. Arsenic species in maize crop in Matehuala (MTA), San Luis Potosi, Mexico (LODAs(V) 55 µg kg^{-1}, LODAs(III) 90 µg kg^{-1}, LODMMA 310 µg kg^{-1}, LOD-DMA 250 µg kg^{-1}).

The As species distribution was evaluated in MTA (zone with the highest As concentration in maize). Inorganic As species were the predominant As form with As(III) and As(V) content varying from 33 to 65% and organic species were not detected (Fig. 1). Although As(V) was the predominant species in water-extractable soil (>90%), As(III)/As_T ratio was the highest in plant. The increment of reduced species can be related to detoxification mechanism which involved complexation and reduction steps (Mallick *et al.*, 2011). Organic As species were not detected in plant and in water-extractable soil fraction. The high content of organic species (DMA and MMA) in plant tissues has been mainly attributed to the bioaccumulation rather than the methylation process (Meharg *et al.*, 2002).

4 CONCLUSIONS

Soil concentration of Ca, Fe, and Mn affect the As transfer in maize. While, Ca favored As accumulation in agricultural soil surfaces, Fe and Mn could reduce the As translocation to the aerial parts of the crops due to As adsorption on (hydr)oxide of these metals. While As(III) was the major species in maize stem, As(V) content was higher in leaf and root. The presence of inorganic As represents a risk to livestock due to its high toxicity.

ACKNOWLEDGEMENTS

This study was supported by the Research fund of PAICyT-UANL and CONACyT through grants UANL-PAICyT-CN885-11 and CONACYT/CB/167372, respectively.

REFERENCES

Cubadda, F. *et al.* 2010. Arsenic contamination of the environment-food chain: a survey on wheat as a test plant to investigate phytoavailable arsenic in Italian

agricultural soils and as a source of inorganic arsenic in the diet. *J Agric Food Chem* 58: 10176–10183.
EPA. Environmental Protection Agency of U.S. [http://www.epa.gov/iris/subst/0278.htm]
Khan, N.I. *et al.*,. 2009. Human arsenic exposure and risk assessment at the landscape level: a review. *Environ Geochem and Health* 31(1): 143–166.
Mallick, S. *et al.* 2011. Study on arsenate tolerant and sensitive cultivars of *Zea mays L.*: differential detoxification mechanism and effect on nutrients status. *Ecotoxicology and Environmental Safety* 74(5): 1316–1324.
Marwa, E.M. *et al.* 2012. Risk assessment of potentially toxic elements in agricultural soils and maize tissues from selected districts in Tanzania. *Sci Total Environ* 416: 180–186.
Meharg, A.A. & Hartley-Whitaker, J. 2002. Arsenic uptake and metabolism in arsenic resistant and nonresistant plant species. *New Phytologist.* 154: 29–43.
Wang, S. & Mulligan, C.N. 2006. Effect of natural organic matter on arsenic release from soils and sediments into groundwater. *Environ Geochem Health* 28(3): 197–214.
Zheng, M.Z. *et al.* 2008. Spatial distribution of arsenic and temporal variation of its concentration in rice. *Environ Sci Technol* 42: 5008–2013.

One Century of the Discovery of Arsenicosis in Latin America (1914–2014) – Litter, Nicolli, Meichtry, Quici, Bundschuh, Bhattacharya & Naidu (Eds)
© 2014 Taylor & Francis Group, London, ISBN 978-1-138-00141-1

Arsenic determination in whole grain industrialized and the persistence after cooking

E.M. Soro, A.B. Livore, C. Liberman, F. Cattaneo & J. Colazo
Instituto Nacional de Tecnología Agropecuaria, Estación Experimental Agropecuaria Concepción del Uruguay (INTA EEA), Prov. de Entre Ríos, Argentina

S.S. Farías, J.A. Londonio & P. Smichowski
Departamento de Química Analítica, Gerencia Química, Comisión Nacional de Energía Atómica, Buenos Aires, Argentina

ABSTRACT: Rice absorbs arsenic and accumulates it in different parts of the plants, including grains, to levels several times higher than the soil content. Determinations of total arsenic contents were done using HG-AAS. Nine samples with total As content higher than the commercial limit were selected, processed and are separated the industrial components. The arsenic persistence was determined in the cooked grain. After its industrialization, grains had a reduction of 36% (average) arsenic concentration. The polishing technique can be used as a tool for reducing the arsenic concentration. Arsenic concentration in cooked grain was on average 87% less than the arsenic in whole grain and 80% less than in polished grain.

1 INTRODUCTION

In rice grains, arsenic is present in inorganic or in methylated form. A wide variation in the proportion that both forms of arsenic are present in the grain from different genotypes has been reported (Norton *et al.*, 2009).

Rice absorbs arsenic and accumulates it in different parts of the plants, including grains, to levels several times higher than the soil content. In regions with high arsenic in soil, rice can contribute substantially to the consumption of this metalloid by human population (Gultz *et al.*, 2005).

2 METHODS

Determinations of total arsenic contents were conducted in the Laboratory of Chemistry of the National Institute of Industrial Technology (INTI) using atomic absorption spectrometry with hydride generation technique (HG-AAS).

Nine samples were selected from the commercial fields of which total arsenic content in whole grain was over 1 mg kg^{-1} and proceeded to its elaboration, separating its industrial components (husk, bran and polished rice). These fractions were analyzed for total arsenic content.

For determining the arsenic persistence, 25 g of polished rice were cooked in 200 mL of water, considering the cooking time previously determined for each variety.

3 RESULTS AND DISCUSSION

The results of the determinations of total arsenic in whole grain and fractions after its preparation are given in Table 1.

After industrialization, grains presented a reduction on arsenic concentration. This reduction was in average of 36%.

Table 1. Arsenic concentration of samples of whole grain of rice and its industrial fractions (in mg kg^{-1}).

Sample Id	Total	As in bran	As in husk	As in polished grain	As in cooked grain
14	1,12	1,43	1,12	0,95	0,13
22	1,01	1,09	0,94	0,60	0,04
25	1,1	1,57	1,04	0,80	0,12
29	1,22	0,69	0,38	0,15	0,04
43	1,28	1,52	1,23	1,24	0,34
1	1,21	1,38	1,09	0,76	0,16
11	1,26	1,16	0,97	0,66	0,09
26	1,01	1,32	1,06	0,67	0,19
15	1,31	1,44	1,37	0,92	0,24

According to the results, the grain polishing technique can be used as a tool for reducing the arsenic concentration in polished grains.

There was no correlation between the arsenic concentration in whole grains and in polished grains.

Arsenic concentration in cooked grain samples was, on average, 87% less than the arsenic content in whole grains and 80% less than the value in polished grains.

Each cooked rice samples analyzed presented As values below the limit marketing.

Arsenic content in polished grains is linearly related to the arsenic content in cooked grains ($R^2 = 0.6675$).

The results are according with previous investigations (Meharg *et al.*, 2009).

4 CONCLUSIONS

Samples whose arsenic concentrations in grain are above the legally established consumption values were detected.

Polishing reduces arsenic concentration values in industrialized grain, although, in some cases, the reduction was not sufficient to achieve values that are below the established consumption limit.

The cooking of rice grains, whose arsenic content are above legal limits established, allows to reduce the concentration of this element in the cooked grain, regardless of initial arsenic concentration.

Arsenic concentration in cooked grain is related to the arsenic in whole grain, and it is independent of the initial concentration in polished grain, provided it is cooked with excess water.

ACKNOWLEDGEMENTS

To the PROARROZ Foundation for their contribution of this work.

REFERENCES

Gulz, P.A., Gupta S.K. & Schulin, R. 2005. Arsenic accumulation of common plants from contaminated soils. *Plant and Soil* 272: 337–347.

Meharg, A.A., Williams, P.N., Adomako, E., Lawgali, Y.Y., Deacon, C., Villada, A., Cambell, R.C., Sun, G., Zhu, Y.G., Feldmann, J., Raab, A., Zhao, F.J., Islam, R., Hossain, S. & Yanai, J. 2009. Geographical variation in total and inorganic arsenic content of polished (white) rice. *Environ. Sci. Technol.* 43: 1612–17.

Norton, G.J., Duan, G.L., Dasgupta, T., Islam, M.R., Lei, M., Zhu, Y., Deacon, C.M., Moran, A.C., Islam, S., Zhao, F.J., Stroud, J.L., McGrath, S.P., Feldmann, J., Price, A.H. & Meharg, A.A. 2009. Environmental and genetic control of arsenic accumulation and speciation in rice grain: comparing a range of common cultivars grown in contaminated sites across Bangladesh, China, and India. *Environ. Sci. Technol.* 43: 8381–8386.

One Century of the Discovery of Arsenicosis in Latin America (1914–2014) –
Litter, Nicolli, Meichtry, Quici, Bundschuh, Bhattacharya & Naidu (Eds)
© 2014 Taylor & Francis Group, London, ISBN 978-1-138-00141-1

Soybean crop exposed to arsenic: A possible risk for the food chain?

A.L. Armendariz, M.A. Talano & E. Agostini
Departamento de Biología Molecular, FCEFQyN, Universidad Nacional de Río Cuarto, Río Cuarto, Córdoba, Argentina

ABSTRACT: Soybean (Glycine max) is often being cultivated in soils with either moderate or high arsenic (As) concentrations or under irrigation with As contaminated groundwater. The purpose of this study was to determine the effect of arsenite (As(III)) and arsenate (As(V)) on growth, antioxidant response and possible accumulation in soybean plants. A decrease in biomass and shoot length was observed when plants were treated with solutions containing As. As induced an antioxidant response, which was more evident in roots treated with As(III). For both treatments, As concentration in roots was higher than in shoots. Plants treated with As(III) accumulated 9% more of As in roots than those exposed to As(V). The results suggest that soybean crops exposed to As would not constitute a source of As entry into the food chain. However, soybean plants could be affected in terms of production, reducing their quality and yield.

1 INTRODUCTION

Soybean (*Glycine max*) is a legume with worldwide economic importance. In Argentina, it is often cultivated in soils with high As levels and, due to the expansion of this crop even to desert regions, it is sometimes irrigated with As-contaminated groundwater. We have previously demonstrated that As produce deleterious effects on soybean germination and seedling development (Talano *et al.*, 2012). In the present study, soybean tolerance and its antioxidant response towards As treatment and As accumulation pattern in roots and aerial parts was explored.

2 MATERIALS AND METHODS

2.1 *Plant growth and As treatments*

Soybean seeds (*Glycine max*) cv. DM 4670 were used. Seeds were germinated at 28 °C in darkness during 3–5 d, then placed in pots containing sterile vermiculite and incubated in a chamber with controlled temperature and under photoperiod regime. To analyze the effect of As on root cell viability, soybean seedlings were exposed during 8 d to 25–200 μM As(V) or As(III) solutions. Cell death was evaluated by Evans blue staining (Suzuki *et al.*, 1999). To evaluate the effect of As on the development of the plants and the oxidative stress response, they were treated with 25 μM of As(V) or As(III) at the beginning of the assay. Plants were harvested after 30 d and Fresh Weight (FW) and root and shoot length were registered for each treatment.

2.2 *Enzymatic and lipid peroxidation analysis*

Frozen roots and shoots (100 mg) were ground with liquid nitrogen, homogenized with 0.2 g of PVP in 1 mL of 50 mM phosphate buffer (pH 7.8) and centrifuged (15,000 rpm during 30 min). The supernatants were used for enzyme assays. Superoxide dismutase (SOD), ascorbate peroxidase (APX) and peroxidase (POD) activities were determined following the methodologies described by Beauchamp & Fridovich (1973), Hossain & Asada (1984) and Sosa Alderete *et al.* (2009), respectively. Lipid peroxidation was estimated by analyzing malondialdehyde (MDA) content as described by Heath & Packer (1968).

2.3 *Total As accumulation analysis*

Root and shoot dried tissues were digested with nitric acid and H_2O_2 (30%), and total As was determined by flame Atomic Absorption Spectrometry (AAS). Results were expressed as mg kg^{-1} of dry weight.

3 RESULTS AND DISCUSSION

3.1 *Soybean growth under As treatment*

As shown in Fig. 1A, the biomass of shoots was significantly decreased under 25 μM As(III) treatment, while shoot length was significantly affected for both treatments (Fig. 1B). Contrarily, root biomass and length did not show significant differences under treatments with solutions containing As compared with control plants (without

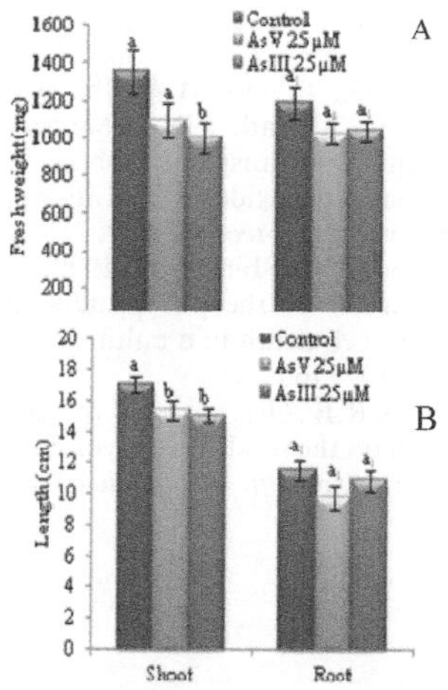

Figure 1. Soybean growth under As treatment. FW (A) and length (B) of roots and shoots of plants treated with 25 µM As(V) and As(III). Data represents means ± SE. Different letters represent significant differences from the control (*Tukey's* test, $p < 0.05$).

As). Further, viability of soybean root cells was examined. The results showed that cell death significantly increased after treatment with 200 µM As(V) and up to 100 µM As(III), showing the higher toxicity of As(III).

3.2 *Antioxidant response to As treatment*

The effect of As on SOD, APX and POD activities in soybean plants was examined. In roots, SOD, APX and POD activity was significantly increased after As(III) treatment compared with control, while in shoots only POD activity was higher compared with control plants for both treatments (Table 1). It is evident that As treatment induced a strong antioxidant response, mainly in roots, which is the organ that has direct contact with the metalloid. Similar effects were found by Shri *et al.* (2009) in rice seedlings. The MDA content in shoots and roots did not showed significantly differences under As(V) and As(III) treatments compared with control plants. Contrarily, other authors found increased MDA values under As treatments although, it was measured in short periods of time (Mészáros *et al.*, 2013).

3.3 *Total As accumulation*

Arsenic concentration in roots was higher than in shoots for both treatments. Arsenic retention in root might be attributed to its efficient compartmentalization in vacuoles, probably associated with phytochelatins (Pickering *et al.*, 2006). However, soybean plants accumulated 9% more As when they were treated with As(III) than when they were treated with As(V). This could be related to an effect of As(V)-P competition since the main route of As(V) absorption in plants is via phosphate transporters. In contrast, As(III) transport is mediated by aquaglyceroporins; they do not have evident competitive effects with other ions, hence As(III) could be more easily incorporated in soybean plants.

Table 1. Antioxidant response to As treatment.

	Treatment	Shoot	Root
SOD	Control	1661 ± 188	257 ± 61
(UI g^{-1})	25 µM As(V)	1793 ± 149	1333 ± 207*
	25 µM As(III)	1494 ± 236	1328 ± 36*
APX	Control	488 ± 60	4679 ± 762
(UI g^{-1})	25 µM As(V)	1058 ± 315	6364 ± 504
	25 µM As(III)	835 ± 36	7335 ± 72*
POD	Control	5947 ± 513	21412 ± 1840
(UI g^{-1})	25 µM As(V)	7772 ± 1209*	22285 ± 385
	25 µM As(III)	8818 ± 1186*	25168 ± 1779*
MDA	Control	39 ± 3	51 ± 3
(µ moles	25 µM As(V)	39 ± 4	59 ± 4
g^{-1})	25 µM As(III)	38 ± 4	56 ± 0.1

Data represents means ± SE. (*) treatments significantly different from the control (*Tukey's* test, $p < 0.05$).

4 CONCLUSIONS

The up-regulation of antioxidant enzyme activities and growth reduction indicated that the presence of As produces stress in soybean plants, that is more pronounced under As(III) treatment. However, the accumulation of As in roots (phytoextraction mechanism) and the low translocation to shoots indicate a low probability that the metalloid can be accumulated in grains during filling. Therefore, soybean crop growing in the presence of As would not constitute a source of As entry into the food chain but they could be affected in terms of production, reducing their quality and yield.

REFERENCES

Beauchamp, C.O. & Fridovich, I. 1973. Isozymes of superoxide dismutase from wheat germ. *Biochem. Biophys.* 317: 50–64.

Heath R.L. & Packer L. 1968. Photoperoxidation in isolated chloroplasts: kinetics and stoichiometry of fatty acid peroxidation. *Arch. Biochem. Biophys.* 125: 189–198.

Hossain, M.A. & Asada, K. 1984. Purification of dehydro-ascorbate reductase from spinach and its char-

acterization as a thiol enzyme. *Plant Cell Phys.* 25: 385–395.
Mészáros, P., Rybanský, L., Hauptvogel, P., Kuna, R., Libantová, J., Moravčíková, J., Piršelová, B., Tirpáková, A. & Matušíková, I. 2013. Cultivar-specific kinetics of chitinase induction in soybean roots during exposure to arsenic. *Mol. Biol. Rep.* 40: 2127–2138.
Pickering, I.J., Gumaelius, L., Harris, H.H., Prince, R.C., Hirsch, G., Banks, J.A., Salt, D.E. & George, G.N. 2006. Localizing the biochemical transformations of arsenate in hyperaccumulating fern. *Environ. Sci. Technol.* 40: 5010–14.
Shri, M., Kumar, S., Chakrabarty, D., Trivedi, P.K., Mallick, S., Misra, P., Shukla, D., Mishra, S., Srivastava, S., Tripathi, R.D. & Tuli, R. 2009. Effect of arsenic on growth, oxidative stress, and antioxidant system in rice seedlings. *Ecotox. Environ. Safe.* 72: 1102–1110.
Sosa Alderete, L.G., Talano, M.A., Ibáñez, S.G., Purro, S., Agostini, E., Milrad, S.R. & Medina, M.I. 2009. Establishment of transgenic tobacco hairy roots expressing basic peroxidases and its application for phenol removal. *J. Biotechnol.* 139: 273–279.
Suzuki, K., Yano, A. & Shinshi, H. 1999. Slow and prolonged activation of the p47 protein kinase during hypersensitive cell death in a culture of tobacco cells. *Plant Phys.* 119: 1465–72.
Talano, M., Cejas, R.B., González, P.S. & Agostini, E. 2012. Arsenic effect on the model crop symbiosis *Bradyrhizobium*-soybean. *Plant Physiol. Bioch.* 63: 8–14.

One Century of the Discovery of Arsenicosis in Latin America (1914–2014) – Litter, Nicolli, Meichtry, Quici, Bundschuh, Bhattacharya & Naidu (Eds)
 ISBN 978-1-138-00141-1

Arsenic in soils, pulses and crops in a Portuguese industrial contaminated site

M. Inácio, V. Pereira & E. Ferreira da Silva
GEOBIOTEC, GeoBiosciences, Geotechnologies and Geoengineering Research Center, University of Aveiro, Portugal

O. Neves
CERENA, Centre for Natural Resources and the Environment, Instituto Superior Técnico, University of Lisbon, Portugal

ABSTRACT: This study was carried out to evaluate the concentrations of As in soils ("near total" and available fraction), forage plants and horticultural crops in the surrounding area of a Portuguese industrial chemical complex. International regulatory guidelines were used to identify contaminated land (>46% of the sampling sites). The spatial distribution of As in the different study samples indicated a similar anthropogenic pattern, the highest values located near the industrial plants and some old sewage outlets. Forage plants (gramineae) present higher As concentration (0.3 to 255 mg kg^{-1} dry weight) than the studied crops (cabbage, tomato, maize, lettuce (0.1 to 3.5 mg kg^{-1} dry weight). Taking into account that some local soils are considered to be contaminated, more data from different environmental medium should be used in the future to investigate possible relations with local health problems.

1 INTRODUCTION

The Estarreja Chemical Complex (ECC), located in North littoral of Portugal (District of Aveiro) is one of the most important centers of the Portuguese chemical industry (Figure 1) and has been operating for more than 7 decades. The most important input of As to the environment is chiefly related to past activities, namely with the production of sulfuric acid from arsenopyrite roasting. The site has a long history of environmental contamination; for many years, the liquid industrial effluents have been discharged in the sewage outlet coming from the factories (Figure 1), and the solid wastes were deposited in the soil without any treatment. At the present time, the situation is different because, according EEC legislation, some environmental geochemical studies have been carried out and some rehabilitation works have been implemented. In the area, the soils (Podzols and Cambisols) are used by the local residents for pasture and agricultural purposes, despite the contamination situation described above.

This study was carried out to evaluate the concentration of As in soils, forage and horticultural crops in the surrounding area of industrial Complex of Estarreja and its transfer from potential contaminated soils to the food chain. The study was performed in two successive phases, the first concerning animal health (forage plants) and the second the local population health (horticultural crops). These studies, focused on the environmental damages to the local population health, were recently started in Estarreja area.

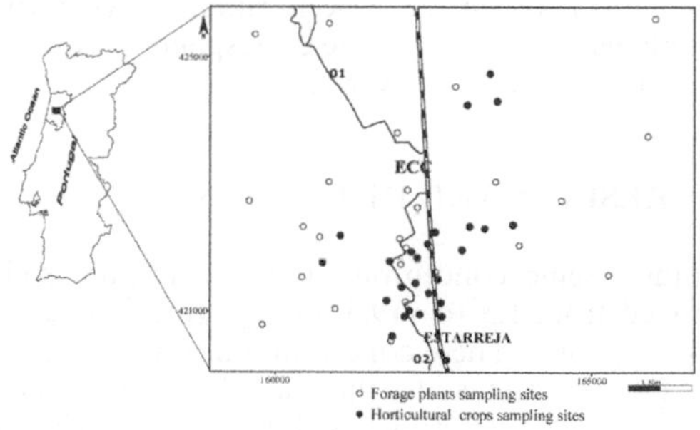

Figure 1. Location of the study area (approximate 60 km^2) and forage plants and horticultural crops sampling sites. O1 and O2 are sewage outlets.

2 MATERIALS & METHODS

In 2010 (1st. phase), composite samples of topsoil (0–15 cm) and forage plants (gramineae) were collected at 26 pasture sites (Figure 1). The species *Avena sativa* L., *Gaudinia fragilis* (L). *Beauv.* and *Holcus lanatus* L. were the ones consumed by the grazing animals. In 2011–2012 (2nd. phase), soil composite samples (rhizosphere area) and horticultural crops were also collected from 26 kitchen gardens and/or small farms (Figure 1) The samples of crops in a total of 26 cabbages (*Brassica oleracea* L.), 25 tomatoes (*Lycopersicon esculentum* Mill.), 12 corns (*Zea mays* L.) and 4 lettuces (*Latucca sativa* L.) were collected according to its availability in the soil sampling sites. All soil samples were air-dried, sieved <2 mm and analyzed for physicochemical parameters following classic methods. After sampling, the edible plant parts (cabbage and lettuces leaves, tomato fruits, grain maize and forage green shoots) were carefully washed several times with tap and distilled water, to remove soil particles and airborne pollutants and then dried at 40 °C until constant mass. All vegetation samples were ground to a fine state of subdivision for analysis. Arsenic analysis of both soils and plants were performed by ICP/ES-MS in an ACME certified laboratory after extraction with *aqua regia* (for "near total" soil and plant concentration) and after soil leaching with ammonium acetate (for mobile and soil As available fraction).

The Bioaccumulation coefficient (BA) is a parameter used to evaluate the capacity of a species to accumulate an element. It was calculated as the ratio between the concentration of As in the edible part plant and in the corresponding soil (all based on dw) for each vegetable.

3 RESULTS AND DISCUSSION

Total arsenic concentration in the pasture soils ranged from 1.5 to 719.3 mg kg^{-1} (median value: 14.5 mg kg^{-1}). These concentrations exceed in 54% of the sampled soils, the Canadian soil quality guideline (CSQG) set for As for all land use classes (12 mg kg^{-1}; CCME, 2011) and also recommended by Portuguese governmental authorities. The available fraction in almost half of soil samples exceed the trigger value of 0.4 mg kg^{-1}, proposed by the German protection law (BBODSCHV, 1999).

In gramineae green shoots, As concentrations ranged from 0.3 to 255 mg kg^{-1} dry weight (median: 2.95 mg kg^{-1} dry weight (dw)). For the forage plants collected, 73% of the samples exceed the As concentration of 1 mg kg^{-1} dw reported as the normal for terrestrial plants (Kabata-Pendias, 2011). Also, 57% present As concentrations higher than the established for undesirable substances in animal feed (2 mg kg^{-1}), according to the values set by European directives.

Concerning As in the agricultural land, its concentrations ranged from 2.9 to 523.3 mg kg^{-1}

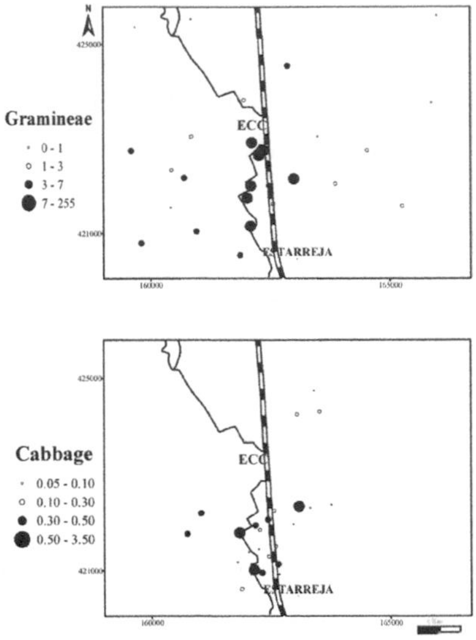

Figure 2. Spatial distribution of As (mg kg^{-1}) in gramineae (green shoots), cabbage leaves.

(median: 11.4 mg kg^{-1}) with 46% of the sampling sites showing concentrations above the CSQG value. Also, the As available fraction in 74% of these soil samples exceeded the trigger value proposed by the German protection law (BBODSCHV, 1999). These As soil fraction was low and ranged between 1.5 and 14.1% (median: 5.6%) of the total.

In the horticultural crops the element concentrations (mg kg^{-1} dw) were as follow: cabbage leaves (<0.1 to 3.5), tomato fruit (<0.1 to 0.4), maize grain and lettuce leaves (<0.1 to 0.3). Except for two cabbage samples (3.5 and 1.1 mg kg^{-1} dw), the results show that As concentration in the studied edible plant tissues is more or less in the same range, even in crops that the presence of the element in the soil in an available form is high. This behavior may suggest an exclusion mechanism. Srivastava & Gupta (1996) reported that in most plants, once As is taken up, it is usually passively absorbed staying in the roots.

The pastures land with the highest As soil concentration (total and available) and forage plants are those located near ECC or sewage outlets (Figure 2).

The BA was higher for forage plants (maximum 1.93). The graminae plants seem to concentrate and translocate As be more efficiently to the edible plant tissues than the studied agricultural crops. In fact, the BA determined for the crops was for all samples below 0.05.

4 CONCLUSIONS

In some locations of the study area, the As levels in the agricultural and pasture soils were above the Canadian soil quality guidelines. Some forage plants presented As concentrations higher than those corresponding to the horticultural crop. Its concentration exceeds the value established in the EU for undesirable substances in animal feed. Although the transfer of As from soil to edible tissues of the different crops was low, in few locations its concentrations were of concern. More research is needed in order to understand the dynamics of As in this particular environment. All data from soils, forage plants and other crops should be used to investigate possible relationships to public health problems.

ACKNOWLEDGEMENTS

The authors thank the financial supports providing from PACOPAR (Community Advisory Panel of the Responsible Action Program) and from CNRS (Centre National de la Recherche Scientifique) of France.

REFERENCES

BBODSCHV, 1999. Federal Soil Protection and Contaminated Sites Ordinance; Action, Trigger and precaution Values—Anexo 2, (BBodSchV) 12 July 1999.

CCME. 2011. Canada Council of Ministers of the Environment (updated 1999, 2001, 2011). Soil Quality Guidelines for the protection of environmental and human health. Winnipeg.

Kabata-Pendias, A. 2011. *Trace Elements in Soils and Plants 4rd ed.* CRC Press, Boca Raton, FL. 505 pp.

Srivastava, P.C. & Grupta, U.C. 1996. *Trace elements in crop production*. Science Publishers, Inc., USA.

One Century of the Discovery of Arsenicosis in Latin America (1914–2014) – Litter, Nicolli, Meichtry, Quici, Bundschuh, Bhattacharya & Naidu (Eds)
© 2014 Taylor & Francis Group, London, ISBN 978-1-138-00141-1

Arsenic levels in grape juice from Mendoza, Argentina

E.M. Martinis & R.G. Wuilloud
QUIANID, Instituto de Ciencias Básicas, Universidad Nacional de Cuyo, Mendoza, Argentina
Consejo Nacional de Investigaciones Científicas y Técnicas (CONICET), Argentina

ABSTRACT: Levels of arsenic in grape juices from Mendoza (Argentina) were investigated. Grapes from different viticulture areas of Mendoza were used for comparison of arsenic levels in grape juices. Elemental concentrations were evaluated with HG-AAS technique. Statistical methods were applied for the interpretation of obtained data. Significant differences between the obtained values for arsenic in grapes from Lavalle compared to grapes from San Martín, Junín and Rivadavia indicate that the geochemical arsenic anomalies present in these regions might contribute to some extent to higher arsenic content in analyzed grapes.

1 INTRODUCTION

Mendoza is the major Argentine viticulture region, and production of high quality grapes is essential for local and international wine and grape juice industries. Lavalle is an area of the Mendoza province where increased concentrations of arsenic in water might represent a potential risk to human health. Groundwater in this area is a source for public water supplies and it is also being used in agriculture for irrigation. High arsenic groundwater originating within the aquifers of this basin reflects mineral composition as well as hydrogeological conditions of the aquifers. In this work, grapes from different productive regions of Mendoza were analyzed for arsenic to evaluate if elevated concentrations of arsenic in groundwater could influence arsenic levels in grapes. Arsenic concentrations were compared with those of grapes from San Martín, Junín and Rivadavia regions.

2 EXPERIMENTAL

2.1 *Instrumentation*

The experiments were performed using a Perkin-Elmer 5100PC atomic absorption spectrometer (Perkin-Elmer, Norwalk, USA) equipped with a flow injection analysis system (FIAS 200) and an AS-90 autosampler. An arsenic electrodeless discharge lamp (Perkin-Elmer, Norwalk, USA) operated at a current of 400 mA and a wavelength of 193.7 nm with a spectral band pass of 0.7 nm was used. A deuterium background corrector was also used. Conditions for HG-AAS analysis were as follows: 2.5% (w/v) sodium borohydride (Merck, Germany) at 5 mL min^{-1}, 10% (v/v) HCl (Merck, Germany) at 9 mL min^{-1}, and 50 mL min^{-1} argon as a carrier. The sample injection volume was 500 µL in all experiments. Tygon type pump tubing was employed to carry sample, reagent and eluent.

2.2 *Sample collection and conditioning*

Grapes were directly collected from producers, covering a vast grape-producing area in the northeast of Mendoza (Argentina). Grape juice was obtained by crushing and blending the grapes into juice. The grape juice (1.0 mL) was transferred into a 10.0 mL volumetric flask, added with 1 mL of 10% (w/v) solution of KI and 1 mL of concentrated HCl, and diluted to 10 mL with ultrapure water. The solution was left for 1 h before arsenic determination by HG-AAS in order to reduce As(V) to As(III).

3 RESULTS AND DISCUSSION

3.1 *Arsenic levels distribution*

Eighty samples of grape juice from four viticulture areas of Mendoza were analyzed: Lavalle, Junín, San Martín and Rivadavia. Arsenic levels distribution (µg L^{-1}) in the samples is shown in Figure 1.

3.2 *Arsenic average concentration in different areas*

Figure 2 shows arsenic average concentrations (µg L^{-1}) found in grape juices from the different viticulture areas of Mendoza under investigation. Significant differences were observed between the

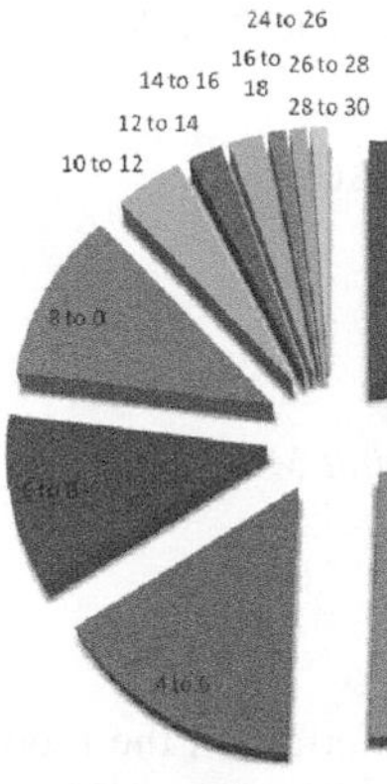

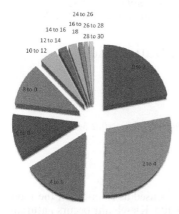

Figure 1. Arsenic levels distribution (μg L^{-1}) and concentrations in grape juices.

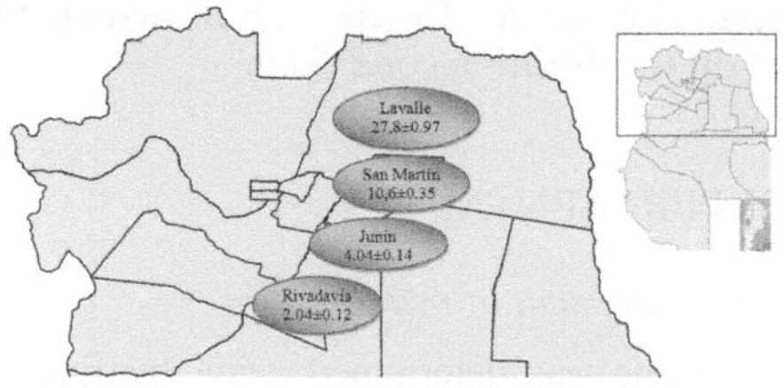

Figure 2. Arsenic concentration media (μg L^{-1}) found in grape juices from different viticulture areas of Mendoza.

obtained values for arsenic in grape juice from Lavalle (27.8 ± 0.97 μg L^{-1}) and others viticulture areas such as Rivadavia and Junín. Lavalle constitutes one of the largest regions of high arsenic levels in groundwaters. Thus, it can be argued that arsenic present in groundwater of Lavalle might contribute to higher arsenic levels in grape juice than that coming from other regions. However, to confirm this, further research is still required.

3.3 *Analytical performance*

The calibration graph was linear with a correlation coefficient of 0.9977 at levels near the detection limits (0.90 μg L^{-1}) and up to at least 180 μg L^{-1}. The limit of detection (LOD), calculated based on three times the standard deviation of the background signal was 0.08 μg L^{-1}. Thus, the methodology offers high sensitivity for determination of low concentrations of the analyte even below the maximum contaminant level (MCL) established by FDA. The relative standard deviation (RSD) for six replicate measurements at 0.2 μg L^{-1} of arsenic were 4.7% for As(III).

A recovery study was developed on the samples containing known additions of As(III) and As(V). The results are shown in Table 1. 100 mL of grape juice (two samples) were divided into portions of 10 mL each. The proposed method was applied to each portions and the average concentration of As(III) and As(V) found were taken as a base value. The remaining aliquots were spiked with 5 μg mL-1 of As(III) and As(V) and were analyzed by the proposed method. The results shown in Table 1 demonstrate that arsenic recoveries were acceptable and in the range of 95–104%.

Table 1. Analyte recovery study in real samples (95% confidence interval; n = 6).

	As(III)		As(V)	
Addition (5 μg L^{-1})	As found (μg L^{-1})	R[a] (%)	As found (μg L^{-1})	R[a] (%)
S1[b]	6.58 ± 0.31	–	6.58 ± 0.31	–
As(III)	11.53 ± 0.29	99	–	
As(V)	–		11.69 ± 0.34	102
As(III) As(V)	16.08 ± 0.32	95	16.54 ± 0.32	95
S2[b]	10.26 ± 0.01	–	10.28 ± 0.01	–
As(III)	15.48 ± 0.03	104	–	–
As(V)	–		15.25 ± 0.01	100
As(III) As(V)	20.25 ± 0.03	99	20.25 ± 0.04	99

[a] R (recovery): 100 × [(Found-base)/added].
[b] S (Sample).

4 CONCLUSIONS

Arsenic levels in grape juice were found to differ among the four investigated viticulture regions of the Mendoza province, with the highest observed in grapes from Lavalle.

Higher concentrations of arsenic measured in grape juice from Lavalle might indicate the influence of groundwater contribution. The extent of this influence must be investigated further.

ACKNOWLEDGEMENTS

This work was supported by Consejo Nacional de Investigaciones Científicas y Técnicas (CONICET), Agencia Nacional de Promoción Científica y Tecnológica (FONCYT) (PICT-BID) and Universidad Nacional de Cuyo (Argentina).

REFERENCES

Cornelis, R., Caruso, J.A., Crews, H. & Heumann, K.G. 2003. *Handbook of elemental speciation: Techniques and methodology*. Chichester: John Wiley & Sons Ltd.

Sanllorente-Méndez, S., Dominguez-Renedo, O. & Arcos-Martínez, M.J. 2010. Experimental design optimization of arsenic speciation in groundwater. *Analytical Letters* 43(12): 1922–1932.

Tašev, K., Karadjova, I. & Stafilov, T. 2005. Determination of inorganic and total arsenic in wines by hydride generation atomic absorption spectrometry. *Microchimica Acta* 149(1–2): 55–60.

*One Century of the Discovery of Arsenicosis in Latin America (1914–2014) –
Litter, Nicolli, Meichtry, Quici, Bundschuh, Bhattacharya & Naidu (Eds)*
 ISBN 978-1-138-00141-1

Release of arsenic from kieselguhr used as filtration aid in the food industry

M. Coelhan
Research Center Weihenstephan for Brewing and Food Quality, Technische Universität München, Freising, Germany

ABSTRACT: Kieselguhr, also known as diatomaceous earth, is used as filter aid in the food industry for clarification of juice, wine, beer, and for extending of shelf life. Kieselguhr occurs naturally in some countries like the USA, Germany, Denmark, France, Mexico, Canada, and Czech Republic. For filtration purposes, kieselguhr is used in different particle size ranges, which are classified as fine, medium, and coarse. Arsenic has been found to be present in some juices and in some beers. Our investigations revealed that kieselguhr may be a significant source of arsenic in beer. The degree of extraction of arsenic from kieselguhr during filtration depends on the type of beer and on the filtration conditions.

1 INTRODUCTION

Heavy metals and arsenic are subject to stringent legislation under law. Beer is a fermented drink which is usually made of water, barley malt and hop. Contaminants like heavy metals or arsenic can be introduced unintentionally into a beverage via raw materials used for production. Although the quality of drinking water is well regulated by law for many parameters worldwide, for food and drinks, a limited number of regulations exist in regard to heavy metals (Martinez *et al.*, 2001; Huang *et al.*, 2012). For instance, U.S. Food and Drug Administration recently proposed an action level of 10 μg L^{-1} for inorganic arsenic in apple juice (FDA). For beer, there is a limit for total arsenic level in some countries, which varies between 100 and 500 μg L^{-1} (Table 1) (Martinez *et al.*, 2001).

Table 1. Action limit for total arsenic in beer in some countries.

Country	Limit (μg L^{-1})*
Bulgarian	200
Hungary	200
Ireland	500
Slovenia	200
Spain	100
Czech Republic	200
UK	500

* Data from Martinez et al.

2 EXPERIMENTAL

2.1 *Measurement of metals*

Total arsenic was determined using hydride generation atomic absorption spectrophotometer (HG-AAS) Aanalyst 600 equipped with a flow injection system FIAS 100 (Perkin-Elmer Corp., Norwalk, CT, USA). Hydride generation was performed using a 0.2 (w/v) $NaBH_4$ in 0.05% NaOH (both Merck Suprapur; E. Merck, Darmstadt, Germany) solution. The radiation source was a hollow cathode lamp of arsenic (Perkin-Elmer) used at a wavelength of 193.7 nm and a spectral slit width of 0.7 nm.

Other metals were measured by Inductively Coupled Plasma Optical Emission Spectrometer (ICP-OES), Optima 5300 DV from Perkin-Elmer (Perkin-Elmer Corp., Norwalk, CT, USA).

Milli-Q water, prepared from deionized water using a Milli-Q Plus UV (Millipore, Billerica, MA) water purification device, was used whenever high purity water was required.

2.2 *Samples*

Beer and juice samples of different origins were purchased from local markets. For arsenic determination, beer and juice samples were digested with a microwave acid digestion unit Mars 6 (CEM, Matthews, NC, USA). 10 mL of the sample was taken into digestion tube and 5 mL of nitric acid (Suprapur, 65%, Merck, Germany) was added to it. The mixture was allowed to stand for 20 minutes. Then, the digestion was performed in the microwave unit at 160 °C for 30 min. Kieselguhr samples were provided by breweries.

Extraction of arsenic from kieselguhr was tested using different beers and potassium hydrogen phthalate solution (1%, w/v). pH of potassium hydrogen phthalate solution was 4.0. For extraction using beer, 3 g of kieselguhr and 150 mL of beer were stirred in an Erlenmeyer flask at 20 °C for 5 minutes. Then, the sample was filtered (0.45 µm) and digested. Alternatively, 1 g kieselguhr and 40 mL of potassium phthalate solution were shaken in a 50 mL plastic (polypropylene) centrifuge tube (Greiner, Austria) on a laboratory shaker at 20 °C for 120 minutes. After centrifugation, samples were analyzed without digestion.

For ICP-OES measurements, the samples were not digested. Prior to measurement, one part of samples was diluted with four parts of HNO_3 (5%). Yttrium was used as an internal standard.

3 RESULTS AND DISCUSSION

3.1 *Measurement of arsenic in beer*

Determination of arsenic in beer provided significantly lower levels when beer samples were analyzed without digestion. In that case, adding the antifoam reagent 1-stearoyl-*rac*-glycerol (Sigma-Aldrich) improved the results; however, they were still too low in regards to digested ones (Table 2).

Some part of arsenic is probably present as bounded to organic compounds in the sample matrix. Without digestion but with antifoam, assumingly only inorganic arsenic is measured.

Table 2. Influence of sample preparation on arsenic results (µg L^{-1}).

Beer	Digested	Not digested	Antifoam/ Not digested
1	8.3	2.5	na
2	12.9	3.7	na
3	10.7	3.4	7.9
4	18.8	4.5	11.2
5	7.4	2.0	5.2

na: not analyzed.

Table 3. Extraction of arsenic and other metals (mg kg^{-1} kieselguhr) from kieselguhr using different beers.

Beer	Arsenic	Aluminum	Calcium	Iron
A	2.9	132.1	203.3	172.0
B	2.8	131.9	211.3	167.5
C	2.8	126.2	252.0	169.4
D	2.4	122.9	204.6	160.1

Table 4. Influence of sample preparation on arsenic results (µg L^{-1}).

Kieselguhr fine	Arsenic	Aluminum	Calcium	Iron
Beer A	2.9	132.1	203.3	172.0
Phthalate	2.6	256.1	494.4	137.4
Kieselguhr Coarse				
Beer A	1.4	69.3	85.3	113.2
Phthalate	1.3	194.7	391.2	151.5

Without antifoam, there is a strong foam formation, which substantially affects the results.

3.2 *Extraction of metals from kieselguhr*

Results from extraction experiments of metals from kieselguhr using different beers are displayed in Table 3. Higher amounts of arsenic, aluminum and iron were extracted when beer A was used. Significantly higher amounts of calcium were found when for extraction beer C was used. Although the beers A-C showed similar results for the extraction of arsenic, aluminum, and iron, with beer D the arsenic level was approx. 20% lower than with the other three beers.

As shown in Table 4, using an 1% potassium hydrogen phthalate solution comparable results for arsenic are produced while, for other metals studied, beer provided a lower extraction efficiency than phthalate. Particularly, calcium and aluminum were extracted by phthalate in much higher levels than with beer. As for iron, beer provided a better extraction medium when fine kieselguhr was used compared to coarse kieselguhr.

4 CONCLUSIONS

For predicting of extractable amounts of arsenic from kieselguhr used for beer filtration an 1% solution (w/v) of potassium hydrogen phthalate at pH 4.0 provides better results in regard to both producibility and efficiency since a digestion of sample is not anymore necessary.

REFERENCES

FDA. Accessed at: http://www.fda.gov/NewsEvents/Newsroom/PressAnnouncements/ucm360466.htm.

Huang, J.H., Hu, K.N., Ilgen, J. & Ilgen, G. 2012. Occurrence and stability of inorganic and organic arsenic species in wines, rice wines and beers from Central European market. *Food Additives and Contaminants, Part A* 29: 85–93.

Martinez, A., Morales-Rubio, A., Cervera, M.L. & de la Guardia, M. 2001. Atomic fluorescence determination of total and inorganic arsenic species in beer. *Journal of Analytical Atomic Spectrometry* 16: 762–766.

One Century of the Discovery of Arsenicosis in Latin America (1914–2014) – Litter, Nicolli, Meichtry, Quici, Bundschuh, Bhattacharya & Naidu (Eds)
 ISBN 978-1-138-00141-1

Determination of arsenic in infusion tea cultivated in north of Iran

A.R. Mesdaghinia & A.H. Mahvi
School of Public Health, Tehran University of Medical Sciences, Tehran, Iran
Center for Solid Waste Research, Institute for Environmental Research, Tehran University of Medical Sciences, Tehran, Iran
National Institute of Health Research, Tehran University of Medical Sciences, Tehran, Iran

S.S. Hosseini & S. Shekoohiyan
School of Public Health, Tehran University of Medical Sciences, Tehran, Iran

ABSTRACT: Tea is one of the most common drinks all over the world. Rapid urbanization and industrialization in recent decades has increased the content of heavy metals and similar pollutants including arsenic (As) in tea and other foods. In this research, As contents were determined in 105 black tea samples cultivated in Guilan and Mazandaran provinces in north of Iran and their tea infusions. The amount of As in black tea infusions was analyzed using Inductively Coupled Plasma Atomic Emission Spectroscopy (ICP—AES). The mean level of As in 5, 15 and 60 min in infusion tea samples was 0.277 ± 0.272, 0.426 ± 0.402 and 0.563 ± 0.454 mg/kg of tea dry weight, respectively. In addition, the results showed that the locations and the infusion times influenced upon the amount of these metals ($p < 0.05$). The level of As was lower than the Iranian standard limit.

1 INTRODUCTION

Approximately 34 thousand hectares of lands in Guilan and Mazandaran provinces have been cultured for tea, almost half of the dry tea interior production, and the remainder comes from imports. The most important source of nutrient uptake by leaf tea is from the medium (Ashraf *et al.*, 2008). The presence of As creates concerns for tea consumers. In addition, fluoride concentration is usually very high in black tea (Mahvi *et al.*, 2006). Arsenic is strongly related with lung and skin cancer in humans, and it may cause other interior cancers and black foot disease (Chen *et al.*, 2006). The main aim of this study was to investigate the As level in tea samples that are cultivated in Guilan and Mazandaran provinces, in north of Iran.

2 MATERIALS AND METHOD

105 black tea samples were obtained randomly from 105 farms in the two different regions of Lahijan and Guilan provinces. The amount of each sample was about 100 g. 5 g of each tea sample were weighed with a digital analytical balance with ± 0.0001 g precision and then were added to 500 mL of boiling tap water and allowed to infuse for 5, 15 and 60 min. Then, samples were filtered under vacuum to eliminate any turbidity or suspended substance. Arsenic contents were determined by Inductively Coupled Plasma Optic Emission Spectrometry End of Plasma (ICP-OES EOP, Spectroacros, Germany). The purity of argon as carrier gas was 99.999% (grade 5), with a flow rate of 0.7 L/min for supplementary and Modified Lichte nebulizer and 13 L/min for coolant flow. The speed of 4 channel peristaltic pump was 60 rpm for 45 s in preflush condition and 30 rpm for analysis. The recovery percentage, detection limits, and % R.S.D for triplicate measurements of the measured elements were 90–95%, 0.3 ppb, and less than 5 %, respectively. Statistical analysis of the obtained results was performed by SPSS 18 and One Way ANOVA test was used; results are expressed by mean ± SD.

3 RESULTS

The mean level of As in 5, 15 and 60 min in infusion tea samples was 0.277 ± 0.27, 0.426 ± 0.40 and 0.563 ± 0.45 mg/kg of tea dry weight, respectively. In addition, results showed that the infusion time influenced upon the amount of As in infusion tea ($p < 0.001$). The maximum level of As was determined in Amlash, where the amount of this metal in infusion times of 5, 15 and 60 min was 0.552 ± 0.3, 0.709 ± 0.35 and 0.771 ± 0.21 mg/kg, respectively. The minimum level of As was determined in Bazkiagorab, where the amount of this metal in infusion times of 5, 15 and 60 min were

0.011 ± 0.00, 0.014 ± 0.00 and 0.022 ± 0.00 mg/kg, respectively. Results also showed that location influenced upon the amount of As in infusion tea samples ($p < 0.001$).

4 DISCUSSION

The results showed that the maximum and minimum level of As was determined in Amlash and Bazkiagorab, respectively. This study showed also that locations and infusion time influenced upon the amount of As in the samples ($p < 0.05$). The concentration of As in tea samples was below than that set as the standard maximum value by the Iranian Ministry of Health (As 150 μg/g). All samples contained As at levels below by set the standard maximum values (4 mg/kg) (Mahmoodi *et al.*, 2006). The arsenic main sources in tea samples are their growth media, i.e., the soil. Since the tea cultivation lands in north of Iran provide much of black tea being consumed in Iran and the exported amount to other countries, it is recommended all elements and compounds in tea infusions to be analyzed for safe consumption of black tea. In order to be in safe side, it is recommended that tea be prepared with heavy metal free and very low fluoride waters, preferably bottled water (Dobaradaran *et al.*, 2008).

5 CONCLUSIONS

After drinking water, tea is the most consumed beverage in Iran. In this regard, health aspects related to tea is very important and therefore consumers should be very confident on the absence of any pollutants in black tea. The presence of any variations in metal contents in this study could be as a matter of the differences in methods of storage, tea leaf processing and the difference in soil metal concentrations. In this study, the levels of arsenic were lower than the standard limits.

REFERENCES

Ashraf, W. & Mian, A. 2008. Levels of selected heavy metals in black tea varieties consumed in Saudi Arabia. *Bull Environ. Contam. Toxicol.* 81:101–104.

Chen, H.W. 2006. Gallium indium, and arsenic pollution of groundwater from a semiconductor manufacturing area of Taiwan. *Environ. Contam. Toxicol.*, 77:289–296.

Dobaradaran, S., Mahvi, A.H., Dehdashti, S. 2008. Fluoride content of bottled drinking water available in Iran. *Fluoride*, 41:93–94.

Mahmoodi, M.R. 2003. *Minerals in Nutrition.* Isfahan: oruj & Esfahan University of Medical Sciences; p. 87–97.

Mahvi, A.H., Zazoli, M.A., Younecian, M. & Esfandiari, Y. 2006. Fluoride content of Iranian black tea and tea liquor. *Fluoride*, 39:266–268.

Section 3: Arsenic and health

3.1 Epidemiological studies

One Century of the Discovery of Arsenicosis in Latin America (1914–2014) – Litter, Nicolli, Meichtry, Quici, Bundschuh, Bhattacharya & Naidu (Eds)
© 2014 Taylor & Francis Group, London, ISBN 978-1-138-00141-1

Arsenic exposure, health effects and biomarker, and treatment of arsenicosis—experience in West Bengal, India

D.N. Guha Mazumder, A. Ghose, D. Deb, A. Biswas & R.N. Guha Mazumder
DNGM Research Foundation, Kolkata, India

C. Saha
Department of Clinical and Experimental Pharmacology, School of Tropical Medicine, Kolkata, India

ABSTRACT: Epidemiological survey on a population of 0.84 million people suspected to be drinking arsenic contaminated water in Nadia, West Bengal, India, showed arsenical skin lesion in 1,616 patients (15.43%) out of 10,469 participants examined. Other major non cancer health effects were chronic lung disease, polyneuropathy and liver fibrosis. Total individual daily arsenic exposure through water and diet were estimated in participants and correlated with arsenic level in urine in Nadia, where arsenic levels in drinking water, rice and vegetables were elevated. In spite of drinking water with arsenic level <50 μg L^{-1} in 94 subjects, dietary arsenic exposure was high (154 ± 83 μg day^{-1}) with raised arsenic level in urine (119 ± 116 μg L^{-1}). Objective assessment of skin lesions in a study for three years showed that drinking arsenic safe water increased the probability of regression of skin lesion in subjects with mild lesions, but not in those with more advanced stage disease.

1 INTRODUCTION

Arsenic pollution in groundwater had been detected in West Bengal, India, as early as 1983. However, reports of large population based study on global assessment of arsenic related health effect are scanty. An epidemiological study was therefore carried out in the whole district of Nadia, West Bengal, having a population of 0.84 million people suspected to be drinking arsenic contaminated water (Guha Mazumder *et al.*, 2010).

Though groundwater arsenic contamination has been considered to be the main source of arsenic exposure in the people, there is increasing evidence of elevated arsenic level in rice grain and vegetables in regions of West Bengal and Bangladesh, where fields are irrigated with arsenic-rich water. Few reports are available that characterize daily arsenic exposure through water and diet and correlate with biomarkers like urine among people living in regions where arsenic level in drinking water, rice and vegetables are elevated. A study has been carried out to ascertain the total individual arsenic exposure and biomarker in such people in West Bengal (Guha Mazumder *et al.*, 2013).

Chronic arsenicosis leads to irreversible damage in several vital organs, and arsenic is an established carcinogen. Various modalities of treatment like chelating agents, antioxidants and retinoids have been used for the treatment of arsenicosis. However, no specific drug has yet been available for altering the natural history of arsenicosis (Guha Mazumder, 2000). Limited information is available regarding the long-term effect of drinking arsenic safe water. The current study was therefore undertaken to assess the effect of drinking arsenic safe water (<50 μg L^{-1}) on disease manifestation of arsenicosis in Nadia.

2 METHODS

2.1 *Arsenic exposure and health effect*

Total individual daily arsenic exposure through water and diet were estimated and correlated with arsenic level in urine in 167 people (Group-1) in Nadia, where arsenic level in drinking water and in rice and vegetables were elevated. Arsenic level in drinking water and each dietary item taken by a participant and arsenic level in urine were estimated by AAS with hydride generation system. Total daily arsenic intake from drinking water and diet were estimated in each participant (Guha Mazumder *et al.*, 2013). A multiple regression model was fitted to urinary arsenic (Biomarker) with average arsenic intake from water and diet as the exposure and age, sex and presence of skin lesions as potential confounders for Group-1A (94) and for Group-1B (72) participants depending on arsenic level in current drinking water source being <50 μg L^{-1} (Safe limit,

India) and <10 μg L^{-1} (Safe limit, WHO) respectively. Out of the 167 (Group-1) participants, 68 participants (Group-1C) were drinking water with arsenic level Below Detection Limit (BDL).

2.2 *Effect of drinking arsenic safe water*

The effect of drinking arsenic safe water on disease manifestation of arsenic exposed people was undertaken in Nadia by carrying on a baseline study on 108 cases (Cohort-I, with skin lesion) and 100 participants (Cohort-II, without skin lesion) using scoring system of skin lesion (Guha Mazumder *et al.*, 2010) and of systemic manifestations (Guha Mazumder *et al.*, 1998) to ascertain their initial scores. Participants having unsafe current source in their households were supplied with arsenic removal filters for getting arsenic safe water (<50 μg L^{-1}) during the follow up period of three years. Severity of disease manifestations in Cohort-I cases was compared objectively every year for a three year follow up period with photographic recording of skin lesions. Assessment of systemic manifestations was also done in both the cohorts with systemic scoring system. Arsenic level in drinking water source was ascertained by AAS with hydride generation system at the beginning of study. The mean (±SD) peak (highest) arsenic level in drinking water source of Cohort-I and Cohort-II participants was 250.56 ± 199.20 μg L^{-1} and 259.53 ± 161.49 μg L^{-1} respectively ($p > 0.05$).

3 RESULTS AND DISCUSSION

3.1 *Arsenic exposure and health effect*

Out of 10,469 participants examined in Nadia, 1,616 patients showed clinical features of arsenical skin lesion with a prevalence rate of 15.43%, while 8853 participants had no skin lesion (controls). Mean arsenic level in tube well water was 103.469 ± 153.289 μg L^{-1} in cases and 73.187 ± 115.105 μg L^{-1} in control ($p < 0.001$).

Out of 1,616 patients with arsenical skin disease 68.8% had only pigmentation and 1.67% had only keratosis and 29.5% had both. Chronic lung disease was found in 207 (12.81%) subjects among cases and 69 (0.78%) in controls ($p < 0.001$). Peripheral neuropathy and pain abdomen were found in 257 (15.9%) and 67 (4.15%) cases and 136 (1.5%) and 67 (0.87%) subjects among controls respectively ($p < 0.001$). Other systemic features found in significantly higher number of cases compared to controls were chronic diarrhea, hepatomegaly, ascites and non pitting edema of the limbs. It was interesting to observe the skin lesions to be mild (skin score-1–2) in largest number (87.56%) of cases (Guha Mazumder *et al.*, 2010).

Various case series reported from West Bengal earlier showed that, over and above skin manifestations, arsenic exposure through drinking water was associated with various systemic manifestations like chronic lung disease (chronic bronchitis, chronic obstructive and/or restrictive pulmonary disease and bronchiectasis), liver disease (non cirrhotic portal fibrosis), polyneuropathy, peripheral vascular disease, hypertension, ischemic heart disease, non-pitting edema of feet/hands, conjunctival congestion, weakness and anemia (Guha Mazumder *et al.*, 1998 2005; Guha Mazumder, 2008). Even in absence of dermatological manifestation, frequency of occurrence of systemic features like weakness, anaemia, chronic lung disease, chronic diarrhea and hepatomegaly were reported to be higher in chronic arsenic exposed people compared to unexposed people in West Bengal (Majumder *et al.*, 2009). High concentrations of arsenic >200 μg L^{-1} during pregnancy were found to be associated with a six-fold increased risk for stillbirth (von Ehrenstein *et al.*, 2006).

3.2 *Arsenic exposure and biomarker*

Mean (±SD) arsenic level in drinking water in 167 participants (Group-1) was 53 ± 63 μg L^{-1}. Total daily arsenic intake was 164 ± 83 μg/day from diet only and 349 ± 261 μg day^{-1} from water and diet in these subjects with arsenic level in urine 124 ± 102 μg L^{-1}. In spite of drinking water with arsenic level <50 μg L^{-1} in 94 Group-1A subjects, dietary arsenic exposure was high (154 ± 83 μg day^{-1}) with raised arsenic level in urine (119 ± 116 μg L^{-1}). Multiple regressions analysis in Group-1A participants showed that daily arsenic dose from water and diet was significantly positively associated with urinary arsenic level. However, daily arsenic dose from diet, but not from water was significantly positively associated with urinary arsenic level in 72 Group-1B participants (arsenic in water <10 μg L^{-1}). Among 68 Group-1C participants (arsenic in drinking water, BDL), significantly increased arsenic excretion occurred in urine with arsenical skin lesion, Figure 1 ($r = 0.573$) while insignificant arsenic excretion occurred in 23 participants without skin lesion, Figure 2 ($r = 0.007$).

A urine sample showing an arsenic concentration above 50 μg L^{-1} may be taken as evidence of recent arsenic exposure (WHO, 2005; Buchet *et al.*, 1981; NRC, 1999). No data are available correlating arsenic intake through diet and arsenic in urine in India and Bangladesh where arsenic contaminated water is used for irrigation of fields for paddy and vegetable cultivation. In the present study, arsenic level in urine was found to be high (>50 μg L^{-1}) in people living in arsenic endemic region but drinking water with arsenic level <50 μg L^{-1}, suggesting contribution of arsenic exposure through diet

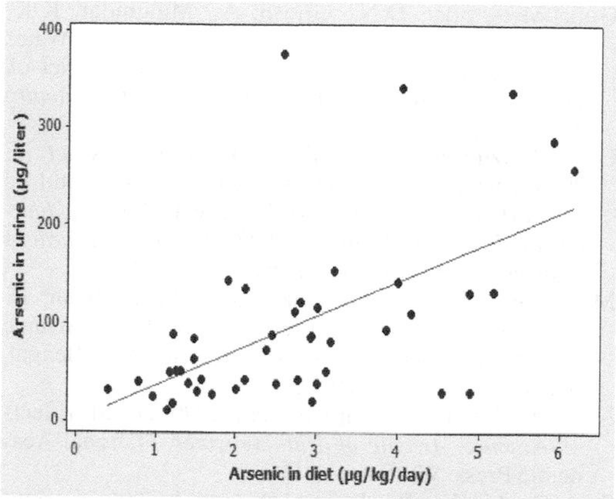

Figure 1. Correlation of urinary arsenic concentration (µg L^{-1}) with daily dietary arsenic intake (µg kg^{-1} day^{-1}) for participants drinking water with arsenic level <0.3 µg L^{-1} having skin lesion ($n = 45$), in West Bengal, India.

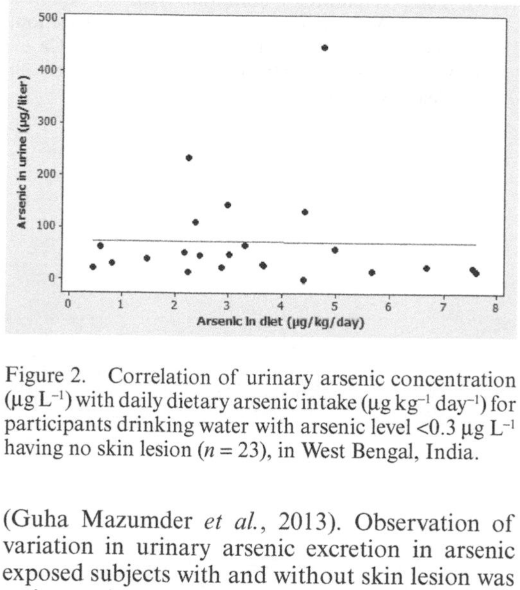

Figure 2. Correlation of urinary arsenic concentration (µg L^{-1}) with daily dietary arsenic intake (µg kg^{-1} day^{-1}) for participants drinking water with arsenic level <0.3 µg L^{-1} having no skin lesion ($n = 23$), in West Bengal, India.

(Guha Mazumder *et al.*, 2013). Observation of variation in urinary arsenic excretion in arsenic exposed subjects with and without skin lesion was an interesting observation and needs further study. Results of our study suggested that lowering of safe limit of drinking water to <10 µg L^{-1} (WHO safe limit) and sustainable agricultural practices that minimize the transfer of arsenic from groundwater to soils to the human food chain need to be propagated for effective reduction of arsenic exposure from water and diet in India and Bangladesh.

3.3 *Effect of drinking arsenic safe water*

Study in Nadia documented with sequential photographic recordings that arsenical skin lesions improve with intake of arsenic safe water for prolonged period. The skin score improved significantly from 2.17 ± 1.09 to 1.23 ± 1.17; $p < 0.001$ at the end of 3 years in Cohort-I participants. The systemic disease symptom score was also found to improve, but less significantly from 2.13 ± 1.58 to 1.64 ± 1.13, $p < 0.01$ in Cohort-I and from 1.45 ± 1.15 to 1.13 ± 0.88, $p < 0.05$ in Cohort-II participants, respectively, at the end of three years. However, persistence of severe symptoms of chronic lung disease and severe skin lesion were observed in spite of drinking arsenic safe water (Ghose *et al.*, 2013). It has also been reported by us (Guha Mazumder, 2003) and by others (Sun *et al.*, 2006), earlier, that drinking predominantly arsenic free water increased the probability of regression of skin lesion in subjects with mild stage lesions but not in those with more advanced stage lesions. However, objective assessment of skin lesion by sequential photographic recording and using scoring system for dermatological lesions and systemic disease manifestations were new in this study.

4 CONCLUSIONS

Large population based study from West Bengal showed that significant non cancer health effects due to chronic arsenic exposure are skin lesions like pigmentation and keratosis, chronic lung disease, polyneuropathy and liver fibrosis. Arsenic exposure from diet plays a significant role for increased arsenic excretion in urine in people in the arsenic endemic region where arsenic contaminated water is used for irrigation of fields. Significantly increased arsenic excretion occurred in urine with arsenical skin lesion while insignificant arsenic excretion occurred in participants without skin lesion. Arsenical skin lesions improve significantly in mild cases with prolonged intake of arsenic safe water.

ACKNOWLEDGEMENTS

The epidemiology study was carried out with financial grant from UNICEF and Government of West Bengal (Ref No. WB/YW/204/1660 dt. 2nd August 2006). Studies on dietary arsenic exposure and treatment were supported by research grant funded by World Bank under National Agricultural Innovative Project 'Arsenic in Food Chain: Cause, Effect and Mitigation' from Indian Council of Agricultural Research (ICAR) Govt. of India (Ref. No. NAIP/C4/C1005, dated 12.6.2007).

REFERENCES

Buchet, J.P., Lauwerys, R. & Roels, H. 1981. Urinary excretion of inorganic arsenic and its metabolites after repeated exposure of sodium meta arsenite by volunteers. *Int. Arch. Occup. Environ. Health* 48: 1111–118.

Ghose, A., Ghose, N., Biswas, A. *et al.* 2012. Changes in severity of arsenical symptoms following intervention with arsenic safe water: A three year follow up study. In *Arsenic contamination in Water and Food Chain, Proceedings of the International workshop on Arsenic in Food Chain: Cause, Effect and Mitigation, 20th February, 2012*. Kolkata, India.

Guha Mazumder, D.N., Chakraborty, A.K., Ghose, A. *et al.* 1988. Chronic arsenic toxicity from drinking tube well water in rural West Bengal. *Bull World Health Organ*. 66(4): 499–506.

Guha Mazumder, D.N., Ghoshal, U.C., Saha, J. *et al.* 1998. Randomized placebo-controlled trial of 2, 3-dimercaptosuccinic acid in therapy of chronic arsenicosis due to drinking arsenic-contaminated subsoil water. *J. Toxicol. Clin. Toxicol*. 36: 683–690.

Guha Mazumder, D.N. 2000. Diagnosis and treatment of chronic arsenic poisoning. *United Nations Synthesis Report on Arsenic in Drinking water*. Available at: www.who.int/water_sanitation_health/dwq/arsenic3/.

Guha Mazumder, D.N. 2003. Chronic Arsenic toxicity: Clinical features, epidemiology and treatment: Experience in West Bengal. *Journal of Environmental Science and Health* 38: 141–163.

Guha Mazumder, D.N., Craig, S., Bhattacharya, P. *et al.* 2005. Bronchiectasis in persons with skin lesions resulting from Arsenic in Drinking Water. *Epidemiology* 16: 760–765.

Guha Mazumder, D.N. 2008. Chronic Arsenic Toxicity and Human Health. *Indian J. Med. Res.* 128: 436–447.

Guha Mazumder, D.N., Ghosh, A., Majumdar, K.K. *et al.* 2010. Arsenic contamination of Ground Water and its Health Impact on Population of District of Nadia, West Bengal, India. *Ind. J. of Community Medicine* 35: 331–338.

Guha Mazumder, D.N., Deb, D., Biswas, A. *et al.* 2013. Evaluation of dietary arsenic exposure and its biomarkers: A case study of West Bengal, India. *Journal of Env. Sci. Health, PART A. Toxic/Hazardous Substance & Environmental Engineering* 49: 1–9.

Majumder, K.K., Guha Mazumder, D.N., Ghosh, N. *et al.* 2009. Systemic manifestations in chronic arsenic toxicity in absence of skin lesions in West Bengal, *Indian J. Med. Res.* 129: 75–82.

NRC (National Research Council). 1999. Health Effects of Arsenic. *Arsenic in drinking water*. National Academic Press: Washington DC.

Sun, G., Li, X., Pi, J., Sun, Y., *et al.* 2006. Current research problems of chronic arsenicosis in China. *J. Health Popul. Nutr.* 24(2): 176–81.

von Ehrenstein, O.S., Guha Mazumder, D.N., Ghosh, N. *et al.* 2006. Pregnancy outcomes, infant mortality and arsenic in drinking water in West Bengal, India. *Am. J. Epidemiol.* 163: 662–669.

WHO, 2005. Technical Publication No. 30. Clinical Aspects of Arsenicosis. A Field Guide for Detection, Management and Surveillance of Arsenicosis Cases. In D. Caussy (ed). WHO Regional Office for South East Asia: New Delhi, 5–9.

One Century of the Discovery of Arsenicosis in Latin America (1914–2014) – Litter, Nicolli, Meichtry, Quici, Bundschuh, Bhattacharya & Naidu (Eds)
© 2014 Taylor & Francis Group, London, ISBN 978-1-138-00141-1

Arsenic environmental and health issues in Uruguay: A multidisciplinary approach

N. Mañay
Cátedra de Toxicología, DEC, Facultad de Química, Universidad de la República (UDELAR), Montevideo, Uruguay

M. Pistón
Química Analítica, DEC, Facultad de Química, Universidad de la República (UDELAR), Montevideo, Uruguay

C. Goso
Departamento de Evolución de Cuencas, Facultad de Ciencias, Universidad de la República (UDELAR), Montevideo, Uruguay

ABSTRACT: Geogenic As in groundwater has been recently studied in different aquifers of Uruguay. However, no background epidemiology studies are available regarding environmental population exposure. As a consequence, it was necessary to conduct scientific research in order to study arsenic as an environmental health issue in the country. This research needs have resulted in joint studies of experts from both the geosciences and biosciences. The groups have coordinated studies to assess the risks of exposure to environmental arsenic in Uruguay with common research objectives and, now, this is a matter of common interest through a multidisciplinary approach. The aim of this work is to present a review of these studies conducted in Uruguay.

1 INTRODUCTION

Medical Geology is a developing discipline in Uruguay since 2005, and Arsenic (As) exposure is one of its major subjects of interest (Mañay, 2010).

Geogenic As in groundwater has been recently studied in different aquifers of the country. However, no background epidemiology studies are available in regards to As environmental health impact Water consumption is led by the state drinking water supplier (OSE) with a population coverage of over 90%. The As maximum acceptable limit value for this official water provider is 20 μg/L (UNIT, 2008).

On the other hand, arsenic exposure at the workplace is now been taken into account to systematically assess workers' health risks, as new occupational legal regulations have been recently established by the Ministry of Public Health (MSP, 2009). Several laws and decrees regulate the quality of water sources and drinking water as well (Poder Legislativo, 2009).

The international recommendations and Uruguayan regulations for urine are: As-Urine <35 μg/L for occupational exposed workers (ACGIH-BEI®, 2011) and 10–20 μg/L for general population (ATSDR, 2007), expressed as inorganic As plus its methylated urine species.

Therefore, there is a special need for conducting research studies on arsenic as an environmental health issue in the country and to develop available analytical tools to assess As levels and its speciation in water and in urine of workers for law-abiding.

All these environmental and health issues are now becoming a matter of concern in Uruguay and need to be undertaken through a multidisciplinary approach. Research teams and experts from both geosciences and biosciences have joined to face those arsenic exposure risks with common research objectives. The present work describes the main studies that have been developed in Uruguay. Several of the ongoing research studies that have been developed focusing on As geological, analytical and toxicological aspects are reviewed.

2 METHODS

2.1 *As analytical and speciation methodologies*

Simple methodologies for routine determinations of total As and its species in water and metabolites in urine were optimized and validated by Atomic Absorption Spectrometry with Hydride Generation (HG-AAS) as available analytical tools in Uruguay. This study developed and optimized an analytical method to determine total As and iAs species in water and determined iAs as well as its methylated metabolites (MMA and DMA) in urine, to be used for routine applications. The methodology was used instead of ICP-MS technique, which is not available in the country for this purposes.

For the screening of toxicologically relevant species iAs—MMA—DMA, separation by means of HPLC-DAD techniques was also optimized (Mañay *et al.*, 2011, Álvarez *et al.*, 2012).

2.2 *Presence of geogenic As in groundwater*

The Uruguayan groundwater uses are mainly agriculture farm, drinking, domestic and thermal—touristic. The main aquifers in the country are Guarani, Mercedes, Raigón and Chuy.

The reviewed studies were performed in sedimentary aquifers sampled from 2007 on. The hydrochemical characterization (major and trace elements) was carried out by both ICP-OES and inductively coupled plasma-mass spectrometry (ICP-MS) in the Laboratory of Earth Sciences Institute "Jaume Almera" (CSIC, Barcelona, Spain) and Laboratory Act Labs (Canada).

2.3 *As unexposed population background levels*

In Uruguay, there is no background of systematic arsenic studies on the unexposed population, although the biological control is done in workers according to the current legislation. For this reason, a pilot study to analyze arsenic in samples of urine of non-exposed population was carried out, to estimate the basal levels and its possible correlation with associate variables. 36 urine samples of voluntary adults were collected according to a designed protocol. The urine was analyzed with a routine developed methodology for toxicologically relevant As species, by means of HG-AAS. The results were statistically assessed and compared with the data previously obtained by a survey (Iaquinta *et al.*, 2012).

2.4 *Predictors of As levels in children*

This study investigates demographic predictors of urinary arsenic in Montevideo school children. Samples of 192 children were analyzed in the Karolinska Institute in Sweden for total urinary As, iAs, and its metabolites (MMA, DMA) using high-pressure liquid chromatography with hydride generation and inductively coupled mass spectrometry, HPLC-ICP-MS (Kordas, 2012).

2.5 *Exposed population*

Urine samples from workers presumably exposed to As, in different worksites, were periodically analyzed for toxicologically relevant species in order to assess health risks according to national regulations.

The total results of this routine biomonitoring were statistically assessed (Iaquinta *et al.*, 2012).

3 RESULTS AND DISCUSSION

3.1 *As analytical and speciation methodologies*

To evaluate the efficiency of the analytical method for determining iAs + MMA + DMA, standard additions were made of these species (in urine), and the recoveries obtained where close to 70%, which was considered adequate for a screening methodology. For urine, the detection limit obtained was 0.3 µg/L, with a precision <5% expressed as RSD. The quantification limit was 1.0 µg/L (Álvarez *et al.*, 2012).

3.2 *Presence of geogenic As in groundwater*

The average and maximum concentration of total As in the Raigón Aquifer System were 14.1 µg/L and 24.19 µg/L, respectively. Out of 37 samples, only six showed concentrations below 10 µg/L of As, according to the WHO recommended limit values for drinking water (WHO, 2004). In the Raigón Aquifer System, distal (deltaic plain and delta front) sediments showed As concentrations higher than those on more proximal (fluvial) sediments in Canelones Department (Manganelli *et al.*, 2007).

All sampled rocks contained some As, typically between 1 and 5 mg/kg. There were several As bearing minerals, including arsenopyrite (AsFeS), realgar (AsS) and orpiment (As_2S_3). Soils, which were formed by the weathering and breakdown of rock to clays, usually contained between 0.1 and 40 mg/kg and on average 5–6 mg/kg.

In this study, a few Cenozoic sediments analyzed in both Raigón and an adjacent formation (called Libertad) show normal As concentrations (between 1–6 mg/kg) (Mañay *et al.*, 2013).

The rocks in the south of the country (specifically in a location called Santa Lucia Basin) were mainly composed of granitic and gneissic suites which do not contain arsenic (Spoturno *et al.*, 2004). This allows us not to consider a geogenic origin in this case (Mañay *et al.*, 2013).

A few sediments were analyzed in both Raigón and Libertad formations, and this allow us to postulate a tentative anthropogenic origin based in several anomalous As concentrations values out of the expected average (>7 mg/kg). Probably, the most significant anthropogenic source of arsenic in this region is from cumulative applications of arsenical pesticides and herbicides used for decades by farmers (Mañay *et al.*, 2013).

3.3 *As unexposed population background levels*

The results of this study can be considered as the first preliminary background of As levels in urine for the not exposed adult population in Uruguay. This study showed average reference levels of 4.96 µg/L, (range 2.67–11.21 µg/L) and allowed to study the influence of several factors that could affect the concentration of As-Urine. The obtained levels were within the reference limits according to ATSDR, 2007, and a significant influence of the age was observed on the concentration of As in urine as the concentration was lower as the age increased (Iaquinta *et al.*, 2012).

3.4 *Predictors of As levels in children*

Preliminary results showed low-level exposure to As from water and low concentrations of inorganic species in urine on a 6–8 years old children population from Montevideo. Several demographic predictors of urinary arsenic, including measures of socioeconomic status, sex, and family consumption of bottled water were identified (Kordas, 2012).

3.5 *Exposed population*

Out of the 100 urine samples analyzed for biomonitoring of As toxicologically relevant species (iAs + MMA + DMA) none showed higher levels than those established by regulations (As-Urine <35 μg/L). The average results of this routine biomonitoring were 3.1 μg/L (range 0.3–18.8 μg/L) (Iaquinta *et al.*, 2012).

4 CONCLUSIONS

As health and environmental issues are being recently studied in Uruguay. The reviewed studies and those that are being conducted show the current status of the problem of arsenic in Uruguay which still have not been addressed as a priority health and environmental issue in the country.

Therefore, it is important to continue developing systematic studies to assess population's chronic exposure to inorganic As through drinking water, food and workplaces, focusing on children health impacts at As low levels. Besides, further studies are necessary to optimize analytical methodologies based on As speciation with coupled techniques available in Uruguay.

We conclude that this multidisciplinary approach in developing As research has been very successful in creating scopes of discussion among researchers and professionals from various disciplines and institutions, to study such a relevant environmental and health topic in Uruguay.

ACKNOWLEDGEMENTS

To DINACYT/PDT—MEC (Uruguay), to Comisión Sectorial de Investigación Científica UDELAR (Uruguay), to Consejo Superior de Investigaciones Científicas (CSIC, Spain) and to National Institute of Environmental Health Science (NIEHS, USA).

REFERENCES

Álvarez, C., Pistón, M., Clavijo, G., Iaquinta, F., Bühl, V. &, Mañay, N. 2012. Avances en la optimización de un método rápido de screening de especies toxicológicamente relevantes de As y As total en orina mediante HG-AAS. *3er Encontro Brasileiro sobre Especiacao Química—EspeQBrasil*, Bento Goncalves, Brasil.

ACGIH, 2011 American Conference of Governmental Industrial Hygienists TLVs© and BEIs©.

ATSDR2007. As Toxicological Profile in: http://www.atsdr.cdc.gov (accessed: October 2013).

Iaquinta, F., Pistón, M., Álvarez, C., Gómez, M.E. & Mañay, N. Evaluación de niveles de referencia de arsénico en orina de población adulta no expuesta ocupacionalmente. Estudio preliminar. 2012. *Aldeq—Anuario Latinoamericano de Educación Química*, XVII: 230–234.

Kordas, K. 2012 Project title: Is low-level arsenic exposure related to neurobehavioral deficits in children? Sponsor: NIEHS, R21 ES019949 (2012–2014).

MSP 2009, Ministerio de Salud Pública. Ordenanza N° 145. Ref. N° 001-3-5137-2008, http://www.msp.gub.uy/sites/default/files/001-3-5137-2008._Vigilancia_sanitaria_de_trabajadores_expuestos_factores_riesgo_laborales.doc (accessed 8-01-2014).

Manganelli, A., Goso, C., Guerequiz, R., Fernández-Turiel, J.L., García-Valles, M., Gimeno, D. & Pérez, C. 2007. Groundwater arsenic distribution in South-western Uruguay. *Environmental Geology*, 53: 827–834.

Mañay, N. 2010. Developing Medical Geology in Uruguay: A Review. *Int. J. Environ. Res. Public Health* 7(5): 1963–1969.

Mañay, N., Pistón, M., Álvarez, C., Clavijo, G. & Gómez, E. 2011. Development of Arsenic Analytical Methodologies In Water And Urine By Hg-AAS For Routine Determinations In Uruguay. *4th International Conference On Medical Geology GEOMED*. Bari—Italia.

Mañay, N., Goso, C., Pistón, M., Fernández-Turiel, J.L., García-Vallés, M., Rejas, M. & Guerequiz, R. 2013. Groundwater Arsenic Content in Raigón Aquifer System (San José, Uruguay). *Revista de la Sociedad Uruguaya de Geología*, 18: 20–38.

Poder Legislativo, 2009, Ley N° 18.610 Política Nacional De Aguas. http://www.ose.com.uy/descargas/documentos/leyes/ley_18_610.pdf

Spoturno, J., Oyhantcabal, P., Aubet, N. & Casaux, S. 2004. Mapa geológico y de recursos minerales del Departamento de San José. DINAMIGE-FCIEN-DINACYT. CD Rom.

UNIT 2010 Instituto Uruguayo de Normas Técnicas (UNIT-BID/Fomin) Referencia 833:2008. Agua Potable: Requisitos. http://www.ose.com.uy/descargas/Clientes/Reglamentos/unit_833_2008_.pdf (accessed 8-01-2014).

WHO 2004. Guidelines for drinking-water quality. Recommendations. World Health Organization, Geneva, 3rd edition. Volume 1, 515.

One Century of the Discovery of Arsenicosis in Latin America (1914–2014) – Litter, Nicolli, Meichtry, Quici, Bundschuh, Bhattacharya & Naidu (Eds)

Arsenic in the environment and its impact on human health: How safe are we?

A.K. Giri
Molecular and Human Genetics Division, CSIR-Indian Institute of Chemical Biology, Calcutta, India

ABSTRACT: Arsenic-induced health effects, genetic damage and genetic variants were analyzed in the population exposed to arsenic through drinking water in West Bengal, India. The incidence of health effects and genetic damage were significantly high in the individuals showing skin lesions when compared to individuals not showing skin lesions and to arsenic unexposed group. Analysis of genetic susceptibility were carried out by studying the prevalence of Single Nucleotide Polymorphisms (SNPs) in the GST group genes, p53, ERCC2 and XRCC3 as they might be involved in arsenic metabolism and detoxification. The minimum threshold dose of arsenic in rice that can induce genetic damage in human has been identified. The multi-facet of arsenic toxicity makes us rethink: "How safe are we?"

1 INTRODUCTION

Arsenic is a major environmental contaminant and more than 137 million people in 70 countries of the world are affected by drinking heavily contaminated groundwater, which includes 26 million individuals from West Bengal, India drinking arsenic-laden water above the WHO threshold limit of 10 μg L^{-1} (Mondal *et al.*, 2010). Some of the hallmarks of chronic arsenic toxicity include noncancerous (raindrop pigmentation and hyperpigmentation), precancerous (palmer and planter hyperkeratosis), and cancerous skin lesions (Basal Cell Carcinoma [BCC], Squamous Cell Carcinoma [SCC], and Bowen's Diseases [BD]). These skin lesions may develop within the latency period of 6 months to 10 years from the first exposure to arsenic. The appearance of skin lesions depends on the concentration of arsenic in drinking water, volume of the intake and health and nutritional status of individuals exposed to arsenic. It also can cause cancer of liver, kidney, bladder, or other internal organs apart from anemia, burning sensation of the eyes, solid edema of the legs, liver fibrosis, chronic lung disease, gangrene of the toes (blackfoot disease) and neuropathy (Guha Mazumder *et al.*, 2001; Paul *et al.*, 2013). During our epidemiological survey, we have found that peripheral neuropathy, respiratory problems and conjunctival irritations in the eyes are the most commonly occurring non-dermatological health effects in the study population. Although arsenic-induced skin lesions are considered as hallmarks of chronic arsenic toxicity, only 15–20% of the total population shows arsenic-induced skin lesions (Banerjee *et al.*, 2011), indicating that genetic variations might play an important role in arsenic susceptibility, toxicity and carcinogenicity.

2 METHODS

2.1 *Recruitment of study participants*

Recruitment of the arsenic exposed individuals was done from three highly arsenic-affected districts of West Bengal, i.e. North 24 Parganas, Nadia and Murshidabad. The control subjects were chosen from East Midnapur district of the same state with little or no history of arsenic contamination in groundwater. The study subjects were matched with respect to age, sex and socio-economic status. Individuals ranging from 18 to 60 years of age with at least 10 years of arsenic exposure were selected as arsenic exposed study participants. Occupationally, the majority of the study participants were farmers and household workers. An interview was performed based on a structured questionnaire that elicited information about demographic factors, life-style, occupation, diet, smoking, medical and residential histories (Ghosh *et al.*, 2007). Expert dermatologists, ophthalmologists, neurologists and pulmonologists, with fifteen years of experience, identified the characteristic arsenic-induced skin lesions and associated non-dermatological health effects required for the recruitment of exposed study participants.

2.2 *Sample collection and analysis*

Urine, water, nail, hair and blood samples were collected only from those subjects who provided informed consent to participate in the study. This study was conducted in accordance with the Helsinki II Declaration and approved by the Institutional Ethics Committee of the CSIR-Indian Institute of Chemical Biology. Blood samples were collected from a total of 1600 (880 males and 720 females) arsenic exposed individuals from three arsenic affected districts of West Bengal, comprising 853 individuals with skin lesions (459 males and 394 females) and 747 individuals without skin lesions (421 males and 326 females). The control blood samples were collected from the 1200 participants (671 males and 529 females) living in the East Midnapur district of the same state. The collected samples included drinking water (approx. 100 mL), urine (100 mL), nails (approx. 250–500 mg) and hair (approx. 300–500 mg). The samples were analyzed by atomic absorption spectrometry, using a Perkin-Elmer Model-3100 (Boston, MA) spectrometer equipped with a Hewlett-Packard (Houston, TX) Vectra computer with GEM software, Perkin-Elmer EDL System-2, arsenic lamp (lamp current 380 mA). Bladder exfoliated urothelial cells were isolated from the urine as well as lymphocyte cultures in RPMI-1640, and were used for analysis of the cytogenetic damage by micronucleus assay. Blood lymphocytes were also used to isolate DNA, RNA and protein for subsequent studies.

3 RESULTS AND DISCUSSION

3.1 *Association of single nucleotide polymorphisms to increased incidence of skin lesions*

PCR, sequence-BLAST and RFLP analysis have found that several single nucleotide polymorphisms (SNPs) of key functional genes of DNA-repair, immunological and tumor suppressor pathways have been associated with increased incidence of arsenic-induced skin lesions.

Table 1 shows our key findings of SNPs in important genes that may be responsible for arsenic induced skin lesions and premalignant form of skin lesions in the population exposed to arsenic through drinking water (Banerjee *et al.*, 2007; Banerjee *et al.*, 2011; De Chaudhuri *et al.*, 2006; De Chaudhuri *et al.*, 2008; Kundu *et al.*, 2011).

Table 1. SNPs in different genes studied in this population.

Gene of interest	Polymorphism	Genotype	Risk/Protection OR [95% C.I.]
p53	Arg72Pro (G > C)	GG	2.02 [1.28–3.15]
ERCC2	Lys751Gln (A > C)	AA	4.77 [2.75–8.23]
XRCC3	Thr241Met (C > T)	CT/TT	0.49 [0.3-.67]
TNF-α	−309 G > A	GA/AA	3.04 [1.78–5.21]
IL-10	−3575 T > A	TA/AA	2.03 [1.26–3.28]

p53: Tumor suppressor protein 53 KDa; *ERCC2*: excision repair cross-complementing rodent repair deficiency, complementation group 2; *XRCC3*: X-ray repair complementing defective repair in Chinese hamster cells 3; *TNF-α*: Tumor Necrosis Factor Alpha; *IL-10*: Interleukin 10.

3.2 *Arsenic induces epigenetic deregulation*

From several associated hypotheses, it has been speculated that altered DNA methylation patterns might contribute to arsenic-induced carcinogenesis. Promoter methylation status was determined by bisulfite conversion of genomic DNA and methylation-specific PCR. Realtime PCR and western blotting determined the expression titer of both the genes. This indicated that significant hypermethylation was found in the promoters of both *DAPK* and *p16* genes in the cases compared with the controls, resulting in down regulation of the genes in the cases. There was a 3.4-fold decrease in the expression of death-associated protein kinase and 2.2-fold decrease in gene expression of *p16* in the cases compared to the controls, the lowest expression being in the cancer tissues. Promoter hypermethylation of the genes was also associated with higher risk of developing arsenic-induced skin lesions, peripheral neuropathy, ocular and respiratory diseases (Banerjee *et al.*, 2013a).

3.3 *Arsenic mitigation strategies provides some relief to chronically exposed individuals*

Several mitigation projects have been adopted by different organizations to ameliorate the effects of arsenic toxicity by reducing arsenic load in drinking water. In a two-wave cross sectional study we found that significant decrease of arsenic exposure (190.1 $\mu g\ L^{-1}$ to 37.94 $\mu g\ L^{-1}$) resulted in significant amelioration of the severity of dermatological disorders ($p < 0.0001$) Micronucleus formation in urothelial cells and lymphocytes also decreased significantly ($p < 0.001$). However, there was a significant ($p < 0.001$) rise in the incidence of each of the non-dermatological diseases like, peripheral neuropathy, conjunctivitis and respiratory distress over the period (Paul *et al.*, 2013). Thus, a complete amelioration calls for better strategies apart from removal of arsenic through drinking water.

3.4 *Rice is a potential source of arsenic exposure to humans*

More than 3,000,000,000 people across the world consume rice as a staple food. Arsenic contents of such rice vary widely, ranging from 20–900 μg kg^{-1}. We therefore designed a study to determine if cooked rice arsenic content on its own is sufficient to give rise to genotoxic effects in humans. We have chosen 417 arsenic exposed individuals from three districts of West Bengal, namely, Murshidabad, Nadia and East Midnapur, where arsenic content in drinking water was <10 μg L^{-1}. The entire study population was divided into 6 exposure groups based on the arsenic content in rice and recorded micronuclei formation in their urothelial cells. Results show that a rice arsenic content of >200 μg kg^{-1} is associated with significant increased genetic damage, as it is evident from the increased micronuclei formation in the urothelial cells of the arsenic exposed individuals. Thus, when arsenic content in rice exceeds this limit it is sufficient to give rise to significant amounts of genetic damage, even when there is little exposure through drinking water (Banerjee *et al.*, 2013).

4 CONCLUSIONS

The facets of arsenic toxicity range from genotypes of an individual (SNPs) to the molecular toxicity (epigenetic alterations) encompass a wide genre of the molecular machinery within the cells. Mitigation strategies to ameliorate the arsenic-induced toxicity in humans have been adopted, by removing arsenic from the drinking water, the largest source of arsenic in humans. However, with the advent of refined technological advances, we find that only water treatment would not suffice to cause amelioration, since arsenic through paddy/rice alone can cause genetic damage in humans without no prior exposure to arsenic through drinking water. Thus, the question still prevails: "*are we safe?*".

ACKNOWLEDGEMENTS

The author acknowledges Council of Scientific and Industrial Research (CSIR), New Delhi, India for funding the research works and field study.

REFERENCES

Banerjee, M., Banerjee, N., Bhattacharjee, P., Mondal, D., Lythgoe, P.R., Martínez, M., Pan, J., Polya, D.A. & Giri, A.K. 2013. High arsenic in rice is associated with elevated genotoxic effects in humans. *Sci Rep.* 3: 2195.

Banerjee, M., Sarkar, J., Das, J.K., Mukherjee, A., Sarkar, A.K, Mondal, L. & Giri, A.K. 2007. Polymorphism in the ERCC2 codon 751 is associated with arsenic-induced premalignant hyperkeratosis and significant chromosome aberrations. *Carcinogenesis* 28(3): 672–676.

Banerjee, N., Nandy, S., Kearns, J.K., Bandyopadhyay, A.K., Das, J.K., Majumder, P., Basu, S., Banerjee, S., Sau, T.J, States, J.C. & Giri, A.K. 2011. Polymorphisms in the TNF-α and IL10 gene promoters and risk of arsenic-induced skin lesions and other nondermatological health effects. *Toxicol Sci.* 121(1): 132–139.

Banerjee, N., Paul, S., Sau, T.J., Das, J.K., Bandyopadhyay, A., Banerjee, S. & Giri, A.K. 2013a. Epigenetic Modifications of DAPK and p16 Genes Contribute to Arsenic-Induced Skin Lesions and Nondermatological Health Effects. *Toxicol Sci.* [Epub ahead of print].

De Chaudhuri, S., Ghosh, P., Sarma, N., Majumdar, P., Sau, T.J., Basu, S., Roychoudhury, S., Ray, K. & Giri, A.K. 2008. Genetic variants associated with arsenic susceptibility: study of purine nucleoside phosphorylase, arsenic (+3) methyltransferase, and glutathione S-transferase omega genes. *Environ Health Perspect* 16(4): 501–505.

De Chaudhuri, S., Mahata, J., Das, J.K., Mukherjee, A., Sarkar, A.K., Mamdal, L.K., Sau, T.J., Giri, A.K. & Roy Chowdhury, S. 2006. Association of specific p53 polymorphisms with keratosis in individuals exposed to arsenic through drinking water In West Bengal, India. *Mutat. Res.* 601: 102–112.

Ghosh, P., Banerjee, M., De Chaudhuri, S., Chowdhuri, R., Das, J.K., Mukherjee, A., Sarkar, A.K., Mondal, L.K., Baidya, K., Sau, T.J., Banerjee, A., Basu, A., Chaudhuri, K., Ray, K. & Giri, A.K. 2007 Comparison of health effects between individuals with and without skin lesions in the population exposed to arsenic through drinking water in West Bengal, India. *J Expo Sci Environ Epidemiol.* 17: 215–223.

Guha Mazumder, D.N., De, B.K., Santra, A., Ghosh, N., Das, S., Lahiri, S. & Das, T. 2001. Randomized placebo-controlled trial of 2,3-dimercapto-1-propanesulfonate (DMPS) in therapy of chronic arsenicosis due to drinking arsenic-contaminated water. *J Toxicol Clin Toxicol.* 39(7): 665–674.

Kundu, M., Ghosh, P., Mitra, S., Das, J.K., Sau, T.J., Banerjee, S., States, J.C. & Giri, A.K. 2011. Precancerous and non-cancer disease endpoints of chronic arsenic exposure: the level of chromosomal damage and XRCC3 T241M polymorphism. *Mutat Res.* 706 (1–2): 7–12.

Mondal, D., Banerjee, M., Kundu, M., Banerjee, N., Bhattacharya, U., Giri, A.K., Ganguli, B., Sen Roy, S. & Polya, D.A. 2010. Comparison of drinking water, raw rice and cooking of rice as arsenic exposure routes in three contrasting areas of West Bengal, India. *Environ. Geochem. Health* 32: 463–477.

Paul, S., Das, N., Bhattacharjee, P., Banerjee, M., Das, J.K., Sarma, N., Sarkar, A., Bandyopadhyay, A.K., Sau, T.J., Basu, S., Banerjee, S., Majumder, P. & Giri, A.K. 2013. Arsenic-induced toxicity and carcinogenicity: a two-wave cross-sectional study in arsenicosis individuals in West Bengal, India *J Expo Sci Environ Epidemiol.* 23(2): 156–162.

One Century of the Discovery of Arsenicosis in Latin America (1914–2014) – Litter, Nicolli, Meichtry, Quici, Bundschuh, Bhattacharya & Naidu (Eds)
© 2014 Taylor & Francis Group, London, ISBN 978-1-138-00141-1

Pharmacodynamic study on the capacity of selenium to promote arsenic excretion in arsenicosis patients in Bangladesh

M. Alauddin & E. Stekolchik
Department of Chemistry, Wagner College, Staten Island, New York, USA

J.E. Spallholz
Division of Nutritional Sciences, Texas Tech University, Lubbock, Texas, USA

S. Ahmed & B. Chakraborty
Institute of Child and Mother Health, Sarkari Hospital, Dhaka, Bangladesh

G.N. George
Department of Geological Sciences, University of Saskatchewan, Saskatchewan, Canada

J. Gailer
Department of Chemistry, University of Calgary, Calgary, Canada

H. Ahsan
Center for Cancer Epidemiology and Prevention, University of Chicago, Chicago, Illinois, USA

P.F. La Porte
Department of Medicine, Emory University School of Medicine, Atlanta, Georgia, USA

ABSTRACT: Several animal model and human studies suggest that dietary selenium can counter toxicity of arsenic through the formation of seleno-bis(S-glutathionyl)arsinium $[(GS)_2\ AsSe]^-$ ion in hepatocytes and excreted through bile. To assess whether oral administration of sodium selenite promotes an increase in fecal excretion of arsenic, a limited pharmacodynamic study was carried out involving four arsenicosis patients in Bangladesh. Patients were selected based on the arsenic level in their drinking water, hair, nail samples and symptoms of diffuse melanosis. In this preliminary study, patients received sodium selenite as oral selenium supplement. Both arsenic and selenium levels have increased in urine and feces of patients in 48 hours after dosing of sodium selenite (800 μg). The enhancement in arsenic and selenium levels in urine and feces after dosing is in agreement with our hypothesis that selenium supplementation promotes excretion of arsenic and selenium through formation of $[(GS)_2\ AsSe]^-$.

1 INTRODUCTION

According to various reports, approximately 70–80 million people in Bangladesh have been exposed to Arsenic (As) above the World Health Organization (WHO) permissible level of 10 μg L^{-1} through drinking waters from aquifers geologically contaminated by arsenic (Mukherjee & Bhattacharya, 2001; Smedley & Kinniburgh, 2002; Ahmed *et al.*, 2006; Naidu *et al.*, 2009). Chronic arsenic exposure causes arsenicosis with pathological conditions that include melanosis and keratosis, vascular and endocrine disorders and several types of cancers such as lungs, liver, kidney and bladder (Kapaj *et al.*, 2006).

1.1 *Arsenic and selenium metabolism*

The two oxyanions of arsenic and Selenium (Se) are arsenite and selenite. If arsenite or selenite is administered separately to animals, they can be toxic. However, if both are administered together, the toxicity is nullified. The arsenic and selenium antagonism has been reported in a number of animal model studies since 1930 (Moxon, 1938). The mechanism of this antagonism has been linked to the formation of a complex ion of arsenic, selenium and glutathione, namely, seleno-bis(S-glutathionyl)arsinium $[(GS)_2\ AsSe]^-$.

Very recently, synchrotron X-ray absorption (XAS) spectroscopy has been utilized to show that the arsenic-selenium antagonism is facilitated through *in vivo* formation of seleno-bis(S-glutathionyl)arsinium anion in hepatocytes of rabbits and excretion through bile (Gailer *et al.*, 2002, 2004). Recent studies also indicate that the Multidrug Resistance Protein 2 (MRP2) transports $[(GS)_2 AsSe]^-$ for excretion through bile (Carew & Leslie, 2010).

1.2 *Arsenic-selenium interaction in humans*

A number of epidemiological studies and clinical trials have suggested benefit to dietary intake of selenium in combating arsenic toxicity. Poor nutrition and protein deficient diet and low intake of selenium have been associated with arsenical melanosis (Wang, 1996). At least, two recent Phase II clinical trials in China and Bangladesh have suggested benefits from oral supplementation of selenomethionine in arsenicosis patients (Yang *et al.*, 2002; Verret *et al.*, 2005). From 2006 to 2009, we have conducted a 48 weeks, 821 patients, randomized, double-blinded placebo controlled Phase III clinical trial in Bangladesh to test the efficacy of oral supplementation of daily dose of 200 µg selenite in combating arsenic toxicity. Post-hoc analyses suggested that the intervention period and dose were not adequate to show a significant decrease in arsenical melanosis. However, in a follow up study by us, selenium administered to arsenic intoxicated hamsters showed marked biliary excretion of arsenic in the form of $[(GS)_2 AsSe]^-$ during the first hour of dosing.

1.3 *Objectives of the current pilot study*

Our recent works in rabbits, rats, hamsters and humans have shown that oral and intravenous selenite administration promote hepatobiliary excretion of arsenic and selenium conjugate $[(GS)_2 AsSe]^-$. Based on these observations, we hypothesize that dietary supplementation of selenite may increase arsenic excretion through feces and urine. In this exploratory pharmacodynamic study, our objectives were to follow the arsenic and selenium excretion in selected patients kept under close supervision for three days when they were provided with meals and drinking water as they would normally consume in their households. The primary objective this study was to determine whether oral sodium selenite supplementation increases in fecal and urinary arsenic excretion.

2 EXPERIMENTAL

2.1 *Study design*

This is a limited study involving four arsenicosis patients from Shahrasty in Bangladesh. The patients were recruited following a selection criteria of arsenic exposure based on the presence of arsenic in their drinking water, dietary habits and all participants' renal and liver function falling in normal ranges. Signed consents were obtained from all participants, and the study protocol was explained to all of them. All participants resided at a private clinic for 3 days where they followed a fixed diet prepared with the drinking water from their home. On the third day, they received 800 µg selenite dose. Two blood samples were collected prior to selenite dose and 20 minutes and 60 minutes after the dose from each patient. All urine and feces were collected from all participants every day during their stay at the clinic. The urine and feces were pooled together separately for each day for each participant. All feces samples were weighed and homogenized in distilled deionized water, and aliquots were preserved at –4 °C in a freezer before further analysis.

2.2 *Chemical analysis, reagents, standards for arsenic and selenium analysis*

Drinking water, blood, urine and feces were collected for each patient. The whole blood samples were digested in high purity (99.999% purity) HNO_3 (ultrex HNO_3; J. T. Baker Co., USA) and diluted with distilled water. Homogenized feces samples were oven dried at 85 °C for several hours and dry weight was obtained. The dried feces samples were digested in high purity HNO_3 and diluted with distilled water to a fixed volume. As and Se analyses in all prepared samples were carried out by graphite furnace atomic absorption spectroscopy (GF-AAS) (Perkin Elmer Model AAnalyst 800) with Zeeman background correction and high energy electrodeless discharge lamps. The detection limits for As and Se in urine were 1.0 µg L^{-1} and for Se in blood was 0.50 µg L^{-1}. The matrix modifier for Se analysis consisted of a solution containing 1% (w/v) $Ni(NO_3)_2.6H_2O$, 2% (w/v) $Mg(NO_3)_2.6H_2O$ and 0.1% (v/v) Triton x–100, while the matrix modifier for As analysis consisted of a solution of 1% (w/v) $Ni(NO_3)_2 \cdot 6H_2O$ and 0.1% Triton X–100. High purity (99.999%) $Ni(NO_3)_2.6H_2O$, 2% (w/v) $Mg(NO_3)_2.6H_2O$ (Sigma-Aldrich Co., USA), and 18 MΩ cm^{-1} water were used throughout the analysis. The AAS instrument was calibrated with As, Se standard solutions (Perkin Elmer Co., USA). The National Institute of Standards and Technology (NIST) Standard Reference Materials (SRM 1640, 1643e, 2670) served as reference material for As, Se analysis in water and urine samples. The method detection limit for As, Se analysis in water and urine was 1.0 µg L^{-1}, with an RSD of 2.5%, and 0.5 µg L^{-1} in blood.

3 RESULTS

3.1 *Blood selenium and arsenic levels*

The patients were selected from areas where their household tube well water arsenic level ranged from 200 to 400 µg L^{-1}. In all patients, blood Se and As increased after dosing of selenite (800 µg). Representative data for blood samples from patient 1 are shown in Figure 1. Samples 1, 2 were collected before dosing, samples 3, 4 were collected 20 minutes and 60 minutes after dosing.

Similar consistent patterns of increase in blood arsenic and selenium levels have been observed for all patients in the group.

3.2 *Urinary and fecal selenium and arsenic data*

Arsenic and selenium in urine from all patients collected before Se dosing and after Se dosing are shown in Table 1.

The dotted line in Figures 2 and 3 represent the point of selenium ingestion by each patient during their stay in the clinic. Urine and fecal samples were collected from all patients before and after ingestion of selenium. While Table 1 shows the As, Se data for pooled sample, Figures 2 and 3 show data for individual samples collected by hours.

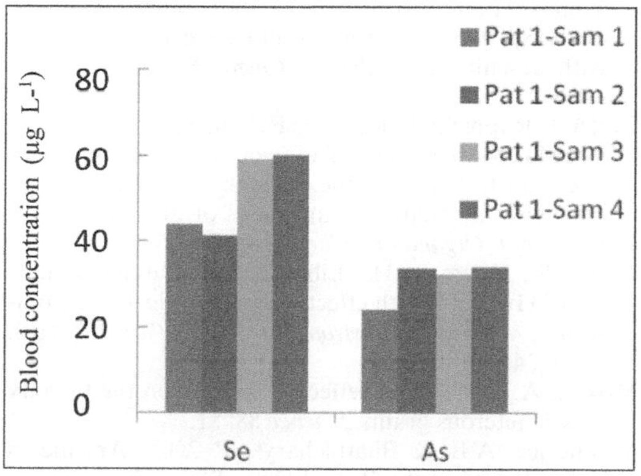

Figure 1. Levels of As and Se in blood for the patients.

Table 1. Creatinine adjusted As, Se in urine samples.

		Before dosing (µg)		After dosing (µg)
Setting	Levels	Day 1	Day 2	Day 3
1	As	13.5 ± 0.5	3.5 ± 0.2	20.6 ± 1.0
	Se	4.3 ± 0.3	2.9 ± 0.2	10.4 ± 0.5
2	As	6.0 ± 0.3	4.8 ± 0.5	6.9 ± 0.4
	Se	4.9 ± 0.3	2.7 ± 0.2	6.6 ± 0.3
3	As	–	33.5 ± 1.8	14.1 ± 0.7
	Se	–	7.8 ± 0.4	12.9 ± 0.6
4	As	–	9.4 ± 0.5	12.3 ± 0.6
	Se	–	4.2 ± 0.2	16.8 ± 0.8

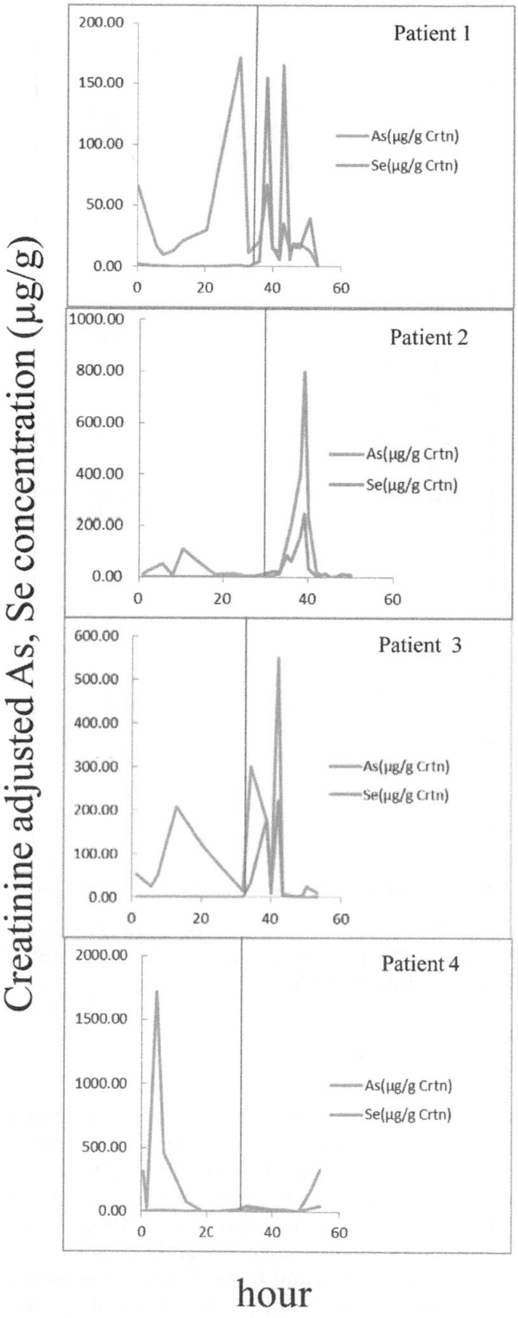

Figure 2. As and Se in urine sample in all the patients.

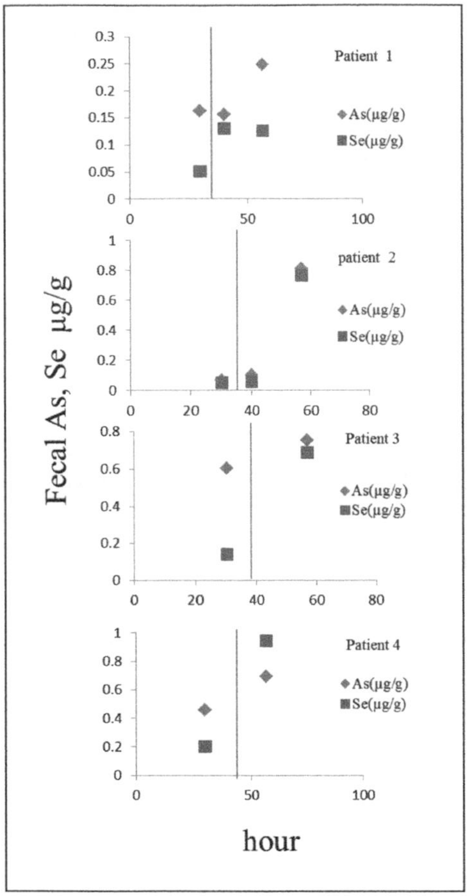

Figure 3. As and Se in feces samples of all patients. The vertical line represents the point of Se ingestion after 33 hour of stay in the clinic.

4 CONCLUSIONS

The fecal As, Se data are quite significant since they show that (a) selenium increases at the appropriate time interval 6–12 hours after Se ingestion, and (b), in every case, fecal arsenic excretion increases after Se ingestion. The parallel increase in As, Se excretion after Se dosing rightly points to the probable formation and hepatobiliary excretion of $[(GS)_2\ AsSe]^-$. The 800 μg dose of selenite closely approximates to 1:1stoichiometric ratio with average daily drinking water arsenic for each patient. Interestingly, the urinary As, Se data also indicate an increase in As, Se excretion after Se dosing. These results, although obtained from a limited number of patients, promises an effective shunting of body burden arsenic by Se supplementation. Whether this increase in As, Se in urine after Se dosing reflects formation of $[(GS)_2\ AsSe]^-$, it warrants more studies. Currently, another pharmacodynamic study involving a larger number of patients is underway in our group.

ACKNOWLEDGEMENTS

We thank Mr. Sanjit Shaha, Mr. Wahidul Hoque and Ms. Taslima Begum from Exonics Technology Center, Dhaka, Bangladesh for assistance in patient recruiting and management during the study.

REFERENCES

Ahmed, M.F., Ahuja, S., Alauddin, M., Hug, S.J., Lloyd, J.R., Pfaff, A., Pichler, T., Saltikov, C., Stute, M. & van Geen, A. 2006. Epidemiology: Ensuring safe drinking water in Bangladesh. *Science* 314: 1687–88.

Carew, M.W. & Leslie, E.M. 2010. Selenium-dependent and independent transport of arsenic by human Multidrug Resistance Protein 2 (MRP2/ABCC2): Implications for the mutual detoxification of arsenic and selenium. *Carcinogenesis* 31: 1450–55.

Gailer, J., George, G.N., Pickering, I.J., Prince, R.C., Younis, H.S & Winzerling, J.J. 2002. Biliary excretion of $[(GS)_2\ AsSe]^-$ after intravenous injection of rabbits with arsenite and selenite. *Chem. Res. Toxicol.* 15: 1466–71.

Gailer, J., Ruprecht, I., Reitmeir, P., Benker, B. & Schramel, P. 2004. Mobilization of exogenous and endogenous selenium to bile after the intravenous administration of environmentally relevant doses of arsenite to rabbits. *Appl. Organomet. Chem.* 18: 670–75.

Kapaj, S., Peterson, H., Liber, K. & Bhattacharya, P. 2006. Human health effects from chronic arsenic poisoning–a review. *J. Environ. Sci. Health Part A* 41 (10): 2399–2428.

Moxon, A.L. 1938. The effect of arsenic on the toxicity of seleneferous grains. *Science* 88: 81.

Mukherjee, A.B. & Bhattacharya, P. 2001. Arsenic in groundwater in the Bengal Delta Plain: slow poisoning in Bangladesh. *Environ. Rev.* 9 (3): 189–220.

Naidu, R. & Bhattachaya, P. 2009. Arsenic in the environment-risks and management strategies. *Environ. Geochem. Health* 31(S1), 1–8.

Smedley, P.L. & Kinniburgh, D.G. 2002. A review of the source, behaviour and distribution of arsenic in natural waters. *Appl. Geochem.* 17: 517–568.

Verret, W.J., Chen, Y., Ahmed, A., Islam, T., Pervez, F., Kibria, M.G., Graziano, J. H. & Ahsan, H. 2005. A randomized double-blind placebo-controlled trial evaluating the effects of vitamin E and selenium on arsenic-induced skin lesions in Bangladesh. *J. Occup. Environ. Med. JT* 47: 1026–1035.

Wang, C.T. 1996. Concentration of arsenic, selenium, zinc, iron and copper in the urine of blackfoot disease patients at different clinical stages. *European J. Clin. Chem. Clin. Biochem.* 34: 493–497.

Yang, L., Wang, W., Hou, S., Peterson, P.J. & William, W.P. 2002. Effects of selenium supplementation on arsenism: an intervention trial in Inner Mongolia. *Environ. Geochem. Health* 24: 359–374.

One Century of the Discovery of Arsenicosis in Latin America (1914–2014) – Litter, Nicolli, Meichtry, Quici, Bundschuh, Bhattacharya & Naidu (Eds)
© 2014 Taylor & Francis Group, London, ISBN 978-1-138-00141-1

Nutrients and genetic polymorphisms modify arsenic metabolism efficiency

C. Hernández-Alcaraz, R.U. Hernández-Ramírez, A. García-Martínez, A.I. Burguete-García & L. López-Carrillo
Instituto Nacional de Salud Pública, Cuernavaca, Morelos, México

M.E. Cebrián & A.J. Gandolfi
Departamento de Toxicología, Centro de Investigación y Estudios Avanzados, México D.F., México
College of Pharmacy, University of Arizona, Tucson, AZ, USA

ABSTRACT: The aim of this study was to examine if five single nucleotide polymorphisms involved in one-carbon metabolism are associated with the efficiency of iAs metabolism and if these potential associations are respectively modified by dietary intake of folate-related nutrients. Most of the nutrients and polymorphisms altered the iAs metabolism. Additionally, vitamin B12 and methionine modified the association between FolH1 T223C and MTR A2756G over the iAs metabolism, respectively. Genetic and dietary differences in one-carbon metabolism may jointly account for differences in iAs metabolism.

1 INTRODUCTION

Nutrients and genetic polymorphisms participating in one carbon metabolism may account for interindividual differences in inorganic arsenic (iAs) metabolism (Hall & Gamble, 2012, Hernández & Marcos, 2008), which in turn may determine susceptibility to iAs-induced disease, including several types of cancer (IARC, 2012).

The aim of this study was to evaluate if single nucleotide polymorphisms in genes *FolH1 T223C*, *MTHFD1 G1958A*, *MTHFR C677T*, *MTR A2756G* and *MTRR A66G*, were associated with the efficiency of iAs metabolism, and if these potential associations were respectively modified by dietary intake of riboflavin, vitamin B6, folate, vitamin B12, choline, betaine and methionine.

2 METHODS

Women (1,027) from an iAs-exposed region in Northern Mexico were interviewed. Blood and urine samples were collected. Micronutrient dietary intake was estimated using a validated food frequency questionnaire. Markers of iAs metabolism included, proportions (%iAs, %MMA, %DMA) and ratios (MMA/iAs, DMA/MMA, DMA/iAs) calculated from urinary As species measured by HPLC-ICP-MS. After DNA extraction using the DNA MidiPrep kit (Quick-gDNA® by Zymo Research) the polymorphisms of *FolH1 T223C* (rs202676), *MTHFD1 G1958A* (rs2236225), *MTHFR C677T* (rs1801133), *MTR A2756G* (rs1805087) y *MTRR A66G* (rs1801394) were analyzed by allelic discrimination using 7900 real-time PCR System (Applied Biosystems®). Linear regression models were used to evaluate the associations between polymorphisms and the efficiency of iAs metabolism, as well as their potential interactions with nutrients.

3 RESULTS AND DISCUSSION

The women had an average age of 54 years, with an intake of 2057.47 kcal day^{-1} and a median of total As (iAs+MMA+DMA+AsB) in urine of 15.68 mcg L^{-1} (p05 = 2.72, p95 = 133.07). The average proportions iAs, MMA and DMA were 11.80%, 10.26% and 77.94%, respectively. Near 80.0% of women were overweighted or obese, 16.0% smoked and 11.0% consumed alcohol at the time of the interview. With increasing age and body, mass index more efficient patterns of iAs metabolism were observed, while the increase in total energy intake per day, smoking and alcohol consumption had the opposite effect.

Our population was depleted of folate (82.67%), choline (95.81%) and betaine (98.00%), whereas the proportion of women below the recommended dietary intake (IDR) for riboflavin, vitamin B6, methionine and vitamin B12 were 21.62%, 37.09%, 28.63% and 66.99%, respectively. The genotype distribution of the polymorphisms of interest was found in Hardy-Weinberg Equilibrium.

The increase in the intake of some of the assessed nutrients was associated with an increase in removal efficiency of iAs. For example, higher intakes of folate were related with a decrease of %iAs. An increase in the intake of choline was associated with higher %DMA and DMA/MMA. Methionine intake was related to a decrease of %iAs and an increase of total DMA/iAs. In addition, for an increase in the intake of vitamin B12, we observed lower %MMA.

Likewise, CC carriers of *FolH1 T223C* showed lower %iAs and DMA/iAs. Furthermore, GG carriers of *MTR A2756G* had lower %MMA and higher DMA/MMA. In contrast, CT carriers of FolH1 T223C were associated with lower DMA/MMA, DMA/iAs, while GA and AA carriers of *MTHFD1 G1958A* had less efficient methylation patterns characterized by the increase of %iAs and %MMA, and the decrease of %DMA as well as MMA/iAs, DMA/MMA and DMA/iAs.

Gene-nutrient adjusted models showed CC carriers of *FolH1 T223C* with an intake above the DRI of vitamin B12 (>2.4 μg day^{-1}) had the most efficient rate of iAs elimination (lower%iAs, higher MMA/iAs, DMA/MMA), apparently indicating that this is a low risk subgroup (Figures 1, 2 and 3).

Furthermore, when methionine intake was <14.0 mg kg^{-1} bw (below DRI), the AA carriers of *MTR A2756G* showed higher %iAs ($\beta = 0.362$, $p = 0.050$;

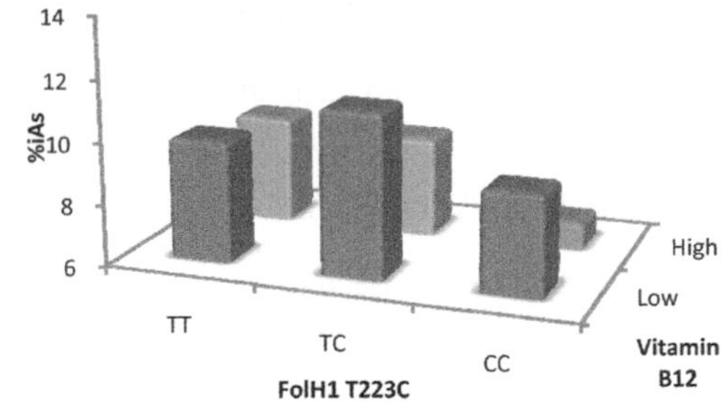

Figure 1. Interaction between FolH1 and Vitamin B12 over %iAs.

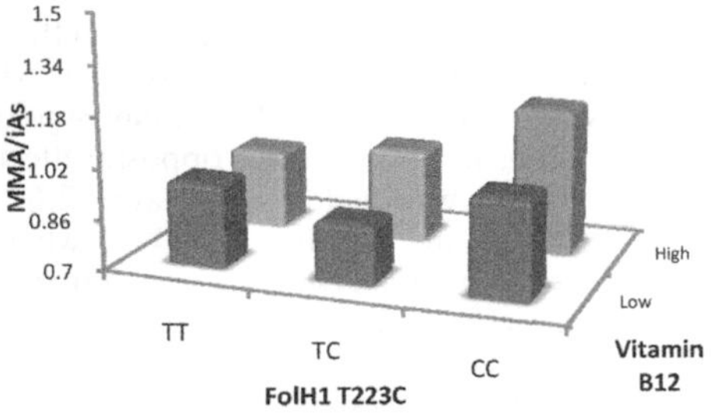

Figure 2. Interaction between FolH1 and Vitamin B12 over first methylation index (MMA/iAs).

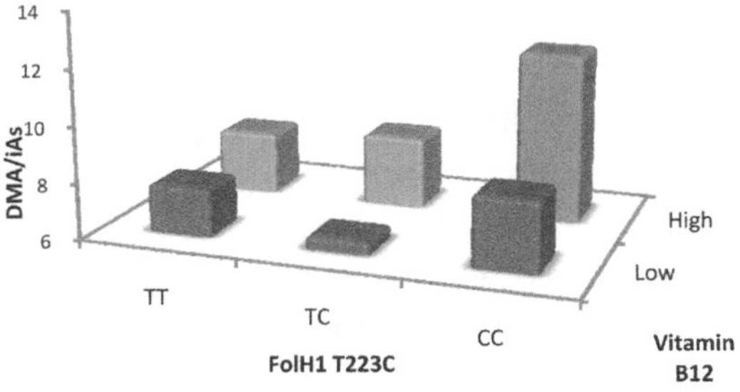

Figure 3. Interaction between FolH1 and Vitamin B12 over total methylation index (DMA/iAs).

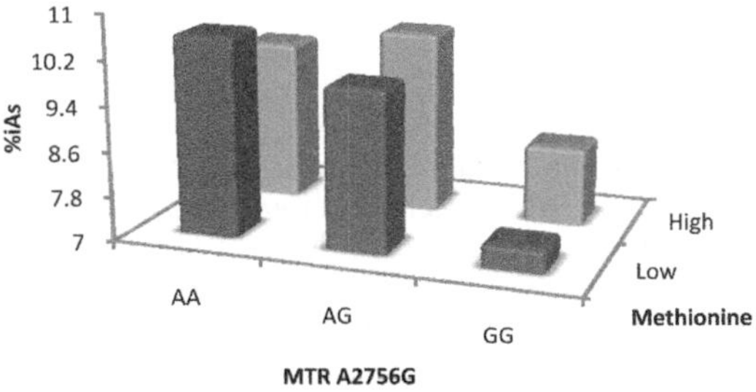

Figure 4. Interaction between MTR and methionine over %iAs.

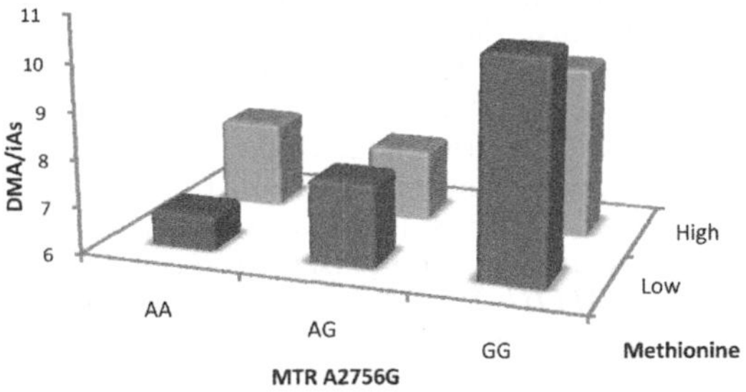

Figure 5. Interaction between MTR and methionine over total methylation index (DMA/iAs).

Figure 4) and lower DMA/iAs ($\beta = -0.457$ $p = 0.081$; Figure 5).

This may suggest the existence of a vulnerable group. These result remained after correcting them by False Discovery Rate test to diminish the probability of type I error, considering a p value for the interaction under 0.2.

4 CONCLUSIONS

Differences in one-carbon polymorphisms and nutrients may affect the availability of methyl group donors and jointly influence As metabolism efficiency. Confirmation of these interactions in other populations is needed.

ACKNOWLEDGEMENTS

This study was supported by CONACyT. Fondo Sectorial de Investigación en Salud y Seguridad Social (2005-2-14373, 2009-1-111384 and 2010-1-140962). Additional support was provided in part by The Dean Carter Binational Center for Environmental Health Sciences (NIH ES04940).

REFERENCES

Hall, M.N. & Gamble, M.V. 2012. Nutritional manipulation of one-carbon metabolism: effects on arsenic methylation and toxicity. *Journal of toxicology*, 2012, p. 595307.

Hernández, A. & Marcos, R., 2008. Genetic variations associated with interindividual sensitivity in the response to arsenic exposure. *Pharmacogenomics*, 9(8), pp. 1113–32.

IARC, 2012. *Arsenic and arsenic compounds*, Lyon, France.

One Century of the Discovery of Arsenicosis in Latin America (1914–2014) – Litter, Nicolli, Meichtry, Quici, Bundschuh, Bhattacharya & Naidu (Eds)

AS3MT genotype in South American populations and their influence on arsenic metabolism

K. Engström
Department of Laboratory Medicine, Section of Occupational and Environmental Medicine, Lund University, Lund, Sweden

K. Broberg
Department of Laboratory Medicine, Section of Occupational and Environmental Medicine, Lund University, Lund, Sweden
Institute of Environmental Medicine, Karolinska Institutet, Stockholm, Sweden

M. Vahter & F. Harari
Institute of Environmental Medicine, Karolinska Institutet, Stockholm, Sweden

A.M. Ronco
Instituto de Nutrición y Tecnología de los Alimentos (INTA), University of Chile, Chile

J. Gardon
IRD—Hydrosciences Montpellier (HSM), France

G. Concha
National Food Administration, Toxicology Division, Uppsala, Sweden

ABSTRACT: High fractions of the arsenic metabolite methylarsonic acid (MMA) in urine is a susceptibility marker for arsenic-related disease. Fraction of MMA is partly determined by genetic factors. The aim of this study was to evaluate genetic variation in the major arsenic-metabolizing gene AS3MT in populations from northern Argentina, northern Chile and highlands of Bolivia and the associations with %MMA in urine. Subjects were genotyped for three AS3MT polymorphisms previously associated with differences in %MMA. Associations between AS3MT and %MMA were in the same directions in all populations and confirm earlier findings of AS3MT genotypes associated with lower %MMA. However, frequencies of alleles associated with lower %MMA varied markedly between study populations (e.g. 25–79% for rs3740393). The highest frequencies were seen in the Bolivian population, showing the highest AS3MT frequencies reported so far for an efficient arsenic-metabolism.

1 INTRODUCTION

High fractions of the arsenic metabolite methylarsonic acid (MMA) in blood and urine is a marker of susceptibility to arsenic-related diseases. There is a marked variability in the metabolism of inorganic arsenic within and between population groups (Vahter, 2002). This is partly explained by polymorphisms in the arsenic (+III) methyltransferase (*AS3MT*) gene. Associations between this gene and arsenic metabolite pattern have been reported for several populations (Sumi *et al.*, 2012) and we have previously found a strong association between *AS3MT* genotypes and arsenic metabolite pattern in a mainly indigenous population in San Antonio de los Cobres (SAC) in the Argentinean Andes (Engström *et al.*, 2011). This population has unusually high frequencies of *AS3MT* alleles associated with a lower percentage of MMA in urine. In this study, we genotyped four additional study groups from South America and investigated their relation between *AS3MT* genotype and fractions of MMA.

2 MATERIALS AND METHODS

2.1 *Study areas and populations*

The study areas were as follows (for levels of urinary arsenic (U-As) and %MMA, see Table 1):

1) Anta, Salta province, Argentina ($N = 27$, of which 14 were women). 2) El Galpon, Salta province, Argentina ($N = 87$; 20 with metabolite data).

Table 1. Total urinary arsenic (U-As), %MMA and %DMA in the different study populations in the different populations.

	U-As (10/90 perc.)	%MMA (10/90 perc.)	%DMA (10/90 perc.)
Bolivia	na	na	na
Arica	23 (17/30)	5.2 (10.1/16.9)	69 (81/90)
Anta	79 (155/1600)	3.4 (9/22)	41 (65/76)
El Galpon	53 (127/678)	5.5 (11.3/17.6)	67 (80/88)
SAC	200 (27/450)	7.9 (4.6/12)	80 (69/89)
Bangladesh	100 (21/390)	10 (5/17)	75 (62/85)

na = Urine samples were not available. There was, however, arsenic in whole blood ($N = 14$, median 2.2 ng/g).

3) Arica, Chile ($N = 25$ women). 4) Oruro, Bolivia ($N = 62$ women). These groups were compared to two previously studied groups (Engstrom *et al.*, 2010): 1) San Antonio de los Cobres (SAC), and surrounding villages, in northern Argentinean Andes ($N = 172$ women). 2) Matlab, Bangladesh ($N = 361$ women). The ethics Committee of the Karolinska Institutet, Sweden and the Health Ministry of Salta, Argentina gave authorization for the sampling.

2.2 *Assessment of arsenic exposure*

Exposure to inorganic arsenic was assessed by the concentration of arsenic metabolites in urine (U-As; sum of inorganic arsenic, MMA and DMA using HPLC hyphened with hydride generation and Inductively Coupled Plasma Mass Spectrometry (ICPMS) (Agilent, Japan and Germany). Arsenic concentrations were adjusted to the mean specific gravity (SG) for each population.

2.3 *Genotyping*

Three non-coding SNPs were genotyped in blood DNA (rs3740400, rs3740393, and rs1046778) either by Sequenom™ (San Diego, CA, USA) technology or by Taqman allelic discrimination assay (ABI).

2.4 *Statistical analyses*

The study groups were analyzed separately. Deviations from Hardy-Weinberg equilibrium were tested using chi-square analysis. For each arsenic metabolite (dependent variables), a multivariable regression model (using the general linear model (GLM) was performed that included genotype and U-As. In consideration of normally distributed residuals, %MMA and U-As were natural log (ln) transformed for all study populations in all analyses.

3 RESULTS AND DISCUSSION

3.1 *Allele frequencies of AS3MT polymorphisms*

The allele frequencies (Table 2) of the protective *AS3MT* alleles were highest in the more indigenous population groups Bolivia and SAC, and lowest in El Galpon, which has mainly inhabitants that are Hispanic descendants. The allele frequencies in El Galpon are similar to these in Bangladesh.

3.2 *AS3MT genotype and %MMA*

A similar pattern for the *AS3MT* genotypes and %MMA was seen among all populations (Table 3);

Table 2. Allele frequencies* in the different populations.

	rs3740400	rs3740393	rs1046778
	C/A	C/G	C/T
Bolivia	0.81/0.19	0.79/0.21	0.81/0.19
Arica	0.52/0.48	0.43/0.57	0.43/0.57
Anta	0.44/0.55	0.37/0.62	0.39/0.61
El Galpon	0.32/0.68	0.25/0.75	0.33/0.66
SAC	0.73/0.27	0.70/0.30	0.71/0.29
Bangladesh	0.45/0.55	0.18/0.82	0.38/0.62

*Alleles associated with lower %MMA are denoted first.

Table 3. Association between *AS3MT* genotype and %MMA (adjusted for U-As).

	%MMA (N)				
SNP, genotype	Arica	SAC	Anta	El Galpon	Bangladesh
rs3740400 CC	6.1 (4)	7.1 (93)	5.5 (3)	*	8.8 (71)
AC	10.9 (16)	8.5 (61)	10.4 (7)	7.6 (6)	10.2 (180)
AA	12.7 (4)	10.3 (15)	12.2 (4)	18.8 (14)	9.7 (103)
p-value	0.002	0.001	0.72	0.52	0.008
rs3740393 CC	5.6 (4)	6.9 (85)	8.7 (2)	*	7.9 (13)
GC	10.3 (14)	8.7 (68)	9.0 (7)	8.3 (6)	9.2 (105)
GG	12.9 (7)	10 (17)	10.7 (5)	18.0 (14)	10.1 (240)
p-value	0.001	<0.001	0.85	0.039	0.13
rs1046788 CC	5.9 (3)	6.9 (87)	6.6 (2)	*	7.9 (45)
TC	9.7 (16)	8.7 (69)	9.1 (8)	10.6 (8)	9.6 (183)
TT	13.9 (6)	10.3 (15)	12.2 (4)	17.4 (12)	10.7 (132)
p-value	0.014	<0.001	0.78	0.24	<0.001

The protective alleles are shown first. * This genotype is pooed with the heterozygotes since only 1 individual had this genotype.

thus the associations between *AS3MT* genotype and %MMA confirm earlier findings of protective *AS3MT* genotypes.

4 CONCLUSIONS

The allele frequencies of the protective *AS3MT* genotypes varied widely in the South American populations. The highest frequencies were found in the more indigenous populations from the Andes Moutains. Arsenic is present naturally in the bedrock around Oruro (Bolivia) (Bundschuh *et al.*, 2012), which may have led to an adaption to arsenic, by a higher frequency of a fast arsenic-metabolizing genotype similar to what have been suggested for SAC (Schlebush *et al.*, 2013). *AS3MT* genotype was confirmed to be associated with %MMA, and in turn, this indicate that some populations might be at higher risk for arsenic-related disease based on their genetic background.

REFERENCES

Bundschuh, J., Litter, M.I., Parvez, F., Román-Ross, G., Nicolli, H.B., Jean, J.S., Liu, C.W., López, D., Armienta, M.A., Guilherme, L.R., Cuevas, A.G., Cornejo, L., Cumbal, L. & Toujaguez, R. 2012. One century of arsenic exposure in Latin America: a review of history and occurrence from 14 countries. *Sci. Total Environ.* 429: 2–35.

Engström, K.S., Hossain, M.B., Lauss, M., Ahmed, S., Raqib, R., Vahter, M. & Broberg, K. 2011. Polymorphisms in AS3MT predict gene expression of AS3MT as well as arsenic metabolism. *Environmental Health Perspectives* 119: 182–188.

Sumi, D. & Himeno, S. 2012. Role of arsenic (+3 oxidation state) methyltransferase in arsenic metabolism and toxicity. *Biol. Pharm. Bull.* 35: 1870–1875.

Schlebusch, C.M., Lewis, C.M. Jr, Vahter, M., Engström, K., Tito, R.Y., Obregón-Tito, A.J., Huerta, D., Polo, S.I., Medina, Á.C., Brutsaert, T.D., Concha, G., Jakobsson, M. & Broberg, K. 2013. Possible positive selection for an arsenic-protective haplotype in humans. *Environ. Health Perspect.* 121: 53–58.

Vahter M. 2002. Mechanisms of arsenic biotransformation. *Toxicology* 181–182: 211–217.

One Century of the Discovery of Arsenicosis in Latin America (1914–2014) –
Litter, Nicolli, Meichtry, Quici, Bundschuh, Bhattacharya & Naidu (Eds)
 ISBN 978-1-138-00141-1

High level of exposure to arsenic and its influence on arsenic urinary methylated metabolites: A study from Argentina

V. Olmos, J.A. Navoni & E.C. Villaamil Lepori
Cátedra de Toxicología y Química Legal, Facultad de Farmacia y Bioquímica, Universidad de Buenos Aires, Argentina

ABSTRACT: The influence of the level of exposure to As on human metabolism was studied in a population of Chaco and Santiago del Estero provinces, northern Argentina. The wide range of levels of urinary arsenic (from 18 to 4103 μg/g of creatinine) showed to correlate positively ($Rs = 0.38$, $p < 0.0001$) with urinary % MMA and negatively ($Rs = -0.32$, $p < 0.001$) with % DMA. The higher the exposure, the greater the percentage of urinary MMA and the lower the urinary DMA. Age, gender and presence of T860C gene polymorphism also showed to influence urinary As metabolite distribution. The existence of populations with such high levels of exposure increases, not linearly, but exponentially, their risk of developing As related diseases.

1 INTRODUCTION

Arsenic (As) is a metalloid with widespread distribution on Earth's crust. Long term human exposure to inorganic arsenic (iAs) through drinking water is an important health problem in Argentina (McClintock *et al.*, 2012; Navoni *et al.*, 2012). Humans metabolize iAs through methylation to monomethylarsonic acid (MMA) and dimethylarsinic acid (DMA), and all metabolites, including InAs, are eliminated in urine in different proportions. Many factors, such as ethnicity, age, gender, pregnancy status, smoking status, presence of some gene polymorphisms and level of As exposure were described to influence As metabolism. In Argentina, there are many regions with current and past high levels of As in water. The Chaco Pampean Plains are known for their elevated levels of As in their groundwater.

The aim of the study was to evaluate the influence of the level of As exposure on urinary As metabolic profile of individuals living in the Chaco-Pampean Plains.

2 EXPERIMENTAL

2.1 *Study population*

A total of 120 inhabitants (61 women and 59 men, 85 children under 13 years old) from Santiago del Estero and Chaco provinces (Chaco-Pampean plain) agreed to participate in the study.

The study was approved by the Ethical Committee of the Hospital de Clínicas "José de San Martín" (Buenos Aires, Argentina) and informed consent was obtained from each participant.

2.2 *Sampling*

All individuals were asked to provide a first morning urine sample, and a buccal swab sample for extraction of desquamation cells. A water sample from each source of drinking water was also obtained to assess the different levels of exposure of the population. Information about age, gender and place of residence was collected by performing a questionnaire.

2.3 *As and metabolites analysis*

Total As (tAs) in water and urinary As (uAs) were quantified by flow injection-HG-AAS (hydride generation-atomic absorption spectrophotometry). Total uAs was expressed in micrograms per gram of creatinine.

Urinary concentration of As(III), As(V), DMA(III+V) and MMA(III+V) was measured using HPLC-HG-AAS. Results were expressed as % of inorganic As (% iAs) (III+V), % MMA and % DMA.

2.4 *Polymorphism analysis*

The presence of the single nucleotide polymorphism Met(287)Thr (T860C) in the As3MT gene was investigated in buccal cells by PCR-RFLP.

3 RESULTS AND DISCUSSION

3.1 *uAs—tAs in water correlation*

Levels of tAs in drinking water were within a wide range of concentrations, from 13 to 1148 µg/L with a mean of 426 µg/L. Urinary As concentrations ranged from 18 to 4103 µg/g of creatinine with a mean of 920 µg/g of creatinine. There was a positive, significant correlation ($Rs = 0.68$, $p < 0.0001$) between uAs and tAs in drinking water.

3.2 *Urinary metabolic profile-uAs correlation*

Urinary % MMA was found to be positively associated to the level of uAs ($Rs = 0.38$, $p < 0.0001$) and % DMA was inversely associated to the level of uAs ($Rs = -0.32$, $p < 0.001$) (Figure 1).

A multiple linear regression analysis was performed considering age, gender, presence of T860C polymorphism and uAs as predictors of As metabolic profile. The results showed a marked influence of age, uAs level, gender and polymorphism, as predictors of urinary % MMA and a marked influence of gender, uAs level and polymorphism as predictors of urinary % DMA and no influence of age (Table 1).

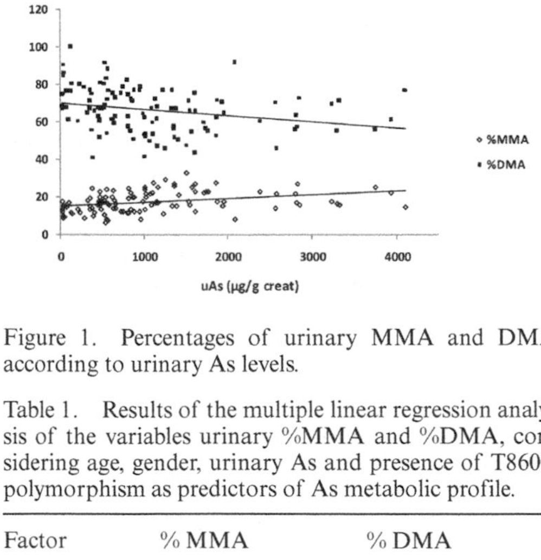

Figure 1. Percentages of urinary MMA and DMA according to urinary As levels.

Table 1. Results of the multiple linear regression analysis of the variables urinary %MMA and %DMA, considering age, gender, urinary As and presence of T860C polymorphism as predictors of As metabolic profile.

Factor	% MMA		% DMA	
	Estimator	*p* value	Estimator	*p* value
Constant	16.57	<0.0001	71.02	<0.0001
Age	−3.54[1]	0.0004*	4.08[1]	0.0588
uAs	0.0014[2]	0.0051*	−0.002[2]	0.0246*
Gender	2.58[3]	0.0037*	−6.82[3]	0.0006*
Polymorphism	3.87[4]	0.0004*	−7.31	0.0026*

Note: [1]with child (under 13 y.o.) as reference group;
[2]per µg of arsenic per gram of creatinine;
[3]with male as reference group; with heterozygous as reference group.
*Statistical significance $p < 0.05$.

3.3 *Presence of T860C polymorphism*

Wild Type (WT) genotype (TT) was found in 94 individuals, Heterozygous (H) genotype (TC) was found in 25 individuals and homozygous genotype (CC) was found in one person.

Urinary As level has been widely used as a biomarker to reflect the exposure to As through drinking water (Meza *et al.*, 2004; Hughes, 2006). However, there is scarce information about higher levels of exposure.

In this study, a positive correlation between uAs level and level of exposure was observed for a wide range of concentrations (from 13 to 1148 µg/L). Although correlation seemed to decrease around the highest level of arsenic in drinking water (1148 µg/L), it was still good enough to consider uAs as an indicator of the level of exposure. Age, gender and presence of the T860C polymorphism showed to be predictors of uAs metabolic profile. A marked influence of the level of exposure was also observed on relative proportions of urinary MMA and DMA in agreement with previously reported studies (Hopenhayn-Rich *et al.*, 1996; Sun *et al.*, 2007). However, in the present study, this influence is more pronounced and also for levels of exposure significantly higher.

4 CONCLUSIONS

Elevated proportions of MMA and a low DMA/MMA ratio are associated with high risks to develop skin lesions (Ahsan *et al.*, 2007), urothelial carcinoma (Huang *et al.*, 2008), bladder cancer (Chen *et al.*, 2003), and lung cancer (Steinmaus *et al.*, 2010). Considering the results of this study, the existence of populations with such high levels of exposure increases, not linearly, but exponentially the risk of developing these diseases.

REFERENCES

Ahsan, H., Chen, Y., Kibriya, M.G., Slavkovich, V., Parvez, F., Jasmine, F., Gamble, M.V. & Graziano, J.H. 2007. Arsenic metabolism, genetic susceptibility, and risk of premalignant skin lesions in Bangladesh. *Cancer Epidemiol. Biomarkers Prev.* 16: 1270–1278.

Chen, Y.C., Su, H.J., Guo, Y.L., Hsueh, Y.M., Smith, T.J., Ryan, L.M., Lee, M.S. & Christiani, D.C. 2003. Arsenic methylation and bladder cancer risk in Taiwan. *Cancer Causes Control.* 14: 303–310.

Hopenhayn-Rich, C., Biggs, M.L., Kalman, D.A., Moore, L.E. & Smith, A.H. 1996. Arsenic methylation patterns before and after changing from high to lower concentrations of arsenic in drinking water. *Environ. Health Perspect.* 104: 1200–1207.

Huang, Y.K., Huang, Y.L., Hsueh, Y.M., Yang, M.H., Wu, M.M., Chen, S.Y., Hsu, L.I., Chen & C.J. 2008. Arsenic exposure, urinary arsenic speciation, and the incidence of urothelial carcinoma: A twelve-year follow-up study. *Cancer Causes Control*. 19: 829–839. Abstract.

Hughes, M.F. 2006. Biomarkers of Exposure: A Case Study with Inorganic Arsenic. *Environ. Health Perspect*. 114: 1790–1796.

McClintock, T.R., Chen, Y., Bundschuh, J., Oliver, J.T., Navoni, J., Olmos, V., Lepori, E.V., Ahsan, H. & Parvez, F. 2012. Arsenic exposure in Latin America: biomarkers, risk assessments and related health effects. *Sci Total Environ*. 1;429: 76–91.

Meza, M.M., Kopplin, M.J., Burgess, J.L. & Gyolfi, A.J. 2004. Arsenic drinking water exposure and urinary excretion among adults in the Yaqui Valley, Sonora, Mexico. *Environ. Res*. 96: 119–126.

Navoni, J.A., De Pietri, D., García, S. & Villaamil Lepori, E.C. 2012. Riesgo sanitario de la población vulnerable expuesta al arsénico en la provincia de Buenos Aires, Argentina. *Rev. Panam. Salud Pública*. 31(1): 1–8.

Steinmaus, C., Yuan, Y., Kalman, D., Rey, O.A., Skibola, C.F., Dauphine, D., Basu, A., Porter, K.E., Hubbard, A., Bates, M.N., Smith, M.T. & Smith, A.H. 2010. Individual differences in arsenic metabolism and lung cancer in a case-control study in Cordoba, Argentina. *Toxicol. Appl. Pharmacol*. 247: 138–45.

Sun, G., Xu, Y., Li, X., Jin Y., Li, B. & Sun, X. 2007. Urinary arsenic metabolites in children and adults exposed to arsenic in drinking water in Inner Mongolia, China. *Environ. Health Perspect*. 115: 648–52.

One Century of the Discovery of Arsenicosis in Latin America (1914–2014) – Litter, Nicolli, Meichtry, Quici, Bundschuh, Bhattacharya & Naidu (Eds)
ISBN 978-1-138-00141-1

Arsenic trioxide induces both apoptosis and autophagy through stimulation of ER stress and inhibition of ubiquitin-proteasome system in human sarcoma cells

H.-W. Chiu, Y.-C. Tseng & Y.-J. Wang
Department of Environmental and Occupational Health, National Cheng Kung University, Medical College, Tainan, Taiwan

ABSTRACT: Osteosarcoma (HOS cells) and fibrosarcoma (HT1080 cells) were used to investigate the anti-cancer effect of arsenic trioxide (ATO) and the underlying mechanisms *in vitro* and *in vivo*. We found that ATO treatment enhanced the percentage of apoptosis and autophagy in HOS and HT1080 cells through increasing the expression of ER stress-associated protein IRE1 and led to induction of LC3-II and cleaved-caspase-3. ATO significantly decreased the phosphorylation of Akt and mTOR and increased the phosphorylation of AMPK, p38, and JNK in HOS and HT1080 cells. Combined treatment of ATO and MG-132 showed synergistic effect through induction of apoptosis and autophagy in HT1080 cells. In *in vivo* study, the combined treatment significantly reduced the tumor volume in SCID mice that had received a subcutaneous injection of HT1080 cells. Thus, a combination of ATO and proteasome inhibitor could be a new potential therapeutic strategy for the treatment of fibrosarcoma.

1 INTRODUCTION

Sarcoma is a rare cancer type with characteristics of aggressive, drug-resistant and highly metastasis (Skubitz & D'Adamo, 2007). In clinical, the efficiency of therapy is rather low. Thus, there is a need to find a new therapeutic strategy and enhance the therapeutic efficiency of sarcoma. Arsenic has long been used as anticancer agent in traditional Chinese medicine. Recently Arsenic Trioxide (ATO) has been successfully employed in the treatment of refractory or relapsed Acute Promyelocytic Leukemia (APL), and its efficacy has been confirmed even in patients resistant to conventional chemotherapy (Shen *et al.*, 2001). Nevertheless, the anticancer effects of ATO on solid tumors remain unclear. It has been reported that Arsenic Trioxide (ATO) not only induced apoptosis but also autophagy in a variety of cancer cells (Kanzawa *et al.*, 2005; Qian *et al.*, 2007). Accumulation of reactive oxygen species (ROS), induction of Endoplasmic Reticulum stress (ER stress), or inhibition of Ubiquitin-Proteasome System (UPS) were suggested to trigger with apoptosis and autophagy (Kim *et al.*, 2008; Salazar *et al.*, 2009).

Autophagy is one of the mechanisms of stress tolerance that maintains cell viability and can lead to tumor dormancy, progression and therapeutic resistance. However, many anticancer drugs could also induce the excessive or prolonged autophagy that triggers tumor cell death. Activation of the PI3 kinase/Akt pathway, a well-known method to inhibit apoptosis, also inhibits autophagy (Mathew *et al.*, 2007). Akt is a serine/threonine protein kinase that plays a critical role in suppressing apoptosis by regulating its downstream pathways (Wang *et al.*, 2008). Akt also phosphorylates mammalian target of rapamycin (mTOR), which has been reported to inhibit the induction of autophagy (Kondo *et al.*, 2005). Both apoptosis and autophagy could be induced in certain tumor cells under the treatment of anti-cancer drugs (Hsu *et al.*, 2009; Qian *et al.*, 2007).

Accumulated evidence indicates that programmed cell death is closely related to anti-cancer therapy. Many studies have shown that tumor cells treated with anti-cancer drugs experience the induction of type I programmed cell death, apoptosis, and type II programmed cell death, autophagy (Hait *et al.*, 2006). The main purpose of this study was to investigate whether the anticancer effect of ATO-induced programmed cell death could be through stimulation of ER stress and inhibition of ubiquitin-proteasome system in human sarcoma cells, and the underlying mechanisms contributing to this effect were also examined.

2 RESULTS

2.1 *Optimal dose and time selection of ATO and MG-132 for treatment of HOS and HT1080 cells*

Cell viability was determined by trypan blue. Treatment with ATO or MG-132 alone reduced the viability of the HOS and HT1080 cells in a time—and dose-dependent manner. Significantly enhanced toxicity was found for the combination treatment compared with ATO and MG-132 treatment alone in HOS and HT1080 cells. The survival curves dramatically shifted downward.

2.2 *Measurement of cell cycle distribution, apoptosis and autophagy in HOS and HT1080 cells treated with ATO and MG-132 alone or in combination*

Cell cycle distribution and early apoptosis with Annexin V-FITC apoptosis detection kit were analyzed by flow cytometry. Quantitative results showed that ATO treatment induced a significantly prolonged G2/M arrest and consequently enhanced the percentage of apoptosis and autophagy in HOS and HT1080 cells. Combined treatment of ATO and MG-132 showed synergistic effect through induction of apoptosis and autophagy in HT1080 cells. By contrast, the occurrence of early apoptosis in HOS cells treated with ATO and/or MG-132 was low. Our results showed that the specific cleavage of PARP and caspase-3 could be found in cells treated with ATO, MG-132 alone or in combination in HT1080 cells. The combined treatment revealed a significant increase in AVOs compared to control groups in HT1080 cells. The ultrastructures of the HT1080 cells for each treatment group were observed by EM photomicrography. The combined treatment also resulted in a large number of autophagic vacuoles and autolysosomes in the cytoplasm. Furthermore, the expression levels of the LC3-II, p62, Atg5 and Atg5–12 proteins also increased with combined treatment in HT1080 cells.

2.3 *Measurement of ER stress and ubiquitin-proteasome degradation and related signaling pathways in HOS and HT1080 cells treated with ATO and MG-132 alone or in combination*

The ER stress-, apoptosis-, and autophagy-associated proteins expression were detected by immunoblotting. In addition, RNA interference technology with IRE1-targeted shRNA was also applied in this study. We found that ATO increased the level of ER stress-associated protein IRE1, whereas, ATO treatment down-regulated p-eIF2α protein levels, an ER stress protein. In addition, ATO caused inhibition of 20S proteasome activity in HOS and HT1080 cells. We also found that ATO significantly decreased the phosphorylation of Akt, mTOR and increased the phosphorylation of AMPK, p38, and JNK in HOS and HT1080 cells. Whereas, ATO inhibited the expression of ERK in HOS cells and increased the expression of ERK in HT1080 cells. Transfection with IRE1 shRNA significantly decreased cytotoxicity, autophagy and apoptosis compared with mock treatment in both sarcoma cells.

2.4 *Tumor growth in SCID mice suppressed by ATO and MG-132*

In *in vivo* study, therapeutic efficacy of ATO and proteasome inhibitor (MG-132) in HT1080 malignant fibrosarcoma xenografts after treatment was assessed. The results demonstrated that none of the treatment regimens produced any loss of body weight, which may be a sign of toxicity. The combined treatment suppressed tumor volume and tumor weight in SCID mice compared with ATO or MG-132 treatments alone. The combination therapy of ATO and MG-132 resulted in a tumor growth inhibition of more than 60%. Thus, the combination treatment possesses anti-tumor growth effect *in vivo*.

3 CONCLUSION

Our results indicate that ATO combined with MG-132 increases the therapeutic efficacy compared to individual treatments in HT1080 human sarcoma cancer cells. Specifically, ATO induced apoptosis and autophagy in HOS cells through inhibition of UPS, induction of ER stress, inhibition of Akt/mTOR and ERK signaling pathway and activation of JNK and p38 signaling pathway. ATO induced apoptosis and autophagy in HT1080 cells through inhibition of UPS, induction of ER stress, inhibition of Akt/mTOR and activation of ERK, JNK and p38 signaling pathway. In conclusion, this study demonstrated that a combination of ATO and proteasome inhibitor could be a new potential therapeutic strategy for the treatment of fibrosarcoma.

REFERENCES

Hait, W.N., Jin, S. & Yang, J.M. 2006. A matter of life or death (or both): understanding autophagy in cancer. *Clin. Cancer Res.* 12(7): 1961–1965.

Hsu, K.F., Wu, C.L., Huang, S.C., Wu, C.M., Hsiao, J.R., Yo, Y.T., Chen, Y.H., Shiau, A.L. & Chou, C.Y. 2009. Cathepsin L mediates resveratrol-induced autophagy and apoptotic cell death in cervical cancer cells. *Autophagy* 5(4): 451–460.

Kim, I., Xu, W. & Reed, J.C. 2008. Cell death and endoplasmic reticulum stress: disease relevance and therapeutic opportunities. *Nat. Rev. Drug Discov.* 7(12): 1013–1030.

Kondo, Y., Kanzawa, T., Sawaya, R. & Kondo, S. 2005. The role of autophagy in cancer development and response to therapy. *Nat. Rev. Cancer* 5(9): 726–734.

Mathew, R., Karantza-Wadsworth, V. & White, E. 2007. Role of autophagy in cancer. *Nat. Rev. Cancer* 7(12): 961–967.

Qian, W., Liu, J., Jin, J., Ni, W. & Xu, W. 2007. Arsenic trioxide induces not only apoptosis but also autophagic cell death in leukemia cell lines via up-regulation of Beclin-1. *Leuk. Res.* 31(3): 329–339.

Skubitz, K.M. & D'Adamo, D.R. 2007. Sarcoma. *Mayo Clin. Proc.* 82(11): 1409–1432.

Wang, J., Yang, L., Yang, J., Kuropatwinski, K., Wang, W., Liu, X.Q., Hauser, J. & Brattain, M.G. 2008. Transforming growth factor beta induces apoptosis through repressing the phosphoinositide 3-kinase/AKT/survivin pathway in colon cancer cells. *Cancer Res.* 68(9): 3152–3160.

One Century of the Discovery of Arsenicosis in Latin America (1914–2014) – Litter, Nicolli, Meichtry, Quici, Bundschuh, Bhattacharya & Naidu (Eds)
 ISBN 978-1-138-00141-1

Pregnancy and arsenic methylation: Results from a population-based mother-child cohort in Argentine Andes

F. Harari, K. Broberg & M. Vahter
Institute of Environmental Medicine, Karolinska Institutet, Stockholm, Sweden

ABSTRACT: Both pregnancy and genetics seem to affect the arsenic methylation phenotype. The present study aims at clarifying the importance of pregnancy in a cohort of women with a genetically determined efficient methylation of inorganic arsenic to dimethylarsinic acid (DMA). We analyzed arsenic metabolites in urine (inorganic arsenic, methylarsonic acid (MMA) and DMA) in 101 women exposed to arsenic through drinking water in northern Argentine Andes, using HPLC-HG-ICPMS. We found a pregnancy-related decrease in urinary %MMA of 1.7% during the second half of the first trimester, a minor decrease in inorganic arsenic, and an increase of 2.3% of %DMA. Three to six months after delivery %MMA had increased by 4.3%, compared to that during the first pregnancy trimester, possibly reflecting pre-pregnancy situations. Thus, pregnancy seemed to up-regulate arsenic methylation, resulting in decreased fetal exposure to MMA, the most toxic metabolite, during pregnancy.

1 INTRODUCTION

Inorganic arsenic (iAs) is metabolized by a series of reduction and methylation reactions, converting iAs to methylarsonic acid (MMA) and dimethylarsinic acid (DMA), which are excreted in urine (Vahter, 2002). There are major differences in the efficiency of arsenic methylation between individuals and population groups. Typically, people excrete 60–80% of the total arsenic in urine in the form of DMA, 10–20% as MMA, and 10–30% as iAs. Poor metabolizers have higher fractions of iAs and MMA (Vahter, 2002), which have been associated with higher risk of toxicity.

It has been observed that pregnancy influences the efficiency of arsenic methylation in Bangladeshi women (Gardner *et al.*, 2012). The %DMA in urine increased and %iAs and %MMA decreased during pregnancy, indicating a more efficient arsenic methylation. This is an important finding as it is likely to imply protection of the fetus against the toxic MMA metabolite.

In particular, arsenic (+III) methyltransferase (AS3MT), but also other methyltransferases, e.g. DNMT1 and N6 AMT1, are capable of methylating arsenic, and polymorphisms in these genes markedly influence arsenic metabolism (Engstrom *et al.*, 2011; Harari *et al.*, 2013).

The aim of this study is to elucidate the arsenic methylation phenotype during pregnancy in a cohort of women with a genetically determined efficient metabolism of iAs to DMA.

2 MATERIALS AND METHODS

2.1 *Study Population*

This study is part of an ongoing population-based mother-child cohort on health effects of early-life exposure to drinking water pollutants (i.e. arsenic, lithium and boron) in the Departamento Los Andes (~3800 m above sea level), Province of Salta, northern Argentina. Pregnant women were recruited (mean gestational week 26; range 7–39 weeks) and followed until six months after delivery. The women were interviewed about their personal characteristics, including age, parity, smoking, dietary preferences, diseases, drinking water sources, and coca chewing. Weight and height were measured in a standardized way and body mass index (BMI; kg/m^2) was calculated. Gestational Age (GA) was calculated based on date of the Last Menstruation Period (LMP). Ultrasound was used to calculate GA in case LMP was uncertain or unknown. A spot-urine sample was collected during each visit and kept frozen at –20° C until analysis.

2.2 *Arsenic exposure and metabolism*

Arsenic methylation efficiency was assessed based on the relative concentrations of arsenic metabolites in urine (%iAs, %MMA and %DMA) measured by high-performance liquid chromatography coupled with hydride generation and inductively coupled plasma mass spectrometry (HPLC-HG-ICPMS) at Karolinska Institutet, Sweden. Briefly, approximately 0.5 mL of each urine sample was filtered (0.20 μm) and transferred to the HPLC-HG-ICPMS system. The sum of urinary metabolites (total urinary arsenic; U-As) was used as exposure measure.

3 RESULTS AND DISCUSSION

3.1 *General characteristics*

One hundred and one women were included in this study. The general characteristics are presented in Table 1. The mean age was 25 years (range 13–41). None reported cigarette smoking, but 52% reported chewing coca.

3.2 *Arsenic metabolite patterns*

We observed a marked decrease in the %MMA in urine (Figure 1) during the second half of the first trimester (gestational weeks 7–13; $\beta = -0.36\%$ per week, 95% CI −0.73; 0.0077; $p = 0.055$), after controlling for total U-As, known to affect the methylation efficiency (Vahter, 2002). Adjusting for other co-variates did not change the estimates.

In the second and third trimesters the %MMA and DMA did not change much (%MMA mean: 3.9, range 1.7–7.4 in second trimester, and 3.7, range 2.0–7.4 in third trimester, $p = 0.37$) (Figure 2). The %MMA was on average 1.7% lower (95% CI −2.5; −0.91; $p < 0.001$) in late pregnancy compared to early pregnancy, while %DMA increased 2.3% (95% CI 0.29; 4.3; $p = 0.025$). However, after delivery the %MMA in urine increased markedly (Figure 2). One week to three months after delivery the %MMA had increased 2.1% (95% CI 1.3–3.0; $p < 0.001$) and 3–6 months after it had increased on average 4.3% (95% CI 3.4–5.3, $p < 0.001$, compared to %MMA in first trimester) (Figure 2). Whether the %MMA after delivery (Figure 2) represents pre-pregnancy data will be further elucidated.

Table 1. Characteristics of the participating women.

Characteristics	Mean, median or %	Range
Age (years, mean)	25	13–41
Parity (*n*, mean)	2.3	0–12
Coca chewing (%, yes)	52	–
BMI (kg/m^2, mean, 1st trimester)	24.8	24.0–25.4
U-As (μg/L, median, 1st trimester)	123	26–421

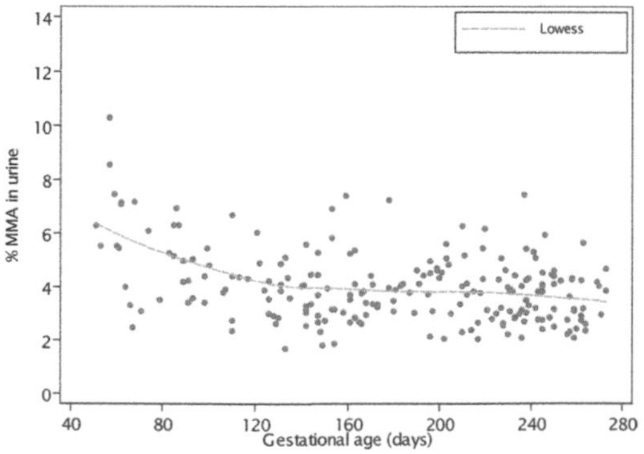

Figure 1. Scatter plot of %MMA in urine and gestational days. Lowess curve is represented as a dashed line.

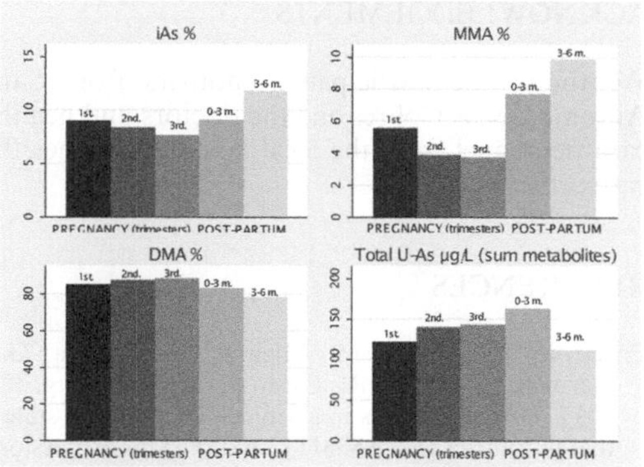

Figure 2. Graph bar showing the %iAs, %MMA, %DMA and total urinary arsenic (μg L^{-1}) during first, second and third trimesters of pregnancy, and 0–3 and 3–6 months after delivery.

The increased arsenic methylation efficiency during pregnancy might be due to regulatory processes of transcription or translation needed during pregnancy, e.g. up-regulation of methyltransferases (Gardner *et al.*, 2012), or up-regulation of one-carbon metabolism. Studies in a mother-child cohort in Bangladesh reported an even stronger decrease of %MMA during pregnancy (5.7% and up to 5.3%) when taking haplotypes of *AS3MT* and *DNMT1* into account (Gardner *et al.*, 2012). We will therefore evaluate the combined effect of pregnancy in the present cohort.

4 CONCLUSIONS

We observed that the efficiency of inorganic arsenic methylation increased markedly during pregnancy, especially in the first trimester. In particular, %MMA, the most toxic metabolite, decreased. These changes might be crucial for adequate fetal growth and development.

ACKNOWLEDGEMENTS

We thank the participating mothers from San Antonio de los Cobres and the doctors and health care personnel from the local hospital and health care centers.

REFERENCES

Engstrom, K., Vahter, M., Mlakar, S.J., Concha, G., Nermell, B., Raqib, R., Cardozo, A. & Broberg, K. 2011. Polymorphisms in arsenic(+III oxidation state) methyltransferase (AS3MT) predict gene expression of AS3MT as well as arsenic metabolism. *Environ Health Perspect* 119(2): 182–188.

Gardner, R.M., Engstrom, K., Bottai, M., Hoque, W.A., Raqib, R., Broberg, K. & Vahter, M. 2012. Pregnancy and the methyltransferase genotype independently influence the arsenic methylation phenotype. *Pharmacogenet Genomics* 22(7): 508–516.

Harari, F., Engstrom, K., Concha, G., Colque, G., Vahter, M. & Broberg, K. 2013. N-6-adenine-specific DNA methyltransferase 1 (N6 AMT1) polymorphisms and arsenic methylation in Andean women. *Environ Health Perspect* 121(7): 797–803.

Vahter, M. 2002. Mechanisms of arsenic biotransformation. *Toxicology* 181–182: 211–217.

One Century of the Discovery of Arsenicosis in Latin America (1914–2014) –
Litter, Nicolli, Meichtry, Quici, Bundschuh, Bhattacharya & Naidu (Eds)
© 2014 Taylor & Francis Group, London, ISBN 978-1-138-00141-1

Candidate single nucleotide polymorphisms associated with arsenic-induced Bowen's disease in Taiwan—a pilot study

Y.-H. Wang
Division of Urology, Department of Surgery, Taipei Medical University, Shuang Ho Hospital, New Taipei City, Taiwan
Graduate Institute of Clinical Medicine, College of Medicine, Taipei Medical University, Taipei, Taiwan

L.-I. Hsu & C.-J. Chen
Genomics Research Center, Academia Sinica, Taipei, Taiwan

ABSTRACT: Arsenic-induced skin cancer may be considered a long-term biomarker of cumulative arsenic exposure. Therefore, we conducted a genome-wide screening to discover novel genes associated with chronic arsenic-induced skin cancers in Taiwan. We included 83 patients with skin cancer and 83 age-, gender—and arsenic exposure level-matched subjects were selected as controls. Then, we conducted a case-control study consisting of 144 skin cancer patients and 631 cancer-free controls for replicating the candidate single nucleotide polymorphisms (SNPs) identified from the genome-wide screening. In a genome-wide screening, we selected four top genes including SYN3(synapsin III), LOXL1(Lysyl Oxidase-Like1), AUH (Au-Specific RNA-binding Protein) and SEC24D (an essential coat protein II) for genotyping. Two SNPs (SYN3 rs137537 G/T and SEC24D rs13127508 G/A) were significantly associated with the risk of skin cancer. In this pilot study, we identify two novel genes including SYN3 and SEC24D which were significantly associated with arsenic-associated skin cancers in Taiwan.

1 INTRODUCTION

Arsenic is a well-known carcinogen of various malignancies such as the skin, lung and bladder cancers. The unique manifestations of chronic arsenic exposure include hyperpigmentation, hyperkeratoses, Bowen's disease, and non–melanoma skin cancers. According to a survey in 1968, the prevalence per 1,000 persons for hyperpigmentation was 183.5; for hyperkeratosis, 71.0; and for skin cancer, 10.6. The significant relationship between arsenic exposure and skin lesions has been reported in Taiwan, Bangladesh, China, and India. Arsenic-induced skin lesions may be considered a long-term biomarker of cumulative arsenic exposure to arsenic. However, the mechanism of arsenic-associated skin lesions was still unclear. Although previous studies have investigated several arsenic-metabolized genes, their results were inconsistent. Therefore, we conducted a genome-wide screening to discover novel genes which may play an important role in chronic arsenic-induced skin cancers in Taiwan.

2 MATERIALS AND METHODS

First, we included a total of 83 patients with skin cancer in both southwestern and northeastern arsenicosis-endemic areas in Taiwan that were followed from 1998 to 2009. In addition, a total of 83 age-, gender—and arsenic exposure level-matched subjects were selected as controls. Second, we conducted a case-control study consisting of 144 skin cancer patients and 631 cancer-free controls for replicating the candidate Single Nucleotide Polymorphisms (SNPs) identified from the genome-wide screening using the Affymetrix Genome-Wide Human SNP Array 6.0. Genotyping of candidate SNPs were performed using Polymerase Chain Reaction-Restriction Fragment Length Polymorphism (PCR-RFLP) and Taqman SNP genotyping assays. The Cumulative Arsenic Exposure (CAE) of each subject was defined as $\Sigma(Ci \times Di)$, where Ci = the median arsenic level (mg/L) in well water of the ith village the subject had lived and Di = the duration (years) of drinking artesian water in the ith village. The effects (odds ratio, OR and 95%

confidence interval, CI) of candidate SNPs on skin cancer risk were evaluated using an unconditional logistic regression analysis adjusted for age, gender, smoking, sun exposure and arsenic exposure. The SAS software (Version 9.1) was applied to all statistical analyses. Values of $p<0.05$ were considered statistically significant.

3 RESULTS

There were 62.7% of skin cancer patients and controls were male. The cumulative arsenic exposure of 10–19 mg/L* years was 30.1% in skin cancer group as well as 37.4% in controls. In addition, the cumulative arsenic exposure of 20–29 mg/L* years was 28.9% in skin cancer group as well as 37.4% in controls. In a genome-wide screening, we identified 13 SNPs in 5 candidate genes. In this pilot study, we selected four top genes including SYN3(synapsin III), LOXL1(Lysyl Oxidase-Like1), AUH (Au-Specific RNA-binding Protein) and SEC24D(an essential coat protein II) for genotyping. Two SNPs (SYN3 rs137537 G/T and SEC24D rs13127508 G/A) were significantly associated with the risk of skin cancer. Subjects carrying G/G and G/T genotypes of SYN3 had a significantly increased risk of skin cancer (OR = 1.78, 95% CI = 1.07–3.10) comparing to those who carried the T/T genotype. As to SEC24D gene, a significantly increased skin cancer risk of 1.87 (95% CI = 1.20–3.01) was found for those with G/G and G/A genotypes.

4 CONCLUSION

In this pilot study, we identify several novel genes including SYN3, LOXL1, AUH and SEC24D which were significantly associated with arsenic-associated skin cancers in Taiwan. Candidate SNPs which located at these genes have to be replicated in a larger study to validate our preliminary findings. In addition, functional assays should be applied to discover the biological effects of these arsenic-related genes.

REFERENCES

Chen, C.J., Hsu, L.I., Wang, C.H., Shih, W.L., Hsu, Y.H., Tseng, M.P., Lin, Y.C., Chou, W.L., Chen, C.Y., Lee, C.Y., Wang, L.H., Cheng, Y.C., Chen, C.L., Chen, S.Y., Wang, Y.H., Hsueh, Y.M., Chiou, H.Y., Wu, M.M. 2005. Biomarkers of exposure, effect, and susceptibility of arsenic-induced health hazards in Taiwan. *Toxicology and Applied Pharmacology* 206(2): 198–206.

Guo, X., Fujino, Y., Kaneko, S., Wu, K., Xia, Y., Yoshimura, T. 2001. Arsenic contamination of groundwater and prevalence of arsenical dermatosis in the Hetao plain area, Inner Mongolia, China. *Molecular and Cell Biochemistry* 222(1–2): 137–140.

Haque, R., Mazumder, D.N., Samanta, S., Ghosh, N., Kalman, D., Smith M.M., Mitra, S., Santra, A., Lahiri, S., Das, S., De, B.K., Smith, A.H. 2003. Arsenic in drinking water and skin lesions: dose-response data from West Bengal, India. *Epidemiology* 14(2): 174–182.

Hsueh, Y.M., Chiou, H.Y., Huang, Y.L., Wu, W.L., Huang, C.C., Yang, M.H., Lue, L.C., Chen, G.S., Chen, C.J. 1997. Serum beta-carotene level, arsenic methylation capability, and incidence of skin cancer *Cancer Epidemiology, Biomarkers & Prevention* 6(8): 589–596.

Tondel, M., Rahman, M., Magnuson, A., Chowdhury, I.A., Faruquee, M.H., Ahmad, S.A. 1999. The relationship of arsenic levels in drinking water and the prevalence rate of skin lesions in Bangladesh. *Environmental Health Perspectives* 107(9): 727–729.

Tseng, W.P. 1977. Effects and dose-response relationships of skin cancer and blackfoot disease with arsenic. *Environmental Health Perspectives* 19: 109–119.

Tseng, W.P., Chu, H.M., How, S.W., Fong, J.M., Lin, C.S., Yeh, S. 1968. Prevalence of skin cancer in an endemic area of chronic arsenicism in Taiwan. *Journal of the National Cancer Institute* 40(3): 453–463.

One Century of the Discovery of Arsenicosis in Latin America (1914–2014) –
Litter, Nicolli, Meichtry, Quici, Bundschuh, Bhattacharya & Naidu (Eds)

The role of an alternative pathway in arsenic metabolism

A.V. Herrera, R.U. Hernández-Ramírez & L. López-Carrillo
Instituto Nacional de Salud Pública, Cuernavaca, Morelos, México

M.E. Cebrián
Departamento de Toxicología Centro de Investigación y Estudios Avanzados, México D.F, México

W. Klimecki & A.J. Gandolfi
College of Pharmacy, University of Arizona, Tucson, AZ, USA

ABSTRACT: Arsenic (As) naturally occurs in water and long-term exposure is linked with health problems, possibly influenced by interindividual variation in nutrient intake and genetic variation. Objective Examine As metabolism and elimination efficiency related to nutrient intake and genetic variants. Methods 415 healthy women from As-exposed regions of Northern Mexico were interviewed, their nutrient intake was estimated, their urinary As species were measured, and DNA genotyped. Results In gene-nutrient adjusted models, methionine was the only nutrient associated with increased As metabolism, particularly in the presence of BHMT. Conclusion Methionine may play a role in one-carbon As metabolism, particularly within certain genetic variant subgroups.

1 INTRODUCTION

Arsenic (As) naturally occurs in water and its long-term exposure to humans is linked with many health problems, including skin lesions and cancer (IARC 2012). As is metabolized through reduction and methylation reactions via one-carbon metabolism, typically through the folate pathway. If an alternative mechanism is needed, the choline pathway is used (Tseng, 2009) and is catalyzed by betaine-homocysteine methyltransferase (*BHMT*), an enzyme derived from a transformation of dietary choline to betaine via choline dehydrogenase (*CHDH*). Varied susceptibility to As-related health outcomes has been shown, and intraindividual variation in nutrient intake and genetic variation may affect As metabolism and elimination, influencing As-related health outcomes (Tseng, 2009). In humans, scant research has studied the nutrients or genetic variants in this alternative pathway. The aim of this study was to assess As elimination, regarding nutrient intake of choline, betaine, and methionine, and genetic variants in genes *CHDH* (rs7626693) and *BHMT* (rs3733890) within Mexican women exposed to As.

2 METHODS

2.1 *Study population*

415 randomly selected healthy women, ≥ age 20, from As-exposed areas of Northern Mexico participated. They were interviewed about demographic and lifestyle characteristics (age, state of residence, education level, smoking status, and alcohol consumption), and their BMI (kg m^{-2}) was calculated.

2.2 *Nutrient intake estimation*

Dietary intake was assessed using a validated semi-quantitative food frequency questionnaire, including both individual foods and regional dishes (Galván-Portillo *et al.*, 2011). Nutrient intake was energy adjusted.

2.3 *Arsenic methylation efficiency*

As species concentration (μg L^{-1}) of arsenite (As^{3+}), arsenate (As^{5+}), monomethylarsonic acid (MMA^{5+}), dimethylarsenic acid (DMA^{5+}), and arsenobetaine

(AsB) were measured with high-performance liquid chromatography with inductively coupled plasma source mass spectrometry from participants' first-morning urine samples. The following As methylation efficiency markers were calculated: iAs (As^{3+} + As^{5+}), total As (TAs; As^{3+} + As^{5+} + MMA^{5+} + DMA^{5+} + AsB), and total As-AsB (TAs-AsB; total inorganic As; As^{3+} + As^{5+} + MMA^{5+} + DMA^{5+}). As species (%iAs, %MMA, %DMA) and methylation ratios, which serve as proxies for methylation capacity, were also calculated (1st = MMA/iAs, 2nd = DMA/MMA, total = DMA/iAs).

2.4 *Polymorphism determination*

Genomic DNA was extracted from participants' blood samples with a semi-automated method. After assessing DNA purity and integrity, allelic discriminations of *CHDH* and *BHMT* were determined. Polymorphic sites were assay developed and genotyped with the 5'-exonuclease-based Taqman assay. Finally, allelic discrimination software designated genotypes. The majority of samples were successfully genotyped (388 for *CHDH* and 387 for *BHMT*).

3 RESULTS

3.1 *Characteristics of the study population*

The average age was 53.7 years and 46.3% of participants were obese (≥30.0 BMI). Average caloric intake was 1,923.2 Kcal day^{-1}. More than half of participants completed primary schooling (60.7%) and the majority of participants (85.4%) did not smoke or consume alcohol (90.8%). Average energy-adjusted intake and percentage of participants who fell short of nutrient RDI for choline, betaine, and methionine (mg/day) was 255.4 (95.4%), 21.9 (79.0%), and 1.2 (29.9%), respectively. Genotype and allele frequencies of *CHDH* were in Hardy-Weinberg's equilibrium (p = 0.12, n = 388 due to missing values), but those for *BHMT* were not (p = 0.01, n = 387 due to missing values), and linkage disequilibrium was not demonstrated between the two genes (D' = 0.0, r = 0.0).

3.2 *Urinary As parameters in study population*

Average total As-AsB (μg L^{-1}) was 22.4 (range: 0.6–223.7). Geometric means (μg/L) of %iAs, % MMA, and %DMA were 8.9, 9.4, and 79.0, respectively. Average creatinine was 62.9 mg dL^{-1}.

3.3 *Multivariate analysis*

In gene-nutrient adjusted models, methionine was the only nutrient associated with increased As metabolism. In the presence of *BHMT*, methionine was significantly associated with decreased %iAs, increased %DMA, and increased total methylation ratio. Only *BHMT* AG genetic variants showed significant associations with As markers: decreased %iAs, decreased %MMA, increased %DMA, increased 2nd methylation ratio, and increased total methylation ratio. For *BHMT* in the dominant genetic model (GG + AG), methionine

Table 1. Urinary As parameters in study population (n = 415).

Urinary As parameter	
Total As-AsB (μg L^{-1}) ($\bar{x}$ ± SD)	22.4 ± 29.3
0.6–5.8 (n = 83)	3.6 ± 1.3
5.8–10.1 (n = 83)	7.7 ± 1.4
10.2–15.9 (n = 83)	13.0 ± 1.7
15.9–29.4 (n = 83)	21.4 ± 3.9
29.4–223.7 (n = 83)	66.3 ± 41.1
Arsenic species (Geometric Mean ± Geometric SD)	
%iAs (As^{5+} + As^{3+})	8.9 ± 1.7
%MMA	9.4 ± 1.6
%DMA	79.0 ± 1.2
Methylation ratios (Geometric Mean ± Geometric SD)	
1st (MMA/iAs)	1.1 ± 1.7
2nd (DMA/MMA)	8.4 ± 1.7
Total (DMA/iAs)	8.8 ± 1.9

Table 2. Multivariate regression for nutrients & genes (n = 415).

	Proportion			Methylation ratios		
*p < 0.05	%iAs	%MMA	%DMA	1st	2nd	Total
Choline (mg/day)						
Model A[1]	−0.0	−0.0	0.0	−0.0	0.0	0.0
Model B[2,4]	−0.0	−0.0	0.0	−0.0	0.0	0.0
Model C[3,5]	−0.0	−0.0	0.0	−0.0	0.0	0.0
Betaine (mg/day)						
Model A[1]	0.2	−0.0	−0.0	−0.2	−0.0	−0.2
Model B[2,4]	0.0	−0.1	−0.0	−0.2	0.1	−0.1
Model C[3,5]	0.2	−0.0	−0.0	−0.2	−0.0	−0.3
Methionine (mg/kg/day)						
Model A[1]	−0.3*	−0.1	0.0	0.2	0.2	0.3*
Model B[2,4]	−0.3	−0.2	0.0	0.1	0.2	0.3
Model C[3,5]	−0.3*	−0.1	0.1*	0.2	0.2	0.4*
Interaction p	0.01	0.04	0.02	0.36	0.03	0.01

[1]Adjusted by energy.
[2]Adjusted by *CHDH*, energy, age, creatinine, TAs-AsB, BMI.
[3]Adjusted by *BHMT*, energy, age, creatinine, TAs-AsB, BMI.

Table 3. Methionine regression coefficients by *BHMT*.

	Proportion			Methylation ratios		
*$p < 0.05$	%iAs	%MMA	%DMA	1st	2nd	Total
BHMT[1]						
GG ($n = 40$)	−0.8	−0.0	0.1	0.8	0.1	0.9
AG ($n = 204$)	−0.6*	−0.5*	0.1*	0.2	0.6*	0.8*
AA ($n = 143$)	0.1	0.2	−0.0	0.1	−0.3	−0.2
GG + AG	−0.6*	−0.4*	0.1*	0.3	0.5*	0.7*
AG + AA	−0.3*	−0.2	0.1	0.2	0.2	0.4*
Energy	0.000*	0.000	−0.000*	−0.000*	−0.000	−0.000*
Age	−0.003	0.001	0.001	0.004	−0.000	0.004
Creatinine	0.001	0.001*	−0.000	0.001	−0.002*	−0.001
TAs-AsB	−0.000	−0.000	−0.000	0.000	−0.000	0.000
BMI	−0.013*	−0.017*	0.004*	−0.004	0.021*	0.017*

[1]Adjusted by energy, age, creatinine, TAs-AsB, BMI.

was significantly associated with decreased %iAs, decreased %MMA, increased %DMA, increased 2nd methylation ratio, and increased total methylation ratio. For *BHMT* in the recessive genetic model (AG + AA), methionine was significantly associated with decreased %iAs and increased total methylation ratio.

4 CONCLUSIONS

Methionine may influence one-carbon As metabolism, particularly within certain genetic variant subgroups. Further studies must evaluate the feasibility of dietary interventions to modulate As metabolism.

ACKNOWLEDGEMENTS

This study was supported by CONACyT, Fondo Sectorial de Investigación en Salud y Seguridad Social (2005-2-14373, 2009-1-111384, 2010-1-140962), and Dean Carter Binational Center for Env Health Sciences (NIH ES04940).

REFERENCES

Galván-Portillo, M.V. *et al.*, 2011. Cuestionario de frecuencia de consumo de alimentos para estimación de ingestión de folato en México, 53(655): 237–246.

IARC (International Agency for Research on Cancer). 2012. *A Review of Human Carcinogens: Arsenic, Metals, Fibres, and Dusts.* Lyon: World Health Organization Press.

Tseng, C.H., 2007. Arsenic methylation, urinary arsenic metabolites and human diseases: current perspective. *J. Environ. Sci. Health C Environ. Carcinog. Ecotoxicol. Rev.* 25: 1–22.

One Century of the Discovery of Arsenicosis in Latin America (1914–2014) – Litter, Nicolli, Meichtry, Quici, Bundschuh, Bhattacharya & Naidu (Eds)
 ISBN 978-1-138-00141-1

Clinical and biochemical features among subjects exposed to arsenic through drinking water

A.G. Soria, R.S. Guber, N.G. Sandoval, M.I. Fernández & C. Baca
Facultad de Bioquímica, Química y Farmacia, Universidad Nacional de Tucumán, Tucumán, Argentina

L.M. Tefaha, M.D. Martínez, R. Toledo, A. Liatto & C. Romero
Facultad de Medicina, Universidad Nacional de Tucumán, Tucumán, Argentina

ABSTRACT: The aim of this study was to determine the association between the chronic exposure to arsenic (As) through drinking water and the clinical and biochemical alterations in the rural population of Tucuman. The sampling included 188 exposed individuals to high As values in drinking water (EG) and 207 individuals no exposed (CG). A physical examination was performed and serum levels of cholesterol, triglyceride, glucose and sialic acid were determined. The 38.4% and the 0% of the studied patients for the EG and CG presented dermatological lesions. The sialic acid was significantly elevated in the EG (mean ± SD) 856.7 ± 232.1 compared to 725.6 ± 134.0 mg L^{-1} in the CG. Our study suggests an association between drinking water with higher arsenic contents and skin lesions. A positive association in the EG with hypertension was found, but not with diabetes or dyslipidemia. Serum sialic acid could serve as an estimate of exposition to high As levels.

1 INTRODUCTION

In the east of the province of Tucuman, the chronic arsenic exposure constitutes a serious problem of public health. The main route of Arsenic (As) exposure is through water contaminated with high concentration of this metalloid, which implies a large population potentially affected to its deleterious effect (Guber *et al.*, 2009). As is metabolized in the liver, where it is transformed to monomethylated and dimethylated arsenicals and then excreted into urine (Marchiset-Ferlay *et al.*, 2012). The chronic As ingestion higher than 0.01 mg/L can lead to trigger the development of Chronic Endemic Regional Hydroarsenicism (in Spanish, *Hidroarsenicismo Crónico Regional Endémico*, HACRE) or arsenicosis, characterized for the progressive cutaneous alterations (hyperhidrosis, hyperkeratosis and melanodhermia) (Sengupta *et al.*, 2008). Hepatic disorders are frequently manifested, like hepatomegaly, cirrhosis with or without hypertension portal, hepatocellular carcinoma and angiosarcoma (Islam *et al.*, 2011). Several epidemiologic studies suggests that the chronic exposition to the As increments the risk of diabetes and cardiac diseases like arterial hypertension, neurologic complication and hematological disorders in the exposed groups (Guha *et al.*, 2008). According to the International agency for Research on Cancer, arsenic is an important carcinogen. The association of water with As consumption and skin, bladder, liver or lung cancers has been well documented (Anetor *et al.*, 2007); the association between prostate cancer and chronic As consumption through drinking water is controverted (Soria *et al.*, 2009, 2011). The aim of this work was to determine the association between the chronic exposure to As in drinking water and HACRE) described in the for the rural population of the Tucuman province (Argentina).

2 METHODS

2.1 *Studied populations*

A descriptive transversal study was performed. As inclusion criteria, the study included 395 rural adults inhabitants aged ≥30 years old and settled in in the area of study for more than 5 years (Graneros department, Tucumán, Argentina). The exposed population (EG) involved 188 exposed volunteers (30–83 years old) and the Control Group (CG), 207 individuals (30–86 years old). Individuals with acute or chronic diseases, e.g. hepatitis or another known hepatic disease or diabetes type 1 were not included in this study. All the objectives of this research were explained, and a consent has to be approved by the population interested to participate.

2.2 *Clinics and demographics variables and biochemical assays*

Population were interviewed following a standard questionnaire aiming to collect information about demographic data such as age, sex, years of residence and drinking water patterns, consumption of tobacco, alcohol and drugs, previous diseases. A physical examination was made by a medical professional, evaluating the dermatology signs of HACRE. A biochemical panel of dyslipidemia that included serum level of cholesterol and triglyceride and glucose serum were done according to the manufacturer's protocol (Wiener Lab). The serum level of sialic acid was determined through a modified Warren method. Arsenic in domestic well water samples was determined by the Gutzeit method. The results were analyzed by SPSS software and $p < 0.05$ was considered significant.

3 RESULTS

The average concentration of As in drinking water consumed by the EG was 0.165 ± 0.313 mg L^{-1}. The concentration range was 0.011 a 3.160 mg L^{-1}. The 38.4% and the 0% of the studied patients for the EG and CG presented dermatological lesions, respectively. In Table 1, the demographic characteristics and the variables in the groups can be observed. The mean ± SD of sialic acid were 856.7 ± 232.1 and 725.6 ± 134.0 mg L^{-1} for EG and CG, respectively. Sialic acid levels were significant ($p < 0.001$). The positive association in the EG with hypertension (Chi-square: $P < 0.001$, RR: 4.6) is observed. A significant association of the EG with diabetes or dyslipidemia was not found.

Table 1. Demographic characteristics and other variables of the study subjects.

Variables	EG % (IC95%)	CG % (IC95%)
Education		
Elementary	87.2	87.5
Middle	11.7	10.1
High	1.1	2.4
Smoking		
No smoker	32.8 (3.6–62)	44.2 (20.0–68.5)
Current smoker	67.2 (38.1–96.3)	55.8 (31.5–80.1)
Alcohol		
No drinker	16.3 (0–39.2)	28.3 (6.7–50)
Regular drinker	83.7 (60.8–100)	71.7 (50.1–93.3)
Dyslipidemia		
No	59.9 (38.4–81.4)	50.0 (28.8–71.2)
Yes	40.1 (18.6–61.6)	50.0 (28.8–71.2)
Hypertension		
No	42.5 (17.2–67.8)	76.3 (57.6–95)
Yes	57.5 (32.2–82.8)	23.7 (5.0–42.4)
Diabetes		
No	67.4 (34.4–100)	76.6 (52–100)
Yes	32.6 (0–65.6)	23.4 (0–47.7)

EG: Exposed Group; CG: Control Group.

4 CONCLUSIONS

The results revealed that the chronic exposure to As in drinking water produced in the study population a high prevalence of dermatological manifestations related to arsenic exposure. An association between exposure and hypertension was found; this could be enhanced with the high percentage of alcohol consumption and smoking habit. The biochemical changes indicate a high percentage of dyslipidemia, but this could not be related with As exposure, as well as an effect of the As in the metabolism of the glucose. Sialic acid has been reported as a marker of acute marker response because increased concentrations have been observed in several pathologies such as diabetes, tumors and alcoholism. Interestingly, serum sialic acid could serve as an As exposure estimator. our study suggests an association between consumption of drinking water with high arsenic levels and skin lesions, hypertension, but not with dyslipidemia and diabetes.

Besides, it must be taken into account that the exposure occurs in depressed areas of the northwest of Argentina, with a low level of instruction, and this explains the lack of knowledge of the harmful effects of the consumption of drinking water with As, due to an inadequate safe water provision. In the east of the province of Tucuman, a failure to understand the arsenic problem takes place by the lack of coordination between the health sector and drinking water supply authorities. This is a public health problem, however, from a public-health perspective. Therefore, effective intervention strategies need to be developed to establish a causal association between As exposure and adverse health effects (skin lesions, chronic diseases as cardiovascular disease, hypertension, diabetes and cancers). Epidemiological, clinical and biochemical studies are needed to diagnose precocious pathologies related to chronic As ingestion. Furthermore, the health sector should be instructed to strengthen rapid diagnostic facilities in endemic rural areas of HACRE. Finally, strategies should be implemented to diminish the risk.

REFERENCES

Anetor, J.I., Wanibuchi, H. & Fukushima, S. 2007. Arsenic exposure and its health effects and risk of cancer in developing countries: micronutrients as host defence. *Asian Pac. J. Cancer Prev.* 8: 13–23.

Guber, R.S., Tefaha, L., Soria, A., Arias, N., Sandoval, N., Toledo, R., Fernandez, M., Bellomio, C., Martinez, M. & Soria de Gonzalez, A. 2009. Levels of arsenic in the drinking water in Leales and Graneros (Tucuman-Argentina). *Acta Bioquimica Clinica Latinoamericana* 43: 201–207.
Guha, M. 2008. Chronic arsenic toxicity & human health. *Indian J. Med. Res.* 128: 436–447.
Islam, K., Haque, A., Karim, R, Fajol, A., Hossain, E., Salam, K.A., Ali, N., Saud, Z.A., Rahman, M., Rahman, M., Karim, R., Sultana, P., Hossain, M., Akhand, A.A., Mandal, A., Miyataka, H., Himeno, S. & Hossain, K. 2011. Dose-response relationship between arsenic exposure and the serum enzymes for liver function tests in the individuals exposed to arsenic: a cross sectional study in Bangladesh. *Environ. Health* 10: 64–75.
Marchiset-Ferlay, N., Savanovitch, C. & Sauvant-Rochat, M.P. 2012. What is the best biomarker to assess arsenic exposure via drinking water? *Environ Int.* 39: 150–171.
Sengupta, S.R., Das, N.K. & Datta, P.K. 2008. Pathogenesis, clinical features and pathology of chronic arsenicosis. *Indian J. Dermatol. Venereol. Leprol.* 74: 559–570.
Soria, A., Guber, R.S. Martínez, M., Arias, N., Tefaha, L., Sandoval, N., Fernandez, M. & Toledo, R. 2009. Biochemical alterations in individuals consuming high levels of arsenic in drinking water in Tucuman, Argentina. *Acta Bioquimica Clinica Latinoamericana* 43: 611–618.
Soria, A., Tefaha, L., Guber, RS., Arias, N., Romero, C., Martínez, M., Valdivia, M., Sandoval, N., Toledo, R. & Czejack, M. 2011. Total prostatic specific antigen levels among subjects exposed and not exposed to arsenic in drinking water. *Rev. Med. Chile* 139: 1592–1598.

One Century of the Discovery of Arsenicosis in Latin America (1914–2014) – Litter, Nicolli, Meichtry, Quici, Bundschuh, Bhattacharya & Naidu (Eds)
© 2014 Taylor & Francis Group, London, ISBN 978-1-138-00141-1

Environmental investigation into a suspected chronic arsenic poisoning case in Far North Queensland, Australia

R. Ware & I. Florence
Environmental Health Officers, Cairns Public Health Unit, Cairns, Queensland, Australia

ABSTRACT: Cairns Public Health Unit, Queensland Health was approached by medical professionals to undertake an environmental assessment of a patient's property after tests confirmed that the patient was suffering from arsenic poisoning. Environmental Health Officers visited the patient's property during the dry season and obtained samples of water, vegetables and soil. The patient utilises water from the river for irrigation and domestic use without any treatment. Sampling results indicated that the levels of arsenic detected within the patient's drinking water and soil were below Australian national guidelines.

1 INTRODUCTION

In November 2011 the Cairns Public Health Unit (CPHU), Queensland Health was advised of a medically confirmed case of arsenic poisoning. CPHU were notified that the diagnosis was based on the patient's symptoms of lethargy, progressive peripheral neuropathy, hyperkeratosis on the forearms and that the patients hair had turned from white to golden. The patient also had an urine arsenic of 0.922 μmol/mmol creatinine. A request was made to the patient to undertake further tests on hair and nail samples; however the patient refused the tests. A request was made by medical professionals to undertake an environmental assessment of the patient's property.

Environmental Health Officers visited the patient's property which is located on the Starcke river, approximately 140 km Northwest of Cooktown, where he spends the majority of his time. It is reported that the patient has lived in this area for an extended period, thought to be in the vicinity of 25 years and has participated in gold mining practices during that time. The patient has always utilised the river water for domestic purposes including consumption and irrigation. The patient occasionally visits larger towns in the area that are on a treated, reticulated water supply. These supplies are regularly tested and no arsenic has been detected.

Far North Queensland has been historically heavily mined for tin, copper, gold, arsenic, bauxite, bismuth, cinnabar, lead, magnesium, molybdenum, nickel, phosphorus, rutile, silver, uranium, wolfram and zinc. Other resources are also mined such as coal, oil, gas and gemstones (Mate, 2010).

The area where the patient lives was a short lived gold mining area which yielded 1108oz of gold bullion from 1139 tons of ore between 1892 and 1896 (www.mines.industry.qld.gov.au).

Inorganic arsenic occurs naturally in soil and in many kinds of rock, especially in minerals and ores that contain copper or lead. Arsenic is mainly obtained as a byproduct of the smelting of copper, lead, cobalt, and gold ores (Anon, 2007).

2 METHODS

Samples of water, soil and vegetables were taken from the patient's property November 2011. The vegetables collected from the patients property included silver beet, rocket, vine pumpkin, bush tomato and tree citrus which are irrigated using water directly from the river.

All environmental samples were stored and transported under temperature control. All environmental samples were sent to the Queensland Health Forensic Scientific Services a NATA accredited laboratory in Brisbane for arsenic analysis. Speciation of any arsenic detected was not requested.

3 RESULTS

The water samples were found to contain traces of arsenic with the highest detection found in the patient's dam at 0.002 mg L^{-1}. A measurement of 0.001 mg L^{-1} was detected in the water at the patients kitchen sink. These measurements were assessed against the Australian Drinking Water Guidelines (ADWG 2011) which set a limit of 0.01 mg L^{-1} for arsenic in drinking water. Another well, located approximately ¾ km east of the patients property was also sampled where the arsenic level was found to be at 0.26 mg L^{-1}. Environmental Health Officers were informed that this well is not utilised.

The results from the vegetables found no detectable levels of arsenic with the exception of the rocket which was found to have an arsenic level of 0.06 mg kg^{-1}. The soil sample was found to have a level of arsenic at 6 mg kg^{-1}. This level was compared to the National Environment Protection (Assessment of Site Contamination) Measure 1999 which has a health investigation level of 100 mg kg^{-1} for a standard residential property.

4 DISCUSSION

The environmental samples taken at that time indicate that the source of the patient's arsenic poisoning could not be directly attributed to the patient's immediate environment. However the results of the water sample taken from a nearby well revealed arsenic levels of 0.26 mg L^{-1} present in groundwater in the vicinity of the property. This suggests that there are levels of arsenic present in groundwater in the area that exceed the ADWG.

The water samples were taken at the end of the dry season and it is possible that higher levels of arsenic may be detected during the wet season due to the movement of water tables. As the patient extracts domestic water directly from the river and stores it in a dam, it must be considered that contaminated water from the previous wet season was being consumed by the patient well into the dry season prior to the patient being tested. Due to this possibility it is intended that further water and environmental samples will be obtained during the current wet season to determine if there are any seasonal variations of river contaminants.

Environmental Health Officers from CPHU will continue to investigate the environmental source of the arsenic and will work with the patient and medical professionals.

REFERENCES

ADWG 2011. *Australian Drinking Water Guidelines - National Water Quality Management Strategy*. The National Health and Medical Research Council Canberra.

Anon 2007. *Toxicological Profile for Arsenic*. Agency for Toxic Substances and Disease Registry U.S. Department of Health and Human Services.

Environment Protection and Heritage Council 1999. NEPM: *Assessment of Site Contamination, Schedule B(1) Investigation Levels for Soil and Groundwater*.

Mate,G., 2010. Mining. www.qhatlas.com.au/content/mining.

National Health and Medical Research Council 2011. *Australian Drinking Water Guidelines 6, Volume 1*. http://mines.industry.qld.gov.au/geoscience Gold Occurrences in far North Queensland: Extracts from Cairns, Atherton, Mossman, Cape Weymouth, Torres Strait, Coen, Cape Melville, Ebagoola, Cooktown, and Hann River Explanatory Notes.

One Century of the Discovery of Arsenicosis in Latin America (1914–2014) – Litter, Nicolli, Meichtry, Quici, Bundschuh, Bhattacharya & Naidu (Eds)
© 2014 Taylor & Francis Group, London, ISBN 978-1-138-00141-1

Cell-type-dependent associations of heme oxygenase-1 GT-repeat polymorphisms with the cancer risk in arsenic-exposed individuals: A preliminary report

M.M. Wu & W.F. Cheng
Graduate Institute of Oncology, National Taiwan University, Taipei, Taiwan

L.-I. Hsu & C.-J. Chen
Genomics Research Center, Academia Sinica, Taipei, Taiwan

H.-Y. Chiou
School of Public Health, Taipei Medical University, Taipei, Taiwan

T.C. Lee
Institute of Biomedical Sciences, Academia Sinica, Taipei, Taiwan

ABSTRACT: Heme oxygenase (HO)-1 is highly up-regulated by many stressful stimuli, including arsenic. A GT-repeat polymorphism in the HO-1 gene promoter can inversely modulate the levels of HO-1 induction. Long-term digestion of arsenic-contaminated drinking water may cause cancers of anatomic variety. We carried out this study to examine the correlation of HO-1 GT-repeat polymorphism with cancer risk in arsenic-exposed individuals during a 10-year follow-up. A total of 1,004 participants who had HO-1 genotyping available from two arsenicosis-endemic areas in Taiwan were included. Baseline characteristics derived from questionnaire interview were collected in 1988–1990 for the subjects of the BFD-endemic area and in 1996–1998 the subjects of the Lanyang Basin area. Allelic GT-repeats were categorized into short (S, <27 GT-repeats) and long (L, ≥27 GT-repeats). Analysis results showed that there were no significant differences in the risk of urinary tract cancer among the three genotype groups after accounting for age, gender, average arsenic exposure, smoking status, and alcohol intake. However, for cancer of trachea bronchus and lung, carriers of L/S genotype versus L/L group had significantly reduced risk [HR 0.34 (95% CI, 0.13–0.95), P = 0.039] while restricting subgroup of study subjects to arsenic levels >150 μg/L-year. Carriers of S/S genotype had a significantly increased risk of non-melanoma skin cancer [HR 3.05 (95% CI, 1.24–7.48), P = 0.015] as compared to those who carried L/L genotype for the entire subjects. Regarding liver and intrahepatic bile ducts cancer, carriers of S/S genotype also had a significantly increased risk [HR 4.14 (95% CI, 1.30–13.21), P = 0.016] when compared to those carried L/S genotype. We provided evidence that HO-1 gene variants may modify the carcinogenic effect of arsenic in a cell-type-dependent pattern.

1 INTRODUCTION

Heme oxygenase (HO)-1 is highly upregulated by many stressful stimuli, including arsenic. A GT-repeat length polymorphism in the HO-1 gene promoter can inversely modulate the levels of HO-1 induction. In contrast to large body of studies in support of cardiovascular protection effect, the results of such investigations for cancer have been inconsistent. Long-term digestion of arsenic-contaminated drinking water may cause vascular-related diseases and cancers of anatomic variety. The objective of this study was to examine the correlation of HO-1 GT-repeat polymorphism to cancer risk in arsenic-exposed individuals. Our study results should have implications in understanding the disease mechanisms and risk assessment of arsenic in drinking water.

2 MATERIALS AND METHODS

The study subjects consisted of 1,004 participants who were enrolled at baseline and had HO-1 genotyping available from two arseniasis-endemic areas in Taiwan: the Blackfoot Disease (BFD)-endemic area and the Lanyang Basin area. Baseline characteristics derived from questionnaire interview were collected in 1988–1990 for the subjects of the BFD-endemic area and in 1996–1998 the subjects of the Lanyang Basin area. Average arsenic exposure was calculated by dividing cumulative arsenic exposure by the years of consuming well water during a subject's lifetime. Users of cigarette smoking or alcohol consumption were defined at a frequency of at least 3 days a week for at least a half year. The number of GT-repeats in the HO-1 gene promoter was determined as described in a previous report.

Identification of cancer cases was determined through record linkage with the Taiwan Cancer Registry profiles. Follow-up person-years for each subject were counted from the date of questionnaire interview to the date of cancer diagnosis, death, or December 31, 2006, whichever came earliest. Identification of cancer site for cases was determined by the presence of ICD ninth revision, clinical modification (ICD-9-CM) for the cancer of urinary tract (188.0–188.9, 189.1–189.9), trachea bronchus and lung (162.0–162.9), liver and intrahepatic bile ducts (155.0–155.2), and skin, non-melanoma (173).

Allelic GT-repeats in the HO-1 gene promoter were categorized into short (S, <27 GT-repeats), and long (L, ≥27 GT-repeats). Subjects were thereafter grouped as carriers of L/L (n = 285), L/S (n = 503), and S/S (n = 216) genotype. Cox proportional hazard method was used to evaluate the independent association of HO-1 genotype with individual cancer risk. To examine the impact of S alleles at high levels of arsenic exposure, the analysis were repeated by restricting study subjects to those who had been exposed to arsenic at levels of >150 μg/L-year (56% of total subjects).

3 RESULTS

During a median 8.9, 10.1, 8.1, and 9.2 years of follow-up, there were 36, 35, 18, and 38 newly diagnosed cases of the urinary tract cancer, trachea bronchus and lung cancer, liver and intrahepatic bile ducts cancer, and non-melanoma skin cancer, respectively. For urinary tract cancer, the respective crude incidence rates (per 100,000 person-years) in carriers of L/L, L/S, and S/S genotype were 293.8, 258.2, and 286.9. The respective incidence rates for trachea bronchus and lung cancer were 291.5, 225.2, and 285.1. The corresponding figures were 158.8, 75.0, and 249.8 for liver and intrahepatic bile ducts cancer, and 212.5, 197.7, and 625.1 for non-melanoma skin cancer. Analysis results showed that there were no significant differences in cancer risk of urinary tract or trachea bronchus and lung cancer among the three genotype groups after the adjustment of age, gender, arsenic exposure, smoking status, and alcohol intake. However, carriers of L/S genotype had significantly reduced risk of trachea bronchus and lung cancer [HR 0.34 (95% CI, 0.13–0.95), P = 0.039] as compared to those of L/L group when the analyses were limited to subjects with arsenic exposure of >150 μg/L-year. In contrast, carriers of S/S genotype had a significantly increased risk of non-melanoma skin cancer [HR 3.05 (95% CI, 1.24–7.48), P = 0.015] when compared to those who carried L/L genotype for the entire study subjects. Regarding liver and intrahepatic bile ducts cancer, carriers of S/S genotype also had a significantly increased risk [HR 4.14 (95% CI, 1.30–13.21), P = 0.016] as compared to those of L/S genotype.

4 CONCLUSIONS

We conclude that heterozygous short GT-repeats of HO-1 gene promoter conferred protection in reducing trachea bronchus and lung cancer risk; homozygous short GT-repeats increased the risk of non-melanoma skin cancer. The HO-1 gene variants may modify the carcinogenic effect of arsenic in a cell-type-dependent pattern.

REFERENCES

Abraham N.G. & Kappas, A. 2008. Pharmacological and clinical aspects of heme oxygenase. *Pharmacological Reviews* 60(1):79–127.

Wu, M.M., Chiou, H.Y., Chen, C.L., Wang, Y.H., Hsieh, Y.C. & Lien, L.M 2010. GT-repeat polymorphism in the heme-oxygenase-1 gene promoter is associated with cardiovasclar mortality risk in an arsenic-exposed population in northeastern Taiwan. *Toxicology and Applied Pharmacology* 248(1):226–233.

Calabrese, E.J. & Baldwin, L.A.. 2001. Hormesis: U-shaped dose responses and their centrality in toxicology. *Trends in Pharmacological Sciences* 22(6):285–291.

3.2 Indicators of exposure and biomarkers

One Century of the Discovery of Arsenicosis in Latin America (1914–2014) –
Litter, Nicolli, Meichtry, Quici, Bundschuh, Bhattacharya & Naidu (Eds)
 ISBN 978-1-138-00141-1

Molecular and genomic biomarkers of arsenic-induced health hazards: Gene-environment interactions

C.-J. Chen
Genomics Research Center, Academia Sinica, Taipei, Taiwan

ABSTRACT: Humans are exposed to ingested and inhaled arsenic through the environment, occupation, diet and medicines. Long-term exposure to arsenic causes systemic health hazards including characteristic skin hyperpigmentation or depigmentation, hyperkeratosis in palms and soles, Bowen's disease, circulatory diseases, goiter, diabetes mellitus, cataract, pterygium, neurological disorder, retarded development, and cancers of the skin, lung, urinary bladder, kidney and liver. Many molecular and genomic biomarkers for internal dose and biologically effective dose of exposure, early biological effects, preclinical lesions, and genetic and acquired susceptibility to arsenic-caused diseases have been developed and validated. These biomarkers are useful for the molecular dosimetry of arsenic exposure from various sources, the prediction and early detection of arsenic-induced diseases, the prevention or intervention of development of end-stage diseases caused by arsenic. Variation in individual susceptibility to chronic arsenic poisoning is determined by gene-environment interactions.

1 INTRODUCTION

Humans are exposed to ingested and inhaled arsenic through environmental, occupational, dietary and medicinal exposures.

2 SPECTRUM OF ARSENIC-CAUSED HEALTH HAZARDS

Long-term exposure to arsenic causes systemic health hazards including characteristic skin hyperpigmentation or depigmentation, hyperkeratosis in palms and soles, Bowen's disease, peripheral vascular disease, ischemic heart disease, cerebral infarction, microvascular diseases, abnormal peripheral microcirculation, carotid atherosclerosis, QT prolongation and increased dispersion in electrocardiography, hypertension, goiter, diabetes mellitus, cataract (specifically posterior subcapsular lens opacity), pterygium, slow neural conduction, retarded neurobehavioral development, erectile dysfunction, and cancers of the skin, lung, urinary bladder, kidney and liver. Dose-response relations have been observed between arsenic exposure levels and risk of arsenic-induced health hazards.

3 MOLECULAR AND GENOMIC BIOMARKERS OF ARSENIC EXPOSURE AND HEALTH EFFECT

Several molecular and genomic biomarkers have been developed and validated to measure the internal dose and biologically effective dose of exposure to arsenic, to assess early biological effects of arsenic, to elucidate acquired and genetic susceptibility to arsenic-induced health hazards. These biomarkers are useful for the molecular dosimetry of arsenic exposure from various sources, the prediction and early detection of arsenic-induced diseases, the prevention or intervention of development of end-stage diseases caused by arsenic.

4 INTERNAL DOSE AND BIOLOGICALLY EFFECTIVE DOSE

Urinary levels of inorganic and inorganic arsenic may be used as the internal dose of exposure to ingested and inhaled arsenic. The urinary level of Monomethylarsonic Acid (MMA) or the percentage of MMA in the total metabolites of inorganic arsenic including arsenite, arsenate, MMA and dimethylarsinic acid (DMA) may be considered as the biologically effective dose of arsenic exposure.

5 BIOMARKERS OF EARLY BIOLOGICAL EFFECTS CAUSED BY ARSENIC

There are several biomarkers of early biological effects of arsenic on various organ systems. Hyperpigmentation and hyperkeratosis are early dermatological signs of arsenic exposure. Carotid atherosclerosis, QT prolongation and increased dispersion in electrocardiography, abnormal microcirculation, and subclinical peripheral vascular disease detected by Doppler ultrasonography are early biological effects of arsenic on the circulatory system. Blood glucose level and hematuria may be considered as the early biomarkers for diabetes and renal disease caused by arsenic, respectively. Micronuclei, chromosomal aberrations and DNA methylation profiles in exfoliated urothelial cells and lymphocytes, serum oxidant level, and urinary 8OHdG level are early biological effects of arsenic on carcinogenesis.

6 BIOMARKERS OF ACQUIRED AND GENETIC SUSCEPTIBILITY

Molecular and genomic biomarkers for acquired and genetic susceptibility to arsenic-induced health hazards include dietary intake of antioxidants and folate, and genetic polymorphisms of enzymes involved in xenobiotic metabolism, DNA repair, and oxidative stress. Variation in susceptibility to chronic arsenic poisoning suggests the arsenic-induced health hazards are determined by gene-environment interactions.

REFERENCES

Chen, C.J., Chuang, Y.C., Lin, T.M. & Wu, H.Y. 1985. Malignant neoplasms among residents of a blackfoot disease-endemic area in Taiwan: high-arsenic artesian well water and cancers. *Cancer Res.* 45: 5895–5899.

Chen, C.J., Chuang, Y.C., You, S.L., Lin, T.M. & Wu, H.Y. 1986. A retrospective study on malignant neoplasms of bladder, lung and liver in blackfoot disease endemic area in Taiwan. *Br. J. Cancer* 53: 399–405.

Chen, C.J., Kuo, T.L. & Wu, M.M. 1988a. Arsenic and cancers. *Lancet* 1: 414–415.

Chen, C.J., Wu, M.M., Lee, S.S., Wang, J.D., Cheng, S.H. & Wu, H.Y. 1988b. Atherogenicity and carcinogenicity of high-arsenic artesian well water: Multiple risk factors and related malignant neoplasms of blackfoot disease. *Arteriosclerosis* 8: 452–460.

Chen, C.J. & Wang, C.J. 1990. Ecological correlation between arsenic level in well water and age-adjusted mortality from malignant neoplasms. *Cancer Res.* 50: 5470–5474.

Chen, C.J. 1990. Blackfoot disease. *Lancet* 336: 442.

Chen, C.J., Chen, C.W., Wu, M.M. & Kuo, T.L. 1992. Cancer potential in liver, lung, bladder and kidney due to ingested inorganic arsenic in drinking water. *Br. J. Cancer* 66: 888–892.

Chen, C.J., Hsueh, Y.M., Lai, M.S., Hsu, M.P., Wu, M.M. & Tai, T.Y. 1995. Increased prevalence of hypertension and long-term arsenic exposure. *Hypertension* 25: 53–60.

Chen, C.J., Chiou, H.Y., Chiang, M.H., 1996. Dose-response relationship between ischemic heart disease mortality and long-term arsenic exposure. Arteriosleer. *Thromb. Vas. Biol.* 16: 504–510.

Chen, C.J., Chiou, H.Y., Huang, W.I., Chen, S.Y., Hsueh, Y.M., Tseng, C.H., Lin, L.J., Shyu, M.P. & Lai, M.S. 1997a. Systemic noncarcinogenic effects and developmental toxicity of inorganic arsenic. In: Abernathy, C.O., Calderon, R.L., Chappell, W.R. (Eds.), *Arsenic: Exposure and Health Effects*. Chapman & Hall, London, pp. 124–134.

Chen, C.J., Hsu, L.I., Shih, W.L., Hsu, Y.H., Tseng, M.P., Lin, Y.C., Chou, W.L., Chen, C.Y., Lee, C.Y., Wang, L.H., Cheng, Y.C., Chen, C.L., Chen, S.Y., Wang, I.H., Hsueh, Y.M., Chiou, H.Y. & Wu, M.M. 2005. Biomarkers of exposure, effect and susceptibility of arsenic-induced health hazards in Taiwan. *Toxicol. Appl. Pharmacol.* 206: 198–206.

Chen, C.J., Hsueh, Y.M., Chiou, H.Y., Hsu, Y.H., Chen, S.Y., Horng, S.F., Liaw, K.F. & Wu, M.M. 1997b. Human carcinogenicity of inorganic arsenic. In: Abernathy, C.O., Calderon, R.L., Chappell, W.R. (Eds.), *Arsenic: Exposure and Health Effects*. Chapman & Hall, London, pp. 232–242.

Chen, C.J., Hsu, L.I., Tseng, C.H., Hsueh, Y.M. & Chiou, H.Y. 1999. Emerging epidemics of arseniasis in Asia. In: Chappell, W.R., Abernathy, C.O., Calderon, R.L. (Eds.), *Arsenic Exposure and Health Effects*. Elsevier, Amsterdam, pp. 113–121.

Chen, K.P. & Wu, H.Y. 1962. Epidemiologic studies on blackfoot disease: 2. A study of source of drinking water in relation to disease. *J. Formosan Med. Assoc.* 61: 611–618.

Chiou, H.Y., Hsueh, Y.M., Liaw, K.F., Horng, S.F., Chiang, M.H., Pu, Y.S., Lin, J.S.N., Huang, C.H. & Chen, C.J. 1995. Incidence of internal cancers and ingested inorganic arsenic: A seven-year follow-up study in Taiwan. *Cancer Res.* 55: 1296–1300.

Chiou, H.Y., Huang, W.I., Su, C.L., Chang, S.F., Hsu, Y.H. & Chen, C.J. 1997b. Dose-response relationship between prevalence of cerebrovascular disease and ingested inorganic arsenic. *Stroke* 28: 1717–1723.

Chiou, H.Y., Chiou, S.T., Hsu, Y.H., Chou, Y.L., Tseng, C.H., Wei, M.L. & Chen, C.J. 2001. Incidence of transitional cell carcinoma and arsenic in drinking water: a follow-up study of 8,102 residents in an arseniasis-endemic area in northeastern Taiwan. *Am. J. Epidemiol.* 153: 411–418.

Hsieh, F.I., Hwang, T.S., Hsieh, Y.C., Lo, H.C., Su, C.T., Hsu, H.S., Chiou, H.Y. & Chen, C.J. 2008. Risk of erectile dysfunction induced by arsenic through well water consumption in Taiwan. *Environ. Health Perspect.* 116: 532–536.

Lai, M.S., Hsueh, Y.M., Chen, C.J., Hsu, M.P., Chen, S.Y., Kuo, T.L., Wu, M.M. & Tai, T.Y. 1994. Ingested inorganic arsenic and prevalence of diabetes mellitus. *Am. J. Epidemiol.* 139: 484–492.

Lin, W., Wang, S.L., Wu, H.J., Chang, K.H., Yeh, P., Chen, C.J. & Guo, H.R. 2008. Associations between arsenic in drinking water and pterygium in southwestern Taiwan. *Environ. Health Perspect.* 116: 952–955.

Morales, K.H., Ryan, L., Brown, K.G., Kuo, T.L., Wu, M.M. & Chen, C.J. 2000. Risk of internal cancers from arsenic in drinking water. *Environ. Health Perspect.* 108: 655–661.

See, L.C., Chiou, H.Y., Lee, J.S., Hsueh, Y.M., Lin, S.M., Tu, M.C., Yang, M.L. & Chen, C.J. 2007. Dose-response relationship between ingested arsenic and cataract among residents in southwestern Taiwan. *J. Environ. Sci. Health Part A* 42: 1843–1851.

Tseng, C.H., Chong, C.K., Chen, C.J., Lin, B.J., Tai, T.Y., 1995. Abnormal peripheral microcirculation in seemingly normal subjects living in blackfoot disease-hyperendemic villages in Taiwan. *Int. J. Microcir.* 15: 21–27.

Tseng, C.H., Chong, C.K., Chen, C.J. & Tai, T.Y. 1996. Dose-response relationship between peripheral vascular disease and ingested inorganic arsenic among residents in blackfoot disease endemic villages in Taiwan. *Atherosclerosis* 120: 125–133.

Tseng, C.H., Tai, T.Y., Chong, C.K., Tseng, C.P., Lai, M.S., Lin, B.J., Chiou, H.Y., Hsueh, Y.M., Hsu, K.H. & Chen, C.J. 2000. Long-term arsenic exposure and incidence of non-insulin-dependent diabetes mellitus: a cohort study in arseniasis-hyperendemic villages in Taiwan. *Environ. Health Perspect.* 108: 847–851.

Tseng, W.P., Chu, H.M., How, S.W., Fong, J.M., Lin, C.S. & Yeh, S. 1968. Prevalence of skin cancer in an endemic area of chronic arsenicism in Taiwan. *J. Natl. Cancer Inst.* 40: 453–463.

Tseng, W.P. 1977. Effects and dose-response relationships of skin cancer and blackfoot disease with arsenic. *Environ. Health Perspect.* 19: 109–119.

Wang, C.H., Jeng, J.S., Yip, P.K., Chen, C.L., Hsu, L.I., Hsueh, Y.M., Chiou, H.Y., Wu, M.M. & Chen, C.J. 2002. Biological gradient between long-term arsenic exposure and carotid atherosclerosis. *Circulation* 105: 1804–1809.

Wang, C.H., Chen, C.L., Hsiao, C.K., Chiang, F.T., Hsu, L.I., Chiou, H.Y., Hsueh, Y.M., Wu, M.M. & Chen, C.J. 2009. Increased risk of QT prolongation associated with atherosclerotic diseases in arseniasis-endemic area in southwestern coast of Taiwan. *Toxicol. Appl. Pharmacol.* 239: 320–324.

Wang, C.H., Chen, C.L., Hsiao, C.K., Hsu, L.I., Chiou, H.Y., Hsueh, Y.M., Wu, M.M. & Chen, C.J. 2010. Arsenic-induced QT dispersion is associated with atherosclerotic diseases and predicts long-term cardiovascular mortality in subjects with previous exposure to arsenic: a 17-year follow-up study. *Cardiovasc. Toxicol.* 10: 17–26.

Wang, S.L., Chiou, J.M., Chen, C.J., Wu, T.N. & Chang, L.W. 2003. Prevalence of diabetes mellitus and its vascular complications in southwestern arseniasis-endemic and non-endemic areas in Taiwan. *Environ. Health Perspect.* 111: 155–159.

Wu, M.M., Kuo, T.L., Hwang, Y.H. & Chen, C.J. 1989. Dose-response relation between arsenic concentration in well water and mortality from cancers and vascular diseases. *Am. J. Epidemiol.* 130: 1123–1132.

One Century of the Discovery of Arsenicosis in Latin America (1914–2014) –
Litter, Nicolli, Meichtry, Quici, Bundschuh, Bhattacharya & Naidu (Eds)
 ISBN 978-1-138-00141-1

Exposure to arsenic and cardiometabolic risk in Chihuahua, Mexico

M.A. Mendez, Z. Drobná, R. Fry, J.B. Buse & M. Stýblo
University of North Carolina at Chapel Hill, North Carolina, USA

C. González-Horta, B. Sánchez-Ramírez, M.L. Ballinas,
M.C. Ishida-Gutiérrez & D.S. Gutiérrez-Torres
Facultad de Ciencias Químicas, Universidad Autónoma de Chihuahua, Chihuahua, México

R. Hernández-Cerón, D. Viniegra-Morales & F. Baeza Terrazas
Colegio de Médicos Cirujanos y Homeópatas del Estado de Chihuahua, A.C., Nayarit

L.M. Del Razo
Departamento de Toxicología, Centro de Investigación y de Estudios Avanzados del IPN, México DF, Mexico

G.G. García-Vargas
Universidad Juárez del Estado de Durango, Gómez Palacio, Durango, México

D. Loomis
International Agency for Research of Cancer, Lyon Cedex, France

ABSTRACT: This study examined associations between Arsenic (As) exposure and Cardiometabolic (CM) risk factors in a cohort of Chihuahua residents who drank water containing 0.05–419.8 μg As L^{-1}. Results show that subjects in the highest exposure tertile (58.3–419.8 μg As L^{-1}) had higher levels of total cholesterol in plasma and were more likely to be diagnosed with dysglycemia (OR 1.60, 95% CI 1.09–2.32), hypertriglyceridemia (OR 1.53, 95% CI 1.07–2.20), and high blood pressure (OR 1.57, 95% CI 1.05–2.35) than subjects in the lowest exposure tertile (≤38.5 μg As L^{-1}). In addition, these CM risk factors were associated with the ratio of dimethyl-As (DMAs) to methyl-As (MAs) in urine. Thus, an increased capacity to convert As to DMAs may be a risk factor for CM disease in individuals chronically exposed to As. Prospective studies are needed to examine causality of the associations of CM risk with As exposure.

1 INTRODUCTION

Cardiometabolic (CM) risk is characterized by a group of factors that serve as indicators of an individual's overall risk for type-2 Diabetes Mellitus (DM) and Cardiovascular Disease (CVD). The traditional CM risk factors include dysglycemia, dyslipidemia, high blood pressure, inflammation and obesity (Del Turco *et al.*, 2013). Inorganic As (iAs) is a potent human carcinogen (ATSDR 2007). However, epidemiologic evidence has also linked iAs exposure to DM (Maull *et al.*, 2012) and several cardiovascular endpoints, including CVD, coronary heart disease, stroke, and peripheral arterial disease (Moon *et al.*, 2012). Here, we present preliminary results of a cross-sectional study in Chihuahua (Mexico) that examined association between exposure to iAs in drinking water and several CM risk factors.

2 METHODS

2.1 *The Chihuahua cohort*

A total of 939 adults (≥18 years old, including 629 women) with a minimum of 5-year uninterrupted residency in the study area were recruited between 2008 and 2012. Pregnant women and subjects with kidney or urinary tract infection were excluded because these conditions affect profiles of iAs metabolites in urine. Individuals with a potential for occupational exposure to As (e.g., those working with pesticides or in mines or smelters) were also excluded.

2.2 *Questionnaire and sample collection*

A study questionnaire recorded data on residency, occupation, drinking water sources, smoking, use of alcohol, and medical history. Samples of drinking water were obtained from subjects' households. Spot urine and fasting venous blood were collected during a medical exam which also included measurement of blood pressure and oral glucose tolerance test with blood drawn 2 h after a 75 g glucose dose.

2.3 *Arsenic and CM risk factor analyses*

Hydride generation-atomic absorption spectrometry coupled with a cryotrap (Hernández-Zavala *et al.*, 2008) was used to determine the concentration of As in drinking water and concentrations of As species in urine, including iAs, methyl-As (MAs) and dimethyl-As (DMAs). Prestige 24i Chemistry Analyzer was used to determine Fasting Plasma Glucose (FPG) and 2-h plasma glucose (2HPG) concentrations, and concentrations of triglycerides (TG), total cholesterol and High-Density Lipoprotein cholesterol (HDL) in fasting plasma. Plasma Low-Density Lipoprotein cholesterol (LDL) was calculated from total cholesterol and HDL values (Oliveira *et al.*, 2013).

2.4 *Statistical analysis*

Logistic regression was used to analyze associations of CM risk factors with iAs exposure (characterized either by As concentration in drinking water or by total speciated As in urine) and with iAs metabolism (characterized by urinary profiles of iAs metabolites). Models adjusted for age, gender, smoking, alcohol consumption, urinary creatinine, elevated waist circumference, and body mass index (BMI).

3 RESULTS

3.1 *The cohort characteristics*

The concentrations of As in drinking water ranged from 0.05 to 419.8 µg L^{-1} with 84% and 42% of samples exceeding 10 µg As L^{-1} and 50 µg As L^{-1}, respectively. All subjects had detectable levels of As in urine. The concentration of total speciated As (iAs + MAs + DMAs) ranged from 0.52 to 375.2 µg L^{-1}. DMAs was the major urinary metabolite representing on average 75.7% of total speciated As, followed by MAs (14.5%) and iAs (9.8%) (Table 1A). Almost 75% of subjects were either overweight or obese, and there was a substantial prevalence of the CM risk factors (Table 1B). In addition, 41% of men and 21% of women reported smoking, and 66% of men and 28% of women consumed alcohol.

Table 1. Basic characteristics of the Chihuahua cohort.

	Distribution
A. Demographics and iAs exposure	
Age (years)	45.7 ± 15.8[a]
Males	33%
Water As (µg L^{-1})	25.9 (22.8–29.5)[b]
Urinary As (µg L^{-1}):	
Total speciated As	46.2 (43.1–49.4)[b]
DMAs	34.8 (32.5–37.2)[b]
MAs	6.3 (5.9–6.8)[b]
iAs	4.0 (3.7–4.4)[b]
B. CM risk factors	
Overweight (25 <BMI <30)	34.5%
Obese (BMI ≥30)	40.0%
Elevated waist circumference (>88/102 cm women/men)	64.9%
Dysglycemia (FPG > 110, 2HPG > 140 mg dL^{-1}, or self-reported DM diagnosis or medication)	32.9%
Hypertriglyceridemia (>150 mg TG dL^{-1})	40.2%
Low HDL (<40/50 mg dL^{-1} men/women)	58.1%
High LDL (>130 mg dL^{-1})	18.2%
High blood pressure (SBP > 140, DBP > 90 or medication)	44.1%
Total cholesterol (mg dL^{-1})	171.2 ± 47.6[a]
LDL (mg dL^{-1})	96.6 ± 37.8[a]

[a]Mean ± SD; [b]geometric mean (95% confidence interval). SBP, systolic blood pressure; DBP, diastolic blood pressure; TG, triglycerides.

3.2 *Associations between CM risk and iAs exposure*

CM risk analysis showed that 39.2% of subjects in the highest exposure tertile (58.3–419.8 µg L^{-1} of drinking water) were diagnosed with dysglycemia as compared to 29.3% in the lowest tertile (≤38.5 µg L^{-1}; $p < 0.05$). Similar differences were also found for hypertriglyceridemia (47.1% vs. 35.4%; $p < 0.05$) and high blood pressure (49.1% vs. 39.7%; $p < 0.10$). In contrast, only 54.1% of subjects in the 3rd tertile had low plasma HDL as compared to 61.4% in the 1st tertile, but this difference was not statistically significant. In addition, subjects in the highest exposure tertile had higher levels (mean ± SD) of total plasma cholesterol (182 ± 44 vs. 160.5 ± 49.5 mg dL^{-1}; $p < 0.05$) and LDL (103.5 ± 34.5 vs. 89.7 ± 39.8 mg dL^{-1}; $p < 0.05$). The Odd Ratios (ORs) for dysglycemia, hypertriglyceridemia, and high blood pressure were significantly higher ($p < 0.05$) for all subjects in the highest exposure tertile as compared with the lowest tertile (Table 2). However, As exposure

Table 2. Multivariate-adjusted odds ratio (95% CI) of CM outcomes associated with highest vs. lowest tertile of iAs exposure characterize either by As concentration in drinking water or by concentration of total speciated As in urine.[a]

CM risk factors[b]	Water As ($n = 876$)[c]	Total speciated As ($n = 933$)
Dysglycemia	1.60 (1.09–2.32)*	1.86 (1.23–2.82)*
Hypertriglyceridemia	1.53 (1.07–2.20)*	2.00 (1.33–2.99)*
Low HDL	0.65 (0.44–0.95)*	0.87 (0.57–1.32)
High LDL	1.28 (0.81–2.04)	1.28 (0.77–2.11)
High blood pressure	1.57 (1.05–2.35)*	1.11 (0.71–1.73)

[a]Models adjusted for age, gender, smoking, alcohol consumption, BMI, waist circumference and urinary creatinine [b]CM risk factors are defined in Table 1. [c]Water As concentration was determined for only 876 subjects. *$p < 0.05$.

was negatively associated with low plasma HDL (OR 0.65, 95% CI 0.44–0.95; $p < 0.05$). Dysglycemia and hypertriglyceridemia were also positively associated with the total speciated As in urine: OR 1.86 (95% CI1.23–2.82; $p < 0.05$) and 2.00 (95% CI1.33–2.99; $p < 0.05$), respectively, for comparison of 3rd tertile (82.4–375.2 μg dL^{-1}) vs. 1st tertile (0.52–32.84 μg dL^{-1}) (Table 2). No statistically significant associations of HDL, LDL, or high blood pressure with total speciated As were found. However, subjects at the 3rd tertile for total speciated As had significantly higher total plasma cholesterol than subjects in the 1st tertile (data not shown). Similar associations were found when total speciated As normalized for urinary creatinine was used for the analysis (data not shown).

3.3 *Associations between CM risk and iAs metabolism*

Regression analysis was also used to examine associations between CM risk factors and indicators of the capacity of the body to methylate iAs, specifically with DMAs/MAs and MAs/iAs ratios in urine (Table 3). DMAs/MAs ratio was positively associated with CM risk factors, including dysglycemia, hypertriglyceridemia, low plasma HDL, and high blood pressure. High MAs/iAs ratio seemed to have a protective effect, at least in case of dysglycemia.

Table 3. Multivariate-adjusted odds ratio (95% CI) of CM outcomes associated with highest vs. lowest tertile of DMAs/MAs and MAs/iAs ratios in urine.[a]

CM risk factors[b]	DMAs/MAs	MAs/iAs
Dysglycemia	2.13 (1.44–3.18)*	0.42 (0.28–0.61)*
Hypertriglyceridemia	2.17 (1.48–3.18)*	1.15 (0.91–1.63)
Low HDL	1.56 (1.05–2.31)*	0.82 (0.56–1.19)
High LDL	1.39 (0.85–2.28)	1.16 (0.74–1.83)
High blood pressure	1.64 (1.07–2.52)*	0.78 (0.52–1.16)

[a]Models adjusted for age, gender, smoking, alcohol consumption, BMI, waist circumference and urinary creatinine. [b]CM risk factors are defined in Table 1. *$p < 0.05$.

4 DISCUSSION

Growing evidence suggests that exposures to some water, air or food contaminants may contribute to CM risk (Thayer *et al.*, 2012). Many of these contaminants are endocrine disruptors that alter either hormone production or hormone-regulated metabolic pathways. Some of the endocrine disruptors may act primarily as obesogens, increasing fat accumulation and, consequently, CM risk; others may increase CM risk by mechanisms that are not associated with obesity. In spite of the growing number of studies focusing on environmental endocrine disruptors, significant knowledge gaps exist that hamper the effort to provide better protection to the public. According to a recent WHO state of the science assessment, "the disease risk associated with exposure to environmental endocrine disruptors may be significantly underestimated" (WHO, 2012).

iAs is a common drinking water contaminant and human carcinogen (ATSDR, 2007). Thus, for many years, the focus of public health and regulatory agencies has been on the carcinogenic effects of iAs exposures. However, epidemiologic evidence suggests that even greater numbers of people exposed to iAs may be at risk of developing non-cancerous diseases, including CVD and DM (Maull *et al.*, 2012; Moon *et al.*, 2012). In addition, laboratory studies have supplied data on potential mechanisms of the CM effects of iAs exposure. We have reported that iAs, and particularly its highly toxic trivalent methylated metabolites, methyhlarsonite (MAs^{III}) and dimethylarsinite ($DMAs^{III}$), can produce dysglycemia by either disrupting the insulin activated signal transduction pathway and glucose uptake (Paul *et al.*, 2007) or by inhibiting the glucose stimulated insulin secretion by pancreatic islets (Douillet *et al.*, 2013). Other laboratories have shown that exposure to iAs in drinking water accelerates or exacerbates atherosclerosis in rats (Cheng *et al.*, 2011) and ApoE-/- mice (Lemaire *et al.*, 2011), possibly by altering cholesterol metabolism or stimulating vascular redox signaling and inflammation in vascular lesions (Druwe *et al.*, 2011). iAs exposure has also been shown to promote hypertension and cardiac hyperthrophy in mice (Sanchez-Soria *et al.*, 2012).

Our present cross-sectional study in Chihuahua provides additional of evinced linking exposure to iAs in drinking water with an increased prevalence of several CM risk factors, including dysglycemia, hypertriglyceridemia, and high blood pressure. Our results also show that iAs metabolism may play a role in determining risk of adverse CM effects of iAs exposure. The DMAs/MAs ratio in urine (also called secondary methylation index) has been used in epidemiologic studies as an indicator of the capacity of the body to metabolize iAs. Both a low DMAs/MAs ratio and a low percentage of DMAs in urine, which are thought to reflect an impaired conversion of MAs to DMAs with possible accumulation of toxic MAs^{III}, have been linked to increased risks of cancer or precancerous skin lesions in populations chronically exposed to iAs (Ahsan *et al.*, 2007; Chen *et al.*, 2003; Chen *et al.*, 2009; Huang *et al.*, 2008; Yu *et al.*, 2000). However, data on the association between DMAs/MAs ratio and risk of non-cancerous disease are inconsistent with several recent reports suggesting that high DMAs/MAs ratio and high percentage of DMAs in urine could be risk factors for DM or hypertension (Del Razo *et al.*, 2011; Kim *et al.*, 2013; Nizam *et al.*, 2013). Results of the present study are in agreement with these reports. We found statistically significant positive associations of high urinary DMAs/MAs ratio with dysglycemia, hypertriglyceridemia, and high blood pressure. Thus, it is possible that an increased capacity to convert iAs to DMAs, including its toxic trivalent form, $DMAs^{III}$, may be a risk factor for CM disease in individuals chronically exposed to As.

Since the present study, as well as most previous studies examining the association between iAs exposure and CM risk used cross-sectional designs, prospective studies are needed to examine causality of this associations and to determine if As metabolism, and specifically production of MAs^{III} and $DMAs^{III}$ can modify this risk.

REFERENCES

Ahsan, H., Chen, Y., Kibriya, M.G., Slavkovich, V., Parvez, F., Jasmine, F., Gamble, M.V. & Graziano J.H. 2007. Arsenic metabolism, genetic susceptibility, and risk of premalignant skin lesions in Bangladesh. *Cancer Epidemiol. Biomarkers Prev.* 16: 1270–1278.

ATSDR. 2007. Toxicological Profile for Arsenic. http://www.atsdr.cdc.gov/toxprofiles/tp.asp?id = 22&tid = 3.

Cannon, C.P. 2007. Cardiovascular disease and modifiable cardiometabolic risk factors. *Clin. Cornerstone* 8: 11–28.

Chen, Y., Parvez, F., Gamble, M., Islam, T., Ahmed, A., Argos, M., Graziano, J.H. & Ahsan, H. 2009. Arsenic exposure at low-to-moderate levels and skin lesions, arsenic metabolism, neurological functions, and biomarkers for respiratory and cardiovascular diseases: review of recent findings from the Health Effects of Arsenic Longitudinal Study (HEALS) in Bangladesh. *Toxicol. Appl. Pharmacol.* 239: 184–192.

Chen, Y.C., Guo, Y.L., Su, H.J., Hsueh, Y.M., Smith, T.J., Ryan, L.M., Lee, M.S., Chao, S.C., Lee, J.Y. & Christiani, D.C. 2003. Arsenic methylation and skin cancer risk in southwestern Taiwan. *J. Occup. Environ. Med.* 45: 241–248.

Cheng, T.J., Chuu, J.J., Chang, C.Y., Tsai, W.C., Chen, K.J. & Guo, H.R. 2011. Atherosclerosis induced by arsenic in drinking water in rats through altering lipid metabolism. *Toxicol. Appl. Pharmacol.* 256: 146–153.

Del Razo, L.M., García-Vargas, G.G., Valenzuela, O.L., Hernandez-Castellanos, E., Sánchez-Peña, L.C., Drobná, Z., Loomis, D. & Stýblo, M. 2011. Exposure to arsenic in drinking water is associated with increased prevalence of diabetes: a cross-sectional study in the Zimapán and Lagunera Regions in Mexico. *Environ. Health.* 10:73.

Del Turco, S., Gaggini, M., Daniele, G., Basta, G., Folli, F., Sicari, R. & Gastaldelli, A. 2013. Insulin resistance and endothelial dysfunction: a mutual relationship in cardiometabolic risk. *Curr. Pharm. Des.* 19: 2420–2431.

Douillet, C., Currier, J.M., Saunders, J., Bodnar, W., Matoušek, T. & Stýblo, M. 2013. Methylated trivalent arsenicals are potent inhibitors of glucose stimulated insulin secretion by murine pancreatic islets. *Toxicol. Appl. Pharmacol.* 267: 11–15.

Druwe, I.L., Sollome, J.J., Sanchez-Soria, P., Hardwick, R.N., Camenisch, T.D. & Vaillancourt, R.R. 2012. Arsenite activates NFκB through induction of C-reactive protein. *Toxicol. Appl. Pharmacol.* 261: 263–270.

Hernández-Zavala, A., Matoušek, T., Drobná, Z., Adair, B.M., Dědina, J., Thomas, D.J. & Stýblo, M. 2008. Speciation of arsenic in biological matrices by automated hydride generation-cryotrapping-atomic absorption spectrometry with multiple microflame quartz tube atomizer (multiatomizer). *J Anal. At. Spectrom.* 23: 342–351.

Huang, Y.K., Pu, Y.S., Chung, C.J., Shiue, H.S., Yang, M.H., Chen, C.J. & Hsueh, Y.M. 2008. Plasma folate level, urinary arsenic methylation profiles, and urothelial carcinoma susceptibility. *Food Chem. Toxicol.* 46: 929–938.

Kim, N.H., Mason, C.C., Nelson, R.G., Afton, S.E., Essader, A.S., Medlin, J.E., Levine, K.E., Hoppin, J.A., Lin, C., Knowler, W.C. & Sandler, D.P. 2013. Arsenic Exposure and Incidence of Type 2 Diabetes in Southwestern American Indians. *Am. J. Epidemiol.* 177: 962–969.

Lemaire, M., Lemarié, C.A., Molina, M.F., Schiffrin, E.L., Lehoux, S., Mann, K.K. 2011. Exposure to moderate arsenic concentrations increases atherosclerosis in ApoE-/- mouse model. *Toxicol. Sci.* 122: 211–221.

Maull, E.A., Ahsan, H., Cooper, G., Edwards, J., Longnecker, M., Navas-Acien, A., Pi, J., Silbergeld, E., Styblo, M., Tseng, C-H., Thayer, K. & Loomis, D. 2012. Evaluation of the association between arsenic and diabetes: a national toxicology program workshop report. *Environ. Health Perspect.* 120: 1658–1670.

Moon, K., Guallar, E. & Navas-Acien, A. 2012. Arsenic exposure and cardiovascular disease: an updated systematic review. *Curr. Atheroscler. Rep.* 14: 542–555.

Nizam, S., Kato, M., Yatsuya, H., Khalequzzaman, M., Ohnuma, S., Naito, H. & Nakajima, T. 2013. Differences in urinary arsenic metabolites between diabetic and non-diabetic subjects in Bangladesh. *Int. J. Environ. Res. Public Health* 10: 1006–1019.

Oliveira, M.J., van Deventer, H.E., Bachmann, L.M., Warnick, G.R., Nakajima, K., Nakamura, M., Sakurabayashi, I., Kimberly, M.M., Shamburek, R.D., Korzun, W.J., Myers, G.L., Miller, W.G. & Remaley, A.T. 2013. Evaluation of four different equations for calculating LDL-C with eight different direct HDL-C assays. *Clin. Chim. Acta.* 423: 135–140.

Paul, D.S., Harmon, A.W., Devesa, V., Thomas, D.J. & Styblo, M. 2007. Molecular mechanisms of diabetogenic effects of arsenic: Inhibition of insulin signaling by arsenite and methylarsonous acid. *Environ. Health Perspect.* 115: 734–742.

Sanchez-Soria, P., Broka, D., Monks, S.L. & Camenisch, T.D. 2012. Chronic low-level arsenite exposure through drinking water increases blood pressure and promotes concentric left ventricular hypertrophy in female mice. *Toxicol. Pathol.* 40: 504–512.

Thayer, K.A., Heindel, J.J., Bucher, J.R. & Gallo, M.A. 2012. Role of environmental chemicals in diabetes and obesity: a National Toxicology Program workshop review. *Environ. Health Perspect.* 120: 779–789.

WHO. 2012. *Endocrine Disrupting Chemicals* – 2012. *An assessment of the state of the science of endocrine disruptors prepared by a group of experts for the United Nations Environment Programme and World Health Organization.* Å. Bergman, J.J. Heindel, S. Jobling, K.A. Kidd & R.T. Zoeller (eds.), WHO, Geneva, Switzerland.

Yu, R.C., Hsu, K.H., Chen, C.J. & Froines, J.R. 2000-Arsenic methylation capacity and skin cancer. *Cancer Epidemiol. Biomarkers Prev.* 9: 1259–1262.

One Century of the Discovery of Arsenicosis in Latin America (1914–2014) – Litter, Nicolli, Meichtry, Quici, Bundschuh, Bhattacharya & Naidu (Eds)
 ISBN 978-1-138-00141-1

Exploring a suitable animal model for chronic arsenic toxicity and carcinogenicity studies

J.C. Ng, J.P. Wang & M. Krishnamohan
National Research Centre for Environmental Toxicology, The University of Queensland, Brisbane, Australia

ABSTRACT: Inorganic arsenic is classified as a human carcinogen based on very strong epidemiological evidence. Surprisingly, there are only few studies available for its carcinogenic effects in animals despite there are many studies for its non-cancer acute and subchronic toxicity in the literature. To date, the urinary arsenic (total or speciation concentrations) is still regarded as the most reliable biomarker for arsenic exposure. In searching for a suitable animal model for arsenic adverse effect studies, we and others have evaluated urinary porphyrins and oxidative indicators for their potential as biomarkers for chronic arsenic exposure. For the carcinogenic effects, there are only few successful studies including research done in our laboratory. In this report, we explore what might be a promising model to employ for further develop our understanding of arsenic chronic toxicity and carcinogenicity.

1 INTRODUCTION

Inorganic arsenic has the potential to cause bodily harm to millions of people who live in endemic areas where arsenic levels are elevated in soils and drinking water (Ng, 2005). More recently, rice arsenic is considered to be an important source contributing to increased cancer risks, particularly when rice is cooked in arsenic-contaminated water (JECFA, 2011). Considering rice is a staple diet for over half of the world population, this has drawn significant attention amongst researchers. The main health concerns for arsenic are its chronic non-cancer and cancer effects. These are backed by strong epidemiological evidence. However, the lack of apparently a suitable animal model(s) has hindered our progress on the understanding of how arsenic induces its chronic effects and particularly its carcinogenicity. In this paper, we summarized our endeavor in exploring for such an animal model(s) addressing certain aspects of arsenic chronic toxicity. For the non-cancer end-points, we studied kidney dysfunction in diabetic rats and porphyrin profile as biomarkers in mice chronically exposed to inorganic arsenate and monomethylarsonous acid (MMA^{III}). Both studies were supported by parallel epidemiological studies. For the cancer end-points, lifetime bioassays were conducted in mice exposing to drinking water containing sodium arsenate and MMA^{III}.

2 METHODS/EXPERIMENTAL

2.1 *Kidney dysfunction and chronic arsenic exposure*

For the animal study, 96 male rats (Sprague-Dawley) were divided into 4 groups and were exposed to different concentrations of arsenic in their drinking water (0, 5, 15 and 30 mg L^{-1}) for up to 8 months. Diabetic conditions were induced in 50% of the rats by a single intravenous injection of STZ (Streptozotocin) (35 mg/kg bw) after 6–8 month. Urinary arsenic and NAG (N-acetyl-β-glucosaminidase) and blood glucose were measured fortnightly and monthly, respectively. Blood glucose tolerance test was carried out by giving rats (5 in each group) 0.2 g glucose as a 20% glucose solution per 100 g body weight by gavage.

For the human study, 235 urine and 233 blood samples were collected from villagers with and without diabetes from the arsenicosis endemic and control areas in Chepaizi, Xinjiang, PR China. Among the 235 urine samples, 111 were from arsenicosis endemic area (57 diabetes and 54 non-diabetes subjects) and 124 from a neighbor control area (72 diabetes and 52 non-diabetes control).

Eighteen drinking water samples were collected from the villages. Water and urinary arsenic levels were measured by ICP-MS. Human and rat urinary NAG (N-acetyl-β-glucosaminidase) concentration analysis were carried out by using a N-acetyl-β-glucosaminidase commercial assay kit (DIAZYME, San Diego, USA) by using ultraviolet spectrophotometry. Details of this study are in Wang *et al.* (2009).

2.2 *Porphyrin profile as biomarker*

Groups of 70 female C57Bl/6 J mice were exposed to drinking water containing MMA^{III} or sodium arsenate (100, 250, 500 μg As L^{-1}) and 105 control mice were given demineralized water (<0.1 μg L^{-1}) *ad lib* for 2 years. Urinary arsenic speciation and porphyrins were measured at 0, 1, 2, 4, 8 weeks and every 8 weeks thereafter until week 104 (Krishnamohan *et al.*, 2007a,b).

For human study, 113 urine samples were collected from villagers who lived in coal-borne arsenic endemic area in Guizhou, PR China and 30 urine samples from the near-by control area for arsenic and porphyrin analyses (Ng *et al.*, 2005). In Xinjiang, 353 urine samples were collected with about equal number from the endemic and control sites (Liu *et al.*, 2013).

2.3 *Animal cancer studies*

The lifetime bioassay dosing regime was that described above by exposing arsenate and MMA^{III} to C57Bl/6 J mice.

3 RESULTS AND DISCUSSION

3.1 *Non-cancer end-points*

Villagers from endemic-area had higher urinary arsenic concentrations compared with the control subject. This couples with elevated NAG suggests that chronic arsenic exposure represents a significant adverse impact on the kidney function. Blood glucose levels of subjects with diabetes were lower than those from the control site. The above observations were validated in rat study (Wang *et al.*, 2009). The rodent model may be a useful tool to study the relationship between chronic arsenic exposure and diabetes.

For the biomarker studies, there was a correlation between urinary arsenic and porphyrin concentrations that demonstrated the effect of arsenic on haem biosynthesis resulting in increased porphyrin excretion particular for uroporphyrin and/or coproporphyrin. The elevation of porphyrin was evident during the earlier phase of exposure in mice. This was also observed in younger age villagers (<20 years) from the endemic-area. The results suggest that porphyrins could be used as early warning biomarkers of chronic arsenic exposure in humans (Ng *et al.*, 2005). Whether alteration of porphyrin profile could predict adverse health effects associated with both cancer and non-cancer end-points in chronically arsenic-exposed populations need further investigation. The mouse model seems to be a suitable model of this type of studies.

3.2 *Cancer end-points*

DMA^{V} was found to be the major urinary metabolite in both As^{V} and MMA^{III} treated mice. Tumors generally occurred 18 months post treatment. Examples of tumors are shown in Figure 1. Tumors-bearing animals appeared to remain healthy until the terminal stages of the experiment. The increased incidences of lymphomas were dose dependent for both As^{V}- and MMA^{III}-treated groups. Other types of tumors and multiplicity of tumors also showed a significant dose response relationship in the treatment groups. Histiocytic sarcoma was higher in both As^{V} and MMA^{III}-treated mice compared to the controls. Plasmacytoid lymphomas were observed only in treated mice. Chronic dermatitis and other degenerative skin lesions were seen only in the As^{V} and MMA^{III}-treated mice and not in the control. Keratoacanthoma of the skin, epidermoid carcinoma and rhabdomyosarcoma were only seen in the MMA^{III} treated mice and not in the As^{V}-treated mice. Immunohistological examination showed that the lymphoma was B cell origin. Significant difference was observed in the incidence of all types of tumors in the treated groups compared with the control ($p < 0.0001$) (Table 1, data for MMA^{III} are not shown).

These studies confirmed the carcinogenic effects of As^{V} in this mouse strain and, for the first time, demonstrated that MMA^{III} is similarly carcinogenic.

Although inorganic arsenic is classified as a human carcinogen, the lack of a proper animal model until recently made it difficult to understand the mechanism of its carcinogenicity.

Previous arsenic carcinogenicity studies could not be taken into account as a proper model

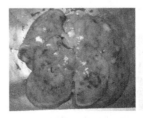

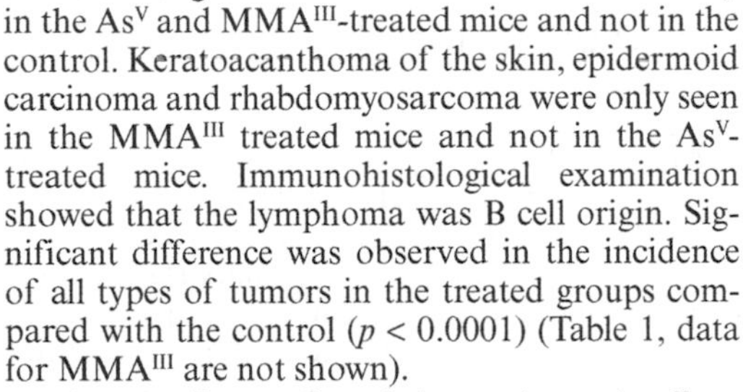

Figure 1. Liver (L) diffusely invaded by tumor tissue. Pale tumor foci were surrounded by hyperemic zones from a mouse dosed with 500 μg L^{-1} As^{V} for two years; Diffuse invasion of the lungs with lymphoid tumor with involvement of the mediastinal lymph node (N) from a mouse dosed with 100 μg/L MMA^{III} for two years.

Table 1. Tumor incidence observed in number of control (0.1 μg L^{-1}) and As^V-treated mice after two years. Brackets indicate percentage.

Incidence	0.1 μg/L	100 μg/L	250 μg/L	500 μg/L
No. of mice exposed for >12 months	85	55	55	55
All types of tumors	13(15)	14(25)	34(62)	31(56)
Lymphoma	10(11.7)	8(15)	17(31)	16(29)
Lymph nodes only	4(4.7)	1(2)	3(5)	4(7)
Organ/lymph nodes	6(7)	7(13)	14(25)	12(22)
Other types of tumors	2(4)	5(9)	13(24)	8(15)
Multiple tumors	1(1)	1(2)	4(7)	7(13)

because of various reasons like unrealistically high dose, route of exposure, short-term exposure and too few animals. Although DMA, a major metabolite of inorganic arsenic, have been shown to cause bladder cancers in rats, inorganic arsenic alone had not been proven to be a carcinogen in any of the known animal models until in a mouse study (Ng *et al.*, 1999) and, subsequently, by another mouse model using a higher concentration of arsenic (Waalkes *et al.*, 2003) were reported.

In an earlier publication of the Environmental Health Criteria, in 1981, it was reported that "at present no definite evidence exists to show that inorganic arsenic compounds are carcinogenic to animals". Not until 1999, Ng *et al.* reported that when female C57Bl/6 J mice were given 500 μg As L^{-1} as sodium arsenate (As^v) in the drinking water for up to 26 months, the tumor incidence in all organs was 41.1%. These findings were reported in a later publication Environmental Health Criteria on Arsenic and Arsenic Compounds (IPCS, 2001), where it was claimed that "this was the first experimental carcinogenicity study in rodents using a relevant route of exposure and relevant exposure level" and that the incidence of tumors was "treatment related".

However, in the Ng's study, no dose response effect was determined. The present study has been designed to investigate whether the human carcinogen inorganic arsenic is carcinogenic in mice, in concentrations that are commonly found in the endemic areas and to confirm Ng *et al.*'s (1999) study. This present study on As^V was carried out using the same strain of female mice (C57Bl/6 J) and recorded the effects of the test doses of 100, 250 and 500 μg As L^{-1} in the drinking water over 2 years. The results from this study have revealed that long-term chronic exposure to As^V causes both cancerous and non-cancerous lesions in the mice. The present study confirms Ng *et al.*'s earlier findings that As^V is carcinogenic to female C57Bl/6 J mice.

In humans, chronic exposure to arsenic was shown to cause cancers in a dose-response relationship mainly in the skin, lungs and bladder in population (JECFA, 2011). Our results have also demonstrated a significant dose-response relationship in all types of tumors observed.

On the other hand, there were *in-vitro* studies which showed that exposure to low concentration of arsenic could be adaptive and protective causing enhanced cell proliferation and viability rather than cytotoxicity. At higher concentrations, toxicity of arsenic can cause an apoptotic response, as was seen in the treatment of promyelocytic leukemia and multiple myeloma. This, in fact, could be a reason for not encountering tumor production in most of earlier animal studies, which employed high doses of arsenicals (cytotoxic).

4 CONCLUSIONS

Arsenic induced kidney dysfunction in rats and humans particularly amongst individuals with diabetes. Our studies also showed long term exposure to As^V or MMA^{III} could affect the haem biosynthetic pathway in mice resulting in the alteration of porphyrin profile. It indicated that the haem metabolic pathway is highly susceptible to alterations induced by As^V or MMA^{III}, and the urinary porphyrin profile can be used as a biological monitoring parameter in addition to urinary arsenic profile prior to the onset of carcinogenesis or clinical manifestations of arsenic toxicity.

Our findings clearly demonstrated that both As^V and MMA^{III} were complete carcinogens to female C57Bl/6 J mice causing tumors in multiple organs at relatively low doses without any promoter.

These two established life-time mouse bioassays and those published by other researchers will contribute significantly to resolve if the argument on whether human carcinogen arsenic is carcinogenic in laboratory animals and dismiss the concept that there are no animal models existing for arsenic carcinogenesis.

The results from this study also suggest that MMA^{III} is a more potent carcinogen than As^V, and support the contention that arsenic methylation is not necessarily a detoxification pathway but, on contrary, it may increase the toxicity of the element and that MMA^{III} could be the cause for arsenic carcinogenicity in both animals and humans. These results can contribute to the risk evaluation of chronic arsenic exposure and the development of arsenic exposure standards.

It is reasonable to conclude that rodent models can be utilized to study chronic arsenic toxicity in both non-cancer and cancer end-points of interest.

As the metabolism of MMA^{III} is extensive with regard to post both acute and chronic exposure, further research is necessary to establish the toxicity of endogenously formed MMA^{III} compared to exogenously administered MMA^{III}. Determination of tissue accumulation of different species of arsenic could be undertaken to facilitate a better understanding of arsenic organ toxicity. It is also important to know more about the effect of arsenic metabolites on the animal and human lymphatic and immune systems. With the current tissue banks established through this current study and that of others, more detailed studies at molecular and genetic level could help to explore the mode of action of chronic arsenic poisoning.

ACKNOWLEDGEMENTS

Entox is a partnership between Queensland Health and the University of Queensland.

REFERENCES

JECFA. 2011. Evaluation of certain contaminants in food. The seventy-second report of Joint FAO/WHO Expert Committee on Food Additives. WHO, 115 pp.

Krishnamohan, M., Qi, L., Lam, P.K.S., Moore, M.R. & Ng, J.C. 2007a. Urinary arsenic and porphyrin profile in C57BL/6 J mice chronically exposed to monomethylarsonous acid (MMAIII) for two years. *Toxicology & Applied Pharmacology* 224: 89–97.

Krishnamohan, M., Wu, H.J. Huang, S.H., Maddalena, R., Lam, P.K.S., Moore, M.R. & Ng, J.C. 2007b. Urinary arsenic methylation and porphyrin profile of C57Bl/6 J mice chronically exposed to sodium arsenate. *Science of the Total Environment.* 379(2–3): 235–243.

Liu, F.F., Wang, J.P., Zheng, Y.J., & Ng, J.C. 2013. Biomarkers for evaluation of population health status 16 years after the intervention of arsenic-contaminated groundwater in Xinjiang, China. *Journal of Hazardous Materials* In Press.

Ng, J.C. 2005. Review: Environmental contamination of arsenic and its toxicological impact on humans. *Environmental Chemistry* 2: 146–160.

Ng, J.C., Seawright, A.A., Qi, L., Garnett, C.M., Chiswell, B. & Moore, M.R. 1999. Tumours in Mice induced by exposure to sodium arsenate in drinking water. In C. Abernathy, R. Calderon & W. Chappell, (Eds), *Arsenic: Exposure and Health effects*: 217–223. Oxford, London, Elsevier Science.

Ng, J.C., Wang, J.P., Zheng, B.S., Zhai C., Maddalena, R., Liu F. & Moore M.R. 2005. Urinary porphyrins as biomarkers for arsenic exposure among susceptible populations in Guizhou province, China. *Toxicology & Applied Pharmacology* 206(2): 176–184.

Waalkes, M.P., Ward, J.M., Liu, J. & Diwan, B.A. 2003. Transplacental carcinogenicity of inorganic arsenic in the drinking water: induction of hepatic, ovarian, pulmonary, and adrenal tumors in mice. *Toxicology & Applied Pharmacology* 186: 7–17.

Wang, J.P., Wang, S.L., Lin, Q., Zhang, L., Huang, D. & Ng, J.C. 2009. Association of arsenic and kidney dysfunction in people with diabetes and validation of its effects in rats. *Environment International* 35(3): 507–511.

One Century of the Discovery of Arsenicosis in Latin America (1914–2014) – Litter, Nicolli, Meichtry, Quici, Bundschuh, Bhattacharya & Naidu (Eds)
© 2014 Taylor & Francis Group, London, ISBN 978-1-138-00141-1

Some cardiovascular effects of inorganic arsenic exposure in children exposed to drinking water

C. Osorio-Yáñez, A. Barrera-Hernández, E. Hernández-Castellanos, L.C. Sánchez-Peña, A. De Vizcaya-Ruiz & L.M. Del Razo
Centro de Investigación y de Estudios Avanzados del IPN, Departamento de Toxicología, México DF, México

J.C. Ayllon-Vergara
Hospital Español, México DF, México

G. Aguilar-Madrid
Unidad de Investigación y Salud en el Trabajo, Instituto Mexicano del Seguro Social, México DF, México

L. Arreola-Mendoza
Departamento de Biociencias e Ingeniería CIIEMAD-IPN, México DF, México

ABSTRACT: The present study investigates associations of inorganic arsenic (iAs) exposure, blood pressure and structural echocardiographic parameters in a cross sectional study of 173 Mexican children residents in an endemic area who presented concentrations of total speciated arsenic (tAs) in urine ranging from 5.7 to 370 ng As/mL. Results show that children in the highest exposure percentile (75th or <678 μg As/L/year) of cumulative As exposure (ΣAsE) were related to increased systolic or diastolic Blood Pressure (DBP), had higher Left Ventricular Mass (LVM) and increased Aortic Root Diameter (ARD) than children in the lowest exposure percentile p25th or <307 μg As/L/year). DBP and LVM were not related to iAs metabolites, but ARD were positively associated with the methyl-As (MAs) level in urine. In conclusion, we observed increased diastolic prehypertension and increased concentric left ventricular hypertrophy prevalence in children exposed to iAs.

1 INTRODUCTION

Exposure to inorganic arsenic (iAs) in drinking water increases the risk and incidence of a number of Cardio-Vascular Diseases (CVD) including ischemic heart disease, acute myocardial infarctions, peripheral vascular disease, atherosclerosis, and hypertension, have been extensively studied in adult populations around the world (Chen *et al.*, 1995; Wang *et al.*, 2007; Hsieh *et al.*, 2011). However, scarce epidemiologic studies have been focused on iAs exposure-related CVD in children. CVD can start from childhood and stays silent until the adulthood, when the clinical manifestations occur. Prehypertensive subjects are at increased risk for developing hypertension and CVD compared with those normotensive (Gupta *et al.*, 2012). Prehypertension may mark the beginning of a progressive remodeling of the left ventricle that may go unnoticed for long time. Increased Left Ventricular Mass (LVM) is an independent predictor of CVD (Cuspidi *et al.*, 2013). Here, we present results of a cross-sectional study in Zimapan (Mexico) that examined association between exposure to iAs in drinking water and Blood Pressure (BP) and structural echocardiographic parameters.

2 METHODS

2.1 *Child recruitment*

A total of 173 children (3–14 years old, 54% boys) with a minimum of 2-year uninterrupted residency in the study area were recruited from two local schools during 2009. Children resided in six local towns close to neighboring school residents. Arsenic levels in drinking water in these locations range from 13 to 263 μg As L^{-1}. Children with diabetes or CVD were not included. We measured body weight and height using standard protocols. We calculated the Body Mass Index (BMI) using the formula weight (kg)/height (m^2). The BMI z—score was calculated and BMI was categorized based on guidelines of the Center of Disease Control and Prevention (http://www.cdc.gov/).

2.2 *Questionnaire, sample collection and clinical evaluation*

Parents of child were interviewed on residential information, child migration, the source of drinking water, and child medical history. First morning void urine samples and fasting venous blood were collected from the children before clinical examination which included measurement of BP and echocardio-graphy measurements using a cardiovascular ultrasound system with transducer frequencies (1–5 MHz) for body size.

2.3 *Arsenic exposure evaluation and structural echocardiographic parameters*

Hydride generation-atomic absorption spectrometry coupled with a cryotrap (Hernández-Zavala *et al.*, 2008) was used to determine the concentration of As species in urine, including iAs, methyl-As (MAs) and dimethyl-As (DMAs). As an indicator of long term iAs exposure, cumulative As exposure (ΣAsE, μg As L^{-1} /year) was calculated by the sum of the products derived by multiplying the iAs drinking water concentration by the water consumption time. LVM (g) was calculated according to the American Society of Echocardiography (Shahn *et al.*, 1978).

2.4 *Statistical analysis*

Percentages were used for categorical variables. Simple and multivariate linear regression models were used to estimate associations of BP and echocardiographic parameters with iAs exposure variables. Models were adjusted for age, gender, BMI, their influence on model fit and the outcomes of interest.

3 RESULTS

3.1 *The study characteristics*

Nearly 70% (n = 173) of the children evaluated were less than 5 years old. Almost 18% of children were overweight and 9% obese. All subjects had detectable levels of As in their urine. The concentration of total speciated As (iAs + MAs + DMAs or tAs) ranged from 5.7 to 370 μg/L. DMAs was the major urinary metabolite re-presenting on average 78.9% of total speciated As, followed by MAs (10.4%) and iAs (9.1%) (Table 1). In total, 79% of the children had total speciated arsenic (tAs) values higher in the urine than the Biological Exposure Index (BEI) of 35 ngAs/mL (ACGIH, 2004) and the ΣAsE values were from 13.95 to 3,601 μg As/L/year.

Table 1. Basic demographic characteristics, As exposure and CVD risk factors of the Mexican children.

	GM[a] or Percentage
A. Demographics and iAs Exposure	
Age *(years)*	5.1 (3–14)
≤5 years	70%
Males	54%
Water As *(μg/L)*	73.8 (13 to 263)
Urinary As *(ng/mL)*:	
iAs	5.4 (0.6–100)
MAs	6.4 (0.2–55.7)
DMAs	46.7 (4.9–236.9)
Total speciated As	59.1 (5.7–369.9)
tAs BEI, 35 *(ng/mL)*	21%
ΣAsE *(μg As/L/year)*	431.1 (13.9–3601)
B. CVD risk factors and structural echocardiographic parameters	
BMI *(kg/m²)*	16.1 (11.8–26)
z-score *(percentile)*	40.0 (1–99)
Plasma glucose *(mg/dL)*	83 (61–125)
Blood pressure (BP)	
Systolic BP *(mmHg)*	86.9 (70–115)
Diastolic BP *(mmHg)*	60.8 (45–80)
Left ventricular mass, LVM *(g)*	55.82 (28.7–109.3)
Aortic root diameter, ARD *(mm)*	14.1 (9–24)

[a]geometric mean (range).

3.2 *Associations between BP and iAs exposure*

Compared with those of normal weight, Systolic Blood Pressure (SBP) was greater in overweight and obese groups (89.9 vs. 85.7 and 93.8 vs. 85.7 mmHg, respectively; $p < 0.001$). The DBP was greater only in the obese groups (67 vs. 60.23 mmHg; $p = 0.000$). SBP and DBP increased with age ($p = 0.000$, Table 2). Although we did not observe systolic or diastolic hypertension, 48% had diastolic prehypertension. Both SBP and DBP were associated with tAs and ΣAsE (Figure 1).

3.3 *Associations between LVM and iAs exposure*

These children presented with a high prevalence of concentric remodeling (61.5%; n = 102) and a small prevalence of concentric hypertrophy (6.6%; 11 children). Concentric remodeling could be a previous stage in concentric hypertrophy, which has also been associated with poor prognosis compared with those that have normal LV geometry, and it is independently related to adverse cardiovascular events (Verdecchia *et al.*, 2007). Notably we observed a significant association between increased LVM, increased ARD and ΣAsE >678 μg/L/year (Figure 2).

The ARD has been proposed to be a subclinical LV dysfunction parameter (Masugata *et al.*, 2011).

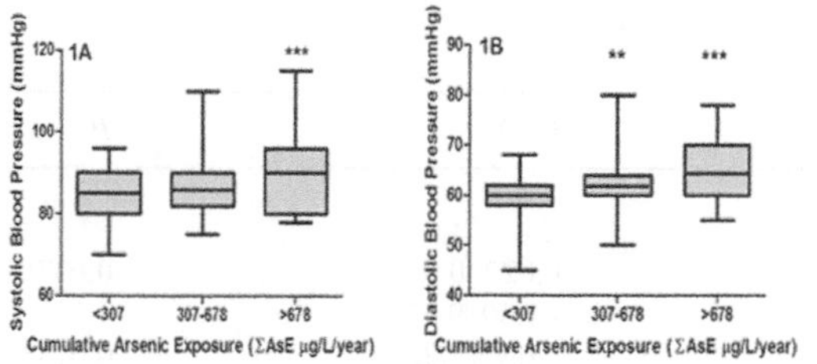

Figure 1. Relationship between cumulative arsenic exposure (ΣAsE) and systolic or diastolic blood pressure (1 A, 1B, res-pectively). ΣAsE was categorized according to percentiles. Simple linear regression analyses were performed; SBP was significantly higher in >678 µg/L/year category compared with <307 µg/L/year category (**$p = 0.009$). Compared with the ΣAsE <307 µg/L/year category, DBP were significantly higher in 307–678 µg/L year (**$p = 0.008$) and >678 µg/L/year ΣAsE categories (***$p < 0.001$), where **$p < 0.01$ and ***$p < 0.001$.

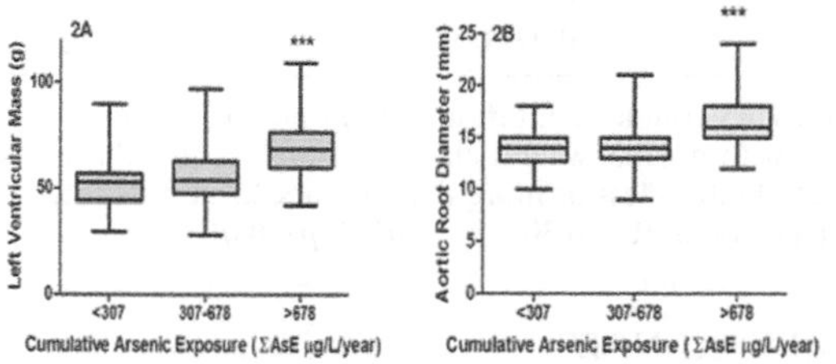

Figure 2. Left ventricular mass and aortic root diameter increase in the ΣAsE >678 µg/L/year category compared with the <307 µg/L/year category (Figure 2 A and 2B, respectively, where ***$p < 0.001$).

Moreover ARD was significantly associated with urinary MAs metabolite (0.043 mm increase per 1 ng/mL; $p = 0.014$) and ΣAsE >678 µg/L/year category ($p = 0.032$; Table 3b). Thus, it is possible that an increased capacity to convert iAs to MAs, including its toxic trivalent form, MAsIII, may be a risk factor for structural heart changes by echocardiography in individuals chronically exposed to iAs. Moreover, MAs and ARD have been related to atherosclerosis in adults (Wu *et al.*, 2006; Jiang *et al.*, 2009).

Urinary tAs values is considered a reliable biomarker of recent exposure (Hughes, 2006) and is used as the main biomarker of exposure. Moreover, ΣAsE provide information on a more long-term exposure. In this study, urinary tAs was not related to ΣAsE (data not shown). A major question in the use of urinary tAs as a biomarker of exposure is how to relate the recent exposure, measured by urinary tAs, to exposures that may have occurred chronically.

In this study, iAs exposure was not related to either obesity or plasma glucose. However, both factors contribute independently and strongly to the increased BP (Table 2).

Table 2. Systolic and diastolic blood pressure multivariate regression analyses.

Explanatory Variable	β Coef.	CI 95%	*P*
Systolic Blood Pressure			
ΣAsE (*µg/L/year*)	0.0055	0.0019–0.009	0.003**
tAs (35–70 *ng/mL*)	1.13	−1.48–3.73	0.394
tAs (>70 *ng/mL*)	3.32	0.58-.05	0.018*
BMI overweight	4.68	1.85–7.50	0.001**
BMI obese	8.34	4.53–12.14	0.000***
Age (>5 years)	2.90	0.765–5.04	0.008**
Gender	−1.09	−2.85–0.68	0.226
Diastolic Blood Pressure			
ΣAsE (*µg/L/year*)	0.0033	0.0013–0.005	0.001**
tAs (*ng/mL*)	0.012	0.0016–0.022	0.023*
Plasma glucose (*mg/dL*)	0.12	0.05–0.19	0.001**
BMI overweight	1.60	−0.22–3.42	0.084
BMI obese	6.88	3.28–10.49	0.000***
Age (years)	0.64	0.12–1.16	0.016*
Gender	−0.68	−2.11–0.75	0.347

Robust linear regression analyses for the SBP model, $R^2 = 0.36$. For the DBP model, $R^2 = 0.34$. The boys were compared with girls. Children more than 5 years old were compared with children less than 5 years old.*$p < 0.05$, **$p < 0.01$,***$p < 0.001$.

In multiple linear regression analyses that were adjusted for gender, BMI and age, LVM was predicted by ΣAsE >678 µg/L/year ($p = 0.014$) and SBP values (0.345 g increase per mmHg increase in SBP; $p = 0.037$, Table 3a).

4 DISCUSSION AND CONCLUSION

We observed BP changes related to urinary tAs and ΣAsE that are consistent with previously reported epidemiological studies showing increased incidence of hypertension in adults exposed to iAs (Abhyankar *et al.*, 2012). Some authors found a more substantial increase in SBP than DBP in healthy women exposed to iAs (Kwok *et al.*, 2007). Hypertension has long been known to be a strong and independent risk factor for the development of major adverse cardiovascular events, whereby the cardiovascular system adopts different abnormal changes as a result of long hypertension. The heart adaptation to these changes includes geometrical reorientation in different patterns notably concentric remodeling, concentric hypertrophy or eccentic hypertrophy (Ganau *et al.*, 1992). The type of geometric pattern is determined mainly by the predominating type of stressor to the myocardium, as volume or pressure overload (Conrady *et al.*, 2004).

Table 3. Multiple linear regression analyses for morphological variables and association with cumulative arsenic exposure.

Explanatory Variable	β	IC 95%	*P*
a. Left Ventricular Mass (g)			
SBP (*mmHg*)	0.345	0.022–0.67	0.037*
BMI overweight (*kg/m²*)	4.16	−1.30–9.62	0.135
BMI obese	9.64	2.05–17.22	0.013*
BMI underweight	−4.52	−11.94–2.9	0.231
ΣAsE (370–678 μg/L/year)	1.73	−3.22–6.69	0.49
ΣAsE (>678 μg/L/year)	10.09	2.05–18.12	0.014*
Age (>5 years)	6.29	1.45–11.12	0.011*
Gender	4.42	0.48–8.35	0.028*
b. Aortic Root Diameter (mm)			
DBP (mmHg)	0.094	0.03–0.16	0.004**
MAs (ng/mL)	0.043	0.009–0.08	0.014*
ΣAsE (370–678 μg/L/year)	−0.19	−0.87–0.48	0.573
ΣAsE (>678 μg/L/year)	1.39	0.12–2.66	0.032*
Age (>5 years)	1.03	0.32–1.74	0.005**
Gender	0.70	0.09–1.32	0.026*

Robust multiple linear regression analyses were conducted for each variable of interest and morphological or systolic function parameters. Increased BMI categories were compared with normal weight children. In left ventricular mass and aortic diameter models, age was categorized into groups of children less or more than 5 years. ΣAsE was categorized by percentiles. For the LVM model, $R^2 = 0.26$; for the ARD model, $R^2 = 0.30$. $^*p < 0.05$, $^{**}p < 0.01$.

Concentric hypertrophy is a cardiac manifestation of pressure overload that results in new sarcomere synthesis in parallel with old sarcomeres, and wall thickness increases without proportional chamber dilation (Lilly, 2011). In our study population, SBP was significantly related to increased LVM, and DBP was significantly related to increased ARD (Tables 3a and 3b, respectively).

With regard to LVM, age of the children SBP and BMI were strong determinants to increased LVM in iAs exposure children. Nearly half of the population presented with diastolic prehypertension. Importantly, prehypertension was concomitant with high concentric remodeling prevalence in these children. It remains to be elucidated whether iAs exposure can produce diastolic prehypertension and/or concentric remodeling in in other human populations. In mice chronically exposed to iAs increased both SBP and DBP, and 43% of LVM demonstrated a concentric hypertrophy pattern (Sanchez-Soria *et al.*, 2012).

The clinical implications of these finding are significant. Children with diastolic prehypertension could present LV adaption to arterial hypertension results in LV geometry responses such as concentric remodeling or concentric hypertrophy. Given the potential of adverse health effects, immediate measures should be taken to provide As safe water.

REFERENCES

American Conference of Governmental Industrial Hygienists (ACGIH) 2004. Arsenic and soluble Inorganic compounds: BEI®. 8th Edition Documentation, Cincinnati, Ohio.

Abhyankar, L.N., Jones, M.R., Guallar, E. & Navas-Acien, A. 2012. Arsenic exposure and hypertension: a systematic review. *Environ. Health Perspect.* 120: 494–500.

Chen, C.J., Hsueh, Y.M., Lai, M.S., Shyu, M.P., Chen, S.Y., Wu, M.M., Kuo, T.L. & Tai, T.Y. 1995. Increased prevalence of hypertension and long-term arsenic exposure. *Hypertension* 25: 53–60.

Conrady, A.O., Rudomanov, O.G., Zaharov, D.V., Krutikov, A.N., Vahrameeva, N.V., Yakovleva, O.I., Alexeeva, N.P. & Shlyakhto, E.V. 2004. Prevalence and determinants of left ventricular hypertrophy and remodelling patterns in hypertensive patients: the St. Petersburg study. *Blood Press* 13:101–109.

Cuspidi, C., Rescaldani, M., Sala, C. 2013. Prevalence of echocardiographic left-atrial enlargement in hypertension: a systematic review of recent clinical studies. *Am. J. Hypertens.* 26:456–464.

Ganau, A., Devereux, R.B., Roman, M.J., de Simone, G., Pickering, T.G., Saba, P.S., Vargiu, P., Simongini, I. & Laragh, J.H. 1992. Patterns of left ventricular hypertrophy and geometric remodeling in essential hypertension. *J. Am. Coll. Cardiol.* 19:1550–1558.

Gupta, P., Nagaraju, S.P., Gupta, A. & Mandya Chikkalingaiah, K.B. 2012. Prehypertension-time to act. *Saudi J. Kidney Dis. Transpl.* 23:223–233.

Hernandez-Zavala, A., Matousek, T., Drobna, Z., Paul, D.S., Walton, F., Adair, B.M., Jiří, D., Thomas, D.J. & Stýblo, M. 2008. Speciation analysis of arsenic in biological matrices by automated hydride generation-cryotrapping-atomic absorption spectrometry with multiple microflame quartz tube atomizer (multiatomizer). *J. Anal. At. Spectrom.* 23:342–351.

Hsieh, Y.C., Lien, L.M., Chung, W.T., Hsieh, F.I., Hsieh, P.F., Wu, M.M., Tseng, H.P., Choir, H.Y. & Chen, C.J. 2011. Significantly increased risk of carotid atherosclerosis with arsenic exposure and polymorphisms in arsenic metabolism genes. *Environ. Res.* 111:804–810.

Hughes, M.F. 2006. Biomarkers of exposure: a case study with inorganic arsenic. *Environ. Health Perspect.* 114:1790–1796.

Jiang, J.J., Chen, X.F., Liu, X.M., Tang, L.J., Lin, X.F., Pu, Z.X., Chen, T.L., Zhang, Y., Wang, Y.P. & Wang, J.A. 2009. Aortic root dilatation is associated with carotid intima-media thickness but not with carotid plaque in hypertensive men. *Acta Cardiol.* 64: 645–651.

Kwok, R.K., Mendola, P., Liu, Z.Y., Savitz, D.A., Heiss, G., Ling, H.L., Xia, Y., Lobdell, D., Zeng, D., Thorp, J.M. Jr, Creason, J.P. & Mumford, J.L. 2007. Drinking water arsenic exposure and blood pressure in healthy women of reproductive age in Inner Mongolia, China. *Toxicol. Appl. Pharmacol.* 222: 337–343.

Lilly, L.S. 2011. *Pathophysiology of Heart Disease*. 5th ed: Wolters Klumer Health Lippincott Williams & Wilkins.

Masugata, H., Senda, S., Murao, K., Okuyama, H., Inukai, M., Hosomi, N., Iwado, Y., Noma, T., Kohno, M., Himoto, T. & Goda, F. 2011. Aortic root dilatation as a marker of subclinical left ventricular diastolic dysfunction in patients with cardiovascular risk factors. *J. Int. Med. Res.* 39: 64–70.

Sahn, D.J., DeMaria, A., Kisslo, J. & Weyman, A. 1978. Recommendations regarding quantitation in M-mode echocardiography: results of a survey of echocardiographic measurements. *Circulation* 58: 1072–1083.

Sanchez-Soria, P., Broka, D., Monks, S.L. & Camenisch, T.D. 2012. Chronic low-level arsenite exposure through drinking water increases blood pressure and promotes concentric left ventricular hypertrophy in female mice. *Toxicol. Pathol.* 40: 504–512.

Verdecchia, P., Angeli, F., Achilli, P., Castellani, C., Broccatelli, A., Gattobigio, R. & Cavallini, C. 2007. Echocardiographic left ventricular hypertrophy in hypertension: marker for future events or mediator of events? *Curr. Opin. Cardiol.* 22: 329–334.

Wang, C.H., Hsiao, C.K., Chen, C.L., Hsu, L.I., Choir, H.Y., Chen, S.Y., Hsueh, Y.M., Wu, M.M. & Chen, C.J. 2007. A review of the epidemiologic literature on the role of environmental arsenic exposure and cardiovascular diseases. *Toxicol Appl Pharmacol* 222: 315–326.

Wu, M.M., Choir, H.Y., Hsueh, Y.M., Hong, C.T., Su, C.L., Chang, S.F., Huang, W.L., Wang, H.T., Wang, Y.H., Hsieh, Y.C. & Chen, C.J. 2006. Effect of plasma homocysteine level and urinary monomethylarsonic acid on the risk of arsenic-associated carotid atherosclerosis. *Toxicol. Appl. Pharmacol.* 216: 168–175.

One Century of the Discovery of Arsenicosis in Latin America (1914–2014) – Litter, Nicolli, Meichtry, Quici, Bundschuh, Bhattacharya & Naidu (Eds)
ISBN 978-1-138-00141-1

Saliva as a biomarker of arsenic exposure

S. Bhowmick
Department of Chemistry, University of Kalyani, Kalyani, West Bengal, India
Faculty of Sciences, University of Girona, Campus de Montilivi, Girona, Spain

D. Halder & D. Chatterjee
Department of Chemistry, University of Kalyani, Kalyani, West Bengal, India

J. Nriagu
Department of Environmental Health Sciences, School of Public Health, University of Michigan, Ann Arbor, MI, USA

D.N. Guha Mazumder
DNGM Research Foundation, Kolkata, India

P. Bhattacharya
KTH-International Groundwater Arsenic Research Group, Department of Sustainable Development, Environmental Science and Engineering, KTH Royal Institute of Technology, Stockholm, Sweden

M. Iglesias
Faculty of Sciences, University of Girona, Campus de Montilivi, Girona, Spain

ABSTRACT: Saliva is a biofluid that has not been used extensively as a biomonitoring tool in epidemiological studies. This study presents the arsenic (As) concentrations in saliva samples collected from populations of West Bengal, India. We found a significant ($p < 0.05$) association between the Log transformed Daily Ingestion of As (μg day^{-1}) and the As concentration in saliva ($r = 0.68$). Additionally, As concentration of saliva and urine also had a significant positive correlation ($r = 0.60$, $p < 0.05$). Male participants, smokers and cases of skin lesion were independently and significantly associated with increase in salivary As. Thus our findings show that saliva is a useful biomarker of As exposure in the study population.

1 INTRODUCTION

Chronic exposure of arsenic (As) manifests as arsenicosis and includes a wide range of cardiovascular, hepatic, hematological, endocrine, renal and dermal related diseases which ultimately leads to cancers (Kapaj *et al.*, 2006). Blood, urine, scalp hair and nail were used as biomarkers in previous studies to monitor exposure of As and its effects on humans but each of these biomarkers has serious limitations.

Salivary glands along with other glands secrete bio-fluids (saliva) which are generated through active transport of water and ion from plasma and has been proved to be useful for detecting a number of natural metabolites, steroids and hormones as well as for Hg concentration in humans (Pesch *et al.*, 2002). Due to non-invasiveness, ease of collection and storage, saliva can also be helpful for studying large population especially involving children. Therefore the aim of our present study is to measure the total concentration of As in the collected saliva samples, and thereby examine the usefulness of these salivary As as a biomarker of As exposure in the study population.

2 METHODS

2.1 *Study population and sample collection*

The present study was carried out in three villages (Chhoto-Itna, Debagram and Tehatta) of Nadia district, West Bengal, India, where we collected saliva (n = 101) and urine (n = 101) samples from participants with age between 18–65 years who have lived in the same locality for a minimum of 10 years prior to the interview. Detailed information about the selected participants was obtained using a standard questionnaire. Participants were characterized as cases of skin lesions and control and were scored accordingly.

Pre-washed polyethylene bottles were used for spot urine sample collection. Prior to saliva sample collection, the participants washed their mouths with Milli-Q water and after 2–3 min, gave saliva samples in 15 mL LDPE bottles. The participants discarded the saliva that was formed in-between the rinsing of mouth and sample collection. Drinking water samples ($n = 16$) were also collected from the sources mentioned by the participants as the primary supplier of their drinking water.

2.2 *Sample preparation and analysis*

Urine samples were filtered with 0.45 µm syringe filter after being brought to room temperature. We adjusted the concentrations of As in the urine samples to the mean specific gravity of the collected samples (1.015 g mL^{-1}). The filtered urine samples were digested with HNO_3 and H_2O_2 (Merck) and the digested sample was cooled and measured for As using HG-AAS (Varian, AA220). The water samples collected during the survey were also measured for total As using HG-AAS.

The saliva samples were thawed to room temperature and were centrifuged. To 1 mL of the sample, appropriate amount of HNO_3 (2% v/v), ethanol (2% v/v) and internal standard (IS) (10 µg L^{-1} Rhodium standard) was mixed in a plastic vial. The resulting solution was analyzed for As in ICP-MS after dilution to 3 mL using Milli-Q water.

3 RESULTS AND DISCUSSION

3.1 *Arsenic exposure and total As concentration in urine and saliva*

The statistical results of the As level in different medium are represented in Table 1. Although the groundwater in our study area have high concentration of As, due to increased social awareness, the participants are now sharing the low As common water sources for drinking purpose. However, the local farmer still uses high As concentration groundwater for irrigation and crop cultivation.

Table 1. Statistical table of the measured As concentration in drinking water, TDI, U_{As} and S_{As} of all the participants.

Medium	N	x ± S.D	Median	Range
C_W (µg L^{-1})	16	120 ± 239	18.0	806–2.50
TDI (µg day^{-1})	101	225 ± 531	102	3168–18.6
U_{As} (µg L^{-1})	101	110 ± 154	67.7	883–0.22
S_{As} (µg L^{-1})	101	7.84 ± 12.6	2.99	84.3–0.22

Thus, there are additional exposures of bioavailable As from foods consumed by the participants (Halder *et al.*, 2013). Therefore, estimation of the total daily intake (TDI) for each participant was calculated as:

$$\text{TDI } (\mu g \text{ day}^{-1}) = (C_W \times V) + (C_R \times W \times 0.92),$$

where, C_W (µg L^{-1}) and C_R (µg kg^{-1}) is the concentration of As in water and rice respectively, V (L) is the volume of water consumed daily, W (kg) is the amount of rice taken daily and 0.92 is the fraction of the inorganic As.

Simple regression analysis between TDI and salivary arsenic (S_{As}) was done to evaluate the viability of the excreted As as a measure of As exposure (Figure 1a). Our study shows that TDI has a positive correlation with S_{As} ($r = 0.68$, $p < 0.05$). This suggests that salivary As can also act as a predictor of As exposure. Simple regression analysis was done between urinary As and S_{As} and there exist a positive, significant correlation between the two parameters ($r = 0.60, p < 0.05$; Figure 1b). This suggests that ingestion of inorganic As is important in determining the As concentration in saliva. Thus, S_{As} can be regarded as biomarker of As exposure and can be used as a surrogate of urine in As epidemiological study.

3.2 *Factors regulating As concentration in urine and saliva*

Influence of age, sex, smoker, BMI and score of skin lesion on S_{As} were tested and male participants and smokers had a higher concentration of S_{As} compared to female and non-smokers respectively, while association of S_{As} with BMI ($p = 0.871$) and age ($p = 0.440$) was not statistically significant. Control had lower concentration of S_{As} and the concentration for severe cases was 2-fold higher than the mild and moderate cases of skin lesion.

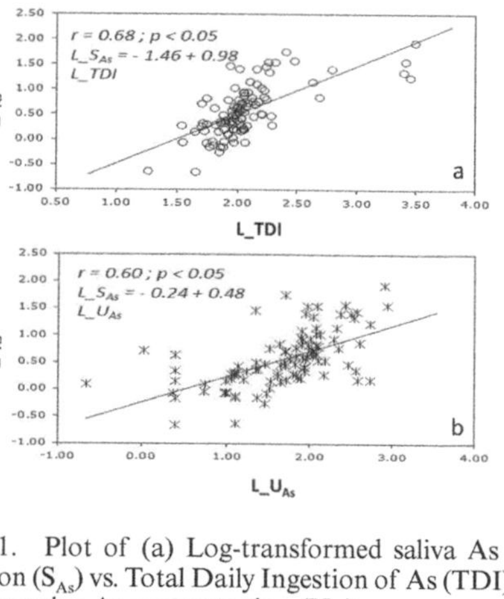

Figure 1. Plot of (a) Log-transformed saliva As concentration (S_{As}) vs. Total Daily Ingestion of As (TDI) and (b) S_{As} vs. urine As concentration (U_{As}).

4 CONCLUSIONS

This study demonstrated S_{As} as a potent biomarker of As exposure in our study population. The use of saliva for exposure assessment has several advantages compared to other already established biomarkers. The metal ions are actively transported from the plasma and thus represent a measure of internal dose. Therefore, monitoring saliva data may provide insight to the As metabolic process.

ACKNOWLEDGEMENTS

The study was funded by Trehan Foundation, University of Michigan, USA.

REFERENCES

Halder, D., Bhowmick, S., Biswas, A., Chatterjee, D., Nriagu, J., Guha Mazumder, D.N., Šlejkovec, Z., Jacks, G. & Bhattacharya, P. 2013. Risk of arsenic in exposure from drinking water and dietary components: Implications for risk management in rural Bengal. *Environmental Science & Technology* 47:1120–1127.

Kapaj, S., Peterson, H., Liber, K. & Bhattacharya, P. 2006. Human health effects from chronic arsenic poisoning: a review. *Journal of Environmental Science and Health* A 41: 2399–2428.

Pesch, A., Wilhelm, L., Rostek, U., Schmitz, N., Weishoff-Houben, M., Ranft, U & Idel, H. 2002. Mercury concentrations in urine, scalp hair, and saliva in children from Germany. *Journal of Exposure Analysis and Environmental Epidemiology* 12: 252–258.

One Century of the Discovery of Arsenicosis in Latin America (1914–2014) – Litter, Nicolli, Meichtry, Quici, Bundschuh, Bhattacharya & Naidu (Eds)
© 2014 Taylor & Francis Group, London, ISBN 978-1-138-00141-1

Exposure to arsenic alters the reproductive axis on offspring rats

N.S. Bourguignon, D. Rodríguez, M.M. Bonaventura, V. Lux Lantos & C. Libertun
Instituto de Biología y Medicina Experimental (IByME, CONICET), CABA, Buenos Aires, Argentina

ABSTRACT: Arsenic is a natural environmental contaminant to which humans are continuously exposed. In some areas, high levels of arsenic are naturally present in drinking water and are a toxicological concern. Arsenic is also postulated as an endocrine disruptor, impacting the metabolic and reproductive axes. In this study, offspring Sprague Dawley rats were exposed to sodium arsenite (As(III)) from conception to adulthood. Neonatal assessment of pup weight, anogenital distance, vaginal opening and testicular descent was carried out, as well as the evaluation of the endocrine status in adulthood. Arsenic exposure determined reduced pup weight at birth in both sexes, which was followed by catch-up growth. Anogenital distance/body weight was decreased on postnatal day 1 in females. Adult females showed normal estrous cycles; nevertheless, they showed elevated serum testosterone and FSH. Therefore, early exposure to As(III) in drinking water alters the reproductive axis mainly in females.

1 INTRODUCTION

In recent years, environmental toxicants have become a serious health concern, and studies on environmental Endocrine Disruptions (EDs) are becoming more prevalent. Arsenic (As) is one of them due to its wide distribution and adverse health effects (Wang *et al.*, 2006). For humans, the primary exposure route to As is through contaminated drinking water, in which inorganic forms of As predominate. Chronic exposure to As during the gestational period has been associated with alteration to the host and fetus, as it can reach the placenta (Concha *et al.*, 1998). Some studies have shown that As may have adverse human reproductive effects including increased rates of fetal loss, congenital malformation, preterm births, and neonatal mortality, as well as decreased birth weight (Ahmad *et al.*, 2001; Kapaj *et al.*, 2006), at fairly low exposure levels. As produces toxic effects on the reproductive system also in rodents (DeSesso *et al.*, 2001), generally evaluated at high concentrations. The aim of the present study was to elucidate the effect of arsenic exposure as an endocrine disruptor in the reproductive axis of male and female offspring, at different maturational ages.

2 MATERIALS AND METHODS

2.1 *Arsenic*

Sodium arsenite ($NaAsO_2$) was dissolved in distilled water to yield 5 mg L^{-1} and 50 mg L^{-1} concentrations in drinking water for rat consumption. The bottles were changed every 2–3 days to avoid oxidation of As(III).

2.2 *Animals and treatment*

All studies on animals were performed according to protocols for animal use approved by the Institutional Animal Care and Use Committee (IByME - CONICET) and by the NIH. All the animals were maintained under a controlled 12 h light/dark cycle and temperature conditions. Rats were allowed standardized pellet food and water ad libitum. Healthy adult Sprague Dawley rats were mated and sperm plug in rat vagina was examined every morning. The day when sperm plug was confirmed was designated as day 1 of gestation (GD 1). Thereafter, rats were randomly housed singly and exposed to sodium arsenite (5 or 50 mg L^{-1}) through drinking water. The control group received distilled water.

At delivery, pups were sexed according to Anogenital Distance (AGD) and each litter was adjusted to eight pups (four males, four females whenever possible). Following birth, which was considered to be postnatal day 1 (PND1), the pups continued to be exposed to As via lactation. Male and female offspring of each dam were weaned on PND21 and divided in two groups: one of them continued to receive the same treatment in drinking water (A5-A5 and A50-A50 group) and the other one received distilled water (A5-C and A50-C group) until sacrifice (adults: 3–4 months). The pups from control dams continued to receive distilled water (C group).

All pups were weighed at PND1 and 21. AGD at PND1 and the age at Vaginal Opening (VO) and Testicular Descent (TD) were recorded as sexual developmental markers. At 2 months of age, estrous cycle of female pups was determined. Finally, serum hormone levels of gonadotropins (LH and FSH), prolactin (PRL), estradiol (E2), testosterone (T) and progesterone (P4) were measured in adulthood.

2.3 *Statistical analysis*

Results are expressed as mean ± SEM, and values are considered significant at $p < 0.05$. Data were analyzed by Chi Square, one-way or two-way analysis of variance (ANOVA) with posttest and transformed when the test for homogeneity of variances so required. Number of animals: $n = 8–25$.

3 RESULTS AND DISCUSSION

3.1 *Body Weights (BW) and sexual developmental markers*

Exposure to 50 mg L-1 As decreased offspring weight at birth, mainly in males. This is followed by catch-up growth at weaning. A5 group increased weight at PND21 on both sexes (Table 1).

As BW on PND1 was different among groups, AGD was relativized to these values. A 50 females offspring showed significantly reduced the AGD/BW at birth (C 0.020 ± 0.002; A5 0.016 ± 0.001; A50 0.014 ± 0.001, *< p 0.05. $n = 9–27$). No differences were found in male offspring (C 0.038 ± 0.001; A5 0.036 ± 0.001; A50 0.038 ± 0.003, $n = 9–27$).

The mean age of VO or TD did not vary with treatment (Table 2).

Table 1. BW (g) from control and As-treated pups.

	Female		Male	
	PND1	PND21	PND1	PND21
C	7.6 ± 0.1	43.6 ± 0.8	8.0 ± 0.1	45.3 ± 0.8
A5	7.6 ± 0.1	47.4 ± 0.8*	7.9 ± 0.1	49.4 ± 0.5*
A50	7.4 ± 0.1*	47.2 ± 0.9*	7.5 ± 0.1*	50.4 ± 1.1*

Two-way ANOVA: interaction: ns, main effect treatment: $p < 0.02$, main effect sex: $p < 0.002$. *: different from *C*: < p 0.05, or less.

Table 2. VO and TD (days) from control and As-treated pups.

	C	A5-A5	A5-C	A50-C	A50-C
VO	37.6 ± 0.56	38.5 ± 1.33	40.9 ± 1.84	38.3 ± 0.94	38.2 ± 1.5
TD	25.2 ± 0.27	25.3 ± 0.25	24.5 ± 0.42	24.9 ± 0.86	27.0 ± 0.0

One- way Anova, ns.

3.2 *Estrous cycle*

The percentage of days in normal cycle or anestrous was calculated for each treatment. No estrous cycles irregularities were observed upon arsenic exposure (Figure 1).

3.3 *Serum Hormones*

In adult females, treatment with 50 mg L^{-1} elevated basal serum levels of T and FSH (Table 3), while LH, PRL, E2 and P4 were unchanged. In adult males, a decrease in FSH levels was also found (Table 4).

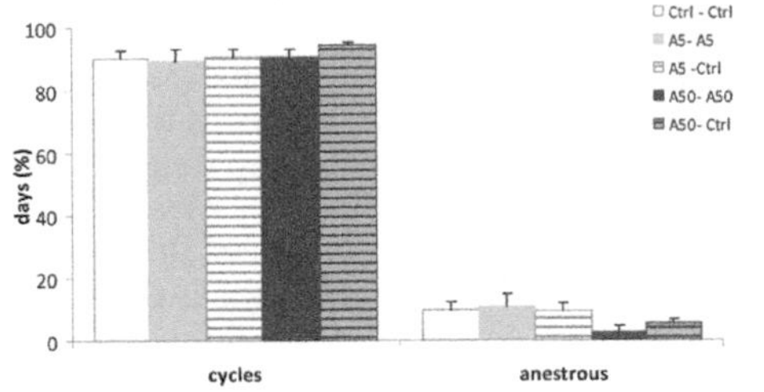

Figure 1. Effects of arsenic exposure on estrous cycles. Results are expressed as percentage of time (days) in normal cycles or anestrous. Chi Square, ns.

Table 3. Hormone dosage (ng mL^{-1}, except E2 pg mL^{-1}) from control and As-treated female offspring.

	C	A5-A5	A5-C	A50-A50	A50-C
LH	0.7 ± 0.1	0.6 ± 0.1	0.7 ± 0.2	0.6 ± 0.1	0.8 ± 0.1
FSH	1.2 ± 0.2	1.2 ± 0.3	1.1 ± 0.2	2.0 ± 0.2*	2.4 ± 0.3*
PRL	14.9 ± 2.6	12.4 ± 4.4	16.7 ± 2.0	10.2 ± 2.2	11.0 ± 2.4
E2	15.6 ± 2.9	15.0 ± 3.8	14.9 ± 3.9	17.5 ± 3.0	17.3 ± 4.5
P4	8.7 ± 1.5	6.8 ± 0.6	5.5 ± 0.6	8.5 ± 1.5	5.2 ± 0.8
T	0.16 ± 0.02	0.18 ± 0.02	0.20 ± 0.01	0.22 ± 0.02*	0.20 ± 0.01

One- way ANOVA, *: different from *C*: p 0.05.

Table 4. Hormone dosage (ng mL^{-1}, except E2 pg mL^{-1}) from control and As-treated male offspring.

	C	A5-A5	A5-C	A50-A50	A50-C
LH	0.43 ± 0.05	0.35 ± 0.	0.23 ± 0.03	0.35 ± 0.07	0.18 ± 0.03
FSH	4.2 ± 0.4	3.9 ± 0.4	3.0 ± 0.4	2.6 ± 0.6*	1.8 ± 0.6*
T	2.8 ± 0.4	2.9 ± 0.6	2.4 ± 0.7	3.8 ± 0.7	4.0 ± 0.7

One- way ANOVA, *: different from *C*: p 0.05.

4 CONCLUSIONS

Arsenic exposure from gestation to adulthood reduces birth weight, alters sexual differentiation and modifies testosterone and FSH levels mainly in female offspring. Further studies are needed to clarify the mechanism of arsenic disruption in the reproductive system.

REFERENCES

Ahmad, S.A., Sayed, M.H., Barua, S., Khan, M.H., Faruquee, M.H., Jalil, A., Hadi, S.A. & Talukder, H.K. 2001. Arsenic in drinking water and pregnancy outcomes. *Environ. Health Perspect.* 109: 629–631.

Concha, G., Vogler, G., Lezcano, D., Nermell, B. & Vahter, M. 1998. Exposure to inorganic arsenic metabolites during early human development. *Toxicol. Sci.* 44: 185–190.

DeSesso, J.M. 2001. Teratogen update: inorganic arsenic. Teratology 64, 170–173.

Kapaj, S., Peterson, H., Liber, K. & Bhattacharya, P. 2006. Human health effects from chronic arsenic poisoning—a review. *J Environ.Sci Health A Tox. Hazard. Subst. Environ. Eng.* 41: 2399–2428.

Wang, A., Holladay, S.D., Wolf, D.C., Ahmed, S.A. & Robertson, J.L. 2006. Reproductive and developmental toxicity of arsenic in rodents: a review. *Int. J. Toxicol.* 25: 319–331.

One Century of the Discovery of Arsenicosis in Latin America (1914–2014) – Litter, Nicolli, Meichtry, Quici, Bundschuh, Bhattacharya & Naidu (Eds)
© 2014 Taylor & Francis Group, London, ISBN 978-1-138-00141-1

The association between arsenic exposure from drinking water and chronic obstructive pulmonary disease in southwestern Taiwan—a 12-year perspective study

L.-I. Hsu & C.-J. Chen
Genomics Research Center, Academia Sinica, Taipei, Taiwan

ABSTRACT: The association of arsenic exposure with respiratory diseases has never been reported in Taiwan. The aim of this study is examining the dose-response relationship between ingested arsenic and Chronic Obstructive Pulmonary Diseases (COPD) in southwestern Taiwan where well arsenic ranged 350–1,140 μg/L. A total of 1,892 residents were followed from 1998 to 2009. The disease status was ascertained through linkage with Taiwan National Health Insurance database. Cox regression analysis was used to determine the hazard ratio of the disease associated with arsenic exposure. The dose-response relationship between Cumulative Arsenic Exposure (CAE) and COPD was observed after considering age, sex, education level, body mass index, cigarette smoking and hypertension. Hazard ratio was 1.00, 1.11 (95%CI, 0.50–2.47). 1.95 (95%CI, 1.06–3.60), 2.14 (95%CI, 1.18–3.87) in relation to CAE<1000, 1000–9999, 10000–19999, 20000+μg/L*years. Bowen's diseases are an independent factor for COPD, associated with 70% increased disease risk after considering individual's arsenic exposure.

1 INTRODUCTION

Arsenic exposure is an important public health issue worldwide. Chronic arsenic ingestion was significantly associated with adverse health effects including skin lesions and Bowen's disease, various cancers, cardiovascular diseases, hypertension, diabetes mellitus and cataract in a dose-response manner. The study in Bangladesh has reported that low or moderate arsenic exposure from drinking water was associated with increased risk of respiratory symptoms including frequent cough, difficulty in breathing and cough with blood (Parvez *et al.*, 2010). *In utero* or early life exposure to arsenic from drinking water greatly increases subsequent mortality in young adults from non-malignant pulmonary disease such as bronchiectasis (Smith *et al.*, 2006). However, the association of this exposure with respiratory diseases has never been reported in Taiwan. The aim of this study is to examine the dose-response relationship between ingested arsenic and Chronic Obstructive Pulmonary diseases (COPD) in arsenicosis area in southwestern Taiwan.

2 MATERIALS AND METHODS

The southwestern cohort was recruited from four Black Foot Disease (BFD)-endemic townships on the southwestern coast of Taiwan. In this area, most residents had used artesian well water for >50 years from 1910 until the early 1970s. In this area, the water from most of these deep wells has arsenic concentrations of 350–1,140 μg/L (median: 780 μg/L). A total of 2,471 subjects aged >30 years were recruited between May 1985 and July 1989, including 257 Black-Foot Disease (BFD) patients at baseline. Skin examination for the study subjects was held by experienced dermatologist during 1989–1996. The Cumulative Arsenic Exposure (CAE) from the artesian well water was defined as $\Sigma(C_i \times D_i)$, where Ci was the median arsenic level of the ith village the subject resided, and Di was the duration of drinking the artesian water in the ith village. The average arsenic concentration of each individual was estimated as the CAE divided by their total years of well water consumption. The disease status of each subject was ascertained through linkage with computerized Taiwan National Health Insurance (NHI) database. In Taiwan, 96.3% of residents were insured for NHI in the first three years when the insurance system was implemented and 99% of residents were insured at the end of 2008. The data including complete outpatients visits, emergency visits, hospital admissions, prescriptions, disease and vital status during the study period were used in this study. Identification of COPD patients was defined including: (1) emphysema: at least one inpatient claim record for ICD-9 code = 492 or at least two refillable prescriptions for emphysema;

(2) bronchiectasis: at least one inpatient claim record for ICD-9 code = 494 or at least two refillable prescriptions for bronchiectasis; (3) chronic airway obstructive, not elsewhere classified: at least one inpatient claim record for ICD-9 code = 496 or at least two refillable prescriptions for COPD. The subjects who died before 1998/01/01, or the prevalent COPD diagnosed before 1998/03/01 were excluded. Finally 1,892 subjects were included for further analysis. Cox regression analysis was used to determine the hazard ratio of the disease associated with arsenic exposure indices including cumulative arsenic exposure and average arsenic concentration. To examine the independent role of arsenical skin lesions, the association of Bowen's disease with COPD was estimated by Cox regression with the adjustment of age, sex, education (illiterate, primary, high school and above), body mass index, cigarette smoking and cumulative arsenic exposure (<1000, 1000–9999, 10000–19999 and 20000+µg/L*years).

3 RESULTS

There were 198 newly diagnosed COPD cases during a follow-up period of 20,756 person-years (mean time: 9.7 years). Old age, male, cigarette smoking and hypertension are significant risk factors for COPD. The crude incidence rate (per 10,000 person-years) was 65.3, 74.3, 80.7, respectively in relation to average arsenic concentrations of ≤ 10.0, 10.1–499 and ≥ 500.0 µg/L; and 65.3, 39.3, 84.0, 121.9 per 10,000 person-years in relation to CAE <1000, 1000–9999, 10000–19999, 20000+ µg/L*years, respectively. Dose-response relationship was observed between two arsenic exposure indices and the COPD risk. With the adjustment of age, sex, education level, cigarette smoking, hypertension and diabetes, the Hazard Ratio (HR) was 1.00, 0.95 (95% Confidence Interval (CI),0.49–1.82), 1.60 (95%CI, 1.00–2.57) in relation to ≤ 10.0, 10.1–499.9 and ≥ 500.0 µg/L; HR was 1.00, 1.11 (0.50–2.47). 1.95 (1.06–3.60), 2.14 (1.18–3.87) in relation to <1000, 1000–9999, 10000–19999, 20000+ µg/L*years. Bowen's disease is an independent risk factor for COPD, with 70% increased disease risk with the consideration of potential confounding factors and arsenic exposure.

4 CONCLUSIONS

There was a significant dose-response trend between ingested arsenic exposure and risk of chronic obstructive pulmonary disease. Arsenic-induced Bowen's diseases are an independent factor associated with increased risk of emphysema, bronchiectasis and unspecified COPD after considering potential risk factors and individual's arsenic exposure.

REFERENCES

Parvez, F., Chen, Y., Brandt-Rauf, P.W., Slavkovich, V., Islam, T., Ahmed, A., Argos, M., Hassan, R., Yunus, M., Haque, S.E., Balac, O., Grazianol, J.H. & Ahsan, H. 2010. A perspective study of respiratory symptoms associated with respiratory symptoms in Bangladesh: findings from the health effects of arsenic longitudinal study (Heals). *Thorax* 65: 528–533.

Smith, A.H., Marshall, G., Yuan, Y., Ferreccio C, Liaw J, von Ehrenstein, O., Steinmaus, C., Bates, M.N. & Selvin, S. 2006. Increased mortality from lung cancer and bronchiectasis in young adults after exposure to arsenic in utero and in early childhood. *Environ. Health Perspect.* 114:1293–1296, 2006.

One Century of the Discovery of Arsenicosis in Latin America (1914–2014) – Litter, Nicolli, Meichtry, Quici, Bundschuh, Bhattacharya & Naidu (Eds)
© 2014 Taylor & Francis Group, London, ISBN 978-1-138-00141-1

Arsenic exposure and blood pressure, homocysteine and haemoglobin concentrations—a study from northern Argentina

S.S. Ameer, K. Engström & K. Broberg
Department of Laboratory Medicine, Section of Occupational and Environmental Medicine, Lund University, Lund, Sweden

F. Harari & M. Vahter
Institute of Environmental Medicine, Section for Metals and Health, Karolinska Institutet, Stockholm, Sweden

G. Concha
National Food Administration, Toxicology Division, Uppsala, Sweden

ABSTRACT: Exposure to inorganic arsenic is a risk factor for cardiovascular disease. Blood pressure was analyzed for 244 women from northern Argentina with varying exposure to inorganic As from drinking water. Blood samples were collected for the measurement of homocysteine, haemoglobin and for genetic analysis; spot urine samples for the measurement of As metabolites. In contrast to what was expected, we found significantly lower concentrations of homocysteine and lower blood pressure with increasing arsenic exposure. However, arsenic seemed to lower haemoglobin concentrations, indicating impairment of the heme synthesis. Data on other risk factors for cardiovascular disease, such as telomere length and markers of oxidative stress, are presented.

1 INTRODUCTION

Inorganic arsenic in drinking water is not only a potent carcinogen but also impairs the cardiovascular system (Moon *et al.*, 2012). Other risk factors/markers of cardiovascular disease include high blood pressure, elevated concentrations of homocysteine in plasma (P-Hcy), chronic kidney disease and possibly anemia (Joseph *et al.*, 2013; Pisaniello *et al.*, 2013). In this study, the effect of exposure to inorganic arsenic in drinking water on the P-tHCy, blood pressure, and hemoglobin was studied, as well as the effect of the *AS3MT* genotype.

2 MATERIALS AND METHODS

2.1 *Study areas and populations*

The study individuals were women recruited in 2008 ($n = 203$) and 2011 ($n = 41$) from the Argentine Andean village San Antonio de los Cobres (~200 μg L^{-1} of As in water, Concha *et al.*, 2010) and from small surrounding villages with lower As values (3.5–70 μg L^{-1}). Urine and blood samples were collected as spot samples (not fasting). Venous blood samples were collected in K_2EDTA (Vacuette®; Greiner Bio-One GmbH) and heparin tubes. The tubes were centrifuged after 10 min of withdrawal to obtain plasma fraction for P-tHcy measurement. Haemoglobin was measured using HemoCue kit (HemoCue, Ängelholm). Blood pressure was measured from the right arm after 5 min rest in a lying position.

The Health Ministry of Salta (Argentina) and the Ethics Committee of the Karolinska Institutet (Sweden) approved the study. Oral and written informed consent was obtained from all participants.

2.2 *Arsenic and genetic analyses*

Exposure to inorganic arsenic was assessed by the sum concentration of arsenic metabolites in urine using HPLC-HG-ICPMS (Agilent 1100 series system; Agilent 7500ce; Agilent Technologies) employing adequate quality control (Schlawicke Engström, 2007). As concentrations were adjusted to the mean specific gravity (SG = 1.020 g mL^{-1}). DNA from peripheral blood was genotyped for one *AS3MT* SNP associated with arsenic metabolism efficiency (Engström *et al.*, 2011) by Taqman allelic discrimination (Applied Biosystems, Life Technologies).

2.3 *Statistical analyses*

Total arsenic in urine and P-tHCy linearized by log-transformation. Linear regression was used, being Ln P-tHCy, blood pressure or haemoglobin the dependent variables and with adjustments for age (model 1) and for age, BMI and coca use (model 2). Stratification for *AS3MT* genotypes was performed to analyse genetic influence. First-degree relatives were excluded from the analyses of genetic effects.

3 RESULTS AND DISCUSSION

3.1 *Arsenic and homocysteine, blood pressure and haemoglobin*

The study participant characteristics are presented in Table 1.

In linear regression analyses, sum of urinary arsenic metabolites was negatively associated with P-tHcy blood pressure and haemoglobin concentrations, after adjusting for influential factors (Table 2).

3.2 *Arsenic and cardiovascular effects in relation to AS3MT genotype*

The genotype for one SNP (rs3740393) was analyzed in *AS3MT*, the major arsenic-metabolising gene, which modified the effects of arsenic. The G allele has previously been associated with less efficient As metabolism (Engström *et al.*, 2011). However, there was a significant ($p = 0.03$) effect

Table 1. General characteristics of the study population.

Variable	*n*	Median	5/95 percentile
Age (years)	244	34	18.2/64
BMI	244	25	18.9/35.0
Parity	239	3	0/11
Coca use (%)	244	51.6	
P-tHcy (homocysteine) (μmol L^{-1})	239	6.1	4.0/10.1
Diastolic blood pressure	244	70	55/85
Systolic blood pressure	244	115	100/140
Haemoglobin (g L^{-1})	234	154	128 to 180
Urinary arsenic metabolites (μg L^{-1})	244	200	21.6 to 537

Table 2. Regression analysis between log-transformed arsenic in urine and log-transformed P-tHcy. Effects of arsenic (β) is presented for each outcome[a].

Outcomes	[a]Model 1		[b]Model 2	
	β (95% CI)	*p*	β (95% CI)	*p*
P-tHCy	–0.043 (–0.082 to –0.005)	0.029	–0.042 (–0.08 to –0.003)	0.033
Diastolic Blood pressure	–1.8 (–2.9 to –0.68)	0.002	–1.8 (–2.9 to –0.70)	0.001
Systolic Blood pressure	–1.6 (–3.1 to –0.057)	0.042	–1.5 (–3.0 to –0.010)	0.049
Haemoglobin	–3.2 (–5.1 to –1.3)	0.001	–3.3 (–5.2 to –1.3)	0.001

[a]Ln P-tHCy/BP (D/S)/Hb = $\alpha + \beta 1 \times$ Ln U-As + $\beta 2 \times$ age. The effect estimate presented for each outcome corresponds to β1 in the model.
[b]Ln P-tHCy/BP (D/S)/Hb = $\alpha + \beta 1 \times$ Ln U-As + $\beta 2 \times$ age + $\beta 3 \times$ BMI + $\beta 4 \times$ coca use. The results were based on 239–244, depending on outcome.

Table 3. Association between arsenic in urine (log-transformed) and outcomes stratified for *AS3MT* genotype.

Variables	Genotype	*n*	β (95% CI)	*p*-Genotype	*p*-interaction
Ln-P-tHCy	CC	93		0.48	0.84
	CG	70	–0.04 (–0.13 to 0.05)		
	GG	17	–0.05 (–0.20 to 0.11)		
Diastolic Blood pressure	CC	94		0.85	0.17
	CG	70	–0.54 (–3.2 to 2.1)		
	GG	19	0.63 (–3.6 to 4.8)		
Systolic Blood pressure	CC	94		0.98	0.87
	CG	70	0.37 (–3.2 to 4.0)		
	GG	19	0.002 (–5.7 to 5.7)		
Haemoglobin	CC	94		0.54	0.03
	CG	70	–0.98 (–5.4 to 3.5)		
	GG	19	–3.9 (–11.0 to 3.1)		

[a]Ln P-tHCy/BP (D/S)/Hb = $\alpha + \beta 1 \times$ Ln U-As + $\beta 2 \times$ age + $\beta 3 \times$ genotype.
[b]Ln P-tHCy/BP (D/S)/Hb = $\alpha + \beta 1 \times$ Ln U-As + $\beta 2 \times$ age + $\beta 3 \times$ genotype + $\beta 4$ (Ln U-As $\times$ genotype).

modification that the slow metabolizers had more severe effect on the hemoglobin levels (Table 3).

4 CONCLUSIONS

Unexpectedly lower concentrations of homocysteine and lower blood pressure with increasing As exposure were found. However, As seemed to lower heamoglobin concentrations, indicating impairment of the heme synthesis by As. Data on other risk factors or mediators for cardiovascular diseases, such as telomere length and markers of oxidatve stress, will be analyzed in future works.

ACKNOWLEDGEMENTS

We thank the study participants and the doctors and health care personnel from the local hospital and health care centers.

REFERENCES

Engström, K., Vahter, M., Mlakar, S.J., Concha, G., Nermell, B., Raqib, R., Cardozo, A. & Broberg, K. 2011. Polymorphisms in arsenic (+III oxidative state) methyltransferase (*AS3MT*) predict gene expression of AS3MT as well as arsenic metabolism. *Environ. Health Perspect.* 19(2): 182–188.

Engström, K.S., Broberg, K., Concha, G., Nermell, B., Warholm, M. & Vahter, M. 2007. Genetic polymorphisms influencing arsenic metabolism: evidence from Argentina. *Environ. Health Perspect.* 115(4): 599–605.

Joseph, J. & Loscalzo, J. 2013. Methoxistasis: Integrating the Roles of Homocysteine and Folic Acid in Cardiovascular Pathobiology. *Nutrients* 5: 3235–3256.

Moon, K.M., Eliseo, G. & Navas-Acien, A. 2012. Arsenic exposure and cardiovascular disease: an updateed systematic review. *Curr. Atheroscler. Rep.* 14: 542–555.

Pisaniello, A.D., Wong, D.T., Kajani, I., Robinson, K. & Shakib, S. 2013. Anemia in chrinic heart failure: more awareness is required. *Intern. Med. J.* 49(9): 999–1004.

One Century of the Discovery of Arsenicosis in Latin America (1914–2014) – Litter, Nicolli, Meichtry, Quici, Bundschuh, Bhattacharya & Naidu (Eds)
 ISBN 978-1-138-00141-1

Antidote effects of plants of Himalayan sub-origin against arsenic induced toxicity

A. Kumar, Md. Ali, R. Kumar, A. Nath & J.K. Singh
Mahavir Cancer Institute and Research Centre, Patna, Bihar, India

ABSTRACT: In India, arsenic poisoning in groundwater in Gangetic basin, especially the districts adjoining the Ganges river, right from Eastern Uttar Pradesh, Bihar to West Bengal are the major problem of concern and where major health related problems are arising. To combat the present problem and because there are no antidotes against arsenic poisoning, a pre-clinical study on Charles foster rats was done. Sodium arsenite was administered to the rats for 60 days and upon this arsenic pretreated rats, *Withania somnifera* and *Pteris longifolia* plant extracts of Himalayan sub-region were selected and administered for 45 days to study the antidote effects. In *in-vivo* conditions, these plants played the vital role to combat the arsenic induced toxicity completely proving to be the best antidote. The plants not only eliminated the effects of arsenic but also reversed the normal physiological activity in the animal, having the best antidote activity against arsenic induced toxicity.

1 INTRODUCTION

Arsenic has become one of the most important global environmental toxicant and has caused health hazards to human population through its contamination of groundwater used for drinking purposes. Recently, in Bihar, India, 16 districts have been found to have groundwater contaminated with arsenic in lower Gangetic plains. The arsenic contamination was also observed in three districts: Ballia, Varanasi and Gazipur of Uttar Pradesh in the upper and middle Ganga plain, India (Ahamed *et al.*, 2006).

Among possible target organs of arsenic, the kidney and central nervous system appear to be the most sensitive ones. Having been absorbed from the alimentary tract, most of the metals and metalloids form durable combination with the protein thionein, forming metallothionein, which plays an important role in the further metabolism of these metals. The kidney and liver are considered to be the most susceptible organs for metals, because they contain most of the metallothionein binding toxic metals, and also produce free radicals such as lipid peroxides and encounters with biomembranes and sub-cellular organelles (Kumar *et al.*, 2013).

Arsenite in *in vivo* conditions are transported into cells through aquaglycoporins 7 and 9, which transport water and glycerol. Arsenite binds to cellular sulfhydryl, especially the vicinal ones and they interfere with high energy generation (Aposhian & Aposhian, 2006).

Since last two decades, the amelioration of various heavy metals borne diseases has gained special attention to researchers (Bhattacharya *et al.*, 2000). In the present investigation, Charles Foster rats were treated with sodium arsenite to observe the ameliorative effect against arsenic induced toxicity. Two medicinal plants, *Withania somnifera* and *Pteris longifolia*, were used for the study.

2 METHODS/EXPERIMENTAL

2.1 *Animals*

Charles Foster rats (24 females), weighing 160 to 180 g, 8 weeks old, were obtained from the animal house of the Mahavir Cancer Institute and Research Centre, Patna, India (CPCSEA Regd-No. 1129/bc/07/CPCSEA). The research work was approved by the IAEC (Institutional Animal Ethics Committee) with IAEC No. IAEC/2012/12/04. Food and water to rats were provided *ad libitum*.

2.2 *Experimental design*

Sodium arsenite at the dose of 8 mg kg^{-1} body weight was administered orally daily for 60 days. Upon these pre-treated groups, *W. somnifera* ethanolic root extract was administered at the dose of 200 mg kg^{-1} body weight while *P. longifolia* ethanolic leaf extracts was administered at the dose of 400 mg kg^{-1} body weight orally daily for 45 days. No treatment was administered to the control group. At end of the treatment, blood samples were collected and serums were extracted. The serums were then assayed for biochemical study as liver function tests, kidney function tests and lipid peroxidation.

3 RESULTS AND DISCUSSION

3.1 *Morbidity and mortality*

The rats after the exposure to arsenic (8 mg kg^{-1} b.w. day^{-1}) for 60 days showed toxicity symptoms such as nausea, nose bleeding, lack of body co-ordination (11 percent of rats showed paralysis like symptoms), blackening of tongue and foot and general body weakness.

3.2 *Biochemical Changes*

The SGPT, SGOT, alkaline phosphatase, total bilirubin, urea, uric acid, creatinine and lipid peroxidation activity showed a significant increase ($p < 0.05$) in the arsenic treated group in comparison to control rat group. However, these values were significantly lowered ($p < 0.05$) in *W. somnifera* and *P. longifolia* treated group. (Table 1).

The abnormally high level of serum SGPT, SGOT, ALP and total bilirubin in the present study are the consequence of arsenic induced liver dysfunction and denotes damage to the hepatic cells. The significant increase in the levels is a direct measure of hepatic injury and they show the status of the liver as there may be cellular leakage and loss of functional integrity of hepatocytes. Furthermore, the significant increase in the levels of the lipid peroxidation denotes oxidative stress produced by arsenic, leading to a high degree of degeneration in the liver cells. Rats treated with *W. somnifera* and *P. longifolia* showed a significant decrease in the serum LFT levels, which denotes their hepatoprotective activity. The hepatoprotective effect of

Table 1. Results are presented as mean ± SD, and total variation present in a set of data was analyzed through one way analysis of variance (ANOVA).

Tests	Control	Arsenic treated	*Withania* treated	*Pteris* treated
SGPT (U mL^{-1})	25.55 ± 1.350	84.50 ± 1.893	37.43 ± 1.386	19.92 ± 0.934
SGOT (U mL^{-1})	27.77 ± 1.307	120.8 ± 1.641	43.33 ± 1.764	28.00 ± 1.915
ALP (K.A Units)	8.66 ± 0.628	29.08 ± 1.519	6.82 ± 0.338	12.73 ± 0.874
Bilirubin (mg dL^{-1})	0.43 ± 0.033	2.54 ± 0.080	0.79 ± 0.034	0.98 ± 0.030
Urea (mg dL^{-1})	25.33 ± 0.881	54.72 ± 1.532	35.17 ± 1.515	36.75 ± 1.778
Uric acid (mg dL^{-1})	3.48 ± 0.250	11.72 ± 0.923	5.13 ± 0.171	6.93 ± 0.296
Creatinine (mg dL^{-1})	0.69 ± 0.034	1.90 ± 0.050	0.88 ± 0.032	0.77 ± 0.029
LPO (nmol mL^{-1})	1.80 ± 0.095	84.13 ± 1.806	7.05 ± 0.303	7.73 ± 0.444

W. somnifera has been well documented and it also proves its antioxidant activity. Lipid peroxidation amelioration by *W. somnifera* and its free radical scavenging activity have been well studied. However, no studies up to date have reported any result of the action of *P. longifolia* on animals.

The significant increase in the serum urea, uric acid and creatinine denotes the high degree of degeneration in the nephrocytes. However, after administration of *W. somnifera* and *P. longifolia* there was a significant decrease in their levels, which denotes their nephroprotective effect. Very few documents support the nephroprotective activity of *W. somnifera*.

4 CONCLUSIONS

The present study is indeed a novel work which deciphers for the first time the action of the antidote against arsenic induced toxicity on liver and kidney, as they are the metabolic organs which show the changes at first level. Thus, these novel plants *W. somnifera* and *P. longifolia* possess hepatoprotective, nephroprotective and antioxidant properties against arsenic induced toxicity.

ACKNOWLEDGEMENTS

The authors extend their appreciation to the Department of Science & Technology (SSTP Division), Ministry of Science & Technology, Government of India, New Delhi, for the financial assistance of this work and to the institute for the entire infrastructural facilities.

REFERENCES

Ahamed, S., Sengupta, M.K., Mukherjee, A., Hossain, M.A., Das, B., Nayak, B., Pal, A., Mukherjee, S.C., Pati, S., Dutta, R.N., Chatterjee, G., Mukherjee, A., Srivastava, R. & Chakraborti, D. 2006. Arsenic groundwater contamination and its health effects in the state of Uttar Pradesh (UP) in upper and middle Ganga plain, India: A severe danger. Sci. *Total Environ.* 370: 310–322.

Aposhian, H.V. & Aposhian, M.M. 2006. Arsenic toxicology: Five questions. *Chem Res Toxicol.* 19(1): 1–15.

Bhattacharya, A., Ramanathan, M., Ghosal, S. & Bhattacharya, S.K. 2000. Effect of *Withania somnifera* glycowithanolides on iron-induced hepatotoxicity in rats. *Phytother Res.* 14(7): 568–570.

Kumar, A., Ali, M., Kumar, R., Suman, S., Kumar, H., Nath, A., Singh, J.K. & Kumar, D. 2013. *Withania somnifera* protects the haematological alterations caused by Sodium Arsenite in Charles Foster rats. *International Journal of Research in Ayurveda & Pharmacy (IJRAP)* 4 (4): 491–494.

One Century of the Discovery of Arsenicosis in Latin America (1914–2014) – Litter, Nicolli, Meichtry, Quici, Bundschuh, Bhattacharya & Naidu (Eds)

Performance on an executive function task among Uruguayan school children exposed to low-level arsenic

G. Barg
Faculty of Psychology, Catholic University of Uruguay, Uruguay

M. Vahter
Institute of Environmental Medicine, Karolinska Institutet, Sweden

E.I. Queirolo
Center for Research, Catholic University of Uruguay, Uruguay

N. Mañay
Faculty of Chemistry, University of the Republic of Uruguay (UDELAR), Uruguay

K. Kordas
School of Social and Community Medicine, University of Bristol, UK

ABSTRACT: Arsenic is a known neurotoxic with demonstrated effects on child IQ, particularly in regions of the world characterized by high-level arsenic exposure. Less is known about the effects of low-level exposure on child cognition and about effects other than general cognitive abilities. We examined the extent to which low-level exposure in school children is related to performance on an executive function task (the Internal/External Dimension Shift task). Sum of inorganic arsenic metabolites in the urine of these children was low (14.4 ± 11.9 μg L^{-1}) and not associated with measures of task performance in adjusted models. The same was true for %MMA. The multivariate models need to be interpreted with caution and replicated in other studies. Based on this study, we find little effects of low-level arsenic exposure on executive function, particularly the ability to learn new rules.

1 INTRODUCTION

Toxic effects of arsenic have been described both in relation to prenatal (Bellinger, 2013) and postnatal exposure (Naujokas *et al.*, 2013). In terms of cognition, most studies have focused on IQ and have been conducted in areas where exposure levels are quite high. A recent meta-analysis of 15 studies concluded that a 50% increase in urinary As concentration in a wide range of body burdens represented by those studies was associated with a 0.4 lower full scale IQ (Rodríguez-Barranco *et al.*, 20123). The association between low-level As exposure and child cognition, and particularly domains other than the IQ, remains under-studied. The aim of this study was to examine potential impairments in executive function—the way children use cognitive capacity to achieve goals—in children with low-level As exposure.

2 METHODS

2.1 *Study sample and setting*

This study was conducted in Montevideo, Uruguay among 1st grade children aged 6–8 y who attended 9 private elementary schools. Low-level exposure to arsenic was previously reported in Montevideo children (Kordas *et al.*, 2010). Children and parents were recruited through schools and attended several study sessions after providing written consent, one for collect biological samples and two cognitive testing sessions. A home visit was conducted to assess the developmental inputs/stimulation in the child's home. Ethical approval was given by Pennsylvania State University, Catholic University of Uruguay, and University of the Republic of Uruguay.

2.2 *Child assessments*

Children provided first morning urine samples in collection cups previously washed with nitric acid. After transport on ice to the laboratory, an aliquot of urine was separated and stored at –20 °C. Samples were transported frozen to the Karolinska Institutet in Sweden and analyzed for inorganic As, MMA, and DMA using HPLC-HG-ICPMS. Children completed the Internal/External Dimension (IED) shift task of the CANTAB testing battery (Cambridge Cognition Inc) at school. The IED

Table 1. Multivariate association between arsenic exposure and IED outcomes.

Outcome	Model 1: SumAs[1]	Model 2: SumAs[1] *w/* covariates[2]	Model 3: SumAs[1] *w/*: covariates[2] and %MMA
Complete stage errors ($\beta \pm SE$)	−0.14 ± 0.09	−0.28 ± 0.18	−0.23 ± 0.18
Complete stage trials ($\beta \pm SE$)	−0.34 ± 0.43*	−0.83 ± 0.43*	−0.73 ± 0.43*
Pre-ED errors above median (OR [95% CI])	0.96 [0.92, 1.00]**	0.92 [0.85, 1.00]**	0.92 [0.84, 1.00]**
IED errors above median (OR [95% CI])	1.00 [0.97, 1.03]	1.00 [0.94, 1.07]	0.99 [0.93, 1.06]
Total errors above median (OR [95% CI])	1.00 [0.96, 1.03]	0.97 [0.90, 1.03]	0.96 [0.89, 1.04]
Total trials above median (OR [95% CI])	1.01 [0.98, 1.04]	1.04 [0.96, 1.11]	1.03 [0.95, 1.12]

[1]SumAs adjusted for specific gravity; [2]Covariates included child IQ, age and sex; maternal IQ; HOME Inventory score, crowding in the household, and SES score; *$p < 0.1$; **$p < 0.05$.

tests rule acquisition and attentional set shifting. The outcomes include pre-shift errors (errors occurring prior to the change of dimension), shift errors (errors occurring at the change of relevant dimension) and total errors, and total number of trials and stages completed.

2.3 *Statistical analysis*

Urinary concentrations of inorganic arsenic, MMA and DMA were summed into the "Sum of arsenic metabolites" (SumAs), and this measure was adjusted for specific gravity to account for hydration. Bivariate and multivariate regression analyses were conducted in STATA. Three types of models were run: Model 1 tested bivariate association between SumAs and IED outcomes, Model 2 included adjustment for covariates, and Model 3 included control for %MMA in addition to covariates. The covariate adjustment included child IQ, age and sex, maternal IQ, HOME Inventory score, crowding index, and socioeconomic score. Overall, statistical significance was determined based on $p < 0.05$.

3 RESULTS AND DISCUSSION

3.1 *Sample characteristics*

Overall, 192 children provided urine samples for analysis. These children had a mean 81.7 ± 6.7 months of age, and 57.3% were boys. Most had working parents (59.4% mothers and 89.3% father) and lived with both parents (70.9%). About 20% of mothers and 27% of fathers only had primary education, and some (3–6%) did not complete it, while ~3% of both mothers and fathers completed university education. Socioeconomic status was low-moderate: 40% of the monthly income (equivalent to \$661 ± 596) was spent on food. This sample was characterized by low concentrations of SumAs (11.4 ± 11.9 $\mu g\ L^{-1}$) and As metabolites in urine. The study children had a mean IQ of 93.0 ± 16.7 points.

3.2 *Association of urinary arsenic with IED*

In the first step of the statistical modeling (Model 1, Table 1), SumAs was associated with fewer completed stage trials ($\beta \pm SE$: −0.3 ± 0.2, $p < 0.1$) and lower likelihood of committing many Pre-ED errors (OR [95% CI]: 0.96 [0.92, 1.00]). Adjustment for covariates (Model 2) strengthened the magnitude of the unadjusted associations. When modeled together with covariates and %MMA, SumAs continued to be associated with these two measures. No other statistically significant associations with SumAs were observed, either in bivariate or adjusted models. In the fully-adjusted model (Model 3), %MMA was associated with more completed stage errors (0.65 ± 0.36, $p = 0.07$), independently of SumAs. The lack of significant associations with arsenic possibly reflects the low-level of exposure and small study group. We plan on recruiting more children to the cohort.

4 CONCLUSIONS

This is the first study to examine the association between very low exposure to inorganic As, measured in urine, and specific domains of cognitive performance in children. Within a range of low-level As exposure, sum of urinary As was generally not associated with performance on the IED Shift Task. %MMA was associated with more completed stage errors and stage trials, but lower likelihood of committing many errors. After covariate adjustment, limited number of associations remained, either because low-level As exposure does not impact on executive functions or because the study was under-powered to detect differences. Additional studies will be needed to confirm these findings.

ACKNOWLEDGEMENTS

D. Ribeiro, G. Yuane and A. Roy for help in sample preparation after collection; J. Deana, M. Pérez, M. Suero, K. Horta and A. Beisso for administering the cognitive task; and M. Granér and B. Palm for U-As analyses.

REFERENCES

Bellinger, D. 2013. Prenatal exposures to environmental chemicals and children's neurodevelopment: an update. *Safety Health Work* 4:1–11.

Kordas, K., Queirolo, E.I., Ettinger, A.S., Wright, R.O. & Stoltzfus, R.J. 2010. Prevalence and predictors of exposure to multiple metals in preschool children from Montevideo, Uruguay. *Sci. Tot Environ.* 408: 4488–4494.

Naujokas, M., Anderson, B., Ahsan, H., Aposhian, H., Graziano, J., Thompson, C. & Suk, A. 2013. The broad scope of health effects from chronic arsenic exposure: an update on a worldwide health problem. *Environ. Health Perspect.* 121(3): 295–302.

Rodríguez-Barranco, M., Lacasaña, M., Aguilar-Garduño, C., Aguilar-Garduño, C., Alguacil, J., Gil, F., González-Alzaga, B. & Rojas-García, A. 2013. Association of arsenic, cadmium and manganese exposure with neurodevelopment and behavioural disorders in children: A systematic review and meta-analysis. *Sci. Tot. Environ.* 454: 562–577.

One Century of the Discovery of Arsenicosis in Latin America (1914–2014) – Litter, Nicolli, Meichtry, Quici, Bundschuh, Bhattacharya & Naidu (Eds)

Human exposure assessment to arsenic and health indicators in Paracatu, Brazil

E.M. De Capitani
State University of Campinas, School of Medicine, Poison Control Center, Campinas, SP, Brazil

I.M. Jesus, K.R.F. Faial, M.O. Lima & K.C.F. Faial
Evandro Chagas Institute, Ministry of Health, Belém, PA, Brazil

A.P. Ferreira
Fundação Oswaldo Cruz, FIOCRUZ, Ministry of Health, Rio de Janeiro, RJ, Brazil

L.M. Domingos & Z.C. Castilhos
Center for Mineral Technology, CETEM, Ministry of Science, Technology and Innovation, RJ, Brazil

ABSTRACT: We conducted a descriptive epidemiological study to assess the human exposure and health effects due to possible arsenic exposure from an open pit golden mine operating in Paracatu. Two subpopulations were assessed regarding hair, urine and blood arsenic concentrations: one living in the region bordering the gold mining operation activities and the other living around 5 km distant from the gold mining process. Statistical significant differences were found in urinary arsenic levels between the two subpopulations, considering plain results in μg L^{-1} and normalized by urinary creatinine concentrations ($p < 0.0001$). However, both means showed to be in the interval of reference values for an urban Brazilian population. Cancer mortality and dermatological morbidity were also evaluated. Results showed that cancer mortality from cancers usually related to arsenic chronic exposure to be in or below the expected background indexes. No typical arsenicosis dermatopathies were found. Hair samples are still under analysis.

1 INTRODUCTION

Several health effects can result from inorganic arsenic chronic exposure, consisting of carcinogenic and non-carcinogenic conditions (Obiri *et al.*, 2006; Pearce *et al.*, 2012). For both kinds of effects, a long period of time is necessary from the start of the exposure and the diagnosis, depending, in this context, on the level of exposure. The most frequent and early non-carcinogenic effect is related to changes in skin (Guha Mazumder, 2000). Also related to the skin is the most prominent carcinogenic effect (multiple basocellular, multiple epidermoid carcinomas, and Bowen's disease).

Studies of health effects due to environmental exposure to As have been published mostly in populations ingesting arsenic contaminated water, like in Chile, Argentina, Bangladesh, West Bengal, and Taiwan. Arsenic inhalation studies are uniquely restricted to the occupational set where deleterious effects have been described in high inhalatory doses. In this respect, related studies have shown that arsenic inhalation is responsible for less than 1% of the total absorbed dose due to non-occupational environmental contamination (ATSDR, 2007).

The objective of this study was to evaluate the human exposure to As in a population living near to an open pit gold mining in Paracatu, MG, Brazil. Results from drinking water sources used by the population showed very low levels of As, all of them below 2 mg L^{-1} (Bidone *et al.*, 2014), indicating potentially no concerns on this exposure pathway. On the other hand, the mine operates in an open pit process involving the production of dust potentially contaminated by As. Results of As in particulate matter in the atmosphere indicate that dusts can reach residential areas of the city, which might contaminate the air and can be an additional source to the surrounded soil (Zamboni *et al.*, 2014).

2 POPULATION AND METHODS

Paracatu is situated in the northwest of MG, with an economy tied to the agricultural and mining sectors (gold, zinc and calcareous). The city had

84,687 inhabitants in 2010 (1.7% annual increase rate) and a Human Development Index of 0.760 (IBGE, 2010).

A descriptive epidemiological study was conducted to assess human exposure and health effects through three complementary approaches: a) population sampling following an epidemiological design which took into account two geographically distinct communities: one living in the region bordering the study area and the other living around 5 km distant from the gold mining process (control area); b) dermatological evaluation of suspected cases previously selected by the public health care system team; c) mortality data analysis regarding cancers known to be associated to chronic arsenic exposure. Urine, blood and hair samples were collected from adults older than 40 years and living in the areas for more than 20 years, a population thought to be the most sensitive to chronic exposure to low arsenic doses. As analyses in urine and blood was done by GF-AAS. Mortality data from lung, urinary bladder, liver and skin cancers from Paracatu of 2000 (not shown here) and 2010 were collected from the Ministry of Health official data bank (DATASUS) and compared with data from cities with similar demographic structure, with gold mining activities (Nova Lima) and without (Três Corações and Itajubá). Comparisons were also made with data from the state of Minas Gerais (MG), and Brazil as a whole. The epidemiological study was approved and registered at the National Commission for Ethics in Research (CONEP/CAE 04328912.0.0000.0019).

3 RESULTS AND DISCUSSION

Cancer mortality was analyzed regarding four kinds of cancers associated to As exposure: lung, liver, urinary bladder and skin, despite the fact that skin cancers, besides melanoma (not involved in cases of As exposure), have very low mortality. Table 1 shows the distribution of mortality rates by 100,000 inhabitants in 2010, and compared with rates from MG and Brazil as a whole. Table 1 shows that the mortality rates in Paracatu regarding lung, liver and urinary bladder cancers are in the same magnitude compared with other cities, MG and Brazil as a whole. There is no indication of increased rates in Paracatu regarding these pathological conditions.

As most cases of lung cancer cases are fatal within five years, relative risks for incidence and mortality are approximately the same.

A preliminary survey of dermatopathies was carried out using the local health care system represented by two of the public Family Medicine Health Services (PSF: Programa de Saúde da Família),

Table 1. Mortality rate by 100,000 in Paracatu compared with 3 cities with similar demographic structure and with MG and Brazil in 2010.

Organ	PCT	NL	TC	ITJ	MG	Brazil
Lung	10.5	19.7	12.4	11.0	9.5	14.0
Liver	3.5	6.1	1.3	6.6	3.5	4.0
UB	1.2	NA	1.3	1.1	1.5	1.6
skin	NA	NA	NA	1.1	0.6	0.8

PCT: Paracatu; NL: Nova Lima; TC: Três Corações; ITJ: Itajubá; MG: Minas Gerais; UB: Urinary Bladder; NA: Data Not Available.

responsible for the care of two geographically distinct communities: one living in the region bordering the mining pit (referring to Amoreiras PSF), and the other living around 5 km distant (Paracatuzinho PSF). Adult patients with any chronic dermatopathy were examined by the two physicians from the researchers group (EMDC and IMJ). The diagnostic approach was based in the peculiar clinical presentation of the possible dermatopathies related to arsenic chronic exposure published elsewhere (Guha Mazumder, 2000). Thirty patients from the study area, and fourteen patients from the control area were minutely examined. No lesions suspected to be associated to arsenic chronic exposure were seen in any of them. Although no significant differences were observed between blood As levels in the two subpopulations studies ($p = 0.42$), they were detected in urinary As levels, indicating that people living close to the mining site are more exposed to As. Nevertheless, urinary As mean and median values, for both groups, showed to be in the interval of reference value for a urban Brazilian population (4.04 $\mu g\ L^{-1}$ SD: 2.01), as well as to the mean (8.10 $\mu g\ L^{-1}$; 7.44–8.83 $\mu g\ L^{-1}$) and median (7.49 $\mu g\ L^{-1}$; 6.90–8.12 $\mu g\ L^{-1}$) of As in urine from US population (CDC, 2012).

4 CONCLUSIONS

Despite the spatial relationship between the gold mine operating in full capacity during the last years, and the communities living in the vicinity, potentially exposed to As contaminated dust, health effects secondary to that exposure showed to be negative regarding specific carcinogenic and skin diseases, considered to be associated to inorganic chronic arsenic exposure. However, the differences between the mean and median urinary arsenic concentrations of the two subpopulations is likely to be related to the proximity of Amoreiras with the mining activities, promoting a heavier As airborne exposure in this locality compared to Paracutuzinho.

REFERENCES

ATSDR (2007). Toxicological Profile for Arsenic (Update). Atlanta, GA, Department of Health and Human Services, Public Health Service.

Brazil. IBGE (2010). Brazilian Institute of Geography and Statistics, www.ibge.gov.br.

Bidone, E., Castilhos, Z.C., Santos, M.C.B., Silva, R.S.V.,. Cesar, R.G & Ferreira, M. 2014. Arsenic levels in natural and drinking water from Paracatu-Brazil. As2014 (unpub.)

CDC (2012) Fourth National Report on Human Exposure to Environmental Chemicals Updated Tables, February, at http://www.cdc.gov/exposurereport

Guha Mazumder, D.N. 2000. Diagnosis and treatment of chronic arsenic poisoning. United Nations synthesis report on arsenic in drinking water. WHO. Geneva.

Obiri, S., Dodoo, D.K., Okai-Sam, F. & Essumang, D.K. 2006. Cancer health risk assessment of exposure to arsenic by workers of Anglo Gold Ashanti-Obuasi gold mine. *Bull Environ Contam Toxicol.* 76(2): 195–201.

Pearce, D.C., Dowling, K. & Sim, M.R. 2012. Cancer incidence and soil arsenic exposure in a historical gold mining area in Victoria, Australia: a geospatial analysis. *J Expo Sci Environ Epidemiol.* 22(3): 248–257.

Zamboni de Mello, W., Matos, J.A., Castilhos, Z.C., da Silva, L.I.D., Carneiro, M.C. & Monteiro, M.I. 2014. As in atmospheric particulate matter at Paractu-Brazil. As2014 (unpub.).

One Century of the Discovery of Arsenicosis in Latin America (1914–2014) – Litter, Nicolli, Meichtry, Quici, Bundschuh, Bhattacharya & Naidu (Eds)
ISBN 978-1-138-00141-1

Toxicity of inorganic thioarsenates for human liver and bladder cells

S. Hinrichsen, R. Lohmayer & B. Planer-Friedrich
Environmental Geochemistry Group, University of Bayreuth, Germany

E. Dopp & R. Zdrenka
Institute of Hygiene and Occupational Medicine, University Hospital Essen, Essen, Germany

ABSTRACT: Arsenic can cause liver and bladder cancer. Its pre-systemic metabolism is influenced by free sulfide in the human gut. We thus investigated the effect of sulfide on arsenite cytotoxicity (S/As molar ratio 4) in human hepatocytes (HepG2) and urothelial cells (UROtsa) focusing on thioarsenate formation and cytotoxicity of individual thioarsenate standards. Cytotoxicity was determined by MTT assay. Arsenic speciation was conducted by anion exchange chromatography coupled to an inductively coupled plasma mass spectrometry. Sulfide reduced arsenite toxicity in both cell lines and thioarsenate formation up to 73% of total arsenic was detected. Testing individual standards, both the order of toxicity and uptake were found to be arsenite > trithioarsenate > monothioarsenate > arsenate. Arsenite was the almost exclusive arsenic species detected intracellularly. It remains unclear whether sulfide regulates the amount of cellular available arsenite by formation of thioarsenates, or if thioarsenates are transferred into the cell and undergo intracellular reduction.

1 INTRODUCTION

Arsenic is a commonly known poison linked to many types of cancer, e.g. liver and bladder cancer. Due to the formation of free sulfide in the human gut lumen, *in vivo* arsenic-sulfur-complexation can significantly influence the pre-systemic arsenic metabolism. Methylated thioarsenates have been detected in urine samples of humans and rodents exposed to arsenic (Naranmandura *et al.*, 2007a; Raml *et al.*, 2007). Microbial arsenate conversion into methylated (thio)arsenicals as well as inorganic thioarsenates was observed both in mouse cecum (Pinyayev *et al.*, 2011) as well as by human fecal microbiota (Van de Wiele *et al.*, 2010). While methylated thioarsenates have been previously investigated for their acute toxicity on different cell types, (Naranmandura *et al.*, 2007b), there are, to the best of our knowledge, no studies on the acute toxicity of the inorganic thioarsenates ($AsS_xO_{4-x}{}^{3-}$) to human cell lines.

A previous study (Planer-Friedrich *et al.*, 2008) used the bioluminescent bacterium *Vibrio fischeri* to test acute toxicity of inorganic thioarsenates, and found increasing toxicity with an increasing number of thio(SH)-groups. While mono—and dithioarsenate were reported to be much less toxic, trithioarsenate was comparable in toxicity to arsenate and arsenite. An increase in toxicity with exposure time was observed for the thioarsenates and interpreted as either lack of detoxification mechanism or conversion into arsenic oxyanions after uptake.

In the present study, we investigated the effect of sulfide on the cytotoxicity of arsenite in human hepatocytes (HepG2) and human urothelial cells (UROtsa) with focus on the intermediate formation of thioarsenates and their individual toxicity.

2 METHODS

HepG2 and UROtsa cells were kindly provided by Prof. Dopp's group, University Hospital Essen, Germany. UROtsa cells were cultivated in minimum essential medium with Earle's Salts (MEM, c·c·pro, Oberdorla, Germany) supplemented with 10% Fetal Bovine Serum (FBS, Gibco), 0.5% gentamycine (c·c·pro, Oberdorla, Germany), and 1% L-glutamine (c·c·pro, Oberdorla, Germany). HepG2 cells were grown in MEM supplemented with 10% FBS, 0.5% gentamycine, 1% L-glutamine, 1% non-essential amino acids, and 1% sodium pyruvate (all c·c·pro, Oberdorla, Germany). HepG2 and UROtsa cells were incubated in a humidified atmosphere at 37 °C and 5% CO_2 (Incubator Galaxy 170 S, New Brunswick Scientific) and exposed to commercially available salts of arsenate, arsenite, and an arsenite-sulfide mixture (S/As molar ratio 4) as well as mono—and trithioarsenate, synthesized in our laboratory. Dithioarsenate could not be synthesized in sufficient purity; tetrathioarsenate was not stable at physiological pH. Cytotoxicity was determined by means of MTT assay (3-(4,5-dimethylthiazol-2-yl)-2,5-diphenyltetrazolium bromide) after 6 and 24 h of exposure. Cellular uptake of the different arsenic compounds at concentrations of 1 to 75 μM in fresh growth medium was determined after 24 h by mechanical cell lysis. Samples for arsenic speciation in cells and medium were flash-frozen immediately after sampling, handled exclusively under nitrogen, and analyzed by Anion Exchange Chromatography coupled to an Inductively Coupled Plasma Mass Spectrometry (AEC-ICP-MS), according to a previously published method for determination of thioarsenates (Planer-Friedrich *et al.*, 2007).

3 RESULTS AND DISCUSSION

Experiments adding excess sulfide to an arsenite solution (molar ratio S/As 4) showed significant reduction of arsenite cytotoxicity for both UROtsa (6 h) and HepG2 cells (6 and 24 h). Speciation analysis of the medium confirmed the expected formation of thioarsenates (up to 73% of total arsenic). Testing individual thioarsenate standards, the order of toxicity was found to be arsenite > trithioarsenate > monothioarsenate > arsenate for both cell lines (6 and 24 h) (Table 1).

The increase of toxicity with a higher degree of thiolation is in accordance with previous results from toxicity tests on bioluminescent bacteria (Planer-Friedrich *et al.*, 2008). In comparison with their chemical structural analogue, arsenate, thioarsenates are more toxic. However, considering their origin as being formed from arsenite in the presence of sulfide, thiolation can be considered a detoxification process. The mediating effect of thioarsenates on arsenite toxicity must, however, be considered in light of their long term instability under changing pH conditions and in the presence of oxygen. This instability with transformation to arsenite might also explain more pronounced increases in cytotoxicity for thioarsenates than for arsenite and arsenate observed in this study (compare IC_{50}s after 6 and 24 h; Table 1) and a previous one (Planer-Friedrich *et al.*, 2007). The amount of cellular arsenic uptake after 24 h was found to reflect the order of toxicity of the four compounds investigated. Arsenite was the dominant to almost exclusive arsenic species detected intracellularly. It currently remains unclear how thioarsenates exactly contribute to arsenite detoxification. They could either regulate the amount of free arsenite available for cellular uptake without entering the cells, or, based on their chemical similarity to arsenate, they could be taken up by similar transporters and reduced rapidly to arsenite in the cell.

Table 1. IC_{50} (Inhibitory concentrations at 50% reduction of cell viability) for arsenite, arsenate, mono-, and trithioarsenate for human liver (HepG2) and bladder (UROtsa) cells.

	HepG2 IC_{50} [μM]		UROtsa	
	6 h	24 h	6 h	24 h
arsenite	287	72	125	4
arsenate	>10,000	3914	552	166
monothioarsenate	1573	371	599	48
trithioarsenate	719	142	162	20

4 CONCLUSIONS

In the present study, we showed that arsenite cytotoxicity was significantly reduced in HepG2 and UROtsa cells in the presence of sulfide, and that formation and transformation of new species—thioarsenates—play an important role in this context. The amount of uptake directly correlates with the cytotoxicity of arsenite, arsenate, mono-, and trithioarsenate, but it is not sufficient to fully explain the observed differences in cytotoxicity. Whether sulfide merely regulates the amount of cellular available, cytotoxic arsenite by extracellular formation of thioarsenates, or thioarsenates are actively transferred into the cell and undergo intracellular transformation, remains subject to future investigation. However, our study clearly showed that experiments focusing on one arsenic species only without considering the presence of interacting species that lead to new species formation such as thioarsenates will yield limited transferability to *in vivo* systems.

ACKNOWLEDGEMENTS

We acknowledge generous funding by the German Research Foundation within the Emmy Noether program (grant # PL 302/3-1) and thank Stefan Will for help with AEC-ICP-MS analyses.

REFERENCES

Naranmandura, H., Suzuki, N., Iwata, K., Hirano, S. & Suzuki, K.T. 2007a. Arsenic metabolism and thioarsenicals in hamsters and rats. *Chemical Research in Toxicology* 20(8): 616–624.

Naranmandura, H., Ibata, K. & Suzuki, K.T. 2007b. Toxicity of dimethylmonothioarsinic acid toward human epidermoid carcinoma A431 cells. *Chemical Research in Toxicology* 20(4): 1120–1125.

Pinyayev, T.S., Kohan, M.J., Herbin-Davis, K., Creed, J.T. & Thomas, D.J., 2011. Preabsorptive Metabolism of Sodium Arsenate by Anaerobic Microbiota of Mouse Cecum Forms a Variety of Methylated and Thiolated Arsenicals. *Chemical Research in Toxicology* 24(4): 475–477.

Planer-Friedrich, B., Franke, D., Merkel, B. & Wallschlaeger, D. 2008. Acute Toxicity of Thioarsenates to Vibrio fischeri. *Environmental Toxicology and Chemistry* 27(10): 2027–2035

Planer-Friedrich, B., London, J., McCleskey, R.B., Nordstrom, D.K. & Wallschläger, D. 2007. Thioarsenates in geothermal waters of Yellowstone National Park: determination, preservation, and geochemical role. *Environmental Science and Technology* 41(15): 5245–5251.

Raml, R., Rumpler, A., Goessler, W., Vahter, M., Li, L., Ochi, T. & Francesconi, K.A. 2007. Thio-dimethylarsinate is a common metabolite in urine samples from arsenic-exposed women in Bangladesh. *Toxicology and Applied Pharmacology* 222(3): 374–380.

Van de Wiele, T., Gallawa, C.M., Kubachka, K.M., Creed, J.T., Basta, N., Dayton, E.A., Whitacre, S., Du Laing, G. & Bradham, K. 2010. Arsenic Metabolism by Human Gut Microbiota upon in Vitro Digestion of Contaminated Soils. *Environmental Health Perspectives* 118(7): 1004–1009.

One Century of the Discovery of Arsenicosis in Latin America (1914–2014) – Litter, Nicolli, Meichtry, Quici, Bundschuh, Bhattacharya & Naidu (Eds)
© 2014 Taylor & Francis Group, London, ISBN 978-1-138-00141-1

Gestational exposure to inorganic arsenic (iAs) modulates cysteine transport in mouse brain

M.E. Gonsebatt, L. Ramos Chávez, C. Valdovinos Flores & J.H. Limón Pacheco
Dep. de Medicina Genómica y Toxicología Ambiental, Instituto de Investigaciones Biomédicas, Universidad Nacional Autónoma de México, México

ABSTRACT: Chronic exposure to iAs through drinking water is associated with learning and memory impairment in children and adults. This exposure can start in utero, damaging neurologic development through oxidative stress. Using CD1 mice, we investigated the effects of gestational exposure to 20 ppm of iAs in drinking water in biomarkers of oxidative stress and the expression of the NMDAR in the cerebral cortex and hippocampus. At birth, exposure was associated with higher levels of oxidized glutathione and expression of essential amino acid transporters for glutathione synthesis. A down regulation of NMDAR NR2B was also observed in 15 days old pups in the same regions. This effect could explain the spatial memory impairment observed mainly in male adults.

1 INTRODUCTION

Human chronic exposure to iAs is associated with different types of cancer and neurotoxicity. Alterations in memory and attention processes have been observed in children exposed to different levels of As whereas adults acutely exposed to high levels of iAs presented encephalopathies and impairments in learning, memory, and concentration. It is possible that exposure can start *in utero*, affecting neurologic development. In the presence of cellular reductants such as thioredoxin (Txn1) or glutathione (GSH), iAs is methylated. Methylated As forms have been found in maternal or cord plasma, suggesting that iAs crosses the placenta. The biomethylation of iAs in the Central Nervous System (CNS) is a significant process because it yields intermediate and final products that are more reactive and toxic than the parent compound. This process also requires S-adenosyl-L-methionine, Txn1 and GSH, which are important intermediate metabolites in the biochemistry of CNS neurotransmission and antioxidant responses to oxidative stress. GSH synthesis in the brain requires cysteine (L-cys), which enters through the xCT (as cystine) or EAAC1 transporters. Glutamate N-methyl-D-aspartate receptor (NMDAR) plays an important role in brain plasticity as well as in spatial learning and memory. NMDAR is a heteromeric complex consisting of subunit 1 (NR1) and 2 (NR2 A, B, C and D). Decreased protein expression of NR2A in hippocampus was associated with iAs exposure in adult rats. Using CD1 mice, we investigated the effects of gestational exposure to iAs in drinking water using biomarkers of oxidative stress and expression of NMDAR NR2 A and B in the cerebral cortex and hippocampus. At the same time, spatial memory behavioral tasks were evaluated.

2 METHODS/EXPERIMENTAL

2.1 *Animals and treatments*

CD1 females and males were exposed to 20 mg/L of sodium arsenite in drinking water one month before mating. Females were exposed during the gestation period until the end of lactation (day 15–16 after birth). Water containing sodium arsenite was prepared daily. The amount of water consumption was recorded daily while weight was recorded weekly. Pups were killed on day 1, 15 or after 3 months, when behavioral tasks were performed.

2.2 *Protein expression and behavior*

After killing, the brain (at day 1) or cortex and hippocampus were obtained (day 15 and 3 months). Protein expression was evaluated by western blotting.

Behavioral tasks were performed in the 3 month old progeny and then animals were killed to obtain hippocampus and cortex brain regions. To evaluate memory for contextual or spatial aspects, we used a novelty-preference paradigm.

3 RESULTS AND DISCUSSION

3.1 *Arsenic exposure did not alter weigh or litter size*

We did not observe significant alterations in weight and litter size between control and treated animals. Water consumption was significantly lower in females during lactation only ($p < 0.05$).

3.2 *GSH and GSSG levels and protein expression*

Exposed one day old pups had higher levels of GSSG and increased expression of xCT in brain homogenates ($p < 0.05$) (Figure 1). At day 15 we did not observed significant differences among GSH or GSSG. Over expression of xCT and EACC1 was observed cortex and hippocampus, which suggests oxidative stress in these structures. At the same time the expression of NMDAR NR2B subunit was significantly down regulated in exposed male and female hippocampus and in male cortex ($p < 0.05$).

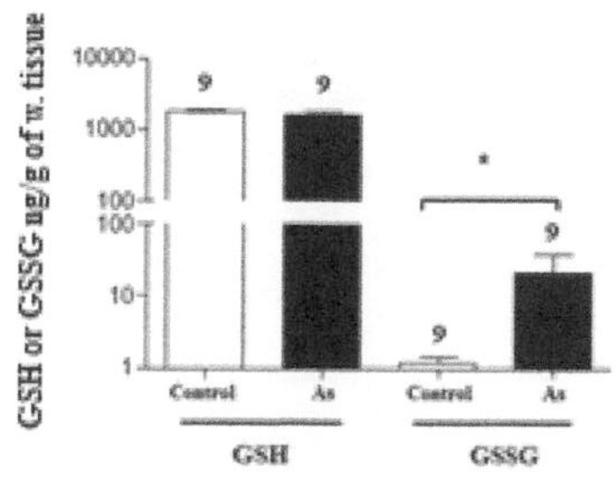

GSH or GSSG levels at day 1

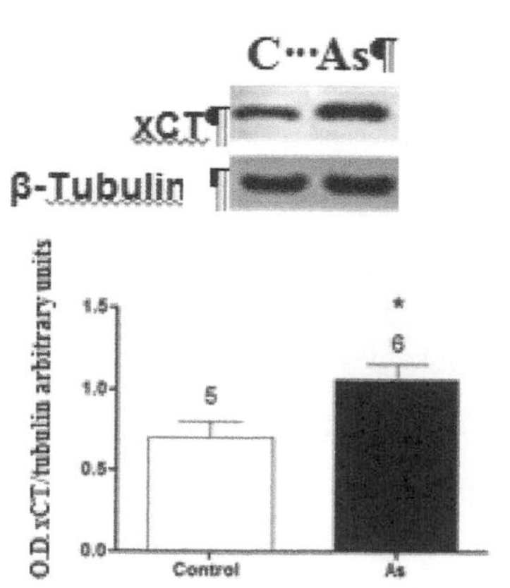

Figure 1. xCT expression in brain homogenates in 1 day old pups.

3.3 *Behavioral task*

The novelty-preference paradigm used to evaluate spatial memory showed that gestational exposure significantly impaired the capacity for spatial memory in 3 month old males ($p < 0.05$). Females were not as affected as males.

4 CONCLUSIONS

Gestational exposure was associated with increased oxidative stress at birth and with increased expression of cystine/cysteine transporters in cortex and hippocampus in both males and females. We also observed an alteration in NMDAR structure and an impairment in the spatial memory mainly in 3 month old males.

ACKNOWLEDGEMENTS

This work was funded by PAPIIT IN207611 and partially by CONACYT 102287.

REFERENCES

Burdo, J., Dargusch R. & Schubert D. (2006) Distribution of the cystine/glutamate antiporter system xc—in the brain, kidney, and duodenum. *J Histochem Cytochem*, 54 (5), 549–557.

Calderon, J., Navarro M.E., Jiménez-Capdeville, M.E., Santos-Díaz, M.A., Golden, A., Rodríguez-Leyva, I. Borja-Aburto, V.H. & Díaz-Barriga, F. (2001) Exposure to arsenic and lead and neuropsychological development in Mexican children. *Environ Res*, 85 (2), 69–76.

Dringen, R. (2000) Glutathione metabolism and oxidative stress in neurodegeneration. Eur J Biochem,. 267 (16), 4903.

Gerr, F., Letz, R., Ryan, P.B. & GreenR.C. (2000) Neurological effects of environmental exposure to arsenic in dust and soil among humans. *Neurotoxicology*, 21 (4), 475–487.

Jin, Y., Xi, S., Lin, X., Lua, C., Lia, G., Xua, Y., Qua, C., Niua, Y. & Suna, G. (2006) Arsenic speciation transported through the placenta from mother mice to their newborn pups. *Environ Res*, 101(3), 349–355.

Lewerenz, J., Klein M. & Methner A. (2006) Cooperative action of glutamate transporters and cystine/glutamate antiporter system Xc—protects from oxidative glutamate toxicity. *J Neurochem*, 98 (3), p. 916–925.

Luo, J.-h., Qiu, Z-q., Zhang, L. & Shu, W.-q. (2012) Arsenite exposure altered the expression of NMDA receptor and postsynaptic signaling proteins in rat hippocampus. *Toxicology Lett*, 211 (2), 39–44.

Luo, J.-h., Qiu, Z-q., Shu, W.-q., Zhang, Y.-y., Zhang, L. & Chen, J.-a. (2009) Effects of arsenic exposure from drinking water on spatial memory, ultra-structures and NMDAR gene expression of hippocampus in rats. *Toxicology Lett*, 184(2), 121–125.

Mukherjee, S.C., Rahman, M.M., Chowdhury, U.K., Sengupta M.K., Lodh, D., Chanda, C.R., Saha, K.C., Chakraborti, D. (2003) Neuropathy in arsenic toxicity from groundwater arsenic contamination in West Bengal, India. *J Environ Sci Health A Tox Hazard Subst Environ Eng*, 38 (1), 165–183.

Rodriguez, V.M., Del Razo L.M., Limón-Pacheco J. H, Giordano M, Sánchez-Peña L.C., Uribe-Querol Eileen, Gutiérrez-Ospina, G. & Gonsebatt M.E. (2005) Glutathione reductase inhibition and methylated arsenic distribution in Cd1 mice brain and liver. *Toxicol Sci*, 84 (1), 157–166.

Sánchez-Peña L.C., Morales M., González N., Gutiérrez-Ospina G, Petrosian P., Del Razo L.M. & Gonsebatt M.E. (2010) Arsenic species, AS3MT amount, and AS3MT gene expression in different brain regions of mouse exposed to arsenite. *Environ Res*, 110 (5), 428–434.

Thomas, D.J., Styblo M. & Lin S. (2001) The cellular metabolism and systemic toxicity of arsenic. *Toxicol Appl Pharmacol*, 176(2), 127–144.

Valdovinos, C. & Gonsebatt M.E. (2013) Nerve growth factor (NGF) exhibits an antioxidant and an autocrine activity in mouse liver that is modulated by buthionine sulfoximine, arsenic and acetaminophen. Free Radical Research, 47, 404–412.

Waalkes, M.P., Liu J. & Diwan B.A. (2007) Transplacental arsenic carcinogenesis in mice. *Toxicol Appl Pharmacol*, 222(3), 271–280.

Xi, S., Guo, L., Qi, R., Sun, W., Jin, Y., Sun, G. (2010) Prenatal and early life arsenic exposure induced oxidative damage and altered activities and mRNA expressions of neurotransmitter metabolic enzymes in offspring rat brain. *J Biochem Mol Toxicol*, 24 (6), 268–378.

One Century of the Discovery of Arsenicosis in Latin America (1914–2014) –
Litter, Nicolli, Meichtry, Quici, Bundschuh, Bhattacharya & Naidu (Eds)
© 2014 Taylor & Francis Group, London, ISBN 978-1-138-00141-1

Prevalence and demographic predictors of low-level arsenic exposure in Uruguayan school children

E.I. Queirolo
Center for Research, Catholic University of Uruguay, Uruguay

M. Vahter
Institute of Environmental Medicine, Karolinska Institutet, Sweden

N. Mañay & M. Guido
Faculty of Chemistry, University of the Republic of Uruguay, Uruguay

K. Kordas
School of Social and Community Medicine, University of Bristol, UK

ABSTRACT: Arsenic exposure is a global health problem but little is known about the prevalence, predictors and effects of low-level exposure, particularly in children. We measured water arsenic and urinary concentrations of inorganic arsenic metabolites in 6–8 years old children from Montevideo, Uruguay. Water As in this group was 0.46 ± 0.28 µg L^{-1}, whereas the sum of inorganic arsenic species in urine was 14.4 ± 11.9 µg L^{-1}. Water arsenic did not predict urinary arsenic levels. Several demographic predictors were identified, including measures of socioeconomic status, sex, and family consumption of bottled water, but these require further investigation. Other sources and health effects of this low-level exposure need to be investigated.

1 INTRODUCTION

Arsenic exposure in children is a public health concern due to its potential effects on growth and development (ATSDR, 2007). The effects of low-level As exposure in children remain under-studied, despite low-level exposure from water and other sources (food) is very prevalent in many countries (FAO, 2011). In Uruguay, pediatric exposure to As has been documented (Kordas *et al.*, 2010), thus it is important to know the predictors of As levels in children. The demographic predictors of urinary As in Montevideo school children were investigated.

2 METHODS

2.1 *Study sample and setting*

The study was conducted in Montevideo among 1st grade children (6–8) attending 9 elementary schools in areas with documented or suspected exposure to metals. Children and parents were recruited via posters and invitation letters sent through schools, and attended several sessions after providing written consent. A home visit was conducted to collect water samples. Ethical approval was given by Pennsylvania State University, Catholic University of Uruguay, and University of the Republic of Uruguay.

2.2 *Assessment of arsenic concentrations*

Children provided first morning urine samples in collection cups previously washed with nitric acid. Samples were stored on ice until transport to the laboratory, and then stored at −20 °C until analysis of urinary As metabolites (MMA, DMA, inorganic As) by HPLC-HG-ICPMS at the Karolinska Institutet. Drinking and cooking water was collected in polyethylene bottles, filtered (45 µm) and acidified to pH <2, then stored at 4 °C until transport and ICPMS analysis at the Pennsylvania State University.

2.3 *Statistical approach*

Sum of As metabolites was adjusted for specific gravity to account for hydration. Descriptive statistics were performed. Child (age, sex), parent (age, education, employment, marital status), and household characteristics (presence of children <5 years, possessions, crowding, home ownership, income, monthly spending on food), as well as household behaviors (maternal and paternal smoking status, water used for drinking and cooking, cultivation and consumption of vegetable in home garden) were used to identify predictors. Water As level was also tested. Bivariate and multivariate linear ordinary least squares regressions were conducted in STATA. Sum of As species and percent inorganic As (%IAs), dimethylarsinic acid (%DMA), and monomethylarsonic acid (%MMA) were modeled as continuous dependent variables. For multivariate models, backward stepwise regression was performed to identify predictors of each measure of urinary As.

3 RESULTS AND DISCUSSION

3.1 *Sample characteristics*

Overall, 192 children provided urine samples for analysis. These children had a mean 81.7 ± 6.7 months of age, and 57.3% were boys. Most had working parents (59.4% mothers and 89.3% father) and lived with both parents (70.9%). About 20% of mothers and 27% of fathers only had primary education, and some (3–6%) did not complete it, while ~3% of both mothers and fathers completed university education. Socioeconomic status was low-moderate: 40% of the monthly income (equivalent to $661 ± 596) was spent on food. As exposure was low, with both the mean and highest water level detected in the study <2.0 μg L^{-1}. This was reflected in low urinary metabolite concentrations and their sum (Table 1), particularly in comparison to many other population groups (Hamadani *et al.*, 2011). Mean %IAs and %MMA were fairly low, while %DMA was high (Table 1).

Table 1. Arsenic exposure and body burden characteristics in 1st-grade children from Uruguay.

Characteristic	Mean ± SD	Range
Environment1,2		
Water arsenic	0.46 ± 0.28	0.10–1.20
Body burden1,3, 4		
Sum of inorganic species	14.4 ± 11.9	1.4–93.9
Inorganic arsenic, IAs	1.2 ± 0.7	0.1–7.0
Monomethylarsonic acid, MMA	1.3 ± 0.9	0.2–7.3
Dimethylarsinic acid, DMA	11.9 ± 8.5	1.7–52.9
%IAs	10.0 ± 4.7	2.5–27.6
%MMA	9.9 ± 3.4	3.2–18.6
%DMA	80.1 ± 7.0	62.9–94.0

[1]Units given as μg L^{-1}; 2N = 165; 3N = 192, 4Total urinary As and metabolite (but not percents) adjusted for specific gravity.

Boys and girls did not differ in sum or percent of inorganic As species in urine.

3.2 *Demographic predictors of body arsenic burden*

In unadjusted regressions, child age was positively associated with %DMA, where for each month of age, DMA was 0.12% higher ($p < 0.1$). Crowding (>2 people/bedroom) was associated with lower %IAs (−1.8 ± 0.8, $p = 0.02$), lower %MMA (−1.0 ± 0.6, $p < 0.1$) but higher %DMA (2.9 ± 1.2, $p = 0.02$). Children in household with higher number of possessions had higher %MMA (1.1 ± 0.5, $p = 0.055$) than the comparison group, possibly due to higher rates of growth. Finally, children consuming bottled water had higher %IAs (1.7 ± 0.8, $p = 0.04$) and lower %DMA (−2.6 ± 1.2, $p = 0.04$), compared to unfiltered tap water. These associations were sex-specific: age was related to lower %MMA in girls, whereas in boys paternal occupation was related to higher concentration of sum of species and crowding was related to lower %MMA and higher %DMA. Finally, associations (higher %IAs and MMA, lower %DMA and sum of species) with bottled drinking water were mostly seen in girls. Stepwise regression modeling did not reveal meaningful predictors of the sum of species; having children <5 years old in the household was the only predictor (3.2 ± 1.8, $p = 0.08$). In turn, maternal smoking (2.7 ± 1.4, $p = 0.06$), crowding (−3.1 ± 1.5 $p = 0.047$), maternal employment in professional jobs (2.7 ± 1.6, $p = 0.09$), and maternal secondary education (−2.9 ± 1.4, $p = 0.04$) were associated with %IAs. In a similar fashion, maternal education and crowding were also associated with %MMA. No predictors of %DMA were identified.

3.3 *Association of water and urinary arsenic*

In bivariate regression, water arsenic concentrations did not predict urinary arsenic concentrations. In multivariate models including the demographic predictors identified above, water arsenic concentrations continued to be unassociated with either sum of inorganic arsenic species or %IAs, %MMA or %DMA, which is contrary to reports in adults (Calderon *et al.*, 2013).

4 CONCLUSIONS

In 6–8 y old children from Montevideo, we found low-level exposure to As from water and low concentrations of inorganic As metabolites in urine. We also identified several demographic predictors of urinary arsenic, including measures of socioeconomic status, sex, and family consumption of bottled water. The associations of socioeconomic status with %MMA need to be investigated in light of children's growth and nutritional status. Finally, water arsenic concentration was not predictive of inorganic arsenic concentrations in this group, suggesting that there may be other, perhaps dietary, sources of exposure.

ACKNOWLEDGEMENTS

We thank Delminda Ribeiro, Graciela Yuane, and Aditi Roy for help during clinic visits and sample preparation after collection.

REFERENCES

ATSDR. *Toxicological profile for arsenic*. In Services DoHaH, (ed.). Atlanta, GA2007.

FAO. 2011. Evaluation of certain contaminants in food: Food and Agriculture Organization and World Health Organization. Contract No.: 959.

Kordas, K., Queirolo, E.I., Ettinger, A.S., Wright, R.O. & Stoltzfus, R.J. 2010. Prevalence and predictors of exposure to multiple metals in preschool children from Montevideo, Uruguay. *Sci. Tot. Environ.* 408: 4488–4494.

Hamadani, J.D., Tofail, F., Nermell, B., Gardner, R., Shiraji, S., Bottai, M., Arifeen, S.E., Huda, S.N. & Vahter, M.2011. Critical windows of exposure for arsenic-associated impairment in cognitive function in pre-school girls and boys: a population based study. *Int. J. Epidemiol.* 40:1593–604.

Calderon, R.L., Hudgens, E.E., Carty, C., He, B., Le, X.C., Rogers, J. & Thomas, E.J. 2013. Biological and behavioral factors modify biomarkers of arsenic exposure in a US population. *Environ. Res.* 126: 134–144.

One Century of the Discovery of Arsenicosis in Latin America (1914–2014) – Litter, Nicolli, Meichtry, Quici, Bundschuh, Bhattacharya & Naidu (Eds)
© 2014 Taylor & Francis Group, London, ISBN 978-1-138-00141-1

Exposure to arsenite alters metabolism and sexual hormones in pregnant rats

M.M. Bonaventura
Instituto de Biología y Medicina Experimental (IByME, CONICET), CABA, Argentina
Facultad de Ingeniería, Universidad de Buenos Aires (UBA), CABA, Argentina

N.S. Bourguignon & D. Rodríguez
Instituto de Biología y Medicina Experimental (IByME, CONICET), CABA, Argentina

C. Libertun
Instituto de Biología y Medicina Experimental (IByME, CONICET), CABA, Argentina
Facultad de Medicina, UBA, CABA, Argentina

V. Lux Lantos
Instituto de Biología y Medicina Experimental (IByME, CONICET), CABA, Argentina

ABSTRACT: Arsenic is a naturally occurring metalloid; the most common source of contamination is through groundwater. In Argentina, population exposure is close to 10%, being pregnant women and children among the most vulnerable. Arsenic impact on the metabolic and reproductive axes is being widely studied; nowadays, it is also considered as an endocrine disruptor. The aim of the present study was to elucidate the effect of arsenic exposure on the metabolic and reproductive axes of pregnant rats. Rats exposed to 5 (A5) or 50 (A50) mg L^{-1} of As(III) in drinking water starting day 1 of pregnancy presented glucose intolerance on Gestational Day 16 (GD16). A50 also presented augmented serum estradiol and testosterone at GD18, and diminished body weight at GD21. These observations are of vital importance since changes during pregnancy in both metabolism and sexual hormones have been widely related to alterations in offspring as well as in dams themselves.

1 INTRODUCTION

Arsenic (As) is a naturally occurring metalloid released into the environment by natural events and human activities. Inorganic As is a worldwide distributed natural contaminant. The most common source of contamination is through groundwater. Population exposed to contaminated ground water is a world concern. In Argentina, the population exposed to As is close to 10%, being pregnant women and children among the most vulnerable groups. Endocrine disruptors are compounds that have the ability to interfere with, mimic or antagonize the function and/or hormone production leading to adverse health effects. Inorganic As is highly toxic and a proved carcinogenic in humans. Recently it has also been described as an endocrine disruptor. Environmental toxicants have become a serious health concern, and studies on environmental Endocrine Disruptions (EDs) are becoming more prevalent. Thus, the impact of As on the metabolic and reproductive axes is being widely studied (Diaz-Villaseñor *et al.*, 2007). Therefore, the aim of the present study was to elucidate the effect of arsenic exposure as an endocrine disruptor on the metabolic and reproductive axes of pregnant rats.

2 MATERIALS AND METHODS

2.1 *Animals and treatment*

Sprague-Dawley rats (200–250 g body weight) from the IByME colony were used. Animals were housed in air-conditioned rooms, with lights on from 0700 to 1900. Studies were performed according to protocols for animal use approved by the Institutional Animal Care and Use Committee, which follows the National Institute of Health (NIH). Animals were treated with sodium arsenite in drinking water (A5: 5 mg L^{-1} or A50: 50 mg L^{-1} in distilled water, or distilled water as control—C-) (Paul *et al.*, 2008) from Gestation Day 1 (GD1) (confirmed by presence of vaginal sperm plug) to sacrifice. Water was given *ad libitum* and changed every 2–3 days to avoid As(III) oxidation, standard chow was given *ad libitum*. Pregnant females were housed singly and Body Weight (BW) was recorded during pregnancy.

2.2 *Glucose Tolerance Test and HOMA-IR index*

On GD16, a Glucose Tolerance Test (GTT) was performed. Briefly, intraperitoneal glucose (2 g kg^{-1} BW) was injected to overnight-fasted rats and blood glucose levels were evaluated at different time points with glucometer (OneTouchUltra). Aditionally, basal insulin was determined from tail blood samples, and HOMA-IR index [(basal glycemia*insulinemia)/22,5] was calculated for these animals. Serum insulin was measured by ELISA (CrystalChem) (Bonaventura *et al.*, 2012).

2.3 *Hormone determinations*

On GD18, animals were sacrificed by decapitation and truncal blood was collected for hormone determination. Testosterone (T), and estradiol (E2) were determined by RIA using specific antisera after ethyl-ether extraction (Bianchi *et al.*, 2004).

2.4 *Statistical analysis*

Results are expressed as mean ± SEM. Differences between means were analyzed by one-way ANOVA or two–way ANOVA with repeated measures design. $P < 0.05$ was considered statistically significant.

3 RESULTS AND DISCUSSION

3.1 *Body Weight*

Rats exposed to 50 mg L^{-1} of As (III) in drinking water showed diminished body weight during pregnancy compared to A5 or C groups, this difference being significant at GD21 (Table 1).

3.2 *Glucose Tolerance Test on GD16*

Pregnant rats intoxicated with both 5 and 50 mg L^{-1} of As(III) in drinking water showed normal

Table 1. Evolution of BW (g) during pregnancy.

	C	A5	A50
GD1	228 ± 5	228 ± 3	231 ± 4
GD7	254 ± 6	261 ± 5	223 ± 6
GD14	283 ± 6	289 ± 5	249 ± 11
GD21	352 ± 8	352 ± 7	301 ± 9*

Values are expressed as means ± SEM,* = A50 different from C and A5 at GD21. $p < 0,01$. $n = 10–14$.

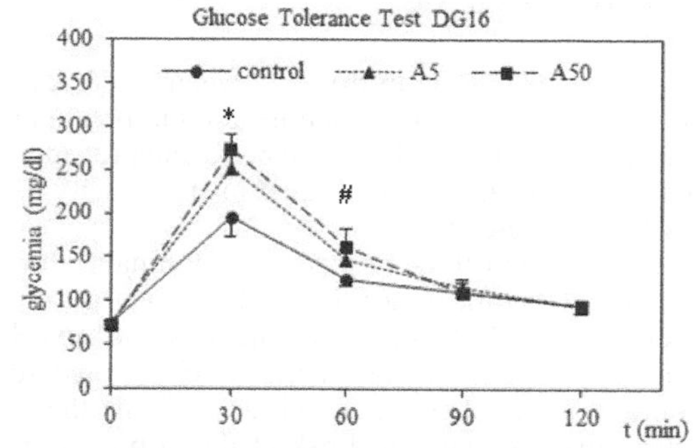

Figure 1. Glucose Tolerance Test on DG16. Two–way ANOVA with repeated measures design, interaction, $p < 0,05$ ($n = 11$). * = A50 y A5 different from C at 30 min. # = A50 different from C at 60 min. $n = 11$.

Table 2. HOMA-IR index of pregnant rats at GD16.

	C	A5	A50
HOMA-IR	5,5 ± 0,5	4,2 ± 0,2*	5,1 ± 0,3

Values are expressed as means ± SEM, * = A5 different from C and A50. $p < 0,05$. ($n = 11–13$).

Table 4. Serum estradiol (E2) and Testosterone (T) in pregnant rats on GD18.

	C	A5	A50
E2 (pg mL^{-1})	16 ± 4	11 ± 2	30 ± 7#
T (ng mL^{-1})	0,36 ± 0,05	0,36 ± 0,1	0,8 ± 0,1*

Values are expressed as means ± SEM,* = A50 different from A5 and C, # = A50 different from A5. $p < 0,05$. $n = 8–13$.

fasting blood glucose at GD16. Nevertheless, they presented altered GTTs, consistent with glucose intolerance, this being more pronounced in the A50 group (Figure 1).

3.3 *HOMA-IR*

HOMA-IR is an index of insulin resistance, being more resistant the higher the HOMA-IR value is. Interestingly, animals treated with 5 mg L^{-1} As(III) present smaller values, indicating augmented insulin sensitivity compared to C and A50 groups.

3.4 *Serum steroid hormone levels*

Increased levels of Testosterone (T) were observed at GD18 in animals receiving 50 mg L^{-1} of As(III), compared to A5 or control. Also, increased estradiol (E2) was determined in the A50 group compared to A5.

4 CONCLUSIONS

Exposure to As(III) in drinking water alters glucose metabolism and sexual hormones in pregnant rats. Further analysis has to be performed to elucidate the mechanisms implicated in the toxicity of As(III) on these axes.

Even though these results are preliminary, they are of vital importance, since alterations during pregnancy on both the metabolic axis and sexual hormone regulation have been widely related to alterations in offspring as well as in dams themselves, like an augmented risk of developing Type II diabetes later in life.

REFERENCES

Bianchi, M.S. *et al.* 2004. Effect of androgens on sexual differentiation of pituitary gamma-aminobutyric acid receptor subunit GABA(B) expression. *Neuroendocrinology* 80: 129–142.

Bonaventura, M.M. *et al.* 2012. Effects of GABA(B) receptor agonists and antagonists on glycemia regulation in mice. *Eur. J. Pharmacol.* 677: 188–196.

Diaz-Villaseñor, A., *et al.* 2007. Arsenic-induced alteration in the expression of genes related to type 2 diabetes mellitus. *Toxicol. Appl. Pharmacol.* 225: 123–133.

Paul, D.S. *et al.* 2008. Environmental arsenic as a disruptor of insulin signaling. *Met. Ions. Biol. Med.* 10, 1–7.

One Century of the Discovery of Arsenicosis in Latin America (1914–2014) – Litter, Nicolli, Meichtry, Quici, Bundschuh, Bhattacharya & Naidu (Eds)
 ISBN 978-1-138-00141-1

Limitations on health affectation assessments due to groundwater arsenic exposure—the Mexican case

R. Rodriguez, I. Morales, M.A. Armienta & I. Rodriguez
Geophysics Institute, Universidad Nacional Autónoma de México, México

ABSTRACT: The presence of As concentrations over the national standards for drinking water, 0.025 mg L^{-1}, implies health risks. Health affectation assessments requires information regarding concentrations, exposure time and water ingestion patterns. The lack of data is the first limitation. The consumption of water of different sources including treated water is the main one. A survey was carried out in two Mexican Cities supplied with water containing As. The main result was that there are not so many cases as was expected.

1 INTRODUCTION

The presence of high concentrations of As in groundwater used as urban water supply is a risk for health population. Arsenic has been detected in many aquifer systems in Mexico, mainly in the Northern part of the country. Groundwater is the main water source in the northern Mexican states. Water scarcity in such regions forces to the uses of groundwater. Rural communities are the groups more at risk. The affectation level is unknown. A few studies to assess the health impact have been carried out (Armienta *et al.*, 1997).

In El Bajío Guanajuatense (Salamanca and Juventino Rosas area), central Mexico, two epidemiological studies were done looking for health affectations and water consumption patterns. Salamanca and Juventino Rosas Cities are supplied with groundwater with As contents over the National standard for drinking water, 0.025 mg L^{-1}.

In Salamanca, there are two potential As and V sources, a refinery and a thermoelectric plant. In J. Rosas there are not potential sources of As.

2 METHODOLOGY

2.1 *Groundwater sampling*

Groundwater sampling was carried out in Salamanca from 1998 to 2004, and in Juventino Rosas in the period 2009–2013. Sampling was realized following international and national standards (APHA, 2005). Chemical determinations were done at the Analytical Laboratory of the Geophysics Institute of the Universidad Nacional Autónoma de México (UNAM). Arsenic was determined by atomic absorption spectrophotometry whereas F with selective electrode potenciomety. In Salamanca, vanadium was also analyzed. In J. Rosas, fluorine analysis was included. In J. Rosas, some urban and agriculture wells were also sampled, at least three times along the study period.

2.2 *Epidemiological survey*

In both cities, a zoning was done considering the As content of the used wells. In Salamanca, the proximity of the well to the detected arsenic potential sources was also considered. In J. Rosas, the urban population was not included, and only the agricultural communities were surveyed. A questionnaire was applied. Non clinical symptomatologies related to As and fluorine exposure were included in the survey. The main objective of the survey was to know the water consumption patterns of such communities.

Standard statistical procedures were applied to calculate the surveyed persons considering an error less than 5% (Rodriguez & Morales, 2013).

Use of water (pipeline water, treated water, drinking water, meal preparation), quantity and frequency of drinking water, residence time, were some of the questions. In the case of As, skin evidences were sought, whereas in the areas with high F concentrations, dental fluorosis was considered.

3 RESULTS

The highest arsenic and fluoride groundwater concentrations detected in Cerro Gordo were 0.039 mg L^{-1} and 3.0 mg L^{-1}, respectively, whereas the maximum F concentration, 2.56 mg L^{-1}, in J. Rosas was measured in the well Valencia. Urban wells of Salamanca have not F, although the aquifer characteristics are very similar. In both cities, there are no correlations between As and cancer incidence (Rodriguez & Morales, 2013).

In Salamanca, there are other exposure routes, as particulate matter inhalation. The use of fuel oil number 6, rich in arsenic and vanadium, provokes emission of As and V (Mejia *et al.*, 2007). This elements fall over aquifer vulnerable areas, facilitating the infiltration to the aquifer.

This survey shows that in Salamanca and J. Rosas more than 40% of the surveyed population consumes pipeline water (Figure 1). People have the strong certainty that urban water has a very bad quality. This idea is reinforced by official campaigns to save water. Beside the health affectations that soft-drink causes, the supplied water has a mean cost of 0.30 U$S per m^3, whereas the cost of a cubic meter of bottled water varies from 770 to 1,100 U$S. This explains why population does not drink pipeline water but they used it for meal preparation. People also believe that boiled water is safe. Children drinks are prepared in general with boiled water.

People belonging to high and middle classes can buy more treated water than poor people. In Salamanca, urban people consume more treated water than urban people in J. Rosas, maybe because there

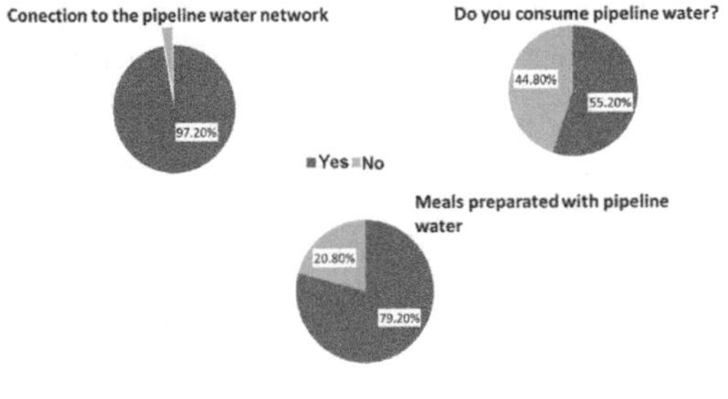

Figure 1. Water consumption pattern, Juventino Rosas, Guanajuato, Mexico, 2012.

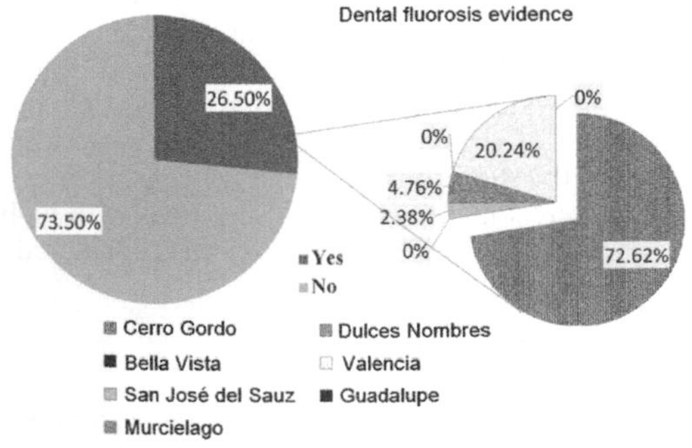

Figure 2. Dental fluorosis detected in Juventino Rosas communities.

are more information regarding the risks associated to the refinery and the thermoelectric plant.

Wastewater in J. Rosas is conducted to the Lerma River by channels. The As and F concentrations in the wastewater are higher than the groundwater values, 0.08–0.15 mg L^{-1} for As and 2.9–3.8 mg L^{-1} for F. This containing As and F water is used for irrigation, part of it infiltrated acting as an additional recharge of the aquifer. J. Rosas produces goat cheese; however, one of the producers found As in the cheese, what put in risk the commercialization of this product (personal communication).

4 DISCUSSION

A health affectation assessment in many Mexican regions indicated that the consumption of treated water becomes an important confusion factor. Population that is not all time at home does not know the quality of water consumed out of home in meals, coffee and infusion preparation.

The non clinical symptomatologies are only an indicator of affectation but not of the seriousness of the health damages. Organic fluid analysis are required (blood, urine, hair, nail). Intake calculation is complicated to evaluate. Persons working out of their communities are exposed to different water qualities. If there are more than one exposure route, total ingestion of As must be considered. Vegetables irrigated with wastewater can be another source of As, as well as goat cheese in this case. In Cerro Gordo, there could be cases of skeletal fluorosis because the high F concentration.

5 CONCLUSIONS

Our results can explain why there are no so many health affectations cases in the area. Population is not continuously exposed to arsenic because they drink treated water. Although there are persons with dental fluorosis evidences, the health affectations due to exposure to F are most evident that the consequences to As exposure. The population at risk is a reduced percentage of the total surveyed population.

ACKNOWLEDGEMENTS

The research was financed by the PAPIIT UNAM grant, Num IN102113. CMAPAJR. Juventino Rosas people helped in the groundwater sampling. Chemical analyses were done by A. Aguayo, N. Ceniceros, and O. Cruz.

REFERENCES

APHA, AWWA, WWF. Standard methods for the Examination of Water and Wastewater. American Public health Association, the American Water Works Association, Association Water Environment Federation, Washington, D.C. 2005.

Armienta, A., Rodriguez, R. & Cruz, O. 1997. Arsenic Content in Hair and People Exposed to Natural Arsenic polluted groundwater at Zimapán, Mexico. *Bulletin of Environmental Contamination and Toxicology* 59(4): 538–589.

Mejia J.A., Rodriguez R., Armienta A., Mata E., & Fiorucci A., 2007. Aquifer vulnerability zoning, an indicator of atmospheric pollutants input? Vanadium in the Salamanca aquifer, Mexico. *Water Air and Soil Poll.*, 185(1–4): 95–100.

Rodriguez, R. & Morales, I. 2013. Water management patterns and health affectations due to groundwater supply with fluorine and arsenic in a village of central Mexico. *Internal Journal of Environmental Contamination*, submitted.

One Century of the Discovery of Arsenicosis in Latin America (1914–2014) – Litter, Nicolli, Meichtry, Quici, Bundschuh, Bhattacharya & Naidu (Eds)
© 2014 Taylor & Francis Group, London, ISBN 978-1-138-00141-1

Nutritional status among the children in arsenic exposed and non-exposed areas

M.R. Karim
Department of Population Dynamics, National Institute of Preventive and Social Medicine (NIPSOM) Mohakhali, Dhaka, Bangladesh

S.A. Ahmad
Department of Occupational and Environmental Health, Bangladesh

ABSTRACT: A study was conducted on 600 children of arsenic exposed ([As] >50 μg L^{-1} of tube well water) and non-exposed (<50 μg L^{-1} of tube well water) areas to find out any difference in the nutritional status. Nutritional status was assessed by z-scores of weight for age, height for age and weight for height and also using 5th and 85th percentiles of the Body Mass Index for age. BMI of the children was found to be strongly associated with the arsenic level of the well water used by the families ($p < 0.01$). Thinness was found more among the children of exposed area (49%) than that of non-exposed one (38%). Comparatively children with normal BMI was found to be more in non-exposed area than in exposed area and the difference was found to be significant. The study suggests that arsenic exposure had negative impact on the nutritional status of children.

1 INTRODUCTION

The threat to public health presented by arsenic contamination of drinking water has attracted much attention since the 1990s, largely due to the scale of the problem in Bangladesh, which was described as "the largest poisoning of a population in history" (Smith, 2006). A WHO report predicted that in most of the southern part of Bangladesh almost 1 in 10 adult deaths will be a result of cancer triggered by arsenic poisoning in the next decade (NGOs 2003). Infants and children are considered more susceptible to the adverse effects of arsenic exposure (Ahamad, 2006). It has been observed that person taking arsenic contaminated water for 4–6 years develop arsenicosis. The youngest reported arsenicosis patient in Bangladesh is 4 years old (Ahmad, 2000). Nutrition plays a decisive role in the prevention of the onset of arsenic related ailments. There is evidence that people in poor socio-economic conditions are more prone to arsenicosis (Jakaria, 2005).

2 MATERIALS AND METHODS

This cross sectional comparative study was carried out among the children of two Upazilla (sub-district) of Bangladesh. A pre-test questionnaire and checklist was used for collection of data. Collected water from the tube wells was analyzed by atomic fluorescence spectrometry method to measure the level of arsenic.

Anthropometric measurements were taken for all the study children. Twenty four hours recalled questionnaire was used for dietary assessment of the study population.

3 RESULTS

The exposed and non-exposed group of this study was found not to be statistically different in terms of age and gender ($p > 0.05$). Furthermore no remarkable difference in overall socio-economic status (e.g. income, education, occupation etc.) between the groups was found. Type of the family and number of the family members no significant difference was observed between the two groups. The nutritional status of children was assessed by z-scores of weight for age, height for age and weight for height and also using 5th and 85th percentiles of the body mass index for age. Using the z-score height for age, weight for age and weight for height the children were grouped as normal or stunted, normal or underweight and normal or wasted respectively. Stunting, underweight and wasting were found to be significantly higher in exposed then in non-exposed group ($p < 0.05$). Thinness (low BMI, *body-mass index,* for age) was found to be more among the children of exposed area 49% than that of non-exposed area 38%. Comparatively children with normal BMI was found to be more in non-exposed area than in exposed area and the difference was found to be significant ($p < 0.05$). The mean BMI of the children was found to be significantly higher ($p = 0.012$) in non-exposed children (14.874 ± 2.167) than that of exposed children (14.423 ± 2.208). When compared after grouping the nutritional status as underweight and normal or overweight (based on BMI) an association between nutritional status and arsenic exposed was observed. In both the group (exposed and non-exposed) monthly income and family size were found to be similar, but in families of similar size, variation in monthly income can influence nutritional status. Similarly in families with similar monthly income, variation in family size may influence nutritional status. Considering these influences of arsenic exposure on nutritional status after adjusting for monthly family income and family size was explore and it was found that under nutrition was still significantly higher ($p < 0.01$) in exposed group then in non-exposed group. To assess the factors influencing the malnutrition among the children binary logistic regression analysis was carried out. The nutritional status of the children was assessed by z-score of weight for age, height for age and weight for height. Children having any of the parameter in terms of underweight, stunting and wasting was consider as malnutrition. The analysis suggests that malnutrition status was 7.2 times higher among the children exposed to arsenic contaminated water.

4 DISCUSSION

This cross sectional study was to find out and compare the body mass index of arsenic exposed and non-exposed children aged 5–14 years in two selected Upazilla (sub-district). In the present study nutritional status was compared between arsenic exposed and non-exposed children, using z-score and 5th and 85th percentile of the body mass index for age. In analyses, nutritional status was found to be significantly lower among the arsenic exposed children. Very few studies have examined nutritional status with reference to arsenic exposure (Mitra. 2002; Smith. 2006). Kabir (2001) in a case control study among adult population of similar socioeconomic status observed that nutritional status of exposed group was significantly lower than that of non-exposed group ($p < 0.001$). In a controlled case study among adult population, (Islam. 2004) showed the 28% of the cases were underweight (malnourished) compared to 15% controls.. However, the study concludes that nutritional status of the exposed children was significantly lower than that of non-exposed children ($p < 0.05$). The study further suggests that arsenic exposure had a negative impact on the nutritional status of children.

5 CONCLUSION

The study findings suggest that there was no remarkable difference in overall socio-economic status (e.g. income, education, occupation etc.) between exposed and non-exposed areas. Dietary consumption also did not show any gross difference between these two groups. However, the study revealed significantly lower number of underweight children in the non-exposed area in comparison to exposed area.

ACKNOWLEDGEMENT

My sincere thanks are due to Dr. Deoraj Caussy (Harry), environmental epidemiologist, WHO; who helped me by quick releasing of the funds and giving me support so that I could conduct the study in time.

REFERENCES

Ahamad, S. 2006.An eight year study report on arsenic contamination in ground water and health effects in Eruani village, Bangladesh and an approach for its mitigation. *Health Popul. Nutr.* 24(2): 129–141.

Islam, LN. 2004. Association of clinical complications with nutritional status and the prevalence of leucopenia among arsenic patients in Bangladesh. *International Journal of environmental research and public health* 1(2): 74–84.

Jakaria, M. 2003. The use of alternative safe water options to migrate the arsenic problem in Bangladesh: community perspective. *BRAC research and evaluation division.* Dhaka, Bangladesh 24:4.

Kabir, MI. 2001.Arsenicosis and body mass index in a selected area of Bangladesh. *Journal of Preventive and Social Medicine* 20(1): 6–12. Smith, A.H. 2006. Do nutritional deficiencies increase susceptibility to arsenic induced health effects? National institute of environmental health science, 1–2.

Mitra, AK. 2002. In arsenic related health problems among hospital patient in southern Bangladesh. *Journal of health, population and nutrition* 20(3): 1–2.

NGOs. 2003. An overview of arsenic issues and mitigation initiatives in Bangladesh: 7–8. In Arsenic information and support unit (NAISU), NGO forum for drinking water supply and sanitation.

Smith, L. 2005. Arsenic in drinking water. Resource centre network for water, sanitation and environmental health: 1–5.

One Century of the Discovery of Arsenicosis in Latin America (1914–2014) – Litter, Nicolli, Meichtry, Quici, Bundschuh, Bhattacharya & Naidu (Eds)

Arsenic levels of residents of Nevşehir province, Turkey

S. Burgaz, N. Ertas, U. Alshana, B. Altun, N.D. Hisarli, E. Asik, G. Cakmak Demircigil & E. Kadioglu
Gazi University, Ankara, Turkey

C.R. Celebi
Akpol Medical Center, Ankara, Turkey

E. Atabey
General Directorate of Mineral Research and Exploration, Ankara, Turkey

O.Y. Ataman
Middle East Technical University, Ankara, Turkey

H. Serce
Urgup Hospital, Nevsehir, Turkey

N. Bilir & A.M. Tuncer
Hacettepe University, Ankara, Turkey

ABSTRACT: Geological studies have recently shown that Arsenic (As) levels in drinking water ranged from 11 to 500 μg L^{-1} in Nevsehir province, Turkey. This study, as a part of ongoing molecular epidemiology research concerning carcinogenic risk related to As exposure through drinking water, was focused on the determination of total As concentration in hair and urine and as well as urinary As speciation of residents collected from six villages with levels of As > 50 μg L^{-1} and between 10 and 50 μg L^{-1} and from four villages with levels of As < 10 μg L^{-1} in drinking water. As levels in hair samples exceeded 1.00 μg/g (ppm) for 71% and 14% of subjects from villages with levels of As > 50 μg L^{-1} and between 10 and 50 μg L^{-1} in drinking water, respectively, indicating the toxic effects.

1 INTRODUCTION

Arsenic (As) is one of the most important carcinogenic pollutants in water. The most important environmental exposure to inorganic As (iAs) occurs by contamination of drinking water from arsenic-rich geogenic areas (IARC, 2004). Arsenic-contaminated waters through mainly geogenic sources were initially recognized in the Kütahya-Emet region in Turkey. However, most of the data were based more on ecologic data than on individual measures of iAs exposure. Recent geological studies in Central Anatolia region reveal that As concentrations in drinking water samples of the Nevsehir province, Turkey, ranged from 11 to 500 μg L^{-1} (Atabey & Unal, 2008). There are no human data on iAs exposure, As–induced adverse health effects and identification of several factors as methylation capacity and genetic polymorphisms in this area yet.

This study as a part of ongoing molecular epidemiology research concerning carcinogenic risk related to As exposure through drinking water was focused on the determination of total As concentration in hair and urine and as well as urinary As speciation of residents (n = 432) collected from six villages with levels of As > 50 μg L^{-1} and between 10 and 50 μg L^{-1} in drinking water. For comparative purposes, hair and urine samples of residents (n = 182) from four villages with levels of As < 10 μg L^{-1} in drinking water were also collected. Detailed information about age, gender, smoking and food habits, occupation, source of drinking water, residence time and medical history were taken through a structured questionnaire. The exposed and control group had similar socioeconomic status. This study was approved by the local ethical committee of the Gazi University, Faculty of Medicine, Ankara, Turkey (15.06.2009).

2 METHODS/EXPERIMENTAL

2.1 *Arsenic determination in urine and hair samples*

Total As concentration was determined using HG-AFS after acid digestion (for urine and hair) while HPLC-HG-AFS was applied for speciation studies (Basu *et al.*, 2002; Lindberg *et al.*, 2007). As(III), As(V), MMA and DMA were separated on a strong anion exchange column. For 100 μL injection loop, the limit of detection (LOD) was in the range of 0.3–0.7 μg L^{-1} with HPLC-HG-AFS and 3 ng L^{-1} with HG-AFS. In order to assess accuracy of both methodologies, Certified Reference Materials (CRMs) for total and speciation studies were also analyzed and the results were statistically not different at 95% confidence level. Urine As levels were not adjusted with creatinine or specific gravity.

3 RESULTS AND DISCUSSION

Results are summarized in Table 1.

Arsenic levels in hair samples exceeded the level; 1.00 μg/g (that may cause skin pathology (Arnold *et al.*, 1990) for 71% and 14% of individuals from villages with levels of As > 50 μg L^{-1} and between 10 and 50 μg L^{-1} in drinking water, respectively, have been measured, suggesting potential health effects to some residents in this area. Due to the drinking water As levels, dose-dependent trend ($p < 0.05$) was observed for urinary total As and As species in our study population.

Table 1. Arsenic concentrations in hair and urine samples.

[As] in drinking water (μg L^{-1})	As > 50 Median (min-max)	10 < As < 50 Median (min-max)	As < 10 Median (min-max)
Total As in hair (μg/g)	1.35 (0.35–6.59) ($n = 175$)	0.57 (0.11–2.37) ($n = 250$)	0.14 (0.02–0.98) ($n = 179$)
Total As in urine (μg L^{-1})	130.5 (8.2–1155.4) ($n = 175$)	55.6 (7.1–441.9) ($n = 257$)	10.5 (0.6–194.6) ($n = 182$)
As species in urine (μg L^{-1})			
As(III)	6.6 (ND-43.4)	4.3 (ND-17.5)	0.8 (ND-1.3)
As(V)	6.3 (ND-62.5)	6.9 (ND-19.7)	ND
DMA	60.6 (ND-935.2)	13.1 (ND-430.8)	1.9 (ND-5.6)
MMA	13.8 (ND-176.6)	4.7 (ND-31.7)	1.0 (ND-2.1)

4 CONCLUSIONS

Our findings suggest that drinking water is the principal factor contributing to As exposure in residents of the Nevsehir province, and this exposure may increase the risk for As-related health effects. Other studies are currently underway to identify arsenic-induced genetic damage, genetic polymorphisms (GSTM1, GSTT1, GSTP1, GST01, AS3MT) and skin lesions in our study population.

ACKNOWLEDGEMENTS

This study was supported by The Scientific and Technological Research Council of Turkey (TUBİTAK), Project No. 109S419.

REFERENCES

Arnold, H.L., Odom, R.B. & James, W.D. (eds) 1990. Diseases of the Skin. In: *Clinical dermatology*. Philadelphia: WB Saunders Company. Atabey, E. & Unal, H. 2008. Batı Anadolu'daki jeolojik unsurlar ve halk sağlığı projesi 2006–2007 yılı Tıbbi Jeoloji ve Etüt Raporu, MTA, No. 11067, Mayıs, Ankara, p. 291.

Basu, A., Mahata, J., Roy, A.K., Sarkar, J.N., Poddar, G., Nandy, A.K., Sarkar, P.K., Dutta, P.K., Banerjeee, A., Das, M., Ray, K., Royehaudhury, S., Natarajan, A.T., Nilsson, R. & Giri, A.K. 2002. Enhanced frequency of micronuclei in individuals exposed to arsenic through drinking water in West Bengal, India. *Mutat. Res.*, 516:29–40.

IARC Monographs on the evaluation of carcinogenic risks to humans. 2004. Some Drinking-water Disinfectants and Contaminants, including Arsenic. Vol. 84.

Lindberg, A.L., Goessler, W., Grander, M., Nermell, B. & Vahter, M. 2007. Evaluation of the three most commonly used analytical methods for determination of inorganic arsenic and its metabolites in urine. *Toxicol. Lett.*, 168: 310–318.

One Century of the Discovery of Arsenicosis in Latin America (1914–2014) – Litter, Nicolli, Meichtry, Quici, Bundschuh, Bhattacharya & Naidu (Eds)
© 2014 Taylor & Francis Group, London, ISBN 978-1-138-00141-1

Lung function and pulmonary biomarkers in children chronically exposed to arsenic in drinking water

E.H. Olivas Calderón & J.A. Espinosa Ferratt
Universidad Juárez del Estado de Durango, Gómez Palacio, Durango, México

R. Recio Vega, G. Ocampo Gómez & T. Cortes González
Salud Ambiental, Facultad de Medicina de la Universidad Autónoma de Coahuila, México

R.C. Lantz & A.J. Gandolfi
Department of Pharmacology and Toxicology, College of Pharmacy, University of Arizona, Tucson, USA

ABSTRACT: The objective of this study was to determine whether arsenic exposure is associated with changes in biomarkers of lung inflammation measured by the ratio of metalloproteinase 9 (MMP-9)/tissue inhibitor of metalloproteinase (TIMP-1), levels of soluble form of Receptor for Advanced Glycation End products (RAGE) in sputum and with decreased lung capacity. Tap water of communities and first morning void urine were analyzed for arsenic. Arsenic levels in this region are 50–120 μg/L in drinking water. Total arsenic urinary (median) was higher in subjects with restrictive spirometric patterns than normal spirometric patterns subjects (123.7 vs. 92.9) ($p < 0.05$). In addition, we observed significantly decreased in ratio MMP-9/TIMP1 and RAGE levels [median (range)] in subjects with restrictive spirometric patterns in comparison with subjects normal [23.21 (17.5, 37.4) vs. 9.0 (3.5,16.4)] and [46.7 (34.4–71.3) vs. 31.2 (20.3–43.2)] respectively. This finding suggests a potential toxic mechanism in lung for arsenic exposure.

1 INTRODUCTION

Intense over-exploitation of the aquifers in the region known as the Comarca Lagunera in Mexico has caused a gradual descent of water levels, which, in turn, has given rise to a serious health issue, hydroarsenicism. Arsenic concentrations in drinking water in this region have surpassed (50–120 μg L^{-1}), the permitted levels by the WHO (10 μg L^{-1}) and the Mexican Official Norms (25 μg L^{-1}) (NOM127). Arsenic exposure through drinking water has been associated with lung function disturbances and respiratory effects (Dauphine, 2011).

Recent studies have reported pathological changes in lungs. Disturbances have also been reported in pulmonary repairing tissue proteins (Josyula, 2006). In addition, clinical human studies have shown as well, diminishment in spirometric values and chronic pulmonary disease frequencies. These findings have been reported in adult populations but not in children. The mains goal of our study were to examine the link between lung function and pulmonary repairing tissue proteins with urinary arsenic levels in children chronically exposed.

2 METHODS

2.1 *Study Subjects and Sample Collection*

A total of 353 healthy children (6–12 age) in Comarca Lagunera, Mexico, were recruited for a cross-sectional study, with a minimum of five-year residency in the Lagunera region. A urine sample was collected from each subject. The urines were aliquoted and snap-frozen in dry ice immediately after collection. Sputum induction was performed as previously described (Burgess, 2002). All sputum supernatant samples were analyzed for levels of pulmonary repairing tissue proteins ratio MMP-9/TIMP-1 and RAGE (n = 269 and 203 respectively).

2.2 *Lung Function test using spirometric proteins biomarkers and urinary arsenic levels*

For lung function testing was used a spirometric test. The tests were repeated 3 times to obtain less of 10% of CV. We used spirometer (Easy One; NDD Medical Technologies) for pulmonary function tests. All sputum supernatant samples were analyzed for levels of RAGE, MMP-9 and TIMP-1 using commercially available ELISA (R&D Systems).

Urine samples were analyzed in the University of Arizona for arsenic determination (HPLC ICP-MS).

3 RESULTS AND DISCUSSION

3.1 *Results*

The mean children age was 8.9 years, 98% of them have been living in the community all their life (mean 8.7 years) and 77.6% of them were conceived at those communities. From the studied population, 6.3% reported chronic cough for more than 2 years and 2.9% for 7–12 years. In addition, 12.1% have been treated for repeated bronchiolitis. The median urinary arsenic level was 113.9 ppb with a range of 62.86–183.38 ppb. In 57.5% of the subjects, the spirometric values such as Forced Vital Capacity (FVC), Forced Expiratory Volume at first second (FEV1) and FEV1/FVC ratio were lower with respect to the reference values for spirometric volumes.

In 205 subjects, restrictive patterns were observed and 148 were normal (Table 1).

3.2 *Lung function and urinary arsenic concentrations*

Table 2 shows lung function and urinary arsenic levels.

Table 1. Spirometric patterns in the population.

Spirometric patterns	Subjects (Total) $n = 353$
Normal n (%)	148 (41.93)
Restrictive n (%)	205 (57.51)

Table 2. Urinary arsenic concentrations by spirometric pattern.

	Subjects (Total) $n = 353$	Subjects with normal spirometric patterns ($n = 148$)	Subjects with restrictive spirometric patterns ($n = 205$)
Total urinary arsenic (ppb)	113.9 (62.86–183.38)	92.89 (57.9–166.1)	123.7* (65.5–188.1)

Median (range). Mann Whitney test * $p < 0.05$.

Table 3. Concentrations of proteins by spirometric patterns.

	Subjects with normal spirometric patterns	Subjects with restrictive spirometric patterns
MMP-9/TIMP-1 $n = (108$ vs. $161)$	23.21 (17.5–37.4)	9.0 (3.5–16.4)*
RAGEs (pg/mL) $n = (90$ vs. $113)$	46.7 (34.4–71.3)	31.2 (20.3–43.2)*

Median (range). Mann Whitney test * $p < 0.05$.

The decrease in FEV1 and FVC (spirometric patterns) identified in this study suggests that exposure to arsenic in drinking water in early life affects lung function.

3.3 *Spirometric patterns and proteins biomarkers*

Sputum RAGE and MMP9/TIMP1 concentrations were significantly different by spirometric patterns (table 3).

4 CONCLUSION

According to our preliminary findings, levels of MMP-9/TIMP-1 and RAGE proteins in sputum could be used as a predictive tool for monitoring lung damage in children exposed to arsenic.

ACKNOWLEDGEMENTS

This work was supported by grants from the University of Arizona (Tucson Az) and UCLA/Fogarty AIDS International Training and Research.

REFERENCES

Burgess, J.L. 2002. Rapid decline in sputum IL-10 concentration following occupational smoke exposure. *Inhal. Toxicol.* 14:133–140.

Dauphine, D.G. 2011. Lung function in adults following in utero and childhood exposure to arsenic in drinking water: Preliminary findings. *Int. Arch. Occup. Environ. Health* 84:591–600.

Josyula, A.B. 2006. Environmental arsenic exposure and sputum metalloproteinase concentrations. *Environ. Research*. 102(3):283–90.

NOM-127-SSA1–1994 Mexico. 1994. *Salud ambiental, agua para uso y consumo humano-Límites permisibles de calidad y tratamientos a que debe someterse el agua para su potabilización.*

One Century of the Discovery of Arsenicosis in Latin America (1914–2014) – Litter, Nicolli, Meichtry, Quici, Bundschuh, Bhattacharya & Naidu (Eds)
 ISBN 978-1-138-00141-1

All-cause and cause-specific mortality and long-term exposure to arsenic in drinking water: A prospective cohort study in northeastern Taiwan

B.-Y. Hsiao, L.-I. Hsu & C.-J. Chen
Genomics Research Center, Academia Sinica, Taipei, Taiwan

Hung-Yi Chiou
School of Public Health, Taipei Medical University, Taipei, Taiwan

ABSTRACT: The association of low-to-moderate arsenic exposure with cause-specific mortality has never been reported in Taiwan. This study aimed to elucidate the dose-response relation of mortality from various causes of death with the arsenic concentration in drinking water and cumulative arsenic exposure. A total of 8,088 residents from northeastern Taiwan were followed from 1991 to 2011. Cox regression analysis was used to determine the hazard ratio of cause-specific mortality associated with two indices of arsenic exposure. A total of 3,106 deaths were ascertained during the follow-up period of 120,768 person-years. The multivariate-adjusted hazard ratio (95% confidence interval) was 1.25 (1.04–1.50), 1.32 (0.96–1.81), 2.41 (1.43–4.05), 4.00 (1.17–13.70), 3.58 (0.92–13.93) and 2.31 (1.43–3.75), respectively, for mortality from all-causes, all cancers, lung cancer, urinary cancer, occlusive stroke, and respiratory disease for arsenic level in drinking water ≥500.0 μg/L compared to <10.0 μg/L after adjustment of age, sex, educational level, and cigarette smoking status.

1 INTRODUCTION

Arsenic is widely distributed in the natural environment and usually transported through water in Taiwan. Areas where residents had exposures to high-arsenic drinking water have been found in Taiwan and many other countries. Previous studies have shown a dose-response relation between arsenic in drinking water and mortality from urinary cancer (Chung *et al.*, 2013), cerebrovascular disease (Cheng *et al.*, 2010), and other diseases (Tsai *et al.*, 1999; Yang, 2006). However, the association of low-to-moderate arsenic level in drinking water with all-causes or cause-specific mortality has never been reported in Taiwan. This study aimed to elucidate the dose-response relation of mortality from various causes of death with the arsenic concentration in drinking water and cumulative arsenic exposure.

2 MATERIALS AND METHODS

2.1 *Northeastern (NE) cohort from Lanyang Basin with low-to-moderate arsenic level in drinking water)*

The northeastern cohort was recruited from 18 villages of Lanyang Basin during 1991–1994. A total of 8,088 residents aged 40 or more years old with complete personal identification information were recruited in the study. The residents in Lanyang Basin had consumed well water (<40 m deep) for more than 50 years since 1940s, and had stopped well water consumption in the 1990s. The water in the wells was found to have an arsenic concentration ranging from undetectable to 3590 μg/L.

2.2 *Well arsenic level and cumulative arsenic level*

Detailed histories of residency and duration of drinking artesian well water were used to derive a cumulative arsenic exposure for each case. For the northeastern cohort, a total of 3,901 well-water samples (one sample from each household) were collected during the home interviews and their arsenic concentrations were estimated by hydride generation combined with flame atomic absorption spectrometry. The Cumulative Arsenic Exposure (CAE) from the artesian well water was defined as ($C \times D$), where C was the arsenic level of the well water from individual's household and D was the duration of drinking the artesian water.

2.3 *Data linkage and statistical analysis*

The vital status of study participants was ascertained through the linkage with the computerized

National death certification registry. Follow-up was from study entry through December, 31, 2011. Cox regression analysis was used to determine the Hazard Ratio (HR) with 95% Confidence Interval (CI) of all-causes or cause-specific mortality in relation to arsenic exposure indices. We calculated the dose-response relation between the arsenic exposure and cause-specific mortality by arsenic concentration in drinking water categorized into four groups (<10.0, 10.0–49.9, 50.0–499.9, and ≥500.0 μg/L) or by cumulative arsenic exposure categorized into three groups (<2000.0, 2000.0–14999.9, and ≥15000.0 μg/L*year) to compare with arsenic concentration in drinking water <10.0 μg/L as the referent group.

3 RESULTS

3.1 *Death events*

A total of 3,106 deaths were ascertained during a follow-up period of 120,768 person-years (mean follow-up period of 14.9 years). Among them, there were 886 cancer deaths, 345 cardiovascular diseases deaths, 390 cerebrovascular diseases deaths, 302 respiratory diseases deaths, and 110 renal disease deaths. Among all cancer cases, 260 died from lung cancer and 29 from urinary cancer. Among cerebrovascular disease deaths, there were 111 hemorrhagic stroke deaths and 44 occlusive stroke deaths. Among respiratory diseases deaths, there were 90 deaths from Chronic Obstructive Pulmonary Disease (COPD, including emphysema, chronic airway obstruction, and bronchiectasis) and 143 pneumonia deaths.

3.2 *Association of arsenic exposure indices and causes of death*

High arsenic concentration in well water (≥500.0 μg/L) was significantly associated with an increased all-causes mortality (HR, 1.25; 95% CI, 1.04–1.50) after the adjustment of entry age, gender, educational level, and cigarette smoking. There was no significant association between all cancer mortality and high arsenic concentration in well water (HR, 1.32; 95% CI, 0.96–1.81). The HR (95% CI) of lung cancer mortality for arsenic concentration in well water of <10.0, 10.0–49.9, 50.0–499.9, ≥500.0 ug/L was 1.00 (referent group), 1.11 (90.77–1.60), 1.56 (1.13–2.23), and 2.41 (1.43–4.05), respectively. The trend was statistically significant (P_{trend} = 0.0003). The HR (95% CI) of urinary cancer mortality for arsenic concentration in well water of <10.0, 10.0–49.9, 50.0–499.9, ≥500.0 ug/L was 1.00 (referent), 0.46 (0.12–1.76), 1.57 (0.60–4.12), and 4.00 (1.17–13.70), respectively. The trend was statistically significant (P_{trend} = 0.0376).

High arsenic concentration in well was associated with the mortality from cerebrovascular disease, specifically for occlusive stroke. The multivariate-adjusted HR (95% CI) of occlusive stroke mortality for arsenic concentration in well water of 50.0–499.9 μg/L was 3.66 (1.55–8.64) compared to arsenic concentration in well water <10.0 ug/L. However, no significant association was observed between hemorrhagic cerebrovascular diseases and arsenic concentration in well water.

High arsenic concentration in well water (≥500.0 ug/L) was significantly associated with an increased mortality of all respiratory diseases and pneumonia, showing HR (95% CI) of 2.31 (1.43–3.75) and 2.85 (1.49–5.46), respectively. No significant association was observed between arsenic concentration in drinking water and mortality of diabetes mellitus, hypertension, cardiovascular disease or renal disease.

The high level of cumulative arsenic exposure (≥15000.0 μg/L*year) was also significantly associated with mortality of all-causes, all cancers, lung cancer, urinary cancer, and respiratory diseases.

4 CONCLUSIONS

This was a 20-year prospective study to examine the relation between low-to-moderate arsenic level in drinking water and mortality of malignant and non-malignant diseases in northeastern Taiwan. The significant associations between arsenic exposure and mortality of urinary and lung cancer was consistent with those reported in previous studies. A significant association with arsenic concentration in drinking water was observed for mortality of respiratory diseases and occlusive stroke.

REFERENCES

Chen, T.J., Ke, D.S., & Guo, H.R. 2010. The association between arsenic exposure from drinking water and cerebrovascular disease mortality in Taiwan. *Water Res.* 44:5770–5776.

Chung, C.J., Huang, Y.L., Huang, Y.K., Wu, M.M., Chen, S.Y., Hsueh, Y.M. & Chen C.J. 2013. Urinary arsenic profiles and the risks of cancer mortality: A population-based 20-year follow-up study in arseniasis-endemic areas in Taiwan. *Environ. Res.* 122:25–30.

Tsai, S.M., Wang, T.N. & Ko, Y.C. 1999 Mortality for certain diseases in areas with high levels of arsenic in drinking water. *Arch. Environ. Health* 54:186–193.

Yang, C.Y. 2006. Does arsenic exposure increase the risk of development of peripheral vascular diseases in humans? J *Toxicol. Environ. Health A.* 69(19):1797–1804.

One Century of the Discovery of Arsenicosis in Latin America (1914–2014) –
Litter, Nicolli, Meichtry, Quici, Bundschuh, Bhattacharya & Naidu (Eds)
© 2014 Taylor & Francis Group, London, ISBN 978-1-138-00141-1

AS3MT Met287Thr polymorphism influences the arsenic-induced DNA damage in environmentally exposed Mexican populations

A. Hernández & R. Marcos
Grup de Mutagènesi, Departament de Genètica i de Microbiologia, Facultat de Biociències, Universitat Autònoma de Barcelona, Bellaterra, Spain
CIBER Epidemiología y Salud Pública, ISCIII, Madrid, Spain

Adriana Sampayo-Reyes
Centro de Investigación Biomédica del Noreste, Instituto Mexicano del Seguro Social (IMSS), Monterrey, Mexico

ABSTRACT: Environmental arsenic contamination has historically been a health concern in the central part of Northern Mexico. To clarify the current scenario of exposure among children and adults living in the Torreón area, total arsenic content in the drinking water and the urine of 124 exposed subjects was measured by HG-AAS. To assess the arsenic-associated risk, the comet assay was used and the DNA damage of the exposed population was evaluated. The study revealed that the urinary arsenic content of the exposed individuals was positively correlated with the arsenic content in their drinking water. A positive association was also found between the level of exposure and the genetic damage measured as percentage of DNA in tail ($p < 0.001$), and *AS3MT* was found to significantly influence the effect ($p < 0.034$) among children carrying the 287Thr variant allele.

1 INTRODUCTION

Arsenic (As) is an environmental carcinogen to which millions of people are chronically exposed worldwide (IARC Working Group on the Evaluation of Carcinogenic Risks to Humans, 2004). Several genetic factors have been proposed to modulate As-associated effects among chronically exposed populations (Hernández & Marcos, 2008). Polymorphisms located in the glutathione S-transferase omega1 (*GSTO1*) and arsenic (3+ oxidation state) methyl transferase (*AS3MT*) genes are among the most explored due to its crucial role in arsenic uptake and biotransformation. Children are considered different from adults in terms of As exposure, metabolism, and response (Tseng, 2009). Studies where both population groups are considered separately are of high interest to better characterize the As-associated risk of future generations. The present study aims to characterize the As-associated risk among children and adults living in the central part of Northern Mexico and to determine whether *GSTO1* and *AS3MT* genetic variation influence the exposure effects.

2 METHODS

2.1 *Arsenic determinations*

Urine and water samples were collected in adequate acid washed (1 nitric acid:1 water) polypropylene bottles and stored at −20 °C. Determination of total As was done by hydride generation-atomic absorption spectrometry using a VP100 Continuous Flow Vapor System with Thermo Electron S Series Atomic Absorption Spectrometer (Thermo Electron Corporation, Cambridge, UK). The standard "Arsenic 1000 mg L^{-1}" (PerkinElmer Life and Analytical Sciences, CO, USA) was used as a quality control, and creatinine in urine for the normalization of urinary As values was measured using the "Creatinine kit" (Merck, DF, Mexico).

2.2 *Comet assay*

Five milliliters of venous blood were collected from subjects using sealed EDTA vacutainer metal-free tubes. Peripheral blood leukocytes were isolated by centrifugation (35 min, 1300 g) in Ficoll-Paque density gradient (Pharmacia LKB Biotechnology, NJ, USA) and Single-cell gel electrophoresis was performed according to Singh *et al.* (1988).

2.3 *Characterization of GSTO1 and AS3MT polymorphisms*

Flinders Technology Associates filter paper matrix cards (FTA cards; Whatman Inc., NJ, USA) were used for nucleic acid storage and preservation.

Real-Time PCR for *GSTO1* Ala140Asp, Glu155-del, Glu208Lys, Ala236Val and *AS3MT* Met287Thr polymorphism detection was carried out in a 96-well plate Applied Biosystems 7300 detector (Applied Biosystems, CA, USA) using TaqMan Universal Master Mix reagent and specific primers and probes. The allelic discrimination was performed using the software system 7300 (Applied Biosystems).

3 RESULTS AND DISCUSSION

3.1 *Exposure assessment*

Urinary As content of the subjects ranged from 0.9 to 147 μg/g creatinine, with a mean value of 34.6 ± 2.3 μg/g creatinine. Individuals were classified into three different groups of exposure according to their urinary As content (see Table 1).

As content in the drinking water of the population was found to be at levels above the WHO permissible exposure limit of 10 μg/L, with concentration values that ranged between 1 and 187 μg/L and a mean concentration value of 16 ± 2.1 μg/L.

A positive correlation was found between the As present in the drinking water and the As present in the urine of the participants (Pearson's coefficient = 0.9, $p < 0.05$), indicating that both biomarkers are a good approach to asses As exposure in our population.

No significant differences in urinary As content were obtained between adults (35.5 ± 3.1 μg/g creatinine) and children (33.5 ± 3.5 μg/g creatinine; *t*-test = 0.03; $p = 0.6$).

3.2 *DNA damage and arsenic exposure*

A positive correlation was found between the urinary As content and the genotoxic damage of the subjects (Pearson's coefficient = 0.48; $p < 0.001$).

Significant differences in DNA damage were obtained between the medium and highly exposed groups when compared with that of the low-exposed group (Table 1).

Table 1. DNA damage of the As-exposed population.

	Number of subjects	% DNA in tail
Overall population	84	28.22 ± 1.18
Low exposed[a]	34	22.90 ± 1.17
Medium exposed[b]	28	32.76 ± 2.55***
High exposed[c]	22	35.80 ± 3.05***

[a]As in urine <30 μg/g creatinine; [b]As in urine 31–60 μg/g creatinine; [c]As in urine >61 μg As/g creatinine; *** Student *t*-test, $p < 0.001$.

Table 2. Influence of As exposure and genetic variation in the genetic damage among children.

DNA damage (% of DNA in tail)	*F*	*P*
Corrected model	5.118	0.000
As exposure level	10.498	0.000
AS3MT Met287Thr	4.992	0.034
Gender	0.319	0.417
Age	0.480	0.793

R^2 for the corrected model = 0.392.

No significant differences in the mean values of genetic damage were observed between adults (29.2 ± 1.6) and children (27.2 ± 1.7, *t*-test = 0.13; $p = 0.4$), indicating a similar As-related response under similar exposure conditions.

The GLM carried out with the informative variables and the % DNA in tail as dependent variable demonstrated that the level of exposure explains the genetic damage after correction by age and gender ($p < 0.001$).

3.3 *SNP analyses*

The allele frequencies for *AS3MT* Met287Thr and *GSTO1* Ala140Asp, Glu155del, Glu208Lys, and Ala236Val variants were 0.09, 0.20, 0.02, 0.02, and 0.05, respectively. Genotype distributions for all polymorphisms were in agreement with the Hardy—Weinberg law.

3.4 *SNPs and As-induced genetic damage*

A linear univariate model was generated to investigate whether the gene variants affecting *AS3MT* or *GSTO1* were associated with the genetic damage induced by As. Out of the five polymorphisms studied, the *AS3MT* Met287Thr SNP was found to significantly modulate the level of genetic damage induced by As, but only among children (Table 2).

4 CONCLUSIONS

The positive association between DNA damage and the level of As-exposure shows that Mexican populations living in the central part of Northern Mexico are at risk. Noteworthy, the As-associated risk is influenced by *AS3MT* genetic variation among children, a group of special concern and poorly studied so far.

ACKNOWLEDGEMENTS

We wish to thank all volunteers that have participated in the study. This work was supported by the grants 2009SGR-725, SAF2008-02933 and APOSTA UAB-2011.

REFERENCES

Hernandez, A. & Marcos, R. 2008. Genetic variations associated with interindividual sensitivity in the response to arsenic exposure. *Pharmacogenomics*, 9:1113–1132

IARC Working Group on the Evaluation of Carcinogenic Risks to Humans. 2004. Some drinking-water disinfectants and contaminants, including arsenic. IARC Monogr Eval Carcinog Risks Hum 84:1–477.

Singh, N.P., McCoy, M.T., Tice, R.R. & Schneider, E.L. 1988. A simple technique for quantization of low levels of DNA damage in individual cells. *Exp. Cell Res.,* 175, 184–191.

Tseng, C.H. 2009. A review on environmental factors regulating arsenic methylation in humans. *Toxicol. Appl. Pharmacol.,* 235: 338–350.

One Century of the Discovery of Arsenicosis in Latin America (1914–2014) – Litter, Nicolli, Meichtry, Quici, Bundschuh, Bhattacharya & Naidu (Eds)
© 2014 Taylor & Francis Group, London, ISBN 978-1-138-00141-1

Human exposure to low arsenic levels in urban dusts: A pilot study carried out in the city of Estarreja, Portugal

A.P. Reis, C. Patinha, S.P. Costa & A.C. Dias
GEOBIOTEC, University of Aveiro, Campus Universitário de Santiago, Aveiro, Portugal

J. Wragg
British Geological Survey, Keyworth, Nottingham, UK

ABSTRACT: This study reports to geochemical and health data of As in urban dusts collected from public areas and households in the city of Estarreja, North of Portugal. This pilot study aims at improving the understanding of the exposure–biomarker relationship between human toenails and low environmental As levels. For the ingestion route, bioaccessible As concentrations were estimated using the Unified BARGE Method (UBM) developed by the Bioaccessibility Research Group of Europe (BARGE). Major fractions of the element are probably in dust phases that are not dissolved by the G fluids, which significantly decrease the health risk.

1 INTRODUCTION

Estarreja is a small city located in the Aveiro district, north of Portugal. The area was classified as a Special Protection Zone (ZPS) because of its salt-marshes that are habitats for a number of animal and plant species. In 2010, the French CNRS creates the Observatoire Hommes-Millieux—Estarreja (OHM.I-Estarreja) and since then it has been funding social, environmental and health studies that aim at increasing the understanding on humans-environment interactions. The urban area is surrounded by agricultural fields but it is also near an important industrial complex, the Chemical Complex of Estarreja (CCE). The plants are producing a number of chemicals, synthetic resins and fertilizers. In the past, huge volumes of sulfides were used to produce sulfuric acid. In the late 90 s, the waste piles were sealed and buried near the complex. The technological upgrades associated to remediation measures allowed reducing the environmental burden of the city.

This study presents and discusses preliminary data on the geochemistry of As in urban dusts and human exposure to low As levels in urban dusts. This pilot study also aims at improving the understanding of the exposure–biomarker relationship between human toenails and low environmental As levels. Ethical approval for this study was provided by the National Committee for Data Protection (CNPD).

2 METHODS

A total of 59 urban dusts were collected in the area: 21 from public areas such as public gardens, playgrounds, urban squares or schoolyards; 19 from outdoor areas and 19 in indoor areas of selected households. Toenail samples were collected from residents living in the city of Estarreja and from a control group living in the countryside. A total of 31 residents from 20 households volunteered to participate. Information was obtained through interviews and a self administered questionnaire.

The questionnaire was designed to provide information relevant to assessing potential exposure to As such as age, gender, whether or not vegetables were grown and eaten from the native soil, and other potential sources such as smoking. Total As concentrations were determined by ICP-MS for dusts and by HR-ICP-MS for the biomarkers of exposure. For the ingestion route, bioaccessible As concentrations were estimated using the Unified BARGE Method (UBM) developed by the Bioaccessibility Research Group of Europe (BARGE). The UBM is a physiologically-based *in vitro* test that simulates the leaching of a solid matrix in the human GI tract (Wragg *et al.*, 2011) and it is a two stage *in vitro* simulation that represents residence times and physicochemical conditions associated with the gastric tract (G phase) and the gastro-intestinal tract (GI phase). The methodology has been validated against a swine model for arsenic (As), cadmium (Cd) and Pb in soils (Denys *et al.*, 2012).

3 RESULTS AND DISCUSSION

Average As concentrations in the samples under study are 27 mg kg^{-1} for outdoor dust collected from public areas, 22 mg kg^{-1} for outdoor household dust (7 mg kg^{-1} for control samples), 11 mg kg^{-1} for indoor

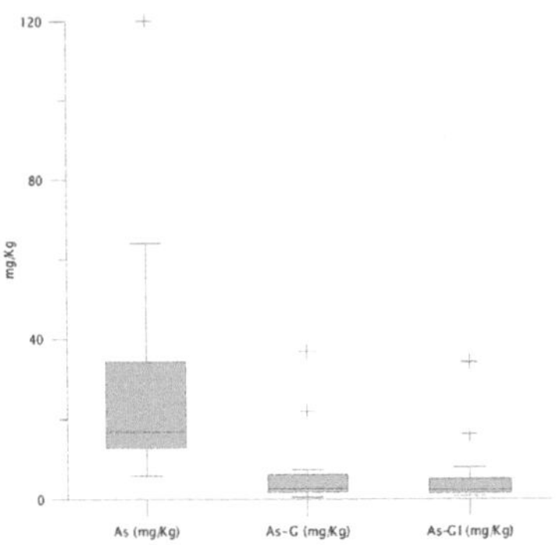

Figure 1. Box-plots for total and bioaccessible (in the Gastric and Gastro-Intestinal phases) As concentrations.

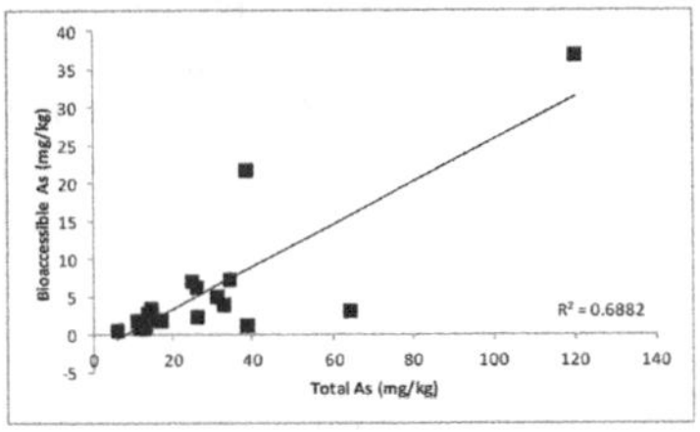

Figure 2. XY plot showing total versus bioaccessible concentrations of As in outdoor dusts.

household dust (4.4 mg kg^{-1} for control samples), and 143 μg kg^{-1} for toenail clippings (125 mg kg^{-1} for control samples). Total As concentrations in outdoor dusts are weakly positively correlated with Th concentrations and negatively correlated with Ga, V and Al concentrations in indoor household dusts ($p < 0.05$). Total As concentrations in indoor dusts are positively correlated with Ga, Tl, Be, Al, V, Cr, Mn, Fe, Sr, Y, Cs and REE ($p < 0.01$), and with As concentrations in toenail clippings ($p < 0.05$). Bioaccessible concentrations of As were determined in a total of 21 dust samples. The solutions extracted were analyzed by ICP-MS. Duplicate samples, blanks and the bioaccessibility guidance material BGS 102 were extracted with every batch of UBM bioaccessibility extractions for quality control. Mean repeatability ($n = 4$), expressed as RSD%, is 1.3% for the G-phase and 4% for the GI phase. Figure 1 shows the box-plots for total concentrations, and bioaccessible concentrations determined in the Gastric (G) and Gastro-Intestinal (GI) phases as determined in outdoor dusts.

A strong decrease is observable from total to bioaccessible As concentrations, indicating that major fractions of the element were not dissolved by the synthetic fluids. However, bioaccessible concentrations in the G and GI phases are similar. The bioaccessible fraction (BAF) was calculated with concentrations determined in the G phase. Bioaccessible concentrations in the G-phase vary between 0.9–36.9 mg kg^{-1} and have an average value of 5.5 mg kg^{-1}. Generally, more elevated total concentrations of As correspond to more elevated bioaccessible concentrations (Figure 2). On average, only 17% of As in the outdoor dusts is solubilized by the G fluids, and is therefore available for intestinal absorption. This indicates that major fractions of As are in dust phases that are not dissolved by the gastric fluids, which significantly decrease the health risk.

4 CONCLUSIONS

Total As concentrations in dusts collected from households and public areas in the city of Estarreja, North of Portugal are not elevated, although slightly more elevated that those of the dusts collected from the control areas. Concentrations in outdoor dusts are more elevated but uncorrelated with indoor concentrations. Therefore, it is probable that other than outdoor dusts are contributors to As indoor dust contents. A weak correlation is observed between concentrations in indoor dusts and in toenail clippings. Therefore, indoor dusts may be a pathway of exposure to low As levels, although other pathways have to be considered. Arsenic concentrations in outdoor dusts are not correlated with other elements, which suggest a different source for this metalloid. Major fractions of the element are probably in dust phases that are not dissolved by the G fluids, which significantly decrease the health risk.

ACKNOWLEDGEMENTS

The authors acknowledge the Labex DRIIHM and the Réseau des Observatoire Hommes-Millieux—Centre National de la Recherche Scientifique (ROHM-CNRS, France) for the support to the project "Human bioaccessibility of potentially harmful elements in indoor dusts from Estarreja" and the Foundation for Science and the Technology (FCT, Portugal) for the supporting the project PEst-C/CTE/UI4035/2011.

REFERENCES

Denys, S., Caboche, J., Tack, K., Rychen, G., Wragg, J., Cave, M., Jondreville, J. & Feidt, C. 2012. In Vivo Validation of the Unified BARGE Method to Assess the Bioaccessibility of Arsenic, Antimony, Cadmium, and Lead in Soils. *Environmental Sciences and Technology* 46: 6252–6260.

Wragg, J., Cave, M., Basta, N., Brandon, E., Casteel, S., Denys, S., Gron, C., Oomen, A., Reimer, K., Tack, K. & Van de Wiele, T. 2011. An inter-laboratory trial of the unified BARGE bioaccessibility method for arsenic, cadmium and lead in soil. *Science of the Total Environment* 409: 4016–4030.

One Century of the Discovery of Arsenicosis in Latin America (1914–2014) – Litter, Nicolli, Meichtry, Quici, Bundschuh, Bhattacharya & Naidu (Eds)
 ISBN 978-1-138-00141-1

Monomethylthioarsenicals are substrates for human arsenic (3+ oxidation state) methyltransferase

P. Alava, K. Bradham, M. Kohan & D.J. Thomas
National Exposure Research Laboratory, NERL, RTP, NC, USA

M. Kohan & D.J. Thomas
NHEERL, RTP, NC, USA

J. Navratilova & M. Stýblo
Department of Nutrition, University of North Chapel Hill, CH, NC, USA

ABSTRACT: In sulfide-rich environments, methylated thioarsenicals form from corresponding methylated oxyarsenicals. The metabolic fate of these methylated thioarsenicals is unknown. Human arsenic (3+ oxidation state) methyltransferase (AS3MT) catalyzes transfer of a methyl group from S-adenosylmethionine (AdoMet) to an arsenical substrate. To date, all identified substrates for AS3MT-catalyzed methylation are oxyarsenicals that contain arsenic in the trivalent oxidation state; thus, MMMTA and MMDTA, which contain arsenic in the pentavalent oxidation state, are unlikely candidates for AS3MT-catalyzed methylation. Using anaerobic conditions to prevent oxidation of thioarsenicals, MMMTA or MMDTA were added to reaction mixtures containing AS3MT, AdoMet, and the non-thiol reductant TCEP. Reaction products were separated by ion-exchange chromatography and detected by ICP-MS. Incubation of MMMTA with AS3MT yields dimethylmonothioarsenate (DMMTA) and dimethyl oxy arsenate (DMAV) as the primary metabolites while having MMAIII as intermediate metabolite. Production of MMAIII suggests that thiolated arsenicals are reduced and are not direct substrates for AS3MT.

1 INTRODUCTION

Enzymatically catalyzed methylation of inorganic arsenic (iAs) is the main pathway for the metabolism of iAs (Styblo *et al.*, 1995). Conversion of iAs to methylated metabolites affects the distribution and retention of arsenic (As) and also produces As species that mediate some of the toxic effects associated with iAs exposure (Chen *et al.*, 2011). Enzymatically catalyzed methylation transfers methyl groups from *S*-adenosylmethionine (AdoMet) to As to produce monomethylarsenic (MAs), dimethylarsenic (DMAs), and trimethylarsenic (TMAs) metabolites that contain either trivalent As (As^{III}) or pentavalent As (As^{V}). Strong evidence suggests that arsenic (3+ oxidation state) methyltransferase (As3mt, EC 2.1.1.137) is the key enzyme catalyzing reactions that form all known methylated oxyarsenical metabolites of iAs (Thomas *et al.*, 2007). These As metabolites include monomethylmonothioarsonic acid (MMMTAV) and dimethylmonothioarsinic acid (DMMTAV) (Raml *et al.*, 2007) and trivalent methylated arsenicals. The role of highly toxic trivalent methylarsenicals and in particular the thiolated methylarsenic species is currently under investigation. Methylated trivalent species, i.e., monomethylarsenous acid (MMAIII), dimethylarsenous acid (DMAIII), and arsenous acid (iAsIII) are two orders of magnitude more cytotoxic than is arsenic acid (iAsV) (Naranmandura *et al.*, 2007). The methylated pentavalent species monomethylarsonic acid (MMAV) and dimethylarsinic acid (DMAV) present a 10-fold lower toxicity than iAsV, whereas trimethylarsine oxide (TMAO) is essentially nontoxic. However, the toxicity profiles of thiolated methylarsenicals are not well characterized. Yet, the most recent findings on cytotoxicity towards human bladder cells point to the following order of toxicity: DMAIII, DMMTAV > iAsIII, iAsV > MMMTAV > MMAV, DMAV, and DMDTAV (Naranmandura *et al.*, 2011).

Sulfur-containing arsenicals have been detected in the urine and feces of experimental animals (Kubachka *et al.*, 2009). More recently, it has been shown that upon *in vitro* digestion of iAsV under gastric and intestinal conditions, followed by the incubation with *in vitro* cultured human colon microbiota, significant methylation and thiolation takes place (Van de Wiele *et al.*, 2010).

To date, we believed that production of thiolated As species is due to interconversion of oxy—and thio-arsenicals by exchange of oxygen and sulfur moieties. This theory is supported to date, as all identified substrates for human arsenic methyl transferase (AS3MT)-catalyzed methylation are oxyarsenicals that contain arsenic in the trivalent oxidation state; oxyarsenicals with arsenic in the pentavalent oxidation state (e.g., MMA) are not substrates for AS3MT-catalyzed methylation. In the present study, we investigated to what extent thiolated arsenicals can act as substrates for human arsenic methyl transferase. This experiment will explain whether thiolated species can be directly methylated (monomethylthiolated species (MMMTAV) to dimethylmonothiolated species (DMMTAV). This can lead to a new metabolic pathway that would work in competition with the pathway where thiolated species are first converted to oxy species to produce a methylated version.

2 METHODS/EXPERIMENTAL

2.1 *In vitro methylation assay*

Catalytic activities of AS3MT was examined in reaction mixtures that contained 100 mM Tris–HCl buffer (pH 7.4), a recombinant enzyme (5–10 μg), 1 mM AdoMet (Sigma), an arsenical substrate, and following reductant: TCEP (Sigma-Aldrich). The final volume of the mixture was 100 μL. The mixtures were incubated in capped 1.5-mL Eppendorf tubes at 25 °C or 37 °C for up to 1 h.

2.2 *Analysis of arsenicals in in-vitro methylation assay mixtures*

Total As content in the mixtures was analyzed by using Thermo scientific X-Series II ICP-MS. Speciation analysis was also carried on these mixtures by using HPLC-ICP-MS. Chromatographic conditions used for the analysis are given in Table 1.

3 RESULTS AND DISCUSSION

3.1 *Primary Results and Discussion*

These are the preliminary results to date. Further experiments are in process. Reaction products were separated by ion-exchange chromatography and detected by ICP-MS. Incubation of MMMTAV with AS3MT yields dimethylmonothioarsenate (DMMTAV) and dimethyl oxy arsenate (DMAV) as the primary metabolites while having MMAIII as intermediate metabolite. Incubation of MMDTAV with AS3MT yields dimethyldithioarsenate (DMDTAV) and dimethyl oxy arsenate (DMAV) as the primary metabolites while having MMAIII as intermediate metabolite. Chromatographs of thiolated species methylation reaction are given in Figures 1 and 2.

Table 1. Chromatographic conditions. Mobile phase A: DI water, Mobile phase B: 0.68% TMAOH. Column: Dionex AS-16 anion exchange.

Gradient (Minutes)	A	B
0–4	100	0
4–6	75	25
6–9	50	50
9–12	0	100
12–15	100	0

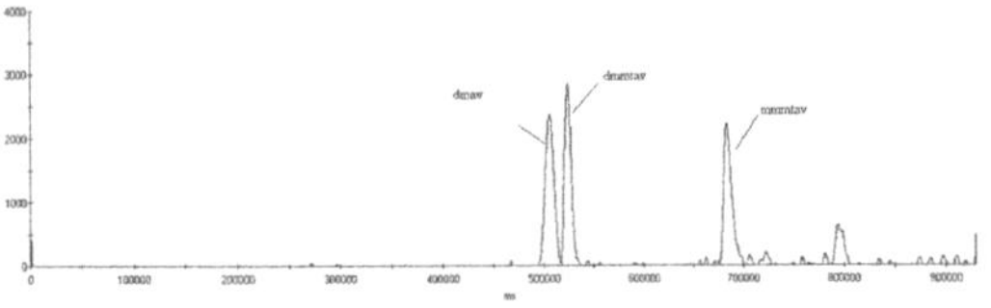

Figure 1. Chromatograph of MMMTAV (parent used substrate).

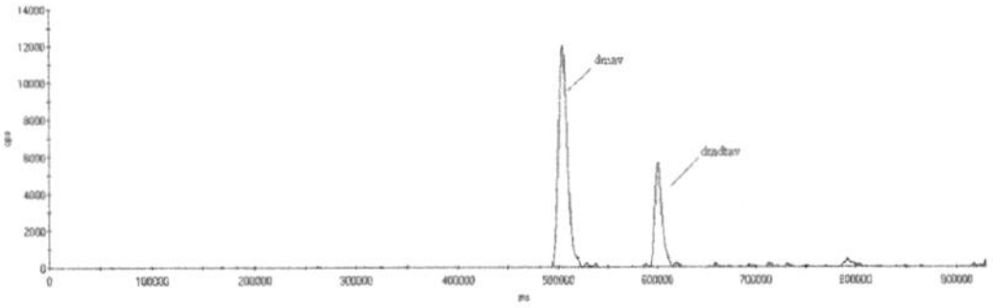

Figure 2. Chromatograph of MMMTAV mixture V after the methylation assay.

4 CONCLUSIONS

Production of MMAIII suggests that thiolated arsenicals are reduced and are first converted to oxy arsenicals. This suggests that pentavalent thiolated arsenic species are not direct substrates for AS3MT.

REFERENCES

Chen, B. 2011. Mouse arsenic (+ 3 oxidation state) methyltransferase genotype affects metabolism and tissue dosimetry of arsenicals after arsenite administration in drinking water. *Toxicological Sciences* 124: 320–326.

Kubachka, K.M. 2009. Exploring the in vitro formation of trimethylarsine sulfide from dimethylthioarsinic acid in anaerobic microflora of mouse cecum using HPLC-ICP MS and HPLC-ESI-MS. *Toxicology and Applied Pharmacology* 239: 137–143.

Naranmandura, H. 2007. Toxicity of Dimethylmonothioarsinic Acid toward Human Epidermoid Carcinoma A431 Cells. *Chemical Research in Toxicology* 20: 1120–1125.

Naranmandura, H. 2011. Comparative Toxicity of Arsenic Metabolites in Human Bladder Cancer EJ-1 Cells. *Chemical Research in Toxicology* 24: 1586–1596.

Raml, R. 2007. Thio-dimethylarsinate is a common metabolite in urine samples from arsenic-exposed women in Bangladesh. *Toxicology and Applied Pharmacology* 222:374–380.

Styblo, M. 1995. Biological mechanisms and toxicological consequences of the methylation of arsenic. M.G. Cherian, R.A. Goyer (eds), *Toxicology of Metals—Biochemical Aspects, Handbook of Experimental Pharmacology* 115: 407–433, Springer Verlag, Berlin.

Thomas, D.J. 2007. Arsenic (+3 oxidation state) methyltransferase and methylation of arsenicals. *Experimental Biology and Medicine* 232: 3–13.

Van de Wiele, T. 2010. Arsenic Metabolism by Human Gut Microbiota upon in Vitro Digestion of Contaminated Soils. *Environmental Health Perspect* 118: 1004–1009.

One Century of the Discovery of Arsenicosis in Latin America (1914–2014) – Litter, Nicolli, Meichtry, Quici, Bundschuh, Bhattacharya & Naidu (Eds)
© 2014 Taylor & Francis Group, London, ISBN 978-1-138-00141-1

Chronic arsenic exposure and type 2 diabetes: A meta-analysis

Abir Tanvir & Hossain Akbar
School of Medicine, University of Western Sydney, Penrith, NSW, Australia

Rahman Bayzidur
The School of Public Health and Community Medicine, Faculty of Medicine, The University of New South Wales, Sydney, NSW, Australia

Islam Rafiqul & Milton Abul Hasnat
Centre for Clinical Epidemiology and Biostatistics (CCEB), The School of Medicine and Public Health, Faculty of Health, The University of Newcastle, Newcastle, NSW, Australia

ABSTRACT: Chronic exposure to inorganic arsenic in drinking water has been associated with Type 2 Diabetes (T2D). However, the association is still inconclusive. We incorporated 33 studies in the analysis including 13 cross-sectional, 6 case-control and 4 cohort studies reporting Odds Ratio (OR), 6 cross-sectional and cohort studies reporting Relative Risk (RR) and 3 mortality studies reporting Standardized Mortality Ratio (SMR). The pooled OR estimate was 1.76 (95% CI: 1.43, 2.17), pooled RR was 1.17 (95% CI: 0.95, 1.45) and pooled SMR was 0.79 (95% CI: 0.59, 1.06), and the overall pooled estimate was 1.55 (95% CI: 1.27, 1.88). The study heterogeneity was very high ($I^2 = 97\%$ for case-control studies, 47% for cohort studies, and overall 98%). Because of the very high heterogeneity and non-significant results from cohort studies, this analysis provides a limited evidence for an association between arsenic and Type 2 Diabetes (T2D) that needs further validation.

1 INTRODUCTION

Chronic arsenic exposure in drinking water has been reported to be associated with type 2 diabetes (T2D) (Rahman *et al.*, 1996, Rahman *et al.*, 1998, Lai *et al.*, 1994). Recently a meta-analysis was published to systematically review the association between chronic arsenic exposure and Type 2 Diabetes. Surprisingly, this meta-analysis did not include all the potential studies (Wang *et al.*, 2013). In our meta-analysis, we included all relevant cross-sectional, case-control, cohort and mortality studies of chronic arsenic exposure and Type 2 Diabetes Mellitus.

2 MATERIALS AND METHODS

2.1 *Literature search*

We used a comprehensive search strategy to identify all relevant studies. The search was carried out in Ovid MEDLINE, EMBASE, PubMed and Google Scholar, without limitation on the time of publication. We also hand-searched for additional relevant studies.

2.2 *Selection of studies*

The included studies (i) cross-sectional (ii) case controls (iii) cohort studies of chronic arsenic exposure reported a Relative Risk (RR) or Odds Ratio (OR) or Standardized Mortality Ratio (SMR). We screened the initial list of articles to identify articles that were irrelevant. We developed a standard data extraction format to extract relevant information from all the included papers.

3 STATISTICAL ANALYSIS

Effect estimates (OR, RR and SMR) were pooled using inverse-variance-weighted random effects method (DerSimonian & Laird, 1986). Heterogeneity was measured using the I^2-statistics (Greenland, 1987). Pooled Effect Measures (EM) were estimated by comparing the highest exposure category to the lowest one.

4 RESULTS

We incorporated 33 studies in the analysis including 13 cross-sectional, 6 case-control and 4 cohort studies reporting OR, 6 cross-sectional and cohort studies reporting RR and three mortality studies based on historical data reporting SMR. The pooled OR estimate was 1.76 (95% CI: 1.43, 2.17), pooled RR was 1.17 (95% CI: 0.95, 1.45) and pooled SMR was 0.79 (95% CI: 0.59, 1.06). After combining all three types of studies the pooled estimate was 1.55 (95% CI: 1.27, 1.88) (Figure 1).

The results of pooled OR reported a strong association but reported RR and SMR reported a weak association. The influence analysis did not detect any single study influencing the pooled estimate. While investigating the publication bias we found the funnel plot highly asymmetric, which demonstrates that several small studies with negative outcomes are missing. This evidence was also supported by the statistical test for publication bias: $p < 0.001$ from Begg's test and $p = 0.04$ from Egger's test.

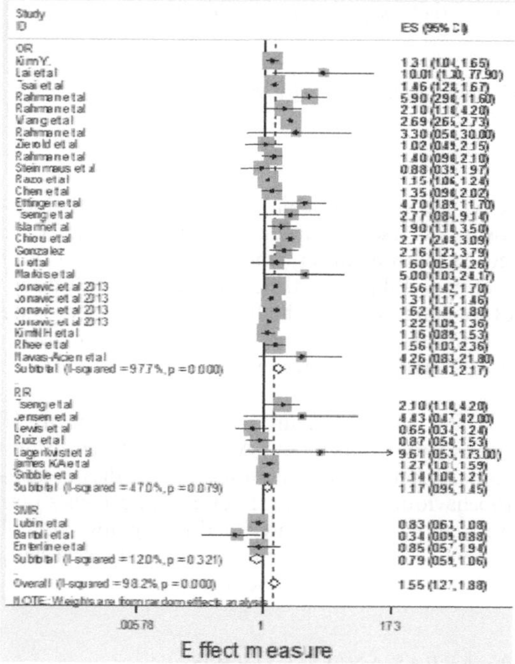

Figure 1. Forest plot of random effects meta-analysis of arsenic exposure and Type 2 Diabetes Mellitus (T2D).

5 CONCLUSIONS AND RECOMMENDATIONS

Because of very high between study heterogeneity and non-significant results from cohort studies, this analysis provides a limited evidence for a relationship between arsenic and Type 2 Diabetes. The hypothesized association is still inconclusive and needs further validation through well conducted prospective cohort studies.

ACKNOWLEDGEMENTS

I am grateful to Professor Prosun Bhattacharya for his valuable comments.

REFERENCES

DerSimonian, R. & Laird, N. 1986. Meta-analysis in clinical trials. *Controlled Clinical Trials* 7(3): 177–188.

Greenland, S. 1987. Quantitative methods in the review of epidemiologic literature. *Epidemiologic Reviews* 9(1): 1–30,

Lai, M.S., Hsueh, Y.M., Chen, C.J., Shyu, M.P., Chen, S.Y., Kuo, T.L., Wu, M,M, & Ta, T.Y. 1994. Ingested inorganic arsenic and prevalence of diabetes mellitus. *Am. Epidemiologist* 139(5): 484–492.

Rahman, M., Tondel, M., Ahmad, S.A. & Axelson, O. 1998. Diabetes mellitus associated with arsenic exposure in Bangladesh. *Am. J. Epidemiology* 148(2):198–203.

Rahman, M., Wingren, G. & Axelson, O. 1996. Diabetes mellitus among Swedish art glass workers–an effect of arsenic exposure? *Scand. J. Work Environ. Health* 22(2):146–149.

Wang, W., Xie, Z., Lin, Y. & Zhang D. 2014. Association of inorganic exposure with type 2 diabetes mellitus: a meta-analysis. *J. Epidemiol. Community Health* 68: 176–184.

One Century of the Discovery of Arsenicosis in Latin America (1914–2014) –
Litter, Nicolli, Meichtry, Quici, Bundschuh, Bhattacharya & Naidu (Eds)
 ISBN 978-1-138-00141-1

Impact of chronic arsenic exposure on neurocognitive functions: First results of the Vietnamese field study

H.J. Kunert
AHG Allgemeine Hospitalgesellschaft Düsseldorf Medical Department University of Göttingen, Germany

D.T.M. Ngoc
National Institute of Occupational and Environmental Health, Hanoi, Vietnam

P.T.K. Trang, V.T.M. Lan & P.H. Viet
Center for Environmental Technology and Sustainable Development, Hanoi University of Science, Vietnam

S. Norra
Institute of Mineralogy and Geochemistry, Karlsruhe Institute of Technology, Karlsruhe, Germany

C. Norra
Department of Psychiatry, Psychotherapy and Preventive Medicine, Ruhr University, Bochum, Germany

ABSTRACT: The health hazards caused by chronic or sub acute arsenic) exposure are well documented. However, the effects of chronic low-level exposure of arsenic on neurocognitive functions are rarely reported. Dysfunctions in the domains of intelligence, learning and memory, attentiveness, as well as visual, spatial and acoustic processing have been described. Taken together, these cognitive dysfunctions point to a comprehensive dysfunctional integrity of the human brain which correlates with exposure to arsenic. First results of a neuropsychological field study in rural areas of the Red River Delta in northern Vietnam demonstrate specific effects on neurocognitive functions regarding to as exposure.

1 INTRODUCTION

Basic research in neuroscience demonstrated disorders in neurobiological and neurofunctional development in animals as a consequence of arsenic exposure. In humans neuropsychological dysfunctions are described in the domains of intelligence, learning and memory, attention, visual, spatial and acoustic processing (Dakeishi et al., 2006; Wasserman et al., 2007) pointing to dysfunctional networks of the human brain. Subjective and social consequences of arsenic exposure regarding cognitive and neurofunctional requirements of daily living (e.g. driving or working abilities) have not been investigated so far.

The WHO recommends a threshold value of 2.1 μg uptake/d of arsenic per kg bodyweight (WHO, 2001). Nevertheless, smaller amounts of arsenic in drinking water and nutrition may also cause negative health effects and a dose-response relationship has been taken into account (Yoshida et al., 2004). Because arsenic in low concentrations is also suspected to determine mental disorders of humans only little knowledge is available with respect to cognitive brain functions in detail. In addition, before first medical damages become visible subtle neurocognitive impairments can occur and potentially interfere with the neurocognitive development of children and adolescents. As there exists hardly any larger systematic population-based studies of neurobehavioural data in arsenics endemic areas so far we were particularly interested in cognitive central nervous effects of arsenic neurotoxicity, especially with low-level chronic arsenic exposure.

2 AREAS AND METHODS

A Vietnamese-German network of experts for the assessment of arsenic in food chain and on the development and optimisation of filter techniques to remove arsenic from contaminated groundwater in rural areas of the Red River Delta, Vietnam (VIGERAS) was initiated. The project was funded by the German International Bureau of the Federal Ministry of Education and Research (BMBF), the German Research Foundation (DFG) and the Vietnamese Ministry of Science and Technology (MOST) from 2008 until 2010 (Norra S. et al. 2009).

In close cooperation with the Vietnamese partners a culture-adapted neuropsychological test battery was developed for the specific Vietnamese open field situation in rural areas of the Red River Delta in order to assess subtle neurocognitive consequences of arsenic neurotoxicity. Previously, part of this neuropsychological test battery has been proven to detect subtle neurocognitive dysfunction in environmental research (Kunert et al., 2004). The neuropsychological test battery included examinations of basic intelligence (fluid intelligence), attentiveness functions (tonic and phasic alertness, divided attention) as well as verbal learning and memory (short term and long term memory) easy to handle in the field.

The study was performed in two rural areas of the Red River Delta near Hanoi, i.e. one area with high As (Mai Dong) and another area with low arsenic in drinking water (Nghia Dan) in spring 2009. The neuropsychological examinations were supported by trained personnel from the Vietnamese institutes who were also taking the human samples (urine, hair, nails). Analyses of human and water samples were carried out at the departments in Hanoi, Vietnam and Karlsruhe, Germany.

3 RESULTS

135 volunteers were examined (m/f: 73/62, age 33.3 ± 16.6 years, range 7–64 years), 66 belonging to one area with high As (190 μg/L, range 15–396 μg/L) and 72 to another area with low arsenic in drinking water (As 3 μg/L, range <1–10 μg/L). Human arsenic concentration in hair was 0.61 ± 0.36 mg/kg in the arsenic high area and 0.21 ± 0.10 mg/kg in the arsenic low area.

Results of the different neuropsychological tests showed diminished capacity especially in higher cognitive functions (i.e. executive functions) according to arsenic concentration in hair ($p < .05$). Interestingly, diminished cognitive capacities could be found in subjects of both study regions with different arsenic contaminated drinking water.

4 DISCUSSION

Neuropsychological examinations revealed cognitive dysfunctions in different cognitive domains but especially in overlapping systems of higher cognitive information processing according to arsenic in hair. More precisely our results suggest dysfunctions in brain related networks which are responsible for superordinated cognitive information processing. Under this perspective, frontal lobe areas are the most important regions for this cluster of cognitive dysfunction (Mesulam 2000).

Dysfunctions in this feed-back control system of the frontal lobe may also explain some of the previously reported neuropsychological peculiarities in subjects with higher arsenic contamination. Apart from that our preliminary results already suggest differences in mental health status in the two arsenic regions.

The further aim of the project is to integrate the individual human findings into the food, household and environmental data of the study for a more comprehensive approach to assess health, esp. neuropsychological consequences of arsenic exposure. Moreover, adapted technologies and prevention strategies have to be developed to limit arsenic entering the food chain as far as possible and to ensure clean water and food resources.

5 CONCLUSION

Chronic exposure to even low levels of arsenic in drinking water may have a negative impact on mental health and neurocognitive functioning as seen in this study with enhancement of different and sometimes subtle symptoms of neurotoxicity. Dysfunctions in frontal lobe related control systems seem responsible for some of our neuropsychological test results. Finally, our neuropsychological test battery has been proven to detect different and subtle cognitive dysfunctions according to arsenic contamination.

REFERENCES

Dakeishi, M., Murata, K. & Grandjean, P. 2006. Long-term consequences of arsenic poisoning during infancy due to contaminated milk powder. *Environ Health* 5: 31.

Kunert, H.J., Wiesmüller, G.A., Schulze-Röbbecke, R., Ebel, H., Müller-Küppers, M. & Podoll, K. 2004. Working memory deficiencies in adults associated with low-level lead exposure: Implications of neuropsychological test results. *Int J Hyg Envir Health* 207: 1–10.

Mesulam, M.M. (ed.). 2000. Principles of behavioral and cognitive neurology. Oxford University Press, New York.

Norra, S., Kunert, H.J., Bahr, C., Berner, Z., Bich, P.T.N., Blömecke, B., Boie, I.; Cat, L.V., Driehaus, W., Eiche, E., Kellermeier, E., Jekel, M., Lan, V.T.M., Vogt, J., Long, D.D., Lübken, M., Sperlich, A., Wegner, A., Wichern, M., Wiesmüller, G.A., Ngoc, D.T.M., Norra, C., Trang, P.T.K. & Viet, P.H. 2009. Development and optimization of measures against the contamination of the food chain by arsenic from polluted groundwater resources in rural areas of the Red River Plain in Vietnam. In: H. Steusloff (ed.), *Integrated Water Resource Management*: 191–199. Karlsruhe: KIT Scientific Publishing.

Wasserman, G.A., Liu, X., Parvez, F., Ahsan, H., Factor-Litvak, P. Kline, J., van Geen, A., Slavkovich, V., Lolacono, N.J., Levy, D., Cheng, Z. & Graziano, J.H. 2007. Water arsenic exposure and intellectual function in 6-year-old children in Araihazar, Bangladesh. *Environmental Health Perspectives* 115: 285–289.

WHO. 2001. Arsenic and arsenic compounds. *Environmental Health Criteria 224*. Geneve.

Yoshida, T., Yamauchi, H. & Sun, G.F. 2004. Chronic health effects in people exposed to arsenic via the drinking water: dose-response relationship in review. *Toxicology and Applied Pharmacology* 198: 243–252.

3.3 Risk assessment of chronic ingestion

One Century of the Discovery of Arsenicosis in Latin America (1914–2014) – Litter, Nicolli, Meichtry, Quici, Bundschuh, Bhattacharya & Naidu (Eds)
© 2014 Taylor & Francis Group, London, ISBN 978-1-138-00141-1

Geospatial human health risk assessment in an Argentine region of hydroarsenicism

D.E. De Pietri & E. de Titto
Dirección Nacional de Determinantes de la Salud e Investigación, Ministerio de Salud de la Nación, Buenos Aires, Argentina

J.A. Navoni, V. Olmos & E.C. Villaamil Lepori
Cátedra de Toxicología y Química Legal, Facultad de Farmacia y Bioquímica, Universidad de Buenos Aires, Argentina

M.C. Giménez
Cátedra Química Analítica I, Universidad Nacional del Chaco Austral, Roque Sáenz Peña, Chaco, Argentina

G. Bovi Mitre
Grupo INQA, Facultad de Ciencias Agrarias, San Salvador de Jujuy, Argentina

ABSTRACT: The aim of the present study was to perform a risk assessment applying spatial analytical approaches to study geogenic Arsenic (As) exposure in a region of Argentine. The study involved inhabitants from Chaco and Santiago del Estero provinces. Arsenic in drinking water and urine was measured by Hydride Generation Atomic Absorption Spectrometry (HG-AAS). Average Daily Dose (ADD), Hazard Quotient (HQ) for dermatologic effects, and Carcinogenic Risk (CR) were calculated, geo-referenced and integrated with demographical data by a Health Composite Index (HCI). As content in drinking water and urine covered a wide range of concentrations: Not Detectable (ND) to 2000 μg L^{-1}, and 11 to 5085 μg g^{-1} creatinine respectively. The time of residence, demographic density and potential health considered outcomes characterized the region by health risk. The geo-spatial approach contributed to delimitate and analyze the changes tendencies in the risk region, broadening the scopes of the results, for decision-making process.

1 INTRODUCTION

Arsenic (As) is an ubiquitous element widely distributed in the environment. Its organoleptic properties make its presence imperceptible. As is transferred from geologic storages to water resources. Consequently, population consuming As contaminated water is chronically exposed to this element, with a high probability to suffer its deleterious effects. Despite of the importance of hydroarsenicism (HACRE) as a problem of health, there is a lack of information about the real risk of the population that consumes contaminated water. The single use of standardized tools to carry on risk assessment is not enough to perform a complete and comprehensive analysis of the risk scenario. The aim of the present study was to perform a risk assessment applying spatial analytical techniques in addition to the use of conventional approaches. In this framework, the specific objectives were: 1) to map water and urine samples for the spatial pattern analysis; 2) to calculate the average daily dose intake to demarcate areas of exposure in the studied region and 3) to characterize different risk scenarios taking into consideration the time of residence and distribution of the population.

2 METHODS

2.1 *Study area*

The study was performed in an area from the center-north region of Argentina in the Santiago del Estero province (SDE, Capital, Banda and Copo departments), and in the Almirante Brown department of the Chaco province. The population was constituted mainly by rural and dispersed inhabitants (54%) and the remaining is settled in urbanized centers. Population were interviewed following a standard questionnaire aimed at collecting information about customs, dietary habits and demographic data such as age, sex, years of residence and drinking water patterns.

2.2 *Sample collection*

The sample collection was performed in the 2010–2011 period. The study population consisted in 650 participants aged between 1 and 96. Water samples were collected in cleaned bottles acidified with HNO_3 (final concentration 0.015% HNO_3 v/v) and stored at 4 °C until arsenic quantification (Standard Methods, 1998). Urine samples (first void) were collected in polyethylene flasks previously soaked in HNO_3 v/v 20%, rinsed with distilled water and dried. As concentration was determined in 192 drinking water samples and in 448 urine samples.

2.3 *Sample analysis*

Arsenic content was quantified in water and urine (UAs) samples using a flow injection hydride generation atomic absorption spectrophotometric method (Navoni *et al.*, 2009; 2010).

2.4 *Exposure assessment*

The individual levels of UAs were compared with the upper end of the normal environmental exposure range of 100 µg g^{-1} creatinine (ATSDR, 2007), to define the exposure status. Correlation analysis between As in drinking water and urine was performed using SPSS 17. The geo-referenced urine and water As levels were used to interpolate the expected As level in the region studied. The tool applied was IDW using Arcgis10 ESRI software. The interpolated As concentration in drinking water was used to calculate and geo-reference the Average Daily Dose (ADD) (EPA, 1998).

2.5 *Human health risk assessment*

Hazard Quotient (HQ) and Carcinogenic Risk (CR) levels were estimated. RfD is the As toxic reference dose of 0.0003 mg kg^{-1} day^{-1} for dermatological manifestations (hyperpigmentation and keratosis). Health risk situation was considered when HQ values were >1 (Khan *et al.*, 2008). Cancer Risk (CR) was calculated using the formula: CR = ADD × CSF, where CSF is the cancer slope factor for As of 1.5 mg kg^{-1} day^{-1}. The population distribution by age group was estimated as a proportion of the total population by censal radius (INDEC, 2001).

3 RESULTS AND DISCUSSION

3.1 *Hazard identification*

The As level in drinking water found in the regions studied showed a wide range of concentrations, from Not Detectable (ND) up to 2000 µg L^{-1}.

Arsenic in drinking water has been described since the early twentieth century, stating natural water contamination and its relationship to As deleterious effect (first description of HACRE). There is an important number of reports that highlights the current situation of geogenic As contamination in a vast region of Argentina (MSAL, 2007).

In these conditions, arsenic oral intake is the most relevant via of exposure, while the others such as dermal or inhalatory ways become negligible (ATSDR, 2007).

The As levels found in this work showed a persistent contamination over the time with maximum values up to 200 times the recommended international standard, in agreement with previously reported data (Navoni *et al.*, 2006; MSAL, 2007).

The high and variable As levels found in water storage systems is linked to their provision from groundwater, because water is an insufficient resource in the region. It is interesting to note that the water supplied by the distribution system in

Table 1. Description of the time of residence and level of As in drinking water of the population interviewed by place of residence. Banda department, Santiago del Estero province: Jumi Pozo (JP); Negra Muerta (NM); Siete Árboles (7 A), Copo department, Santiago del Estero province: San José del Boquerón (SJB); Urutaú (U); Monte Quemado (MQ); Santos Lugares (SL); Venado Solo (VS); La Firmeza (LF); Malvinas (M); San Bernardo (SB); Lujan (L); Las Termas; (LT) and Almirante Brown department, Chaco province: Taco Pozo (TP); Santa Teresa de Carballo (STC); Pozo Hondo (PH); El Rosillo (ER); San Telmo (ST); Brasil (BR); El Quinto (EQ); Kilómetro 27 (Km27).

	Population interviewed		Time of residence (years)		As drinking water (µg L^{-1})
Region	Location	N	Mean	Range	Mean value
Banda (SDE)	7 A	12	9,6	7–13	11
	NM	3	38	10–56	14
	JP	38	19	3–70	20
Copo (SDE)	U	114	23	0,2–75	10
	MQ	14	27	5–64	10
	M	5	32	6–63	33
	LT	23	26	1,5–63	3
	LF	41	39	3–96	90
	SL	19	29	0,6–60	387
	VS	28	19	0,6–78	813
A Brown (Chaco)	SJB	95	9	1–16	160
	TP	75	12	5–44	230
	ST	20	12	6–51	262
	BR	20	12	6–38	435
	STC	28	23	2–54	668
	Km27	13	12	6–32	656
	ER	48	16	2–80	382
	EQ	34	17	4–67	512
	PH	20	34	8–51	997

Monte Quemado and Urutú (SDE) comes from a narrow aqueduct, with an As content near to 1 mg L^{-1}. The As content is reduced to acceptable values (10 μg L^{-1}) by a treatment based on the precipitation/flocculation of As with ferrous chloride, in a plant installed in the sixties. Opposite to this situation was that found in Taco Pozo (Chaco). The treatment applied to reduce the As content (based on the precipitation/flocculation with aluminum salts) is insufficient to get the required levels.

The level of As in drinking water was comparable with those reported previously in other endemic regions of hydroarsenicism such as Chile, México, India and Bangladesh (Rahman *et al.*, 2005) describing a highly contaminated region.

3.2 *Exposure assessment*

As is mainly excreted through the renal system. Therefore, urine is the best biological specimen for assessing recent exposure to this element (ATSDR, 2007). The degree of exposure covered a wide range of concentration, from 11 to 5085 μg g^{-1} creatinine. More than 90% of people presented values higher than the considered intervention value of 100 μg g^{-1} creatinine (ATSDR, 2007). The observed biomonitoring of As revealed a exposure degree comparable with those found in other areas from Latin America, such as Chile, and quite similar to those reported in other endemic regions of hydroarsenicism, such as India and Bangladesh.

A statistically significant correlation was observed between drinking water and UAs levels $Y = 1.502x + 233.5$ (Pearson's correlation coefficient: 0.66; $P < 0.00001$). The estimated level of UAs (using the correlation equation) and the observed values (average and maximum) were compared, showing that the observed value was higher than the expected one.

Figure 1 describes the spatial variation of As in drinking water (surface) and UAs content (iso-lines). In both representations (color graduation) are shown the relative values on As concentration. This approach indicates the areas where there was a geographical overlap between the As concentrations in water and urine.

The analysis of UAs confirmed As exposure. The effective exposure level clearly exceeded the proposed intervention cut off, depicting an effectively exposed population. The curve fitting ratified that the level of exposure was as high as water source indicated. The difference between the expected and observed UAs levels, suggested the possibility that other sources were involved in the exposure to this toxic, such as the food prepared and or cultivated with groundwater (Navoni *et al.*, 2007; Uchino *et al.*, 2006). Besides, children face additional risk from their common recreational activities, being exposed to other sources such as dust and/or soil sticked to hands or toys, and inadvertently ingested.

3.3 *Risk characterization*

In Figure 2, HQ is described by age group. Map series showed the increasing extent of the risk zone as time goes by. People living all the time in the study region and aged 40 or older are at risk for development of skin manifestations.

Skin is considered the most sensitive organ to As exposure. Mosaferi *et al.* (2007) found a clear dose-response relationship between skin disorders and As in drinking water below 150 μg L^{-1}. Therefore, most of the locations included in this study have suitable conditions for the population to be affected by As deleterious effects. Skin tumors are the most cited carcinogenic effect (Mosaferi *et al.*, 2007; Kazi *et al.*, 2009). Besides, the malignization of skin lesions is considered a possible alert for future internal organ cancers (Bates *et al.*, 1992). Thus, dermal manifestations could be suggesting us the future development of other oncologic effects.

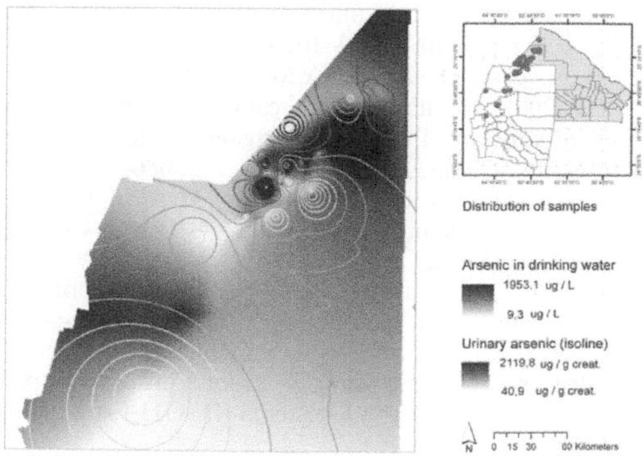

Figure 1. Geo-spatial distribution of the As level in drinking water and urinary As content in the region.

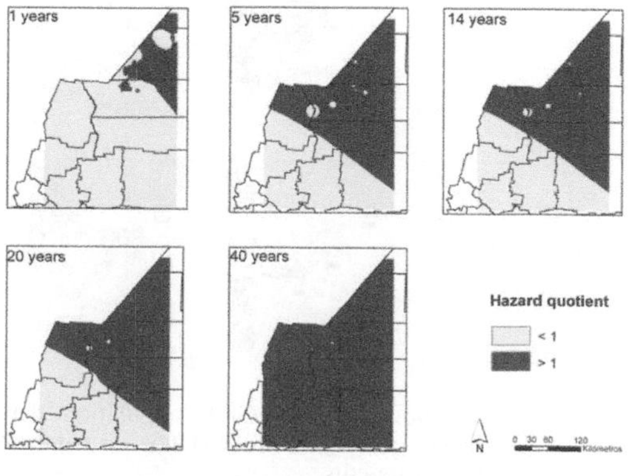

Figure 2. Health risk (HQ) areas vs. time of residence.

People from all the locations in Almirante Brown department (Chaco) and more than half locations in Copo department (SDE) should be expected to develop dermatological manifestations. A high number of As related cases of cancer should be expected in these departments, especially from urbanized centers.

A Health Composite Index (HCI) by age group was performed for the identification of Carcinogenic Risk (CR) areas, which is represented in Figure 3a. The classification of the geo-referenced information by regular intervals facilitated the identification and the "expansion" of the risk area related to age. The CR index characterized the sites of the study area, based on the probability that a resident becomes ill as a result of ingestion of As through drinking water. In Figure 3b, the distribution (density) of the population by age groups of 0–14 and 0–40 years old is shown, measured as a proportion of the total population and represented in quintiles. This information was integrated through a HCI that included the value of the probability of developing cancer and the probability that a person of 14 or 40 years old were settled in a specific area according to the considered population group.

The risk area increased with age in the same way as HQ did, from a covered area of 411.935 ha (8%) to 710.589 ha (13.8%) for the population of 14 years and 40 years, respectively.

Thus, the construction of maps of average daily intake of As containing water gave a regional vision of the risk, together with the time of exposure (as suggested by the time of residence) (Figure 3a). This description linked to population density (Figure 3b) with the above variables through the HCI, that showed the time and spatial tendency of progress expected of the regional health risk.

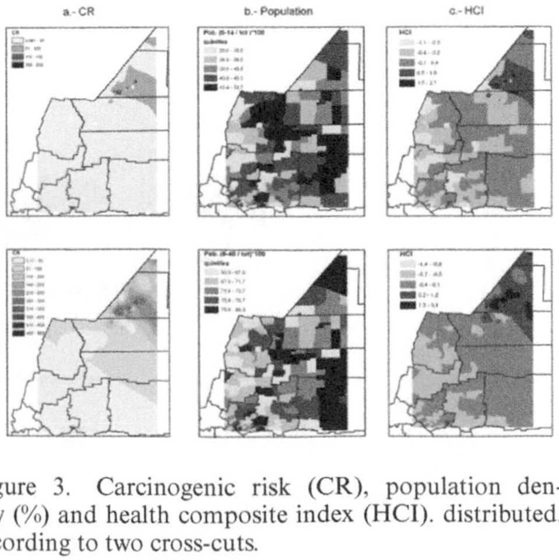

Figure 3. Carcinogenic risk (CR), population density (%) and health composite index (HCI). distributed, according to two cross-cuts.

4 CONCLUSIONS

The spatial and temporal analysis improved the scope of the results. The tools applied helped to make a diagnosis on the risk status of the region related to As. The characterization of space-time changes described particular scenarios. The interpretation of these observations indicated a relationship between the distance to the source of contamination and the residence time becoming a new approach for estimating the risk.

HCI integrated both concepts and compared the sites each other. This gave a relative measurement of the regional human risk when considering the hazard (arsenic in water) and vulnerability (proportion of people settled in a particular zone). HCI compared with CR included the distribution of the population, becoming a more reliable tool in the risk characterization, defining intervention priorities, especially relevant in rural and/or urban areas with low population density, that are not usually accompanied neither by risk planning programs nor by a warning criterion, facing to improve risk management.

Therefore, the applied tools had advantages over the standardized methods of risk assessment, including space-time risk variables broadening the scopes of the results and becoming a better approach for decision-making process.

REFERENCES

ATSDR. 2007. Agency for Toxic Substances and Disease Registry. *Toxicological Profile for Arsenic.* [on line] available at: http://www.atsdr.cdc.gov/toxprofiles/tp2.html. Accessed: September 18th, 2013.

Bates, M.N., Smith, A.H. & Hopenhayn-Rich, C. 1992. Arsenic ingestion and internal cancers a review. *Am. J. Epidemiol.* 1;135: 452–476.

Kazi, T.G., Arain, M.B., Baig, M.K., Afridi, H.I., Jalbani, N., Sarfraz, R.A., Shah, A.Q. & Niaz, A. 2009. The correlation of arsenic levels in drinking water with the biological samples of skin disorders. *Sci. Total Environ.* 407: 1019–1026.

Khan, S., Cao, Q., Zheng, Y.M., Huang, Y.Z. & Zhu, Y.G. 2008. Health risk of heavy metals in contaminated soils and food crops irrigated with wastewater in Beijing China. *Environ. Pollut.* 152: 686–692.

Instituto Nacional de Estadística y Censos. INDEC, 2001. Censo Nacional de Población, Hogares y Viviendas, Argentina.

Navoni, J.A., Olivera, N.M. & Villaamil Lepori, E. 2009. Validación metodológica de la cuantificación de arsénico por inyección en flujo-generación de hidruros—espectrometría de absorción atómica previa derivatización con L-cisteína (IF-GH-EAA). *Acta Toxicol. Argent.* 17(2): 45–54.

Navoni, J.A., Olivera, N.M. & Villaamil Lepori E. 2010. Cuantificación de arsénico por inyección en

flujo-generación de hidruros—espectrometría de absorción atómica (IF-GH-EAA) previa derivatización con L-cisteína. Validación y comparación intermetodológica utilizando dos técnicas de referencia. *Acta Toxicol. Argent*. 18(2): 29–38.
Navoni, J.A., Olivera, N., Garcia, S. & Villaamil Lepori, E. 2007. Evaluación de riesgo por ingesta de arsénico inorgánico en poblaciones de zonas endémicas argentinas. *Alimentación Latinoamericana* 270: 66–70.
Navoni, J.A., González Cid, M., Olivera, M., Tschambler, J., Bovi Mitre, G., Larripa, I. & Villaamil Lepori, E. 2006. Daño al ADN asociado al contenido de arsénico urinario en una población de jóvenes expuesta al arsénico por el agua de bebida. *Acta Toxicol. Argent.* 14(sup): 48–51.
Ministerio de Salud (MSAL). Epidemiología del Hidroarsenismo Crónico Regional Endémico—Estudio Colaborativo Multicéntrico. Villaamil Lepori *et al.* 2007 [en línea]. Disponible en http://www.ambiente.gov.ar/archivos/web/UniDA/File/libro_hidroarsenicismo _completo.pdf (Last accessed: 1/10/2013).
Mosaferi, M., Yunesian, M., Dastgiri, S., Mesdsaghinia, A. & Esmailnasab, N. 2007. Prevalence of skin lesions and exposure to arsenic in drinking water in Iran. *Sci. Total Environ*. 390: 69–76.
Rahman, M., Sengupta, M., Ahamed, S., Chowdhury, U., Hossain, A., Das, B., Lodh, D., Saha, K., Pati, S., Kie,s I., Barua, A. & Chakraborti, F. 2005. The magnitude of arsenic contamination in groundwater and its health effects to the inhabitants of the Jalangi—one of the 85 arsenic affected blocks in West Bengal, India. *Sci. Total Environ*. 338: 189–200.
Uchino, T., Roychowdhury, T., Ando, M. & Tokunaga, H. 2006. Intake of arsenic from water, food composites and excretion through urine, hair from a studied population in West Bengal, India. *Food Chem Toxicol*.; 44:455–461.

One Century of the Discovery of Arsenicosis in Latin America (1914–2014) – Litter, Nicolli, Meichtry, Quici, Bundschuh, Bhattacharya & Naidu (Eds)
© 2014 Taylor & Francis Group, London, ISBN 978-1-138-00141-1

Health risk assessment of arsenic in a residential area adjoining to a gold mine in Brazil

J.C. Ng
National Research Centre for Environmental Toxicology (Entox), The University of Queensland, Brisbane, Queensland, Australia

M. Gasparon
Earth Sciences, The University of Queensland, Brisbane, Australia
National Institutes of Science and Technology Acqua, Universidade Federal de Minas Gerais—UFMG, Belo Horizonte, Brazil

G. Duarte, A.M. Oliveira & V.S.T. Ciminelli
National Institutes of Science and Technology Acqua, Universidade Federal de Minas Gerais—UFMG, Belo Horizonte, Brazil

ABSTRACT: Paracatu is a town adjoining to a gold mine in Minas Gerais, Brazil where we collected drinking water and soil/dust samples were collected from seven residential locations. Fifteen common food categories were also collected from local vendors and stores. Arsenic was analysed by ICP-MS. A physiologically based extraction test for bioaccessibility (BAc) of arsenic was conducted on the soil/dust samples. Water arsenic ranged from non-detected to 3.9 μg/L; soil arsenic 6–461 mg/kg in the <250 μm fraction; food arsenic were all below the level of quantification; and BAc ranged 1.8–7.5%. For the health risk assessment, Brazilian exposure parameters or international standards were applied together with site specific data obtained from this study to derive daily intake of arsenic. When the soil intake was adjusted for BAc the potential risk estimate to local residents is thought to be insignificant.

1 INTRODUCTION

Arsenic is a well known environmental toxicant and it's often found as a common contaminant associated with mine waste materials generated from gold mining activities. Some of the houses in the Paracatu township are in close proximity of a large gold mine operation in Brazil. The fugitive dust and its potential impact upon the environment and health has been an ongoing concern by Paracatu residents.

Bioavailability is an important factor for risk assessment and is assumed to be 100% in the absence of site specific data. This conservative approach could result in excessive estimation of potential risk. BA is usually measured using an *in-vivo* animal model. However its *in-vivo* measurement can be expensive and time consuming. Various physiologically based extraction tests for the measurement of bioaccessibility (BAc) have been developed as a surrogate test of BA (Ng *et al.*, 2010). It is the aim of this study to measure BAc of As in order to give a more refined risk assessment of this element upon the health of Paracatu residents.

2 EXPERIMENTAL

In May 2011 (dry season), drinking water samples and geogenic materials (top 20 mm surface soil/dust) were collected from 7 locations in Paracatu in close proximity of the mine site. ICP-MS (Agilent 7500 cs, Agilent technologies, Japan) was used to measure the elemental concentrations together with appropriate certified reference materials (TM24.3 and BCSS-1from NRC) as the QA/QC check. Water samples were filtered through 0.45 μm filters and preserved in 2% nitric acid until analyses; and geogenic samples were dried at 40°C, sieved to obtain the ≤250 μm fraction followed by Aqua Regia acid digestion before analyses. BAc was measured based on a Physiologically Based Extraction Test (PBET) simulating the gastrointestinal tract pH conditions fasting, semi-fed, and fully-fed stomach (stomach phase I—pH 1.5, 2.5 and 4.0), and the small intestine (phase 2 pH 7.0) respectively (Bruce *et al.*, 2007). BAC was done in duplicate.

3 RESULTS AND DISCUSSION

Arsenic concentrations of water and ≤250 μm fraction of surface geogenic samples; and BAc are shown in table 1. BAc data shown represent the averaged value across all pHs and time intervals from the duplicate.

The results for limited set of household drinking water samples indicated that they were all below available Brazilian water guideline of 10 μg L^{-1} for As. None of the water samples tested represents any health concern. These results confirmed that the mining operation had not impacted on the drinking water quality in Paracatu. This is consistent with historical data bases as supplied by the local water supplier company (COPASA).

A recent review of As by the Joint FAO/WHO Expert Committee on Food Additives (JECFA, 2011) has established a BMDL0.5 of 2.1 to 7.0 μg kg^{-1} d^{-1}. The Bench Mark Dose was set at 0.5% incremental lung cancer risk. The lower BMD of 2.1 μg kg^{-1} d^{-1} is similar to the previous Provisional Tolerable Weekly Intake (PTWI) of 2.14 μg kg^{-1} d^{-1} (WHO, 1989) and is used as a guide here for the health risk assessment (acknowledging the margin of exposure might be about 30). Hence, the Acceptable Daily Intake (ADI) threshold of arsenic is 0.15 mg d^{-1} and 0.034 mg d^{-1} for a 70 kg adult and 16 kg child in accordance with Brazilian exposure factors. The daily soil ingestion rates via hand-to-mouth are 50 mg and 200 mg for adults and children; and the daily water consumptions were 2 L and 1 L respectively.

A local food survey is yet to be validated by independent laboratories although the initial data by a local commercial Brazilian laboratory suggested that the As in 15 common food categories were negligible. In a previous study in Catarina 1 region (Fávaro *et al.*, 1994) the As intake is similar to that of the Australian population at 63 μg d^{-1} for adults. After adjusting for 75% of the total arsenic in the form of organic arsenic the inorganic arsenic intakes for adults and young children are 0.01575 mg d^{-1} and 0.0036 mg d^{-1} in accordance with 70 kg and 16 kg body weight respectively. These values were used for the calculations of dietary intake (Table 2).

Table 1. Arsenic concentrations in water (μg L^{-1}) and surface geogenic (soil/dust) samples (mg kg^{-1}); and percent BAc.

House	Water As	Soil/Dust As	Soil/Dust BAc (%)
1	<0.1	6	7.5
2	<0.1	461	3.2
3	<3.9	186	2.9
4	<0.1	321	1.8
5	<0.1	53	3.6
6	<0.1	166	5.1
7	<0.1	211	1.8

Table 2. Daily arsenic ingestion (mg d^{-1}) with and without BAC adjustment for health risk assessment.

	As Intake		BAc Adjusted As Intake*	
House	Adult	Child	Adult	Child
1	0.0003	0.0012	0.0236	0.0076
2	0.0231	**0.0922**	0.0243	0.0105
3	0.0093	**0.0372**	0.0238	0.0086
4	0.0161	**0.0642**	0.0238	0.0087
5	0.0027	0.0106	0.0236	0.0079
6	0.0083	0.0332	0.0240	0.0092
7	0.0106	**0.0422**	0.0237	0.0083

Bold value indicates intake from soil/dust ingestion only would exceed ADI; *ingestion from soil/dust adjusted for BAc and dietary intake from water and food.

All soil/dust samples had levels of As one to two orders of magnitude lower than the sources (data not shown). This indicates that these values might have been a natural geological nature and that fugitive dust dispersion from the mining operation was limited. This is supported by the fact that the samples have a relatively low BAc of mostly less than 10% (Table 1). When As ingestion from soil was BAc adjusted all surveyed houses did not exceed the ADI.

4 CONCLUSIONS

The water results have confirmed the good quality of drinking water in Paracatu. Fugitive dispersion of dust from the mining operation upon Paracatu was not apparent and that the BAc of residential geogenic materials was relatively low. Hence the potential risk due to As in Paracatu residential area is limited. A second survey conducted in the wet season (December) is being completed.

ACKNOWLEDGMENTS

Logistic and field support by Juliana M. Esper, Marcos A. Morais and other staff of Kinross Paracatu Mining are acknowledged. Entox is a partnership between Queensland Health and the University of Queensland.

REFERENCES

Bruce, S., Noller, B.N., Matanitobua, V. & Ng, J.C. 2007. In-vitro physiologically-based extraction test (PBET)

and bioaccessibility of arsenic and lead from various mine waste materials. *Journal of Toxicology and Environmental Health* Part A. 70: 1700–1711.
Fávaro, D.I.T., Maihara, V.A., Armelin, M.J.A. Vasconcellos, M.B.A. & Cozzolino, S.M. 1994.. Determination of As, Cd, Cr, Cu, Hg, Sb and Se concentrations by radiochemical neutron activation Analysis in different Brazilian regional diets. *J. Radioanalytical Nuclear Chemistry* 181 (2): 385–394.
JECFA 2011. *Evaluation of certain contaminants in food.* The seventy-second report of Joint FAO/WHO Expert Committee on Food Additives. WHO: 1–115.
Ng, J.C., Juhasz, A.L., Smith, E. & Naidu, R. 2010. *Contaminant bioavailability and bioaccessibility: Part 1. Scientific and Technical Review.* CRC CARE Technical Report 14. CRC for Contamination Assessment and Remediation of the Environment, Adelaide, Australia: 1–74.

One Century of the Discovery of Arsenicosis in Latin America (1914–2014) – Litter, Nicolli, Meichtry, Quici, Bundschuh, Bhattacharya & Naidu (Eds)
 ISBN 978-1-138-00141-1

Chronic ingestion of arsenic—fluorine and its repercussion in the peripheral nervous system of inhabitants of the Córdoba province (Argentina)

A. Villagra Cocco, V. Goldaracena, N. del Valle Gait, L. Quinteros & C. Mondino
Nuevo Hospital San Roque, Córdoba, Argentina

ABSTRACT: In accordance with the edaphoclimatic characteristics, the southwest region of Córdoba (Argentina) has excellent agricultural aptitudes, but suffers limitations due to the quality of the available hydric resources. In this investigation, the quality of the water consumed in Chajan (33° 33′ 25.45″ S–65° 00′ 15.08″W), Department of Rio Cuarto, was evaluated. The main source of water is subterranean and comes from perforations in the phreatic stratum (3–15 m depth, 40%) or from spring water perforations (80–150 m depth, 44%). The levels of arsenic and fluorine in subterranean water were evaluated together with clinical manifestations.

1 INTRODUCTION

The high levels of toxicity of arsenic and fluorine, and its compounds demand a rigorous control of water and food. It accumulates in the organism in low doses and causes chronic intoxication. In Argentina, fluorosis in the southwest region of Córdoba becomes really important, with clinical manifestations like mottled dental enamel, bone structures, joints, and cognitive fluency in the population.

In this investigation, we evaluated the quality of the water in Chajan, a rural town located in the Department of Rio Cuarto, Córdoba province, Argentina. It is situated on the national route N° 8, 30 km from Sampacho, 95 km to the west of Rio Cuarto and 320 km from the capital city of Córdoba and a population of 1000 habitants. 154 native inhabitants, 15 to 80 years old, were clinically studied, with Electroencephalogram (EEG) and electromyography.

2 OBJECTIVES

2.1 *General*

The main purpose of the investigation is to prove the influence of arsenic and fluorine on the population of Chajan, and the implication in clinical manifestations.

2.2 *Specific*

- To evaluate the concentration of arsenic and fluorine in drinking water of this population.
- To determine the ingest source, and the fluorine total consumption by this population.
- To define the arsenic and fluorine base concentration in the source of water.
- To make a clinic- neurological/neurophysiological exam in the population.

3 METHODS AND MATERIALS

The developed procedures respected the established guidelines of Helsinki Declaration (1975 and revised in 1983). Besides, an informed consent was requested to the involved patients.

The population study included a complete clinic neurologic exam, cognitive evaluation and complementary exams:

- Electroencephalogram.
- Electromyography.
- Electroneurography.
- Toxicological exam of urine and blood samples (from people who partake in the investigation).
- Toxicological analysis of groundwater samples from different sources (consumed by the studied population).

The neurophysiological exams (electromyography and electroneurography) were performed with the ModelecSynergy device, with NCS, EMG and EEG options of 20 channels, contact electrodes to achieve sensitive and motor nervous stimulation. The results were interpreted by a specialized neurologist. The descriptive analysis of 154 patients from Chajan, between 15 to 80 years old was done. The diagnosis criteria of The American Association of Neuropathy of DYCK, World Health Organization (WHO) and the American Academy of Neurology (AAN), were followed in this investigation.

3.1 *Statistic methodology*

The statistic program "Demo instat" was used to analyze the results. The concentrations of fluorine and arsenic in water supplies were analyzed and compared with international standards. The fluorine analysis of the groundwater of 32 different wells, and of the rivers of Rio Aji and Chajan were performed by the Department of Hydric Resources of the Córdoba province by y the method of ion selective electrode, and alizarin Y methodology. Arsenic (mg L^{-1}) was analyzed by the Merck test methodology. The blood samples of the population were analyzed by the Toxicology Environmental Unit of Córdoba. The analysis of arsenic and fluorine in water and urine was performed by CEPROCOR.

4 CRITERIA

4.1 *Inclusion*

- Patients in the age between 15 and 80 years.
- Patients who have been residents in the place for 15 years or more.
- Patients who agreed and signed the informed consent.
- Patients who accomplished the neurologic and neurophysiologic criteria proposed by AAN and DYCK.
- 100 healthy patients without pathologic antecedents or intoxicated by ingestion were evaluated.

4.2 *Exclusion criteria*

- Patients 40 years old or above without life expectation.
- Patients with several motor or functional brain damage, where the language and speech are compromised.
- Ill patients, compromised by a serious systemic illness or neoplasia.
- Patients who had consumed endogenous or exogenous toxics.
- Patients with metabolic pathologies.
- People who refuse to sign the informed consent.

5 ANALYSIS

Fluorine concentration in examined wells was from 4 to 14 mg L^{-1}, when the limiting value in drinking water is 1 mg L^{-1}. Arsenic concentration of the examined wells was 0.50 mg L^{-1}, when the limiting value is 0.10 mg L^{-1}. A descriptive analysis of 154 patients of Chajan (15 to 80 years old), 86 women (56.8%) and 68 men (44.2%) was performed. 96 patients were normal (63.3%), 39 (8.4%) showed axonal manifestation (25.3%, 16 feminine and 23 masculine) and 13 patients showed demyelinising expression (1 feminine and 12 masculine). 6 patients were not included in the results because they could not achieve the inclusion criteria.

6 DISCUSSION

The clinical manifestations of chronic ingestion of arsenic and fluorine are fatigue, intolerance to physical activities, mental slowness, social complacency, obsequiousness to authority, neglect ion of critical thinking, arthrogryposis, decalcification, mottling on teeth and dermis (Gallará *et al.*, 2011; Pizzo *et al.*, 2007; Driscoll *et al.*, 1983).

This study manifests that 33.7% of population suffers neuropathies and axonal as the most prevalent (25.3%) and 8.4%, demyelinising. In both cases, the masculine sex was the most affected.

Other ions were evaluated in this revision, including selenium, which sometimes combines with fluorine in place of arsenic.

7 CONCLUSIONS

It is really important to control the concentration of fluorine and arsenic in water. This work indicates the need of providing good water treatment plants in the zone (Conaughton, 2007).

This would maintain a population without bone diseases and mottling teeth. Regarding to neuropathies, some manifestations such as the neglection of critical thinking, hyperkinesia and anxiety would be less common.

ACKNOWLEDGEMENTS

To Gladys Montachini and Pilar Cebollada (Ministerio de Agua, Ambiente y Energía, Córdoba).

REFERENCES

Conaughton, G. 2007. Southern California Water Supplies to Be Fluoridated Starting October, *North County Times*, August 1, 2007.

Driscoll, W.S., Horowitz, H.S., Meyers, R.J., Heifetz, S.B., Kingman, A. & Zimmerman, E.R. 1983. Prevalence of dental caries and dental fluorosis in areas with optimal and above optimal water fluoride concentration. *J. Am. Dent. Assoc.* 107: 42–47.

Gallará, R., Piazza, L. & Piñas, M. 2011. Fluorosis endémica en zonas rurales del norte y noroeste de Córdoba. Revista de Salud Pública 26(5): 40–48.

Pizzo, G., Piscopo, M.R., Pizzo, I. & Giuliana, G. 2007. Community water fluoridation and caries prevention: a critical review. *Clin. Oral Investig.* 11(3): 189–193.

One Century of the Discovery of Arsenicosis in Latin America (1914–2014) –
Litter, Nicolli, Meichtry, Quici, Bundschuh, Bhattacharya & Naidu (Eds)
 ISBN 978-1-138-00141-1

Health risk assessment of arsenic near a gold mine in Brazil

J.C. Ng
National Research Centre for Environmental Toxicology (Entox), The University of Queensland, Brisbane, Queensland, Australia

M. Gasparon
School of Earth Sciences, The University of Queensland, Brisbane, Australia
National Institute of Science and Technology Acqua, Universidade Federal de Minas Gerais—UFMG, Belo Horizonte, Brazil

G.C. Silva & V.S.T. Ciminelli
National Institutes of Science and Technology Acqua, Universidade Federal de Minas Gerais—UFMG, Belo Horizonte, Brazil

ABSTRACT: In addition to a food survey conducted in 2011–12, soil, dust and water were collected in 2010 and 2011 to generate input parameters for the health risk assessment of arsenic (As) in the residential area near a gold mine in Paracatu, Brazil. For a better estimate of As intake, bioaccessibility data for As in geogenic materials were obtained. Inhaled dust As contributed to only about 2 and 7% to the total exposure for adults and children, respectively. Ingestion is the major exposure route; however the As contribution from soil and dust is much lower than that from food intake. Dietary As contributed to 91 and 56% of the total exposure for adults and children, respectively. Bioaccessibility (BAC) data are essential for a more refined health risk assessment. Without this, overestimate of risk could result.

1 INTRODUCTION

Historically, there have been many townships established in close proximity of mine sites worldwide. Despite best practice technology and stringent environmental control, houses built in mineral-rich regions could face elevated natural and anthropogenic concentrations of potentially harmful elements, including arsenic, in their residential area. The Morro do Ouro Mine in Paracatu, Brazil, is currently operated by Kinross Brasil Mineração (KBM). This study was aimed at conducting a detailed investigation on associated occurrence of As in the residential area of Paracatu considering all possible exposure pathways that are relevant for the health risk assessment of As, and to address the public health concern.

2 METHODS

2.1 *Risk assessment approach*

The approach for Health Risk Assessment was based on the determination of the population exposure to As considering all pathways, such as ingestion (water + food + soil/dust) and inhalation (dust). The measured As exposure is compared to the Benchmark dose $BMDL_{0.5}$. Site specific bioaccessibility data were obtained for a more accurate estimate of As intake via the ingestion route. Local Brazilian exposure parameters including body weight, soil ingestion rate, air inhalation volume, water and food consumption were used for total daily intake calculation. A conceptual exposure scenario is shown in Figure 1.

2.2 *Sample collection and bioavailability*

In May 2011, at the end of the wet season, seven surface soil/dust from seven local properties and with

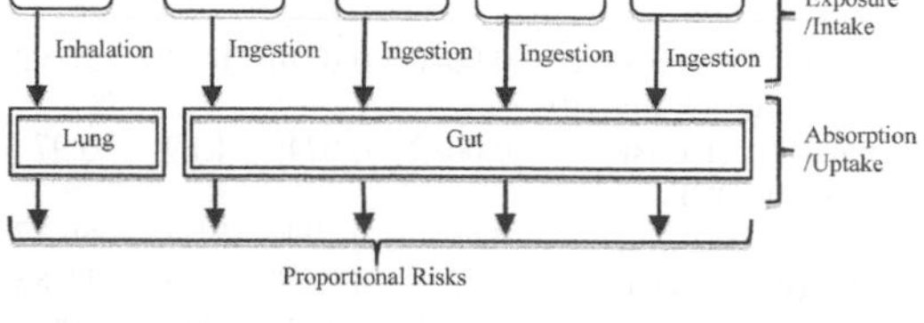

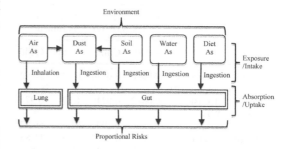

Figure 1. Conceptual exposure pathways of arsenic to derive proportional risks. Dermal As exposure is considered insignificant.

seven matching water samples were collected. In November 2011, at the end of the dry season, soil/dust from the same seven properties and another three properties were collected together with twenty water samples including replicates. In February 2012 (wet season), nine more water samples were collected. Dry soil samples were sieved to <250 μm. Water was preserved with HNO_3 (1%) until analysis. Ongoing air monitoring from 8 stations from August 2010 to June 2012 provided 387 Total Suspended Particles (TSP) data sets. A food survey was conducted in 2011/12 (see Silva *et al.*, this book).

A Physiological Based Extraction Test (PBET) was used for the measurement of bioaccessibility (BAC) as a surrogate for the prediction of bioavailability (BA) (Ng *et al.*, 2013). As concentrations were measured after appropriate acid digestion for food and geogenic materials, and water samples were measured without further treatment.

3 RESULTS AND DISCUSSION

3.1 *Arsenic levels in various matrices and health risk assessment*

Total daily inorganic As intake is shown in Table 1.

The mean (0.89 μg/L) of the As in water is 11 times lower than the Brazilian and the WHO limit of 10 μg/L.

The mean of BAC of soil/dust was (3.4 ± 2%) with a median of 2.9%. When adjusted for BAC (3.4 ± 2%), the mean bioaccessible As concentration of soil/dust was 5.9 mg/kg with a median of 3.1 mg/kg. The BAC value found in the present work is in agreement with that of 4.2% reported by Ono *et al.* (2011) in samples collected near KBM. The BAC adjusted mean As was used to calculate for its uptake upon ingestion.

TSP gave a median air arsenic concentration of 12.5 ng/m^3, a 75th percentile of 23.6 ng/m^3, and a mean of 26.1 ng/m^3. Under the reasonable assumption that the BAC of air As is similar (3.4%) to that of the residential geogenic materials, then the adjusted As medium daily intake from the inhalation route would be 0.00011 μg/kg bw/d for a 70 kg adult; 0.00034 μg/kg bw/d for a 16 kg child.

Table 1. Total daily inorganic As intake from various exposure pathways.

	IAs Intake (μg/kg bw/day)		% of Total As Intake	
Pathway	Adult	Child	Adult	Child
Ingestion of geogenic material (5.9 mg/kg)	0.0040	0.0740	0.92	20.92
Inhalation of dust (26.1 ng/m^3)	0.0082	0.0246	1.89	6.97
Food	0.3980	0.1990	91.45	56.27
Water (0.89 μg/L)	0.0250	0.0560	5.74	15.83
Total	0.4352	0.3536	100	100
$BMDL_{0.5}$	3	3		

The contribution of inhaled As to total intake is 1.89% for adults and 6.97% for children, even under the conservative assumptions of mean As concentration in the air and BAC = 100% in the calculations.

The air As represents limited risk in terms of respiratory cancer mortality. Respiratory cancers had only been associated with very high levels of air arsenic, such as those found in occupational exposure without personal protective equipment.

Average concentrations of As in food samples coupled with their respective consumption rates of the local population gave a daily intake of 0.398 μg/kg bw/d. Food contributed to 91 and 55% of the total exposure to As for adults and children, respectively. The highest As concentrations were found in rice and beans (not locally produced) that contributed to 90% of the total dietary As intake.

4 CONCLUSIONS

The overall results indicate that local geology and mining activities do not appear to have a significant impact on the total As exposure in Paracatu. The major risk contributor is the dietary source, but geogenic sources are relatively minor. Furthermore, the risk associated with the exposure—0.4352 μg/kg/day for adults and 0.3536 μg/kg/day for children—can be considered low by comparison to the $BMDL_{0.5}$ of 3 μg/kg/day (JECFA, 2011). This conclusion is supported by the results from another investigation (CEMEA, 2012), which showed that the incidence of diseases including cancers in Paracatu is similar to that from other cities in the region and to the average values in Minas Gerais (MG). This study demonstrates that it is necessary to consider bioavailability and all exposure pathways including dietary intake in order to provide a meaningful risk assessment. Without BAC assessment, the risk is likely to be over-estimated.

ACKNOWLEDGEMENTS

Logistic and field support by Juliana M. Esper, Marcos A. Morais and other staff of KBM are acknowledged. Entox is a partnership between Queensland Health and the University of Queensland.

REFERENCES

CEMEA 2012. Centro Mineiro de Estudos Epidemiológicos e Ambientais - CEMEA. *Perfil de morbimortalidade*

em municípios da região de Paracatu. Relatório Técnico, 1–26.
JECFA, 2011. Evaluation of certain contaminants in food. The seventy-second report of Joint FAO/WHO Expert Committee on Food Additives. WHO, pp. 1–115.
NEPC, 2013. Amendment of the National Environmental Protection (Assessment of Site Contamination) Measure. National Environmental Protection Council, Australia.
Ng, J.C., Juhasz, A., Smith, E. & Naidu, R. 2013. Assessing bioavailability and bioaccessibility of metals and metalloids. *Environ. Sci. Pollut. Res.*, in press.
Ono, F.B., Guilherme, L.R.G., Penido, E.S., Carvalho, G.S., Hale, B., Toujaguez, R. & Bundschuh, J. 2011. Arsenic bioaccessibility in a gold mining area: a health risk assessment for children, *Environ. Geochem. Health*, 34(4):457–65.

One Century of the Discovery of Arsenicosis in Latin America (1914–2014) – Litter, Nicolli, Meichtry, Quici, Bundschuh, Bhattacharya & Naidu (Eds)
 ISBN 978-1-138-00141-1

Gene-environment interactions between arsenic exposure from drinking water and genetic susceptibility in cardiovascular disease risk and carotid artery intima-media thickness in Bangladesh

F. Wu, M.L. Liu, X. Cheng, J.Y. Jiang & Y. Chen
New York University, New York, NY, USA

F. Parvez, V. Slavkovich, D. Levy & J.H. Graziano
Columbia University, New York, NY, USA

F. Jasmine, M.G. Kibriya & H. Ahsan
The University of Chicago, Chicago, IL, USA

T. Islam, A. Ahmed, M. Rakibuz-Zaman, S. Roy & R. Paul-Brutus
U-Chicago Research Bangladesh, Ltd., Dhaka, Bangladesh

ABSTRACT: Arsenic exposure from drinking water has been linked to subclinical and clinical outcomes of cardiovascular disease (CVD). However, no large-scale studies have evaluated whether the cardiovascular effects of arsenic exposure could be modified by genetic factors. Using data from the Health Effects of Arsenic Longitudinal Study (HEALS) in Bangladesh, we evaluated whether the association between arsenic exposure and CVD risk or carotid artery Intima-Media Thickness (cIMT) differs by single-nucleotide polymorphisms (SNPs) in 18 genes related to arsenic metabolism, oxidative stress, and inflammation and endothelial dysfunction. We found significant interactions between well-water As and two SNPs in ICAM1 and VCAM1 in CVD risk, and of both well-water As and urinary As with 3 SNPs in AS3MT in cIMT. Our data provide novel evidence that the cardiovascular effects of arsenic exposure may vary with some common genetic variants in genes related to arsenic metabolism and endothelial dysfunction.

1 INTRODUCTION

Arsenic (As) is a naturally occurring element primarily encountered in drinking water and foods, exposing millions of people worldwide to this toxic agent. Chronic exposure to As from drinking water has been linked to subclinical and clinical outcomes of Cardiovascular Disease (CVD) (Moon et al., 2012). Carotid artery intima-media thickness (cIMT) is a widely accepted indicator of subclinical atherosclerosis and a valid surrogate marker for clinical endpoints. The literature suggests that As-related health effects may be modified by genetic factors. However, no large-scale studies have evaluated whether the cardiovascular effects of As exposure could be modified by genetic factors.

2 METHODS

2.1 *Gene-environment interactions in CVD risk*

Using the resources from the Health Effects of As Longitudinal Study (HEALS) in Bangladesh, a prospective cohort of more than 20,000 participants recruited in 2000 (Ahsan *et al.*, 2006), was conducted a case-cohort study of 447 incident fatal and nonfatal cases of CVD, including 165 stroke cases and 238 cases of Coronary Heart Disease (CHD), and a subcohort of 1,375 subjects randomly selected from the HEALS to evaluate whether the association between As exposure and risk of CVD, CHD, and stroke differs by single-nucleotide polymorphisms (SNPs) in 18 genes related to As metabolism (*GSTM1*, *GSTT1*, *GSTO1*, *GSTP1*, *MTHFR*, *CBS*, *PNP*, and *AS3MT*), oxidative stress (*HMOX1*, *NOS3*, *SOD2*, and *CYBA*), and inflammation and endothelial dysfunction (*APOE*, *TNF*, *IL6*, *ICAM1*, *S1PR1*, and *VCAM1*).

Candidate genes were selected if they 1) are involved in As metabolism, or 2) have been related to As exposure in animal, *in vitro*, or epidemiologic studies and are known to play a key role in CVD risk in epidemiologic studies. A comprehensive approach was used to select SNPs in the candidate genes of interest. First was selected tag SNPs from International Hapmap Project and

Seattle SNPs using the r^2-based Tagger program with a pairwise $r^2 \geq 0.80$ and a minor allele frequency (MAF) ≥5%. The selection was performed for each ethnic group in the Hapmap/SeattleSNPs data separately to compile a list that includes all the tag SNPs. We also selected validated, non-synonymous SNPs with a MAF ≥ 5% from SeattleSNPs and potentially functional SNPs from the F-SNP database (Lee & Shatkay, 2008). In addition, we included SNPs that have been related to CVD risk and/or phenotypic markers of interest in the literature. After removing SNPs with genotyping efficiency <95%, monomorphic genotype data, Hardy-Weinberg equilibrium <0.0001, and an MAF <5% in the study population, a total of 170 SNPs in 17 genes were remained for analysis.

The multiplicative interaction was assessed by the cross-product term of As exposure and each candidate SNP using the Cox proportional hazards models. Interaction was also assessed on the additive scale (synergy) by testing whether the joint effect of As exposure and a SNP was greater than the sum of their independent effects.

2.2 *Gene-environment interactions in cIMT*

A cross-sectional study of 1078 participants in the subcohort was conducted to evaluate gene-environment interactions between As exposure and the abovementioned SNPs in cIMT. We used the mean of the near and far walls of the maximum common carotid artery IMT from both sides of the neck (mean of 4 sites) as the main outcome variable. The multiplicative interaction between As exposure and each SNP was assessed using multiple linear regression models. cIMT (β) in relation to every standard deviation (SD) increase in As exposure alone, presence of each SNP alone, and the joint effect of As exposure and each SNP was assessed.

3 RESULTS

3.1 *Interactions between As exposure and SNPs in risk of CVD, CHD, and stroke*

The multiplicative interactions between well-water As and two SNPs, rs281432 in *ICAM1* ($P_{adj}=0.0002$) and rs3176867 in *VCAM1* ($P_{adj} = 0.035$) in CVD risk, were significant after adjustment for multiple testing of all 170 tests. The hazard ration (HR) for CVD risk was 1.82 [95% confidence interval (CI): 1.31, 2.54] for every SD increase (101.3 μg/L) in well-water As among those with GG genotype of rs281432, much greater than the HR of 0.96 (95% CI: 0.65, 1.42) associated with GG genotype alone or the HR of 1.08 (95% CI: 0.94, 1.25) associated with one SD increase in well-water As alone. The magnitude of the interaction between well-water As and rs3176867 in *VCAM1* was weaker; the HR for CVD risk was 1.34 (95% CI: 0.95, 1.87) for one SD increase in well-water As among those with CC genotype of rs3176867. These associations were similar for stroke risk but weaker for CHD risk.

For main effects of SNPs, AA genotype of *NOS3* rs2853792 and GG genotype of *SOD2* rs5746088 were associated with a significantly reduced risk of CVD (HR = 0.51, 95% CI: 0.38, 0.69; and 0.30, 95% CI: 0.17, 0.51, respectively) and CHD (HR = 0.49, 95% CI: 0.34, 0.70; and 0.29, 95% CI: 0.15, 0.58, respectively). CC genotype of *MTHFR* rs1801133 was related to a significantly increased risk of stroke (HR = 2.33; 95% CI: 1.51, 3.61).

3.2 *Interactions between As exposure and SNPs in cIMT*

Although not significant after correcting for multiple testing, nine SNPs in *APOE*, *AS3MT*, *PNP*, and *TNF* genes had a nominally significant interaction with well-water As in cIMT. cIMT in relation to the joint effect of both higher well-water As exposure and CT or TT genotype of *APOE* rs7256173 (β = 46.0 μm, 95% CI = 10.7, 81.3), CC genotype of *AS3MT* rs10883790 (β = 35.8 μm, 95% CI = 9.2, 62.4), AA genotype of rs11191442 (β = 38.3 μm, 95% CI = 12.1, 64.5), and GG genotype of rs3740392 (β = 40.9 μm, 95% CI = 14.4, 67.5), was greater than the cIMT in relation to the genotype alone (β = −8.1 μm, −5.4 μm, −5.8 μm, and −5.1 μm, respectively) or As exposure alone (β = 8.7 μm, 8.0 μm, 7.8 μm, and 7.2 μm, respectively). These SNPs also showed similar pattern of interactions with urinary As. Additionally, the at-risk genotypes of the *AS3MT* SNPs were positively related to increased proportion of monomethylarsonic acid (MMA) and negatively related to proportion of dimethylarsinic acid (DMA) in urine, indicators of suboptimal As methylation capacity.

4 CONCLUSIONS

This data provide novel evidence that genetic variants in genes related to endothelial dysfunction may modify the risk of CVD associated with As exposure whereas genetic variants in genes involved in As metabolism may play a more important role in As-induced subclinical atherosclerosis.

ACKNOWLEDGEMENTS

This work is supported by grants R01ES017541, P42ES010349, P30ES000260, and R01CA107431 from the National Institutes of Health.

REFERENCES

Ahsan, H., Chen, Y., Parvez, F., Zablotska, L., Argos, M., Hussain, I., Momotaj, H., Levy, D., Cheng, Z., Slavkovich, V., van Geen, A., Howe, G.R. & Graziano, J.H. 2006. Arsenic exposure from drinking water and risk of premalignant skin lesions in Bangladesh: baseline results from the Health Effects of Arsenic Longitudinal Study. *Am J Epidemiol* 163(12): 1138–1148.

Lee, P.H. & Shatkay, H. 2008. F-SNP: computationally predicted functional SNPs for disease association studies. *Nucleic acids research* 36: D820–D824.

Moon, K., Guallar, E. & Navas-Acien, A. 2012. Arsenic exposure and cardiovascular disease: an updated systematic review. *Current atherosclerosis reports* 14(6): 542–555.

One Century of the Discovery of Arsenicosis in Latin America (1914–2014) – Litter, Nicolli, Meichtry, Quici, Bundschuh, Bhattacharya & Naidu (Eds)
 ISBN 978-1-138-00141-1

Characterization of Human hepatocellular carcinoma cells (HepG2), for experiments of effects the sex steroids hormones in arsenic metabolism

C.O. Puente-Valenzuela, J.J. Duarte-Sustaita, E. Sierra-Campos & G.G. García-Vargas
Universidad Juárez del Estado de Durango, Gomez Palacio Dgo, México

ABSTRACT: Population studies showed that arsenic metabolism is more efficient in women, suggesting possible regulations by sex steroid hormones. This study evaluates the effect of testosterone, progesterone and estradiol on the profile of arsenic species in the cell line HepG2, which is able to metabolize 0.2 μM arsenite, to generate: 36.8% of iAs, 21.2% MMA and 42.0% DMA. This concentration of arsenic did not active the antioxidant system GSH/GSSG. Finally, the concentrations of steroid hormones were settled to no decrease cell viability below 80%. The effects of Testosterone 0.3–10 μg/dL, Progesterone 0.75–12 μg/dL and Estradiol 6.25–100.0 ng/dL were characterized.

1 INTRODUCTION

Arsenic metabolism is performed by the As3MT, which requires the s-adenosylmethionine (SAM) as a donor of methyl groups and the presence of glutathione (GSH) a reducing agent to carry out this catalytic activity. This makes arsenic metabolism to interact with synthetic and recycling routes of GSH and SAM, and its possible regulation by sexual steroid hormones, which is consistent with the work of Linderberg *et al.* (2008) and Hopenhayn *et al.* (2003), who suggest an influence of sex steroid hormones on arsenic metabolism, resulting in an increase of metabolic efficiency in women during gestation. The cell line HepG2 has been used for the study of the metabolism of arsenic; in addition, it has been documented that this line expresses receptors for hormones (testosterone, estradiol and progesterone). The objective of this study was to establish the experimental conditions to evaluate the effect of the sex steroid hormones in the profile of arsenic species (As^{3+} + As^{5+}) in the cell line HepG2.

2 METHODS/EXPERIMENTAL

2.1 *Cells and sample preparation*

HepG2 liver cells (ATCC® HB-8065TM) were grown in Eagle's Minimum Essential Medium, and supplemented with 10% fetal bovine serum as recommended by the supplier. For hormonal experiments, free medium fetal bovine serum was used for quantification of arsenic species, 30×10^6 cells were used, which were placed in 0.5 mL of PBS pH 7 (4 °C). For the quantification of GSH and GSSG, 30×10^6 cells were also used, which were placed in 0.5 mL of 2% phosphoric acid (4 °C). In both cases, the cells were lysed in the cell disruptors Genie, using glass spheres of 0.5 mm. Then, lysates were centrifuged at 10,000 rpm for 10 minutes, recovering the supernatant and storing them at −70 °C until analysis.

2.2 *Arsenic species determination and GSH/GSSG measures*

For quantification of total arsenic species we used HG-CT-AAS method, adapted from that reported by Hernández-Zavala *et al.* (2008). The reduced and oxidized glutathione were measured using the technique of HPLC reverse phase in C18 column reported by Spadaro *et al.* (2005).

3 RESULTS AND DISCUSSION

3.1 *Dose-response curve—As vs. GSH and GSSG concentrations*

This experiment was addressed to determine the concentrations of intracellular GSH and GSSG as response of arsenite additions, to reflect the degree of GSH oxidation induced by As(III). The arsenite concentrations were in the range 0.2–5.4 μM. The GSH was increased in function to arsenite concentration, but not the GSSG. This can be explained by the generation of Reactive Oxygen Species (ROS) by the arsenic, which induced the synthesis of the glutamate-cysteine ligase (EC. 6.3.2.2) enzyme. through the signaling of NrF2. Figure 1A shows that the GSH increase is only significant at a concentration of 1.8 μM As(III), and Table 1 shows the data.

3.2 *Intracellular arsenic species*

This experiment was addressed to determine the ability to metabolize sodium arsenite by HepG2 cells. Table 1 and Figure 1B show the profiles of arsenic species obtained in each case (0.0, 0.2, 0.6 and 1.8 um). It can be seen that the methylation capacity decreases as the arsenic concentration increases. Metabolic efficiency is lost despite the increase of GSH concentration (Figure 1A) These results are consistent with those found for L. Ding *et al.* (2012), who suggested a modulatory role for the second arsenic methylation by glutathione, through a possible glutathionylation of the cysteine residues of AS3MT (Ding *et al.*, 2012).

3.3 *Dose—response of cell viability vs. arsenite and hormone concentrations*

Cell viability was assessed using the MTT assay; the goal of this experiment was to test if arsenic or hormones are capable of decreasing the viability below 80% in the physiological range of the total content. Figures 1C and D show that viability is not significantly affected by the concentration of sex steroid

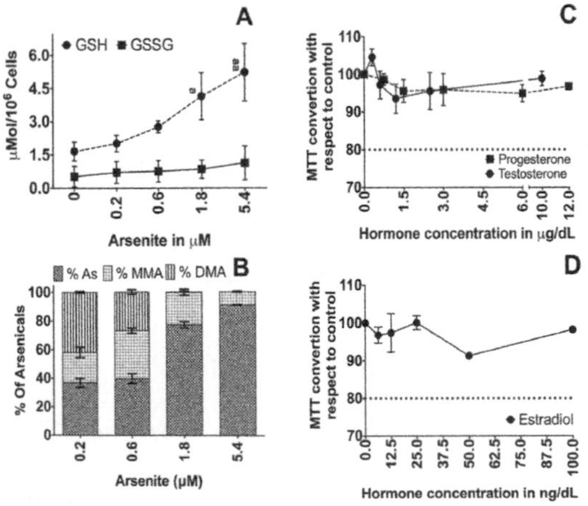

Figure 1. Graphs obtained from three independent experiments. (A) GSH and GSSG concentrations relative to arsenite concentrations in culture media. (B) Profiles of intracellular arsenic species obtained in the different treatments, the control treatment contained no arsenite (data not shown). (C) and (D) Curve dose—response of cell viability (MTT assay) against the concentration sex steroid hormone concentrations. Points plotted are means and SF. One-way ANOVA test/Dunnett.

Table 1. Intracellular concentrations of As(V) species, GSH and GSSG.

	Arsenite treatment (μM, as As(III))			
	0.0 x ± SD (%)	0.2 x ± SD (%)	0.6 x ± SD (%)	1.8 x ± SD (%)
Cell viability* (% ± SD)	100 ± 0.0	102.5 ± 2.3	101.9 ± 4.5	103.2 ± 4.1
iAs (μM/10⁶ cells)	–	0.19 ± 0.01 (36.8)	0.83 ± 0.40 (39.6)	2.68 ± 0.68 (77.0)
MMA (μM/10⁶ cells)	–	0.11 ± 0.04 (21.2)	0.74 ± 0.48 (33.5)	0.81 ± 0.34 (22.8)
DMA (μM/10⁶ cells)	–	0.21 ± 0.03 (42.0)	0.60 ± 0.39 (26.9)	0.01 ± 0.02 (0.30)
GSH (μM/10⁶ cells)	1.66 ± 0.43 (75.9)	2.01 ± 0.38 (73.9)	2.77 ± 0.27 (78.6)	4.15 ± 1.06 (83.0)
GSSG (μM/10⁶ cells)	0.53 ± 0.47 (24.1)	0.71 ± 0.49 (26.1)	0.76 ± 0.49 (21.6)	0.85 ± 0.41 (17.0)

The data are the mean of three independents experiment. *Cell viability was measured with MTT assay using the range of concentration of 0.2–48.6 with increases of 3X; the percentages were calculated with respect to control.

hormones (testosterone 0.3–10.0 μg/L, progesterone 0.75–12.0 μg/L and estradiol 0–100 ng/dL). Table 1 shows the cell viability when the cells were exposed to arsenite. The cell viability decreased to 74.5 ± 12.4 (% ± SD), when the concentration of arsenic was 15.6 μM (data not shown).

4 CONCLUSIONS

HepG2 hepatocytes model is capable of producing a metabolic profile model for total arsenic species at a concentration of 0.2 μM of arsenite. At this concentration, arsenic does not increase GSH levels, and the sexual hormones addition did not impact on cell viability. Finally, the HepG2 cells can be used to evaluate the effect of sex steroid hormones in the arsenic metabolism.

REFERENCES

Ding, L., Saunders, R.J., Drobná, Z., Walton, F.S., Xun, P., Thomas, D.J. & Stýblo, M. 2012. Methylation of Arsenic by Recombinant Human Wild-type Arsenic (+3 Oxidation State) Methyltransferase and Its Methionine 287 Threonine (M287T) Polymorph: Role of Glutathione. *Toxicol. Appl. Pharmacol.* 264(1): 121–130.

Hernández-Zavala, A., Valenzuela, O.L., Matousek, T., Drobná, Z., Dědina, J., García-Vargas, G.G., Thomas, D.J., Del Razo, L.M. & Stýblo, M. 2008. Speciation of Arsenic in Exfoliated Urinary Bladder Epithelial Cells From Individuals Exposed to Arsenic in Drinking Water. *Environ. Health Persp.* 116(12): 1656–1660.

Hopenhayn, C., Huang, B., Christian, J., Peralta, C., Ferreccio, C., Atallah, R. & Kalman, D. 2003. Profile of Urinary Arsenic Metabolites During Pregnancy. *Environ. Health Persp.* 111(16): 1888–1891.

Lindberg, A.L., Ekström, E.C., Nermell, B., Rahman, M., Lönnerdal, B., Persson, L.A. & Vahter, M. 2008. Gender and Age Differences in the Metabolism of Inorganic Arsenic in a Highly Exposed Population in Bangladesh. *Environ. Res.* 106: 110–120.

Spadaro, A., Bousquet. E., Santagati, N.A, Vittorio, F. & Ronsisvalle, G. 2005. Simple Analysis of Glutathione in Human Colon Carcinoma Cells and Epidermoid Human Larynx Carcinoma Cells by HPLC with Electrochemical Detection. *Chromatography* 62: 11–15.

One Century of the Discovery of Arsenicosis in Latin America (1914–2014) – Litter, Nicolli, Meichtry, Quici, Bundschuh, Bhattacharya & Naidu (Eds)
© 2014 Taylor & Francis Group, London, ISBN 978-1-138-00141-1

Synchrotron-based analytical techniques for elemental characterization of tumors in arsenic exposed patients

C.A. Pérez & E.X. Miqueles
Laboratório Nacional de Luz Sincrotron (LNLS), Campinas, Brazil

G.A. Mirabelli
Instituto Bac Spinoza, Córdoba, Argentina

D. Cremonezzi
Universidad Nacional de Córdoba, Córdoba, Argentina

G.A. Bongiovanni
CONICET-Universidad Nacional del Comahue (IDEPA), Neuquén-Cinco Saltos, Argentina

ABSTRACT: The populations of some areas of Argentina are exposed to arsenic (As) through drinking water and there has been a high correlation with bladder and lung cancer. Two of the greatest medical challenges are the early diagnosis and localized treatment, without damaging healthy tissue. Tumor samples were obtained from two inhabitants exposed to arsenic (As) and were analyzed by X-ray fluorescence microtomography and microX-ray fluorescence spectrometry, using Synchrotron Radiation (SR-XFCT and SR-μXRF). These technologies provide valuable information related to the distribution and accumulation of elements that can be used in the diagnosis and localization of malignant formations.

1 INTRODUCTION

Arsenic (As) is an environmental pollutant that causes several types of cancer and non-cancer diseases. It is known that up to 10% of Argentine population drinks As contaminated water with a high arsenic-associated cancer mortality (mainly bladder) (Nicolli *et al.*, 2012). The application new tumor biomarkers and diagnostic techniques may benefit from preventive interventions and treatment. It is also necessary to locate the cancerous lump, in as painless manner as possible. A technique to detect cancer at a higher sensitivity has thus been much awaited within the medical community and people at risk. Previous studies showed that Trace Elements (TEs) determined by X-Ray Fluorescence (XRF) techniques are found in significantly higher concentrations in neoplastic tissues when compared with normal tissues (Farah *et al.*, 2010; Silva *et al.*, 2012). X-rays have been used for non-invasive high-resolution imaging since their discovery in 1895. Now, the primary beam intensity needed in fluorescence x-ray analysis is on a lower order of magnitude than that used in mammography (Hayashi & Okuyama, 2010). Thus, the originated hypothesis was that XRF may be advantageous tools for cancer diagnostic in the future. In order to study the applicability of XRF on detection of arsenic-associated tumor, the aim of this work was to analyze the chemical composition of tumors from patients exposed to arsenic by drinking water. Elemental concentration was determined by micro X-Ray Fluorescence analysis using Synchrotron Radiation (SR-μXRF) and the 3D reconstruction of elemental distribution was determined by XRF-microtomography (SR-XFCT), both techniques available at the DO09B beamline of the Brazilian Synchrotron Light Laboratory (LNLS).

2 METHODS

2.1 *Areas under study*

In order to determine level of exposition, arsenic in the household water network from Villa del Rosario, Province of Córdoba, Argentina, was analyzed by the Arsenic Merck Test (cat. N°1.17927.0001), with 0.005 mg L^{-1} as detection limit. Accuracy of the method was determined by a calibration curve using TraceCERT® arsenic standard from Sigma-Aldrich.

2.2 *X-ray micro-fluorescence using synchrotron radiation (SR-μXRF)*

Tumor sections of 2 × 2 mm were obtained from biopsy material and confirmed by histopathological diagnostic. Multielemental analysis of tumor and Certified Reference Materials was done at the DO09B beamline of the LNLS where the XRF setup is placed. In order to carry out elemental determinations, *Kα* X-ray fluorescent intensities of S, Cl, K, Ca, Ti, Cr, Mn, Fe, Ni, Cu, Zn, As, Br and Sr were measured in conventional geometry (SR-μXRF) according to previously reported setup (Pérez *et al.*, 2006) using a collimated white beam of 300 × 300 μm and 300 sec as counting time.

2.3 *XRF tomographic setup*

X-ray fluorescence microtomography using synchrotron radiation was performed at the DO09B beamline at the LNLS by exciting the sample pixel by pixel with a collimated white beam of 100 × 100 μm. Samples were mounted onto a holder, which was in turn attached to translation and rotation stages. Acquisition time was 1 sec/pixel (Pérez *et al.*, 2014).

3 RESULTS AND DISCUSSION

3.1 *Arsenic in drinking water*

Since there is a high reporting of cancer cases in Villa del Rosario city (personal communication head physician), local drinking water was then analyzed. Drinking water in Villa del Rosario is obtained from five different depths underground sources. Ten sample waters from different sectors were analyzed and As around 0.025–0.050 mg L^{-1} was measured. These results indicate that population is chronically exposed to low arsenic doses.

3.2 *Multielemental characterization of tumors*

One bladder tumor from a patient exposed to 0.025 mg L^{-1} and two tumors (lung and adrenal gland) from patients exposed to 0.025 mg L^{-1} in Villa del Rosario city, were analyzed by SR-μXRF. Figure 1 shows the multielemental composition of tumors, being Cl, K, Ca, Cr, Mn, Fe, Cu, Zn and Br the major TEs in tumoral tissues. The results suggest that Fe and Zn are in highest proportions. These elements may be easily detectable and measurable by XRF analysis. 3D distribution of specific elements from SR-XFCT data were done by using a dedicated code recently developed by Miqueles & De Pierro (2010).

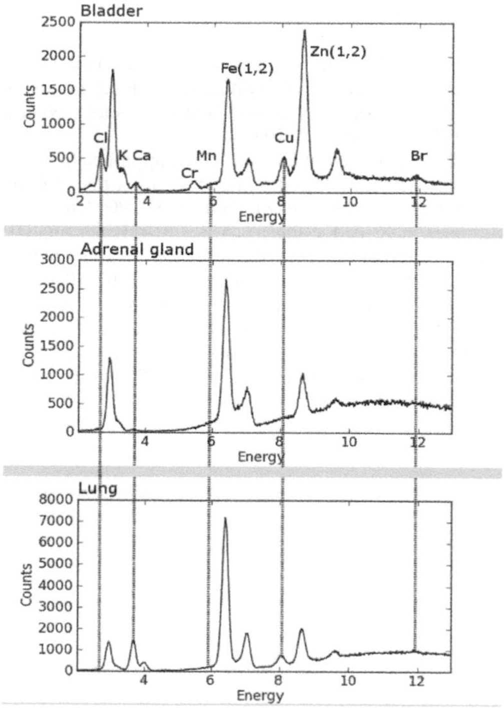

Figure 1. XRF spectra of tumors. The energies are expressed in keV. For comparisons, some elements were marked with dotted line.

4 CONCLUSIONS

Noninvasive and fast visualization of tumors is an important issue for physicians, biologists, medical physicists, etc. The XRF technique has enabled us to determine the distribution of the elements inside of the tumors of patients from hydroarsenical areas without destructing it and without toxic contrast agents. We expect that the ability to study local chemical concentrations such as Fe/Zn by XRF will be useful in early detection of arsenic-related tumors.

ACKNOWLEDGEMENTS

The authors would like to thank Brazilian Synchrotron Light Laboratory (LNLS). This work was supported by Brazilian Center for Research in Energy and Materials (CNPEM) (grants D09B-XRF-13421/2012, D09B-XRF-13495/2012 and D09B-XRF-15256/2013), CONICET, National University of Comahue (grant 04-A110) and FONCyT (grants PICT97-PRH33 and PICT214), Argentina.

REFERENCES

Farah, I.O., Trimble, Q., Ndebele, K. & Mawson, A. 2010. Significance of Differential Metal Loads in Normal Versus Cancerous Cadaver Tissues. *Biomed. Sci. Instrum.* 46: 404–409.

Hayashi, Y. & Okuyama, F. 2010. New approach to breast tumor detection based on fluorescence x-ray analysis. *GMS Ger Med Sci.* 8(18). DOI: 10.3205/000107.

Miqueles, E.X. & De Pierro, A.R. 2010. Exact analytic reconstruction in x-ray fluorescence CT and approximate versions. *Phys. Med. Biology* 55: 1007–1024.

Nicolli, H.B., Bundschuh, J., Blanco, M.C., Tujchneider, O.C., Panarello, H.O., Dapeña, C. & Rusansky, J.E. 2012. Arsenic and associated trace-elements in groundwater from the Chaco-Pampean plain, Argentina: Results from 100 years of research. *Science of the Total Environment* 429: 36–56.

Pérez, C.A., Miqueles, E. & Bongiovanni, G.A. 2014. Analysis of As bioaccumulation by XRF-microtomography at the DO09B beamline of the Laboratorio Nacional de Luz Sincrotron (LNLS). *LNLS Activity Report* 2012 (in press)

Pérez, R.D., Rubio, M., Pérez, C.A., Eynard, A.R. & Bongiovanni, G.A. 2006 Study of the effects of chronic arsenic poisoning on rat kidney by means of synchrotron microscopic X ray fluorescence analysis. *X Ray Spectrom.* 35(6): 352–358.

Silva, M.P., Soave, D.V., Ribeiro-Silva, A. & Poletti, M.E. 2012. Trace elements as tumor biomarkers and prognostic factors in breast cancer: a study through energy dispersive x-ray fluorescence *BMC Research Notes* 5: 194.

One Century of the Discovery of Arsenicosis in Latin America (1914–2014) – Litter, Nicolli, Meichtry, Quici, Bundschuh, Bhattacharya & Naidu (Eds)
© 2014 Taylor & Francis Group, London, ISBN 978-1-138-00141-1

Arsenic contamination in groundwater and its associated health effects in Maner and Shahpur blocks of Bihar, India

B.K. Thakur & V. Gupta
National Institute of Industrial Engineering (NITIE), Mumbai, India

ABSTRACT: This study tries to understand the associated effects on health of arsenic (As) contaminated drinking water. We conducted a primary survey of 388 households from two blocks in Bihar, India. A field test kit was used for testing the arsenic contamination level of household drinking water. From the water sample analysis, we found that 82.2% of households in the study area have poor water quality, while 72.4% of households have excess As concentration (more than 10 μg L^{-1}) in their drinking water. The mean As concentration was found 89.31 μg L^{-1}. We investigated the arsenic symptoms prevailing at the households and found that arsenicosis cases (mainly skin diseases, thickening and roughness of palms and soles, and lever problems) are more severe among children and women compared to men.

1 INTRODUCTION

Bihar, along with Bangladesh, Nepal and the state of West Bengal in India, are facing a major problem due to arsenic (As) groundwater contamination. Groundwater As contamination and its associated human health impacts have already been reported in different regions around the globe (Kapaj *et al.*, 2006). The chronic effects of inorganic As exposure through drinking water include skin lesions, such as hyperpigmentation and black foot disease, and respiratory symptoms, such as cough and bronchitis. Arsenic exposure during pregnancy can adversely affect several reproductive outcomes (Mukherjee, 2006). Given the background, the objectives of the work were: a) to measure the As contamination status in groundwater of the Maner and Shahpur blocks, b) to estimate the number of vulnerable population exposed to As contamination in drinking water, and c) to understand the health problem associated to the presence of As in drinking water.

2 METHODS AND MATERIALS

2.1 *Study area description*

The state of Bihar, India consists of 38 districts. Each district is divided into subdivisions and several blocks, and blocks are divided into Gram Panchayats (GPs), GPs having several villages. Out of 38 districts, 13 districts, which consists of 65 blocks and more than 892 habitations above have As concentration in groundwater above the permissible limit of 50 μg L^{-1}. The Maner block has 19 GPs and 49 villages, with a total population of 2,66,457 (Census 2011), and Shahpur block contains 20 GPs and 86 villages with a total population of 2,12,170 (Census 2011). Maner and Shahpur blocks were chosen for our study area because these two blocks are among the most As affected from Bihar.

2.2 *Water sample collection and instrumentation*

A field test kit was used to test As and Iron concentrations in the household hand tube well. 388 households water samples were collected from the household and tested immediately through the kit. The concentration of As was identified on the basis of the color of the test kit from 10 to 500 μg L^{-1}. In few samples, the results were contrasted against an As laboratory test and almost similar results were found.

3 RESULTS AND DISCUSSION

3.1 *Arsenic concentration in hand tube wells*

We tested 388 hand tube wells from the study area consisting of 215 and 173 water samples from Maner and Shahpur blocks, respectively. From the results, 23.45% of water samples contained As concentrations less than 10 μg L^{-1}, 18.30% contained between 10 and 50 μg L^{-1}, 18.04% contained concentrations in the range of 50–100 μg L^{-1}, 28.10% contained between 100 and 300 μg L^{-1} and 8.24% between 300–500 μg L^{-1}. 3.87% of water samples contained above 500 μg L^{-1}, the concentration in drinking water that in few years causes As skin lesions and cancer (Chakraborti *et al.* 2005).

From the analysis of water samples of the Maner block. It appears that 11.6% of water samples contained As concentrations below 10 μg L^{-1}, 15.8% contained between 10–50 μg L^{-1}, 23.7% in the range 50–100 μg L^{-1}, 34.4% between 100–300 μg L^{-1}, 9.3% in the range 300–500 μg L^{-1} and 5.1% above 500 μg L^{-1}. Results from the Shahpur block indicated that 6.94% households contain As concentrations in the range 300–500 μg L^{-1}, and 2.31% samples contained more than 500 μg L^{-1}.

3.2 *Vulnerable population*

This study tries to estimate the population vulnerable to As contamination based on the water sample results. Based on studies by Chakraborti *et al.* (2001, 2002, 2003, and 2005) and Ghosh *et al.* (2007), we tried to estimate the population drinking As contaminated water in both blocks. Ghosh *et al.* (2007, 2012) estimated the population at risk in the areas of As contaminated aquifers from four district of Bihar and suggested that aquifers of three districts are more contaminated. The study found that 3724 sources of drinking water were contaminated, with 19 and 30 villages at high risk from Patna and Bhojpur districts, respectively. From the analysis of water, the majority of the population from the study area drinks water containing above 50 μg L^{-1} of As 82.22% of households from the study area do not use safe water for drinking, and 72.42% of households have excess of iron in waters. The total population drinking water with more than 500 μg L^{-1} As is 13263, which is around 3.87% of the total rural population of study area.

3.3 *Patients with skin lesions*

Children and infants are more vulnerable and more susceptible to the adverse effects of As and toxic metals than adult males and females (NRC, 1993). Out of the 215 households surveyed from the Maner block, we found 75 (3.87% of the surveyed population) suspected cases of male suffering from skin lesions, 79 (4.08%) cases of female suffering skin lesions, and 46 (2.37%) cases of children suspected of suffering skin lesions. In the Shahpur block, out of the 161 households we surveyed, we found 115 (8.15%) cases with persons having skin lesion problems, 31 (2.19%) cases of male patients, 41 (2.90%) cases of female patients, and 43 (3.04%) cases of children suffering from skin lesion problems. In the Dudhghat village of Semariya GPs and Dudhaila and Haldi Chapda villages from Kitta Chauhattar GPs, a group of children, drinking highly contaminated water, is facing skin lesion problems. It was also found that the children presented black spots and black and white spot problems on their entire body, although earlier studies suggested that children are less vulnerable to melanosis, spotted melanosis, keratosis and spotted keratosis (Chakraborti *et al.*, 2005).

4 CONCLUSIONS

From the field survey, groundwater As levels in groundwater aquifers are found to be alarming in the study area. Around 76.7% and 64% hand tube wells analyzed from Maner and Shahpur blocks contain As concentrations above 10 μg L^{-1}. A team of villagers and doctors informed us that the incidences of cancer and skin lesions are increasing in these blocks.

A few plans have been initiated by the government, but due to bad implementation, no achievement has been obtained.

ACKNOWLEDGEMENTS

The authors thank the director of CIMP, Patna, for providing support in field work. Prof. A.K. Ghosh, Dr. D. Saha, and Dr. B. Sahoo helped us to understand the technical issues of arsenic contamination. We thank the villagers for their participation.

REFERENCES

Chakraborti, D., Basu, G.K., Biswas, B.K., Chowdhury, U.K., Rahman, M.M., Paul, K., Chowdhury, T.R., Chanda, C.R., Lodh, D. & Ray, S.L. 2001. Characterization of arsenic bearing sediments in Gangetic delta of West Bengal, India. *Arsenic Exposure and Health Effects* Elsevier New York pp. 27–52.

Chakraborti, D., Mukherjee, S.C., Pati, S., Sengupta, M.K, Rahman, M.M., Chowdhury, U.K., Lodh, D., Chanda, R.C., Chakraborti, A.K. & Basu, G.K. 2003. Arsenic groundwater contamination in middle Ganga plain, Bihar India: A future danger, *Environmental Health Perspectives* 119 (9): 1194–1201.

Chowdhury, U.K., Biswas, B.K., Chowdhury, T.R., Samanta, G., Mandal, B.K., Basu, G.C., Chanda, C.R., Lodh, D., Saha, K.C., Mukherjee, S.K., Roy, S., Kabri, S., Quamruzzaman, Q., Chakraborti, D. 2000. Groundwater arsenic contamination in Bangladesh and West Bengal, India, *Environmental Health Perspective* 108(5): 393–396.

Chowdhury, U.K., Biswas, B.K, Chowdhury, T.R., Mandal, B.K., Samanta, G., Basu, G.K., Chanda, C.R., Lodh, D., Saha, K.C., Chakraborti, D., Mukherjee, S.C., Roy, S., Kabir, S., Quamruzzaman, Q. 2001. Groundwater arsenic contamination and human sufferings in West Bengal, India and Bangladesh, *Environmental* Science 8(5): 393–415.

Ghosh, A.K., Singh, S.K., Bose, N. & Chaudhary, S. 2007. Arsenic contaminated aquifers: a study of the Ganga levee zones in Bihar, India. *Symposium on Arsenic: The Geography of a global problem*, Royal Geographical Society, London.

Ghosh, A.K., Bose, N., Kumar, R., Bruining, H., Lourma, S., Donselaar, M.E. & Bhatt, A.G. 2012. Geological origin of arsenic groundwater contamination in Bihar, India. In.: J.C. Ng, B.N. Noller, R. Naidu, J. Bundschuh & P. Bhattacharya (eds.) *Understanding the Geological and Medical Interface of Arsenic, As 2012*: 85-87. Leiden, CRC Press/Taylor and Francis (ISBN-13: 978-0-415-63763-3), pp. 522–525.

Kapaj, S., Peterson, H., Liber, K. & Bhattacharya, P. 2006. Human health effects from chronic arsenic poisoning: a review. *Journal of Environmental Science and Health* A 41: 2399–2428.

Mukherjee, A., 2006. Arsenic contamination in groundwater: a global perspective with emphasis on the Asian scenario, *International centre for diarrheal research centre* Bangladesh.

One Century of the Discovery of Arsenicosis in Latin America (1914–2014) – Litter, Nicolli, Meichtry, Quici, Bundschuh, Bhattacharya & Naidu (Eds)
© 2014 Taylor & Francis Group, London, ISBN 978-1-138-00141-1

Arsenic, obesity, and inflammation cytokines in Mexican adolescents

M. Rubio-Andrade & G.G. García-Vargas
Facultad de Medicina, UJED, Gómez Palacio, México

E.K. Silbergeld, R. Zamoiski, C. Resnick, V. Weaver, A. Navas-Acien & E. Guallar
Department of Environmental Health Sciences, BSPH, Johns Hopkins University, Baltimore, USA

S.J. Rothenberg
Instituto Nacional de Salud Pública, Cuernavaca, México

A.J. Steuerwald & P. Parsons
New York State Department of Health, Albany NY, USA

ABSTRACT: The aim of this work was to determine the associations among arsenic (As) exposure, adiposity, and inflammatory cytokines in 384 adolescents aged 12–15 years in a cross-sectional study. As was measured by total As concentration in urine adjusted by creatinine, adiposity was assessed using Body Mass Index (BMI) and bioimpedance segmental body composition, cytokines were determined by standard procedures. Our results show prevalence of overweight and obesity of 39.5%. Median (interquartile range) of total As in urine was 36.2 (26.01, 46.8) μg/g creatinine. We found an inverse association between As exposure and the percentage of total body fat, which was no dependent from inflammatory cytokines.

1 INTRODUCTION

There is a growing interest in the role of environmental pollutants and obesity, but the existing literature of an association of arsenic (As) exposures with obesity is considered as insufficient (Maull *et al.*, 2012).

In 2011, Gomez Rubio *et al.*, found a negative association between Body Mass Index (BMI) and the efficiency of As methylation (Gómez Rubio *et al.*, 2011), and this association was more robust in women (Gomez Rubio *et al.*, 2012). Arsenic exposure was not associated with BMI in women from Santiago city, Chile (Ronco, 2010) In adolescents from Taiwan, however, participant with normal weight had higher urinary arsenic levels compared to participants with overweight (Su *et al.*, 2012).

Lagunera Region is located in northern Mexico and is site of documented endemic arsenic exposure. The problem has been mitigated in part in the recent years, but arsenic in tap water has still been reported at concentrations that surpass the WHO guideline of 10 μg/L. The aim of this work was to assess the association of arsenic exposure, obesity and inflammatory cytokines in adolescents exposed to arsenic via drinking water.

2 METHODS

2.1 *Population recruitment*

We recruited 627 adolescents 12–15 years of age who were living in Torreón Coahuila, México, without acute or chronic illness or chronic medication. We explained the study procedures and asked each adolescent and their parents to provide written consent to participate in the study. Subjects answered a questionnaire, and only 512 completed a clinic visit for a physical examination, and biological specimen collection. 63 participants were excluded due to fish consumption the previous week, and 65 participants because the questionnaire or anthropometric measurements were incomplete.

2.2 *Arsenic determinations*

Total arsenic concentrations were measured in a urine spot sample. Subjects were asked to collect fresh urine at morning, in a 100 mL plastic flask. Urine samples were stored frozen (–70 °C) until analyses. Arsenic was measured using a PE DRCII inductively coupled plasma-mass spectrometer in the DRC mode (Perkin Elmer Life and Analytical Sciences, Shelton, Connecticut, USA).

2.3 *Anthropometric measurements*

Body Mass Index (ratio of weight and height squared) was calculated, and classified by Z-score (ZBMI) according to the World Health Organization (WHO). Additionally we categorized in obese or non obese according to the International Obesity Task Force (IOTF) by BMI, sex and age Segmental body composition was measured by bio impedance procedures using impedance electronic scale Tanita® 418.

2.4 *Cytokines*

Interleukins, IL-6, IL-1β, and TNF-α were measured in serum at the Biochemistry Laboratory of the Johns Hopkins Bayview Medical Center General Clinical Research Center, using ELISA method.

2.5 *Statistical analysis*

Urine concentrations of total urinary arsenic and cytokines were right skewed and log transformed for the analyses. Tertiles were generated based on the distribution of urine arsenic concentrations in the overall study sample.

Linear models were used to estimate adjusted ratios of the geometric mean values for fat mass and ZBMI by urine arsenic concentrations. We performed linear regression in models with progressive degrees of adjustment. Initially, we adjusted by sex, age, parental education, birth weight and family income. Then we further adjusted for fat mass for BMI models, and for ZBMI for fat mass models. Finally, we further adjusted for interleukin levels.

3 RESULTS AND DISCUSSION

3.1 *Population description*

The study sample was 384 adolescents, the prevalence of obesity and overweight was 14.8% and 21.1% respectively. The mean ± Standard Deviation (SD) of ZBMI was 0.43 ± 1.5 for male and 0.64 ± 1.4 for females. The median fat mass was 26.4% (19.9–32.75), which was significantly higher in female than males.

The median (IQR) urinary arsenic level was 40.1 μg/L (25.1, 61.4). We found higher urinary arsenic concentrations in males than in females.

The median of IL-6 was 0.69 ng/mL (0.46, 1.05) ng/mL, IL-1beta was 0.07 (0.0003–0.14) ng/mL and level TNF-α was 5.4 ng/mL (4.6,-6.5) ng/mL. Only TNF-α levels were higher in male adolescents.

3.2 *Obesity associated variables*

The median (IQR) of arsenic in urine in non-obese subjects was 42.5 (25.8, 62.3) μg/L and in adolescents with overweight and obesity it was 37.8 (24.4, 61.1) μg/L.

Table 1. Spearman correlation coefficient for arsenic corrected by creatinine, ZBMI and body composition variables. ($n = 384$).

	Arsenic corrected by creatinine	ZBMI	Fat mass
Arsenic corrected by creatinine	1.00		
ZBMI (kg/m^2)	–0.005	1.00	
Fat mass (%)	–0.15**	0.78***	1.00
Interlukin 6	0.16**	0.35***	0.33***
TNF-α	0.19***	0.13**	–0.06
Interleukin 1β	–0.07	0.07	0.08

* $p < 0.05$; ** $p < 0.01$; *** $p < 0.001$.

In a Spearman's correlation analysis, Interleukin 1β correlated only with As exposure, but not with obesity. IL-6 and TNF-α correlates with ZBMI (Table 1).

Arsenic was negatively associated with BMI and fat body mass percentage (Table 1). However, there was a positive association with IL-6 and TNFα cytokines.

3.3 *Discussion*

Our results showed a negative association of adiposity with arsenic exposure. These findings are in agreement with those studies done in human populations exposed to low levels of As in water (Su *et al.*, 2012). Obesity is considered a disorder related to an inflammatory process, while arsenic also has an effect on the immune system. Our results suggest the association of IL-6 and TNF-α with ZBMI, but only IL-6 was associated with body fat percentage.

ACKNOWLEDGEMENTS

This project was supported by NIH grant No. R01 ES015597.

REFERENCES

Gomez-Rubio, P., Roberge, J., Arendell, L., Harris, R.B., O'Rourke, M.K., Chen, Z., Cantu-Soto, E., Meza-Montenegro, M.M., Billheimer, D., Lu, Z. & Klimecki, W.T. 2011. Association between body mass index and arsenic methylation efficiency in adult women from southwest U.S. and northwest Mexico, *Toxicol. Appl. Pharmacol.* 252(2):176–82.

Gomez-Rubio, P., Klimentidis, Y.C., Cantu-Soto, E., Meza-Montenegro, M.M., Billheimer, D., Lu, Z., Chen, Z., & Klimecki, W.T. 2012. Indigenous

American ancestry is associated with arsenic methylation efficiency in an admixed population of northwest Mexico, *J. Toxicol. Environ. Health A*, 75(1):36–49.
Maull, E.A., Ahsan, H., Edwards, J., Longnecker, M.P., Navas-Acien, A., Pi, J., Silbergeld, E.K., Styblo, M., Tseng, C.H., Thayer, K.A. & Loomis, D. 2012. Evaluation of the association between arsenic and diabetes: a National Toxicology Program workshop review, *Environ. Health Perspect.*, 120(12):1658–1670.
Ronco, A.M. 2010. Lead and arsenic levels in women with different body mass composition. *Biol. Trace Elem. Res.* 136(3):269–278.
Su, C.T., Lin, H.-C., Choy, C.-S., Huang, Y.-K., Huang, S.-R., Hsueh, Y.-M. 2012. The relationship between obesity, insulin and arsenic methylation capability in Taiwan adolescents. *Sci. Total Environ.* 414:152–158.

One Century of the Discovery of Arsenicosis in Latin America (1914–2014) – Litter, Nicolli, Meichtry, Quici, Bundschuh, Bhattacharya & Naidu (Eds)
© 2014 Taylor & Francis Group, London, ISBN 978-1-138-00141-1

ABCA1 is another ABC sub-family gene which contributes to arsenic efflux besides ABCB1 and ABCC1

X.H. Tan, X.B. Wang & C.H. Di
School of Medicine, Hangzhou Normal University, Hangzhou, Zhejiang, China

L. Yang
School of Medicine, Hangzhou Normal University, Hangzhou, Zhejiang, China
Ministry of Education Key Laboratory of Xinjiang Endemic and Ethnic Disease, Shihezi University, China

S.L. Guo
Ministry of Education Key Laboratory of Xinjiang Endemic and Ethnic Disease, Shihezi University, China

ABSTRACT: Increasing of arsenic efflux has been regarded as the most important mechanism for arsenic resistance. The ATP-Binding Cassette (ABC) subfamily members, ABCB1 (MDR1) and ABCC1 (MRP1), have been reported increased in arsenic resistance cells. In our previous studies, we found that another ABC gene, ABCA1, was increased in the arsenic resistant ECV-304 (AsRE) cells. The results from this study showed that ABCA1 expression level was associated with the survive rate and intracellular arsenic accumulation after arsenic treatments. Interestingly, silence all the ABCA1, ABCB1 and ABCC1 genes deplete the arsenic resistance feature compared with the ECV304 cells. Our results suggested that ABCA1 is another ABC gene which contributes arsenic efflux besides ABCB1 and ABCC1.

1 INTRODUCTION

Arsenic is a toxic element widely distributed in nature, and nearly all organisms develop strategies to tolerate arsenic toxicity in some degrees. Increase arsenic efflux (Rappa *et al.*, 1997; Vernhet *et al.*, 2001) and a decrease arsenic intake (Leung *et al.*, 2007) have been regarded as the most important mechanisms for arsenic resistance. The ATP-Binding Cassette (ABC) genes, such as ABCB1 and ABCC1 have been reported in some arsenic resistance cell lines (Lee *et al.*, 2006; Miller *et al.*, 2007). In previous studies, we constructed a human arsenic-resistant ECV-304 cell line (AsRE) by chronic exposure the ECV304 cells in low dose arsenic, and used the Suppression Subtractive Hybridization (SSH) and microarray analysis to identify arsenic-resistant genes in these cells. ABCA1, another gene belonging to the ABC super family, was found increased in AsRE cells. In the present study, results demonstrate that ABCA1 also contributes to arsenic tolerance by reducing cellular arsenic accumulation.

2 METHODS/EXPERIMENTAL

Silence ABCA1, ABCB1 and ABCC1 in AsRE cells by RNA interference. The ABCA1 was cloned into the pCDNA3.1 expression vector, and transfected into Hela cells. The ABCA1, ABCB1 and ABCC1 mRNA and protein production were determined by Real-time quantitative PCR and Western blot analyses. The survival rate and IC_{50} of the ECV-304, AsRE and Hela cells after arsenic treatment were determined by MTT (3-(4, 5-dimethylthiazol-2-yl)-2, 5-diphenyltetrazolium bromide, a tetrazole) assay. Cellular arsenic accumulation was determined by an atomic fluorescence spectrophotometer.

3 RESULTS AND DISCUSSION

Silence ABCA1 in AsRE cell by RNAi (Figure 1A) decreased cells survive rate (13.22 μM vs. 7.01 μM, $p < 0.01$) (Figure 1B), and enhanced the intracellular arsenic accumulation (Figure 1C). Overexpression ABCA1 in Hela cell (Figure 1D), increased its survival rate (36.74 μM vs. 20.72 μM, $p < 0.01$) (Figure 1E), and lowered the intracellular arsenic accumulation (Figure 1F). These results indicated that ABCA1expression level is associated with arsenic resistance and intracellular arsenic accumulation.

After adjusting the concentrations of siRNAs against ABCA1, ABCB1 and ABCC1, we got the similar RNAi efficiency (>90%) of each gene

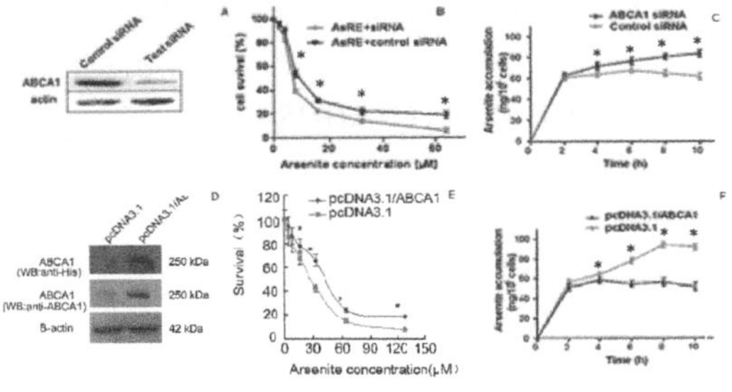

Figure 1. Silence ABCA1 in AsRE cells lowered the survive rate and elevated intracellular arsenic contents after arsenic treatment (A–C); overexpression ABCA1 in Hela cells increased the survive rate and decreased arsenic contents after arsenic treatment (D–F). Within each figure, bar represent mean (±SEM), *: $p < 0.05$.

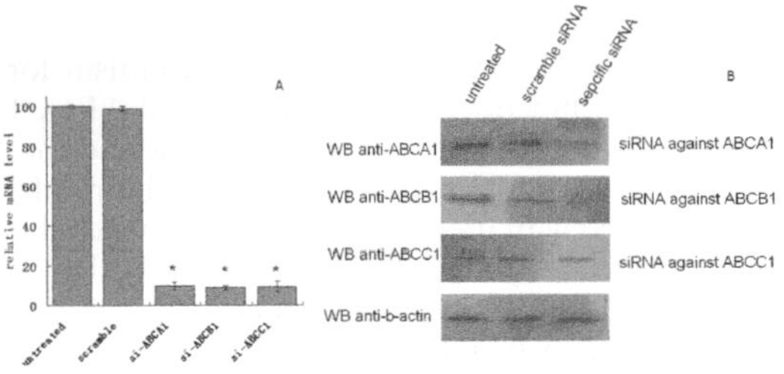

Figure 2. mRNA and protein levels of ABCA1, ABCB1 and ABCC1 were knocked down by siRNAs. Expression of mRNA level of ABCA1, ABCB1 and ABCC1 in AsRE cells by real-time quantitative PCR after RNA treated with scramble siRNA or specific siRNA against ABCA1, ABCB1 and ABCC1 for 48 h (A). Bar represent mean (±SEM) of relative expression of target mRNA. Western blots showing relative protein levels in AsRE cells after RNA interference (B).

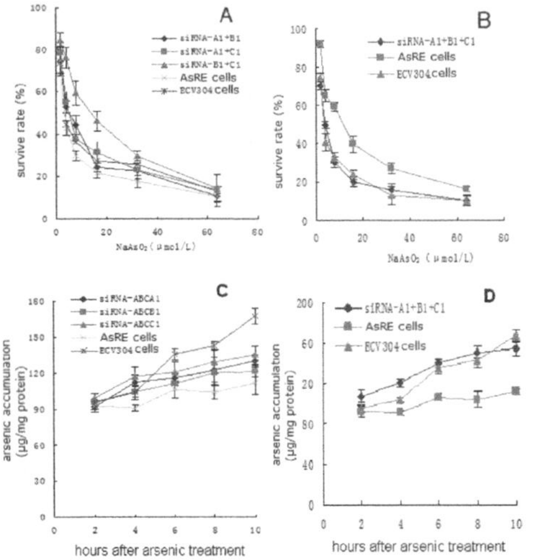

Figure 3. Decrease of the arsenic resistance by silence ABCA1, ABCB1 and ABCC1. The survive rate of the AsRE cells treated with scramble siRNA or the target gene-specific siRNA separately (A), or all the 3 genes(B). The arsenic accumulation of the AsRE cells treated with scramble siRNA or the target gene-specific siRNA separately (C), or all the 3 genes(D). The untreated ECV304 cell as a negative control and untreated AsRE cells as a positive control.

compared with the controls in mRNA level after 48 h treatment with the target gene-specific siRNAs. The mRNA and protein levels are shown in Figure 2.

Silence ABCA1 decreased the arsenic resistance in AsRE cells, and when all the 3 ABC genes were silenced, the AsRE cells lost its arsenic resistance feature compared with the ECV304 cells (Figure 3A–B). The IC50 value for the AsRE cells treated with siRNA against ABCA1, ABCB1, ABCC1 or all the 3 genes were 5.68 ± 0.2, 6.38 ± 0.1, 4.87 ± 0.1 and 4.08 ± 0.2, respectively; that of the cells treated with scramble siRNA and the ECV304 cells were 12.2 ± 0.2 and 4.08 ± 0.1, respectively. AsRE cells transfected siRNA(s) then treated with 10 μM arsenic were harvested after indicated time points of arsenic treatment. Compared with AsRE cells treated scramble siRNA, silence ABCA1 increased cellular arsenic concentration. When all the ABCA1, ABCB1 and ABCC1 genes were silenced, the arsenic accumulation was similar with ECV304 cells at all studied time points (Figure 3C–D). These results suggested that ABCA1 will synergy with ABCB1 and ABCC1 to pump arsenic out from intracellular.

Studies have found that MDR, MRP and other genes belonging to the ABC family are associated with arsenic resistance. In the present study, results showed that silence ABCA1 dramatically lower arsenic resistance in AsRE cells, and silence all the ABCA1, ABCB1 and ABCC1 deplete theirs arsenic resistance. The results suggested that ABCA1 can synergy with ABCB1 and ABCC1 to pump out the intracellular arsenic, therefore to resist its toxicity.

4 CONCLUSIONS

As ABCB1 and ABCC1, ABCA1 also is an arsenic resistance gene, which contributes to arsenic resistance.

ACKNOWLEDGEMENTS

This work was part supported by grants from the Natural Science Foundation of China (30060074; 30560129).

REFERENCES

Lee, T.C., Ho, I.C., Lu, W.J. & Huang, J.D. 2006. Enhanced expression of multidrug resistance-associated protein 2 and reduced expression of aquaglyceroporin 3 in an arsenic-resistant human cell line. *J. Biol. Chem.* 281(27): 18401–18407.

Leung, J., Pang, A., Yuen, W.H., Kwong, Y.L. & Tse, E.W. 2007. Relationship of expression of aquaglyceroporin 9 with arsenic uptake and sensitivity in leukemia cells. *Blood.* 109(2): 740–746.

Miller, D.S., Shaw, J.R., Stanton, C.R., Barnaby, R., Karlson, K.H., Hamilton, J.W. & Stantonet, B.A. 2007. MRP2 and acquired tolerance to inorganic arsenic in the kidney of killifish (*Fundulus heteroclitus*). *Toxicol. Sci.* 97(1): 103–110.

Rappa, G., Lorico, A., Flavell, R.A. & Sartorelli, A.C. 1997. Evidence that the multidrug resistance protein (MRP) functions as a co-transporter of glutathione and natural product toxins. *Cancer Research* 57(23): 5232–5237.

Vernhet, L., Allain, N., Payen, L., Anger, J.P., Guillouzo, A. & Fardel, O. 2001. Resistance of human multidrug resistance-associated protein 1-overexpressing lung tumor cells to the anticancer drug arsenic trioxide. *Biochem. Pharmacol.* 61(11): 1387–1391.

One Century of the Discovery of Arsenicosis in Latin America (1914–2014) – Litter, Nicolli, Meichtry, Quici, Bundschuh, Bhattacharya & Naidu (Eds)
© 2014 Taylor & Francis Group, London, ISBN 978-1-138-00141-1

Basic study for the establishment of criteria and priorities in health coverage and water quality for La Pampa province, Argentina, 2012–2015

C.G. Elorza
Ministerio de Salud, La Pampa, Argentina

E.B. Buitrón, L.L. Sereno & P.B. Pratts
Ministerio de Obras y Servicios Públicos, Ministerio de Salud & Secretaría de Recursos Hídricos, La Pampa, Argentina

ABSTRACT: La Pampa is one of the Argentine provinces with the highest contents of arsenic (As) in water. For this reason, the Ministry of Health and the Water Resources Secretary of the Nation performed studies to evaluate the water quality in Argentina. These studies will take into account that the WHO recommendation (provisionally setting the maximum acceptable value of As in water in 10 μg/L) implies to our country a significant economic cost. The goal is to conduct an epidemiological assessment of the health impact of the consumption of water with As in all locations of the country, in order to give better supports on acceptable values of As in water and to provide the financial resources to improve the access to water throughout the country. This study refers to La Pampa province.

1 INTRODUCTION

Arsenic is a chemical element with toxic recognized ability. The presence of arsenic in La Pampa groundwater is related to geological events in the eastern region of the province (Figure 1). The presence of the element is differential and inconsistent. Quaternary loess sediments, reworked by aeolian processes, are the main source of As and associated trace elements. Two factors, space and time, determine the presence or absence in water of the element and others, such as selenium, uranium and vanadium (Nicolli *et al.*, 1997). Regarding space, distribution depends on: a) climatic factors, b) geological factors (structural and lithological), c) geomorphology. Regarding time, climatological conditions, mainly rains, are the most important (Schulz *et al.*, 2002).

Direct relationship was found about arsenic and fluoride concentrations, representing a high degree of complexity in order to define areas with different values, since their behavior is different in the horizontal or vertical layout.

The situation in La Pampa province concerning elevated concentrations of arsenic in water has been studied for decades. The presence of As is constant in groundwater of the Pampean Formation or Cerro Azul, in concentration as high as tenths of mg/L.

The Provincial Law No. 1027 of 1980 regulates the maximum range of this element in water for human consumption in 0.15 to 0.18 mg/L, while the Argentine Food Code (CAA) amended in May 2007 the limit

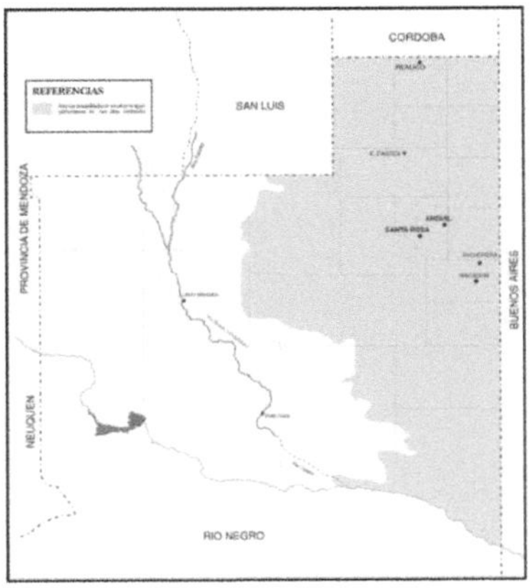

Figure 1. Presence of arsenic associated with the formation pampas. Taken from Schulz, 2005.

from 0.05 to 0.01 mg/L, same limits recommended before by the WHO (2004) and the USEPA (2005).

According to the existing database in the Laboratory of Provincial Water Management (APA), in a record of 83 villages with water service provision by network distribution, the concentrations tested between 1994–2012 ranged from 0.03 to 0.417 mg/L.

In the Ministry of Health of La Pampa Province there are no available studies demonstrating the existence of morbidity and mortality from diseases associated with the risk of exposure to arsenic consumption. There are no either records about background cases of Chronic Regional Endemic Hydroarsenicism (HACRE).

2 METHODS FOR THE WORK PLAN

Under the Framework Agreement concluded in 2011 between the Ministry of Water Resources, Public Investment and Services, and the Argentine provinces, including La Pampa, a work plan was proposed, comprising the following modules.

2.1 *Module 0: Investigations about the presence of arsenic in water in the province of La Pampa*

The goal is to collect and analyze information from various sources (virtual libraries, documentation centers, officers, etc.) on the hydrogeology of the arsenic in La Pampa.

2.2 *Module 1: Diagnosis of the current sanitation situation in the Province of La Pampa*

The objective is to evaluate the coverage of population by potable water and sanitation, identifying collection and treatment systems, and obtaining health coverage maps of the province of La Pampa.

Additionally, towns without centralized water and sanitation will be identified.

2.3 *Module 2: Arsenic risk map for the population covered by centralized water services in the province of La Pampa*

The provincial profile of the population exposed to arsenic through drinking water will be updated and identified from available data in the competent agency, on the historical context of arsenic in water provided by centralized water services in the Province of La Pampa and data of the population covered by the service. To collect environmental data for exposure records, work will be done with arsenic concentration in elevated tanks. Sampling will be conducted in terms of drinking water quality to check against existing records in the localities under study.

Arsenic determinations will be performed in reference laboratories using standard methods.

2.4 *Module 3: Epidemiological study of morbidity and mortality from cancers associated with exposure to arsenic in drinking water*

An ecological study of cancer in departments/locations and groups exposed in La Pampa, based on previous studies and the availability of official data sources, will be done.

Exposure data will come from the Risk Map built earlier in the project. If possible, taking into account the number of cancer deaths in La Pampa, we will analyze the mortality for different levels of exposure to arsenic.

Standardized Mortality Ratios (SMR) for kidney, lung, liver and bladder cancers for each location/department, separately by sex, will be calculated using the same methodology to calculate the SMR departments/locations of high, medium and low exposure (determined levels).

2.5 *Module 4: Study of water-borne diseases in populations arsenical not lacking sanitation*

A descriptive cross-sectional study of the effects of the deficit in sanitation, socio-economic and environmental factors that, by themselves or in combination, increase the risk of suffering from waterborne diseases in towns/departments of the Province of La Pampa will be done.

The study population will consist of the consultants of waterborne diseases occurring in public health who attend selected populations.

This study will focus on two areas: the group of pathogens and the chemicals transmitted primarily by water.

3 RESULTS

The project, intersectorial and multidisciplinary unprecedented, will be conducted in populations that consume water between 10 and 50 µg/L, where there is interest in assessing the risk of morbidity and mortality associated to cancers after risk map processing, depending on the concentrations of arsenic in drinking water received through centralized services.

REFERENCES

Corey, G., Tomasini, R. & Pagura, J. 2005. Estudio Epidemiológico de la exposición al Arsénico a través del consumo de agua. 3° Parte: Mortalidad por cánceres asociados a Arsénico. Santa Fe, República Argentina. ISBN 987-23193-0-8. http://www.sertox.com.ar/img/item_full/3a%2PARTE%20–0MORTALIDAD%20POR%20CANCER.pdf.

Nicolli, H. Smedley, P. & Tullio, J. 1997. Aguas subterráneas con altos contenidos de F, As y otros oligoelementos en el norte de La Pampa. En: Actas del Congreso Internacional del Agua. Buenos Aires, Argentina.

Schulz, C. Castro, E. Mariño, E. & Dalmaso, G. 2002. El agua potable en la Provincia de La Pampa. Consecuencias por presencia de Arsénico. Taller sobre As en aguas subterráneas, XXXII Congreso Internacional de Hidrología Subterránea y V Congreso Latinoamericano de Aguas Subterráneas, Mar del Plata.

WHO (World Health Organization). 2004. Guidelines for drinking-water quality. Vol. 1, Recommendations. Geneva. 3rd edition. 515 pp.

One Century of the Discovery of Arsenicosis in Latin America (1914–2014) – Litter, Nicolli, Meichtry, Quici, Bundschuh, Bhattacharya & Naidu (Eds)
© 2014 Taylor & Francis Group, London, ISBN 978-1-138-00141-1

Recent findings on arsenic exposure and respiratory outcomes from the Health Effects of Arsenic Longitudinal Study (HEALS)

F. Parvez, V. Slavkovich & J.H. Graziano
Department of Environmental Health Sciences, Mailman School of Public Health, Columbia University, New York, NY, USA

Y. Chen & F. Wu
Departments of Environmental Medicine, New York University School of Medicine, New York, NY, USA

M. Yunus, R. Hasan, A. Ahmed & T. Islam
Columbia University Arsenic Research Project, Dhaka, Bangladesh

ABSTRACT: Relatively little is known about the effects of water arsenic (As) and non-malignant respiratory outcomes among individuals exposed to low-to-moderate levels of As or without skin lesion. In a population-based study, using follow-up data in the Health Effects of Arsenic Longitudinal Study (HEALS) cohort, we found that As exposure from drinking water was significantly related to pulmonary infections, impaired lung function, incident and mortality from obstructive lung diseases.

1 INTRODUCTION

Chronic exposure to arsenic (As) in drinking water has been associated with a number of non-malignant respiratory outcomes including clinical respiratory symptoms, respiratory infections, impaired lung function, and obstructive lung diseases in population living in arsenic endemic areas of Bangladesh, China, India and Pakistan. A number of studies from an As endemic area of Chile have also reported increased mortality from pulmonary tuberculosis and chronic obstructive lung disease including chronic bronchitis and bronchiectasis. However, evidence on non-malignant respiratory effects of As are largely based on studies with methodological limitations including ecological nature of the study design, retrospectively collected exposure and outcome data with limited sample size. Additionally, most of the studies were conducted in population exposed to very high levels of water with high As levels or with As induced skin lesions. Limited evidence exists on the effects of As induced pulmonary outcomes, particularly those exposed to low-to-moderate levels of As exposure or without skin lesion.

We prospectively evaluated the effects of As exposure on respiratory outcomes in a population exposed to wide range of water with As concentrations (0.1–1,517 μg/L) and includes a large number of study participants exposed to low-to-moderate levels of exposure in the Health Effects of Arsenic Longitudinal Study (HEALS) cohort.

2 MATERIALS AND METHODS

The HEALS was established in Araihazar in 2000, Bangladesh, to prospectively examine health effects of As exposure from drinking water among 30,000 adults and their children. Since baseline recruitment, we have been following our study participants for their health status using a number of procedures including active quarterly visits by research assistants, bi-annual visits by physicians, and surveillance by village health workers. During our follow-up visits all cohort members were evaluated by trained physicians using protocol including pulmonary function tests to ascertain lung disorders.

We conducted population-based studies to evaluate the association between As exposure, measured in well water and urine samples, and pulmonary infection, lung function, incident of lung diseases and mortality in 20,033 HEALS participants.

3 RESULTS AND DISCUSSION

In a prospective analysis, we observed a dose-response association between baseline As exposure and respiratory infections among 784 individuals. As compared to those at the lowest quintile of well As level (< = 6 μg/L), the HRs (hazard ratios) for having respiratory infections were 1.2 (1.0–1.5), 1.2 (0.9–1.4) and 1.4 (1.1–1.9) for the 2nd-4th quartiles of baseline water As concentration (6–46, 46–116, and >116 μg/L), respectively, in model adjusted for age, gender, BMI, smoking, and socioeconomic status and skin lesion status. A slightly stronger relationship was observed between baselines urinary As. In a sub-sample of HEALS participants (N = 950) with *any respiratory symptoms*, our prospective analysis revealed that individuals in the highest tertile of baseline well water As concentration (>97 μg/L) had a significantly lower FEV1 (–80.6 ml, $p < 0.01$) and FVC (–97.3 ml, $p < 0.05$) compared with individuals with baseline well As <19 μg/L in adjusted models. We also observed an increased risk of obstructive lung disease [HR: 1.01 (0.95–1.09) and 1.07 (1.02–1.11)] associated with 1 Standard Deviation (SD) of baseline water and urinary As, with significantly higher risk for CT-diagnosed bronchiectasis in adjusted models. We have not observed any increased risk associated with emphysema. Importantly, our analysis revealed an increased mortality from obstructive lung diseases, with an HR of 1.04 (0.9–1.1) and 1.4 (1.05–1.89) associated with one SD increase in baseline water and urinary As respectively. The risk was much higher among males and smokers.

4 CONCLUSION

To our knowledge, our study is the first to report a positive association between low-to-moderate levels of As exposure and respiratory infections, impaired pulmonary function, incident and mortality from obstructive lung disease. Our study has strengths including a large sample size, prospectively-collected data on As exposure measured in water and urine samples, and objective measure of infection, pulmonary function and obstructive lung disease. We also have found that the associations for impaired lung function particularly the obstructive pattern and mortality were stronger among smokers. This finding was consistent with studies that found adverse effects of As on bladder and lung cancer.

REFERENCES

Ahsan, H., Chen, Y., Parvez, F., Argos, M., Hussain, A.I., Momotaj, H., Levy, D., van Geen, A., Howe, G., Graziano, J. 2006. Health Effects of Arsenic Longitudinal Study (HEALS): description of a multidisciplinary epidemiologic investigation. *J. Expo. Sci. Environ. Epidemiol.*, 16(2): 191–205.

Ahsan, H., Chen, Y., Kibriya, M.G., Slavkovich, V., Parvez, F., Jasmine, F., Gamble, M.V., Graziano, J.H. 2007. Arsenic metabolism, genetic susceptibility, and risk of premalignant skin lesions in Bangladesh. *Cancer Epidemiol. Biomarkers Prev.* 16(6): 1270–1278.

Dauphine, D.C., Ferreccio, C., Guntur, S., Yuan, Y., Hammond, S.K., Balmes, J., Smith, A.H, Steinmaus, C. 2011. Lung function in adults following in utero and childhood exposure to arsenic in drinking water: preliminary findings. *Int. Arch. Occup. Environ. Health,* 84(6): 591–600.

De, B.K., Majumdar, D., Sen, S., Guru, S., Kundu, S. 2004. Pulmonary involvement in chronic arsenic poisoning from drinking contaminated ground-water. *J. Assoc. Physicians India.* 52: 395–400.

Mazumder, D.N., Haque, R., Ghosh, N., De, B.K., Santra, A., Chakraborti, D., Smith, A.H. 2000. Arsenic in drinking water and the prevalence of respiratory effects in West Bengal, India. *In.t J. Epidemiol.,* 29(6): 1047–1052.

Mazumder, D.N., Steinmaus, C., Bhattacharya, P., von Ehrenstein, O.S., Ghosh, N., Gotway, M., Sil, A., Balmes, J.R., Haque, R., Hira-Smith, M.M. Smith, A.H. 2005. Bronchiectasis in persons with skin lesions resulting from arsenic in drinking water. *Epidemiology,* 16(6): 760–765.

Nafees, A.A., Kazi, A., Fatmi, Z., Irfan, M., Ali, A., Kayama, F. 2011. Lung function decrement with arsenic exposure to drinking groundwater along River Indus: a comparative cross-sectional study. *Environ. Geochem. Health,* 33(2): 203–216.

Parvez, F., Chen, Y., Brandt-Rauf, P.W., Slavkovich, V., Islam, T., Ahmed, A., Argos, M., Hassan, R., Yunus, M., Haque, S.E., Balac, O., Graziano, J.H., Ahsan, H., 2010. A prospective study of respiratory symptoms associated with chronic arsenic exposure in Bangladesh: findings from the Health Effects of Arsenic Longitudinal Study (HEALS). *Thorax, 65*(6): 528–533.

Parvez. F., Chen, Y., Yunus, M., Olopade, C., Segers, S., Slavkovich, V., Argos, M., Hasan, R., Ahmed, A., Islam, T., Akter, M.M., Graziano, J.H., Ahsan, H. 2013. Arsenic Exposure and Impaired Lung Function: Findings from a Large Population-based Prospective Cohort Study. *Am. J. Respir. Crit. Care Med.*, 188(7): 813–819.

Steinmaus, C., Yuan, Y., Kalman, D., Rey, O.A., Skibola, C.F., Dauphine, D., Basu, A., Porter, K., Hubbard, A., Bates, M., Smith, M.T., Smith, A.H. 2010. Individual differences in arsenic metabolism and lung cancer in a case-control study in Cordoba, Argentina. *Toxicol. Appl. Pharmacol.*,247(2): 138–145.

Smith, A.H., Marshall, G., Yuan, Y., Ferreccio, C., Liaw, J., von Ehrenstein, O. Steimaus, C., Bates, M.N., Selvin, S. 2006. Increased mortality from lung cancer and bronchiectasis in young adults after exposure to arsenic in utero and in early childhood. *Environ. Health Perspect.*, 114(8): 1293–1296.

von Ehrenstein, O.S., Mazumder, D.N., Yuan, Y., Samanta, S., Balmes, J., Sil, A., Gosh, N., Hira-Smith, M., Haque, R., Purushothamam, R., Lahiri, S., Das, S., Smith, A.H. 2005. Decrements in lung function related to arsenic in drinking water in West Bengal, India. *Am. J. Epidemiol.*, 162(6): 533–541.

One Century of the Discovery of Arsenicosis in Latin America (1914–2014) – Litter, Nicolli, Meichtry, Quici, Bundschuh, Bhattacharya & Naidu (Eds)
© 2014 Taylor & Francis Group, London, ISBN 978-1-138-00141-1

Arsenic in drinking water supply systems in the Vojvodina Region, Serbia: A public health challenge

T. Knezevic & C.D. Dragana Jovanovic
Institute of Public Health of Serbia "Dr Milan Jovanovic Batut", Belgrade, Serbia

B. Jakovljevic
Institute of Hygiene and Medical Ecology, School of Medicine, Belgrade, Serbia

ABSTRACT: Vojvodina, a northern region of Serbia, belongs to the Pannonian Basin, whose aquifers contain high concentrations of arsenic. This study represents arsenic levels in drinking water in ten municipalities in Vojvodina, Serbia. Around 63% of all water samples exceeded Serbian and European standards for arsenic in drink water. Large variations in arsenic were observed among supply systems.

1 INTRODUCTION

Vojvodina is a northern region in Serbia geographically belonging to the southern part of Pannonian Basin in which quaternary sedimentary aquifers contain high concentration of naturally occurring arsenic. The magmatic rock contains 3.1 mg As kg^{-1}, whereas the soil contains 10 mg As kg^{-1}. Arsenic concentrations are lower in the Danube River water and shallow riparian groundwaters (0.1–8 μg L^{-1}) (Dangic, 2007). However, arsenic has been a known water pollutant in Vojvodina for many decades and recognized as a major public health challenge.

2 MATERIALS AND METHODS

2.1 *Materials*

The National Monitoring Program of Drinking Water Quality in Serbia shows public water supply systems in 20 municipalities in Vojvodina Region that cover 44.41% of all municipalities. Ten municipalities provide all data. Total arsenic concentration was measured in water samples from public water supply systems in 2008 and 2009. No statistical differences was observed between the samples taken in various years and, therefore, the most recent data were included in the analysis. Sampling and analyses of drinking water were performed at the IPHs laboratories. All laboratories were accredited and authorized according to SRPS ISO/IEC 17025 and SRPS ISO 9001 standards.

Laboratory procedures for sample management, analytical methods and quality control measures (accuracy, precision and detection limits) were standardized by Serbian law. Current Serbian regulations limit arsenic levels at 10 mg L^{-1}.

2.2 *Methods*

HG-AAS technique was used for total arsenic water determination. Hydride generation was performed using a 3% (w/v) $NaBH_4$ in 1% NaOH solution. The radiation source was a hollow cathode lamp of arsenic. With this method, a limit of detection of 0.5 μg L^{-1} and limit of quantification of 2 μg L^{-1} were obtained by analyzing 4 series of 10 repeated analysis of blank samples and calculated the three standard deviations of these responses. Recovery of standards was 80–120%. As a descriptive statistics methods, median values, 10th. and 90th. percentiles for numeric variables and percentage for categorical variables were used. Differences between several samples were analyzed using Kruscal Walis test for non parametric data.

3 RESULTS AND DISCUSSION

3.1 *Results*

In total, 577 water samples from public water supply systems in 10 municipalities in Vojvodina were analyzed for arsenic. Almost 63% of all samples exceeded the current standard. Arsenic was detected in all water samples from Senta and Kanjiza (northeast of Vojvodina) and in 94% of samples in Zrenjanin (eastern part of Vojvodina) (Table 1). The distribution of total arsenic in drinking water by public water supply systems in the investigated municipalities is shown in Table 2.

Large variations in arsenic concentration were observed among the water supply systems (median ranged from 2 to 250 mg L^{-1}; maximum value ranged from 5 to 349 mg L^{-1}). The highest concentrations were reported in the city of Zrenjanin and

Table 1. Number of analyzed samples, number and percent of samples with exceeded arsenic levels in public water supply systems in Vojvodina Region, according to Serbian maximum allowed value.[a]

Municipality	No. of analyzed samples	No. of samples exceeding	% of samples exceeding
Zrenjanin	144	135	93.8
Bačka Palanka	174	126	72.4
Bački Petrovac	60	24	40.0
Bač	8	3	37.5
Srbobran	42	3	7.1
Titel	18	16	88.9
Temerin	55	27	49.1
Subotica	52	5	9.6
Kanjiza	12	12	100.0
Senta	12	12	100.0
Total	577	363	62.9

[a]Over 10 μg L^{-1}

Table 2. Total inorganic arsenic concentrations (mg L^{-1}) in water samples from public water supply systems in Vojvodina Region by municipalities and villages.

Municipality/ Village	Median As Level	10th percentile of As level	90th percentile of As level	Maximum As level
Zrenjanin	77.0	61.6	99.0	200.0
Melenci	232.5	73.0	346.3	349.0
Taras	250.0	76.0	336.8	344.0
Jankov Most	60.0	41.6	83.0	85.0
Mihajlovo	82.5	60.0	95.5	110.0
Elemir	112.5	35.4	209.8	224.0
Klek	70.0	51.0	89.2	96.0
Aradac	90.0	42.4	130.0	155.0
Perlez	2.0	2.0	4.7	5.0
Farkadzin	13.0	10.0	29.0	30.0
Lazarevo	5.0	4.1	8.0	8.0
Bačka Palanka	32.0	15.0	49.0	98.0
Despotovo	21.5	11.0	30.7	53.0
Obrovac	43.0	34.0	63.9	98.0
Čelarevo	87.0	86.0	87.0	88.0
Silbas	6.0	2.0	11.0	11.0
Pivnice	2.0	1.0	12.0	15.0
Tovariševo	20.0	16.0	33.8	36.0
Mladenovo	38.0	31.0	52.0	70.0
Backi Petrova	14.5	11.0	17.5	18.0
Bač	15.0	12.0	45.0	52.0
Srbobran	13.0	12.0	15.0	15.0
Titel	14.0	11.0	23.0	25.0
Temerin	46.0	23.2	230.2	420.0
Subotic	46.0	16.0	69.0	71.0
Kanjiza	17.0	16.0	20.0	20.0

most of its villages (eastern part of Vojvodina), in municipalities Temerin (central to northeast), Subotica (north) and village Celarevo (west).

In total, only 11% of all water supply systems in Vojvodina had arsenic concentrations below the limit (10 mg L^{-1}). About half (50.4%) of them reported arsenic concentrations ranging from 11 to 50 mg L^{-1} and further 38.6% systems had mean arsenic levels over 50 mg L^{-1}. The differences among the investigated municipalities were highly statistically significant (Kruskal Wallis test: w2j137.039; po0.001).

3.2 *Discussion*

This is a first study on the distribution of arsenic in public water supply systems in Vojvodina and Serbia. The results show a large variation in arsenic levels in the region. Mean arsenic levels were much higher than those reported in other countries in the Pannonian Basin. The arsenic content in drinking water in some parts of Vojvodina (primarily Zrenjanin) is higher than in other geologically independent parts of the world, such as Spain (ranging from 1 to 118 mg/L). The main objective for the future research will be to estimate individual exposure to arsenic which have to be based on arsenic levels in drinking water. At the same time, other exposure pathways, such as the use of bottled water or water from individual wells, must be taken into consideration, because they could increase the size of the exposed population.

4 CONCLUSIONS

This study has reported very high concentrations of arsenic in public water supply systems across ten municipalities in Vojvodina Region, Serbia. Elevated levels of arsenic should initiate social, economic and technological actions in order to reduce it to acceptable limits. Furthermore, this should raise public health concern and produce further environmental and epidemiological study.

REFERENCES

Dangic, A. 2007. Arsenic in surface—and groundwater in central parts of the Balkan Peninsula (SE Europe). In: P. Bhattacharya, A.B.B. Mukherjee, J. Bundschuh, R. Zevenhoven & R.H. Loeppert (eds.), *Arsenic in Soil and Groundwater Environment—Biogeochemical Interactions, Health Effects and Remediation.*

Trace Metals and other Contaminants in the Environment 9, pp. 207–236.

One Century of the Discovery of Arsenicosis in Latin America (1914–2014) – Litter, Nicolli, Meichtry, Quici, Bundschuh, Bhattacharya & Naidu (Eds)
© 2014 Taylor & Francis Group, London, ISBN 978-1-138-00141-1

Effect of sodium arsenite in a model of experimental carcinogenesis in mouse skin

M.A. Palmieri
Biodiversity and Experimental Biology Department, School of Exact and Natural Sciences, University of Buenos Aires, Buenos Aires, Argentina

B.L. Molinari
National Atomic Energy Commission (CNEA), National Scientific and Research Council (CONICET), Buenos Aires, Argentina

ABSTRACT: Environmental exposure to arsenic is a public health problem. It affects a large number of populations worldwide and arsenic is carcinogenic to human beings. The aim of the present study was to assess the influence of arsenical compounds on potential carcinogenic processes, both pre-existing and concomitant with the intake of water containing arsenite. Experimental protocols were designed to evaluate the effect of chronic administration of sodium arsenite (As^{3+}) in the drinking water, in Sencar mice. Increasing doses of As^{3+} were administered with a variable overlapping with a classic carcinogenesis process. The effect of As^{3+} on the process of experimental carcinogenesis resulted in a reduction of the number of tumors in all groups exposed to As^{3+} compared to control. However, the number of malignant tumors was higher in the experimental groups than in the control group. This response was consistent with the different aspects of the effect of As^{3+} described in the literature.

1 INTRODUCTION

In certain geographical areas, there are arsenic-rich layers and this can result in higher levels of As^{3+} in the water (EPA, 1988; Pfeifer *et al.*, 2002) because natural lixiviation of arsenic depends on the mineral composition of the environment, the acidity of the medium and the availability of O_2 (Méndez & Armienta, 2003; Zhu *et al.*, 2003). The major routes of entry of the environmental arsenic in living organisms are through drinking water with high contents of inorganic arsenic as well as through the intake of food from contaminated areas. Arsenic exposure is associated with deleterious effects on human health and is extensively documented (Tsuchiya, 1977; Yoshida *et al.*, 2004).

Germolec *et al.* (1997) used transgenic mice (Tg. AC) to study the effect of As^{3+} in drinking water. These mice have genetically initiated skin and were used as an experimental carcinogenesis model. After few applications of a promotor and fourteen weeks later, an increase in the number of papillomas was observed in mice receiving As^{3+} in the drinking water.

The aim of this work was to study the effects of sodium arsenite (As^{3+}) in the drinking water in mice submitted to an experimental carcinogenesis protocol in skin. The intention was to replicate scenarios similar to those that might occur in human populations living in areas with a high content of As^{3+} in water.

For this purpose, the analysis of the effect of As^{3+} on the model of two-stage carcinogenesis in SenCar mice skin was carried out. The oral route was selected as chronic exposure, and two possible situations were considered: increasing doses of As^{3+} and different exposure periods.

2 MATERIALS AND METHODS

2.1 *Drugs and animals*

Sodium arsenite (As^{3+}), 7,12-dimethylbenz (a) antracene (DMBA) and, 12-O-tetradecanoilforbol-13-acetate (TPA) were purchased from Sigma Chemical Co.

Female SenCar mice, 7–9 weeks old were bred at CNEA and housed in a controlled environment.

2.2 *Two-stage skin carcinogenesis protocol*

Mouse skin tumors were induced in dorsal skin area following the initiation-promotion schedule:

Initiation stage: a single topical application of DMBA (5.13 μg/200 μL acetone).

Promotion stage started ten days later: TPA solution (2 μg/200 μL acetone) were topically applied twice a week during 9 months.

2.3 *Statistical analysis*

The tumor number data were well-fitted by the Gompertz model, whose mathematical function is:

$$Y_t = \alpha e^{-\beta e^{-\gamma t}}$$

The Alpha parameter is the load capacity and the Gamma parameter is the growth rate. Thus, estimation of the parameters of the number of tumors curve was obtained for each animal (Nelder & Mead, 1965). ANOVA was performed on the calculated parameters. Comparisons were made using the Tukey test.

2.4 *Histology*

Skin tumors were prepared using conventional protocols for paraffin sections and hematoxylin-eosin staining. They were histopathologically evaluated and categorized as previously described (Klein-Szanto, 1989)

2.5 *Experimental design in vivo*

Exp#1). Different As^{3+} doses were administered before (two months) and during TPA promotion (nine months). Treatment groups were nominated as Group A (As^{3+} 0, control), Group B (As^{3+} 2), Group C (As^{3+} 20) and Group D (As^{3+} 200). The number correspond to As^{3+} concentration in mg L^{-1}.

Exp#2). A single dose of 20 mg L^{-1} As^{3+} was administered or not during four months before and/or during the TPA promotion period (nine months). Experimental groups were nominated as

Group E: water-water: without As^{3+} (control).
Group F: water-As^{3+}: As^{3+} only during promotion.
Group G: As^{3+}-water: As^{3+} only before promotion.
Group H: As^{3+}-As^{3+}: As^{3+} before-during promotion.

3 RESULTS

For each experiment, the number of tumors/mouse was counted (mean ± standard error, Figures 1-A and 1-B). The number of tumors for each mouse as a function of time was adjusted to Gompertz growth curve and values for alpha and gamma parameters were calculated.

The tumor growth rate (gamma parameter) was higher in all experimental conditions in which water contained As^{3+} vs. control (Exp#1 and Exp#2). The asymptotic value (alpha parameter) was lower than the control for all As^{3+} doses delivered (Exp#1). Conditions F and G reached an intermediate value between control and H (Exp#2).

In both experiments, differences in tumor size were found only in the first stage of cancerization.

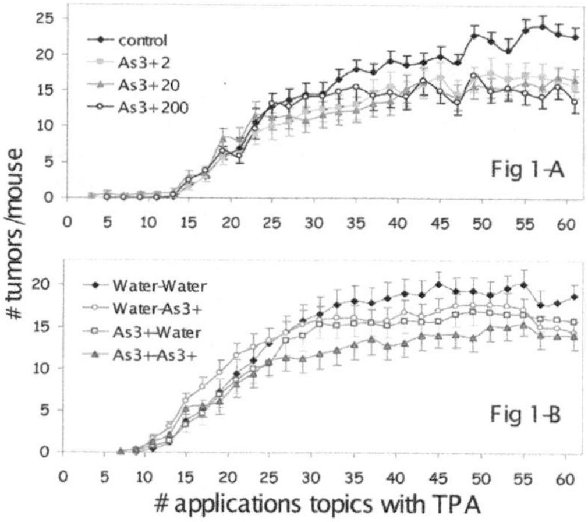

Figure 1. Average tumor number ± standard error vs. number of topical TPA application. 1-A Exp#1. Different As^{3+} doses. 1-B Exp#2. A single dose before or during the TPA period.

However a tendency to the development of larger tumors in As^{3+} conditions in Exp#2 or D condition in Exp#1 was observed at later stages. Concerning histopathologic findings, in all experimental conditions a higher rate of malignant tumors was found when compared to controls (Exp#2), particularly at low As^{3+} doses (Exp#1). On the other hand, mice of D condition consumed less arsenical water from the beginning of the experiment, associated with less feed intake compared to all other conditions.

4 CONCLUSIONS

The effects of As^{3+} in the drinking water in a model of experimental carcinogenesis in mouse skin were a) a lower average number of tumors per mouse from thirty weeks of promotion in all animals that drank water with As^{3+} was observed, irrespective of the protocol used, and b) a marked tendency to an increased rate of malignant tumors.

REFERENCES

Germolec, D.R., Spalding, J., Boorman, G.A., Wilmer, J.L., Yoshida, T., Simeonova, P.P., Bruccoleri, A., Kayama, F., Gaido, K., Tennant, R., Burleson, F., Dong, W., Lang, R.W. & Luster, M.I. 1997. Arsenic can mediate skin neoplasia by chronic stimulation of keratinocyte-derived growth factors. *Mutation Research* 386: 209–218.

Klein-Szanto, A.J.P. 1989 Pathology of human and experimental skin tumors. In: *Carcinogenesis, a comprehensive survey*. Conti C, Slaga TJ, Klein-Szanto AJP. (eds) Raven press, New York, Vol 11, pp36–41.

Méndez, M. & Armienta, M.A. 2003. Arsenic phase distribution in Zimapán mine tailings, Mexico. *Geofísica Internacional* 42(1): 131–140.
Nelder, J.A. & Mead, R. 1965. Downhill simplex method in multidimensions. *Computer Journal* 7: 308–315.
Pfeifer, H.R., Beatrizotti, G., Berthoud, J., De Rossa, M., Girardet, A., Jäggli, M., Lavanchy, J.C., Reymond, D., Righetti, G., Schlegel, C., Schmit, V. & Temgoua, E. 2002. Natural arsenic-contamination of surface and ground waters in Southern Switzerland (Ticino). *Bulletin for Applied Geology* 7(1): 81–103.
Tsuchiya, K. 1977. Various effects of arsenic in Japan depending on type of exposure. *Environmental Health Perspectives* 19: 35–42.
U.S. EPA. 1988. Special Report on Ingested Inorganic Arsenic; Skin Cancer; Nutritional Essentiality Risk Assessment Forum. July 1988. EPA/625/3–87/013.
Yoshida, T., Yamauchi, H. & Fan Sun, G. 2004. Chronic health effects in people exposed to arsenic via the drinking water: Dose–response relationships in review. *Toxicology and Applied Pharmacology* 198(3): 243–252.
Zhu, Y., Merkel, B.J., Stober, I. & Bucher, K. 2003. The hydrology of arsenic in the Clara mine, Germany. *Mine Water and the Environment* 22: 110–117.

One Century of the Discovery of Arsenicosis in Latin America (1914–2014) – Litter, Nicolli, Meichtry, Quici, Bundschuh, Bhattacharya & Naidu (Eds)
© 2014 Taylor & Francis Group, London, ISBN 978-1-138-00141-1

Plant extracts as biopharmaceutical products and analysis of their activity against arsenicosis

V. Sotomayor & G.A. Bongiovanni
IDEPA (CONICET-UNComa), Neuquén, Argentina

Cecilia S. Roldán
F.A.C.A. (UNComa), Cinco Saltos, Río Negro, Argentina

Ignasi Queralt
Jaume Almera Institute, (ICTJA-CSIC), Barcelona, Spain

ABSTRACT: Arsenic (As) has been classified by the US Environmental Protection Agency as a human carcinogen, being inorganic As the natural contaminating element present mostly in water sources because of its high solubility. Almost 10% of the Argentine population is exposed to As with high association to cancer. In the present work, we aimed to determine plant extracts able to counteract As oxidative stress effect, the main cellular response triggered by exposure. We analyzed *Araucaria araucana* used in popular medicine for inflammatory processes. We studied aqueous and organic extracts of seeds and leaves (and stems) in order to determine antioxidant activities in an *in vitro* model. We also analyzed elemental composition by X-ray fluorescence by dispersive energy (EDXRF). The results suggest that seeds of *A. araucana* are a suitable source of antioxidants and oligoelements as magnesium, potassium and calcium, worth to be studied as a potential pharmaceutical drug.

1 INTRODUCTION

Inorganic arsenic (As) is widely distributed in the environment being a water contaminant considered to have carcinogenic effects, and associated to incidence of diabetes mellitus, nephrotoxicity and cerebral ischemia (Meliker *et al.*, 2007). As(V) is absorbed at intestinal level being transformed to As(III), which is known to cause oxidative stress implicated in arsenical toxicity and carcinogenicity (Shi *et al.*, 2004; Soria *et al.*, 2014). Plants exposed to natural pressure of the surroundings have led to the development of antioxidant compounds use in adaptation to the environmental conditions and hence, survival. Accordingly, the aim of this work is to evaluate the antioxidant capacity of plant extracts, particularly of *Araucaria araucana* seeds, which had been being used in Argentine traditional medicine. We also analyzed the elemental composition of the seeds by X-ray fluorescence by dispersive energy (*EDXRF*).

2 METHODS/EXPERIMENTAL

2.1 *Cell culture*

African green monkey (*Cercopithecus aethiops*) kidney cells (Vero cell line ATCC *n*° CCL-81) were cultured in Dulbecco's Modified Eagle's medium with 10% fetal bovine serum, 100 U/mL penicillin-G, and 40 μg/mL gentamycin sulfate, incubated at 37°C in a 5% CO_2 atmosphere. This line comes from normal mammalian renal epithelia, with kidneys being target organs of arsenic (Pérez *et al.*, 2006).

2.2 *Plant material and extraction*

A. araucana is a conifer tree endemic of Chile and Argentina whose seed (Piñon) is mainly constituted by starch. Seeds were collected in Caviahue, Argentina (latitude of S 37° 53' 12.48" and a longitude of W 71° 4' 16.8594"). To obtain different extracts, air-dried powdered seeds (S), hull (H) and leaves/stem (L&S) (12 g) were macerated with 50 ml of solvent (water, *n*-hexane and ethyl ether) at RT in darkness for 24 h. Extracts were dried and resuspended in dimethylsulfoxide (DMSO) (Soria *et al.*, 2008).

2.3 *Treatment*

Cells were incubated for 2 h with 200 μM $NaAsO_2$, 48 h after seeding in a medium containing 0.2% DMSO (Positive control) or 200 μM $NaAsO_2$ plus plant extract (200 μg/mL) (Soria *et al.*, 2008).

2.4 *Lipid hydroperoxides measurement*

After treatments, culture medium was removed and cells were lysed by 1% SDS. The PeroxiDetect™ Kit (Sigma Co) was used to determine cellular lipid hydroperoxides (LHP, μM) (Soria *et al.*, 2008).

2.5 *X-ray fluorescence by dispersive energy (EDXRF)*

Elemental characterization of samples (L&S and H) and Certified Reference Materials by EDXRF spectrometry were carried out at the X-ray Analytical Applications Laboratory (ICTJA-CSIC), Barcelona, Spain. Briefly, X-ray fluorescence analysis in pellets was performed using an EDXRF Fischer XDV-SD spectrometer with a 50 kV tungsten-anode X-ray tube as a radiation source (50 W). The characteristic Kα X-ray lines of elements were used for calculations of the net peak area and mass concentrations by WinFTM®—v.6.20 software (Helmut Fisher GmbH Co., KG).

3 RESULTS AND DISCUSSION

3.1 *Cellular lipid hydroperoxides*

Protective effect of *A. araucana* (L&S and S extracts) against arsenic-induced oxidative response was assessed by measuring LHP concentration. Of all the tested extracts, aqueous extracts of the seeds were able to decrease LHP levels hence showing the highest antioxidant capacity (Figure 1). L&S induced cell death after 2 h incubation (data not shown).

3.2 *Multielemental composition of seeds*

In this report, Na, Mg, Al, Si, P, S, Cl, K, Ca, Mn, Fe, Cu, Zn, Br, Rb, Sr, Zr were detected in seeds of *A. araucana,* being Na, Mg, Cl, K, Ca the elements found in the highest proportion (Table 1).

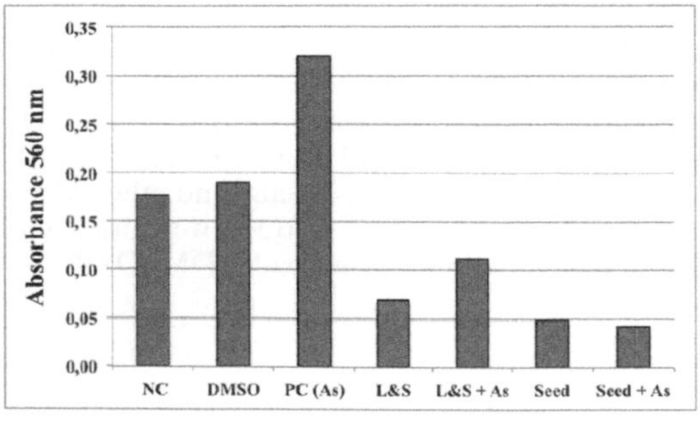

Figure 1. Cellular lipid hydroperoxides (LHP, μM) after aqueous Leaves and Steam (L&S) and seeds extracts or As treatments. NC: Negative Control (DMSO), PC: Positive Control (As).

Table 1. Major elements found in seeds.

	EDXRF	
	Hull (H)	Seed (S)
Main elements	μg/g dry weight (ppm)	
Na	1185.00	1178.00
Mg	253.50	408.33
Cl	2345.00	1806.67
K	3680.00	7550.00
Ca	1111.50	308.33

Values are expressed as mean of two independent determinations and data represent μg of element per g of dry weight (ppm) obtained by EDXRF.

4 CONCLUSIONS

Although molecular carcinogenic mechanism is unknown, arsenite is bioaccumulated and induces tissue damage and oxidative stress (Rubato Birri *et al.*, 2010; Soria *et al.*, 2014). *A. araucana* aqueous seed extracts were protective against As-induced oxidative response. Analysis by the XRF method is especially advantageous, because it is non-destructive and components can undergo multielemental analysis without prior sample preparation (Marguí *et al.*, 2009). Elemental analyses reveled that seeds of *A. araucana* are a reliable source of oligoelements as Mg, K and Ca. Using Vero cells treated with arsenite under the above-described conditions allowed us to reinforce our model in order to search for plant antioxidants as an strategy to arsenical exposure. Further analysis including antioxidant activity in embryos and endosperm from *A. araucana* seeds are under study.

ACKNOWLEDGEMENTS

This work was supported by ANPCyT, PICT97-PRH33 and PICT214, and CONICET, Res. 863/10 of Argentina.

REFERENCES

Marguí, E., Queralt, I., & Hidalgo, M. 2009. Application of X-ray fluorescence spectrometry to determination and quantitation of metals in vegetal material. *Trends Anal. Chem.* 28(3): 362–372.

Meliker, J.R., Wahl, R.L., Cameron, L.L., & Nriagu, J.O. 2007. Arsenic in drinking water and cerebrovascular disease, diabetes mellitus and kidney disease in Michigan: A standardized mortality ratio analysis. *Environ. Health* 6:4.

Pérez, R.D., Rubio, M, Pérez, C.A., Eynard, A.R. & Bongiovanni, G.A. 2006. Study of the effects of chronic arsenic poisoning on rat kidney by means of synchrotron microscopic X ray fluorescence analysis. *X Ray Spectrom.* 35(6): 352–358.

Rubatto Birri, N., Pérez, R.D., Cremonezzi, D., Pérez, C.A., Rubio, M., Bongiovanni, G.A. 2010. Association between Cu and As renal cortex accumulation and physiological and histological alterations after chronic arsenic intake. *Environ. Res.* 110: 417–423.

Shi, H., Shi, X. & Liu, K.J. 2004. Oxidative mechanism of arsenic toxicity and carcinogenesis. *Mol. and Cel. Biochem.* 255:67–78.

Soria, E.A., Bongiovanni, G.A., Díaz Luján, C., Eynard, A.R. 2014. Effect of arsenic on nitrosative stress in human breast cancer cells and its modulation by flavonoids. *Curr. Pharm. Design*, (http://www.ncbi.nlm.nih.gov/pubmed/24138720).

Soria E.A., Goleniowski, M.E., Cantero, J.J., Bongiovanni, G.A. 2008. Antioxidant activity of different extracts of Argentinean medicinal plants against arsenic-induced toxicity in renal cells. *Human Exp. Toxicol.* 27: 341–346.

*One Century of the Discovery of Arsenicosis in Latin America (1914–2014) –
Litter, Nicolli, Meichtry, Quici, Bundschuh, Bhattacharya & Naidu (Eds)*

Arsenic, diet and occupational exposure on prostate cancer risk in Córdoba, Argentina

M.D. Román, S.E. Muñoz, C. Niclis & M.P. Díaz
Escuela de Nutrición e Instituto de Investigaciones en Ciencias de la Salud, CONICET, Facultad de Ciencias Médicas, Universidad Nacional de Córdoba, Córdoba, Argentina

ABSTRACT: Arsenic in drinking water was associated to higher cancer incidence. To analyze the risk of prostate cancer in relation to arsenic and occupational exposure in subjects with different dietary patterns, a case control study was conducted in Córdoba, Argentina during 2008–2013. Subjects were interviewed about lifestyle and, sociodemographic characteristics. A two level logistic regression model was fitted using prostate cancer status as the outcome. Age, occupational exposure, dietary patterns and arsenic in drinking water were included as covariates in the first level taking into account family history of cancer as the variance component in the second level. Exposure to high levels of arsenic combined to occupational exposure in rural workers constitute a significant risk increase in prostate cancer occurrence. High levels of arsenic content in drinking water and adhesion to unhealthy dietary patterns have an important role on prostate cancer risk.

1 INTRODUCTION

Ecologic studies in Córdoba, Argentina, have found increased cancer incidence rates in some areas where water wells are contaminated with arsenic concentrations higher than the value suggested by international regulations (Aballay *et al.*, 2012). Prostate cancer is a second most frequent type of cancer in this Province with Age Standardized Incidence Rate (ASIR) of 35.04 preceded by lung cancer (Díaz *et al.*, 2010).

Several lifestyle characteristics of population have been related with prostate cancer risk, however it is not know whether diet effect on prostate cancer risk is modified by different exposure to arsenic through drinking water (World Cancer Research Fund, 2007).

The objective of this study was to analyze arsenic and occupational exposure in subjects with different dietary patterns on prostate cancer risk.

2 METHODS

2.1 *Case-control study*

A case-control study was carried out in Córdoba, Argentina, between 2008 and 2013. Cases were individuals with incident histologically confirmed prostate cancer identified by the Córdoba Cancer Registry. Controls were of identical sex, age (± 5 years) and place of residence and were interviewed in the same period than the cases. A total of 138 cases and 288 controls were included. All participants signed an informed consent according to established bioethical norms. Cases and controls were interviewed about socio-demographic characteristics, anthropometric variables, lifestyle factors, personal medical history and family history of cancer. To assess dietary exposure, a validated food frequency questionnaire was used (Navarro *et al.*, 2001). For several subjects in different areas, determinations on the arsenic content in samples of drinking water were performed.

2.2 *Statistical analysis*

Factor analysis was performed to identify dietary patterns using the information of the control group. After estimation, an individual score for each pattern was calculated, and used as covariate in the risk analysis. A two level logistic regression model was fitted using prostate cancer status as the outcome. Age, occupational exposure (without, rural, transport, industry), dietary patterns (traditional, sweet beverages) and arsenic in drinking water were included as covariates in the first level taking into account family history of cancer as the variance component in the second level (Skrondal *et al.*, 2004). Odds Ratios (OR) and the corresponding 95% Confidence Intervals (CI) were computed. All the analyses were performed using STATA 12.1 (StataCorp, 2011).

3 RESULTS AND DISCUSSION

3.1 *Descriptive analysis*

The demographic and lifestyle characteristics of the case-control study are summarized in Table 1.

3.2 *Risk analysis*

Adjusting by age, drinking water with different arsenic content and occupational exposure, it can be seen that the adherence to a traditional dietary pattern (red fat meat, eggs, nonstarchy vegetables and refined grains) and to sweet drinks pattern (sweet soft drinks and juices) increases risk of prostate cancer.

In this model, it can be seen that exposure to high levels of arsenic, combined to occupational exposure in rural workers constitutes a significant risk increase in prostate cancer occurrence (OR 3.55; 95% CI 1.59–7.94) (Table 2).

Precision levels of estimated OR increase when variability is captured within cancer family history of cancer categories. Thus, a 12.5% of variability of those estimated risks is captured when grouping by cancer family history (intraclass correlation—ICC-), The Median Odd Ratios (MOR) of those people having cancer family history, respect who do not, is of 2.20 (Table 2).

Table 1. Distribution of selected demographic and lifestyle characteristics. Prostate cancer case-control study, Córdoba 2008–2013.

	Cases $n = 138$	Controls $n = 288$
Age	71 years (±8.46) *n* (%)	70 years (±8.61) *n* (%)
Social status		
High	34 (25)	77 (27)
Medium	55 (40)	103 (36)
Low	49 (35)	108 (37)
Risk occupations		
Low risk worker	70 (51)	189 (66)
Rural worker	31(23)	30 (11)
Transport worker	11 (8)	23 (8)
Industry worker	24 (18)	42 (15)
Family history of prostate cancer		
No	112 (82)	261 (95)
Yes	24 (17)	14 (5)
Body Mass Index		
≤24.9	34 (25)	78 (27)
25–29.9	80 (56)	146 (51)
≤30	24 (17)	64 (22)
Caloric intake		
<2800 calories	38 (28)	96 (33)
2800 to 3800 calories	47 (34)	96 (33)
>3800 calories	53 (38)	96 (33)
Smoking habit		
No	45 (33)	88 (31)
Yes	93 (67)	200 (69)
Arsenic in drinking water		
<10 μg/L	51 (37)	103 (36)
>10 μg/L	49 (63)	185 (64)

Table 2. Prostate cancer risk (OR, 95% CI) according to to dietary pattern adherence, arsenic in drinking water and occupational exposure. Prostate cancer case-control study, Córdoba, 2008–2013.

	OR* (95% CI)	p-value
Dietary pattern		
Traditional	1.20 (0.97–1.48)	0.091
Sweet drinks	1.35 (1.05–1.74)	0.018
Arsenic in drinking water >10 μg/L and Rural worker	3.55 (1.59–7.94)	0.002
Second level variance component		
Cancer family history	**Var (SE)****	**CI 95%**
	0.69 (0.39)	0.22–2.133

*Multiple model OR, including term for age ** $p < 0.0015$.

4 CONCLUSIONS

High levels of arsenic content in drinking water and adhesion to unhealthy dietary patterns have an important role on prostate cancer risk. It is also noteworthy the combined effect of arsenic coupled to occupational exposure in our population.

ACKNOWLEDGEMENTS

This work was supported by fellowships provided by the National Scientific and Technical Research Council (CONICET) and by the Science and Technology Secretary of the National University of Córdoba. This research was supported by the National Science and Technology Agency, FONCyT Grant PICT 2008–1814, PICT-O 2005–36035, and Science and Technical Secretariat of the University of Córdoba (SECyT-UNC) Res. 162/12.

REFERENCES

Aballay, L.R., Díaz, M.P., Francisca, F.M. & Muñoz, S.E. 2012. Cancer incidence and pattern of arsenic concentration in drinking water wells in Córdoba, Argentina. *Int. J. Environ. Health Res.*, 22(3):220–31.

Díaz, M.P., Corrente, J.E., Osella, A.R., Muñoz, S.E. & Aballay, L.R. 2010. Modeling spatial distribution of cancer incidence in Córdoba, Argentina. *Applied Cancer Research*, 30(2):245–252.

Navarro, A., Osella, A.R., Guerra, V., Muñoz, S.E., Lantieri, M.J. & Eynard, A.R. 2001. Reproducibility and validity of a food-frequency questionnaire in assessing dietary intakes and food habits in epidemiological cancer studies in Argentina. *J. Exp. Clin. Cancer Res.*, 20(Suppl 3):203–208.

Skrondal, A. & Rabe-Hesketh, S. 2004. Generalized latent variable modeling: multilevel, longitudinal and structural equation models. Boca Raton, FL: Chapman & Hall.

StataCorp. Stata Statistical Software: versión 12.1. College Station, TX: StataCorp LP. 2011.

World Cancer Research Fund/American Institute for Cancer. 2007. Research Food, nutrition, physical activity, and the prevention of cancer: a global perspective. Washington, DC: AICR.

One Century of the Discovery of Arsenicosis in Latin America (1914–2014) – Litter, Nicolli, Meichtry, Quici, Bundschuh, Bhattacharya & Naidu (Eds)

DNA methylation profiles in arsenic-induced urothelial carcinomas

T.-Y. Yang
Molecular and Genomic Epidemiology Center, China Medical University Hospital, China Medical University, Taichung, Taiwan

L.-I. Hsu & C.-J. Chen
Genomics Research Center, Academia Sinica, Taipei, Taiwan

ABSTRACT: Arsenic is a well-documented carcinogen of Urothelial Carcinoma (UC) with incompletely understood mechanisms. This study aimed to compare DNA methylation profiles of arsenic-induced UC (AsUC) and non-arsenic-induced UC (Non-AsUC), and to assess associations between site-specific methylation levels and Cumulative Arsenic Exposure (CAE). DNA methylation profiles in 14 AsUC and 14 non-AsUC were analyzed by Illumina methylation array and validated by pyrosequencing. Mean methylation levels ($\bar{\beta}$) in AsUC and non-AsUC were compared by their ratio ($\bar{\beta}$ ratio) and difference ($\Delta\bar{\beta}$). Among overall methylation sites analyzed, 231 sites had $\bar{\beta}$ ratio >2 or <0.5 and 45 sites had $\Delta\bar{\beta}$ >0.2 or <–0.2. There were 13 sites showing statistically significant differences in $\bar{\beta}$ between AsUC and non-AsUC. Significant associations between CAE and DNA methylation of 28 patients were observed in 11 CpG sites (11 genes). Arsenic exposure may cause UC through the hypermethylation of genes involved in cell adhesion, transcriptional regulation, and ion transport.

1 INTRODUCTION

Arsenic has been well documented as a Group 1 human carcinogen by the International Agency for Research on Cancer (IARC, 2012). A significant dose-response relation exists between arsenic in drinking water and risk of non-melanoma skin cancers and Urothelial Carcinoma (UC) (Tseng, 1977; Chen *et al.*, 1985, 1988, 1990, 1992).

A relation between arsenic and DNA methylation has been suggested based on the observation that arsenic biotransformation and DNA methylation share the same methyl group donor, S-adenosyl methionine (SAM). Competition between arsenic biotransformation and DNA methylation for available methyl groups can lead to differential DNA methylation distribution in arsenic-induced diseases including urothelial carcinoma (Wilhelm-Benartzi *et al.*, 2010).

This study aimed to compare the DNA methylation patterns between Arsenic-induced Urothelial Carcinomas (AsUC) and non-AsUC.

2 MATERIALS AND METHODS

2.1 *Enrollment of patients affected with AsUC and non-AsUC*

28 urothelial carcinomas were obtained from 14 matched pairs of patients with and without exposure to arsenic through drinking artesian well water. They were enrolled from two medical centers, Chi-Mei Hospital and National Taiwan University Hospital. DNA Methylation analysis Illumina Infinium Methylation27 BeadChip (symbols as Methylation BeadChip) (Illumina Inc., San Diego, CA, USA) containing 27,578 methylation sites, was used for the analysis of genome-wide DNA methylation. Bisulfite conversion of DNA specimens was performed using the EZ DNA Methylation kit (Zymo Research, Irvine, CA, USA) in accordance with the manufacturer's recommended protocol. The methylation level per site of the urothelial carcinoma from each participant was compared to negative controls from both the methylated and unmethylated signals. The ratio of the methylated signal to the sum of both methylated and unmethylated signals was calculated and defined as the β-value. The β-value is a continuous variable between 0 and 1 (Bibikova *et al.*, 2009).

2.2 *Bisulfite pyrosequencing*

Bisulfite-converted DNA (1 μL) was amplified using Hot-Start Taq-polymerase. Amplicons were analyzed on the PyroMark Q24 pyrosequencer as specified by the manufacturer, and the percentage of methylation was quantified as a ratio of C (methylated C) to C + T (methylated C + unmethylated C) using PyroMark Q24 software. PCR amplification of target sequences were included with these significant CpG sites from Methylation BeadChip.

2.3 *Statistical analysis*

Based on the literature review of epigenetic studies using Methylation BeadChip, two criteria were used to identify the methylation sites with differential DNA methylation patterns between AsUC and non-AsUC. First criterion was the ratio of mean β-values between AsUC and non-AsUC indicated as $\bar{\beta}$ ratio. Second criterion was the difference in mean β-values between the AsUC and non-AsUC indicated as $\Delta\bar{\beta}$. The methylated sites with a $\bar{\beta}$ ratio >2 or <0.5 and a $\Delta\bar{\beta}$ >0.2 or <-0.2 were considered the differential methylation sites between AsUC and non-AsUC.

The statistical significance of the difference in $\bar{\beta}$ at each site between AsUC and non-ASUC was further assessed by the Wilcoxon signed-rank test using SAS/JMP 10 (SAS Institute Inc., Cary, NC, USA). The q-values were derived using the q-value package of R software (R Development CT, Vienna, Austria) in order to obtain an estimated false discovery rate. The consistency of methylation levels detected by both BeadChips and pyrosequencing methods at these arsenic-associated sites were assessed by pairwise correlation coefficients.

3 RESULTS

3.1 *Differential DNA methylation patterns between AsUC and non-AsUC*

Using the Wilcoxon signed-rank test to examine the statistical significance of the differences in mean methylation levels between AsUC and non-AsUC (multiple comparison *q* value <0.05), we found AsUC and non-AsUC had significantly different mean methylation levels at 34 methylation sites in 33 genes with $\Delta\bar{\beta}$ >0.2 or <-0.2 and at 75 methylation sites in 70 genes with $\bar{\beta}$ ratio >2 or <0.5. In combination of both criteria of $\bar{\beta}$ ratio and $\Delta\bar{\beta}$, there were 13 methylation sites in 13 genes showing significantly different $\bar{\beta}$ between AsUC and non-AsUC (q value <0.05).

3.2 *Bisulfite pyrosequencing for validation of methylation levels detected by Illumina Infinium Methylation27*

The DNA methylation levels of 13 sites with significant differences between AsUC and non-AsUC were further validated using bisulfite pyrosequencing. The methylation levels of specific sites detected by the Methylation BeadChip and bisulfite pyrosequencing were compared in 28 DNA samples. Bisulfite pyrosequencing data were very consistent with the Methylation BeadChip data. Their pairwise correlation coefficients were above 0.85 at 11 sites (*PCDHB2, CTNNA2, KCNK17, ZNF132, PDGFD, NPY2R, KLK7, HSPA2, FBXO39, DCDC2*, and *CYP1B1*). There were only two sites with lower correlation coefficients (<0.7) in *SIPA1* and *ATP5G2*.

4 CONCLUSION

We compared the genome-wide DNA methylation patterns in AsUC and non-AsUC, and identified 11 sites with differential methylation levels which involved in cell adhesion, transcriptional regulation, and ion transport.

REFERENCES

Bibikova, M., Le, J., Barnes, B., Saedinia-Melnyk, S., Zhou, L., Shen, R., & Gunderson, K.L., 2009. Genome-wide DNA methylation profiling using Infinium® assay. *Epigenomics* 1: 177–200.

Chen, C.J., Chen, C.W., Wu, M.M., & Kuo, T.L., 1992. Cancer potential in liver, lung, bladder and kidney due to ingested inorganic arsenic in drinking water. *Brit J Cancer* 66: 888–892.

Chen, C.J., Chuang, Y.C., Lin, T.M., & Wu, H.Y., 1985. Malignant neoplasms among residents of a Blackfoot disease-endemic area in Taiwan: High-arsenic artesian well water and cancers. *Cancer Res.* 45: 5895–5899.

Chen, C.J., Kuo, T.L. & Wu, M.M., 1988. Arsenic and cancers. *Lancet* 331: 414–415.

Chen, C.J. & Wang, C.J., 1990. Ecological correlation between arsenic level in well water and age-adjusted mortality from malignant neoplasms. *Cancer Res.* 50: 5470–5474.

Hsu, L.I., Chiu, A.W., Pu, Y.S., Wang, Y.H., Huan, S.K., Hsiao, C.H., Hsieh, F.I. & Chen, C.J., 2008. Comparative genomic hybridization study of arsenic-exposed and non-arsenic-exposed urinary transitional cell carcinoma. *Toxicol. Appl. Pharmacol.* 227: 229–238.

International Agency for Research on Cancer, 2012. IARC Monographs on the Evaluation of Carcinogenic Risks to Humans. A Review of Human Carcinogens: Arsenic, Metals, Fibres, and Dusts. Lyon: IARC, 1–469.

Tseng, W.P., 1977. Effects and dose—response relationships of skin cancer and Blackfoot disease with arsenic. *Environ. Health Perspect.* 19: 109–119.

Wilhelm-Benartzi, C.S., Koestler, D.C., Houseman, E.A., Christensen, B.C., Wiencke, J.K., Schned, A.R., Karagas, M.R., Kelsey, K.T. & Marsit, C.J. 2010. DNA methylation profiles delineate etiologic heterogeneity and clinically important subgroups of bladder cancer. *Carcinogenesis* 31: 1972–1976.

Yang, T.Y., Hsu, L.I., Chiu, A.W., Pu, Y.S., Wang, S.H., Liao, Y.T., Wu, M.M., Wang, Y.H., Chang, C.H., Lee, T.C. & Chen, C.J., 2013. Comparison of genome-wide DNA methylation in urothelial carcinomas of patients with and without arsenic exposure. *Environmental Research* (In press).

One Century of the Discovery of Arsenicosis in Latin America (1914–2014) –
Litter, Nicolli, Meichtry, Quici, Bundschuh, Bhattacharya & Naidu (Eds)

Association between arsenic in groundwater and malignant tumors in La Pampa, Argentina

R. Molina
Population Record of Tumors, La Pampa Province, Argentina

C. Schulz, J. Bernardos & M. Dalmaso
Faculty of Exact and Natural Sciences, UNLPam, Argentina

ABSTRACT: In the province of La Pampa, groundwater is extensively used for human consumption, where the high prevalence of tumors in the population has historically been associated with arsenic. In 2003, this led to provincial and national Health Department authorities to implement recording of tumors in the province, with the primary objective of providing statistics of high international quality on this pathology. It is vitally important for reliable records to be obtained in order to develop health policies based on accurate data. However, in this work, no link was found between arsenic in drinking water and the incidence of the most frequent tumors in the population of the province of La Pampa.

1 INTRODUCTION

The presence of arsenic (As) in groundwater in the east of La Pampa province, Argentina (Figure 1) limits the water usage, primarily restricting the use as drinking water. The concentrations of As in groundwater in this region vary widely, and are often erratic and difficult to predict.

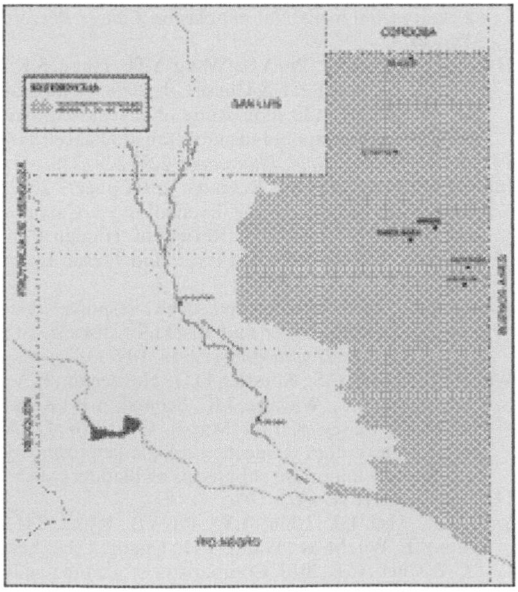

Figure 1. Location of the problem area.

Arsenic is generally heterogeneously distributed, and concentrations within aquifers vary both vertically and horizontally within the aquifer. The vertical distribution results in a pronounced hydrochemical stratification and gives the greatest uncertainty. Consistent with this, reoccurrence occurs in areas of rapid infiltration (dune areas), where As occurs in low amounts, increasing in the direction of flow and towards the lower levels of the main aquifer (Pampeano).

The presence of As has no relationship with the regional hydrogeological characteristics, but it rather, seems to be related to particularities of local type. Likewise, the mechanism that controls the dissolution of As is a combination of factors including speed of flow (hydraulics) as well as physical and chemical reactions with the mineral surface (Schulz *et al.*, 2002).

Likewise, other ions present in these waters cannot easily be related with any scientific rigor to hydrogeochemical patterns.

2 MATERIALS AND METHODS

Spearman correlation coefficients were determined between As concentrations in wells supplying drinking water in different localities, with the incidence of the most common malignant tumors in these locations for the period 2003–2007 (Molina *et al.*, 2010).

The prevalent tumors for women are those of breast, colon, cervix, endometrium while, for men, the most frequently reported tumors are those of prostate, lung, colon and stomach.

Average arsenic concentration in the groundwater water for the studied departments varied between 0.035 and 0.39 mg L^{-1} (Table 1).

For both women and men, the association between As and the most prevalent tumors in the province of La Pampa was not significant (Table 2 and 3).

Table 1. Concentration of arsenic in drinking water in La Pampa province (Argentina).

Department	Arsenic (mg L^{-1})
Atreucó	0.045
Maracó	0.065
Rancul	0.17
Chapaleufú	0.12
La Pampa city	0.12
Conhelo	0.235
Realicó	0.06
Utracán	0.045
Caleu Caleu	0.04
Catriló	0.145
Guatraché	0.07
Hucal	0.035
Toay	0.22
Trenel	0.25
Loventué	0.39
Quemú	0.04

Table 2. Association between the concentration of arsenic in drinking water and the prevalent tumors in women for the province of La Pampa (2003–2007).

Tumor	Spearman correlation	
Mamma	–0.32	($p > 0.22$)
Colon	0.07	($p > 0.77$)
Cervix	0.30	($p > 0.25$)
Endometrium	–0.17	($p > 0.52$)

Table 3. Association between the concentration of arsenic in drinking water and the prevalent tumors in men for the province of La Pampa (2003–2007).

Tumor	Spearman correlation	
Prostate	0.19	($p > 0.46$)
Lung	0.01	($p > 0.96$)
Colon	–0.29	($p > 0.26$)
Stomach	0.05	($p > 0.84$)

3 CONCLUSIONS

While high groundwater As content is a distinctive characteristic of many of the aquifers of La Pampa, no association was found between the concentration of As in drinking water and the incidence of the most common tumors in the population. This can be explained because probably their etiologies have another origin. Nevertheless, it is still necessary to maintain the monitoring of both tumors as well as the water quality to provide early warnings of potential problems.

REFERENCES

Schulz, C., Castro, E., Mariño, E. & Dalmaso, G. 2002. El agua potable en la provincia de La Pampa. Consecuencias por presencia de arsénico. In *Taller sobre Arsénico en Aguas Subterráneas*. XXXII Congreso Internacional de Hidrología Subterránea y V Congreso Latinoamericano de Aguas Subterráneas. Asociación Internacional de Hidrogeólogos y Asociación Latinoamericana de Hidrología Subterránea para el Desarrollo, Mar del Plata, 21–25 October 2002.

Molina, R. & Giménez, P. 2010. *Resultados del registro poblacional de tumores, Provincia de La Pampa*. Datos del Quinquenio 2003–2007. Inédito. Gobierno de La Pampa. 44 pp.

One Century of the Discovery of Arsenicosis in Latin America (1914–2014) – Litter, Nicolli, Meichtry, Quici, Bundschuh, Bhattacharya & Naidu (Eds)

Chemical and male infertility oriented metabolomics in population based health risk research: The case of arsenic

H. Shen, J. Zhang & W. Xu
Key Lab of Urban Environment and Health, Institute of Urban Environment, Chinese Academy of Science, Xiamen, China

ABSTRACT: Biomarkers indicate the major responses of the molecular network to environmental risk factors such as arsenic exposure. The direct linkage of arsenic exposure to biomarkers based on population bio-monitoring data has not been widely used for risk assessment or epidemiological studies to identify a chemical as a disease risk factor. In the present work, we have examined both arsenic and male infertility specific metabolite biomarkers in urine, in combination with arsenic biomonitoring data to assess their roles in risk assessment and molecular epidemiology.

1 INTRODUCTION

Different strategies are commonly applied for the chemical risk assessment and for identification of factors associated to health disease. For chemical risk assessment, the exposure assessment and the dose-response assessment are initially conducted using exposure data that may come from top-down environmental monitoring or from bottom-up human bio-monitoring. The Lowest-Observed-Adverse-Effect Level (LOAEL) and No-Observed-Adverse-Effect Level (NOAEL) criteria are derived from dose-response curves, and are usually the results of different tiers of *in vitro* and *in vivo* animal studies. However, uncertainty and assumptions exist in the process of linking the indirectly determined toxic criteria to human health. In comparison, in health risk assessment, epidemiological studies use biomonitoring data to identify the role of chemicals in disease risk, in which the relationship between the exposure data and health outcome is established solely on the basis of biostatistical analysis.

In this work, to address the direct correlations between exposure (biomonitoring data) and observed effects in health risk assessment for arsenic, and to understand the toxicity behind the statistic correlation of exposure data and disease incidence, biomarkers were mined from metabolomic data, i.e., arsenic (As) risk analysis and epidemiological study.

2 METHODS

Human biomonitoring data of arsenic in urine with health controls and male infertile cases were measured by HPLC-ICP-MS. Metabolomic analysis was conducted on the same urine samples using HPLC-TOF-MS. The screening of both of the arsenic and male infertile oriented biomarkers was conducted by using urinary metabolite profiling acquisition and the established OPLS-DA model.

The screening of arsenic oriented biomarkers was conducted using urinary metabolite profiling acquisition and established OPLS-DA models, and the dose-response curves between arsenic and biomarkers, such as estrone, testosterone, and some alkylcarnitines, were established based on arsenic concentrations and biomarker occurrence odds ratios.

3 RESULTS AND DISCUSSION

In risk analysis, the biological hypothesis is that arsenic exposure can cause dose-dependent responses in potential urinary metabolites, and the response curves can indicate the risk criteria of arsenic directly.

Elevated arsenate exposure was associated with infertility. Then the male infertility oriented to biomarker mining was run by OPLS-DA model, in which the male infertility disease was characterized by poor semen quality. Levels of urinary biomarkers, such as acylcarnitines, aspartic acid, hydroxyestrone correlated with both male infertility and arsenic exposure.

4 CONCLUSIONS

Chemical oriented metabolomic approaches may be used directly in population based risk assessment in combination of bio-monitoring data. The disease oriented metabolomic approach may be used to illustrate the inherit toxicology underlying the observed epidemiological association of the disease incidence.

ACKNOWLEDGEMENTS

This work was supported by Hundred Talent Program of Chinese Academy of Sciences (CAS) for 2010 on Human Exposure to Environmental Pollutant and Health Effect, NSFC 2011 research foundation (21177123) and CAS/SAFEA International Partnership Program for Creative Research Teams (KZCX2-YW-T08).

REFERENCES

Albertini, R., Bird, M., Doerrer, N., Needham, L., Robison, S., Sheldon, L. & Zenick, H. 2006. The Use of Biomonitoring Data in Exposure and Human Health Risk Assessments. *Environ. Health Perspect.* 114: 1755–1762.

Shen, H., Xu, W., Zhang, J., Chen, M., Martin, F.L., Xia, Y., Liu, L., Dong, S. & Zhu, Y.-G. 2013. Urinary Metabolic biomarkers link oxidative stress indicators associated with general arsenic exposure to male infertility in a Han Chinese population. *Environ. Sci. Technol.* 47: 8843–8851.

One Century of the Discovery of Arsenicosis in Latin America (1914–2014) – Litter, Nicolli, Meichtry, Quici, Bundschuh, Bhattacharya & Naidu (Eds)

Historical exposure to arsenic in drinking water and risk of late fetal and infant mortality—Chile 1950–2005

P. Ríos
National Direction of Epidemiological surveillance, Ministry of Health, Ecuador

S. Cortés, L. Villarroel & C. Ferreccio
Faculty of Medicine, Department of Public Health, Pontifical Catholic University of Chile, Chile

ABSTRACT: A previous study comparing two Chilean cities demonstrated significant increase in fetal mortality in relation to increase in arsenic (As) exposure. Our aim was to confirm this association increasing size and variation of the comparison group, extending the time interval and measuring additional potential confounders. We used time series analysis of infant mortality indicators. We confirmed the clear excess of risk of late fetal death following the increase of the As concentration in drinking water, and the rapid return to the baseline mortality rate in response to the removal of arsenic after 1973. The effect was lower for neonatal and post-neonatal mortality. This study confirms that fetus is more sensitive to As and that regulations should consider additional protection for pregnant women and small children.

1 INTRODUCTION

Arsenic crosses the placenta barrier increasing the risk of spontaneous abortion, low birth weight, Late Fetal Mortality (LFM) and Infant Mortality (IM), both neonatal-NM—and Post-Neonatal Mortality—PNM. In the northern city of Antofagasta (32°–27° latitude), levels of As in drinking water averaged 860 µg L^{-1} from 1958–1970, dropping rapidly to 50 µg L^{-1} since 1971, after the installation of arsenic treatment plants, until 2003 when they reached the current levels of 10 µg L^{-1} (Borgoño and Griber, 1971; Ferreccio and Sancha, 2006). At the national level, infant mortality (deaths/1000 live births) dropped from 120 in the 50's to 70 in the 70's and continued to fall to the current rate of 7 (Barría *et al.*, 1992).

In a previous study comparing only two cities (Antofagasta and Valparaiso) from 1950 to 1996, we showed that during the high As exposure period in Antofagasta fetal mortality increased significantly returning to normal after the treatment plants were installed (Hopenhayn *et al.*, 2000). We decided to confirm this finding including a larger control sample, extending the study to 2005 and including new potential confounders to confirm the validity of our previous finding. Our aim was to study the association of As exposure in drinking water with infant mortality, in particular LFM, NM and PNM, by comparing the trend of these measures in five geographic regions with different As exposure: Antofagasta (860 µg L^{-1}), Iquique (60 µg L^{-1}) Copiapó (15 µg L^{-1}), La Serena (6 µg L^{-1}) and Valparaíso (2 µg L^{-1}). The study was approved by the Institutional Review Board (IRB) of the Pontificia Universidad Católica de Chile (Project #: 10–180) on November 2, 2010.

2 METHODS

We used time series analysis of infant mortality indicators in relation to As exposure adjusting for socio-demographic factors. Data on As exposure was provided by water companies obtained from the distribution networks in each region.

We used different metrics to analyze As: mean 4-years As concentration (µg L^{-1}) from 1952–2005, level of As in six categories based on 4-years averages (860 µg L^{-1}, 636 µg L^{-1}, 250 µg L^{-1}, 110 µg L^{-1}, 60 µg L^{-1} and 15 µg L^{-1}). Infant mortality indicators were based on national vital statistics and standard WHO definitions. Covariates considered were rate of hospital delivery (RTD), regional Gross Domestic Product (GDP)-Central Bank of Chile. All study variables were compiled for each region from 1952 to 2005. We used Poisson and Quasi-Poisson regression to model infant mortality in association to arsenic exposure adjusting for covariates ("region", "calendar time", "hospital delivery" and "GDP"). We used R statistical program version 2.11 and SPSS 17. Subsection 2.2.

3 RESULTS AND DISCUSSION

We present preliminary results showing the excess infant mortality rates for Antofagasta compared with each of the five cities for the period 1952–2004. The rate difference clearly shows that during the high exposure period until 1973 (860 μg L^{-1}) Antofagasta presented excess of deaths in comparison with any of the other cities, reaching more than 20% in some moments. This excess of deaths dropped abruptly soon after 1973, and from then till now, Antofagasta presents similar or lower infant mortality rate than any of the other comparison cities. The risk difference is particularly noticeable for late fetal mortality (Figures 1–3).

Others results not shown here using Poison regression model gave a relative risk of 1,007 (IC 95% 1,006–1,008) for late fetal mortality for each 10 μg L^{-1} of As in drinking water.

The risk for neonatal-NM—and post-neonatal mortality—PNM were 1,004 (IC 95% 1,002–1,005) and 1,002 (IC 95% 1,001–1,004), respectively.

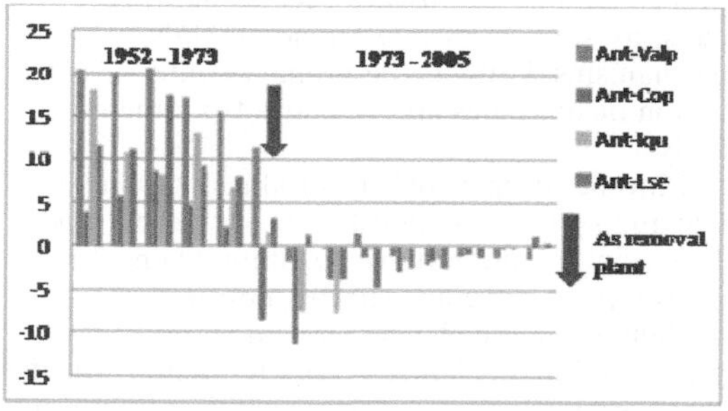

Figure 1. Risk difference Late Fetal Mortality rate of Antofagasta vs. 4 unexposed cities. Chile, 1953–2005. Ant: Antofagasta; Cop: Copiapó; Iqu: Iquique; Lse: La Serena.

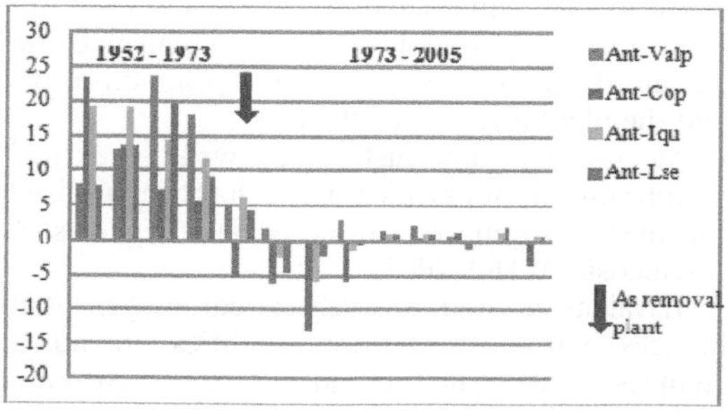

Figure 2. Risk difference Neonatal Mortality rate of Antofagasta vs. 4 unexposed cities. Chile, 1953–2005. Ant: Antofagasta; Cop: Copiapó; Iqu: Iquique; Lse: La Serena.

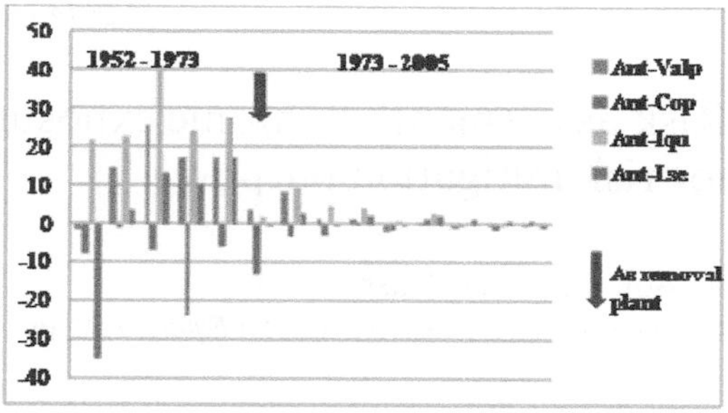

Figure 3. Risk difference **Post Neonatal Mortality** rate of Antofagasta vs. 4 unexposed cities. Chile, 1953–2005. Ant: Antofagasta; Cop: Copiapó; Iqu: Iquique; Lse: La Serena.

4 CONCLUSIONS

Arsenic exposure caused increase in infant mortality in Antofagasta when average As in drinking water was 860 μg L^{-1}. The risk difference between Antofagasta and the rest of evaluated cities is particularly noticeable for late fetal mortality. This finding confirms our previous report and confirms that fetuses are more sensitive to As and that regulations should consider additional protection for pregnant women and small children.

ACKNOWLEDGEMENTS

This study was funded by internal funds of Department of Public Health of the Pontificia Universidad Católica de Chile.

REFERENCES

Borgoño, J. & Griber R. 1971. Epidemiologic study of arsenic poisoning in the city of Antofagasta. *Rev. Med. Chil.* 99: 702–707.

Barría Concha, M. & Mardones Azocar, G. 1992. Mortalidad infantil por causas de muerte según legitimidad en Chile. *Rev. Chil. Pediatr*. 63(6):332–341.

Ferreccio, C. & Sancha, A.M. 2006. Arsenic exposure and its impact in health in Chile. *J. Health Popul. Nutr.* 24(2): 164–175.

Hopenhayn-Rich, C., Browning, S.R., Hertz-Picciotto, I., Ferreccio, C., Peralta, C. & Gibb, H. 2000. Chronic arsenic exposure and risk of infant mortality in two areas of Chile. *Environ. Health Perspect.* 108(7):667–673.

One Century of the Discovery of Arsenicosis in Latin America (1914–2014) – Litter, Nicolli, Meichtry, Quici, Bundschuh, Bhattacharya & Naidu (Eds)
© 2014 Taylor & Francis Group, London, ISBN 978-1-138-00141-1

Risk assessment of chronic exposure to arsenic—regional mapping for risk mitigation purposes

A. Boischio & H. Nasr
Sustainable Development and Health Equity, Pan American Health Organization, USA

ABSTRACT: Risk assessments are recommended for mapping arsenic exposures and attributable health outcomes in the Americas. To conduct such assessments, a review to addressing the Lowest Observable Adverse Effect Level (LOAEL) is combined with potential arsenic concentrations in water exposure levels. This regional mapping of arsenic exposure may also be utilized for policy impact purposes.

1 INTRODUCTION

Risk assessment methods include the identification of hazards, exposures, and health effects of one or more chemicals. By comparing the Lowest Observable Adverse Effect Levels (LOAELs) within a given exposed population, it is possible to summarize figures into hazard quotients for the purpose of decision making on mitigation measures. In the case of arsenic, LOAELs have not been easy to establish due to the various sub-clinical and multi-causal aspects of signs and symptoms. It is imperative to recognize the history of exposure levels for early diagnosis.

Arsenic is often referred to as a metal, although it is in fact a metalloid, with properties of both metals and nonmetals. Elemental arsenic, or metallic arsenic, has zero oxidation whereas in inorganic compounds, arsenic exists in one of two different oxidation states—arsenites (which are trivalent) and arsenates (which are pentavalent). Like other elements, organic arsenic contains a carbon bond; however, toxic effects mostly arise from inorganic arsenic, and trivalent arsenicals tend to be more potent toxicants than the pentavalent ones. After absorption of arsenic by the body, arsenic is rapidly cleared from the blood thus entering the liver where it is detoxified by conversion into methylated arsenical compounds. The metabolism of arsenic is characterized by a series of sequential reactions in which the reduction of pentavalent arsenic to trivalent arsenic in the presence of gluthathione occurs. This is followed by an oxidative methylation reaction in which the trivalent forms of arsenic are sequentially methylated to form mono-, di-and trimethylated products using methionine as a methyl donor and glutathione as an essential co-factor (ATSDR, 2007).

Arsenate interferes in the oxidative phosphorylation in the mitochondria by replacing inorganic phosphate in the process of ATP synthesis (i.e., in the process of cellular respiration). Arsenite has a high affinity to sulfhydryl groups of many essential enzymes and proteins. This leads to its accumulation in keratin-rich tissues such as skin, hair and nails, as well as its interference in DNA related mechanisms. Urine is the primary route for elimination of arsenates and arsenites from the human body.

Chronic arsenic toxicity produces various dermal and systemic manifestations including cancer. The skin manifestations resulting from chronic ingestion of arsenic can be non-malignant or malignant. Dermal changes include increased pigmentation (melanosis) and hardening of the skin (keratosis) that can occur within 6 to 9 years depending on the exposure dose and other factors. The characteristics and patterns of dermal effects have been clinically well established. Melanosis is pronounced on the trunk and extremities in a bilaterally symmetric manner; melanosis is considered an early and common manifestation of arsenicosis. Keratosis is considered to be a more sensitive and advanced stage of arsenicosis; it appears on palms and the plantar region of the feet.

Arsenic exposure history, as well as dermal manifestations of arsenic toxicity, has been utilized together as strong indicators for the diagnosis of arsenicosis (WHO, 2005).

Trivalent arsenic is likely a carcinogen that induces chromosomal abnormalities including changes in the structure and number of chromosomes and sister chromatid exchanges rate.

Other systems affected by arsenicals include neurological (such as paresthesia and neurodevelopment delay and impairment), hematological (Blackfoot disease—arteriosclerosis and gangrene; leukopenia, anemia and splenomegaly); gastrointestinal (anorexia, abdominal pain or chronic diarrhea; liver enlargement with or without non-cirrhotic portal fibrosis) and respiratory (chronic cough or bronchitis).

The purpose of this paper is to use risk assessment methods for mapping current arsenic exposures in water for risk characterization and mitigation measures recommendations.

2 MATERIALS AND METHODS

A long list of studies with the identification of the LOAELs and health effects in different systems (dermal, hematological, gastrological, respiratory, hepatic, etc.) has been compiled by ATSDR (2007). Taking into account exposure duration and frequency, human studies were identified with low LOAELs ranging from 50 to 200 $\mu g\ kg^{-1}$ body weight per day for both less serious and serious effects. These figures are to be compared with daily intakes of around 0.30–1.40 $\mu g\ kg^{-1}$ bw day^{-1} for the As water content in the range of 10 to 50 $\mu g\ L^{-1}$, adult water consumption of 2 liters of water, and body weight of 70 kilos. Similar intake figures for children consuming 1 liter of water per day and body weight of 25 kilos would result in the range of 0.4–2 $\mu g\ kg^{-1}$ bw day^{-1}. Comparisons are made with previous provisional tolerable intake of 2.1 $\mu g\ kg^{-1}$ bw day^{-1} used to the benchmark dose of 0.5% increased incidence of lung cancer, has been considered recently no longer appropriate by the joint FAO/WHO expert committee on food additives (2010).

Based on different sources of information, including literature reviews, exposure data regarding arsenic contents in water, are compiled to be compared to a set of LOAELs related to different health outcomes.

3 RESULTS AND CONCLUSIONS

Risk assessments allow for a better, more wholesome understanding of the issue at hand, thus honing in on a more centralized solution. This method allows for the associations of exposures and health outcomes to be summarized in figures (hazard quotients) that can be plotted in a map for regional intervention policy drafts. According to the identified hazard quotients, small scale mitigation options are provided. In the Americas, there are countries (such as Argentina and Chile) for which exposure data is available, whereas in various other countries, information is incomplete or missing. This poses great difficulty in attaining respective hazard quotients, thus interventional policies may more generalized,, than policies derived from risk assessments in which hazard quotients were available.

REFERENCES

FAO/WHO Expert Committee On Food Additives 2010. Available at: http://www.who.int/foodsafety/chem/summary72_rev.pdf

U.S. Department Of Health And Human Services Public Health Service Agency for Toxic Substances and Disease Registry 2007. Toxicological Profile For Arsenic.

World Health Organization 2005. A Field Guide for Detection, Management and Surveillance of Arsenicosis Cases. Deoraj Caussy (ed.), New Delhi, India.

One Century of the Discovery of Arsenicosis in Latin America (1914–2014) – Litter, Nicolli, Meichtry, Quici, Bundschuh, Bhattacharya & Naidu (Eds)
© 2014 Taylor & Francis Group, London, ISBN 978-1-138-00141-1

Human health risk assessment by As environmental exposure in Paracatu: An integrated approach

C.A. Lima
Independent Risk Assessor

Z.C. Castilhos
Center for Mineral Technology, Ministry of Science, Technology and Innovation, Rio de Janeiro, Brazil

ABSTRACT: The objective of this study was to estimate the potential risk on the health of residents of Paracatu (adults and children) due exposure to arsenic present in soil, water, air and food. For the assessed receptors, the risk due to the present contamination scenario was higher than the limit considered acceptable. The main pathways were ingestion of water while swimming and inhalation of particulates. It is important to highlight that human health risk assessment is a very conservative modeling, trying to protect the human health including critical subpopulations. Epidemiological study carried out at Paracatu city will bring new elements to the uncertainties in the human health risk assessment.

1 INTRODUCTION

Arsenic (As) is a naturally occurring element widely distributed in the Earth's crust. However, it is one of the most toxic elements to humans, whose exposure usually occurs through food, soil, water and air. The relationship between the intensity of the environmental contamination and the potential risks to human health can be assessed by the human health risk assessment methodology proposed by USEPA (1989). This methodology allows expressing risk as a comparable numeric estimates, which permits to establish priority of impacted areas, as well as the evaluation of remediation techniques. Therefore, the objective of this study was to estimate the potential risk on the health of residents of Paracatu (adults and children) due exposure to arsenic present in soil, water, air and food. This study is part of the environmental and health assessment conducted by Brazilian research institutions under the general coordination of Center for Mineral Technology (CETEM).

2 METHODS

2.1 *Study area*

Paracatu (8,229.6 km^2) has a population of *ca.* 90,000 inhabitants, of which about 95% live in the urban area. The economy is centered on cattle raising, agriculture (mainly soybean, corn, rice and beans) and gold mining. The gold-mine operations in Morro do Ouro, a low-grade gold deposit located close to the northern border of Paracatu city, began in the late 80 s. Nowadays, it is the largest open pit gold mine in the world.

2.2 *Sampling and analysis*

Surface water, sediment and soil from the sub-basins of Córrego Rico and Ribeirão-Entre-Ribeiros, which may be impacted by gold mining, were collected. Samples were also collected in the sub-basin of the Escuro river (reference area). Arsenic levels were analyzed in all samples (Bidone *et al.*, 2014). In addition, arsenic levels were accessed in atmospheric PM (Zamboni *et al.*, 2014). Moreover, tap water samples provided by the water supply company were collected. Arsenic concentration resulted not only below the drinking-water quality criteria (10 μg/L) but also, below the limit of detection (<0.5 μg/L), except one sample (1.6 μg/L) (Bidone *et al.*, 2014). Thus, ingestion of tap water was not considered an exposure pathway in this risk assessment and the "water ingestion" in this work is related only to the ingestion of water while swimming.

2.3 *Human Health Risk Assessment*

The human health risk assessment was accomplished according to the methodology proposed by USEPA (1989) and the guidelines suggested by CONAMA 420 (2009). This methodology is composed of four stages: 1) elaboration of a qualitative conceptual model concerning to arsenic contamination and transference among environmental multimedia, exposure pathways and potential receptors: adults and children; 2) chronic exposure assessment (magnitude, frequency and duration of the exposure) for the receptors and exposure pathways previously selected (by using estimated or measured As levels in soils, freshwater, tap water and atmospheric PM); 3) Toxicity assessment: the weight of evidence for As carcinogenicity (Group A) exists and a carcinogenic unit risk (which depends on the exposure pathway and is related to the likelihood of developing cancer as a result of that specific exposure) is provided and the reference of dose (RfD) for As non-carcinogenic effects and; 4) risk characterization, which summarizes and combines outputs of the exposure and toxicity assessments to characterize baseline risks, both in quantitative expressions and qualitative statements.

3 RESULTS AND DISCUSSION

Figure 1 shows the conceptual model developed for the study area after a qualitative risk assessment.

For adults, the hazard quotient regarding the exposure to non-carcinogenic contaminants for all assessed exposure pathways was less than 1, resulting in an acceptable risk (Figure 2).

For children, the hazard quotient to non-carcinogenic contaminants for all assessed exposure pathways was less than 1. However, the hazard index (sum of all exposure pathways HQs) was higher than 1, resulting in potential hazard (Figure 4). For carcinogenic effects, the Incremental Lifetime Cancer Risk (ILCR) was higher than 10^{-5} (for the ingestion of water and inhalation of particulates exposure pathway), resulting in an unacceptable risk (Figure 5).

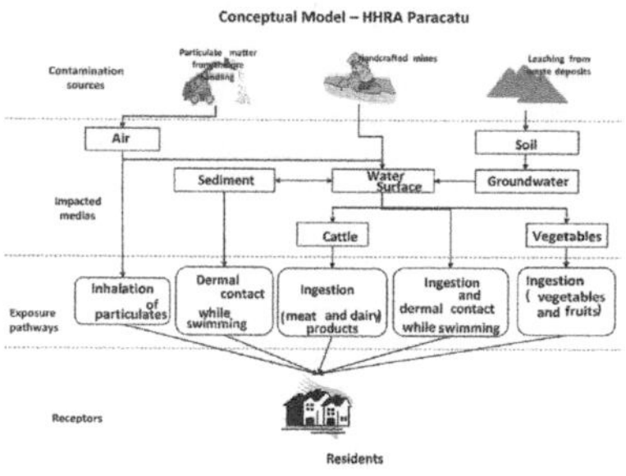

Figure 1. Conceptual model for human health risk assessment.

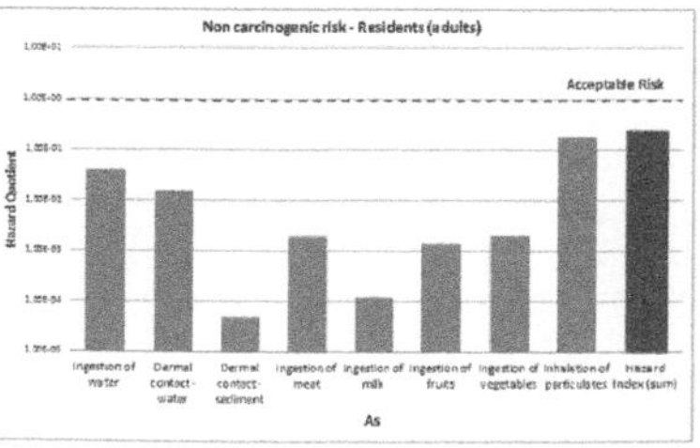

Figure 2. Noncarcinogenic hazard for adult residents.

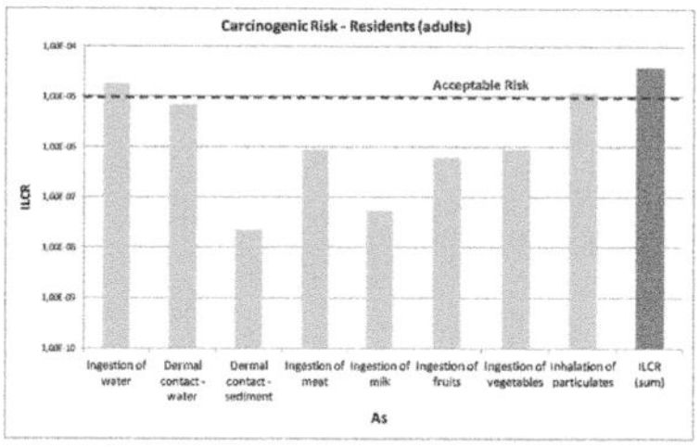

Figure 3. Carcinogenic risk for adult residents.

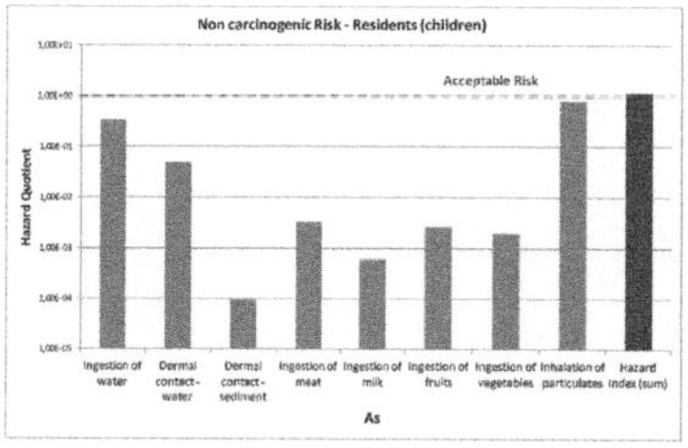

Figure 4. Non carcinogenic risk—Residents (children).

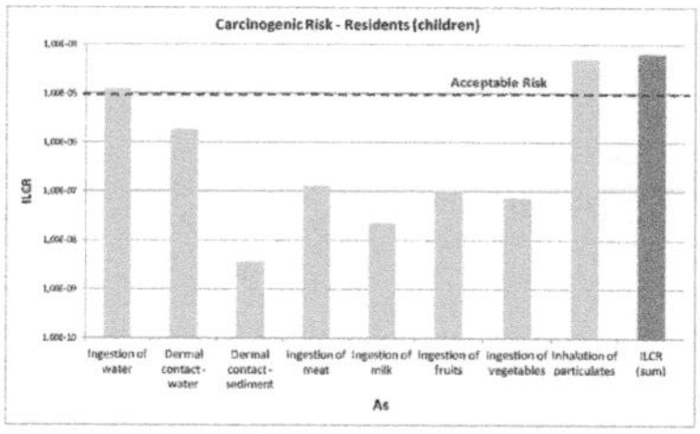

Figure 5. Carcinogenic risk—Residents (children).

4 CONCLUSIONS

The results showed that the environmental exposure to As does not represent potential hazard of non-carcinogenic effects for adults, but children are at risk. For carcinogenic effects, children and adults are at risk. The pathways that contributed most to this result were ingestion of water (while swimming) and inhalation of particulates.

REFERENCES

Bidone, E., Castilhos, Z., Santos, M., Silva, R., Cesar, R. & Ferreira, M. 2014. Arsenic in soils and sediments from Paracatu, Brazil. This issue.

CONAMA. 2009. Resolução nº 420. In *DOU, 249*: 81–84.

USEPA. 1989. Risk Assessment Guidance for Superfund: Human Health Evaluation Manual. EPA/540/1–89/002.

Zamboni, W., Mattos, J., Silva, L., Carneiro, M., Monteiro, M. & Castilhos, Z. 2014. As in atmospheric particulate matter at Paracatu-Brazil. This issue.

Section 4: Arsenic remediation and removal technologies

4.1 Technologies based on adsorption and co-precipitation

One Century of the Discovery of Arsenicosis in Latin America (1914–2014) – Litter, Nicolli, Meichtry, Quici, Bundschuh, Bhattacharya & Naidu (Eds)
© 2014 Taylor & Francis Group, London, ISBN 978-1-138-00141-1

Arsenic sorption by Iron Based Sorbents (IBS)

A. Fiúza & A. Futuro
CIGAR, University of Porto, Faculty of Engineering, Portugal

M. Guimarães
LNEG, National Laboratory for Geology and Energy, S. Mamede de Infesta, Portugal

ABSTRACT: Iron products have been used in the treatment of As in the groundwater, especially Zerovalent Iron (ZVI), Zerovalent Iron Nanoparticles (nZVI) and furnace slags. The potentialities of other iron based sorbents, such as oxides and hydroxides, commercially known as ARM 300 and GEH 102, have not been thoroughly examined. The behavior of these materials using laboratory studies at batch scale and in columns and their performance is compared with nZVI. A similar loading capacity was found for both sorbents: 20.6 g As kg^{-1} for GEH 102 and 20.8 g As kg^{-1} for ARM 300. The spatial distribution of As in the sorbent was analyzed using an electron microprobe that produced backscattered electron images and elemental mapping, evidencing that the deposition of As occurs on exposed surfaces and coincides with a net depletion in oxygen.

1 INTRODUCTION

Anomalous arsenic concentrations in groundwater are generally caused by natural dissolution of minerals. Arsenic is always present in natural waters either in the reduced As(III) form or in the oxidized As(V) form. Toxicity depends upon the chemical species and, in general, As(III) is 25 to 60 times more toxic than As(V). Arsenic concentrations in clean superficial waters and in groundwater are normally in the range 1–10 μg L^{-1}. In Portugal and Spain, it is possible to find concentrations of As in former mining areas ranging from 100 to 2000 μg L^{-1}.

Conventional technologies for the treatment of groundwater like pump and treat have important disadvantages: very high costs, especially if the treatment is long, and the efficiency is moderated. For this reason, the in-situ technologies have received greater attention in the last years. Some available alternatives are bioremediation, permeable reactive barriers, in-situ chemical treatment, enhanced natural attenuation and electrokinetics. Permeable reactive barriers is the in-situ technology that has received more attention, and some applications were implemented in several countries. The main objective of our research was to assess the capacity of using Iron Based Sorbents (IBS), of the oxide-hydroxide type, as active reagents for those barriers and compare their performance with Zerovalent Iron Nanoparticles (nZVI).

2 METHODS AND MATERIAL

All chemicals were reagent grade and were used without any further purification. All solutions were prepared with deionized water. All glassware was cleaned by soaking in 10% HNO_3 and rinsed with deionized water. Arsenic solutions were prepared by direct dilution of 1000-mg/L standard solutions from Fluka for As(V) and from Panreac for As(III), and were kept in dark glass flasks for the preparation of all the required diluted solutions. nZVI used in this work was manufactured by Toda Kogyo Corporation (Japan).

Total As determinations were carried by inductively coupled plasma optical emission spectroscopy with coupled hydride generation. As(III) was determined by cathodic stripping voltammetry with square wave using a hanging mercury drop electrode as described by Ferreira & Barros (2002). Samples that were not analyzed immediately were preserved by acidification to a pH lower than 2 with concentrated HCl and stored in acid washed high-density polyethylene containers (ISO, 1994). All samples were analyzed within seven days after collection. The pH values were obtained by immediate measurement using a pH meter (Crison, GLP22). The sand used was pure quartz, which was washed, dried and sieved to grain sizes between 710 and 1000 μm. The ARM 300 commercial sorbent was supplied by BASF and GEH 102 by Wasserchemie GmbH & Co.

Column tests were performed at 20°C. The solutions contaminated with As were artificially prepared and were pumped from a large volume container using a peristaltic multi-channel Ismatec BV-GES pump into an Omnifit column with an internal diameter of 2.5 cm and a total height of 15 cm. Later, larger columns (3.1 cm in diameter and 40 cm in height) were also used. Cumulative effluent volumes were periodically measured and samples were simultaneously collected for chemical analysis.

Particle size distribution of nZVI was determined by a Mastersizer 2000 laser diffraction particle size analyzer from Malvern Instruments.

3 RESULTS AND DISCUSSION FOR ARM 300 AND GEH 102 SORBENTS

3.1 *Preliminary tests*

Tests were performed using: a) As(III) solutions; b) As(V) solutions and c) solutions with both As(III) and As(V). The sorption isotherms were obtained using flasks with 100 mL of a 5 mg L^{-1} As solution and adding different masses of sorbent. Control tests were simultaneously performed in the same way but without sorbent. The flasks were incubated in an orbital shaker (Ritabit, Selecta) during 24 hours. Then, pH was measured and the content was filtered under pressure; a sample of 10 mL was collected from each flask and preserved by adding 100 μL of 37% HCl.

This procedure was used for both As(III) and As(V).

The continuous tests through a fixed bed sorption column, filled with the reactive medium mixed with quartz sand, were programmed in order to quantify the influence of the interstitial water velocity through the pores. Three different flow rates were used, 5, 10 and 20 mL min^{-1}, corresponding respectively to interstitial velocities of 1.7, 3.4 and 5.1×10^{-4} m s^{-1}. The last value was the highest that could be safely used taking into account the size and the characteristics of the column. The cumulative effluent volume was periodically measured and samples were collected for immediate measurement of pH and conductivity.

General conclusions could be drawn from these experiments:

- pH, starting from neutral water, decreases slightly with time;
- hydrogen ions are generated as a consequence of the As sorption on ARM 300; this proves that the sorption is not only of physical origin but that there are simultaneous chemical reactions involved.

Figure 2 represents the values for the As concentrations in the effluent as a function of the cumulative effluent volume. The uptake capacities for the flow rates of 5 and 10 mL min^{-1} are similar, with a value of 20 g As kg^{-1} of sorbent.

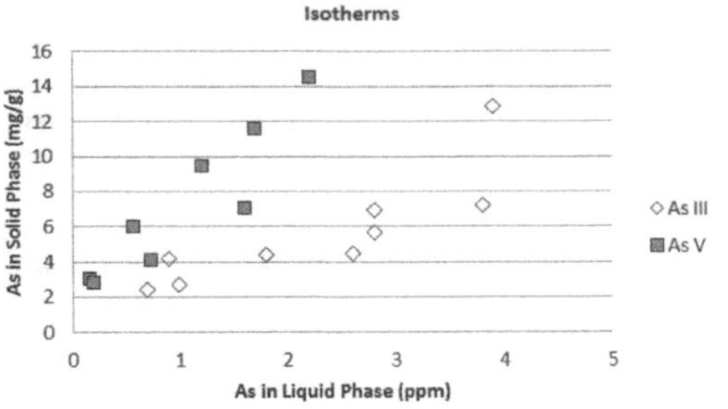

Figure 1. Sorption isotherms for As(III) and As(V) using ARM 300.

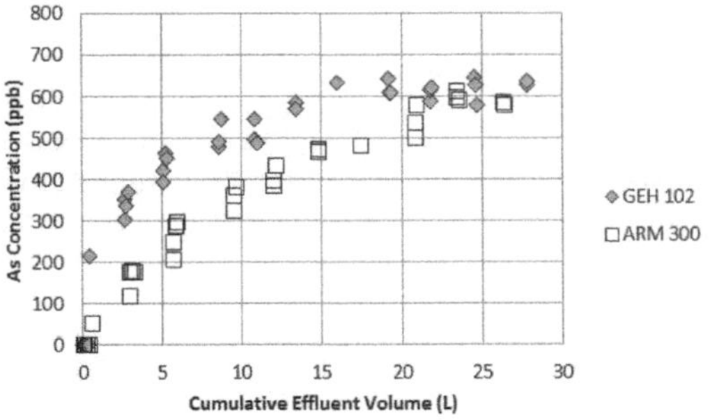

Figure 2. Total arsenic breakthrough curves for different flow rates.

3.2 *Comparison tests between the sorbents*

Comparison tests between ARM 300 and GEH 102 were performed using two columns fed in parallel by the same As solution (around 1000 μg L^{-1}) using a multi-channel peristaltic pump. Tests were also performed for each sorbent using the same operating conditions: same amount of sorbent (1.5 g) and sand, and same flow rate (around 0.110 L h^{-1}).

The procedures for collecting samples and measuring the cumulative effluent volumes were similar.

The most relevant conclusions are the following:

- ARM 300: the residence time was too short in relation with the kinetics demand. For this reason, not all the As had enough contact time to be adsorbed. This result was not expected because the used water velocity was lower than the recommended for operation in water treatment columns.
- GEH 102 demonstrates a slightly better performance.

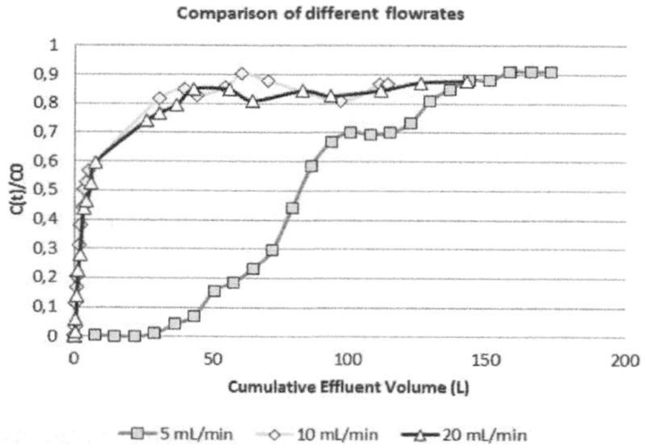

Figure 3. Comparison between ARM 300 and GEH 102.

- In both cases, the saturation of the uptake capacity was not reached and the volume of water that passed through the column was insufficient.
- An extrapolation method was developed to estimate the uptake capacities of both sorbents, based on the sigmoidal shape of the loading curve, and using a normalized logistic equation with two parameters to calculate the mass of As captured by the sorbent. The following results were obtained: 20.6 g of As/kg sorbent for GEH 102 and 20.8 for ARM 300. In a previous study, Ipsen (2005) found an uptake capacity of 36 g As/kg for the GEH sorbent.
- In a potential field application, it would be required to build two permeable reactive columns, with a reasonable distance between them, in a way that the second barrier would allow the attainment of environmentally acceptable As concentrations.
- Alternatively, an excess of at least 100% of the sorbent in relation to the necessities estimated by the uptake capacity would be required.

3.3 *Study of the sorption mechanisms using an Electron Microprobe Analyzer*

In order to better understand the mechanisms of sorption, samples of two sorbents were collected after their usage in fixed bed columns tests, and were analyzed using an Electron Microprobe Analyzer.

The samples were impregnated in resin and polished with diamond paste for examination and analysis by a Field Emission Electron Microprobe (EPMA), model Jeol JXA-8500F. Electron backscattered images were obtained from the sorbent particles.

The observation of Figure 1 allows to extract the following conclusions:

- ARM 300: the deposition of As was predominantly occurring at the lower face of the particle, indicating that it is a superficial reaction. Although the sorbent particle has several fractures, As has not penetrated into the inner part of the particle through those fractures. It can be concluded that the reaction occurs only at the exposed surface in contact with the solution.
- GEH 102: As is captured all around the external surface of the sorbent particle constituting a layer with variable thickness.
- The increase in As concentrations coincides with a net depletion in O_2.
- The same effect takes place with iron, although the thickness of the depletion layer is thinner.

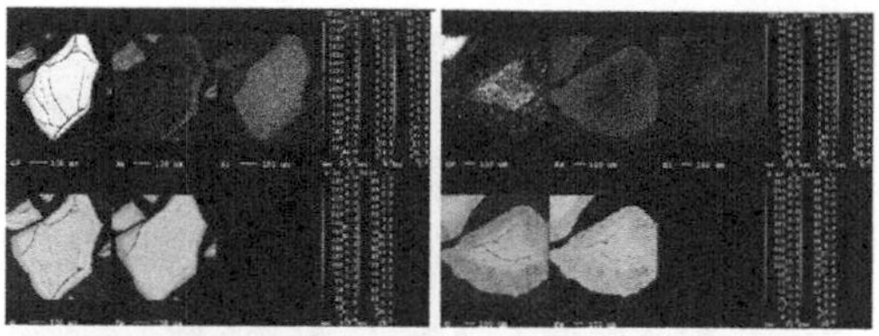

Figure 4. Mapping of concentrations in a particle of ARM 300 (left) and a particle of GEH 102 (right).

Quantitative analysis at selected points and Wavelength-Dispersive X-Ray Spectroscopy (WDS) intensity dot maps were obtained with the following operating conditions: 15 kV, 60 nA and 1 μm beam diameter. A dwell time of 10 ms was used for X-ray dot maps.

The observations in Figure 5 and Table 1 led to the following conclusions:

- ARM 300: the concentration of adsorbed As is higher at the surface of the sorbent, although As has apparently penetrated to the inner part of the particle, even if the particle has a large size (scale is indicated at the picture). Nevertheless, concentrations decrease from the outer to the inner side of the particle. Notice that the highest As concentration coincides with the lowest Fe concentration, suggesting that part of Fe is reduced to a soluble Fe(II) form. A reductive precipitation is possible, and probably coexists with other removal mechanisms.
- GEH 102: there is no As in the inner part of the particle; the highest As concentration is found at point 10 and belongs to the external sorption layer.
- In the sorption layer located at the opposite side of the particle (points 6 and 7), concentrations are relatively homogeneous but lower than at the opposite face. This suggests that chemisorption prevails over the other possible precipitation mechanisms.

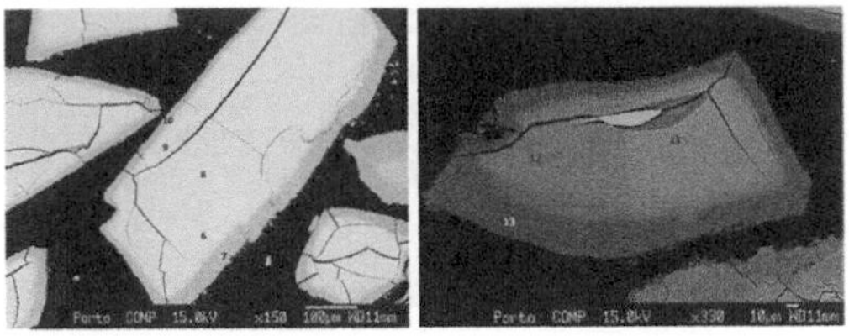

Figure 5. Points selected for chemical analysis in an ARM 300 particle (right) and GEH 102 particle (left).

Table 1. Spatial distribution of concentrations in a particle of GEH 102 and a particle of ARM 300 (percentage).

	ARM 300			GEH 102			
Element	11	12	13	6	7	9	10
As_2O_5	2.82	3.07	3.59	1.08	0.97	1.92	2.47
FeO	64.51	64.91	54.84	64.08	51.01	67.16	56.78
SiO_2	1.38	1.14	0.68	1.80	0.99	1.52	0.84

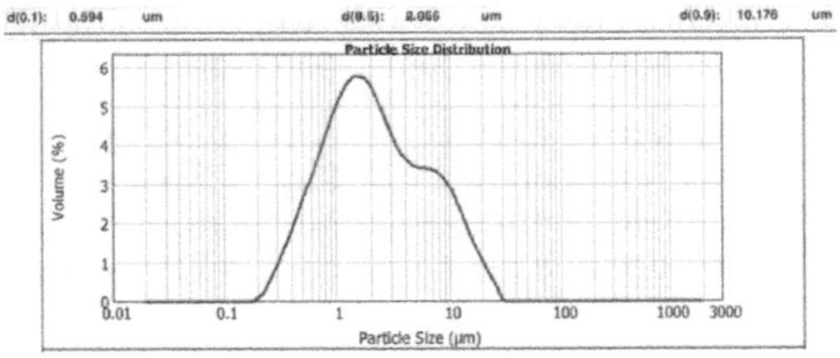

Figure 6. Particle size distribution of nZVI.

4 RESULTS AND DISCUSSION FOR ZEROVALENT IRON NANOPARTICLES

In order to compare the performance of IBS with nZVI, we performed some tests with a suspension of this material, manufactured by Toda Kogyo Corporation (Japan).

Figure 6 shows the particle size distribution of the nZVI used in the experiments.

Kinetics and tests were performed using flasks with 500 mL of an 1.1 mg L^{-1} As solution adding different nZVI masses: 0.1, 0.2 and 0.55 g. The flasks were stirred in an orbital shaker for 48 h. During this time, 5 samples of 10 mL were collected from each flask and preserved by adding 100 µL of 37% HCl.

The observation of Figure 7 led to extract several conclusions:

- As concentration decreases rapidly up to 4 h and then the sorption decreases.
- Test with 1 mL of nZVI: after 24 h, the As concentration stabilized at 270 µg L^{-1}, suggesting that saturation of the uptake capacity was reached.
- Tests with 2 and 5 ml of nZVI: despite the occurrence of a drastic decrease in the sorption, in both cases the saturation of the uptake capacity was not reached.

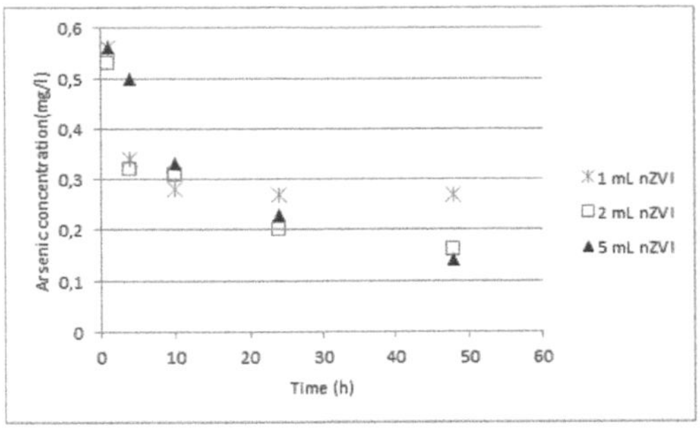

Figure 7. Time evolution of total As concentrations using different nZVI masses.

5 CONCLUSIONS

This research proves that IBS may be used as reactive media in PRB. The two commercial sorbents tested, ARM 300 and GEH 102, remove efficiently both As(III), As(V) and their mixtures. Breakthrough curves obtained in different operating conditions allowed comparison of both sorbents: the loading capacity is similar, around 20 g of As kg^{-1}, and GEH has a slightly more favorable performance. Particles of the sorbents, after submitted to adsorption tests in fixed bed columns during 30 days, were analyzed using an electron microprobe analyzer. Spatial mapping of the concentrations allows to infer that the removal mechanisms are different: while for ARM 300 reductive precipitation is relevant, for GEH 102, the adsorption mechanisms with complex formation are prevalent.

Preliminary tests with nZVI show that this sorbent has a faster kinetics, mainly due to its high specific surface. Comparing both types of materials, nZVI evidences faster kinetics than IBS; nevertheless, the physical implementation of IBS in a permeable barrier is easier and the usage of polymers to fix nVZI probably reduces drastically its loading capacity.

ACKNOWLEDGEMENTS

The authors would like to thank The National Foundation for Science and Technology (Fundação para a Ciência e Tecnologia FCT, Portugal) for the financial support of this research, that is part of the PTDC/ECM/70216 project.

REFERENCES

Ferreira, M.A. & Barros, A.A. 2002. Determination of As(III) and arsenic(V) in natural waters by cathodic stripping voltammetry at a hanging mercury drop electrode. *Anal. Chim. Acta* 459: 151–159.

Ipsen, S.-O., Gerth, J. & Förstner, U. 2005. Identifying and testing materials for arsenic removal by permeable reactive barriers. *Consoil* 2005: 1815–1817.

Silva, A., Freitas, O., Figueiredo, S., Vandervliet, B., Ferreira, A. & Fiúza, A. 2009. Arsenic Removal Using Synthetic Adsorbents: Kinetics, Equilibrium and Column Study. In *12th EuCheMS International Conference on Chemistry and the* Environment. 7 June 2009, Stockholm, Sweden.

One Century of the Discovery of Arsenicosis in Latin America (1914–2014) – Litter, Nicolli, Meichtry, Quici, Bundschuh, Bhattacharya & Naidu (Eds)
© 2014 Taylor & Francis Group, London, ISBN 978-1-138-00141-1

Evaluation of arsenic removal process in water using zerovalent iron oxidation

I.D. Garcés Mendoza, M. Vaca Mier & R. López Callejas
Universidad Autónoma Metropolitana, Azcapotzalco, México

ABSTRACT: Sequential batch experiments were conducted to study the efficiency of removal of arsenic from water by means of an oxidation process using zerovalent iron fillings. Triplicate experiments were done in three sequential steps with acid-arsenic mists recirculation, using a iron-fillings/water relation of 0.4 (w/v), at pH values of 3, 5, and 7 and different concentrations of arsenic (2000 mg L^{-1}, 4000 L^{-1} and 8000 mg L^{-1}). Reactors were run for a total period of 72 hours. Oxidation of iron took place with an air flow of 950 mL min^{-1}. Efficiencies greater than 70% removal of As in water were observed in the first batch step. At concentrations of 2000 and 4000 mg L^{-1}, 99% removal efficiencies were obtained in the second batch reactor. Therefore, it was inferred that a third phase in the sequential reactors could be redundant.

1 INTRODUCTION

The pollution of drinking water caused by arsenic is a very important problem that has won the world's attention. Arsenic has been classified as a carcinogen of group 1 (IARC, 2004). The World Health Organization (WHO) sets 10 µg L^{-1} of arsenic in drinking water as maximum allowed limit. Drinking arsenic-rich water for a long period leads to arsenic poisoning, a disease with high incidence in Asia and Latin America (Litter *et al.*, 2010).

The iron-oxidation process consists in the utilization of zerovalent iron to remove arsenic found in water. This mechanism of elimination of arsenic (III) and (V) is due to the formation of rust products, a mixture of amorphous Fe(III), iron oxide/hydroxide, among others, creating adsorption sites for As (III) and As (V) (Tyrovola *et al.*, 2007; Farrell *et al.*, 2001).

Although other authors have already proven that oxidation of arsenic in the presence of zerovalent iron is an efficient removal method, we aimed to study the use of sequential reactors, in acid solutions containing arsenic, recirculating the formed acid mists which contained part of the arsenic, and determining the efficiency of applying industrial zerovalent Fe shavings.

2 METHODS

The iron shavings from a metalworking industry were used, with particle sizes in the range of 1.0–17.0 mm. The average BET adsorption surface, estimated by means of nitrogen adsorption was 0.83 m^2 g^{-1}. Synthetic polluted water was prepared using sodium arsenite ($NaAsO_2$) (J.T. Baker), sulfuric acid (H_2SO_4) at 98% (Sigma-Aldrich) and deionized water.

Three aerated-sequential batch reactors (800 mL) were set. Fe shavings/water relation was kept at 0.4 (w/v) in each reactor. The air flow was in the range of 950–1000 ml min^{-1}. The reactor reaction time was 24 hours. Three different values of pH (3, 5 and 7) with three different concentrations of arsenic (2000, 4000 and 8000 mg L^{-1}) were tested.

3 RESULTS AND DISCUSSION

3.1 *Effect of pH*

The efficiency of arsenic removal process by oxidation of zerovalent iron was studied as a function of pH and the initial concentration of arsenic.

The oxidation of iron shavings was performed in aqueous medium in the presence of oxygen; OH^-ions were released, and iron(II) and iron(III) were formed; further reaction with water molecules formed iron hydroxide, $Fe(OH)_3$; in the complex process of adsorption and/or precipitation of these oxides, arsenic is removed. The formed OH^- promotes the increment of pH in almost 2 units (Ramaswami *et al.*, 2001; Hug & Leupin, 2003).

3.2 *Arsenic removal efficiency*

Elimination of arsenic using the iron oxidation was efficient for the lowest initial concentration,

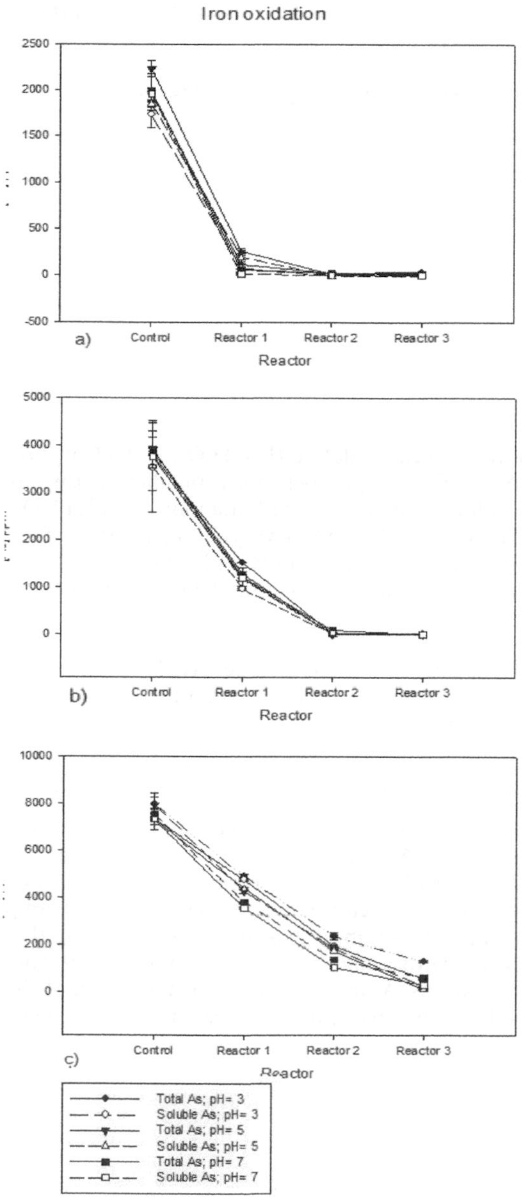

Figure 1. Iron oxidation, a) As 2000 mg L^{-1}, b) As 4000 mg L^{-1}, c) As 8000 mg L^{-1}.

2000 mg L^{-1}, up to 99% in the second stage, while at higher concentrations arsenic removal decreased, the process of oxidation being slightly affected by pH. The best arsenic removals were observed at pH close to 7, for all the tested concentrations.

Results are summarized Figure 1. It is observed that arsenic removal is efficient starting from the second reactor in concentrations up to 4000 mg L^{-1}, but at 8000 mg L^{-1} this efficiency is only 85.6% and the third reactor is required to obtained a significant efficiency.

All results were statistically significant.

4 CONCLUSIONS

The results obtained in this study were found to be optimal for low arsenic concentrations at pH values near neutrality, although the oxidation of iron in the three tests proves efficient, reporting end efficiencies higher than 70%. The pH had a direct influence on the efficiency of the process.

ACKNOWLEDGEMENTS

To CONACyT and the Autonomous Metropolitan University, Azcapotzalco unit for provided the resources and space to perform this work.

REFERENCES

Hug, S.J. & Leupin, O. 2003. Iron-Catalyzed Oxidation of Arsenic(III) by Oxygen and by Hydrogen Peroxide: pH-Dependent Formation of Oxidants in the Fenton Reaction. *Environmental science*, 2734–2742.

IARC 2004. *International Agency for Research on Cancer. World Health Organization.* Recuperado el 28 de Febrero de 2013, de http://monographs.iarc.fr/ENG/Monographs/vol84/volume84.pdf

Litter M.I., Morgada M.E. & Bundschuh J. 2010. Possible treatments for arsenic removal in Latin American waters for human consumption. *Environmental Pollution, 158*, 1105–1118.

Ramaswani, A., Tawachsupa, S. & Isleyen, M. 2001. Batch-mixed iron treatment of high arsenic waters. *Research note*, 4474–4479.

Tyrovola K., Peroulaki E. & Nikolaidis N. P. 2007. Modeling of arsenic immobilization by zero valent iron. *European Journal of Soil Biology, 43*, 356–367.

One Century of the Discovery of Arsenicosis in Latin America (1914–2014) – Litter, Nicolli, Meichtry, Quici, Bundschuh, Bhattacharya & Naidu (Eds)

A combined system for arsenic removal from water by photochemical oxidation and adsorption technology

M.R. Lescano
INTEC (UNL-CONICET), Güemes, Santa Fe, Argentina

C.S. Zalazar & R.J. Brandi
INTEC (UNL-CONICET), Güemes, Santa Fe, Argentina
FICH (UNL), Ciudad Universitaria, Santa Fe, Argentina

ABSTRACT: A combined technology employing photochemical oxidation (UV/H_2O_2) and adsorption process for arsenic removal from water was evaluated. A middle scale photochemical annular reactor was developed being connected alternately to a pair of adsorption columns filled with titanium dioxide (TiO_2) and Granular Ferric Hydroxide (GFH). The experiments were performed by varying the relation of As concentration (As(III)/As(V) ratio) at constant hydrogen peroxide concentration and incident radiation. The combined technology results effective and promising for arsenic removal at small and medium scale.

1 INTRODUCTION

Arsenic pollution is a worldwide problem being one of the most important environmental issues in last years and has been extensively reported by literature (Mohan & Pittman, 2007; Litter *et al.*, 2010). In few publications, a combination of oxidation and adsorption processes for arsenic removal from water was discussed (Nakajima *et al.*, 2005; Yoon & Lee, 2007). An oxidation step is required according to previous publications of our group (Lescano *et al.*, 2010; Lescano *et al.*, 2011) and also because of the important percentage of As(III) present in the treated water. The aim of this work is to evaluate and study the effectiveness of the combination of UV/H_2O_2 process for As(III) oxidation with an adsorption technology, employing titanium dioxide and granular ferric hydroxide as adsorbent materials

2 METHODS/EXPERIMENTAL

2.1 *Chemical and materials*

Arsenic stock solutions containing 1000 mg L^{-1} of As(III) and As(V) were performed employing sodium (meta) arsenite ($AsNaO_2$, ≥ 99%, Sigma-Aldrich p.a.) and sodium dibasic heptahydrated ($Na_2HAsO_4.7H_2O$, 99–102%, Sigma-Aldrich, ACS reagent) respectively. Also, a stock solution of 500 mg L^{-1} of H_2O_2 was prepared from hydrogen peroxide (30% w/v, Ciccarelli p.a.). Two commercially available adsorbents for arsenic removal, based on titanium dioxide (Adsorbsia® As500) and Granular Ferric Hydroxide (GFH), were purchased from Dow Chemical and Pro H_2O companies, respectively.

2.2 *Experimental device and operation conditions*

A photograph of the combined system is shown in Figure 1. Arsenic contaminated water that comes from a 10 L glass tank was pumped to the inlet of an annular reactor with a tubular germicidal lamp (Philips 15 W) in its axis. A centrifugal pump that works with a high flow rate assures good mixing conditions. At the outlet of the reactor, part of the solution was recirculated to the same reactor, while

Figure 1. Combined system for As removal.

the other portion was forced to let into the adsorption columns. The whole system was operated in a continuous mode.

Experimental conditions are detailed in Table 1. As concentration was measured by Atomic Absorption Spectroscopy (ASS) (Perkin Elmer AAnalyst 800 equipment with a TGHA graphite furnace). Prior to the As(III) determination, a speciation process was performed employing Solid Phase Extraction (SPE). H_2O_2 concentration was analyzed by a spectrophotometric method. The spectral fluence rate at the inner wall of the reactor was experimentally measured by ferrioxalate actinometry.

3 RESULTS AND DISCUSSION

3.1 *Photochemical oxidation of As(III)*

Initially, steady state conditions for the oxidation process were studied. A concentration lower than 4 μg L^{-1} and 7 μg L^{-1} could be reached in 20 min (steady state) after the UV/H_2O_2 process for As(III)/As(V) ratios: 25/125–50/150 and 75/125–100/100, respectively. In addition, conversions of hydrogen peroxide varied between 8.47 and 9.3%. Experimental results for As(III) oxidation at the outlet of the annular system reactor-recycle are presented in Table 2.

3.2 *Removal of arsenic by UV/H_2O_2—TiO_2/GFH system*

As(V) concentration was measured at the outlet of the adsorption column, and the results indicate a complete removal of the contaminant for every ratio studied for both TiO_2 and GFH (As(V) concentration lower than 4 μg L^{-1}). H_2O_2 concentration was also measured at the outlet of the adsorption columns. The presence of H_2O_2 was not detected at the outlet of both columns, indicating the adsorption of hydrogen peroxide onto the materials. This finding adds an advantage to the designed system because it removes in one step arsenic and also, the remaining H_2O_2.

Table 1. Experimental Conditions.

Variable and units	Value
Arsenic inlet Conc. (μg L^{-1})	200
As(III)/As(V) ratio	25/175 50/50 75/125 100/100
Hydrogen peroxide inlet conc. (mg L^{-1})	3
Spectral fluence rate at the inner wall of the reactor (einstein cm^{-2} s^{-1} × 10^8)	2.17
Initial pH	5.19
Recirculation flow rate (cm^3 s^{-1})	133
Flow rate at the outlet of the column (cm^3 s^{-1})	6
Column/Reactor high (cm)	41
Column/Reactor volume (cm^3)	870
Residence time column (min)	2.42
Reaction Time (min)	20

Table 2. Experimental data for the As(III) oxidation reaction in the annular reactor.

	As(III)	H_2O_2
$C_{As(III)}/C_{As(V)}$	X_{Exp}	X_{Exp}
50/150	93.01	9.30
25/175	93.56	9.84
75/125	90.05	8.95
100/100	91.26	8.47

3.3 *Groundwater assays*

In order to evaluate the feasibility of the combined system, the process was carried out employing a real groundwater sample. The sample was spiked with 150 and 50 μg L^{-1} of As(V) and As(III), respectively. Experimental conditions were: H_2O_2 concentration = 3 mg L^{-1}, spectral fluence = 2.17×10^{-8} einstein cm^{-2} s^{-1}, adsorbent: TiO_2. Conversions for 11 min were compared with the results previously obtained employing ultrapure water. As(III) conversion obtained at the outlet of the reactor for groundwater sample and ultrapure water sample were 60 and 93%, respectively. Some components in the groundwater sample probably affect the activity of hydroxyl radicals generated by H_2O_2 photolysis, resulting in a decrease of the oxidation rate (especially because of the presence of carbonates and bicarbonate ions evidenced by the high alkalinity (762 mg L^{-1})). The concentration of As(V) was also measured at the outlet of the adsorption column, and a complete removal of the contaminant was also achieved. This result shows that in spite that As(III) conversion at the outlet of the reactor is not complete, the lowest concentration obtained (18 μg L^{-1}) could be easily adsorbed by the solid adsorbent. Therefore, the steady state, for treating real waters, should be determined taking into account the particular physicochemical composition in each case.

4 CONCLUSIONS

A middle scale photochemical reactor for oxidation process and an adsorption column filled with the adsorbents TiO_2 and GFH were designed. From the obtained results, it can be concluded that the combined process UV/H_2O_2 – adsorption onto TiO_2/GFH is a feasible and efficient technique for complete arsenic removal from water with high concentration of this contaminant at small and middle scale.

ACKNOWLEDGEMENTS

The authors are grateful to UNL, CONICET and ANPCyT, Argentina, for financial support.

REFERENCES

Lescano, M., Zalazar, C., Cassano, A. & Brandi, R. 2011. Arsenic(III) oxidation of water applying a combination of hydrogen peroxide and UVC radiation. *Photochem. Photobiol. Sci.* 19: 1797–1803.

Lescano, M., Zalazar, C., Cassano, A. & Brandi, R. 2012. Kinetic modelling of arsenic (III) oxidation in water employing the UV/H_2O_2 process, *Chem. Eng.* J. 211–212: 360–368.

Litter, M., Morgada, M. & Bundschuh, J. 2010. Possible treatments for arsenic removal in Latin American waters for human consumption, *Environ. Poll.* 158(5), 1105–1118.

Mohan, D. & Pittman Jr., C. 2007. Arsenic removal from water/wastewater using adsorbents-a critical review, *J. Hazard. Mater*. 142: 1–53.

Nakajima, T., Xu, Y., Mori, Y., Kishita, M., Takanashi, H., Maeda, S. & Ohki, A. 2005. Combined use of photocatalyst and adsorbent for the removal of inorganic arsenic (III) and organoarsenic compounds from aqueous media, *J. Hazard. Mater* B120, 75–80.

Yoon, S. & Lee, J. 2007. Combined use of photochemical reaction and activated alumina for the oxidation and removal of As(III). *J. Ind. Eng. Chem.* 13(1): 97–104.

One Century of the Discovery of Arsenicosis in Latin America (1914–2014) – Litter, Nicolli, Meichtry, Quici, Bundschuh, Bhattacharya & Naidu (Eds)
© 2014 Taylor & Francis Group, London, ISBN 978-1-138-00141-1

Drinking water dearsenification with granular ferric hydroxide: Comparison of field experience with waterworks units in Germany and POU units in Argentina

C. Bahr
GEH Wasserchemie GmbH & Co. KG, Osnabrück, Germany

W. Ewy
ProH2O S.A., Buenos Aires, Argentina

ABSTRACT: Fixed-bed adsorption has proven to be the simplest and most reliable dearsenification process for groundwater in drinking water production. Synthetic granular ferric hydroxide has a highly successful track record as an arsenic adsorbent both in large-scale units in water treatment plants as well as in compact point-of-use (POU) units. Adsorption capacities attained in practice are highly dependent on the composition of the raw water processed, in particular its pH level and concentrations of competitive anions. While adsorber beds in German waterworks routinely attain lifetime throughput capacities of up to 300,000 bed volumes, lifetime throughput capacities attained with POU treatment units in use in Argentina are in the order of roughly 4,000–12,000 bed volumes due to the less favorable groundwater composition found in these regions.

1 INTRODUCTION

Dearsenification of groundwater is the most urgent drinking water treatment problem facing many regions worldwide. As compared with various other methods used for dearsenification, such as flocculation, ion exchange and reverse osmosis, adsorber beds have proven to be the most reliable and easiest method to implement from a technical standpoint. Granular Ferric Hydroxide (abbreviated in this presentation as "GFH") was developed in Germany in the 1990's at Technische Universität Berlin, providing the world's first ferric hydroxide-based granular adsorbent for drinking water treatment (Driehaus *et al.*, 1998). GFH has official certification for use in drinking water treatment (e.g. ANSI/NSF Standard 61) in many countries and has been used successfully for many years worldwide.

2 TREATMENT AGENTS/METHODS

2.1 *Granular ferric hydroxide GFH*

GFH is a synthetically manufactured ferric hydroxyde composed of akaganeite (β-FeOOH) with a ferrihydrite ($Fe(OH)_3$) component. The granular adsorbent has a particle size distribution ranging from 0.2 to 2.0 mm and a high specific surface area for adsorption, approx. 300 $m^2\ g^{-1}$, as determined by the BET method (Saha *et al.*, 2005). GFH complies with all quality requirements of European Standard EN DIN 15029.

2.2 *GFH treatment units in waterworks*

In regions with centralized drinking water treatment and supply, GFH is used in adsorber beds housed in large-scale plastic or stainless steel pressure vessels. These beds generally process water throughputs in the range of 10 to 100 $m^3\ h^{-1}$, for the most part in continuous operation (Driehaus, 2002). GFH adsorber beds are normally dimensioned for operation at a water flow speed of 10–15 $m^3\ h^{-1}$ with an Empty Bed Contact Time (EBCT) of 3–5 min. Three typical water treatment units located in different regions in Germany (waterworks "B" in Hessia, "N" in Bavaria and "S" in Thuringia) were studied in detail with regard to their arsenic breakthrough behavior. The relevant analysis data of the raw water treated at each site is summarized in Figure 1.

2.3 *POU units in Argentina*

The 10 $\mu g\ L^{-1}$ arsenic limit specified for drinking water by the WHO Guidelines is not complied with in all areas in Argentina to date. As a result, great demand has arisen from end users in highly

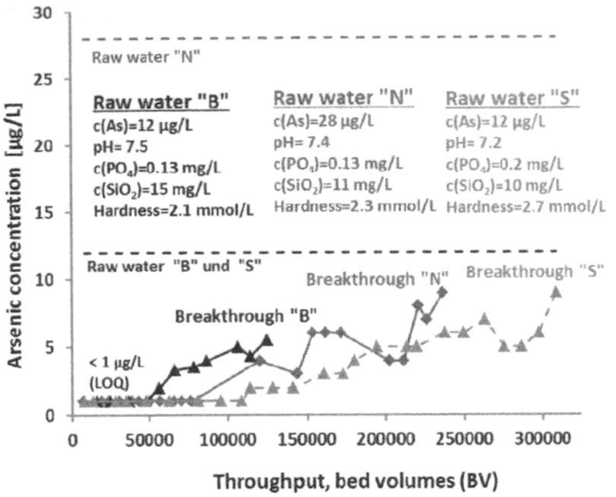

Figure 1. Arsenic breakthrough plots for GFH adsorber beds in three typical waterworks in Germany.

contaminated regions (e.g. La Pampa province) for point-of-use (POU) drinking water treatment units. Unlike the large-scale GFH adsorber beds used in centralized water works, POU units provide lower water throughput rates and are operated only intermittently in accordance with the users' demand. POU units are self-operated by end users and therefore simplicity of operation and minimum maintenance are prime requirements. In recent years, roughly 400 POU units consisting of one or more GFH adsorber beds and a downstream activated-carbon filter bed (average pore size: 5 µm) have been installed in Argentina alone. The 6.4 cm diameter adsorber beds hold approx. 550 g GFH and are normally operated at a throughput of 1 L min^{-1}. These POU treatment units are inspected on site at regular intervals, including determination of outlet arsenic concentration by commercially available rapid testing systems.

3 RESULTS AND DISCUSSION

3.1 *Field experience with large-scale treatment units*

Evaluation of data gathered at the three different waterworks in Germany shows that GFH adsorbers provide lifetime throughput capacities (lifetime = operating life before outlet arsenic concentration reaches 10 µg L^{-1} limit) of up to 300,000 bed volumes (Figure 1). In field practice, this translates to an adsorbent replacement interval of 3–4 years.

In general, the adsorption capacity of GFH beds is highly dependent on the chemistry of the inlet water treated, e.g. its arsenic concentration and pH level. Other factors have also been identified as adsorption inhibitors, primarily phosphate and silicate concentration as confirmed by literature sources (Sperlich & Werner, 2005; Guan *et al.*, 2008). The only maintenance work required with GFH adsorbers in waterworks is occasional backwashing of the bed. This is required in the event that suspended particulates from the inlet raw water accumulate in the adsorbent, causing increased pressure drop across the bed. The spent GFH has an arsenic content of roughly 2–10 g kg^{-1}. Leaching tests (e.g. TCLP from the United States EPA) show that the arsenic is securely bonded with the GFH and is not elutable under normal conditions. As a result, exhausted GFH can be disposed of as common municipal refuse.

3.2 *Field experience with POU treatment units*

POU units in Argentina are primarily used in residences; however, a number of systems are also in use in schools and other public institutions. The units are dimensioned to provide water supply for a family of 4 with a combined consumption of roughly 10 liters of treated water daily. Depending on the arsenic content of the water treated, the service lifetimes attained in field practice are in the order of 0.5–1.5 years, corresponding to a lifetime throughput capacity of 4,000–12,000 bed volumes. These capacity figures, much lower than those achieved with the larger-sized beds described in Section 3.1 above, are attributable to the relatively high arsenic concentrations and high pH values of the treated groundwater. In addition, the POU units, due to their compact design, have lower Empty Bed Contact Times (EBCT < 1 min). Table 1 provides an overview of the POU units and their typical operating conditions—including bed replacement interval—in four different regions in Argentina.

Table 1. Overview of the installed POU systems in Argentina.

City/ Province	P.O.U. units installed	Arsenic in water [µg L^{-1}]	pH in water [–]	Replacement interval
Junín/ Buenos Aires	200	80–120	8.2	2× yearly
Suipacha/ Buenos Aires	100	100	8,0	2× yearly
Villa María/ Córdoba	20	25	7.9	yearly
Chivilcoy/ Buenos Aires	10	50	7.8	yearly

In groundwater regions with extremely high arsenic concentrations, two or more GFH adsorbers can be used in series to enhance the treatment effectiveness (lead/lag configuration). Regular backwashing of POU units is normally not required.

4 CONCLUSIONS

Field experience in recent years shows that GFH adsorbers realize highly efficient water dearsenification both in waterworks and in POU units, providing a reliable water treatment technology requiring only minimum maintenance.

REFERENCES

Driehaus, W., Jekel, M. & Hildebrandt, U. 1998. Granular ferric hydroxide - a new adsorbent for the removal of arsenic from natural water. *Journal of Water Supply: Research and Technology - AQUA* 47(1): 30–35.

Driehaus, W. 2002. Arsenic removal - Experience with the GEH process in Germany. *Water Science and Technology: Water Supply* 2(2): 275–280.

Guan, X.-H., Wang, J. & Chusuei, C.C. 2008. Removal of arsenic from water using granular ferric hydroxide: Macroscopic and microscopic studies. *Journal of Hazardous Materials* 156 (1–3): 178–185.

Saha, B., Bains, R. & Greenwood, F. 2005. Physicochemical characterization of granular ferric hydroxide (GFH) for arsenic(V) sorption from water. *Separation Science and Technology* 40(14): 2909–2932.

Sperlich, A. & Werner, A. 2005. Breakthrough behaviour of granular ferric hydroxide (GFH) fixed–bed adsorption filters: Modeling and experimental approaches. *Water Res.* 39: 1190–1198.

One Century of the Discovery of Arsenicosis in Latin America (1914–2014) –
Litter, Nicolli, Meichtry, Quici, Bundschuh, Bhattacharya & Naidu (Eds)
 ISBN 978-1-138-00141-1

Arsenic removal in polluted water using pillared clays

A. Iriel & A. Fernández Cirelli
Instituto de Investigación en Producción Animal (INPA-UBA-CONICET), Facultad de Ciencias Veterinarias, Buenos Aires, Argentina

J.L. Marco-Brown
Instituto de Investigación e Ingeniería Ambiental (3IA), Escuela de Ciencia y Tecnología, UNSAM, San Martín, Argentina

M.A. Trinelli, A.L. Pérez & M. dos Santos Afonso
INQUIMAE/DQIAQF, Facultad de Ciencias Exactas, UBA, Buenos Aires, Argentina

ABSTRACT: Iron oxide pillared clays synthesized by a conventional method was developed as a new material that is capable and useful for retaining As in drinking water. Montmorillonite (Mt) was used as clay mineral starting material. Solid materials were characterized by X-ray Diffraction (XRD), N_2 adsorption/desorption isotherms and the BET area was determined. Dispersions were prepared with solid material and arsenic solutions in a ratio of 8 g L^{-1} and the adsorption capacity was evaluated at several pH values. As(V) was completely removed from aqueous solution at pH 4.0 and 6.5 using the modified clays, while low adsorption was observed using Mt. Adsorption equilibrium was reached at reaction time of one hour approximately.

1 INTRODUCTION

Arsenic (As) is an element widely spread in Argentina, which has its origin in the Andes volcanic activity. The distribution region includes the provinces of Córdoba, La Pampa, Santiago del Estero, San Luis, Santa Fe, Buenos Aires, Chaco, Salta and Tucumán. Its high toxicity is responsible for a disease caused by ingestion of arsenic over long periods of time, known as Chronic Regional Hydroarsenicism Endemic (HACRE, McClintock *et al.*, 2012).

There are emerging and innovative technologies suitable to be employed at small scale in rural and urban areas (Bundschuh *et al.*, 2010). Pillared clays (PILCs) are interesting materials to be used as adsorbents, due to the abundance and low cost of clays that are used as raw material, as well as to the fact of being tunable in terms of properties. PILCs exhibit multi-charged centers, large area, high interlayer space and thermal stability. PILCs have been previously used in studies of adsorption of organic compounds and toxic elements (Lenoble *et al.*, 2002; Marco-Brown *et al.*, 2012).

In view of this background, the aim of the present work is to evaluate the use of modified clays to remove As(V) from drinking water.

2 METHODS/EXPERIMENTAL

2.1 *Preparation and characterization of pillared clays*

The pillared clays were prepared following a method previously reported (Marco-Brown *et al.*, 2012). The pillaring solution was prepared by dissolution of $Fe(NO_3)_3.9H_2O$ (Merck) in a 2 M KOH solution, until reaching a ratio of $OH^-/Fe = 2$. Afterwards, the solution was kept at room temperature for 12 h with continuous stirring for aging.

Furthermore, the natural clay mineral, a montmorillonite (Mt), previously milled, was added to the pillaring solution in order to obtain a Fe/Mt ratio of 6 mmol Fe/g Mt. The dispersion was allowed to react, stirring for 24 h. Later on, the solid was separated by centrifugation and washed by deionized water through successive redispersions/centrifugation cycles until a conductivity value lower than 10 μS was attained. Then, the product was air-dried, and separated in two portions. One of them was kept and denoted as FeMt. The other portion was heated at 300 °C for 2 h and denoted as CFeMt.

The X-ray diffraction patterns (reflection (001), counting time of 10 s/step collected from 2 to

12° (2θ)) were obtained using a Philips X'Pert diffractometer with CuKα radiation source ($\lambda K\alpha 1$ = 1.54060 Å). Samples were X-rayed as semioriented.

Nitrogen adsorption-desorption isotherms were recorded at 77 K using a Micromeritics ASAP 2010 instrument. All samples were degassed at 150 °C prior to analysis.

2.2 *Adsorption experiments*

Experiments were carried out with a sorbent concentration of 8 g solid L^{-1}. Each solid was mixed with As(V) (Merck) solutions of concentration ranging from 50 to 160 μg L^{-1} using an orbital shaker. Ionic strength was constant at 1 mM KNO_3. Arsenic removal by different solid materials were determined as a function of pH between 3 and 8 adding 0.1 M HNO_3 or KOH drops until equilibrium was reached. For kinetic studies, aliquot of samples were obtained at regular time intervals and filtered through membranes (pore size of 0.45 μm). The adsorption capacity of material can be obtained according eq. 1:

$$Q = (C_i - C_f)V/(1000W) \quad (1)$$

where Q is the adsorption capacity (mg g^{-1}); C_i and C_f are the initial and final concentrations of As (V) (mg L^{-1}), W is the solid mass (g) and V is the aliquot volume.

2.3 *Arsenic analysis*

Arsenic analysis was carried out using an ICP-OES (Perking Elmer, Optima 2000). All measurements were performed at 193.7 nm using an external calibration. Reference material from Lake Ontario water was provided by Certified Reference Materials & Quality Assurance.

3 RESULTS AND DISCUSSION

3.1 *Characterization of pillared clays*

Incorporation of Fe oxide in the montmorillonite interlayer resulted in the disappearance of the (001) reflection indicating that pillaring produces mainly a delamination of the raw clay mineral (inset in Figure 1). Similar results were previously reported (Marco-Brown *et al.*, 2012). All materials presented a type II nitrogen adsorption-desorption isotherms with a type H3 hysteresis associated with low porosity materials formed by sheet agglomerations (Figure 1). The specific surface areas obtained by BET method for Mt, FeMt and CFeMt were 24, 107 and 83 m^2/g respectively.

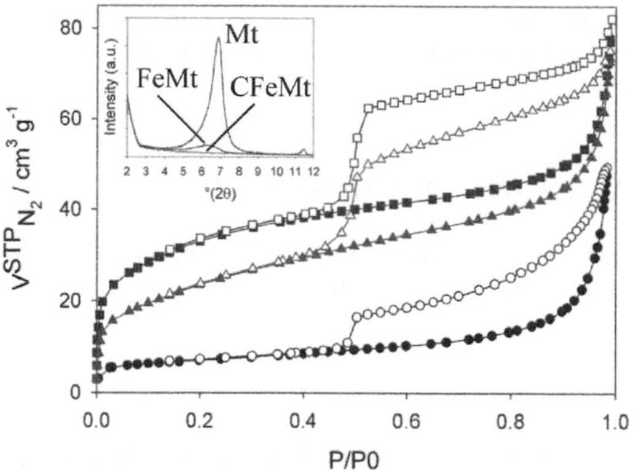

Figure 1. N_2 adsorption/desorption isotherms of Mt (•), FeMt (■) and CFeMt (▲). Inset: XRD of indicated samples.

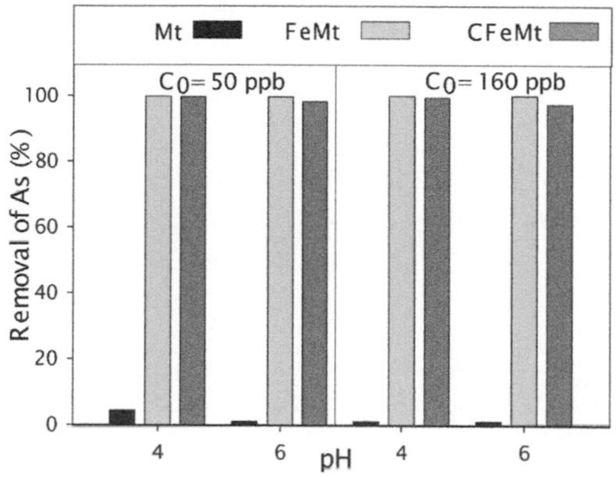

Figure 2. Arsenic removal at pH 4 and 6.5.

3.2 *Effect of pH*

The pH value is one of the most important parameters controlling the adsorption process of ions. Usually, As removal increases with decreasing pH. However, a neutral pH is preferred due to the applicability of the method. In this work, adsorption experiments were carried out in Mt, FeMt and CFeMt adjusting the pH at 4 and 6.5. In all cases, equilibrium concentration was reached in the first hour. Results are showed in Figure 2. Mt is not able to remove As(V) while FeMt and CFeMt remove around 99% of the total arsenic (Figure 2). The adsorption capacity of FeMt was higher than 19 μg g^{-1}.

4 CONCLUSIONS

FeMt and CFeMt showed an extraordinary removal power of As(V) from contaminated water at neutral pH. Further analysis is required in order to determine desorbed process and material effectiveness.

ACKNOWLEDGEMENTS

To University of Buenos Aires and Consejo Nacional de Investigaciones Científicas y Tecnológicas.

REFERENCES

Bundschuh, J., Litter, M.I., Parvez, F., Román-Ross, G., Nicolli, H.B., Jean, J., Liu, C., López, D., Armienta, M.A. Guilherme, L.R.G., Gomez Cuevas, A., Cornejo, L., Cumbal, L. and Toujaguez, R. 2012. One century of arsenic exposure in Latin America: A review of history and occurrence from 14 countries, *Science of Total Environment* 429: 2–35.

Lenoble, V., Bouras, O, Deluchat, V., Serpaud, B. & Bollinger, J-C. 2002. Arsenic Adsorption onto Pillared Clays and Iron Oxides, *Journal of Colloid and Interface Science* 255: 52–58.

McClintock, T.R., Chen, Y., Bundschuh, J. Oliver, J.T. Navoni, J., Olmos, V. Villaamil Lepori, E. Ahsan, H., Parvez, F. 2012. Arsenic exposure in Latin America: Biomarkers, risk assessments and related health effects, *Science of Total Environment* 429: 76–91.

Marco-Brown, J.L., Barbosa-Lema, C.M., Torres Sánchez, R.M., Mercader, R.C. & dos Santos Afonso, M. 2012. Adsorption of picloram herbicide on iron oxide pillared montmorillonite. *Applied Clay Science* 58: 25–33.

One Century of the Discovery of Arsenicosis in Latin America (1914–2014) – Litter, Nicolli, Meichtry, Quici, Bundschuh, Bhattacharya & Naidu (Eds)
© 2014 Taylor & Francis Group, London, ISBN 978-1-138-00141-1

Arsenic reduction levels in drinking water using granular ferric hydroxide oxide

O.J. Graieb & J.C. Lujan
Universidad Tecnológica Nacional/CEDIA, Tucumán, Argentina

ABSTRACT: This paper aims at testing the arsenic removal efficiency in drinking water using granular ferric hydroxide oxide in a pilot trial using a mobile plant in three different sources from the towns of Romera Pozo and Agua Azul (Tucumán province, Argentina) with varied contents of initial arsenic. The results of the carried out studies were adjusted when the effluent reached a level of 10 μg L^{-1}. A great capacity treatment for certain types of water, where the pH and silica content are key determinants for treated water matrixes, was observed.

1 INTRODUCTION

In the eastern part of the province of Tucumán (Argentina), the Regional Chronic Endemic Hydroarsenicism (HACRE) is mainly due to the use of drinking water drilled from groundwater wells (Pal, 2001). The depth of these wells varies between 2 and 20 meters. The drinking water from the first ground wells has high contents of As(V) and As(III), with prevalence of As(V). In Ranchillos and influence zones, the values range from 20 to 700 μg L^{-1}, with an average of 279 μg L^{-1} in both arsenic species, As(III) and As(V) (García *et al.*, 2006).

pH varies between 6.8 and 8.6 with a mean of 7.8. According to (Viapiano, 1996), most of the water wells in the departments of Cruz Alta, Leales and Graneros have As contents above 10 μg L^{-1}, with peak values of 1000–1200 μg L^{-1} (Fernandez and Graieb, 2006), determining average values of 500 μg L^{-1} for the same region.

The presence of arsenic in concentrations that exceed the limits established by the Argentine Food Code (CAA, 2013) in drinking water would affect the population that drinks water from wells of the first ground in the whole eastern rural area and some deep water in the South West of the Province of Tucumán (Nicolli *et al.*, 2007).

2 METHODS/EXPERIMENTAL

2.1 *Choosing a treatment method*

Studies indicate that with conventional methods only As(V) can be efficiently removed from water. If As(III) is present in the source water, previous oxidation is essential in the treatment methods (Lujan, 2001). Arsenic removal with ferric salts, followed by coagulation and filtration are known techniques that allow to use relatively low doses of ferric salts, attaining very low final arsenic concentrations. Osmosis and nanofiltration can be also satisfactory for the treatment, but with high rejection of salt waters and high cost for the user. Arsenic adsorption with activated alumina results in a simple and effective method to reduce the arsenic content (Luján, 2001; Graieb *et al.*, 2006). In the process of coagulation-filtration, ferric salts show better removal efficiency than aluminum salts used in equal dose (Pierce & Moore, 1982). Therefore, the use of a granular activated iron oxide or ferric hydroxide should have a great capacity for arsenic adsorption from water using a fixed bed. In all arsenic removal tests, a stream of water was passed through the bed and the presence of arsenic up to 10 μg L^{-1} was evaluated.

2.2 *Granular iron hydroxide*

According to Pal (2011), the granular iron hydroxide oxide has a high porosity (75%) and a specific surface area of approx. 300 m^2/g. Arsenic (III and V) is adsorbed on a wide range of pH (7 to 8.5), competing with phosphate. Also, iron, manganese and silica contents have a strong influence on the removal efficiency. The residual product from the treatment process is a small amount of compact grain that can be securely handled by authorized companies of hazardous waste disposal.

2.3 *Matrix water to be treated*

Arsenic in groundwater before any treatment may be accompanied by chemical species that may hinder the removal effectiveness. Taking into account the previous knowledge of the composition of the drinking water matrix from underground sources (especially the first ground wells mainly used in dispersed households), it will be possible to evaluate the use of the granular iron hydroxide as an accessible low-cost technology.

Our experience was focused on the removal of arsenic from water from Romera Pozo and Agua Azul in the Tucumán province (Argentina) with various initial arsenic contents (65, 110 and 920 μg L^{-1}) and pH between 8 and 8.6. Analysis of phosphates, silica and iron content were emphasized.

3 RESULTS AND DISCUSSION

Arsenic content was analyzed by colorimetry with silver diethyldithiocarbamate. Beds with 465 cm^3 of HHG (granular iron hydroxide) were prepared. The results show that the initial iron content diminishes from between 130–160 μg L^{-1} to less than 20 μg L^{-1}. The conductivity decreases slightly to 10%. The hardness expressed in carbonate increased slightly to a percentage between 5 and 7%. The initial content of silica was significantly reduced by 80%. The initial content of phosphates was not relevant (0.25 μg L^{-1}) and no phosphates were detected in the outlet water. In the plant operation with groundwater with an initial arsenic content of 65 μg L^{-1}, 10 μg L^{-1} were attained in the treated water after 10,000 L of Bed Volumes (BV). For water with an initial content of arsenic of 110 μg L^{-1}, the value was overcome after 8,000 L BV and for water with an initial content of 920 μg L^{-1}, with 1,000 L.

4 CONCLUSIONS

The results showed studies carried out with natural groundwater in a small mobile plant installed in Romera Pozo and Agua Azul in Tucumán, Argentina for three water sources with different initial arsenic concentrations, and pH between 6.8 and 8.6. The content of arsenic reached the value of 10 μg L^{-1} after 10,000 BV for groundwater with an initial arsenic content of 65 μg L^{-1}, 8,000 for water with 110 μg L^{-1} and 1,000 for water with 920 μg L^{-1}.

ACKNOWLEDGEMENTS

We thank to the community of Romera Pozo and Agua Azul. We also thank to Sociedad Aguas de Tucumán (S.A.T.) laboratories, S.I.P.R.O.S.A. and Research and Development Center of Environmental Engineering (CEDIA) UTN—FRT.

REFERENCES

Codigo Alimentario Argentino (CAA) 2013.

Falcón, C.M., Graieb, O.J., García, J.W. & Sayago, J.M. 2006. La distribución del arsénico en los materiales loessicos en la llanura oriental tucumana, factores de su movilidad—*Asauee 2006*.

Fernández, R.I. & Graieb, O.J. 2006. La presencia de arsénico como factor de riesgo geoambiental en la provincia de Tucumán. Argentina—*Asauee 2006*.

Ferrari R.R. & Graieb O.J. 2006. Selección de tecnologías para el tratamiento de aguas contaminadas con arsénico en países en desarrollo—*Asauee 2006*.

García, MG., Sracek, O., Fernández, D.S. & Hidalgo, M.D.V. 2007. Factors affecting arsenic concentration in groundwaters4 from Northwestern Chaco-Pampean Plain, Argentina. *Environmental Geology,* 52(7):, 1261–1275.

Graieb, O.J. & Lujan, J.C. 2006. Química y determinación analítica del arsénico—*Asauee 2006*.

Graieb, O.J., Graieb V.C. & Tintilay S.C. 2006. Tecnología para el abordaje del HACRE en pobladores rurales dispersos—*Asauee 2006*.

Graieb, O.J., Iturre, A.E. & Moreno, P.G. 2006. Intervenciones comunitarias sustentables: estrategias metodológicas para el tratamiento del arsénico en el agua de consumo humano—*Asauee 2006*.

Lujan, J.C. 2001. Un hidrogel de hidróxido de aluminio para eliminar el arsénico del agua. *Revista OPS ISSN 1020–4989*

Nicolli, H., Garcia, W., Falcon, C. & Tineo, A. 2007. Contaminación con arsénico de fuentes de aguas subterráneas en la Cuenca del Rio Salí—Argentina. *Water Rock Interaction—Kunming Yunnan (China.)*

Pal, B.N. 2001. Granular Ferric Hydroxide for Elimination of Arsenic from Drinking Water. *In: Technology for Arsenic Removal From Drinking Water,* M.F. Ahmed, M.A. Ali & Z. Adeel (eds) Dhaka,: pp. 59–68.

Pierce, M.L. & Moore, C.B. 1982. Adsorption of Arsenite and arsenate on amorphous iron hydroxide. *Water Res. 16: 1247–1253*.

Standard Methods 2002, American Public Health Association (APHA), The American Water Works Association (AWWA) and the Water Environment Federation (WEF).

Viapiano, J.S. 1996. Informe breve sobre hidroarsenicismo (HACRE). *Departamento Básico. Documento SIPROSA. Tucumán.*

One Century of the Discovery of Arsenicosis in Latin America (1914–2014) – Litter, Nicolli, Meichtry, Quici, Bundschuh, Bhattacharya & Naidu (Eds)
© 2014 Taylor & Francis Group, London, ISBN 978-1-138-00141-1

Effects of changes in operating conditions on the iron corrosion rate in zerovalent iron columns

E. Berardozzi & J.M. Galindez
Dpto. Hidráulica, Facultad de Ingeniería, UNLP, Argentina

F.S. García Einschlag
Dpto. de Química, Facultad de Ciencias Exactas, Instituto de Investigaciones Fisicoquímica Teóricas y Aplicadas (INIFTA), UNLP, Argentina

ABSTRACT: The Zerovalent Iron (ZVI) technique has proven to be an effective method in arsenic removal from drinking water. Continuous studies using packed columns with iron wool were carried out in order to evaluate the most favorable operating conditions and to estimate the reaction rates of the most important steps involved in the heterogeneous process. The results suggested that the flow rate and the iron amount are important parameters for optimization since they determine the corrosion rates. Adjusting these parameters it is possible to obtain high efficiencies of arsenic removal and long service times of the fixed ZVI-beds.

1 INTRODUCTION

Arsenic is a contaminant widely distributed in groundwater in many regions of Argentina where the concentration levels significantly exceed the values recommended by the WHO (MCL = 10 μg L^{-1}) (Castro de Esparza and Wong, 1999).

Many different technologies have been developed to remove arsenic from water in order to comply with the WHO. Zerovalent Iron (ZVI) technique has gained attention during the last years for the removal of arsenic due to its applicability under different geochemical conditions, operational simplicity, low cost maintenance and minimal environmental impact. In addition, since this methodology is feasible for continuous treatment drinking water, scale-up processes are simplified.

The removal mechanism of arsenic by ZVI involves complex physicochemical process of adsorption and/or coprecipitation, which are closely related to the corrosion products generated by iron oxidation (Biterna *et al.*, 2010; Noubactep and Caré, 2010; Melitas *et al.*, 2002; Triszcz *et al.*, 2009; Kundu and Gupta, 2007). The following reaction scheme depicts the main steps associated with the overall process:

$$Fe(0) + 2\,H_2O \rightarrow Fe(II) + H_2 + 2\,HO^- \qquad (R\,1)$$

$$Fe(0) + H_2O + \tfrac{1}{2}\,O_2 \rightarrow Fe(II) + 2\,HO^- \qquad (R\,2)$$

$$Fe(0) + 2\,Fe(III) \rightarrow 3\,Fe(II) \qquad (R\,3)$$

$$Fe(II) + \tfrac{1}{2}\,H_2O + \tfrac{1}{4}\,O_2 \rightarrow Fe(III) + HO^- \qquad (R\,4)$$

$$Fe(III) + 3\,H_2O \rightarrow Fe(HO)_3 + 3\,H^+ \qquad (R\,5)$$

$$Fe(II) + 2\,H_2O \rightarrow Fe(HO)_2 + 2\,H^+ \qquad (R\,6)$$

$$As(III/V) + {\equiv}FeOH \rightarrow As_{Adsorbed} \qquad (R\,7)^1$$

$$As(III/V) + Fe(II/III) + HO^- \rightarrow As_{Coprecipited} \qquad (R\,8)$$

In the first steps, metallic iron oxidation to Fe(II) takes place through reactions R1 or R2. Then, if dissolved oxygen is present, Fe(II) species are rapidly oxidized to Fe(III) by R4. Finally, since Fe(II) and, especially, Fe(III) species are not soluble at the typical pH values of natural drinking waters (between 5.5 and 9), colloidal phases are rapidly generated (R5–R6). These species, can remove As (R7–R8), depending on the operating conditions.

The purpose of this work is to understand the behavior of the iron species formed during the flow of contaminated water through a column filled with steel wool. The effects of changing the main operational variables on the removal capacity of columns packed with (ZVI) are discussed.

2 METHODS/EXPERIMENTAL

The experimental part of the study was comprised of a series of a small-scale short-duration column

[1] (≡ denotes surface)

laboratory experiments, conducted to evaluate the effect of flow rate and iron loading on iron corrosion rates.

2.1 *Materials*

Tests were run in continuous mode using glass columns of 3.8 cm inner diameter and 25 cm length packed with commercial steel wool (Mapavirulana®). The steel wool was used as received. The arsenic concentration of the contaminated water was 400 μg L^{-1} as As(V). The working solutions were prepared using tap water from La Plata artificially contaminated with $Na_2HAsO_4.7H_2O$.

2.2 *Analytical methods*

With the aim of characterizing the production of iron species two operationally defined fractions were used, namely "total" and "filterable" iron based on physical separation through 0.45 μm syringe filters. The samples were collected over water acidified to pH 4 with HCl. This pH value retards the O_2-mediated oxidation of Fe(II) species and also prevents the dissolution of precipitated species.

The concentrations of Fe(III) and Fe(II) in both fractions were quantified spectrophotometrically, through the complexes formed with KSCN and o-phenantroline, respectively. The absorbance of the colored complex formed with KSCN was recorded at 525 nm and that of the complex formed with o-phenantroline at 510 nm.

2.3 *Experimental conditions*

The operating flow rate and ZVI content were selected as study variables. The columns were designed with upward water flow between 5 and 30 mL/min and iron loadings between 10 and 100 g/L. The experimental runs lasted approximately 5 hours. Oxygen content, pH and flow rate were periodically monitored at the reactor outlet. Samples were taken every half hour from the column effluent and Fe(II), soluble Fe(II), Fe(III) and soluble Fe(III) profiles were obtained.

2.4 *Numerical model*

A multicomponent reactive transport model was set up by means of the computer program CrunchFlow in order to simulate the corrosion of ZVI, the further oxidation of Fe(II) in solution and the precipitation of iron oxides under the different experimental conditions described above (Steefel, 2009). Results obtained in terms of the evolution of the concentration of both the relevant species in solution and the volume fractions of the minerals involved were found to be in good agreement with the experimental measurements.

3 RESULTS AND DISCUSSION

The experimental and theoretical analysis of the profiles obtained for iron species and oxygen concentrations at the column outlet allows the estimation of the rate constants associated with the main physico-chemical processes involved inside the column. The experimental results showed that the two operating variables analyzed are of major importance for the fixed bed lifetime and the efficiency of As removal.

At high water flow rates and/or low iron loads, the reactions that take place along the path of the fixed bed are not fast enough to exhaust the oxygen at the column outlet, favoring the premature formation of oxidized species. Under these conditions, the oxides formed precipitate on the iron surface inside the column thereby reducing both the contact area and the column porosity. The reduction in the area of the interface ZVI/solution results in a decrease of the corrosion rate, whereas the reduction in the column porosity increases the pressure drop through the column that decreases the flow rate over time. Consequently, the overall efficiency is reduced since the ZVI corrosion rate is significantly diminished and clogging of the fixed bed is accelerated.

On the other hand, at low flow rates and/or high iron loading, it is possible to both obtain sustained production of Fe(II) and retard column clogging processes. Under these conditions, the Fe(II) species eluted from the column can be rapidly and easily oxidized by gently bubbling oxygen or even air. Thus, the resulting Fe(III) species precipitate and efficiently remove the As species from the water matrix.

4 CONCLUSIONS

Results of the present study show that it is necessary to adjust the column design and operating conditions in order to achieve a sustained Fe(II) production and a minimized formation of Fe(III) species. This mode of operation allows optimizing the performance of the entire prototype since it ensures the release of the amount of iron required to remove the arsenic and prevents the clogging of the column over time.

ACKNOWLEDGEMENTS

The authors wish to thank the Departamento de Hidráulica of the Facultad de Ingeniería of the Universidad Nacional de La Plata (Argentina) for the financial support.

REFERENCES

Biterna, M., Antonoglou, L., Lazou E. & Voutsa, D. 2010. Arsenite removal from waters by zero valent iron: Batch and column tests. *Chemosphere* 78: 7–12.

Castro de Esparza, M.L. & Wong, M. 1999. Remoción de arsénico a nivel domiciliario. *CEPIS* HDT N° 74.

Kundu, S. & Gupta, A.K. 2007. As(III) removal from aqueous medium in fixed bed using iron oxide-coated cement (IOCC): Experimental and modeling studies. *Chemical Engineering Journal* 129: 123–131.

Melitas, N., Wang, J.P., Conklin, M., O'Day, P. & Farrell, J. 2002. Understanding soluble arsenate removal kinetics by zerovalent iron. *Environ. Sci. Technol.* 36: 2074–2081.

Noubactep, C. & Caré, S. 2010. Dimensioning metallic iron beds for efficient removal. *Chemical Engineering Journal* 163: 454–460 (and references cited there).

Steefel, C.I. 2009. CrunchFlow—software for modeling multicomponent reactive flow and transport. Lawrence Berkeley National Laboratory.

Triszcz, J.M., Porta, A. & García Einschlag, F.S. 2009. Effect of operating conditions on iron corrosion rates in zero-valent iron systems for arsenic removal. *Chemical Engineering Journal* 150: 431–439 (and references cited there).

One Century of the Discovery of Arsenicosis in Latin America (1914–2014) – Litter, Nicolli, Meichtry, Quici, Bundschuh, Bhattacharya & Naidu (Eds)
© 2014 Taylor & Francis Group, London, ISBN 978-1-138-00141-1

Treatment of As contaminated water from mining industry by precipitation of Al-Fe (hydr)oxides

J.W.V. de Mello
Department of Soils, Federal University of Viçosa, Viçosa-MG, Brazil
National Institute of Science and Technology, INCT-Acqua, Belo Horizonte-MG, Brazil

M. Gasparon
School of Earth Sciences, The University of Queensland, Brisbane, Queensland, Australia

J. Silva
EMBRAPA Vegetables, Brasília, Federal District, Brazil

ABSTRACT: Treatment of water contaminated with arsenic is a worldwide problem. Al-Fe (hydr) oxides were precipitated from Al and ferrous sulfates at different Fe:Al ratios in order to remove arsenic from water. The resulting suspensions were aged for three months and As was periodically analyzed in the supernatant during this period. At the end of the aging period, the precipitates were collected and oven dried at 50 °C. TCLP procedure and discrete extractions with BCR sequential extraction solutions were used in order to evaluate the stability of As in the precipitates. Results showed that all treatments were efficient to remove As from water, and the resulting precipitates were considered non-toxic and stable residues under acidic or reducing conditions.

1 INTRODUCTION

Treatment of wastewater is a critical problem faced by mining industry. Effluents generated during ore processing may present high concentration of trace elements including arsenic (As). The problem is even more critical when sulfide-bearing ores are exposed to atmospheric conditions, producing Acid Mine Drainage (AMD) with high polluting potential.

Precipitation of Fe and Al (hydr)oxides have been widely considered in methods to treat wastewater contaminated with As. Even these methods have been proved efficient to remove As from water, there is some concern about the stability of the resulting muds. It is well known that Al hydroxides are less efficient than Fe (hydr)oxides in retaining inorganic pollutants, but Fe is less stable than Al under reducing conditions. Nevertheless, Al substituting Fe enhances the redox stability and the adsorption capacity of Fe (hydr)oxides. Therefore, the purpose of this study was to evaluate the efficiency of the water treatment by coprecipitating Al-Fe (hydr)oxides in the presence of As, as well as the stability of the resulting phases.

2 METHODS/EXPERIMENTAL

Al-Fe (hydr)oxides were synthesized following Schwertmann & Cornell (2000) report, but As was added together with Fe and Al salts, before precipitating Al-Fe (hydr)oxides. Therefore, it is expected As coprecipitation and not only adsorption. Different Fe-Al (hydr)oxides were synthesized from ferrous and Al sulfates, in three Fe:Al ratios (1:0.7; 1:0.3; 1:0). A Merck standard solution (10,000 mg L^{-1}) was added to deionized water in order to obtain three different concentrations of As (5.0; 1.0; 0.2 mg L^{-1}). Then, the ferrous (1 M $FeSO_4.7H_2O$) and Al sulfates (0.5 M $Al_2(SO_4)_3.18H_2O$) were added to the solutions. Precipitation was achieved by adding 5 M KOH solution to pH 11.7, and the suspensions were aged during 3 months. Oxidation of Fe^{2+} and incorporation of Al^{3+} in the Fe (hydr)oxides structure were achieved by daily stirring the suspensions during some minutes. All treatments were conducted in duplicate.

Aliquots of the supernatants solutions were collected 1, 7, 14, 21 and 90 days after precipitation in order to measure As concentrations in equilibrium with precipitates. These analyses were performed by HG-AFS. At the end of the aging period, the precipitates were collected and oven dried at 50 °C. Toxicity Characteristic Leaching Procedure (TCLP) and discrete extractions with Biological Contaminant Removal (BCR) sequential extraction solutions were used in order to evaluate the stability of As in the precipitates. Analyses in these extracts were performed by ICP-OES. Arsenic analyses refer to total contents, as speciation was not performed in this study.

3 RESULTS AND DISCUSSION

Soluble As in equilibrium with precipitates drastically dropped in the first day after precipitation (Figure 1). All treatments were effective to remove arsenic from water, as the concentrations decreased to less than the WHO recommended limit for drinking water. It is worth of note that the detection limit was 0.50 μg L^{-1}, as these analyses were performed by HG-AFS. For the lower initial As concentration (200 μg L^{-1}) that limit was attained in only 24 hours, but at higher initial concentrations (5,000 μg L^{-1}) it was necessary more than a week to reach that threshold for treatments containing more Al. On the other hand, the As concentrations dropped below the detection limit in the treatments containing only Fe in as fast as 24 h, independently of the initial As concentration. These results agree with previous results showing that Al is less effective than Fe to treat As contaminated water.

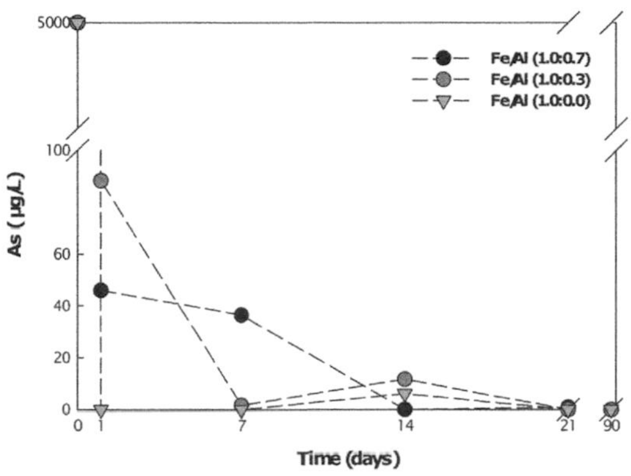

Figure 1. Concentration of soluble As in equilibrium with precipitates during the aging period.

Table 1. Concentrations of arsenic, aluminum and iron by acetic acid solution after TCLP. (Detected by ICP-OES[1]).

Fe:Al Ratio	Initial As	As	Al	Fe
		μg L^{-1}		
1:0.7	5.0	<DL	35.3 (18.9)[2]	56.7 (28.5)
1:0.3	5.0	<DL	22.2 (5.4)	43.7 (14.1)
1:0.0	5.0	<DL	<DL	75.3 (70.4)
1:0.7	1.0	<DL	29.8 (9.8)	63.3 (15.5)
1:0.3	1.0	<DL	32.3 (0.1)	66.1 (9.3)
1:0.0	1.0	<DL	<DL	219.4 (15.3)
1:0.7	0.2	<DL	31.5 (24.3)	69.2 (51.8)
1:0.3	0.2	<DL	44.1 (19.4)	47.9 (6.8)
1:0.0	0.2	<DL	<DL	110.0 (113.1)

[1]Detection Limits: As = 14.5 μg/L; Al = 5.6 μg/L; Fe = 1.4 μg/L; [2]Standard deviations.

Results for TCLP analyzes showed that the leached As was below the ICP-OES detection limit (Table 1) and the materials can be considered "non toxic" independent on the initial As or Fe:Al ratio.

The analysis following the BCR discrete extractions showed As concentrations below the detec-

Table 2. Concentrations of arsenic, aluminum and iron on BCR discrete extractions. (Detected by ICP-OES[1]).

Fe:Al Ratio	Initial As (mg/L)	As	Al	Fe
			mg kg^{-1}	
Acid leaching fraction[3]				
1:0.7	5.0	<DL	264 (16)[2]	24 (1.4)
1:0.3	5.0	<DL	284 (29)	7.6 (2.4)
1:0.0	5.0	<DL	DL	415 (41)
1:0.7	1.0	<DL	257 (5)	2.9 (0.1)
1:0.3	1.0	<DL	330 (12)	8.0 (0.1)
1:0.0	1.0	<DL	DL	300 (37)
1:0.7	0.2	<DL	235 (5)	2.4 (0.9)
1:0.3	0.2	<DL	460 (109)	9.1 (5.7)
1:0.0	0.2	<DL	DL	604 (410)
Reducible fraction[4]				
1:0.7	5.0	<DL	207 (30)	3278 (324)
1:0.3	5.0	<DL	203 (23)	3237 (313)
1:0.0	5.0	<DL	3 (0.56)	9877 (532)
1:0.7	1.0	<DL	183 (2)	3319 (95)
1:0.3	1.0	<DL	220 (29)	2942 (55)
1:0.0	1.0	<DL	3 (0,56)	9878 (190)
1:0.7	0.2	<DL	182 (26)	2851 (28)
1:0.3	0.2	<DL	380 (215)	3260 (381)
1:0.0	0.2	<DL	3 (0.56)	9177 (932)
Oxidizable fraction[5]				
1:0.7	5.0	2.00 (1.06)	<DL	0.36 (0.07)
1:0.3	5.0	<DL	<DL	0.51 (0.20)
1:0.0	5.0	<DL	<DL	0.49 (0.13)
1:0.7	1.0	1.04 (0.04)	<DL	0.43 (0.12)
1:0.3	1.0	<DL	<DL	0.48 (0.04)
1:0.0	1.0	<DL	<DL	0.57 (0.01)
1:0.7	0.2	<DL	<DL	0.47 (0.27)
1:0.3	0.2	<DL	<DL	0.69 (0.07)
1:0.0	0.2	<DL	<DL	0.61 (0.05)

[1] Detection Limits: acid leaching fraction (As = 1.50 mg kg^{-1}; Al = 2.37 mg kg^{-1}; Fe = 0.15 mg kg^{-1}). Easily reducible fraction (As = 0.36 mg/kg; Al = 3.20 mg kg^{-1}; Fe = 4.00 mg kg^{-1}). Reducible fraction (As = 0.38 mg kg^{-1}; Al = 1.92 mg kg^{-1}; Fe = 1.28 mg kg^{-1}). Oxidizable fraction (As = 0.95 mg kg^{-1}; Al = 0.45 mg kg^{-1}; Fe = 0.27 mg kg^{-1}).
[2] Standard deviations.
[3] Extracted by acetic acid 0.11 mol L^{-1}; solid to solution ratio = 1:40.
[4] Extracted by hydroxylammonium chloride 0.5 mol L^{-1} pH 2.0; solid to solution ratio = 1:40.
[5] Extracted by hydrogen peroxide 8.8 mol L^{-1} digestion.

tion limits (Table 2), implying that the residues are stable under acidic or reducing conditions.

4 CONCLUSIONS

1. Fe-Al (hydr)oxides precipitation was efficient to clean up the water contaminated with As, but the lower the Al content the faster was the WHO threshold attained.
2. The resulting precipitates can be considered non-toxic and stable residues under acidic or reducing conditions.

ACKNOWLEDGEMENTS

To the funding Brazilian agencies CNPq, CAPES and FAPEMIG.

REFERENCE

Schwertmann, U. and Cornell, R.M. (2000). Iron oxides in the laboratory: preparation and characterization' Wiley-VCH, Weinheim; Chichester. pp. 727–738.

One Century of the Discovery of Arsenicosis in Latin America (1914–2014) –
Litter, Nicolli, Meichtry, Quici, Bundschuh, Bhattacharya & Naidu (Eds)
 ISBN 978-1-138-00141-1

Final disposal of clay wastes from arsenic removal processes: Studies of leaching and mechanical properties

E.G. De Seta
Departamento de Ingeniería Química, Facultad Regional Buenos Aires, Universidad Tecnológica Nacional, Argentina
Departamento de Ingeniería Civil, Facultad Regional Buenos Aires, Universidad Tecnológica Nacional, Argentina
UDB-Química, Facultad Regional Buenos Aires, Universidad Tecnológica Nacional, Argentina

J.M. Meichtry
Departamento de Ingeniería Química, Facultad Regional Buenos Aires, Universidad Tecnológica Nacional, Argentina

A.R. López
Departamento de Ingeniería Química, Facultad Regional Buenos Aires, Universidad Tecnológica Nacional, Argentina
Departamento de Ingeniería Civil, Facultad Regional Buenos Aires, Universidad Tecnológica Nacional, Argentina

F.D. Reina & F.I. Mugrabi
Departamento de Ingeniería Química, Facultad Regional Buenos Aires, Universidad Tecnológica Nacional, Argentina

L.E. Lan & E. Lavezzari
Departamento de Ingeniería Química, Facultad Regional Buenos Aires, Universidad Tecnológica Nacional, Argentina
UDB-Química, Facultad Regional Buenos Aires, Universidad Tecnológica Nacional, Argentina

E.J. Domingo & N. Tojo
Departamento de Ingeniería Civil, Facultad Regional Buenos Aires, Universidad Tecnológica Nacional, Argentina

ABSTRACT: Leaching and mechanical tests were performed on wastes generated by arsenic removal from water using natural clay. The clay wastes showed negligible As(V) leaching when used for As(V) removal, but rather high As leaching (1.5 to 6.7%, mainly as As(V)) was observed for clay used to remove As(III). Higher As leaching was observed for the calcined samples. In addition, an As-containing clay (9.73 mg g^{-1}) was mixed with cement (25%) and sand (67.5%) to make probes that were submitted to mechanical and leaching tests. The compression and bending mechanical tests showed a decreased performance when compared with the material without clay. The leaching tests showed a good stability of As until a HCl concentration of 0.2 M of higher, where a significant As leaching was observed. This study helps to define possible alternatives for the safe disposal of As-containing solid wastes.

1 INTRODUCTION

Clay is an almost ready-to-use material that has a very low cost and can remove As efficiently from water (Litter *et al.*, 2010; Meichtry *et al.*, 2012). However, few studies have been reported related to the stability of the solid and semisolid wastes generated by clay treatment of As containing water. In this work, the stability of the As in clay is studied by leaching tests. In addition, an alternative disposal of As-containing clay by using it in construction, mixed with cement and sand, is also evaluated.

2 METHODS/EXPERIMENTAL

2.1 *Materials*

The natural clay used was from Misiones province (Argentina); its chemical composition was previously described (Meichtry *et al.*, 2012). Silica sand was oriental type, appropriate for using in sand filters. Normal portland cement (CP40) was used to make the probes. All other reagents used were of analytical grade. Distilled water was used in the leaching experiments and in the analytical determinations.

2.2 *Analytical determinations*

As(V) and As(III) determination was carried out by the spectrophotometric method established by Lenoble *et al.* (2003), using quartz cells of 1 cm of pathlength. The detection limit was 0.05 mg L^{-1}. pH was measured with a digital pH meter.

2.3 *Experimental setup*

The cement probes were prepared according to the IRAM 1622 norm. The composition was: 25% cement, 67.5% sand and 7.5% clay (As probe); control probes (T) were also prepared (25% cement, 75% sand). The clay used contained a 50:50 mixture of As(III):As(V) (9.73 mg g^{-1}, total As concentration in the clay); the concentration of total As in the As probe was 730 mg kg^{-1}.

The mixture was poured into steel molds of $4 \times 4 \times 16$ cm, and compacted according to IRAM 1570 norm. The vertical movement of the mold falling from 12.7 mm was conducted 40 times, removing the entrained air while mixing the mortar. The probes were demoulded after 72 hours and aged for 28 days before the tests.

Leaching experiments were done by adaptations of EPA TCLP 1311 (1992) and LSP 1313 (2012) methods using an orbital shaker (Fercea) at 100 rpm and at constant temperature (16–20 °C). Mechanical tests of the cement probes were done according to the IRAM 1622 norm for portland cement strength determination.

3 RESULTS AND DISCUSSION

3.1 *As leaching from clay*

As-containing clay wastes were obtained from the treatment of aqueous As solutions at pH 7. Then, the samples were dried under three different conditions: at 30 °C for 7 days, at 100 °C for 7 days, and calcined at 700 °C for 3 h. Two different As concentration in clay were tested, 0.1 and 0.2 mg As g^{-1} clay. For the leaching experiments, 8 g of clay were mixed with 200 mL of distilled water and stirred for 7 days. The results are shown in Table 1.

As(V) leaching from dried clay used for As(V) removal was almost negligible. In the calcined samples, a higher amount of As(V) leached was observed, probably due to structural changes in the clay and/or to the leaching of the interfering phosphate (Lenoble *et al.*, 2003) released by the organic matter consituting the clay (Meichtry *et al.*, 2012). As leaching was higher in clay samples used for As(III) removal, being As released mainly as As(V); this behavior can be explained considering that As(III) oxidation to As(V) takes places during the drying process of the clay.

Table 1. Leaching test of As-containing clay. Reaction time: 7 days. Brackets indicates % of As desorbed, found as As(V).

Treatment	Sample	% of total As leached
30 °C, 7 days	0.2 mg g^{-1} As(III)	3.7 (77% As(V))
	0.1 mg g^{-1} As(III)	6.7 (87% As(V))
	0.2 mg g^{-1} As(V)	0.4
	0.1 mg g^{-1} As(V)	0.4
100 °C, 7 days	0.2 mg g^{-1} As(III)	1.5 (76% As(V))
	0.1 mg g^{-1} As(III)	3.2 (94% As(V))
	0.2 mg g^{-1} As(V)	0.2
	0.1 mg g^{-1} As(V)	0.4
700 °C, 3 hours	0.2 mg g^{-1} As(III)	9.1 (55% As(V))
	0.1 mg g^{-1} As(III)	11.1 (98% As(V))
	0.2 mg g^{-1} As(V)	4.4
	0.1 mg g^{-1} As(V)	21

Table 2. Leaching tests of the cement probes. Total As leached represents the sum of As(III) and As(V) leached respect to total As contained in the probes.

Treatment	final pH	[As(III)] leached, mg L^{-1}	[As(V)] leached, mg L^{-1}	% of total As leached
Water	11.6	≤ 0.05	≤ 0.05	≤ 0.05
Method 1311	9.5	≤ 0.05	≤ 0.05	0.1
Method 1313	10.42[a]	≤ 0.05	≤ 0.05	≤ 0.05
	8.15[a]	0.10	≤ 0.05	0.2
	7.37[a]	0.18	≤ 0.05	0.3
	3.25[a]	2.07	0.05	2.9
	2.96[a]	3.92	0.22	5.7
	1.05[a]	30.3	1.48	43.6
	0.92[a]	31.3	2.34	46.1

[a] HCl concentration of 0.05, 0.20, 0.30, 0.40, 0.50, 0.75 and 1.00 M, respectively.

3.2 *Tests on cement probes*

The mechanical tests showed that the As probes have a decrease of 21% in bending (from 210 to 165 kg cm^{-2}) and 12% in compressive strength (from 391 to 345 kg cm^{-2}) when compared with the T probes. The decrease can be attributed to the effect of hydration water, related with the presence of clay. Arsenic should not be detrimental for the properties of cement at concentrations up to 500 mg kg^{-1} (Minoche & Bhatnagar, 2007).

Three leaching tests were performed on the probes:

a. Water: 10 g of the probes were stirred with 200 mL of water at pH 7 for 18 h.
b. Method 1311: 10 g of the probes were stirred with 200 mL of an aqueous acetic acid solution for 18 h.
c. Method 1313: 20 g of the probes were stirred with 200 mL of an aqueous HCl solution 24 h.

Arsenic leaching was only significant in the samples treated with HCl concentrations higher than 0.2 M. Almost all As leached corresponded to As(III), indicating that this species is more mobile than As(V) in the probe.

4 CONCLUSIONS

As(V) leaching from clay used to remove As(V) was negligible, while clay samples used for As(III) removal showed significant As leaching, mainly as As(V). More studies are necessary to evaluate whether the leached species desorbed from the calcined clay is As(V) or P(V).

The disposal of As-containing clay combined with portland cement as construction material is a viable alternative, as As leaching is not significant under normal conditions (HCl ≤ 0.2 M). However, the samples with clay showed decrease in mechanical strength.

REFERENCES

EPA. 1992. Method 1311. *Toxicity characteristic leaching procedure*.

EPA. 2012. Method 1313. *Liquid-solid partitioning as a function of extract pH using a parallel batch extraction procedure*.

Lenoble, V., Deluchat, V., Serpaud, B. & Bollinger, J.C. 2003. Arsenite oxidation and arsenate determination by the molybdene blue method. *Talanta* 61: 267–276.

Litter, M.I., Sancha, A.M. & Ingallinella, A.M. 2010. *Tecnologías Económicas para el Abatimiento de Arsénico en Aguas*. IBEROARSEN, CYTED (ed.), Argentina.

Meichtry, J.M., Castiglia, M.D., Mugrabi, F., Reina, F.D., De Seta, E.G., Bressan, S., López, A.R. & Domingo, E. 2012. Remoción de Arsénico en Agua Mediante Materiales de Bajo Costo y Segura Disposición Final. In M. Dos Santos Afonso & R.M. Torres Sánchez (eds.), *1er Congreso Argentina y Ambiente, 28 May–1 June 2012*. Asociación Argentina para el Progreso de las Ciencias. pp. 980–-985.

Minocha, A.K & Bhatnagar, A. 2007. Immobilization of Arsenate (As^{5}+) Ions in Ordinary Portland Cement: Influence on the Setting Time and Compressive Strength of Cement. *Res. J. Environ. Toxicol.* 1: 45–50.

One Century of the Discovery of Arsenicosis in Latin America (1914–2014) – Litter, Nicolli, Meichtry, Quici, Bundschuh, Bhattacharya & Naidu (Eds)
© 2014 Taylor & Francis Group, London, ISBN 978-1-138-00141-1

Arsenic removal using low-cost materials

J.M. Meichtry, E.G. De Seta, A.R. López, F.D. Reina, F.I. Mugrabi, L.E. Lan, E. Lavezzari, E.J. Domingo & N. Tojo
Facultad Regional Buenos Aires, UTN, Argentina

ABSTRACT: Low-cost materials for As removal from water were evaluated: a natural clay, coagulation-flocculation with $Al_2(SO_4)_3$ and zerovalent iron, as micrometric powder (µFe(0)) or as commercial iron wool (wFe(0)). An initial As concentration of 5 mg L^{-1} at pH 7 was used in all experiments, with 99% of removal as the target. For As(V) removal, the most efficient material was $Al_2(SO_4)_3$, requiring the smallest dosage (0.5 g L^{-1}) and reaction time (1 h); for As(III), the smallest dosage corresponded to µFe(0) (1 g L^{-1}). As(V) and As(III) removal with the clay and As(V) removal with µFe(0)/wFe(0) were adjusted to simple kinetic models. $Al_2(SO_4)_3$ coagulation-flocculation was the best material for As(V) removal, but it is not efficient for As(III); µFe(0) provides a good efficiency for both As species, but the removal rate is very slow.

1 INTRODUCTION

Arsenic is a well known toxic agent, responsible for human diseases such as *Hidroarsenicismo Crónico Regional Endémico* (HACRE) and cancer. The consumption of water containing arsenic is the most frequent exposure way; in Argentina, over 2 million people are potentially affected, especially in the countryside or small villages with no tap water access. Therefore, the need of searching simple and low-cost systems for arsenic removal is very clear.

Natural clay is an almost ready-to-use material for As removal with a very low cost (Litter *et al.*, 2010; Sharma *et al.*, 2009), but it exhibits a low removal capacity. Zerovalent iron is a very studied material due to their low cost and high As removal capacity (Litter *et al.*, 2010; Noubactep *et al.*, 2009). Both natural clay and zerovalent iron can be used for household level treatment units and also for higher scale applications, but previous capacity and removal kinetic tests should be performed. In this work, these materials were compared with the use of $Al_2(SO_4)_3$ coagulation-flocculation methods (Baskan & Pala, 2010).

2 METHODS/EXPERIMENTAL

2.1 *Materials*

A natural clay was obtained from the Misiones province, Argentina, with mesh 20 grinding as the only treatment. The chemical composition was determined by sample digestion according to US-EPA 3051 (2007), followed by ICP-OES measurement.

Table 1. Chemical composition of the clay used.

Element	Measurement λ (nm)	% m/m of the sample
Al	396.153	9.26
Fe	238.204	5.94
Mn	257.610	0.09

Micrometric zerovalent iron (µFe(0), mesh 200) was of high purity (99,8%). Zerovalent iron wool was comercial (wFe(0)). The coagulant-flocculants and all other reactives were of analytical grade. NaOH or HCl 1 M solutions were used for pH adjustment. Low conductivity distilled water was used for the solutions and in the experiments.

2.2 *Analytical determinations*

As(V) and As(III) determination was carried out by the spectrophotometric method established by Lenoble *et al.* (2003), using quartz cells of 1 cm of pathlength. The detection limit was 0.05 mg L^{-1}.

2.3 *Experimental setup*

200 mL of aqueous solution (5 mg L^{-1} of As(III) or As(V), pH 7) was poured into a 250 mL glass flask open to air. Then, the desired amount of material was added under strong orbital stirring. The reaction temperature was fixed at 18 °C. 1 mL samples were taken and filtered with a 0.22 µm cellulose nitrate filter.

3 RESULTS AND DISCUSSION

3.1 *Removal capacity of As*

The removal capacity of the different materials was tested for As(III)/As(V), using different initial concentration of materials (0–50 g L^{-1}) and leaving the system until no changes in As concentration in solution ($[As]_S$) were observed; the time required to reach this pseudo-equilibrium was 2 h for $Al_2(SO_4)_3$, 168 h for the clay, 240 h for µFe(0)) and 300 h for wFe(0). The smaller dosage for a 99% As(III)/As(V) removal can be observed in Figure 1.

For all the studied materials, As(V) was removed in preference to As(III); this is especially noticeable with $Al_2(SO_4)_3$, as reported by Litter *et al.* (2010).

The smallest dosage for As(V) removal corresponds to $Al_2(SO_4)_3$, closely followed by µFe(0). For As(III), a removal higher than around 33% could not be obtained with $Al_2(SO_4)_3$, while for µFe(0) the removal was similar to that obtained for As(V). Partial As(III) oxidation by dissolved oxygen is appreciated for the control, clay and µFe(0). The combination of clay with µFe(0) did not improve the efficiency of the latter.

3.2 *As removal kinetics*

In Figure 2, the kinetic experiments of As removal using the materials at their minimal dosage, as determined in Section 3.1, can be observed. The results for $Al_2(SO_4)_3$ were not included, but as indicated before, the process was much faster.

In the experiments shown in Figure 2, the temporal evolution of $[As]_S$ could be fitted to a simple biexponential model, as represented by eq. (1).

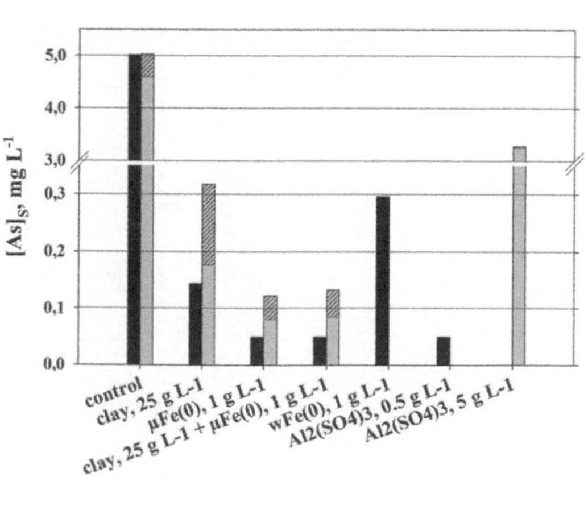

Figure 1. As removal capacity of the different studied materials. Conditions: $[As]_0 = 5$ mg L^{-1}, pH 7, T = 18 °C. Blank bars: experiments with As(V). Gray bars: experiments with As(III), indicating As(III) (smooth) and As(V) (stripped).

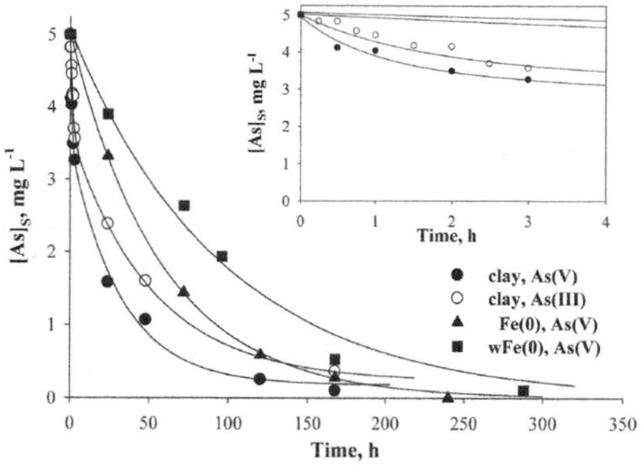

Figure 2. As(V) and As(III) adsorption kinetics on clay and on Fe(0). Conditions: $[As]_0 = 5$ mg L^{-1}, pH 7, T = 18 °C. Insert: short reaction times. Lines: fittings to eq. (1).

Table 2. Fitting parameters to eq. (1) of results from Figure 2.

Material/As species	A_1, mg L^{-1}	t_1, *h*	A_2, mg L^{-1}	t_2, *h*	$[As]_\infty$, mg L^{-1}
clay, 25 g L^{-1}/As(V)	1.46	0.96	3.31	32	0.18
clay, 25 g L^{-1}/As(III)	1.26	1.53	3.53	48	0.21
µFe(0), 1 g L^{-1}/As(V)	–	–	5.0	57	0.0
wFe(0), 1 g L^{-1}/As(V)	–	–	5.0	95	0.0

$$[As]_S = A_1 \times \exp^{-t/t_1} + A_2 \times \exp^{-t/t_2} + [As]_\infty \qquad (1)$$

A_1 and A_2 are the concentration of As removed by the processes 1 and 2, t_1 and t_2 are the characteristic time of these processes, and $[As]_\infty$ represent the As that remains in solution. The parameters are displayed in Table 2.

For clay, a biexponential behavior with similar parameters can be observed for As(V) and As(III) (somewhat slower for As(III)), which can be related to two different adsorption sites. For both µFe(0) and wFe(0) with As(V), only one simple process can be observed, with a characteristic time larger than that corresponding to the clay. The higher value of t_2 for wFe(0) reflects its slower removal rate.

4 CONCLUSIONS

For four different materials, minimum dosage and removal rates were determined. $Al_2(SO_4)_3$ was the fastest and more efficient material for As(V) removal, but it is not useful for As(III). µFe(0) has a very good capacity for both As(V) and As(III), but the small removal rate can be a limitation. The clay requires a much higher dosage, but it is efficient for both As species, and the small value of t_1 indicates that As removal could be achieved within hours.

REFERENCES

Baskan M.B. & Pala, A. 2010. A statistical experiment design approach for arsenic removal by coagulation process using aluminum sulfate. *Desalination* 254: 42–48.

Lenoble, V., Deluchat, V., Serpaud, B. & Bollinger, J.C. 2003. Arsenite oxidation and arsenate determination by the molybdene blue method. *Talanta* 61: 267–276.

Litter, M.I., Sancha, A.M. & Ingallinella, A.M. 2010. *Tecnologías Económicas para el Abatimiento de Arsénico en Aguas*. IBEROARSEN, CYTED (ed.), Argentina.

Sharma, V.K. & Sohn, M. 2009. Aquatic arsenic: Toxicity, speciation, transformations, and remediation. *Environ. Intern.* 35: 743–759.

Noubactep, C., Caré, S., Togue-Kamga, F., Schöner, A. & Woafo, P. 2010. Extending Service Life of Household Water Filters by Mixing Metallic Iron with Sand. *Clean—Soil, Air, Water* 38 (10): 951–959.

US-EPA Method 3051. 2007. *Microwave assisted acid digestion of sediments, sludges, soils and oils.* Available at: http://www.epa.gov/osw/hazard/testmethods/sw846/pdfs/3051a.pdf (accessed 15/12/2013).

One Century of the Discovery of Arsenicosis in Latin America (1914–2014) – Litter, Nicolli, Meichtry, Quici, Bundschuh, Bhattacharya & Naidu (Eds)
 ISBN 978-1-138-00141-1

Using of rehydrated kaolinites for decontamination of areas polluted by arsenic, antimony and selenium

M. Lhotka, B. Doušová, V. Machovič & P. Houdková
Institute of Chemical Technology in Prague, Czech Republic

ABSTRACT: Anionactive sorbents were prepared from rehydrated kaolinite using the modification of raw clay with 0.6 M $FeCl_2$. New sorbents were used for the removal of arsenic, selenium and antimony, which are human carcinogens that, in small amounts, are widely distributed in water. The natural kaolinite was calcined to metakaolinite (dehydroxylation) at 650 °C and then rehydrated at different temperature. To improve sorption properties to anionic particles, prepared metakaolinite was treated by rehydration in autoclave with Fe^{2+} ions. The best adsorption properties were found at the kaolinite rehydrated for 7 days. Adsorption parameters for As/Se/Sb oxyanions were calculated according to the Langmuir model. The adsorption affinity of toxic oxyanions decreased in the order of Sb(V) > As(V) > Se(IV) > As(III), at the theoretical adsorption capacities range of 0.08–0.16.

1 INTRODUCTION

Under the hydrothermal treatment, kaolinite is transformed into semi-crystalline very reactive metakaolinite, and the rehydroxylation of metakaolinite to kaolinite is possible. Rocha *et al.* (1990) studied the rehydration of metakaolinite to kaolinite which was heated at 155–250 °C for 1–14 days, and concluded that the amorphous metakaolinite can be transformed to a crystalline form. The specific surface area highly increased during the rehydration. Lhotka *et al.* (2012) studied the rehydration of metakaolinite to kaolinite which was heated at 150–250 °C for 1–14 days and concluded that the rehydration of metakaolinite to kaolinite was strongly dependent on temperature and time of the hydrothermal process. The optimum transformation from the point of view of the surface properties was observed after longer-term autoclaving (4–7 days) at 175 °C, when the specific surface S_{BET} of raw kaolinite increased more than three times. Cationic particles were better adsorbed on the longer-term rehydrated raw sorbents in the order of Cd^{2+} > Zn^{2+} \$ Pb^{2+}, while oxyanions showed a higher adsorption affinity for Fe-modified sorbents. Pb^{2+} and SeO_3^{2-} particles exhibited the best adsorption properties. During the interaction of raw clay and Fe salt solution, ion-exchangeable Fe^{3+} particles in amorphous and/or poorly crystalline form have been fixed on the clay surface forming active adsorption sites. Substantial variability in growing Fe phases (hydrated Fe_2O_3, non-specific Fe^{3+} species, ferrihydrite) resulted from using different types of aluminosilicate carrier and the treatment conditions (Dousova *et al.*, 2009). The aim of this work was to prepare kaoline-based sorbents using rehydration method and to compare sorption efficiency and sorption capacity of modified sorbents from the point of view of rehydration time and the type of oxyanionic particle.

2 METHODS/EXPERIMENTAL

2.1 *Preparation of rehydrated kaolinites*

The crystalline kaolinite from West Bohemia was used for the preparation of modified sorbents. This kaolinite was calcined at 650 °C for 3 h and converted to metakaolinite. For the preparation of rehydrat-ed kaolinite, 8 g of metakaolinite was mixed with 30 g of water. The suspension was stirred for 2 min at room temperature and then was inserted into an auto-clave. The autoclave was heated at 175 °C for 1, 4, 7 and 14 days. In next experiments, 30 g of 0.6 M $FeCl_2$ were used instead of water. After removal from the auto-clave, the suspension was filtered off, washed with distilled water and dried at 100 °C for 24 h.

2.2 *Arsenic and selenium decontamination and analytical methods*

The As(V), As(III), Sb(V) and Se(IV) model solutions were prepared from KH_2AsO_4, $NaAsO_2$, $NaSb(OH)_6$ and Na_2SeO_3 of analytical quality and distilled water in the concentration about 40 mg/L and pH ≈ 6.4–7.0. The suspension of model solution and sorbent (6 g L^{-1}) was shaken in sealed polyethylene bottle at room temperature for 24 hours. The product was filtered off; the filtrate was analyzed for residual As concentration and pH value. Equilibrium adsorption isotherms of nitrogen were measured at 77 K using static volumetric adsorption systems (ASAP 2020 analyzer, Micromeritics). The adsorption isotherms were fitted in the BET specific surface area and the pore size distribution by the DFT and BJH method. The concentration of As, Sb and Se in aqueous solutions was determined by HG-AFS using PSA 10.055 Millennium Excalibur.

3 RESULTS AND DISCUSSION

3.1 *Characterization of rehydrated kaolinites*

The samples of sorbents were prepared under different reaction conditions. The surface area of newly prepared sorbents was much larger than those of raw kaolinite and metakaolinite (from 15.8 to ~103.1 $m^2\ g^{-1}$). At the rehydration temperature of 175 °C, the maximum surface area was achieved. Using the maximum S_{BET} at 175 °C (see Table 1), this temperature was used for the Fe treatment of samples. The comparison of S_{BET} of sorbents prepared at 175 °C with and without Fe ions is shown in Table 1.

Fe treated kaolinites reached the maximum of S_{BET} after four days reaction time. The decrease of their S_{BET} value compared with Fe-free rehydrated samples corresponded to the binding Fe^{3+} particles on the surface, forming active adsorption sites. It was observed that the median pore size of the samples is similar (pore diameter of 3.8 nm) (Lhotka *et al.*, 2012).

3.2 *Arsenic and selenium decontamination*

The experimental series for the characterization of As/Sb/Se adsorption on Fe-modified kaolinite (175 °C, 4–14 days) were performed with 8 samples of varying sorbent dosage for each reaction time. This arrangement enabled to evaluate sorption parameters according to the Langmuir model.

Table 1. Specific surface area of raw kaolinite, metakaolinite and rehydrated sorbents (temperature of rehydration—175 °C).

Sample	S_{BET} (m^2/g)
Kaolinite (K)	15.84
Metakaolinite (MK)	17.81
K4 (4 days)	98.57
K7 (7 days)	103.10
K4 (4 day-Fe)	70.21
K7 (7 days-Fe)	73.81
K10 (10 days-Fe)	71.05
K14 (14 days-Fe)	69.71

Table 3. Theoretical sorption capacities Q of Fe-modified kaolinite.

	Theoretical sorption capacity Q (mmol/g)			
Oxyanion	4 days	7 days	10 days	14 days
As(III)	–	–	–	–
As(V)	0.09	0.10	0.08	0.08
Se(IV)	0.06	0.08	0.06	0.07
Sb(V)	0.11	0.16	0.10	0.14

Table 2. As(V)/Sb(V)/Se(IV) adsorption on Fe-modified rehydrated kaolinite under the various rehydration time.

	As(III)		As(V)		Sb(V)		Se(IV)	
Sample	q*	ε**	q	ε	q	ε	q	ε
K4	0.07	57	0.14	99	0.18	96	0.07	100
K7	0.08	59	0.14	99	0.24	97	0.09	99
K10	0.09	56	0.12	100	0.10	98	0.07	99
K14	0.08	61	0.08	99	0.23	98	0.11	99

* maximum adsorption capacity (mmol/g).
** maximum sorption efficiency (%).

The comparison of adsorption efficiency ε and maximum sorption capacities q for investigated oxyanions are summarized in Table 2. Sorption parameters evaluated according to the Langmuir model are shown in Table 3.

The results in Tables 2 and 3 demonstrated a high selectivity of Fe-modified rehydrated kaolinite to all oxyanions except As(III), where only 60% was removed. This phenomenon resulted from the structure of AsO_3^{3-} tetrahedrons, which bind mostly via less stable outer-sphere complexes due to the presence of free electron pair (Dousova *et al.*, 2009). In terms of that, adsorption of As(III) oxyanions did not run according to Langmuir or Freundlich adsorption models. The kaolinite rehydrated for 7 days proved the best sorption properties in all sorption series. The adsorption affinity of investigated oxyanions to modified 7-days kaolinite declined in the following order:

Sb(V) > As(V) >Se(IV) > As(III)

ACKNOWLEDGEMENTS

This work was the part of projects 13–24155S and P210/10/0938 (Grant Agency of Czech Republic).

REFERENCES

Doušová, B., Fuitová, L., Grygar, T., Machovič, V., Koloušek, D., Herzogová, L. & Lhotka, M. 2009. Modified aluminosilicates as low-cost sorbents of As(III) from anoxic groundwater. *J. Hazard. Mater.* 165: 134–140.

Lhotka, M., Doušova, B. & Machovič, V. 2012. Preparation of modified sorbents from rehydrated clay minerals. *Clay Minerals* 47: 251–258

Rocha, J., Adams, J.M. & Klinowski J. 1990. The Rehydration of Metakaolinite to Kaolinite: Evidence from Solid-State NMR and Cognate techniques. *J. Solid State Chem.* 89: 260–274.

One Century of the Discovery of Arsenicosis in Latin America (1914–2014) –
Litter, Nicolli, Meichtry, Quici, Bundschuh, Bhattacharya & Naidu (Eds)
© 2014 Taylor & Francis Group, London, ISBN 978-1-138-00141-1

Field assessment of applicability of adsorptive arsenic removal using IOCS and in-situ regeneration for treatment of anoxic groundwater with high arsenic concentration

B. Petrusevski & Y.M. Slokar
Unesco-IHE Institute for Water Education, Delft, The Netherlands

M. Tasic
Water Supply Company of Subotica, Serbia

C. van Ommen & J. van Paassen
Water Supply Company Vitens, The Netherlands

ABSTRACT: Field assessment of applicability of adsorptive arsenic removal based on adsorption on Iron Oxide Coated Sand (IOCS), a by-product from iron removal treatment plants, and in-situ regeneration, for treatment of anaerobic groundwater with high arsenic concentration was conducted during 10 months using an industrial scale pilot (capacity 10–20 m^3 h^{-1}). Pre-treatment based on conventional aeration and Rapid Sand Filtration (RSF) highly effectively removed ammonia, iron, and manganese present in groundwater. Two adsorptive filters with IOCS, installed in series, demonstrated potential to consistently and completely remove high concentration of arsenic ($\leq$ 275 µg L^{-1}) present in groundwater. The 2nd adsorptive filter can be omitted if more frequent regeneration of the 1st. adsorptive filter is applied. Ten months of continuous pilot plant operation demonstrated that adsorptive arsenic removal with IOCS and in-situ regeneration, supported by conventional aeration-RSF, is very robust treatment approach for treatment of anoxic groundwater with very high arsenic concentration.

1 INTRODUCTION

The UNESCO-IHE Institute for Water Education (Delft, The Netherlands) in cooperation with Vitens, the largest Dutch water supply company, has been developing an arsenic removal technology based on adsorption on Iron Oxide Coated Sand (IOCS), and in-situ regeneration of the saturated adsorbent. The regeneration process, which is based on in-situ treatment of arsenic saturated IOCS with ferrous solution, forms new nanolayers of iron (hydro)oxides above the previously adsorbed arsenic (Figure 1). IOCS is a by-product of iron removal treatment plants utilizing aeration—Rapid Sand Filtration (RSF), and is consequently very cheap. Adsorptive arsenic removal with IOCS demonstrated potential to very effectively remove both As(V) and As(III) under field conditions using groundwater with relatively low arsenic concentration (Petrusevski *et al.*, 2007). To achieve consistent removal of arsenic < 10 µg L^{-1}, under field conditions, two stage adsorptive filtration with IOCS was required in combination with monthly regeneration cycles.

The goal of the study presented in this paper was to assess the suitability of the adsorptive arsenic removal with IOCS and in-situ regeneration for treatment of complex groundwaters with very high arsenic concentrations in combination with other impurities typical of anoxic groundwater (e.g. iron, manganese, ammonia).

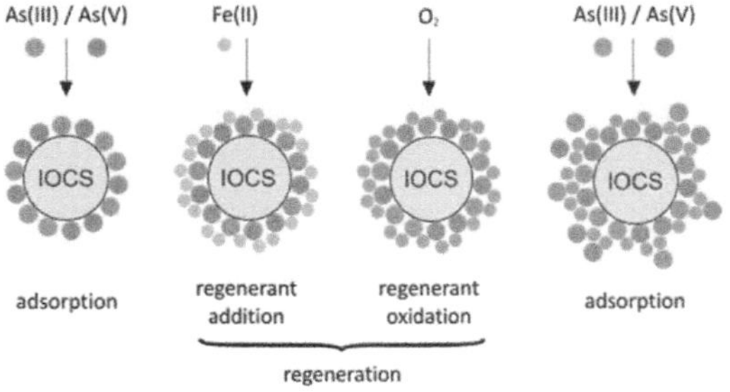

Figure 1. Arsenic removal with IOCS and in-situ regeneration.

2 METHODS/EXPERIMENTAL

A pilot plant with capacity of $\leq$ 20 m^3 h^{-1} was continuously in operation during 10 months in the village of Backi Vinogradi, Serbia. Untreated groundwater with the highest arsenic concentration in the region (≤ 275 μg L^{-1}) is supplied at present to population of this village as "drinking water" without treatment. This groundwater has also elevated ammonia (0.6–1.1 mg L^{-1}), iron (0.6–2.2 mg L^{-1}), manganese (≤ 0.10 mg L^{-1}) and phosphate (≤ 0.75 mg L^{-1}). The groundwater is slightly alkaline (pH 7.5–7.7), and has high HCO_3^- concentration (340–380 mg L^{-1}). The process scheme of the pilot plant includes a pre-treatment based on conventional aeration and RSF, and two adsorptive filters with IOCS installed in series, which operate at a filtration rate of 2.5 m h^{-1}. A relatively low filtration rate was applied, and the 2nd. adsorptive filter was included in the process scheme as a safety measure given the very high arsenic concentration in the groundwater. The pilot plant performance was intensively monitored on the site with test kits (Arsenator), and by laboratories of the water supply company and the Public Health Authorities.

3 RESULTS AND DISCUSSION

Highly effective iron removal was achieved from the very start of the pilot operation with residual iron concentrations after RSF consistently ≤ 0.03 mg L^{-1}. Removal of ammonia was highly effective (NH_4^+ ≤ 0.10 mg L^{-1}), after the initial ripening period of 4 weeks. RSF that was provided as pretreatment, achieved only partial arsenic removal, mainly through process of coagulation with iron present in the groundwater.

Arsenic was below the Detection Limit (DL) of analytical techniques applied (~2 μg L^{-1}) after the 1st adsorptive filter during the first 90 days of operation (Figure 2).

After approximately 3 months of continuous operation, residual arsenic was detected in the filtrate of the 1st. adsorptive filter, with arsenic concentrations ≤ 9 μg L^{-1}. The 2nd. adsorptive filter completely removed arsenic escaping from the 1st. adsorptive filter throughout the whole testing period. At this stage, biweekly regeneration cycles of the 1st. adsorptive filter were introduced. Prolonged continuous operation of the plant under these conditions resulted in slow increase of arsenic levels after the 1st. filtration step to approximately 13 μg L^{-1}. It was obvious that biweekly regeneration of the 1st. adsorptive filter was insufficient to keep the residual arsenic < 10 μg L^{-1}. An increase of the regeneration frequency that was introduced from mid of January 2013 to twice per week resulted in reducing arsenic levels that after a few weeks dropped to a level <DL (Figure 3). The much higher required regeneration frequency than that applied earlier (Petrusevski

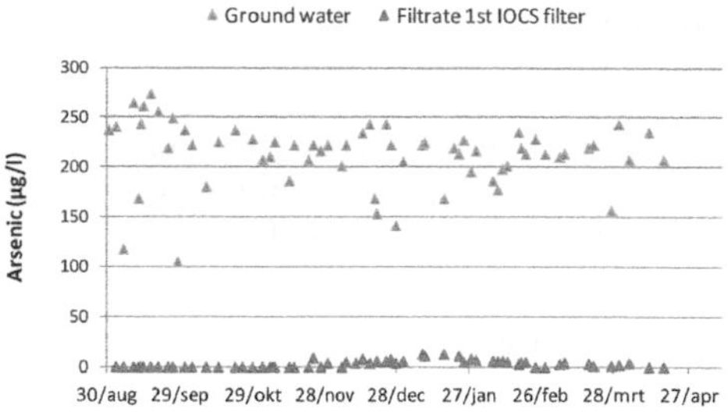

Figure 2. Arsenic concentration in groundwater and filtrate of the 1st. adsorptive filter; pilot plant in Backi Vinogradi (Serbia).

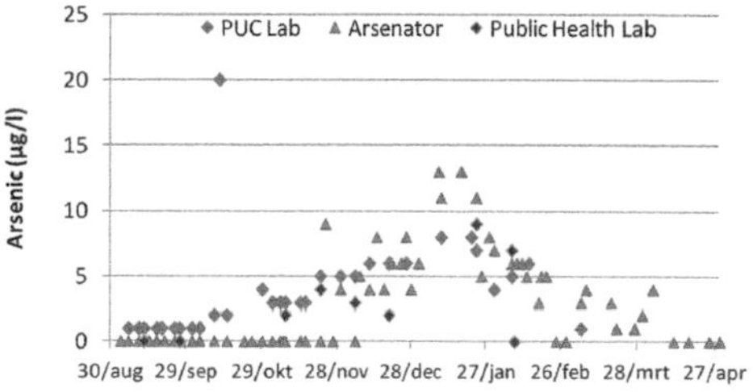

Figure 3. Effect of increased regeneration frequency applied from the mid of January 2013 on arsenic concentration in the filtrate of the 1st. adsorptive filter.

et al., 2007) could be attributed to the significantly higher arsenic concentration in the groundwater and to the presence of phosphate, which is known to compete with arsenic for adsorption sites on IOCS (Dzombak & Morel, 1990; Katsoyiannis & Zouboulis, 2002).

Results have shown that the 2nd. adsorptive filter is not strictly necessary, and that arsenic concentration after the 1st. adsorptive filter can be kept very low by frequent regeneration. Head loss development in IOCS filters was very low due to coarse filter media (d_{avg} = 2.6 mm), low filtration rate, and complete adsorption of ferrous iron used for regeneration, with associated backwashing frequency of once in several months.

Field investigation demonstrated that the technology is simple, reliable and easy to operate.

4 CONCLUSIONS

Adsorptive arsenic removal based on the use of IOCS and in-situ regeneration procedure, in combination with conventional aeration-RSF, demonstrated to be highly suitable for treatment of anaerobic groundwater with very high arsenic concentration. Arsenic concentration in the filtrate can be controlled by the frequency of regeneration.

ACKNOWLEDGEMENTS

Research presented in this paper was financed by Agentschap, an agency of the Dutch Government. Foundation Water is Our World provided the pilot plant used in field investigation.

REFERENCES

Dzombak, D.A. & Morel, F.M.M. 1990. *Surface complexation modeling: Hydrous ferric oxide*, John Willey & Sons Inc. Wiley-Interscience, New York USA.

Katsoyiannis, I.A. & Zouboulis, A.I. 2002. Removal of arsenic from contaminated water sources by sorption onto iron-oxide-coated polymeric materials. *Wat. Res.* 36: 5141–5155.

Petrusevski, B., van der Meer, W.G.J., Baker, J., Kruis, F., Sharma, S.K. & Shippers, J.C. 2007. Innovative approach for treatment of arsenic contaminated groundwater in Central Europe. *Water Sci. & Technol.: Water Supply* 7(3): 131–138.

One Century of the Discovery of Arsenicosis in Latin America (1914–2014) – Litter, Nicolli, Meichtry, Quici, Bundschuh, Bhattacharya & Naidu (Eds)
 ISBN 978-1-138-00141-1

Synthesis of adsorbents with iron oxide and hydroxide contents for the removal of arsenic from water for human consumption

S.E. Garrido
Instituto Mexicano de Tecnología del Agua, Jiutepec, Morelos, Mexico

L. Romero
Universidad Politécnica de Morelos, Jiutepec, Morelos, Mexico

ABSTRACT: Two types of adsorption media, synthetic goethite and Fe(III)-coated silica sand, were developed to remove arsenic (As) present in water for human consumption. A surface area of 43.04 m^2 g^{-1} was determined for the synthetic goethite and 2.44 m^2 g^{-1} for Fe(III)-coated silica sand. A 2^3 factorial design was applied to evaluate concentration of As, pH and mass of the adsorbent to obtain the optimal conditions. The experimental data were fitted to a pseudo-second order kinetic model with a rate constant of adsorption (k_{ad}) of 4.019 g mg^{-1} min^{-1} for goethite and 0.745 g mg^{-1} min^{-1} for Fe(III)-coated silica sand. The equilibrium values, with an adsorption capacity (q_m) of 0.4822 mg g^{-1} for goethite and 0.2494 mg g^{-1} for Fe(III)-coated silica sand, were fit to the Langmuir II model. The most efficient medium was synthetic goethite with 98.61% removal of arsenic and a concentration in treated water of 0.005 mg L^{-1}.

1 INTRODUCTION

Over time, the presence of many contaminants has become permanent in diverse environmental matrices. Arsenic is primarily found in groundwater. Nowadays, new processes of adsorption have been developed to remove arsenic from drinking water. When an adsorption medium is used, it is necessary that it contain oxides/hydroxides (iron oxides), with a specific high porosity to facilitate the adsorption process (Glibota, 2005). As adsorption is a surface phenomenon, only adsorbents with high internal surface can be used (Sans and Paul, 1989).

In Mexico, several cases of contamination of groundwater with As have occurred, one of which has occurred in the Huautla mountains, in the municipality of Tlaquiltenango, Morelos. The objective of this work was to obtain two synthetic adsorption media for arsenic removal from drinking water.

2 EXPERIMENTAL METHODS

Two adsorbent media were prepared: Fe(III)-coated silica sand and synthetic goethite. The media were selected according to the research conducted by Thirunavukkarasu *et al.* (2001) for the case of Fe(III)-coated silica sand and, for the case of goethite, conducted by Garrido *et al.* (2008). The particle size was determined using granulometric techniques CAPT6–01 used in the Instituto Mexicano de Tecnología del Agua (Garrido *et al.*, 2008). The media were analyzed in a Micrometrics Model ASAP 202 surface area and porosity analyzer. The experiments were performed using a 2^3 factorial design for each medium (Statgraphics Centurion XVI version 2012) and the factors evaluated were: As concentration, pH and mass of the adsorbent. The water used for batch tests came from the locality of Sierra de Huautla, from the Pájaro Verde Tlaquiltenango mine shaft, Morelos. The water was characterized: electrical conductivity: 452 µS cm^{-1}; turbidity: 19 UTN; true color: 2 UPt-Co; pH 8.57; redox potential: 233 mV; HCO_3^-: 233 mg L^{-1}; PO_4^{3-}: 0.45 mg L^{-1}; NO_3^-: 6.3 mg L^{-1} and As 0.196 mg L^{-1}. Ten experimental tests were conducted for each medium under different conditions (Table 1).

Batch tests were conducted to obtain the adsorption kinetics in Jar Tests (PHIPPS & BIRD STIRRER PB700) with an agitation of 120 rpm for synthetic goethite and 200 rpm for Fe(III)-coated

Table 1. Experimental Design Matrix.

	Silica Sand + Fe(III)			Synthetic goethite		
Factors	+1	0	−1	+1	0	−1
As (mg L^{-1})	0.39	0.29	0.19	0.36	0.27	0.19
pH	8.5	8.0	7.5	8.5	8.0	7.5
Mass (g)	4.0	3.0	2.0	2.0	1.5	1.0

silica sand. The pH was maintained by adjusting with NaOH and HCl. After reaching the optimal conditions, the experimental values were adjusted to a pseudo-second order adsorption kinetics.

$$\frac{t}{q_t} = \frac{1}{k_{ad} q_e^2} + \frac{1}{q_e} t$$

where: q_e is the maximum adsorption capacity, q_t is the adsorption capacity with respect to time, and k_{ad}, the adsorption rate constant.

Langmuir adsorption isotherms (when the solute is adsorbed in the monolayer on the surface adsorbent) and Freundlich adsorption isotherms were obtained based on the concentration of arsenic in equilibrium. Arsenic was quantified according to the Wagtech Arsenator 2005, a photometric method that is more precise than Merck and Hach kits, measuring concentrations of 1–0.5 mg L^{-1} (Garrido *et al.*, 2009).

3 RESULTS AND DISCUSSION

Table 2 describes the characteristics of each medium. The goethite sample has 17 times more surface area than the Fe(III)-coated silica sand, i.e. goethite has a larger number of active sites for retaining As.

Based on the analysis of the 2^3 factorial design, the optimal conditions were obtained for: As concentration: 0.360 mg L^{-1}, pH: 7.5 and mass of goethite: 2.0 g L^{-1} (Figure 1). For Fe(III)-coated silica sand: As concentration: 0.392 mg L^{-1}, pH: 8.5 and mass of sorbent: 4 g L^{-1}

Table 2. Characterization of the media.

Analysis	Units	Silica sand +Fe(III)	Synthetic goethite
Surface area	m2 g^{-1}	2.44	43.04
Porous volume	cm^3 g^{-1}	0.002	0.12
Pore diameter	Å	53.60	119.05
Particle size	mm	0.3–0.6	0.1–0.6

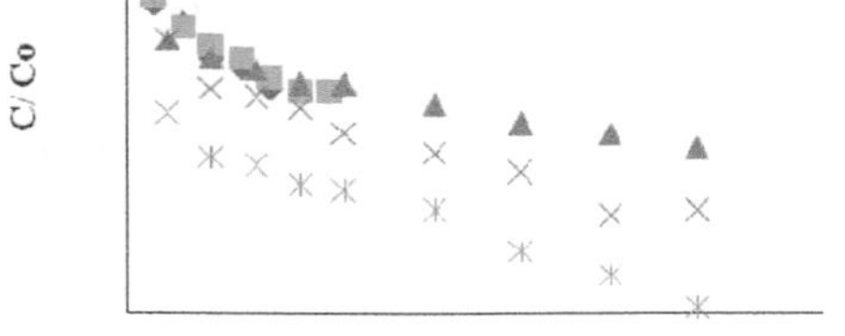

Figure 1. Adsorption kinetics of synthetic goethite.

Table 3. Adsorption kinetic constants.

Constants	Units	Silica sand +Fe(III)	Synthetic goethite
Mass	(g)	4.0	2.0
kad	(g mg^{-1} min^{-1})	0.745	4.019
h	(mg g^{-1} min^{-1})	0.0024	0.155

Table 4. Langmuir adsorption isotherm constants.

Constants	Unit	Silica Sand +Fe (III)	Synthetic goethite
q_m	(mg g^{-1})	0.2494	0.4822
B	–	0.5102	5.0841
R^2	–	0.7795	0.9085

In previous studies (Garrido, 2008), the greatest amount of arsenic found in water in Huautla was $HASO_42^-$, which has a radius of 3.97 Å. Therefore, the pore diameter of both media was enough to allow the As(V) species to be adsorbed.

Table 3 shows the adsorption rate constant (k_{ad}) for goethite (4.019 g mg^{-1} min^{-1}), which was 5.39 times higher than that for Fe(III)-coated silica sand (0.745 g mg^{-1} min^{-1}). This is attributed to the fact that the surface area of the synthetic goethite is larger than that of the Fe(III)-coated silica sand.

The adsorption isotherms and constants were obtained based on the concentrations of As in equilibrium (Table 4). It is observed that the results of both media were better fitted to the Langmuir II model. The number of moles of the adsorbed solute per unit weight of the adsorbent on the surface (q_m) in the present work was greater than that obtained by Thirunavukkarasu *et al.* (2001) for both adsorbents, i.e.: $q_m = 0.183$ mg g^{-1} for iron oxide-coated silica sand and $q_m = 0.285$ mg g^{-1} for ferrihydrite.

However, when comparing the results reported by Paredes (2012) for natural goethite, the q_m obtained was 9.30 mg g^{-1}, a higher value than the found for synthetic goethite, 0.4822 mg g^{-1}.

4 CONCLUSIONS

When fitting the experimental data to a pseudo-second order equation, the kinetic adsorption study shows that the goethite has a higher adsorption rate constant (kad) compared with other investigations. The average arsenic removal was 98.61% and a concentration in treated water of 0.005 mg L^{-1}, which complies with NOM-127-SSA-1994.

REFERENCES

Garrido, S.E. 2008. Origen Hidrogeológico del arsénico, metodos alternativos innovativos para cuantificación y tratamiento. Informe final. CONACYT-2004-C01-47076. México.

Garrido, S., Avilés, M., Ramírez, A., Calderón, C., Ramírez-Orozco, A., Nieto, A., Shelp,G., Seed, L., Cebrian, M. & Vera, E. 2009. Arsenic removal from water of Huautla, Morelos, Mexico using Capacitive deionization. Natural arsenic in Groundwaters of Latin America. Ed. Taylor & Francis. London UK. pp. 665–676.

Glibota, G.S. (2005). Avances en opciones para eliminación de arsénico en aguas. Argentina: Universidad Nacional del Nordeste Comunicaciones Científicas y Tecnológicas, Argentina.

Paredes, J.L. 2012. Remoción de arsénico del agua para uso y consumo humano mediante diferentes materiales de adsorción. Tesis de Maestría. UNAM, México, pp. 106.

Sans, F.R. & De Pablo, R.J. (1989). Ingeniería ambiental contaminación y tratamientos, Colombia: MARCOMBO, S.A.

Thirunavukkarasu, O.S, Viraraghavan, T. & Subramanian, K.S. 2001. Removal of arsenic in drinking water by iron oxide coated sand and ferrihydrite batch studies. Water Qual. Res. J. Canada 36:55–70.

One Century of the Discovery of Arsenicosis in Latin America (1914–2014) – Litter, Nicolli, Meichtry, Quici, Bundschuh, Bhattacharya & Naidu (Eds)
© 2014 Taylor & Francis Group, London, ISBN 978-1-138-00141-1

Study of arsenic adsorption in columns of Activated Natural Siderite (ANS) and its application to real groundwater with high arsenic contents

K. Zhao, H.M. Guo, Y.S. Cao & X.Q. Zhou
School of Water Resources and Environment, China University of Geosciences, Beijing, China

ABSTRACT: Column tests were performed utilizing Activated Natural Siderite (ANS) to remove As from artificial solutions. Results indicated that ANS-packed column reactor removed As(III) more efficiently than As(V). The presence of background electrolytes and coexisting anions within certain levels in tap water had no significant effect on the removal of As(III) by ANS-packed column in the early stages. Field pilot test using ANS as media material was conducted to remove As from real groundwater with high As contents. Though As(III) was the dominant As species in the natural groundwater, removal efficiency of ANS on real groundwater was relatively lower compared to As(III)-spiked DI water, which was attributed to oxidation of As(III) to As(V) occurring when natural groundwater flowed through the ANS-packed filter.

1 INTRODUCTION

Arsenic removal from drinking water is of serious concern and has become an important global problem. It is an urgent and challenging issue to develop novel materials for As removal from As-contaminated drinking water. Batch experiments have demonstrated that Activated Natural Siderite (ANS) was quite effective in adsorbing As from aqueous solution, the material presenting improved kinetic rate and adsorption capacity compared with natural minerals. The maximum adsorption capacities of ANS at 25 °C, estimated from the Langmuir isotherm, were 2.66 and 2.19 mg g^{-1} for As(III) and As(V), respectively (Zhao & Guo, 2013). It is considered that by the use of the novel adsorbent a balance between cost and adsorption performance can be achieved. However, the performances of fixed bed adsorption of arsenic on ANS are still unknown, which are vital parameters for obtaining basic engineering data and predicting the column breakthrough.

2 MATERIALS AND METHODS

Natural siderite particles with particle size fraction of 0.5–1.0 mm, in which siderite, clay minerals, and quartz accounted for 64.7, 22.9, and 7.2 wt.%, respectively, were calcined in a muffle furnace at 350 °C for about 2 h; this material will be designated as ANS (Zhao & Guo, 2013).

Plexiglass columns, with an internal diameter of 30 mm, a height of 115 mm, were used in the column study with an up-flow rate of 1.8 mL min^{-1} obtained with a peristaltic pump fixed-bed reactor. The Empty Bed Contact Time (EBCT) was about 45 min. ANS (118 g) was packed in each column. Column A was used for treating 1.0 mg L^{-1} As(III)-spiked DI water, Column B for 1.0 mg L^{-1} As(III)-spiked tap water, and Column C for 1.0 mg L^{-1} As(V)-spiked DI water. Field pilot test (D) was conducted using a filter of 60 cm diameter and 87 cm bed depth (416 kg), with an up-flow rate of 4.0 L min^{-1}, to on-site treat groundwater containing high As concentrations taken from the Hetao basin, Inner Mongolia, China. The EBCT of the field pilot test was about 60 min. The chemical composition of tap water and real groundwater are shown in Table 1. Samples were collected at certain time intervals and then analyzed for residual As concentration.

3 RESULTS AND DISCUSSION

Breakthrough curves of ANS-packed columns for As removal are illustrated in Figure 1a. The Figure shows that the removal efficiency of Column A for treating 1.0 mg L^{-1} As(III)-spiked DI water appeared to be approximately equal to that of Column B for treating 1.0 mg L^{-1} As(III)-spiked tap water, although tap water contained some coexisting anions (Table 1), demonstrating that effects of background electrolytes and coexisting anions on As(III) adsorption onto ANS were not significant within certain levels. Both Columns A and B treated approximately 1200 bed volumes (BV) of 1.0 mg L^{-1} As(III) solutions prior to breakthrough of 10 µg L^{-1} As, and over 1600 BV before the Chinese drinking water standard in rural areas of 50 µg L^{-1} As was broken through. However, effluent As concentration of Column B increased rapidly after about 2200 BV compared with that of Column A. The ANS presented a much better performance than Fe oxide-coated sand (1403 BV with 0.5 mg L^{-1} As(III)-spiked Regina tap water) (Thirunavukkarasu *et al.*, 2003) and GAC-based Fe-containing adsorbent (7500 BV with influent As(III) concentration of 56 µg L^{-1}) (Gu *et al.*, 2005).

Column C, treating 1.0 mg/L As(V)-spiked DI water, was setup to investigate the effect of As species on As removal by ANS. As shown in Figure 1a, the drinking water standard of 10 µg L^{-1} was exceeded after 500 BV and the 50 µg L^{-1} MCL was surpassed after 700 BV, which was relatively lower than with Column A.

A mass balance calculation indicated that total As load in the Column C was 0.48 mg/g before 50 µg L^{-1} of As breakthrough, while in Columns A and B the values were around 1.10 mg g^{-1}. The results indicated that ANS-packed column reactor removed As(III) more efficiently than As(V).

The breakthrough curve of field pilot test is shown in Figure 1b. The As load in the filter was 289 µg g^{-1} before 50 µg L-1 of As breakthrough. Although its EBCT was slightly greater than that in column study, removal efficiency from real high-As groundwater was relatively lower in comparison with Column A. Meanwhile, As(V) was the dominant As species in the effluents. It was inferred that As(III), the dominant As species in the influents, was oxidized to As(V) when flowing through ANS-packed filter, a different behavior

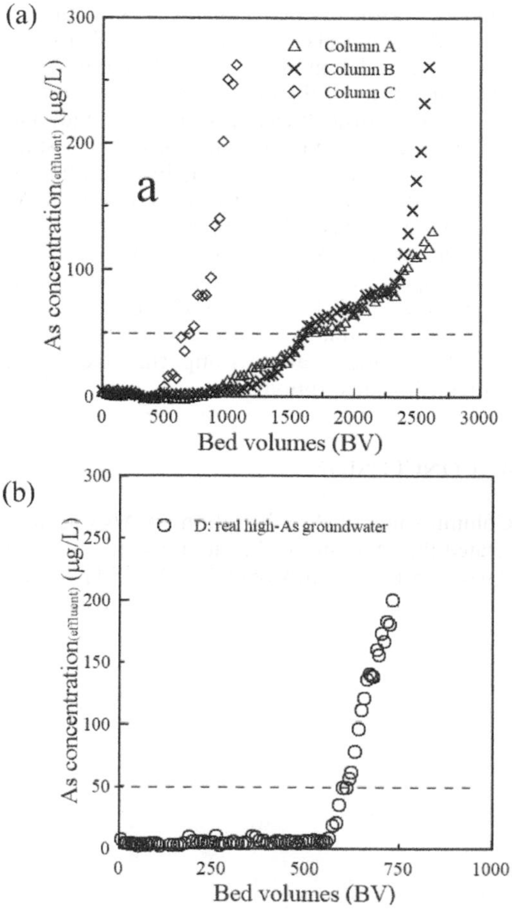

Figure 1. Breakthrough curves of ANS-packed columns (a) and field pilot test (b).

Table 1. Chemical composition of natural As-bearing groundwater and tap water used in batch and column experiments.

	Total As (µg)	pH	Na^+ (mg L^{-1})	Mg^{2+} (mg L^{-1})	K^+ (mg L^{-1})	Ca^{2+} (mg L^{-1})	Fe (µg L^{-1})	Mn (µg L^{-1})
Natural groundwater	793.3	8.38	314	21.0	3.89	16.0	564	37.1
Tap water	0.35	7.31	14.3	14.1	2.42	39.8	8.40	4.35
	As(III) (µg L^{-1})	As(V) (µg L^{-1})	Cl^- (mg L^{-1})	HCO_3^- (mg L^{-1})	NO_3^- (mg L-1)	SO_4^{2-} (mg L^{-1})	Si (mg L^{-1})	P (mg L^{-1})
Natural groundwater	668.3	107.1	232	617	N.D.	N.D.	2.63	N.D.
Tap water	N.D.	N.D.	31.1	26.2	66.0	70.9	N.D.	N.D.

N.D. = non-detectable.

compared with that of the artificial As solutions, due to the chemical complexity of natural groundwater. The oxidation process from As(III) to As(V) was believed to be the most important reason for the lower removal efficiency in the pilot study. On the other hand, high concentrations of coexisting anions in natural groundwater, like bicarbonate and silicate (Table 1), had adverse effect on As adsorption on ANS. Besides, because interaction between As and ANS is endothermic in nature, high temperature would strengthen As adsorption on ANS (Zhao & Guo, 2013). However, the temperature of natural groundwater was only about 10 oC, which could be another important factor that depressed As adsorption.

4 CONCLUSION

Column study of As adsorption on ANS demonstrated that ANS-packed column reactor removed As(III) more efficiently than As(V). Field pilot test indicated that removal efficiency on ANS from real high-As groundwater was relatively lower compared to As(III)-spiked DI water, even though As(III) was the dominant As species in the natural groundwater.

REFERENCES

Gu, Z.M., Fang, J. & Deng, B.L. 2005. Preparation and evaluation of GAC-based iron-containing adsorbents for arsenic removal. *Environ. Sci. Technol.* 39: 3833–3843.

Thirunavukkarasu, O.S., Viraraghavan, T., Subramanian, K.S., 2003. Arsenic removal from drinking water using iron oxide-coated sand. *Water, Air, Soil Pollut.* 142: 95–111.

Zhao, K. & Guo, H.M. 2013. Behavior and mechanism of arsenate adsorption on activated natural siderite: evidences from FTIR and XANES analysis. *Environ. Sci. Pollut. Res.* DOI 10.1007/s11356-013-2097-8.

One Century of the Discovery of Arsenicosis in Latin America (1914–2014) –
Litter, Nicolli, Meichtry, Quici, Bundschuh, Bhattacharya & Naidu (Eds)
 ISBN 978-1-138-00141-1

Influence of the Al-for-Fe substitution in the adsorption of As(V) onto goethites—morphological surface characterization

A.E. Tufo
INQUIMAE, Facultad de Ciencias Exactas, Universidad de Buenos Aires, Buenos Aires, Argentina

M.C. Marchi
INQUIMAE, Facultad de Ciencias Exactas, Universidad de Buenos Aires, Buenos Aires, Argentina
Centro de Microscopías Avanzadas, Facultad de Ciencias Exactas, Universidad de Buenos Aires, Buenos Aires, Argentina

M. dos Santos Afonso & E.E. Sileo
INQUIMAE, Facultad de Ciencias Exactas, Universidad de Buenos Aires, Buenos Aires, Argentina

S.M. Landi
Instituto Nacional de Metrologia, Qualidade e Tecnologia, Divisão de Metrologia de Materiais, RJ, Brasil

ABSTRACT: Arsenic derived from natural and anthropogenic sources occurs in groundwater, and its mobility is highly influenced by goethite and ferrihydrite. As the substitution of Al-for-Fe in natural goethites is well documented, our aim was to explore the adsorption of As(V) onto several Al-substituted goethites, and to analyze by STEM micrographs the possible formation of $FeAsO_4$ as a surface precipitate that blocks ulterior sorption. Adsorption isotherms ([As]adsorbed vs. [As]solution) of all samples were obtained, and an uncommon bell shaped curve with a maximum at 0.25 mM was found for all goethites. The STEM micrographs of the As-adsorbed Al-goethites were obtained at [As] > 0.25 mM and confirmed that no surface precipitates were formed. Based on these results, the formation of bidentated complexes were proposed to explain the decrease of the adsorption by As(V) coordination to two metal sites.

1 INTRODUCTION

Natural As mobilization results from weathering reactions, biological activity and volcanic emissions, followed by transportation through the environment by water (Abdullah *et al.*, 1998). In the volcanic Andes area, Argentina has one of the world largest aquifers polluted with high As concentrations of natural origin. The two common species found in natural environments are arsenite (As(III)) and arsenate (As(V)), which are the most widespread in surface waters. In soils, under oxidizing conditions, As(V) is the predominant form, and previous studies revealed that As is associated primarily with the Fe(III) oxyhydroxide coatings on soil particles (Wang *et al.*, 2006) or it is strongly adsorbed onto clays, Mn oxides/hydroxides and organic matter. The minerals goethite and ferrihydrite, commonly found in soils, present a strong affinity for both As species, highly influencing the mobility behavior of As (Manning *et al.*, 1996; Aguilar *et al.*, 2007).

2 METHODS

2.1 *Adsorbents*

Three samples of pure and Al-substituted goethites were prepared following the method of Schwertmann and Cornell (2000). Samples containing μ_{Al} values of 0.0, 3.78 ± 0.02 and 7.61 ± 0.02 were prepared (μ_{Al} = [Al] × 100/[Al] + [Fe], [Me] M) and aged for 24 days at 70 °C. The samples (GAl_0, $GAl_{3.78}$ and $GAl_{7.61}$) were extracted to remove poorly crystalline compounds. The final precipitates were washed, dialyzed until the conductivity of the solution was similar to that of doubly distilled water, and dried at 50 °C for 48 hours.

2.2 *Adsorption isotherms*

Batch experiments were carried out to obtain adsorption isotherms at pH 5.5 ± 0.2 and 25.00 ± 0.02 °C. The As(V) concentration was varied between 0.03 and 0.67 mM.

2.3 *Scanning/Transmission Electron Microscopy (S/TEM)*

The morphological and chemical characterization of the samples was performed using a FEI scanning/transmission electron microscope (S/TEM) model Titan 80–300, equipped with FEG, Cs probe corrector and an energy dispersive X-ray (EDS) analyzer. Z-contrast images were acquired using a high angle annular dark-field detector in STEM mode. TEM samples were prepared depositing a drop of the particles suspension on a Holey Carbon film on 300 mesh Copper grids.

3 RESULTS

3.1 *Adsorption isotherms*

Arsenic(V) adsorption isotherms showed that $GAl_{3.78}$ was the sample with the higher adsorption capacity. In addition, all the isotherms displayed a maximum value at ~ 0.25 mM (Figure 1), which represents the saturation of the available sorption sites. At higher As(V) concentrations, the sorption decreased, indicating the blocking of adsorption sites. The form of the isotherm is unusual, and its bell shaped curve was already reported by Dixit *et al.* (2003) when studying the adsorption of As(V) onto amorphous Fe oxide (HFO).

3.2 *HRTEM and STEM microscopy*

As the decrease in adsorption detected at [As(V)]~0.25 mM could be ascribed to the formation of a surface precipitate of $FeAsO_4$ that blocks subsequent sorption, several micrographs of the surfaces of the As-adsorbed particles were analyzed by High Resolution TEM (HRTEM) and STEM microscopy. Figures 2 a) and b) show the HRTEM micrographs of $GAl_{3.78}$ after the adsorption in a solution containing $[As]_{solution} = 0.53$ mM.

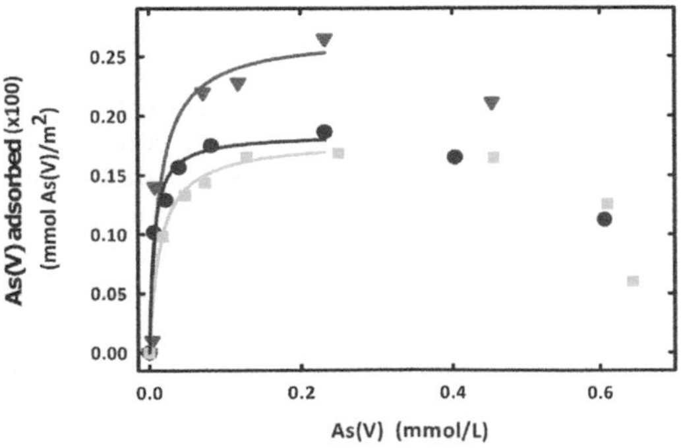

Figure 1. Adsorption isotherm of As(V) normalized by surface area. (●) GAl_0, (▼) $GAl_{3.78}$ and (■) $GAl_{7.61}$.

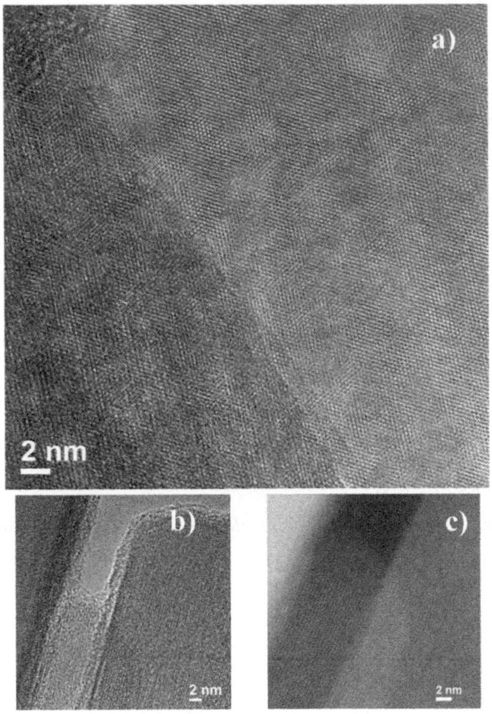

Figure 2. (a-b) HRTEM and (c) STEM images of As adsorbed onto $GAl_{3.78}$.

Figure 2a) displays two overlapping platelets with clear surfaces. No precipitate clusters are detected. Figure 2b) shows the crystallite domains of two almost parallel particles. Figure 2c) shows the STEM micrograph of the same particle at a high-Z contrast. An interplanar distance of 4.32 Å is detected and corresponds to (111) planes. The micrographs presented in Figure 2 are only a selection from a large series of micrographs with increasing magnification, none of them showing a surface precipitate.

4 CONCLUSIONS

The HRTEM and STEM analyses indicated that the surface of the sample with higher adsorption capacity is free of As precipitates. Based on these results, the formation of bidentated complexes, with As(V) bonded to two metal sites, was proposed. This coordination decreases the adsorption efficiency, and the form of the isotherm may be ascribed to the formation of monodentated mononuclear, and bidentated mono- and bi-nuclear species. These species have been formerly reported by

Fendorf *et al.* (1997) and Farrell *et al.* (2013), who observed that the type of the formed complex was dependent on the surface loading, with the bidentate complexes dominating at high surface loadings. We conclude that the decrease in adsorption is due to an increased formation of bidentated strong complexes that inhibit further As mobilization.

ACKNOWLEDGEMENTS

Electron microscopy analyses were carried out at INMETRO, Brazil. We thank MINCYT (Argentina) by PICT 2008–0780 grant.

REFERENCES

Abdullah, M. & Reis, M. 1998. Probable role of nutrition on arsenic toxicity. In International Conf. On Arsenic Pollution of ground water in Bangladesh: causes, Effects and Remedies; pp 8–12.

Aguilar, J., Dorronsoro, C., Fernández, E., Fernández, J., García, I., Martín, F., Sierra, M. & Simón, M. 2007. Remediation of As-Contaminated Soils in the Guadiamar River Basin. *Water, Air & Soil Pollution*, 180: 271–281.

Dixit, S. & Hering, J.G. 2003. Comparison of arsenic(V) and arsenic(III) sorption onto iron oxide minerals: implications for arsenic mobility. *Environ. Sci. Technol.*, 37: 4182–4189.

Farrell J. & Chaudhary B.K. 2013. Understanding Arsenate Reaction Kinetics with Ferric Hydroxides. *Environ. Sci. Technol.*, 47: 8342–8347.

Fendorf, S., Eick, M.J., Grossl, P. & Sparks, D.L. 1997. Arsenate and chromate retention mechanisms on goethite. 1. Surface structure. *Environ. Sci. Technol.*, 31(2): 315–320.

Manning, B. & Goldberg, S. 1996. Modeling Competitive Adsorption of Arsenate with Phosphate and Molybdate on Oxide Minerals. *Soil Sci. Soc. Am. J.* 60: 121–131.

Schwertmann, U. & Cornell, R.M. 2000. Iron Oxides in the Laboratory, Preparation and Characterization, 2nd Ed., *Wiley-VCH, Weinheim, Germany.*

Wang, S. & Mulligan, C.N. 2006. Occurrence of arsenic contamination in Canada: Sources, behavior and distribution. *J. Hazard. Mater.*, 138: 459–470.

One Century of the Discovery of Arsenicosis in Latin America (1914–2014) –
Litter, Nicolli, Meichtry, Quici, Bundschuh, Bhattacharya & Naidu (Eds)
 ISBN 978-1-138-00141-1

Effect of pH on As^{3+} sorption from aqueous solution by red clays

J.R. Rivera-Hernández
Posgrado en Ciencias del Mar y Limnología, Universidad Nacional Autónoma de México, Sin., México

C.R. Green-Ruiz
Unidad Académica Mazatlán, Instituto de Ciencias del Mar y Limnología, Universidad Nacional Autónoma de México

ABSTRACT: There is a global interest in As removal from aquatic environments. High As concentration can cause serious effects on the health of many living organisms, including humans. In this experimental work, the removal of As^{3+} from aqueous solution by red clays at different pH values (4, 6, 8 and 10), different initial concentrations (0, 0.1, 0.5, 1, 5 and 10 mg L^{-1}) and constant temperature (25 °C) was studied. An effect of pH on the sorption of As^{3+} was observed and the highest As(III) sorption occurred at pH 8 in 2 hours. These experiments showed a strong sorption of As^{3+} onto red clays at the beginning of the process (30 minutes) and the highest sorption at pH 4 for high concentrations, pH 8 for intermediate concentrations and pH 10 for low concentrations. There was not a significant effect of temperature on As(III) adsorption.

1 INTRODUCTION

The high concentrations of Arsenic (As) found in aquatic environment, mainly in groundwater, is the product of both natural and industrial processes. These concentrations have caused several serious health problems for many livings organisms, including the human, around the world (Litter *et al.*, 2010). There are many methods for As removal, such as adsorption, electrooxidation, precipitation, ion exchange, etc. The successful application of red clays as sorbent materials is due to their large surface area and porosity; in addition, these clays have negative surface charges, high availability in the environment and great capacity for regeneration (Carvalho *et al.*, 2008).

The main objective of this experimental work was to evaluate the effect of the initial concentration and pH on the sorption of As^{3+} by red clays.

2 METHODS/EXPERIMENTAL

0.25 g of red clays (from Huayacocotla, Hidalgo-Veracruz region, Mexico) were put into different polyethylene flasks (250 mL) containing 100 mL of As^{3+} solutions at five different concentrations (0.1, 0.5, 1, 5 and 10 mg L^{-1}). Prior to the experiments, these solutions were adjusted at pH values of 4, 6, 8, and 10 with 0.5 N HCl or NaOH. Aliquots of 5 mL were taken at different time intervals and centrifuged at 3000 rpm. Arsenic concentration in the residual solution was quantified in the supernatant by Atomic Fluorescence Spectroscopy (AFS). The quantity of As removed per gram of clay was estimated from the differences between initial (t_0) and final (t_n) concentrations, considering the volume and sorbent mass employed. The experiments were performed by quadruplicate. Kinetics (first, second and third order reaction; pseudo-first and pseudo-second order reaction; parabolic diffusion and Elovich reaction) and adsorption isotherms (Langmuir & Freundlich) were applied to evaluate the behavior and natural process. Averages and standard deviations were obtained for each one. Blanks were run under the same As concentrations but without red clays, in order to quantify the metal lost due to other than adsorption process. All materials used in this work were washed following the standards for metal analysis, and the reagents were analytical grade

3 RESULTS AND DISCUSSION

Figure 1 shows the contact time and pH effects on the As^{3+} sorption by red clays. The As^{3+} sorption was observed early after starting the experiments. The equilibrium was reached in 2 hours, except for pH 4, where equilibrium was not reached during the experimental time. A quick As^{3+} sorption was also observed by López *et al.* (1998) and Genç-Fuhrman (2004). They explained this by: 1)

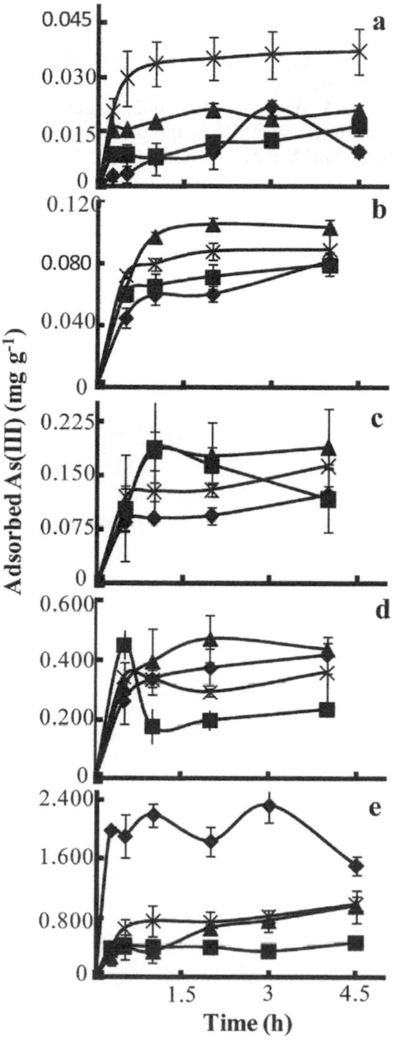

Figure 1. Kinetic of the sorption of As^{3+} by red clays (0.25 g), under different pH values (4[●], 6[■], 8[▲] and 10[X]), constant temperature (25 °C), different initial concentrations (0.1[a], 0.5[b], 1[c], 5[d] and 10[e] mg L^{-1}).

the high As^{3+} concentration in the solution and, 2) the high availability of sorption sites on the surface of the minerals, which are susceptible to be occupied by As^{3+} ions until equilibrium is reached.

For low concentration (0.1 mg L^{-1}), alkaline conditions (pH 10) favored As^{3+} sorption (Figure 1a). At intermediate concentrations (1 and 5 mg L^{-1}), the highest As^{3+} sorption occurred at pH 8 (Figures 1b and 1c). When the concentration was increased (10 mg L^{-1}), this process became more efficient under acid conditions (pH 4; Figure 1d). Rubinos *et al.* (2005) observed that at pH 4 there is a dissolution of red mud releasing Si, Al, Fe and Ca as well as As(III) previously sorbed into the experimental solutions. This fact can explain why the sorption process at this pH value did not reach the equilibrium in our experiments.

As regards to the pH effect at high As^{3+} concentration, several authors (Waltham & Eick, 2002; Dixit & Hering, 2003) mentioned that this behavior is due to the sorbent surface acquiring a net negative charge at high pH values, which generates a repulsion effect of the H_3AsO_3 deprotonated species ($H_2AsO_3^-$ and $HAsO_3^{2-}$).

Pseudo-second order reaction (R^2 above 0.95) and Langmuir models were the models that better fitted the experimental points of the As^{3+} adsorption process (Table 1). This suggests that the process takes place on a homogeneous surface, and that the adsorbate only interacts with one active site on the adsorbent.

Table 1. Parameters for As^{3+} sorption of by red clays.

	Freundlich		Thermodynamic		
T (°C)	K_f (L mg^{-1})	R^2	$\Delta G°$ (KJ mol)	$\Delta S°$ (J mol^{-1} K^{-1})	$\Delta H°$ (KJ mol^{-1})
10	1.232	0.9895	54972	−92.518	28776
25	1.313	0.9857	56360		
40	1.387	0.9662	57748		

4 CONCLUSIONS

Results of this experimental work demonstrated that As^{3+} was strongly sorbed on red clays in the first 30 minutes of contact time. In addition, the pH value for the maximum sorption capacity varied from high (10) to low pH (4) when the As^{3+} concentration was increased from 0.1 to 10 mg/L. The red clays from the Huayacocotla region are good As^{3+} geosorbent agents because they have a large specific surface area and suitable electrostatic charge.

REFERENCES

Carvalho, W.A., Vignado, C. & Fontana, J. 2008. Ni(II) removal from aqueous effluents by silylated Clays. *J. Hazard. Mater.* 153: 1240–1247.

Dixit, S. & Hering, J.G. 2003. Comparison of arsenic(V) and arsenic(III) sorption onto iron oxide minerals: Implications for Arsenic Mobility. *Environ. Sci. Technol.* 37: 4182–4189.

Genç-Fuhrman, H. 2004. *Arsenic removal from water using seawater-neutralized red mud (Bauxsol)*. PhD

Thesis. Environment & Resources DTU. Technical University of Denmark, 1–67.

Litter, M.I., Morgada, M.E. & Bundschuh, J. 2010. Possible treatments for arsenic removal in Latin American waters for human consumption. *Environ. Pollut.* 158: 1105–1118.

López, E., Soto, B., Arias, M., Núñez, A., Rubinos, D. & Barral, M.T. 1998. Adsorbent properties of red mud and its use for wastewater treatment. *Water Res.* 32: 314–1322.

Rubinos, D.A., Arias, M., Diaz-Fierro, F. & Barral, M.T. 2005. Speciation of adsorbed arsenic(V) on red mud using a sequential extraction procedure. *Mineral. Mag.* 69: 591–600.

Waltham, C.A. & Eick, M.J. 2002. Kinetics of arsenic adsorption on goethite in the presence of sorbed silicic acid. *Soil Sci. Soc. Am. J.* 66: 818–825.

One Century of the Discovery of Arsenicosis in Latin America (1914–2014) – Litter, Nicolli, Meichtry, Quici, Bundschuh, Bhattacharya & Naidu (Eds)
© 2014 Taylor & Francis Group, London, ISBN 978-1-138-00141-1

Enhancing arsenic removal by means of coagulation—adsorption—filtration processes from groundwater containing phosphates

A. González, A.M. Ingallinella, G.S. Sanguinetti, V.A. Pacini, R.G. Fernández & H. Quevedo
Centro de Ingeniería Sanitaria, Facultad de Ciencias Exactas, Ingeniería y Agrimensura, Universidad Nacional de Rosario, Argentina

ABSTRACT: The objective of this project was to improve Arsenic (As) removal from groundwater containing phosphates (PO_4^{3-}) in full-scale plants operating with the ArCIS-UNR® process. This process consists of a coagulation step with polyaluminum chloride (PACl), at pH 6.9 and double filtration. Bench-scale tests were conducted to evaluate the efficiency of As and PO_4^{3-} removal under different coagulant doses and pH conditions. The tests were performed using natural waters from two locations with As concentration ranging 0.15–0.20 mg/L and PO_4^{3-} ranging 0.5–2.5 mg/L. An acceptable PO_4^{3-} removal (70–80%) was reached at pH 6.9 and with 80 mg/L of coagulant. At this pH, another series of trials adding PACl coagulant in two stages was performed to evaluate the improvement in As removal. Preliminary results indicated that the addition of the appropriate dose of coagulant (PACl) in two stages could be a suitable alternative to increase As removal when PO_4^{3-} is present.

1 INTRODUCTION

The matrix of the water to be treated is an important factor in the process of selection of As removal technologies. Despite the importance of the presence of competing ions, not enough attention was given to the matrix when coagulation- adsorption—filtration processes were applied. The information available in the bibliography (Meng *et al.*, 2000, 2002; Holm, 2002; Roberts *et al.*, 2004; Jeong *et al.*, 2007) is diverse, because the studies were done either using a model water or natural waters with very different characteristics. However, the mentioned authors conclude that the most important competing ions are silicate (SiO_4^{4-}) and phosphates (PO_4^{3-}). The aim of the present project was to minimize the influence of the phosphate interference on As removal from natural waters treated in full-scale plants operating under the ArCIS-UNR® process (Ingallinella *et al.*, 2003). This process consists of a coagulation step with polyaluminum chloride (PACl), at pH 6.9 and double filtration. This study began with a compilation of the results in As and PO_4^{3-} removal, either at full-scale plants or at pilot-scale plants using natural waters, both of them based in the ArCIS-UNR® process. By the analysis of the results, it was detected that, with doses of 100–120 mg/L PACl, not only As had been removed (between 70–80%) but also had been PO_4^{3} (between 50–70%). In order to evaluate the behavior of PO_4^{3-}, at different pH and PACl doses, a series of trials at laboratory scale were carried out. The results of these assays led to try the addition of the coagulant in two stages in order to assess whether in such way an increase in As removal could be achieved. As hypothesis, it was proposed that the PO_4^{3-} removal in the first stage would lead to "less interference" to remove the residual As in the second stage.

2 EXPERIMENTAL METHODS

Two sets of Jar Tests using groundwater of different matrices obtained from two locations in the Santa Fe Province (Argentina), Villa Cañas and Pueblo Andino, were performed. Villa Cañas water has As concentrations between 0.15 and 0.20 mg/L, and PO_4^{3-} concentrations between 0.50 and 0.90 mg/L. Water from Pueblo Andino has similar concentrations of As, but higher concentrations of PO_4^{3-}, between 2.0 and 2.5 mg/L. The modifications of pH from natural groundwater were carried out by the addition of 1 N sulfuric acid, H_2SO_4 (96%). The used coagulant was PACl (16.9 m/m % Al_2O_3).

2.1 *Trials under different coagulant dose and pH*

Two sets of trials were performed in duplicate, with each of the natural waters, using doses of PACl of 40, 60, 80, 100, 120 and 140 mg/L, and different initial pH values: Trial 1 (natural pH 8.0) and Trial 2 (pH 6.9). Once the stage of sedimentation was completed, the supernatant of each jar was filtered by 0.45 μm pore cellulose nitrate membranes (Millipore).

2.2 *Trials with the addition of coagulant in two stages*

Trials at pH 6.9 and applying coagulant dose in two stages, with natural waters of Villa Cañas and Pueblo Andino, were carried out. In the first stage, a dose of 80 mg/L of PACl was used. This dose was chosen according to the results of the first series of tests. In a second stage, a dose of 40 or 60 mg/L of PACl was added, and thus the total doses of PACl used in this series of trials were 120 mg/l and 140 mg/L, respectively. As control, 120 or 140 mg/L of PACl were added in a single stage in one of the jars in order to compare results.

2.3 *Analytical methods*

In the filtered samples pH (Electrometric Method), turbidity (nephelometric method, Turbidimeter HACH 2100), As (silver diethyldithiocarbamate method, and atomic absorption spectrophotometry method) and PO_4^{3-} (molibdovanadate method) were determined.

3 RESULTS AND DISCUSSION

3.1 *Results using different dose of coagulant and pH*

Figures 1 and 2 show the average efficiency of PO_4^{3-} and As removal from groundwater at different PACl doses and pH.

The previous graphs show that, in both cases, with doses of 80 mg/L of PACl and at pH 6.9, a PO_4^{3-} removal between 70 and 80% was achieved. On the other hand, As removal increased with the increase of the coagulant dose. In addition, removal was greater at pH 6.9 than that at natural pH (8.0); this result matched with those reported in previous works (Ingallinella *et al.*, 2003).

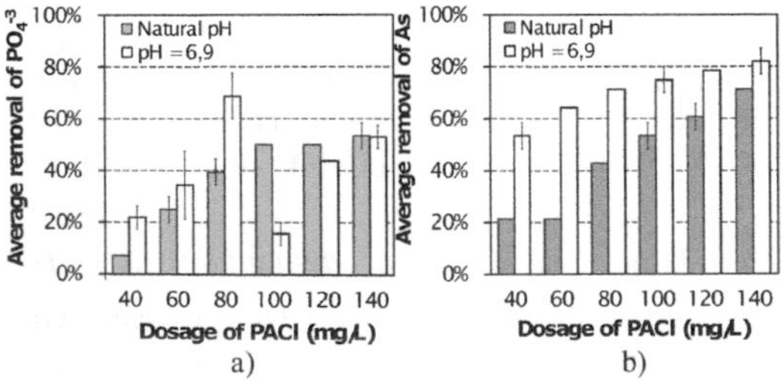

Figure 1. PO_4^{3-} (a) and As (b) removal from Villa Cañás waters at different pH and PACl doses. Initial As concentration: 0.14 mg/L. Initial PO_4^{3-}concentration: 0.80 to 0.95 mg/L.

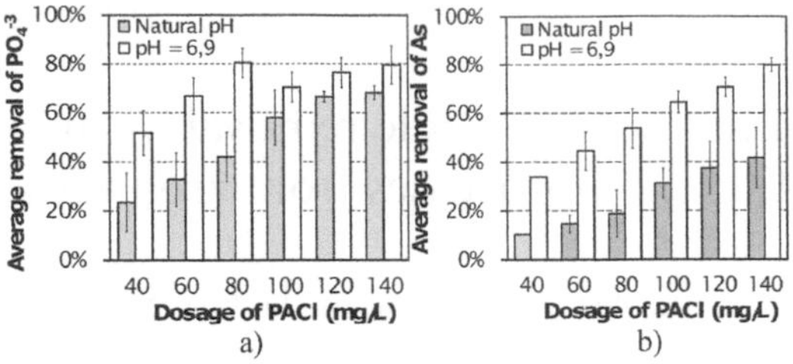

Figure 2. PO_4^{3-} (a) and As (b) removal from Pueblo Andino waters at different pH and PACl doses. Initial As concentration: 0.16 mg/L. Initial PO_4^{3-} concentration: 1.8 to 2.3 mg/L.

Table 1. As and PO_4^{3-} removal for Villa Cañás and Pueblo Andino waters, at pH 6.9, using 120 mg/L and 140 mg/L of PACl in a single dose or applied by two stages.

Dose (mg/L PACl)	Mean removal (%) Villa Cañas		Mean removal (%) Pueblo Andino	
	PO_4^{3-}	As	PO_4^{3-}	As
120	42	79	78	72
140	–	–	79	80
80 + 40	58	86	69	81
80 + 60	–	–	87	88

3.2 *Results with addition of coagulants with dosage in two stages*

Table 1 shows the results of As and PO_4^{3-} removal in the trials applying the addition of coagulant in one and in two stages at pH 6.9. With the addition of PACl in two stages, higher As removal was achieved than with the addition in only one stage for both natural sources of groundwater studied.

4 CONCLUSIONS

The addition of the appropriate dose of coagulant (PACl) in two stages could be a suitable alternative to increase As removal when PO_4^{3-} is present in groundwater and would be feasible to be applied at full-scale plants.

Further research work should be performed with different combinations of coagulant dose to be applied in two stages. In addition, the influence of other ions commonly found in groundwater of Argentina as silicates and carbonates should be studied in the future.

REFERENCES

Holm, T.R. 2002. Effects of CO_3^{2-}/bicarbonate, Si, and PO_4^{3-} on Arsenic sorption to HFO. *Journal AWWA*, 174–180.

Ingallinella, A. M., Fernández, R.G. & Stecca, L.M. 2003. Proceso ArCIS-UNR para la remoción de As y F -en aguas subterráneas: una experiencia de aplicación. *Revista Ingeniería Sanitaria y Ambiental, Edición* 66: 36–40 / 67: 61–65, AIDIS Argentina.

Jeong, Y., Maohong, F., Van Leeuwen, J. & Belczyk, J.F. 2007. Effect of competing solutes on As(V) adsorption using iron and aluminum oxides. *J. Environ. Sci.* 19(8): 910–919.

Meng, X., Bang, S. & Korfiatis, G.P. 2000. Effects of silicate, sulfate, and carbonate on arsenic removal by ferric chloride [J]. *Water Res.* 34(4): 1255–1261.

Meng, X., Korfiatis, G.P. & Bang, S., 2002. Combined effects of anions on arsenic removal by iron hydroxides. *Toxicol. Lett.* 133(1): 103–111.

Roberts, L., Hug, S., Ruettimann, T., Billah, M., Wahab Khan, A. & Rahman, M. 2004. *Environ. Sci. Technol.* 38(1): 307–315.

Sancha, A.M. 2010. In *Tecnologías económicas para el abatimiento de As en aguas*, M.I. Litter, A.M. Sancha & A.M. Ingallinella (eds), Editorial Programa Iberoamericano de Ciencia y Tecnologia para el Desarrollo, pp. 145–153.

One Century of the Discovery of Arsenicosis in Latin America (1914–2014) – Litter, Nicolli, Meichtry, Quici, Bundschuh, Bhattacharya & Naidu (Eds)
© 2014 Taylor & Francis Group, London, ISBN 978-1-138-00141-1

Adsorptive filter for removal of arsenic from drinking water

F.M. Francisca & M.E. Carro Perez
Universidad Nacional de Córdoba, Córdoba, CONICET, Argentina

A. Krapp & D.R. Panique Lazcano
Universidad Nacional de Córdoba, Córdoba, Argentina

ABSTRACT: The purpose of this paper is to design a novel module for the removal of Arsenic (As) in drinking water. The main removal mechanism of the proposed device consists on adsorption on a geomaterial. The adsorptive material is a granular media from Sierras Chicas in Cordoba Province, Argentine. Design of the removal device is based on previous batch and soil column tests performed to evaluate the capability of this material for removal of As from water. The removal efficiencies ranged from 12 to 92% depending on the initial As concentration and flow rate. Obtained results show that the developed filters successfully reduced As concentration in drinking water during more than 100 days.

1 INTRODUCTION

Available alternatives for arsenic removal include chemical oxidation, precipitation, coagulation, reverse osmosis, adsorption, biological degradation and electrokinetics (Litter *et al.*, 2010). However, adsorption was preferred in many cases due to its low cost and the wide range of reactive materials that may be used for arsenic retention (Mohan & Pittman, 2007; Maji *et al.*, 2008; Jovanović & Rajaković, 2010; Carro Perez & Francisca, 2013a). The adsorbing surfaces or reactive media usually include activated alumina, iron-based media or other oxides, bauxite, hematite, feldspar, laterite, clay minerals (e.g., bentonite and kaolin), activated carbon, cellulosic material, blast furnace slag, surfactant modified zeolite and ion exchange resin, among others.

The purpose of this work is to develop a low cost adsorptive filter for the removal of As from drinking water. A natural material is characterized with the purpose of assessing its sorbent capacity. Available batch and column tests experimental results are used for the development and calibration of a domestic filter for removal of As in drinking water.

2 EXPERIMENTAL METHODS

2.1 *Materials*

The sorbent material used in this research is a granular media obtained from Cuesta Colorada, La Calera, in Córdoba Province, Argentina. The tested geomaterial is composed of 26% of graves, 66% of sand and 8% of fines identified as montmorillonite by XRD analysis. Significant presence of iron oxides was also identified. Arsenic solutions were prepared from the dissolution of arsenic trioxide (As_2O_3) in an alkaline sodium hydroxide (NaOH) medium. Used arsenic concentrations were 0.1, 0.5, 1, 5, 10 and 15 mg L^{-1} to cover the registered As concentration in most aquifers in the center of Argentine (OSN, 1942). A natural water with As concentration of 0.05 mg L^{-1} was also included.

2.2 *As detection method*

Determinations of arsenic in water were performed by means of the Quantofix Arsen10 test kits using the optimized detection method developed by Carro Perez & Francisca (2013b).

2.3 *Filter development*

Two different filters for As removal from water were tested in laboratory. A transparent acrylic cell having 5 cm in diameter and 10 cm in length was used to calibrate a 1D mass transport model. The main characteristics of the column test are as follows: porosity = 0.48, bulk density = 15 KN m^{-3}, flow rate = 1.7×10^{-7} m^3 m^{-2} s^{-1}. These results allow obtaining calibration parameters to simulate the behavior of filters. On the other hand, a commercial 25 L water container was modified to include a reactive filter for As removal. The reactive media was placed between two layers of fine gravel and separated by geotextiles to avoid migration of small particles (Figure 1).

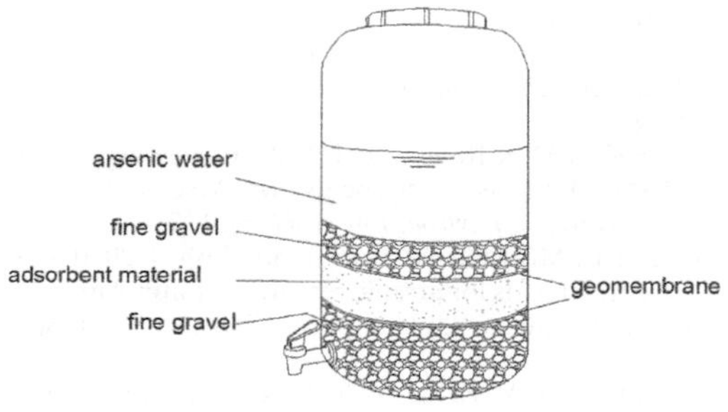

Figure 1. Developed filter.

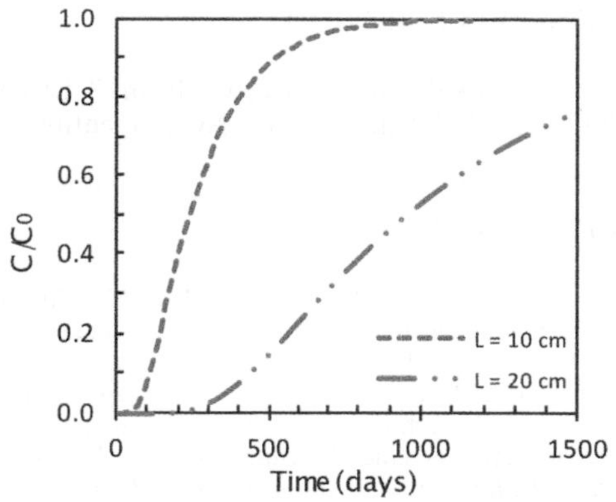

Figure 2. Life time of filters with different reactive length.

2.4 *Test procedures*

Filters were permeated with distilled water following the constant head technique (ASTM D2434, ASTM 2007) until obtaining constant hydraulic conductivity and then were permeated with arsenic solutions of known initial concentration. Arsenic in the effluent was periodically measured.

3 RESULTS AND DISCUSSION

3.1 *Laboratory soil column*

A 1D mass transport model was fitted to experimental results by means of least square fitting of the advection-dispersion-retardation equation (Fetter, 1993). Once calibrated, this model was used to obtain life-time charts for different filters with different lengths and initial As concentrations. As removal is mainly related to the presence of iron oxides, as suggested by Carro Perez & Francisca (2013a). Figure 2 presents the change in As concentration ($C_o = 0.3$ mg L^{-1}) with time for two reactive columns, 25 cm diameter and 10 cm and 20 cm height, hydraulic head 0.25 m. This chart was used to determine the time required for the As concentration to surpass 0.01 mg L^{-1} in the outlet port, the maximum allowed As concentration in drinking water according to the World Health Organization (WHO). Results are presented in Table 1. Once the As concentration surpasses 0.01 mg L^{-1}, the reactive materials have to be replaced.

Table 1. Filter lifetime for a 25 cm in diameter and 10 cm in length filters.

Initial concentration (mg L^{-1})	Lifetime (days)
0.05	107
0.1	104
0.3	80
0.5	60

3.2 *Case study: Colonia Las Pichanas*

Colonia Las Pichanas is located in San Justo County in the east of the Córdoba province. Rural and dispersed population at this location is near 500 people.

This community is settled in a region with aquifers having high arsenic concentrations according to historical data. In the past decade, a 350 meters deep groundwater well was installed for water provision. As concentration in these aquifers is in the 0.04–0.05 mg L^{-1} range, which is in good agreement with past local regulations. All other chemical parameters make this source of water compatible to be used as drinking water without any further treatment. However, at present, international regulations recommends As concentrations lower than 0.01 mg L^{-1} for drinking water. Then, the use of filters as the one developed in this work may be necessary at any time. From the As concentration measured in this aquifer and according to Table 1, the lifetime of filters developed in this work will be close to 100 days.

4 CONCLUSIONS

A low-cost module for As removal in drinking water was developed. The proposed filter could be used to provide practical solutions to rural and dispersed population without any other source of water. The tested filters were efficient in removing As from drinking water during more than 100 days. The efficiency of the filter for As removal from real groundwater has to be proved before use.

ACKNOWLEDGEMENTS

Authors thank the support received from FCEFyN-UNC, CONICET and ISEA-UNC (Argentina).

REFERENCES

Carro Pérez, M.E. & Francisca, F.M. 2013a. Arsenic entrapment in reactive columns of residual soils. *J. Environ. Eng.* 139(6): 788–795.

Carro Perez, M.E., & Francisca, F.M. 2013b. Digital analysis technique for uncertainty reduction in colorimetric arsenic detection method. *J. Environ. Sci. Health. Part A: Environ. Sci. Eng. Toxic Hazard. Subst. Control,* 48(2): 191–196.

Fetter, C.W. 1993. Contaminant hydrogeology, Prentice Hall, Upper Saddle River, New Jersey.

Francisca, F.M. & Carro Pérez, M.E. 2013. Remoción de arsénico en agua mediante procesos de coagulación-floculación, *Rev. Int. de Contaminación Ambiental* (in press).

Jovanović, B.M. & Rajaković, L.V. 2010. New Approach: Waste Materials as Sorbents for Arsenic Removal from Water. *J. Environ. Eng.* 136(11): 1277–1286.

Litter, M.I., Morgada, M.E. & Bundschuh, J. 2010. Possible treatments for arsenic removal in Latin American waters for human consumption. *Environ. Pol.* 158(5): 1105–1118.

Maji, S.K., Pal, A., & Pal, T. 2008. Arsenic removal from real-life groundwater by adsorption on laterite soil. *J. Hazard. Mat.* 151: 811–820.

Mohan, D. & Pittman Jr., C.U. 2007. Arsenic removal from water/wastewater using adsorbents—A critical review. *J. Hazard. Mat.* 142: 1–53.

OSN, 1942. El problema del agua potable en el interior del país. Tomo II. Buenos Aires.

One Century of the Discovery of Arsenicosis in Latin America (1914–2014) – Litter, Nicolli, Meichtry, Quici, Bundschuh, Bhattacharya & Naidu (Eds)
© 2014 Taylor & Francis Group, London, ISBN 978-1-138-00141-1

Adsorptive removal of arsenic from waters using natural and modified Argentine zeolites

P.A. Camerotto Andreani & M.A. Trinelli
Instituto de Investigación e Ingeniería Ambiental (3iA), UNSAM, Buenos Aires, Argentina
Departamento de Ciencias Geológicas, Facultad de Ciencias Exactas y Naturales, UBA, CABA, Argentina
INQUIMAE/DQIAQF, Facultad de Ciencias Exactas y Naturales, UBA, CABA, Argentina

P.R. Leal
Departamento de Ciencias Geológicas, Facultad de Ciencias Exactas y Naturales, UBA, CABA, Argentina

A.V. Rodriguez, A.I. Rodriguez, J. Llano, & M. dos Santos Afonso
INQUIMAE/DQIAQF, Facultad de Ciencias Exactas y Naturales, UBA, CABA, Argentina

ABSTRACT: A removal method for arsenic in polluted water using zeolites was evaluated. Natural and commercial zeolites were used; the second one was also treated with iron (T-CZ). Adsorption isotherms were evaluated. Pure natural zeolites had more adsorption capacity than commercial ones, and T-CZ presented a significantly higher adsorption. Also, experiments at pH 5.5 showed better adsorption capacities compared with those at lower pH values. T-CZ represents a novel material that may be used as filter for arsenic-contaminated waters.

1 INTRODUCTION

According to the WHO, in Latin America, the main affected zones by the presence of Arsenic (As) in drinking water are Argentina, Chile and Mexico. In Argentina, the allowable limit according to the Argentine Food Code (CAA) is 10 μg L^{-1}.

Zeolites possess cavities of molecular dimensions that allow them to work as ionic exchangers and as true molecular sieves. These properties make zeolites good adsorbents for several pollutants. They can also be specifically modified to increase their detoxification capacity (Chmielewská & Pilchowski, 2006). The aim of the present work was to study the removal of As from water using zeolites.

2 METHODS/EXPERIMENTAL

2.1 *Zeolite samples*

Zeolite samples were collected from different sites in Argentina. Cli-PZ, Heu-St-PZ, St-PZ, Ana-PZ, Lau-b-PZ and Lau-PZ samples were obtained from the Neuquén province, whereas Cha-PZ was collected from the Misiones province. In most samples, zeolite assemblages are intergrown with other minerals, such as calcite, chalcedony and Fe and Mg phyllosilicates. Cli-PZ is mainly composed by clinoptilolite with minor amount of heulandite, barrerite and stilbite; Lau-PZ sample mainly contains laumontite with very low quantities of epidote and chlorite (Depine *et al.*, 2003). Samples were concentrated in order to reduce the occurrence of accessories species. A commercial sample from the San Juan province was also used (CZ).

2.2 *Preparation of iron treated zeolite*

Zeolites were treated with iron following the same method used for synthesis of pillared clays (Marco-Brown *et al.*, 2012). CZ, previously milled or not, was added to the treating solution. A part of the solid product was heated at 300 °C for 2 hours (TC-CZ). The other no calcined portion was kept and will be denoted as T-CZ.

2.3 *Adsorption experiments*

Solids were mixed with As(V) (Merck) solutions with concentrations ranging from 0 to 500 μg L^{-1} on an orbital shaker. Ionic strength was maintained constant at the value of 1 mM with KNO_3. Arsenic-zeolite adsorption studies were determined at two different pH values, 4.0 and 5.5, adjusted with HNO_3 and KOH 0.1 M until equilibrium was reached. The As uptake was calculated using the following mass balance equation (eq. 1):

$$q = (C_i - C_{eq})V/(10^6 \times m) \quad (1)$$

where q is the adsorption capacity (mg g^{-1}), C_i and C_f are the initial and final concentrations of As(V) (μg L^{-1}), V is the aliquot volume (mL) and m is the mass used of the solid (g). All experiments were performed by duplicate. Relative standard deviations were below 5% in all cases. The statistical data analysis was performed using the Sigma Plot software package.

2.4 *Arsenic analysis*

Arsenic analysis was carried out using an ICP-MS (Agilent, 7500 series).

3 RESULTS AND DISCUSSION

3.1 *Preliminary tests*

Preliminary tests, starting with an As concentration of 160 μg L^{-1}, showed that Cha-PZ, Lau-b-PZ, St-PZ, Heu-St-PZ and Ana-PZ adsorbed As in the range of 10% to 48%. In contrast, Cli-PZ and Lau-PZ adsorbed 71% and 65%, respectively. According to these results, the Lau-PZ sample and the commercial one were selected to obtain the adsorption isotherms.

3.2 *Effect of pH*

CZ and Lau-PZ presented lower efficiency at pH 4 than at pH 5.5. This can be explained taking into account the As speciation diagram: at pH 5.5 there is a larger proportion of $HAsO_4^{2-}$ than at pH 4 were the main species is $H_2AsO_4^-$. The stronger affinity of the As species at pH 5.5 for the zeolite surface containing positive charges is then expected, being consistent with the experimental results.

3.3 *Adsorption isotherms*

CZ shows an increasing monotonous isotherm at pH 5.5 (Figure 1). Lau-PZ isotherm at pH 5.5 shows a novel tendency with a bell form (Figure 1). This behavior can be explained by the formation of bidentate complexes between the zeolite surface and aqueous As.

Similar results were found by Tufo *et al.* (unpubl.). The maximum adsorption capacity reached by Lau-PZ doubles the one obtained by CZ (as it can be also observed in Fig. 2). This implies that the main active sites for As adsorption reside on the zeolite surface. In CZ material, besides zeolite, other substances are present, making the product impure, and this explains the form of the isotherm as the ones obtained for clays (Damonte *et al.*, 2007).

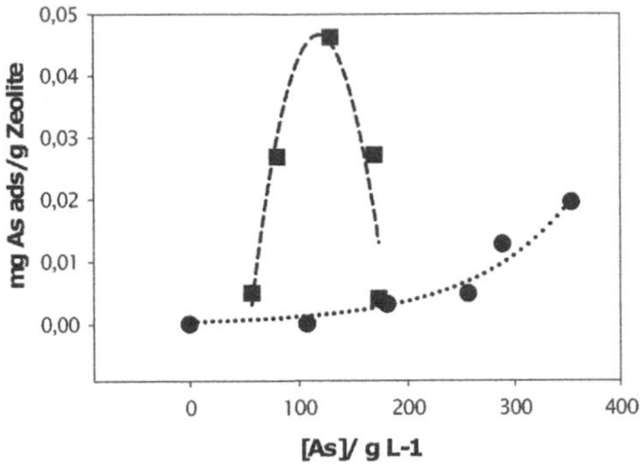

Figure 1. Adsorption isotherms for CZ (•) and Lau-PZ (■) at pH 5.5. Dotted lines are guide to the eyes.

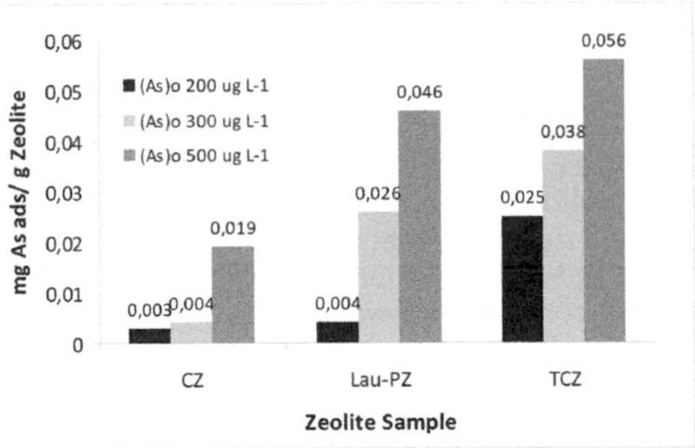

Figure 2. Arsenic removal capacities for different materials at pH 5.5.

3.4 *Treated zeolites*

The treatment with iron was very effective. Figure 2 shows, for different initial As concentrations, that, among the three zeolite samples, the best performance is reached by T-CZ. Grained and milled treated zeolites did not show significant differences for the As removal efficiency. Besides, the calcination process did not improve the removal efficiency. The efficiency for all the iron-treated samples was 0.056 mg g^{-1}. For all As initial concentrations tested, the remaining As concentration found in solution was lower than 3 μg L^{-1}.

4 CONCLUSIONS

The iron treatment in the commercial zeolite (T-CZ) highly increased aqueous arsenic removal capacity up to levels lower than the CAA limit (10 μg L^{-1}). In consequence, this new material is strongly promising for the development of filters for treating arsenic contaminated waters.

ACKNOWLEDGEMENTS

UBA, UNSAM, CONICET (Argentina), J.L. Marco-Brown.

REFERENCES

Chmielewská, E. & Pilchowski, K. 2006. Surface modifications of natural clinoptilolite-dominated zeolite for phenolic pollutant mitigation. *Chem. Papers*, 60: 98.

Damonte, M., Torres Sánchez, R.M. & dos Santos Afonso, M. 2007. Some aspects of the glyphosate adsorption on montmorillonite and its calcined form. *Appl. Clay Sci.* 36: 86–94.

Depine, G., Gargiulo, M.F., Leal, P.R., Scaricabarozzi, N., Spagnuolo, C. & Vattuone, M.E. 2003. Paragénesis de ceolitas en rocas volcánicas de la Cordillera Patagónica Septentrional, Villa La Angostura, Neuquén, Argentina. *10th Chilean Geologycal Congress.* 1–11.

Marco-Brown, J.L., Barbosa-Lema, C.M., Torres Sánchez, R.M., Mercader, R.C. & dos Santos Afonso, M. 2012. Adsorption of picloram herbicide on iron oxide pillared montmorillonite. *Appl. Clay Sci.* 58: 25–33.

Tufo, A., dos Santos Afonso, M. & Sileo, E. unpubl. Influence of Aluminium Incorporation in the Arsenic Adsorption onto Goethite. *Enviromental Science and Technology*.

One Century of the Discovery of Arsenicosis in Latin America (1914–2014) – Litter, Nicolli, Meichtry, Quici, Bundschuh, Bhattacharya & Naidu (Eds)
© 2014 Taylor & Francis Group, London, ISBN 978-1-138-00141-1

Sorption and chemical stability of As(V) onto tin doped goethites

A.L. Larralde, A.E. Tufo & E.E. Sileo
INQUIMAE, DQIADF, Facultad de Ciencias Exactas y Naturales, Universidad de Buenos Aires, Buenos Aires, Argentina

ABSTRACT: The effect of Sn(IV)-for-Fe(III) substitution on the adsorption of As(V) and on the reactivity in acid media were studied for pure and Sn-substituted goethites. Adsorption isotherms indicated that the As adsorption increases in the Sn-goethites almost duplicating the value of pure goethite. Dissolution experiments showed sigmoidal curves with K rate values that decreased with the Sn(IV) incorporation. The results indicate that the Sn-for-Fe substitution enhances the adsorption of As(V) and stabilizes the oxide towards dissolution in acid media, confirming the usefulness of Sn-goethites in removal technologies.

1 INTRODUCTION

Goethite (α-FeOOH) is a common mineral widely found in nature. It is rarely stoichiometric and may contain trace concentration of foreign cations. The mineral exhibits a highly reactive surface and a strong affinity for inorganic anions, organic ligands, Metal cations (Me) and As. The sorption properties of the oxide are altered by the Me-for-Fe(III) substitution, and when the incorporating cation is in a tetra oxidation state (Me(IV), the superficial charge is substantially changed modifying significantly the adsorption efficiency of the oxy(hydr)oxide.

2 METHODS

2.1 *Adsorbents*

Pure and tin-doped goethites were prepared following Schwertmann and Cornell (2000), from initial solutions containing different volumes of 1 M $Fe(NO_3)_3$ and 1 M $SnCl_2$ solutions, in the appropriate fractions to yield the desired μSn ratio (μSn [mol mol^{-1}] = [Sn] × 100/[Sn] + [Fe], [Me]: mol L^{-1}). Three suspensions were aged in a basic media for 20 days at 70 °C. The amorphous materials were removed with 0.4 M HCl at 80 °C. The final products were filtered, washed and dialyzed. Final μSn contents in the samples were 0.0, 2.1 and 5.5.

2.2 *Adsorption isotherms*

Batch experiments were carried out to obtain adsorption isotherms at pH 4.0, 5.5 and 7.0 ± 0.2 and 25.00 ± 0.02 °C. The As(V) concentration was 35 mM.

2.3 *Dissolution experiments*

Batch measurements were carried out to obtain dissolution rates in acidic media (HCl 6 M and oxalic acid 0.1 M/Fe(II), 5.12×10^{-4} M and $NaClO_4$ to control the ionic strength).

2.4 *Characterization of samples*

The Fe content in the samples was determined spectrophotometrically and the tin content was determined by ICP measurements. Specific surface area measurements (BET method) were obtained in a Micromeritics AccuSorb 2100 equipment. Diffraction patterns were recorded in a Siemens D5000 diffractometer and the data were analyzed using the GSAS system with the EXPGUI interface. Thermogravimetric (TGA) analysis was performed in a TGA-51 equipment in a N_2 atmosphere in the temperature range 30–450 °C. Previous Mössbauer experiments have shown that Sn is incorporated as Sn(IV) creating vacancies in the adsorber (Larralde *et al.*, 2012).

3 RESULTS AND DISCUSSION

3.1 *Chemical analysis*

Table 1 shows the nominal and final concentrations of Sn in the prepared samples.

3.2 *Surface area*

Figure 1 shows the BET isotherms for all samples. The three isotherms belong to *Type IV*, presenting a very thin hysteresis cycle. The almost constant values observed at low relative pressures indicate a strong interaction between adsorbent and adsorbate caused by the adsorption of N_2 onto the micropores.

Table 1. Molar Sn content and particle mean sizes for the prepared goethites.

	GSn0	GSn2.1	GSn5.5
µSn nominal	0.0	5.0 ± 0.2	10.0 ± 0.2
µSn incorporated	0.0	2.1 ± 0.1	5.5 ± 0.1
Pperp. (nm)	36	96	175
Pparal. (nm)	116	178	186

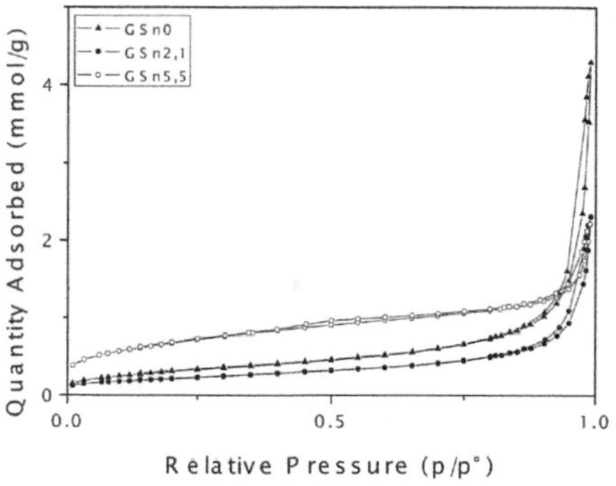

Figure 1. N_2 adsorption isotherms for the tin-doped goethites.

Specific Surface Areas (SSA) follow the trend GSn2.1 (17.19) < GSn0 (25.75) < GSn5.5 (54.90 $m^2\ g^{-1}$), indicating that Sn-incorporation modifies the SSA, and that the sample with higher Sn-content (GSn5.5) presents the largest superficial area.

3.3 *Rietveld simulation*

Rietveld simulation indicated that the crystallite size increases along tin incorporation. See Table 1 for Pperp and Pparal values.

3.4 *TG analyses*

The weight losses indicated that the formed goethites are metal deficient and present the following stoichiometry:

GSn-0: α-$Fe_{0.983}O_{0.949}(OH)_{1.051}$
GSn-2.1: α-$Fe_{0.979}Sn_{0.021}O_{0.906}(OH)_{1.094}$
GSn-5.5: α-$Fe_{0.934}Sn_{0.044}O_{1.023}(OH)_{0.977}$

3.5 *As(V) adsorption*

As(V) adsorbed by the samples was measured at pH = 4.0; 5.5 and 7.0 (see Figure 2). The adsorption expressed as mmol As per g follows the trend GSn2.1 (2.80×10^{-2}) < GSn0 (3.97×10^{-2}) < GSn5.5 (7.21×10^{-2} mmol As g^{-1}), indicating that the As-adsorption onto GSn5.5 duplicates the As adsorbed onto pure goethite. The adsorption per m^2 follows the same trend.

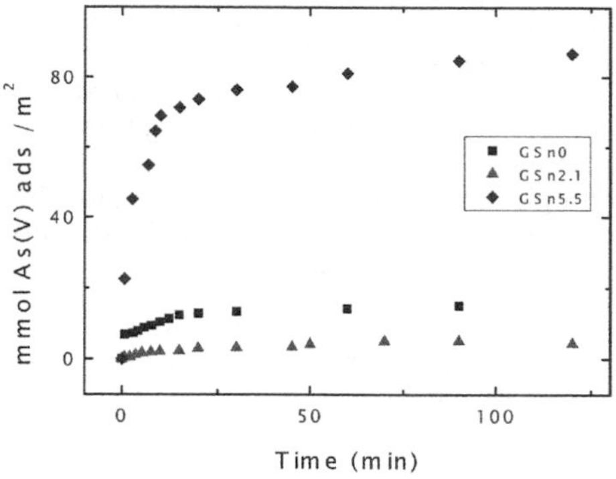

Figure 2. Adsorption of As(V) onto pure and tin-doped goethites.

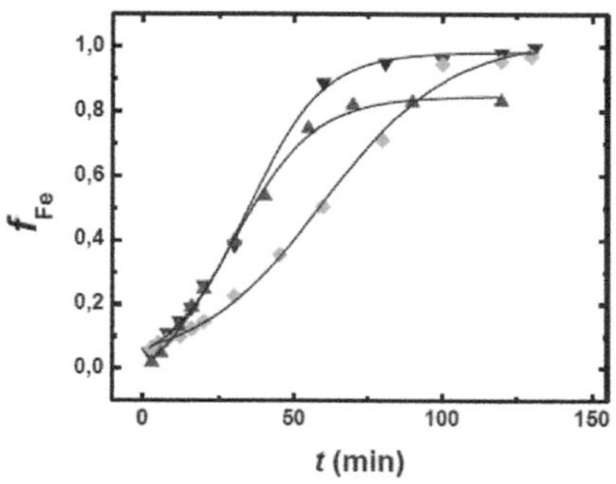

Figure 3. f_{Fe} vs. t (min) for (▲) GSn0; (▼) GSn2.1 and (■) GSn5.5.

The removal of As(V) after 2 hours at 25.0 °C was also measured and follows: GSn5.5 (90.00%) > GSn0 (49.54%) > GSn2.1 (34,95%), highlighting the good efficiency of GSn5.5 for As-removal.

3.6 *Dissolution experiments*

In order to evaluate the stability of the prepared samples in acid media, we have measured the dissolution in HCl 6 M. Figure 3 shows the results. The curves are more sigmoidal as the Sn(IV) content increases. The experimental data have been modeled using the Avrami Erofe´ev approach The obtained dissolution rate values are: GSn0 (0.0110) > GSn2.1 (0.108) > GSn5.5 (0.0089).

The results indicate that sample GSn5.5 dissolves more slowly than pure goethite.

CONCLUSIONS

We have obtained Sn(IV)-doped samples that:

- are non-stoichiometric goethites,
- the doped-samples present vacancies because of the Sn(IV)-for Fe(III) substitution,

- Sample GSn5.5 presents the highest SSA value and the highest crystallinity,
- Sample GSn5.5 shows the greatest As(V) adsorption, duplicating the adsorption of pure goethite,
- Sample GSn5.5 present the lowest dissolution constant in HCl 6 M, indicating the stability of the Sn-goethite in solution.

We are currently performing other kinetics runs in oxalic acid plus Fe(II), and measuring also the reactivity of the As-adsorbed samples.

ACKNOWLEDGEMENTS

We thank MINCYT by PICT 2008–0780 grant.

REFERENCES

Larralde, A. Ramos, C.P., Arcondo, B. Tufo, A.E. Saragovi, C. & Sileo E.E. 2012. *Materials Chemistry and Physics*, 133: 735–740.

Schwertmann, U. & Cornell, R.M. 2000. Iron Oxides in the Laboratory, Preparation and Characterization, 2nd Ed., Wiley-VCH, Weinheim, Germany.

*One Century of the Discovery of Arsenicosis in Latin America (1914–2014) –
Litter, Nicolli, Meichtry, Quici, Bundschuh, Bhattacharya & Naidu (Eds)*
© 2014 Taylor & Francis Group, London, ISBN 978-1-138-00141-1

Arsenic removal with compressed laterite

Eduardo Hryczyñski, W.G. López & J. Alegre
Laboratorio de Investigación y Desarrollo, UNCAUS, P.R. Saenz Peña, Chaco, Argentina

E. Soria
Laboratorio de Farmacotecnia, UNCAUS, P.R. Sáenz Peña, Chaco, Argentina

M.C. Giménez
Laboratorio de Química Analítica, UNCAUS, P.R. Sáenz Peña, Chaco, Argentina

ABSTRACT: In rural areas of the Chaco province, the provision of drinking water is difficult, so groundwater is used from water table aquifers. The quality of these waters varies considerably in terms of elements, having high salt concentrations and, in other cases, the presence of trace elements such as arsenic. An economical alternative, valid and easy to apply is the removal of arsenic using compressed laterite. The land laterites, with a high content of iron from Posadas, Misiones, Argentina, in the form of tablets, were used, showing excellent results. The amount of arsenic that can be removed with this method can be up to 90%.

1 INTRODUCTION

In Latin America, more than four million people consume water with concentrations greater than 0.05 mg As L^{-1} (Bundschuh *et al.*, 2006). In the province of Chaco, a large population consumes waters with high content of salinity and hardness, and also with the presence of other elements such as vanadium, selenium, fluorine, among others (Blanes *et al.*, 2006).

An economical and easy to handle arsenic removal method, which uses regional and laterite soils as a filter medium, were employed used in previous studies by Storniolo *et al.* (2007). Analysis performed demonstrated the ability of materials to remove arsenic due to the high content of metal oxides and hydroxides (Fe, Al, Mn).

In previous studies (Hryczyñski *et al.*, 2011), it was shown that arsenic removal after 30 minutes contact time was within the range of 95–99% when the ground laterite in the form of spheres were used. However, a remarkable disintegration was observed, with significant obstructions in the system. Accordingly, the aim of this work was to use tablets of laterite in order to provide better mechanical strength, contact area and compactness in the filter material.

2 METHODS/EXPERIMENTAL

2.1 *Materials*

The experiment was performed using water from underground boreholes of President Roque Saenz Peña, Chaco Province, containing 0.25 mg As L^{-1} and land laterite from Posadas, Misiones Province, Argentina.

The material, laterite, was thermally treated and sifted, mixed with alloying elements, and then pressed to a tablet, in order to give sufficient mechanical strength, a high contact area and compactness. Sand of the Paraná river as siliceous support material, was also used.

2.2 *Methods*

The laterite tablets were made sterilizing earth at 140 °C for 18 hours. Then, the powder was pulverized, sieved using a mesh N° 12 and mixed with ethyl cellulose. Subsequently, it was moisturized with alcohol 96° GL until consistency of wet mass. The powder was then granulated and heat in oven at 50 °C for 12 hours. Slip agents (talc and magnesium stearate) were added, and the mixture was compressed by a Model 3914 MP4 Minipress machine. The dimensions of the compressed laterite were 1.2 cm in diameter and 0.3 cm in thickness.

The filtration system was constituted by two filter elements of the same dimensions. The dimensions of the filter cartridges are 21 centimeters long and 7 centimeters in diameter with a capacity of 800 cubic centimeters. In the first filter, layers of sand and layers of laterite tablets were alternatively distributed. In the second filter element, in series, 1600 grams of sand were placed. The filters were arranged to have a continuous system.

Groundwater to be treated was previously chlorinated at a concentration of 3 mg L^{-1}. Then, water was passed (flow rate: 1 L h^{-1}) first through the cartridge with tablet laterite and sand, and then by the sand cartridge.

The analytical determination of As in the samples was performed by HGAAS (Hydride Generation Atomic Absorption Spectroscopy) using hollow cathode lamps at 193.7 nm wavelength. The instrument quantification limit was 5 mg L^{-1}, and an intermediate accuracy of less than ± 10% was achieved.

3 RESULTS AND DISCUSSION

In previous works, high soil-water ratios, such as 10%, were needed for a good performance (Storniolo *et al.*, 2007). In contrast, in our experiences, the soil-water ratio was only of 0.14%.

Samples of 105, 210 and 420 g of laterite, distributed in uniform layers, were used in layers alternating with sand in the filter. Fifteen runs were made in each case, except with the sample of 420 g, because the flow was interrupted to reach 50 L in 5 trials.

The average results obtained in each experiment are gathered in Table 1. It can be seen that best results were obtained when 210 g of laterite was used in the filter. For example, after passing 150 L of groundwater, arsenic concentration decreased from 0.25 to 0.05 mg As L^{-1}, whereas, when passing 20 L, this concentration decreased from 0.25 to 0.005 mg As L^{-1}, reaching an arsenic removal greater than 90%. However, further research is needed to establish the reasons why the filter is clogging after 50 liters when 420 g of laterite is used.

Table 1. Residual arsenic (mg L^{-1}) in treated water with compressed laterite at different flows.

Amount of laterite	Flow of groundwater (L)						
(g)	5	20	50	75	100	125	150
105	0,02	0,01	0,03	0,05	0,1	0,2	0,25
210	0,02	0,005	0,01	0,01	0,025	0,03	0,05
420	0,02	0,05	0,01	–	–	–	–

Taking in account the amount of tablets of laterite used, a good volume of water could be treated.

4 CONCLUSIONS

The results show that alternative arsenic removal by water-soil contact with compressed laterite is feasible reaching an arsenic removal greater than 90%.

In order to establish the reasons why the filter is clogging after 50 liters when 420 g of laterite is used, it is necessary carried out more experiments. This work is under development in our laboratory.

ACKNOWLEDGEMENTS

The authors acknowledge Universidad Nacional del Chaco Austral (PI 36/00005) for financial support.

REFERENCES

Bundschuh, J., García, M.E., Bhattacharya, P. & Guérèquiz, A.R. 2006. Latinoamérica Rural—Una parte olvidada del problema global. *II Seminario hispano-Latinoamericano sobre temas actuales de Hidrología subterránea -de arsénico en agua subterránea.*

Blanes, P.S., Herrera Aguad, C.E. & Giménez, M.C. 2006. Arsénico y otros oligoelementos asociados en aguas subterráneas de la provincia del Chaco. Preliminar study.

Hryczyñski, E., Lopez, W.G., Alegre, J. & Giménez, C. 2011. Comunicaciones Científicas y Técnicas, Universidad Nacional del Chaco Austral. Eliminación de arsénico en aguas subterráneas utilizando arcillas lateríticas.

Storniolo, A., Martín, R., Thir, M., Cortes, J., Ramírez, A., Mellano, F., Bundschuh, J. & Bhattacharya, P. 2005. Disminución del contenido de Arsénico en el agua mediante el uso de material geológico natural. Taller de Arsénico en aguas: origen, movilidad y tratamiento: 173–182. *II Seminario Hispano-Latinoamericano sobre temas actuales de Hidrología Subterránea y IV Congreso Hidrogeológico Argentino. Río Cuarto—Argentina.*

Storniolo, A., Martín, R., Thir, M., Cortes, J., Ramírez, A., Terribile, M., Bejarano, R. & Bundschuh, J. 2007. La laterita en la disminución del contenido de arsénico en el agua subterránea.

One Century of the Discovery of Arsenicosis in Latin America (1914–2014) – Litter, Nicolli, Meichtry, Quici, Bundschuh, Bhattacharya & Naidu (Eds)
 ISBN 978-1-138-00141-1

Granular Ferric Hydroxide (GFH) shows promise for removal of natural arsenic from water

A.H. Mahvi
School of Public Health, Tehran University of Medical Sciences, Tehran, Iran
Center for Solid Waste Research, Institute for Environmental Research, Tehran University of Medical Sciences, Tehran, Iran
National Institute of Health Research, Tehran University of Medical Sciences, Tehran, Iran

A.R. Asgari & N. Yousefi
School of Public Health, Tehran University of Medical Sciences, Tehran, Iran
Center for Solid Waste Research, Institute for Environmental Research, Tehran University of Medical Sciences, Tehran, Iran

S.S. Hosseini, S.A. Hosseini, H. Boudaghi Malidareh & S. Namavar
School of Public Health, Tehran University of Medical Sciences, Tehran, Iran

ABSTRACT: Granular Ferric Hydroxide (GFH) has been used as a synthetic adsorbent for removal of natural As in three concentrations (0.5, 1.0 and 2.0 mg L^{-1}). Description of adsorption isotherms has also been accomplished. This adsorbent showed good performance when used in low dose of 0.5 g L^{-1}, and no considerable interference was observed by other anions (SO_4^{2-} and Cl^-). It appeared that absorbability of both arsenate and arsenite by GFH could be expressed by Freundlich isotherm with $R^2 > 0.96$, whereas arsenate adsorption (with $R^2 > 0.94$) can be better described by a Langmuir isotherm than arsenite ($R^2 > 0.92$). Results also indicate that the amount of iron added to water was much higher than the standard value of 0.3 mg L^{-1} set for drinking water. Nevertheless, this method is more advantageous than others and it can be easily accomplished without need of pH modification.

1 INTRODUCTION

Reports of arsenic contamination of western parts of Iran, especially in the Kurdistan province, were published (Van *et al.*, 2003). Arsenic can be removed from surface and groundwater supplies by iron or alum coagulation and filtration, reduced iron oxidation system, and lime softening, among other methods (Chen *et al.*, 2002). Many iron oxides such as Granular Ferric Hydroxide (GFH) have also shown a high adsorption capacity for arsenic, particularly in the As(V) form (Jakel & Seith, 2000). GFH is an adsorbent, developed especially for arsenic removal from natural water. This product was first manufactured at the Technical University of Berlin, Germany, Department of Water Quality Control, for selective removal pollutants (Driehaus, 1998). The objective of this project was to determine the effectiveness of this new adsorbent in removal of As compounds (arsenate and arsenite) from drinking water.

2 EXPERIMENTAL

2.1 *Preparation of the adsorbent*

Before use, Granular Ferric Hydroxide (GFH, Wasserchemie Gmb H and Co. KG.) was dehydrated in oven (105 °C) for 90 minutes and placed in desiccators for cooling (Van *et al.*, 2003).

2.2 *Preparation of water samples*

Salts of arsenic used for making of synthetic water samples were analytical grade sodium arsenate ($Na_2HAsO_4.7H_2O$) and sodium arsenite ($NaAsO_2$), and the required solution concentrations of these salts (0.5, 1 and 2 mg L^{-1}) were daily prepared in order to prevent the possible oxidation of arsenite to arsenate.

2.3 *Experimental procedure*

The doses of GFH used in these testes were 0.25, 0.5, 1.0 and 1.5 g L^{-1}. Arsenic adsorption was examined at different contact times of 5, 10, 15, 30 and 60 minutes, and detection of residual concentration of As(III) and As(V) was accomplished by the use of an appropriate arsenic test kit (As Test EZ Kit) produced by American Hach Company. pH adjustment of water samples in the range of 5.0, 5.5, 6.0, 6.5, 7.0, 7.5 and 8.0 had been carried out by addition of 6 N nitric acid or 6 N sodium hydroxide solution, and pH analysis was performed by a pHmeter model E520. Development of adsorption isotherms for this study has been done by exposing known concentrations of arsenate and arsenite (0.25. 0.5, 1.0, 1.5, 2.0 and 3.0 mg L^{-1}) to a fixed dosage of adsorbent (5 g L^{-1}) at pH 7.5. These solutions as well as a blank were then allowed to equilibrate in a rotating apparatus at 10 rpm (at a constant temperature of 25°C). Equation 1 was used for the calculation of the milligrams of arsenate or arsenite adsorbed on one gram of adsorbent (qe):

$$qe = \frac{(C_0 - C_e)V}{M} \qquad (1)$$

where qe = mg adsorbate per gram of adsorbent (mg g^{-1}), C_0 = initial concentration of adsorbate (mg L^{-1}), C_e = equilibrium concentration of adsorbate (mg L^{-1}), V = volume of solution (L) and M = mass of adsorbent (g).

3 RESULTS AND DISCUSSION

Figure 1 demonstrates the effect of changing contact time on the adsorption of arsenate and arsenite onto GFH. The minimum removal efficiencies of arsenate and arsenite were obtained in 5 minutes contact time, but these efficiencies were considerably increased at higher contact times and maximized at 30 minutes.

The effects of GFH different dosages on removal efficiency of arsenate compounds are shown in Figures 2 and 3. According to Figure 2, it can be concluded that arsenate adsorption was enhanced by an increase in the adsorbent dosage. However, 0.5 g GFH per liter of solution was sufficient to maximize this treatment.

In Figure 4, the results of developing adsorption isotherms for arsenate and arsenite adsorption by GFH can be considered. By comparison of correlation coefficients obtained from Freundlich isotherm it could be concluded that the absorbability of both arsenate and arsenite may be expressed by this isotherm with about the same R^2 values (> 0.96).

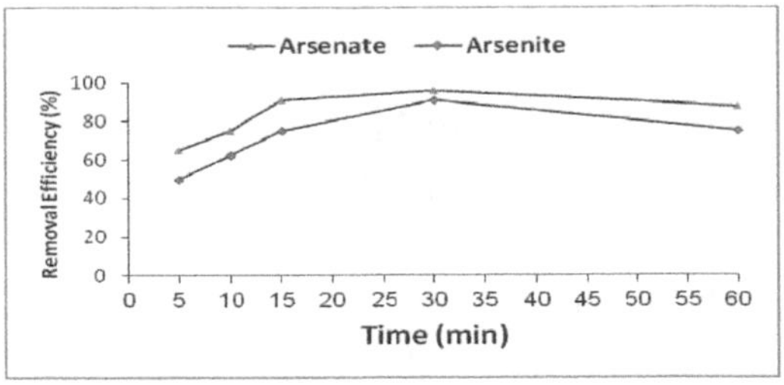

Figure 1. Contact time effect on removal efficiency of arsenate and arsenite (initial concentration: 2 mg L^{-1}, dose adsorbent: 1 g L^{-1}, pH 7.5).

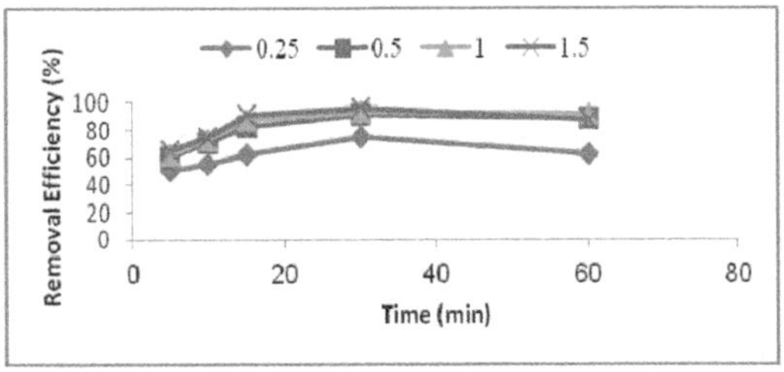

Figure 2. Adsorbent dose effect on removal efficiency of arsenate (initial concentration: 2 mg L^{-1}, pH 7.5).

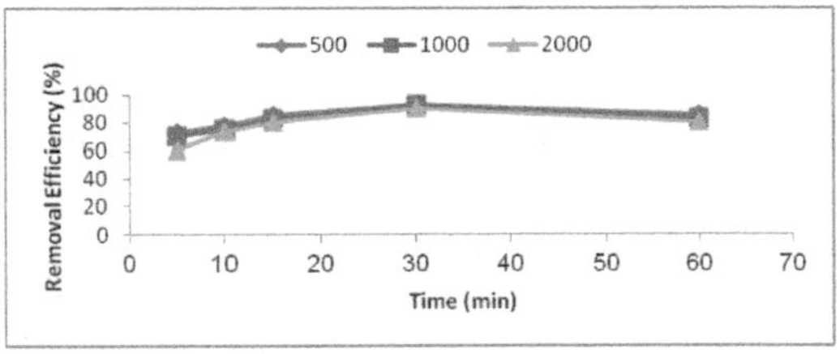

Figure 3. Initial concentration effect on removal efficiency of arsenate (adsorbent dose: 1 mg L^{-1}, pH 7.5).

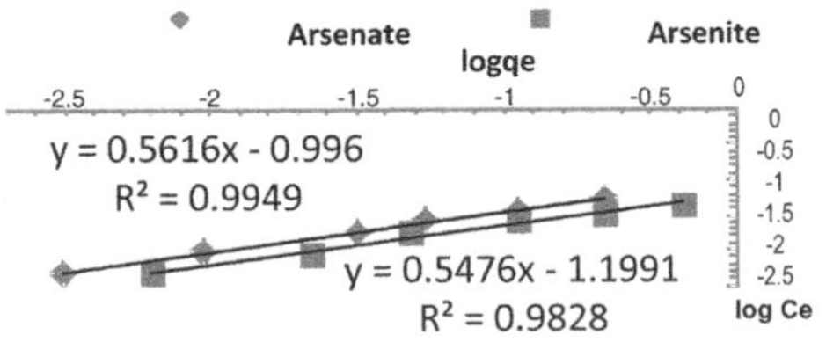

Figure 4. Freundlich isotherm for arsenate and arsenite removal by GFH.

4 CONCLUSIONS

The optimum contact time for the treatment of GFH for As removal is estimated to be 30 minutes. Thus, it can be recommended the use of this adsorbent in the sedimentation tank or preferably in the rapid mix chamber of water treatment plants. The best treatment efficiency was observed at pH 7.5, which is in the pH range of natural water samples. Therefore, there is no need to modify pH which would increase the costs of operation. Although it would be possible to improve the efficiency of treatment by increasing the GFH dosage, this practice is not recommended because GFH is in fact quite efficient even at low doses. Analysis of the used adsorbents by Toxic Characteristic Leaching Producer (TCLP) is part of forthcoming studies.

REFERENCES

Asgari, A.R., Vaezi, F., Nasseri, S., Dördelmann, O., Mahvi, A.H. & Fard, E.D. 2008. Removal of hexavalent chromium from drinking water by granular ferric hydroxide. *Iranian J. Environ Health Sci. Eng.* 5(4): 277–282.

Chen, A.S.C., Fields, K.A., Sorg, T.J. & Wang, L.L. 2002. Field evaluation of As removal by conventional plants, *J. Am. Water Works Assoc.* 94(9): 64–77.

Driehaus, W., Jakel, M. & Hildeberandt, U. 1998. Granular ferric hydroxide- a new adsorbent for the removal of arsenic from natural water, Blackwell Science Ltd. *J Water SRT- Aqua* 47: 30–35.

Jakel, M. & Seith, R. 2000. The removal of the arsenic: comparison of conventional and new techniques for the removal of arsenic in a full-scale water treatment plant. *Water Supply* 18(1): 628–631.

Van, G., Ahmad, K.M., Seddique, A.A. & Ahamsudduha, M. 2003. Community wells to mitigate the current arsenic crisis in Bangladesh. *Bull. WHO* 81: 632–638.

One Century of the Discovery of Arsenicosis in Latin America (1914–2014) – Litter, Nicolli, Meichtry, Quici, Bundschuh, Bhattacharya & Naidu (Eds)
© 2014 Taylor & Francis Group, London, ISBN 978-1-138-00141-1

Adsorption technology for removal of arsenic in water

G. Borneo & L. Barrionuevo
AISA IONIC S.A., Buenos Aires, Argentina

ABSTRACT: About two million people in the Southern Cone are potentially exposed to water with arsenic concentrations above 50 mg/L, and therefore have a high risk for arsenicosis. The aim of this work was to study the effectiveness of adsorption technology for removal of arsenic in drinking water. We used seven samples from different wells of Buenos Aires City. The method used for the determination of arsenic was the silver diethyldithiocarbamate method (3500-As). It was concluded that, within the range of technologies used for the removal of arsenic in drinking water, studies show that the adsorption provides 80% removal of arsenic in water in 60% of cases, thus obtaining water quality within the parameters required by existing legislation in Argentina.

1 INTRODUCTION

About two million people in an area of 1.7×10^6 km^2 in the South American Cone are potentially exposed to drinking water with arsenic concentrations exceeding 50 mg L^{-1} and consequently have an elevated risk of arsenicosis (Fernandez-Turiel *et al.*, 2005).

Epidemiological studies show that the potential health hazards of exposure to natural arsenic in drinking water are serious. Moreover, it is known that only a few chemicals cause health effects at high scale through exposure to drinking water (WHO, 2004).

Effects in humans from chronic exposure to drinking water with high concentrations of inorganic arsenic may be multiple and are grouped under the names of arsenicosis or chronic endemic regional hydroarsenicism (HACRE), including internal (lung and bladder) and external (skin) cancers plus dermatitis, involvement of central and peripheral nervous system, hypertension, peripheral vascular disease, cardiovascular diseases, respiratory diseases and diabetes mellitus (Besuschio *et al.*, 1980; BEST, 2001, Yoshida *et al.*, 2004). It could also have an effect on reproduction, increasing the mortality rate of late gestation fetuses and children (Hopenhayn-Rich *et al.*, 2000). It is estimated that these effects appear after exposure to 5–15 years doses of 0.01 mg kg^{-1} per day or 0.5–3 years doses above 0.04 mg kg^{-1} per day of inorganic arsenic (Bhattachariya *et al.*, 2003).

In conventional water treatment plants, the applied processes are basically oxidation, coagulation-flocculation, filtration and post-chlorination. With regards to arsenic, the first three processes can be active. Various oxidation products (ozone, chlorine, ClO_2, H_2O_2, $KMnO_4$) can oxidize As(III) to As(V), which is easier to remove from water than the reduced form (Bissen & Frimmel, 2003).

A coagulation-flocculation stage can be used to allow different compounds to coprecipitate with arsenic. The most commonly used are iron, aluminum, manganese, calcium and magnesium compounds. (Karcher *et al.*, 1999; Muñoz *et al.*, 2005). The filtration step in a fixed bed also offers interesting alternatives to lower arsenic levels in water. Arsenic adsorption is achieved with activated carbon filters, iron compounds, metallic iron or activated alumina (Bissen & Frimmel, 2003). Filter materials, natural zeolites or bentonites may also be useful in some waters (Elizalde-González *et al.*, 2001). These processes can be very effective treatments for the removal of other undesirable compounds in drinking water, as in the case of fluorine (Muñoz *et al.*, 2005).

2 METHODS/EXPERIMENTAL

2.1 *Materials*

Seven samples of drinking water from different wells belonging to the Autonomous City of Buenos Aires were analyzed.

2.2 *Methods*

Silver diethyldithiocarbamate method (3500-As C of Standard Methods for the Examination of Water and Wastewater APHA-AWWA-WPCF). Arsen QUANTOFIX 10 (from 0.01 to 0.05 mg L^{-1} As^{3+}/As^{5+}) MACHEREY-N (as a first approximation to the result).

Table 1. Arsenic concentration present in samples of well water without filtration (S/F) and with filtration (C/F).

Sample	Sample S/F (ppm)	Sample C/F (ppm)
1	0.05	0.01
2	0.08	0.02
3	0.07	0.01
4	0.08	0.01
5	0.04	0.01
6	0.05	0.02
7	0.09	0.02

Table 2. % Residual arsenic and removed arsenic in water filtration adsorption technology.

Sample	% Residual arsenic	% Removed arsenic
1	20	80
2	25	75
3	14	86
4	13	88
5	50	50
6	20	80
7	22	78

3 RESULTS AND DISCUSSION

3.1 *Results*

Table 1 shows the results obtained from the study of the drinking water samples from different areas of the Autonomous City of Buenos Aires.

The experimental results, Table 1, show that, in 57% of the studied cases, arsenic concentrations in filtered water have reached the limit set by the Argentine Food Code (CAA), 10 $\mu g\ L^{-1}$.

The 43% of the remaining cases have decreased concentration values between 18 and 20 $\mu g\ L^{-1}$.

In Table 2, the effectiveness of removal of arsenic in water studied can be verified.

In 57% of the studied cases, arsenic removal of water by 80% by adsorption filtration processes took place.

In 43% of cases, remaining percentage of arsenic removal of water was in a range between 50 and 78%.

4 CONCLUSIONS

It was concluded that, within the range of technologies used for the removal of arsenic in drinking water, studies show that the adsorption provides 80% removal of arsenic in water in 60% of cases, thus obtaining water quality within the parameters required by existing legislation in Argentina.

ACKNOWLEDGEMENTS

To Aisa Ionic S.A. Trad. Micaela Bocchio. National University of the Center of Buenos Aires Province, College of Agronomy Azul.

REFERENCES

BEST (Board on Environmental Studies and Toxicology), 2001. *Arsenic in drinking water: 2001 update*. National Academy Press, Washington DC, 225.

Besuschio, S.C., Desanzo, A.C., Perez, A. & Croci, M. 1980. Between Epidemiological associations arsenic and cancer in Argentina. *Biol. Trace Element Res.* 2: 41–55.

Bhattacharyya, R., Chatterjee, D., Nath, B., Jana, J., Jacks, G. & Vahter, M. 2003. High arsenic groundwater: Mobilization, metabolism and mitigation—an overview in the Bengal Delta Plain. *Molec. Cell. Biochem.* 253 (1–2): 347–355.

Bissen, M. & Frimmel, F.H. 2003. Arsenic—a review. Part II: Oxidation of arsenic and its removal in water treatment. *Acta Hydrochim. Hydrobiol.* 31 (2): 97–107.

Elizalde-González, M.P., Mattusch, J., Einicke W.-D. & Wennrich, R. 2001. Sorption on Natural solids for arsenic removal. *Chem. Eng. J.* 81 (1–3): 187–195.

Fernandez-Turiel, J.L., Galindo, G., Parada, M.A., Gimeno, D.M., García-Valles, M. & Saavedra, J. 2005. *Current State of Knowledge about arsenic in water from Argentina and Chile: Origin, Mobility and Treatment*, 1–4.

Hopenhayn-Rich, C., Browning, S.R., Hertz-Picciotto, I., Ferreccio, C., Peralta, C. & Gibb, H. 2000. Chronic arsenic exposure and risk of infant mortality in two areas of Chile. *Environ. Health Persp.* 108 (7): 667–673.

Karcher, S., Caceres, L., Jekel, M. & Contreras, R. 1999. Arsenic removal from water supplies in northern Chile using ferric chloride coagulation. *J. Chart. Inst Water Environ. Manag.* 13 (3): 164–169.

Mead, M.N. 2005. Arsenic: In search of an antidote to a Global Poison. *Environ. Health Persp.* 6: A378–A386.

Muñoz, M.A., Buitrón, J.A. & De Ormaechea, B. 2005. Removal of arsenic and fluorine, a case study in the town of Eduardo Castex, La Pampa Arsenic in water: origin, mobility and tratamiento. Taller. *II Spanish-Latin American Seminar on current issues in groundwater hydrology -IV Hydrogeological Congress Argentino.* Rio Cuarto, Argentina.

WHO (World Health Organization), 2004. *Guidelines for drinking-water quality*. Volume 1 Recommendations. Geneva. 3rd edition.

Yoshida, T., Yamauchi, H. & Sun, G.F. 2004. Chronic health effects in people exposed to arsenic via the drinking water: dose-response relationships in review. *Toxicol. Appl. Pharmacol.* 198 (3): 243–252.

4.2 Arsenic removal based on biomaterials

One Century of the Discovery of Arsenicosis in Latin America (1914–2014) – Litter, Nicolli, Meichtry, Quici, Bundschuh, Bhattacharya & Naidu (Eds)

Biomaterial based sorbents for arsenic removal

L. Ansone & M. Klavins
Department of Environmental Science, University of Latvia, Riga, Latvia

ABSTRACT: Due to toxicity and wide distribution, arsenic is the most studied metalloid. Arsenic contamination of waters is a global problem and one of the possible solutions could be the development of new low-cost sorbents based on natural materials for its removal. Biomaterials (peat, shingles, straw, canes, moss and sand) modified with iron compounds as well as a Fe-modified peat graft polymer were prepared. Sorption of different arsenic forms onto iron-modified peat sorbents was investigated as a function of pH and temperature, and the impact of competing anions was determined. Sorption capacity increased with temperature, and was spontaneous and endothermic.

1 INTRODUCTION

Arsenic is mobilized in natural water systems through the range of anthropogenic as well as natural processes. For example, geochemical interactions that occur between arsenic-containing rocks and minerals and water, biological activity and volcanic emission, as well as soil erosion and leaching are some of natural sources of arsenic. Anthropogenic sources of arsenic include discharges from various industries, such as smelting, petroleum refinery, glass manufacturing, fertilizer production and intensive application of arsenic insecticides, herbicides and crop desiccants in past (Henke, 2009; Mohan & Pittman, 2007).

Arsenic can be found in both inorganic and organic speciation forms, but inorganic species are more common and more toxic in comparison to organic species. In addition, inorganic species are the predominant forms in polluted waters and they exist in two oxidation states—As(III) and As(V), depending on pH and redox conditions (Ansari *et al.*, 2007). In the pH range 3–9, the neutral H_3AsO_3 is the predominant specie, and those of As(V) are the negatively charged $HAsO_4^{2-}$ and $H_2AsO_4^-$ (Nemade *et al.*, 2009).

Nowadays, contamination of natural water sources with As is a global problem, especially in Southeast Asia, South America, United States and Europe. Drinking water supplies in polluted areas contain dissolved arsenic in quantities higher than 10 $\mu g\ L^{-1}$, which is the threshold value recommended by the WHO (Henke, 2009).

In comparison to other widely used technologies for metalloid removal, adsorption is considered one of the best methods due to its simplicity, potential for regeneration; it is easy to set up and it is economical.

Although many different sorbents have been used for metal and metalloid removal so far, due to unsatisfactory efficiency and high costs of these sorbents, opportunities are still open for finding new environmentally friendly and cost effective sorbents. Recently, great attention has been paid to the sorbents based on natural materials. Thus, as arsenic can strongly bind to Fe containing materials, metalloid removal using new Fe-modified sorbents based on common natural materials was investigated in this study. A novel peat-based anion exchanger was synthesized by graft polymerization and its application in sorption of As(V) from aqueous media was investigated.

2 EXPERIMENTAL

Fe modified biomaterial (peat, shingles, straw, moss, canes, sand) sorbents were synthesized. The synthesis method was based on impregnation of the material with Fe oxyhydroxide, followed by thermal treatment.

A novel sorbent—peat graft polymer was also used for arsenic removal. The peat graft polymer was synthesized by graft polymerization reaction of peat and glycidylmethacrylate using N,N-methylene-bis-acryl-amide as a cross linker and benzoyl peroxide as an initiator followed by amination and acid treatment (Anirudhan & Jalajamony, 2010).

Scanning electron microscopy (SEM), specific surface area measurements, Fourier transformation infrared spectra (FT-IR), as well as moisture content, organic substances content and Fe_2O_3 analyses were used to characterize the sorbents and to assess the success of the modification methods.

Sorption experiments were carried out using Fe-modified biomaterials as well as peat graft polymer.

Inorganic forms of arsenic—sodium arsenate heptahydrate and sodium meta-arsenite as well as organic form—cacodylic acid were used. As concentrations were determined using FAAS (flame atomic adsorption spectrometry) and ETAAS (electro thermal atomic adsorption spectrometry).

The obtained sorption data were correlated using the Langmuir and Freundlich isotherm models.

To investigate the impact of physicochemical conditions on arsenic removal, sorption experiments were carried out at different temperatures (275, 283, 298 and 313 K); to investigate the impact of pH, 0.1 M HCl and 0.1 M NaOH were used to obtain the desired solution pH (pH interval 3–9), and the impact of competing anions—chloride, nitrate, phosphate, sulfate, oxalate, tartrate, carbonate as well as humic substances on As (V) sorption capacity were investigated as well.

3 RESULTS AND DISCUSSION

Sorbent characterization demonstrates changes that have arisen after modification. SEM figures show surface morphology changes, while FT-IR spectra show different signal intensity changes that may be related to the formation of Fe phenolates and carboxylates. However, although FT-IR spectra are useful for sorbent characterization, it cannot give complete information regarding the modification results. Considering that specific surface area is one of the parameters that affect sorption capacity and due to arsenic high affinity to iron containing materials, results of specific surface area along with iron oxide content allow predicting what sorbents will have the highest sorption capacity. In this case, the best was Fe-modified peat.

Sorption experiments indicated that material modification with iron compounds significantly enhances the sorption capacity of the obtained sorbents in comparison to the raw (unmodified) material. The possible reason could be the formation of As-O-Fe bonds in the case of Fe-modified sorbents. Arsenic sorption capacity is dependent on the biomass sorbent used, although reaction conditions for the material modification with Fe compounds were similar. The highest sorption capacity for As(V) using Fe-modified sorbents was observed for the Fe-modified peat, whereas Fe-modified moss, Fe-modified straw and Fe-modified shingles showed similar but not as high results (Figure 1).

In contrast to Fe-modified sorbents, As(V) sorption capacity is significantly higher using peat graft polymer as a sorbent (Figure 1), thus indicating that peat graft polymer could be useful for remediation of As contaminated waters. Due to the relatively high As(V) sorption capacity peat graft polymer is one of the promising sorbent materials. As(V) sorption capacity using peat graft polymer

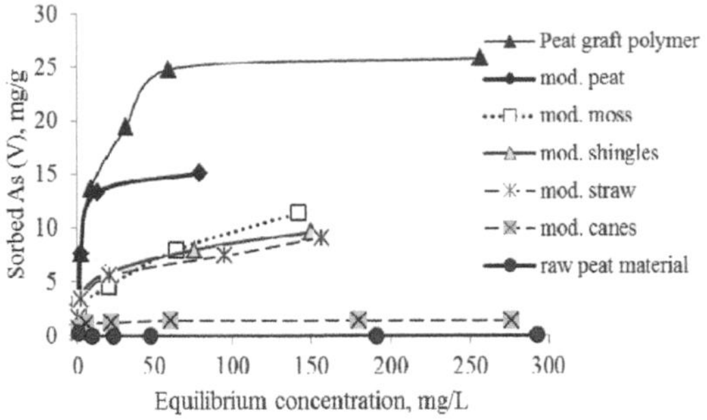

Figure 1. As (V) sorption isotherms using peat graft polymer, Fe-modified sorbents and raw peat material.

reached 25 mg/g, i.e. As removal exceeded 95%, even if the initial As concentration was 200 mg L^{-1}, decreasing to 83%, if the initial As concentration exceeded 370 mg L^{-1}. The possible removal mechanism could be related to hydrogen bonding of arsenate with a tertiary amine functional group.

The environmental fate and behavior of arsenic is significantly influenced by physicochemical conditions such as pH. Maximum binding of As(V) was observed at pH interval 3–6.5 for the Fe-modified peat.

Arsenic sorption in the presence of competing chemical species indicate that sulfate, nitrate, chloride and tartrate had practically no influence on As(V) sorption onto Fe-modified peat, whereas the presence of phosphate ions and humic acid significantly lowered the arsenic removal rate.

Variation in arsenate sorption onto modified peat at four temperatures (275, 283, 298 and 313 K) showed that the sorption capacity for both sorbents increased with temperature; this allowed the calculation of thermodynamic sorption parameters. The sorption process was spontaneous ($\Delta G° < 0$) and endothermic ($\Delta H° = 41.43$ kJ mol^{-1}).

4 CONCLUSIONS

Modification of materials with Fe compounds significantly enhanced the sorption capacities of the sorbents used for arsenic (III, V) sorption.

Due to relatively high As(V) sorption capacity peat graft polymer is one of the promising sorbent materials and it is the best sorbent for arsenic removal found in this study.

The best sorption conditions of As(V) are in the pH interval 3–6.5 for Fe-modified peat sample. The results of arsenic sorption in the presence of competing substances indicate that the presence of phosphate ions and humic acid significantly lowers the arsenic removal rate. Arsenate sorption capacity increased with temperature and arsenic sorption was spontaneous and endothermic.

REFERENCES

Anirudhan, T.S. & Jalajamony, S. 2010. Cellulose-based anion exchanger with tertiary amine functionality for the extraction of arsenic (V) from aqueous media. *Journal of Environmental Management* 91: 2201–2207.

Ansari, R. & Sadegh, M. 2007. Application of activated carbon for removal of arsenic ions from aqueous solutions 0973-4945; CODEN ECJHAO http://www.e-journals.net *E-Journal of Chemistry* 4(1): 103–108.

Henke, K.R. 2009. *Arsenic: environmental chemistry, health threats, and waste treatment*. Great Britain, Wiltshire, Chippenham, John Wiley and Sons.

Mohan, D. & Pittman, A.U. Jr. 2007. Arsenic removal from water/wastewater using adsorbents—A critical review. *Journal of Hazardous Materials* 142: 1–53.

Nemade, P.D., Kadam, A.M. & Shankar, H.S. 2009. Adsorption of arsenic from aqueous solution on naturally available red soil. *Journal of Environmental Biology* 30(4): 499–504.

One Century of the Discovery of Arsenicosis in Latin America (1914–2014) – Litter, Nicolli, Meichtry, Quici, Bundschuh, Bhattacharya & Naidu (Eds)
© 2014 Taylor & Francis Group, London, ISBN 978-1-138-00141-1

Development of a bio-adsorbent medium to remove arsenic from water

J. Paredes
Posgrado en Ingeniería UNAM, Jiutepec, Morelos, Mexico

S.E. Garrido
Instituto Mexicano de Tecnología del Agua, Mexico

ABSTRACT: The objective of this work was to study the removal of arsenic from water for human consumption by developing a bio-adsorbent using passion fruit, able to meet the maximum allowable limits of 0.025 mg L^{-1} established by Mexican NOM-127-SSA-1994. This was performed with synthetic water ([As] = 0.22 mg L^{-1}). The pectin was extracted using two methods, conventional acid extraction and alkali treatment with steam drag. After the pectin was extracted and characterized, experiments were performed to obtain adsorption kinetics and isotherms. The results showed adsorption kinetics of pseudo-second order with a rate constant of adsorption, k_{ad}, of 2.107 g mg^{-1} min^{-1} for the acidic pectin and of 0.550 g mg^{-1} min^{-1} for the alkaline pectin. With regard to adsorption isotherms, the experimental data fitted the Freundlich model, the K_F being 2.08 for acidic pectin and 2.032 for alkaline pectin. Arsenic removal percentages were 91% for acidic pectin and 81% for alkaline pectin.

1 INTRODUCTION

Epidemiological studies report potentially serious risks to health caused by exposure to natural arsenic in water for human consumption (WHO, 2004). Studies by Ghimire *et al.* (2003) used cellulose and orange waste coated with Fe^{3+}, which were chemically modified by phosphorylation to create a chelate environment suitable for the removal of As. The orange waste gel had a capacity of 1.21 mmol g^{-1} to adsorb iron ions, while the cellulose had a capacity of 0.96 mmol g^{-1}. This result is likely due to the presence of pectin in the orange peel. The orange waste gel was observed to have greater capacity under these conditions, as well as under neutral pH conditions. On the other hand, Biswas *et al.* (2008) reported that gels obtained from orange waste enriched with Zr^{4+} ions had a greater capacity to remove arsenate and arsenite, 88 and 130 mg g^{-1} of material, respectively. The suitable pH range was 2 to 6 for removing As^{5+} and 9 to 10 for removing As^{3+}. Iliná *et al.* (2009) showed that the materials obtained from the skin and fiber of passion fruit peel (solid waste from the national food industry) using the proposed technique can be considered biosorbents of arsenite and arsenate ions.

The present work was conducted using the adsorption process as a method for arsenic removal, in which a bio-adsorbent was produced with pectin from passion fruit.

2 METHODS/EXPERIMENTAL

A stock solution was prepared with sodium arsenate heptahydrate ($Na_2HAsO_4.7H_2O$, 98% purity). The pectin was extracted from the *Flavicarpia degener* species of passion fruit by means of two methods: the conventional technique using precipitation with alcohol (Jasme & Poirrier, 1993) and that using steam drag (Estrada, 1998). The adsorbent material was characterized by an analysis of elements, surface area and porosity.

Batch bioadsorption kinetics was performed in a Jar Tester. One gram of bioadsorbent material was used for each 1 L and agitated at 120 rpm (G: 65 s^{-1}) (Paredes, 2012). The adsorption tests were conducted with an initial arsenic concentration of 0.22 mg L^{-1}. The pH was maintained at 5.25. Arsenic was measured with a digital photometer (Garrido *et al.*, 2009), in the filtered samples.

3 RESULTS AND DISCUSSION

3.1 *Characterization of adsorbents*

Table 1 summarizes the results of the percentage, by weight, of the elements present. The mineral analysis indicates that carbon, hydrogen and iron were the primary elements present in the samples analyzed.

The pectin surface area (BET) was 0.144 mg^2 g^{-1} with a micropore volume of 0.0157 cm^3 g^{-1}.

Table 1. Elemental analysis of percentage by weight of pectin from passion fruit.

Element*	Percentage by weight (%)
C	29.93
N	0.38
P	0.086
S	0.057
H	3.520
Fe	1.075

* Elements reported without oxidation state.

Table 2. Constants for the pseudo-second order kinetics of the adsorbents in the study.

		Type of pectin	
Parameters	Units	Acidic	Alkaline
Mass*	(g)	1.0	1.0
q_e	(mg g^{-1})	0.192	0.185
k_{ad}	(g mg^{-1} min^{-1})	2.107	0.550
H	(mg g^{-1} min^{-1})	0.078	0.019

*In 1 L of solution.

3.2 *Adsorption Kinetics*

To quantify the changes in As concentration by adsorption with respect to the time required by a suitable kinetic model, the pseudo-second order equation was used:

$$\frac{t}{q_t} = \frac{1}{k_{ad} q_e^2} + \frac{1}{q_e} t$$

Where: q_e is the adsorption capacity, k_{ad}, the adsorption rate constant and h the initial adsorption rate as $t \to 0$. Table 2 shows the kinetic constants.

3.3 *Adsorption isotherms*

After performing the kinetic experiments, the adsorption isotherms were obtained for each of the adsorbents, Figure 1, Figure 2 and Table 3. Out of the models evaluated based on adsorption isotherms, the Freundlich 1/n<1 model resulted in a better fit of the data than the Langmuir model.

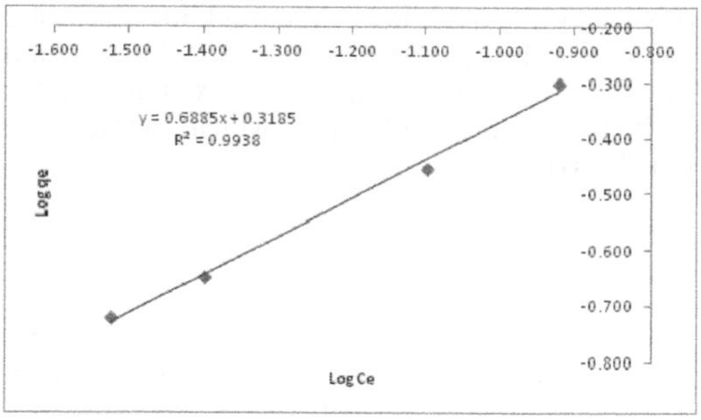

Figure 1. Freundlich isotherms for acidic pectin.

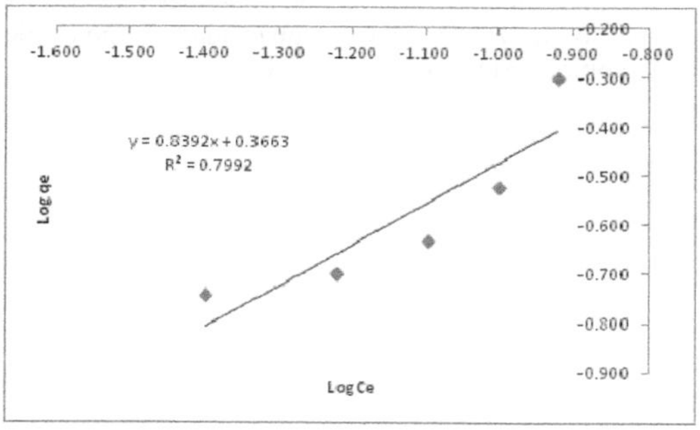

Figure 2. Freundlich isotherms for alkaline pectin.

Table 3. Freundlich adsorption isotherm constants for the different adsorbents.

	Type of pectin	
Parameters	Acidic	Alkaline
K_F	2.08	2.32
n	1.45	1.19
R^2	0.9938	0.7992

4 CONCLUSIONS

The best arsenic removal, 91%, was achieved by means of acidic pectin.

A natural adsorbent was produced, which achieved better results in terms of contact time (a decrease from 24 to 4.5 hours) compared with those obtained by Iliná *et al.* (2009).

The Freundlich 1/n<1 model resulted in a better fit of the data than the Langmuir model. On the other hand, adsorption presented in the acidic pectin enriched Fe(III) (creating a chelator) could be associated in part with the ion exchange of the carboxyl groups or other ions which interact with ions Fe(III).

REFERENCES

Biswas, B.K., Inoue, J.C., Inoue, K., Ghimire, K.N., Harada, H., Ohto, K & Kawakita, H. 2008. Adsorptive removal of As(V) and As(III) from water by a Zr(IV)–loaded orange waste gel. *J. Hazard. Mater.* 154: 1066–1074.

Estrada, A. 1998. Pectinas cítricas. Efecto del arrastre de vapor en la extracción y de diferentes métodos de secado. In *Revista Noos*. Universidad Nacional de Colombia. Manizales; pp. 23–34.

Garrido, S.E. 2008. Origen Hidrogeológico del arsénico, métodos alternativos innovativos para cuantificación y tratamiento. Informe final. CONACYT SEP-2004-C01–47076. México.

Ghimire, K.N., Inoue, K., Yamagchi, H., Makino, K. & Miyajima, T. 2003. Adsortive separation of arsenate and arsenite anions from aqueous medium by using orange waste. *Water Res.* (37):4945–4953.

Iliná, A., Martínez–Hernández, J.L., Segura–Ceniceros, E.P., Villarreal–Sánchez, J.A. & Gregorio–Jáuregui, K.M. 2009. Biosorción de arsénico en materiales derivados de maracuyá. *Rev. Int. Contam. Ambient* 25(4).

Jasme Miranda, M.E. & Poirrier González, P. 1993. Características, producción y utilización de pectinas. *Revista Alimentación, Equipos y Tecnología* 12 (9): 61–66.

One Century of the Discovery of Arsenicosis in Latin America (1914–2014) – Litter, Nicolli, Meichtry, Quici, Bundschuh, Bhattacharya & Naidu (Eds)
 ISBN 978-1-138-00141-1

Phytofiltration mechanism of arsenic from drinking water using *Micranthemum umbrosum*

Md.S. Islam & M. Kurasaki
Hokkaido University, Sapporo, Hokkaido, Japan

ABSTRACT: Phytofiltration is an emerging eco-friend technology to remediate pollutants from waters and soils. Arsenic (As) is one of the most noxious and carcinogenic contaminant, causing severe pollution in water followed by developing different kinds of diseases in human beings like dermatological disorders, cancer, etc. Previous studies showed that *Micranthemum umbrosum* plant is capable to remove 79.3–89.5% As from 0.2–1 μg As mL^{-1} solution. After Sephadex G-50 gel filtration of the soluble fractions of leaf of the plant grown in As contaminated water, fractions of As and- SH contents were determined. The results indicated that low molecular weight substances having thiol groups may berelated to the detoxification of As within the plant body. This research demonstrates that *M. umbrosum* has high potentiality as phytofiltrator for As. This As phytofiltration or detoxification procedure increases the scope of phytoremediation research for environmental contaminants.

1 INTRODUCTION

Arsenic (As) contamination in drinking water has been recognized as a serious global problem, threatened the health of millions of people, for example, in Bangladesh, China, Argentina, West Bengal, Vietnam, etc. Different forms of As exist in the environment, e.g., inorganic (arsine, arsenious acid, arsenite, arsenic acids, arsenate, etc.), organic (monomethylarsonic acid or dimethylarsinic acid, etc.), biological and other forms (Rahman & Hasegawa, 2011). Groundwater in Bangladesh is currently contaminated with up to 2,000 μg/L As, mainly inorganic As (arsenite) comprising about 50% of the total (Abedin *et al.*, 2002), whereas the As standard in drinking water is 10 μg/L (WHO, 2011). Thus, remediation of As from drinking water is a burning issue all over the world. Among different remediation process, phytofiltration is an eco-friend and emerging technology to remediate toxic heavy metals or contaminants from the environment using plant species (Vamerail *et. al.*, 2010). *Micranthemum umbrosum* can remove 79.3 to 89.5% As from 0.2–1 mg As L^{-1} solution (Islam *et al.*, 2013). Phytofiltration procedures of arsenite contaminated drinking water by green *M. umbrosum* plants were investigated in the present study.

2 METHODS/EXPERIMENTAL

2.1 *Plant Culture*

M. umbrosum plants were grown in glass pots containing 0, 0.2, 0.45 and 1 mg As L^{-1} as $NaAsO_2$(III)). Seven days after a hydroponic experiment, *M. umbrosum* was harvested and separated into leaf, stem and root and preserved in refrigerator for measurement of arsenic content and arsenic binding species within the plant.

2.2 *Analysis of plant samples*

Exactly 0.5 g leaf samples were homogenized with 5 mL Tris-HCl buffer at pH 7.4 and 0.5 mL 0.1% SDS solution. Then, the homogenized sample was centrifuged at 10,000 rpm for 30 min at 4 °C. 3 mL of supernatant were applied on a Sephadex G 50 column (1.0 cm × 110 cm) and 60 fractions (each of 2 mL eluent, flow rate 5.47 min/2 mL) were collected. Then, As was measured in each fraction by ICP-MS and the 280 nm absorbance was recorded by UV/VIS spectrophotometry for protein determination (Layne, 1957). -SH (thiol) contents were measured by the absorbance at 480 nm after treating samples with Ellman's Reagent (Ellman, 1959). The accuracy of the analysis was checked by the use of an As certified standard reference material (013–15481, Lot ALK 9912, 1000 ppm: Wako Pure Chemical, Japan). Control plant samples were also treated by same procedure.

3 RESULTS AND DISCUSSION

After Sephadex G-50 gel filtration of the leaf extract, the maximum amount of arsenic was found at 94–106 mL of eluents (Figures 1 and 2). As shown in Figure 1, it can be easily observed that As bonded substance(s) other than proteins because they gave different peaks at the G-50 column. Figure 2 showed that the maximum amount of thiol (11 μM) was found at the fraction having the maximum amount of As. On the contrary, the eluent of the control leaf

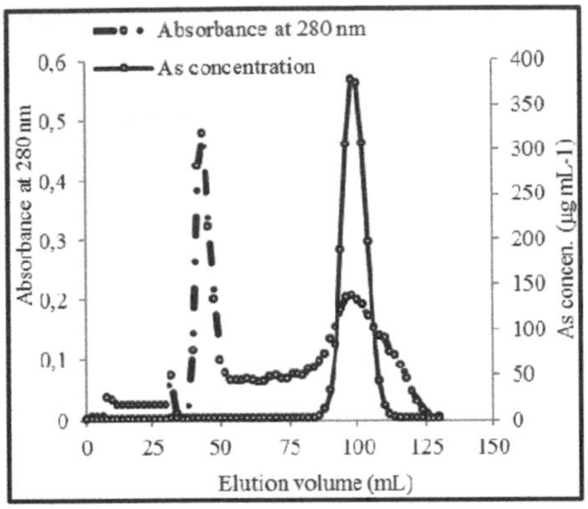

Figure 1. Arsenic concentration (µg mL^{-1}) and absorbance at 280 nm of each 2 mL eluents obtained from the Sephadex G-50 gel filtration column using *Micranthemum umbrosum* leaf treated with 1 mg As/L solution.

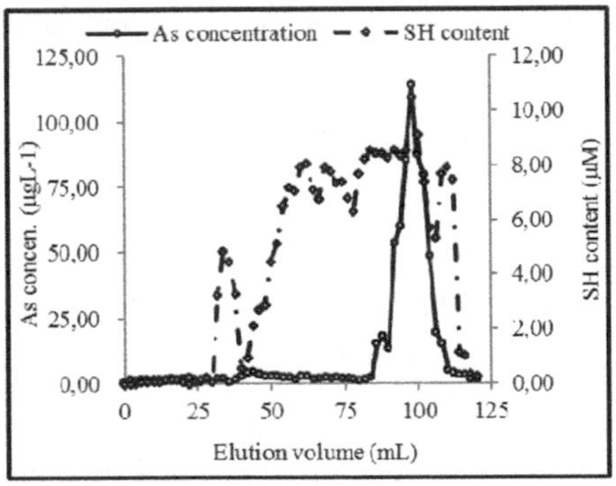

Figure 2. Arsenic concentration (µg mL^{-1}) and- SH content (µM) of each 2 mL eluents obtained from the Sephadex G-50 gel filtration column using *Micranthemum umbrosum* leaf treated with 1 mg As/L solution.

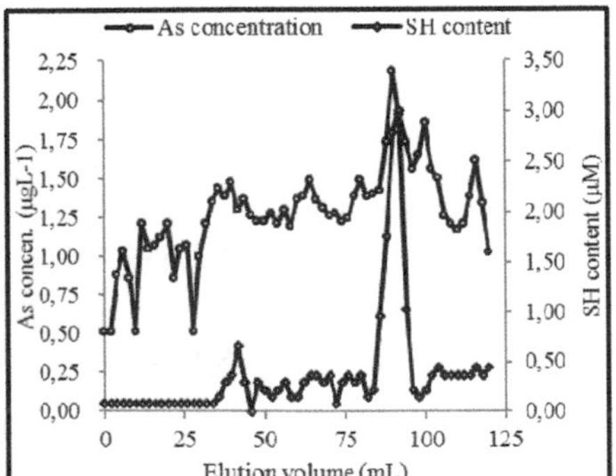

Figure 3. Arsenic concentration (µg mL^{-1}) and- SH content (µM) of each 2 mL eluents obtained from Sephadex G-50 gel filtration column using *Micranthemum umbrosum* leaf treated with 0 mg As/L solution (Control).

sample had a lower thiol content (2.8 µM) when treated with 0 mg As L^{-1} solution (Figure 3).

From this study, it can be concluded that As can induce the formation of low molecular thiol compound(s) within the plant body to detoxify or enhance accumulation of As from water. This result was also supported by Cai *et al.* (2004), who observed an unidentified thiol in addition to cysteine and glutathione, which might be responsible for the As detoxification mechanism in *Pteris vittata.*

4 CONCLUSIONS

This preliminary study showed that As can induce *thiol* formation within the plant body, which is responsible for As detoxification or enhanced As uptake within the plant body. More biochemical analysis will be performed to identify the specific thiol compounds which can bind with As within *Micranthemum umbrosum* plant.

ACKNOWLEDGEMENTS

This research was supported by GCOE research fund, Graduate School of Environmental Science, Hokkaido University, Japan, and the authors would like to thank to Aqua Friend Hokusui, Japan, for providing plants.

REFERENCES

Abedin, M.D.J. Cresser, M.S. Meharg, A.A. Feldmann, J. & Cotter-Howells, J. 2002. Arsenic accumulation and metabolism in rice. *Environ. Sci. Technol.* 36:962–968.

Cai, Y., Jinhui, S. & Ma, L.Q. 2004. Low molecular weight thiols in arsenic hyperaccumulator *Pteris vittata* upon exposure to arsenic and other trace elements, *Environmental pollution*, 129:69–78.

Ellman, G.L. (1959) *Arch. Biochem. Biophys.* 82: 70–77.

Islam, M.S., Ueno, Y., Sikder, Md.T. & Kurasaki, M. 2013. Phytofiltration of Arsenic and Cadmium From the Water Environment Using Micranthemum Umbrosum (J.F. Gmel) S.F. Blake As A Hyperaccumulator, *International Journal of Phytoremediation*, 15: 1010–1021.

Layne, E. 1957. Spectrophotometric and Turbidimetric Methods for Measuring Proteins. *Methods in Enzymology 3:* 447–455.

Rahman, M.A. & Hasegawa, H. 2011. Aquatic arsenic: Phytoremediation using floating macrophytes. *Chemosphere* 83:633–646.

Vamerali, T., Bandiera, M. & Mosca, G. 2010. Field crops for phytoremediation of metal-contaminated land: A review. *Environ. Chem. Lett.* 8: 1–17.

WHO. 2011. Guidelines for drinking-water quality, 4th ed. Chemical fact sheets-chemical contaminants in drinking water. Geneva, *World Health Organization.*

One Century of the Discovery of Arsenicosis in Latin America (1914–2014) – Litter, Nicolli, Meichtry, Quici, Bundschuh, Bhattacharya & Naidu (Eds)
© 2014 Taylor & Francis Group, London, ISBN 978-1-138-00141-1

Removal of arsenic from water for human consumption using a bioadsorbent obtained from orange wastes

V. Gómez
Instituto Tecnológico de Zacatepec, Morelos, México

S.E. Garrido
Instituto Mexicano de Tecnología del Agua, Jiutepec, Morelos, Mexico

ABSTRACT: Arsenic (As) is particularly known for its toxic effect, especially due to its carcinogenic action and significant effects on the skin. Food and water are the primary sources of ingesting As. The search for low-cost adsorbents that are easily available has led to investigations on organic materials. Thus, a bioadsorbent medium was developed to remove arsenic from water for human consumption, using orange albedo coated with Fe(III). The bioadsorbent medium had a surface area of 0.0216 $m^2\ g^{-1}$ for the case of uncoated pectin and 0.0894 $m^2\ g^{-1}$ for pectin coated with Fe(III) ions. The active groups present in the surface of the solid adsorbent were identified using infrared spectroscopy with Fourier Transform (FTIR). A composition analysis was conducted by X Ray Diffraction (XRD) and sweep electronic microscopy. Adsorption capacity obtained was 0.5498 $g\ mg^{-1}\ min^{-1}$ for Fe(III)-coated pectin.

1 INTRODUCTION

Epidemiological studies report potentially serious risks to health caused by exposure to natural arsenic in water for human consumption (WHO, 2004). Therefore, water should be treated to remove this metalloid. Among the treatments proposed are bioadsorbents of different origins, such as passion fruit and Fe(III)-coated orange pectin. The cell walls of bioadsorbent materials contain polysaccharides, proteins and lipids, and therefore, many functional groups are able to link heavy metals and attach them on the surface of bioadsorbent. Among these functional groups, amino, carboxylic, hydroxylic, phosphate and thiol groups are possible, which differ in their affinity and specificity with respect to their ability to link to different metallic ions (Ghimire *et al.*, 2003). Orange wastes contain cellulose or pectin (Figure 1), hemicellulose and other low-weight molecular compounds such as limonene. Therefore, they possess active functional groups such as carboxylic (pectin) and hydroxylic (cellulose) groups that can bind with dissolved metallic ions (Ghimire *et al.*, 2003). Ions of heavy metals and metalloids bind to the sorption active centers of biological material through complex formation, chelators, ion exchange, microprecipitation in the inner part of the material, etc.

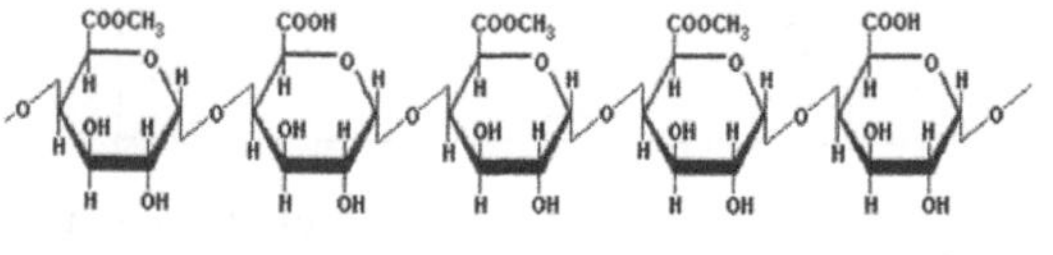

Figure 1. Pectin structure.

2 EXPERIMENTAL METHODS

2.1 *Obtaining pectin from orange wastes*

To obtain the material of interest, the peel of a fresh waste orange (albedo) was used, and underwent several stages of extractopm in the laboratory, Figures 2 and 3 (Paredes, 2012).

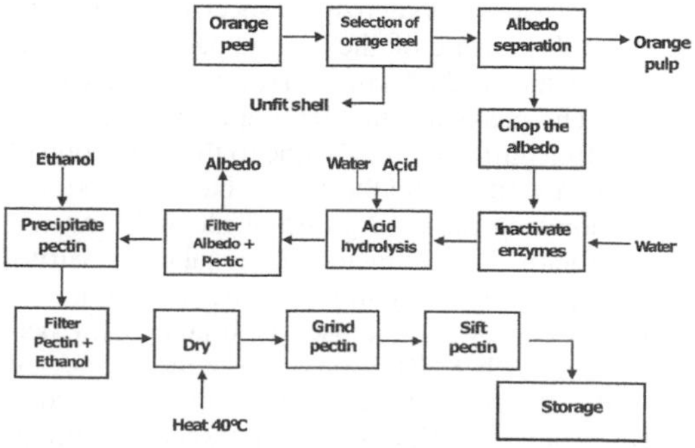

Figure 2. Block diagram of the extraction of orange pectin.

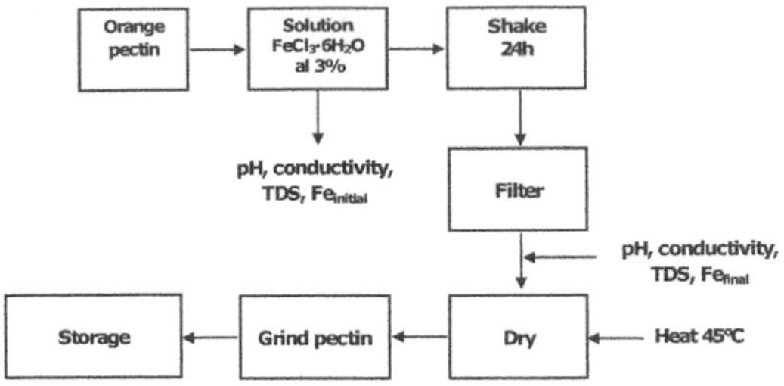

Figure 3. Block diagram of Fe(III) enrichment of pectin.

2.2 *Characterization of the pectin medium*

Surface area was obtained using a Micrometrics Model ASAP 2020 analyzer. The isoelectric point was determined with a Zeta Plus equipment. Functional groups on the pectin, responsible for the adsorption of As, were determined using infrared spectroscopy with Fourier Transform (FTIR), Thermo Sci. equipment, Nicolet 6700 spectrometry brand.

2.3 *Bioadsorption kinetics*

Batch bioadsorption kinetics was performed in a Jar Tester. One gram of the bioadsorbent material was used for each 1 L of solution and agitated at 220 rpm (Paredes, 2012). The adsorption tests were conducted with an initial arsenic concentration of 0.2 mg L^{-1}. pH was maintained at 5.25. Arsenic was measured with digital photometry (Garrido *et al.*, 2009).

3 RESULTS AND DISCUSSION

3.1 *Characterization of the pectin*

The BET surface area for the orange pectin without Fe(III) was 0.0216 m^2 g^{-1} and with Fe(III) ion coating, 0.0894 m^2 g^{-1}. The isoelectric point of the Fe(III)-coated pectin was found to be at pH 5.25. As shown in Figure 4, the infrared spectrum contains some very wide peaks across the spectrum, indicating the complex nature of the orange pectin. The peak at approximately 2900 cm^{-1} corresponds to the C–H group. The peaks around 1600 cm^{-1} correspond to C = C stretching that may be due to the C–C bond of the aromatic rings, and the intense band at 1000 cm^{-1} corresponds to the C–O group of alcohols and carboxylic acids. The characteristic signal of the carbonyl group is not clearly shown in the spectrum.

Figure 5 shows a microscopic image of the pectin coated with Fe(III) ions. The zone in the box indicates the fragment corresponding to the composition analysis that shows the spectrogram of the pectin in Figure 6. Percentage values in weight for each element are shown.

3.2 *Adsorption Kinetics*

Figure 7 shows the pseudo-second order constants kinetics for As removal. The equilibrium was reached after 120 min.

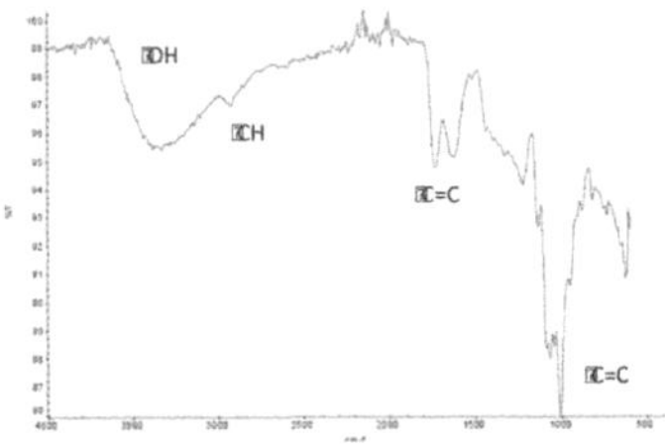

Figure 4. FTIR spectrum of acidic orange pectin coated with Fe(III) ions.

Figure 5. View with electronic microscope of the orange pectin coated with Fe(III) ions, at 1 000x amplification.

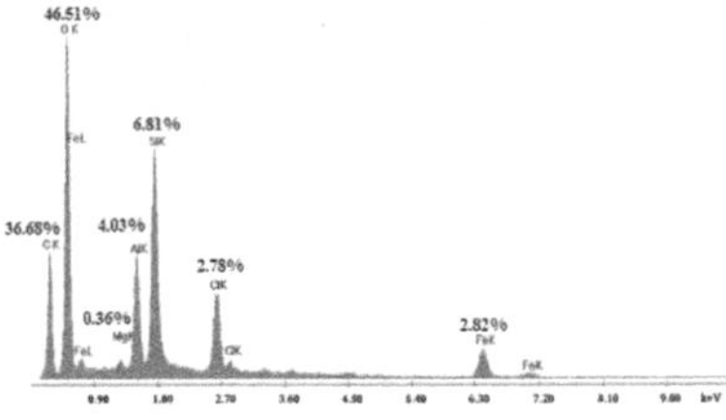

Figure 6. Spectrogram of the pectin and percentage in weight of the composition analysis.

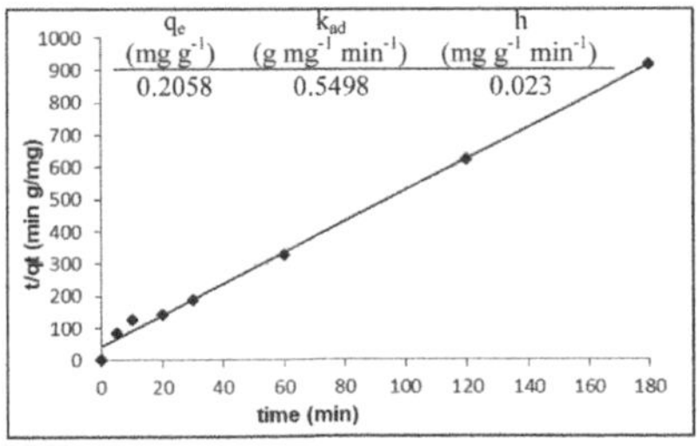

Figure 7. Pseudo-second order kinetics of As on pectin, R^2: 0.9952.

4 CONCLUSIONS

The adsorption rate constant, k_{ad}, for this bioabsorbent, 0.5498 g mg^{-1} min^{-1}, is lower than that reported by Paredes (2012) for pectin- Fe(III) coated (2.107 g mg^{-1} min^{-1}).

REFERENCES

Garrido, S., Avilés, M., Ramírez, A., Calderón, C., Ramírez-Orozco, A., Nieto, A., Shelp, G., Seed, L., Cebrian, M. & Vera, E. 2009. Arsenic removal from water of Huautla, Morelos, Mexico using Capacitive deionization. *Natural arsenic in Groundwaters of Latin America*. Ed. Taylor & Francis. London UK. pp. 665–676.

Ghimire, K.N. Inoue, K., Yamaguchi, H., Makino, K. & Miyajima, T. 2003. Adsorptive separation of arsenate and arsenite anions from aqueous medium by using orange waste. *Water Research* 37: 4945–4953.

Paredes, J.L. 2012. Remoción de arsénico del agua para uso y consumo humano mediante diferentes materiales de adsorción. Tesis de Maestría. UNAM, México.

WHO, 2004. World Health Organization. Guidelines for drinking Water Quality. Volume 1, Geneva, 3erd edition.

One Century of the Discovery of Arsenicosis in Latin America (1914–2014) – Litter, Nicolli, Meichtry, Quici, Bundschuh, Bhattacharya & Naidu (Eds)
© 2014 Taylor & Francis Group, London, ISBN 978-1-138-00141-1

Arsenic removal feasibility from aqueous solutions using shells and bone derivatives as adsorbents: A preliminary study

J.I. García, S. Boeykens, A. Saralegui & A. Gobbi
Laboratorio de Química de Sistemas Heterogéneos (LaQuíSiHe), Departamento de Química, Universidad de Buenos Aires, Buenos Aires, Argentina

C. Fernandez & A. Ozols
Instituto de Ingeniería Biomédica, Universidad de Buenos Aires, Buenos Aires, Argentina

ABSTRACT: In this work, feasibility of Arsenic (As) removal from aqueous solutions employing low cost adsorbents was analyzed. Solutions of 10 mg/L As were treated with 1 g of adsorbents based on hydroxyapatite, three derivatives, nacre and external surface of shells, for 24 h to ensure saturation. As concentration was measured by Atomic Absorption Spectroscopy. Hydroxyapatite showed the highest adsorption performance removing approximately 43%.

1 INTRODUCTION

Groundwater geochemical arsenic contamination has turned a big concern in Argentina. Research and development of low-cost adsorbents have become very important in the last years, due to the fact that potable water availability for low-income people represents a great challenge in some regions. Moreover, heavy metal removal from contaminated water is a worldwide problem nowadays (Litter *et al.*, 2010).

Natural, abundant and renewable resources seem to be ideal candidates for the development of As adsorption materials. Bovine bones and mollusk shells represent an option. Bones are primarily composed of calcium phosphate, hydroxyapatite (HA) and collagen; and shells, by calcium carbonate and chitin (a polysaccharide) in the External Surface (ES), and nacre in the inner side (NC). HA was reported as an effective heavy metals adsorbent (El Asri *et al.*, 2010; Czerniczyniec *et al.*, 2007).

This work shows preliminary studies on As adsorption capability of NC, ES, HA and three of their derivatives.

2 EXPERIMENTAL AND MEASUREMENT METHODS

2.1 *Experimental method*

The method was based on the immersion of adsorbent material in AsO_4^{3-} aqueous solutions. 1 g of each adsorbent was put into a beaker with 50 mL of a 10 mg/L AsO_4^{3-} solution, and was left under continuous shaking during one day. The samples were filtered, and the As concentration was measured by Flame Atomic Absorption.

2.2 *Adsorbent preparation*

Six adsorbents from two different groups (bones and shells) were tested. HA, obtained from fresh bovine bone, was treated by chemical washing (in soft acid solutions), pyrolyzed under oxidant atmosphere (900 °C for 2 h) and milled to particle size less than 200 μm (HA1). This powder was used to prepare blends: one with 25% (in wt.) of TiO_2 (HATi) and another one with 40% (in wt.) of tricalcium phosphate ($Ca_3(PO_4)_2$) (HATCP). The procedure comprised wet ball milling (in ethylic alcohol), drying followed by pressing (200 MPa), slow heating (10 °C/min), sintering (1200 °C by 2 h), and final milling. Also, HA1 was blended with isocyanate and polyol to polymerize as polyurethane foam with 40% (in wt.) of polymer (HAPU). Regarding the other group, mechanical separation of external and internal surfaces of shells resulted in two different adsorbents: ES and NC. All materials were milled to less than 45 μm particle size, except for HAPU (< 200 μm).

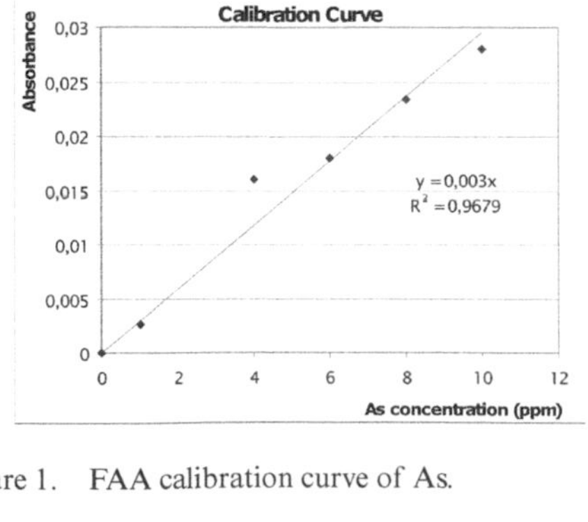

Figure 1. FAA calibration curve of As.

2.3 *Measurement method*

A flame Atomic Absorption Spectrophotometer with acetylene combustion was used in all measurements (*Buck Scientific* model 210VGP), and an As lamp at 1937 nm was employed.

The calibration curve (Figure 1) was determined using several dilutions of the original 10 mg/L solution (1, 4, 6 and 8 mg/L). To rectify errors introduced by equipment fluctuation and shifts in the zero baseline, five measurements were done in each case.

Sensitivity was not very good for low As concentrations since the maximum value of absorbance was 0.028 for a 10 mg/L concentration, and this is a small value in comparison to signals of other heavy metals that were tested (e.g. 0.390 for a 10 mg/L Cu solution). Besides, an As detection limit of 1 mg/L was determined, which is bigger than the maximum As content recommended in potable water (0.01 mg/L). However, the method permitted us to make a first selection of adsorbent material, which was the primary objective of this work.

3 RESULTS AND DISCUSSION

Table 1 shows the results obtained from all experiments for each adsorbent. A sample of the initial solution was preserved in order to be used as a reference when measuring. The percentage of As removed for each adsorbent was determined and it is shown in Figure 2. As it can be seen, none of the adsorbents had a resulting concentration below the detection limit or near it, and this is in concordance with our previous argument about the selected measurement method.

Results showed that HA1 had the highest performance removing up to 43% of the initial As content. The second place was occupied by HAPU, removing up to 33.5%. HATi and HATCP were moderate As adsorbents, they retained up to 31% and 22% respectively. The lowest performance was showed by NC and ES with removals close to 11%, and 3.4%.

Table 1. Results obtained from all experiments.

Exp.	Sample	Absorbance	As concentration (mg/L)
–	Initial sn	0.030	10,382
1	HATi	0.021	7,152
2	HA1	0.018	5,910
3	HATCP	0.024	8.075
4	HAPU	0.021	6.904
5	NC	0.027	9.282
6	ES	0.029	10.027

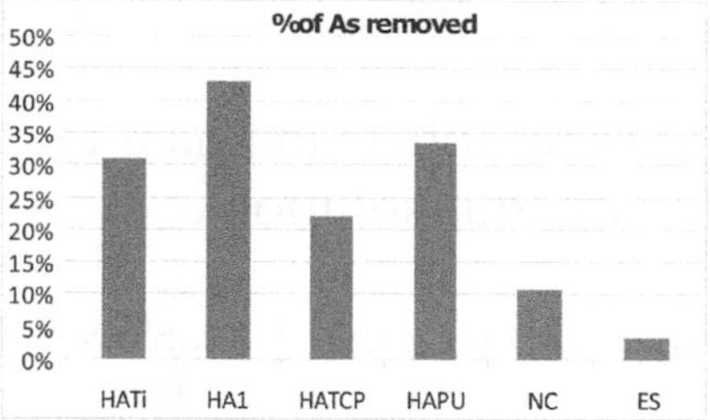

Figure 2. Percentage of As removed by each adsorbent.

4 CONCLUSIONS

From our results, we can conclude that HA can be a possible low-cost adsorbent of As since it showed the most satisfactory performance. Indeed, it was revealed that a derivative process is not necessary because it does not increase the performance. The fact that it removed approximately 43% allows us to believe that by designing a proper process for water treatment we will be able to reduce the As concentration in water for drinking purposes. Forthcoming studies on the effect of other variables (such as relationship of adsorption with time, adsorbent saturation and specific area analysis) are needed in order to design new technologies. This work also leaves some questions that could guide us in the future: Is it possible to reduce As concentration below 0,01 mg/L with HA? Can HA be recycled or reused? Can we obtain As from adsorption with HA? How do other components present in the solution interfere with As adsorption? Is HA selective? Does HA saturate very easily?

ACKNOWLEDGEMENTS

Projects UBACyT 20020090100102 and 200201201 00201.

REFERENCES

Czerniczyniec, M., Farías, S., Magallanes, J. & Cicerone, D. 2007. Arsenic(V) Adsorption onto Biogenic Hydroxyapatite: Solution Composition Effects. *Water Air Soil Pollut.* 180: 75–82.

Litter, M., Morgada, M., Bundschuh, J. 2010. Possible treatments for arsenic removal in Latin American waters for human consumption. *Environmental Pollution* 158: 1105–1118.

El Asri, S., Laghzizil, A., Coradinb, T., Saoiabi, A., Alaoui, A. & M'hamedi, R. 2010. Conversion of natural phosphate rock into mesoporous hydroxyapatite for heavy metals removal from aqueous solution. *Colloids and Surfaces A: Physicochem. Eng. Aspects* 362: 33–38.

One Century of the Discovery of Arsenicosis in Latin America (1914–2014) –
Litter, Nicolli, Meichtry, Quici, Bundschuh, Bhattacharya & Naidu (Eds)
 ISBN 978-1-138-00141-1

Dead macrophyte biomass utilization for the adsorption of arsenic from aqueous solutions

A. Saralegui, J.I. García, A. Gobbi & S. Boeykens
Laboratorio de Química de Sistemas Heterogéneos (LaQuiSiHe), Departamento de Química, Universidad de Buenos Aires, Buenos Aires, Argentina

ABSTRACT: The development of low-cost methods for the abatement of arsenic in natural waters is a matter of interest because of concern in many regions of our country and worldwide. In this paper, we present a preliminary study on the use of untreated biomass for adsorption of arsenate ions from aqueous solutions. The biomass materials coming from dead aquatic macrophytes: *Pistia stratiotes, Limnobium* Sp., *Azolla pinnata and Lemna* Sp. were tested. The better adsorption capacity was obtained for the *Azolla pinnata* (0.16 mg As/g biomass). Results of these experiments with biomass derived from various species of macrophytes demonstrate the feasibility of their use for water purification.

1 INTRODUCTION

Non-anthropogenic contamination with arsenic of groundwaters in the Pampas plain has been reported in numerous scientific works (Bundschuh *et al.*, 2008; Farías *et al.*, 2003). The presence of As represents a problem for the provision of drinking water, especially in rural communities. Thus, many efforts have been made to develop low-cost technologies for the abatement of this element. Among the new technologies, biosorption has a great potential, characterized by the use of inexpensive, non-toxic and biodegradable materials (Elifantz & Tel-Or, 2002). In a previous work, we have studied the adsorption of metal cations by non-living macrophyte biomass with promising results (Miretzky *et al.*, 2006).

The objective of this work was to test biomass materials coming from different aquatic macrophytes indigenous of the Pampas plain, *Pistia stratiotes*, *Limnobium* Sp., *Azolla pinnata* and *Lemna* Sp. to evaluate their performance in the adsorption of arsenate ions from aqueous solutions without any pretreatment.

2 MATERIALS AND METHODS

2.1 *Reagents*

For adsorption tests a 10 mg L^{-1} As solution was prepared from a dilution of a 1000 mg L^{-1} As standard MERCK. All chemicals used were of analytical-reagent grade. Ultrapure water was used to prepare all solutions.

2.2 *Biosorbents*

The four selected aquatic macrophyte species were collected in plastic bags of different natural aquatic environments in the Buenos Aires province. At the laboratory, the plants were separated from potentially invasive species, washed with pure water, dried in an oven at 60 °C for 24 hours, milled and sieved (< 500 μm). Processed biomass was stored in containers for subsequent adsorption tests.

2.3 *Adsorption tests*

1 g of each biomass material was put into a beaker with 50 mL of 10 mg L^{-1} As solution, and left for 24 h. Subsequently, the solutions were filtered through a 45 μm membrane. As concentrations in aqueous solutions before and after the experiment were measured. A blank test for each experiment was done in order to discard other types of adsorption (as due to the glass material).

2.4 *Measurement method*

An acetylene flame atomic absorption spectrophotometer (210VGP Buck Scientific) with an As lamp (λ = 193.7 nm) was used. The calibration curve (Figure 1) was developed using four dilutions of the original 10 mg L^{-1} solution (1, 4, 6 and 8 mg L^{-1}).

To eliminate random errors introduced by oscillation of the spectrophotometer, we took the average of five consecutive values, we subtracted from it the average of ten values (five before and five after the sample) obtained from ultrapure water.

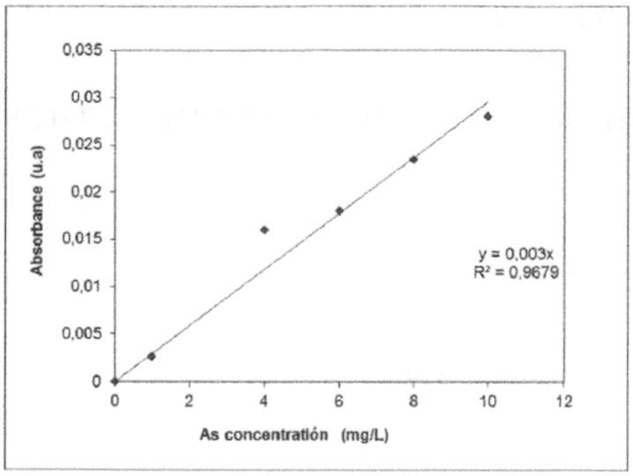

Figure 1. Calibration curve. Absorbance vs. as concentration.

The detection limit for As was determined to be 1 mg L^{-1}, and the sensibility was 0.003 mg L^{-1}. Even though it was high in relation to the maximum concentration recommended by the Argentinean Alimentary Code (CAA), the equipment was useful to determine which of the solids tested represent a viable alternative for removal of this heavy metal.

3 RESULTS AND DISCUSSION

Table 1 shows the results from the experiments with the four different biomass materials tested. During the experiment, a sample of the initial solution was preserved to be used as a reference when measuring. As it can be observed, none of the adsorbents had a resulting concentration below the detection limit or near it, and this is in concordance with our previous argument about the selected measurement method.

The absorption capacity (q) can be calculated by:

$$q = (C_f - C_i)\ V/m$$

where C_f and C_i are the final and initial concentrations, respectively, V is the volume of the solution and m is the macrophyte mass.

Out of the four adsorbents tested in this work, none showed a great performance. However, these results show that *Azolla* has the highest As adsorbent performance with an adsorption capacity of 0.16 mg g^{-1}.

Taking into account that the adsorption of anionic species by untreated dead macrophyte biomass has been reported previously only in few works (Pennesi *et al.*, 2012), this preliminary study turns out encouraging.

Table 1. Result from experiments with different biomass materials.

Macrophyte biomass	As concentration (mg L^{-1})	As removed from water	
	Initial	Final	%
Pistia	10.382	9.637	7.17
Limnobium	10.382	9.779	5.08
Azolla	10.382	7.152	31.1
Lemna Sp.	10.382	9,992	3.75

4 CONCLUSIONS

Forthcoming studies should be undertaken to address the design of suitable purification technologies, considering recirculation of the contaminated flow, macrophyte reuse, and recycling or disposal of arsenic.

From our preliminary results we can conclude that the dead macrophyte biomass material can be a possible low-cost adsorbent of As.

ACKNOWLEDGEMENTS

Work performed with UBACyT 20020090100102 and 20020120100201 projects (Argentina).

REFERENCES

Bundschuh, J., Perez Carrera, A. & Litter, M. 2008. Distribución de As en las regiones Ibérica e Iberoamericana. IBEROARSEN. CYTED, Argentina.

Elifantz, H. & Tel-Or, E. 2002. Heavy metal biosorption by plant biomass of the macrophyte *Ludwigia stolonifera*. *Water Air Soil Pollut*. 141: 207–218.

Farias, S., Casa, V.A., Vázquez, C., Ferpozzi, L., Pucci, G.N. & Cohen, I.M. 2003. Natural contamination with arsenic and other trace elements in ground waters of Argentine Pampean Plain, *The Science of the Total Environment* 309: 187–199.

Miretzky, P., Saralegui, A. & Fernández Cirelli, A. 2006. Simultaneous heavy metal removal mechanism by dead macrophytes. *Chemosphere* 62: 247–254.

Pennesi, C., Veglio, F., Totti, C., Romagnoli, T. & Beolchini, F. 2012. Nonliving biomass of marine macrophytes as arsenic(V) biosorbents. *J. Appl. Phycol.* 24(6): 1495–1502.

One Century of the Discovery of Arsenicosis in Latin America (1914–2014) – Litter, Nicolli, Meichtry, Quici, Bundschuh, Bhattacharya & Naidu (Eds)
 ISBN 978-1-138-00141-1

Removal of arsenic using a phosphonium ionic liquid impregnated-resin

L.B. Escudero
Chemistry Institute of San Luis (INQUISAL), San Luis, Argentina

A. Castro Grijalba & R.G. Wuilloud
Laboratory of Analytical Chemistry for Research and Development (QUIANID), ICB-UNCuyo, Mendoza, Argentina

ABSTRACT: A preliminary approach based on statics and dynamics studies involving removal of Arsenic (As) -as As(V)- from aqueous solutions is presented in this work. Raw and trihexyltetradecylphosphonium chloride ionic liquid-impregnated polymeric resins were evaluated for removal of As(V). Measurements of As(V) were made by Electrothermal Atomic Absorption Spectrometry (ETAAS). A 95% extraction efficiency was achieved for As(V) when the procedure was developed under optimal experimental conditions. This work leads to the possible development of simple and green remediation processes involving As species uptake.

1 INTRODUCTION

Arsenic (As) is a ubiquitous and potentially toxic element in the environment (Majidi & Shemirani, 2010). It can cause multiple adverse health effects, including dermal, respiratory, cardiovascular, and carcinogenic effects (Mandal & Suzuki, 2002). In order to study removal of As -as As(V)- from aqueous solutions, preliminary studies in batch and dynamic mode using macroreticular crosslinked aromatic polymeric resins impregnated with a quaternary phosphonium Ionic Liquid (IL) were developed. Variables such as the type of polymeric resin, amount of IL for resin impregnation, time of contact between the impregnated resin and the As solution were studied. This work reports the first application of IL-impregnated polymeric resins for the successful removal of As(V) from aqueous solutions.

2 EXPERIMENTAL

2.1 *Resin impregnation*

Three non-ionic, hydrophobic and polymeric resins were tested, each of them showing different physical properties (porosity, pore diameter, surface area, etc.). Before impregnation, polymeric resins were purified with aliquots of water, 10% (v/v) HCl solution and ethanol, and finally dried at room temperature. The impregnation of the resin was performed by contact of 1.0 g resin with a solution containing 0.5 g of trihexyltetradecylphosphonium chloride IL in 5 mL of ethanol for 7 h. The solvent was removed by filtration with 0.45 μm pore size PTFE membrane filters. The IL-impregnated resin was dried at 50 °C for 30 minutes.

2.2 *Procedure for As(V) removal*

A standard solution containing 50 mL of 2 mg L^{-1} As(V) was placed in a beaker with 500 μL hydrochloric acid (37% w/w), and 6 mL 0.32 mol L^{-1} $(NH_4)_6Mo_7O_{24}$. A final ratio for complex formation of 1:1 × 10^4 (As:Mo) was chosen following a previous work (Matsumiya *et al.*, 2009). In a static mode, 1 g of impregnated resin was added to the previous solution for 10 min. In a dynamic mode, an on-line system was developed using a cylindrical column packed with IL impregnated polymeric resin. The column was 6 cm long with an inner diameter of 2 mm. The initial As(V) solution was loaded on the column at a flow rate of 5 ml min^{-1} with a peristaltic pump.

3 RESULTS AND DISCUSSION

3.1 *Impregnated resins and As(V) retention*

Retention capacity of IL-impregnated polymeric resins was studied under static and dynamic conditions using three different polymeric resins. For the static study, the procedure for As removal was performed as described previously and the supernatant was analyzed by Electrothermal Atomic Absorption Spectrometry (ETAAS). Under dynamic conditions, the effluent was collected and subsequently analyzed by ETAAS.

Results were different for each resin studied. Table 1 shows the physical properties of the resins. The best results were obtained using the polymeric resin that exhibited the lowest pore diameter (140 Å). Thus, 95% As(V) retention onto the solid material under optimal experimental conditions was achieved.

Due to the differences between physical properties of the resins under study, the amount of IL immobilized on the resin was evaluated following the procedure described previously (Arias *et al.*, 2011). Figure 1 shows the amount of IL impregnated for each resin. Moreover, it could be observed that As(V) retention was linearly related to the amount of IL impregnated on the resin ($R^2 = 0.9954$). Thus, the resin with the highest As retention capacity showed also the highest amount of impregnated IL.

3.2 *Recovery As(V)*

Once As(V) was removed from the aqueous solution, different compounds including acids and alkalis were tested in order to reach the recovery of the analyte. Organic solvents could not be

Table 1. Physical properties of the resins under study.

	Resins		
Properties	A	B	C
Particle size (mm)	0.49–0.69	0.56–0.71	0.35–0.60
Surface area (m^2 g^{-1})	≤ 725	≤ 800	≤ 500
Porosity (mL mL^{-1})	≤ 0.50	≤ 0.55	≤ 0.60
Pore diameter (Å)	40	100	140
Pore volume (mL g^{-1})	0.98	1.82	1.68

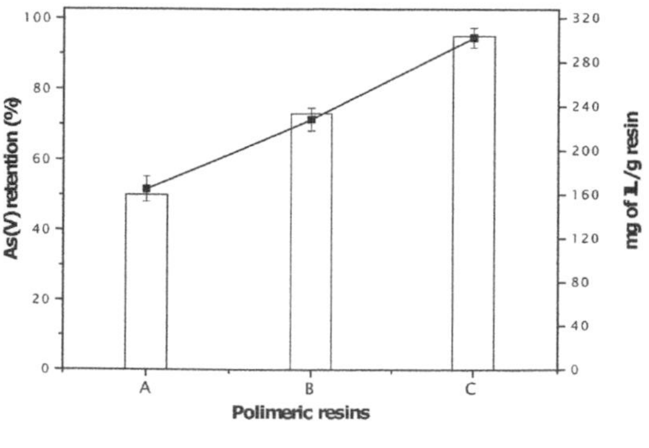

Figure 1. As(V) retention (%) using IL-impregnated resins (bar graph). Amount of IL per gram of solid material (line graph). A: resin with a pore diameter of 40 Å; B: 100 Å; C: 140 Å.

evaluated because of the solubilization of the IL in these media. Thus, 10% (v/v) nitric acid, 10% (v/v) sulfuric acid, 1 mol L^{-1} ammonium hydroxide and 1 mol L^{-1} sodium hydroxide were evaluated as possible recovery agents. It could be observed that the acid solutions were not useful as recovery agents for As. In contrast, the use of alkalis was the most effective agent for the recovery step. Concentrations of sodium hydroxide in the range 0.1–3 M were studied not only for the static mode but also for the dynamic mode. Results showed that the best As recovery (60%) was using 1.5 M sodium hydroxide.

However, future plans for this work aim to study other recovery agents, such as citric acid and potassium iodide.

3.3 *Study of As detection by ETAAS*

The graphite furnace program (pyrolysis and atomization temperatures) was carefully optimized to obtain the highest absorbance-to-background signal ratio. In order to reduce a potential matrix interference and to increase the accuracy, a chemical modifier or a modifier mixture is essential for ETAAS measurements. Thus, two modifiers, palladium nitrate solution and magnesium nitrate solution, were investigated. When a mixture containing 5 µg Pd and 3 µg Mg was employed as a matrix modifier, the arsenic absorption signal was well shaped, i.e. narrow, sharp and symmetric peaks were observed, which led to select this mixture in these quantities for the analysis.

Finally, As(V) measurements were performed with this mixture. Optimal pyrolysis and atomization temperatures were 1200 °C and 2300 °C, respectively.

4 CONCLUSIONS

A simple procedure for As (V) removal from aqueous solutions using green solvents has been presented in this work. The results demonstrated the possibility of using a quaternary phosphonium IL-impregnated polymeric resin for efficient recovery of As (V). Despite sodium hydroxide was the best analyte recovery agent, more studies could be needed in order to optimize the results already obtained. The method could be useful for selective retention of As(V).

ACKNOWLEDGEMENTS

This work was supported by Consejo Nacional de Investigaciones Científicas y Técnicas (CONICET), Agencia Nacional de Promoción Científica y Tecnológica (FONCYT)(PICT-BID) and Universidad Nacional de Cuyo (Argentina).

REFERENCES

Arias, A., Saucedo, I., Navarro, R., Gallardo, V., Martinez, M. & Guibal, E. 2011. Cadmium(II) recovery from hydrochloric acid solutions using Amberlite XAD-7 impregnated with a tetraalkyl phosphonium ionic liquid. *Reactive Functional Polymer* 71: 1059–1070.

Majidi, B. & Shemirani, F. 2010. In situ solvent formation microextraction in the presence of ionic liquid for preconcentration and speciation of arsenic in saline samples and total arsenic in biological samples by electrothermal atomic absorption spectrometry. *Biological trace elements research* 143: 579–590.

Mandal, B.K. & Suzuki K.T. 2002. Arsenic around the world: a review. *Talanta*, 58: 201–235.

Matsumiya, H., Kikai, T. & Hiraide, M. 2009. Solid-phase extraction followed by HPLC for the determination of Phosphorus(V) and Arsenic(V). *Analytical sciences*. 25: 207–210.

One Century of the Discovery of Arsenicosis in Latin America (1914–2014) – Litter, Nicolli, Meichtry, Quici, Bundschuh, Bhattacharya & Naidu (Eds)
© 2014 Taylor & Francis Group, London, ISBN 978-1-138-00141-1

Characterization of a chitosan biopolymer and arsenate removal for drinking water treatment

M. Annaduzzaman & P. Bhattacharya
KTH-International Groundwater Arsenic Research Group, Department of Sustainable Development, Environmental Science and Engineering, KTH Royal Institute of Technology, Stockholm, Sweden

M. Ersoz
Advanced Technology Research and Application Center, Selcuk University, Turkey

Z. Lazarova
Department of Health and Environment, AIT Austrian Institute of Technology, Vienna, Austria

ABSTRACT: Chitosan biopolymer with a deacetylation degree of 85%, was assessed for its capability to adsorb As(V) from drinking water by batch experiments. To characterize the chitosan biopolymer, chitosan was analyzed by FTIR and SEM. The results showed that chitosan is an effective and promising sorbent for As(V) from drinking water. From the batch tests, results showed a maximum adsorption of 355 μg/L of As(V) with 1.18 μg g^{-1} adsorption capacity at pH 6. The kinetic data, obtained at pH 6 could be fitted with pseudo-second order equation (adsorption capacity: 0.923 μg g^{-1}) and the process was suitably described by a Freundlich ($R^2 = 0.9933$) model than by a Langmuir model ($R^2 = 0.9741$). The results above indicated that chitosan is a very favorable sorbent for As(V) removal from aqueous solution.

1 INTRODUCTION

Drinking water, contaminated with Arsenic (As), pose serious health risk for humans all over the world. Millions of people have severe As-related diseases through consumption of As rich water, which increases the importance of treating water contaminated with As. Several methods are available for this purpose, namely, lime softening, iron coagulation, filtration, Reverse Osmosis (RO), ion exchange, membrane process, colloid floatation, etc. (Bhattacharya *et al.*, 2002). These chemical methods may be easy for operation; however, the generated residues after the treatment processes are highly toxic to human health and not environmentally suitable. By considering these aspects, chitosan is a low cost and effective biopolymer proper for treating As(V) contaminated water, since it is easily available, inexpensive, non-toxic, renewable and biodegradable.

In this study the adsorption capacity and the filtration properties of chitosan with a deacetylation degree of 85% for removing As(V) from water was evaluated. Thus, this study aims: (1) to characterize the chitosan biopolymer (2) to evaluate the adsorption capacities by Langmuir and Freundlich isotherms with kinetic models. The final objective was to evaluate the applicability of the chitosan as a low cost adsorbent for As(V) removal.

2 EXPERIMENTAL DESIGN

Chitosan with a 85% deacetylation degree was manufactured by BioLog Biotechnologie und Logistik GmbH (Germany) for the batch experiments and evaluation of adsorption characteristics. Arsenic(V) standard (1000 μg mL^{-1}) was purchased from ThermoFisher Scientific, USA. Liquid parafilm, nitric acid, sodium hydroxide, and other reagents used for the experiments were of analytical grade. Characterization of chitosan biopolymer was performed by FTIR and SEM analysis performed in Selcuk University, Turkey. Arsenic(V) concentration in aqueous solutions and pH were determined by ICP-OES-6000 and standard digital Ion-pH meter. The adsorption study was investigated by considering different parameters as pH, effects of contact time, concentration and chitosan dosages (Benavente, 2008). The kinetic studies was conducted at the optimum condition, and As(V) concentrations of 50, 100, 150, 200 and 250 μg L^{-1} were considered.

3 RESULTS AND DISCUSSION

3.1 *Characterization of the chitosan structure*

The chitosan structure was confirmed by FTIR analysis (Figure 1). At 3000–3500 cm^{-1}, the bands indicated NH and O-H stretching vibrations and at 1400–1650 cm^{-1}, the bands corresponded to C = O bond. The peaks around 2950, 2347, 1641, 1578, 1425, 1300 to 1025 cm^{-1} in the spectrum were due to stretching vibrations of 3-methylbenzyl chloride, ethyl sorbet, chlorogenic acid hemihydrate, 3-nitrophenol, 3-methoxypropionitrile, phenylethyleneglycol, 4-chlorobenzophen-one, respectively, which are the characteristic bands of a chitosan polysaccharide. The SEM-EDS analysis showed that chitosan consisted of O (50.25%), C (30.85%), N (10.39%), and As (0.89%) by weight (Figure 2).

3.2 *Effect of pH on As(V) adsorption*

The effect of pH on As(V) adsorption on chitosan adsorbent was studied in the range 6–9 (considering the drinking water values). At pH 6, removal was about 100%, decreasing with increasing pH. At the higher pH values, the ionization of both the adsorbate and adsorbent caused repulsion at the chitosan surface, decreasing net As(V) adsorption.

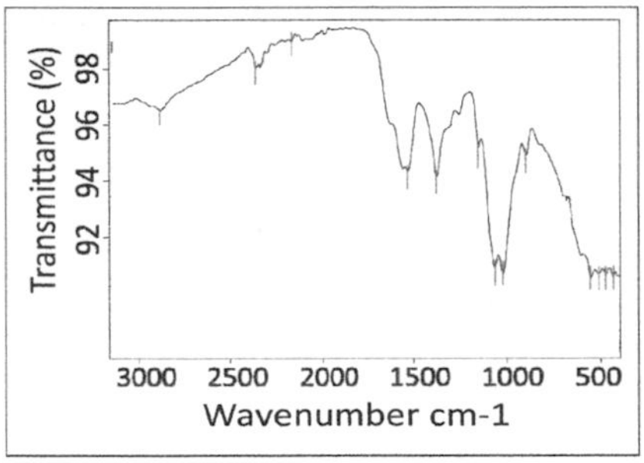

Figure 1. FTIR spectrum of chitosan used in experiments.

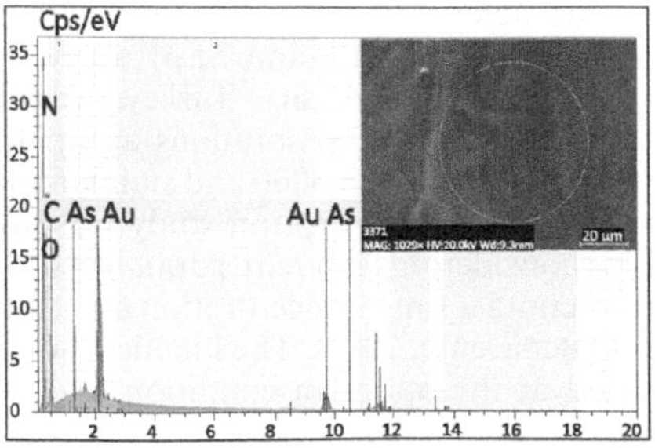

Figure 2. SEM and EDS analysis of the chitosan biopolymer.

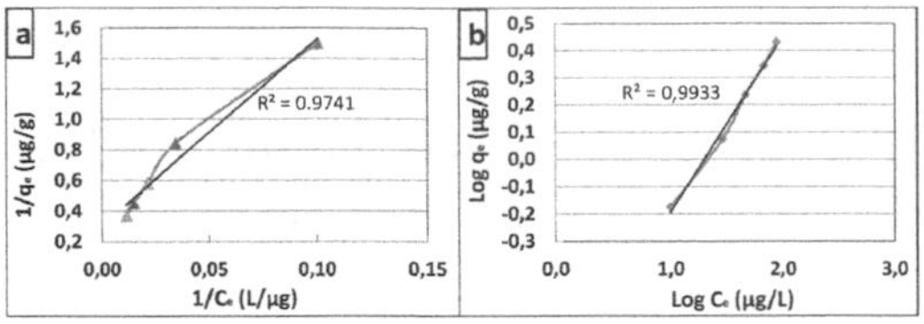

Figure 3. a) Langmuir fitting and b) Freundlich fitting for As(V) adsorption on the chitosan biopolymer at the optimum removal condition (pH 6, [As(V)] = 100 µg L^{-1} and [chitosan] = 70 g L^{-1}).

3.3 *Arsenic adsorption isotherm*

The maximum As(V) adsorption capacity of chitosan was determined by Langmuir and Freundlich adsorption isotherms, at an initial As(V) concentration of 100 µg L^{-1} and pH 6 (optimum condition). Based on the regression coefficient results of the Langmuir (0.9741) and Freundlich (0.9933) model plots (Figures 3a and 3b, respectively), the results showed that the adsorption follows both models, with sorption coefficients of 0.025 and 0.147 L μg^{-1}, correspondingly. This isotherm results clearly indicates that As(V) adsorption takes place on a homogeneous monolayer surface of chitosan and that the As(V) species do not interact with each other during the sorption process.

3.4 *Kinetics of arsenate adsorption*

The validity of pseudo-first and pseudo-second order models were assessed with various kinetic experiments at different As(V) initial concentrations. The adsorption was suitably described by a pseudo-second order equation, and the amount of As(V) adsorbed was estimated for various As(V) concentrations. A quite good correlation within the experimental results was obtained from the pseudo-first fitting (qe = 1.18 µg g^{-1}) and from pseudo-second-order (qe = 0.923 µg g^{-1}). These results indicate that the rate constant decreases with an increases in the initial concentration of As(V).

4 CONCLUSIONS

The results revealed a better correlation of the experimental data with the Freundlich adsorption isotherm, in comparison with that with the Langmuir adsorption isotherm. This also indicates that the adsorption of As(V) takes place on a homogeneous monolayer of chitosan with a maximum adsorption of 1.18 µg/g. The adsorption kinetics of chitosan followed a pseudo-second order model. The chitosan with a deacetylation degree of 85%, has good potentiality for removing of As(V) from drinking water.

ACKNOWLEDGEMENTS

The authors are thankful to the ChitoClean project (FP7-SME) for financial support and Selcuk University (Turkey), for providing some of the experimental facilities during this study.

REFERENCES

Benavente, M. (2008). Adsorption of Metallic Ions onto Chitosan: Equilibrium and Kinetic Studies. TRITA CHE Report 2008:44.

Bhattacharya, P., Jacks, G., Frisbie, S.H., Smith, E., Naidu, R. & Sarkar, B. 2002. Arsenic in the Environment: A Global Perspective in Sarkar, B. (ed), *Heavy Metals in the Environment; New York, USA, 2002.*

One Century of the Discovery of Arsenicosis in Latin America (1914–2014) – Litter, Nicolli, Meichtry, Quici, Bundschuh, Bhattacharya & Naidu (Eds)
© 2014 Taylor & Francis Group, London, ISBN 978-1-138-00141-1

Seasonal variation of redox potential and arsenic removal from water in a constructed wetland mesocosm

M.C. Valles-Aragón & M.T. Alarcón-Herrera
Advanced Materials Research Center, CIMAV, Chihuahua, Chih, Mexico

ABSTRACT: Arsenic retention in constructed wetlands is influenced by environmental, physicochemical and biological conditions. The study was conducted in three constructed wetland prototypes. Two planted with *E. macrostachya* and *S. americanus*, respectively; a third unplanted one was used as control. The system was continuously operated for 343 days, fed by groundwater with As concentration at 90.66 ± 14.95 μg/L, and a hydraulic retention time of 2 days. Redox potential, environmental temperature, and inlet and outlet As concentration were constantly monitored. Oxidizing conditions (87 to 516 mV) were recorded during most of hot and warm seasons (84–90%), while reducing conditions (up to -539 mV) were recorded the rest of the time. On the cold season, only oxidizing conditions were detected. Plants translocate O_2 in the mesocosm, inducing a proper media for As precipitation. Planted prototypes had higher As retention than the unplanted, as plants play an important role on As retention.

1 INTRODUCTION

Constructed wetlands represent a low cost alternative for As removal from drinking water supply in less developed communities (Zurita *et al.*, 2012). Environmental physicochemical conditions have a great influence on the As retention mechanisms in constructed wetlands, which can be distinguish by *Eh* variations, in addition to water temperature dependence (Lizama *et al.*, 2012). Therefore, the main objective of the present research was to analyze the seasonal variation of *Eh* and As retention in a constructed wetland mesocosm.

2 METHODS/EXPERIMENTAL

2.1 *Experimental conditions*

The experiment was performed in three constructed wetland prototypes (large: 1.5 m, wide: 0.5 m, height: 0.5 m, 2.5% slope) located inside of a greenhouse without temperature control to identify seasonal effects in the system. Prototypes were evenly filled with 300 kg of silty sand soil (ρ = 1.4 g/cm^3, porosity: 38%, hydraulic conductivity: 18.53 cm/h). Coarse gravel (2.5 up to 4.0 cm) was used at the inlet as well at the outlet (wide: 10 cm). Water level was adjusted 5 cm below the surface of the sand bed in order to preserve submerged flow conditions.

Previous research reported a successful retention of As using *Eleocharis macrostachya* and *Schoenoplectus americanus* [23]. Therefore, a field collection of these plants was held at 53 km from Chihuahua City (28°35′05″N, 105°34′22″ W). Prototypes denominated HA and HB were planted (with *E. macrostachya* and *S. americanus*, respectively), whereas HC remained unplanted as a control reference. HA, HB, HC were operated simultaneously during 343 days. The system was fed with synthetic water, prepared with groundwater added with sodium arsenite ($NaAsO_2$) in order to reach As values of 90.66 ± 14.95 μg/L, considering a theoretical hydraulic retention time of 2 days.

2.2 *Monitored parameters*

Redox potential (*Eh*): monitored automatically every hour at the entrance, middle and exit section of wetlands, by digital equipment Hach, PC SC and RC model: Sc 1000.

Environmental temperature: monitored every 15 minutes using temperature sensors HOBO, Light Logger UA-002-64.

Analytic determination of As concentration: samples were taken weekly from water inflow and outflow. These were properly prepared for analysis of total As, with an acid digestion in a MARSX microwave (EPA method 3015). Analytical As determination was carried out using an atomic absorption spectrophotometer with hydride generator, GBC Avanta Sigma. Duplicate samples, certified standards (As in 2% HNO_3 of 1000 μg/L High-Purity 10003-1, traceable at National Institute of Standards and Technology, NIST), and blanks were taken. As recovery was 96% ± 3% for all samples. As quantification limit was 5 μg/L.

3 RESULTS AND DISCUSSION

3.1 *Temperature*

Considering environmental conditions, classification of operational time was: *June, July, August, September, April* and *May*; classified as hot season, with a range of 10 to 48°C. *October, February* and *March*; classified as warm season, with a variation range of 4 to 33°C. *November, December* and *January*; classified as cold season with a range of –1 up 26°C.

3.2 *Redox potential (Eh)*

Hot season in planted (HA, HB) and unplanted (HC) wetlands, recorded oxidizing conditions 85% of the time, in a range of 87 to 501 mV. *Eh* presented negative values up to –513 mV. During *warm season* (4 to 33° C) oxidizing conditions of 161 to 516 mV were recorded 90% of the time on HA and HB, whereas 100% of the time on HC. Negative values of *Eh* up to –539 mV were recorded. In *cold season* (–1 up to 26 °C) only oxidizing conditions in a range of 139 up to 492 mV were recorded for HB and HC whereas 99.5% of the time for HA, with negative values of *Eh* up to –161 mV.

Eh values on wetlands system present a correlation with regard to water pore temperature. Oxidizing conditions were maintained most of the time. Reducing conditions were recorded during peak temperature hours, at the beginning of the experiment, increasing the duration and incidence towards the end of the experiment (day 282–343). There was a higher frequency for oxidizing conditions and a lower frequency for reducing conditions in warm season than in hot season. As a result of low plant development at cold temperature, theoretically the rhizosphere had a lower microbial activity. Consequently, microbial growth and metabolism rate (O_2 consumption) was lower during cold season (Faulwetter *et al.*, 2009), having a direct impact on the frequency of reducing conditions, reporting mainly oxidizing conditions.

Further, redox conditions on the rhizosphere changed rapidly during hot season from 500 to -500 and viceversa. Similar researches reported redox values from 800 to –400 mV (Dusek *et al.*, 2008) and from 500 to –400 mV (Wießner *et al.*, 2005). This behavior was related to the influence of O_2 release from plants, which is consumed by microorganisms present in the rhizosphere.

3.3 *As retention*

During the first 178 days of the operation of the wetlands, the reported As retention in HA was 92% and in HB was 81%, while HC had a much lower retention, of 59%. After 343 operation days, HA and HB retention percentage dropped to 76 and 67%, respectively, while HC dropped to only 35% (Figure 1).

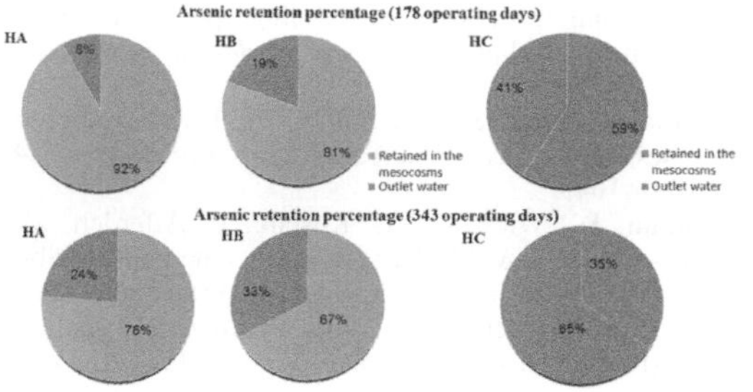

Figure 1. As retention in the constructed wetland mesocosms.

At the end of the experiment, As immobilization on planted wetlands was 32 to 41% higher than on the unplanted. In addition, this percentage was higher than found in other researches, in which 15% (Rahman *et al.*, 2011) and 20% (Zurita *et al.*, 2012) of As retention were reported from planted wetlands compared with unplanted ones.

Although plants have a secondary role in arsenic absorption, they are indirectly essential for the retention of this metalloid, owing that plants oxygenate the culture bed through roots, providing an extended zone of aerobic and anaerobic conditions (Lizama *et al.*, 2012). Plants also bring carbon which provides an optimum surface for the development of microorganisms in wetlands, along with roots, as a proper media for precipitation of Fe and As hydroxides (Rahman *et al.*, 2011).

4 CONCLUSIONS

A direct influence of seasonal variation on the *Eh* of constructed wetlands was detected, related to the prevalence of oxidizing conditions and As retention. Planted prototypes had higher As retention than unplanted ones. Plants contribute to induce more oxidizing conditions in the mesocosm of a constructed wetland, and play an important role on As retention on the system.

REFERENCES

Dusek, J., Picek, T. & Cizcova, H. 2008. Redox potential dynamics in a horizontal subsurface flow constructed wetland for wastewater treatment: Diel, seasonal and spatial fluctuations. *Ecological Engineering* 34: 223–232.

Faulwetter, J., Gagnon, V., Sundberg, C., Chazarenc, F., Burr, M., Brisson, J., Camper, A. & Stein, O. 2009.

Microbial processes influencing performance of treatment wetlands: A review. *Ecological Engineering* 35: 987–1004.
Lizama, K., Fletcher, T. & Sun, G. 2011. Removal processes for As in constructed wetlands. *Chemosphere* 84: 1032–1043.
Rahman, K., Wiessner, A., Kuschk, P., Afferden, M., Mattuschc, J. & Müllera, R. 2011. Fate and distribution of As in laboratory-scale subsurface horizontal -flow constructed wetlands treating an artificial wastewater. *Ecological Engineering* 37: 1214–1224.
Wießner, A., Kappelmeyer, U., Kuschk, P. & Ka¨ stn, M. 2005. Influence of the redox condition dynamics on the removal efficiency of a laboratory-scale constructed wetland. *Water Research* 39: 248–256.
Zurita, F., Del Toro-Sánchez, C., Gutierrez-Lomelí, M., Rodríguez-Sahagún, A., Castellanos-Hernández, O., Ramírez-Martínez, G. & White, J. 2012. Preliminary study on the potential of As removal by subsurface flow constructed mesocosms. *Ecological Engineering* 47: 101–104.

One Century of the Discovery of Arsenicosis in Latin America (1914–2014) – Litter, Nicolli, Meichtry, Quici, Bundschuh, Bhattacharya & Naidu (Eds)

ISBN 978-1-138-00141-1

Arsenic removal by phytoremediation technique in a greenhouse using water hyacinth (*Eichhornia crassipes Mart. Solms*) at varying growth stages

T.A. Laniyan
Department of Earth Sciences, Olabisi Onabanjo University, Ago-Iwoye, Ogun State, Nigeria

A.F. Abimbola
Department of Geology, University of Ibadan, Ibadan, Nigeria

M.K.C. Sridhar
Department of Environmental Health Sciences, Faculty of Public Health, University of Ibadan, Ibadan, Nigeria

ABSTRACT: An experimental study was performed on arsenic removal by water hyacinth (*Eichhornia crassipes Mart. Solms*) in a greenhouse. Matured water hyacinth was tested for its tolerance to As at 10.0, 20.0, 50.0 and 100.0 mg L^{-1}. Water hyacinth at different stages of maturity (sprouting, flowering, and matured) were also placed in contact with As at a maximum tolerance level of 100.0 mg L^{-1} for 0, 2, 12, 24, 48 and 120 hours, and the As accumulation was assessed. The highest arsenic uptake was recorded in the leaves of the sprouting plant (12 h) with a transfer factor of 200 as well as in roots of both flowering (48 h) and matured (120 h) with a transfer factor of 5400 and 6500, respectively. Matured water hyacinth plants were found to be the best for As remediation. This technique may be used to treat effluents rich in arsenic.

1 INTRODUCTION

In Nigerian environment, arsenic (As) occurs in water mostly from natural minerals and certain industrial effluents (Laniyan, 2012). More recent studies showed that water hyacinth is a valuable bioremediation plant capable of removing large amounts of organic and inorganic pollutants from water bodies in Nigeria (Talabi, 2008). For the possible utilization of aquatic plants to treat polluted waters and industrial effluents, Sridhar (1988) made use of water lettuce (*Pistia stratiotes L*) obtained from an eutrophic lake in Ibadan, Nigeria, to examine its mineral and trace the potential uptake of elements. The present study aimed at evaluating As removal using water hyacinth under greenhouse conditions in order to test the ability to be applied as an economic method for As removal from contaminated waters on a large scale.

2 METHODS/EXPERIMENTAL

2.1 *Study location*

The study was conducted in Ibadan, capital of Oyo State (southwestern Nigeria with latitude 7°15′–7°30′ and longitude 3°45′–4°00′E N). The location of the experimental study was on latitude 7°20.828′N and longitude 3°53.851′E. Water hyacinth used for the experimental study was harvested at the Awba Dam at the University of Ibadan at latitude 7°2639.6′N and longitude 003°5400.5′E.

A. Arsenic acid solutions at different concentrations were added to various buckets in the greenhouse: 0 (control with no As), 10, 20, 50 and 100 mg L^{-1} before cultivation of water hyacinth. Samples of water were collected at 0, 5, 10, and 15 days and analyzed for pH, Electrical Conductivity (EC), and Total Dissolved Solids (TDS).
   As content was measured using inductively coupled-ion chromatography.
B. The experiment was designed to determine the best age of the plant (sprouting, flowering, and matured stage of growth) and duration of the treatment (at start and at the end of 2, 12, 24, 48 and 120 h). Arsenic dose was maintained at the maximum tolerance level of 100 mg L^{-1} in each of the buckets.

2.2 *Data analysis*

The Transfer Function (TF) is a measure of bioaccumulation or uptake of metals by plants from the soil or water-based substrate, expressed as:

$TF = CAs_{plant}/CM_{substrate}$

where CAsplant is the measured concentration of As in the plant and CMsubstrate is the measured concentration of the soil or water substrate that the plant was taken from (Uchida & Tagami, 2005). Transfer uptake is interpreted when uptake or bio-accumulation is greater than 1 (TF > 1).

3 RESULTS AND DISCUSSION

3.1 *Physical features of the water where water hyacinth was cultivated*

pH values increase with the number of days (5th, 10th and 15th day). EC and TDS decreased with increase in contact period (5th, 10th and 15th day). Results showed that water hyacinth has the capacity to increase alkalinity as it grows.

3.2 *Effect of varying the arsenic concentration on accumulation of As in various parts of the plant*

Increase of As uptake was observed as As concentration increased from 0 to 50 mg L^{-1}; the progressive increase started on the fifth (5th) day, the 10th day and the 15th day. A similar increase was observed for plants cultivated on 100 mg L^{-1} arsenic. Calculated transfer uptake (Table 1) indicates values > 1. This implies that arsenic bioaccumulation metal occurred in the plants, but the highest bioaccumulation occurred in the roots and leaves of the plant, with the highest As concentration being observed in the 50 and 100 mg L^{-1} solutions.

3.3 *Effect of age of water hyacinth on the As uptake*

Calculated transfer uptake indicates values > 1. The revealed highest uptake for arsenic in the leaves of sprouting plants occurred at 12 h and at the roots between 12 and 48 h. The uptake for flowering plants by leaves and roots were between the 48 and 120 h, respectively, while that of the matured leaves and roots were between the 120 and 48 h, respectively. The transfer factor, therefore, shows the highest rate of absorption between 12 and 120 h.

Table 1. Uptake function of water hyacinth in concentrates.

As mg L^{-1}	leaf	stem	root
0	118.42	118.42	118.42
10	157.89	197.37	1105.26
20	118.42	118.42	1697.37
50	1263.16	394.74	14526.32
100	1302.63	473.68	18986.84

CONCLUSIONS

Water Hyacinth (*Eichhornia crassipes*) grown in a greenhouse was found to remove arsenic from water. The transfer factor indicated that the highest uptake was in the root system, followed by leaves and stolon in decreasing order. In the duration and maturity experiment, the highest uptake of arsenic was found to be between the 12 and 120 hours on the leaves of the young plant and the root of the old plant.

In conclusion, the study showed that arsenic contaminated waters can be remediated and that water-hyacinth can bioaccumulate arsenic.

REFERENCES

Laniyan, T.A. 2012. Geochemical evaluation and possible remediation methods of arsenic in surface and groundwater in Ibadan metropolis. PhD thesis submitted to the postgraduate school of University of Ibadan. pp. 122–211.

Sridhar, M.K.C. 1988. Uptake of trace elements by water lettuce (*Pistia stratiotes*). *Acta Hydrochimica et Hydrobiologica* 16(3): 293–297.

Talabi, L. 2008. Nigerian scientists unveils threats, benefits from Water Hyacinth. *The Guardian: Agrocare*. July 20: A85.

Uchida, S. & Tagami, K. 2005. Concentrations of rare—earth elements, Th and U in paddy field soils and rice and their behavior in soil—rice plant system. In *ICOBTE-Book of Abstract of conference on Biogeochemistry of trace elements*, pp. 112–113.

4.3 Ion exchange and membrane technologies

One Century of the Discovery of Arsenicosis in Latin America (1914–2014) – Litter, Nicolli, Meichtry, Quici, Bundschuh, Bhattacharya & Naidu (Eds)
© 2014 Taylor & Francis Group, London, ISBN 978-1-138-00141-1

Small-scale membrane-based desalinators—developing a viable concept for sustainable arsenic removal from groundwater

J. Hoinkis, S.A. Deowan, M. Herrmann, S. Schmidt, E. Gukelberger, F. Fiedler & B. Großmann
Karlsruhe University of Applied Sciences, Germany

A.K. Ghosh
Anugrah Narayan College, Patna, Bihar, India

D. Chatterjee
University of Kalyani, India

J. Bundschuh
University of Southern Queensland, Toowoomba, Australia

ABSTRACT: Arsenic poisoning in drinking water is a health issue in many countries worldwide including Bangladesh, India, Argentina, China and Vietnam. In areas where the drinking water supply contains unsafe levels of arsenic, technologies to remove arsenic are of prime importance. Many technologies have been developed for the removal of arsenic. Among those, Reverse Osmosis (RO) and NanoFiltration (NF) are very promising techniques because they have the advantage of removing dissolved arsenic along with other dissolved and particulate compounds. So far, however, this kind of membrane filtration has needed bulky and sophisticated units, which are not suitable for application in the rural areas of developing and newly industrializing countries. Therefore, the purpose of this work was to develop and test a simple and viable concept for membrane based arsenic removal in rural areas of developing countries. For this purpose, laboratory work was done at Karlsruhe University of Applied Sciences and pilot-scale experiments were conducted in rural India.

1 INTRODUCTION

Arsenic is a naturally-occurring contaminant of geological origin caused by weathering of arsenic-bearing rocks, minerals and ores. Arsenic in drinking water can cause a wide range of health problems such as skin lesions, circulatory disorders, diabetes, cancers, etc. Groundwater contamination with arsenic is a global issue, reported in 36 countries in Asia, North and South America, Europe and the Pacific, and the most serious pollution has been found in Bangladesh and West Bengal, India (Petrusevski *et al.*, 2007). The less developed the country is the less they are able to deal with this issue. The Maximum Contaminant Level (MCL) for arsenic in drinking water in most of the countries worldwide follows the recommended guideline limit of WHO, which is 10 $\mu g\ L^{-1}$ (WHO, 2013).

In areas where drinking water supply contains unsafe arsenic levels, arsenic removal technologies are of prime importance. To this day, many technologies for the removal of arsenic have been developed. The most commonly used technologies include oxidation, co-precipitation and adsorption onto coagulated flocks, lime treatment, adsorption onto sorption media, ion exchange resin and membrane techniques (Chwirka *et al.*, 2000; Mohan & Pittman, 2007; Song *et al.*, 2006; Uddin *et al.*, 2007; Shih, 2005). All of these technologies have their advantages and disadvantages; however, membrane technology, like Reverse Osmosis (RO) and NanoFiltration (NF), are very promising techniques due the advantage of removing dissolved arsenic along with other dissolved and particulate species (bacteria, viruses, dissolved iron, etc.). The benefits of membrane processes compared to other treatment options (e.g., ion exchange, precipitation) can be summarized as:

- high efficiency
- easy operation
- high effluent water quality
- modularity and flexibility.

Therefore, research on membrane-based arsenic removal has attracted great interest for many years (Shih, 2005; Figoli *et al.*, 2010).

To this day, RO and NF membrane filtration have mostly used bulky and sophisticated units with high-energy consumption, which are not suitable for application in rural areas of developing countries. As a consequence, most of the documented experience in drinking water production has been with large treatment plants in developed countries. Some time ago, however, also small-scale marine reverse osmosis units (known as watermakers) came onto the market. They are meanwhile widely used to produce drinking water from seawater on boats (for an overview, see Watermakers, 2013). It is now a proven technology, which works reliably at remote locations under mechanically, climatically and chemically rough conditions. Most of these systems have been optimized in terms of energy efficiency and productivity. Some of them can be powered by photovoltaics or can be operated manually. They are off-grid and can be operated in remote rural areas at the seashore by solar and wind energy (see, e.g., Moerk, 2013).

The purpose of this work was to develop a small-scale membrane-based membrane unit with regard to arsenic removal from drinking water to be used in developing countries.

2 METHODS

2.1 *Water quality*

The experimental work conducted by the Karlsruhe University of Applied Sciences (KUAS) comprised laboratory work as well as pilot trials on the ground in India. The experiments at KUAS were carried out with arsenic-spiked local tap water (Geucke *et al.*, 2009), whereas the pilot trials were done near Patna, Bihar, India, using local groundwater (Hoinkis *et al.*, 2012).

2.2 *Laboratory work*

The experiments were carried out with a small RO desalinator type "Power Survivor™ 160E", supplied by the Swiss company Katadyn (Katadyn, 2013). This system was initially designed for marine drinking water production on small boats. It is a compact unit, which is simple to install, to operate and to maintain. The membrane housing is fitted with a 2521 spiral wound module. Furthermore, a 20 L vessel was used as feed tank. The permeate and the concentrate solution were recirculated back to the feed tank. The pressure was read out at pressure gauges in the inlet and outlet of the membrane housing. The RO pump makes use of an energy recovery system taking advantage of stored energy in the high-pressure reject water that is typically wasted in conventional systems. The water recovery rate of the unit was fixed at 10%.

All the membranes that were used during the experiments were thin-film polyamide composite membranes supplied by the company Dow. XLE is an extra low-energy brackish water element for municipal and industrial applications. TW is an element for home drinking water, while SW is an element for desalination in marine applications (Dow Water and Process Solutions, 2013). The tests were carried out with tri - and pentavalent arsenic solutions.

2.3 *Pilot trials*

The experiments were carried out at two sites, namely Bind Toli and Ramnagar near Patna, Bihar, India, with a technical RO desalinator type "Smart 60", supplied by the company Schenker (Schenker, 2011). This desalinator is similar to the Katadyn unit typically used for marine drinking water production. However, here, it is based on a double piston pump and needs significantly lower energy than that needed for seawater desalination (ca. 4 Wh L^{-1}) compared to the Katadyn desalinator. Prior to the RO treatment, the groundwater was aerated and passed through a granular medium filter (GMF). First, a simple sand filter and, in a separate test, a granular medium filter based on anthracite (Evers, 2013) were applied in order to remove precipitates of Fe(III) (hydr)oxides. Similarly to sand filter, GMF based on anthracite has also the ability of effectively removing iron precipitates. However, GMF proved to be better backflushable in the cleaning procedure due to the lower weight of anthracite compared to sand.

The commercially available one pass unit was revamped to a two pass unit in order to increase the arsenic rejection. The membrane housing of the Schenker unit was fitted with 2540 (first pass) and 2521 (second pass) spiral wound modules (Dow Water and Process Solutions, 2013) and with sensors for pressure, permeate flow, electrical conductivity, power consumption. All data were recorded on a notebook by the software LabView (National Instruments). The permeate recovery rate of the first pass unit was kept at 10% in order to prevent membrane scaling, and the permeate recovery of the second pass has been adjusted between 50–90%. The pilot plant was fed with groundwater provided by local hand pumps and has been operated for 2–4 hours on each test day.

3 RESULTS AND DISCUSSION

3.1 *Laboratory work*

As can be seen from Figures 1 and 2, the rejection rate for trivalent arsenic is generally much lower than that for pentavalent arsenic. This agrees with preliminary lab scale results using flat sheet membranes and other publications (Deowan, 2008; Shih, 2005).

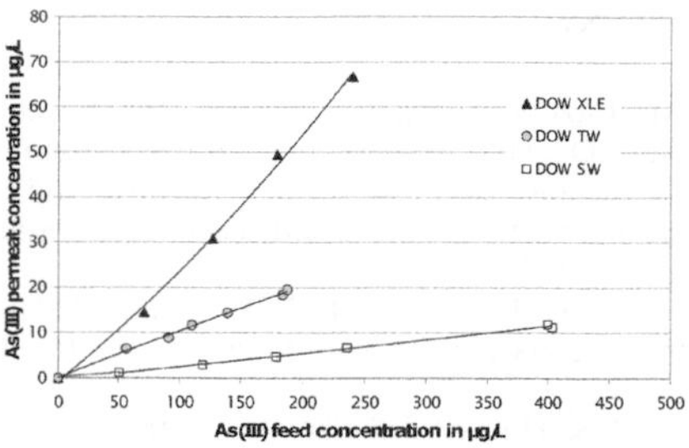

Figure 1. Arsenic (V) removal for the tested RO membranes.

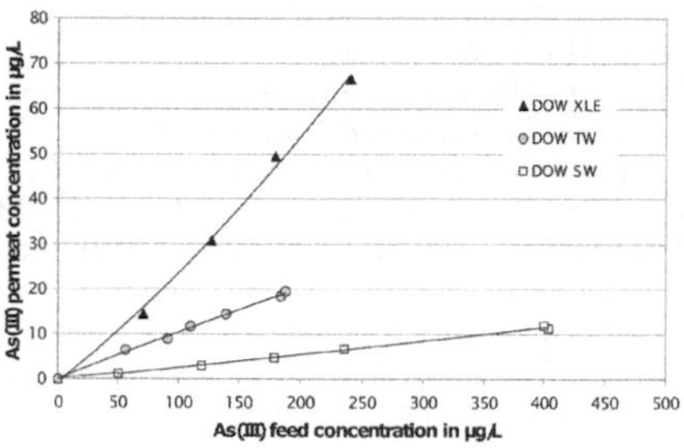

Figure 2. Arsenic (III) removal for the tested RO membranes.

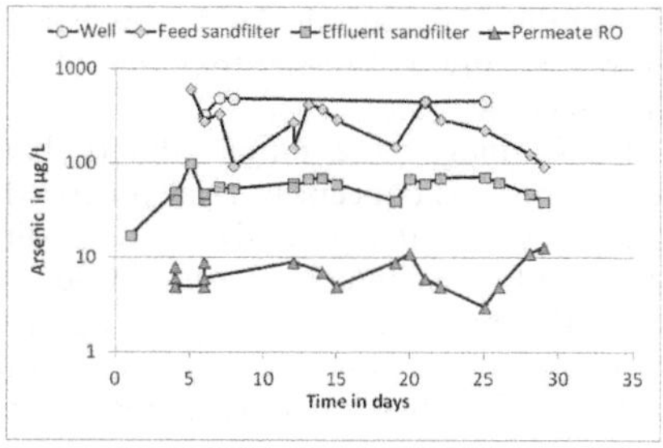

Figure 3. Arsenic (III) removal for pilot trials in Bind Toli.

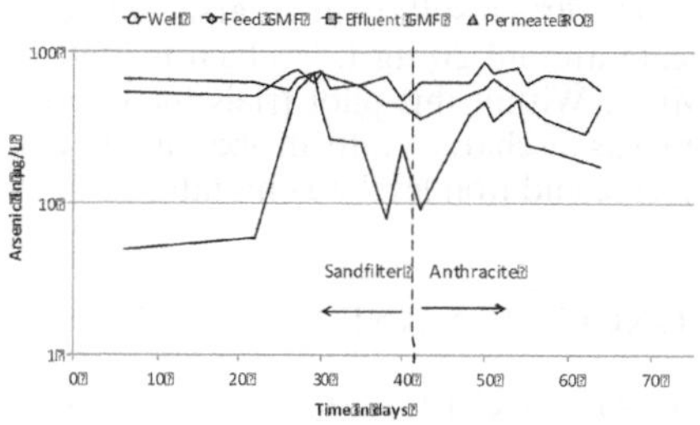

Figure 4. Arsenic (III) removal for pilot trials in Ramnagar.

For the "dense" TW and SW brackish and seawater membranes, As(V) rejection was so high, that the permeate water quality complied with the WHO recommended Maximum Contaminant Level (MCL) of 10 µg L^{-1} even at a feed concentration of 2400 mg L^{-1}. In the case of As(III), only feed concentrations below 350 µg L^{-1} resulted in permeate concentration lower than the MCL. This fact can be explained as follows. All the membranes consist of polymers with negatively charged groups. A charge exclusion effect enhances the rejection of negatively charged ions such as As(V). In the observed pH range, As(III) is only neutrally charged, hence it is less rejected. The power consumption was in the same range for all employed membranes 8–9 kW L^{-1}.

3.2 *Pilot trials*

Figures 3 and 4 show results of the pilot trials at the two different pilot sites using a single pass RO treatment over a period of about 3 month in total. More than 99% of iron in groundwater can be removed by aeration and subsequent sand filter; this means that iron is not affecting the downstream RO in terms of scaling. The iron level in the permeate of the RO unit is below the detection limit. The findings of these experiments further indicate that up to more than 80% of the arsenic occurring in groundwater can be removed by co-precipitation and filtration by sand filter. This is in line with previously published results stating that high natural iron concentration can facilitate arsenic removal (Berg, 2006). However, the arsenic level in the effluent of the GMF cannot comply with the stringent drinking water standards for arsenic (National Indian Standard: 10 µg L^{-1}). Only after having passed the effluent of the GMF through the RO desalinator (single pass system), arsenic in permeate was mostly below 10 µg L^{-1} (average As removal by RO: 89%) and hence the pilot trials verified the suitability of the system as a viable option for arsenic remediation in rural India. Eventually, the total removal rate of arsenic in the groundwater sums up to more to 99%.

The average permeate flow was relatively constant between 70–60 L h^{-1} at transmembrane pressures ranging from 9 to 12 bar. The pressure loss between inlet and outlet of the membrane module is less than 1 bar. The energy consumption per liter permeate was only ~3 4 W h L^{-1}.

Within the Ramnagar pilot tests, the RO unit has been also operated in the two pass mode at a recovery of 80% for the second pass. However, no significant reduction in arsenic concentration has been noticed since the As level of the single pass mode has been already below the detection limit.

Two wastestreams occurred, which need to be treated or safely disposed, namely the backwash water of the GMF and the concentrate of the RO desalinator. Whereas the arsenic concentration of the backwash water may be considerably above the initial groundwater value, the level of arsenic in the concentrate is only 10% higher than that of the RO feed (RO recovery rate: 10%).

At the Ramnagar site, initial pilot trials using a modified process concept were also carried out. With this process, the groundwater was directly extracted with a submerged groundwater pump replacing the previously used hand pump, and was subsequently fed under anaerobic conditions through a small particle filter into the RO unit. The advantage of this concept is that no aeration and GMF is needed, and, in addition, only one waste-stream has to be disposed of (RO concentrate). However, the first findings showed an arsenic rejection of 60–70% resulting in an arsenic level in the permeate around 20 μg L^{-1}, which is higher than the MCL. Within this pilot trials the concentrate stream was discharged into an abandoned well and the arsenic and iron level was monitored.

4 CONCLUSION AND OUTLOOK

Reverse Osmosis (RO) offers a viable option for treatment of arsenic rich groundwater. In general, As(V) can be much better removed compared with As(III). However, the majority of As(III) groundwaters in India and Bangladesh contain high levels of Fe and, hence, most of the arsenic can be removed by oxidation, precipitation as FeO(OH) and a simple granular media filter (GMF). This has been successfully demonstrated within pilot scale findings conducted at two locations in rural Bihar, India. The total arsenic reduction on both grounds was around 99% and, in most of the cases, the arsenic concentration in the permeate was in compliance with the National Indian Standard of 10 μg L^{-1}. However, two wastestreams needed to be disposed of, namely sludge of GMF and RO concentrate. In order to simplify the treatment process, a one step treatment was studied. For this purpose, the groundwater has been extracted under anaerobic conditions and subsequently split in permeate and concentrate by RO treatment. Consequently, only the concentrate stream needed to be disposed of. However, the initial data showed that the arsenic level in the permeate (around 20 μg L^{-1}) was higher than the MCL. Hence, this concept needs to be improved. The findings will eventually contribute to developing a simple and cost effective membrane RO water filter for developing and newly industrializing countries.

ACKNOWLEDGEMENTS

We are grateful to the State Level Water Testing Laboratory, PHED, Government of Bihar, India, for conducting laboratory analysis of arsenic and iron.

REFERENCES

Berg, M., Luzi, S., Trang, P.T.K., Viet, P.H., Giger, W. & Stuben, D. 2006. Arsenic removal from groundwater by household sand filters: Comparative field study, model calculations, and health benefits. *Environ. Sci. Technol.* 40: 5567–5573.

Chwirka, J.D., Thomson, B.M. & Stomp, J.M. 2000. Removing arsenic from groundwater, *J. Am. Water Works Assoc.* 2:3: 79–88.

Deowan, A.S., Hoinkis, J. & Pätzold, C. 2008. Low-energy reverse osmosis membranes for arsenic removal from groundwater. In P. Bhattacharya, A.L. Ramanathan, J. Bundschuh, A.K. Keshari & D. Chandrasekharam (eds), *Groundwater for Sustainable Development—Problems, Perspectives and* Challenges: 375–386, Balkema/Taylor & Francis.

Dow Water and Process Solutions. 2013. www.dowwater andprocess.com (accessed 10th Sep. 2013).

Evers. 2013. www.evers.de (accessed 10th Sep. 2013).

Figoli, A., Cassano, A., Criscuoli, A., Mozumder, M.S.I., Uddin, M.T., Islam, M.A., Drioli, E. 2010. Influence of operating parameters on the arsenic removal by nanofiltration *Water Res.* 44(1): 97–104.

Geucke, T., Deowan, S.A., Hoinkis, J. & Pätzold, Ch. 2009. Performance of a small scale RO desalinator for arsenic removal. *Desalination* 239: 198–206.

Hoinkis, J., Hermann, M., Schmidt, S., Gukelberger, E., Ghosh, A.D. Chatterjee, D. & Bundschuh, J. 2012. Arsenic removal from groundwater by small-scale reverse osmosis unit in rural Bihar, India. In J.C Ng, B.N. Noller, R. Niadu, J. Bundschuh & P. Battacharya (eds), *Understanding the Geological and Medical Interface of Arsenic—As 2012: Proceedings of the 4th International Congress on Arsenic in the Environment, 22–27 July 2012*, CRC, Cairns.

Katadyn. 2013. www.katadyn.com (accessed 10th Sep. 2013).

Moerk. 2013. www.moerkwater.com (accessed 10th Sep. 2013).

Mohan, D. & Pittman, C.U. 2007. Arsenic removal from water/wastewater using adsorbents—A critical review. *J. Hazard. Mater.* 142(1–2): 1–53.

Petrusevski, B., Sharma, S., Schippers, J.C. & Shordt, K. 2007. Arsenic in Drinking Water, IRC International Water and Sanitation Centre. http://www.irc.nl/page/33113 (accessed 10th Sep. 2013).

Shih, M.C. 2005. An overview of arsenic removal by pressure-driven membrane processes. *Desalination* 172(1): 85–97.

Song, S., Lopez-Valdivieso, A., Hernandez-Campos, D.J., Peng, C., Monroy-Fernandez, M.G. & Razo-Soto, I. 2006. Arsenic removal from high-arsenic water by enhanced coagulation with ferric ions and coarse calcite, *Water Res.* 40(2): 364–372.

Uddin, M.T., Mozumder, M.S.I., Figoli, A., Islam, M.A. & Drioli, E. 2007. Arsenic removal by conventional and membrane technology: An overview. *Indian J. Chem. Techn.* 14(5): 441–450.

WHO. 2013. www.who.int/mediacentre/factsheets/fs372/en/ (accessed 10th Sep. 2013).

Watermakers. 2013. www.nauticexpo.com/boat-manufacturer/watermaker-1223.html.

One Century of the Discovery of Arsenicosis in Latin America (1914–2014) – Litter, Nicolli, Meichtry, Quici, Bundschuh, Bhattacharya & Naidu (Eds)
© 2014 Taylor & Francis Group, London, ISBN 978-1-138-00141-1

Reduction of arsenic from groundwater by using a coupling of RO and Ultra-Osmosis (URO) processes

A. Armas, V. Olivieri, J.J. Mauricci & V. Silva
Cooperativa Obras Sanitarias de Venado Tuerto, Venado Tuerto, Santa Fe, Argentina

ABSTRACT: Nowadays the high arsenic concentration levels in Argentina affects around 7% of the total population, with the highest levels in the north and middle-east part of the country. In this project, an industrial scale water treatment plant was installed in Venado Tuerto city (Santa Fe province, Argentina) with the aims of reducing the As content below 10 µg/L and to make the process more sustainable by the concentrated flow reduction without compromising the membrane lifetime. The arsenic level in the wells for the water intake is around 325 µg/L and the final product for distribution contains less than 10 µg/L. The membrane system is using a RO processes with special membranes developed by Dow Chemical which have very high arsenic rejection while letting SiO_2 to go through, which is an important fouling source in membrane processes. The plant has been successfully running for more than a year with approximately 4800 m^3/day of water intake.

1 INTRODUCTION

Drinking water contaminated with arsenic (As) has been recognized as a human threat. Therefore, the World Health Organization (WHO, 2008) has recommended that the concentration in drinking water should be at least below 10 µg/L.

There are several technologies available for arsenic reduction in water like adsorption, precipitation / coagulation, ion-exchange and membrane technology. Lately, membrane processes have shown important advantages when compared to conventional processes since they do not use large amount of chemicals or energy and they do not generate hazardous waste as ion-exchange does. For As removal, membrane separation technologies like Reverse Osmosis (RO), Nanofiltration (NF) and Ultrafiltration are the most efficient to reduce arsenate (As(V)) between 85–99%, depending on the membrane properties, water characteristic, contaminants, etc. (Dutta, 2012).

Some membranes have thin layers with a chemical structure and active groups which interact with the water itself and selectively with the present ions. Moreover, the membrane rejection for a specific ion (as arsenate) can be enhanced by the presence of a coion for which the membrane has preferential affinity.

In Argentina, several RO water treatment plants have been installed in order to reduce As concentration for drinking purposes. The principal challenge in these cases is the high silica content since it interferes in the good membrane performance by forcing the system to work under low water recovery rates and, at the same time, the amount of salts in the permeate is very low and the final product is often mixed with a quota of underground raw water (only microfiltered) to increase the salinity and water final taste. To improve the situation, a new membrane developed by Dow Chemical was tested at pilot scale showing a low silica and high arsenic rejection. Even better, the membrane is able to permeate some common coions in groundwater as HCO_3^-, Cl^-, and SO_4^{2-}, allowing higher salinity in the product. To differentiate this particular membrane from common NF or RO conventional membranes, it was internally called Ultra-Osmosis membrane (URO).

In this project, an industrial scale RO plant was installed and further optimized over the years in order to accomplish the following objectives:

1. To reduce the As content below 10 µg/L according to the legislation.
2. To make the process more sustainable by reducing the concentrated flow without compromising the membrane lifetime by silica fouling.

2 METHODS/EXPERIMENTAL

2.1 *Venado Tuerto case*

Venado Tuerto is one of the mayor cities in the Santa Fe province in Argentina. The well water characteristics are indicated in Table 1. It is a mid-salinity water, containing high bicarbonate levels, moderated silica concentration and high arsenic content. The water after treatment is distributed to the community by the Cooperativa de Obras Sanitarias y Servicio Anexos de Venado Tuerto (COSVT), who also performs the sewage water treatment.

The As concentration in this region is around 325 μg/L; consequently, As concentration needs to be reduced in around 80%. For this purpose, the RO water treatment plant was installed with approximately 4800 m^3/day of water intake. It has 22 water wells that work 12 h/day with an average water flow less than 18 m^3/h.

Originally, the plant was built in with three RO modules, according to Table 1.

The permeate resulted in a conductivity of around 10 μS/cm and As concentration lower than 5 μg/L. This stream was mixed with row groundwater to increase the conductivity to about 300 μS/cm, and the resulting As content was less than 50 μg/L. Later on, the COSVT needed to further reduce the As concentration to around 10 μg/L while maintaining the conductivity between 300 and 600 μS/cm and increasing the water recovery.

2.2 *RO-URO system*

In order to accomplish the aims, some improvements were introduced and the final configuration is presented in Figure 1. In this scheme, the water coming from the wells is collected in a 100 m^3 buffer tank from where it is pumped to the modules according Table 2. In module 1.2 and 1.3, conventional RO membranes remained installed with 60% of water recovery adopted in order to avoid the silica precipitation and consequently to avoid to use of SiO_2 specific antiscalants. After optimization, module 1.1 is operating with the URO membranes in order to reach the target conductivity of around 300 μS/cm.

Table 1. Design modules flow for old configuration.

Flow (m^3/h)	Feed	Permeate	Concentrate
RO module 1.1	46	30	16
RO module 1.2	100	65	35
RO module 1.3	100	65	35

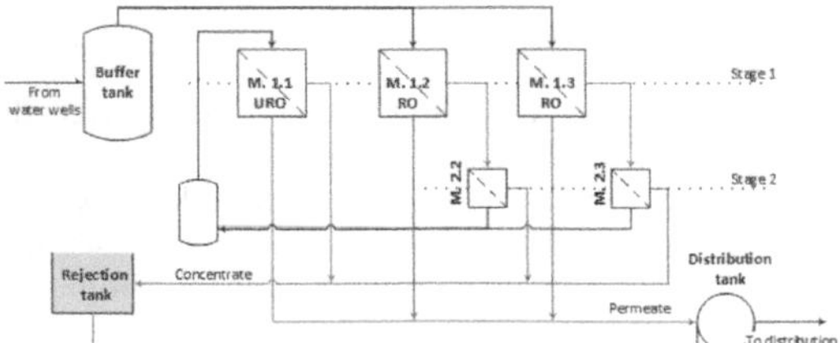

Figure 1. Flow diagram in Venado Tuerto treatment plant in final configuration including the RO-URO system.

Table 2. Design modules flow for new configuration.

Flow (m^3/h)	Feed	Permeate	Concentrate
UO module 1.1	34	27	7
RO module 1.2	100	60	40
RO module 1.3	100	60	40
UO module 2.2	40	23	17
UO module 2.3	40	23	17

Table 3. Water recuperation from concentrate of RO plant.

Property	Well	RO conc. modules 1.2 and 1.3	URO perm. module 2.2	URO-Conc. module 2.2
pH	7.84	8.02	7.38	8.05
TDS (mg/L)	1,263	3,000	649	6,061
As (mg/L)	0.325	0.8	0.005	1,900
SiO_2 (mg/L)	48	115	87	151
Recovery (%)	–	60	57	

To achieve the second goal of reducing the retentate, a second concentration stage was incorporated in modules 1.2 and 1.3, which are also using the URO membranes. By doing this, more than 50% of concentrate flow coming from those modules is recovered. The permeate from the second stage is mixed and used as feed for module 1.1. The permeated fractions in modules 1.1, 1.2 and 1.3 are collected for further chlorination and distribution. On the other hand, the concentrated in modules 1.1, 2.2 and 2.3 are mixed and pumped to the sewage collection system. Under these conditions, the As concentration remains in the limits allowed to spill into sewage collection of around 500 μg/L.

3 RESULTS AND DISCUSSION

The water treatment plant has been running with a water quality that matches the required standards for drinking water for more than a year. The final drinking water quality has: As concentration lower than 10 μg/l, conductivity of 150 μS/cm and pH of 7.9. As mentioned before, the rejected water is conducted to the sewage collection system, where approximately an 88% of the distributed water is returned for further treatment and discharge. Consequently, the As concentration disposed into the lake remains similar to the one taken from the wells.

The water properties and ion concentrations for the water well, URO 2.2 concentrate and permeate are shown in Table 3 in order to evaluate membrane performance. The silica rejection in module 1.2 and 1.3 is around 93% compared to 25% in modules 2.2 and 2.3, which confirms the good performance of this membranes to avoid fouling.

4 CONCLUSIONS

The drinking water supply system is working according to the expectations and the URO membranes have shown an excellent performance allowing a higher water recovery in the system. The plant is currently expanding to overcome the average consumption by connection which is in continuous growing. New modules are planned to be installed in the RO-URO system.

REFERENCES

Dutta T. et al., (2012). Removal of arsenic using membrane technology—A review. Int. J. Eng. Res. and Tech. Vol 1 (9). ISSN: 2278–0181.

World Health Organization (2008). Guidelines for drinking-water quality: second addendum. Vol. 1, Recommendations. 3rd ed.

One Century of the Discovery of Arsenicosis in Latin America (1914–2014) – Litter, Nicolli, Meichtry, Quici, Bundschuh, Bhattacharya & Naidu (Eds)
© 2014 Taylor & Francis Group, London, ISBN 978-1-138-00141-1

Arsenic decrease using *Pseudomonas aeruginosa* on mineral matrix for bioremediation

E.E. Pellizzari
Laboratorio de Microbiología, Universidad Nacional del Chaco Austral, P.R. Sáenz Peña, Chaco, Argentina

M.C. Giménez
Laboratorio de Química Analítica, Univ. Nacional del Chaco Austral, P.R. Sáenz Peña, Chaco, Argentina

ABSTRACT: Groundwater contamination with Arsenic (As) is a problem that has received considerable worldwide attention in the last years. Bacteria have the metabolic capability of degrading chemical compounds. They may use the bioavailability of heavy metals such as cadmium, mercury, vanadium and arsenic, among others. The aim of this paper was to study As degradation using Pseudomonas aeruginosa isolated from groundwater containing 0.25 mg As L^{-1} As at an stimulating temperature of 23 °C from Presidencia Roque Sáenz Peña, Chaco province. The strains were immobilized on the matrix of natural stone and covered with broth formulated with salts, and a concentration of 0.5 mg L^{-1} As. We observed biofilm formation on the matrix, obtaining higher cell concentration. This facilitated the interaction with the solution to be removed. The percentage of arsenic removed in a period of three months reached 50% from the initial concentration.

1 INTRODUCTION

Arsenic is a contaminant in many aquifers in the Chaco Province, where the population consumes this water (Giménez *et al*,. 2000). All life on Earth is connected through four-dimensional space-time (Guerrero, 2002). Living systems have developed the availability of tolerating different chemical elements, including arsenic. Therefore, some organisms are able to tolerate this metalloid. In previous studies (Pellizzari *et al.*, 2011), we observed the ability of certain microorganisms to remove arsenic from water, and this fact could be used in biotechnological processes for the purification of water. Banerjee *et al.* (2011) observed a good growth of *Pseudomonas aeruginosa* in the presence of As(V). Studies like this give a good perspective of use of these bacteria in bioremediation of arsenic.

2 METHODS/EXPERIMENTAL

2.1 *Materials*

The experiments were performed using *Pseudomonas aeruginosa* isolated from groundwater containing 0.25 mg As L^{-1} at a temperature of 23 °C, in Presidencia Roque Sáenz Peña, Chaco province.

Solutions of 1 mg As/L were prepared with distiller water and As(V) stock solutions by dissolvingsodium arsenate ($Na_2HAsO_4.7H_2O$) in deionized water. All prepared solutions were sterilized by autoclave at 120 °C for 15 min.

The used reactor was a glass container of 1 L volume, with top loading mouth and bottom discharge. Natural rock stones were used as immobilizing material; as they are very uneven and porous, they allow the adhesion of bacteria and the formation of a biofilm. The stones were selected by size and were 4 cm to facilitate its introduction in the reactor.

2.2 *Methods*

The removal of arsenic was performed using *Pseudomonas aeruginosa* immobilized on the matrix of natural stone. In the scientific literature, the very dynamic nature of the *P. aeruginosa* biofilm was demonstrated (Klausen *et al.*, 2003, van Hullebusch, 2003). To prove the ability of biofilm formation by *Pseudomonas aeruginosa* two methods were used: the first, where bioreactor stones were removed and plated on nutrient agar, and the second, where sterile swabs rubbed on stones were plated in nutrient agar. Cultured obtained in both case were incubated at 35 °C for 24 hours.

The mineral salts minimal medium (MN) used in the tests comprised (g/L): 0,5 KH_2PO_4, 1.0 NH_4Cl, 2.0 Na_2SO_4, 2.0 KNO_3, 0.001$CaCl_2$.6H2O, 1.0 $MgSO_4.7H_2O$, 0.0004 $FeSO_4$ (Chitiva, 2003).

Two trials were conducted on the matrix. The first stage of the test was performed with *Pseudomonas* in 50 mL MN, incubated at 35 °C for 24 hours, this culture was transferred to 1000 mL of the same solution, incubated for 24 hours and transferred to the reactor. The second solution used was prepared with MN salts and arsenic solution at 1 mg/L ionic strength. The reactor was empty and filled with the second solution of arsenic to 1 mg L^{-1}. The bacteria were grown on the stone media.

Samples were taken to analyze the concentration of bacteria and arsenic. The bacteria sample was prepared in nutritive agar, inoculating 1 mL of sample at 37 °C ± 2 for 48 hours. All samples were analyzed for total As(V).

The analytical determination of As in the extracted samples from the reactor was performed by HGAAS (hydride generation atomic absorption spectroscopy) using hollow cathode lamps at 193.7 nm. The instrument quantification limit was 5 μg L^{-1}; an intermediate precision of less than ± 10% was achieved.

3 RESULTS AND DISCUSSION

3.1 *Results*

Pseudomonas grew very well in culture media made with 1 mg As L^{-1}. The highest rate of growth was reached in 24 h. The growth of bacteria on the surface of the stones in the reactor was also very good. The stones allowed having a porous media with enough oxygen for the bacteria development. While conditions were for optimal growth, the decrease on As concentration in the solution was 50%.

After 24 h, incubation biofilm formation on the stones was observed.

3.2 *Discussion*

The stones as a supported media in the reactor permitted adhesion of the bacteria, allowing a significant increase in the biomass in direct contact with the water to be treated. The biofilm was formed very fast. Furthermore, the gaps left by the stone allowed having oxygen, promoting a fast metabolism of Pseudomonas aeruginosa.

It is proposed that all these factors enhanced the As removal rate.

4 CONCLUSIONS

The following conclusions were drawn based on the experimental study:

- Arsenic can be removed at 1 mg As L^{-1} in a stone filtration column.
- *Pseudomonas aeruginosa* performed well for removal. They are easy to handle and do not require strict temperature and pH controls.
- The reactors are easy to assemble and of low maintenance.
- Few bacteria were observed in the reactor discharge. Therefore, it was concluded that the interaction between microorganisms and stones is very good.

ACKNOWLEDGEMENTS

This work was supported with funds from PI 36/00005 UNCAus (Universidad Nacional del Chaco Austral).

REFERENCES

Banerjee, S. 2011. Arsenic accumulating and transforming bacteria isolated from contaminated soil for potential use in bioremediation. *Journal of Environmental Science and Health* part A. 46: 1736–1747.

Chitiva Urbina, L. & Dussán, J. 2003 Evaluación de matrices para la inmovilización de Pseudomonas spp. en biorremediación de fenol. Revista colombiana de biotecnología. Vol. 2, 5–10. Edificio Manuel Ancizar. Instituto de Biotecnología.

Gimenez, M.C., Benítez M.E., Osicka R.M. & Castro M.P. 2000. Arsénico en aguas naturales subterráneas en la región central de la provincia del Chaco. *Comunicaciones cientificas y tecnologicas*. Chaco. Universidad Nacional del Nordeste.

Guerrero, R. 2002. La simbiosis como mecanismo de evolución. Sociedad Española de Microbiología (*SEM*), 33:10–14.

Klausen, M., Heydorn, A., Ragas, P., Lambertsen, L., Aaes-Jørgensen,A., Molin, S. & Tolker-Nielsen, T. 2003. Biofilm formation by *Pseudomonas aeruginosa* wild type, flagella and type IV pili mutants. *Molecular Microbiology* 48 (6): 1511–1524

Pellizzari, E.E., Buchhamer, E. & Giménez, M.C. 2011. Selección de bacterias autóctonas para ensayos de crecimiento en medios con alto contenidos de As(V) aisladas de aguas subterráneas de la provincia del Chaco. VIII Congreso Iberoamericano de Ingeniería en Alimentos. Lima. Perú.

van Hullebusch, E.D., Marcel H. Zandvoor, M.H. & Lens, P.N.L. 2003. Metal immobilisation by biofilms: Mechanisms and analytical tolos. *Re/Views in Environmental Science & Bio/Technology* 2: 9–33.

One Century of the Discovery of Arsenicosis in Latin America (1914–2014) –
Litter, Nicolli, Meichtry, Quici, Bundschuh, Bhattacharya & Naidu (Eds)
© 2014 Taylor & Francis Group, London, ISBN 978-1-138-00141-1

The removal of arsenic by horizontal subsurface flow constructed wetlands

K. Lizama Allende
Universidad de Chile, Santiago, Chile

D.T. McCarthy
Monash University, Melbourne, VIC, Australia

T.D. Fletcher
The University of Melbourne, Melbourne, VIC, Australia

ABSTRACT: Constructed wetlands are a cost-effective, environmental-friendly treatment technology to remove As from water. In this study, lab-scale horizontal flow wetlands were used to verify the effectiveness of two media types, one being zeolite and the other a mix of limestone and cocopeat. Inflow water had [As] = 2.6 mg L^{-1}, [Fe] = 97.3 mg L^{-1} and [B] = 30.8 mg L^{-1} at pH 2 ± 0.2. Both types were highly effective: zeolite removed 99.9% and 96.1% of As and Fe respectively, whereas the limestone/cocopeat wetlands removed 99.8% and 87.3%. Results confirmed the key role of the wetland media in fostering specific removal processes: As co-precipitation with Fe in limestone wetlands, and As and Fe removal by cation exchange capacity of zeolite. Limestone/cocopeat wetlands offer a more suitable treatment, given the neutral pH achieved, but zeolite wetlands were able to achieve lower concentrations of Fe, despite the acidic pH in the treated effluent.

1 INTRODUCTION

Constructed wetlands are low-cost green systems successfully applied in water treatment. They are able to remove different pollutants including As (Lizama Allende *et al.*, 2010), but very little is known about their efficiency and reliability for this purpose. They have been increasingly applied for the treatment of metal-contaminated water, which is often acidic (e.g. acid mine drainage). Previous studies have shown than vertical flow wetlands with alternative media could be employed (Lizama Allende *et al.*, 2012). Little is known about the performance of horizontal flow wetlands.

The aim of this study was to assess the effectiveness of horizontal flow constructed wetlands in the removal of arsenic (As) and iron (Fe) from acidic contaminated water, using different wetland media. The Azufre River, Northern Chile, was selected as the case study given its elevated concentrations of As and metals under acidic conditions (Ríos *et al.*, 2011).

2 MATERIALS AND METHODS

2.1 *The wetland system*

Two types of wetland cells were built, using zeolite and limestone/cocopeat as the main media.

Each wetland group had three replicates, which were operated individually. Young *Phragmites australis* were harvested and planted five months before the experiments started. Synthetic water resembling the As, Fe and B levels in the Azufre River was prepared using tap water and reagents (2.6 ± 0.5 mg L^{-1} As, 30.8 ± 6.2 mg L^{-1} B, 97.3 ± 14.0 mg L^{-1} Fe, pH 2.0 ± 0.2). The system was located in a greenhouse and was pseudocontinuosly dosed daily during 22 weeks. The hydraulic loading rate was 30 mm d^{-1} resulting in a dosing rate of 150 mL h^{-1}.

2.2 *Sampling and analysis*

Water samples were filtered immediately using 0.45 μm filters and acidified to pH < 2 using HNO_3.

Water quality parameters (pH, Eh) were also measured. Media samples were collected after the experiments finished and were dried at 40 °C until constant weight was achieved. Plants from the inlet and outlet of every cell were divided in shoots and roots, dried at 55 °C and the metal concentrations were analyzed by USEPA methods 3051A and 3060A. Total and dissolved metal fraction in all samples was analyzed by ICP-OES and ICP-MS.

3 RESULTS AND DISCUSSION

3.1 *Pollutants removal and water quality*

Arsenic removal rates were 99.8% and 99.9% in limestone/cocopeat and zeolite wetlands, respectively; whereas iron removal rates were 87.3% and 96.1%. pH and Eh were the most affected parameters: limestone/cocopeat raised the pH to 6.95, while zeolite raised it to 4.1. Redox potential decreased to negative values in limestone/cocopeat wetlands (from 475 to −37 mV), whereas it was still positive in zeolite wetlands (315 mV). Thus, alkaline and reducing conditions provided in limestone/cocopeat wetlands were not sufficient to achieve better removal as in zeolite wetlands, although both performed better than in vertical flow.

3.2 *Metal accumulation in media*

Zeolite presented higher As and Fe levels than did limestone and cocopeat. Furthermore, these levels tended to decrease towards the outlet, and towards the bottom: arsenic fluctuated between 230 and 8 mg kg^{-1}, iron fluctuated between 11,000 and 4,500 mg kg^{-1}. This suggests that the media plays the main role in accumulating the target pollutants.

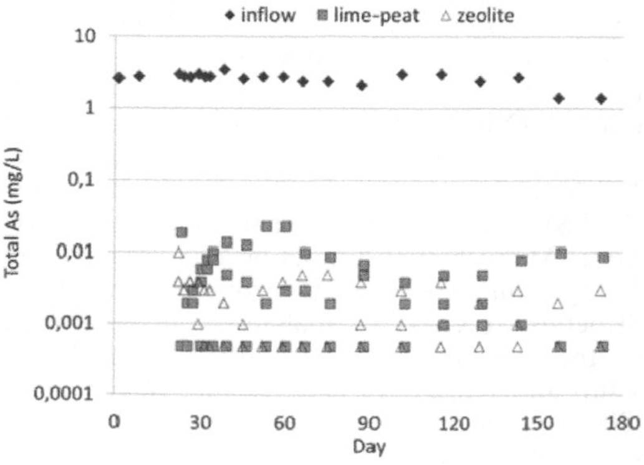

Figure 1. Total As concentrations in the inflow and outflow from the system.

3.3 *Metal accumulation in P. australis*

Plants located in the inlet uptook higher As and Fe levels than did plants in the outlet, in both wetland types: arsenic concentrations reached 130 mg kg^{-1} in zeolite and 40 mg kg^{-1} in limestone/cocopeat, whereas iron concentrations reached 10,000 and 22,000 mg kg^{-1} respectively. Roots accumulated higher As and Fe levels than did shoots, as in previous studies (Kadlec & Wallace, 2009). Despite these high concentrations, the accumulation of pollutants was < 3% of the total mass loading, thus confirming the minor role of plant uptake as a metal removal mechanism in constructed wetlands (e.g. Ye *et al.*, 2003).

4 CONCLUSIONS

Alternative media may be used since high As and Fe removal efficiencies were observed for both wetland types. Sorption capacity of zeolite was slightly more effective than coprecipitation with iron oxides—triggered by the pH adjustment of limestone—for this case study. Metals were mainly retained in the media rather than in the plants.

ACKNOWLEDGEMENTS

We thank the Chilean Government (Becas Chile) for sponsoring Katherine Lizama Allende's Ph.D. studies and the staff from the Department of Civil Engineering, Monash University (Australia), for assistance with the experimental setup.

REFERENCES

Kadlec, R.H. & Wallace, S.D. 2009. *Treatment Wetland (2nd ed.)*, Boca Raton. CRC Press.
Lizama Allende, K., Fletcher, T.D. & Sun G. 2010. Removal processes for arsenic in constructed wetlands. *Chemosphere* 84(8): 1032–1043
Lizama Allende, K., Fletcher, T.D. & Sun G. 2012. The effect of substrate media on the removal of arsenic, boron and iron from an acidic wastewater in planted column reactors. *Chemical Engineering Journal* 179:119–130.
Ríos, P.L, Guerra, P.A., Bonilla, C.A., Escauriaza, C.R., Pizarro, G.E. & Pastén, P.A. 2011. Arsenic occurrence in fluvial sediments: Challenges for planning sustainable water infrastructure in the Lluta river basin. *Abstracts of Papers of the American Chemical Society*. 404-ENVR.
Ye, Z.H., Lin, Z.Q., Whiting, S.N., de Souza, M.P. & Terry, N. 2003. Possible use of constructed wetland to remove selenocyanate, arsenic, and boron from electric utility wastewater. *Chemosphere* 52(9):1571–1579.

One Century of the Discovery of Arsenicosis in Latin America (1914–2014) – Litter, Nicolli, Meichtry, Quici, Bundschuh, Bhattacharya & Naidu (Eds)
 ISBN 978-1-138-00141-1

Arsenic immobilization from industrial effluents

F.K. Opio, J. Peacey & H.E. Jamieson
Queen's University, Canada

ABSTRACT: In this study, tooeleite was investigated as an alternative potential disposal option for As(III) immobilization from weak acid effluents. Batch coprecipitation experiments were performed in order to evaluate the precipitation of tooeleite under varying conditions of pH and neutralizing agent, and to determine the effect of these conditions on the As removal efficiency, and the precipitate characteristics and stability. The As removal efficiency was enhanced with increasing pH over the pH range 2–10. The use of lime as opposed to sodium hydroxide as neutralizing agent enhanced the As removal efficiency, with > 85% As(III) being removed at pH > 2.7 in the lime neutralized system. Tooeleite formed at pH 2–3.5, but rapidly transformed at > pH 4 to form a poorly crystalline equimolar ferric arsenite which was stable at pH 6 to 10. Toxicity Characteristic Leaching Procedure tests indicated that tooeleite and ferric arsenite have relatively high As solubilities.

1 INTRODUCTION

Various treatment practices are currently used for the removal and immobilization of arsenic from weak acid effluents, of which the preferred method involves coprecipitation of As(V) with ferric iron (Fe:As molar ratio > 3) upon neutralization with the addition of lime to form arsenical ferrihydrite (As(V)-bearing ferrihydrite). However, this practice has a few drawbacks including high ferric iron requirements, low As content (< 10%), the generation of large volumes of sludge with poor filtration properties and the pretreatment requirement to oxidize As(III) to As(V) prior to precipitation, resulting in high operating and disposal costs (Riveros *et al.*, 2001).

Tooeleite, a ferric arsenite sulfate hydrate ($Fe_6(AsO_3)_4(SO_4)(OH)_4 \cdot 4H_2O$), has been proposed as a potential disposal option for the fixation of the As(III) species from weak acid effluents due to its high As content (25%), high As removal efficiency, low Fe:As molar ratio of 1.2, and because it readily forms under ambient conditions (Nishimura & Robins, 2008). Nishimura & Robins (2008) investigated the precipitation of nano-crystalline tooeleite from sulfate-based Fe(III)-As(III) bearing weak acid solutions (Fe:As = 1.0), and conducted preliminary characterization tests (XRD, SEM, TG/DTA) on the synthetic precipitates. The authors reported that tooeleite is stable in the pH range 2 to 3.5, and hypothesized that it is likely to transform to FeOOH(As) at >pH 3.5. However, their study did not evaluate the stability of tooeleite. The main objective of this study was to investigate tooeleite as a potential disposal option for As(III) immobilization from weak acid effluents by: (i) investigating the precipitation of tooeleite from Fe(III)-As(III) bearing weak acid solutions under varying conditions of pH and neutralizing agent; (ii) determining the mineralogical characteristics of the synthetic precipitates using conventional and synchrotron-based techniques; and (iii) evaluating the stability of the synthetic precipitates using the US EPA Toxicity Characteristic Leaching Procedure (TCLP) (Method 1311).

2 METHODS/EXPERIMENTAL

2.1 *Batch coprecipitation*

Tooeleite was synthesized using the procedure that is described by Nishimura and Robins (2008), and sodium hydroxide (NaOH) and lime as the bases. The tooeleite and gypsum-bearing tooeleite precipitates were subjected to powder X-Ray Diffraction (XRD) analysis, and synchrotron-based techniques using the hard X-ray microprobe, X26A, at the National Synchrotron Light Source (NSLS), Brookhaven National Laboratory, New York. Micro-X-Ray diffraction (μXRD) and As K-edge micro-X-Ray Absorption Near Edge Structure (μXANES) analyses were conducted on the synthetic precipitates, using the analytical setup and procedures described in Walker *et al.* (2005), in order to determine their precise mineralogical composition and As speciation. The stability of the synthetic precipitates was evaluated using a modified US EPA Method 3111 Toxicity Characteristic Leaching Procedure (TCLP) (US EPA, 1992). The TCLP test was modified from the original method described by the US EPA (1992) in that an Orbit Environ Shaker operated at 200 rpm was used instead of an end-over-end agitation device operated at 30 ± 2 rpm.

3 RESULTS AND DISCUSSION

3.1 *Effect of pH and neutralizing agent*

The use of lime as a base as opposed to NaOH enhanced the As uptake of the initial Fe:As = 1.0 system, resulting in > 90% As(III) removal in the form of tooeleite at pH > 3. For both NaOH and lime neutralized systems, the As and Fe uptake from solution was enhanced with increasing pH over the pH range 2 to 10, and the maximum As removal was achieved in the pH range 6 to 8 (> 97%).

3.2 *Characterization of solids*

Bulk XRD and synchrotron-based μXRD analytical results of the precipitates are presented in Tables 1 and 2, respectively.

Bulk XRD analyses of the precipitates formed from the NaOH neutralized system in the pH range 2 to 4 revealed the presence of tooeleite, and a poorly crystalline As-bearing phase was identified in the precipitates formed at pH 6 to 10. Tooeleite and gypsum were identified in the precipitates formed from the lime-neutralized system in the pH range 2 to 4, while gypsum was the only phase revealed in the precipitates formed at pH 6 to 10.

Table 1. Mineral phases identified by bulk XRD analysis.

Samples	Precipitation pH range	Mineral phases
Initial Fe: As molar ratio = 1.0 [NaOH]	2–4	Tooeleite
	6–10	Poorly crystalline As-bearing phase
Initial Fe: As molar ratio = 1.0 [$Ca(OH)_2$]	2–4	Tooeleite and gypsum
	6–10	Gypsum

Table 2. Mineral phases identified by μXRD analysis.

Samples	Precipitation pH range	Mineral phases
Initial Fe: As molar ratio = 1.0 [NaOH]	2–4	Tooeleite, arsenolite
	6–10	Ferric arsenite
Initial Fe: As molar ratio = 1.0 [$Ca(OH)_2$]	2–4	Tooeleite, bassinite and gypsum
	6–10	Gypsum and ferric arsenite

μXRD analyses of the precipitates formed from NaOH neutralized system in the pH range 2 to 4 revealed the presence of tooeleite and trace amounts of arsenolite. Arsenolite possibly formed as a result of high As concentrations and a shortage of Fe. Tooeleite, gypsum and bassinite were identified in the precipitates formed from the lime neutralized system at pH 2 to 4. Bassinite probably formed during the preparation of the thin section. μXRD analysis of precipitates formed from the NaOH and lime neutralized systems at pH 6 to 10 revealed the presence of a poorly crystalline phase with a μXRD pattern with 2 broad peaks at d-spacings of ~0.29 nm and ~0.17 nm. These broad peaks are presumed to indicate the presence of an amorphous ferric arsenite phase formed from the transformation of tooeleite. These results suggest that tooeleite is formed in the pH range 2 to 3.5, and subsequently begins to transform at > pH 4 to form an equimolar ferric arsenite. The As K-edge μXANES analysis confirmed that As(III) was prevalent in both tooeleite and ferric arsenite.

3.3 *Short term leaching tests (U.S. EPA TCLP)*

TCLP stability analysis of the precipitates formed from NaOH and $Ca(OH)_2$ neutralized systems indicated that both tooeleite and ferric arsenite have a relatively high As solubility at pH 5.

4 CONCLUSIONS

In summary, this study indicated that tooeleite can effectively be used to immobilize As(III) in As-bearing weak acid effluents. Tooeleite was found to form within the pH range 2 to 3.5, but rapidly transformed at > pH 4 to form a poorly crystalline equimolar ferric arsenite phase which was most stable in the pH range 6 to 8. TCLP stability analysis indicated that both tooeleite and ferric arsenite have a relatively high As solubilities. Tooeleite and ferric arsenite could be considered as viable low-cost disposal options for the treatment of As(III)-bearing drinking water due to their low Fe demand, high As content and high As removal efficiency, but the solid residues would need to be disposed of as hazardous waste in a lined pond.

ACKNOWLEDGEMENTS

The authors gratefully acknowledge financial support from Natural Sciences and Engineering Research Council (NSERC) and Xstrata through grants to NSERC-Xstrata Executive Industrial Chair in Minerals and Metals Processing at Queen's University. Portions of this work were performed at Beamline 26A, National Synchrotron Light Source (NSLS), Brookhaven National Laboratory. X26A is supported by the Department of Energy (DOE)—Geosciences (DE-FG02-92ER14244 to The University of Chicago—CARS). Use of the NSLS was supported by DOE under Contract No. DE-AC02-98CH10886.

REFERENCES

Nishimura, T. & Robins, R.G. 2008. Confirmation that tooeleite is a ferric arsenite sulfate hydrate and is relevant to arsenic stabilization. *Minerals Engineering* 21: 246–251.

Riveros, P.A., Dutrizac, J.E. & Spencer, P. 2001. Arsenic disposal practices in the metallurgical industry. *Canadian Metallurgical Quarterly* 40: 395–420.

U.S. Environmental Protection Agency. 1992. Toxicity Characteristic Leaching Procedure (Method 1311). In SW-846: Test methods for evaluating solid waste, physical/chemical methods. Office of Solid Waste, Washington DC.

Walker, S.R., Jamieson, H.E., Lanzarotti, A., Andrade, C.F. & Hall, G.E.M. 2005. Determining arsenic speciation in iron oxides: Application of synchrotron micro-XRD and micro-XANES at the grain scale. *The Canadian Mineralogist* 43: 1205–1224.

One Century of the Discovery of Arsenicosis in Latin America (1914–2014) – Litter, Nicolli, Meichtry, Quici, Bundschuh, Bhattacharya & Naidu (Eds)
© 2014 Taylor & Francis Group, London, ISBN 978-1-138-00141-1

Removal of arsenic from aqueous systems using Fe/MnO_2 3-dimensional nanostructure

M. Lei, P. Qin, L. Peng, Y. Ren & Q. Zeng
College of Resource and Environment, Hunan Agricultural University, Changsha, P.R. China

ABSTRACT: Fe/MnO_2 flower-like 3-dimensional nanostructure synthesized by hydrothermal procedure was used to remove As from aqueous solution. The Fe/MnO_2 dispersing flower-like sphere with the diameter of 500~800 nm and the width of petal of 20~30 nm was characterized by SEM. Sorption of arsenic to Fe/MnO_2 reached equilibrium in less than 180 s, which was much faster than that of pure MnO_2 and amorphous iron manganese binary oxide. The adsorption isotherm agreed well (R^2 = 0.99) to the Freundlich adsorption model, with adsorption capacities of 26.5 mg/g, at an equilibrium concentration of 34.5 mg/L at room temperature. As adsorption was exothermic, and the adsorption capacity decreased by increasing the temperature from 20 to 70 °C. The removal percentage of As by Fe/MnO_2 reached 99% from aqueous solution at pH 4.0 to 6.0. The Fe/MnO_2 nano-flowers are a potential highly efficient nanomaterial for removal of arsenic from water.

1 INTRODUCTION

The As-poisoned drinking water events have been often reported worldwide. As removal from drinking water is mandatory to warrant human health. Among removal methods, adsorption has been deemed to be the best available technology. MnO_2 is an important adsorbent for arsenic removal (Li *et al.*, 2010), although its adsorption capacity is low (Deschamps *et al.*, 2005; Lenoble *et al.*, 2004), what limits its application. In order to enhance the surface-to-volume ratio, the adsorbents are usually prepared as nanoparticles. In our study, the novel Fe/MnO_2 flower-like 3-dimensional nanostructure (nanoflower) has been prepared by a hydrothermal method. The adsorbent with 3-dimensional nano-structure was more easily separated than nanoparticles through filtration or membrane process. Furthermore, this structure can keep the required high surface-to-volume ratio. The Fe/MnO_2 large surface induces a quick arsenic removal rate, and would be a potential useful adsorbent.

2 METHODS/EXPERIMENTAL

2.1 *Preparation Fe/MnO_2 nanoflowers*

Fe/MnO_2 nanoflowers were prepared by modification of a previous method (Camacho *et al.*, 2011). In a typical procedure, $MnSO_4 \cdot H_2O$ (0.6830 g), $Fe(NO_3)_3 \cdot 9H_2O$ (1.6406 g), $K_2S_2O_8$ (1.0868 g) and 4 mL H_2SO_4 were mixed in 76 mL of deionized (DI) water and stirred with a magnetic stirrer for 10 min to form a homogeneous solution at room temperature. The precipitates were collected, washed with DI water and absolute ethanol for several times to remove impurities, and then dried at 60 °C for 8 h and cooled down in air. Scan Electron Microscopy (SEM) was JSM-6380LV (Japan). The crystal forms of materials were measured with a Rigaku-TTRIII X-ray diffractometer (Japan). The IR-spectrum was obtained with a PerkinElmer Spectrum (USA) apparatus. The zeta potential of nanoflowers was measured at various pH values with a DELSA 440SX (USA) equipment.

2.2 *Experimental design*

The effect of pH: As adsorption was measured in batch experiments. 0.01 g sorbent and 50 mL of test solution (sodium arsenite) were stirred at a 140 rpm for 10 min at room temperature, and the sorbent was collected by centrifugation. Removal efficiency was calculated by the following equation:

$$\text{As removal} = \frac{C_0 - C_f}{C_0} \times 100$$

(C_0 and C_f are the initial and final As concentrations, respectively).

Sorption Capacity: The initial arsenic concentration was varied among 0.05, 0.50, 1.00, 20.00 and 60.00 mg L^{-1}.

Sorption Kinetics: 0.01 g of Fe/MnO_2 were added to the solution containing 1.00 mg L^{-1} As; 1.00 mL samples of solution were taken to be analyzed at 3, 5, 10, 20, 60 and 120 min at room temperature.

Sorption thermodynamics: The bottle containing 0.01 g sorbent and 50 mL of 1.00 mg L^{-1} As solution was kept at different temperatures (22, 30, 40, 50 and 70 °C), and then 1.00 mL of each solution was extracted to be analyzed after 10 min.

3 RESULTS AND DISCUSSION

3.1 *Characterization of Fe/mno₂*

The SEM images of the as-prepared Fe/MnO_2 were shown in Figure 1. The Fe/MnO_2 nanoflowers were aggregated as a sphere with the diameter of 500–800 nm, and the width of petal was of 20–30 nm. This nanoflower structure with high surface-to-volume ratio induced a high adsorption efficiency to remove arsenic from water. A powder X-ray Diffraction (XRD) pattern of the resulting product is shown in Figure 1d. Some diffraction peaks in Figure 1d can be assigned to the tetragonal phase of α-MnO_2 (JCPDS 44-0141, a = 9.784, c = 2.863 Å), and some diffraction peaks corresponded to birnessite and β-MnOOH. This result showed that the Fe/MnO_2 was a composite, and the low intensity of XRD indicated that the material was of poor crystalline nature. There was no peak corresponding to an iron oxide, implying that the iron oxide phase was amorphous or that the Fe ions replaced Mn ions in part of MnO_2.

3.2 *Effect of pH*

The highest arsenic removal efficiency (95%) was found at pH 4.0. At pH 2.0, the removal efficiency

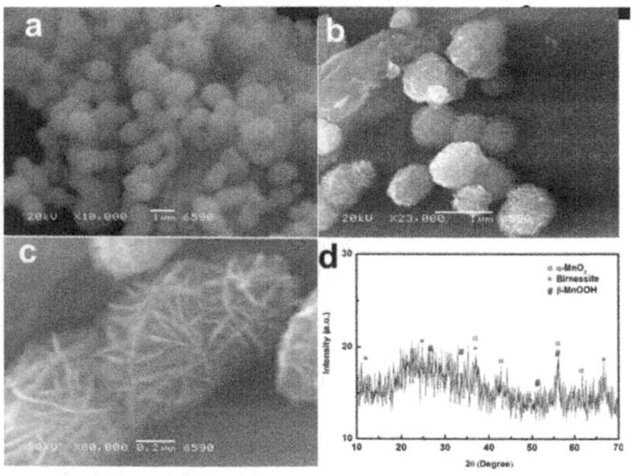

Figure 1. a,b) SEM of Fe/MnO_2, scale 1 μm, c) SEM of Fe/MnO_2, scale 200 nm. d) XRD pattern of Fe/MnO_2.

was 88%, while it was 73% at pH 8.0, indicating that arsenic sorption on the Fe/MnO_2 surface was significantly influenced by the pH. The change in pH of the solution results in the formation of different ionic species and different surface charge on Fe/MnO_2. Fe/MnO_2 can oxidize As (III) to As (V) and then adsorb As(V) on the surface of material. When pH is higher than the pH_{PZC} of Fe/MnO_2 (~7.10), the surface of Fe/MnO_2 is negative and it has weak interaction with arsenic anions ($H_2AsO_4^-$). When the solution pH is lower than that of Fe/MnO_2, the surface of Fe/MnO_2 is positive, with a strong interaction with arsenic ions. The removal efficiency of arsenic increased at lower pH; however, at solution pH values lower than 2, the form of As in water is H_3AsO_4 (pKa = 2.2), and the removal efficiency decreases.

3.3 *Sorption kinetics*

The sorption equilibrium was reached in ~3 min, which was the shortest equilibrium time of all tested materials, as the sorption equilibrium time of MnO_2 modified natural clinoptilolite was 48 h (Maliyekkal *et al.*, 2009), and those of common birnessite MnO_2 (Li *et al.*, 2010) and MnO_2 coated-alumina was ~2 h (Zhang *et al.*, 2007). The faster sorption kinetics is likely due to the presence of the iron oxide, because pure α-MnO_2 nano-needles prepared by a similar method present a low arsenic adsorption rate.

The As(III) sorption process of MnO_2 can be divided in two steps: first, As(III) is oxidized to As (V), and then it is adsorbed by both the iron and the manganese oxide. The presence of the iron oxide increases the adsorption rate. However, slow adsorption kinetics of the amorphous iron manganese binary oxide has been reported (Zhang *et al.*, 2007). The high crystallinity of our Fe/MnO_2, presenting higher oxidization ability than that of an amorphous structure, may be the responsible for the highr adsorption rate.

4 CONCLUSIONS

Fe/MnO_2 prepared from hydrothermal procedure is dispersing flower-like sphere with the diameter of 500–800 nm and the width of petal is of 20–30 nm. Sorption of the arsenic to Fe/MnO_2 reached equilibrium in less than 180 s, and agreed well to the Freundlich adsorption model with adsorption capacities of 26.5 mg g^{-1} at equilibrium concentration of 34.5 mg L^{-1}. The effect of temperature revealed that the adsorption of the arsenic is exothermic, and the adsorption is decreased as increasing temperature form 20 to 70 °C. The Fe/MnO_2 was able to remove 99% of arsenic in water at pH 4.0 to 6.0.

ACKNOWLEDGEMENTS

This work was supported by the National Natural Science Foundation of China (No.21007014), the National Environmental Protection Public Welfare Program (No.201009047), Changsha S&E project (K1301103–11).

REFERENCES

Camacho, L.M. Parra & R.R. Deng, S. 2011. Arsenic removal from groundwater by MnO_2-modified natural clinoptilolite zeolite: Effects of pH and initial feed concentration, *J. Hazard. Mater.* 189, 286–293.

Deschamps, E. Ciminelli, V.S.T. Holl, W.H. 2005. Removal of As(III) and As(V) from water using a natural Fe and Mn enriched sample, *Water Res.* 39, 5212–5220.

Lenoble, V. Laclautre, C. Serpaud, B. Deluchat, V. Bollinger, J.C. 2004. As(V) retention and As(III) simultaneous oxidation and removal on a MnO_2-loaded polystyrene resin. *Sci. Total Environ.* 326, 197–207.

Li, X. Liu, C. Li, F. Li, Y. Zhang, L. Liu, C. Zhou, Y. 2010. The oxidative transformation of sodium arsenite at the interface of MnO_2 and water, *J. Hazard. Mater.* 173, 675–681.

Maliyekkal, S.M. Philip, L. Pradeep, T. 2009. As(III) removal from drinking water using manganese oxide-coated-alumina: Performance evaluation and mechanistic details of surface binding, *Chem. Eng. J.* 153, 101–107.

Zhang, G.S. Qu, J.H. Liu, H.J. Liu, R.P. Li, G.T. 2007. Removal Mechanism of As(III) by a Novel Fe-Mn Binary Oxide Adsorbent: Oxidation and Sorption, *Environ. Sci. Technol.* 41, 4613–4619.

One Century of the Discovery of Arsenicosis in Latin America (1914–2014) – Litter, Nicolli, Meichtry, Quici, Bundschuh, Bhattacharya & Naidu (Eds)
© 2014 Taylor & Francis Group, London, ISBN 978-1-138-00141-1

Recovery of elemental arsenic using electrodeposition with rejection water from capacitive deionization and reverse osmosis

J.A. Ramírez Alvarado
Posgrado en Ingeniería Universidad Nacional Autónoma de México, Campus Morelos-IMTA, México

S.E. Garrido
Instituto Mexicano de Tecnología del Agua, México

A. Baeza Reyes
Universidad Nacional Autónoma de México, México

ABSTRACT: An electrolytic cell was developed at a microscale with copper working electrodes to evaluate the efficiency of the electrodeposition of arsenic as an alternative for recovering this metalloid from the rejection water of capacitive deionization and reverse osmosis processes. Electrodeposits tests were made with a 2^3 factorial design. Factors included As concentration, pH and current density. Maximum arsenic removal efficiencies of 96.06%, 82.76% and 67.30% were obtained with synthetic water, and rejection waters of capacitive deionization and reverse osmosis processes, in concentrations of total As 1.0 mg L^{-1}, 5.0 mg L^{-1} and 0.198 mg L^{-1} respectively. In electrodeposits with maximum As removal, a higher removal of nitrates (70.36%) was observed compared to SO_4^{2-} and PO_4^{3-} anions. Scanning electron microscopy showed the presence of As deposits on the surface of the electrode with synthetic water, while As deposition was not observed in tests with rejection water.

1 INTRODUCTION

Arsenic exists in the environment in different organic and inorganic chemical forms and in four oxidation states—As(V), As(III), As(0) and As(-III) (Ballantyne et. al., 2000). Generally, As(III) is predominant in groundwater because of reducing conditions, while As(V) is found in oxidizing environments. Treatment and disposal of rejection water of processes like capacitive deionization or reverse osmosis processes is currently in decline because of the high treatment costs. Rejection waters resulting from arsenic removal processes can be more toxic than the initial ones because of the concentration of residual arsenic. The development of electrodeposition for recovering elements is very well known and is therefore considered a positive alternative for recovering arsenic from those rejection waters. Recovery of arsenic is of great interest particularly for pharmaceutical, metallurgical and advanced electronics industries. The objective of this work was to evaluate the efficiency of electrodeposition of arsenic as an alternative to recovering this metalloid.

2 EXPERIMENTAL METHODS

An electrolytic glass cell with a volume of 1 L was developed at an experimental scale, with the adaptation of a hermetically sealed gate with three orifices to adapt the electrical current connections to the electrodes, exhaust gases and sampling. The cathode was used as a working electrode and the stainless steel anode as an auxiliary electrode. A connection was designed for collecting the gases. Three types of water were used: synthetic, rejection from capacitive deionization and from reverse osmosis. The synthetic rejection water was prepared with distilled water and sodium arsenite ($NaAsO_2$), CAS 7784-46-5, anions like as nitrates, sulfates, phosphates and chlorides. The rejection waters from capacitive deionization were obtained with DesEl Model 400 capacitive deionization equipment. The rejection water from osmosis was obtained from three reverse osmosis systems located in the states of Durango and Coahuila, Mexico (total As between 0.198–0.073 mg L^{-1}).

Table 1. 2^3 Factorial design, from electrode-position tests with three types of water.

	Water		
Factors	+1	0	−1
As (mg L^{-1})	5.00	3.00	1.00
pH	3.0	2.0	1.0
CD (A cm^{-2})	0.033	0.0245	0.016

The electrodeposits were conducted with a 2^3 factorial design using the three types of water, with a convective regime. Factors included arsenic concentrations of 1.00, 3.00, and 5.00 mg L^{-1}, pH of 1.0, 2.0 and 3.0 and a current density of 0.016, 0.0245 and 0.033 A cm^{-2} (Table 1).

The runs with the greatest removal efficiencies for arsenic in solution were selected to analyze anions that could present interferences with the electrodeposition, as well as to analyze working electrodes. This was performed with Scanning Electron Microscopy (SEM) in order to observe the existence of an arsenic deposit on the surface of the electrode.

Total initial and final arsenic concentrations were determined with the chronoamperometric technique using a Wagtech arsenator digital photometer, approved by UNICEF (Garrido *et al.*, 2009), and with atomic absorption. The electrolytic solution was characterized according to the following parameters: oxide reduction potential (ORP), pH, total dissolved solids (TDS) and electrical conductivity (EC), with electrodeposition times of 0, 30, 60 and 120 min.

3 RESULTS AND DISCUSSION

Table 2 shows the main results from the experiments with the electrodeposition process.

First, the highest As removal occurred with the tests using synthetic water and an initial As concentration of 1 mg/L, resulting in an efficiency of 96.06%. The removal efficiencies for EC and TDS decreased 7.69 and 7.56%, respectively. In the next test, with 3 mg As L^{-1}, a greater decrease was observed for TDS (16.45%) and conductivity (17.99%). The last test, with 5 mg/L using synthetic water, also registered decreases in removal efficiencies for TDS and conductivity, roughly 32%, although reducing conditions were not attained.

In tests with rejection water from capacitive deionization, an average decrease of 12.72% was observed for As removal efficiencies with respect to the efficiencies obtained previously with synthetic water. The As removal efficiencies for the three concentrations with rejection water from capacitive deionization were very similar, with an average of 81.37% and a standard deviation of ±2.34. The variation in removal of TDS did not exceed 20%, while the oxide reducing potentials registered were 148.0, −40.0 and −40.0 mV for As concentrations of 1.0, 3.0 and 5.0 mg L^{-1}, respectively. Finally, in the tests with rejection water from reverse osmosis, a decrease of 20% was observed in As removal efficiencies with respect to those obtained with deionization water.

With respect to the influence of different anions present in rejection water, in the electrodeposition of As, nitrates showed the highest removal percentages, 70%, followed by sulfates and phosphates (Table 3).

Figure 2 shows the surface of the copper electrode, where an irregular and laminar distribution of As can be seen.

Table 2. Maximum as removal efficiencies, EC and ORP.

	As		Removal As	Conductivity			ORP	
	(mg L^{-1})			(mS)			(mV)	
Water	t_0	t_{120}	(%)	t_0	t_{120}	(%)	t_0	t_{120}
Synthetic	1.00	0.04	96.06	15.60	14.40	7.69	220.80	275.90
	3.00	0.28	90.80	11.34	9.30	17.99	502.00	−60.00
	5.00	0.02	95.40	25.00	17.06	31.76	507.00	30.00
Deionization rejection	1.00	0.18	82.69	8.90	7.35	17.42	525.00	148.00
	3.00	0.64	78.67	6.61	4.79	27.53	475.00	−40.00
	5.00	0.86	82.76	42.90	40.30	6.06	544.00	−40.00
Reverse osmosis	0.07	0.03	59.12	25.70	14.06	45.29	586.00	−59.00
	0.17	0.07	57.73	31.20	27.90	10.58	499.00	−128.0
	0.20	0.07	67.30	52.50	46.50	11.43	601.00	40.00

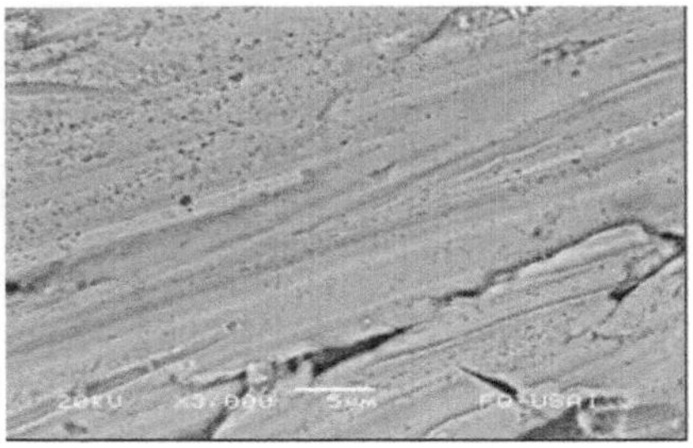

Figure 2. 3000X photography of the Copper Electrode Surface during electrodeposition tests with synthetic rejection water, As removal of 96.06%.

Table 3. Analysis of anions removed in electrodeposition tests with highest removal of as.

Type of water	As (mg/L)	Removal (%)				Cr_{total} (mg/L)
		As	NO_3^-	SO_4^{2-}	PO_4^{3-}	
Synthetic	1.00	96.06	66.19	49.80	35.24	4.0
Capacitive deionization	5.00	82.76	70.36	72.66	17.86	16.0
Reverse osmosis	0.073	59.13	70.00	59.97	50.00	31.0
	0.168	57.73	69.17	53.27	52.00	15.0
	0.198	67.30	42.0	50.0	60.00	18.5

4 CONCLUSIONS

The electrodeposition process is viable for samples of real waters. Nevertheless, more studies are needed to identify the best operating conditions for the process in order to obtain more stable arsenic forms and to prevent the formation of gaseous As species.

ACKNOWLEDGEMENTS

The authors thank the Drinking Water Laboratory of the Instituto Mexicano de Tecnología del Agua, and the Laboratory of Analytical Electrochemistry and Microanalytical Chemistry of the Universidad Autónoma Metropolitana de México.

REFERENCES

Ballantyne B., Marrs T. & Syversen T. 2000. *General and applied toxicology.* Second edition. Mc-Graw Hill. USA.

Garrido S., Avilés M., Ramírez A., Calderón C., Ramírez-Orozco A., Nieto A., Shelp G., Seed L., Cebrian M. & Vera E. 2009. Arsenic removal from water of Huautla, Morelos, Mexico using Capacitive deionization. *Natural arsenic in Groundwaters of Latin America*. Ed. Taylor & Francis. London UK. 665–676.

One Century of the Discovery of Arsenicosis in Latin America (1914–2014) – Litter, Nicolli, Meichtry, Quici, Bundschuh, Bhattacharya & Naidu (Eds)
© 2014 Taylor & Francis Group, London, ISBN 978-1-138-00141-1

Arsenic removal enhanced by the use of ultrafiltration

F. de Carvalho, N. Masetto, A. Lima, J. Arbelaez & R.G. Ramos
Dow Water and Process Solutions

C. Avalos
MIT Sloan School of Management, USA

O. Quiñones
Aquatech Internacional SA de CV, México

ABSTRACT: The presence of arsenic in drinking water is a serious public health problem and approximately 137 million people around the world face problems in water supply systems with high toxic concentrations of arsenic. This problem has caught the attention of different organizations and water companies that are trying to find a solution for the high arsenic concentrations in water. In Mexico, the government started to build plants for arsenic removal using different technologies and one of the technologies used is coagulation -flocculation-filtration. One alternative to this process is the use of ultrafiltration (UF) instead of a general filtration media. A case study is presented to demonstrate the feasibility of ultrafiltration technology in a municipal water treatment plant in Celaya, Mexico. The application of UF for arsenic removal proved to be adequate to produce high quality water proper to distribution for the community.

1 ARSENIC OVERVIEW

1.1 *Toxicity*

Arsenic (As) is a poisonous element for the multicellular life and it has been classified as toxic. Exposure to arsenic can cause health effects such as irritation of the stomach, decreased production of red and white blood cells, DNA damage and cancer. Exposure to arsenic concentrations in drinking water in excess to 300 μg L^{-1} is associated with various diseases; however, little is known about the health consequences of exposure from low-to-moderate levels of arsenic (10–100 μg L^{-1}) (Soto *et al.*, 2004).

1.2 *Arsenic in drinking water*

Arsenic may be found in some drinking water supplies including wells and can get into the water naturally through weathering reactions and microbiological activity or by human action such as metal mining and groundwater abstraction, use of arsenical pesticides and wood preservation. Drinking water constitutes an important pathway of exposure to arsenic in humans. The World Health Organization (WHO) establishes a guideline value for As in drinking water of 10 μg L^{-1}. Some examples of the documented and severe cases of contaminated groundwater by arsenic are located in Asia (Bangladesh, China, India, Nepal) and America (Chile, Argentina, Mexico, Peru) (Smedley, 2008).

There are different technologies available to remove arsenic from water and the efficiency of each technology depends on the ionic form in which arsenic is present and the water chemistry. Common technologies are coagulation/flocculation, iron/manganese oxidation, lime softening, activated alumina, iron oxide coated sand, ion exchange resins, reverse osmosis and nanofiltration.

1.3 *Arsenic in Mexico*

In Mexico, 65% of the population relies on groundwater as a potable supply, and the presence of arsenic is affecting around 450,000 persons just in the north region of Mexico. The main form of arsenic present in water is As(V) and the concentrations are between 8 to 624 μg L^{-1}. The Mexican law (NOM-127-SSA1–1994, 2000) establishes a maximum concentration of 25 μg L^{-1} of arsenic in drinking water, and the government is looking to remove arsenic from tap water.

2 ARSENIC REMOVAL CELAYA PLANT: COAGULATION—FLOCCULATION—UF

2.1 *The Celaya problem*

Celaya is a small city in the state of Guanajuato with a population of 340,400 inhabitants. Celaya relies on groundwater to supply drinking water to the population. Even though most of the wells are under the As limit, the ones located in the north-east part of the city present arsenic levels above 25 μg L^{-1}.

2.2 *Plant overview*

The Celaya plant has a production capacity of 108 m^3/h, based on 32 ultrafiltration modules (2 skids with 16 modules each), a system recovery higher than 98% and design flux of 66 L m^{-2} h^{-1}.

Ultrafiltration technology is based on:

- pressurized vertical modules (51 m^2 active area) with hydrophilic PVDF polymeric hollow fiber membranes providing high strength and chemical resistance;
- 0.03 μm nominal pore diameter for removal of bacteria, viruses, and particulates;
- outside-in flow configuration for high tolerance to feed solids;
- dead-end flow which offers higher recoveries.

2.3 *Removal process*

Figure 1 shows a flow diagram of the Celaya plant. Well water (turbidity 3–5 NTU and total As 49 μg L-1) is oxidized and cooled. Afterwards, coagulation with ferric chloride allows the formation of arsenic flocks that will be removed from water by ultrafiltration modules. High quality water produced is finally sent to a municipal water storage tank to be pumped as tap water to the community. All water used for backwash and cleanings is entirely separated from slurry. Water then is recirculated to the initial storage tank to be treated and the slurry is sent to disposal by a third party. Table 1 shows the operational parameters of plant.

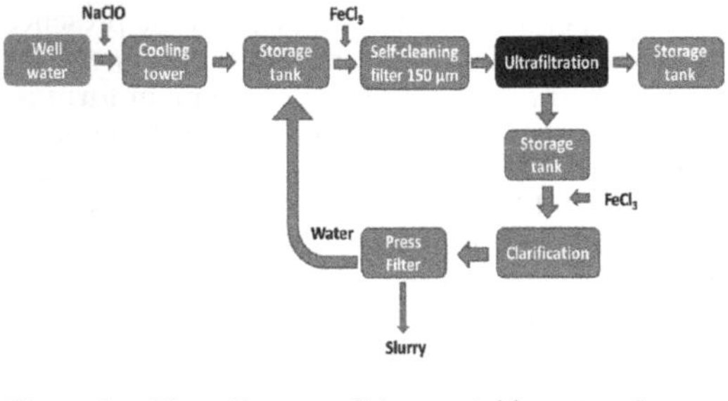

Figure 1. Flow diagram: Celaya potable water plant.

Table 1. Operational parameters.

Parameter	Frequency	Duration	Chemical consumption
Filtration	–	30 min	–
Air Scour	30 min	40 s	–
Backwash	30 min	4 min	–
Forward Flush	30 min	60 s	–
CEB acid	48 h	14 min	0.1% HCl
CEB alkaline	48 h	14 min	0.05% NaOH, 0.1% NaOCl
CIP acid	1 month	3 h	0.2% HCl
CIP alkaline	1 month	3 h	0.1% NaOH, 0.2% NaOCl

3 CELAYA PLANT PERFORMANCE

Performance data was collected from three months of operation. Figure 2 shows arsenic level in the feed stream and in the product. Huge variations are shown in arsenic level in the feed, mainly due to salt variability monitored in the deep well and some deviations coming from plant instrumentation. However, the arsenic level in product was always below 25 μg L-1, which was the value required from regulation.

In Figure 3, the behavior of turbidity in feed water and in product water is shown. It can be seen that turbidity in water product was constantly lower than 0.2 NTU, below of the regulation limit of 5 NTU. On the other hand, feed water turbidity showed variability over time going from values closer to 1 NTU up to values closer to 7 NTU.

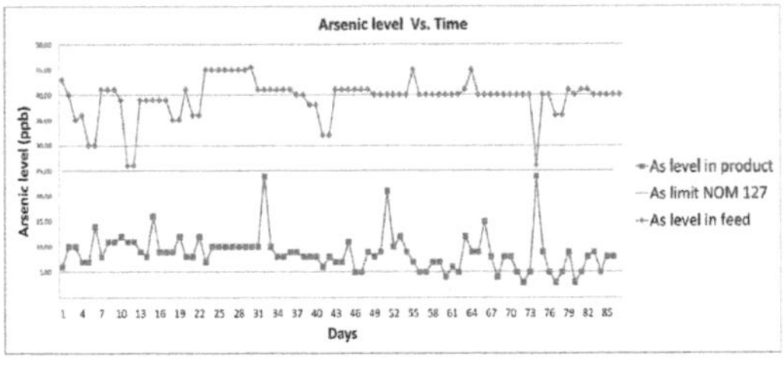

Figure 2. Product and feed water As concentration versus time.

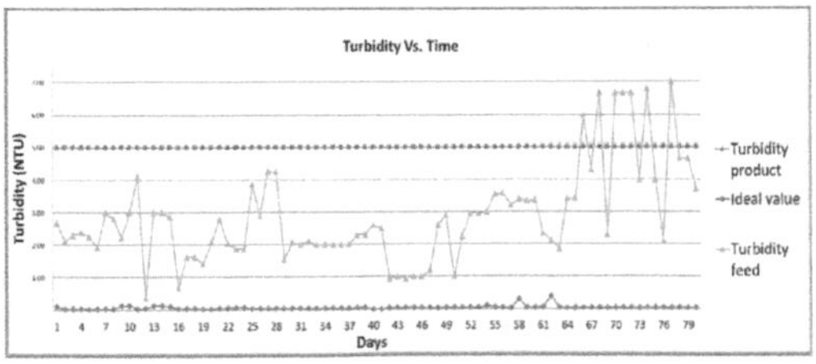

Figure 3. Product and feed water turbidity versus time.

4 CONCLUSIONS

Celaya plant demonstrates application feasibility of ultrafiltration for arsenic removal. The UF pore size allows the removal of the arsenic flocks providing high quality water and reducing the amount of coagulant needed in the process. Additionally, the method is capable of removing bacteria and virus without using a secondary step. The use of ultrafiltration decreases the amount of waste to be disposed, being more environmentally friendly. Furthermore, the water used in the backwash is recycled and the slurry is sent to the adequate disposal. Future studies should seek to increase the arsenic removal, making UF technology becoming more widespread.

REFERENCES

NOM-127-SSA1–1994, 2000, *Norma Oficial Mexicana sobre Salud ambiental, agua para uso y consumo humano-límites permisibles de calidad y tratamiento que debe someterse al agua para su potabilización*, México D.F.

Soto, P., Lara F., Portillo L. and Cianca A. 2004, *An Overview of Arsenic's groundwater occurrence in Mexico*; Gerencia de aguas subterráneas; Comision Nacional del Agua; Mexico DF.

Smedley, Pauline L.. 2008 Sources and distribution of arsenic in groundwater and aquifers. In: Appelo, Tony, (ed.) *Arsenic in Groundwater: a World Problem.* Utrecht, the Netherlands, IAH, 4–32.

One Century of the Discovery of Arsenicosis in Latin America (1914–2014) –
Litter, Nicolli, Meichtry, Quici, Bundschuh, Bhattacharya & Naidu (Eds)
 ISBN 978-1-138-00141-1

Polymer inclusion membranes: A promising approach for arsenic detection and removal from waters

C. Fontàs & E. Anticó
Department of Química, University of Girona, Girona, Spain

S.D. Kolev
Faculty of Science, School of Chemistry, The University of Melbourne, Australia

A. Illa
Ingesco, Terrassa (Barcelona), Spain

ABSTRACT: A novel polymer inclusion membrane (PIM) based system was developed and successfully applied to the removal of As(V) from polluted natural waters and to its detection at μg L^{-1} levels. The latter application involved membrane separation of As(V) followed by its colorimetric detection based on the molybdenum blue reaction.

1 INTRODUCTION

Chemical membranes normally contain a specific reagent (carrier) that interacts with the target species to be transported. In the case of polymer inclusion membranes (PIMs), the carrier is entrapped into a polymer matrix to provide mechanical strength and, thus, to ensure better membrane stability. Sometimes, a plasticizer can be used to provide elasticity or a modifier to vary the properties of the membranes. In our previous work (Güell *et al.*, 2011), a PIM made of cellulose triacetate (CTA) as the polymer and Aliquat 336 as the carrier exhibited a very good performance in terms of As(V) transport. As it was described, the transport mechanism is based on the fact that anionic species of As(V) present in waters at neutral pH (mainly $H_2AsO_4^-$ and $HAsO_4^{2-}$) can be extracted by anion-exchange carriers, such as Aliquat 336.

Taking advantage of the versatility of this separation system, both an As detection method and a water remediation treatment technology can be designed.

The PIM incorporated in an especial device has allowed the preconcentration of As(V), which has been subsequently determined colorimetrically using the molybdenum blue reaction (Fontàs *et al.*, 2013). Moreover, the same PIM has shown to effectively remove As from several water samples from the region of Girona (Spain), which naturally contain As at < 100 μg L^{-1} levels.

2 METHODS/EXPERIMENTAL

2.1 *Reagents*

Stock solutions (100 mg L^{-1}) of As(V) and As(III) were prepared from $Na_2HAsO_4.7H_2O$ and $NaAsO_2$, respectively. PIMs were prepared with the extractant Aliquat 336, which is a mixture of quaternary alkylammonium chlorides, and the polymer polyvinyl chloride (PVC). A 0.5 M Aliquat 336 solution in tetrahydrofuran (THF) was used to prepare the polymeric films. For the molybdenum blue method, the colorimetric reagent solutions were prepared in accordance with the latest improved version of the method (Tsang *et al.*, 2007).

2.2 *Preparation of PIMs and transport experiments*

PIMs were prepared by dissolving 400 mg of PVC and Aliquat 336 (150–400 mg) in 20 mL THF. This solution was stirred during 4 hours, and then, the resulting mixture was poured into a 9.0 cm in diameter flat bottom glass Petri dish, which was set horizontally and loosely covered. The solution was allowed to evaporate over 24 h at room temperature. The resulting film was then carefully peeled of the bottom of the Petri dish and discs of appropriate diameter were cut for the experiments. The device designed for detection purposes consists of a glass tube with two openings, one at the top (0.9 cm diameter) and another one at the bottom (1.8 cm diameter).

The PIM was fixed at the bottom opening. A fixed volume of 0.1 or 2 M NaCl stripping solution was placed inside the tube. The transport cell for As removal is a spiral type cell (V feed = 100 mL; V stripping = 5 mL; membrane area = 64 cm^2).

2.3 *Water samples*

Water samples with naturally occurring arsenic were collected in Northern Catalonia (Spain) from three different wells in the village of Lles and two in the village of Setcases. Both sites are located close to the Pyrenees Mountains.

3 RESULTS AND DISCUSSION

3.1 *As removal with PIMs*

Preliminary experiments to optimize the transport system were performed in a two compartment membrane cell where 190 mL of both feed and stripping solutions (0.1 M NaCl) were used. The transient concentration curves corresponding to a PIM with 31 wt.% Aliquat 336 (Figure 1) show that a quantitative transport of As is achieved in only 5 h.

Thus, this membrane system was implemented in a spiral cell that allowed the continuous flow of the natural water to be treated. Water samples from the villages of Lles and Setcases, with different chemical compositions, and As concentrations between 50–100 μg L^{-1}, were processed in the membrane system and removal efficiencies of 100% were achieved.

3.2 *PIM-based device for As detection*

Previous studies have demonstrated the possibility of arsenate preconcentration using a PIM-based device contacted with a 100 mL solution of 100 μg L^{-1} As(V) in Milli-Q water for 24 h under constant agitation, and 5 mL of 0.1 M NaCl stripping solution (Fontàs *et al.*, 2013). Under these conditions, a 70% transport efficiency was achieved

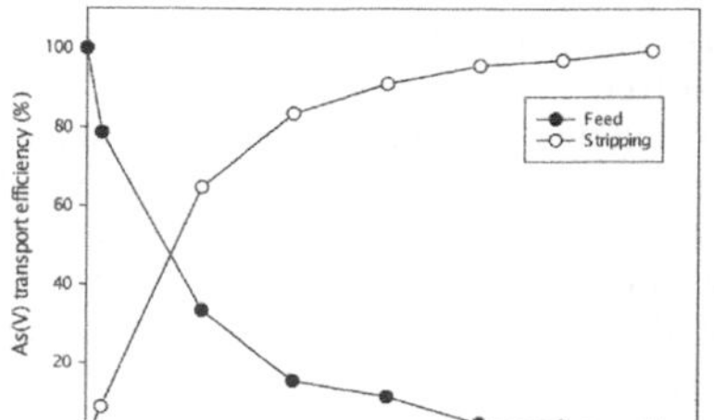

Figure 1. Transient concentration curves of As(V) in transport experiments involving a PIM made of 69 wt.% PVC and 31 wt.% Aliquat 336. Feed solution: 10 mg L^{-1} As(V), pH 7; stripping solution: 0.1 M NaCl.

Table 1. Comparison of As concentrations measured by ICP-MS and using the proposed method (PIM: 31 wt.% Aliquat 336).

	Arsenic concentration (μg L^{-1})	
Sample	ICP-MS	PIM-based device (SD)
Lles 1	70	65 (8)
Lles 2	54	45 (8)
Lles 3	48	42 (8)
Setcases 2	55	50 (8)
Setcases 3	22	30 (8)

and a good reproducibility was obtained (RSD <5%). Moreover, the coupling of the molybdenum blue method and the PIM-based device allowed the measurement of the As concentration with a limit of detection of 5 μg L^{-1}. The validation of the PIM-based device for As detection was done by comparing results with those obtained with the well-established ICP-MS technique (Table 1). Different water samples were analyzed as shown in this table.

In order to improve the performance of the detection device in view of its application for *in-situ* As monitoring, and to account for different matrix compositions, several studies have been conducted. It has been shown that the use of a 2 M NaCl stripping solution allowed the detection of As in waters even in the presence of a high content of other anions. Besides, the transport can be improved using a thinner PIM containing 50 wt.% of Aliquat 336. Under these conditions, the transport efficiency increased three times after 2 h contact.

4 CONCLUSIONS

In this study, we have presented a novel green detection method for As screening in natural waters. Moreover, the results obtained indicate that PIMs can be used for the development of membrane-based systems for the removal of As from natural waters since these membrane systems do not need high operational or capital costs.

REFERENCES

Fontàs, C., Vera, R., Batalla, A., Kolev, S.D. & Anticó, E. 2013. Screening of arsenic in groundwater: a novel and green detection method. *Environm. Sci. Poll. Res. (submitted).*

Güell, R., Anticó, E., Kolev, S.D., Benavente, J., Salvadó, V. & Fontàs, C. 2011. Development and characterization of polymer inclusion membranes for the separation and speciation of inorganic As species. *J. Membr. Sci.* 383: 88–95.

Tsang, S., Phu, F., Baum, M.M. & Poskrebyshev, G.A. 2007. Determination of phosphate/arsenate by a modified molybdenum blue method and reduction of arsenate by $S_2O_4 2^-$. *Talanta* 71: 1560–1568.

One Century of the Discovery of Arsenicosis in Latin America (1914–2014) – Litter, Nicolli, Meichtry, Quici, Bundschuh, Bhattacharya & Naidu (Eds)
© 2014 Taylor & Francis Group, London, ISBN 978-1-138-00141-1

Developing a viable concept for sustainable arsenic removal from groundwater in remote area—findings of pilot trials in rural Bihar, India

S. Schmidt, M. Herrmann, E. Gukelberger, F. Fiedler, B. Großmann, S.A. Deowan & J. Hoinkis
Karlsruhe University of Applied Sciences, Germany

S.A. Ghosh
Anugrah Narayan College, Patna, Bihar, India

D. Chatterjee
University of Kalyani, India

J. Bundschuh
University of Southern Queensland, Toowoomba, Australia

ABSTRACT: Arsenic contamination of groundwater is posing a serious challenge to the drinking water supply, particularly in the large alluvial and deltaic tracts of south-eastern Asia, which are densely populated, and rely on shallow aquifers for portable water. In order to find a solution for this issue, an arsenic treatment unit was developed based on the reverse osmosis principle, including an energy recovery unit to lower the energy consumption for the system. This unit promises big advantages in handling arsenic contamination in groundwater. Initial tests have been successfully conducted under laboratory conditions and, subsequent to them, pilot studies have been carried out in Bind Toli, a village located in the rural area of Bihar, India.

1 INTRODUCTION

The most critical elements in the groundwater of Bihar are arsenic, iron and fluoride. These three contaminants are found in many cases to be beyond the permissible WHO limits of 10 μg L^{-1}, 1.0 mg L^{-1} and 1.5 mg L^{-1} (WHO, 2012). In 2010, the Indian Ministry of Water Resources published that about 10.4 million people are exposed to the high arsenic loaded groundwater in Bihar. Arsenic concentration reported so far goes up to a value of 1810 μg L^{-1} (Govt. of India, 2010). Typical indicators for high arsenic loaded water to the human body are skin lesions, diabetes, cancers, etc. Groundwater contamination with arsenic above the permissible limit is a worldwide issue challenging inhabitants of more than 36 countries in Asia, America, Europe and the Pacific area (Petrusevski *et al.*, 2007). As arsenic is a naturally occurring geological contaminant in the rock beds in the underground, it is a real challenge to find sustainable concepts to treat water that is pumped from this soils. Until now, many different arsenic removal technologies have been developed and tested. The most simple and commonly used technologies work with the principles of oxidation, co-precipitation and adsorption (Chwirka *et al.*, 2000; Mohan & Pittman, 2007; Uddin *et al.*, 2007; Shih, 2005). Despite the big advantages of these cheap and simple solutions, there is one disadvantage that all of them have in common, which is the problem of disposal of the spent materials or sludge. In contrast, membrane technologies promise the possibility of a safe concentrate disposal back to its origin, together with the advantage of removing dissolved arsenic along with other dissolved and particulate compounds (viruses, bacteria, dissolved iron, etc.). The benefits of the membrane technology are high efficiency, easy operation, high effluent water quality, modularity and flexibility. To date, Reverse Osmosis (RO) and Nanofiltration (NF) have the reputation of being highly sophisticated units with an enormous energy consumption, being not suitable for their application in rural areas of developing countries.

The tested pilot unit presented in this work, makes use of a RO desalinator unit for sailing boats with a very low recovery rate of 10% and an integrated energy recovery rate that lowers the energy consumption significantly (for overview, see Watermakers, 2013). By running the system with this low recovery rate, usually there is no need of additional use of chemicals.

2 EXPERIMENTAL WORK

Prior to the pilot trials in the field, numerous tests under laboratory conditions have been conducted using both a small RO desalinator type "Power SurvivorTM 160E", supplied by the Swiss company Katadyn (Katadyn, 2013) and "Watermaker" supplied by Schenker, Italia. Different membrane types from DOW Chemicals have been tested, and it has been found that the reaction rate for trivalent arsenic (As(III)) is generally much lower than that for pentavalent arsenic (As(V)). For trivalent arsenic, a feed solution containing up to 2400 mg L^{-1} As(V) resulted in a permeate quality below the Maximum Contaminant Level (MCL) of 10 μg L^{-1}. For As(III), only a feed concentration below 350 μg L^{-1} resulted in a permeate quality below the MCL (Geucke *et al.*, 2009). This effect occurs because all the membranes consist of polymers with negatively charged groups (Geucke *et al.*, 2009). A charge exclusion effect enhances the rejection of negatively charged ions such as As(V). In the observed pH range, As(III) is only neutrally charged, hence it is less rejected. Figure 1 shows the system sketch of the complete arsenic mitigation system installed in the field (Bind Toli, Bihar, India).

Firstly, groundwater was extracted by a handpump into tank 1. Afterwards, it was pumped, by using an electrical pump, into tank 2, equipped with showerheads in order to support precipitation of iron hydroxides occurring in the groundwater. After having passed the sand filter, all precipitated iron is removed, and the clear water flows into the feed tank of the RO-facility. It has been demonstrated that 80–90% of the total arsenic occurring in groundwater can be removed by co-precipitation and filtration in the sand filter. Hence, the RO-filter downstream acts as polisher in order to comply with

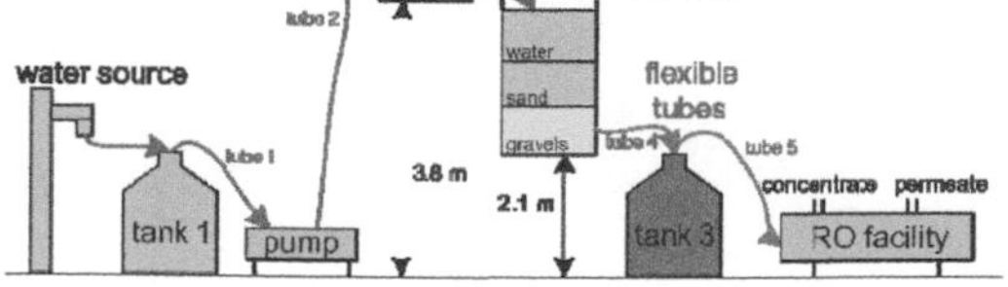

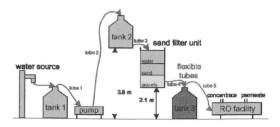

Figure 1. System setup in Bind Toli.

the stringent drinking water standards for arsenic (10 μg L^{-1}, Govt. of India, 2010). After having passed the sand filter, all the precipitated iron was removed and only a very low concentration of dissolved iron is left. The permeate flux shows values between 20 and 26 L h m^{-1} during the pilot studies. The initial values of 26 L h m^{-2} decreased slightly, probably due to some fouling on the membrane surface when testing alternative pre-filtering methods.

3 SOCIO-ECONOMIC ASPECTS

General apathy exists among the affected villagers towards already adopted mitigation practices. Hence, the socio-economic conditions of the test site have been also studied within this study. The highlights found out within this study are:

1. Very low literacy level, about 46% are illiterate.
2. Monthly income per person in average 7.33 €.
3. Very low awareness of arsenic contamination (14%). During interviews with the local population it has been reported that there is one golden rule for safe drinking water: "As long as there is no bad smell, taste and as long as it looks clear, everything is fine with the water". However, since this basic rule for safe drinking water cannot be applied for arsenic loaded water, it has been shown that the awareness of arsenic occurrence in the water and its impact to the human body is an underestimated challenge in the awareness creation of the inhabitants. On the other hand, it has been demonstrated that the awareness grew during the piloting phase of the treatment unit and increasing interest in the tested unit and the ongoing interviews have been observed as well within the 3 months of testing's in the field.
4. Community based water supply is preferred.
5. 71% are not able to pay for safe drinking water.

4 CONCLUSIONS AND OUTLOOK

- Additional long-term pilot trials in Ramnagar Bihar, India, will be done, with different groundwater qualities.
- Further studies on options for sludge and brine treatment as well as for their disposal.
- Continuation of socio-economic investigations.
- Development and test a commercial system powered by renewable energies in order to have an autonomous, easy to operate and self-sufficient system.
- Testing of the unit under anaerobic conditions in order to avoid the effect of precipitation and directly discharge of the concentrated water back to the groundwater.

REFERENCES

Chwirka, J.D., Thomson, B.M. & Stomp, J.M. 2000. Removing arsenic from groundwater. *J. Am. Water Works Assoc.* 2–3: 79–88.

Geucke, T., Deowan, S.A., Hoinkis, J. & Pätzold, Ch. 2009. Performance of a small-scale RO desalinator for arsenic removal. *Desalination* 239: 198–206.

Govt. of India. 2010 http://india.gov.in/knowindia/stbihar.php 14.08.2010.

Katadyn. 2013. www.katadyn.com

Mohan, D. & Pittman, C.U. 2007. Arsenic removal from water/wastewater using adsorbents—A critical review. *J. Hazard Mater.* 142(1–2): 1–53.

Petrusevski, B., Sharma, S., Schippers, J.C. & Shordt, K. 2007. Arsenic in Drinking Water, IRC International Water and Sanitation Centre, http://www.irc.nl/page/33113 (accessed 10th Sep. 2013).

Shih, M.C. 2005. An overview of arsenic removal by pressure-driven membrane processes. *Desalination* 172(1): 85–97.

Uddin, M.T., Mozumder, M.S.I., Figoli, A., Islam, M.A. & Drioli, E. 2007. Arsenic removal by conventional and membrane technology: An overview. *Indian J. Chem. Techn.* 14(5): 441–450.

WHO. 2012. //www.who.int/mediacentre/factsheets/fs372/en/ (accessed 10th Sep. 2013).

Watermakers. 2013. www.nauticexpo.com.

One Century of the Discovery of Arsenicosis in Latin America (1914–2014) –
Litter, Nicolli, Meichtry, Quici, Bundschuh, Bhattacharya & Naidu (Eds)

Influence of household reverse-osmosis systems and its neutralizer accessory for arsenic removal in drinking water

A. Maleki
Kurdistan Environmental Health Research Center, Kurdistan University of Medical Sciences, Kurdistan, Iran

A.H. Mahvi
School of Public Health and Center for Environmental Research Medical Sciences, University of Tehran, Iran

H. Daraei
Kurdistan Environmental Health Research Center, Kurdistan University of Medical Sciences, Kurdistan, Iran

ABSTRACT: In this study, the efficiency of household reverse-osmosis systems (HRO) with and without a neutralizer accessory as well as the output water quality and stability were investigated in both real and synthetic samples. The real samples were drinking water samples collected from rural and urban areas in the Kurdistan province, Iran, served by public systems with and without a primary treatment. The HRO model RO100GPD (Luna Water Co., Canada) with and without a neutralizer accessory was used in all experiments.

1 INTRODUCTION

A significant part of the water consumed by the public in Iran as well as in other countries is treated by household reverse-osmosis systems (HRO). The results of this treatment on the quality of standard and contaminated feed waters should be precisely probed.

Therefore, the purpose of this study was the characterization of the quality of output water treated by HRO and the evaluation of a neutralizer accessory used in this equipment.

2 MATERIALS AND METHODS

2.1 *Study area and samples*

The Kurdistan province of Iran was selected as the study area. Two urban drinking water samples collected from Sanandaj (Kurdistan, Iran) and Sarvabad (Kurdistan, Iran) and three rural drinking water samples collected from Daraki (Sarvabad, Kurdistan, Iran), Goor-Baba-Ali (Divandare, Kurdistan, Iran), and Gharakhlar (Bijar, Kurdistan, Iran) were analyzed.

2.2 *Equipment*

A household reverse-osmosis system (HRO) model RO100GPD (Luna Water Co., Canada) with and without a neutralizer accessory was used. In addition to the neutralizer accessory, the RO100GPD equipment contains three prefilters (a 25 μm sediment pre-filter, a 10 μm micron active carbon prefilter and a 1 μm sediment pre-filter) and an osmosis membrane. In the filtration stage, feed water passes first through the sediment filters using a primary pump where silt, sediment, sand and clay particles are removed.

2.3 *Sampling and procedure*

Samples (20 L) were collected as feed water from the study areas. After 10 L treatment by HRO, output samples were collected from the output port. Washing by 10 L of distilled water was applied to prevent the memory effect on HRO. Two repeated experiments were done for Sanandaj and Sarvabad samples to investigate the precision of the experiment and sampling.

3 RESULTS AND DISCUSSION

3.1 *Results*

Physicochemical tests were done for both the feed and the output water samples. The experimental data are summarized in Figure 1. The Iran national drinking water standards levels, the Maximum Acceptable Concentration (MAC) and the Maximum Desirable Concentrations (MDC) are presented for each parameter in the same table.

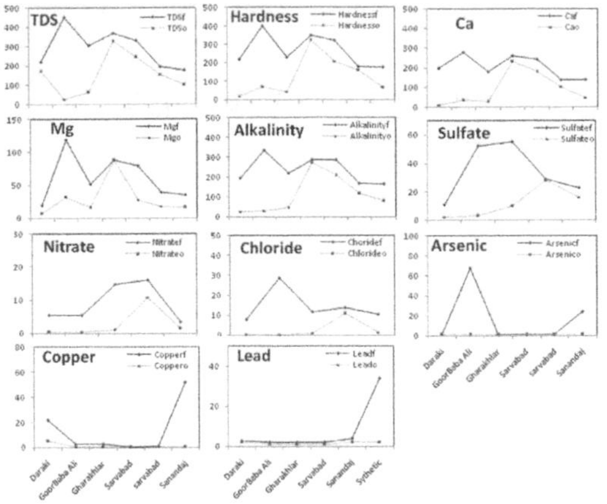

Figure 1. Physicochemical parameters for feed and output water samples.

Table 1. T-test results for evaluation of neutralizer accessory and repeated samples.

T-test objects	P-Values
T-test for HRO output with and without neutralizer accessory	0.01
T-test for repeated output of Sanandaj	0.03
T-test for repeated feed HRO samples for Sanandaj	0.11
T-test for repeated output of Sarvabad	0.00
T-test for repeated feed HRO samples for Sarvabad	0.06

Arsenic content, hardness, Ca^{2+}, Mg^{2+}, TDS, EC, Alkalinity, Cl^-, Br^-, SO_4^{2-}, PO_4^{3-}, NO_3^-, NO_2^-, copper and lead were analyzed by standard methods for the feed (f) and output (o) samples.

3.2 *Discussion*

The pair-sample T-test was applied for statistical evaluation of differences between the HRO output with and without the neutralizer accessory. In addition, T-test was applied to evaluate statistically the differences between the repeated input/output samples (Table 1).

The results show that the output water completely meets the standard levels of MAC and even MDC regardless to the feed water quality. However, the output water did not show the proper repeatability. It can be concluded that for both polluted and standard feed waters, the application of HRO produces an increase in the quality of the output water. However, the method is more applicable to polluted feed waters, especially those containing heavy metals.

Concerning the neutralizer accessory, the use of this device is not necessary, as the output water contains the proper amount of TDS. This could indicate that that the membranes of these systems are not reverse osmosis membranes but, probably, they are nanofilters, according to the results of TDS and hardness in output. Thus, in real reverse osmosis membrane base systems, the use of the neutralizer could be more applicable.

4 CONCLUSIONS

In this study, a household HRO was evaluated for the quality of feed and output water regarding standard MAC and MDC levels, considering 1) the influence of the use of a new neutralizer accessory, and 2) the stability (repeatability) of the output water. Several rural and urban samples with different initial physicochemical characterization were used as feed water. It can be concluded that the output water is completely meets the standard levels of MAC and even MDC regardless to the feed water quality but did not show the proper repeatability in the output water. The use of a neutralizer accessory is not necessary, as the output water contains a the proper amount of TDS.

ACKNOWLEDGEMENTS

This work was financially supported by Kurdistan Environmental Health Research Center, Kurdistan University of Medical Sciences.

REFERENCES

Choi, J.S., Lee, S., Kim, J.M., & Choi, S. 2009. Small-scale desalination plants in Korea: Technical challenges. *Desalination* 247(1): 222–232.

Jackson, P.E. 2001. Determination of inorganic ions in drinking water by ion chromatography. *Trends in Analytical Chemistry* 20(6–7): 320–329.

Lashkaripour, G.R., & Zivdar, M. 2005. Desalination of brackish groundwater in Zahedan city in Iran. *Desalination* 177(1): 1–5.

Mahmood, Q., Baig, S.A., Nawab, B., Shafqat, M.N., Pervez, A. & Zeb, B.S. 2011. Development of low cost household drinking water treatment system for the earthquake affected communities in Northern Pakistan. *Desalination* 273(2): 316–320.

Manjikian, S. 1974. Reverse osmosis water purifying system for household use. U.S. Patent No. 3,849,305, 4 P, 1 Fig, 8 Ref.

Walker, M., Seiler, R.L., & Meinert, M. 2008. Effectiveness of household reverse-osmosis systems in a Western US region with high arsenic in groundwater. *Science of the Total Environment* 389(2): 245–252.

4.4 Emerging technologies

One Century of the Discovery of Arsenicosis in Latin America (1914–2014) – Litter, Nicolli, Meichtry, Quici, Bundschuh, Bhattacharya & Naidu (Eds)
© 2014 Taylor & Francis Group, London, ISBN 978-1-138-00141-1

From a modified naturally occurring material to the design of novel receptors for the removal of arsenic from water

A.F. Danil de Namor
Department of Chemistry, Thermochemistry Laboratory, University of Surrey, UK
Instituto Nacional de Tecnología Industrial, Argentina

J.R. Cartwright, W. Abou Hamdan & O.A. Webb
Department of Chemistry, Thermochemistry Laboratory, University of Surrey, UK

N. Al Hakawati
Department of Chemistry, Thermochemistry Laboratory, University of Surrey, UK
Department of Chemistry, Lebanese American University, Lebanon

L. Valiente
Instituto Nacional de Tecnología Industrial, Argentina

S. Korfali
Department of Chemistry, Lebanese American University, Lebanon

ABSTRACT: Two approaches have been investigated for the removal of arsenic speciation from water. The first involved the modification of diatomaceous earth (diatomite) using ferric oxide, which increased the uptake of arsenite and arsenate compared to raw diatomite. The kinetics of arsenic sorption was studied and increased through heat treatment prior to Fe modification. Temperature was found to have minimal effect on arsenic kinetics while significant changes on As uptake were observed with increasing pH. Additionally, high matrix waters (tap & spring) have minimal effect on arsenate uptake but a larger effect on arsenite sorption was identified. The three forms of interest; raw, calcined and modified diatomite were successfully characterized by multiple techniques. The second approach included the application of macrocyclic receptors based on calix[4]pyrrole and its dimer for the removal of arsenic speciation from water. The optimum conditions for the extraction process were investigated and compared to polypyrroles.

1 INTRODUCTION

The presence of arsenic speciation in water exceeding the limits established by the Wealth Health Organization (WHO) is a serious problem affecting many regions of the world. This is mainly due to the fact that arsenic compounds are classified as carcinogenic. Arsenic speciation in water is arsenite, arsenate and organic arsenic. The toxicity of arsenite is the highest due to the enhanced cellular uptake and the difficulties involved in its removal (Mahata 2003a,b, Harisha *et al.*, 2010).

Technological approaches for arsenic removal include precipitation and coagulation as well as adsorption and ion exchange processes. Precipitation and coagulation with iron (III) (Roberts *et al.*, 2004a,b, Wilkie & Hering 1996), aluminum (III) (Wilkie & Hering 1996a,b, Gregor 2001, Giles *et al.*, 2011), calcium (Bothe & Brown 1993a,b, Moon *et al.*, 2004) and lanthanum salts (Tokunaga *et al.*, 1999), which convert anions into insoluble forms, have been widely used. Although these methods are simple and economical, the processes lead to a wet bulky sludge disposal problem and require final filtration for secondary treatment. Comparatively, the ion exchange and adsorption processes seem to be the most promising for the removal of arsenic speciation from water. Sorbents most often used for the removal of arsenic speciation from water are activated alumina (Lin & Wu, 2001), activated carbon (Payne & Fattah, 2005), diatomite and modified diatomite (Danil de Namor *et al.*, 2012) and other materials (Demarco *et al.*, 2003a,b; Chen *et al.*, Bissen & Frimmel, 2003). However, as a result of the low removal capacity and poor kinetics, these materials cannot be widely used. Recent research has demonstrated that rare earth and Zr(IV) elements have affinity for arsenate; thus, various rare earth compounds (Danil de Namor & Abbas, 2007a,b; Danil de Namor, 2007c; Danil de Namor, 2007d; Danil de Namor *et al.*, 2007) and Zr(IV) loaded polymers (Iwamoto *et al.*, 2002a,b; Seko, 2004) have been developed to remove arsenate ions from water. Although the capacity of these materials for this purpose has been improved, the kinetics of the process is still slow, as several hours are required to reach equilibrium. This results in operational, control and maintenance problems, which limit the commercial application of these materials. Given that the viability of these technologies depends on the development of suitable materials for the efficient removal of arsenic speciation from water, in this paper we report results recently obtained for the removal of arsenic speciation from water using a modified naturally occurring material (diatomite) and decontaminating agents based on supramolecular chemistry. For the latter, calix[4]pyrrole derivatives are used and the results are compared with those based on polypyrroles (Danil de Namor *et al.*, 2006a,b; Danil de Namor & Khalife, 2008; Danil de Namor & Abbas, 2010). In this form, the optimum conditions for arsenic removal are investigated taking into account several factors such as: the solid solution ratio, the pH of the aqueous solution, the temperature, as well as the kinetics of the process.

2 METHODS & EXPERIMENTAL

2.1 *Diatomaceous earth modification*

Modification of diatomite was achieved using diatomaceous earth (Sigma-Aldrich, Gillingham, UK) in a 0.5 M iron(III) chloride (Lancaster, Worecambe, UK) solution. Three separate samples (10 g) were weighed and further washed three times with distilled water (100 mL) using vacuum filtration. Diatomite was then placed in a furnace at 105 °C for 12 hours and kept in a desiccator in-between use. Two different modifications of the diatomite were tested: one was the calcined material prior to ferric oxide sorption, and the other involved the non-calcined material treated with ferric oxide. Also, one sample of diatomaceous earth was left unmodified as a control. The calcined and the non-calcined samples of diatomite were placed into two separate solutions of iron(III) chloride (0.5 M), at a solid/solution concentration ratio of 200 g/L. Magnetic stirrers were introduced to both suspensions with a rotational speed of 600 rpm and both were adjusted to the basic range using a sodium hydroxide (1 M) (Fisher Scientific, Loughborough, UK) solution. Diatomite was left for 24 hours in order to reach full saturation of ferric oxide. After the sorption process, both samples were washed again with distilled water followed by vacuum filtration, to wash away any excess $FeCl_3$, then was dried at 105 °C for 12 hours.

2.2 *Characterization of the materials*

Energy dispersive X-Ray (EDS) and Scanning Electron Microscopy (SEM)—Operation for both areas were performed on a Hitachi S-3200N SEM fitted with an Oxford Instruments X-act EDS. SEM was performed in a Variable Pressure SEM (VP-SEM) and electron strength of 20 kV. The data for EDS analysis is given in percentage weight of present elements. Brunauer Emmett Teller (BET) method analysis—Samples for nitrogen sorption tests at 77 K were performed on a Micromeritics Gemini V Surface Area and Pore Size Analyzer. Samples were outgassed at 150 °C for 12 hours on a Micromeritics Flowprep 060 unit prior to analysis.

2.3 *Arsenic uptake tests*

Arsenic uptake tests were performed with the two most abundant inorganic forms of arsenic: arsenate and arsenite. Stock solutions of sodium arsenate dibasic heptahydrate ($Na_2HAsO_4.7H_2O$) (Sigma-Aldrich, Gillingham, UK) and sodium (meta)arsenite ($NaAsO_2$) (Sigma-Aldrich, Gillingham, UK) were made to 1 g/L total arsenic.

All analyses were performed in HDPE bottles using an arsenic solution (10 mL), with diatomite, 0.1 g (10 g/L). Apart from matrix testing, all arsenic solutions were spiked into distilled water, and apart from temperature controlled assessment, samples were contained in a 25 °C controlled water bath throughout testing. All samples (unless otherwise stated) were allowed to interact for 24 hours. For pH studies, the pH of the arsenate and arsenite solutions were adjusted using concentrated NaOH solution and concentrated HNO_3 (Sigma-Aldrich, Gillingham, UK). For water matrices analysis, the major anion content was measured of different waters (spring, tap and distilled water) using an ICS-5000 Dionex DC (Detector/Chromatography) equipped with suppressed conductivity. The column used was an IonPac® AS18 Analytical (2 mm width × 25 mm length), a flow rate of 0.25 mL/min coupled with a 3.3×10^{-2} mol KOH eluent.

Samples were analyzed by i) Inductively Coupled Plasma—Mass Spectrometry (ICP-MS) using an Agilent 7700 series with an ASX-500 series auto sampler (Agilent, Wokingham, United Kingdom), ii) spectrophotometrically at a wavelength of 650 nm using a UV-Visible spectrophotometer (Cecyl 8000). This method involves the formation of molybdoarsenic acid which is reduced to a colored complex, molybdenum blue and iii) Atomic

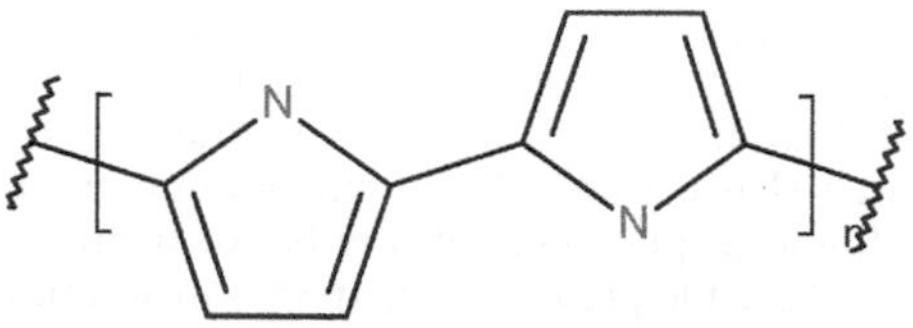

Figure 1. The structure of polypyrrole.

Absorption Spectroscopy (AAS) using the AA 6300 Shimadzu instrument.

2.4 *Materials based on calix[4]pyrroles and polypyrroles*

These materials were synthesized and characterized as described elsewhere (Abbas 2007a,b; Kałędkowski & Trochimczuk, 2006; Saville, 2005). Polypyrrole (Figure 1) was synthesized by the method described by Danil de Namor & Abou Hamdan (unpubl.) and characterized by several analytical techniques.

2.5 *Extraction experiments*

Typically, a known amount of the appropriate material (0.1 g) was equilibrated with an aqueous solution of arsenate (10 mL) at 25 °C. The uptake isotherms were studied by varying the concentration of arsenate anions (1.00×10^{-5} to 1×10^{-2} mol L^{-1}) at a fixed dose of the material, and the effect of the solution pH on the uptake of this anion by these materials was observed in the 2–12 pH range. After the equilibrium was achieved, aliquots were taken and carefully analyzed. The kinetics of the extraction process was investigated using the optimum conditions for extraction, analyzing the samples at different periods of time.

3 RESULTS & DISCUSSION

3.1 *Elemental composition of diatomites*

Following the characterization of (raw) diatomite, and the two Fe modified diatomite (one with pre-calcination treatment and the other without), the elemental composition is shown in Table 1.

The elemental composition shows a loss of carbon between natural and calcined diatomite. This could be largely due to any volatile organic compounds trapped in the porous diatomite material desorbing and evaporating from the surface, as was also seen by Ibrahim & Selim (2012). It should be noted that aluminum percentage should be taken with caution. The medium used to analyze the diatomite was aluminum based, and a large portion would be due to a signal from this. Hydrogen cannot be quantified by EDS, so analysis of the silanol chemistry is not so obvious. A more effective method for this would have been through Fourier Transform Infrared Spectroscopy (FTIR), which could show the presence of Si-O bonds and O-H bonds in the sample. The dehydration of silanols on the silica surface has been shown by Yuan (Yuan *et al.*, 2004). It shows how the silanols are hydrogen bonded initially to water, and upon heat treatment loses water to leave siloxane on the silica surface. Subsequently, loss of oxygen would be expected which would be reflected in the EDS analysis. However, the EDS showed a gain in oxygen, whilst the silicon weight percentage remained unchanged. This could be due to a low concentration of silanols on the surface in the raw diatomite; again, FTIR would be required to confirm this.

Table 1. Elemental compositions in weight (%) of diatomite forms.

Element	Natural diatomite Weight (%)	Calcined diatomite Weight (%)	Fe-modified diatomite Weight (%)
Carbon	3.6	1.7	5.2
Oxygen	53.3	55.6	48.3
Aluminum	2.6	1.7	2.1
Silicon	39.0	39.1	28.9
Iron	0.8	0.9	13.7
Other	0.9	1.0	1.8

3.2 *Arsenic speciation uptake by raw and modified diatomite*

Figure 2 shows the comparison of the uptake of arsenite by raw (natural) and two modified diatomites, one with pre-calcined treatment and one without. It can be visualized that modification of diatomite dramatically increases as uptake by the material. The calcined Fe-diatomite reaches maximum extraction within 2 hours.

This was done to a concentration of solution being 2 mg/g arsenic to diatomite, to study the trends between the differing diatomite. Figure 3 gives the sorption of arsenite and arsenate by diatomite (10 mg/g) respectively. Figure 3 shows that diatomite removes 9.87 mg/g arsenate and 9.90 mg/g arsenite.

It is concluded that about 10 mg/g arsenic of both species is taken up by the modified diatomite, which was over 50 times higher than for raw diatomite. It was also confirmed that calcination of diatomite prior modification can increase the kinetics due to an increase in permeability of the diatomite surface.

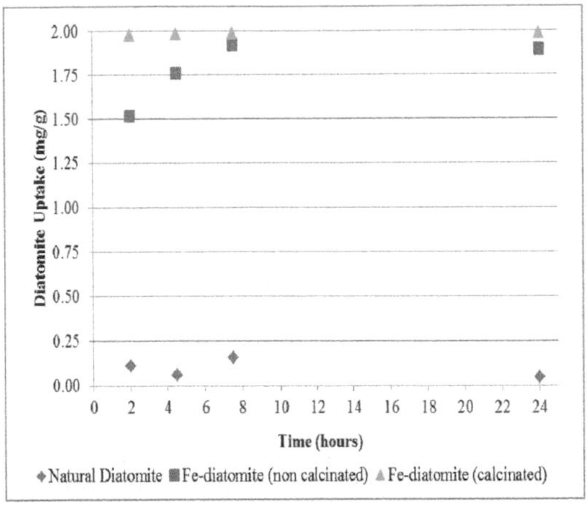

Figure 2. Comparison of diatomite arsenite uptake.

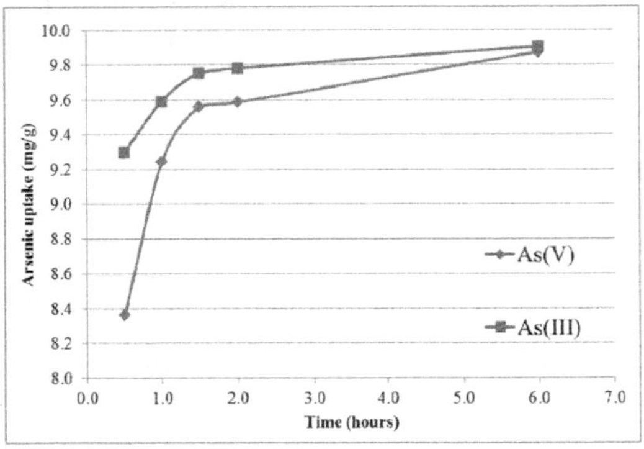

Figure 3. Arsenate and arsenite adsorption by diatomite.

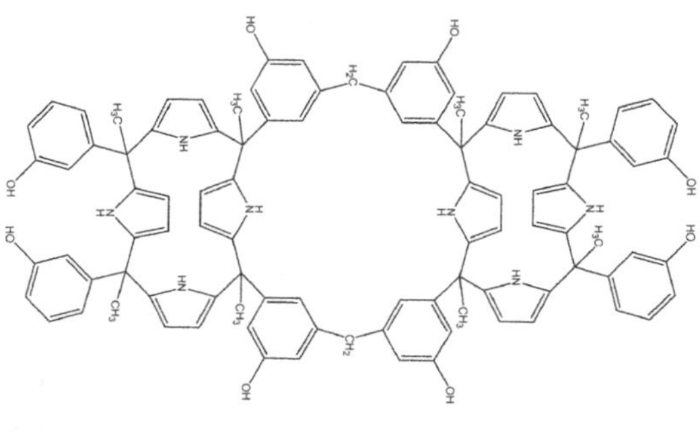

Figure 4. The proposed structure of the calix[4]pyrrole dimer.

3.3 *Arsenic uptake by calix[4]pyrrole dimer and polypyrrole*

A suggested structure for the calix[4]pyrrole dimer is shown in Figure 4. The optimum conditions for the extraction of arsenate by the dimer of calix[4] pyrrole, which showed maximum extraction (90%) for a concentration of 1×10^{-4} mol L^{-1}, was found to be 2 grams of resin per liter of solution. As far as the pH is concerned, the maximum extraction was found at pH 4.66, halfway between pK_{a1} and pK_{a2}, where the predominant species in solution is the monovalent anion $H_2PO_4^-$. Experimental work is in progress to determine the capacity of the dimer for removal of arsenate and arsenite from water. Regarding polypyrrole, experimental results so far obtained indicate that, among the materials considered, its capacity for the removal of arsenate is lower than that for the dimer.

4 CONCLUSIONS

It is concluded that approximately 10 mg/g arsenic of both species is taken up by the modified diatomite, which was over 50 times higher than for raw diatomite. It was also confirmed that calcination of diatomite prior to modification can increase the kinetics due to an increase in permeability of the diatomite surface. As far as the calixpyrrole dimer is concerned, the optimum conditions for the removal of arsenate from water have been established. The same applies to polypyrrole. For the latter, studies have been extended to involve the addition of surfactants.

REFERENCES

Abbas, I. 2007. Anion and cation binding studies of calix[4]pyrrole receptors. Detailed thermodynamic studies and environmental applications. *PhD thesis, University of Surrey*.

Bissen, M. & Frimmel, F.H. 2003. *Acta Hydrochimica et Hydrobiologica* 31: 97.

Bothe, J.V. & Brown, P.W. 1993. *Environmental Science & Technology* 33: 3806.

Chen, A.S.C. *et al., EPA 68-C-00-185*: USA.

Danil de Namor, A.F. & Abbas, I. 2007. *Journal of Physical Chemistry B* 111: 5803.

Danil de Namor, A.F. & Abbas, I. 2010. *Analytical Methods* 2: 63.

Danil de Namor, A.F. & Abou Hamdan, W., unpublished work.

Danil de Namor, A.F. & Khalife, R. 2008. *Journal of Physical Chemistry B* 112: 15766.

Danil de Namor, A.F. 2007. *Journal of Thermal Analysis and Calorimetry* 87: 7.

Danil de Namor, A.F. 2007. *Water Science Technology: Water Supply* 7: 33.

Danil de Namor, A.F. *et al.*, 2006. *Journal of Physical Chemistry B* 110: 12653.

Danil de Namor, A.F. *et al.*, 2007. *Journal of Physical Chemistry B* 111: 12177.

Danil de Namor, A.F. et al., 2012. *Journal of Hazardous Materials* 14: 241–242.

Demarco, M.J. *et al.*, 2003. *Water Research* 37: 164.

Giles, D.E. *et al.*, 2011. *Journal of Environmental Management* 92: 3011.
Gregor, J. 2001. *Water Research* 26: 1659.
Harisha, R.S. *et al.*, 2010. *Desalination* 252: 75.
Ibrahim, S.S. & Selim, A.Q. 2012. *Physicochemical Problems of Mineral Processing* 48: 413.
Iwamoto, M. *et al.*, 2002. *Chemistry Letters* 31: 814.
Kałędkowski, A. & Trochimczuk, A.W. 2006. *Reactive & Functional Polymers* 66: 740.
Lin, T.F. & Wu, J.K. 2001. *Water Research* 35: 2049.
Mahata, J. *et al.*, 2003. *Mutation Research/Genetic Toxicology and Environmental Mutagenesis* 534: 133.
Moon, D.H. *et al.*, 2004. *Science of the Total Environment* 330: 171.
Payne, K.B. & Fattah, A. 2005. *Journal of Environmental Science and Health, Part A: Toxic/Hazardous Substances and Environmental Engineering* 40: 723.
Roberts, L.C. *et al.*, 2004. *Environmental Science and Technology* 38: 307.
Saville, P. 2005. *Polypyrrole Formation and Use, Defence Research and Development*: Canada.
Seko, N. 2004. *Reactive and Functional Polymers* 59: 235.
Tokunaga, S. *et al.*, 1999. *Water Environment Research* 71: 299.
Wilkie, J.A. & Hering, J.G. 1996. *Colloids and Surfaces: Physicochemical and Engineering Aspects* 107: 97.
Yuan, P. *et al.*, 2004. *Applied Surface Science* 227: 30.

One Century of the Discovery of Arsenicosis in Latin America (1914–2014) – Litter, Nicolli, Meichtry, Quici, Bundschuh, Bhattacharya & Naidu (Eds)
© 2014 Taylor & Francis Group, London, ISBN 978-1-138-00141-1

Arsenic removal from groundwater using magnetic nanoparticles

M.T. Alarcón-Herrera, M.Z. López Paraguay & I.R. Martín Domínguez
Centro de Investigación en Materiales Avanzados (CIMAV), Chihuahua, Chihuahua, México

ABSTRACT: The objective of this study was to investigate the technical feasibility of using nanoparticles obtained from oxidation of metallic wool, to remove arsenite (As^{3+}) from water. The obtained magnetic, microporous-nanomaterial had a superficial area of 88.30 m^2/g and was composed mainly of lepidocrocite (γ-FeO(OH)). As(III) was removed from water until it reached undetectable levels (<5 μg L^{-1}) in a retention time of seven minutes by using 0.55 g L^{-1} of γ-FeO(OH) under operating conditions (pH 7.8 ± 0.2 and temperature 23 ± 0.3°C). The removal capacity of the sorbent is found to be 2.9 10^{-2} mmol As g^{-1}. The magnetic field intensity for the magnetic filtration was 0.24 Teslas. Lepidocrocite nanoparticles sorption, in conjunction with magnetic filtration constitutes an innovative and highly efficient technique for the removal of arsenic from water.

1 INTRODUCTION

One of the biggest problems that affect the quality of drinking water is the presence of arsenic in drinking water sources. In our time, to develop new and economic alternatives for arsenic (As) removal from water is a relevant issue. In Mexico, the NOM-127-SSA1–1994 has established a maximum concentration of arsenic in drinking water of 25 μg L^{-1} (2005). To approach the problem, there are different technologies such as the conventional water treatments (oxidation, ion exchange, coagulation-precipitation, reverse osmosis, adsorption) and the so-called emerging technologies (phytoremediation, electrocoagulation, use of nanomaterials as adsorbents) (Prasenjit *et al.*, 2012). All of these have advantages and limitations; therefore, the search for new technologies is required. Recently, there has been a growing interest in the development and application of nanomaterials for water treatment. Considering that micro- and macro-iron oxide particles are effective adsorbent media and economically affordable for arsenic removal (Maiti *et al.*, 2007), the present research investigates the technical feasibility of using iron nanoparticles produced from oxidation of metallic wool for arsenic removal from water, taking advantage of the magnetic properties of the material for their separation from the aqueous medium by using a magnetic field.

2 METHODS/EXPERIMENTAL

Arsenic (III) stock solutions were prepared from $NaAsO_2$ reagent grade (Fisher Scientific Laboratories). In order to work under conditions closer to reality, solutions were prepared with groundwater containing arsenic. The arsenic concentrations of the prepared working solutions were 113, 313, 513, 613, 713 and 913 μg L^{-1} respectively.

2.1 *Nanomaterial preparation*

In the present research, steel wool was used to produce nanoparticles with high iron content and good magnetic properties. The material was subjected to an extensive washing process and then it was periodically moistened with water to induce its oxidation. The oxidized particles were collected and screened to remove coarse material (# 400 mesh). Following this, a suspension was prepared with screened material using deionized water. This slurry was allowed to precipitate. Sedimentation of particles was accelerated by the use of permanent magnet plates. Finally, the settled material was dried and ready for use in the experimentation process.

2.2 *Experimental procedure*

The sorption process and magnetic filtration were performed at room temperature (23 ± 0.3°C) and the natural pH of the water (pH 7.8 ± 0.2). Different amounts of nanoparticles (0.2 to 1.2 g L^{-1}) were put in contact with the prepared As(III) solutions for 0.5 to 10 minutes. The solution containing As and nanoparticles was then fed through the magnetic filtration process. The flow rate was controlled at 1 mL min^{-1}. The filtered samples were prepared for total As quantification.

2.3 *Magnetic filtration*

The separation of the nanomaterial with arsenic from water was performed through magnetic filtration using an electromagnet at different magnetic field strengths. The electromagnet was calibrated with a Walker Scientific MG-3DP Gaussmeter. The voltage ranged between 0 and 25 volts and the current varied from 0 to 0.16 A. The column used for magnetic filtration was filled with stainless steel fine wool (magnetic grade) purchased from SoBo Distribution Inc. The glass column was 3.6 cm high and had a diameter of 6 mm. The mass of fine wool was 0.7 g.

2.4 *Characterization of nanoparticles*

The surface area of the nanoparticles was measured by the Brunauer-Emmett-Teller (BET) method using a Quantachrome Nova Corporation (Series 1000). The elemental analysis and the particle size were determined using a Transmission Electron Microscope (TEM) JEOL, JEM-2200FS model, with STEM+Cs corrector. The resolution of the device was 0.187 nm. The magnetic properties of nanoparticles and that of the fine stainless steel wool (hysteresis curve, coercive field, saturation magnetization) were determined with a Vibrating Sample Magnetometer (VSM) LDJ brand, model 9600.

2.5 *Analytical determinations*

Total As concentrations were determined using a GBC Atomic Absorption Spectrophotometer (AAS) (model Avanta Sigma) coupled to a Hydride Generator (HG). This equipment was calibrated with reference standard solutions (High Purity Control Standards) traceable by The U.S. National Institute of Standards and Technology (NIST). The average percentage of analyte recovery was 99.8 ± 1.8%. The lower limit of detection of the equipment was 5 μg L^{-1}. pH, temperature and conductivity of the solutions were measured by Orion pH-meter, model 1260.

3 RESULTS AND DISCUSSION

3.1 *Characterization of the nanoparticles*

The physicochemical characteristics of the nanoparticles are shown in Table 1. The pore size distribution obtained by the Dubinin Astakhov and BJH (Barrett-Joyner-Halenda) method suggests a mainly microporous material. X-ray diffraction analysis shows that the sorbent is composed mainly of ferric oxyhydroxide known as lepidocrocite (γ-FeO(OH)) and Fe_2O_3.

Table 1. Characteristics of the adsorbent.

Characteristic	Adsorbent
BET surface area (m^2 g^{-1})	88.30
Average particle size range (nm)	5–150
Density (g cm^{-3}),	3.96 (aprox.)
Porosity	Microporous (mainly)
Pore size (nm)	<2.0
Composition	γ-FeO(OH), Fe_2O_3, $MnMoO_4.H_2O$

The inorganic composition of the sorbent was Fe (50.7%), O (42.5%), Na (4.3%), Cl (0.9%) and Mn (1.6%).

3.2 *Effect of the sorbent on As(III) sorption*

The As removal efficiency in relation to time and dosage of nanoparticles is presented in Figure 1. At dosages higher than 0.7 g L^{-1}, practically a total As removal was achieved in less than four minutes. For dosages from 0.4 to 0.6 g L^{-1}, the removal rate was moderate. This may be due to the coverage of the sorbent surface by the arsenic molecules until the equilibrium state has been reached. At the dosage of 0.4 g L^{-1} and 4 minutes of contact time, the treated water met the water quality required by the international standards (<10 μg As L^{-1}).

Table 2 shows the removal efficiencies and conditions reported by other authors and those obtained in this study. The amounts of sorbent used are similar to those used in this study; however, their contact times make the process technically unfeasible for a practical water treatment process.

3.3 *Effect of pH on As(III) sorption*

To determine the optimum pH for As(III) removal, the effects of pH were evaluated. For pH values between 6 and 7, the sorption process was less effective, while removal efficiencies higher than 93% were obtained between pH 8 and 9. According to the Eh-pH diagram, the predominant arsenic species at pH 7 is H_3 AsO_3; this neutral form would be another possible reason for the decrease in removal efficiency at this pH. A study of arsenic adsorption on clay minerals by Henke (2009) concluded that the As(V) sorption decreased at pH 7.5, while the As(III) sorption increased. That difference could be attributed to the dissociation of H_3 AsO_3 and H_3 AsO_4. This phenomenon could also have happened in the current study.

3.4 *Adsorption kinetic study*

The kinetic experiments were carried out with 0.5 g L^{-1} of nanoparticles, an initial concentration

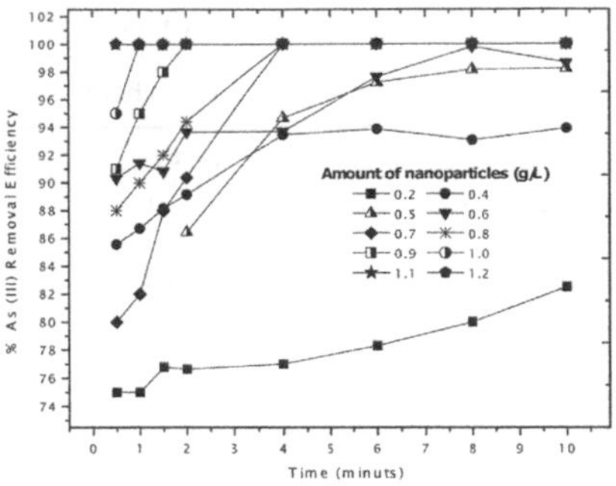

Figure 1. As(III) removal for different amounts of nanoparticles (C_o = 113 μg L^{-1}, pH 7.8 ± 0.2, T = 23 ± 3 °C, maximum time = 10 min).

Table 2. Comparison of the results obtained with those reported in the literature.

Author	Nano-particles	Dose ($g L^{-1}$)	Time (hour)	pH	Efficiency
Mayo *et al.* (2007)	Magnetite (Fe_3O_4)	0.5	24	8	99.2% As(III) 98.4% As(V)
Rahman *et al.* (2010)	Magnetite-maghemite	0.4	24	2	96% As(III)
Lin *et al.* (2012)	Magnetic Fe_2O_3	0.8	0.5	6	61.2% As(III)
Song *et al.* (2013)	Iron oxide	1.0	24	6	2.9 mg g^{-1} As(III)
Present research	FeO(OH)	0.55	0.12	7.8	99.8% As(III)

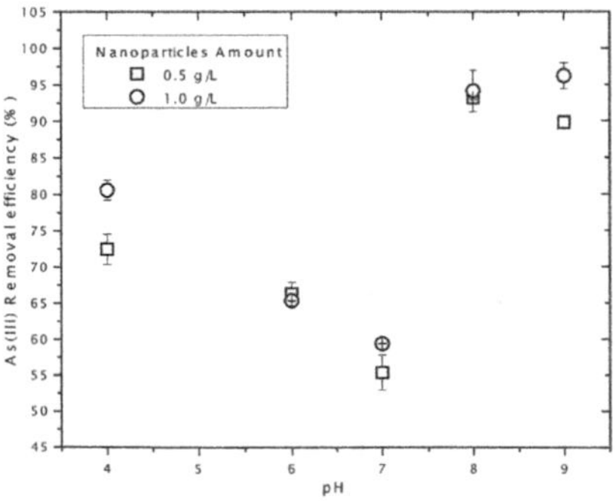

Figure 2. pH effect on the As(III) removal efficiency (C_o: 113 μg L^{-1}; pH 7.8 ± 0.2; [adsorbent]: 0.5 and 1.0 g L^{-1}, Time: 10 min, T: 23 ± 3°C).

of 713 μg As L^{-1} and 30 minutes of contact time. The experimental results were analyzed by four kinetic models (Maji *et al.*, 2008, Guo *et al.*, 2007, Yuh Shan Ho., 2006, Jiang *et al.*, 2013).

Table 3. Parameters of the analyzed kinetic models.

Kinetic model	Equation	Constant	Value
First order	$\ln C_t = \ln C_o - K.t$	K (1 h^{-1})	0.34
		R^2	0.78
Pseudo first order	$\log(qe - qt) = \log q_e - K.t/2.303$	K (1 h^{-1})	0.64
		R^2	0.76
Second order	$1/C_t = K.t + 1/C_o$	K (L μg^{-1} h^{-1})	0.003
		R^2	0.79
Pseudo second order	$t/q_t = 1/(K.q_e^2) + t/q_e$	K (g μg^{-1} h^{-1})	0.50
		R^2	0.99

The kinetic data fitted well with the pseudo-second order kinetic model. Similar results were reported by Repo et al., 2012, where the kinetics of As(III) and As(V) adsorption by synthetic lepidocrocite was best described also by the pseudo-second order model, and the initial sorption rate was 2.2 mg g^{-1} min^{-1}. The coefficients and constants of the four applied models are presented in Table 3.

3.5 *Adsorption isotherms*

The adsorption parameters were determined applying the isotherms of Langmuir, Freundlich, Dubinin Radushkevich and BET. The constants of the isotherm models applied are presented in Table 4.

According to Langmuir, the sorption capacity of the nanomaterial was 2.1 mg g^{-1} (Q_o), which should represent the removal capacity of a monolayer. The Dubinnin model suggests a 2.12 mg/g as the maximum sorption capacity of the material, which is near to the amount (2.18 mg g^{-1}) calculated with the Langmuir model. On the other hand, the Freundlich model suggests that there is interaction between the sorbed molecules (physisorption). This may mean that there are existing active sites with heterogeneous sorption energies. The value of the Freundlich constant (n > 1) indicates also that arsenite ions are favorably sorbed under the presented experimental conditions (Loukidou *et al.*, 2004). The nanomaterial was able to adsorb around 2.2 mg g^{-1}.

3.6 *Thermodynamic adsorption parameters*

The free energy of adsorption (E) represents the change of energy when an ion mole is transferred from an infinite distance/point in the aqueous solution to the surface of the sorbent. Majia *et al.* (2007) classified the adsorption process even further. According to them, when the value of E > 8 kJ mol^{-1}, it is termed as chemisorption and

Table 4. Langmuir, Freundlich, BET and Dubinnin adsorption parameters.

Isotherm	Constants	Units	Value
Langmuir	K	L mg^{-1}	2.18
	Q_o	mg g^{-1}	9.94
	R^2	–	0.99
Freundlich	Kf	(mg g^{-1}) (L $mg^{-1})^1/n$	62.18
	n	adimensional	1.61
	R^2	–	0.99
BET	B	adimensional	–1.86
	Q_o	mg g^{-1}	1.69
	R^2	–	0.89
Dubinin Radushkevich	K	mol^2 kJ^{-2}	0.019
	q_m	mg g^{-1}	2.12
	R^2	–	0.99

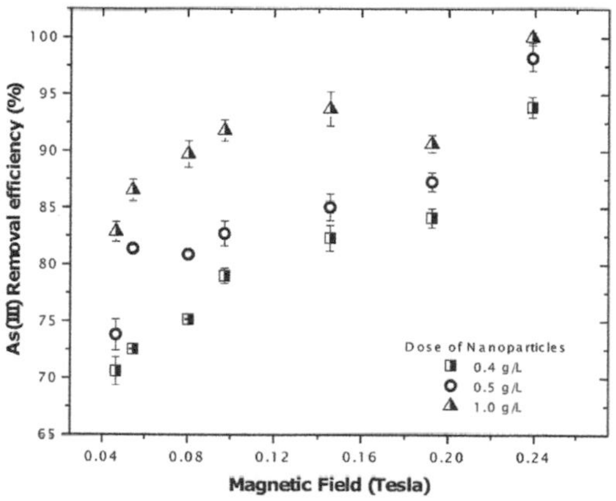

Figure 3. As(III) removal efficiency at different magnetic field (Co: 113 μg L^{-1}, pH 7.8 ± 0.2, agitation time: 10 min, magnetic filtration time: 40 min, T: 23 ± 3°C).

for $E < 8$ kJ mol^{-1} it is classified as physisorption. On the other hand, $-\Delta G^o$ (kJ mol^{-1}) represents part of the total energy of the system that is available to do useful work to displace the sorbate from the adsorbent surface. The value of E (-5.12 kJ mol^{-1}) was estimated by the relationship $E = -1/\sqrt{2K}$, where K is the constant of Dubinin. According to the literature, the sorption mechanism is a physisorption. The negative sign indicates that the process is exothermic. The energy released by transferring a mol of adsorbate until the solid surface was 5.12 kJ.

3.7 *Influence of the magnetic field*

At higher magnetic field intensity, the efficiency of As(III) removal was also higher (Figure 3). It is possible that at the strongest field there is an increased retention of nanomaterial inside the magnetic column. Under the conditions carried out in this study, it was possible to obtain a treated water that met the quality required by the international standards (As <10 μg L^{-1}) at 0.24 T. Also with a higher amount of nanoparticles, the removal efficiency was higher. Efficiencies above 93% were observed for 0.24 T, and doses of 0.4, 0.5 and 1.0 g L^{-1}, respectively.

4 CONCLUSIONS

Arsenic was adsorbed onto iron nanoparticles and separated efficiently from water by applying an external magnetic field. The surface area of the microporous material, identified as lepidocrocite (γ-FeO (OH)), was 88.3 m^2 g^{-1}. As(III) removal was complete using 0.55 g L^{-1} of γ-FeO(OH) in seven minutes. The kinetics of the sorption process fitted well with a pseudo-second order model. The maximum As removal capacity of the material was 2.9 10^{-2} mmol g^{-1}. The multilayer adsorption is consistent with the conventional Freundlich model and corresponds to a physisorption process. Under the experimental conditions, at 0.24 Tesla, treated water met the quality required by the international standards (As < 10 μg L^{-1}).

REFERENCES

Guo, Z., Wang, L.F., Gao, Z. & Zhang, W.W. 2007. Equilibrium and kinetic studies on the adsorption of VB12 onto CMK-3. *Chinese Chemical Letters* 18: 233–236.

Henke, K.R. 2009. Arsenic: Environmental chemistry, health, threats and waste treatment. 1st ed. John Wiley and Sons Ltda. Great Britain, England.

Ho, Y.-S. 2006. Review of second-order models for adsorption systems. *Journal of Hazardous Materials* B136: 681–689.

Jiang, W., Pelaez. M., Dionysiou, D.D., Entezari, M.H., Tsoutsou, D. & O'Shea, K. 2013. Chromium (VI) removal by maghemite nanoparticles. *Chemical Engineering Journal* 222: 527–533.

Loukidou, M.X., Zouboulis, A.I., Karapantsios, T.D. & Matis, K.A. 2004. Equilibrium and kinetic modeling of chromium(VI) biosorption by Aeromonas caviae. Colloids and Surfaces A: *Physicochem. Eng. Aspects* 242: 93–104

Maiti, A., Das Gupta, S., Basu, K.J. & De, S. 2007. Adsorption of arsenite using natural laterite as adsorbent. *Separation and Purification Technology* 55: 350–359.

Maji, S.K., Pal, A. & Pal, T. 2008. Arsenic removal from real-life groundwater by adsorption on laterite soil. *Journal of Hazardous Materials* 151: 811–820.

Majia S.K., Pal, A., Pal, T. & Adak, A. 2007. Modeling and fixed bed column adsorption of As(V) on laterite soil. *Journal of Environmental Science and Health, Part A* 42(11): 1585–1593.

Mayo, J.T., Yavuz, C., Yean, S., Cong, L., Shipley, H., Yu, W., Falkner, J., Kan, A., Tomson, M. & Colvin, V.L. 2007. The effect of nanocrystalline magnetite size on arsenic removal. *Science and Technology of Advanced Materials* 8: 71–75.

Mondal, P., Mohanty, B. & Majumder, C.B. 2012. Removal of Arsenic from Simulated Groundwater Using GAC-Ca in Batch Reactor: Kinetics and Equilibrium Studies. *Clean—Soil, Air, Water* 40(5): 506–514.

Rahman, C.S. & Yanful, E.K. 2010. Arsenic and chromium removal by mixed magnetite emaghemite nanoparticles and the effect of phosphate on removal. *Journal of Environmental Management* 91: 2238–2247.

Repo, E., Makinen, M., Rengaraj, S., Natarajan, G., Bhatnagar, A. & Sillanpa, M. 2012. Lepidocrocite and its heat-treated forms as effective arsenic adsorbents in aqueous medium. *Chemical Engineering Journal* 180: 159–169.

Song, K., Kim, W., Suh, C.-Y., Shin, D., Ko, K.-S. & Ha, K. 2013. Magnetic iron oxide nanoparticles prepared by electrical wire explosion for arsenic removal. *Powder Technology* 246: 572–574.

One Century of the Discovery of Arsenicosis in Latin America (1914–2014) – Litter, Nicolli, Meichtry, Quici, Bundschuh, Bhattacharya & Naidu (Eds)
© 2014 Taylor & Francis Group, London, ISBN 978-1-138-00141-1

Robust and reusable hybrid nanosorbent to mitigate arsenic crisis: From laboratory to masses in the field

L. Cumbal
Centro de Nanociencia y Nanotecnologia, Universidad de las Fuerzas Armadas, Quito, Ecuador

M. German & A. SenGupta
Department of Civil and Environmental Engineering, Lehigh University, Bethlehem, USA

H. Shengheng
Institute of Technology of Cambodia, Phnom Penh, Cambodia

ABSTRACT: Although there have been many research projects, awards and publications, appropriate treatment technology has not been matched to ground level realities and water solutions have not scaled to reach millions of people. However, for thousands of people from Nepal to India to Cambodia, Hybrid Anion Exchange (HAIX) resins have provided arsenic-safe water for up to nine years. Synthesis of HAIX resins has been commercialized and they are now available globally. Robust, reusable and arsenic-selective, HAIX has been in operation in rural communities over numerous cycles of exhaustion-regeneration. Removed arsenic is safely stored in a scientifically and environmentally appropriate manner to prevent future hazards to animals or people. Recent installations have shown the profitability of HAIX-based arsenic treatment, with capital payback periods of only two years in ideal locations. With an appropriate implementation model, HAIX-based treatment can rapidly scale and provide arsenic-safe water to at-risk populations.

1 INTRODUCTION

Arsenic groundwater contamination across the Gangetic delta extends over a large area of Bangladesh and India and is one of the worst calamities of the world in recent times. Over 100 million people are at-risk by drinking water well above the WHO recommended limit of 10 µg As/L (Chatterjee *et al.*, 1995; Bearak, 1998). The crisis also affects several countries in South Asia, including Nepal, Burma, Vietnam, Cambodia, Laos, etc (Berg *et al.*, 2007, Stanger *et al.*, 2005; Christen, 2001). As in drinking water also affects Latin America countries such as Mexico, Argentina, Peru, Colombia, Uruguay, Guatemala, Honduras, etc. (Bundschuh *et al.*, 2012). As in groundwater has been a focus for public health scientists and engineers over the last twenty years (Bagla & Kaiser 1996, Sancha & Castro de Esparza, 2000, Ng *et al.*, 2003, Ravenscroft *et al.*, 2005, Bhattacharya *et al.*, 2007, Levy *et al.*, 2012). Nevertheless, in Bangladesh, 20% of deaths were related to As consumption during the previous decade (Argos *et al.*, 2010).

It is common for groundwater to be potable and palatable, with only minimal treatment beyond trace contaminant removal. Thus, a highly specific adsorbent is needed. A significant proportion of As adsorption technologies, if not all, use innocuous Hydrated Metal Oxides (HMOs) with high As affinity, the metals being iron, aluminum, titanium and zirconium (Bang *et al.*, 2005, Driehaus *et al.*, 1998, DeMarco *et al.*, 2003, Dutta *et al.*, 2004, Suzuki *et al.*, 2000).

Aluminum oxides, such as activated alumina or AA (Al_2O_3), have high capacity at certain conditions for fluoride and arsenate or As(V). However, AA has poor arsenite or As(III) capacity (Cumbal, 2004). Hydrated Ferric Oxides (HFOs) have high affinity for both As(III) and As(V) because of the functional surface groups. HFOs form monodentate or bidentate inner-sphere complexes where Fe(III), a transition metal, serves as the electron-pair acceptor or Lewis acid (Dzombak and Morel, 1990). However, landfill disposal of iron adsorbents is inadvisable because, in a landfill, Fe(III) can reduce to Fe(II), become aqueous and leach As, although the EPA TCLP indicates landfill disposal is safe (Ghosh *et al.*, 2004).

Maintaining the morphology of the desired metal oxide nanoparticles in a continuous process, with low head losses, is optimal for efficient long-term adsorption of trace contaminants.

When metal oxide nanoparticles are impregnated in quaternary ammonium-functionalized polymer beads, the non-diffusible, positive functional groups enhances the local concentration of anions, e.g., As(V), near the hydrated metal oxides. In this Hybrid Anion Exchange (HAIX) resin, there is a synergy of intraparticle diffusion, contaminant selectivity and mechanical strength (Donnan, 1911, 1995, Cumbal & SenGupta, 2005). The specific objective of this study is to provide evidence of community-scale HAIX installations as effective long-term solutions for As contaminated groundwater.

2 METHODS

2.1 *HIAX synthesis*

Impregnation or dispersion of hydrated Fe(III) oxide or HFO nanoparticles within an anion exchanger posed major challenges because of the associated difficulty in introducing positively charged ferric ions (Fe^{3+}) within a positively charged anion exchange resin with quaternary ammonium functional groups. Through several iterations, a reproducible, operationally simple protocol was developed and described in detail in United States Patent US 7291578 (SenGupta & Cumbal 2007, Sarkar *et al.*, 2007). HAIX is now commercially available globally.

2.2 *Field work*

Over the past 15 years, about 200 Sustainable As Removal Systems For Affected Communities (SARSACs) have been installed and they continue to operate in rural communities of West Bengal, adjacent to Bangladesh, Cambodia, Laos and Nepal (Sarkar *et al.*, 2010, 2012). Over time, these efforts have been assisted and coordinated by several organizations including Bengal Engineering Science University (BESU), Institute of Technology in Cambodia (ITC), Technology with a Human Face (THF), and the Tagore-SenGupta Foundation.

3 RESULTS AND DISCUSSION

3.1 *Long-term performance in West Bengal, India*

Amorphous HFOs of size 20–100 nm were precipitated on the gel-phase of macroporous anion exchange resins (HIAX) (DeMarco *et al.*, 2003, Cumbal *et al.*, 2003). HAIX loaded with high surface area HFOs has similar capacity to Granular Ferric Hydroxide (GFH) for both As(III) and As(V), even though the iron content of HAIX is much less (Sarkar *et al.*, 2007). The mechanical integrity of polymer resins allow for several cycles of use and regeneration with minimal changes in metal oxide morphology and content; GFH is a single-use adsorbent (Sarkar *et al.*, 2008).

In early 2004, the first public use of HAIX was in a community-scale installation for a rural community in Ashok Nagar, West Bengal, India. In the design, groundwater is first pumped by hand to the top of the column. Upon entering the column, water is aerated, Fe(II) is oxidized and Fe(III) is filtered by the adsorbent bed. Iron removal, caused primarily by oxidation of Fe(II) into Fe(III) by oxygen, is a simple, but a key step for aesthetic appeal and consumer confidence in water purity. Incipient precipitation in the top portion of the column, leading to the formation of HFO nanoparticles, also renders removal of both dissolved As(III) and As(V) (Sarkar *et al.*, 2007). Remaining As is then removed by HAIX. The transparency of water is greatly improved following iron removal.

Many rural communities in West Bengal are electrified for limited daily use. While electricity is available, As contaminated groundwater can be pumped to an elevated water storage tank (Figure 1). At minimal elevation, water can be fed by gravity through aeration and adsorption treatment. Treated water is available to people for on-site pick-up or rickshaw delivery. Today, communities in Ashok Nagar continue to purchase As-safe water from HAIX treatment nine years after installation. The most recent water test indicates the need for regeneration of Column No. 1 in Nabarun Sangha is upcoming an annual requirement (Figure 2). From July 2004 to March 2013, Column No. 1 has produced over 6.3 million L of safe water. Consistent operation in Ashok Nagar

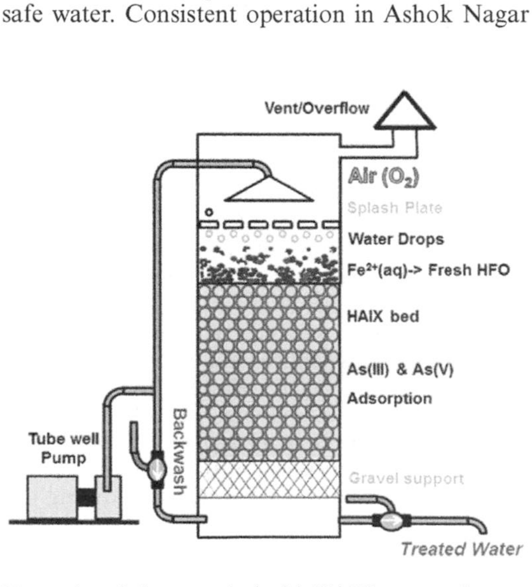

Figure 1. Column packed with HAIX to treat As contaminated water in West Bengal, India.

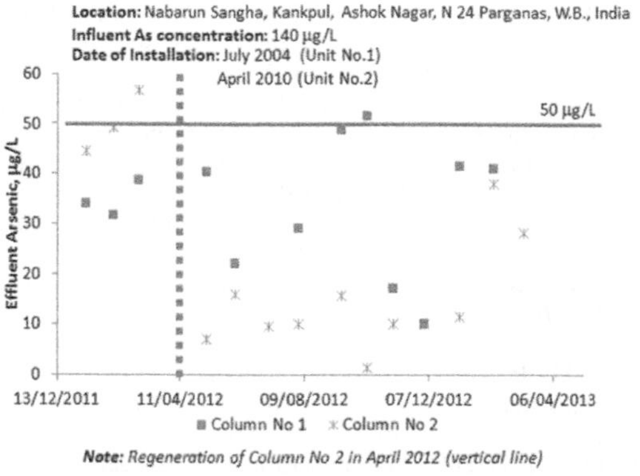

Figure 2. Arsenic profile in two columns loaded with HAIX vs. time.

shows the long-term durability, and high As capacity of HAIX treatment has been proven in both lab-scale (Sarkar *et al.*, 2007) and community-scale systems.

3.2 *On-going efforts in Cambodia*

Cambodia has very high As levels (600–1600 ppb) with the majority of impacted people near the Phnom Penh province (Buschman *et al.*, 2007, Berg *et al.*, 2007). Currently, there are tens of HAIX As treatment systems in operation from Nepal to Cambodia that produce safe water. In the Sambour District of Kratie Province, Cambodia, local women have operated SARSACs using HAIX since October 2009. Over the years, As-safe water has been provided throughout the community at an affordable rate of 500 riel (U$S 0.12) per 20 L or at 1000 riel (U$S 0.25) per 20 L for door-to-door delivery of water. In Cambodia, rain water is preferred when available during the wet season, and a SARSAC may be underutilized for several months, but it still operates as designed upon restart. Such reliability is an important, non-quantitative operational consideration when systems are remote, technical expertise is distant and maintenance can be costly, e.g., RO membrane replacement.

3.3 *Economic sustainability: From water crisis to water business*

In Ashok Nagar, there are over 1300 families who purchase safe water from three HAIX based As treatment systems. The families pay 20–30 Rs (U$S 0.40-U$S0.60) per month for daily pick-up of 20 L potable water; people can pay extra for cycle rickshaw delivery. Both communities in Nabarun Sangha and Binimaypara have had HAIX installations selling safe water since 2005. At the new Sakthi Sadhana Community Club installation, in Ashok Nagar (2010-present), there are over 600 families currently purchasing water at 30 Rs per month. Revenue was great enough for the club to purchase a truck for the delivery of water. Now, if a family desires daily water delivery, they pay the local truck drivers an additional 60–150 Rs per month, depending on the distance.

The most recent installation was commissioned on April 15, 2013. The cost of resin is only 15% of the overall costs, a minor portion of the overall expense. Many past systems have used AA instead of HAIX because of the lower initial costs. Using AA would have saved U$S 450 in initial costs, or 10%, but this cost is insignificant when looking at the overall revenue during operation and the improvement in treatment performance (Sarkar *et al.*, 2010).

4 CONCLUSIONS

Community-scale As removal with HAIX, referred to as SARSAC, is a robust process that has been effectively used for several years in rural communities of India and Cambodia. Regular operation, maintenance and testing of the system are managed locally by a committee of representative villagers. The As data points presented were tested by laboratories with funds collected from the local consumers. Upon exhaustion, HAIX is able to paid be repeatedly regenerated by removing the adsorbed As because HFOs are chemically resilient and the parent anion exchange resin is physically robust (Cumbal, 2004, Sarkar *et al.*, 2010).

During regeneration, the removed As is concentrated in waste regenerant. Preventing concentrated As from re-entering the ecosystem is important for the well-being of all life in the community; improper disposal could lead to wildlife consumption, acute poisoning and death. Concentrated As is safely stored at the community-level in a scientifically sound manner that is environmentally benign. A central regeneration facility in rural West Bengal has safely stored the As waste from over one hundred regeneration cycles; a similar central regeneration facility is currently under construction in Cambodia under the auspices of ITC. The underlying chemistry has been presented and substantiated through field data in the literature (Blaney and SenGupta, 2006, Delemos *et al.*, 2006, Ghosh *et al.*, 2004, Sarkar *et al.*, 2005, 2010).

Robustness, reusability and excellent As removal capacity of HAIX have catalyzed community participation in tens of rural communities in As affected countries in South and Southeast Asia. Not only is As-safe water provision possible in remote communities, it can also be profitable, as demonstrated in many communities. Simple system operation, low operational costs and high water quality have been possible with HAIX.

ACKNOWLEDGEMENTS

Financial support received from USEPA P3 Phase II, NCIIA, Tagore-SenGupta Foundation, Water for People and private donors are gratefully acknowledged. M. German is grateful for one-year Fulbright-Nehru Fellowship from the USA State Department and the India government. We are also thankful for our past and continued association with and assistance from Technology with a Human Face (THF), Resource Development International-Cambodia (RDIC), Bengal Engineering and Science University (BESU) and Anugrah Narayan (A.N.) College.

REFERENCES

Argos, M., Kalra, T., Rathouz, P.J., Chen, Y., Pierce, B., Parvez, F., Islam, T., Ahmed, A., Rakibuz-Zaman, M., Dasan, R., Sarwar, G., Slavkovich, V., van Geen, A., Grazaino, G. & Ahsan, H. 2010. Arsenic Exposure from Drinking Water, and All-cause and Chronic-disease Mortalities in Bangladesh (HEALS): A Prospective Cohort Study. *The Lancet*. 376(9737): 252–258.

Bagla, P. & Kaiser, J. 1996. India's spreading health crisis draws global arsenic experts. *Science* 274: 174–175.

Bang, S., Patel, P., Lippincott, L. & Meng, X. 2005. Removal of arsenic from groundwater by granular titanium dioxide adsorbent. *Chemosphere* 60(3): 389–397.

Bearak, D. 1998. New Bangladesh Disaster: Wells that Pump Poison. The New York Times, New York, USA.

Berg, M., Stengel, C., Trang, P.T., Viet, P.H., Sampson, M.L., Leng, M., Samreth, S. & Fredericks, D. 2007. Magnitude of arsenic pollution in the Mekong and Red River deltas-Cambodia and Vietnam. *Sci. Total Environ.* 372(2–3): 413–425.

Bhattacharya, P., Welch, A.H., Stollenwerk, K.G., McLaughlin, M.J., Bundschuh, J. & Panaullah, G. 2007. Arsenic in the environment: biology and chemistry. *Sci. Total Environ.* 379(2–3): 109–120.

Blaney, L.M. & SenGupta, A.K. 2006. Comment on "Landfill-Stimulated Iron Reduction and Arsenic Release at the Coakley Superfund Site (NH)." *Environ Sci. Technol.* 40(12): 4037–4038.

Bundschuh, J., Litter, M., Parvez, F., Román-Ross, G., Nicolli, H.B., Jean, J.-S., Liu, C.-W., López, D. Armienta, M.A., Guilherme, L.R.G., Gomez-Cuevas, A., Cornejo, L., Cumbal, L. & Toujaguez, R. 2012. One century of arsenic exposure in Latin America: A review of history and occurrence from 14 countries. *Sci. Total Environ.* 429: 2–35.

Buschmann, J., Berg, M., Stengel, C. & Sampson, M.L. 2007. Arsenic and Manganese Contamination of Drinking Water Resources in Cambodia: Coincidence of Risk Areas with Low Relief Topography. *Environ. Sci. Technol.* 41(7): 2146–2152.

Chatterjee, A., Das, D., Mandal, B.K., Chowdhuri, T.R., Samanta, G. & Chakraborti, D. 1995. Arsenic in groundwater in six districts of West Bengal, India: the biggest arsenic calamity in the world. Part 1. Arsenic species in drinking water and urine of the affected people. *Analyst 120*: 643–650.

Christen, K. 2001. The arsenic threat worsens. *Environ. Sci. Technol.* 35(13): 286A–291A.

Cumbal, L., Greenleaf, J., Leun, D. & SenGupta, A.K. 2003. Polymer supported inorganic nanoparticles: characterization and environmental applications. *React. Funct. Polym.* 54: 167–180.

Cumbal, L.H. 2004. Polymer-Supported Hydrated Ferric Oxide nanoparticles: Characterization and environmental applications [dissertation]. Bethlehem (PA): Lehigh University.

Cumbal, L. & SenGupta A.K. 2005. Arsenic removal using polymer supported hydrated iron (III) oxide nanoparticles: role of Donnan membrane effect. *Environ. Sci. Technol.* 39: 6508–6515.

Delemos, J.L., Bostick, B.C., Stürup, S. & Feng, X. 2006. Landfill-Stimulated Iron Reduction and Arsenic Release at the Coakley Superfund Site (NH). *Environ. Sci. Technol.* 40: 67–73.

DeMarco, M.J., SenGupta A.K. & Greenleaf, J.E. 2003. Arsenic removal using a polymeric/inorganic hybrid sorbent. *Water Res.* 37: 164–176.

Donnan, F.G. 1911. Theorie der Membrangleichgewichte und Membranpotentiale bei Vorhandensein von nicht dialysierenden Elektrolyten. Ein Beitrag zur physikalisch-chemischen Physiologie. *Z. Elektrochem. Ang. Phys. Chem* 17: 572–581.

Donnan, F.G. 1995. Theory of membrane equilibria and membrane potentials in the presence of non-dialysing electrolytes. A contribution to physical-chemical physiology *J. Membr. Sci.* 100: 45–55.

Driehaus, W., Jekel, M. & Hildebrandt, U. 1998. Granular ferric hydroxide: a new adsorbent for the removal of arsenic from natural water. *J. Water SRT-Aqua* 47: 30–35.

Dutta, P.K., Ray, A.K., Sharma, V.K. & Millero, F.J. 2004. Adsorption of arsenate and arsenite on titanium dioxide suspensions. *J. Colloid Interface Sci.* 278(2): 270–275.

Dzombak, D.A. & Morel, F.M. 1990. Surface Complexation Modeling: Hydrous Ferric Oxide. 1st ed. Hoboken: Wiley-Interscience, New York, USA.

Ghosh, A., Mukiibi, M. & Ela, W. 2004. TCLP underestimates leaching of arsenic from solid residuals under landfill conditions. *Environ. Sci. Technol.* 38: 4677–4682.

Hossain, M.A., Sengupta, M.K., Ahamed, S., Rahman, M.M., Mondal, D., Lodh, D., Das, B., Nayak, B., Roy, B.K., Mukherjee, D. & Chakraborti, D. 2005. Ineffectiveness and poor reliability of arsenic removal plants in West Bengal, India. *Environ. Sci. Technol.* 39(11): 4300–4306.

LayneRT [Internet]. Kansas: Layne Christensen Company; 2000 - [updated 20132 Oct 29; cited 2013 Aug 21]. Available from: http://www.layne.com/en/technologies/laynert.aspx.

Levy, I., Mizrahi, M., Ruano, G., Zampieri, G., Requejo, F.G. & Litter, M. 2012. TiO_2-photocatalytic reduction of pentavalent and trivalent arsenic: production of elemental arsenic and arsine. *Environ. Sci. Technol.* 46(4): 2299–2308.

Mukherjee, R. 2010. Providing Safe Drinking Water for the Poor in India. *Enterprise Development and Microfinance* 21(3): 205–215.

Ng, J.C., Wang, J. & Shraim, A. 2003. Global health problem caused by arsenic from natural sources. *Chemosphere* 52: 1353–1359.

Ravenscroft, P., Burgess, W.G., Ahmed, K.M., Burren, M. & Perrin, J. 2005. Arsenic in groundwater of the Bengal Basin, Bangladesh: distribution, field relations, and hydrogeological setting. *Earth Environ. Sci.* 13(5–6): 727–751.

Ravenscroft, P., Brammer, H. & Richards, K.S. 2009. Arsenic Pollution a Global Synthesis. Chichester: Wiley-Blackwell.

Sancha, A.M. & Castro de Esparza, M.L. 2000. Arsenic status and handling in Latin America. Univ. Chile, Grupo As de AIDIS/DIAGUA, CEPIS/OPS, Lima, Peru.

Sarkar. S., Gupta, A., Biswas, R.K., Deb, A.K, Greenleaf, J.E. & SenGupta, A.K. 2005. Well-head arsenic removal units in remote villages of Indian subcontinent: field results and performance evaluation. *Water Res.* 39(10): 2196–2206.

Sarkar, S., Blaney, L.M., Gupta, A., Ghosh, D. & SenGupta, A.K. 2007. Use of ArsenXnp, a hybrid anion exchanger, for arsenic removal in remote. *React. Funct. Poly. 27*: 1599–1611.

Sarkar, S., Blaney, L.M., Gupta, A., Ghosh, D. & SenGupta, A.K. 2008. Arsenic Removal from Groundwater and Its Safe Containment in a Rural Environment: Validation of a Sustainable Approach. *Environ. Sci. Technol.* 42(12): 4268–4273.

Sarkar, S., Greenleaf, J.E., Gupta, A., Ghosh, D., Blaney, L.M., Bandyopadhyay, P., Biswas, R.K., Dutta, A.K. & SenGupta, A.K. 2010. Evolution of Community-Based Arsenic Removal Systems in Remote Villages: Assessment of Decade-Long Operation. *Water Res.* 44: 5813–5822.

Sarkar, S., Greenleaf, J.E., Gupta, A., Uy, D. & SenGupta, A.K. 2012. Sustainable Engineered Processes to Mitigate the Global Arsenic Crisis in Drinking Water: Challenges and Progress. *Annual Review of Chemical and Biomolecular Engineering* 3: 497–512.

SenGupta, A.K. & Cumbal, L.H., inventors; SenGupta AK, assignee. Hybrid anion exchanger for selective removal of contaminating ligands from fluids and method of manufacture thereof. U.S. patent US 7291578, 2007, Nov 6.

Stanger, G., Truong, T.V., Ngoc, K.S., Luyen, T.V. & Thanh, T.T. 2005. Arsenic in groundwaters of the lower Mekong. *Environ. Geochem. Health* 27(4): 341–357.

Suzuki, T.M., Bomani, J.O., Matsunaga, H. & Yokoyama, Y. 2000. Preparation of porous resin loaded with crystalline hydrous zirconium oxide and its application to removal of arsenic. *React. Funct. Polym.* 43(1–2): 165–172.

One Century of the Discovery of Arsenicosis in Latin America (1914–2014) – Litter, Nicolli, Meichtry, Quici, Bundschuh, Bhattacharya & Naidu (Eds)
 ISBN 978-1-138-00141-1

Dark reduction of As(V) by accumulated electrons stored in alcoholic TiO_2 nanoparticles

I.K. Levy, N. Quici, G. Custo & M.I. Litter
Gerencia Química, Comisión Nacional de Energía Atómica, San Martín, Prov. de Buenos Aires, Argentina

M. Brusa, M.E. Aguirre & M.A. Grela
Departamento de Química, Facultad de Ciencias Exactas y Naturales, Universidad Nacional de Mar del Plata, Prov. de Buenos Aires, Argentina

E. San Román
INQUIMAE/DQIAyQF, Facultad de Ciencias Exactas y Naturales, Universidad de Buenos Aires, Argentina

ABSTRACT: Direct evidence is presented that the otherwise thermodynamically forbidden reduction of As(V) by conduction band electrons of the semiconductor can be driven in the dark by accumulated electrons stored in alcoholic TiO_2 nanoparticles. Accumulation of electrons and participation in As(V) reduction was directly followed by UV-vis spectrophotometry and by Electron Paramagnetic Resonance, detected as Ti(III) species.

1 INTRODUCTION

As(III) and As(V) reduction to solid As(0) has been recently investigated by our group using TiO_2 as photocatalyst and UV light (Levy *et al.*, 2012). Studies demonstrated that As(V) direct reduction to As(0) through the conduction band (CB) electrons of TiO_2 (e_{CB}^-) is not possible. This is consistent with the fact that the one electron reduction potential of As(V) to As(IV) in solution ($E^0_{As(V)/As(IV)} = -1.2$ V vs. NHE (Kläning *et al*,. 1989)) lies well above the CB. To overcome this limitation, it is possible to use sacrificial donors that generate intermediates of strong reducing power like the $^{\bullet}CH_2OH$ radical, formed by oxidation of CH_3OH. Another less explored strategy is the modification of the Fermi level of the electrons of the semiconductor (SC) by electron accumulation. TiO_2 colloids prepared by the HCl catalyzed hydrolysis of a titanium alkoxide at a low H_2O/Ti ratio display a high electron storage capacity (Di Iorio *et al.*, 2012), which can promote reactions that occur at potentials more negative than the CB.

The results reported here and already published (Levy *et al.*, 2013) show that As(V) can be reduced in the dark by electrons stored in alcoholic TiO_2 nanoparticles.

2 EXPERIMENTAL

2.1 *Reagents*

Phenylglyoxylic acid (Sigma) was recrystallized from carbon tetrachloride (Merck). 3,7-bis(dimethylamino)-phenothiazin-5-ium chloride, Methylene Blue, (MB, Merck, for microscopy), sodium meta-arsenite ($NaAsO_2$, Baker), sodium arsenate dibasic 7-hydrate ($Na_2HAsO_4.7H_2O$, Baker), ammonium molybdate ($(NH_4)_6Mo_7O_{24}$, Stanton), potassium antimonyl tartrate (K(SbO) $C_4H_4O_6$.1/2 H_2O, Baker) were used as received. All other chemicals were of the highest purity.

All solutions were prepared using ultrapure water (resistivity >18 MΩ cm). HCl (Merck, p.a. 37%) was used to adjust the pH. Titanium(IV) ethoxide (Aldrich) and absolute ethanol (Cicarelli, p.a. 99.5%) were employed in the synthesis of TiO_2 sols.

The preparation of the TiO_2 sols was made according to Di Iorio *et al.* (2012).

2.2 *Irradiation procedure*

Photolysis was carried out in 3 mL square quartz prismatic cells (path length, l = 1 cm). A high pressure Hg-Xe lamp coupled to a Kratos-Schoeffel monochromator was used to select the 303 ±

10 nm irradiation wavelength. Photon flux determinations were performed by actinometry using phenylglyoxylic acid (Di Iorio *et al.*, 2008), giving $P_a = P_0 \times (1-10^{-A}{}_{303})$ using the incident photon flux, $P_0 = 5.24 \times 10^{-6}$ M^{-1} s^{-1} and the measured absorbance, A_{303}. Most of the experiments were performed using 3 mM TiO_2 sols, which gives $A_{303} \sim 4.5$.

2.3 *Reduction of arsenic species by stored charges*

As(V) concentration in solution was measured by UV-vis spectrophotometry (Lenoble *et al.*, 2003). Conversion of As(III) was determined by TXRF spectroscopy (Juvonen *et al.*, 2009) after removing TiO_2 nanoparticles by precipitation and filtration. Analysis of total Ti in solution was also performed by the TXRF technique. An X-ray fluorescence spectrometer with total reflection geometry (TXRF) S2-PICOFOX was used.

3 RESULTS AND DISCUSSION

The anaerobic irradiation of the ethanolic TiO_2 sol resulted in a typical blue coloration, detected as a broad band in the UV-vis spectrum between 400 and 800 nm(Duonhong *et al.*, 1982; Kormann *et al.*, 1988), and to a blue shift of the band edge (Burnstein-Moss effect, Kamat *et al.*, 1994), indicative of accumulation of electrons. After irradiation, a grey-bluish solid was clearly discernible onto the SC surface but solely in the samples treated with accumulated electrons. These results suggest the formation of As(0).

The irradiated anaerobic samples of colloids showed an intense EPR signal at $g = 1.9551$ that can be ascribed to a Ti(III) species. The quantification of electrons was performed by titration with MB. When 20–50 μL of a degassed concentrated aqueous As(V) solution was injected in the 3 mL cell containing the sol of nanoparticles previously irradiated, and the mixture was allowed to react in the dark for at least 30 min, the loss of the Ti(III) paramagnetic signal was observed, together with the decrease of the [As(V)]. Particularly, we observed that nearly 5 electrons are consumed in the disappearance of a single As(V). The extent of reaction was evaluated from the difference between the [As(V)] determined after injecting the same aliquot of the purged As(V) solution in the 3 mL cell containing the original and the pre-irradiated TiO_2 sol (after electron accumulation). In a typical experiment, 80 ± 8 μM ($(3.3 \pm 0.3) \times 10^{-4}$ mol As/g TiO_2) were consumed from a 210 μM As(V) solution. After reaction with As(V), the UV-vis and the EPR spectra indicated that no trapped electrons remained in the sample.

Separate experiments demonstrated that As(III) can also be reduced by accumulated charges.

These experimental findings can be discussed on the basis of the position of the potential energy bands of the semiconductor and the aqueous standard reduction potentials for As(V) transformations at pH 0 (Levy *et al.*, 2013). The successive one electron reduction of As(V) to As(III) must overcome the high-energy of the As(IV) intermediate. According to the reports, the As(V)/As(IV) redox couple lies nearly 0.9 V above the TiO_2 CB (Kläning *et al*,. 1989). If As(IV) can be formed, less energetic stored electrons, if present, could be used to complete the reduction of As(III) to As(0). Thus, as soon as the high energy electrons are depleted, As(V) reduction should become a forbidden process.

In the present approach, two experimental conditions favor electron accumulation and allow As(V) reduction. First, alkoxide and chloride coordination probably impede electron discharge and charge recombination by promoting a high degree of surface dehydroxylation (Di Iorio *et al.*, 2012). Secondly, the small size of TiO_2 nanoparticles dictates that a low electron concentration per particle may cause the semiconductor to become degenerate due to the quantization of the CB levels. This fact, in turn, results in the increase of the electron Fermi level.

4 CONCLUSIONS

The results demonstrate that electrons stored in TiO_2 nanoparticles can, under controlled conditions, drive the dark reduction of As(V), not possible by e_{CB}^-, and contributes to the understanding of electron transfer processes occurring at the SC interface as well as As(V) photocatalytic reduction.

ACKNOWLEDGEMENTS

This work was founded by CONICET (PIP 319).

REFERENCES

Di Iorio, Y., San Román, E., Litter, M.I. & Grela, M.A. 2008. Photoinduced reactivity of strongly coupled TiO_2 ligands under visible irradiation: An examination of an alizarin red@TiO_2 nanoparticulate system. *J. Phys. Chem. C* 112: 16532–16538.

Di Iorio, Y.D., Aguirre, M.E., Brusa, M.A. & Grela, M.A. 2012. Surface chemistry determines electron storage capabilities in alcoholic sols of titanium dioxide nanoparticles. A combined FTIR and room temperature EPR investigation. *J. Phys. Chem. C* 116: 9646–9652.

Duonghong, D., Ramsden, J. & Grätzel, M. 1982. Dynamics of interfacial electron-transfer processes in colloidal semiconductor systems. *J. Am. Chem. Soc.* 104: 2977–2985.

Juvonen, R., Parviainen, A. & Loukola-Ruskeeniemi, K. 2009. Evaluation of a total reflection X-ray fluorescence spectrometer in the determination of arsenic and trace metals in environmental samples. *Geochem. Explor. Environ. Anal.* 9: 173–178.

Kamat, P.V., Bedja, I. & Hotchandani, S. 1994. Photoinduced charge transfer between carbon and semiconductor clusters. One-electron reduction of C60 in colloidal TiO_2 semiconductor suspensions. *J. Phys. Chem.* 98: 9137–9142.

Kläning, U.K., Bielski, B.H.J. & Sehesteds, K. 1989. Arsenic(IV). A pulse-radiolysis study. *Inorg. Chem.* 28: 2717–2724.

Kormann, C., Bahnemann, D.W. & Hoffmann, M.R. 1988. Preparation and characterization of quantum-size titanium dioxide. *J. Phys. Chem.* 92: 5196–5201.

Lenoble, V., Deluchat, V., Serpaud, B., & Bollinger, J.-C. 2003. Arsenite oxidation and arsenate determination by the molybdene blue method. *Talanta* 61: 267–276.

Levy, I.K., Mizrahi, M., Ruano, G., Zampieri, G., Requejo, F.G. & Litter, M.I. 2012. TiO_2-photocatalytic reduction of pentavalent and trivalent arsenic: production of elemental arsenic and arsine. *Environ. Sci. Technol.* 46: 2299–2308.

Levy, I.K., Brusa, M.A., Aguirre, M.E., Custo, G., San Román, E., Litter, M.I. & Grela, M.A. 2013. Exploiting electron storage in TiO_2 nanoparticles for dark reduction of As(V) by accumulated electrons, *Phys. Chem. Chem. Phys.* 15: 10335–10338.

One Century of the Discovery of Arsenicosis in Latin America (1914–2014) – Litter, Nicolli, Meichtry, Quici, Bundschuh, Bhattacharya & Naidu (Eds)
© 2014 Taylor & Francis Group, London, ISBN 978-1-138-00141-1

Removal of arsenic with zerovalent iron nanoparticles in the dark and under UV-vis light

I.K. Levy, C. Ramos, G. Custo & M.I. Litter
Gerencia Química, Comisión Nacional de Energía Atómica, San Martín, Prov. de Buenos Aires, Argentina

M. Mizrahi & F. Requejo
Instituto de Investigaciones Fisicoquímicas Teóricas y Aplicadas, La Plata, Argentina

ABSTRACT: Removal of As(III) from aqueous solutions using zerovalent iron nanoparticles (nZVI) was studied at different As/Fe ratios, in the dark or under UV-vis irradiation, and in the presence of O_2 or anoxic conditions. Solid products were analyzed by Mössbauer and XANES spectroscopy correlating As and Fe speciation with removal efficiency.

1 INTRODUCTION

The use of nanoparticulated ZVI for pollutant removal from water is currently under investigation, (Pradeep & Anschup, 2009; Crane & Scott 2012; O'Carroll *et al.*, 2013). As(V) removal studies using commercial iron nanoparticles were recently initiated by our group (Morgada *et al.*, 2009).

2 EXPERIMENTAL

2.1 *Reagents*

The zerovalent iron nanoparticles (nZVI) were provided by NANO IRON (NANOFER 25®). Sodium metaarsenite ($NaAsO_2$, Baker), sodium arsenate dibasic 7-hydrate ($Na_2HAsO_4.7H_2O$, Baker) and all other chemicals were of the highest purity.

2.2 *Experiments of As removal*

A glass cylindrical cell thermostatic at 25 °C was used. For irradiation, a UV-vis Philips HPA 400S lamp ($q^0_{n,p}/V = 65.9$ μeinstein s^{-1} L^{-1}, with ferrioxalate) was placed at 5 cm from the front of the cell.

For As(III) experiments, 250 mL of a 10 mg L^{-1} solution (pH 9) were used. Experiments of As(V) removal were performed at 1 mg L^{-1} (pH 7). The solutions were stirred in contact with nZVI for 1 h at different As:Fe mass concentration ratios, CR. Changes of [As(V)] in solution were measured spectrophotometrically (Lenoble *et al.*, 2003). Total As in solution was measured by TXRF.

2.3 *Characterization of initial and final solids*

A ^{57}Fe Mossbauer spectrometer was used in transmission mode. XANES measurements at As K-edge (11867 eV) and Fe K-edge (7112 eV) were performed using the synchrotron facilities at the D08B-XAFS2 beamline at LNLS (Campinas, Brazil).

3 RESULTS

3.1 *Experiments of As removal*

Experiments with As(III) were performed with different CR (1:3, 1:10, 1:100). The highest removal (>70%, 1 h) was attained with CR 1:100; removal was faster under UV-vis, reaching 90% in the same time. Removal under N_2 (CR 1:100) was also important, reaching 76 and 85% in the dark and under irradiation. In all cases, the concentration of As(V) in solution was low, but higher under irradiation, indicating that light promotes As(III) oxidation.

In the case of As(V), experiments were performed with CR 1:50, 1:100 and 1:10 000 in the presence of O_2. With CR1:100 and 1:10000, removal was total, reaching As values in accordance with the WHO (2011) regulation (10 μg L^{-1}). The same result was obtained either under irradiation or in the dark, indicating that for As(V) the effect of light is not relevant, at least for CR 1:100. Removal took place without changes in As oxidation state.

3.2 *Characterization of initial and final solids*

Mössbauer and XANES spectra of the original NANOFER 25 allowed describing the material as composed of an α-Fe core (68%) and a shell of γ-Fe_2O_3 and Fe_3O_4. These spectra were compared with those of the solid products obtained after treatment with As (CR 1:100). The sample obtained in the presence of O_2 in the dark was more similar in composition to the original NANOFER 25, in agreement with the lower removal efficiency obtained under this condition. The sample obtained with O_2 under UV-vis light indicated a lower amount of Fe(0) together with a higher amount of iron oxides. This trend was also observed for the samples treated under N_2 with UV-vis irradiation and in the dark. Results of analysis of hyperfine parameters from Mössbauer spectra suggest that the mechanism of As removal takes place by interaction of As with the iron oxides and not with Fe(0).

The results of XANES at the As-K edge after the treatment showed that As is retained in the solid phase as both As(III) and As(V), with predominance of As(V), particularly under O_2.

4 DISCUSSION

In oxic media, nZVI oxidation by water and O_2 begins with ferrous production (Kanel *et al.*, 2005), followed by reaction with dissolved O_2, generating Reactive Oxygen Species (ROS, e.g. $HO^{\bullet}$, $O_2^{\bullet -}$/$HO_2^{\bullet}$) and oxidation to Fe(III) and Fe(IV) species (Morgada *et al.*, 2009).

The presence of Fe(II) leads to Fe_3O_4, $Fe(OH)_2$, and/or $(Fe(OH)_3$, etc. as fresh corrosion products ("FeOx"). In the present work, formation of fresh Fe_3O_4 and γ-Fe_2O_3 after treatment with As was proved by Mössbauer and XANES.

Arsenic removal occurs by direct As(III) adsorption on iron oxides and oxidation to As(V), which is also adsorbed on FeOx (Morgada *et al.*, 2009). Fresh oxides provide a high surface area for adsorption of As(III) and As(V) *via* surface complexes. As(III) is oxidized to As(V) *via* $HO^{\bullet}$ and O_2 (Hug and Leupin, 2003) and via Fe^{3+} on the surface of iron oxides (Greenleaf *et al.*, 2003).

In anoxic media, Fe(0) oxidation by water generates Fe^{2+} and, consequently, iron oxides *in situ*. Then, the removal process would continue in a similar way as the mechanism under oxic conditions.

In all conditions, the removal process was increased under irradiation. Iron oxides initially present in NANOFER 25 and iron oxides formed by corrosion might be responsible of this effect due to the fact that of γ-Fe_2O_3 and Fe_3O_4 absorb radiation in the UV-vis range and can act as semiconductors.

The signal of As(0), around 11867.4 eV (Levy *et al.*, 2010) was never observed. While other studies report As-Fe(0) interactions, with reduction to As(0) (Yan et al. 2012), we found that removal occurs through As-FeOx interactions where As is effectively retained, and *via* an oxidative mechanism. Differences could be attributed to the nature of the layer of oxides surrounding the Fe core.

In the case of As(V), removal occurs without changes in the As oxidation state, and mainly by adsorption on fresh FeOx. No relevant differences in results of removal were observed under irradiation and in the dark. This fact can be explained because As(V) can be adsorbed on iron oxides surfaces at a very fast rate in comparison with the rate of generation of fresh FeOx, which is improved by light.

5 CONCLUSIONS

The advantage of the use of nZVI compared with the direct use of nanoparticles of iron oxides lies in the occurrence of a continuous corrosion process leading to the *in situ* formation of FeOx, which may be highly reactive due to their higher surface area. From the point of view of the application, the most favorable conditions for As(III) removal with NANOFER 25 are in O_2 under UV-Vis irradiation, whereas, for As(V) removal, irradiation would not be necessary to achieve a high removal extent.

ACKNOWLEDGEMENTS

Work founded by ANPCyT PICT 463. I.K.L, C.R., M.I.L., M.M., F.R. are members of CONICET. To LNLS for their facilities. To NANO IRON, s.r.o., Czech Republic for zerovalent iron nanoparticles.

REFERENCES

Crane, R.A. & Scott, T.B. 2012. Nanoscale zero-valent iron: Future prospects for an emerging water treatment technology. *J. Hazard. Mater.* 211: 112–125.

Greenleaf, J.E., Cumbal, L, Staina, I. & Sengupta, A.K. 2003. Abiotic As(III) Oxidation by Hydrated Fe(III) Oxide (HFO) Microparticles in a Plug Flow Columnar Configuration. *Process Safety and Environ. Protection* 81:2: 87–98.

Hug, S. & Leupin, O. 2003. Iron-Catalyzed Oxidation of Arsenic(III) by Oxygen and by Hydrogen Peroxide: pH-Dependent Formation of Oxidants in the Fenton Reaction. *Environ. Sci. Technol.* 37: 2734–2742.

Kanel, S.R., Manning, B., Charlet, L. & Choi, H. 2005. Removal of Arsenic(III) from Groundwater by Nanoscale Zero-Valent Iron. *Environ. Sci. Technol.* 39: 1291–1298.

Lenoble, V., Deluchat, V., Serpaud, B. & Bollinger, J.-C. 2003. Arsenite oxidation and arsenate determination by the molybdene blue method. *Talanta* 61: 267–276.

Levy, I.K., Mizrahi, M., Ruano, G., Zampieri, G., Requejo, F.G. & Litter, M.I. 2012. TiO_2-photocatalytic reduction of pentavalent and trivalent arsenic: production of elemental arsenic and arsine. *Env. Sci. Technol.* 46: 2299–2308.

Morgada, M.E., Levy, I.K., Salomone,V., Farías, S., López, G. & Litter, M.I. 2009. Arsenic(V) removal with nanoparticulate zerovalent iron: effect of UV light and humic acids. *Catal. Today* 143: 261–268.

O' Carroll D., Sleep B., Krol M., Boparai H. & Kocur C. 2013. Nanoscale zero valent iron and bimetallic particles for contaminated site remediation. *Adv. Water Resources* 51: 104–122.

Pradeep, T. & Anshup. 2009. Noble metal nanoparticles for water purification: A critical review. *Thin Solid Films* 517: 6441–6478.

World Health Organization. Arsenic in drinking-water, background document for development of WHO - Guidelines for drinking water quality. WHO/SDE/WSH/03.04/75/Rev/1. 2011.

Yan, Y.W., Vasic, R., Frenkel, A.I. & Koel, B.E. 2012. Intraparticle Reduction of Arsenite (As(III)) by Nanoscale Zerovalent Iron (nZVI) Investigated with In Situ X-ray Absorption Spectroscopy. *Environ. Sci. Technol.* 46: 7018–7026.

One Century of the Discovery of Arsenicosis in Latin America (1914–2014) – Litter, Nicolli, Meichtry, Quici, Bundschuh, Bhattacharya & Naidu (Eds)

TiO_2 photocatalytic oxidation of As(III) in the presence of Hg(II)

D.J. Rodríguez
Departamento de Ciencias Básicas, Universidad Nacional de Lujan, Luján, Prov. de Buenos Aires, Argentina

N. Quici, G. Custo, L. Cherchiet, M. Ortiz & M.I. Litter
Gerencia Química, Comisión Nacional de Energía Atómica, San Martín, Prov. de Buenos Aires, Argentina

ABSTRACT: As(III) and Hg(II) are known by their toxicity and environmental damage. Heterogeneous photocatalysis (HP) appears as a convenient alternative for As and Hg removal and, in this work, the HP oxidation of As(III) was studied in the presence of Hg(II) in equimolar ratio. Straight evidence is presented of enhanced As(III) oxidation using TiO_2 photocatalysis in the presence of Hg(II) in the absence of oxygen.

1 INTRODUCTION

Arsenic and mercury are highly toxic and associated with many health problems. They can be found together in soil (Liu *et al.*, 2010), in watersheds coming from atmospheric deposition or from power plants and other industrial sources (Lawson & Mason, 2001); these wastewaters arrive to wells used for irrigation (Mousavi *et al.*, 2013) constituting a serious environmental problem.

Removal of arsenite, As(III), from water is difficult, and oxidation to As(V) is generally needed to improve the treatment in most of the technologies.

Heterogeneous Photocatalysis (HP) with titanium dioxide (TiO_2) is an effective technology for removal of a variety of metal and metalloid ions in aqueous effluents (Litter and Quici, in press). After UV irradiation of a TiO_2 aqueous suspension, conduction band electrons (e_{CB}^-) and valence band holes (h_{VB}^+) are created, which can reduce or oxidize the metal or metalloid species present in the system:

$$SC + h\nu \rightarrow e_{CB}^- + h_{VB}^+ \quad (1)$$

TiO_2 HP has been shown as a good alternative for As(III) oxidation and for reduction of Hg(II) (Litter and Quici, in press and references therein).

In this work, the HP oxidation of As(III) in the presence of Hg(II) (equimolar concentrations, 5×10^{-4} M) in different experimental conditions has been studied as a possible method for enhanced As(III) oxidation in water with simultaneous Hg(II) removal.

2 METHODS/EXPERIMENTAL

2.1 *Reagents*

Mercuric chloride ($HgCl_2$, Bromfield), sodium meta-arsenite ($NaAsO_2$, Baker), sodium arsenate dibasic 7-hydrate ($Na_2HAsO_4.7H_2O$, Baker), ammonium molybdate ($(NH_4)_6Mo_7O_{24}$, Stanton), potassium antimonyl tartrate ($K(SbO)C_4H_4O_6.1/2\ H_2O$, Baker) were used as received. Sodium hydroxide (NaOH, Carlo Erba) was used for adjusting the initial pH. All other chemicals were of the highest purity. TiO_2 Degussa P25 (now Aeroxide P25, Evonik) was used.

2.2 *Irradiation procedure and analytical techniques*

Batch experiments were carried out in a cylindrical jacketed Pyrex glass reactor with magnetic stirring at 25 °C. Starting concentrations were and 0.5 mM $NaAsO_2$ and 0.5 mM $HgCl_2$, pH 7, and 1 g L^{-1} TiO_2 in HP experiments. Experiments in the absence of O_2 were performed by bubbling N_2 (1 L min^{-1}) during all the experiment. For irradiation, a Xe lamp (450 W) was employed with a band-pass filter (300 nm < λ < 400 nm). Aliquots were taken at different times for analysis.

As(V) concentration in solution was measured by UV-vis spectrophotometry (Lenoble *et al.* 2003). Total mercury was determined by TXRF, using an X-ray fluorescence spectrometer with total reflection geometry (TXRF) S2-PICOFOX.

3 RESULTS AND DISCUSSION

Figure 1 shows the results of As(III) oxidation experiments in different conditions. According to thermodynamical data (Bard & Parsons, 1985), oxidation of As(III) to As(V) at pH 7 in the presence of an equimolar of Hg(II) concentration is possible in dark ambient conditions (open to the air). In accordance, after mixing equimolar solutions of As(III) and Hg(II) in the above conditions, 28% of As(III) was converted to As(V) in ca. 90 min. Addition of TiO_2 led to the same conversion, indicating not significant adsorption of As(V) on TiO_2.

Near UV-light irradiation of the As(III) system in the presence of TiO_2 and O_2 (absence of Hg(II)) showed a 40% of As(III) conversion to As(V) in 90 min. This is due to the reaction with photocatalytic formed Reactive Oxygen Species (Litter and Quici, in press):

$$As(III) + HO^{\bullet}/h_{VB}^{+}/HO_2^{\bullet}/O_2^{\bullet -}/O_2 \rightarrow As(V) + \ldots \quad (1)$$

In the absence of O_2 (N_2 bubbling), only 18% of As(III) was photooxidized to As(V) in similar conditions, the reaction taking place due probably to traces of O_2 adsorbed or occluded in the pores of the solid. When Hg(II) was equimolarly added to the same system, As(III) conversion was much higher (74%), and a grey-pale solid was found onto the photocatalyst surface. Hg(II) is photocatalytically reduced by e_{CB}^{-}, according to:

$$Hg(II) + e_{CB}^{-} \rightarrow Hg^{+} + e_{CB}^{-} \rightarrow Hg(0) \quad (2)$$

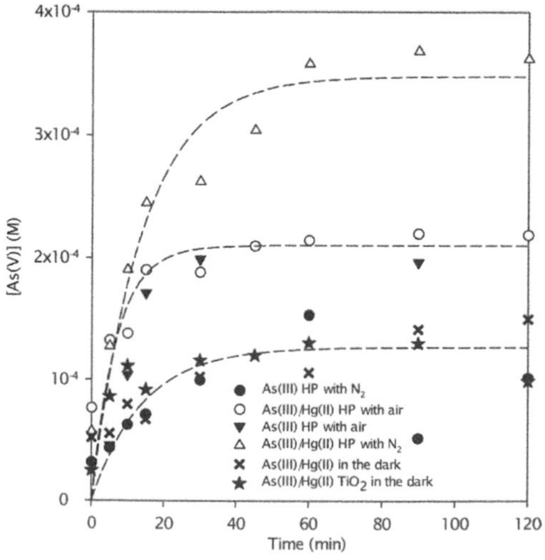

Figure 1. As(V) generation vs. time in different experimental conditions.

The grey deposit would correspond to a mixture of Hg(0) and Hg_2Cl_2, as previously obtained in the $HgCl_2$ TiO_2-photocatalytic reduction in similar conditions (Botta *et al.*, 2002). These results evidence that a synergy takes place for As(III) oxidation in the presence of Hg(II) in the absence of O_2. Although reoxidation of mercury by h_{VB}^{+} or $HO^{\bullet}$ (Equation (3)) is possible, it is here prevented by As(III) oxidation (Equation (1)). Also, formation of HgO by Equation (4) would be hindered in the absence of O_2.

$$Hg(0)/(I) + h_{VB}^{+} (HO^{\bullet}) \rightarrow Hg(I)/(II) + (OH^{-}) \quad (3)$$

$$Hg(0) + O_2 \rightarrow HgO \quad (4)$$

In agreement, in the presence of O_2, TiO_2 and UV-irradiation, the extent of As(III) oxidation was the same either in the presence or in the absence of Hg(II), indicating a competence between O_2 and Hg(II).

Further experiments are underway in order to understand the mechanisms involved in this complex system, to complete these preliminary results.

4 CONCLUSIONS

The results here presented demonstrate that HP with TiO_2 nanoparticles, under controlled conditions, drives the oxidation of As(III) simultaneously with Hg(II) reduction. As(V) could be then removed by other techniques (adsorption, addition of zerovalent iron, etc.) and, at the same time, mercury, deposited as solid, could be separated by subsequent physical step. On the other hand, the system is relevant to understand mechanistic implications of the presence of simultaneous metals and metalloids in photocatalytic systems.

ACKNOWLEDGEMENTS

Work funded by PICT-2011-0463 (Argentina). N.Q. and M.I.L. are members of CONICET (Argentina).

REFERENCES

Bard, J.A. & Parsons, R. (eds). 1985. *Standard potentials in aqueous solutions*: 162–171 and 265–286. New York: Marcel Dekker Inc.

Botta, S.G., Rodríguez, D.J., Leyva, A.G & Litter, M.I. 2002. Features of the transformation of Hg^{II} by heterogeneous photocatalysis over TiO_2. *Catalysis Today* 76: 247–258.

Lawson, N.M & Mason, R.P. 2001. Concentration of mercury, methylmercury, cadmium, lead, arsenic, and

selenium in the rain and stream water of two contrasting watersheds in western Maryland. *Water Research* 35: 4039–4052.
Lenoble, V., Deluchat, V., Serpaud, V. & Bollinger, J.C. 2003. Arsenite oxidation and arsenate determination by the molybdene blue method. *Talanta* 61: 267–276.
Liu, Y., McDermott, S., Lawson, A. & Aelion, C.M. 2010. The relationship between mental retardation and developmental delays in children and the levels of arsenic, mercury and lead in soil samples taken near their mother's residence during pregnancy. *International Journal of Hygiene and Environmental Health* 213: 116–123.
Litter, M.I. & Quici, N. 2013 (in press). New advances of heterogeneous photocatalysis for treatment of toxic metals and arsenic. In B.I. Kharisov, O.V. Kharissova & H.V. Rasika Dias (eds), *Nanomaterials for environmental protection*. Hoboken: John Wiley & Sons.
Mousavi, S.R., Balali-Mood, M., Riahi-Zanjani, B., Yousefzadeh, H. & Sadeghi, M. 2013. Concentrations of mercury, lead, chromium, cadmium, arsenic and aluminium in irrigation water wells and waste waters used for agriculture in Mashad, northeastern Iran. *International Journal of Occupational and Environmental Medicine* 4: 80–86.

One Century of the Discovery of Arsenicosis in Latin America (1914–2014) – Litter, Nicolli, Meichtry, Quici, Bundschuh, Bhattacharya & Naidu (Eds)
© 2014 Taylor & Francis Group, London, ISBN 978-1-138-00141-1

Photocatalytic oxidation of As(III) using TiO_2-Cr_2O_3 semiconductors

Luis A. May-Ix & R. Gómez
Universidad Autónoma Metropolitana – Iztapalapa, Depto. de Química, Grupo ECOCATAL. México, D.F. México

R. López
Universidad Juárez Autónoma de Tabasco, Div. de Ingeniería y Arquitectura, Cunduacán, Tabasco, México

ABSTRACT: In this work, the synthesis and characterization of TiO_2 and TiO_2-Cr_2O_3 mixed oxide prepared by the sol-gel method is reported and the role of Cr_2O_3 in the photocatalyzed oxidation of As(III) to As(V), in basic pH medium under UV-Vis light irradiation is discussed. Characterization of these solids by XRD show the presence of nanostructured TiO_2 in which the presence of Cr ions stabilized the TiO_2 anatase structure. BET results revealed a pore size in the range of 93–114 Å, with a specific surface area between 75–148 m²/g. The E_g band calculated by the Kubelka–Munk method was of 2.95 and 1.92 eV for TiO_2 and TiO_2-Cr_2O_3, respectively. Thermal analysis (DSC and TGA) of these materials showed phase transitions around 500 °C. In the evaluation of the photooxidation of As(III), improved photoactivity was found with the TiO_2-Cr_2O_3 photocatalyst in comparing with the bare TiO_2.

1 INTRODUCTION

Arsenic contamination in groundwater has been recognized as a great threat to the public health worldwide because high levels of arsenic, ranging from tenths to several thousands of $\mu g\ L^{-1}$, have been found in groundwater in many regions around the world (Smedley & Kinniburg, 2002). In this way, extensive efforts have been made to develop effective methods to remove arsenic from drinking water. The toxicity and mobility of As depend on its oxidation state where arsenite (As(III)) is 50–100 times more toxic than arsenate (As(V)) and oxidation of arsenite, As(III), to arsenate, As(V), is recommended by many water treatment technologies. To carry out photooxidation process, the use of TiO_2 as photocatalyst has been successfully used for the abatement of polluted water (Agrios & Pichat, 2005). With the aim to improve the photoactivity of TiO_2, in the present work TiO_2-Cr_2O_3 was used as photocatalyst for the oxidation of As(III) to As(V). The material was characterized structural and texturally by powder X-ray diffraction (XRD), N_2 adsorption-desorption, UV-Vis-DRS spectra and thermogravimetric analyses.

2 METHODS/EXPERIMENTAL

2.1 *Synthesis of TiO_2-Cr_2O_3 photocatalyst*

The TiO_2 photocatalyst modified with Cr_2O_3 (10 wt.%) was synthesized by the sol-gel method. An appropriate amount of chromium nitrate nonahydrated was added to a flask containing 18 mL of distilled-deionized water (18 mΩ.cm) and 44 mL of 1-butanol. Subsequently, 44 mL of titanium (IV) butoxide were added dropwise to the solution during 4 h. The gelling solution was then heated at 70 °C under reflux and maintained under constant stirring for 24 h until the gel was formed. The obtained xerogel was dried at 70 °C for 24 h. The dried solid was ground in an agate mortar until a fine and homogeneous powder mixture was obtained. Finally, the solid was annealed in air at 500 °C for 5 h, using a heating rate of 2 °C/min. As a reference, undoped TiO_2 sol-gel sample was prepared according to the protocol described above without the addition of the corresponding chromium precursor.

2.2 *Photocatalytic experiments*

The photocatalytic reaction was carried out in a glass reactor containing 0.2 g of photocatalyst and 200 mL of aqueous solution with initial As(III) concentrations in the range of 30–210 μmol. The suspension was stirred for 20 min in dark to allow the adsorption equilibrium of As(III) on TiO_2-Cr_2O_3. The photoirradiation was carried out using a high pressure mercury lamp (UV lamp, emitting at 254 nm, 2.16 watts, 18 mA) protected with a quartz tube and immersed in the center of vessel. The pH of the initial solution was 9 with no variations in the irradiated solution. The oxidation of As(III) was monitored by collecting samples at intervals of 10 min. To avoid interferences in the UV-Vis analysis associated with suspended solids, each sample was filtered through a nylon membrane (0.45 μm, Millipore) to remove the particles before the analyses.

2.3 *Solution preparation and analytical measurements*

As(III) stock solution (13.3 mmol L^{-1}, 1000 mg L^{-1}) was prepared by dissolving the appropriate amounts of sodium metaarsenite ($NaAsO_2$-Aldrich) in deionized water. A stock solution of 1 mmol/L was prepared by further dilution. As(III) concentration was measured by analyzing the sample using a modified colorimetric molybdate-blue method (Dhar *et al.*, 2004). In brief, the method is based in the formation of a complex of As(V) with reduced molybdate which strongly absorbs at 880 nm. A calibration curve was constructed by measuring the absorbance of As(V) complex derived from As(III) solutions of known concentration.

3 RESULTS AND DISCUSSION

XRD patterns of the bare TiO_2 and TiO_2-Cr_2O_3 (10 wt.%) calcined sample exhibit peaks at 2θ = 25.38° (100), 38.14° (004), 48.08° (200) and 53.89° (105) assigned to the anatase phase TiO_2 (JCPDS code: 21–1272), indicating that the incorporation of Cr(III) does not change the TiO_2 network structure. Cr_2O_3 reflections at 2θ = 36.19° (JCPDS code: 04–0765) are not visible probably because the oxide was found as highly dispersed on the TiO_2 surface forming very small crystallites no detected by XRD technique.

Nitrogen adsorption-desorption isotherms of the solids correspond to type IV, which is characteristic of mesoporous materials; they present an H1 type hysteresis loop according to IUPAC classification. This type of hysteresis loop is characteristic of solids formed by agglomerates of spheroidal particles, with pore size and uniform shape.

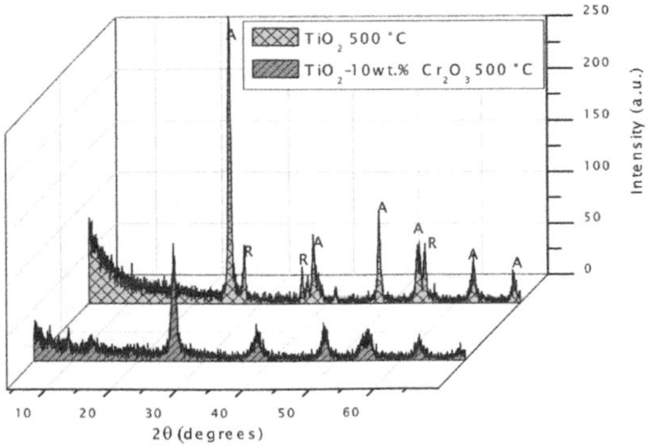

Figure 1. XRD patterns of TiO_2 and TiO_2-Cr_2O_3 (10 wt.%) calcined at 500 °C.

Table 1. Textural properties and E_g values of the photocatalysts.

Photocatalyst (500 °C)	TiO_2-Cr_2O_3-10 wt.%	TiO_2
Pore size (Å)	114.5	93.70
Pore volume (cm^3/g)	0.085	0.020
Specific surface area (m^2/g)	148.8	75.10
E_g (eV)	1.920	2.950

The synthesized materials containing chromium show a significant increase in the specific surface area (Table 1); therefore, other factors, such as a chromium oxidation state or crystallite size, must be taken into account in order to explain the enhancement in photoactivity. Additionally, the E_g value obtained considering indirect transitions of the TiO_2-Cr_2O_3 semiconductor is lower than the value obtained for the bare TiO_2. Thermal analysis (DSC and TGA) of this material showed a phase transition around 500 °C, and at this temperature was calcined the synthetic material to produce the corresponding mixed oxide.

3.1 *Photocatalytic oxidation of As(III)/ TiO_2-Cr_2O_3 system*

As(III) photocatalytic oxidation proceeds relatively fast even for the highest As(III) concentration tested. When the initial As(III) concentration was 30 μM L^{-1}, complete As(III) oxidation was accomplished in 10 min, while at initial As(III) concentration of 75 and 120 μM L^{-1} complete photocatalytic oxidation achieves in 30 and 60 min, respectively. Further increase of initial As(III) concentration from 120 to 210 μM L^{-1} resulted in practically the same concentration-time profile for As(III) oxidation. The pure TiO_2 catalyst required 60 min for complete oxidation of 120 μM L^{-1} As(III). Nevertheless, concentrations above 120 μM L^{-1} were not completely oxidized at all after 120 min. It can be concluded that for initial As(III) concentration in the range of 30–210 μM L^{-1}, photocatalytic oxidation using TiO_2-Cr_2O_3 (10 wt.%) can be accomplished within 10–60 min.

4 CONCLUSIONS

TiO_2-Cr_2O_3 photocatalyst prepared by the sol-gel method with high specific surface area and E_g band in the visible region was obtained. The incorporation of Cr_2O_3 (10 wt.%) to the sol-gel titania stabilize the anatase phase. The results showed that the addition of Cr_2O_3 significantly improves the As(III) photooxidation, in which the Cr(III)/Cr(VI) pair play and important role as electron acceptor in the process.

ACKNOWLEDGEMENTS

Luis A. May-Ix acknowledges the scholarship given by CONACyT México 252088 grant supporting this research.

REFERENCES

Agrios, A. & Pichat, P. 2005. State of the art and perspectives on materials and applications of photocatalysis over TiO_2. *J. Appl. Electrochem.* 35: 655–663.

Dhar, R.K., Zheng, Y., Rubenstone, J. & van Geen, A. 2004. A rapid colorimetric method for measuring arsenic concentration in groundwater. *Anal. Chim. Acta* 526: 203–209.

Smedley, P.L. & Kinniburg, D.G. 2002. A review of the source, behaviour, and distribution of arsenic in natural waters. *Appl. Geochem.* 17: 517–568.

One Century of the Discovery of Arsenicosis in Latin America (1914–2014) – Litter, Nicolli, Meichtry, Quici, Bundschuh, Bhattacharya & Naidu (Eds)
© 2014 Taylor & Francis Group, London, ISBN 978-1-138-00141-1

Removing arsenic from water in West Bengal, India using CuO nanoparticles

K.J. McDonald & K.J. Reddy
Department of Ecosystem Science and Management, University of Wyoming, USA

ABSTRACT: Arsenic contamination in drinking water is a worldwide health crisis. Treatment of arsenic laden water in areas of the world such as West Bengal, India has proven to be an extremely difficult task. Cupric Oxide (CuO) nanoparticles have shown promising characteristics as a sorbent to remove arsenic from water. Presented in this study is, to our knowledge, the first time CuO nanoparticles have been used to treat groundwater from West Bengal that is naturally high in arsenic. Batch experiments were conducted by reacting CuO nanoparticles with 16 groundwater samples from West Bengal that exceed 10 µg L^{-1}. All samples showed near complete removal of arsenic following the treatment with CuO nanoparticles. The removal of arsenic was unaffected by the presence of high concentrations of competing ions such as bicarbonate (HCO_3^-), phosphate (PO_4^{3-}), and sulfate (SO_4^{2-}).

1 INTRODUCTION

The widespread nature and severity of health issues associated with arsenic contamination in drinking water is a chief global health concern. Existing arsenic removal techniques are hindered by many factors including pre- and post-treatment requirements, the disposal of byproducts, operational expertise, lack of onsite arsenic analysis techniques, and socioeconomic barriers (Johnston *et al.*, 2010). Recent studies have found that CuO is a novel and effective arsenic adsorbent that circumvents many of the obstacles other treatment techniques may encounter. However, these studies have been conducted only with groundwater samples collected from the greater Rocky Mountain Region of the United States. Studying the performance of CuO in areas where different geochemical conditions exist and where arsenic related health issues are more prevalent, such as West Bengal, India will establish a better understanding of the capabilities of this remediation technology. (Chakraborti *et al.*, 2009).

2 MATERIALS AND METHODS

2.1 *Sampling and Analytical techniques*

Sampling locations were determined based on spectral signatures of various geomorphologic landform features inferred from the satellite imagery conducted by Singh *et al.* (2013). These groundwater samples were collected at the wellhead and stored in polypropylene bottles. All bottles were filled with zero headspace and were transported back to the lab and stored at 4 °C to avoid chemical alteration. A portion of sample aliquots were collected and acidified in the field with diluted HNO_3 for analysis.

The analysis of cations and anions was performed as described by the American Public Health Association (APHA, 2005). pH measurements were taken using a Hanna pH meter.

2.2 *CuO nanoparticles and batch experiments*

CuO nanoparticles were synthesized using the procedure described by Martinson and Reddy (2009) with ethanolic solutions of copper chloride and sodium hydroxide with polyethyleneglycol as a dispersant. The nanoparticles were dried, ground and analyzed by TEM and BET analysis.

50 mL of unacidified groundwater samples were combined with 0.2 g of CuO nanoparticles in 50 mL centrifuge tubes. The solution was allowed to react on an orbital shaker table for 30 minutes at 250 rpm. Following the reaction, the samples were filtered using 0.45 µm filter paper and plastic syringes.

3 RESULTS AND DISCUSSION

3.1 *Sample characterization*

TEM analysis shows that the nanoparticles formed spherical and cylindrical shapes. The BET surface area of the CuO nanoparticles gave a specific surface area of 62 m^2 g^{-1}. These nanoparticles are similar in size and surface area to the nanoparticles prepared by Martinson & Reddy (2009).

Of the 38 total groundwater samples collected, sixteen contained concentrations of arsenic above the United States Environmental Protection Agency (USEPA) and World Health Organization (WHO) limit of 10 μg L^{-1} (Table 1). Concentrations of arsenic ranged from 10 to nearly 70 μg L^{-1}. The pH of these samples ranged from 8.1 to 8.8. Further, many samples showed significant concentrations of PO_4^{3-}, ranging from 1 to greater than 500 mg L^{-1}, and SO_4^{2-}, ranging from Below Detection Limits (BDL) to greater than 40 mg L^{-1}.

Similarly high concentrations of phosphate were found by the British Geological Survey in the groundwaters of comparable regions in Bangladesh (BGS & DPHE, 2001). Copper concentrations ranged from BDL to almost 0.2 mg L^{-1}. Ca levels were found to range between approximately 50 to 170 mg L^{-1}, Na concentrations ranged from 40 to 150 mg L^{-1} and Mg from ca. 10 to 60 mg L^{-1}.

3.2 *Arsenic removal by CuO nanoparticles*

Following the treatment with CuO nanoparticles, the water samples showed slight decreases in pH and minor changes in major and trace element concentrations after the reaction with CuO nanoparticles. All samples showed significant reduction in arsenic concentrations to levels well below the USEPA and WHO standard value of 10 μg L^{-1}. Results of arsenic removal are shown in Table 1. The effectiveness of many arsenic sorbents is significantly limited by the presence of competing ions such as phosphate and sulfate.

Results shown in Figure 1 show no discernible effect on the rate of arsenic removal by CuO nanoparticles due to high concentrations of potential competing ions such as HCO_3^-, PO_4^{3-} or SO_4^{2-}. Further, the reaction with CuO nanoparticles did not significantly change the concentrations of any other chemical constituents of the water samples that were measured. Slight changes are reported in concentrations of NO_3^- and SO_4^{2-} following treatment with CuO nanoparticles. However, these changes are insignificant and do not exceed 2.5 mg L^{-1} for NO_3^- and 2 mg L^{-1} for SO_4^{2-}. Concentrations of copper before and after treatment with CuO nanoparticles are of particular concern. Most of these samples show an increase in copper concentrations following the batch experiment process ranging from BDL to almost 0.2 mg L^{-1}. However, all treated samples remained well below the EPA drinking water maximum contaminant level goal of 1.3 mg L^{-1}.

Table 1. Concentration of arsenic before and after treatment with CuO nanoparticles for all 16 groundwater samples.

Sample no.	Pre-CuO (As)	Post-CuO (As)	Sample no.	Pre-CuO (As)	Post-CuO (As)
1	62	1.62	9	70	1.85
2	58	0.18	10	13	BDL
3	20	BDL	11	58	0.06
4	39	0.02	12	17	BDL
5	17	BDL	13	48	0.04
6	36	BDL	14	42	0.04
7	15	0.04	15	15	BDL
8	14	BDL	16	14	BDL

* Units for As in μg L^{-1}, BDL = below detection limit.

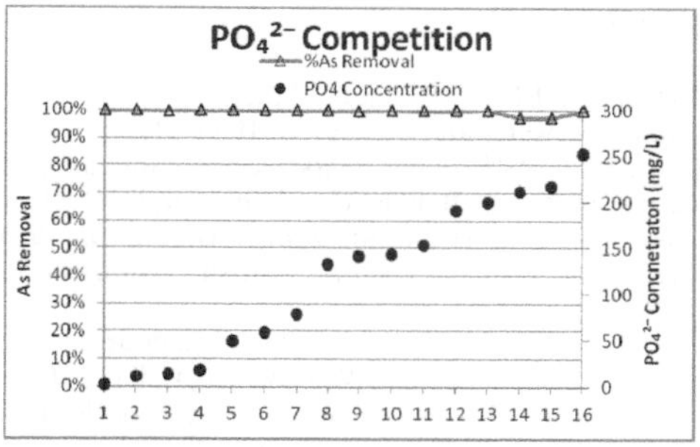

Figure 1. Concentration of PO_4^{3-} vs. removal of arsenic for all 16 wells.

4 CONCLUSIONS

CuO nanoparticles effectively removed arsenic from sixteen groundwater samples collected in West Bengal, India. All samples contained naturally high levels of arsenic reaching concentrations as high as 70 μg L^{-1} seven times more than the drinking water guideline held by the USEPA and WHO. Furthermore, many samples contained high concentrations of potential competing ions such as HCO_3^-, PO_4^{3-}, and SO_4^{2-}. The effectiveness of CuO nanoparticles in removing arsenic was unchanged by the presence of these potential competing ions. The treatment of natural groundwater samples suggests that CuO nanoparticles are effective across a wide range of geochemical conditions. The combination of these characteristics lend to the viability of CuO nanoparticles as an effective adsorbent to remove arsenic from groundwater that merits further study and development for field application.

ACKNOWLEDGEMENTS

To the University of Wyoming and Jawaharlal Nehru University for providing the facilities and resources used in this research.

REFERENCES

APHA, A. WEF. 2005. Standard methods for the examination of water and wastewater 1965, 21.

BGS and DPHE. 2001. Arsenic contamination of groundwater in Bangladesh. Kinniburgh, D G and Smedley, P L (Editors). *British Geological Survey Technical Report WC/00/19.* British Geological Survey: Keyworth.

Chakraborti, D. *et al.* Status of groundwater arsenic contamination in the state of West Bengal, India: A 20-year study report. Molecular Nutrition & Food Research 2009, 53(5), 542–551; 10.1002/mnfr.200700517.

EPA, U. 40 CFR Parts 9, 141, 142: Final rule. Fed. Regist. 2001, 66 (14), 6976–7066.

Johnston, R.B. *et al.* The socio-economics of arsenic removal. Nature Geoscience 2010, 3 (1), 2–3.

Martinson, C.A. & Reddy, K. 2009. Adsorption of arsenic (III) and arsenic (V) by cupric oxide nanoparticles. *J. Colloid Interface Sci.* 336 (2): 406–411.

Singh, N. *et al.* Hydro-geological processes controlling the release of arsenic in part of 24 Parganas district, West Bengal. Environmental Earth Sciences 2013, (in review).

One Century of the Discovery of Arsenicosis in Latin America (1914–2014) – Litter, Nicolli, Meichtry, Quici, Bundschuh, Bhattacharya & Naidu (Eds)
© 2014 Taylor & Francis Group, London, ISBN 978-1-138-00141-1

Advanced Oxidation-Coagulation-Filtration (AOCF)—an innovative treatment technology for targeting drinking water with <1 μg/L of arsenic

A. Ahmad
Hoofd Ingenieursbureau, Brabant Water N.V., 's-Hertogenbosch, The Netherlands
KTH-International Groundwater Arsenic Research Group, Department of Sustainable Development, Environmental Science and Engineering, KTH Royal Institute of Technology, Stockholm, Sweden

S. van de Wetering & M. Groenendijk
Hoofd Ingenieursbureau, Brabant Water N.V., 's-Hertogenbosch, The Netherlands

P. Bhattacharya
KTH-International Groundwater Arsenic Research Group, Department of Sustainable Development, Environmental Science and Engineering, KTH Royal Institute of Technology, Stockholm, Sweden

ABSTRACT: Advanced Oxidation-Coagulation-Filtration (AOCF) has been investigated for producing drinking water with less than 1 μg L^{-1} of As through a series of bench scale and pilot scale experiments. At bench scale, the suitable coagulant, its combination dose with $KMnO_4$ oxidant, the optimum process pH and kinetics of As removal were determined. The optimized AOCF technique was capable of consistently reducing the As concentration to below 1 μg L^{-1} when implemented at pilot scale and did not adversely affect the already existing removal processes of Fe, Mn and NH_4^+. Dual media filter solved the filter run time reduction issue.

1 INTRODUCTION

Arsenic (As) is an toxic element which causes contamination of groundwater in many parts of the world. Although, the World Health Organization (WHO) currently recommends 10 μg/L as the guideline value for drinking water, it can pose serious threat to human health even at very low concentrations (Kapaj *et al.*, 2006). The US Environmental Protection Agency (USEPA, 1998) and the US Natural Resources Defense Council (NRDC, 2000) have recommended As guidelines below 1 μg L^{-1} to attain an acceptable lifetime cancer risk. Optimizing the available treatment techniques and developing new methods for As removal is currently of great urgency and high priority. The main goal of this research was to develop an efficient As removal technique that would be able to produce drinking water with As concentration of less than 1 μg L^{-1}. For this purpose, an innovative three step technique, Advanced Oxidation-Coagulation-Filtration (AOCF), was investigated through a series of bench scale and pilot scale experiments using As contaminated source water from one of the drinking water production plants in the Netherlands. AOCF is an efficient As removal technique, comprising of an advanced oxidation step to convert As(III) to As(V) with potassium permanganate ($KMnO_4$), followed by the sorption of As(V) onto/into the precipitating coagulations (flocs) formed after a suitable coagulant is added to the aqueous system and, finally, by the removal of the floc-As matrix through granular media filtration.

2 METHODS/EXPERIMENTAL

2.1 *Bench scale optimization of AOCF*

The bench scale experiments were conducted using a jar test apparatus in the controlled conditions of the treatment plant laboratory. In order to develop an efficient As removal method, firstly the most suitable type of coagulant was determined which could achieve high As removal from the source water in hand. Three commonly used coagulants in the water industry, i.e., $FeSO_4$, $FeCl_3$ and alum were investigated for this purpose. After the selection of the suitable coagulant, the combination doses of the oxidant ($KMnO_4$) and the selected coagulant were determined, which could achieve a residual As concentration of less than 1 μg L^{-1}. The oxidation and adsorption kinetics of As(III) and As(V) were studied to determine the dosing points in the pilot setup for the further evaluation of AOCF. The third and final step of bench scale optimization was to determine the optimum process pH. The optimized technique was then implemented at plot scale.

2.2 *Pilot scale evaluation of AOCF*

The pilot plant consisted of a pair of aeration cascades, a pair of filtration columns (300 mm diameter, 1.8 m bed height), a chemical dosing setup, and a network of pipes, sampling points and drains. In one of the columns, virgin sand media (VS media) was filled and, in the other column, metal oxide coated sand extracted from the filters of the full scale treatment plant was placed (MOCS media). The effective grain size for both the media was 1–1.6 mm. The primary reason to use VS besides MOCS was to investigate the influence of sand media age/coating on the efficiency of AOCF treatment. Before the implementation of AOCF on the pilot plant, both the filters were ripened at a filtration velocity of 4 m h^{-1} for a period of 8 weeks to start-up the removal processes of Fe, Mn and NH_4^+. During the ripening phase and after the implementation of AOCF, filters were operated continuously and backwashed when required with air and water. The dosing of chemicals was increased in a step-wise manner to optimize the As removal and to evaluate the effect of AOCF on the removal of Fe, Mn and NH_4^+.

3 RESULTS AND DISCUSSION

3.1 *Bench scale optimization of AOCF*

Ferric chloride, $FeCl_3$, showed higher As removal efficiency than its competitors, i.e., $FeSO_4$ and alum, not only at all the pH values between 5 and 8.5—but also at all the investigated coagulant doses between 2 and 10 mg/L (coagulant dose in terms of the concentration of the total metal added). Alum showed the lowest As removal efficiency. Although, in general, an increase in As removal was noticed with the increase of coagulant dose, the removal of As to the desired level of less than 1 μg L^{-1} was not achieved with any of the coagulant alone in the range of 2–10 mg L^{-1} metal dose. It was due to the presence of As in the reduced arsenite form, i.e., As(III). When a pre-oxidation step through the use of $KMnO_4$ was combined with the $FeCl_3$ treatment, a significant increase in As removal was noticed and residual As levels lower than 1 μg L^{-1} were achieved with various $KMnO_4$-$FeCl_3$ dosing combinations (Table 1).

Table 1. Optimized dosing combinations at different pH values obtained from the bench scale study.

pH	$KMnO_4$ (mg L^{-1})	$FeCl_3$ (mg L^{-1})
7.0	1.0	1.5
7.5	1.0	2.0
8.0	1.5	2.0

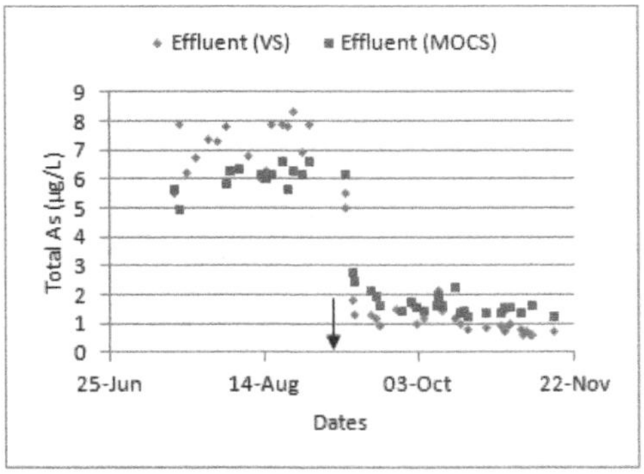

Figure 1. Total As concentration in the effluents of Virgin Sand filter (VS) and Metal Oxide Coated Sand filter (MOCS). The arrow mark indicates implementation of AOCF.

3.2 *Pilot scale evaluation of AOCF*

The As removal efficiency of AOCF was evaluated by applying an optimized combinations of $KMnO_4$ and $FeCl_3$ doses at the suitable dosing points in the pilot system. Variations over time in the effluent As concentrations of VS and MOCS media are provided (Figure 1). As soon as the AOCF was implemented, levels of As in both the effluents significantly decreased. In the effluent of VS media residual As level of lower than 1 μg/L was recorded consistently for several weeks; however, the effluent from MOCS contained a slightly higher concentration of As (1–1.5 μg/L). The relatively lower As removal efficiency of MOCS was due to the grain size growth over the period of sand media use in the full scale filter. It should be noted that before the implementation of AOCF, the MOCS media was producing better quality water in terms of total As content (Figure 1). However, the situation reversed after the new technology was implemented. It is also worth-mentioning that the application of AOCF did not disturb the pre-existing removal processes of Fe, Mn and NH_4^+. However, a decrease in average filter run time from 96 to 24 h was noticed for both the filters. In order to optimize the filter run time, dual media/double layer filtration with anthracite (1–1.6 mm) and finer sand (0.5–0.8 mm) was evaluated with the optimum chemical dosing combination. Average filter run time increased to more than 48 h.

4 CONCLUSIONS

The AOCF technology was capable of consistently reducing effluent As concentrations below 1 $\mu g\ L^{-1}$ both during bench scale and pilot scale investigations. No adverse effect was noticed on the removal of Fe, Mn and NH_4^+. The new technology showed the potential to be easily implemented and integrated with any existing water treatment system. The optimum pH range for AOCF is 7–8. In order to avoid shorter filter run times, dual media filtration may be considered.

ACKNOWLEDGEMENTS

The authors would like to acknowledge the Engineering Department at Brabant Water for the financial support for this study. Thanks to Jink Gude (Process Engineer at Brabant Water) and Tim van Dijk (Process Technologist at Brabant Water) for continuous support for the research project.

REFERENCES

Kapaj, S., Peterson, H., Liber K. & Bhattacharya, P. 2006. Human health effects from chronic arsenic poisoning—A Review. *J. Environ. Sci. Health, Part A* 41(10): 2399–2428.

Natural Recourses Defense Council 2000. Arsenic and old laws: A scientific and public health analysis of arsenic occurrence in drinking water, its health effects, and EPA's outdated arsenic tap water standard, 2000: http://www.nrdc.org/water/drinking/arsenic/aolinx.asp.

US Environmental Protection Agency. 1998. Integrated Risk Information System, Arsenic, inorganic; CASRN 7440-38-2. www.epa.gov/NCEA/iris/subst/0278.htm.

One Century of the Discovery of Arsenicosis in Latin America (1914–2014) –
Litter, Nicolli, Meichtry, Quici, Bundschuh, Bhattacharya & Naidu (Eds)
© 2014 Taylor & Francis Group, London, ISBN 978-1-138-00141-1

Titanium dioxide-photocatalytic reduction of pentavalent and trivalent arsenic: Production of elemental arsenic and arsine

M.I. Litter & I.K. Levy
Comisión Nacional de Energía Atómica, Buenos Aires, Argentina

F. Requejo & M. Mizrahi
Instituto de Investigaciones Fisicoquímicas Teóricas y Aplicadas, La Plata, Argentina

G. Ruano & G. Zampieri
Comisión Nacional de Energía Atómica, Bariloche, Argentina

ABSTRACT: Heterogeneous photocatalytic reduction of Arsenic (As V) and arsenic (As III) at different concentrations over titanium dioxide (TiO_2) under ultraviolet light in deoxygenated aqueous suspensions is described. For the first time, arsenic (As 0) was unambiguously identified together with arsine (AsH_3) as reaction products. Arsenic (V) reduction requires the presence of an electron donor (methanol in the present case) and takes place through the hydroxymethyl radical formed from methanol oxidation by holes or hydroxyl radicals while arsenic (III) reduction takes place through direct reduction by the TiO_2-conduction band electrons. Detailed mechanisms for the photocatalytic processes are proposed. Although reduction to solid As(0) is convenient for As removal from water as a deposit on TiO_2, attention must be paid to formation of AsH_3, one of the most toxic forms of As, and strategies for AsH_3 treatment should be considered.

1 INTRODUCTION

Heterogeneous Photocatalysis (HP) with titanium dioxide (TiO_2) is one of the most studied Advanced Oxidation Processes (AOP) for water treatment. After irradiation with photons, electrons in the conduction band (e_{CB}^-) and holes in the valence band (h_{VB}^+) are produced, followed by redox reactions with solution species (Litter, 1999, 2009).

$$TiO_2 + h\nu \rightarrow e_{CB}^- + h_{VB}^+ \quad (1)$$

Oxidative HP reactions of As(III) to As(V) over TiO_2 have been thoroughly studied (see e.g. Litter, 2009) and proposed to occur through oxidation to As(IV), easily driven by hydroxyl radical ($HO^•$) produced from water or by h_{VB}^+, taking into account the reduction potential of h_{VB}^+ generated from P25 (+2.9 V, all potentials vs. NHE), the value of the reduction potential of the $HO^•/H_2O$ couple ($E^0 \approx +2.7$ V) and that of the As(IV)/As(III) couple ($E^0 \approx +2.4$ V) (Kläning *et al.*, 1989).

$$H_2O + h_{VB}^+ \rightarrow HO^• + H^+ \quad (2)$$

$$As(III) + h_{VB}^+ \{HO^• + H^+\} \rightarrow As(IV) \{H_2O\} \quad (3)$$

On the contrary, HP reduction of As (V) or As (III) has been scarcely studied. Actually, reduction of As (V) to As (IV) by e_{CB}^- (eq. (4)) is not thermodynamically possible ($E^0 \approx -1.2$ V, Kläning et al., 1989) in relation with the reduction level of P25 e_{CB}^- (≈ -0.3 V):

$$As(V) + e_{CB}^- \rightarrow As(IV) \quad (4)$$

However, an indirect reductive mechanism might take place in the presence of sacrificial electron donors like alcohols or carboxylic acids, able to produce strongly reductive radicals. In agreement, negligible As (V) HP removal was observed in the absence of electron donors (anoxic conditions), but removal was actually possible in the presence of different electron donors (Choi et al., 2010, Yang et al., 1999). On the other hand, As (III) reduction was never proposed and, generally, only oxidation to As (V) in solution was evaluated (Jayaweera *et al.*, 2003, Ferguson *et al.*, 2005).

In this paper, reduction of As (V)/(III) species by TiO_2-HP under anoxic conditions is revisited and thoroughly analyzed. As (III) effective reduction in the absence of donors, and clear identification of As (0) and arsine (AsH_3) as products are presented for the first time (Levy *et al.*, 2012).

2 EXPERIMENTAL SECTION

2.1 *Materials and chemicals*

TiO_2 (AEROXIDE® TiO_2 P25, Evonik) was used as received. Sodium meta-arsenite (Baker), sodium arsenate dibasic 7-hydrate (Baker), methanol (MeOH, 99.9%, Carlo Erba) and silver diethyldithiocarbamate (Merck) were used. $HClO_4$ (70–72%, Merck) was employed for pH adjustments. All reagents were of the highest purity. Solutions and suspensions were prepared with deionized water (Apema Osmoion, resistivity = 18 MΩ cm).

2.2 *Photocatalytic experiments*

Irradiations were performed in a quartz photoreactor well, provided with a medium pressure mercury lamp (125 W, maximum emission at 366 nm), surrounded by a thermostatic jacket at 25°C acting as IR filter.

TiO_2 suspensions containing As (V) or As (III) at fixed concentrations were ultrasonicated for 30 s, poured in the outer jacket of the well, and magnetically stirred and bubbled with N_2 all throughout the reaction period. An initial stirring for 30 min in the dark was performed to reach the adsorption equilibrium of species onto TiO_2; the decrease of As concentration before switching on the lamp was discounted to establish the initial concentration and take into account changes due only to light irradiation. No changes in As concentration were observed under irradiation in the absence of photocatalyst.

Initial As (V) concentrations were 0.525, 0.065 and 0.013 mM, while those of As (III) were 0.525 and 0.013 mM; pH was initially adjusted to 3, and left to vary freely during the runs. TiO_2 concentration was always 1 g L^{-1}. MeOH was added at 0.4 M. Periodically, samples were taken from the suspensions and filtered through 0.2 μm Millipore membranes before analysis.

Actinometry with ferrioxalate indicated a photon flow per unit volume incident on the cell wall of 127 μeinstein s^{-1} L^{-1}.

2.3 *Analytical determinations in filtered samples*

Changes in As (V) concentration in solution were measured using the arsenomolybdate technique (Lenoble et al., 2003). Total As was determined similarly by previous oxidation of As (III) with $KMnO_4$. Quantify® Arsen 10 strips (Macherey-Nagel) allowed semi quantitative measurement of AsH_3. Quantitative AsH_3 determination was performed by the silver diethyldithiocarbamate spectrophotometric method (AgDDC, modified Gutzeit method, Rand et al., 1976).

2.4 *Analysis of solid residues*

For XRD, a Philips PW-3710 diffractometer was used. For SEM-EDS, a Fei Company Quanta 200 apparatus was employed. TEM images were obtained with a TEM-EM 301 Philips apparatus (60 kV). X-ray Photoemission Spectra (XPS) were taken with a hemispherical electrostatic energy analyzer (r = 10 cm) using Al K_α radiation (hν= 1486.6 eV). X-ray absorption near edge spectroscopy (XANES) analyses was performed using the X-ray absorption spectrometer Rigaku R-XAS Looper.

3 RESULTS

3.1 *As (V) photocatalytic experiments*

Figure 1 shows results of HP experiments with 0.525 mM As (V). In the absence of MeOH, As (V) concentration did not change during the run (not shown), but the decay was evident when MeOH was added, reaching an almost total As (V) removal at ca. 150 min. At the end of the run, a gray-bluish solid layer was found onto the TiO_2 surface, suggesting the formation of As (0) in the α form. In same Figure 1, evolution of total As in solution is plotted, from which As (III) was calculated. As(III) increased steadily, but around 90 min, when As(V) was almost totally depleted, As(III) began to decay, indicating photocatalytic transformation of this species. As (0) evolution was calculated by mass balance. Data at 270 min indicated still the presence of 68% of As in solution, in the form of As (III).

Analysis with the Quantify® strips allowed AsH_3 detection in the headspace of the photoreactor. In a separate quantitative experiment, AsH_3

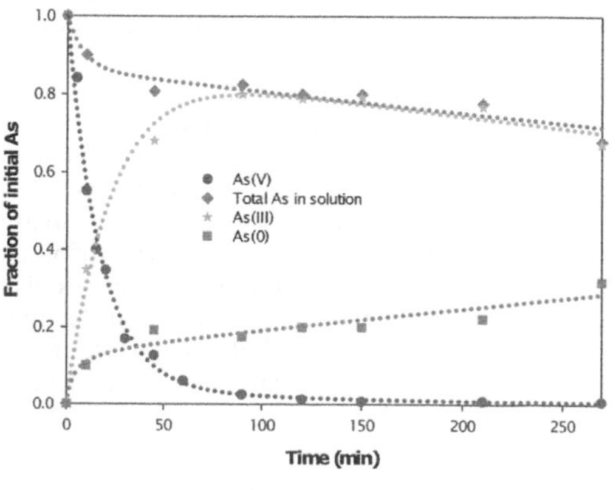

Figure 1. Temporal evolution of As species under light irradiation over TiO_2 in the presence of MeOH starting from As (V).

began to be detected at 135 min, and 3.9×10^{-4} mM were measured at 160 min.

Similar HP experiments starting from lower As (V) concentration (0.065 mM) gave 82% removal after 30 min in the dark, without darkening of the photocatalyst, indicating that the decrease was due only to adsorption. However, after 30 min more under irradiation, 90% As (V) removal was achieved, and the gray deposit of As (0) was observed, together with AsH_3 evolution. When starting from 0.013 mM As V), 82% removal was observed after 30 min in the dark. After 30 min more under irradiation, As (0) appearance and AsH_3 evolution also took place, while As remaining in solution was below 10 μg L^{-1}. These results are clearly indicative that reductive pathways took place under irradiation.

3.2 *As (III) photocatalytic experiments*

Experiments starting from 0.525 mM As (III) and 0.4 M MeOH are shown in Figure 2(a). After 195 min of irradiation, 38% As (III) was removed. As (V) was not found in solution but was detected in traces in the solid residue after desorption with $NaHCO_3$. Figure 2(a) also shows that As (III) decays at a lower rate than As (V) in Figure 1, with a totally different kinetic profile. As (0) was deposited on the catalyst; a mass balance indicates 38% at 195 min. AsH_3 evolved initially at 30 min; at 135 min, Quantofix® strips indicated a rather higher production than in the As (V) system.

An experiment starting from 0.013 mM As (III) yielded more than 90% As removal in the dark after initial contact with the photocatalyst, without evidences of gray deposits. After 30 min more under irradiation, As(0) and AsH_3 formation were observed, leaving less than 10 μg As L^{-1} in solution.

Figure 2(b) shows similar experiments but in the absence of MeOH. In contrast with the total lack of reaction of As (V) in the absence of donor, a rather important decay of As (III) (similar to that of total As) was observed in these conditions, although somewhat lower than in the presence of MeOH. In this case, As (V) was actually detected in solution: the maximum quantity, ca. 2%, was found at 10 min, followed by a decrease (Figure 2(b), inset), which could be attributed to adsorption onto TiO_2. Figure 2(b) shows also As (0) formation. AsH_3 was detected at 45 min. A separate experiment for AgDDC measurement indicated that at 60 min 0.06% of As had been transformed into AsH_3.

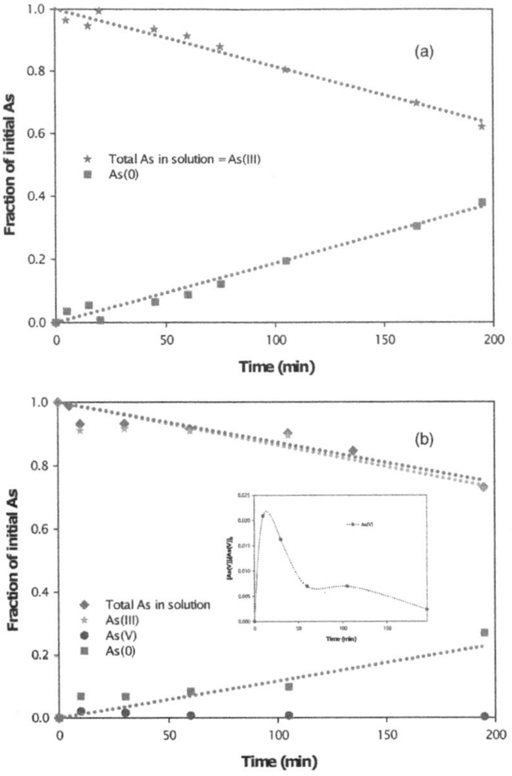

Figure 2. Temporal evolution of As species under light irradiation over TiO_2 starting from As (III), (a) in the presence of 0.4 M MeOH, (b) in the absence of MeOH.

3.3 *Analysis of solid residues*

Dark residues, attributed to As (0), were deposited on TiO_2 after the HP reactions and, when exposed to air, whitened gradually with time, indicating oxidation to arsenic oxides. XRD patterns corresponded always to P25, without As signals. SEM/EDS showed that the deposit was composed of nanoparticles, formed by As in a TiO_2 matrix. TEM images indicate a particle size between 10 and 15 nm, disposed in chain form.

XPS and XAS studies also demonstrated unambiguously the presence of As (0) on the solids after HP experiments starting from both As (V) and As (III) (Levy et al., 2012).

4 DISCUSSION

As said before, direct As(V) reduction by P25 e_{CB}^- is not possible, but an indirect reductive mechanism is possible by the reducing hydroxymethyl radicals ($^{\bullet}CH_2OH$) formed by $h_{VB}^+/HO^{\bullet}$ attack to MeOH, (eqs. (5)). In the absence of O_2, $^{\bullet}CH_2OH$ can donate electrons to the CB (current doubling effect, eq. (6)) or be the effective As (V) reductant, with formaldehyde generation (eq. (7)).

$$CH_3OH + h_{VB}^+ \{HO^\bullet\} \rightarrow {}^\bullet CH_2OH + H^+ \{H_2O\} \quad (5)$$

$$^\bullet CH_2OH \rightarrow CH_2O + H^+ + e_{CB}^- \quad (6)$$

$$^\bullet CH_2OH + As(V) \rightarrow CH_2O + As(IV) + H^+ \quad (7)$$

In the HP system, formaldehyde can be transformed to Formic Acid (FA) and finally to CO_2. FA generates a much stronger reducing agent, $CO_2^{\bullet-}$, which can contribute to the reducing process of As (V).

$$CH_2O \rightarrow HCOOH \rightarrow CO_2 \quad (8)$$

$$HCOOH + h_{vb}^+ \{HO^\bullet\} \rightarrow CO_2^{\bullet-} + 2\,H^+ \{H_3O^+\} \quad (9)$$

$$CO_2^{\bullet-} + As\,(V) \rightarrow CO_2 + As\,(IV) \quad (10)$$

As (IV) is easily reduced to As (III) by CB or trapped electrons (11), by $^\bullet CH_2OH$ (11) or by $CO_2^{\bullet-}$ (not shown):

$$As(IV) + e_{CB}^- / e_{trapp}^- \{^\bullet CH_2OH\} \rightarrow As(III) \{CH_2O + H^+\} \quad (11)$$

At the high MeOH concentrations used (0.4 M), (5) would be preferred over competing As (IV) reoxidation:

$$As(IV) + h_{VB}^+ \{HO^\bullet + H^+\} \rightarrow As(V) \{H_2O\} \quad (12)$$

As (IV) can also rapidly disproportionate:

$$2\,As(IV) \rightarrow As(III) + As(V) \quad (13)$$

Once formed, As (III) can be also photocatalytically reduced (Figures 1 and 2(a)). However, As (III) reduction can actually take place in the absence of MeOH (Figure 2(b)), oppositely to As (V). As (III) reduction would lead to the unstable As (II) form:

$$As(III) + e_{CB}^- / e_{trapp}^- \{^\bullet CH_2OH\} \rightarrow As(II) \{CH_2O + H^+\} \quad (14)$$

(or equivalent reaction with $CO_2^{\bullet-}$).

Successive reactions would lead to As (0) and AsH_3.

In the absence of MeOH, As (III) HP reduction is somewhat less efficient (cf. Figures 2(a) and (b)) because the anodic reaction is the sluggish oxidation of water by h_{VB}^+:

$$2\,H_2O + 4\,h_{VB}^+ \rightarrow O_2 + 4\,H^+ \quad (15)$$

Competition of As (II) for the charge carriers would lead to unproductive short-circuiting and reoxidation to As (III):

$$As(II) + h_{VB}^+ \{HO^\bullet + H^+\} \rightarrow As(III) \{H_2O\} \quad (16)$$

In the absence of MeOH, As(III) can be not only reduced through (11), but also oxidized to As(IV) through (3), helped by traces of O_2 produced in (15). This leads ultimately to As (V) formation through (12), (13) or by injection of electrons to the CB (eq. (18)), explaining the formation of low amounts of As (V) in the system.

$$As(IV) \rightarrow As(V) + e_{CB}^- \quad (17)$$

H^+ reduction to $H^\bullet$, ending in H_2, can be another reducing source. $H^\bullet$ and H_2 would be rapidly consumed, forming e.g. AsH_3, or reducing As (V) and As (III) to As (0). H^+ reduction can be assisted by small spots of nanosized As (0), with formation of a Schottky barrier at the As/TiO_2 interface, enabling electrons to flow from TiO_2 to As (0):

$$H^+(As(0)) + e_{CB}^- \rightarrow H^\bullet(As(0)) \quad (18)$$

$$2\,H^\bullet(As(0)) \rightarrow H_2 \quad (19)$$

$$H^\bullet(As\,(0)) + CH_3OH \rightarrow As\,(0) + H_2 + {}^\bullet CH_2OH \quad (20)$$

5 CONCLUSIONS

As (III) HP reduction is possible in deoxygenated suspensions. As (V) can be also reduced, but only in the presence of an electron donor. As (0) and AsH_3 are formed. The reductive process is very efficient at the lowest As concentrations (e.g., 0.013 mM ≡ 1 mg L^{-1}, a common value found in natural polluted groundwaters) attaining As levels in agreement with drinking water regulations (<10 μg L^{-1}).

As(0) formation results a definite way to make less mobile As(V) and As(III) species present in aqueous media. However, it is important to remark that AsH_3 can be formed and should be removed or contained.

ACKNOWLEDGMENTS

This work was carried out as part of Agencia Nacional de Promoción Científica y Tecnológica PICT-512, PICT-2008-00038, PAE-PME-2007-00039, PICT 33432 and CYTED 406RT0282.

REFERENCES

Choi, W., Yeo, J., Ryu, J., Tachikawa, T. & Majima, T. 2010. Photocatalytic oxidation mechanism of As (III) on TiO_2: unique role of As (III) as a charge recombinant species. *Environmental Science and Technology* 44: 9099–9104.

Ferguson, M.A., Hoffmann, M.R. & Hering, J.G. 2005. TiO_2-photocatalyzed As (III) oxidation in aqueous suspensions: reaction kinetics and effects of adsorption. *Environmental Science and Technology* 39: 1880–1886.

Jayaweera, P.M., Godakumbra, P.I. & Pathiartne, K.A.S. 2003. Photocatalytic oxidation of As (III) to As (V) in aqueous solutions: A low cost pre-oxidative treatment for total removal of arsenic from water. *Current Science* 84: 541–543.

Kläning, U.K., Bielski, B.H.J. & Sehesteds, K. 1989. Arsenic (IV). Pulse-radiolysis study. *Inorganic Chemistry* 28: 2717–2724.

Lenoble, V., Deluchat, V., Serpaud, B. & Bollinger, J.-C. 2003. Arsenite oxidation and arsenate determination by the molybdene blue method. *Talanta* 61: 267–276.

Levy, I.K., Mizrahi, M., Ruano, G., Zampieri, G., Requejo, F.G. & Litter, M.I. 2012. TiO_2-photocatalytic reduction of pentavalent and trivalent arsenic: production of elemental arsenic and arsine, *Environmental Science and Technology* 46(4): 2299–2308.

Litter, M.I. 1999. Heterogeneous Photocatalysis. Transition metal ions in photocatalytic systems. *Applied Catalysis B: Environmental* 23: 89–114.

Litter, M.I. 2009. Treatment of chromium, mercury, lead, uranium and arsenic in water by heterogeneous photocatalysis. *Advances in Chemical Engineering* 36: 37–67.

Rand, M.C., Greenberg, A.E. & Taras, M.J. (eds). 1976. *Standard methods for the examination of water and wastewater*, APHA-AWWA-WPCF: Washington D.C., pp. 283–284.

Yang, H., Lin, W.-Y. & Rajeshwar, K. 1999. Homogeneous and heterogeneous photocatalytic reactions involving As (III) and As(V) species in aqueous media. *Journal of Photochemistry and Photobiology A: Chemistry* 123: 137–143.

4.5 Remediation of arsenic contaminated soil environments

*One Century of the Discovery of Arsenicosis in Latin America (1914–2014) –
Litter, Nicolli, Meichtry, Quici, Bundschuh, Bhattacharya & Naidu (Eds)*
 ISBN 978-1-138-00141-1

Characterization of the soil porosity for a hierarchy of pore-length scales, from nano-pores to macro-pores: Implications to the efficiency of soil remediation technologies

C.D. Tsakiroglou
Foundation for Research and Technology Hellas, Institute of Chemical Engineering Sciences, Platani, Patras, Greece

ABSTRACT: Mineral soils are heterogeneous at multiple scales with pore sizes spanning several orders of magnitude from nm to mm. A methodology is suggested for the complete characterization of such porous media by combining the autocorrelation function of Back-scattered Scanning Electron Microscopy (BSEM) images with inverse modeling of mercury intrusion porosimetry data. The method is demonstrated with application to samples of a low permeability and heterogeneous soil contaminated with jet fuel, and its results are used to interpret the remediation efficiency of an in situ bioventing pilot experiment. The pore space characteristics of soils are of key importance for the efficient in situ immobilization or removal of arsenic from vadose or saturated zone by injecting nanoparticle suspensions.

1 INTRODUCTION

Unlike many synthetic porous materials designed to have narrow pore-size distributions (Tsakiroglou *et al.*, 2004) mineral soils comprise networks of interconnected pores, the sizes of which span several orders of magnitude from nm to mm. Back-scattered Scanning Electron Microscopy (BSEM) is suitable to provide statistically significant micro-structure data at length scales from several mm to a few μm (Tsakiroglou & Ioannidis, 2008). Indirect imaging methods like the Small-Angle Neutron Scattering and X-ray Scattering (SANS and SAXS) yield the volume-averaged Fourier transform of the two-point correlation function at length scales 1 nm–10 μm (Radlinski *et al.*, 2004). Mercury porosimetry is widely used to probe the pore space in the range 20 nm-200 μm, but instead of the pore size distribution, this method provides the distribution of pore volume that is accessible to Hg through pore throats of different size. The statistical fusion of SANS and BSEM data and their subsequent interpretation in terms of a polydispersed spherical pore model was proposed as a potential method to characterize the pore structure of sandstones and detect pore sizes in the range 1 nm–1 μm, including the fractal and Euclidean pore space, as well (Radlinski *et al.*, 2004). The method was modified and information normally provided by SANS measurements has been substituted approximately by a fractal scaling law using estimates of the surface fractal dimension from mercury intrusion porosimetry (Tsakiroglou & Ioannidis, 2008; Tsakiroglou *et al.*, 2009). The single—and multiphase transport properties, measured with flow tests in undisturbed soil columns, are governed by preferential flow paths (Aggelopoulos & Tsakiroglou, 2009) and multi-scale models are required for their numerical prediction (Tsakiroglou, 2012).

In this work, a method is presented for the detailed characterization of the microporous structure of highly heterogeneous porous media in terms of dual pore networks. The method is demonstrated with application to soil samples, and its results are employed to interpret the main mechanisms of a bioventing pilot test applied to the vadose zone of a low permeability and heterogeneous soil contaminated with jet fuel.

2 MATERIALS AND METHODS

2.1 *In situ soil remediation*

In the context of STRESOIL project (2004–2007), the abandoned Kluczewo airport in Poland was used as experimental site to develop an integrated strategy for the enhanced treatment of the vadose zone of fractured low permeable soils by combining hydraulic fracturing with two in situ remediation technologies: steam injection (Nilsson *et al.*, 2011), and bioventing. The subsurface (Figure 1) is characterized by a low permeable glacial till matrix intersected by desiccation fractures in upper layers, sand lenses in intermediate layers, and glacio-tectonic fractures in lower layers (Tzovolou *et al.*, 2009).

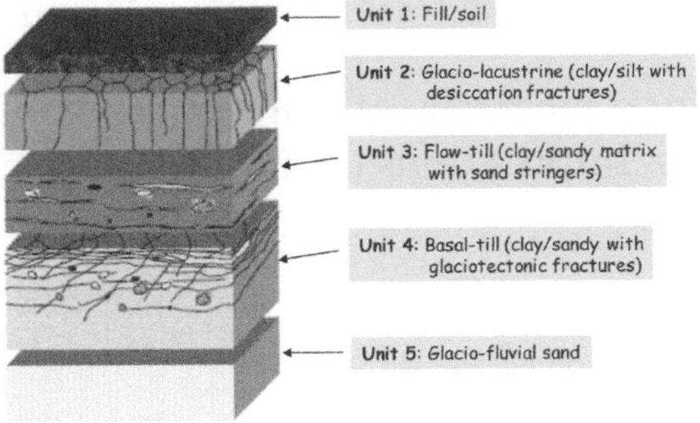

Figure 1. Geological conceptual model of Kluczewo site.

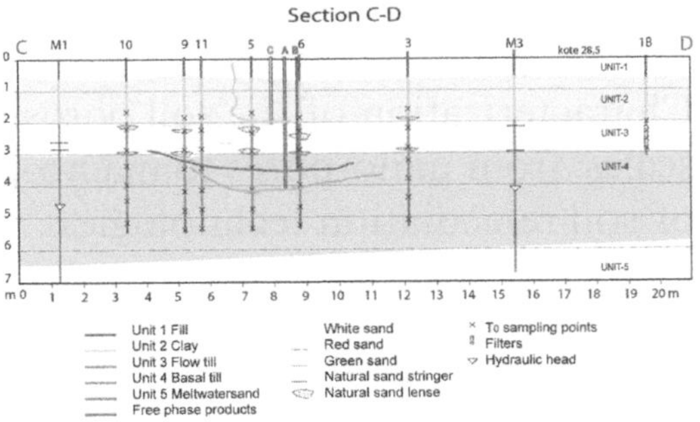

Figure 2. Cross-sections C-D of the fracture distribution profile on Cell 1a (bio-venting) from excavation data.

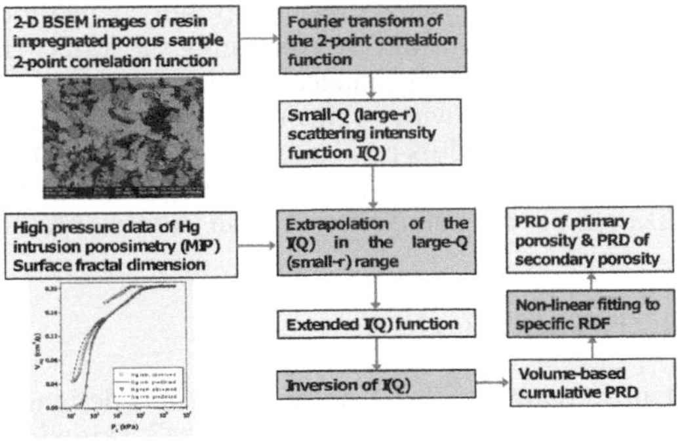

Figure 3. Methodology for the estimation of the complete Pore Radius Distribution (PRD) from BSEM images and high pressure MIP data.

Bioventing was applied to three Cells (1a, 1b, 1c) of dimensions 10 m × 10 m × 5 m. In Cell 1a, three hydraulic fractures were opened at various depths (Figure 2) by injecting jets of a slurry (sand gravels mixed with guar gum) at high pressure (Nilsson *et al.*, 2011). The air was injected in the green (lower) fracture without applying any vacuum in extraction well connected to the red (upper) fracture (Figure 2). Due to the presence of tectonic fractures in Unit 4 (Figure 1), the air flow, oxygen diffusion, and Volatile Organic Compounds (VOCs) extraction were facilitated. The remediation efficiency was monitored by collecting samples during three campaigns (t = t0, t1, t2) from seven (7) depths of 12 wells, providing thus a total number of eighty four (84) samples for off-site chemical analyses with GC-MS.

Batch biodegradation essays were performed in penicillin flasks by varying the soil mass and monitoring the CO_2 and O_2 concentrations in the gas phase. The limiting factors related to the presence of nutrients (N, K and P) were evaluated and the concentration of residual hydrocarbons were measured with GC-FID.

Soil column biodegradation tests were conducted by introducing 100 g of distributed soil or undisturbed soil aggregates of Units 3 and 4 inside a sand matrix. The experiments were conducted at 20 °C and lasted 90 days. The oxygen consumption, particularly with reference to nutrient addition, was significant and monitored by a respirometer.

2.2 *Pore structure analysis*

A new method was developed to characterize the complicated pore structure of mineral soils (e.g. clayey/silty sands) by combining data of the analysis of BSEM images of resin-impregnated pore-casts and mercury intrusion/retraction curves with analytical percolation models and inverse modeling numerical algorithms (Tsakiroglou *et al.*, 2009).

The pore space is regarded as a dual pore network consisting of a primary Euclidean pore-and-throat network and a secondary fractal pore system that is accessed through primary pores. First, well-polished thick slices of resin-impregnated soil samples are prepared. Back-scattered Scanning Electron Microscope (BSEM) images of the 2-D area of the slices are captured and employed to determine the AutoCorrelation Function (ACF). The Fourier transform of this function provides the Small-Angle Neutron Scattering (SANS) intensity function, which is extended by using the surface fractal dimension obtained from high pressure Hg intrusion (MIP) data (Figure 3). Then, the inverse Fourier transform of Scattering Intensity Function (SIF) produces the volume-based radius distribution function of equivalent spherical pores (PRD) (Figure 3).

The complete volume based PRD is fitted with a composite PRD composed of a lognormal primary PRD and a power (fractal) secondary PRD with upper and lower cut-offs. The complete PRD and Throat Radius Distribution (TRD) of the primary network along with the Drainage Accessibility Functions (DAFs) of the primary and secondary pore networks are estimated with inverse modeling

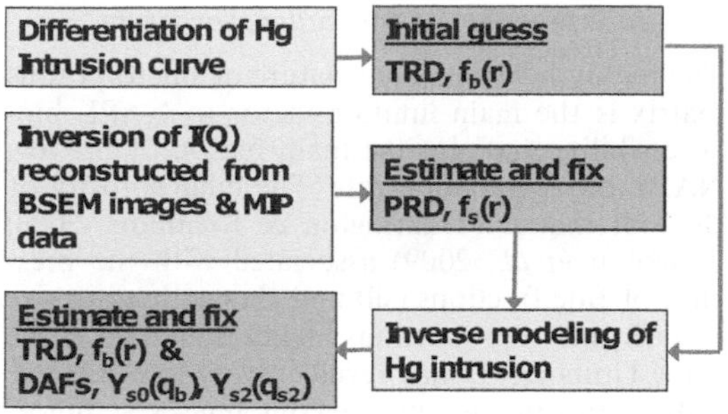

Figure 4. Methodology for the estimation of Throat Radius Distribution (TRD) and Drainage Accessibility Functions (DAFs) from Mercury Intrusion Porosimetry (MIP) data.

Table 1. Estimated total NAPL mass in cell C1a before, during and after bioventing.

Depth(m)/ Unit	Total NAPL mass (kg) t0 = 0 days $\langle M\rangle \pm \sigma_M$	t1 = 60 days $\langle M\rangle \pm \sigma_M$	t2 = 269 days $\langle M\rangle \pm \sigma_M$	Efficiency (%)
1.9/2	264 ± 392	471 ± 320	253 ± 156	4
2.8/3	134 ± 78	442 ± 411	164 ± 87	–23
3.2/3	152 ± 172	259 ± 183	122 ± 45	19
3.6/4	96 ± 91	185 ± 118	172 ± 338	–79
4.2/4	315 ± 275	223 ± 146	96 ± 25	70
4.8/4	1471 ± 984	319 ± 205	137 ± 56	91
5.4/4	1802 ± 901	605 ± 249	241 ± 147	87
Sum	4233 ± 984	2504 ± 411	1186 ± 338	72

(non-linear optimization) of the Hg intrusion curve by using analytic percolation models (Figure 4).

The aforementioned methodology was demonstrated with application to the characterization of the pore space of soil samples collected from all Units of Kluczewo site (Figure 1), so that the efficiency of bioventing pilot experiment was interpreted.

3 RESULTS AND DISCUSSION

3.1 *Assessing the efficiency of bioventing*

The concentrations of residual Non-Aqueous Phase Liquids (NAPL) measured with GC-MS in soil samples collected from the various wells during the bioventing pilot experiment were used to estimate the total mass of accumulated NAPL in the various layers of Cell 1a (Table 1). The highest NAPL removal efficiency (70–90%) was observed in the lower layers of Unit 4 where the highest initial NAPL concentrations were measured. No significant NAPL removal efficiency occurred in Units 2 and 3; in contrast, in some cases and for intermediate times, the residual NAPL mass in upper layers was increased (Table 1).

The batch tests revealed that NAPL biodegradability increases with the mass of soil (or NAPL) increasing and the presence of nutrients, whereas the residual NAPL mass is governed by pore-scale soil heterogeneities (Tsakiroglou & Ioannidids, 2008) and subsequent non-uniform distribution of NAPL in soil samples. For both disturbed and undisturbed soil columns without nutrients, no positive biodegradation rate was evident. The biodegradation conditions were changed after 90 days by adding salt solution in the columns to enhance the gas/water and oil/water interfacial areas, and facilitate the O_2 and hydrocarbon transfer between the various phases, so that the biodegradation rate was increased.

3.2 *Pore structure properties of soil samples*

In massive clay (Unit 2/sample C2), the pore and grain sizes are very fine exhibiting a relatively small variability (Figure 5a). In contrast, in sandy soils (Unit 4/sample C15) the pore and grain sizes vary over a very broad range of length scales (Figure 5b). In this manner, C2 might be modeled as a single pore network with fractal pore-wall roughness representing the pore corners, nook and crannies. In contrast, four samples from Units 3 and 4 were modeled as dual pore networks.

Evidently, the void space of sandy soils is a mixture of pore systems dominated by different pore length scales that cover several orders of magnitude (Figure 5b). This is attributed to the very broad grain size distribution of the sandy material and the high percentage (~20–30%) of clay and silt content (Figure 5). The surface fractal dimension was determined from the high pressure MIP data, and was employed to extend the scattering intensity function (derived from the Fourier transform of the average ACF) to low pore radii (Figure 1). The inversion of the extended SIF produced the volume-based cumulative pore radius distribution (Figure 6).

The volume-based PRD resulting from the analysis of BSEM images was decomposed into the number-based component PRDs of primary and secondary porosity by fitting the observed datasets with log-normal distribution functions. By fixing the so-estimated component PRDs, introducing the analytical percolation model into the ATHENA Visual Studio software package, and inverting the Hg intrusion dataset (Figure 2), the TRD (Figure 7) and DAFs were estimated (Tsakiroglou & Ioannidis, 2008).

Some general comments on the pore structure of sandy soils (Units 3 & 4) of Kluczewo site follow: (i) the TRD of primary porosity is very broad extending to pore length scales that differ by 4–5

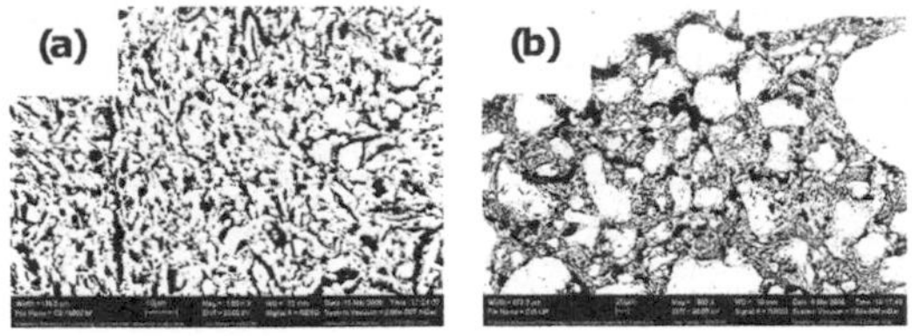

Figure 5. 2-D BSEM binary images of (a) sample C2 (Unit 2, 1500x), (b) sample C15 (Unit 4, 300x).

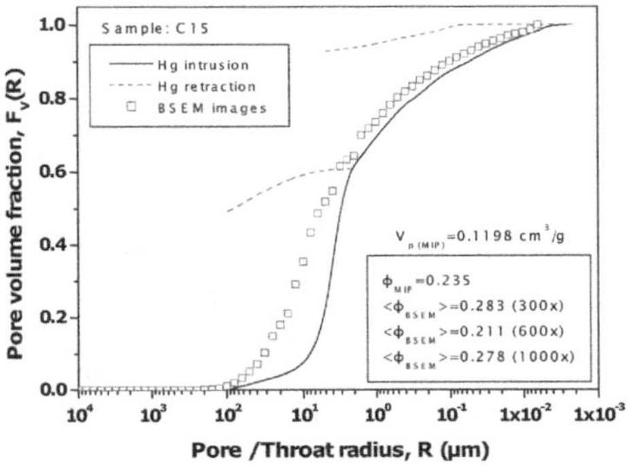

Figure 6. Cumulative volume-based spherical pore radius distribution derived from BSEM images and MIP data (surface fractal dimension $D_s = 2.7$).

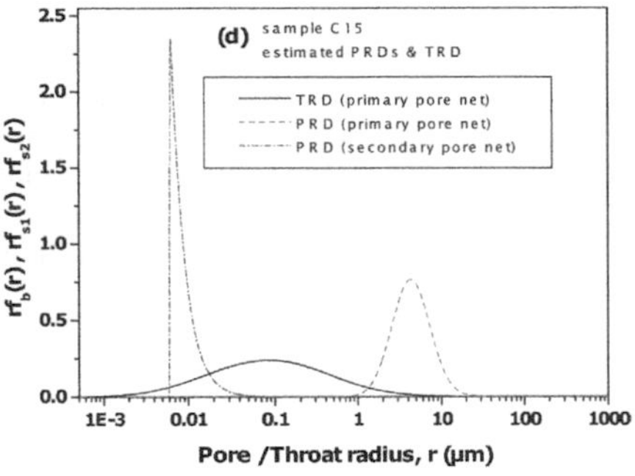

Figure 7. Estimated PRD & TRD of primary (macroporosity) and PRD of secondary (microporosity) network.

orders of magnitude (Figure 7); (ii) the PRD of primary porosity extends to large pore sizes (Figure 7), it is a very small percentage ($<10^{-7}$) of the total number-based PRD but represents more than 50% of the total pore volume; (iii) the PRD of secondary porosity extends to nm scale (Figure 7), it is fractal, and represents a significant percentage of the total pore volume (30–40%); (iv) the percolation threshold is very low (~0.01–0.03), and hence the hydraulic conductivity is governed by preferential flow paths (Aggelopoulos & Tsakiroglou, 2009).

3.3 *Interpretation of bioventing experiment*

The highly heterogeneous nature of micro-porous matrix is the main limiting factor to NAPL bio-accessibility and also the main reason of the low NAPL biodegradation rate. The dual porosity of the soil matrix (Tsakiroglou & Ioannidis, 2008; Tzovolou *et al.*, 2009) associated with the presence of fine fractions (silt and clay with grain size <0.063 μm) at proportions higher than 25% in the three Units 2/3/4, may result in very low effective diffusivities in gas and aqueous phases acting as another limiting factor to the jet fuel biodegradation kinetics. Under unsaturated conditions, the wetting aqueous phase occupies the sub-networks of finest pores so that the access of indigenous bacteria to gas phase (oxygen) and NAPL constituents, and subsequently the bio-accessibility of NAPL may be limited.

It is evident that the main mechanism of NAPL (jet fuel) removal during bioventing was ventilation. (Figure 8). The NAPL mass trapped in micro-porous matrix is bypassed by the injected air that moves along preferential flow pathways. The vapors of hydrocarbons created above NAPL blobs have the tendency to move via diffusion toward preferential flow paths. Due to the forced convective air flow, the concentration of NAPL compounds is kept close to zero along these pathways, and diffusion rate is enhanced. In this manner, the vaporization rate of volatile and semi-volatile NAPL compounds increases.

3.4 *Implications to arsenic removal from soils*

During the last years, attention has also been paid on the in situ remediation of contaminated soils by injecting suspensions of reactive/adsorptive nanoparticles (Kanel *et al.*, 2007). Among the various nano-materials explored for remediation, nanoscale zerovalent iron (nZVI), iron-oxide and nanocomposites have been suggested for the in situ immobilization or removal of As(III) and As(V) from contaminated soils (An & Zhao, 2012). In order to assess the efficiency of relevant processes (e.g., As sorption in nanoparticles, As(III) oxidation to As(V)), information is required not only for the mobility/longevity/reactivity/sorption capacity

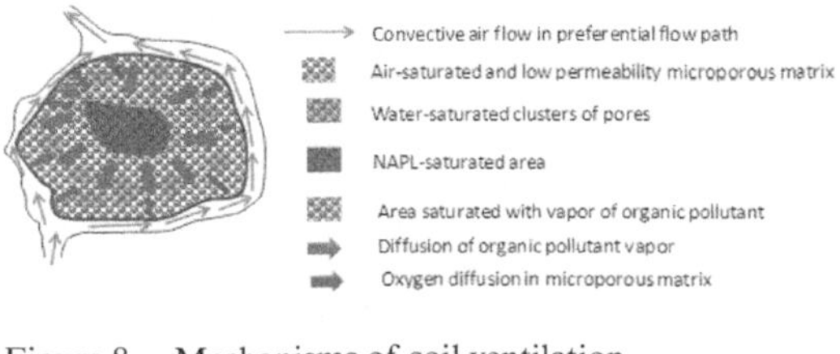

Figure 8. Mechanisms of soil ventilation.

of nanoparticles, but also for the soil properties quantifying the pore structure and heterogeneity.

REFERENCES

Aggelopoulos, C.A. & Tsakiroglou, C.D. 2009. A multi-flowpath model for the interpretation of immiscible displacement experiments in heterogeneous soil columns. *J. Contam. Hydrol.* 105: 146–160.

An, B. & Zhao, D. 2012. Immobilization of As(III) in soil and groundwater using a new class of polysaccharide stabilized Fe–Mn oxide nanoparticles. *J. Hazard. Mat.* 211–212: 332–341.

Kanel, S.R., Nepal, D., Manning, B. & Choi, H. 2007. Transport of surface-modified iron nanoparticle in porous media and application to arsenic(III) remediation. *J. Nanoparticle Res.* 9:725–735.

Nilsson B., Tzovolou, D., Jeczalik, M., Kasela, T., Slack, W., Klint, K.E., Haeseler, F. & Tsakiroglou, C.D. 2011. Combining steam injection with hydraulic fracturing for the in situ remediation of the unsaturated zone of a fractured soil polluted by jet fuel. *J. Environ. Manag.* 92: 695–707.

Radlinski, A.P., Ioannidis, M.A., Hinde, A.L., Hainbuchner, M., Baron, M., Rauch, H. & Kline, S.R. 2004. Angstrom-to-millimeter characterization of sedimentary rock microstructure. *J. Colloid Interface Sci.* 274: 607–612.

Tsakiroglou, C.D., V.N. Burganos, V.N. & Jacobsen, J. 2004. Pore structure analysis by using nitrogen sorption and mercury intrusion data. *AIChE J.* 50: 489–510.

Tsakiroglou, C.D. & Ioannidis, M.A. 2008. Dual porosity modeling of the pore structure and transport properties of a contaminated soil. *Eur. J. Soil Sci.* 59: 744–761.

Tsakiroglou, C.D., Ioannidis, M.A., Amirtharaj, E. & Vizika, O. 2009. A new approach for the characterization of the pore structure of dual porosity rocks. *Chem. Eng. Sci.* 64: 847–859.

Tsakiroglou, C.D. 2012. A multi-scale approach to model two-phase flow in heterogeneous porous media. *Transp. Porous Media* 94: 525–536.

Tzovolou, D.N., Benoit, Y., Haeseler, F., Klint, K.E. & Tsakiroglou, C.D. 2009. Spatial distribution of jet fuel in the vadoze zone of a heterogeneous and fractured soil. *Sci. Total Environ.* 407: 3044–3054.

One Century of the Discovery of Arsenicosis in Latin America (1914–2014) – Litter, Nicolli, Meichtry, Quici, Bundschuh, Bhattacharya & Naidu (Eds)

Transport of arsenic species in geological media: Laboratory and field studies

Q.H. Hu
China University of Geosciences, Wuhan, China
The University of Texas at Arlington, Arlington, Texas, USA

S.Q. Kong, X.B. Gao & Y.X. Wang
China University of Geosciences, Wuhan, China

ABSTRACT: Using the laboratory and field approaches, as well as sensitive quantification of As(III) and As(V) with LC-ICP-MS, this work investigates the sorption, transport, and removal of arsenic species in the Datong Basin of China. It is found that Datong sediments sorb As(V) much strongly than As(III), and no inter–conversion of arsenic species is detected under the experimental conditions. A new nanoscale iron-manganese binary oxides sorbent, which can oxidize As(III) to As(V), is prepared and applied for effective ground arsenate removal in the field.

1 INTRODUCTION

Arsenic (As) has at least four oxidation states: −3, 0, +3 and +5, and is present in a variety of inorganic and organic forms in the environment (Smedley and Kinniburgh, 2002); the most common chemical forms of arsenic present in groundwater are the inorganic forms of As(V) and As(III). Arsenate is known to sorb to many mineral phases, thus exhibiting a lesser mobility in the environment. However, reduction of the sorbed As(V) results in the mobilization of the more mobile, toxic and bioavailable As(III).

Arsenic is strongly adsorbed by iron or manganese oxides, and the adsorption of As(III) is generally less significant than that of As(V). Therefore, oxidation of As(III) to As(V) is proposed as an effective option to increase the adsorption potential and to decrease the environmental risk.

Proven to be an effective, low-cost and easily-operative technology for removing arsenic from groundwater, the permeable reactive barriers packed with an arsenic-removal adsorbent is the most popular technique (Mohan and Pittman 2007). Common sorbents include activated carbons, natural minerals, metal-based methods such as zerovalent iron, bi-metallic sorbent and metal oxides or hydroxides (Mohan and Pittman 2007). We will use iron–manganese binary oxides that have received special attention from their unique characteristics for arsenic removal (Manning *et al.*, 2002; Zhang *et al.*, 2007a and 2007b).

Using integrated laboratory and field experiments, as well as advanced analytical tools, this work studies the sorption, transport, and removal of arsenic species in the sediments of Datong Basin, China. In addition, iron–manganese binary oxide nanomaterials are developed and utilized in the field to treat arsenic-laced groundwater in Datong Basin.

2 METHODS/EXPERIMENTAL

2.1 *Datong Basin sediment samples*

The Datong Basin in China is a Cenozoic basin of the Shanxi rift system (Figure 1). The evolution of the Datong Basin began in the early Pleistocene. The thickness of Cenozoic sediments in the basin ranges from 50 to 2,500 m. Shallow groundwater occurs in the Quaternary alluvial, alluvial–pluvial and alluvial–lacustrine aquifers. The shallow aquifer (<60 m) sediments are usually made up of lacustrine and alluvial–lacustrine medium–fine sand, silty clay and clay; the color is gray to blackish. The depth of the water table is generally less than 2 to 5 m in the study area.

Based on previous hydrogeochemical studies and field investigation in the Datong Basin, a site for the sediment sampling was chosen where the groundwater arsenic content was as high as up to 1,060 μg/L (Gao *et al.*, 2011). The sediment samples were collected in May 2006 from one borehole inside high-arsenic area (named as DY in Figure 1)

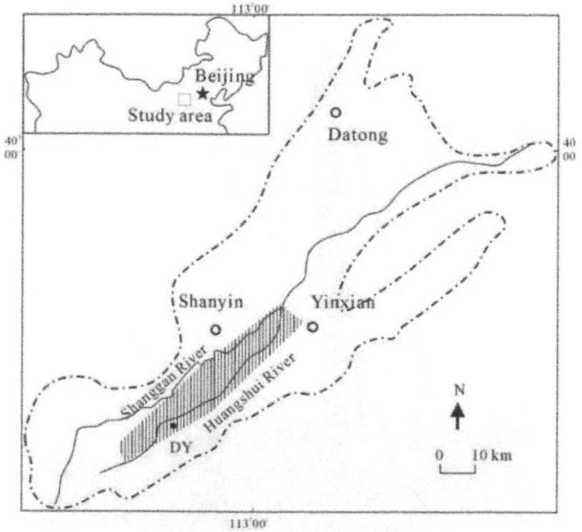

Figure 1. Location of the study area and the drilling site (DY) for sediment sampling. The shadow area indicates where high arsenic groundwater occurs.

at depths of 24.1–24.5 m (DY-12) and 50.2–50.6 m (DY-22). The samples were capped immediately with a PVC pipe and wax–sealed, thereby minimizing the exposure to the atmosphere. After collection, the samples were stored at 4 °C in the dark, air–dried in the dark at room temperature (23 °C), grinded and screened through a 2 mm sieve and stored in desiccators. The field sediment samples consist of 39% sand, 32% silt, and 29% clay. The major mineralogical phases in the sediment are quartz (40%), feldspar (20%), calcite (10%), chlorite (10%), and illite (10%); minor contents of dolomite and amphibole are noted. The sediments are alkaline, with pH values of about 7.9 for a solid: water ratio of 1:4. The total organic carbon in the sediment is 0.24%. A significant fraction of oxalate extractable As and Fe, representing As in oxyhydroxide fraction, was noticed in the sediment samples (Gao *et al.*, 2012).

2.2 *Batch sorption tests*

In batch tests, a solution of As(V) or As(III) was placed in contact with Datong sediments at a solution to solid ratio of 4:1 (Hu *et al.*, 2012). For each sediment, we conducted triplicate blank treatments (adding only 1 mM $Ca(NO_3)_2$ as the electrolyte solution), and triplicate treatments of either As(V) or As(III) at the concentration of 1 mg/L (1.3×10^{-5} M). Control samples, with As(V) or As(III) solution but no sediment, were also included to discern the potential conversion of As(V) and As(III). After 24-h equilibration, the mixture was centrifuged at 6,000 rpm for 30 min and then filtered through 0.25 μm Supor membranes with the help of vacuum.

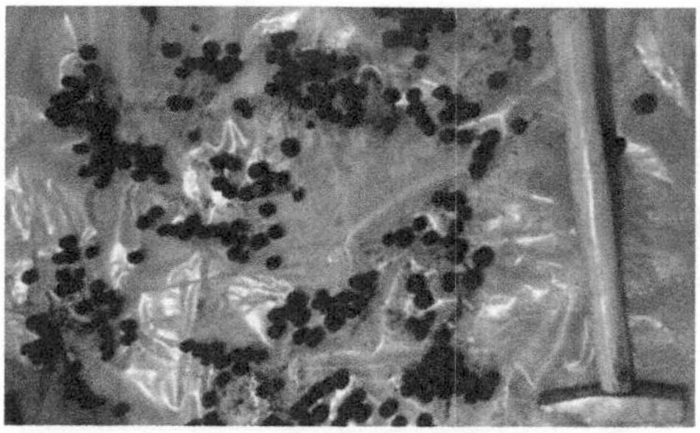

Figure 2. Bimetallic nanoscale iron–manganese oxides used for groundwater arsenic removal.

The liquid samples were analyzed for pH with a standard glass electrode, and for the concentration of As species using LC-ICP-MS, where arsenic species were separated by an anion exchange column (Hamilton PRP–X100, 4.1 mm i.d. × 250 mm long, 10 μm) and quantified with a PerkinElmer/SCIEX ELAN DRC II ICP-MS (Hu *et al.*, 2012).

2.3 *Bimetallic iron-manganese oxides*

The bimetallic nanoscale iron–manganese oxides loaded on zeolite were prepared in the laboratory to test its application of removing arsenic in groundwater in the Datong Basin (Kong *et al.*, 2013). The prepared nanoscale materials were painted on the surface of wet cement balls, which were used after drying for groundwater arsenic removal (Figure 2).

3 RESULTS AND DISCUSSION

3.1 *Batch sorption*

In the batch sorption tests, the apparent distribution coefficient values of either As(III) or As(V) include the concentration decrease both from its sorption onto solid phases and the possible interconversion; such information can only be garnered from the simultaneous measurements of As species (Figure 3). Two sediments at different vertical depths from Datong Basin of Shanxi in China (Datong DY-12 and DY-22) show much higher sorption capacity for As(V) than As(III). For sediment DY-12, the distribution coefficients are 149 ± 5.42 mL/g and 1,060 ± 238 mL/g for As(III) and As(V) species, respectively; the values are 275 ± 9.95 mL/g and 1,434 ± 552 mL/g for DF-22. The results are consistent with those of numerous reports citing stronger sorption of As(V) than As(III) in minerals and soils at the concentration ranges typical of natural systems. In addition, no inter-conversion of arsenic species was noticed under the experimental (atmospheric) conditions.

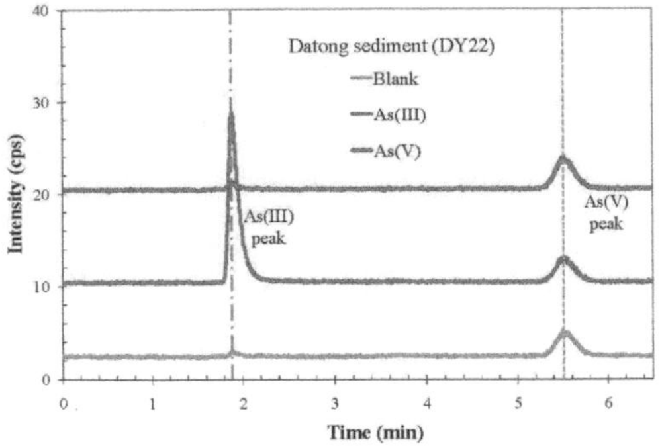

Figure 3. LC chromatograms with ICP-MS detection for Datong sediment DY-22 with three treatments (blank: 1 mM $Ca(NO_3)_2$ electrolyte solution; As(V): 1.3×10^{-5} M As(V) in the electrolyte solution; and As(III):1.3×10^{-5} M As(III) in the electrolyte solution.

Figure 4. Well–drilling in the field.

3.2 *Field deployment of bimetallic nano-materials*

To treat the contaminated groundwater in Datong Basin, we emplaced the nylon-mesh bags, filled with lab–prepared bimetallic nanoscale iron–manganese oxides, into a 26 m depth drilled well (Figure 4) with high arsenic concentrations (about 652 μg/L). The pumping rate was 0.25 L/s. The results showed that total arsenic was efficiently removed from groundwater. In addition, no other constituents (Fe, Mn, Si and Al) were observed to increase. In the first day of the nanomaterial application with three groundwater pumping cycles, arsenic was detected to be less than 10 μg/L. In the second day of groundwater pumping, arsenic concentration was still below 10 μg/L in the first cycle of pumping groundwater for about 4 h. Concentration of arsenic then increased to 25 μg/L (but still below the Chinese national standard for arsenic in drinking water of 50 μg/L) when groundwater was kept on pumping for several cycles (Figure 5), indicating that the nanomaterial starts to reach its capacity of arsenic removal from continuous usage.

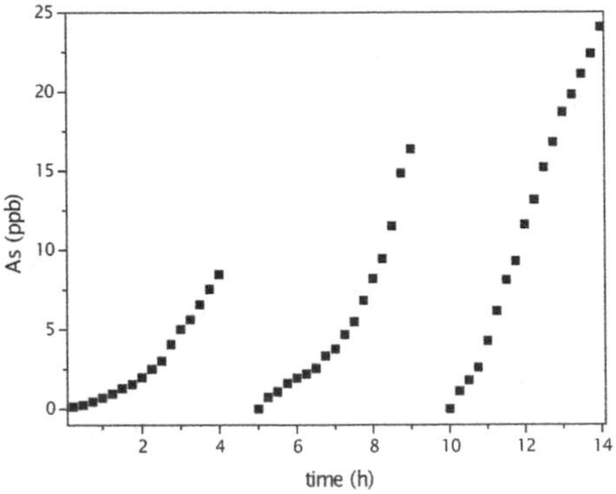

Figure 5. Arsenic removal by the bimetallic nanoscale iron–manganese oxides in field work.

3.3 *Arsenic removal mechanism using bimetallic nano-materials*

To further investigate the adsorption and oxidation of As(III) on nanomaterials, the variation of As(III) and As(V) in aqueous and solid phases were determined (Figure 6). The nanomaterials were dissolved in 2 M HCl, with the aqueous phase measured for both As(III) and As(V) to calculate the concentration changes of arsenic species on solid surfaces. The concentration of As(III) adsorbed in solid phase peaked within initial 10 min and decreased afterwards. Correspondingly, the concentration of As(V) increased in the whole course

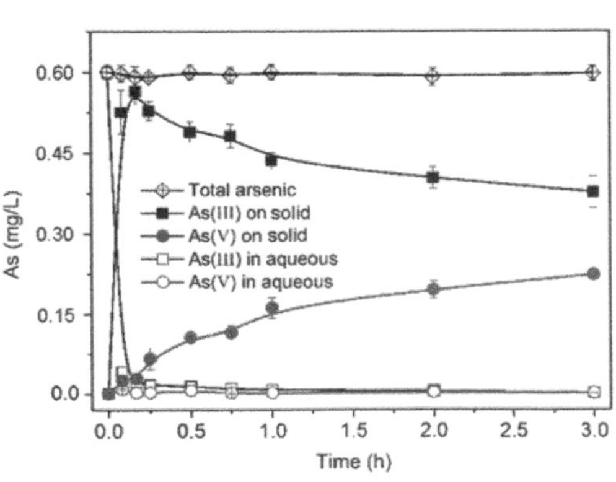

Figure 6. Adsorption/oxidation of arsenic species on bimetallic nano-material surfaces.

of adsorption. At 3 h, As(III) adsorbed on the solid phase increased to about 86% in first 15 min and then declined to about 62.5% of the total arsenic, while As(V) on solid phase achieved 37.5% of the total arsenic. The gradual conversion of As(III) to As(V) indicates that the adsorbed As(III) can be oxidized to As(V) by the bimetallic nanomaterials. The appearance of an extremely low concentration of As(V) in the aqueous phase (below 30 μg/L) proves that the As(V) produced from As(III) oxidation on solid surface can be quickly adsorbed by the co-existing Fe-oxides. Thus, the mechanism of As(III) removal by the bimetallic nanomaterial surfaces can be preliminarily proposed as the adsorption of As(III) by the Fe-oxides, the oxidation of adsorbed As(III) to As(V) by the Mn–oxides, and finally the adsorption of As(V) by the Fe–oxides. A certain fraction of As(V) can be released to the solution due to the reductive release of Mn–oxides accompanied by As(III) oxidation. At the end, As(V) and Mn^{2+} were adsorbed to the surface of iron–manganese binary oxides adsorbent.

4 CONCLUSIONS

Using advanced analytical tools as well as integrated laboratory and field experimental approaches, this work has demonstrated different extent of sorption for inorganic arsenic species in the sediments of Datong Basin, where high groundwater arsenic is prevalent. The field work on the utility of prepared bimetallic nanoscale iron-manganese in removing arsenic shows that the bimetallic nanomaterial is efficient in arsenic removal by reducing its concentration for 96%.

ACKNOWLEDGEMENTS

This work was supported by the University of Texas at Arlington (USA), China University of Geosciences (Wuhan), and National Natural Science Foundation of China (No. 40830748, No. 41120124003).

REFERENCES

Gao, X.B., Wang, Y.X., Hu, Q.H. & Su, C.L. 2011. Effects of anion competitive sorption on arsenic enrichment in groundwater. *Journal of Environmental Science and Health, Part A-Toxic/Hazardous Substance and Environmental Engineering* 46(5): 471–479.

Gao, X.B., Su, C.L., Wang, Y.X., & Hu, Q.H. 2012. Mobility of arsenic in aquifer sediments at Datong Basin, northern China: Effect of bicarbonate and phosphate. *Journal of Geochemical Exploration*, http://dx.doi.org/10.1016/j.gexplo.2012.09.001.

Hu, Q.H., Sun, G.X., Gao, X.B., and Zhu, Y.G. 2012. Conversion, sorption, and transport of arsenic species in geological media. *Applied Geochemistry* 27: 2197–2203.

Kong, S.Q., Wang, Y.X., Zhan, H.B., Liu, M.L., Liang, L.L., & Hu, Q.H. 2013. Competitive adsorption of humic acid and arsenate on nanoscale iron-manganese binary oxides-loaded zeolite in groundwater. *Journal of Geochemical Exploration* (submitted).

Manning, B.A., Fendorf, S.E., Bostick, B., & Suarez, D.L. 2002. Arsenic(III) oxidation and arsenic(V) adsorption reactions on synthetic birnessite. *Environmental Science and Technology* 36: 976–981.

Mohan, D., & Pittman, Jr. C.U. 2007. Arsenic removal from water/wastewater using adsorbents—A critical review. *Journal of Hazardous Materials* 142(1–2): 1–53.

Smedley, P.L., & Kinniburgh, D.G. 2002. A review of the source, behaviour and distribution of arsenic in natural waters. *Applied Geochemistry* 17: 517–568.

Zhang, G.S., Qu, J.H., Liu, H.J., Liu, R.P., & Wu, R.C. 2007a. Preparation and evaluation of a novel Fe-Mn binary oxide adsorbent for effective arsenite removal. *Water Research* 41: 1921–1928.

Zhang, G.S., Qu, J.H., Liu, H.J., Liu, R.P., & Li, G.T. 2007b. Removal mechanism of As(III) by a novel Fe-Mn binary oxide adsorbent: Oxidation and sorption. *Environmental Science and Technology* 41(13): 4613–4619.

One Century of the Discovery of Arsenicosis in Latin America (1914–2014) –
Litter, Nicolli, Meichtry, Quici, Bundschuh, Bhattacharya & Naidu (Eds)
 ISBN 978-1-138-00141-1

Arsenic phytoextraction by *Pteris vittata* L. and frond conversion by solvolysis: An integrated gentle remediation option for restoring ecosystem services in line with the biorefinery and the bioeconomy

M. Mench, M. Galende, L. Marchand & F. Kechit
INRA, UMR BIOGECO, University of Bordeaux, Talence, France
INRA Cestas, France

M. Carrier & A. Loppinet-Serani
CNRS, University of Bordeaux, ICMCB, IPB-ENSCBP, Pessac, France

N. Caille & F.-J. Zhao
Rothamsted Research, Harpenden, Hertfordshire, UK

J. Vangronsveld
Environmental biology, Hasselt University, Diepenbeek, Belgium

ABSTRACT: Since 2004, *Pteris vittata* L. was cultivated for stripping bioavailable As in a Belgian soil. Fronds were annually harvested in spring and autumn. Soil treatments, i.e. Beringite (B), iron grit (Z) and their combination (BZ), and season generally did not influence frond yield. On the 2006–2013 period, leachate As concentration remained lower in the Z-treated soils than in the Unt and B soils. Frond As concentrations (in mg As kg^{-1}) varied in the 970–2870 range for the contaminated soils. Frond As removal varied from 3894 to 2278 mg As m^{-2} in the decreasing order: Unt, B > BZ, Z. Shoot As concentration was lower in BZ-, B- and Z-lettuces cultivated after the fern. Fronds were converted with either sub-or supercritical water conditions. At 300 °C, major organic compounds were guaiacols. Cyclopentenones prevailed at 400 °C and phenol concentrations raised while those of guaiacols and other compounds decreased for both control and As-contaminated fronds.

1 INTRODUCTION

One challenge of the ecological restoration of metal (loid)-contaminated soils through phytomanagement options is to both reduce pollutant linkages and promote ecosystem services, notably the provisioning of non-food crops in line with biorefineries, i.e. the co-production of a spectrum of biobased products (food, feed, materials, chemicals) and energy (fuels, power, and heat) from biomass (IEA Bioenergy task42, Mench *et al.*, 2010). Pilot-scale experiments are needed to investigate this challenge. One small-scale field experiment dedicated to study As in-situ stabilization (Mench *et al.*, 2006) was redirected in 2004 to explore bioavailable As stripping. The aims were to phytoextract the remaining labile As in the soils using the As-hyperaccumulator *Pteris vittata* L., to regularly assess frond As removals, the root-to-shoot As transfer in several vegetables (e.g. lettuce) after fern cultures, and As leaching from the root-zone, and to use the frond biomass for producing biobased chemicals.

This abstract reported the influence of previous in-situ As stabilization treatments on the annual frond yields and ionome of *P. vittata*, frond As removal depending on soil treatments, changes in root and shoot DW yields and ionome of lettuces cultivated after several fern cuts, and changes in leachates As concentrations. Frond biomass was converted with either sub- or supercritical water conditions. This aims at demonstrating As phytoextraction in line with frond conversion to platform chemicals and the restoration of ecosystem services, notably safety of following crop productions.

2 METHODS/EXPERIMENTAL

2.1 *Small-scale field experiment*

The Reppel small-scale field experiment was set up in 1997, i.e. 10 lysimeters (1 m^2, 0.5 m depth) filled with a Belgian soil polluted by atmospheric fallout from a former As smelter and two lysimeters with an uncontaminated soil, all placed in a greenhouse

(Mench *et al.*, 2006). Beringite (B, 5% w/w), iron grit (Z, 1% w/w) and their combination (BZ) were incorporated into the soil for immobilizing metals and As.

2.2 *Plants*

Since 2004, *Pteris vittata* L. was cultivated for stripping bioavailable As. Fronds were harvested in spring and autumn. Restoration of the safety of crop productions was investigated with lettuce, radish and French bean. Metal and As concentrations in initial biomass and solid phase after biomass conversion were determined by ICP-OES analysis after wet-digestion in HNO_3/H_2O_2 (Carrier *et al.*, 2011).

2.3 *Leachates from the root-zone*

Leachates were analyzed for trace elements by ICP-AES and ICP-MS; leachate pH and EC were measured (WTR). Leachate phytotoxicity was assessed by radish germination.

2.4 *Sub- and super critical water treatments*

The biomass was converted by sub-and supercritical water treatments at 300 and 400 °C with 25 MPa; material and methods are detailed in Carrier *et al.* (2011).

3 RESULTS AND DISCUSSION

3.1 *Ferns*

Frond DW yield was doubled in the contaminated soils compared to the control soil. Soil treatments, i.e. Beringite (B, 5% w/w), iron grit (Z, 1% w/w) and their combination (BZ), and season did not influence annual frond yield, except differences between B and BZ in November and between November and May for the untreated (Unt) and B soils. Mean values of frond As concentrations (in mg As kg^{-1}) varied in the 60–171 range for the control soil and in the 970–2870 range for the contaminated soils. In May 2006, frond As removal varied from 3894 to 2278 mg As m^{-2} in the decreasing order: Unt, B > BZ, Z.

3.2 *Lettuce testing*

Root DW yield of lettuce was higher in the BZ soils and lower in the B ones. Shoot DW yield did not differ across soil treatments. Shoot As concentration (in $\mu g\ g^{-1}$ DW) varied from 1.28 ± 0.25 to 2.5 ± 0.5 and was lower in the BZ-, B-, and Z-lettuces.

3.3 *Leachates from the root-zone*

Mean values of leachate As concentrations varied between 83 and 836 $\mu g\ L^{-1}$. On the 2006–2013 period, leachate As concentration remained lower in Z-treated soils than in the Unt and B soils.

3.4 *Treatment of the fronds in sub- and supercritical water conditions*

Frond biomass was reduced between 70 and 77%. Compared to subcritical conditions, supercritical conditions decreased C and inorganic contents in both the solid and liquid phases for uncontaminated and contaminated fronds and promoted CH_4 formation. Higher As, Fe and Zn contents in contaminated fronds promoted decreasing C contents and the formations of cyclopentenones and benzenediols in the liquid phase. Al, Fe, P, Zn and Ca mainly remained in the solid phase whereas As and S were transferred to the liquid phase for both phytomasses. As the temperature increased from 300 to 400 °C, the concentrations of cyclopentenones and phenols in the liquid phase rose while those of guaiacols and other compounds decreased for both phytomasses. Arsenic in the liquid phase was removed by sorption on hydrous iron oxide.

4 CONCLUSIONS

Disposal of plant biomass produced in the phytoremediation of metal(loid)-contaminated soils and water is a key-point. This abstract reported a long-term integrated study exploring bioavailable As strapping in a polluted soil in line with decrease in pollutant linkages and frond conversion by either sub- or supercritical water conditions. Increase in temperature slightly reduced inorganic concentrations in solid and liquid phases. At 300 °C, major organic compounds were guaiacols whereas cyclopentenones prevailed at 400 °C. No furfurals were detected. A main difference probably due to presence of inorganics was the formations of cyclopentenones, benzenes and diols in the liquid phase of As-contaminated fronds. Some main products from sub- and supercritical water treatments of fronds, e.g. methylcyclopentenones, phenols and guaiacols, are platform chemicals used by industries: material of fragrance, herbicides precursor and antiseptic, respectively.

ACKNOWLEDGEMENTS

This work was supported by ADEME, Angers, France, the Aquitaine Region Council, Bordeaux, France, and the European Commission under the Seventh Framework Programme for Research (FP7-KBBE-266124, GREENLAND). Dr. Galende is grateful to Euskampus Fundazioa, Leioa Bizkaia, Spain for post-doctoral fellowship.

REFERENCES

Carrier, M., Loppinet-Serani, A., Absalon, B.C., Marias, F., Arias, F., Aymonier, C. & Mench M. 2011. Conversion of fern (*Pteris vittata* L.) biomass from a phytoremediation trial in sub- and supercritical water conditions. *Biomass and Bioenergy* 35: 872–883.

Mench, M., Vangronsveld, J., Beckx, C. & Ruttens, A. 2006. Progress in assisted natural remediation of an arsenic contaminated agricultural soil. *Environmental Pollution* 144: 51–61.

Mench, M., Bert, V., Schwitzguébel, J.P., Lepp, N., Schröder, P., Gawronski, S. & Vangronsveld, J, 2010. Successes and limitations of phytotechnologies at field scale: Outcomes, assessment and outlook from COST Action 859. *Journal Soils Sediments* 10: 1039–1070.

One Century of the Discovery of Arsenicosis in Latin America (1914–2014) – Litter, Nicolli, Meichtry, Quici, Bundschuh, Bhattacharya & Naidu (Eds)
© 2014 Taylor & Francis Group, London, ISBN 978-1-138-00141-1

The use of Eucalyptus and other plants species biomass for removal of arsenic from fly ash leachate

Z. Khamseh Safa & C. McRae
Macquarie University, Sydney, NSW, Australia

ABSTRACT: The outer bark of the plants, *Eucalyptus deanei* (Mountain Blue Gum), *Lophostemon confertus* (Brush Box) and especially *Melaleuca quinquenervia* (Paperbark) have been found to be capable adsorbents of total arsenic from fly ash leachate. The leachability of arsenic from two acidic and one alkaline class F fly ashes was investigated using batch leaching at room temperature. The effects of pH, solid:liquid ratio and leaching time were examined. Batch adsorption experiments were carried out using the *E. deanei*, *L. Confertus* and *M. quinquenervia* bark as the adsorbent for total As removal from synthetic solutions under various conditions. It was observed that *M. quinquenervia* bark has the highest capacity for arsenic removal compared to the other species. Up to 100% of arsenic was removed from fly ash leachates after 24 hours using 20 g L^{-1} *M. quinquenervia* bark at pH 11.

1 INTRODUCTION

Coal fly ash, which is generated during the combustion of coal for energy production, is an industrial by-product of coal-fired power stations. The current annual production of coal fly ash worldwide is estimated at around 750 million tons. On average 25% of fly ash produced is utilized mostly in cement and concrete. However, up to 75% of fly ash is either stored temporarily in stockpiles, or disposed of in ash landfills, ponds or lagoons. Fly ash is recognized as an environmental pollutant because it contains high concentrations of toxic elements with the potential for the leaching of those hazardous elements into the surrounding environment. In this study, the use of outer bark of three different tree species as biosorbent for removal of total As from Fly Ash Leachate (FAL) was investigated.

2 METHODS/EXPERIMENTAL

2.1 *Fly ash sample and preparation of the leachates*

Two acidic (called as MP and WW) and one alkaline (called as VP) class F fly ash were collected from three different coal power stations located in central west of NSW, Australia in 2012 and 2013. Leaching tests from all samples were carried out in plastic beakers at room temperature (23 ± 2 °C) under batch mode with 300 rpm magnetic stirring at different experimental conditions, Concentrations of total As in all leachates were analyzed using hydride generation atomic absorption spectroscopy (HG-AAS model GBC 908/909 and HG-3000) and Agilent 4100 Microwave Plasma-Atomic Emission Spectrometer (MP-AES). All the chemicals used were of AR grade obtained from Sigma-Aldrich, Australia and Agilent, USA.

2.2 *Bark samples and biosorption from synthetic solutions and FALs*

Eucalyptus deanei (Mountain Blue Gum, Ed), *Lophostemon confertus* (Brush Box, Lc) and *Melaleuca quinquenervia* (Paperbark, Mq) bark were collected from the campus grounds of Macquarie University, Sydney, NSW, Australia. After washing with tap and distilled water, bark was dried (60 °C, 12 h), ground, sieved to a particle size of 150–710 μm and stored in sealed plastic bags. To find the optimum conditions for total arsenic removal from synthetic solutions at room temperature, batch sorption experiments were carried on 100 μg L^{-1} arsenic (V) solutions at pH 2–12, bark dosage 0.0125–1.000 g and contact time 5 min to 24 h. After sorption equilibrium, the filtrate was collected and the initial and final arsenic concentrations were determined. Later, batch sorption experiments were carried out to remove total arsenic from actual FALs using optimum biosorption conditions such as pH, bark dosage and contact time.

3 RESULTS AND DISCUSSION

3.1 *Effect of pH, time and solid:liquid ratio on FAL*

The effect of initial pH 4, 7, and 10, contact time 1 h and 24 h and solid:liquid (S/L) ratio 1:3.5 and 1:10 on mobility of total As from fly ash was investigated by leaching experiments according to EPA procedure 1313, U.S. 2010. Table 1 shows a small part of the results for one acidic FAL. Overall, regardless of the initial pH of the solutions used in leaching tests, the final pH equalizes at approximately the natural pH level of the FALs (Ward *et al.*, 2009). The most notable result is that the leachable amount of As from the acidic fly ash moderately increases with increasing initial pH in earlier contact time (Izquierdo, 2012; Ward *et al.*, 2009). However, the alkaline FAL did not follow the same pattern. The maximum amount of As was found around neutral pH, especially after 24 hours.

In addition, by decreasing the S/L ratio from 1:3.5 to 1:10, the amount of As in all leachate at all pH values moderately increased, independent of contact time. A possible reason for this phenomenon could be an initial release of iron into the leachate (Ward *et al.*, 2009) which then oxidizes and precipitates out of solution, thereby removing As from the leachate.

3.2 *Effect of pH, time and S/L ratio on biosorption*

The effect of initial pH (2–12) on the percentage removal of total As for three biosorbents is shown in Figure 1. The maximum percentage adsorption of total As occurred at pH 5 for Ed, 11 for Lc and 12 for Mq. During the adsorption process, the pH of the solution increased possibly due to acid catalyzed hydrolysis of the bark matrix material (proteins and carbohydrates). Similarly, a decrease of pH for pH 11 sample was observed due to the same mechanism.

Table 1. Initial pH, final pH and concentration of As in the acidic FALs at S/L ratio of 1:3.5 and 1:10 after 1 h and 24 h leaching time for MP.

Leaching time	Initial pH	Mean of Final pH	As (μg/kg dry ash) 1/3.5	As (μg/kg dry ash) 1/10
1 hour	4	4.33	16.99	239.08
	7	4.43	31.35	416.56
	10	4.42	106.3	651.04
24 hour	4	4.33	13.18	558.36
	7	4.36	49.06	395.33
	10	4.49	27.61	260.75

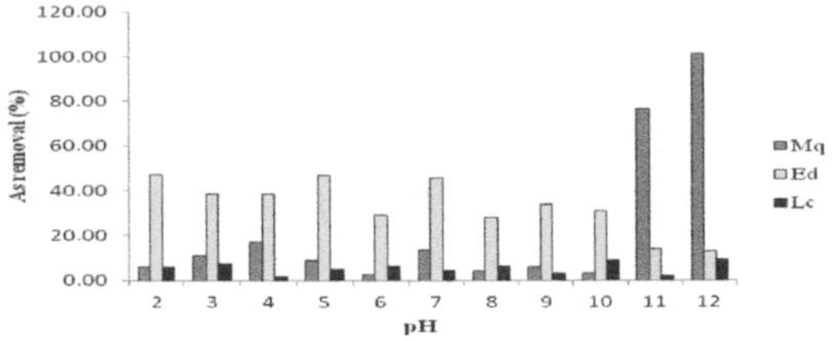

Figure 1. Effect of pH on As removal from a 100 μg L^{-1} solution containing As using 10 g L^{-1} *E. deanei* (Ed), *L. Confertus* (Lc) and *M. quinquenervia* (Mq) bark after 2 h contact time at 23 ± 2 °C.

Additionally, the effect of contact time (5 min to 24 hour) and biosorbent dosage (2.5–20 g L^{-1}) on adsorption of As were studied using aqueous solutions of 100 μg L^{-1} As at optimum pH values. It was found that, although the removal of As occurred rapidly in early stages, the rate of removal slowly decreased due to decreasing active sites on bark reaching a constant value within 24 hour. As expected, therefore, biosorption of arsenic gradually increased with the increasing dosage of bark as a result of the larger number of available active sites for adsorption of arsenic.

3.3 *Application of biomass for removal of total arsenic from FALs*

The biosorption of total As from acidic and alkaline FALs with the highest amount of total arsenic at optimum conditions of pH (5 for Ed and 11 for Mq), bark dosage 20 g L^{-1} and contact time 24 hours were examined. The results indicated that since the amount of arsenic in all leachate was in the range of bark biosorption efficiency, the percentage of total arsenic removal was up to 100%.

4 CONCLUSIONS

To assess the potential use of *Eucalyptus deanei, Lophostemon confertus* and especially *Melaleuca quinquenervia* bark as a biosorbent for arsenic removal from acidic and alkaline FALs, adsorption of this element from 100 μg L^{-1} synthetic solutions was confirmed. It was found that up to 98% arsenic could be bioadsorbed depending on pH, contact time, initial ion concentration and bark dosage. Using optimized conditions of the biosorption process, pH 5 and 11 for Ed and Mq, respectively, bark dosage 20 g L^{-1} and contact time 24 hours for total arsenic, it was found that it was possible to achieve the removal of almost all total As from actual acidic and alkaline FALs.

ACKNOWLEDGEMENTS

The authors would like to express their sincere gratitude to the Macquarie University for the Macquarie University Research Excellence Scholarship (MQRES) Scheme. We would also like to thank Delta Energy for fly ash samples supplied without cost.

REFERENCES

Izquierdo, M.X.Q., 2012. Leaching behaviour of elements from coal combustion fly ash: An overview. *Int. J. Coal Geol.* 94: 54–66.

Ward, C.R., French, D., Jankowski, J., Dubikova, M., Li, Z., Riley, K.W., 2009. Element mobility from fresh and long-stored acidic fly ashes associated with an Australian power station. *Int. J. Coal Geol.* 80: 224–236.

One Century of the Discovery of Arsenicosis in Latin America (1914–2014) – Litter, Nicolli, Meichtry, Quici, Bundschuh, Bhattacharya & Naidu (Eds)
© 2014 Taylor & Francis Group, London, ISBN 978-1-138-00141-1

Chemically stabilized arsenic-contaminated soil for landfill covers

J. Kumpiene
Waste Science and Technology, Luleå University of Technology, Sweden

L. Niero
Waste Science and Technology, Luleå University of Technology, Sweden
Padua University, Italy

ABSTRACT: Arsenic (As) stabilization using zerovalent iron (Fe^0) and its combination with peat was investigated in soil used as a pilot scale landfill cover in Northern Sweden. Leachate percolating through a 2 m thick layer of treated and untreated soil was collected in field. Chemical fractionation using sequential extraction, phytotoxicity test with dwarf beans and bioaccessibility tests simulating gastric solution were performed to assess the residual risks to the environment and human health. The results show that the exchangeable As-fraction in stabilized soils decreased when compared to the untreated soil, while other fractions remained unaffected. All the morphological parameters of plants improved and the bioaccessible As-fraction significantly decreased in the Fe-peat treated soil. The analysis of the leachates collected in field showed a substantially decreased As concentration in the Fe-peat amended soil. Further sampling is on-going in order to determine whether or not the treatment is successful in a long-term.

1 INTRODUCTION

Iron minerals and iron-containing industrial by-products show a great potential for in situ remediation of contaminated soil (Komarek *et al.*, 2013). Due to their strong binding capacities, Fe oxides have been extensively evaluated as potential stabilization amendments in soils contaminated with metals and Arsenic (As). However, it has been shown that the As-stabilization using Fe is effective only in the upper soil layer, where oxidizing condition prevail (Kumpiene *et al.*, 2013). Peat has been shown to improve the soil texture, which can facilitate the air penetration into the deeper soil layers, which is necessary to keep Fe-As complexes stable (Kumpiene *et al.*, 2013).

In this study, As stabilization using zerovalent iron (Fe^0) and its combination with peat was investigated in soil used as a pilot scale landfill cover in northern Sweden.

2 MATERIALS AND METHODS

2.1 *Pilot scale landfill cover*

Soil (800 tons) from a former wood impregnation site in Northern Sweden, contaminated with Chromated Copper Arsenate (CCA) chemical K33, was transported to a landfill for treatment. The soil contained 190 ± 6 mg/kg As, pH was 7.8 ± 0.1 and electrical conductivity was 403 ± 10 μS/cm. One-third of the soil volume was mixed with 1 wt.% powdered Fe by-product from Swedish Steel AB, containing 98.3% zerovalent Fe and some impurities (mainly C); one-third was mixed using a tractor scoop with a combination of 1 wt.% Fe and 5 wt.% peat, while the last one-third was not treated and was used as a control. The soils were placed in a 2 m layer on the top of a pilot scale landfill cover as a vegetation/protection layer.

2.2 *Evaluation of soil treatment*

1 m² glass fiber lysimeters were placed below the soil layers in field (three for each area) and used to collect leachate over one year. Chemical fractionation using sequential extraction was performed (Kumpiene *et al.*, 2012) to evaluate distribution of As between six soil fractions and to better understand the changes in As binding caused by the soil treatment. Phytotoxicity test with dwarf beans (*Phaseolus vulgaris*) (Vangronsveld & Clijsters, 1992) and bioaccessibility tests simulating gastric solution (Juhasz *et al.*, 2009) were performed to assess the residual risks to the environment and human health. All laboratory tests were done with the samples after one year of the treatment with the amendments.

3 RESULTS AND DISCUSSION

3.1 *Changes in As availability*

The results of the sequential extraction show that the exchangeable As fraction in Fe-amended soil decreased by 30% and in Fe-peat treatment by 47%

when compared to the untreated soil, while other fractions remained unaffected.

All the morphological parameters of plants (primary leaf area, fresh biomass, dry weight of shoots and roots) improved on average in the Fe-peat treated soil, especially the shoot biomass (Figure 1), indicating an improved the soil quality.

The bioaccessibility test showed a small (20%) but statistically significant (at 95% confidence level) decrease in extractable As in the Fe-peat treated soil. In the soil amended with only Fe, the decrease was smaller than in the Fe-peat containing soil (Figure 2).

3.2 *As leaching in field*

The analysis of the leachates collected in field showed a substantially decreased As concentration in the Fe-peat amended soil. The values varied quite markedly during the second and third sampling and on average were higher than in the first sample. Further sampling is on-going in order to determine whether or not the treatment is successful in a long-term.

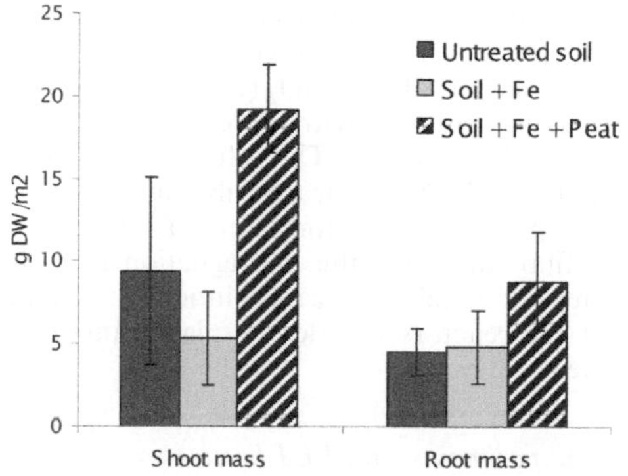

Figure 1. Dry weight of shoots and roots of dwarf beans grown on untreated and treated soil, n = 4.

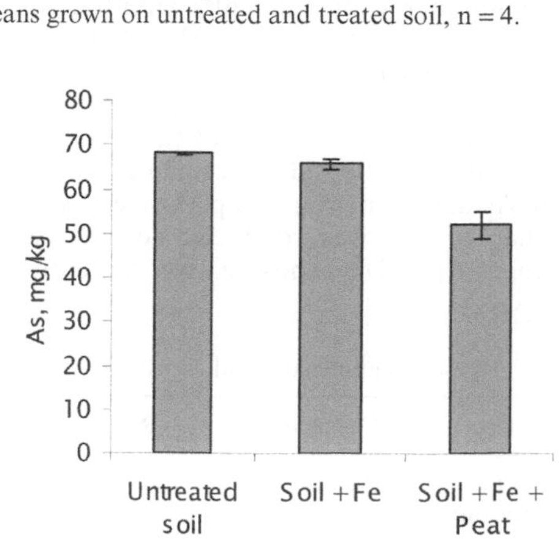

Figure 2. Bioaccessible concentration of As in untreated and treated soil.

Soil amended with Fe alone had a reverse effect on As leaching, i.e. As solubility increased substantially exceeding that of the untreated soil. This is in agreement with the previous results, which showed that As solubility in Fe-amended soil dramatically increases in layers deeper than 0.5 m (Kumpiene *et al.*, 2013).

4 CONCLUSIONS

All evaluation tests showed an improved soil quality and As stability in soil amended with the combination of zerovalent iron and peat. The positive effects are due to the reduced availability of As through its binding to iron oxides and supply of nutrients by the addition of peat. The negligible effect of applying Fe alone on biological responses is most likely due to the nutrient deficiency in soil.

The As leachate in field however varied substantially and it needs to be monitored for a longer period to evaluate the long-term effectiveness of this treatment for As stabilization.

ACKNOWLEDGEMENTS

The authors thank the Swedish Research Council Formas, Swedish Foundation for Strategic Research and the EU Regional Development Fund Objective 2 project North Waste Infrastructure for the financial support of the project.

REFERENCES

Juhasz, A.L.,Weber, J., Smith, E., Naidu, R., Marschner, B., Rees, M., Rofe, A., Kuchel, T. & Sansom, L. 2009. Evaluation of SBRC-gastric and SBRC-intestinal methods for the prediction of in vivo relative lead bioavailability in contaminated soils. *Environmental Science & Technology* 43: 4503–4509.

Komarek, M., Vanek A. & Ettler, V. 2013. Chemical stabilization of metals and arsenic in contaminated soils using oxides—a review. *Environ. Pollution* 172: 9–22.

Kumpiene, J., Desogus, P., Schulenburg, S., Arenella, M., Renella, G., Brännvall, E., Lagerkvist, A., Andreas, L. & Sjöblom, R. 2013. Utilisation of chemically stabilized arsenic-contaminated soil in a landfill cover. *Environmental Science and Pollution Research* 20: 8649–8662.

Kumpiene J., Fitts J.P. & Mench, M. 2012. Arsenic fractionation in mine spoils 10 years after aided phytostabilization. *Environmental Pollution* 166: 82–88.

Vangronsveld, J. & Clijsters, H. 1994. Toxic effects of metals. In M. Farago (ed.), *Plants and the chemical elements*. pp. 149–177. VCH Verlagsgesellshaft, Weinheim, Germany.

One Century of the Discovery of Arsenicosis in Latin America (1914–2014) – Litter, Nicolli, Meichtry, Quici, Bundschuh, Bhattacharya & Naidu (Eds)
 ISBN 978-1-138-00141-1

Use of heterogeneous photocatalysis in water for arsenic removal and disinfection

L.G. Córdova-Villegas & M.T. Alarcón-Herrera
Advanced Materials Research Center (CIMAV), Chihuahua, Mexico

N. Biswas
Civil and Environmental Engineering, University of Windsor, Windsor, Ontario, Canada

ABSTRACT: Heterogeneous photocatalysis with UV (solar radiation) and ferric chloride was used in the simultaneous removal of arsenic as well as pathogens from water. Effects of pH, exposure time and Fe(III) concentration were studied. With the optimized conditions determined in a batch process, a continuous flow system was developed. It was demonstrated that the process of heterogeneous photocatalysis in the presence of iron salts is a viable method for oxidizing and removing arsenic while achieving a high level of disinfection.

1 INTRODUCTION

According to the World Health Organization (WHO, 2012), in 2012, 783 millions of people were without access to an adequate water source. Because of this, more than 3000 children die every day of diarrheal diseases (WHO, 2012). In addition to the problem that is brought on by bacteriologically contaminated water source, the presence of other pollutants, like arsenic, also adds a serious public health problem. In Latin America, more than 14 million people are exposed to arsenic daily, usually in low-income areas, including Mexico (Bundschuh *et al.*, 2011). This is a serious health issue, and therefore, it is important to develop effective low-cost treatment techniques that are practical and economically feasible. The heterogeneous photocatalysis process allows inactivation of bacteria and, at the same time, propitiates the oxidation and adsorption of As onto iron salts and its precipitation (Malato *et al.*, 2009). This process has the advantage of using sunlight, which is an abundant source of energy. The aim of this work was to achieve the disinfection and removal of arsenic through TiO_2 immobilized in polyethyleneterephthalate (PET) bottles and a plastic tube, under the action of sunlight and in the presence of iron salts, followed by a continuous flow system.

2 EXPERIMENTAL METHODS

2.1 *Impregnation process*

The experimental process was divided in two phases. PET bottles (600 mL) were used first, and, in the second phase, a plastic tube (29.8 m length and 8 cm diameter) was used. For the impregnation, a solution of 5% TiO_2 (P25 Aeroxide Acros Organics), acidified with perchloric acid (analytical grade) was used. The solution (pH 2.5) was poured into the bottle or the tube, and was stirred until it produced a uniform film. To increase the TiO_2 film thickness, the impregnation procedure was repeated and the excess solution was drained. Then the reactors (bottles or plastic tube) were allowed to dry.

2.2 *Experiments with PET bottles*

The impregnated bottles were filled with 400 mL of groundwater with different concentrations of As(III) (sodium arsenite), Fe(III) (ferric salts), and coliform bacteria (Table 1). Bottles were then exposed to sunlight in a horizontal position in a solar concentrator. At the end of each experiment, the contents were allowed to precipitate for 24 hours before As determination was performed. A factorial analysis (2k) was conducted to analyze the influence of pH, UV exposure time, concentration

Table 1. Experimental values and variables.

Fixed values	Variables	Range
TiO_2 concentration	[Fe(III)]	0–21 mg L^{-1}
	[As(III)]	25–160 μg L^{-1}
Impregnation procedure	Irradiation time	0–180 min
	Coliform	150–1000 NMP/100 mL

of Fe(III) and the initial concentration of As(III) on the overall removal efficiency.

2.3 *Analytical methods*

As and Fe concentrations were determined by atomic absorption spectrophotometry (GBC model AVANTA P) after acid digestion in a microwave. Coliform analyses were done before and after treatment using the standardized Colilert procedure.

3 RESULTS AND DISCUSSION

3.1 *Influence of pH and Fe (III)*

Figure 1 shows the influence of pH at various concentrations of Fe(III) with a reaction time of 3 hours. Three levels of pH (5.5, 6.8 and 8.0) were selected. Iron concentration ranged from 0 to 20 mg L^{-1}. For the same iron content, lower pH (5.5) resulted in higher As removal. Arsenic removal was as high as 95% with 4.8 mg L^{-1} (initial) of Fe(III) and the final concentrations of As in treated water ranged from 8 µg L^{-1} to 11.2. At pH of 6.8, increased level (13 mg L^{-1}) of iron was required to achieve As concentrations below 25 µg L^{-1} in the treated effluent. To achieve the As concentration recommended by WHO for drinking water (< 10 µg L^{-1}), 15.5 mg L^{-1} of Fe(III) was required in this study.

The pH of groundwater in arid regions like Chihuahua ranges from 6.5 to 8.5. According to this study, it is possible to get the required quality of drinking water, either by increasing the iron concentration (preferred) or by lowering the pH of water, which would be a costlier alternative.

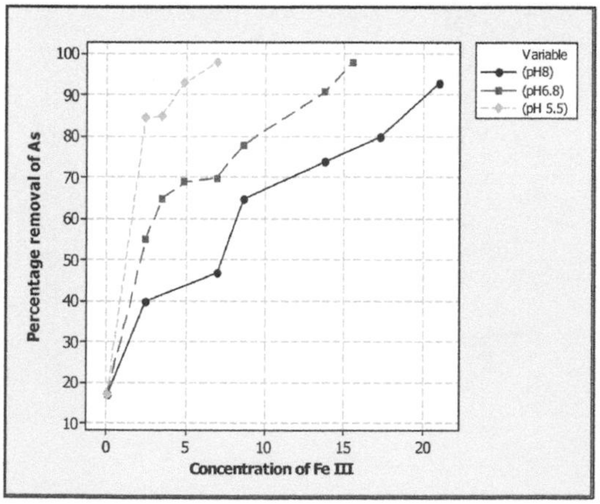

Figure 1. Arsenic removal at different pH values and concentrations of Fe(III). Exposure time: 180 minutes. [TiO_2] = 5%. [As] = 160 µ L^{-1}.

3.2 *Influence of the reaction time and initial arsenic concentration*

For this part of the study, the reaction time was varied between 120 to 180 minutes and the initial concentration of As ranged from 25 µg L^{-1} to 160 µg L^{-1}. For a higher As level (160 µg L^{-1}), the increase in the reaction time from 120 minutes to 180 minutes resulted only in 4% increase in the As removal efficiency. For a lower As content (25 µg L^{-1}), the increase was only 2%, indicating that 120 minutes would be a preferred reaction time. The As removal efficiency increased slightly with increasing concentration of Fe(III).

3.3 *Disinfection with TiO_2 and UV*

According to the results obtained with the TiO_2 and a level of UV radiation of 8.205 mW cm^{-2} (average of a typical day in Chihuahua), 15 minutes of UV radiation were required to reduce the concentration of Total Coliform (TC) in bacteriologically polluted groundwater. The initial values of TC and *E. coli* (150 MPN/100 mL), were reduced to <1 MPN/100 mL. Without the presence of the titanium dioxide film, the required time to achieve a similar level of disinfection was 30 minutes. For polluted water with high initial TC concentrations (up to 1000 MPN/100 mL), the exposure times required for disinfection were well within the range of the time (120 min) required for the removal of As. This confirms that the process was effective for both water disinfection as well as for the removal of As.

Once the batch system (bottles) was optimized for arsenic removal and disinfection, the processes was evaluated in a continuous flow prototype; the pH was 6.8, initial concentrations of As(III) and Fe(III) were 160 mg L^{-1}, and 15.5 mg L^{-1}, respectively, and the flow rate was 0.5 L/h. Results showed that As concentration in treated water was near the detectable level, which was around 5 µg L^{-1}. Likewise, for raw water, with initial concentration of TC of 724.7 MPN/100 mL, the disinfection efficiency was acceptable, since the treated water contained less than 1 MPN/100 mL of TC after treatment.

4 CONCLUSIONS

Heterogeneous photocatalysis process with TiO_2 in the presence of iron salts is effective for arsenic oxidation and precipitation as well as for disinfection of bacteriologically contaminated water. The national and international standards were met and the treated water was suitable for human consumption. This can be scaled to a full-scale continuous flow system to treat larger volumes of water.

REFERENCES

Bundschuh, J., Litter, M.I., Parvez, F., Román-Ross, G., Nicolli, H.B., Jiin-Shuh, J., Chen-Wuing, L., López, D., Armienta, M.A., Guilherme, L.R.G., Gomez-Cuevas, A., Cornejo, L., Cumbal, L. & Toujaguez, R. 2011. One century of arsenic exposure in Latin America: A review of history and occurrence from 14 countries. *Sci. Tot. Environ. 429: 2–35.*

Fostier, A.H., Pereira, M.S.S., Rath, S. & Guimaraes, J.R. 2008. Arsenic removal from water employing heterogeneous photocatalysis with TiO_2 immobilized in PET bottles. *Chemosphere* 72: 319–324.

Malato, S., Fernández-Ibañez, P., Maldonado, M.I., Blanco J. & Gernjak, W. 2009. Decontamination and disinfection of water by solar photocatalysis: Recent overview and trends. *Catal. Today* 147: 1–59.

WHO UNICEF, 2012. Progress on drinking water and sanitation. EUA, UNICEF and World Health Organization http://whqlibdoc.who.int/publications/2012/9789280646320_eng_full_text.pdf.

One Century of the Discovery of Arsenicosis in Latin America (1914–2014) – Litter, Nicolli, Meichtry, Quici, Bundschuh, Bhattacharya & Naidu (Eds)
© 2014 Taylor & Francis Group, London, ISBN 978-1-138-00141-1

Stabilization of As-contaminated soil with fly ashes

I. Travar
Luleå University of Technology, Luleå, Sweden
Ragn Sells AB, Stockholm, Sweden

J. Kumpiene
Luleå University of Technology, Luleå, Sweden

A. Kihl
Ragn Sells AB, Stockholm, Sweden

ABSTRACT: The main aim of this study is to evaluate the stabilization of As-contaminated soil with Fly Ash (FA) and the environmental impact assessment of a soil/ash mixture after treatment. Two soil samples heavily contaminated with As were stabilized with two FA that originate from incineration of biofuels (BFA) and mixture of coal and crushed olive kernel (CFA). The As solubility was reduced between 39 and 93% in soil/ash mixtures. Leaching of As from soils decreased with increased amount of ash. Immobilization of As was positively correlated to pH and negatively correlated to the amount of organic matter in soil. Geochemical modeling showed that the release of As was controlled by $Ca_3(AsO_4)_2{:}4H_2O$ in soil/ash mixtures that can be a possible explanation for the reduced solubility of As from soil/ash mixtures. A negative effect of the treatment was the mobilization of Cu from ashes and Dissolved Organic Carbon (DOC) from soil.

1 INTRODUCTION

There are many industrial sites heavily contaminated with As in Sweden. Remediation of these sites most often involves soil excavation and disposal at landfills. Excavated soil has to be treated before its disposal or reuse. Stabilization of As-contaminated soil with amendments implies a reduction of As mobility by mixing soil with e.g. Fe salts, zerovalent Fe, synthetic $Al(OH)_3$, Mn-oxides, alkaline materials such as cement, clay minerals, etc. (Kumpiene *et al.*, 2008). Fly Ash (FA) is rich in elements such as Ca, Al, Si, Fe and Mn, which could reduce leaching of As. The main aim of this work is to evaluate the efficiency of different FA to immobilize As in contaminated soil.

2 METHODS/EXPERIMENTAL

2.1 *Materials*

The FA originated from incineration of biofuels such as woo chips (BFA) and a mixture of coal and crushed olive kernel (CFA). Soil samples, LS (containing 4.8 ± 0.5% organic matter) and LB (containing 14 ± 1.5% organic matter), originated from a sawmill and wood impregnation site at Lessebo, Sweden. Two material mixtures consisting of 30% ash + 70% soil and 50% ash + 50% soil on dry base were used to evaluate the influence of ash/soil ratio on immobilization of As.

2.2 *Methods*

A one step batch leaching test according to Swedish standard SS-EN 12457–3 at liquid to solid ratio (L/S) of 10 L/kg was used to study the immobilization of As in the soil/ash mixtures. Deionized water was used as leaching medium. Leaching test was performed in triplicates. The released concentrations were compared to the Leaching Limit Values (LLV) for waste acceptance at EU landfills (2003/33/EC).

Multivariate Data Analysis (MVDA) was performed. The score and loading plots are produced as a result of the Principal Component Analysis (PCA). The score plot is used to identify groups and differences/similarities between samples. The loading plot shows variation and correlation between variables.

Visual Minteq ver. 2.61 was used to identify minerals that might precipitate and control release of As from soil/ash mixtures.

3 RESULTS AND DISCUSSION

Leachable amounts of As were 61 and 64 mg/kgTS for LB and LS respectively. The LLV for As for granular waste acceptable at landfills for hazardous waste (LHW) is 25 mg/kgTS (2003/33/EC). Both LS and LB soils cannot be landfilled without pretreatment, since amounts of leachable As are higher than LLV for the waste to be disposed of at LHW. Results from the leaching test shows that soil, ash and their mixtures are clearly grouped according to the element composition of the eluate (Figure 1). Soil samples showed the highest leached amounts of As, PO_4-P, Mn, Fe, Al and Sb compared with other samples. Ashes released more Pb, Mo, Cd, Cr, Zn, salt forming elements (Cl, Na, K, Ca) and SO_4 compared to soils. Soil/ash mixtures showed reduced leaching of As from soils and Pb and Ni from FA. A negative effect of the treatment was the mobilization of Cu from ashes and Dissolved Organic Carbon (DOC) from soil (Figure 1).

Figure 2 shows that the treatment of LS soil with BFA showed the best results in immobilization of As (reduction between 79 and 93%). CFA was the most efficient amendment in stabilization of As in LB soil when 30% of ash was added (reduction 57%). In 50%ash/50%LB mixtures, both ashes showed about the same efficiency in stabilization

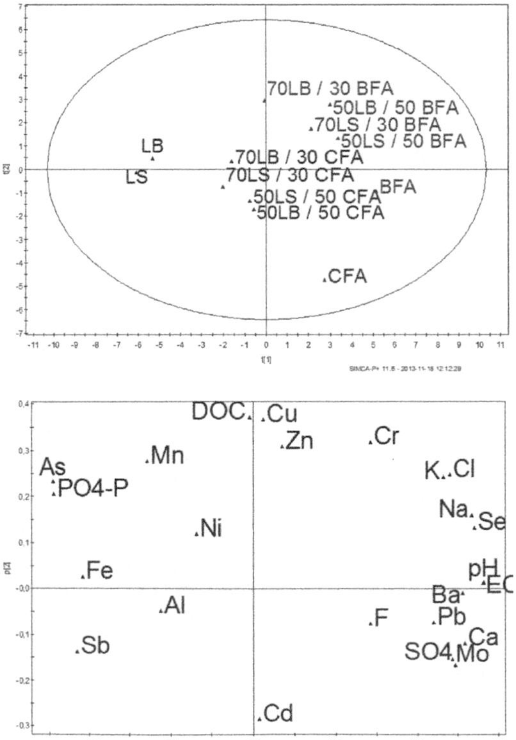

Figure 1. PCA score plot (t[1]/t[2]) and loading plot (p[1]/p[2]) of the first and second principal components showing results of the leaching test. Element concentrations in eluates were recalculated to mg kg^{-1} TS and then modeled.

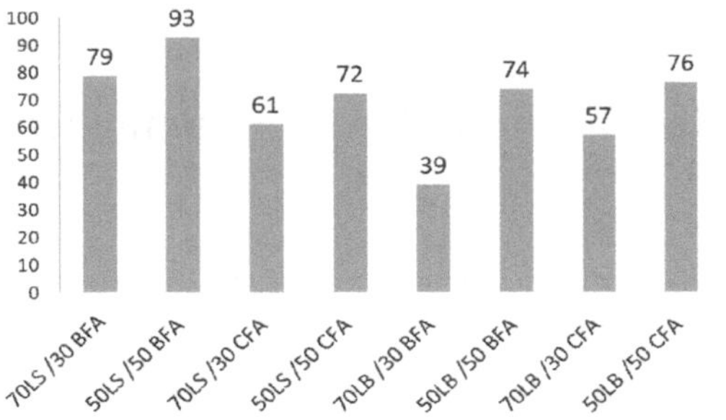

Figure 2. Immobilization of As in soil/ash mixtures expressed as% of leachable fraction.

of As (reduction by 74–76%). Decrease of As release from the soil was positively related with the amount of ash in the mixture, i.e. higher amount of ash in the mixture resulted in higher reduction of leachable As. Moreover, immobilization of As was positively correlated to pH and negatively correlated to organic matter (Figure 1).

Geochemical modeling showed that the release of As was controlled by $Ca_3(AsO_4)_2{:}4H_2O$ in soil/ash mixtures. Formation of insoluble Ca-As precipitates can be an explanation for the lower release of As from contaminated soil after stabilization with fly ashes (Moon *et al.*, 2004).

Increased release of DOC from soils after stabilization with ashes was probably attributed to change in pH from neutral (in soils) to alkaline in soil/ash mixtures (pH = 10–13). Cu was probably leached from FA due to complex formation with organic matter from soil as indicated by strong correlation between these two variables (Figure 1). It is expected that DOC and Cu will be less soluble when pH decreases due to carbonization of ashes (Meima *et. al.*, 2002).

All analyzed elements in eluates of soil/ash mixtures except for As in 70%LB/30%BFA meets leaching limit values at LHW. Hence, soil/ash mixtures can be deposited after treatment at landfills for hazardous waste.

4 CONCLUSIONS

FA can be beneficially used as amendment to stabilize As-contaminated soil. Immobilization of As was positively related to amount of FA in the mixture and pH and negatively related to organic content in soil. Amount of FA amendment should be optimized between positive effects on immobilization of As, Pb and Ni and negative effects such as increased release of Cu and DOC.

As-contaminated soil can be landfilled at LHW after treatment with FA. Future work will be conducted to study the influence of the landfill environment on the stability of As in soil/ash mixtures over time.

ACKNOWLEDGEMENTS

This investigation was financially supported by Ragn Sells AB and the Swedish Research Council FORMAS.

REFERENCES

EU Council. 2002. Council decision of 19 December 2002 establishing criteria and procedures for the acceptance of waste at landfills pursuant to Article 16 of and Annex II to directive 1999/31/EC (2003/33/EC).

Kumpiene, J., Lagerkvist, A. & Maurice, C. 2008. Stabilization of As, Cr, Cu, Pb and Zn in soil using amendments—A review, *Waste Management* 28: 215–225.

Meima, J.A. & Comans, R.N.J. 2002. Carbonation processes in municipal solid waste incinerator bottom ash and their effect on the leaching of copper and molybdenum. *Applied Geochemistry* 17: 1503–1513.

Moon, D.H., Dermatas, D. & Menounou, N. 2004. Arsenic immobilization by calcium-arsenic precipitates in lime treated soils. *Science of the Total Environment* 330: 171–185.

One Century of the Discovery of Arsenicosis in Latin America (1914–2014) – Litter, Nicolli, Meichtry, Quici, Bundschuh, Bhattacharya & Naidu (Eds)
© 2014 Taylor & Francis Group, London, ISBN 978-1-138-00141-1

Arsenic removal from contaminated soils by Pteris plants

P. Tlustos, J. Szakova, D. Pavlikova, J. Najmanova & K. Brendova
Department of Agroenvironmental Chemistry and Plant Nutrition, Czech University of Life Sciences, Prague, Czech Republic

ABSTRACT: Two Pteris species were tested for their ability to accumulate As and other elements from three highly contaminated soils for two growing periods. Results show sufficient growth of both Pteris species at all three soils. Accumulation of As was significantly higher by *P. vittata* than by *P. cretica* at all soils. The content of As in the soil did not substantially affect As accumulation due to sufficient plant availability. The bioaccumulation factor of individual species was relatively low, mainly affected by the capacity of fronds for As accumulation.

1 INTRODUCTION

Geogenic origin as well as former mining activities led to contamination of top soil layer by arsenic and other toxic elements in several locations of the Czech Republic. Due to high density of population, the majority of the land is used for crop production. Therefore, much effort has to be taken for the remediation to produce healthy food. Phytoremediation belongs to environmentally friendly technologies; however, the accumulation of As by plants is very limited (Tlustos *et al.*, 2007). Only a few plant species are known to accumulate As at reasonable levels (Baroni *et al.*, 2004; Ma *et al.*, 2001).

The objective of the present study was to compare the suitability of two Pteris species to grow at contaminated soils of different origin and to accumulate As and other toxic elements in two growing periods.

2 METHODS/EXPERIMENTAL

2.1 *Experimental design*

Soils investigated in the experiment were sampled at three sites. The samples differed in the content of the element and the soil properties (Table 1). Arable soil from Mokrsko, contained gold and high amounts of As. Other two soils were from Kutná Hora, a former silver mining town with ore tailings mixed either with fertile arable soil or with weakly fertile grassland. Five kg of air-dried soil was mixed with 0.5 g N, 0.16 g P and 0.4 g K and one plant was inserted into the pot. The biomass was harvested four months later. During the second year, a nutrient solution was applied at the beginning of the plant regrowth and the biomass was harvested in July. Two Pteris species, *P. cretica* and *P. vittata,* were grown. Afterwards, the harvested biomass was dried, grounded and analyzed for element content.

2.2 *Analyses*

Plant samples were decomposed using the dry ashing procedure as follows: an aliquot (~1 g) of the dried and powdered above ground biomass or roots were weighed into a borosilicate glass test-tube and decomposed in a mixture of oxidizing gases ($O_2 + O_3 + NO_x$) at 400 °C for 10 hours in an Apion Dry Mode Mineralizer (Tessek, Czech Republic). The ash was dissolved in 20 mL of 1.5% HNO_3 and kept in glass tubes until the analysis. Aliquots of the certified reference material RM NCS DC 73350 Poplar leaves (Analytika, CZ) were processed under the same conditions for quality assurance.

Table 1. Total content of elements (mg kg^{-1}) in contaminated soils.

	As	Cd	Zn
Soil	x ± sd	x ± sd	x ± sd
Mok*	1028 ± 42	0.367 ± 0.028	46.3 ± 0.4
KH -a**	1434 ± 127	13.8 ± 0	1399 ± 50
KH -g***	289 ± 3	1.13 ± 0.02	148 ± 3

x: mean, sd: standard deviation, *Mokrsko, **Kutna Hora—arable land, *** Kutna Hora—grass.

3 RESULTS AND DISCUSSION

3.1 *Biomass production*

Total amount of dry biomass per plant and treatment produced within two growing periods is displayed in Table 2. The mean biomass mass varied substantially among years and treatments.

The largest total differences were found on Mok. Soil. *P. cretica* produced significantly higher biomass amount than *P. vittata*. Biomass production was similar on the other two soils. The presence of a high As content as well of other elements did not affect plant growth.

3.2 *Arsenic accumulation in plants*

The As accumulation was mainly affected by the tested species. *P. vittata* accumulated three to six fold more arsenic than *P. cretica*. Wang *et al.* (2007) also found higher As accumulation by *P. vittata* than by two *P. cretica* species. Differences grew with growing As content in soil. Different As accumulation did not correspond with the results of Zhao *et al.* (2002), who determined similar As accumulation of both Pteris species in As spiked substrates. The As accumulation of both species grew up at the second harvest. The soil As content did not properly correspond to the As accumulation in fronds of Pterises. The As highest content in soil at KH-a led to the highest As accumulation in *P. vittata*, not in *P. cretica*.

3.3 *Pteris phytoextraction capacity for arsenic*

For the expression of phytoextraction capacity, total uptake of As presented by yield and As content plays an important role. The higher As removal after the two harvests was determined by *P. vittata* fronds at the three tested soils, due to significantly higher element accumulation (Table 4). The lower As removal determined at Mok. soil was mainly due to the low biomass production of *P. vittata* at the second harvest.

The clearest results of phytoextraction ability of the tested plants are given by the bioaccumulation coefficient representing the relative removal of elements from certain amounts of soil by individual species of plants (Table 5). Due to the extremely high content of As in the tested soils, the total bioaccumulation coefficient was low in the majority of the treatments. The only sufficient amount of removed arsenic by plants was found in soil the with 289 mg kg-1 of As which is a 10 fold higher content than the national limit for As agricultural land contamination in the Czech Republic. Relative values were low, except for *P. vittata* treatment at the least contaminated soil, where the mean removal reached 1%.

Table 2. Total biomass yield (g) of above ground plant (two years experiment).

	P. cretica L.	*P. vittata* L.
Soil	x ± sd	x ± sd
Mok.*	41.4 ± 7.25	19.9 ± 4.5
KH -a*	26.6 ± 5.55	22.5 ± 3.85
KH -g*	24.8 ± 8.45	27.7 ± 5.65

*see Table 1.

Table 3. The mean accumulation of As (mg kg^{-1}) in above ground biomass of plants.

	P. cretica L.	*P. vittata* L.
Soil	x ± sd	x ± sd
Mok.*	658 ± 41	2075 ± 886
KH -a*	535 ± 148	2554 ± 418
KH -g*	340 ± 107	2200 ± 480

Table 4. Total uptake of As (mg/plant) by plants (two years experiment).

	P. cretica L.	*P. vittata* L.
Soil	x ± sd	x ± sd
Mok.*	24.88 ± 2.54	31.4 ± 7.51
KH -a*	16.07 ± 2.40	56.17 ± 10.60
KH -g*	9.048 ± 3.17	51.71 ± 14.55

*see Table 1.

Table 5. The bioaccumulation factor (%) of plants.

	P. cretica L.	*P. vittata* L.
Soil	x ± sd	x ± sd
Mok.*	0.242 ± 0.06	0.306 ± 0.06
KH -a*	0.112 ± 0.07	0.392 ± 0.08
KH -g*	0.314 ± 0.11	1.790 ± 0.83

*see Table 1.

4 CONCLUSIONS

Results of our study show sufficient growth of both Pteris species at extremely As contaminated soils, as well as no effect of other contaminants on plant development. Accumulation of As was significantly higher by *P. vittata* than by *P. cretica* at all tested soils. As content in soil did not substantially affect As accumulation due to sufficient plant availability. The bioaccumulation factor of individual species was mainly affected by the As accumulation capacity by fronds. Relative values were low, except for *P. vittata* treatment at the least contaminated soil, where the mean removal reached 1%.

ACKNOWLEDGEMENTS

Supp. by project *TACR BROZEN č. TA01020366* (Czech Republic).

REFERENCES

Baroni, F., Boscagli, A., DiLella, L.A., Protano, G., Riccobono, F. 2004. Arsenic in soil and vegetation of contaminated areas in southern Tuscany (Italy). *J. Geochem. Expl.* 81: 1–14.

Ma, L.Q., Komar, K.M., Tu, C., Zhang, W.H., Cai Y. & Kennelley, E.D. 2001. A fern that hyperaccumulates arsenic—A hardy, versatile, fast-growing plant helps to remove arsenic from contaminated soils. *Nature* 409: 579.

Tlustos, P., Szakova, J., Vyslouzilova, M., Pavlikova, D., Weger, J. & Javorska, H. 2007. Variation in the uptake of Arsenic, Cadmium, Lead, and Zinc by different species of willows *Salix spp. CEJB* 2: 254–275.

Wang, H.B., Wong, M.H., Lan, C.Y., Baker, A.J.M., Qin, Y.R., Shu, W.S., Chen, G.Z. & Ye, Z.H. 2007. Uptake and accumulation arsenic by 11 Pteris taxa from southern China. *Environmental Pollution* 145: 225–233.

Zhao, F.J., Dunham, S.J. & McGrath, S.P. 2002. Arsenic hyperaccumulation by different fern species. *New Phytologist* 156: 27–31.

One Century of the Discovery of Arsenicosis in Latin America (1914–2014) – Litter, Nicolli, Meichtry, Quici, Bundschuh, Bhattacharya & Naidu (Eds)
© 2014 Taylor & Francis Group, London, ISBN 978-1-138-00141-1

Exploring low-cost arsenic removal alternatives in Costa Rica

L.G. Romero, J. Valverde, P. Rojas, M.J. Vargas & J.A. Araya
Instituto Tecnológico de Costa Rica, Cartago, Costa Rica

ABSTRACT: Costa Rica was considered as a country without arsenic in drinking water. However, polluted groundwater resources (up to 200 μg L^{-1}) in the North and North Pacific regions were detected in 2008. This document presents preliminary coagulation results of experiments of arsenic removal with iron chloride and a synthetic or a natural flocculant, Solar Oxidation and Removal of Arsenic (SORAS), and local adsorbents. The coagulation experiments showed 96% removal from around 200 μg L^{-1} arsenic spiked water. The SORAS experiments showed similar results with a residual arsenic concentration below 10 μg L^{-1} after 4 h of sunlight. In both cases, coagulation and SORAS, a further sand filtration step to reduce turbidity and color is needed. Among the local adsorbents tested, a soil rich biotite (iron(II)-bearing silicates) looks promising as powder application. More than 94% removal was obtained in 200 μg L^{-1} spiked water dosing 8 g L^{-1} of the soil. The systems evaluated look promising as family system.

1 INTRODUCTION

In Latin America it was estimated that around 14 million people were at risk because of drinking water polluted with arsenic (Litter *et al.*, 2012). Until few years ago, Costa Rica was considered as a country with no arsenic in drinking water. However, recent groundwater analysis detected naturally occurring arsenic up to 200 μg L^{-1} in several groundwater sources of the North and North Pacific regions of the country. Those sources supply water for rural communities after chlorination as unique treatment. As a primary effort, the Costa Rican Water and Sanitation Institute (AyA) started to close some wells and the interconnection of polluted and clean ones. AyA is going to contract adsorption removal systems for small communities up to 10000 people in 2014. However, the population with no connection to those systems will need a source of clean water. Besides, the AyA systems are based on imported adsorbent materials and technologies. Hence, it is necessary to look for alternatives to provide safe drinking water to individual houses and the use of low cost adsorbents. Coagulation with iron chloride followed by a synthetic and a natural flocculant and Solar Oxidation and Removal of Arsenic (SORAS) could be good alternatives for individual houses. The use of local adsorbents would reduce the costs for AyA community systems. Therefore, the aim of this study is to look for local and low cost solutions for individual houses and local adsorbent alternatives for community systems. This document presents preliminary results of the research.

2 METHODS/EXPERIMENTAL

2.1 *Coagulation processes*

To determine the effectiveness of iron chloride and a natural flocculant "Mozote" (*Triumfetta semitriloba*), jar tests experiments were conducted. Optimum conditions were determined using a natural and a commercial flocculant (FK-930 S di-allyl-di-methylammonium chloride). The initial As(V) concentration was 170 μg L^{-1}. The jar test consisted in the addition of the coagulant and flocculant with 1 min rapid mixing and 10 min slow mixing at 20 rpm, and filtration through a 0.8 μm filter paper.

2.2 *SORAS*

Commercial fine steel wool fiber for dishwashing was evaluated for SORAS treatment. The procedure did not involve the addition of lemon juice as was evaluated by Litter *et al.* (2010). PET bottles were filled with a model water containing 160 μg L^{-1} of As(V) and 0.6 g L^{-1} of steel wool fiber. Residual arsenic concentration of filtered samples (0.8 μm filter paper) was determined after 1, 2, 4 and 7 h of sunlight. Turbidity, color, pH and residual iron was also evaluated. The experiments were conducted during cloudy days.

2.3 *Evaluation of local adsorbents*

Four different local materials were evaluated as As adsorbents: rich iron beach dark sand, a residual material from a steel wiredrawing local company, natural biotite (iron(II)-bearing silicates) rich rocks from the north pacific region, and calcite from the same region. The adsorption tests were performed by adding in a plastic container 1 to 100 g of the adsorbent to 200 mL of synthetic water containing 200 μg L^{-1} As. The materials were shaken in an orbital shaker for 24 h at 27 ± 2 °C before As analysis.

2.4 *Synthetic water and analytical procedures*

The model water was prepared by adding inorganic salts (ACS grade) to type I grade of reagent water to simulate NSF (National Sanitation Foundation) model water accordingly to Amy *et al.* (2005). The synthetic water was spiked with 150–200 μg L^{-1} As(V) and pH was adjusted to 6.0–7.0.

As was determined following the USEPA method No. 206.3 AA, Gaseous-Hydride, in a Perking Elmer apparatus model AAnalyst 800. The turbidity, color and iron tests were performed using a La Motte colorimeter. The corresponding methods were absorptimetric, platinum-cobalt and bipyridyl, respectively.

3 RESULTS AND DISCUSSION

The coagulation experiments showed 96% arsenic removal from 174 to below 10 μg L^{-1} using 12 mg L^{-1} $FeCl_3$ and 1 mg L^{-1} of commercial flocculant. The maximum removal was obtained at pH 6. The local natural flocculant behaved similarly, reaching 96% removal from 211 μg L^{-1} initial arsenic concentration. The coagulant dose was 14 mg L^{-1} $FeCl_{3,}$ pH was 6 and the "mozote" dosage, 300 mg L^{-1}. This high dosage of natural flocculant is necessary because the calculation is based on the weight of the whole pieces of the plant, not only of the active ingredient dosage. A better performance of the natural flocculant is expected at optimized conditions. More experiments are underway to determine the optimum iron chloride and natural flocculant dosages. Further experiments include applying a small sand filter to quickly remove the flocs via direct filtration.

SORAS experiments during cloudy days demonstrated good removal within 4 h (Figure 1). The final arsenic concentration was well below 10 μg L^{-1}. Similar results have been reported by Litter *et al.* (2010) after 2 h of exposure to artificial UV-light (800 μW cm^{-2}) using an initial arsenic concentration of 1000 μg L^{-1} in synthetic water. The same authors reported the reduction from 340 μg L^{-1} initial concentration to 97 μg L^{-1} using natural water. Therefore, it is necessary to evaluate the system with natural local waters. Another evaluated factor was the aesthetic conditions of the water produced. Iron oxidation generated iron hydroxides that affect the turbidity, color and iron content of the water. Values up to 40 FTU, 400 UC and 0.58 mg L^{-1}, respectively, were observed in the treated water. Turbidity and color were removed to below the drinking water guidelines after filtration through 0.8 μm filter paper. Therefore, sand filter experiments are needed. Experiments using lower steel wool amount are also needed in order to reduce the remaining iron in the water.

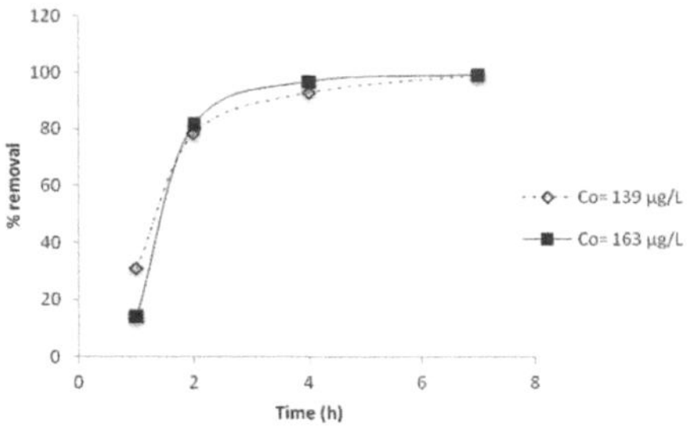

Figure 1. Arsenic removal during SORAS experiments.

The adsorbents studied showed little adsorption of arsenate. The dark sand and the steel wiredrawing materials did not fit the linear, Freundlich or Langmuir isotherms. The calcite match well the linear isotherm; however, the adsorption capacity was very poor (linear coefficient 0.018 L kg^{-1}). The most promising adsorbent was the biotite rich soil. That one did not match any isotherms, but showed 94% removal after a dosage of 8 g L^{-1}. The biotite rich soil is a soft material; therefore, it is not likely to work as a filter material, but could be applied as powder. According to Appelo & Postma (2010), iron(III) hydroxide is one of the weathering products of biotite. Therefore, adsorption onto iron hydroxides might be the responsible of the material efficiency.

4 CONCLUSIONS

Removal of arsenic through coagulation with iron chloride and the SORAS process looks promising as alternatives for water treatment in individual houses. Remaining suspended iron hydroxides should be removed via sand filtration. Thus, future experiments include the development of a simple sand filter system. Another possibility for the local people could be the use of biotite as powder adsorbent. In all cases, it is necessary to evaluate the capacity to remove arsenic in natural waters and to evaluate the perception of the local community.

ACKNOWLEDGEMENTS

The authors thank Tom Reinmann and Fabian Conrad from the University of Applied Sciences, Dresden for their support in the adsorption tests.

REFERENCES

Amy, G., Chen, H.C., Drizo, A., von Gunten, U., Brandhuber, P., Hund, R. & Banerjee, K. 2005. *Adsorbent treatment technologies for arsenic removal* (p. 164). Denver.

Appelo, C.A.J. & Postma, D. 2010. *Geochemistry, groundwater and pollution* (2th ed., p. 649). London, CRC Press.

Litter, M.I., Alarcón-Herrera, M.T., Arenas, M.J., Armienta, M.A., Avilés, M., Cáceres, R.E. & Pérez-Carrera, A. 2012. Small-scale and household methods to remove arsenic from water for drinking purposes in Latin America. *The Science of the total environment* 429: 107–22. doi:10.1016/j.scitotenv.2011.05.004

Litter, M.I., Morgada, M.E., Lin, H., García, M.G., Hidalgo, M. del V, D'Hiriart, J. & Mateu, M. 2010. Experiencias de remoción de arsénico por tecnologías fotoquímicas y solares en Argentina. In M.I. Litter, A.M. Sancha & A.M. Ingallinella (eds.), *Iberoarsen Tecnologías económicas para el abatimiento del arsénico en aguas* (pp. 191–208). CYTED.

One Century of the Discovery of Arsenicosis in Latin America (1914–2014) – Litter, Nicolli, Meichtry, Quici, Bundschuh, Bhattacharya & Naidu (Eds)
© 2014 Taylor & Francis Group, London, ISBN 978-1-138-00141-1

Arsenic retention and distribution in treatment wetlands

M.T. Alarcón-Herrera
Advanced Materials Research Center (CIMAV), Chihuahua, Mexico

E. Llorens i Ribes
ICRA, Catalan Institute for Water Research, Spain

M.A. Olmos-Márquez & I.R. Martín-Domínguez
Advanced Materials Research Center (CIMAV), Chihuahua, Mexico

ABSTRACT: Arsenic (As) can be removed from water through its retention in constructed wetlands. The aim of this study was to analyze the arsenic retention in treatment wetland prototypes, as well as its distribution along the flow gradient in a treatment wetland mesocosmos. Experiments were carried out in laboratory-scale wetland prototypes, two planted with *E. macrostachya* and one without plants. Samples of water were taken at the inlet and outlet of the prototype during the testing period. At the end of the experiment, plants and soil (silty-sand) from each prototype were divided into three equal segments (entrance, middle and exit) and analyzed for their arsenic content. Results revealed that the planted wetlands have a higher As-mass retention capacity (87–90% of the total As inflow) than prototypes without plants (27%). As-mass balance in the planted wetlands revealed that 78% of the total inflowing As was retained in the soil bed. Nearly 2% was absorbed in the plant roots. In the prototype without plants, the soil retained only 16% of As-mass, 72% of the arsenic was accounted for in the outflow. The plants retained only 2% of the total arsenic mass in their roots; however its presence was determinant for arsenic retention in the wetland soil medium. Therefore, it can be assumed that treatment wetlands are a suitable option for treating As-contaminated water.

1 INTRODUCTION

Treatment wetlands (TWs) are an innovative technology in water treatment, which has proven to be both effective and affordable. Some species of aquatic macrophytes can be used for water treatment, due to the great extent to which they accumulate arsenic from water. The behavior of metals and metalloids in aquatic systems is complex and may include interactions among or between major wetland compartments: above-ground plant parts, roots, litter, biofilms, soil, and water (Kadlec and Knight,1996). Chemical reactions, such as acid–base, precipitation, complexation, oxidation/reduction, and sorption, all play a role in removing metal ions from the water column, resulting in a metal-ion complex more or less rapidly settling in the sediments (Yong, 1995). To understand As-retention capacities, it is necessary to analyze the specific behavior of arsenic during uptake by plants, as well as the influence of media (soil) retention (Frohne et al. 2011). The mass balance of As in treatment wetlands, both in the presence and absence of wetland plants, could yield insights on the capacity of treatment wetlands (TWs) to retain arsenic from water. Therefore, the objectives of this study were: a) to investigate the arsenic retention capacity of subsurface-flow TWs prototypes with and without plants, and b) to investigate the distribution of total arsenic in main wetland compartments (plants and soil bed) and in three segments (entrance, middle and exit) of each prototype along the flow path.

2 MATERIALS AND METHODS

Laboratory-scale prototypes; Three wetland prototypes built with acrylic (length: 150 cm, width: 50 cm, height: 50 cm) were used in this study. The prototypes were uniformly filled with 200 kg of silt sand (porosity of 31%, and hydraulic conductivity of 6.89×10^{-4} cm/s). Two prototype wetlands (W1 and W2) were planted with *E. macrostachya*, and one prototype remained unplanted (W3) as a control.

2.1 *Arsenic mass balance*

After 4 months of operation, the experiment was ended. An arsenic mass balance was performed in each prototype unit and within the three segments by considering the total As-mass input, the total As-mass output, and the total As retained in the soil and plant biomass. The remaining (loss or gain) of arsenic from the mass balance was considered to be unknown. The total As-mass input and output in each unit was calculated from the cumulative total As-mass inflow and outflow during the whole operation time period.

2.2 *Sample collection and analysis*

Throughout the experiment, water samples were taken before the entrance and after the exit of the three units. Samples were preserved for later digestion and measurement of arsenic content. At the end of the experiment, plant biomass samples were collected. For plant sampling, each prototype (W1 and W2) was divided into three equal segments (entrance, middle and exit). The plant samples were prepared for As content analysis in accordance to the EPA method 200.2. For soil sampling, the wetland units (W1, W2 and W3) were divided into three equal segments. Soil samples were digested according to the SW-3051 EPA method and analyzed for arsenic content.

3 RESULTS

3.1 *Arsenic retention by plants*

The prototypes with plants showed the greatest arsenic retention in the exit section of the prototype; reaching an As concentration in the roots (dry weight, dw) of 33 ± 1.37 and 47 ± 1.42 mg/kg. The entrance sections retained 17 ± 0.70 and 27± 1.01 mg/kg in W1 and W2, respectively.

3.2 *Arsenic retention in soil*

As concentration in soil decreased between the entrance and exit segments from 13.05 ± 0.75 to 5.86 ± 0.26 mg/kg (dw). In the un-planted prototype W3, the arsenic concentration values presented the same trend but also significantly lower concentrations in all three segments (4.91 ± 0.30, 1.86 ± 0.20 and 0.14 ± 0.05 mg/kg (dw) for the entrance, middle, and exit segments, respectively). As retention in the soil of the prototypes showed an inverse behavior to that of the plants. A better retention of Arsenic was obtained at the entrance section of each unit. The obtained arsenic concentrations show a great difference between the retention of arsenic from the soil of the units with and without plants.

3.3 *Arsenic mass balance and distribution*

The total mass balance showed that the soil with the plants is the main As-retention compartment in units W1 and W2. The mass of As that was retained in the units with plants was 1210 and 1044 mg for W1 and W2, respectively, while the unit without plants (W3) retained only 230 mg. The units with plants (W1, W2) presented a smaller quantity of As in the exit water (182 and 140 mg, respectively) than the unit without plants W3 (1034 mg). The plants' retention of As was very small (2%) in comparison to that obtained from the soil-bed of units W1 and W2, which retained 84 and 72%, respectively. Like in natural wetlands, a TW's environment is very complex. It is known that plants add to the wetland environment a high capacity of As retention through different physical-chemical mechanisms like precipitation, adsorption, chelation, and complexation among others.

4 CONCLUSIONS

In a treatment wetland, As is mainly retained by soil; however, plants play an important role in the As retention capacities of wetland systems, making the whole system able to retain As better than unplanted wetland systems.

ACKNOWLEDGEMENTS

The present study was developed with the support of AECID (Agencia Española de Cooperación Internacional para el Desarrollo).

REFERENCES

Frohne, T., Rinklebe, J., Diaz-Bone, R., Du Laing, G. (2011). Controlled variation of redox conditions in a floodplain soil: impact on metal mobilization and biomethylation of arsenic and antimony, Geoderma. 160, 414–424.

Kadlec, R.H., Knight, R.L. (1996). Treatment Wetlands. CRC Press, Boca Raton, FL, pp. 893.

Yong, R.N. (1995). The fate of toxic pollutants in contaminated sediments. In: Demars, K.R., Richardson, G.N., Yong, R.N., Chaney, R.C. (Eds.), Dredging, Remediation, and Containment of Contaminated Sediments, ASTM STP, vol. 1293. American Society for Testing and Materials, Philadelphia, pp. 13–39.

Section 5: Mitigation management and policy

5.1 Arsenic mitigation aspects

One Century of the Discovery of Arsenicosis in Latin America (1914–2014) – Litter, Nicolli, Meichtry, Quici, Bundschuh, Bhattacharya & Naidu (Eds)
© 2014 Taylor & Francis Group, London, ISBN 978-1-138-00141-1

From the presence of arsenic in natural waters to HACRE

M.A. Hernández, N. González & M.M. Trovatto
Cátedra de Hidogeología, Facultad de Ciencias Naturales y Museo, Universidad Nacional de La Plata, La Plata, Argentina

M. del P. Álvarez
CONICET, Argentina

ABSTRACT: An update on a proposal put forward by the authors, which up to date has not been discussed in depth, is provided. The problem of the different arsenic tolerance thresholds in drinking water and their validity is analyzed in the face of the uncertainty as to the concentrations above which chronic endemic regional hydroarsenicism (known as 'HACRE' by its initials in Spanish) is present. Once the current situation and the changes in the standards have been discussed, the focus is shifted to the need for local regulation with ecotoxicological criteria and the provision of truly viable treatments for water conditioning.

1 INTRODUCTION

Ever since 1913/1914, when the health problems caused by the intake of water with high arsenic content (Litter, 2010; Auge, 2013) were disseminated, the regulations on drinking water quality for Argentina have evolved as the guidelines of the World Health Organization (WHO) were incorporated, as well as the consequent standards of the Código Alimentario Argentino (CAA). As Argentina is a federal state, the CAA can only be enforced in the provinces that have adopted the regulations, many of which have standards of their own, causing a considerable diversity in guideline values, as this is the case of arsenic. A previous contribution of the authors (Hernández *et al.*, 2005a) proposes a series of criteria to set standards, which in general have not been taken into consideration up to date. Subsequently, the current situation and the perspectives are reviewed, as well as the lack of a clear limit between the presence of arsenic in water and the conditions in which HACRE develops, against the backdrop of the persistent absence of ecotoxicological studies in Argentina. Finally, the problems deriving from the possible need for the treatment of water, to conform the potability standards, are discussed. The contribution focuses on natural occurrence, not on high values caused by environmental disasters and/or anthropogenic presence.

2 DISCUSSION

In 2007, by means of a joint resolution (No. 68 and 196/2007), the 982 and 983 articles of the CAA were modified, decreasing the arsenic standard in water for human consumption from 0.05 to 0.01 mg L^{-1}, and setting a 5-year period—whose deadline was in June 2012—to adapt to the new values. Subsequently, the Comisión Nacional de Alimentos (CONAL; National Food Commission) (2011), decided to extend the original deadline until the results of the study "Hidroarsenicismo y Saneamiento Básico en la República Argentina. Estudios básicos para el establecimiento de criterios y prioridades sanitarias en cobertura y calidad de aguas" ("Hydroarsenicism and Basic Sanitation in Argentina. Basic studies to establish guidelines and sanitary priorities in water coverage and quality"), whose terms were set by the Subsecretaría de Recursos Hídricos de la Nación (National Undersecretariat of Water Resources). Neither task had set a deadline, nor does the validity of the mentioned standard. However, an interesting prospect has emerged in line with the above-mentioned proposal of the authors, if the principles formulated are taken into consideration.

2.1 *Arsenic concentrations and HACRE*

In Argentina, arsenic concentrations in groundwater vary "between 4 and 5,300 µg L^{-1}, and an extreme value of 14,969 µg L^{-1} was measured in Santiago del Estero," according to Litter (2010). Nevertheless, there is uncertainty regarding the value over which the symptoms make it possible to identify the incidence of HACRE; in certain cases, it has been determined as being above 10 µg L^{-1} (Litter, 2010). However, the context of certain conditioning factors remains to be studied, such as the ethnobiological factors, the volume of daily intake, the local climate (in relation to the previous factor), whether the population lives in an open or closed community, time of exposure, age of the individual, body weight, time of residence in the locality and other related factors. The lack of more precise etiological knowledge of HACRE in general leads to the common tendency towards asserting its existence when the arsenic content in natural water is at a minimum and there is no medical evidence, as it happens in a large section of the humid Pampas in the Province of Buenos Aires (Pelusso *et al.*, 2007). No cases that can be attributed to HACRE were reported until occasionally public concern was aroused for different reasons; it was combined with poor social communication with the population, as was the case in Junín (Buenos Aires) due to legal action, which was then dismissed (Hernández *et al.*, 2005b). In such cases, the situation has reached an extreme situation: the population has the habit of replacing the consumption of supply network water by bottled water coming from the same aquifer and the same location. If the validity of the 10 $\mu g\ L^{-1}$ limit for arsenic were enforced today in the Province of Buenos Aires—which adheres to the CAA—and considering the map of arsenic distribution (Auge, 2013), it would be easy to confirm that most of the territory with groundwater supply would be unfit for domestic use.

2.2 *Treatment*

Considering such a possible alternative, the need arises for a treatment that would make the water of the supply network to meet the standard, and it would require the distinction between rural localities and middle to large cities. In the first case, methods of abatement of the arsenic concentrations have been developed, ranging from small gadgets of individual use to water treatment microplants, all of which are generally based on the fixation of arsenic by means of sorption. The use of treatment units of reverse osmosis with energy supplied by solar panels, which was implemented several years ago in educational institutions, has the same disadvantages that will be discussed below at a different scale. When the urban areas are of a certain size, from small to large cities, the problem mainly becomes more complicated due to two factors, depending on the volumes to be treated: the essentially economic aspect and the environmental one. If sequestration treatments are chosen, such as reverse osmosis, economic drawbacks arise, deriving from the considerable cost of energy at present and of the replacement of osmotic membranes when dealing with such large volumes. Regarding the environmental aspect, a safe destination should be found for solid wastes, not so much due to its quantity as to its hazardous quality, which requires that they be disposed in high security repositories with the consequent special transportation and cost of disposal. On the positive side, it should be mentioned that these techniques would bring about other improvements in water quality, due to the decrease of other inconvenient solutes.

2.3 *Need for ecotoxicological studies*

Apart from the modification and/or adoption of new standards, it is essential to perform the test for arsenic content on the main parts of the body that accumulate it (i.e., skin, hair and nails) in a population that has continuously inhabited each area for over thirty years, as already proposed by the authors (Hernández *et al.*, 2005a). It would be advisable for the universe sampled to be supplied both from the water supply network and from individual sources, and for a regional target area with little or no arsenic content to be available. In the studies, public and private healthcare providers, professional associations of physicians and biochemists, the university and other research centers should participate in order to compile an etiological record of disorders derived from the consumption of water with arsenic and other related pollutants. A recommendation in this sense was aptly put forward by Vazquez *et al.* (1999), when they state, "The adoption of standards in water quality regulations for parameters that have an impact on health, on the basis of the guideline values proposed by the WHO whenever there is a certain measure of uncertainty and difficulty to reach them, must be preceded by local epidemiological studies. These studies will allow the authorities and monitoring bodies to support the revision, update and adoption of standards for drinking water, taking acceptable risks depending on the different exposure times compatible with the hydrogeological, social and economic reality of the region...". The suspension implemented by the CONAL (Comisión Nacional de Alimentos, 2011) and its intervention in the letter of the CAA regulations are excellent opportunities to carry out such studies.

3 CONCLUSIONS

In Argentina, there are still problems regarding the acceptable limits of arsenic in groundwater for human consumption, even though the national regulations (CAA) have extended the current threshold, while a basic study supporting the new standard is awaited. Should a limit such as the one previously set by the CAA be adopted, vast regions would have As concentrations above the threshold, which is why water treatment would be required.

In this respect, the case of the rural localities (where small units or individual low-tech solutions would be suitable) should be differentiated from the case of the supply networks of middle to large cities, which require large volumes to provide their service. Regardless of the decision, the new value should arise from ecotoxicological studies based on determinations performed on individuals meeting certain characteristics, such as residence in the locality and source of supply. The participation of healthcare providers, public health authorities, professional associations and researchers from different fields of knowledge is necessary.

REFERENCES

Auge, M., Espinoza Viale, G. & Sierra, L. 2013. Arsénico en el agua subterránea de la Provincia de Buenos Aires. In: N. González, E. Kruse, M.M. Trovatto & P. Laurencena (eds), *Agua Subterránea recurso estratégico: T. II*, 58–63. La Plata: Edulp.

Comisión Nacional de Alimentos. 2011. Acta N° 93. CONAL. Buenos Aires: unpublished.

Hernández, M.A., González, N., Trovatto, M.M., Ceci, J.H. & Hernández, L. 2005a. Sobre los criterios para el establecimiento de umbrales de tolerancia de arsénico en aguas de bebida. In: G. Galindo, J.L. Fernández Turiel, M.A. Parada & D.G. Torrente (eds), *Arsénico en aguas. Origen, movilidad y tratamiento*: 167–172. Río Cuarto.

Hernández, M.A., González, N., Trovatto, M.M., Ceci, J.H. & Hernández, L. 2005b. Ocurrencia de arsénico en aguas de los acuíferos Pampeano y Puelche. Junín. Provincia de Buenos Aires. *XVI Congreso Geológico Argentino: T III*, 687–694. La Plata.

Litter, M.I. 2010. La problemática del arsénico en Argentina: el HACRE. Revista de la Sociedad Argentina de Endocrinología Ginecológica y Reproductiva (SAEGRE), 17: 5–10.

Pelusso, F., Othax, N. & Usunoff, E. 2007. Efectos sobre el análisis de riesgo sanitario por arsénico a partir de cambios en los valores referenciales toxicológicos. In: G. Galindo & H. Nicolli (comp), *Hacia una integración de las investigaciones*: 13–22. UNER. Paraná.

Vazquez, H., Ortolani, V., Rizzo, G., Bachur, J. & Pidustwa, V. 1999. Arsénico en Aguas Subterráneas. Criterios para la adopción de límites tolerables. Documento ENRESS. Rosario, Santa Fe.

One Century of the Discovery of Arsenicosis in Latin America (1914–2014) – Litter, Nicolli, Meichtry, Quici, Bundschuh, Bhattacharya & Naidu (Eds)

Arsenic exposure through well water and household behavior in a rural Maine community: Implications for mitigation

S.V. Flanagan & Y. Zheng
Columbia University Lamont-Doherty, NY, USA
University of New York, USA

ABSTRACT: In the US, private well water is unregulated by federal drinking water standards and is solely the responsibility of the owner to ensure quality. The Columbia University Superfund Research Program found that 31% of domestic wells in its 17-town project area in Maine exceed the EPA MCL for As, resulting in an estimated 13,300 population at risk of drinking As-contaminated water. Analysis of new household survey data on water testing and treatment practices reveals an estimated 69% of those at risk likely remain exposed at present. The different categories of exposed will require unique strategies for mitigation. exposure.

1 INTRODUCTION

Elevated arsenic (As) concentrations (>10 μg L^{-1}) in well water affect an estimated 140 million people in 70 countries (Ravenscroft *et al.*, 2009) with increased risks of cancer, cardiovascular disease, and neuropathy. About 15 percent of the U.S. population, over 43 million people, relies on private wells for their drinking water (Huston *et al.*, 2004). Although national standards for As have been in place since the Safe Drinking Water Act in 1974, domestic wells are private, so it is entirely the responsibility of the well owner to have their water tested and treated as necessary. No authority is tasked to ensure that private drinking water is brought into compliance with federal regulations; some states have taken steps to alert their residents of risks and have in some cases tried to fill the gap with further regulations, but it ultimately falls to the homeowner to take action. The result is that significant portions of communities in at-risk areas remain exposed to elevated As through drinking their well water.

The greater-Augusta area of Maine is such an area with frequent natural As groundwater contamination and high rates of private well water supply. Seventeen towns of Kennebec County comprise the project area of the Columbia University Superfund Research Program (SRP), where well water was sampled 2006–2011 and household well testing and treatment surveys were conducted in Jan 2013. These recent surveys give insight into the water testing and treatment behaviors of the local community and the stages of behavior that result in continued arsenic exposure among this study population, important for developing any mitigation action plan.

2 DATA SOURCES

2.1 *Water quality*

Between 2006 and 2011, the SRP of Columbia University and the Maine Geological Survey tested 1,428 domestic well water samples in 17 towns of Kennebec County and found that 31% of domestic wells exceeded the EPA Maximum Contaminant Level (MCL) for As. All participating households received a letter informing testing results of 40 water quality parameters including As, and an educational brochure provided by Maine CDC and other contact information for appropriate action.

A total of 131 households in known high-arsenic clusters throughout the state of Maine participated in a treatment system assessment conducted by the US Geological Survey selected because of known high arsenic concentrations (>10 μg L^{-1}) in well water, and previously installed water-purification systems designed specifically for arsenic removal, either at the point-of-entry (POE) or at the point-of-use (POU). Water samples were collected over two time periods: 2001–2002 (n = 31) and 2006–2007 (n = 100) and analyzed for As concentration; "influent" well-water samples (pre-water-treatment system) were collected at the point-of-entry to the house and compared to "effluent" samples (post-water-treatment system) which were collected at the kitchen faucet.

2.2 *Behavior surveys*

Two household surveys were conducted by mail in January 2013. Survey 1 was mailed to 900 random addresses in the 13 towns of the project area with the highest rates of well-water supply, and achieved a response rate of 58.3%. Participants completed a 10-page questionnaire on their water testing and treatment practices, preferences, and opinions.

Survey 2 was targeted to the 466 households identified by the previous testing program to have well water exceeding the EPA MCL, containing >10 μg L^{-1} As. The survey response rate was 73.1% among the 386 successfully mailed surveys. Participants completed an 8-page questionnaire similar to Survey 1, although focusing more on treatment actions taken in response to As specifically. The study protocol and survey instruments were approved by the IRB of Columbia University.

3 RESULTS AND DISCUSSION

3.1 *Total population at risk*

The 17-town project area (Augusta, Belgrade, Chelsea, China, Farmingdale, Hallowell, Litchfield, Manchester, Monmouth, Mount Vernon, Readfield, Sidney, Vassalboro, Waterville, West Gardiner, Windsor, and Winthrop) have a combined population of 85,668, of whom about 43,100 drink private well water (US Census 2010; Nielsen *et al.*, 2010). Based on the well sampling conducted by Columbia, an estimated 5411 households, or 13,300 people, are at risk of As exposure through drinking water due to wells above the MCL. This population at risk is based on natural As occurrence but does not take into account any steps the households have taken to reduce their exposure risks. For that, it is needed to know the testing and treatment practices of this population.

3.2 *Group 1: untested population*

The random household survey found that up to 59% of well households may have ever tested for arsenic, although the majority cannot remember the results. Therefore, at least 41% of households in this area have never tested their well water for As. Among the 13,300 population already at risk of As exposure, testing rates suggest that 7847 have potentially received results of >10 μg/L As in their well water, but the remaining 5453 people are likely still exposed to As in their well water and unaware of it.

3.3 *Group 2: tested but not treating*

The results of Survey 2 of households known to have received test results of >10 μg L^{-1} As indicate homeowner behavior following testing. 24% report "seldom or never" using their well water for drinking, likely relying on bottled water or another source. Although there may be residual exposures

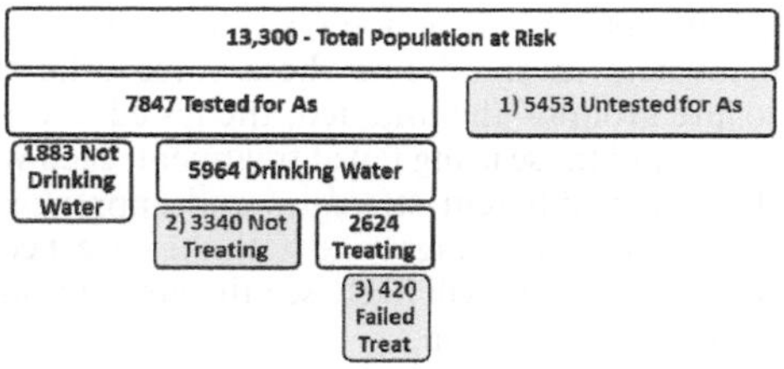

Figure 1. Estimated population exposed to As in drinking water out of total population at risk in Maine project area.

among those households, it is primarily the 76% of households that report drinking their well water at least sometimes that remain at exposure risk. Of these, only 44% report using a water treatment system capable of removing As. Applied to our at risk population estimates, this means another 3340 people at risk of arsenic exposure are likely still exposed because they have decided not to treat the well water that they drink, despite their water test results.

3.4 *Group 3: using failing treatment*

Based on comparison of untreated and treated well water samples, the USGS assessment of household As removal systems revealed treatment failure as a serious risk. The overall rate of failure for the 131 treatment systems, defined as failing to produce treated water below the MCL of 10 μg L^{-1} As, was 16%. Among the 47 systems treating well water with >100 μg L^{-1}, failure rates jumped to 32%. The danger here is that motivated homeowners who have successfully tested their well water and made the decision to take action to mitigate their exposure to As may yet remain exposed due to an ineffective or inefficient treatment system. If we apply this failure rate to our population estimates, out of the 2624 people we estimate have tested their well water and have taken action to treat it for As, about 420 may still be drinking water above MCL.

4 CONCLUSIONS

Our analysis applied water quality data and survey findings on household water testing and treatment behavior to estimate that 9213 people (69%) out of our total at risk population of 13,300 likely remain exposed to As in their drinking water. Although a portion have made the conscious decision not to treat As in their well water, the majority is likely unaware of the exposure. Mitigation of As from private well water in the US remains challenging due to responsibility falling primarily on homeowners, but any action planning to motivate homeowners must address these three different exposure groups—the untested, the tested but not treating, and those using failed treatment; each will likely require different strategies and approaches. A precursor to later stages and with the lowest cost barrier, well testing will likely see the largest initial impact from promotion.

ACKNOWLEDGEMENTS

Thanks to Maine Geological Survey, Maine CDC, and US Geological Survey Maine.

REFERENCES

Huston, S.S., Barber, N.L., Kenny, J.F., Linsey, K.S., Lumia, D.S. & Maupin, M.A. 2004. *Estimated use of water in the United States in 2000. Circular 1268*. US Geol. Survey.

Nielsen, M.G., Lombard, P.J. & Schalk, L.F. 2010. Assessment of arsenic concentrations in domestic wells by town in Maine 2005–2009. U.S. Geol. Survey Report 2010–5199.

Ravenscroft, P., Brammer, H. & Richards, K. 2009. *Arsenic Pollution: A Global Synthesis*. John-Wiley & Sons, Oxford.

One Century of the Discovery of Arsenicosis in Latin America (1914–2014) – Litter, Nicolli, Meichtry, Quici, Bundschuh, Bhattacharya & Naidu (Eds)
 ISBN 978-1-138-00141-1

Demonstration of the genetically modified ARSOlux biosensor in Jyot Sujan, West Bengal, India

K. Siegfried, S. Hahn-Tomer & E. Osterwalder
Helmholtz Centre for Environmental Research, UFZ, Department of Environmental Microbiology, Leipzig, Saxony, Germany

S. Pal & A. Mazumdar
School of Water Resources Engineering, Jadavpur University, Kolkata, India

R. Chakrabarti
Adelphi, Berlin, Germany

ABSTRACT: Arsenic concentrations in groundwater were analyzed with the ARSOlux biosensor directly in the field near Murshidabad, in West Bengal, India as part of the EU funded ECO-India project. The focus of this particular study was to demonstrate the functionality, practicality and biologic safety of the biosensors to stakeholders and researchers responsible for biosafety aspects. The non-pathogenic, genetically modified biosensor was imported with the permission of the Institutional Biosafety Committee of the Jadavpur University, Kolkata. The biosensors were deactivated after usage; they pose no danger to the environment or health, which is stated in a risk assessment report prepared by the Central Commission on Biologic Safety (ZKBS) in Germany, which was published in 2013. It is intended to commercialize the biosensor in due course to enable governments, companies and NGOs working in the water sector internationally to use this technology, which could be of great benefit to prevent arsenicosis.

1 INTRODUCTION

As part of the European Union funded ECO-India project, arsenic concentrations in a limited number of groundwater samples were analyzed with the ARSOlux biosensor, which was developed in Switzerland and Germany by scientists of the University Lausanne and the Helmholtz Centre for Environmental Research—UFZ (Stocker *et al.*, 2003, Harms *et al.*, 2005, 2006; Siegfried *et al.*, 2012). According to the guidelines of the Cartagena Protocol on Biosafety, the deployment of the genetically modified bioreporter bacteria included in the test kit requires the permission of ministries and committees in the country where the bioreporters are utilized.

2 PREPARATIONS AND METHODOLOGY

2.1 *Application for research and development purpose*

An application was submitted to the Institutional Biosafety Committee of the Jadavpur University (IBSC) in Kolkata in order to get the approval for the demonstration of the non-pathogenic, genetically modified ARSOlux biosensor in West Bengal. The IBSC members approved field tests with a limited number of biosensors to observe the performance of the technology and the fulfillment of safety requirements under real field and laboratory settings. If the ARSOlux biosensor adheres to the given requirements regarding accuracy, practicality and biosafety, the IBSC will compose a recommendation to officially apply for the overall distribution of the ARSOlux biosensor for research and development purpose in India. This request must be submitted to the Review Committee on Genetic Manipulation, Ministry of Science and Technology in Delhi.

2.2 *Study site and methodology of field testing*

The testing sites were located in the village Jyot Sujan, Murshidabad Jiaganj Gram Panchayat, Murshidabad district in West Bengal state in India (24°9'11.23"N, 88°15'24.06"E). Numerous studies and arsenic mitigation projects have been implemented in the surrounding areas. The area is prone to moderate to severe arsenic contamination of shallow groundwater aquifers used for drinking purpose (Chakraborti *et al.*, 1987; Farooq *et al.*, 2011).

3 RESULTS AND DISCUSSION

3.1 *Performance of the ARSOlux biosensor*

The results of the ARSOlux biosensor were in good agreement with the results of ICP-OES (data not shown) and the Arsenator chemical test kit (Table 1). Handling and measurement procedures posed no problem for a student, who was trained by a member of the ARSOlux team. An independent scientist observed the testing procedure. Arsenic concentrations in most of the tested tube wells were very high, while arsenic above the WHO threshold in drinking water of 10 μg L^{-1} was even detected in some of the surrounding surface water ponds as well as in a deep tube well.

3.2 *Biosafety measures*

The ARSOlux biosensor contains genetically modified, non-pathogenic *E. coli* K12 bacteria, which by following the manual instructions never enter the open environment before, during and after the measurement process. Nevertheless, the bacteria have to be deactivated and properly disposed after usage according to the guidelines of the Cartagena Protocol on Biosafety. Adding a disinfectant to each biosensor vial after the measurement, ensures deactivation of the bacteria. After testing campaigns, all biosensor vials are transported back to a laboratory where they are autoclaved. The remaining glass vials and biomass are considered normal waste. They, however, can also be cleaned and refilled with freeze-dried bio-reporter bacteria.

3.3 *Implementation and market transfer*

Currently, the ARSOlux biosensor is used in the framework of various research projects and evaluation campaigns. It is not yet available on the free market. The German government permitted the use of the sensors in the field in mobile labs in the German federal state of Saxony. After a request by the Mongolian government, a risk assessment report of the ARSOlux test system was prepared by the Central Commission on Biologic Safety of Germany (ZKBS), which is part of the Federal Ministry of Consumer Protection and Food Safety. The risk assessment report, including a detailed description, was published on the website of the ZKBS and the Biosafety Clearing House in 2013. The report states that the ARSOlux biosensor poses no danger to the environment and health. The Mongolian government authorized the import of the ARSOlux biosensor to Mongolia in May 2013. It is planned to commercialize the sensor to make it available to stakeholders working in the field of arsenicosis prevention and arsenic mitigation worldwide. Therefore, applications for import and commercialization of the patented and certified test system have to be submitted to selected governments to enable the transfer of the biosensor to the markets in arsenic affected countries.

Table 1. Percentage of contaminated wells in Syot Juan determined with different field-testing methods (n = 20).

Method	As >10 μg L^{-1}	As ≥ 50 μg L^{-1}	As ≥ 200 μg L^{-1}
	%	%	%
ARSOlux	95	75	15
Arsenator	95	90	15

4 CONCLUSIONS

The demonstration of the ARSOlux biosensor made clear that ARSOlux is highly beneficial specifically for fast and low cost screening campaigns. With ARSOlux, less material and financial resources are needed to quickly achieve accurate results. Constraints regarding biosafety and market transfer can be overcome while following recommendations of local governments and ministries in the affected countries. Utilizing biosensors for field tests, is an innovative and environmentally friendly opportunity to introduce sustainable progress to the area of mobile water analysis.

ACKNOWLEDGEMENTS

We are thankful to the students and scientists of the School of Water Resource Engineering, the inhabitants of the village Jyot Sujan and to Aidan Quinn of Tyndall Institute, Cork, Ireland. This particular study is part of the ECO-India project, which is funded by the Seventh Framework Program of the European Union.

REFERENCES

Chakraborty, A.K. & Saha, K.C. 1987. Arsenical dermatosis from tubewell water in West-Bengal. *Indian J. Med. Res.* 85: 326–334.

Farooq, S.H., Chandrasekharam, D., Norra, S., Berner, Z., Eiche, E., Thambidurai, P. & Stueben, D. 2011. Temporal variations in arsenic concentration in the groundwater of Murshidabad District, West Bengal, India. *Environmental Earth Sciences* 62(2): 223–232.

Harms, H., Rime, J., Leupin, O., S.J. Hug, S.J. & van der Meer, J.R. 2005. Effect of groundwater composition on arsenic detection by bacterial biosensors. *Microchim. Acta* 151: 217–222.

Harms, H., Wells, M.C. & van der Meer, J.R. 2006. Whole-cell living biosensors—are they ready for environmental application? *Appl. Microbiol. Biotechnol.* 70: 273–280.

Siegfried, K., Endes, C., Bhuiyan, A.F.M.K., Kuppardt, A., Mattusch, J., van der Meer, J.R., Chatzinotas, A. & Harms, H. 2012. Field testing of arsenic in groundwater samples of Bangladesh using a test kit based on lyophilized bioreporter bacteria. *Environmental Science & Technology* 46(6): 3281–3287.

Stocker, J., Balluch, D., Gsell, M., Harms, H., Feliciano, J.S., Daunert, S., Malik, K.A. & van der Meer, J.R. 2003. Development of a set of simple bacterial biosensors for quantitative and rapid field measurements of arsenite and arsenate in potable water. *Environmental Science & Technology* 37: 4743–4750.

One Century of the Discovery of Arsenicosis in Latin America (1914–2014) –
Litter, Nicolli, Meichtry, Quici, Bundschuh, Bhattacharya & Naidu (Eds)
 ISBN 978-1-138-00141-1

Is the arsenic toxicity in soils dependent on the adsorptive properties of soils?

M.L.B. de Moraes, J.W.V. de Mello, W.A.P. Abrahão & M.C. Melo
Soils Department, Federal University of Viçosa, Viçosa, MG, Brazil
National Institute of Science and Technology INCT-Acqua, Belo Horizonte, MG, Brazil

ABSTRACT: Guiding values are important tools in the management of contaminated areas to support decisions, not only to protect soil and groundwater quality, but also for pollution control and remediation. For arsenic, these values are related to the toxicity characteristics of this element to living organisms and, therefore, depend on the EC_{50} values. This study aimed at evaluating the influence of some adsorptive characteristics of soils and the age of contamination on the toxicity of As in terrestrial organisms. Experiments were then conducted to evaluate the acute toxicity of As in terrestrial organisms using different types of soils. Results showed that EC_{50} and the prevention values depended on the aging period of contamination as well as on the adsorptive properties of the soils. The higher the contamination aging period and the higher the adsorption capacity, the lower the As toxicity in soils.

1 INTRODUCTION

Management of contaminated areas is an issue relatively new in Brazil. The tools used in the management of such areas include the use of guiding values. In Minas Gerais, Brazil, according to COPAM Normative Resolution N° 166, dated 29 June 2011, three guiding values are considered: 1) quality reference value (VRQ), which is related to the regional background; 2) prevention value (VP), that is the concentration of a substance in soil above which detrimental changes may occur to soil quality and groundwater, 3) investigation value (VI), which is a concentration in the soil that represents risks to human health. The VP for arsenic adopted in Brazil was obtained from an international bibliographical survey and equals to 15 mg kg^{-1}. This study aimed at evaluating the influence of some adsorptive characteristics of the soils and the age of contamination on the toxicity of As in terrestrial organisms.

2 METHODS/EXPERIMENTAL

Samples of soils with low contents of arsenic (a Ferralsol, an Arenosol and a Cambissol) were contaminated with increasing doses of sodium arsenate. Two experiments were conducted for the evaluation of acute toxicity of As in terrestrial organisms, according to the ISO-11.269-2:2005 and EPA 712-C-96-167: 1996 standard methods.

Aiming at evaluating the age of contamination, the experiments were performed for 24 h and 6 weeks after the contamination (incubation time) of the soils with increasing doses of As (0, 15, 150, 1500 and 3000 mg kg^{-1}). Soils were incubated at 80% of moisture.

We also tested the toxicity of As in *Glycine max* and *Sorghum bicolor* cultivated in three Leptosols from gold mining sites with high background for As (28, 2285 and 8542 mg kg^{-1}, respectively). The biomass of these plants at the 15th day after germination was then compared with the biomass of the plants cultivated in the other three soils with low background of arsenic. Tukey-Kramer test ($\alpha = 0.05$) was used to compare the data.

The adsorptive characteristics of soils were evaluated including analyses of Arsenic Maximum Adsorption Capacity (MAC-AS) by Langmuir Isotherms and clay mineralogy by X-ray diffraction. The Wenzel (2001) fractionation method was used to assess the As phases after incubation.

The concentrations of As in soils that reduced the growth in 50% (EC_{50}) were determined for all organisms tested. The VP were determined according to Verbuggen *et. al.* (2001), considering a background of 8 mg kg^{-1}, which is the VRQ adopted for As in Minas Gerais soils (COPAM, 2011): $VP = 8 + (EC_{50}/10)$.

3 RESULTS AND DISCUSSION

In general, the longer the arsenic incubation period, the higher was the EC_{50}. It means that toxicity of As decreases over the age of contamination (Table 1). Such results can be ascribed to

Table 1. Arsenic Maximum Adsorption Capacity (MAC-As), EC_{50} and prevention values (VP) for soils, incubation times and organisms.

Soil	MAC-As (mg g^{-1})	Organism	Incubation time	EC^{50} (mg kg^{-1})	VP
Arenosol	1,522	Glicine max	24 h	120	20
			6 weeks	233	31
		Sorghum bicolor	24 h	452	53
			6 weeks	797	87
Cambisol	1,782	Glicine max	24 h	651	73
			6 weeks	1207	128
		Sorghum bicolor	24 h	194	27
			6 weeks	907	98
Ferralsol	2,414	Glicine max	24 h	1602	168
			6 weeks	2502	258
		Sorghum bicolor	24 h	637	71
			6 weeks	2193	227
Cambisol	1,782	Eisenia andrei	24 h	984	106
Ferralsol	2,414	Eisenia andrei	24 h	1277	135

the stabilization of As phases in soils with the incubation time. This was demonstrated by the As fractionation analyses. It can be seen that the labile forms of arsenic in soils decreased and the most stable forms increased with the incubation time (Figure 1).

Our results are in line with findings from Woolson *et. al.* (1973). These authors also showed that the As toxicity for *Zea mays* and *Avena sativa* decreased along the incubation time as the labile forms of As in soils trended to stabilize after 4–6 weeks of incubation.

Another obvious finding was that the higher the MAC-As, the higher the EC_{50}. Soils with higher clay content also had higher MAC-As. However, it can be pointed out that the clay type also interferes in As adsorption. In general, Fe and Al (hydr) oxides have higher As adsorption capacity than aluminosilicates. Thus, different VP can be found depending on the type of soil used as reference, in spite of the organisms adopted as indicators.

There was no significant differences (Tukey-Kramer test $\alpha = 0.05$) between plants cultivated in soils with high background compared with that cultivated in low background uncontaminated soils (data not presented). These results imply again that the As toxicity in soils depends on contamination aging. We can ascribe such results to the stability of As phases in soil, as the fractionation analyses showed that the high background soils have almost the totality of their arsenic content in less labile forms.

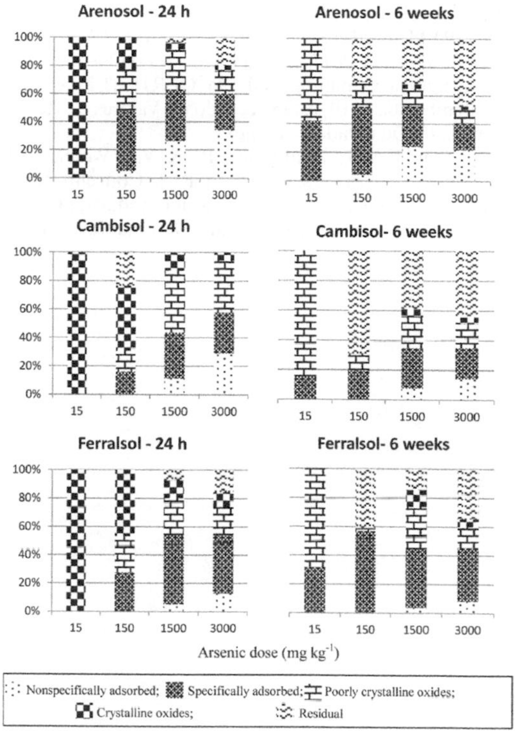

Figure 1. Arsenic fractioning results, expressed in total relative percentage.

4 CONCLUSIONS

The EC_{50} and so the prevention values for arsenic in soils, depend on the aging period of contamination as well as the adsorptive properties of the soils. The higher the contamination aging period, as well as the higher the adsorption capacity, the lower the As toxicity in soils.

ACKNOWLEDGEMENTS

To the funding Brazilian agencies CNPq (National Council for Scientific and Technological Development), CAPES (Coordination of Improvement of Higher Education Personnel) and FAPEMIG (Foundation for Research of the State of Minas Gerais).

REFERENCES

EPA. 712-C-96-167: 1996 – *Ecological Effects Test Guidelines* – OPPTS 850.6200.

ISO—International Standard Organization - ISO-11.269-2:2005 (Determination of the effects of pollutants on soil flora—Part 2: Effects of chemicals on the emergence and growth of higher plants).

Minas Gerais, Conselho Estadual de Política Ambiental—COPAM. *Deliberação Normativa COPAM N° 166, de 29 de Junho de 2011.* Altera o Anexo I da Deliberação Normativa Conjunta COPAM CERH nº 2 de 6 de setembro de 2010, estabelecendo os Valores de Referência de Qualidade dos Solos.

Verbuggen, E.M.J., Posthumus, R. & Van Wezel, A.P. 2010. Ecotoxicological Serious Risk Concentrations for soil, sediment and (ground)water: updated proposals for first series of compounds. National Institute of Public Health and the Environment, Bilthoven, The Netherlands. RIVM Report N° 711 701 020. 263 p.

Wenzel, W.W. 2001. Arsenic fractionation in soils using an improved sequential extraction procedure. *Analytica Chimica Acta* 436: 309–323.

Woolson, E.A., Axley, J.H. & Kearney, P.C. 1973. The chemistry and Phytotoxicity of Arsenic in Soils: II. Effects of Time and Phosphorus. *Soil Sci. Amer. Proc.* 37: 254–259.

One Century of the Discovery of Arsenicosis in Latin America (1914–2014) – Litter, Nicolli, Meichtry, Quici, Bundschuh, Bhattacharya & Naidu (Eds)
© 2014 Taylor & Francis Group, London, ISBN 978-1-138-00141-1

Arsenic content in cover materials used for revegetation of mining areas

L.E. Dias, S.C. Silva & I.R. Assis
Soil Department, Federal University of Vicosa, Vicosa, MG, Brazil

A.S. Araújo & O. Ferreira
Kinross Gold Corporation, Paracatu, MG, Brazil

ABSTRACT: The gold extraction is frequently associated with the generation of Acid Mine Drainage (AMD). This process may promote the contamination of areas near to mining, making the land reclamation process more complex. Moreover, it is notable the occurrence of the mobilization of heavy metals and metalloids such as Arsenic (As), resulting from AMD in these areas. This study aimed to evaluate As content in different materials used to cover sulfide substrate remaining of gold exploration. The capillary break, sealant and cover layers were used as cover of the remaining areas. These layers were composed by limestone, laterite, soil and/or B1-substrate (original rock, more weathering). The average As concentration available in the soil (cover layer) was significantly lower than B1 substrate. Along the evaluation time, an increase in the As concentration available in the soil of the cover layer was observed, showing that As is rising from the lower layers by capillarity. The use of laterite in capillary break layer was more favorable for As retention, preventing it from reaching the cover layers.

1 INTRODUCTION

The mining activity in areas that contain sulfide minerals such as pyrite (FeS_2) and arsenopyrite (FeSAs) promotes the oxidation of these minerals due to exposure to water and air (Blowes *et al.*, 1998). The products of oxidation of sulfides, in addition to being highly soluble, show strongly acidic reaction, so they are easily dissolved in the liquid phase by acidifying the drain water (Mello & Abrahão, 1998). Due to the low pH, toxic elements such as arsenic, if present in the material, are solubilized and mobilized in drainage waters. Therefore, these elements can be absorbed by plants to toxic levels and incorporated into the food chain.

In this sense, recovery techniques aimed at immobilization of this element should be used. Among the techniques, there are cover layers placed above the sulfide substrate to avoid contact with water and oxygen, besides to promote plant establishing and growth.

This study aimed to evaluate available arsenic content in different materials used to cover sulfide substrate remaining from gold exploration in order to mitigate the occurrence of acid drainage and provide favorable conditions for vegetation establishment and growth.

2 MATERIAL AND METHODS

The study area is located in the Paracatu city, Minas Gerais State; known as Morro do Ouro, where gold and silver is currently mined.

The experiment was performed on the top of a sulfide substrate layer that has considerable concentrations of sulfide, iron and arsenic. Above this layer, a three layers model was adopted. The first layer is a capillary breaking layer—CBL (10 cm depth) consisting of crushed limestone or laterite. The sealant layer—SL (20 cm), when present, was over the break capillary layer, consisting of soil or B1-substrate (substrate oxidized with lower levels of sulfides, in the order of 3.0 g kg^{-1}). Finally, the cover layer—CL (20 cm) was constructed with soil or B1-substrate (Table 1). Above the cover layer, five different plant species were used: *Melinis minutiflora* (grass fat), *Mucuna pruriens* (velvet bean), *Lolium multiflorum* Lam (ryegrass), *Crotalaria spectabilis* (crotalaria), *Stylosanthes spp.*—*S. capitata* and *S. macrocephala*—(estilosante).

Samples of soil and B1 substrate were collected in the cover layer. These samples were dried in the shade and passed through a sieve 2 mm mesh to form the Air-Dried Soil (ADS).

Table 1. Treatments description implemented in field experiment about revegetation of sulfite substrate.

Treatment	Layer Break capillarity	Sealant (20 cm)	Cover (20 cm)
1	B1	B1	B1
2	Limestone	Soil	B1
3	Limestone	Soil	Soil
4	Limestone	B1	B1
5	Limestone	B1	Soil
6	Limestone	–	B1
7	Limestone	–	Soil
8	Laterite	Soil	B1
9	Laterite	Soil	Soil
10	Laterite	B1	B1
11	Laterite	B1	Soil
12	Laterite	–	B1
13	Laterite	–	Soil

The As available content was determined in triplicate by the Mehlich-III extractant (Mehlich, 1984).

The results were submitted to variance analysis for the different treatments. The freedom degrees of the treatments were deployed in orthogonal contrasts in order to test the effect of treatments.

3 RESULTS AND DISCUSSION

In the first evaluation, carried out in the dry period, the As available contents in B1-substrate layers (Table 2) were significantly higher than in soil layers ($p < 0.01$) (contrast C_4, C_5, C_6, C_9, C_{10} and C_{11}) due to the natural presence of this element (B1-substrate). On the other hand, no significant difference was observed for the As available content regarding the presence or absence of sealant layer (C_3, C_8).

The As content average was 1.09 mg dm^{-3} ($p < 0.10$) greater in the treatments containing crushed limestone in the layer of capillary break than those containing laterite (C_2). Thus, it is expected that the laterite can better retain the element as it is formed of iron oxides. These minerals promotes adsorption due to the presence of charges on its surface (high specific surface area), which plays an important role in the adsorption of heavy metals and anions. The crushed limestone, in turn, with increasing pH, should have promoted As co-precipitation with Fe and S. However, this effect was smaller than the laterite adsorption, thus resulting in lower levels of this element when this was present in CBL.

Table 2. Average contrasts and its significance for As available contents in cover layers in the 5th and 13th month after planting

Contrast	Contrast Medium As (5th month)	As (13th month)
C1	–6,18**	–6,55**
C2	–1,09°	–0,29
C3	0,31	0,11
C4	–6,67**	–4,45**
C5	–13,96**	–10,73**
C6	–2,89**	–2,01**
C7	1,03	0,74
C8	3,33	3,73
C9	–7,54**	–6,20**
C10	–6,99**	–5,84**
C11	–13,49**	–12,61**
C12	6,71*	3,74
CV (%)	33,96	39,27

°, *, **: Significant at 10, 5 and 1 % by F test, respectively. C1 (T1 vs. T2+T3+T4+T5+T6+T7+T8+T9+T10+T11+T12+T13); C2 (T2+T3+T4+T5+T6+T7 vs. T8+T9+T10+T11+T12+T13); C3 (T2+T3+T4+T5 vs. T6+T7); C4 (T2 vs. T3); C5 (T4 vs. T5); C6 (T6 vs. T7); C7 (T2+T3 vs. T4+T5); C8 (T8+T9+ T10+T11 vs. T12+T13); C9 (T8 vs. T9); C10 (T10 vs. T11); C11 (T12 vs. T13) and C12 (T8+T9 vs. T10+T11). C: contrast; T: treatment.

An increased As content in the soil cover layers over time was observed. The average for the treatments with soil layer covering in the fifth month was 0.30 mg dm^{-3}, while in the thirteenth month, it was 0.74 mg dm^{-3}. The presence of plants in this layer may have contributed to the rise of this element due to evapotranspiration that would have caused a moisture gradient in the profile. Moreover, the presence of the vegetation layer above this material may have contributed to the element absorption by plants. As these plants are of relatively short cycle, its disposal in the soil after the death of some individuals may have favored soil incorporation.

4 CONCLUSIONS

Along evaluation time, an increase in the As available content in the soil from cover layer was observed, possibly due to a rise from the lower layers.

The use of laterite in capillary break layer was more favorable to the retention of arsenic, preventing it from reaching the cover layers. This occurred because the laterite is rich in iron oxides. However, other evaluations over time are necessary to observe the behavior of this element and the efficiency of this layer because of climate variations.

ACKNOWLEDGEMENTS

We thank Kinross Brazil Mining by the opportunity to carry out this study.

REFERENCES

Blowes, D.W., Jambor, J.L., Hanton-Fong, C.J., Lortie, L. & Gould, W.D. 1998. Geochemical, mineralogical and microbiological characterization of sulphide-bearing carbonate-rich gold-mine tailings impoundment, Joutel, Québec. *Applied Geochemistry* 13(6): 687–705.

Mehlich, A. 1984. Soil test extractant: a modification of Mehlich 2 extractant. *Comm. Soil Science Plant Analysis*, 15(2): 1409–1416.

Mello, J.W.V. & Abrahão, W.A.P. 1998. Geoquímica da drenagem ácida. In: Dias, L.E., Mello, J.W.V. *Recuperação de áreas degradadas*. Viçosa, MG: Folha de Viçosa.

One Century of the Discovery of Arsenicosis in Latin America (1914–2014) – Litter, Nicolli, Meichtry, Quici, Bundschuh, Bhattacharya & Naidu (Eds)
© 2014 Taylor & Francis Group, London, ISBN 978-1-138-00141-1

Arsenic investigation and mitigation trials in Nepal

S.K. Shakya
Environment and Public Health Organization, New Baneshwor, Kathmandu, Nepal

ABSTRACT: Water source for all purposes in lowland Nepal with more than half of the total population of the country is primarily groundwater extracted from shallow or deep aquifers. According to blanket arsenic testing results, 7.5% of 1,120,912 samples exceeded the WHO guideline value (10 μg L^{-1}) with 1.8% samples above the national standard (50 μg L^{-1}). Studies in some of the arsenic affected communities have identified several hundred arsenicosis cases. Government, national and international organizations have implemented arsenic mitigation programs aiming to prevent health damage and have provided different arsenic safe water alternatives to high arsenic exposed households. Thus, the integrated approach for wider awareness generation, local capacity building on arsenic mitigation, resources mobilization and proper management with active community involvement is necessary for addressing the arsenic problem in a sustainable way in lowland Nepal.

1 INTRODUCTION

The acute toxicity of arsenic at high concentrations has been known about for centuries. It was only relatively and recently that a strong adverse effect on health was discovered to be associated with long-term exposure to even very low arsenic concentrations.

The first major investigation of arsenic in groundwater was completed in 2001 by DWSS/UNICEF, testing 4,000 tube wells in all 20 Terai districts of Nepal. The study found over 3% tested tube wells above 50 μg L^{-1} and over 10% exceeded the WHO guideline of 10 μg L^{-1}. The "State of Arsenic in Nepal-2003" study by NASC/ENPHO in 2003 tested 18,635 tube wells in 20 Terai districts and reported 23.7% tube wells above the WHO guideline of 10 μg L^{-1} and 7.4% exceeding the Nepal Interim Standard guideline of 50 μg L^{-1} (NASC/ENPHO, 2003 and 2010).

The database includes 11,01,536 blanket tested wells from the twenty districts of the Terai region. The distribution of these tube wells over the twenty districts of Terai is not uniform. The highest number of tube wells is located in the Jhapa district with 10.50% (113,829), followed by Morang 10.30% (111,688), while the lowest number is located in the Banke district with 2.26% (24,530) of total number of tube wells in Terai district.

Although the percentage of tube wells exceeding the Nepal Interim Standard is 1.73% for the twenty districts, this percentage varies strongly between the districts from 0.05% in Jhapa to 11.69% in the Nawalparasi district. Similarly, the percentage of tube wells exceeding the WHO Guideline ranges from 0.18% in Chitawan to 17.04% in Rautahat district.

2 METHODS/EXPERIMENTAL

2.1 *Health survey*

A study on 'health impact survey in arsenic affected areas in three VDCs of Kailali district' was conducted to identify arsenicosis cases. VDC (village development district) is the lower administrative division of district. The study took into account the extent of manifestations and status of arsenic exposure among risk population in three arsenic affected VDCs, namely Chaumala, Lalbojhi and Kota Tulsipur, in the Kailali district. Identification of arsenicosis cases was done according to a WHO Guide (World Health Organization, 2005). Arsenic concentrations were measured in 50 tube well water, 150 spot urine (male-76 and female-74), 26 hair (male-13 and female-13) and 26 nail (male-13 and female-13) samples. Despite using the same tube well water, the mean urinary arsenic level for male (34.4 μg kg^{-1}) was lower than that for female (44.8 μg kg^{-1}).

2.2 *Water quality testing and questionnaire survey*

The water quality testing on 'Prevalence of Arsenicosis in Terai, Nepal' was conducted during 2001–2004 in six arsenic contaminated districts of Terai, namely Nawalparasi, Bara, Parsa, Rautahat, Rupandehi and Kapilbastu.

3 RESULTS AND DISCUSSION

Of the total studied population ($n = 18{,}288$) of the six districts, 9,015 (49.3%) were male and 9,273 (50.7%) were female. Of them, 400 were identified as cases of arsenicosis. The prevalence of arsenicosis in these districts, on average, was 2.2%, but ranged from 0.7% to 3.6%. The highest prevalence of arsenicosis was found in the Nawalparasi district (3.6%), which was also reported to be a highly arsenic contaminated district. The prevalence of arsenicosis was lower in Rupandehi (0.9%) and Kapilbastu (0.7%) among people aged 50 years and above. In the age group of 5–14 years, five (0.1%) cases of arsenicosis were reported and in the age group below five years, only one (0.3%) person had arsenicosis. Of the total arsenicosis patients identified in the six districts, 268 (67%) were male and 132 (33%) were female. In the community based studies in four VDCs, namely Goini, Thulo Kunwar, Sano Kunwar and Patkhouli, 1,864 (79.7%) of the total cases were examined for skin manifestations. Of them, 190 were found diagnosed as arsenicosis cases. The prevalence of arsenicosis was found highest (18.6%) in the Patkhouli village, where 95.8% of the tube wells were arsenic contaminated, i.e. having more than 50 $\mu g\ L^{-1}$, while none of the tube wells in Goini village was arsenic safe. It is also reported that, among the cases, the most common manifestation was melanosis (95.7%), followed by leukomelanosis (57.7%) and keratosis (55.9%). Arsenic contamination was high in Nawalparasi and, in this study, the prevalence of arsenicosis was also highest. Next to Nawalparasi, the prevalence of arsenicosis was found to be high in Rautahat.

Thus, a project '*Local Capacity Building for Arsenic Mitigation in Nawalparasi district, Nepal*' funded by JICA and implemented by the Kyushu University, Japan and Environment and ENPHO has been undertaken in the Nawlaparasi district, Nepal, from January 2011 to March 2013. It covers high arsenic exposed 59 communities from 32 wards of 15 VDCs and 12 wards of the Ramgram Municipality, which have been selected based on the blanket arsenic testing results.

4 CONCLUSIONS

Groundwater in lowland Terai (Nepal) is arsenic contaminated and health effects as skin manifestations have been detected. Both governmental and non-governmental organizations have made continuous efforts for arsenic mitigation. In context of mitigation on arsenic, practical implementation in the actual field was found conducted by different stakeholders in order to reduce the concentration of arsenic in the groundwater and vulnerability of population exposed to arsenic. More than 7000 of improved tube wells, Biosand filters, Kanchan Arsenic filters have been installed in the arsenic affected areas. Lack of monitoring and awareness on arsenic related problems have recently raised the concerns on the sustainable use of the provided arsenic safe water alternatives. The experience has suggested the need of the integrated approach for a wider awareness generation, local capacity building, resources mobilization and proper management with active community involvement for sustainable arsenic mitigation in lowland Nepal.

REFERENCES

NASC/ENPHO. 2003. *The state of arsenic in Nepal-2003*, Kathmandu, Nepal: National Arsenic Steering Committee/Environment Public Health Organization.

Shakya, S.K. & Maharjan, M. 2012. Arsenic mitigation initiatives and sustainability issues in lowland Nepal. In The 3rd International Symposium on Health Hazards of Arsenic Contamination of Groundwater and its Countermeasures, 23–25 November 2012, Miyazaki, Japan.

Shakya, S.K. & Maharjan, M. 2012. Local capacity building for sustainable arsenic mitigation in lowland, Nepal. In 18th Japan Arsenic Symposium, *24–25 November 2012*, Miyazaki, Japan.

WHO. Arsenic in drinking water (Fact sheet No. 210, Rev. ed) 2001, http://www.who.int/mediacentre/factsheets/fs210/en

World Health Organization. 2006. A Field Guide for Detection, Management and Surveillance of Arsenicosis Cases, D. Caussy (ed), WHO Technical Publication No. 30, Regional Office for South-East Asia, New Delhi.

5.2 Sustainable mitigation and policy directions

One Century of the Discovery of Arsenicosis in Latin America (1914–2014) – Litter, Nicolli, Meichtry, Quici, Bundschuh, Bhattacharya & Naidu (Eds)
 ISBN 978-1-138-00141-1

Groundwater arsenic pollution: A conceptual framework for sustainable mitigation strategy

P. Bhattacharya, M. Hossain & G. Jacks
KTH-International Groundwater Arsenic Research Group, Department of Sustainable Development, Environmental Science and Engineering, KTH Royal Institute of Technology, Stockholm, Sweden

K.M. Ahmed & M.A. Hasan
Department of Geology, University of Dhaka, Dhaka, Bangladesh

M. von Brömssen
Ramböll Sweden AB, Stockholm, Sweden

ABSTRACT: Tubewells installed by local drillers, provide access of drinking water in rural Bangladesh. Significant proportion of these wells contains arsenic (As) above the WHO guideline and the Bangladesh Drinking Water Standard. Various attempts for mitigation at household and community scale have resulted in limited success, but through the local driller's initiatives, the tubewells are the source of priority drinking water supply. We have developed a concept of Sustainable Arsenic Mitigation (SASMIT) to identify and target the safe aquifers through detailed hydrogeological studies for scientific validation of the water quality with respect to the color of the shallow sediments as perceived by local drillers. Together with water quality monitoring, we have also targeted the Intermediate Depth Aquifers (IDA) for providing As-safe and low manganese (Mn) water. SASMIT intervention logic also considered the relevant socio-economic scenario, such as household distribution, poverty issues and available safe water access for prioritizing safe well installation.

1 INTRODUCTION

Drinking water supply in rural Bangladesh mostly depends upon manually operated hand tubewells installed by the local community drillers. The occurrence of natural arsenic (As) in groundwater and its scale of exposure drastically reduced the safe water access across the country (Figure 1) and rendered tens of millions of people under health risk (Ahmed *et al.*, 2004; von Brömssen *et al.*, 2007). Tens of millions of people are exposed to concentrations above the national drinking water standard (BDWS: 50 µg L^{-1}) and the WHO guideline (10 µg L^{-1}) with visible manifestations of the toxic effects of long-term exposure to As, a well known carcinogen. The magnitude of the human tragedy will depend on the rate at which mitigation programs are implemented and, presently, the key challenge is to adopt a cost efficient and sustainable mitigation strategy that could be implemented for scaling up the safe water access.

Notwithstanding the current knowledge on the source and distribution of As, and its mobilization in groundwater since the discovery of As in the

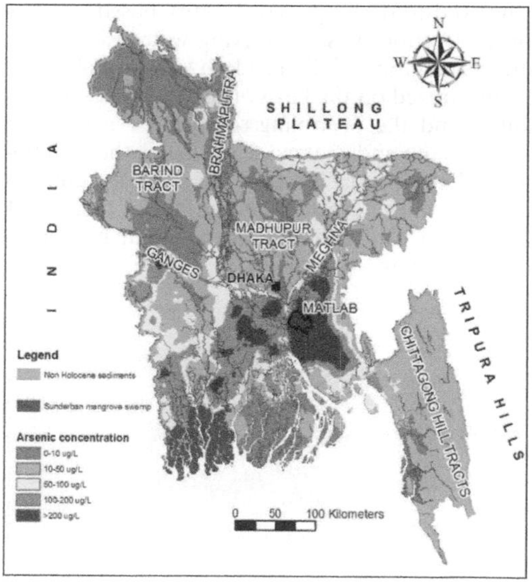

Figure 1. Occurrence of arsenic in groundwater of Bangladesh and location of the SASMIT intervention area.

drinking water wells in 1993, there has been limited success in mitigation attempts in Bangladesh. Over a period of two decades, various options have been implemented at household and community scale, including As-removal filters, rainwater harvesters (RWH), pond sand filters (PSF), dug wells (DW), and deep tube wells (DTW). Among these options, DTW (>150 m) offer an alternative source of As-safe drinking water over the major part of the country. These options have been assessed on several criteria, such as community acceptability, technical viability and their socio-economic implications (Jakariya *et al.*, 2004). It has been found that community acceptance of many of the options is low as people do not find them as convenient as tubewells (Figure 2).

This has led to the creation of a huge gap between the magnitude of arsenic exposure and the pace of mitigation, which is primarily attributable to:

- Lack of recognition and capacity building of local drillers community and their indigenous practice to identify the safe aquifers for well installation;
- Dependence on government and other NGOs for deep wells, being promoted as a safe water wells; and
- Unplanned installation of wells by different agencies overlooking safe water access at community levels and sustainability.

In order to bridge this inadequacy, the project on Sustainable Arsenic Mitigation (SASMIT) was conceived to develop a community based strategy for installation of safe drinking water wells in As affected regions of Bangladesh. The installations are optimized on the basis of hydrogeological suitability and the prevailing scenario of safe water access to the underserved segments of the communities. This approach can be scaled up for improving safe water access in other affected regions of Bangladesh and elsewhere with similar hydrogeological settings.

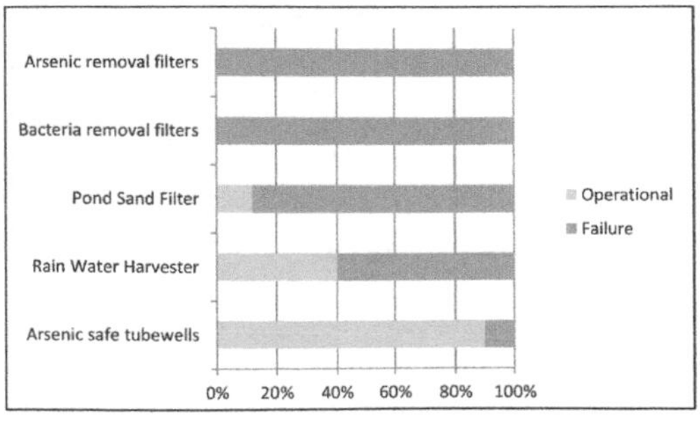

Figure 2. Performance analysis of the different options adopted for arsenic mitigation in Bangladesh.

2 APPROACH FOR SUSTAINABLE ARSENIC MITIGATION (SASMIT)

The approach for developing a strategy for Sutainable Arsenic Mitigation (SASMIT) was developed through multidisciplinary action research on geoscientific and social aspects. A combination of both hydrogeological suitability and social mapping to optimize the sites for well installation to improve the safe water access for rural and disadvantaged communities exposed to elevated As in drinking water from groundwater sources. The strategy is based on the following considerations:

- Assessing the available safe water options;
- Developing an implementation strategy based on hydrogeological suitability and social mapping through:
  - Identifying the intervention areas based on mapping of socio-economic status of the communities;
  - Prioritization of underserved cluster of households with respect to safe water access;
- Site-selection and optimization through dialogue with villagers and local drillers and GIS mapping;
- Capacity building of the local drillers in rural communities to enhance their indigenous skills to target safe aquifers for well installations;
- Baseline water quality investigation followed by systematic monitoring of the newly installed wells.

3 LOCAL DRILLERS—DRIVERS OF SASMIT

The increased dependence on groundwater for drinking in rural and urban areas of Bangladesh pose a challenge to the local drillers for targeting aquifers for installation of safe drinking water wells. Nearly 90% of the estimated more than 10 million tubewells are installed privately by local drillers in Bangladesh and elsewhere in the developing world.

The awareness of local drillers on elevated As concentrations in tubewell water at shallow depths have made them shift their practice of installation of tubewells. Using the visual color attributes of the shallow sediments (<50 m) and content of dissolved Fe, which in general are associated with high As concentrations, the local drillers install community tubewells at depths targeting off-white (buff) or red/brownish sediments. The drillers' perception and local knowledge of the sediments are often commendable and there is a good potential for capacity building of the local drillers to target safe aquifers. Involving the local drillers in this process would contribute to enhance their awareness and knowledge that will help them target safe aquifers on a country-wide scale and thereby reduces the exposure to As through drinking water.

4 THE MILESTONES OF SUSTAINABLE ARSENIC MITIGATION (SASMIT)

4.1 *Assessing the available safe water options*

Based on the evaluation of the various alternative drinking water options that were provided in Matlab area during Sida-AsMat project (2001–2006) and social survey in the project intervention area, tubewells emerged as the most preferred and accepted safe drinking water option. Tubewells were ranked as the most feasible and viable option for safe drinking water supply, mainly due to technical suitability in terms of installation and operation, almost negligible cost of maintenance and availability of good quality water throughout the year. Moreover, the tubewells are also ranked high due to their user-friendliness, especially to women and children. It is important to note that as compared to safe water demand, the number of safe wells is very low, as the cost of installation for deeper tubewells is beyond affordability of the local community especially for the poor section of the society.

4.2 *Assessing hydrogeological suitability*

As a part of the SASMIT action research for targeting safe aquifers in Matlab, subsurface geology and hydrogeological suitability was explored and assessed through installation of 17 piezometer nests (groundwater observation well-nests) at shallow, intermediate and deep aquifers (Figure 3). For a comprehensive assessment of the groundwater system, detailed monitoring of groundwater level fluctuations and water quality were carried out over a period of three consecutive years twice a year during the pre-monsoon and post-monsoon periods. In addition, monitoring of water quality also included a number of drinking water wells installed privately by the local drillers, and other stakeholders and all these information were merged and synthesized for the decision making process for new well installations. In addition, the hydrology of the aquifer systems were evaluated through groundwater modeling to assess the suitability of targeting intermediate depth aquifers (IDA).

4.3 *Implementation strategy based on social mapping*

Social mapping was used as a tool which is essential to identify the sites for tubewell installation in underserved areas. This strategy is recommended for the stakeholders during the planning and implementation of safe tubewell installation, which involves village and focus group discussions with the local communities to maximize the safe water coverage and the beneficiaries in a consultative manner. Additionally these meetings also serve as a platform for generating awareness on safe drinking water access as well as the community initiatives to monitor and maintain the installed tubewells. Mapping and interpretation techniques have been developed through the application of GIS for presenting and analyzing the social data generated from the implementation areas prior to installation.

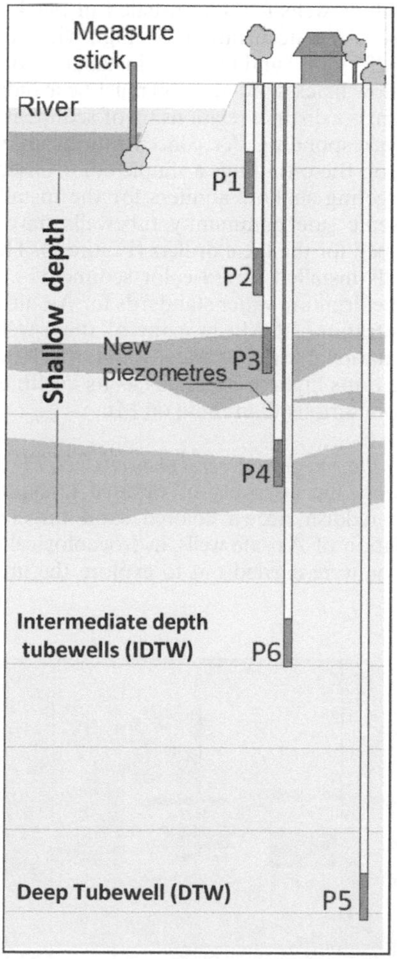

Figure 3. Assessment of hydrogeological suitability for drinking water wells installation of the piezometer nests (groundwater observation well-nests) at shallow, intermediate and deep aquifers in Matlab, Bangladesh.

4.4 *Development of mitigation approach*

4.4.1 *Sediment color tool for targeting As-safe aquifers at shallow depths*

Local drillers are installing shallow tubewells in reddish brown sediment with low concentrations of As with average and median values lower than the WHO drinking water guideline (10 μg L^{-1}). The levels of As in the off-white sediments were also similar; however, targeting off-white sands could be limited due to uncertainty of proper identification of color, specifically when day-light is a factor. Elevated Mn in red, off-white and white sands is a constraint for installation of safe tubewells, which warrants a better understanding on the impacts of elevated Mn on human health. White colored sediments at shallow depths were rare and apparently less important for well installations. In most of the shallow wells (>90%) installed in black sands (n = 64), As concentration was high with an average of 235 μg L^{-1} and therefore installation of wells in shallow black sand aquifers must be avoided.

There is a distinct relationship of sediment color and corresponding As concentrations in water. Based on these findings a simple color based tool for targeting shallow aquifers for the installation of arsenic safe community tubewells have been developed for the local drillers (Figure 3). The low As wells installed in red color sediments comply with the drinking water standards for As, although concentrations of Mn in many of these wells are above national drinking water standards. However, As warrants highest attentions as its health effects are more serious than those on Mn.

4.4.2 *Intermediate Deep Tubewells (IDTW)*

Following the discovery of elevated levels of Mn in the reddish brown colored sand targeted for installation of As safe wells, hydrogeological investigations were carried out to explore the intermediate deep aquifers (designated as P6, between 100–130 m, Figure 4) to avoid the risk for both As and Mn. These aquifers were found to be low both in As and Mn, and the SASMIT project successfully installed 268 IDTW during the project period. Among the IDTW installed, 96 % were found to be As-safe according to BDWS. In all of the IDTW, Mn concentrations were found to be within and/or very close to WHO values of 0.4 mg L^{-1} (currently there is no health based limit for Mn in WHO Drinking Water Quality Standards). These newly explored intermediate depth aquifers could be a potential source for As- and Mn-safe water supply at a reasonable cost. Replication trials in neighboring areas validated the wider applicability of the IDTW strategy of As mitigation (Ahmed and Bhattacharya, 2014).

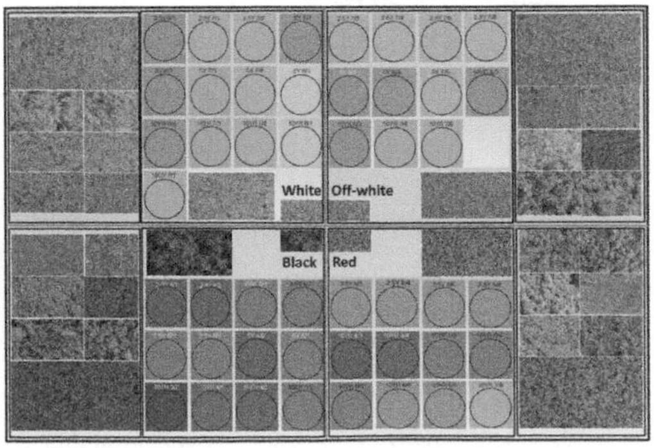

Figure 4. A simplified color tool for the local drillers.

4.5 *Follow-up on water quality monitoring and surveillance*

Sustainability of the provided safe water option needs to be ascertained by conducting various field and laboratory investigations and using these data to run predictive models. This would give essential information for risk assessments for cross-contamination and sustainability of these safe aquifers. The present focus is on As only and other parameters are not considered critically. Since groundwater may contain many other types of contaminants such as pathogens and other elements of health concern, it is important to focus on safe water rather than As safe water. Water Safety Plan (WSP) has been included for the provided option (targeting As safe aquifer) for SASMIT project implemented wells through a detailed baseline water quality survey as a primary step towards the achievement of the WSP. However, there is a need for continuous monitoring and surveillance of the installations for compliance with the WSP from holistic perspectives.

5 SASMIT—AN INTEGRATED CONCEPT FOR OPTIMIZED SAFE WATER ACCESS

The uniqueness of SASMIT strategy based on technical (hydrogeological) suitability and social aspects of arsenic mitigation to develop a cost-effective solution which involves and strengthens the initiative and capacity building of the local community. Implementation of future arsenic mitigation project, based on the SASMIT strategy would contribute to the capacity building of the local drillers for safe tubewell installations, besides raising the awareness among the affected communities in rural areas of Bangladesh and elsewhere in the developing countries. Focus on the human resource development at all stages of the project implementation, especially on the training of the local drillers for capacity building and disseminate the invented method of installing safe tube wells. The SASMIT concept has thus generated a vision to scale up safe water access in the Bangladesh, where tubewell is the widely accepted and most dependable drinking water option. SASMIT strategy could be applied in areas with similar geology and, considering our knowledge base, we certainly expect that it would be globally applicable for aquifers with similar hydrogeologic setting. Thus, bringing this strategy to policy makers would aid the feasibility of wider applications.

ACKNOWLEDGEMENTS

We thankfully acknowledge the financial support from the Swedish International Development Cooperation Agency (Sida) for the financial support throught the grant 75000854.

REFERENCES

Ahmed, K.M. & Bhattacharya, P. 2014. Groundwater arsenic mitigation in Bangladesh: two decades of advancements in scientific research and policy instruments. (This Volume).

Ahmed, K.M., Bhattacharya, P., Hasan, M.A., Akhter, S.H., Alam, S.M.M., Bhuyian, M.A.H., Imam, M.B., Khan, A.A. & Sracek, O. 2004. Arsenic contamination in groundwater of alluvial aquifers in Bangladesh: An overview. *Appl. Geochem.* 19(2): 181–200.

Bundschuh, J., Litter, M. & Bhattacharya, P. 2010. Targeting arsenic-safe aquifers for drinking water supplies. *Environ. Geochem. & Health* 32: 307–315.

Jakariya, M., von Brömssen, M., Jacks, G., Chowdhury, A.M.R., Ahmed, K.M. & Bhattacharya, P. 2007. Searching for sustainable arsenic mitigation strategy in Bangladesh: experience from two upazilas. *Int. J. Environment and Pollution* 31(3/4): 415–430.

von Brömssen, M., Jakariya, Md., Bhattacharya, P., Ahmed, K.M., Hasan, M.A., Sracek, O., Jonsson, L., Lundell, L. & Jacks G. 2007. Targeting low-arsenic aquifers in groundwater of Matlab Upazila, Southeastern Bangladesh. *Science of the Total Environment* 379: 121–132.

One Century of the Discovery of Arsenicosis in Latin America (1914–2014) – Litter, Nicolli, Meichtry, Quici, Bundschuh, Bhattacharya & Naidu (Eds)
 ISBN 978-1-138-00141-1

Groundwater arsenic mitigation in Bangladesh: Two decades of advancements in scientific research and policy instruments

K.M. Ahmed
Department of Geology, University of Dhaka, Dhaka, Bangladesh

P. Bhattacharya
KTH-International Groundwater Arsenic Research Group, Department of Sustainable Development, Environmental Science and Engineering, KTH Royal Institute of Technology, Stockholm, Sweden

ABSTRACT: Two decades have passed since the first detection of arsenic above allowable limits in groundwater of Bangladesh. A good number of scientific research and mitigation projects have so far been completed but still today more than 22 million people are exposed to arsenic leaves of 50 μg L^{-1} or more. As there are many untested new wells, it is not precisely known how many people are exposed to what level. Scientific knowledge about occurrences, distribution and release mechanisms have enhanced significantly. Although deep tube wells have emerged as the most effective mitigation measure over most of the country, still there are areas where this does not work. Recent studies reported effectiveness of alternative options like intermediate deep wells and subsurface arsenic removal. There has been a major paradigm shift in the policy arena regarding arsenic mitigation.

1 INTRODUCTION

Two decades have passed since first detection of arsenic in Bangladesh in 1993. The first national surveys reported exposure of millions of people above national drinking water limit for arsenic. The occurrence and distribution arsenic in the shallow alluvial aquifer have been well studied and areas with arsenic hot spots have been identified. Lot of efforts and resources has been spent in order to reduce the arsenic exposure. Sadly, until today, a large number of people are exposed to high As levels in different parts of the country. Lot of scientific studies have been carried out which significantly enhanced the knowledge base. New policy instruments have been formulated and introduced to deal with the issue of serious public health concerns. This paper reviews the scientific researches and policy instruments in the field of arsenic mitigation in order to provide a pragmatic strategy for As mitigation.

2 SCIENTIFIC KNOWLEDGE

2.1 *Mitigation options*

Many different mitigation options have been tested in the country so far. Effectiveness of household arsenic removal units have been found to be insignificant in providing safe water although community units have been found effective. Role of surface water is also very low in providing safe water. Groundwater remains the main source of safe water despite wide spread occurrences of arsenic. Deep tube wells (DTW) (>150 m deep) have become the most sustainable and community liked mitigation measure. However, DTWs are found to be ineffective in certain parts of the country, constrained mainly by hydrostratigraphy. In addition, DTWs in many parts are not effective in providing safe water due to the occurrence of high concentrations of Mn. High salinity is also becoming a reason for failure of some deep tube wells. Occurrence of high Mn in deep aquifers is linked to the nature of the sediments whereas occurrence of salinity is linked to entrapped seawater from the geological past.

2.2 *Newer options*

Although DTWs have been very useful in providing safe water in many parts of the country, these are expensive and cannot be installed using the conventional hand drilling methods. Recent investigations in Matlab and surrounding areas in southern Bangladesh highlighted the possibility of using the intermediate aquifers as a source of water safe from both As and Mn (SASMIT, 2012;

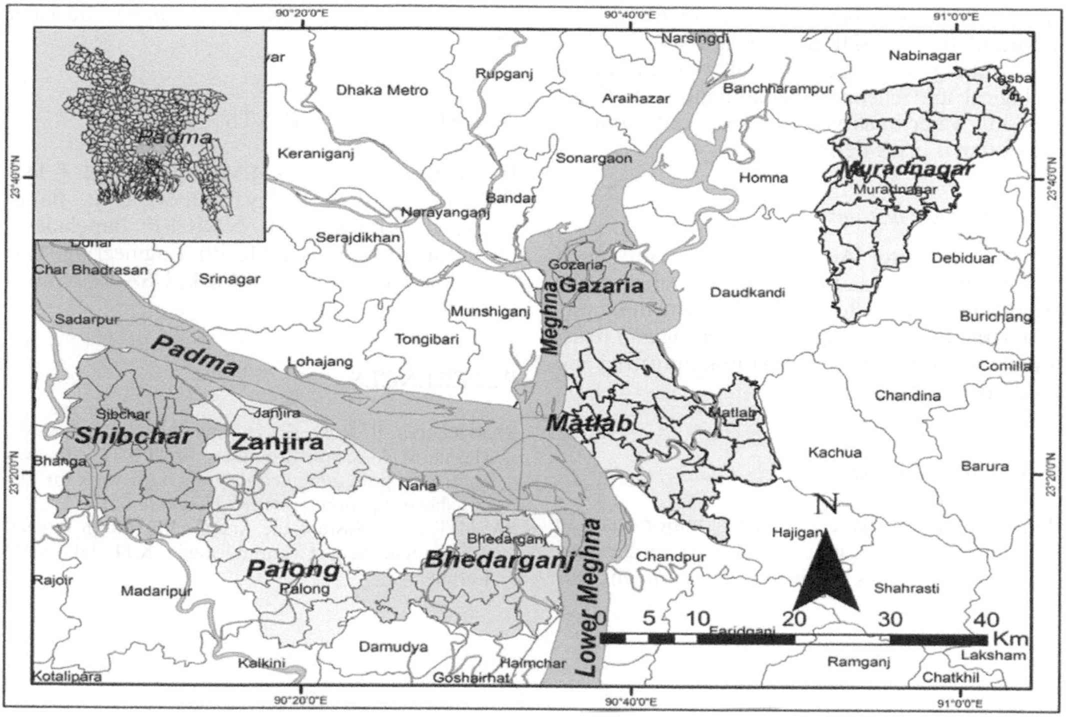

Figure 1. Areas in Bangladesh where Intermediate Deep Tube Well (IDTW) feasibility has been tested under the SASMIT project.

Bhattacharya *et al.*, 2014). The intermediate deep wells (IDW) are less expensive and can be even drilled by local hand percussion method.

Limited testing of Subsurface Arsenic Removal (SAR) in the shallow alluvial aquifers has also been successful in reducing arsenic concentrations below national drinking water limits. Managed Aquifer Recharge (MAR) in the shallow costal aquifers also provide evidences of lowering in As concentrations.

3 POLICY INSTRUMENTS

3.1 *Short term emergency measures*

The issue of arsenic mitigation was taken as a public health emergency and government mobilized resources, and introduced policy and action plan for arsenic mitigation. However, the initial mitigation action plan was not substantiated by scientific and factual evidences. As a result, huge amounts of investment were not effective in reducing arsenic exposure. Hundreds of thousands of household arsenic removal units were provided to the affected communities and most of them became nonoperational without making an impact on arsenic mitigation. Similarly, many dugwells and pond sand filters were installed in different parts of the country without considering the issue of overall water safety. Most of them are now nonfunctional and do not play any role in providing safe water. Investments in such mitigation options have been stopped.

3.2 *Long-term policy and plans*

There has been effort in revising the arsenic mitigation action plan. Instead of considering As mitigation in the field of water supply only, suggestions have been made to take action plans highlighting the irrigation and health aspects. Instead of installing As-safe water options, efforts have been made to install safe water options. Other parameters considered include microbiological and chemical contaminants. Water Safety Plan (WSP) has been introduced for all existing water supply options in order to protect quality from source to end user. A twenty year long Sector Development Plan (SDP) has been introduced by the government in order to provide safe water along with proper sanitation.

Water quality issues have been given high emphasis in formulating the SDP with allocation for providing drinking water safe from all type of microbiological and chemical contaminants. National drinking water quality surveillance and monitoring system is being designed.

3.3 *New legal instrument*

Bangladesh Water Act 2013 has recently been ratified by the parliament. This has created the opportunity for stricter water management by way of abstraction control through licensing and registration of water wells for irrigation, industrial and drinking purposes.

4 CONCLUSIONS

Reducing As exposure remains a major issue in Bangladesh. New scientific findings and policy instruments creates a better environment for bringing all people in the country under safe water coverage.

ACKNOWLEDGEMENTS

We acknowledge the SASMIT project of KTH-DU-NGO Forum; Policy Support Unit, Bangladesh of Government; UNICEF Bangladesh; Department of Public Health Engineering; and Columbia University, New York, USA for various supports.

REFERENCES

Bhattacharya, P., Jacks, G., Hossain, M., Ahmed, K.M., Hasan, M.A. & von Brömssen, M. 2014. Groundwater arsenic pollution: a conceptual framework for sustainable mitigation strategy. (This Volume).

SASMIT. 2012. Sustainable Arsenic Mitigation Project. 4th Annual Report (unpublished), KTH-DU-NGO Forum.

One Century of the Discovery of Arsenicosis in Latin America (1914–2014) – Litter, Nicolli, Meichtry, Quici, Bundschuh, Bhattacharya & Naidu (Eds)
© 2014 Taylor & Francis Group, London, ISBN 978-1-138-00141-1

Arsenic safe drinking water in rural Bangladesh: Perceptions of households and willingness to pay

N.I. Khan
Fenner School of Environment and Society, The Australian National University, Canberra, Australia

ABSTRACT: This study examined the Willingness To Pay (WTP) for Arsenic (As) free drinking water of rural Bangladeshi households across three different districts by applying a double bound discrete choice value elicitation approach. The majority of the households (87%) were willing to pay on average about 5% of their disposable average annual household income for As-free drinking water. However, the amount they were willing to pay for As safe water was very low compared to other household expenses. Factors which influenced WTP included the As contamination level, household income, water consumption, awareness of water source contamination, whether household members were affected by As contamination and whether they already had taken any mitigation measures.

1 INTRODUCTION

Arsenic exposure to humans is a major public health concern, especially in Bangladesh, where it is estimated that between 30–40 million people out of a population of 129 million are potentially at risk of As poisoning from drinking water sources (Ahmad *et al.*, 2006). The estimated annual health costs associated with As contamination in tubewell water in Bangladesh is USD 2.7 billion (Maddison *et al.*, 2005). In a recent study, Khan & Haque (2010) estimate that the cost of illness due to As exposure, including mitigation expenses, was USD 51 per household per year.

Bringing the tens of millions of As exposed people under safe water coverage is an immensely complex and expensive task. Therefore, mitigation of As contamination in Bangladesh is also a complex task requiring sound remediation technologies which are both socially and institutionally acceptable. Understanding this complex issue from the household level stakeholders' perspective and their views on Willingness To Pay (WTP) is crucial for a sustainable arsenic mitigation in Bangladesh. Thus, the main objective of this study was to examine WTP for As-free drinking water across different risk zones of Bangladeshi rural households.

2 METHODS

An economic evaluation of As safe drinking water options was conducted using a questionnaire survey to determine the key factors which influenced rural Bangladeshi households WTP. Double bound dichotomous choice WTP questions (e.g., Hanemann and Kanninen, 1999) for the one-off capital and monthly operation and maintenance costs were asked. The Contingent Valuation (CV) method was then applied to determine WTP in 650 households, located in three districts (Comilla, Munshiganj and Pabna) of Bangladesh. Rural household knowledge of Arsenic (As) contamination and preference for various mitigation measures were also identified. To estimate WTP for communal DTW bi-variate probit regression models were developed.

3 RESULTS AND DISCUSSION

Households consisted of an average of five family members with an average household monthly income slightly over 9,000 BDT (USD 132). Based on a poverty income threshold value of USD 125 per capita per year (BBS, 2005), 53% of the households lived below the poverty threshold. Household monthly food expenditure (59%) was higher than other non-food related expenses (41%) and, at the time of survey, some of the households were spending only 0.5% of the total monthly expenditure on water. At present, household expenditure on water is very minor compared with other expenditures, indicating that the concept of "paying for water" has not been developed in the rural community but they valued good quality drinking water.

Only 7% of the households solely (100% usage) used Deep TubeWells (DTW) and 82% households solely (100% usage) used shallow tubewells (STW) as their drinking water source. The use of Household As Removal Filters (HAsRF) were not a popular option amongst the Bangladeshi households being used by only 18% of the households.

The overall As concentration in the tubewell water ranged between 5 and 750 μg L^{-1} (unpublished data). Overall 18% of the As concentration in drinking water of the households were below the WHO guideline value of 10 μg L^{-1}, and only 4% was below the Bangladesh guideline value of 50 μg L^{-1}. The majority of the As concentrations (78%) were much higher than the Bangladesh guideline value.

Amongst the currently available As mitigation measures, the preference for particular mitigation options decreased in the order DTW (49%) > piped water system (22%) > rain water harvesting (20%) > dugwell (14%) > HAsRF (9%) > community filter (4%). Results of this study are thus in agreement with previous studies by Shafiquzzaman *et al.* (2009), Ravenscroft *et al.* (2009) and Khan & Hong (20122), which also found that DTW was the most preferred option amongst the users and option providers.

Most respondents (87%) expressed in principle WTP for As safe water, indicating rural populations valued water. However, the bi-variate probit regression model indicated that the amount that households were willing to allocate for safe drinking water without As was only 3 and 5% of their average monthly income for the payment of one-time capital cost and operation and maintenance cost, respectively. The annuity of the more conservative double bound WTP was only 5.4% of the annual household income. The estimated median WTP for the capital investment costs to construct an As safe community DTW was USD 4.0 and 2.6 based on the single bound and double bound WTP questions respectively. Also, estimated median WTP for the monthly operation and maintenance costs was USD 7.0 and 6.7 per household per year for the single and double bound WTP question respectively. Based on the self-reported cost of illness (COI) the estimated mean cost of illness was USD 24 per household per year. This is substantially lower than the cost of illness per household per year estimated by Khan and Haque (2009), that was USD 51. According to the respondents, medical treatment for arsenicosis was provided free of charge; therefore, a reliable estimation of household level COI was difficult. A higher level of WTP for As safe water options were estimated from those respondents who were already exposed to high levels of As contamination through drinking water, who were aware that their tubewell was contaminated with As and who had arsenicosis patients in their household. Factors that influenced WTP were mainly household income, education, exposure level of As, knowledge on adverse effect of As contamination and amount of daily water consumption. Household respondents located in a medium or high As risk zone were willing to contribute significantly more towards the capital investment costs for a community As DTW than respondents living in a risk free zone.

4 CONCLUSIONS

The positive attitude towards WTP for As safe water indicated that rural households have started to value good quality water for ensuring good health, and avoiding As exposure. However, the amount of money they were willing to pay for As safe water was very low compared to other household expenses. Allocation of lower amounts for water can be attributed mainly to the latent onset of detrimental health effects from As exposure, to water currently being considered as a free commodity, to the basic right to have water and also to the absence of the concept of "paying for water" among the rural communities in Bangladesh.

ACKNOWLEDGEMENTS

The author thanks EAWAG, the Swiss Federal Institute of Aquatic Science and Technology, for providing financial support and Dhaka Community Hospital (DCH) for logistical support during fieldwork in Bangladesh.

REFERENCES

Ahmad, J. Goldar, B. & Misra, S. 2005. Value of Arsenic-Free Drinking Water to Rural Households in Bangladesh. *Journal of Environmental Management* 74, 173–185.

Ahmad, J., Misra, S. & Goldar, B. 2006. Rural Communities' Preferences for Arsenic Mitigation Options in Bangladesh. *Journal of Water Health* 4: 463–478.

Bangladesh Bureau of Statistics (BBS). 2005. *Statistical year book 2005*. Available at: www.bbs.gov.bd

Hanemann, W.M. & Kannienen, B. 1999. *The Statistical Analysis of Discrete-Response CV Data*. Oxford University Press, Oxford.

Khan, M.Z.H. & Haque, A.K.E. 2010. Economic Cost of Arsenic Disaster: Policy Choices for Bangladesh, Erg Occasional Paper, 1/2010. Economic Research Group, Bangladesh.

Khan, N.I. & Yang, H. 2012. An analysis of institutional stakeholders' opinion on arsenic mitigation in Bangladesh, In Ng, J. Noller, B. Naidu, R. Bundschuh, J. & Bhattacharya, P. (eds.), Understanding the Geological and Medical Interface of Arsenic, Taylor and Francis Group, London, ISBN 978–0–415–63763–3.

Maddison, D., Catala-Luque, R. & Pearce, D. 2005. Valuing the Arsenic Contamination of Groundwater in Bangladesh. *Environmental Resource Economics* 31: 459–476.

Ravenscroft, P., Brammer, H. & Richards, K. 2009. Water Supply Mitigation Arsenic Pollution: A Global Synthesis. Wiley-Blackwell, Oxford, UK. doi: 10.1002/9781444308785.

Shafiquzzaman, M., Azam, M.S., Mishima, I. & Nakajima, J. 2009. Technical and Social Evaluation of Arsenic Mitigation in Rural Bangladesh. *Journal of Population Health and Nutrition* 27: 674–683.

One Century of the Discovery of Arsenicosis in Latin America (1914–2014) – Litter, Nicolli, Meichtry, Quici, Bundschuh, Bhattacharya & Naidu (Eds)

Security of deep groundwater against ingress of arsenic and salinity in Bangladesh: Policy aspects

M. Shamsudduha
Institute for Risk and Disaster Reduction, University College London, London, UK

A. Zahid
Ground Water Hydrology, Bangladesh Water Development Board, Dhaka, Bangladesh

W.G. Burgess
Department of Earth Sciences, University College London, London, UK

K.M. Ahmed
Department of Geology, University of Dhaka, Curzon Hall Campus, Dhaka, Bangladesh

ABSTRACT: Over the past two decades, the use of deep tubewells has become the most widely used mitigation response to the groundwater Arsenic (As) crisis in Bangladesh. Also over this time, much has been learned about the deeper levels of the Bengal Aquifer System through field investigation, experiments and modeling. The deep wells are vulnerable to contamination by As drawn down from its shallow source, but the magnitude and timing of the As ingress and the security of alternative groundwater pumping strategies are all uncertain. Vulnerability to salinity in some areas is an additional concern. Policy direction and monitoring strategies are urgently needed as a basis for effective management of deep groundwater abstraction. Both requirements may be guided by predictive modeling of alternative groundwater abstraction strategies and scenarios, accounting for the heterogeneous character of the aquifer, the spatial distribution of deep pumping, and the likely scale of increasing demand.

1 INTRODUCTION

At the first formal discussion on deep groundwater as a mitigation option for the Arsenic (As) crisis in Bangladesh (DPHE/UNICEF/WB, 2000) little was known about the aquifer conditions at depth. Over ensuing years, the concept 'deep groundwater' has become preferred to 'deep aquifer', as hydraulic continuity within the 'Bengal Aquifer System' has been recognized at whole-basin and regional scales (Michael & Voss, 2008; Burgess *et al.*, 2010; Hoque, 2010; Hoque & Burgess, 2012; Zahid *et al.*, 2012). Much has been learned from field investigations, experimentation and modeling. Many tens of thousands of deep tubewells (>150 m) have been installed (Figure 1), including for hand-pumped domestic supplies, rural piped systems, and municipal and commercial supplies (DPHE/JICA, 2009). High-yielding deep wells have been installed in over 100 rural water supply schemes and at more than 20 towns. Deep tubewells now account for over 70% of the mitigation response across Bangladesh (DPHE/JICA, 2009). The vulnerability of these deep wells to contamination by As drawn down over time from its shallow source is uncertain, however, and contamination by salinity in some areas is an additional, but unquantified, risk (Burgess *et al.*, 2010) (Figure 2). Also, irrigation tubewells at intermediate depth (75–100 m) are ultimately vulnerable to As contamination and, hence, there are pressures for irrigation water to be derived from deeper wells, which will exacerbate the deep groundwater vulnerability (Michael & Voss, 2008).

2 POLICY ASPECTS

Policy direction and monitoring strategies urgently need to catch up with this de-facto situation in which deep groundwater pumping has become the most popular, practical and economic mitigation response to the As crisis. Caution against excessive exploitation of the deep groundwater resource with reference to a 1000 year timeframe for judging sustainability using a basin-scale model of groundwater flow (Michael & Voss, 2008) is counter-balanced by the outcomes of short-term re-sampling of deep wells which shows no adverse impact on

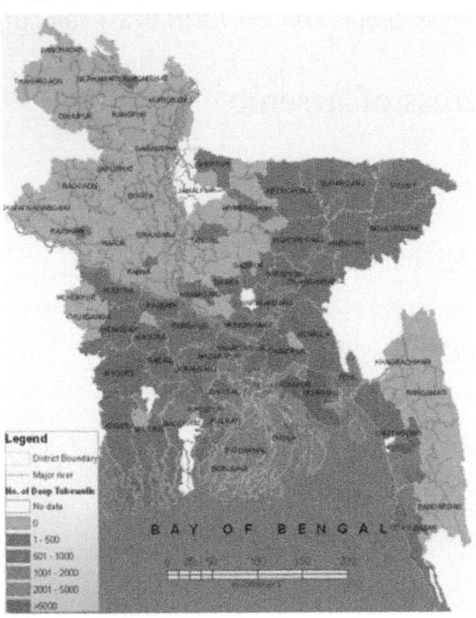

Figure 1. Spatial distribution of deep tubewells in Bangladesh (DPHE/JICA, 2009).

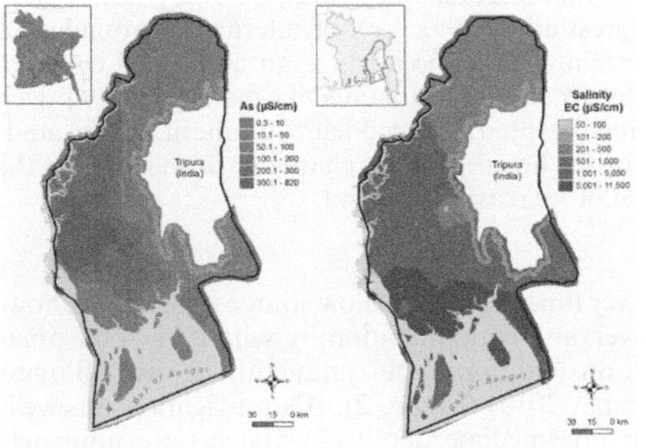

Figure 2. Spatial distribution of As (left) and salinity (right) in shallow groundwaters in south-east Bangladesh (UCL, 2013).

water quality attributable to deep groundwater pumping over a period of 13 years (Ravenscroft *et al.*, 2013). Policy development must consider the options presented by deep groundwater and its security over an intermediate period of time, decades or longer. These options may be strategically valuable even though ultimately unsustainable, and address the questions: (1) Could deep wells supply water unaffected by As and salinity for a period of time sufficient to be of strategic value, while also not excessively depleting the shallow water table? (2) How does the amount and allocation of deep groundwater pumping influence the security of the deep resource?

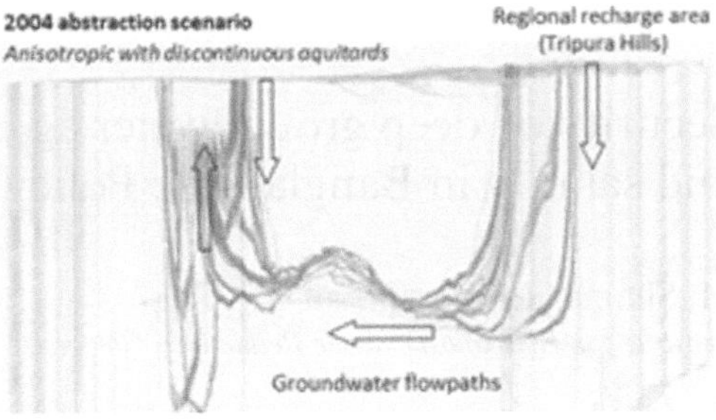

Figure 3. Multiple flow paths converging to points of deep groundwater abstraction (Zahid *et al.*, 2012).

3 INDICATIONS FROM MODELING

Recognizing that deep groundwater has enormous strategic value for water supply, health protection and development in Bangladesh, we suggest that for water supply planning, timeframes of 100 years be applied. Policy and the development of the deep groundwater resource over this time period should be guided by predictive modeling of the impacts of a range of groundwater abstraction strategies and scenarios, accounting for the heterogeneous character of hydraulic conductivity of the Bengal Aquifer System to depths of 300–400 m, the spatial distribution of deep pumping, and the likely scale of increasing demand. Groundwater flowpath (Figure 3) and travel-time statistics under a variety of abstraction scenarios have been applied to map deep groundwater security against As and salinity invasion across the region.

4 CONCLUSIONS

Deep groundwater abstraction for public water supply in southern Bangladesh may, in general, be secure against ingress of As for at least 100 years, but recognizes local vulnerability in some areas. Development should be backed up by an effective system of abstraction management and monitoring.

ACKNOWLEDGEMENTS

We acknowledge the Arsenic Policy Support Unit, Department of Public Health Engineering, Bangladesh Water development Board, Bangladesh Agricultural Development Corporation, JICA Bangladesh, UNICEF Bangladesh and Dhaka University. We acknowledge the UK Engineering and Physical Sciences Research Council (EPSRC) for financial support (KTA 40461).

REFERENCES

Burgess, W.G., Hoque, M.A., Michael, H.A., Voss, C.I., Breit, G.N. & Ahmed, K.M. 2010. Vulnerability of deep groundwater in the Bengal Aquifer System to contamination by arsenic. *Nature Geosci.* 3(2): 83–87.

DPHE/JICA 2009. Situation analysis of arsenic mitigation. Department of Public Health Engineering, Japan International Cooperation Agency Bangladesh, Dhaka.

DPHE/UNICEF/WB 2000. Deeper Aquifers of Bangladesh—A Review. Meeting, 20–22 August 2000. Local Government Division, Ministry of LGRD and Cooperatives, Bangladesh, Dhaka.

Hoque, M.A. 2010. Models for managing the deep aquifer in Bangladesh (PhD thesis). London University College, London.

Hoque, M.A. & Burgess, W.G. 2012. 14-C dating of deep groundwater in the Bengal Aquifer System, Bangladesh: implications for aquifer anisotropy, recharge sources and sustainability. *Journal of Hydrology* 444–445: 209–220.

Michael, H.A. & Voss, C.I. 2008. Evaluation of the sustainability of deep groundwater as an arsenic-safe resource in the Bengal Basin. *PNAS* 105(25): 8531–8536.

Ravenscroft, P., McArthur, J.M. & Hoque, M.A. 2013. Stable groundwater quality in deep aquifers of Southern Bangladesh: The case against sustainable abstraction. *Sci Total Environ.* 454–455: 627–638.

UCL 2013. The security of deep groundwater in southeast Bangladesh: recommendations for policy to safeguard against arsenic and salinity invasion. University College London, London, p. 78.

Zahid, A., Hassan, M.Q. & Ahmed, K.M. 2012. Model simulation on groundwater development stresses to determine sustainable zones of multi-layered aquifer system in Bengal Basin. In A. Zahid, M.Q. Hassan, A. Rahman, M.S. Khan, M.A. Hashem & L. Hassan (eds.), *Impact of climate change on water resources and food security of Bangladesh*: 147–168. Dhaka:Alumni Association of German Universities in Bangladesh.

One Century of the Discovery of Arsenicosis in Latin America (1914–2014) – Litter, Nicolli, Meichtry, Quici, Bundschuh, Bhattacharya & Naidu (Eds)
© 2014 Taylor & Francis Group, London, ISBN 978-1-138-00141-1

Intervention model on removal of arsenic from drinking water

M.S. Frangie, A. Hernández & M.P. Orsini
Instituto Nacional de Tecnología Industrial, San Martín, Buenos Aires, Argentina

ABSTRACT: The National Institute of Industrial Technology (INTI, Argentina) developed an intervention model on removal of arsenic from drinking waters". INTI also designed a rural equipment for arsenic removal. Twenty-five equipments were installed in rural schools from the town of Taco Pozo, Chaco, Argentina. Physical and chemical analyses were carried out on about fifty samples of water. Teachers were trained to build understanding and expertise on the operation of the equipment. Education and community engagement are the key factors for ensuring successful interventions. INTI will continue monitoring control measurements and verifying the effectiveness of the water treatment.

1 INTRODUCTION

Provision of water is limited in many regions of Argentina. In agreement with the new Código Alimentario Argentino regulation, removal of arsenic from groundwater is a great challenge, due to the toxicological, economic and infrastructure implications.

In 2009, INTI—Química (Chemical Center) developed an intervention model on removal of arsenic from drinking water, which provides solutions for:

- communities supplied by public water distribution in piped systems.
- communities supplied by private wells that are not controlled by any regulatory institution. This may be the case of communities located in the peripheral urban areas not reached yet by public water network distribution, remote and rural populations, etc. These communities have been classified in communities with or without energy supply.

INTI—Quimica developed a rural equipment, called INTI-Q-DR, which removes arsenic from water and does not need the use of electric power.

Several people approached INTI looking for solutions to the problem of arsenic in water. This paper shows INTI´s intervention in rural schools from the Impenetrable Chaqueño, near Taco Pozo city, Chaco, Argentina.

2 INTERVENTION IN TACO POZO: DEVELOPMENT AND IMPLEMENTATION OF ARSENIC REMOVAL EQUIPMENTS

2.1 *Activities*

INTI carried out five interventions in Taco Pozo (Table 1). During the third intervention, a temporary laboratory was installed in order to analyze the drinking water consumed by the community.

Figure 1. Equipment installed in Lebreton's school.

Table 1. Intervention / Activities.

Intervention no	Activities
1-April 2011	Three equipments were installed. The intervention was communicated to Taco Pozo community.
2-August 2011	Twenty- two equipments were installed. Equipments installed in April 2011 were checked.
3- November 2011	All the installed equipments were checked. A temporary laboratory (arsenic and microbiological analysis) was installed.
4-August / September 2012	All the installed equipments were checked. A temporary laboratory was installed. New teachers were trained on the operation of the equipments.
5-June 2013	All the installed equipments were checked. A temporary laboratory was installed. Preliminary actions to install the laboratory in Taco Pozo hospital.

A list of relevant activities undertaken during each intervention are described below:

- samples of water were taken from wells and cisterns of 25 rural schools.
- physical and chemical analyses were carried out on about fifty samples of water.
- A study of natural settling, sedimentation and a study of coagulation and flocculation were carried out in INTI—Química.
- twenty five equipments were installed from April to August 2011.
- delivery of coagulants and oxidizing agents
- manuals on installing and operation instructions in DVDs were delivered.
- teachers were trained to build understanding and expertise on the operation of the equipments.
- Taco Pozo community was educated on the impacts on health caused by the consumption of water with arsenic and on mitigation measures.

2.2 *INTI-Q-DR Equipment*

Technology: coagulation—filtration
Coagulants: $FeCl_3/Al_2(SO_4)_3$
Oxidizing agent: sodium hypochlorite
Capacity: 30 liters/batch
Filter: sand filter
Batch time: 120 minutes

3 RESULTS

In Tables 2–4, different aspects of the interventions are indicated.

Seven equipments were taken away from the schools with good systems of rainwater harvesting

Table 2. Performance of the equipment INTI-Q-DR 2 installed in El Pintado school.

Parameter	Well water	Treated water
pH	8,7	7,5
Fluoride (mg L^{-1})	1,9	1,0
Arsenic (μg L^{-1})	500	<2

Table 3. Performance of the equipment INTI-Q-DR 1 installed in Brasil school.

Parameter	Well water	Treated water
pH	7,6	6,2
Chloride (mg L^{-1})	45	135
Fluoride (mg L^{-1})	2,9	2,0
Arsenic (μg L^{-1})	490	<2

Table 4. Effectiveness of operation and use of equipments.

Intervention	Checked Schools	Operation		
		Good	Regular	No use
1	3			
2	25	1	1	1
3	21	9	2	10
4	18	7	7	4
5	18	7	7	4

4 CONCLUSIONS

The most important action in affected communities is the prevention of further exposure to arsenic by the provision of a safe water supply. This supply must substitute a groundwater source with high arsenic contents. Rain water and treated surface water, microbiologically safe, can be also used.

All members of the communities should understand the risks of high arsenic exposure through drinking contaminated water.

Education and community engagement are the key factors for ensuring successful interventions.

INTI will continue monitoring control measures and verifying the effectiveness of the water treatment.

ACKNOWLEDGEMENTS

The authors are grateful to INTI—Quimica and the following INTI laboratories: Química del Agua, Análisis de Trazas, Micro y Eotoxicología and Planta Piloto.

REFERENCES

Degremont, Water Treatment Handbook, Ed. 6, Vol. 1 and 2.

EPA, 2000. Technologies and Cost for Removal of Arsenic from Drinking Water.

EPA, 2003. Arsenic Treatment Technology Evaluation Handbook for Small Systems.

IBEROARSEN (CNEA, CYTED). Taller de distribución de As en Iberoamérica, Red Temática 406RTO282.

Lesikar, B., Arsenic Agrilife Extension; Drinking Water Problems (11–05).

One Century of the Discovery of Arsenicosis in Latin America (1914–2014) – Litter, Nicolli, Meichtry, Quici, Bundschuh, Bhattacharya & Naidu (Eds)

Sustainable mitigation of arsenic contaminated groundwater in India

A.K. Ghosh, N. Bose & R. Kumar
A.N. College, Patna, India

M. German
Lehigh University, Bethlehem, USA

ABSTRACT: The "water surplus" state of Bihar is faced with the serious problem of arsenic contaminated aquifers whose water is used both for drinking and irrigation purposes. A large number of mitigation strategies are being adopted by the authorities, without obtaining the desired results of clean water supplies to the arsenic affected rural population. The objectives of this study were to identify the gaps in such mitigation techniques and to establish a holistic, innovative technology that is integrated with the socio-economic milieu of the study area. An adsorbent based arsenic removal technology has been tested and is being operated with the community participation. The results revealed a marked reduction in iron and arsenic concentration, effective operational processes and a financially viable clean water production for a community of 25 families, with scope for upscaling this mitigation model.

1 INTRODUCTION

The Holocene sediments of the river basins of the state of Bihar, India, are presently recognized as having the world most extensive areas of arsenic contaminated underground water (Acharyya *et al.*, 2007). Arsenic is a slow bio-accumulative genotoxin (Faita *et al.*, 2013). The desirable limit of arsenic in drinking water is 10 μg L^{-1} according to WHO guidelines (WHO, 2004), and Bihar aquifers have registered up to eighteen times more than this limit (Ghosh *et al.*, 2010). This has exposed approximately 30 million rural inhabitants residing in the Gangetic Plains of Bihar to be vulnerable to the carcinogenic properties and other serious symptoms of arsenic poisoning. Furthermore, in Bihar, the situation is further aggravated by acute poverty, malnutrition and illiteracy and a public/community participation becomes essential for any sustainable arsenic mitigation strategy.

The objectives of this study were to test an energy-efficient, adsorption-based arsenic filter under private-public ownership, and generate a cost-effective design for providing clean water to the community. The principles of participation sought to be the guidelines of this initiative encompass inclusion, equal partnership, transparency, sharing power, sharing responsibility, empowerment and cooperation.

2 METHODS/EXPERIMENTAL

The experiments were carried out with an easy-to-operate single filtration unit to provide arsenic-safe drinking water to 30 families using regenerable arsenic adsorbents. When the filtering capacity of unit was exhausted, the absorbent material was regenerated. The water of the selected hand pump had more than 90 μg L^{-1} As, with seasonal dilution to 30 μg L^{-1} in the monsoon season. A filter, with an upper tank having activated alumina and a lower tank having hybrid anion exchange resin as media, was installed near the hand pump and connected with the same, so as to yield the arsenic (As) and iron (Fe) safe water upon manually operating the hand pump. Waste backwash was made to pass through a coarse sand filter to reduce the arsenic concentration before its release to the environment. Both raw and filtered water were tested for As and Fe concentration by atomic absorption spectrophotometer (AAnalyst 200-PerkinElmer) with a graphite furnace. The precision of analyses was better than 10%. Community involvement was obtained by interactions and interviews with 30 households in the village, and training of selected persons on operation and maintenance of the filter unit. Regeneration of adsorbent media was done by passing high pH alkali through the adsorbing resin. The hydrated ferric oxides present in the resin became negatively charged and the previously adsorbed arsenic released from the resin into the bulk solution due to Donnan co-ion exclusion. To return the resin to its previous function, the resin was then rinsed with acid to protonate the metal oxide surface groups. Community involvement was achieved by training selected users for operation and maintenance of the filter unit.

3 RESULTS AND DISCUSSION

The pilot trials verified the suitability of the system as a viable and sustainable option for arsenic remediation in rural India. The system was successful in reducing both iron and arsenic concentrations within the WHO permissible limit. This technology conserved scarce power supply in the study area. The filtration process reduced iron to within the permissible limit of 0.3 mg $L^{-}1$ of water (WHO, 2011) over a span of 28 months, after which the efficacy of the media declined (Table 1, Figure 1).

Table 1. Iron removal by adsorption filter (BVs = bed volumes).

Date	Flow (kL)	BVs	Raw water (iron (mg/L))	Treated water (iron (mg/L))
09-06-2011	22	220	1.26	0.04
09-10-2011	38	380	1.63	0.06
09-12-2011	52	520	1.54	0.08
16-03-2012	71	710	1.64	0.05
20-05-2012	92	920	1.52	0.05
18-09-2012	112	1120	1.39	0.03
15/12/2012	136	1360	1.28	0.06
10-02-2013	161	1610	1.41	0.05
12-05-2013	186	1860	1.3	0.01

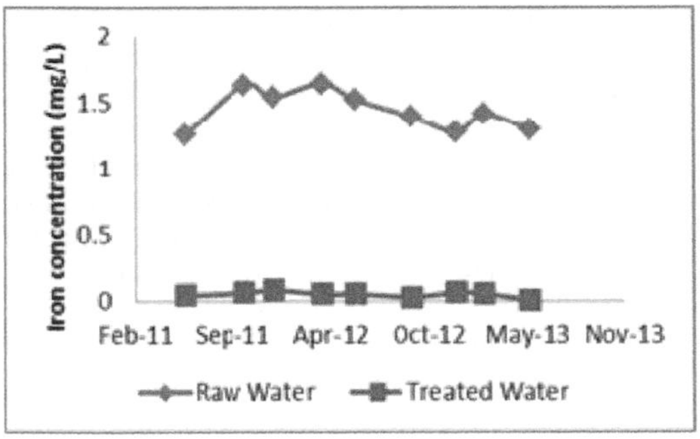

Figure 1. Iron removal by adsorption filter. (BVs = Bed volumes).

Safe arsenic levels were also maintained in the filtered water (10 ± 6 μg $L^{-}1$), while the backwash after being subject to sand filtration, had lower arsenic levels than raw water (Table 2, Figure 2).

The arsenic reduction was up to 93% and, throughout the study period, the arsenic concentration in filtered water ranged 10 ± 6 μg L-1. The present approach of creating demand for arsenic and iron free water, individual ownership and leadership development was backed by collective decision-making. Making such unaware arsenic affected community change over from one traditional drinking water source to alternate options immediately was very difficult. This filter was accepted by the villagers as it was perceived as an extension of the hand pump in place. Secondly, the issue of granting ownership of the filter to one family was mutually decided by the community and social conflicts avoided. A feasible water pricing mechanism based upon cost-effectiveness of the filter operations was calculated at an affordable sum of $ 0.002 per liter of arsenic free water. It was proved that sustainable technologies in developing countries greatly depend upon their integration with local socio-economic parameters.

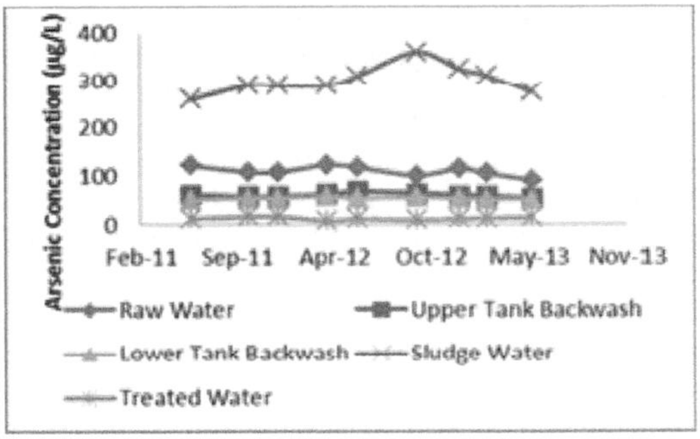

Figure 2. Arsenic removal by adsorption filter.

Table 2. Arsenic removal by adsorption filter.

Date	Flow (kL)	BVs	Raw Water (As Conc (μg/L))	Upper Tank Backwash (As Conc (μg/L))	Lower Tank Backwash (As Conc (μg/L))	Sludge Water (As Conc (μg/L))	Treated Water (As Conc (μg/L))
09-06-2011	22	220	122	62	51	260	12
09-10-2011	38	380	109	58	55	290	16
09-12-2011	52	520	109	58	55	290	16
16-03-2012	71	710	123	63	58	290	8
20-05-2012	92	920	118	68	54	310	11
18-09-2012	112	1120	101	65	56	360	9
15/12/2012	136	1360	115	60	53	325	11
10-02-2013	161	1610	105	59	51	310	12
12-05-2013	186	1860	91	54	52	275	14

4 CONCLUSIONS

The study confirms that simple and low cost technology for arsenic removal is sustainable through community participation in developing economies like India. The pilot-scale findings conducted in rural Bihar, India demonstrate that a granular media community filter can be a viable option to treat arsenic and iron rich groundwater and supply safer water to poverty stricken rural population at low cost.

ACKNOWLEDGEMENTS

This work has been supported by Dept. of Science and Technology (DST), Government of India and The Taagore-SenGupta Foundation, USA.

REFERENCES

Acharyya, S.K. & Shah, B.A. 2007. Groundwater arsenic contamination affecting different geologic domains in India—a review: influence of geological setting, fluvial geomorphology and Quaternary stratigraphy, *J. Environ. Sci. Health, Part A*, 42(12): 1795–1805.

Faita, F., Cori, L., Bianci, F. & Andreassi, M.G. 2013. *Int. J. Environ. Res. Public Health*. 2013; 10(4): 1527–1546.

Ghosh, A.K., Bose, N., Bhatt, A.G. & Kumar, R. 2010. New dimensions of groundwater arsenic contamination in Mid Ganga Plain, India. In: *Arsenic in Geosphere and Human Diseases*, Bundschuh, J. & Bhattacharya, P. (ed), Taylor & Francis Group, London.

World Health Organization. Guidelines for drinking-water quality. 2011, incorporating first addendum. Vol. 1, Recommendations. - 4th ed.

One Century of the Discovery of Arsenicosis in Latin America (1914–2014) – Litter, Nicolli, Meichtry, Quici, Bundschuh, Bhattacharya & Naidu (Eds)

Access, use and quality of water: Water economy and policy in rural populations with arsenic problems in Santiago del Estero, Argentina

S. Pereyra
Instituto de Altos Estudios Sociales, Universidad Nacional de San Martín, Campus Miguelete, San Martín, Prov. de Buenos Aires, Argentina

M.I. Litter
Gerencia Química, Comisión Nacional de Energía Atómica, San Martín, Prov. de Buenos Aires, Argentina

C.E. López Pasquali
Universidad Nacional de Santiago del Estero, Santiago del Estero, Argentina

ABSTRACT: A diagnosis about access, uses and quality of water with arsenic problems in the rural periphery of Clodomira (Dept. of Banda—Santiago del Estero Province, Argentina), followed by an intervention in the zone, is performed by a transdisciplinary group involving chemical and social scientists. The implementation of technologies to provide safe water to selected populations not connected to water networks is foreseen. Further replicates of this project can be performed in other similar regions.

1 INTRODUCTION

In many isolated rural and periurban places of Latin America and particularly in Argentina, access, use and quality of water is rather poor, in agreement with the difficult socioeconomic life conditions and low incomes of the population. Waters are generally affected by the presence of biological and chemical pollution including organics and toxic metals or metalloids such as As. Fatal endemic diseases (hepatitis, typhus or cholera) are current and represent a dangerous health risk. As known, the presence of As in groundwaters causes the incidence of arsenicosis (HACRE, chronic regional endemic hydroarsenicism). Thus, ways to facilitate the access to safe water to the populations are imperative.

2 OBJECTIVES AND STUDY AREA

The project links development of remediation technologies for As polluted water with social studies on life conditions and conflicts related to the access to water (Neiburg & Nicaise, 2009) in populations of low socioeconomic resources where HACRE cases have been detected.

For the project, the rural periphery of the locality of Clodomira, Banda Department, Santiago del Estero province, Argentina, has been chosen. In this area, around 3000 disperse inhabitants live in rural populations, hamlets, villages or isolated settlements of few houses, with an economy of subsistence and important difficulties of access to water. According to the 2010 Census (Censo 2010, INDEC) more than 41% of households of the Province have no access to water network inside the house. The area is located in the Río Dulce basin, where the presence of As in water has been amply documented and where 95% of the inhabitants could be affected by consuming well waters containing As (0.05–2 mg L^{-1}, Bundschuh *et al.*, 2010, Litter *et al.*, 2010). In the study region, the average intensity of solar light is around 5–6 kWh m^{-2} (Litter & Mansilla, 2001).

The project comprises two stages. In the first diagnosis stage, the daily practices of the study population related to the access, uses and habits of water will be collected. The social mediations involved, such as present and historical conflicts related to water, map of collective actors, related cultural traditions and water circuits in the domestic economy and in the productive activities will be analyzed. In parallel, water samples will be taken in the zone together with measurement of solar radiation to evaluate the implementation of water remediation strategies. In the second stage, an intervention in the zone will be made with implementation of methods for safe water provision, especially free of As. For this purpose, low-cost technologies will be validated, with emphasis in those using bottles under solar irradiation.

3 A TRANSDISCIPLINARY DIAGNOSIS

For the transdisciplinary diagnosis, a team of chemists, engineers, sociologists and anthropologists designed strategies of data collection, involving analysis of qualitative and quantitative data.

The diagnosis stage is presently ongoing. The statistical sources available for a socioeconomic and demographic characterization of Clodomira, with data of the 2010 Census (Censo 2010, INDEC) were used. The objective is the elaboration and analysis of some indicators of economic, social and human development that allow defining the more relevant characteristics and issues of the zone (employment and working market, education, health, poverty, etc.). However, these data are not enough for a good characterization because the project is referred specifically to the peripheral rural population, and the census data have a strong bias towards the urban population (which is the majority). Thus, a qualitative field work has been designed, combining ethnographic observations in environments and situations significant for the research with the implementation of 40 interviews to people, selected on the basis of a previous exploratory work including a visit to the zone and interviews with key informants.

A first screening campaign was performed in August 2013, where four schools of the rural peripheral zone of Clodomira have been selected and visited, serving as nodes for the future interviews and for establishing contacts. This exploratory work indicated that most of the population has now access to safe water, free of As, coming from a network recently provided by the new government of the Province. Based on this first campaign, the field work will be made in isolated settlements of 5 or 6 houses close to the schools, without water network connection or with a precarious one. Water samples will be also collected from water points of use of the houses for microbiological and As measurements. Rapid As measurements with field kits will be performed to select suitable samples with appreciable pollution to merit treatment (higher than 50 $\mu g\ L^{-1}$). A more precise analysis will be made later in the laboratory before the treatment. The solar intensity in several points of the region and in different times of the year and of the day will be measured with an appropriate radiometer and contrasted against existing data.

4 REMEDIATION STRATEGIES

Innovative procedures to provide safe water should be accessible, environmentally friendly and adapted to the cultural practices and rules of the population. A variety of technologies to improve the microbial quality of water and reduce the hydric diseases, including physical and chemical methods that can be used at the level of settlement clusters (schools, community centers, buildings and villages) are available (Meichtry & Litter, 2012).

In last times, the use of solar radiation has been amply investigated. Solar energy is costless and can be used *in situ*. Three low-cost technologies for providing safe water to isolated households are proposed in this project: Solar Disinfection (SODIS, Acra *et al.*, 1990), Arsenic Removal by Solar Oxidation (SORAS, Wegelin *et al.*, 2000) and heterogeneous photocatalysis (HP, Bundschuh *et al.*, 2010, Litter *et al.*, 2010). The three technologies can be used alone or combined, and they are visualized as ideal sociocultural processes to be accepted by rural and periurban communities. For the intervention stage, the Project foresees the design and elaboration of devices for households which use some of these remediation technologies. At the same time, the design should accomplish certain requirements related to the ways of life of the people object of this work. The goal is that the technologies could be used and appropriated by the people, with the less possible alteration of uses and habits related to water.

5 CONCLUSIONS

The scope and first advances of a transdisciplinary project about access, uses and quality of water in the rural zone of Clodomira are presented. The daily practices of the population related to the use and habits of water are considered. Samples of water will be collected and their analysis included in the diagnosis, while low-cost technologies in bottles such as SODIS, SORAS and HP will be validated. Finally, an intervention in the zone will be performed in few selected places with the implementation of devices for disinfection and arsenic removal. This methodology can be replicated in other similar zones.

ACKNOWLEDGEMENTS

This work was funded by UNSAM Diálogo entre las Ciencias Project, Argentina. The authors are members of CONICET.

REFERENCES

Acra, A., Jurdi, M., Mu'allem, H., Karahagopian, Y. & Raffoul, Z. *Water Disinfection by Solar Radiation, Assessment and Application Technical Study 66e* (1990). http://www.idrc.ca/library/ document/041882/.

IDRC Library: Documents, Ottawa, Canada, 1998, accessed 8-10-2013.
Bundschuh, J. Litter, M., Ciminelli, V, Morgada, M.E., Cornejo, L. Garrido Hoyos, S., Hoinkis, J. Alarcón-Herrera, M.T. Armienta M.A., Bhattacharya P. 2010. Emerging mitigation needs and sustainable options for solving the arsenic problems of rural and isolated urban areas in Iberoamerica—A critical analysis. *Water Research* 44: 5828–5845.
Censo INDEC 2010, http://www.censo2010.indec.gov.ar/CuadrosDefinitivos/H2-P_Santiago_del_estero.pdf. accessed 8-10-2013.
Litter, M.I. Morgada, M.E. & Bundschuh, J. 2010. Possible treatments for arsenic removal in Latin American waters for human consumption, *Environmental Pollution* 158: 1105–1118.
Meichtry, J.M. & Litter M.I. 2012. Solar disinfection as low-cost technologies for clean water production. In J. Bundschuh & J. Hoinkis (eds), *Renewable Energy Applications for Freshwater Production*: 207–238. London: CRC Press/Balkema, Taylor & Francis Group.
Neiburg, F. & Nicaise, N. 2009. *The social life of water*, Río de Janeiro, Viva Rio.
Wegelin, M., Getcher, D., Hug, S. & Mahmud, A. 2000. *SORAS—a simple arsenic removal process*. In Water, Sanitation and Hygiene: challenges of the millennium, 26th WEDC Conference, Dhaka, Bangladesh.

One Century of the Discovery of Arsenicosis in Latin America (1914–2014) – Litter, Nicolli, Meichtry, Quici, Bundschuh, Bhattacharya & Naidu (Eds)
© 2014 Taylor & Francis Group, London, ISBN 978-1-138-00141-1

Eradication of HACRE, a National Program Project, its rationale and feasibility study

J.P. Nigrele, N. Derderián, F. Fort & R.M. Manrique Carrera
Licenciatura en Gestión Ambiental, Universidad CAECE, Buenos Aires, Argentina

ABSTRACT: It is estimated that in Argentina the population living in areas with arsenical waters is at least 2,500,000 inhabitants, which constitutes almost 7% of the population of the country. It is a fact that arsenic concentrations above the maximum permissible values in drinking water (0.01 mg/L according to the WHO) can cause serious damage to human health. The main objective of this project is to create a formal framework for the progressive eradication of HACRE (Hidroarsenicismo Crónico Regional Endémico or arsenicosis) nationwide and to develop the necessary tools for decision making by national and local authorities. We propose the gradual implementation of measures sustained over time to provide the affected population with safe drinking water with arsenic levels below the standard set by the WHO.

1 INTRODUCTION

Water pollution by naturally occurring arsenic is a worldwide problem. It is estimated that in Argentina, the population living in areas with arsenical waters is at least 2,500,000 inhabitants, almost 7% of the population of the country (MINSAL, UnIDA and ATA, 2006). A remarkable feature of this problem is the difficulty in the early detection of the disease by health care providers, because when symptoms appear, there is no chance of being reversed. For this reason, it is essential to ensure preventive measures, mainly to avoid prolonged intake of water with arsenic concentrations above the WHO guidance level (0.01 mg/L).

Considering the present scenario, the creation of a national regulation that promotes the coordination of the efforts of different stakeholders, in order to take preventive measures, to develop intervention models for arsenical water treatment and to provide feasible solutions is proposed, mainly in regions where infrastructure and economic resources are scarce.

2 METHODS

2.1 *Analysis of the intervening factors*

Based on an analysis of different intervention scenarios, we developed a checklist of various factors to include in a SWOT (strengths, weaknesses, opportunities, and threats) matrix to categorize and assess their impact on a scale of 1–10. The assessment of the factors was performed individually by team members and then the results were averaged in order to minimize the subjectivity. This allowed us to know the favorable scenarios and those that may adversely affect the feasibility of the project.

2.2 *Operational feasibility analysis*

As part of the feasibility analysis, brainstorming was performed to identify and describe the relations between the different aspects of the problem and the stakeholders. In this way, we arrived at a first global understanding of the problem and identified the key points (current legal regulations, technologies on the market, professional training of parties involved, of groundwater extraction methods used in the different regions, etc.).

2.3 *Economic prefeasibility analysis*

For each of the steps described in the previous section, we have calculated the costs associated with each task necessary for the allocation of both human and material resources. We considered the costs involved in implementing the program, the incentives for research of new treatment technologies and the diffusion of the program.

3 RESULTS AND DISCUSSION

3.1 *Technologies*

The selection of treatment methods depends on the physicochemical characteristics of the aquifers and socioeconomic factors of the affected populations. We also contemplated the handling and disposal of generated wastes. In Argentina, conventional and emerging technologies for arsenic removal are available, with multiple positive experiences, such as oxidative photochemical processes, electrochemical corrosion, use of iron fixed beds, etc. This project sets the foundations to promote research, registration and validation of various technologies in order to improve technology accessibility.

3.2 *Social feasibility*

The social feasibility of the project is positive from all points of view, since it provides an improvement in the quality of life of affected people, providing a sustainable solution to the HACRE (*Hidroarsenicismo Crónico Regional Endémico or arsenicosis*) problem.

Employment generation, both direct and indirect, in the areas of health care, science, construction, transport, etc. is another important positive social aspect. This project would create jobs from research and manufacturing to maintenance and ongoing operation of the treatment plants.

The HACRE Project has the support of relevant voices that played an important role in the formulation of this project. Among them, Senator María Esther Lavado, Dr. Susana Isabel Garcia (National Ministry of Health), Engineer Graciela De Seta (National Technological University) can be mentioned.

As part of the project, we developed a methodology for the objective prioritization of affected areas, based on quantitative indicators.

3.3 *Results of the financial and economic feasibility analysis*

The gradual implementation structure of this project allows a phased funding, in accordance to the economic reality of Argentina. To demonstrate the economic and financial feasibility, a comparative analysis was made between the annual cost of implementation of the project and the national budget allocated for the year 2013 under the item "Drinking Water and Sewage".

Table 1. Cost Analysis, AR $ values.

Year	Cost of all project stages	Percentage of national budget (Drinking Water and Sewage Item)
2013	2,000,000.00	0.02%
2014	6,000,000.00	0.06%
2015	6,000,000.00	0.05%
2016	40,188,533.84	0.27%
2017	74,377,067.68	0.42%
2018	108,565,601.52	0.51%
2019	138,754,135.36	0.55%
2020	172,942,669.20	0.57%
2021	207,131,203.04	0.57%
2022	241,319,736.88	0.55%
2023	275,508,270.72	0.52%
2024	309,696,804.56	0.49%
2025	343,885,338.40	0.45%

This comparison demonstrates the feasibility of one of the essential components of this project, consisting in the creation of a National Fund designed to support HACRE eradication actions.

4 CONCLUSIONS

At present, there are no national projects aiming at integration of legal, technological, economic and management criteria. The HACRE Project envisages a network of smart and sustainable solutions, with technical, financial and regulatory support, aiming at reversing this issue in the medium term, with the active participation of the society.

The technical and legal variables analyzed have shown that the project has a high degree of feasibility. Furthermore, the political support (expressed by means of an agreement of a national senator) and the recognition received in the last Congress of Environmental Sciences COPIME 2013 (First Place) show that this project results viable in its multidimensional extension.

This initiative will result in a significant contribution to the improvement of the quality of life of individuals and to the reduction of direct costs in the health system.

ACKNOWLEDGEMENTS

This work was supported by CAECE University, members of federal agencies, national senators and Lic. Leonardo Pflüger as tutor.

REFERENCES

Höll, W. & Litter, M. 2010. Ocurrencia y química del arsénico en aguas. Sumario de tecnologías de remoción de arsénico de aguas. In: M.I. Litter, A.M. Sancha, A.M. Ingallinella (eds.) *Tecnologías económicas para el abatimiento de arsénico en aguas.* Buenos Aires, Argentina: CYTED, pp. 17–27.

Ingallinella, A.M. & Fernández, R. 2010. Experiencia argentina en la remoción de arsénico por diversas tecnologías. In: M.I. Litter, A.M. Sancha, A.M. Ingallinella (eds.) *Tecnologías económicas para el abatimiento de arsénico en aguas.* CYTED, Buenos Aires, Argentina, pp. 155–166.

INTI Química. 2009. Poblaciones con sistema de distribución de agua por red. In: *Modelo de Intervención para el Abatimiento de Arsénico en Aguas de Consumo.* INTI, Buenos Aires, Argentina, pp. 14–30.

Litter, M.I. & Morgada, M.E. 2011. Formas arsenicales en agua y suelos. In: M.I. Litter, M.A. Armienta, S.S. Farias (eds). *Metodologías analíticas para la determinación y especiación de arsénico en aguas y suelo*. Buenos Aires, Argentina: CYTED. pp. 19–27.

Litter, M.I. & Morgada, M.E. 2011. Tratamiento de las muestras de agua, suelos y sedimentos para determinación de arsénico. In: M.A. Armienta, S.S. Farias (eds). *Metodologías analíticas para la determinación y especiación de arsénico en aguas y suelo*. CYTED, Buenos Aires, Argentina, pp. 29–32.

Ministerio de Salud, UnIDA. & Asociación Toxicológica Argentina. 2006. *Epidemiología del Hidroarsenicismo Crónico Regional Endémico en la República Argentina*. Estudio Colaborativo Multicéntrico. Buenos Aires; SAyDS; pp. 2–81.

Navoni, J.A., De Pietri, D., García, S. & Villaamil Lepori, E.C. 2012. *Riesgo sanitario de la población vulnerable expuesta al arsénico en la provincia de Buenos Aires, Argentina*. Rev. Panam. Salud Pública: pp. 1–8.

Sancha, A.M. (2010). Importancia de la matriz de agua a tratar en la selección de las tecnologías de abatimiento de arsénico. In: M.I. Litter, A.M. Sancha, A.M. Ingallinella (eds.) *Tecnologías económicas para el abatimiento de arsénico en aguas*. CYTED, Buenos Aires, Argentina, pp. 145–151.

One Century of the Discovery of Arsenicosis in Latin America (1914–2014) – Litter, Nicolli, Meichtry, Quici, Bundschuh, Bhattacharya & Naidu (Eds)
© 2014 Taylor & Francis Group, London, ISBN 978-1-138-00141-1

Regulatory strategy to mitigate effects of arsenic on the population served by water systems in the province of Santa Fe

H.P. Vázquez, V. Ortolani & V. Pidustwa
Quality Control Management, "Ente Regulador de Servicios Sanitarios", Province of Santa Fe, Argentina

ABSTRACT: In Santa Fe, Argentina, in 1995, a new regulatory framework for the Sanitary Services was established. It created a new agency for control and regulation (ENRESS) and defined new Water Quality Standards that supports a maximum level of arsenic of 50 µg/L. The ENRESS approach to achieve compliance with the arsenic standard within a Water Safety Plan is based on quality management, hazard analysis and risk assessment to promote continuous improvement. The results are satisfactory and experience has shown that the establishment of new quality standards should be accompanied by an integral sanitary policy to harmonize scientific, technical and economic factors and human resources.

1 INTRODUCTION

The province of Santa Fe is located in east central Argentine, in the Chaco-Pampean plain, with a population of nearly 3,200,000 inhabitants and an area of 133.007 km^2. Currently, there are 392 water systems that provide water to the 91% of the people. The *Ente Regulador de Servicios Sanitarios* (ENRESS), created by the provincial law N° 11.220 in 1995, is responsible for regulating and controlling the Sanitary Services. Nowadays, 352 waters systems use groundwater sources with natural arsenic level between 2 and 300 50 µg/L. Arsenic in drinking water has been the subject of permanent monitoring and study by the Agency (ENRESS).

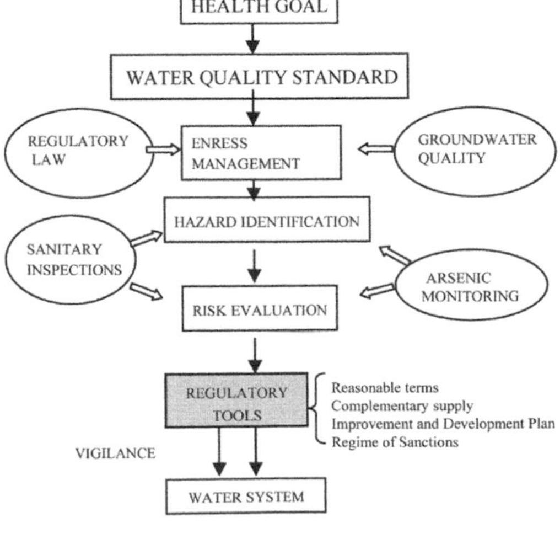

Figure 1. Scheme of water safety plan.

This works aims at presenting to the sanitary community the ENRESS regulatory and controlling actions in order to achieve compliance with the arsenic standard within the current legislation in the province of Santa Fe.

2 WATER SAFETY PLAN IN RELATIONSHIP WITH THE ARSENIC

When ENRESS began in 1995, 196 waters systems in the province used groundwater as a source, and 60 out of them did not comply with the level of arsenic established by the Quality Standards.

At the end of the three year period admitted by the Law 11.220 for adjustments to the new system, most of these services had not been improved for technical or economic reasons. Under this adverse sanitary situation, the Agency encouraged the development of an epidemiological study and adopted regulatory measures tending to achieve the gradual improvement of water quality within a Water Safety Plan (Figure 1).

3 HAZARD IDENTIFICATION AND RISK EVALUATION

Hazard identification is performed by sanitary inspections including sampling throughout the water supply chain, with a frequency set according to the population supplied. The samples are analyzed by Atomic Spectrometry in the ENRESS laboratories. The analytical quality is guaranteed by inter-laboratory and intra-laboratory controls. A complete and dynamic diagnosis is obtained from these activities. It reflects the quality of the

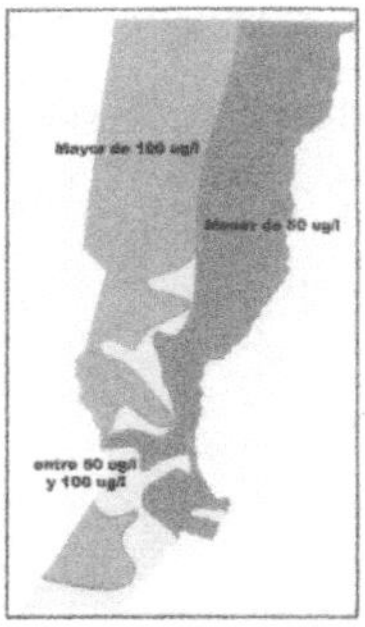

Figure 2. Arsenical risk map Santa Fe province (Argentina).

supplies from the provided centralized services, particularly arsenic level at sources of supply, water distribution network and complementary services.

Once the hazard profile was finished, ENRESS established agreements with the Pan American Health Organization in order to develop an epidemiological study. Finally, between 2000 and 2002, a research of "Risk evaluation study of consumption of water with arsenic" was carried out. The general objective was to evaluate the health impact of consuming water with different arsenic levels. Based on historical arsenic concentrations and population sizes, an Arsenical Risk Map was elaborated and the most exposed population was identified (Figure 2).

The prevalence of HACRE (arsenicosis) was determined in populations exposed to levels above the arsenic standard and the Mortality Risk Relative to cancers associated to arsenic was estimated. Even though in Argentine there is plentiful clinical information about HACRE, there is scarcity of representative epidemiological studies.

Regulatory policies have been established in the province based on the results of this research, supported by local measurements of risk indicators.

4 REGULATORY ACTIONS

Among the activities carried out by the Agency, emphasis was put on the dictation of resolutions to establish a regime of gradual improvements in water services in reasonable terms based on an acceptable risk evaluation. The suppliers that for economic reasons cannot achieve the required level have the possibility of providing a complementary service of drinking water that complies with the standard, with 2 L per capita per day as a minimum. The Agency carries out the evaluation following a "Development Improvement Plan" of Water Services, together with a "Regime of Sanctions".

5 RESULTS: ACTUAL SITUATION OF THE WATER SUPPLY

The sanitary policy carried out by the Agency has given satisfactory results.

The number of water systems has increased to 392, and 98 arsenic removal plants have been installed and operated to achieve the standard level.

6 DISCUSSION ON THE NORMATIVE

The Argentine "National Commission of Food" (CONAL) proposed in 2005 to reduce the standard of arsenic content in drinking water to 10 μg/L, set in the Argentine Code Food (CAA).

During a public consultation, the ENRESS Quality Control Management presented a scientific report regarding risk assessment and proposed: "Not to adopt a limit of arsenic below 30 μg/L before carrying out local epidemiology studies that assess with a better accuracy the risk of consuming water with arsenic levels below 50 μg/L, and to estimate the required investment based on a National Risk Map thus as to reach an acceptable risk level".

The Assessor Council of CONAL agreed and recommended "Not to modify the maximal arsenic limit until studies justify the need of modification".

Nevertheless, the modification of the drinking water arsenic standard to 10 μg/L with a term of 5 year to compliance was confirmed. Currently, this period is suspended until the conclusion of a national epidemiological study, in which the ENRESS will be in charge of updating the risk map and selecting the locations involved in the Santa Fe province.

7 CONCLUSIONS AND RECOMMENDATIONS

The ENRESS strategy to achieve compliance with the arsenic standard within a Water Safety Plan is based on quality management, hazard analysis and risk assessment. It intends to achieve a gradual improvement of the water quality in order to protect the public health.

The results are satisfactory and the experience has shown that the establishment of quality standards should be set in the context of an integral sanitary policy to harmonize scientific, technical and economic factors and human resources.

REFERENCES

Bartram, J., Corrales, L., Davison, A., Deere, D., Drury, D., Gordon, B., Howard, G., Rinehold, A. &

Stevens, M. 2009. Manual para el desarrollo de planes de seguridad del agua. Metodología pormenorizada de gestión de riesgos para proveedores de agua de consumo. Ginebra: OMS/IWA.

Corey, G. 2005. *Estudio Epidemiológico de la exposición al arsénico a través del agua de consumo*. Imprenta Oficial de la Provincia de Santa Fe. ISBN 987-23193-0-8. Argentina.

Hantke–Domas, M. & Jouravlev, A. 2011. *Lineamientos de politicas públicas para el sector agua potable y saneamiento.* CEPAL, Comisión Económica para América Latina y el Caribe, Naciones Unidas. Santiago de Chile.

One Century of the Discovery of Arsenicosis in Latin America (1914–2014) – Litter, Nicolli, Meichtry, Quici, Bundschuh, Bhattacharya & Naidu (Eds)
© 2014 Taylor & Francis Group, London, ISBN 978-1-138-00141-1

Management and arsenic mitigation policies in Costa Rica: The case of Bagaces, Guanacaste

R. Blanco
Environmental Management Unit, Caja Costarricense de Seguro Social, Retired Professor, Universidad de Costa Rica, Costa Rica

L. Arce
Life Quality Unit, Defensoría de los Habitantes, Costa Rica

M. Picado
Member, La Voz de Bagaces Association, Costa Rica

ABSTRACT: Costa Rica is facing a serious problem of arsenicosis, which was ignored for a considerable time and affects more than 30,000 people in 20 communities. The lack of experience has caused delays and solution approaches do not seem to solve the problem in the medium and long term. What has been achieved is the result of community organizing and legal advocacy and media. Current evidence is indicative of a possible chronic endemic regional hydroarsenicism (HACRE), which has not been recognized by health authorities.

1 INTRODUCTION

North part of Costa Rica is a rich area in geothermal and volcanic activity. There is also an intensive and extensive crops and livestock exploitation. Most of the 2 million tourists that visit Costa Rica each year choose such zone to visit due to its beautiful nature.

During a routine physical chemistry analysis of the quality of potable water, in the city of Cañas, Guanacaste, it was found that the levels of As in water used by the Integral Center for Health Attention were between 26 and 36 μg/L, over the Permissible Maximum Value (PMV) of 10 μg/L.

Subsequently, it was detected that other northern cities were also receiving water with As levels above the PMV, and that the water supply and quality control authorities knew of the presence of arsenic since 2005–2007.

This presentation will describe the actions of misinformation and demerit of the health significance of the detected levels of As and risk of the exposed population (more than 30,000 inhabitants) conducted by official bodies, regardless of the amount of risk level. Also presented laboratory results were used to justify the "drinkability" of water supplied to the population, with averages impossible or giving a misconception to the PMV. Furthermore, it shows how an organized community faces a serious environmental pollution situation and manages press authorities to seek sustainable solutions and not just palliative solutions.

2 RESULTS AND DISCUSSION

This gave rise to an unprecedented situation in Costa Rica, in which one of the affected communities, Bagaces city, given the inaction of local authorities, organized and conducted activities of reclamation to the national Costa Rica health authorities (Ministry of Health, MS) and to the body responsible for the supply and control of the quality of water (National Institute of Aqueducts and Sewers, AyA) which were neglected and forced to submit to the Costa Rican Constitutional Court several accusations of violation of fundamental rights, which culminated in a conviction for AyA and MS, with a deadline of six months to resolve the situation and to determine the environmental source of arsenic.

From the political point of view, it is interesting to note that the treatment given by AyA to this serious water pollution, affecting 22 communities and nearly 40,000 inhabitants, was always to belittle the seriousness of it and try to minimize it. At first, it was said that As was a micronutrient and that the amounts found were good for the health of people. Then AyA tried to raise the maximum allowable value, using a legal regulation for health services and raising the PMV to 50 μg/L, which made most of the aqueducts were at right.

3 CONCLUSIONS

This response of an organized community who confronted serious environmental pollution and managed to press authorities to seek sustainable solutions, not merely palliative panaceas, illustrates how an informed community was able to organize and to provide technical and legal arguments to safeguard their right to a healthy environment. In this case, the As problem in Bagaces, Guanacaste, Costa Rica has been explained.

ACKNOWLEDGEMENTS

The authors wish to thank the community of Bagaces, Guanacaste, and neighboring sites affected by the presence of arsenic in the drinking water, by the example of dedication and courage in defending their right to a healthy environment for themselves and their children.

One Century of the Discovery of Arsenicosis in Latin America (1914–2014) – Litter, Nicolli, Meichtry, Quici, Bundschuh, Bhattacharya & Naidu (Eds)
© 2014 Taylor & Francis Group, London, ISBN 978-1-138-00141-1

Arsenic in Paracatu: A conceptual model for environmental and epidemiological assessment and political contextualization

Z.C. Castilhos
Center for Mineral Technology, Rio de Janeiro, Brazil
Environmental Geochemistry Department, Fluminense Federal University, Niteroi, RJ, Brazil

E. Mello De Capitani
School of Medicine, State University of Campinas, Campinas, SP, Brazil

I.M. de Jesus, M.O. Lima & K.C.F. Faial
Evandro Chagas Institute (IEC/MS), Belém, Pará, Brazil

S. Patchineelam, W. Zamboni & E.D. Bidone
Environmental Geochemistry Department, Fluminense Federal University, Niteroi, RJ, Brazil

ABSTRACT: The objective of this work was to share the methodological bases applied in this research project, starting from a conceptual model for As environmental and epidemiological assessment in Paracatu city, from which a basic framework was elaborated, herein called Plan of Action (PoA).

1 INTRODUCTION

When it comes to perform an environmental contamination assessment and epidemiological study, a multidisciplinary research team is essential. Additionally, many local technicians and supporting staff are fundamental. However, isolated human resources are not enough. More than recognized experts from many fields, this sort of research needs a priceless integration space, where creativity can emerge to solve unusual problems. The objective of this work was to share the methodological bases applied in this research project, starting from a conceptual model for As environmental and epidemiological assessment in Paracatu city, from which a basic framework was elaborated, herein called Plan of Action (PoA). The team of researchers has also considered distinct aspects, such as the socioeconomic and political contextualization in Paracatu, the local population health vulnerability by additional diseases, financial and educational concerns that may become more difficult to deeply and/or to clearly understand the hypothesis of this project and its results. The results should return to the communities and they are not, unavoidably, in complete agreement with populations′ expectations. This study is part of the 3-years environmental and health assessment research project performed by Brazilian institutions under the general coordination of CETEM.

2 METHODS

2.1 *Rationale and plan of action*

This research project was designed after it was demanded by the Municipality of Paracatu and lunched more than one year after the first contacts. The basic question of the local Government was on human health risks by Arsenic (As) environmental exposure in Paracatu city. The concern about this issue is increasing locally due to the expansion of gold mining activities (from 18 Mtpa to 61 Mtpa) on distinct mineral deposits (harder—B2- sulfide was found as the mine goes deeper) and, also, as a consequence of notes in some local newspapers and websites about the As potential exposure by the Paracatu population and its associated risks to the human health, i.e., its carcinogenic effects and connected deaths.

The PoA proposed was guided by the toxicological aspects of As environmental exposure. Most cases of human toxicity from As have been associated with exposure to inorganic As. There is good evidence that inorganic As is carcinogenic to humans by both oral and inhalation routes. Oral uptake is generally the most important route of exposure, whereas inhalation normally contributes less than 1% to the total dose. Non-cancer effects observed after inhalation of air, with high As levels at workplaces, are increased mortality from cardiovascular diseases, neuropathy and gangrene of the extremities. Changes in skin are also indicated as the early non-carcinogenic effect and the most prominent carcinogenic effect. In addition, several organic arsenicals accumulate in fish and shellfish, but these derivatives (mainly arsenobetaine and arsenocholine, also referred to as "fish arsenic") have been found to be essentially nontoxic. Because of this fact, the epidemiological study included a fish and shellfish ingestion question. PoA also considered the Paracatu specific environmental and its political situation. Thus, geographic data, hydrography, land uses, soil types, economic development (present and past activities, historical gold mining development since artisanal to companies; other mineral exploitation, economic crisis, employment rate, etc.) and future plans, expressed in "Plano Diretor do Município", were evaluated.

2.2 *Paracatu city and gold mining general aspects*

Paracatu city is located at the northeast of the Minas Gerais State. The gold mining has been developed in Paracatu since the 1700's with the discovery of placer gold in the creeks. Small scale gold mining peaked in the mid 1800's until the 1980's. The industrial open pit gold mine activities began in 1976–77 and it is estimated to continue until 2042. Major mining-related features at the Paracatu mine includes an open pit mine, two process plants, two tailings facilities and related surface infrastructure. Estimated site restoration costs are around 200 million dollars.

Paracatu has approximately 85,000 inhabitants (2010) and more than 85% of population lives in urban areas. Although the mining site area is classified as a rural area, there are neighborhoods bordering the site and, in some places, no more than few meters separates mining from homes.

2.3 *Performed steps of Plan of Action*

The following steps were performed to evaluate environmental aspects: (i) As levels in soils and sediments from the three watershed, sub-basins of Rio Paracatu; (ii) As levels in freshwater; (iii) As levels in drinking water (from sources as far as tap water, based on Municipality approved plan for water supply), and, (iv) As levels in groundwater for human consumption and from monitoring wells; (v) As levels in atmosphere. After that, using the (vi) human health risk assessment methodology (US EPA, 1989), all the environmental data were integrated to estimate risks of As environmental exposure. Additional studies on (vii) mineralogy and soil characterization, (viii) As ecotoxicology in soils and in freshwater and (ix) microbial activities in soil and sediments were performed in order to better understand the mobility/retention and ecological significance of As levels measured in environmental matrices. In parallel, human health aspects were evaluated: (x) carcinogenic and skin problems statistics in Paracatu and (xi) mortality data cancers type linked to As exposure from official data bank. In addition, it was performed a (xii) social network analysis aimed to create a strategy to communicate the main results to the focal population. Finally, the epidemiological strategy (xiii) was established after environmental data availability and was performed after approval by the National Commission for Ethics in Research. Socioeconomic and quality of life indicators were also evaluated (xviii).

2.4 *Political contextualization*

One should declare that support by local government as well as by the Public Attorney Office at state level were essential. This support is time-demanding because it depends on the comprehension the scientific bases of PoA by managers. However, it was fundamental to access the environmental reports of the mining companies that have been sent to the Environment Municipal Secretary, as well as to access to municipality health data bases from the Health Municipal Secretary. This closeness with local municipalities and their administrative background, on the other hand, brings their reality, which includes personal efforts, technical gaps, overcharge of work, low income, etc. In consequence, technical reports on environmental monitoring of several mining companies may not be adequately evaluated regarding to the sufficiency. Additionally, the local health data base is complex to search information, but worst situation takes place considering information available under notification and/or quality control. This situation is general in municipalities, unfortunately, with obvious exceptions. A very close relationship with local health care workers is needed and recommended, as they have known local population for a long period, they are welcome at homes and act as the link between population and researchers, revealing unique aspects from a specific population. Overall, a chronic problem is the discontinuity that naturally occurs after changes in the public government, which may happen within each 4 years. Unfortunately, this research project passed through by this situation. Fortunately, the local social links were strong enough to overcome any concern.

3 CONCLUSIONS

The model of action discussed here tend to promote and establish a better background for cooperation between local and external institutions, either in the technical and scientific fields, as in the local political field in the Brazilian mining city actual context.

ACKNOWLEDGEMENTS

The authors thank Rejane Lopes, Rosimar Fonseca, Abgail Soares, Leila Ruela, Cristiane Rabelo, Jussara Silva, Vânia Simão and Maria J.F. Gonçalves; Luciana T. de Oliveira, Raquel Macedo, Simone Soares, Cleone Martins, Antonieta Oliveira, Geni Brito, Patrícia Amorim, Muriene Xavier, Marilene Xavier, Amanda Costa, Emilene Serra, Tatiany Santos, Edson Lopes, Vanuza Campos, Thalita Gonçalves, Sayure Kayashima, Claudia Peres, Rosalba Cassuci, and TECSOMA students: Nilda Aragão, Jéssica Silva, Bianca Silva, Juliana Alves, Thassa Barros, Jaynne Guimarães, Camilla Cambronio, Thays Costa, Geisiane Silva, Samira Rabelo, Yara Soares and UNB students: Juliana Assis, Lucas Silva, Marcela Britto, Marina Alves and Sheila Silva.

One Century of the Discovery of Arsenicosis in Latin America (1914–2014) – Litter, Nicolli, Meichtry, Quici, Bundschuh, Bhattacharya & Naidu (Eds)
© 2014 Taylor & Francis Group, London, ISBN 978-1-138-00141-1

Author index

Printed by CPI Group (UK) Ltd, Croydon

Printed and bound by CPI Group (UK) Ltd, Croydon, CR0 4YY
21/10/2024
01777100-0004